Third Edition

Wine
Science

Principles

and

Applications

Third Edition

Wine Science

Principles

and

Applications

Ronald S. Jackson, PhD

ELSEVIER

AMSTERDAM • BOSTON • HEIDELBERG
LONDON • NEW YORK • OXFORD • PARIS
SAN DIEGO • SAN FRANCISCO
SYDNEY • TOKYO
Academic Press is an imprint of Elsevier

Academic Press is an imprint of Elsevier
30 Corporate Drive, Suite 400, Burlington, MA 01803, USA
84 Theobald's Road, London WC1X 8RR, UK
525 B Street, Suite 1900, San Diego, California 92101-4495, USA

First edition 1994
Second edition 2000
Third edition 2008

Notice
No responsibility is assumed by the publisher for any injury and/or damage to persons or property
as a matter of products liability, negligence or otherwise, or from any use or operation of any
methods, products, instructions or ideas contained in the material herein. Because of rapid
advances in the medical sciences, in particular, independent verification of diagnoses and drug
dosages should be made

Library of Congress Cataloging in Publication Data
A catalog record for this book is available from the Library of Congress

British Library Cataloguing in Publication Data
A catalogue record for this book is available from the British Library

ISBN: 978-0-12-373646-8

For information on all Academic Press publications
visit our web site at books.elsevier.com

Typeset by Charontec Ltd (A Macmillan Company), Chennai, India
www.charontec.com
Printed and bound in Canada

08 09 10 11 12   10 9 8 7 6 5 4 3 2 1

The book is dedicated to the
miraculous microbes that can turn a
marvelous fruit into a seraphic beverage,
to God who has given us the ability to
savor its finest qualities and pleasures,
and to my mother and father, to whom
I will eternally owe a debt of gratitude
for their unwavering support.

Contents

7

Fermentation

8

Postfermentation Treatments and Related Topics

12

Wine and Health

About the Author

The author received his bachelor's and maser's degrees from Queen's University and doctorate from the University of Toronto. His time in Vineland, Ontario, and subsequently on a sabbatical at Cornell University, redirected his interest in plant disease toward viticulture and ecology. As part of his regular teaching duties, he developed the first wine techonology course in Canada at Brandon University. For Many years he was a technical advisor to the Manitoba Liquor Control Commission, developed sensory tests to assess candidates of its Sensory Panel, and was a member of its External Tasting Panel. In addition he is author of *Wine Tasting: A Professional Handbook, Conserve Water Drink Wine*, numerous technical reviews, and an annual section in Tom Stevenson's Wine Report. Dr Jackson is retired from university activity and now concentrates his on writing. To contact the author send correspondence to his attention at Elsevier, 525 B Street, Suite 1900, San Diego, CA 92101–4495 USA.

Preface

There are three pillars of wine science – grape culture, wine production, and sensory analysis. Although it is traditional to cover these topics separately, a joint discussion is valuable and reinforces their natural interrelationships.

Consistent with present biological thought, much of wine science is expressed in terms of chemistry. Because of the botanical nature of the raw materials and its microbial transformation into wine, the physiology and genetics of the vine, yeasts, and bacteria are crucial to an understanding of the origins of wine quality. Similarly, microclimatology and soil physicochemistry are revealing the vineyard origins of grape quality. Finally, a knowledge of human sensory psychophysiology is essential for interpreting wine quality data. For those more interested in applications, much of the scientific discussion has been placed so that the practical aspects can be accessed without necessarily reading and understanding the scientific explanations.

Much of the data used in the book is derived from a few cultivars that originated in the cooler central regions of Western Europe. Thus, caution must be taken in extrapolating much of the information to warmer climates. The value of challenging established wisdom is evident from the success of Australian wine produced from cultivars grown in regions quite different from their European birthplace. In addition, the oft-quoted value of cooler mesoclimates must be qualified because it is derived from cultivars that arose in moderate climates. Cultivars that originated in cold climates generally are considered to develop best in the warmest sites of the ancestral region. Thus, for varieties derived in hot regions, the most favorable conditions for flavor accumulation are likely to be considerably different from those commonly quoted for moderate and cool climates.

Specific recommendations are avoided because of the international scope of the work. Even books with a regional focus find it difficult to give precise directions due to the variability in regional and site specific conditions. Science can suggest guidelines and reasons for good practice, and enunciate the potential advantages and disadvantages of particular options. However, it is the grape grower and winemaker who knows the subtleties of his or her sites, cultivars, and fermentation conditions. Individual experimentation and data recording are the only certain way for them to maximize grape potential.

One of the negative side-effects of our rapidly advancing (changing) state of knowledge is the confusion created as to what is the "truth." Too often non-scientists get annoyed with the inconstant recommendation from "experts." There is the misconception that scientists *know*, rather than are *searchers* for the truth. For some, this has resulted in their discarding technological advances for ancient techniques. This certainly facilitates many vinivicultural decisions, and can be used profitably in the "back to nature" philosophy of winemaking. While I cannot deny the commercial success some producers have with this approach, it is not the route by which quality wine will fill the supermarket shelf.

It is hoped that this book will help place our present knowledge in perspective and illustrate where further study is needed. It is not possible in a book to provide a detailed treatment of all diverging views. I have chosen those views that in my opinion have the greatest support, practical importance, or potential for significance. In addition, several topics are quite contentious among grape growers and winemakers. For some issues, further study will clarify the topic; for others, personal preference will always be the deciding factor. I extend my apologies to those who may feel that their views have been inadequately represented.

The effects of global warming on viticulture is increasingly coming under investigation. However, its true influence is only speculation at the moment. Thus, these have not been included. If some of the scenarios suggested come to fruition, the effect will be horrific. Although some famous vineyards may be under water, and grape adaptation to site be seriously dislocated, the more devastating effects are likely to result from the extreme and destructive disruption of world agriculture, trade, and economy, and the political and social strife that will follow.

Where no common chemical name is available or preferred, I.U.P.A.C. terminology has been used. In conformity with the International Code of Botanical Nomenclature, grape cultivar names are noted by single quotes (i.e. 'Pinot noir'), in lieu of the other accepted practice, placing *cv.* before the name. Except in tables, the present-day practice of naming rootstock cultivars with a number and the originator's name is used, *in lieu* of the number and a contraction of the originator's name (i.e. 3309 Couderc vs. 3309 C).

A list of Suggested Readings is given at the end of each chapter to guide further study. Although several are in languages other than English, they are excellent sources of precise information. To have omitted them would have done a disservice to those wishing to pursue the topics concerned. In addition, References are given in the book if the information is very specific or not readily available in the Suggested Readings. Further details can be obtained from sources given for the figures and tables.

Samuel Johnson made a cogent observation about the subject of this book:

This is one of the disadvantages of wine; it makes man mistake words for thoughts.

Ronald S. Jackson

Acknowledgments

Without the astute observations of generations of wine-makers and grape growers, and the dedicated research of countless enologists and viticulturalists, this work would have been impossible. Thus, acknowledgment is given to those whose work has not been specifically cited. Appreciation also is given to those who read and provided constructive criticism of various chapters of the manuscript. Credit must also go to the various editors who have helped over the years in the preparation of various editions of the text. However, special thanks goes to Nancy Maragioglio. She has facilitated every aspect of the preparation of the third edition. Her constant encouragement and creativity has not only provided considerable improvements, but made its preparation a joy.

Gratitude is also expressed to the many researchers, companies, institutes, and publishers who freely donated the photographs, data, diagrams or figures reproduced in the book.

Finally, but not least, I must express my deepest appreciation to my wife, Suzanne Ouellet, for her unshakable support in the preparation of the various editions of this work.

1

Introduction

Grapevine and Wine Origin

Wine has an archeological record dating back more than 7.5 thousand years. The earliest suspected wine residues come from the early to mid-fifth millennium B.C. – Hajji Firuz Tepe, in the northern Zagros Mountains of Iran (McGovern *et al.*, 1996). Evidence from Neolithic pottery from Georgia suggests that contemporaneous wine production was dispersed throughout the region (McGovern, in preparation). Older examples of fermented beverages have been discovered (McGovern *et al.*, 2004), but they appear to have been produced from rice, honey, and fruit (hawthorn and/or grape). Such beverages were being produced in China as early as 7000 B.C. The presence of wine residues is usually identified by the presence of tartaric acid residues, although additional procedures for identifying grape tannin residues are in development (Garnier *et al.*, 2003).

Other than the technical problems associated with identifying wine residues, there is the thorny issue of what constitutes wine – does spontaneously fermented grape juice qualify as wine, or should the term be restricted to juice fermented and stored in a manner to retain its wine-like properties?

The first unequivocal evidence of intentional winemaking appears in the representations of wine presses from the reign of Udimu (Egypt), some 5000 years ago (Petrie, 1923). Wine residues also have been found in clearly identified wine amphoras in many ancient Egyptian tombs, beginning at least with King Semerkhet – 1st Dynasty, 2920–2770 B.C. (Guasch-Jané *et al.*, 2004). They have also discovered evidence for both white and red wine in amphorae found in King Tutankhamun's tomb (1325 B.C.). Identification of red wine was made by the presence of syringic acid, an alkaline breakdown product of malvidin-3-glycoside. The same technique was used to establish the red grape origin of the ancient Egyptian drink – *Shedeh* (Guasch-Jané *et al.*, 2006).

Most researchers believe that winemaking was discovered, or at least evolved, in southern Caucasia. This area includes parts of present-day northwestern Turkey, northern Iraq, Azerbaijan, and Georgia. It is also generally thought that the domestication of the wine grape (*Vitis vinifera*) ensued in the same area. Remains of what appear to be domesticated grapes have been found in a Neolithic village in the Transcaucasian region of Georgia (Ramishvili, 1983). It is in this region that the natural distribution of *V. vinifera* most closely approaches the probable origins of Western agriculture – along the Tigris and Euphrates Rivers (Zohary and Hopf, 2000). Grapevine domestication also may have occurred independently in Spain (Núñez and Walker, 1989).

Although grapes readily ferment, due to the prevalence of fermentable sugars, the wine yeast (*Saccharomyces cerevisiae*) is not a major, indigenous member of the grape flora. The natural habitat of the ancestral strains of *S. cerevisiae* appears to be the bark and sap exudate of oak trees (Phaff, 1986). If so, the habit of grapevines climbing trees, such as oak, and the joint harvesting of grapes and acorns, may have encouraged the inoculation of grapes and grape juice with *S. cerevisiae*. The fortuitous overlap in the distribution of the progenitors of both *S. cerevisiae* and *V. vinifera* with the northern spread of agriculture into Anatolia may have fostered the discovery of winemaking, as well as its subsequent development and spread. It may not be pure coincidence that most major yeast-fermented beverages and foods (wine, beer, mead, and bread) have their origins in the Near East.

The earliest evidence of the connection between wine and *Saccharomyces cerevisiae* comes from an amphora found in the tomb of Narmer, the Scorpion King (ca. 3150 B.C.). *S. cerevisiae* was confirmed by the extraction of DNA from one of the amphoras. The DNA showed more similarity with modern strains of *S. cerevisiae* than closely related species, *S. bayanus* and

S. paradoxus (Cavalieri *et al.*, 2003). The latter is considered to be the progenitor of *S. cerevisiae*. Specific words referring to yeast action (ferment) begin to appear about 2000 B.C. (Forbes, 1965).

Other yeasts indigenous to grapes, such as *Kloeckera apiculata* and various *Candida* spp., can readily initiate fermentation. However, they seldom complete fermentation. Their sensitivity to the accumulating alcohol content and limited fermentative metabolism curtails their activity. In contrast, beer with its lower alcohol content may have initially been fermented by yeasts other than *S. cerevisiae*.

The Near Eastern origin and spread of winemaking are supported by the remarkable similarity between the words meaning wine in most Indo-European languages (Table 2.1). The spread of agriculture into Europe appears to be associated with the dispersion of Proto-Indo-European-speaking Caucasians (or their language and culture) (Renfrew, 1989). In addition, most eastern Mediterranean myths locate the origin of winemaking in northeastern Asia Minor (Stanislawski, 1975).

Unlike the major cereal crops of the Near East (wheat and barley), cultivated grapes develop an extensive yeast population by maturity, although rarely including the wine yeast (*Saccharomyces cerevisiae*). Piled unattended for several days, grape cells begin to self-ferment as oxygen becomes limiting. When the berries rupture, juice from the fruit is rapidly colonized by the yeast flora. These continue the conversion of fruit sugars into alcohol (ethanol). Unless *S. cerevisiae* is present to continue the fermentation, fermentation usually ceases before all the sugars are converted to alcohol. Unlike the native yeast population, *S. cerevisiae* can completely metabolize fermentable sugars.

The fermentation of grape juice into wine is greatly facilitated if the fruit is first crushed. Crushing releases and mixes the juice with yeasts on the grape skins (and associated equipment). Although yeast fermentation is more rapid in contact with slight amounts of oxygen, continued exposure to air favors the growth of a wide range of yeasts and bacteria. The latter can quickly turn the nascent wine into vinegar. Although unacceptable as a beverage, the vinegar so produced was probably valuable in its own right. As a source of acetic acid, vinegar expedited pottery production and the preservation (pickling) of perishable foods.

Of the many fruits gathered by ancient man, only grapes store carbohydrates predominantly in the form of soluble sugars. Thus, the major caloric source in grapes is in a form readily metabolized by wine yeasts. Most other fleshy fruits store carbohydrates as starch and pectins, nutrients not fermentable by wine yeasts. The rapid and extensive production of ethanol by *S. cerevisiae* quickly limits the growth of most bacteria and other yeasts in grape juice. Consequently, wine yeasts generate conditions that rapidly give them almost exclusive access to grape nutrients. Subsequent yeast growth is possible after the sugars are metabolized, if oxygen becomes available. An example is the respiration of ethanol by *flor* yeasts (see Chapter 9).

Another unique property of grapes concerns the acids they contain. The major acid found in mature grapes is tartaric acid. This acid occurs in small quantities in the vegetative parts of some other plants (Stafford, 1959), but rarely in fruit. Because tartaric acid is metabolized by few microbes, wine remains sufficiently acidic to limit the growth of most bacteria and fungi. In addition, the acidity gives wine much of its fresh taste. The combined action of grape acidity and the accumulation of ethanol suppresses the growth and metabolism of most potential wine-spoilage organisms. This property is enhanced in the absence of air (oxygen). For ancient man, the result of grape fermentation was the transformation of a perishable, periodically available fruit, into a relatively stable beverage with novel and potentially intoxicating properties.

Unlike many crop plants, the grapevine has required little genetic modification to adapt it to cultivation. Its mineral and water requirements are low, permitting it to flourish on soils and hillsides unsuitable for other food crops. Its ability to grow up trees and other supports meant it could be grown with little tending in association with other crops. In addition, its immense regenerative potential has allowed it to permit intense pruning. Intense pruning turned a trailing climber into a short shrub-like plant suitable for monoculture. The short stature of the shrubby vine minimized the need for supports and may have decreased water stress in semiarid environments by shading the soil. The regenerative powers and woody structure of the vine also have permitted it to withstand considerable winterkill and still possess the potential to produce commercially acceptable yields in cool climates. This favored the spread of viticulture into central Europe and the subsequent selection of, or hybridization with, indigenous grapevines.

The major change that converted "wild" vines into a "domesticated" crop was the selection of bisexual mutants. The vast majority of wild vines are functionally unisexual, despite usually possessing both male and female parts. In several cultivars, conversion to functional bisexuality has involved the inactivation of a single gene. However, the complexity of sexual differentiation in some cultivars (Carbonneau, 1983) suggests the involvement of mutations in several genes.

How ancient peoples domesticated the grapevine will probably never be known. However, two scenarios seem likely. Several Neolithic sites show significant collections of grape seeds in refuse piles, indicating the importance of grapes to the local inhabitants. Although most of these seed remains indicate charring, seed escaping the heating process could have found conditions ideal for growth among the ashes. Were any of these progeny rare bisexual (self-fertile) vines, they could have produced a crop, despite being isolated from feral vines. More likely, functional bisexual vines were unintentionally selected when feral vines were planted adjacent to settlements, and away from wild populations. Self-fertile vines would have become conspicuous by their fruitfulness, especially if unfruitful (male) vines were rogued. Cuttings from such vines could have provided plants appropriate for the initiation of nascent viticulture.

Although other modifications may characterize domesticated strains, changes in seed and leaf shape are not of viticultural value. The lower acidity and higher sugar content that characterize cultivated varieties are not the exclusive attributes of domesticated vines. These properties may reflect more cultural conditions than genetic modifications.

Because canes lying on the ground root easily when covered by soil, layering probably developed as the first method of vegetative propagation. Success with layering would have ultimately led to propagation by cuttings. Early viticulturalists, if they did not already know from other perennial crops, would have come to realize that to retain desirable traits, vegetative propagation was preferable to sowing seed. Vegetative propagation retains desirable combinations of genetic traits unmodified.

In drier regions, the limited growth of vines could be left to trail on the ground. However, in moister regions, it would have been better to plant vines next to trees for support. This technique is still used in some parts of Portugal, and was, until comparatively recently, fairly common in parts of Italy. It had the advantage of leaving arable land free for annual food crops. One of the major problems with training up trees is that most of the fruit is soon located out of easy reach. Some inventive cultivator probably found that staking and trimming restricted growth to a convenient height, facilitating fruit gathering. In addition, pruning off excess growth at the end of the season would have been discovered to benefit fruit maturation. The combination of easier harvesting and improved ripening probably spurred further experimentation with pruning and training systems. Combined with advances in wine production and storage, the stage would have been set for the development of wine trade.

The evolution of winemaking from a periodic, haphazard event to a common cultural occurrence presupposes the development of a settled lifestyle. A nomadic habit is incompatible with harvesting a sufficient quantity of grapes to produce steady supply of wine. In addition,

unlike major field crops, grapevines provide significantly less yield, have a shorter harvest period, and
produce a perishable fruit (unless dried or converted
into wine). A dependable supply of grapes would have
become important when wine developed an association with religious rites. To assure a reliable supply of
wine required the planting of grapevines in or around
human settlements. Because grapevines begin to bear a
significant crop after only 3–5 years, and require several additional years to reach full productivity, such an
investment in time and effort would be reasonable only
if the planter resided nearby. Under such conditions,
grape collection for winemaking could have initiated
the beginnings of viticulture. If, as seems reasonable,
wine production is dependent on a settled agricultural
existence, then significant wine production cannot pre-
date the agricultural revolution. Because grapevines
are not indigenous to the Fertile Crescent (the origin
of Western agriculture), the beginnings of winemaking
probably occurred after the knowledge of agricultural
skills moved into southern Caucasia.

From Caucasia, grape growing and winemaking
probably spread southward toward Palestine, Syria,
Egypt, and Mesopotamia. From this base, wine consumption, and its socioreligious connections, spread
winemaking around the Mediterranean. Despite this,
Stevenson (1985) has provided evidence for an extensive system of grape culture in southern Spain, several
centuries before the Phoenicians established colonies in
the region. Nevertheless, colonization from the eastern
Mediterranean is still viewed as the predominant source
of early grape-growing and winemaking knowledge. In
more recent times, European exploration and colonization has spread grapevine cultivation into most of the
temperate climatic regions of the globe.

Throughout much of this period, contemporary wine
styles either did not exist or occurred in forms considerably different from their present form. Most ancient
and medieval wines probably resembled dry to semidry
table wines, turning vinegary by spring. Protection from
oxidation was generally poor, and the use and value of
sulfur dioxide apparently unknown. Thus, prolonged
storage of wine would probably have been avoided.

Nonetheless, various techniques were available in
ancient times that could extend the drinkable life of a
wine. A lining of pitch (1–2 mm thick) was often used
to waterproof amphoras, the majority of which were
unglazed and otherwise porous. Resins, dissolving into
the wine from the pitch, may have had the added benefit of acting as a mild antimicrobial agent, retarding
spoilage. It also supplied a flavorant that could partially mask the beginnings of spoilage. However, the
ancients eventually developed a process for generating
pottery with an impervious inner layer, termed Type A

amphoras. This was achieved by adding a flux of potash to an illitic clay. A rapid reductive firing (~1000°C)
produced an inner, gray, vitreous lining (Vandiver and
Koehler, 1986). The typical red surface coloration of
amphoras comes from the oxidation of iron oxide in
the outermost layers. This occurs after the introduction
of air near the end of firing.

Wine amphoras were normally sealed with cork in
Roman times. Underwater archeology has supplied
amble support for cork use (Cousteau, 1954; Frey
et al., 1978). Pitch was used to affix a circular cap of
cork to the rim of the amphora. An overlay cap of
pozzuolana (volcanic clay) subsequently protected the
cork seal. The procedure is documented in ancient
Roman writings.

Amphoras seem to have been stored on their sides
or upside down, thus keeping the cork wet (Addeo
et al., 1979; Grace, 1979, photo 63; Jashemski, 1979,
plate 256; Koehler, 1986). Thus, the minimum conditions for extended wine aging were met. That they
were met is suggested from ancient literary sources. For
example, there is frequent mention of quality distinc-
tions between vintages, specific vineyards, and different regions. Aged wine was highly prized. Horace, the
famous Roman poet, praises a wine aged for forty-six
years in a cork-stoppered amphora. In addition, ancient
authors such as Athenaeus and Hermippos employed
wine descriptors that sound surprisingly modern (white
flowers, violets, roses, hyacinths, and apples) (see
Henderson, 1824, p. 62; Stanley, 1999). Thus, there
seems little doubt that ancient Greeks and Romans produced wine that, were we able to taste them, we would
probably rate highly (Henderson, 1824; Allen, 1961;
Tchernia, 1986; Stanley, 1999).

Although wine production techniques were primitive, compared to today, some modern procedures have
ancient counterparts. For example, Cato recommends
storing amphoras of wine in the sun, having added to
them a portion of boiled must. Pliny the Elder (23–79
A.D.) makes the same suggestion. This could be the origin of wines such as *vin santo*. Several procedures for
the production of sweet wine are noted, the simplest
being the addition of boiled-down must, or leaving the
grapes to partially dry in the sun (see Stanley, 1999).
More demanding were procedures involving the fermentation of juice that oozed out of the grapes under
their own weight, for example Priam and Saprias wines
(see Stanley, 1999). The latter was apparently made
from molded (botrytized?) grapes. However, whether
wines stored in a room through which smoke and
heat rose would resemble modern madeira is a moot
point. It was recommended by Columella to achieve
early wine maturity. Although appreciated by some,
Columella also notes that the technique was open to

abuse. Pliny clearly felt that wine should be aged naturally, not smoked.

Despite the probability some Roman wine would please modern palates, the majority clearly would not. Grape yields were often high, leading to acidic juice, low in sugar content. Most wines were also stored in amphoras coated with pitch. This probably would have masked any subtle flavors the wine might have possessed. Pitch was also frequently added to the wine, possibly to mask wine defects. Lower grade wines were also often treated with heat-concentrated must, honey, flavored with herbs, or perfumed with myrrh. Many of the formulae in ancient texts seem to refer to wine used as solvent for medical herbs and assorted potions. Inferior quality wine was also made from boiled must or grape pressings soaked in water. However, it was for the poor and slaves that the wine was definitely doctored, usually with sea water and/or vinegar. An example of a recipe from Cato (234–149 B.C.) gives a clear indication of its probable quality:

Combine 10 quadrantals of must, 2 quadrantals of sharp vinegar, 2 quadrantals of boiled must with 50 quadrantals of fresh water. Stir with a stick three times daily for five days. Add 64 sextarii of aged sea-water, seal, and let set for 10 days. The wine should last until the summer. The remainder will be excellent, sharp vinegar.

Wines began to take on their modern expression during the seventeenth century. The use of sulfur in barrel treatment seems to have become fairly common in Western Europe about this time. This would have greatly increased the likelihood of producing better-quality wines and extending their aging potential. Stable sweet wines able to age for decades or centuries also started to be appear in the mid-1600s, commencing with the Tokaj wines of Hungary.

For the commercial production of sparkling wine, a prerequisite was necessary – the production of strong glass bottles. This occurred in England in the mid-1600s. The reintroduction of cork as a wine closure, and the production of bottles able to withstand the high pressures generated by carbon dioxide, set the stage for the commercial development of sparkling wines.

The development of vintage port also depended on the ready availability of inexpensive bottles, made possible by the industrial revolution. The evolution in bottle shape, from bulbous to cylindrical, permitted bottles to be laid on their sides. Because the cork stayed wet in this position, the wine remained isolated from oxygen and had the opportunity to develop a smooth character and complex bouquet. The development of modern port also depended on the perfection of wine distillation.

Distilled spirits are added to the fermenting juice to prematurely stop fermentation. As a consequence, grape sugars are retained, along with the extraction of sufficient pigments, to produce a sweet, dark-red wine. Modern sherries also depend on the addition of grape spirits. Although alcohol distillation was first developed by the Arabs, the adoption of the technique in medieval Europe was slow. Thus, fortified wines are of relatively recent origin.

With mechanization, glass bottles became the standard container for both wine maturation and transport. The reintroduction and widespread adoption of cork as a bottle closure in the seventeenth century provided conditions favorable for the production of modern wine. The discovery by Pasteur in the 1860s of the central importance of yeasts and bacteria to fermentation set in motion a chain of events that has produced the incredible range of wines that typify modern commerce.

Commercial Importance of Grapes and Wine

From its humble origins, grape production has developed into the world's most important fresh fruit crop. Worldwide grape production in 2002 was about 62 million metric tons. This compares with roughly 57, 50, and 43 million metric tons for oranges, bananas, and apples, respectively. The area planted under grapevines in 2002 is estimated at about 7.9 million hectares, down from a maximum of 10.2 million in the late 1970s. Approximately 66% of the production was fermented into wine, 18.7% consumed as a fresh fruit crop, and the remaining 7.7% dried for raisins (OIV, 2005). The use varies widely from country to country, often depending on the physical and politicoreligious (wine prohibition) dictates of the region. Despite its world importance, vines only cover about 0.5% of agriculture land, and its produce constitute but 0.4% of global household expenditure (Anderson, 2004).

Grape production is largely restricted to climatic regions similar to those of the indigenous range of *Vitis vinifera*. This zone approximates the area between the 10 and 20°C annual isotherms (Fig. 1.1). Grape culture is further largely restricted to regions characterized by Mediterranean-type climates. Extension into cooler, warmer, or more moist environs is possible when local conditions modify the climate or viticultural practice compensates for less than ideal conditions. Commercial production even occurs in subtropical regions, where severe pruning stimulates nearly year-round vine growth.

In Europe, where 61% of the world's vineyards are located, about 77% of the crop is fermented into wine. The latter percentage is slightly less for world

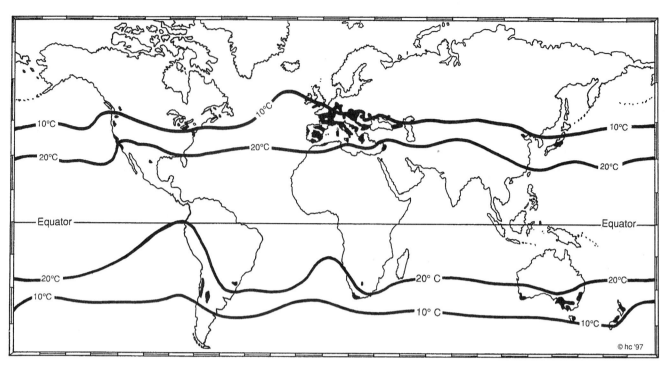

Figure 1.1 Association between the major viticultural regions of the world, with the 10 and 20 °C annual isotherms. (Drawing courtesy of H. Casteleyn, reproduced by permission)

production (71%), owing to the predominant use of grapes as a table or raisin crop in Islamic countries. Since the 1970s, wine production has ranged from about 250 to 330 million hl (66 to 87 million gallons), with recent production levels being about 270 million hl. Although Spain has the largest vineyard hectarage, France and Italy produce the largest volumes of wine. Together, France and Italy produce about 50% of the world's wine, but supply about 60% of world wine exports. The increasing economic significance of wine export is partially reflected in the marked increase in research conducted throughout the world (Glänzel and Veugelers, 2006). Statistics on wine production and export for several countries are given in Fig. 1.2. Several major wine-producing nations, such as Argentina and the United States, export a relatively small proportion of their production. In contrast, countries such as Chile and Portugal export the majority of their production.

Although Europe is the most important wine producing and exporting region, in terms of volume, it is also the primary wine-consuming region. For centuries, wine has been a significant caloric food source in the daily diets of many workers in France, Italy, Spain and other Christian Mediterranean nations. Because wine was an integral part of daily food consumption, heavy drinking did not have the tacit acceptance found in some northern European countries. Alcohol abuse, especially in the United States, spawned the prohibitionist and current neoprohibitionist movements. Their views that

consuming beverages containing alcohol is detrimental to human health are in marked conflict with evidence supporting the healthful benefits of moderate wine consumption (see Chapter 12). The reticence of some governments to acknowledge the beneficial consequences of moderate wine consumption does injustice to the long, extensive, and efficacious use of wine in medicine (Lucia, 1963).

The trend toward reduced or stabilized per capita wine consumption is noted in Fig. 1.3. Additional data is available in Anonymous (1999). The reasons for these changes are complex and often region-specific. Occasionally, the decline in per capita consumption has been optimistically interpreted as a shift toward the use of less, but better-quality wine. Although possibly true in some cases, in the traditional wine-consuming regions of Europe, the decline in wine consumption appears to be associated with a rise in the use of distilled spirits.

Wine Classification

Except in the broadest sense, there is no generally accepted system of classifying wines. They may be grouped by carbon dioxide or alcohol content, color, or stylistic, varietal or geographic origin. Each has its advantages and disadvantages. For taxation purposes, wines often are divided into three general categories: still, sparkling, and fortified – the latter two typically being taxed at a higher

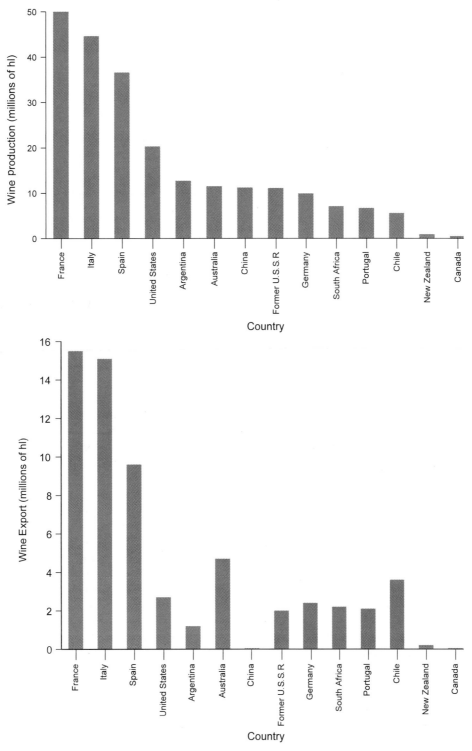

Figure 1.2 Wine production and export statistics (2002) for several wine-producing countries. (Data from OIV, 2005)

rate. This division recognizes significant differences, not only in production, but also use. In addition, classification by color provides the purchaser with a rough indication of the wine's flavor intensity. Stylistic and geographical origin often go hand-in-hand, at least for many European appellations, supplying additional information about the wine's likely characteristics. Varietal origin furnishes further clues as to the potential flavor attributes of the wine. Although useful, these sources of information do not provide consumers with adequate

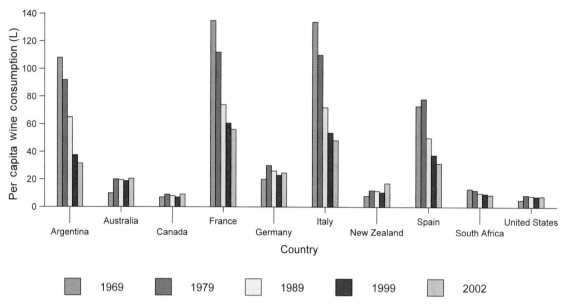

Figure 1.3 Changes in per capita wine consumption in several countries over about 30 years. (Data from OIV, 2005)

information on which to confidently base wine purchases. However, without classification, precise language would be nonexistent, and most thought impossible. Thus, no matter how inadequate, codifying the eclectic range of wines into categories is a necessity, in spite of there being no precise sensory system on which classification can be rationalized. Of classification systems, those based on geographic origin are the most common. For consumers, it gives the impression of being concrete and permits wine selection based on regional bias. Regrettably, it seldom provides sufficient or detailed information on potential flavor characteristics.

The arrangement presented below is traditional, being based primarily on stylistic features. Wines are initially grouped based on alcohol concentration. This commonly is indicated by the terms "table" (alcohol contents ranging between 9 and 14% by volume) and "fortified" (alcohol contents ranging between 17 and 22% by volume). Table wines are subdivided into "still" and "sparkling" categories, depending on the wine's carbon dioxide content.

Still Table Wines

Because most wines fall into the category of still table wines, it requires the largest number of subcategories (Table 1.1). The oldest division, based on color, separates wines into white, red, and rosé subgroups. Not only does this have the benefit of long acceptance, it reflects distinct differences in flavor, use, and production methods. For example, red wines are more flavorful, typically drier, and more astringent than white wines. In contrast, white wines are generally more acidic, floral in nature, and come in a wide range of sweetness styles. Rosés fall in between, being lighter than red wines, but more astringent than whites.

Because most white wines are intended to be consumed with meals, they typically are produced to possess an acidic character. Combined with food proteins, the acidic aspect of the wine becomes balanced and can both accentuate and harmonize with food flavors. Most white wines are given little if any maturation in oak cooperage. Only wines with distinct varietal aromas tend to benefit from an association with oak flavors. Those with a sweet finish generally are intended to be consumed alone – as a "sipping" wine, to accompany or replace dessert. Most botrytized (late-harvest) wines and icewines fall into this category.

Modern red wines are almost exclusively dry. The absence of a detectable sweet taste is consistent with their intended use as a food beverage. The bitter and astringent compounds that characterize most red wines bind with food proteins, producing a balance that otherwise would not develop. Occasionally, well-aged red wines are saved for enjoyment after the meal. Their diminished tannin content obviates the need for food to develop smoothness. Also, the complex subtle bouquet of aged wines often can be more fully appreciated in the absence of competing food flavors.

Most red wines that age well are given the benefit of some maturation in oak. Storage in small oak cooperage (~225-liter barrels) usually speeds maturation and adds subtle flavors. Following in-barrel maturation, the wines typically receive further in-bottle aging at the winery before release. When less oak character is desired, cooperage over 1000-liter capacities are used. Alternately, the wine may be matured in inert tanks to avoid oxidation and the uptake of accessory flavors.

Table 1.1 Classification of still wines based on stylistic differences[a]

White[b]			
Long-aging (often matured and occasionally fermented in oak cooperage)		Short-aging (seldom exposed to oak)	
Typically little varietal aroma	Varietal aroma commonly detectable	Typically little varietal aroma	Varietal aroma commonly detectable
Botrytized wines Vernaccia di San Gimignano Vin Santo	'Riesling' 'Chardonnay' 'Sauvignon blanc' 'Parellada' 'Sémillon'	'Trebbiano' 'Muscadet' 'Folle blanche' 'Chasselas' 'Aligoté'	'Müller-Thurgau' 'Kerner' 'Pinot blanc' 'Chenin blanc' 'Seyval blanc'

Red[b]			
Long-aging		Short-aging	
Tank oak-aging (many European wines, except those from France)	Barrel oak-aging (most French, "new" European, and New World wines)	Little varietal aroma detectable	Varietal aroma often detectable
'Tempranillo' 'Sangiovese' 'Nebbiolo' 'Garrafeira'	'Cabernet Sauvignon' 'Pinot noir' 'Syrah' 'Zinfandel'	'Gamay' 'Grenache' 'Carignan' 'Barbera'	'Dolcetto' 'Grignolino' 'Baco noir' 'Lambrusco'

Rosé	
Sweet	Dry
Mateus Pink Chablis Rosato Some blush wines	Tavel Cabernet rosé White zinfandel Some blush wines

[a] Although predominantly dry, many have a sweet finish. These include both light "sipping" wines and the classic botrytized wines.
[b] Representative examples in single quotes refer to the names of grape cultivars used in the wine's production.

One of the more common differences between red wines depends on the consumer market for which they are intended. Wines processed for early consumption have lighter, more fruity flavors, whereas those processed to enhance aging potential often do so at the expense of early enjoyment and are initially excessively tannic. Beaujolais *nouveau* is a prime example of a wine designed for early consumption. In contrast, premium 'Cabernet Sauvignon' and 'Nebbiolo' wines illustrate the other end of the spectrum, in which long aging is typically required for the development of their finest qualities.

Rosé wines are the most maligned of table wines. To achieve the light rosé color, the juice of red grapes is often left in contact with the skins for only a short period. This limits not only anthocyanin extraction, but also flavor uptake. In addition, rosé wines soon lose their initial fruity character and fresh pink coloration (turning orangish). Many rosé wines are also finished with a slight sparkle and sweet taste. This has made many connoisseurs view rosés with disdain, considering them to possess the faults of both white and red

wines, but none of their benefits. To counter the stigma attached to the term rosé, many North American versions are called "blush" wines or "white" renderings of red cultivars.

Sparkling Wines

Sparkling wines often are classified by method of production (see Table 1.2). The three principal techniques are the traditional (champagne), transfer, and bulk (Charmat). They all employ yeasts to generate the carbon dioxide that produces the effervescence. Although precise, classification based on production method need not reflect significant differences in sensory characteristics. For example, the traditional and transfer methods typically aim to produce dry to semidry wines that accentuate subtlety, limit varietal aroma, and possess a "toasty" bouquet. Sparkling wines differ more due to duration of yeast contact and grape variety than method of production. Although most bulk-method wines tend to be sweet and aromatic (i.e., Asti Spumante), some are dry with subtle fragrances.

Table 1.2 Classification of sparkling wines with some representative examples

| With added flavors, coolers (low alcohol) | Natural (without flavors added) | |
	Highly aromatic (sweet)	Subtly aromatic (dry or sweet)

Fruit-flavored, carbonated wines	Asti-style Muscat-based wines	Traditional-style
		  Champagne
		  Vin Mousseux
		  Cava
		  Sekt
		  Spumante

		Crackling/carbonated
		  Perlwein
		  Lambrusco
		  Vinho Verde

Carbonated sparkling wines (deriving their sparkle from carbon dioxide incorporated under pressure) show an even wider range of styles. These include dry white wines, such as vinho verde (historically obtaining its sparkle from malolactic fermentation); sweet sparkling red wines, such as lambrusco; most crackling rosés; and fruit-flavored "coolers."

Fortified Wines (Dessert and Appetizer Wines)

All terms applied to this category (see Table 1.3) are somewhat misleading. For example, some subcategories achieve their elevated alcohol contents without the addition of distilled spirits (e.g., the sherry-like wines from Montilla, Spain). Thus, they are technically not fortified. The alternative designation of aperitif and dessert wines also has problems. Although most are used as aperitif or dessert wines, many table wines are used similarly. For example, sparkling wines are often viewed as the ultimate aperitif, whereas botrytized wines can be a numinous dessert wine.

Regardless of designation, wines in this category typically are consumed in small amounts, and are seldom completely consumed shortly after opening. Their high alcohol content limits microbial spoilage, and their marked flavor and resistance to oxidization often allow them to remain stable for weeks after opening. These are desirable properties for wines consumed in small amounts. The exceptions are *fino* sherries and vintage ports. Both lose their distinctive properties several months after bottling, or several hours after opening, respectively.

Fortified wines are produced in a wide range of styles. Dry or bitter-tasting forms are normally consumed as aperitifs before meals. They stimulate the appetite and

Table 1.3 Classification of fortified wines with some representative examples

With added flavors	Without added flavors
Vermouth	Sherry-like
Byrrh	Jerez-xerès-sherry
Marsala (some)	Malaga (some)
Dubonnet	Montilla
	Marsala
	Château-chalon
	New World solera and submerged sherries
	Port-like
	Porto
	New World ports
	Madeira-like
	Madeira
	Baked New World sherries and ports
	Muscatel
	Muscat-based wines
	Setúbal
	Samos (some of)
	Muscat de Beaunes de Venise
	Communion wine

activate the release of digestive juices. Examples are *fino*-style sherries and dry vermouths. The latter are flavored with a variety of herbs and spices. More commonly, fortified wines possess a sweet attribute. Major examples are *oloroso* sherries, ports, madeiras, and marsalas. These wines are consumed after meals, or as a dessert substitute.

Wine Quality

What constitutes wine quality often changes with experience. It is also affected by the genetic makeup of the individual. Nevertheless, quality does have components that can be more or less quantified. Negative quality factors, such as off-odors, are generally easier to identify and control. Positive quality factors tend to be more elusive.

Wine quality often is defined in incredibly diverse ways. It may be evaluated in terms of subtlety and complexity, aging potential, stylistic purity, varietal expression, ranking by experts, or consumer acceptance. Each has its justification and limitations. Nevertheless, the views of experts (either self-proclaimed or panels of trained tasters) have had the greatest influence on winemakers. Premium wine sales constitute only a small fraction of world wine production, but have had a profound influence on the direction of enologic and viticultural research. This has resulted in the marked improvement of wine quality during the last half of the twentieth century. Its influence has been felt all the way down to bulk-wine production. It has also brought fine-quality wine to a broader selection of people than ever before.

Occasionally, this change has been viewed as potentially bringing "fine wine to all, on the supermarket shelf" (NOVA, 1978). However, this view confuses availability with acceptance. It is unlikely that simple economic availability will increase the appreciation of premium wine, any more than opera on television has generated higher consumer demand. Those psychophysical features that make premium-quality wine appealing to a small group of connoisseurs are still poorly understood. This means that for the majority of wine producers, understanding the desires of the majority of consumers is far more lucrative than a select group of connoisseurs. Understanding how a target group perceives quality and value-for-money is particularly important (Cardello, 1995; Lawless, 1995). A clear example of the importance of perception was the marked increase in red wine sales following airing of the "French Paradox" on *60 Minutes*. Pretorius *et al.* (2006) have clearly enunciated the marketing view of quality by stating that "quality is defined as *sustainable customer and consumer satisfaction.*" However, accurately measuring this aspect is fraught with difficulty. Consumer loyalty is often fickle. It is also uncertain whether purchase is based on opinions expressed in questionnaires (Köster, 2003; Jover *et al.*, 2004).

Perceived quality is the principal driving force among connoisseurs. How else can one explain the continuing importance of a quality ranking developed in the mid-1800s – the *cru classé* system for Bordeaux wines? In few other areas of commerce is historical ranking still considered of any significance.

For the occasional wine drinker, knowledge of geographic or varietal origin tends to be secondary – ease of availability, price, and previous experience being the overriding factors in selection. Pleasure on consumption is usually assessed on subjective, highly idiosyncratic criteria. In contrast, geographic origin and reputation strongly influence the purchases of, and presumably appreciation by, wine connoisseurs. For the connoisseur, whether and how well a wine reflects expectations can be crucial to perceived quality. Historical or traditional expectations are central to the quality percepts embodied in most appellation control laws.

In addition to the purely subjective and historical views of quality, esthetic quality is the most highly prized attribute possessed by premium wines. Esthetic quality is defined similarly, and uses the same language as artistic endeavors such as sculpture, architecture, and literature. Aspects of esthetic quality include balance, harmony, symmetry, development, duration, complexity, subtlety, interest, and uniqueness. Defining these terms precisely is impossible, owing to human variability in perception. Nevertheless, balance and harmony in wine commonly refer to a smooth taste and mouth-feel, without any aspect interfering with the overall pleasurable sensation. Symmetry refers to the perception of compatibility between sapid (taste and mouth-feel) and olfactory (fragrant) sensations. Development typically refers to the changes in intensity and aromatic character after pouring. When pleasurable, development is important in maintaining interest. Fragrance duration is also essential to the esthetic perception of wine quality. Complexity and subtlety are additional highly valued attributes of fragrance and flavor. The impact of these factors on memory is probably the most significant determinant of overall wine quality.

Health-related Aspects of Wine Consumption

Until the 1900s, wine was used in the treatment for several human afflictions (Lucia, 1963). It also acted as an important solvent for medications. In the twentieth century, well-meaning but misguided prohibitionists waged war against all beverages containing alcohol. They succeeded in persuading several national governments, and the medical establishment in general, that consuming alcohol-containing beverages was undesirable. The only exception, grudgingly permitted, was use in religious services.

Since the 1990s, there has been a marked renewal in interest among the medical profession in the health benefits of moderate wine consumption (see Chapter 12). One of the more widely documented benefits relates to cardiovascular disease. However, wine also can reduce the undesirable influences of stress, enhance sociability, lower rates of clinical depression, and improve self-esteem and appetite in the elderly (Baum-Baicker, 1985; Delin and Lee, 1992). Imperative in all such studies is the need to minimize the potential influence of cultural, environmental, and individual factors on the results. For example, in one study, wine consumers were found to purchase more "healthy" foods than their beer-drinking counterparts (Johansen et al., 2006). Thus, the importance of studies such as that of Mukamal et al. (2006). It compared the alcohol consumption and incidence of coronary heart disease in men with comparable healthy lifestyles. Neoprohibitionists are all too quick to point out both real and imaginary faults in any study that presents findings contrary to their established beliefs.

The benefits of moderate wine consumption on favoring a healthful balance of low- and high-density lipoprotein in the plasma are now well established (Rimm et al., 1991; Kinsella et al., 1993; Soleas et al., 1997). Even the National Institute on Alcohol Abuse and Alcoholism (1992) was moved to record that "there is a considerable body of evidence that lower levels of drinking decrease the risk of death from coronary heart disease."

In an intriguing study by Lindman and Lang (1986), wine was the only beverage containing alcohol associated with positive social expectations. Thus, wine appears unique in its being identified with happiness, contentment, and romance. Additional studies also have found that wine is associated with more socially desirable stereotypes than other alcohol-containing beverages (Klein and Pittman, 1990a; Delin and Lee, 1992; Duncan et al., 1995). Wine consumption is also rarely associated with intoxication and other alcohol-related problems (Smart and Walsh, 1999) (Fig. 1.4). Wine is also the alcoholic beverage most associated in the mind of consumers with food consumption (Pettigrew and Charters, 2006).

In addition to revealing the potential benefits of wine consumption, researchers are also beginning to investigate the occasionally unpleasant consequences of moderate wine use. For example, the induction of headaches by red wine has been correlated with insufficient production of a platelet phenolsulphotransferase (Littlewood et al., 1988). Also, headache prevention has been associated with the prior use of acetylsalicylic acid (Kaufman, 1992) and other prostaglandin synthesis inhibitors.

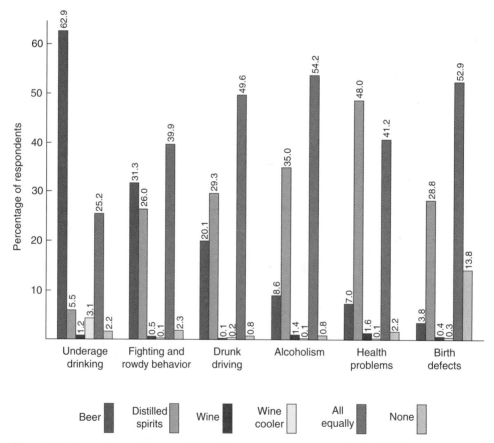

Figure 1.4 Comparison of the perception of adverse consequences associated with the consumption of different beverages containing alcohol. (Reprinted from Klein and Pittman, 1990b, p. 481 by courtesy of Marcel Dekker, Inc.)

Although wine consumption is contraindicated in a few medical instances, such as gastrointestinal ulcerations and cancers, in most situations, the daily consumption of wine in moderate amounts (between one and two glasses of wine) is beneficial to human health.

Suggested readings

Wine History

Allen, H. W. (1961) *A History of Wine.* Faber and Faber, London.

Fleming, S. J. (2001) *Vinum: The Story of Roman Wine.* Art Flair, Glen Mills, PA.

Fregoni, M. (1991) *Origines de la vigne et de la viticulture.* Musumeci Editeur, Quart (Vale d'Aosta), Italy.

Henderson, A. (1824) *The History of Ancient and Modern Wines.* Baldwin, Craddock and Joy, London.

Hyams, E. (1987) *Dionysus – A Social History of the Wine Vine.* Sidgwick & Jackson, London.

Johnson, H. (1989) *Vintage: The Story of Wine.* Simon & Schuster, New York.

Lesko, L. H. (1977) *King Tut's Wine Cellar.* Albany Press, Albany, NY.

Loubère, L. A. (1978) *The Red and the White – The History of Wine in France and Italy in the Nineteenth Century.* State University of New York Press, Albany, NY.

McGovern, P. E. (2003) *Ancient Wine: The Search for the Origins of Viniculture.* Princeton University Press, Princeton, NJ.

McGovern, P. E., Fleming, S. J., and Katz, S. H. (eds.) (1995) *The Origins and Ancient History of Wine.* Gordon and Breach Publishers, Luxembourg.

Redding, C. (1851) *A History and Description of Modern Wines,* 3rd edn. Henry G. Bohn, London.

Soleas, G. J., Diamandis, E. P., and Goldberg, D. M. (1997) Wine as a biological fluid: History, production, and role in disease prevention. *J. Clin. Lab. Anal.* 11, 287–313.

Unwin, T. (1991) *Wine and the Vine. An Historical Geography of Viticulture and the Wine Trade.* Routledge, London.

Weinbold, R. (1978) *Vivat Bacchus – A History of the Vine and Its Wine.* Argus Books, Watford, UK.

Younger, W. (1966) *Gods, Men and Wine.* George Rainbird, London.

Wine and Culture

Burton, B. J., and Jacobsen, J. P. (2001) The rate of return on investment in wine. *Econ. Inquiry* 39, 337–350.

Heath, D. B. (1995) *International Handbook on Alcohol and Culture.* Greenwood Press, Westport, CT.

Oczkowski, E. (2001) Hedonic wine price function and measurement of error. *Econ. Record* 77, 374–382.

References

Addeo, F., Barlotti, L., Boffa, G., Di Luccia, A., Malorni, A., and Piccioli, G. (1979) Constituenti acidi di una oleoresina di conifere rinvenuta in anfore vinarie durante gli scavi archeologici di Oplonti. *Ann. Fac. Sci. Agrarie Studi Napoli, Portici* 13, 144–148.

Allen, H. W. (1961) *A History of Wine.* Faber and Faber, London.

Anderson, K. (2004) *The World's Wine Markets. Globalization at Work.* Edward Elgar, Cheltenham, UK.

Anonymous (1999) *World Drink Trends – 1998.* Produktschap voor Gedistilleerde Dranken, Schiedam, Netherlands.

Baum-Baicker, C. (1985) The psychological benefits of moderate alcohol consumption: A review of the literature. *Drug Alcohol Depend.* 15, 305–322.

Beck, C. W., and Borromeo, C. (1991) Ancient pine pitch: Technological perspectives from a Hellenistic wreck. *MASCA Res. Papers Sci. Archaeol.* 7, 51–58.

Carbonneau, A. (1983) Stérilités mâle et femelle dans le genre *Vitis.* II. Conséquences en génétique et sélection. *Agronomie* 3, 645–649.

Cardello, A. V. (1995) Food quality: Relativity, context and consumer expectations. *Food Qual. Pref.* 6, 163–168.

Cavalieri, D., McGovern, P. E., Hartl, D. L., Mortimer, R., and Polsinelli, M. (2003) Evidence for *S. cerevisiae* fermentation in ancient wine. *J. Mol. Evol.* 57, S226–S232.

Cousteau, J.-Y. (1954) Fish men discover a 2,200-year-old Greek ship. *Natl. Geogr.* 105 (1), 1–34.

de Blij, H. J. (1983) *Wine: A Geographic Appreciation.* Rowman and Allanheld, Totowa, NJ.

Delin, C. R., and Lee, T. L. (1992) Psychological concomitant of the moderate consumption of alcohol. *J. Wine Res.* 3, 5–23.

Duncan, B. B., Chambless, L. E., Schmidt, M. I., Folsom, A. R., and Szklo, M. (1995) Association of the waist-to-hip ratio is different with wine than with beer or hard liquor consumption. *Am. J. Epidemiol.* 142, 1034–1038.

Forbes, R. J. (1965) *Studies in Ancient Technology,* Vol. III, 2nd edn. E. J. Brill, Leiden, p. 83, n. 17.

Frey, D., Hentschel, F. D., and Keith, D. H. (1978) Deepwater archaeology: The Capistello wreck excavation, Lipari, Aeolian Islands. *Intl J. Naut. Archaeol. Underwater Explor.* 7, 279–300.

Garnier, N., Richardin, P., Cheynier, V., and Regert, M. (2003) Characterization of thermally assisted hydrolysis and methylation products of polyphenols from modern and archaeological vine derivatives using gas chromatography-mass spectrometry. *Anal. Chim. Acta* 493, 137–157.

Glänzel, W., and Veugelers, R. (2006) Science for wine: A bibliometric assessment of wine and grape research for wine-producing and consuming countries. *Am. J. Enol. Vitic.* 57, 23–32.

Grace, V. R. (1979) *Amphoras and the Ancient Wine Trade.* American School Classical Studies, Athens, Princeton, NJ.

Guasch-Jané, M. R., Ibern-Gómez, M., Andrés-Lacueva, C., Jáurequi, O., and Lamuela-Raventós, R. M. (2004) Liquid chromatography with mass spectrometry in tandem mode applied for the identification of wine markers in residues from Ancient Egyptian vessels. *Anal. Chem.* 76, 1672–1677.

Guasch-Jané, M. R., Andrés-Lacueva, C., Jáuregui, O., and Lamuela-Raventós, R. M. (2006) The origin of the ancient Egyptian drink *Shedeh* revealed using LC/MS/MS. *J. Archaeol. Sci.* 33, 98–101.

Henderson, A. (1824) *The History of Ancient and Modern Wines.* Baldwin, Craddock and Joy, London.

Jashemski, W. F. (1979, 1993) *The Gardens of Pompeii, Herculaneum and the Villas Destroyed by Vesuvius.* Vols I & II. Caratzas Brothers, New Rochelle, NY.

Johansen, D., Friis, K., Skovenborg, E., and Grønbæk, M. (2006) Food buying habits of people who buy wine or beer: Cross sectional study. *Br. Med. J.* 332, 519–522.

Jover, A. J. V., Montes, F. J. L., and Fuentes, M. M. F. (2004) Measuring perceptions of quality in food products: The case of red wine. *Food Qual. Pref.* 15, 453–469.

Kaufman, H. S. (1992) The red wine headache and prostaglandin synthetase inhibitors: A blind controlled study. *J. Wine Res.* 3, 43–46.

Kinsella, J. E., Frankel, E., German, J. B., and Kanner, J. (1993) Possible mechanisms for the protective role of antioxidants in wine and plant foods. *Food Technol.* 47, 85–89.

Klein, H., and Pittman, D. (1990a) Drinker prototypes in American society. *J. Substance Abuse* **2**, 299–316.

Klein, H., and Pittman, D. (1990b) Perceived consequences of alcohol use. *Intl J. Addictions* **25**, 471–493.

Koehler, C. G. (1986) Handling of Greek transport amphoras. In: Recherches sur les Amphores Greques (J.-Y. Empereur and Y. Garlan, eds.), *Bull. Correspondance Hellénique*, suppl. 13. École française d'Athènes. Paris, pp. 49–67.

Köster, E. P. (2003) The psychology of food choice: Some often encountered fallacies. *Food Qual. Pref.* **14**, 359–373.

Lawless, H. (1995) Dimensions of sensory quality: A critique. *Food Qual. Pref.* **6**, 191–200.

Lindman, R., and Lang, A. R. (1986) Anticipated effects of alcohol consumption as a function of beverage type: A cross-cultural replication. *Int. J. Psychol.* **21**, 671–678.

Littlewood, J. T., Glover, W., Davies, P. T. G., Gibb, C., Sandler, M., and Rose, F. C. (1988) Red wine as a cause for migraine. *Lancet* **1**, 558–559.

Littlewood, J. T., Glover, W., Sandler, M., Peatfield, R., and Rose, F. C. (1982) Platelet phenolsulphotransferase deficiency in dietary migraine. *Lancet* **1**, 983–986.

Lucia, S. P. (1963) *A History of Wine as Therapy*. Lippincott, Philadelphia.

Masquelier, J. (1988) Effets physiologiques du vin. Sa part dans l'alcoolisme. *Bull. O.I.V.* **61**, 555–577.

McGovern, P. E., Glusker, D. L., Exner, L. J., and Voigt, M. M. (1996) Neolithic resinated wine. *Nature* **381**, 480–481.

McGovern, P. E., Zhang, J., Tang, J., Zhang, Z., Hall, G. R., Moreau, R. A., Nuñez, A., Butrym, E. D., Richards, M. P., Wang, C-S., Cheng, G., Zhao, Z., and Wang, C. (2004) Fermented beverages of pre- and proto-historic China. *Proc. Natl Acad. Sci.* **101**, 17593–17598.

Mukamal, K. J., Chiuve, S. E., and Rimm, E. B. (2006) Alcohol consumption and risk for coronary heart disease in men with healthy lifestyles. *Arch. Intern. Med.* **166**, 2145–2150.

National Institute on Alcohol Abuse and Alcoholism (1992) Moderate Drinking. *Alcohol Alert* **16**, 315.

NOVA (1978) *The Great Wine Revolution*. BBC/WBGH (PBS) production.

Núñez, D. R., and Walker, M. J. (1989) A review of paleobotanical findings of early *Vitis* in the Mediterranean and of the origins of cultivated grape-vines, with special reference to prehistoric exploitation in the western Mediterranean. *Rev. Paleobot. Palynol.* **61**, 205–237.

OIV. (2005) The world viticultural situation in 2002. http://www.oiv.int/uk/statistiques/index.htm

Petrie, W. M. F. (1923) *Social Life in Ancient Egypt*. Methuen, London.

Pettigrew, S., and Charters, S. (2006) Consumers' expectations of food and alcohol pairing. *Br. Food J.* **108**, 169–180.

Phaff, H. J. (1986) Ecology of yeasts with actual and potential value in biotechnology. *Microb. Ecol.* **12**, 31–42.

Pretorius, I. S., Bartowsky, E. J., de Barros Lopes, M., Bauer, F. F., du Toit, M., van Rensburg, P., and Vivier, M. A. (2006) The tailoring of designer grapevines and microbial starter strains for a market-directed and quality-focussed wine industry. In: *Handbook of Food Science, Technology and Engineering*, Vol. 4 (H. Y. Hui and F. Sherkat, eds.), pp. 174-1–174-24. CRC Press, New York.

Ramishvili, R. (1983) New material on the history of viniculture in Georgia (in Georgian, with Russian summary) *Matsne (Hist. Archaeol. Ethnol. Art Hist. Ser.)* **2**, 125–140.

Renfrew, C. (1989) The origins of Indo-European languages. *Sci. Am.* **261** (4), 106–116.

Shackley, M. (1982) Gas chromatographic identification of a resinous deposit from a 6th century storage gar and its possible identification. *J. Archaeol. Sci.* **9**, 305–306.

Smart, R. G., and Walsh, G. (1999) Heavy drinking and problems among wine drinkers. *J. Stud. Alcohol* **60**, 467–471.

Stafford, H. A. (1959) Distribution of tartaric acid in the leaves of certain angiosperms. *Am. J. Bot.* **46**, 347–352.

Stanislawski, D. (1975) Dionysus westward: Early religion and the economic geography of wine. *Geogr. Rev.* **65**, 427–444.

Stanley, P. V. (1999) Gradation and quality of wine in the Greek and Roman worlds. *J. Wine Res.* **10**, 105–114.

Stevenson, A. C. (1985) Studies in the vegetational history of S.W. Spain. II. Palynological in investigations at Laguna de los Madres, Spain. *J. Biogeogr.* **12**, 293–314.

Tchernia, A. (1986) *Le Vin de l'Italie Romaine: Essai d'Histoire Economique d'après les Amphores*. École Française de Rome, Rome.

Vandiver, P., and Koehler, C. G. (1986) Structure, processing, properties, and style of Corinthian amphoras. In: *Ceramics and Civilization*, Vol. 2. *Technology and Style* (W. D. Kingery, ed.), pp. 173–215. American Ceramics Society, Columbus, OH.

White, K. D. (1970) *Roman Farming*. Thames and Hudson, London.

White, K. D. (1984) *Greek and Roman Technology*. Cornell University Press, Ithaca, NY.

Younger, W. (1966) *Gods, Men and Wine*. George Rainbird, London.

Zohary, D., and Hopf, M. (2000) *Domestication of Plants in the Old World*, 3rd edn. Oxford University Press (Clarendon), Oxford.

2

Grape Species and Varieties

Introduction

Grapevines are classified in the genus *Vitis*, family Vitaceae. Other well-known members of the family are the Boston Ivy (*Parthenocissus tricuspidata*) and Virginia Creeper (*P. quinquefolia*). Members of the Vitaceae typically show a climbing habit, have leaves that develop alternately on shoots (Fig. 2.1), and possess swollen or jointed nodes. These may generate tendrils or flower clusters opposite the leaves. The flowers are minute, uni- or bisexual, and occur in large clusters. Most flower parts appear in groups of fours and fives, with the stamens developing opposite the petals. The ovary consists of two carpels, partially enclosed by a receptacle that develops into a two-compartment berry. The fruit contains up to four seeds.

The Vitaceae is predominantly a tropical to subtropical family, containing possibly more than a thousand species, placed in some 15 to 16 genera (Galet, 1988). In contrast, *Vitis* is primarily a temperate-zone genus, occurring indigenously only in the Northern Hemisphere. Related

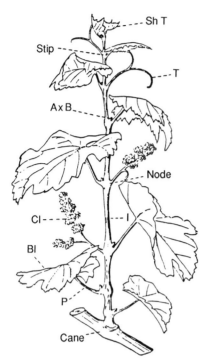

Figure 2.1 *Vitis vinifera* shoot, showing the arrangement of leaves, clusters (Cl), and tendrils (T); Ax B, axillary buds; Bl, blade; I, internode; P, petiole; Sh T, shoot tip; Stip, stipule. (After von Babo and Mach, 1923, from Pratt, 1988, reproduced by permission of the American Phytopathological Society)

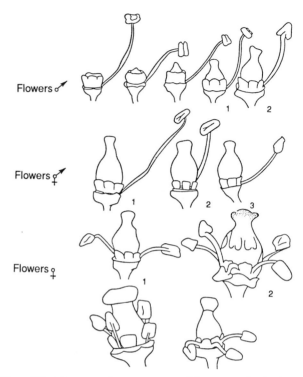

Figure 2.2 Diagrammatic representation of the variety of male, female, and bisexual flowers produced by *Vitis vinifera*. (After Levadoux, 1946, reproduced by permission)

genera are *Acareosperma, Ampelocissus, Ampelopsis, Cayratia, Cissus, Clematicissus, Cyphostemma, Nothocissus, Parthenocissus, Pterisanthes, Pterocissus, Rhoicissus, Tetrastigma,* and *Yua*.

The Genus *Vitis*

Grapevines are distinguished from related genera primarily by floral characteristics. The flowers are typically functionally unisexual, being either male (possessing erect functional anthers and lacking a fully developed pistil) or female (containing a functional pistil and either producing recurved stamens and sterile pollen, or lacking anthers) (Fig. 2.2). The fused petals, called a **calyptra** or cap, remain connected at the apex, while splitting along the base from the receptacle (see Plate 3.6). The petals are shed at maturity. Occasionally, though, the petals separate at the top, while remaining attached at the base (Longbottom *et al.*, 2004). These 'star' flowers possess an appearance more typical of angiosperms, a situation found in some members of the Vitaceae, such as *Cissus*. Swollen nectaries occur at the base of the ovary (see Fig. 3.19). They generate a mild fragrance that attracts pollinating insects. The sepals of the calyx form only as vestiges and degenerate early in flower development. The fruit is juicy and acidic.

The genus has typically been divided into two subgenera, *Vitis*[1] and *Muscadinia*. *Vitis* (bunch grapes) is the larger of the two subgenera, containing all species except *V. rotundifolia* and *V. popenoei*. The latter two species are placed in the subgenus *Muscadinia*. The two subgenera are sufficiently distinct to have induced some taxonomists to separate the muscadine grapes into their own genus, *Muscadinia*.

Members of the subgenus *Vitis* are characterized by having shredding bark without prominent lenticels, a pith interrupted at nodes by woody tissue (diaphragm), tangentially positioned phloem fibers, branched tendrils, elongated flower clusters, fruit that adheres to the fruit stalk at maturity, and pear-shaped seeds possessing a prominent beak and smooth chalaza. The chalaza is the pronounced, circular, depressed region on the dorsal (back) side of the seed (Fig. 2.3C). In contrast, species in the subgenus *Muscadinia* possess a tight, nonshredding bark, prominent lenticels, no diaphragm interrupting the pith at nodes, radially arranged phloem fibers, unbranched tendrils, small floral clusters, berries that separate individually from the cluster at maturity, and boat-shaped seeds with a wrinkled chalaza. Some of these characteristics are

---

[1] According to the International Code of Botanical Nomenclature, the prefix *eu-* is to be no longer applied to the main subgenus of *Vitis* (it was formerly designated *Euvitis*).

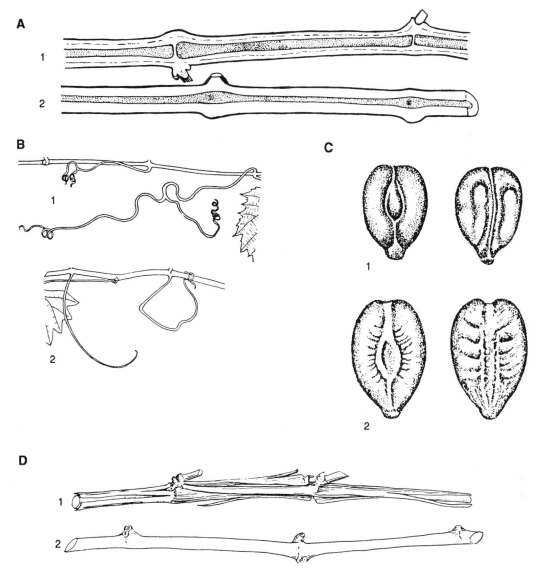

Figure 2.3 Properties of the *Vitis* (1) and *Muscadinia* (2) subgenera of *Vitis*. (**A**) Internal cane morphology; (**B**) tendrils; (**C**) front and back seed morphology; (**D**) bark shredding. (A, B, and D from Bailey, 1933; C from Rives, 1975, reproduced by permission)

diagrammatically represented in Fig. 2.3. Plate 2.1 illustrates the appearance of *Muscadinia* grapes and leaves.

The two subgenera also differ in chromosomal composition. *Vitis* species contain 38 chromosomes (2n = 6x = 38), whereas *Muscadinia* species possess 40 chromosomes (2n = 6x = 40). The symbol *n* refers to the number of chromosome pairs formed during meiosis, and *x* refers to the number of chromosome complements (genomes). Grapevines are thought to be ancestrally hexaploids (Patel and Olmo, 1955). Successful crosses can be made experimentally between species of the two subgenera, primarily when *V. rotundifolia* is used as the pollen source. When *V. vinifera* is used as the male plant, the pollen germinates well, but does not effectively penetrate the style of the *V. rotundifolia* flower (Lu and

Lamikanra, 1996). Although generally showing vigorous growth, the progeny frequently are infertile. This probably results from imprecise pairing of the unequal number of chromosomes (19 + 20), and imbalanced separation of the chromosomes during meiosis. The genetic instability so produced disrupts pollen growth and results in infertility. This may result from the synthesis of inhibitors, such as quercetin glycosides in the pistil (Okamoto *et al.*, 1995).

The evolution of the genus *Vitis* (and possibly other genera in the Vitaceae) is thought to have involved the crossing of diploid species, followed by a later crossing of their tetraploid offspring with one of several diploids. In each instance, accidental chromosome doubling in the hybrids could have imparted fertility. In the absence of chromosome doubling, offspring of inter-species crosses are usually infertile. This results from improper chromosome pairing and unbalanced separation during meiosis (the same problem noted in *Vitis* × *uscadinia* crosses).

In *Vitis*, the ancestral progenitors are hypothesized to have possessed six and seven chromosome pairs, respectively. Their crossing would have given rise to hybrids possessing 13 univalents (Fig. 2.4). Chromosome doubling could have regenerated fertile tetraploids (4x = 26). Subsequent crossing of the tetraploids with separate diploids could have produced progenitors of the two *Vitis* subgenera. Depending on whether the diploid had six or seven chromosome pairs, respectively, the subgenus *Vitis* (6x = 38) or the subgenus *Muscadinia* (6x = 40) could have arisen.

Chromosome numbers for other members of the Vitaceae are *Cyphostemma* (22), *Tetrastigma* (22, and occasionally 44), *Cissus* (24, and occasionally 22 or 26), *Cayratia* (32, 72, or 98), and *Ampelocissus* and *Ampelopsis* (40). Although no existing members of the Vitaceae are known to possess six or seven chromosome pairs, *Cissus vitiginea* possesses a chromosome number appropriate for a potential tetraploid ancestor (Shetty, 1959). Only about half of the genera and a small fraction of species in the Vitaceae have been investigated cytogenetically.

Evidence that polyploidy has been involved includes the presence of four nucleosome-related chromosomes in species possessing 24 and 26 chromosomes, and six nucleosome-related chromosomes in species possessing 38 or 40 chromosomes. Typically, diploid species possess two nucleosome-related chromosomes. In addition, observations of meiotic figures of *Vitis* × *Muscadinia* crosses show chromosome pairings with 13 bivalents (similar chromosomes) and 13 (7 + 6) dissimilar univalents (Patel and Olmo, 1954). These data are consistent with the evolutionary scheme provided in Fig. 2.4. Regrettably, confirmation of such a hypothesis is not possible by standard cytogenetic means, due to the very small size of the chromosomes and their morphological similarity.

Although the genus *Vitis* probably descended from hexaploid progenitors, it probably has undergone diploidization similar to that of cultivated wheats and other polyploid crops (Briggs and Walters, 1986; Wang *et al.*, 2005). This may explain why there are only two prominent nucleoli per cell (Haas *et al.*, 1994), in contrast to the six that would be expected. Diploidization can result from the inactivation (or structural modification) of excess duplicate genes, turning a polyploid into a functional diploid. An important consequence of diploidization is prevention, or regulation, of multivalent crossovers between the multiple sets of similar chromosomes. It is critical that chromosome separation occurs evenly during meiosis. Otherwise, unequal chromosome complements will occur in pollen and egg cells, leading to partial or complete sterility.

In contrast to the relative genetic isolation imposed by the differing chromosome complements of *Vitis* and *Muscadinia*, crossing between species of each subgenus is comparatively easy. Additionally, the progeny are often fertile and vigorous. The ease with which inter-species crossing occurs complicates the task of delineating species boundaries. Many of the criteria commonly used to differentiate species are not applicable to grapevines. Most *Vitis* species have similar chromosome numbers, are cross-fertile, often sympatric (overlap in geographic distribution) (Fig. 2.5), and show few distinctive morphological differences. The quantitative differences that do exist between species, such as shoot and leaf hairiness, are often strongly influenced by environmental conditions. Evolution into distinct species appears to be incomplete, and some local populations may be more appropriately viewed as ecospecies or ecotypes rather than biological species.

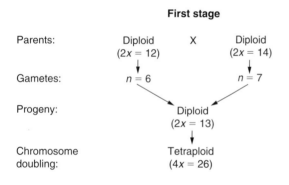

First stage

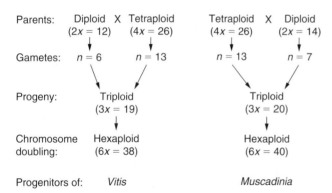

Second stage

Figure 2.4 Hypothesized evolution of the *Vitis* and *Muscadinia* subgenera of *Vitis*, involving sequential hybridization and chromosome doubling of the progeny. (Based on the work of Patel and Olmo, 1955)

The establishment of a definitive taxonomic classification of *Vitis* spp. may require a genetic analysis of the morphological features on which species delimitation is currently based. Nevertheless, the properties of some species are sufficiently distinct to be of use as sources of genetic variation in grapevine breeding. A recent classification of the eastern North American species of *Vitis* is given by Moore (1991).

Geographic Origin and Distribution of *Vitis* and *Vitis vinifera*

Where and when the genus *Vitis* evolved is unclear. The current distribution of *Vitis* species includes northern South America (the Andean highlands of Colombia and Venezuela), Central and North America, Asia, and Europe. In contrast, species in the subgenus *Muscadinia* are restricted to the southeastern United States and northeastern Mexico. The distribution of the North American species of *Vitis* is shown in Fig. 2.5.

In the nineteenth century, many extinct species of *Vitis* were proposed, based on fossil leaf impressions (Jongmans, 1939). These are no longer accepted as valid designations, due to the dubious nature of the evidence. Not only do several unrelated plants possess leaves of similar outline, but individual grapevines may show remarkable variation in leaf shape, lobbing, and dentation (Zapriagaeva, 1964). Of greater value is seed morphology, even though interspecies variation exists (Fig. 2.6). On the basis of seed morphology, two groups of fossilized grapes have been distinguished, namely those of the *Vitis ludwigii* and *V. teutonica* types. Seeds of the *V. ludwigii* type, resembling those of muscadine grapes, have been found in Europe from the Pliocene (2 to 10 million years B.P.). Those of the *V. teutonica* type, resembling those of bunch grapes, have been discovered as far back as the Eocene (40 to 55 million years B.P.). However, these identifications are based on comparatively few specimens, and thus any conclusions remain tenuous. In addition, related genera, such as *Ampelocissus* and *Tetrastigma* produce seed similar to those of *Vitis*. Although most grape fossils have been found in Europe, this may reflect more the distribution of appropriate sedimentary deposits (or paleobotanical interest) than *Vitis*.

Baranov (in Zukovskij, 1950) suggests that the ancestral forms of *Vitis* were bushy and inhabited sunny locations. As forests expanded during the more humid Eocene, the development of a climbing growth habit allowed *Vitis* to retain its preference for sunny conditions. This may have involved mutations modifying some floral clusters into tendrils, thus improving climbing ability. This hypothesis is not unreasonable since differentiation of bud tissue into flower clusters or tendrils is based simply on the balance of gibberellins and cytokinins (Srinivasan and Mullins, 1981; Martinez and Mantilla, 1993).

Regardless of the manner and geographic origin of *Vitis*, the genus established its present range by the end of the last major glacial period (~8000 B.C.). It is believed that periodic advances and retreats of the last glacial period markedly affected the evolution of *Vitis*, notably *V. vinifera*. The alignment of the major mountain ranges in the Americas, versus Eurasia, also appears to have had an important bearing on its evolution. In the Americas and eastern China, the mountain ranges run predominantly north–south, whereas in Europe and western Asia they run principally east–west. This would have permitted North American and eastern Chinese species to move south or north, relative to movement of the ice sheets. The southward movement of grapevines in Europe and western Asia would have been largely restricted by the east–west mountain ranges (Pyrenees, Alps, Caucasus, and Himalayas). This may explain the existence of only one *Vitis* sp. (*V. vinifera*) from the Atlantic coast of Europe to the western Himalayas, whereas China possesses about 30 plus species (Fengqin *et al.*, 1990) and North and Central America some 34 species (Rogers and Rogers, 1978).

Although glaciation and cold destroyed most of the favorable habitats in the Northern Hemisphere, major southward displacement was not the only option open for survival. In certain areas, favorable sites (refuges) permitted the continued existence of grapevines throughout glacial periods. In Europe, refuges occurred around the Mediterranean basin and south of the Black and Caspian Seas (Fig. 2.7). For example, grape seeds have been found associated with anthropogenic remains in caves in southern Greece (Renfrew, 1995) and southern France (Vaquer *et al.*, 1985) near the end of the last glacial advance. These refuges may have played a role in the evolution of the various varietal groups of *Vitis vinifera*.

Although periodically displaced during the various Quaternary interglacial periods (Fig. 2.8), *V. vinifera* was again inhabiting southern regions of France some 10,000 years ago (Planchais and Vergara, 1984). For the next several thousand years, the climate slowly improved to an isotherm about 2–3°C warmer than presently (Dorf, 1960). The preferred habitats of wild *V. vinifera* were in the mild humid forests south of the Caspian and Black Seas and adjacent Transcaucasia, along the fringes of the cooler mesic forests of the northern Mediterranean, and into the heartland of Europe along the Danube, Rhine, and Rhone rivers. The current situation of wild *Vitis vinifera* is discussed in Arnold *et al.* (1998).

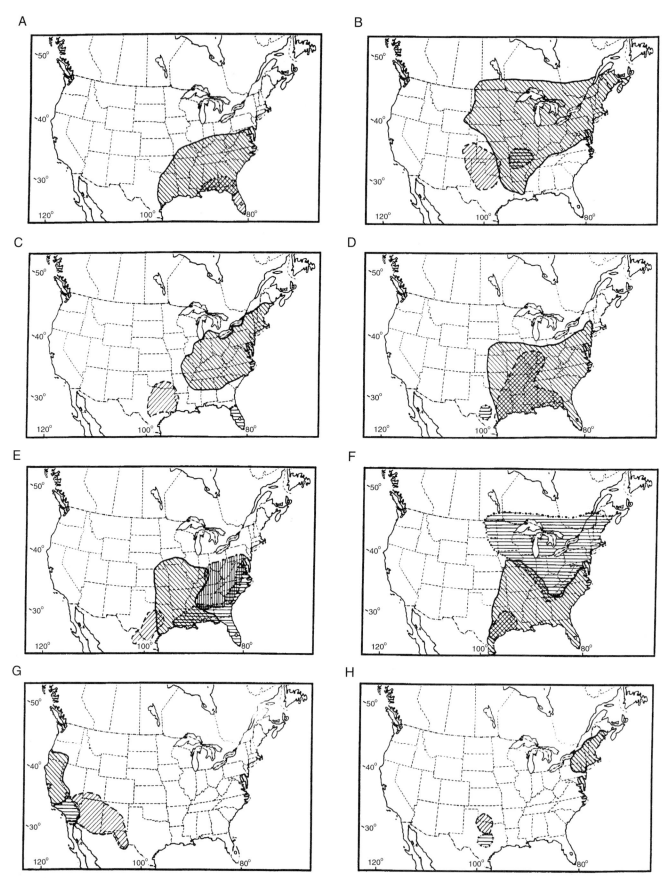

Figure 2.5 Geographic distribution of *Vitis* species in North America: subgenus *Muscadinia*: (**A**) *V. rotundifolia* var. *rotundifolia* (———), and *Vitis rotundifolia* var. *munsoniana* (· · ·); subgenus *Vitis*: (**B**) Series *Labruscae*, *V. labrusca* (———), *V. shuttleworthii* (· · ·), and *V. mustangensis* (- - -); (**C**) series *Ripariae*, *V. riparia* (———), *V. rupestris* (· · ·), and *V. acerifolia* (- - -); (**D**) series *Cordifoliae*, *V. vulpina* (———), *V. monticola* (· · ·), and *V. palmata* (- - -); (**E**) series *Cinerescentes*, *V. cinerea* var. *cinerea* (———), *V. cinerea* var. *floridana* (· · ·), *V. cinerea* var. *helleri* (- - -), and *V. cinerea* var. *baileyana* (- · -); (**F**) series *Aestivalis*, *V. aestivalis* var. *aestivalis* (———), *V. aestivalis* var. *bicolor* (· · ·), and *V. aestivalis* var. *lincecumii* (- - -); (**G**) series *Occidentales*, *V. californica* (———), *V. girdiana* (· · ·), and *arizonica* (- - -); and (**H**) hybrids, *V.* × *novae-angliae* (———), *V.* × *champinii* (· · ·), and *V.* × *doaniana* (- - -). (Data supplied by M. Moore)

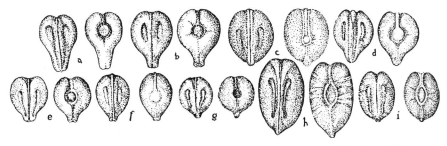

Figure 2.6 Lateral and dorsal view of grape seed (× 2.5): a, *Vitis vinifera*; b, *V. labrusca*; c, *V. vulpina*; d, *V. cinerea* var. *helleri*; e, *V. cinerea* var. *baileyana*; f, *V. illex*; g, *V. vulpina* var. *praecox*; h, *V. rotundifolia*; i, *V. rotundifolia* var. *munsoniana*. (From Bailey, 1933, reproduced by permission)

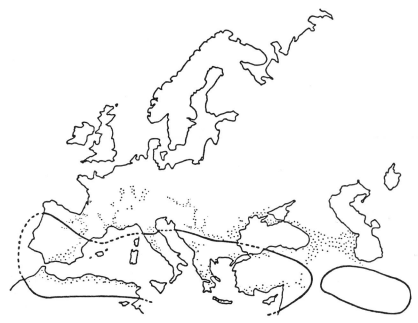

Figure 2.7 Distribution of wild *Vitis vinifera* vines about 1850 (dots) superimposed on forest refuges in the Mediterranean and Caucasian regions during the last ice age (line). (After Levadoux, 1956, reproduced by permission)

Domestication of *Vitis vinifera*

Grapevine cultivars show few of the standard signs of plant domestication (Baker, 1972; de Wet and Harlan, 1975). There views could be summarized as follows: cross- to self-fertilization, no need for seed and bud vernalization, phenologic plasticity (loss of regulation by the photoperiod), fruit or seed dehiscence upon maturation, parthenocissus (fruit production independent of seed development), increase in shoot to root ratio, increase in fruit (or seed) size, enhanced crop yield, reduction in phytotoxin production (if any), conversion to annual habit.

Of these, only conversion to self-fertility is characteristic of domesticated grapevines. Other domesticated attributes are less marked. For example, slight reduction in photoperiod sensitivity and need for vernalization; easier fruit dehiscence; increased fruit size (notably

table grapes); and seedlessness in some table and raisin cultivars. Other features that tend to differentiate wild and domesticated grapevines are a shift from small round berries to larger elongated fruit; bark separating in wider, core-coherent strips (vs. bark separating in long thin strips); larger elongated seeds (vs. small rounded seed); and large leaves with entire or with shallow sinuses (vs. small, usually deeply three lobed leaves) (Olmo, 1976; Fig. 2.9). Plate 2.2 illustrates the grape cluster of wild (*sylvestris*) grapevine.

The principal indicator of domestication in archeological finds has been the seed index – the ratio of seed width to length. Although of no known selective advantage, seed index appears to correlate with a shift from cross- to self-fertilization. Seeds from wild (*sylvestris*) vines are rounder, possess a nonprominent beak, and show an average seed index of averaging about 0.64 (ranging from 0.54–0.82). In contrast, seeds from

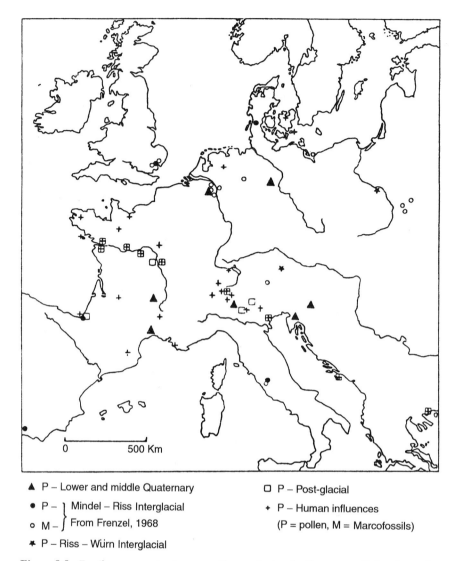

Legend:

▲ P – Lower and middle Quaternary   □ P – Post-glacial

● P – ⎫ Mindel – Riss Interglacial   + P – Human influences
○ M – ⎭ From Frenzel, 1968   (P = pollen, M = Marcofossils)

✦ P – Riss – Würn Interglacial

Figure 2.8 Fossil grape remains in eastern Europe during the Quaternary. Pollen data refer
to all but the Würm period (occurring only around the Mediterranean basin) and macrofossils
for the Mindel-Riss interglacial. (From Planchais, 1972–1973, reproduced with permission)

domesticated (*sativa*) vines are more elongated, possess a prominent beak and have a seed index averaging about 0.55 (often ranging from 0.44–0.75) (Renfrew, 1973; Fig. 2.10). However, evidence based on carbonated seed remains may be of uncertain value – charring appears to increase the relative length of grape seeds and, thus, decrease the seed index. In addition, considerable variation in seed size and shape (Zapriagaeva, 1964) makes conclusions based on small sample sizes unreliable.

Seed index data suggest a slow domestication of the grapevine in northeastern Iran and Anatolia (see Miller, 1991). However, clear evidence of domestication appears only in the late fourth millennium B.C. – Jericho about 3200 B.C. (Hopf, 1983) and northeastern Iran; and after 2800 B.C. in Macedonia and Greece (Renfrew, 1995). This is almost two millennia after the first archeological evidence of wine production (McGovern *et al.*, 1996).

However, the latter is consistent with the predicted spread of agriculture out of the Fertile Crescent into northern Iran and Anatolia (Ammerman and Cavalli-Sporza, 1971). Thus, the production of wine may have developed concurrently with agriculture and may have predated preserved morphological evidence of grapevine domestication (Zohary, 1995).

In Europe, evidence of wild grapevine use has been found in a Neolithic village near Paris (about 4000 B.C.) (Dietsch, 1996). Semidomesticated grape seed remains (2700 B.C.) have been discovered in England (Jones and Legge, 1987), and in several Neolithic and Bronze Age lake dwellings in northern Italy, Switzerland, and the former Yugoslavia (see Renfrew, 1973). In addition, grape seed and pollen remains have been identified from Neolithic sites in southern Sweden and Denmark (see Rausing, 1990). Evidence consistent with grape culture

Cultivated   Wild

L

F

B

S

0
10 cm

0
5 mm

0
10 cm

0
5 nm

Figure 2.9 Morphological differences between the leaves, flowers, fruit clusters and seeds of cultivated (*sativa*) and wild (*sylvestris*) grapevines are illustrated. (From This *et al.*, 2006, reproduced by permission)

in southwestern Spain has been found dating from about 2500 B.C. (Stevenson, 1985). The presence of grapevine pollen has also been used as an indicator of ancient viticultural sites (Turner and Brown, 2004).

Differences in floral sexual expression and leaf shape also occur between wild and domesticated grapevines (Levadoux, 1956). For example, male vines have larger inflorescences, flower earlier and for a longer period, whereas bisexual flowers show poorly developed nectaries. These features are, however, unlikely to be preserved or recognizable in vine remains. Additional differences have been noted (Olmo, 1995), but most of these are properties readily influenced by cultivation. Thus, until studies have been conducted under equivalent situations, acceptance of the validity of such differences must be held in reserve.

Another archeologically useful property involves changes in the fruit pedicel. In domesticated *Vitis vinifera*, the pedicel frequently breaks off the main stem (rachis) when the berries are harvested. In contrast, the stems are stout in wild vines and the pedicel seldom breaks easily

from the peduncle. Thus, the relative frequency of seeds attached to the pedicel is an indicator of domestication in archeological remains.

Of all the fossil evidence, the greatest confidence may be placed in pollen data. Pollen found in sedimentary deposits is often sufficiently distinctive and well preserved to establish the presence of a species or subspecies at a prehistoric site (Fig. 2.8). In *Vitis vinifera*, differences exist between fertile pollen (produced by male and bisexual flowers) and sterile pollen (produced by some female flowers). Fertile pollen is tricolporate (containing three distinct ridges) and produces germ pores, whereas sterile pollen generally is acolporate (possessing no ridges) and produces no germ pores (Fig. 2.11). Thus far, differences in the fertile/infertile pollen ratio have not been used in indicating the relative frequency of dioecious (wild) and bisexual (cultivated) vines at prehistoric sites. This may result from the poorer release of pollen from bisexual flowers and, corresponding, limited collection in sedimentary deposits. Nevertheless, pollen differences have been used in distinguishing the presence, or absence of domesticated vines near ancient lake and river sediments (Planchais, 1972–1973).

It is generally believed that domestication of *Vitis vinifera* occurred in or around Transcaucasia, or neighboring Anatolia (~4000 B.C.). Because the climate then was similar to that of today, the distribution of *Vitis vinifera* was probably comparable to that in the mid-1850s (Fig. 2.7), before the decimation caused by phylloxera. In the region between and below the Black and Caspian Seas, the indigenous distribution of *Vitis vinifera* most closely approaches the Near Eastern origin(s) of Western agriculture. Therefore, it is likely that, when agricultural technologies spread into the region, domestication was the eventual result. However, direct archeological support for this hypothesis is lacking. In contrast, Núñez and Walker (1989) present evidence that strongly suggests domestication of the grapevine in southern Spain, long before Phoenician and Phocaean colonization. The view of at least two independent sites of grapevine domestication has received support from molecular evidence on cultivars around the Mediterranean (Arroyo-García *et al.*, 2006).

That cultivars were carried westward, associated with colonization, seem probable. The importation of grape cultivars into Italy by ancient Greek colonists, and into France, Spain, and Germany by Roman settlers is suggested from ancient writings. What is more certain is the implantation of a grape-growing culture. However, that the vine was transported into areas such as Israel, Egypt and ancient Babylonia is indisputable. The vine is indigenous in neither of these regions.

Despite human migration and its consequences, changes in seed morphology and pollen shape indicate that local

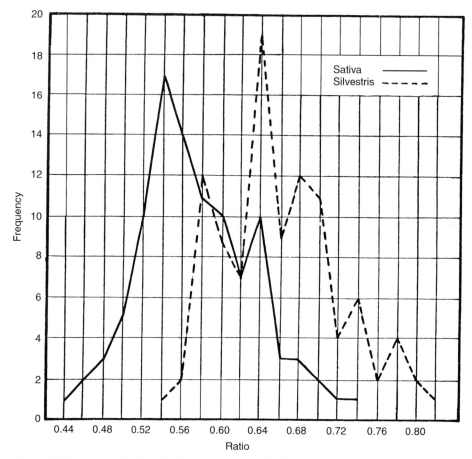

Figure 2.10 Variation in the width/length index of seeds of wild (*sylvestris*) and cultivated (*sativa*) subspecies of *Vitis vinifera*. (From Stummer, 1911, reproduced by permission)

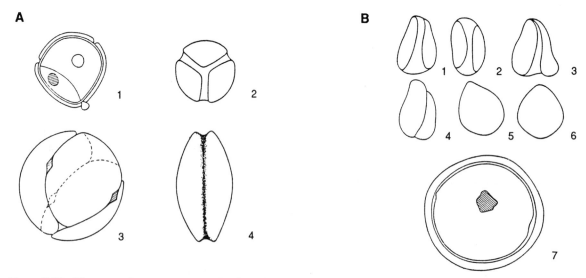

Figure 2.11 Diagrammatic representations of pollen of *Vitis vinifera*: (**A**) fertile pollen: 1, swollen grain; 2, dry grain, top view; 3, swollen grain in water; 4, dry grain, side view; (**B**) infertile pollen: 1–6, dry grains; 7, swollen grain in water. (After Levadoux, 1946, reproduced by permission)

grape domestication was in progress long before the agricultural revolution reached Europe. Traits such as self-fertilization and large fruit size often have been attributed to varieties brought in by human activity. Although fruit size often increases under cultivation, there is evidence that genetic control is also involved. Fernandez *et al.* (2006) suggest that mutation of the *flb* gene (or possibly other similar genes) may have been involved in the increase in berry size during domestication. In addition, bisexuality occurs, albeit rarely, in *Vitis* species unexposed to human selective influences (Fengqin *et al.*, 1990). Thus, the occurrence of the bisexual habit in indigenous wild (*sylvestris*) populations (Anzani *et al.*, 1990) need not presuppose crossing between wild and domesticated vines. Locally derived cultivars would have had the advantage of being adapted to prevailing soil and climatic conditions. In addition, most data suggest that many cultivars fall into ecogeographic groups (Bourquin *et al.*, 1993), consistent with local origin. The presence of numerous traits (such as rounder seeds), reminiscent of *V. vinifera* f. *sylvestris*, in cultivars such as 'Traminer,' 'Pinot noir,' and 'Riesling,' has been taken as evidence of indigenous origin. This, of course, does not preclude possible crossing and introgression of traits from imported domesticated varieties into local vines. Data on a genetic comparison of long-established cultivars with indigenous *sylvestris* vines in Italy are, however, not compatible with their being the source of most Italian (*sativa*) cultivars (Scienza *et al.*, 1994). Contradictory data from a range of studies is most compatible with a combination of cultivar importation, limited introgression with local strains, and *de novo* domestication of cultivars derived from indigenous wild strains.

Until devastated by foreign pests and disease in the middle of the nineteenth century, wild grapevines grew extensively from Spain to Turkmenistan. Although markedly diminished at present, wild vines still occur in significant numbers in some portions of its original range, notably in the Caucasus.

Domestication of the grapevine probably occurred progressively, with the slow accumulation of agronomically valuable mutations. The early association of wine with religious rites in the Near East (Stanislawski, 1975) could have provided the incentive for selection, and the beginnings of concerted viticulture. For example, grapevine cultivation would have initially required the planting of both fruit-bearing (female) and non-fruit-bearing (male) vines. Propagating only fruit-bearing cuttings would have resulted in a reduction or cessation in productivity, by isolating female vines from pollen-bearing vines (growing wild in open woodlands). This would have highlighted the productivity of self-fertile vines and presumable led to their selection. Viticulture also would have favored the selection of vines showing increased fruit size, as well as improved visual, taste, and aroma characteristics.

Support for the connection between winemaking and grapevine domestication comes from the remarkable similarity between the words for *wine* and *vine* in Indo-European languages (Table 2.1). In contrast, little resemblance exists between words for *grape*. The persistence of local terms for grape in European languages suggests that grapes were used (and verbally recognized) long before the introduction of viticulture and winemaking, and the dispersion of Indo-European languages into Europe (Renfrew, 1989). This is consistent with the demic diffusion of Neolithic farmers into Europe (Chicki *et al.*, 2002). In contrast, the adjacent Semitic languages, which evolved outside the indigenous habitat of grapevines, possess terms for grape and vine that are similar to that for wine (Forbes, 1965). These terms are also related to what may be the ancestral term for wine, *woi-no* (Gamkrelidze and Ivanov, 1990). This suggests that wine preceded or coincided with the appearance of the grapevine in Semitic cultures.

One view of the diffusion routes of viticulture into Europe is given in Fig. 2.12. The beginnings of agriculture are thought to have spread to, or developed in, the southern Caucasus (Anatolian) region of Turkey between 6000 and 5000 B.C. (Renfrew, 1989). Another line of evidence suggesting the antiquity of grape culture in the Caucasus comes from the advanced state of cultivar evolution in the region. Local varieties possess many recessive mutants, such as smooth leaves, large branched grape clusters, and medium-size juicy fruit. The number

Table 2.1 Comparison of synonyms for "wine," "vine," and "grape" in the principal Indo-European languages

Language	Wine	Vine	Grape
Greek	οἶνος	ἄμπελος	βότρυς
Latin	*vinum*	*vitis*	*uva*
Italian	*vino*	*vite, vigne (pl.)*	*uva*
French	*vin*	*vigne*	*raisin*
Spanish	*vino*	*vid*	*uva*
Rumanian	*vin*	*vita*	*strugure*
Irish	*fin*	*finemain*	*fin*
Breton	*gwin*	*gwinienn*	*rezinenn*
Gothic	*wein*	*weinatriu*	*weinabasi*
Danish	*vin*	*vinranke(-stok)*	*drue*
Swedish	*vin*	*vinranka(-stock)*	*druva*
English	wine	vine	grape
Dutch	*wijn*	*wijnstok*	*druif*
German	*Wein*	*Weinstock Rebe*	*Traube, Weinbeere*
Lithuanian	*vynas*	*vynmedis*	*kekė, vynuogė*
Lettic	*vins*	*vina kuoks*	*k'eke', vinuogé*
Lettic	*vīns*	*vīna kuoks*	*grozd*
Czech	*víno*	*réva, vinný keř*	*hrozen*
Polish	*wino*	*winorośl*	*winogrono*
Russian	*vino*	*vinograd*	*vinograd*
Sanskrit	*drākṛṣarasa*	*drākṣā-*	*drākṣā-*

Source: Information from Buck, 1949

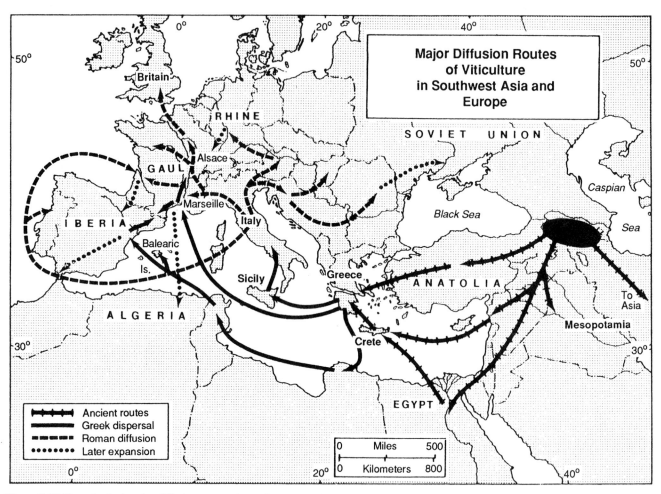

Figure 2.12 Hypothesized major diffusion routes of viticulture in southwest Asia and Europe. (From de Blij, 1983, reproduced by permission)

of accumulated recessive traits is often taken as an indicator of cultivar age. Varieties showing these traits were classified by Negrul (1938) in the cultivar grouping *proles orientalis*. Also included in this division are the genetically distinct *vinifera* cultivars of China and Japan (Goto-Yamamoto *et al.*, 2006). In contrast, cultivars found along the northern Mediterranean and in central Europe were considered to be of relatively recent origin (possessing few recessive traits) and closely resemble wild vines. They were placed in the *proles occidentalis* grouping. Varieties found in the regions between the former two (Georgia and the Balkans) show properties intermediate between *proles orientalis* and *proles occidentalis*. These Negrul called *proles pontica*. Characteristics of the three groups are given in Table 2.2.

Additional studies along these lines were conducted by Levadoux (1956). Extensions of Negrul's classification system can be found in Tsertsvadze (1986) and Gramotenko and Troshin (1988). The division of cultivars, and the probability of distinct centers of origin, have received modern support (Aradhya *et al.*, 2003; Arroyo-Garcia *et al.*, 2006). In addition, table grapes

(more common in the Near East) show greater genetic divergence from wild grapes than do wine grapes. Table grapes have also diverged significantly from wine grapes.

Cultivar Origins

Except for cultivars of recent origin, for which there is ample record, the origin of most cultivars is shrouded in mystery. Archeological finds are insufficiently detailed, while ancient writing do not discuss the issue. The written record of intentional breeding begins in the late 1600s, with the development of hybrid hyacinths in Holland. However, selection of improved strains (often due to accidental if not intentional crossing) clearly goes back to the origins of agriculture.

For annual crops, seed collection for the next year probably functioned as the principal agent by which cultivars developed. In contrast, vegetative propagation has acted as the principal vehicle for selecting perennial crops, such as grapevines. In both instances, the

Table 2.2 Classification of varieties of *Vitis vinifera* according to Negrul (1938)

Proles orientalis	Proles pontica	Proles occidentalis
Regions		
Central Asia, Afghanistan, Iran, Armenia, Azerbaijan	Georgia, Asia Minor, Greece, Bulgaria, Hungary, Romania	France, Germany, Spain, Portugal
Vine properties		
Buds glabrous, shiny	Buds velvety, ash-gray to white	Buds weakly velvety
Lower leaf surfaces glabrous to setaceous pubescent	Lower leaf surface with mixed pubescence (webbed and setaceous)	Lower leaf surfaces with webbed pubescence
Leaf edges recurved toward the tip	Leaf edges variously recurved	Leaf edges recurved toward the base
Grape clusters large, loose, often branching	Grape clusters medium size, compact, rarely loose	Grape clusters generally very large, compact
Fruit generally oval, ovoid, or elongated, medium to large, pulpy	Fruit typically round, medium to small, juicy	Fruit often round, more rarely oval, small to medium, juicy
Varieties mostly white with about 30% rosés	About equal numbers of white, rosé, and red varieties	Varieties commonly white or red
Seeds medium to large with an elongated beak	Seeds small, medium, or large (table grapes)	Seeds small with a marked beak
Fruiting properties		
Many varieties partially seedless, some seedless	Many varieties partially seedless, some completely so	Seedless varieties rare
Varieties produce few, low-yielding fruiting shoots	Varieties often produce several, highly productive fruiting shoots	Varieties typically produce several, highly productive fruiting shoots
Varieties short-day plants with long growing periods, not cold-hardy	Varieties relatively cold-hardy	Varieties long-day plants with short growing periods, cold-hardy
Most varieties are table grapes, few possess good winemaking properties	Many varieties are good winemaking cultivars, a few are table grapes	Most varieties possessing good winemaking properties
Grapes low in acidity (0.3–0.6%), sugar content commonly 18–20%	Grapes acidic (0.6–1.0%), sugar content commonly 18–20%	Grapes acidic (0.6–1%), sugar content commonly 18–20%
Self-crossed seedling of certain varieties possessing simple leaves	Self-crossed seedings of certain varieties with dwarfed shoots and rounded form	Self-crossed seedlings of certain varieties having mottled colored leaves

Source: After Levadoux, 1956, reproduced by permission

cultivar evolution undoubtedly took many generations and occurred surreptitiously. For grapevines, it was both simpler and quicker to propagate grapevines by layering (and subsequently by cuttings) than by seed. Had anyone planted the seeds, the new vines would have had properties different from those of the obvious (female) parent. In contrast, astute growers would have quickly realized that layering maintained and multiplied vines possessing the parent's traits. This is likely the origin of the ancient cultivars noted by Roman authors such a Columella and Pliny.

During Roman times, it is clear that the benefits of breeding were well known, at least for animals. However, if evidence from existing European grape cultivars is correct (see below), crossing played a major role in their development. Crossing could have occurred accidentally between cultivated vines in a vineyard, or by introgression with indigenous feral vines. While possible, the fortuitous germination and growth of seeding to maturity in established vineyards seem unlikely. For such vines to have been selected and propagated, they must also have been clearly different from the surrounding vines. More probable is intentional breeding

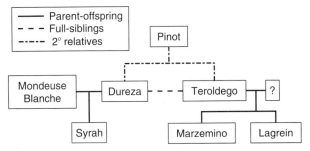

Figure 2.13 Most likely pedigree reconstruction from an analysis of 60 microsatellite markers of cultivars from northern Italy and southern France. (Reprinted by permission from Macmillan Publishers Ltd: *Heredity* 97, 102–110. Vouillamoz and Grando, 2006)

and selection. Regrettably, no records exist of such activity. The most likely sites for such developments are abbeys associated with vineyards. The monks not only could read ancient texts, but also had the time and writing skills to collect data on the performance of superior vines. That deliberate breeding was involved seems supported by molecular evidence that permits the construction of pedigree lines, linking the parentage for several Italian and French cultivars (Fig. 2.13).

Many European cultivars are considered ancient, but evidence is either absent or circumstantial. Many named cultivars were known to the Romans, but during the Middle Ages, varietal designation was largely abandoned. Wine was principally differentiated only into *vinum hunicum* (poor quality) and *vinum francicum* (high quality). These terms subsequently were used for groups of cultivars ('Heunisch' and 'Frankisch') (Schumann, 1997). Precise naming of cultivars reappeared in the fourteenth and fifteenth centuries, examples being 'Traminer' (1349), 'Ruländer' (1375), and 'Riesling' (1435) (Ambrosi *et al.*, 1994).

Name derivation has occasionally been used to suggest local origin, for example 'Sémillon' from *semis* (seed), and 'Sauvignon' from *sauvage* (wild) (Levadoux, 1956). However, the existence of multiple and unrelated names for many European cultivars does not lend credence to name derivation as an important line of argument. However, DNA fingerprinting provides data supporting the local origin of most well-known European cultivars.

Up until the development of DNA techniques, the best evidence for varietal origin came from morphological (ampelographic) comparisons. Such data were particularly useful when cultivars diverged from a common ancestor, via somatic mutation. Examples are the color mutants of 'Pinot noir' – 'Pinot gris,' and 'Pinot blanc.' Vegetative propagation maintains the traits of the progenitor, except where modified by subsequent somatic mutation. In contrast, seed propagation produces progeny with characteristics different from their parents (Bronner and Oliveira, 1990). This tends to blur differences in morphologic traits and makes such attributes unreliable as indicators of origin.

Researchers also used phenotypic chemical indicators to investigate varietal origin and classification. Isozyme and phenolic distributions are examples. However, the growing number of DNA fingerprinting techniques, such as amplified fragment-polymorphism (AFLP) and microsatellite allele (simple sequence repeat) (SSR) analysis are far more powerful and universally applicable. They circumvent some of the difficulties associated with the interpretation of restriction fragment-length polymorphism (RFLP) patterns and standardization with the random amplified polymorphic DNA (RAPD) procedure. Microsatellite markers also have the advantage of high polymorphism. In addition, they are inherited co-dominantly (valuable in parentage studies) and occur as relatively short repeats of one to five base pairs.

Different studies have provided support for views that European cultivars evolved from indigenous vines, expressing little evidence of introgression with introduced cultivars (Sefc *et al.*, 2003) and that cultivars show little genetic similarity with sympatric wild vines, and thus were probably introduced (Grassi *et al.*, 2003; Snoussi *et al.*,

2004). In addition, DNA investigations indicate that, for some 164 European cultivars, genetic divergence among their gene pools roughly correlates to their geographic distance apart (Fig. 2.14). Data from chloroplast DNA (inherited maternally) also supports the view that many cultivars evolved locally (Arroyo-García *et al.*, 2002).

Initially, sequencing techniques were thought to be independent of the tissues used – in contrast to isozymes and pigments that are produced only in certain tissues or at particular developmental stages. However, tissue-specific DNA banding is now known (Donini, 1997). This is not surprising due to the selective amplification or mutation of genes during tissue development. Thus, some of the divergence in data interpretation may originate with the tissues used in DNA collection.

One of the major advantages of DNA fingerprinting (see Caetano-Anolles and Gresshoff, 1997) is the number of differences that can be resolved. For example, AFLP can resolve potentially more than 100,000 DNA markers. Microsatellite allele analysis (SSR) also provides a direct means of assessing genetic relationships among cultivars (Thomas *et al.*, 1994; Sefc *et al.*, 1998). With the aid of DNA fingerprinting, it has been possible to establish the identity of 'Zinfandel' (California) and 'Primativo' (Italy) with 'Crljenak kastelanski' (Croatia) (Maletić *et al.*, 2004) and the identity of most 'Petit Sirah' in California as 'Durif' (Meredith *et al.*, 1999). In addition, DNA (plus additional) techniques have indicated that 'Cabernet Sauvignon' is the likely result of a crossing between 'Cabernet Franc' and 'Sauvignon blanc' (Bowers and Meredith, 1997). 'Chardonnay,' 'Gamay noir,' Aligoté,' and 'Melon' appear to be the progeny of separate crossings between 'Pinot noir' and

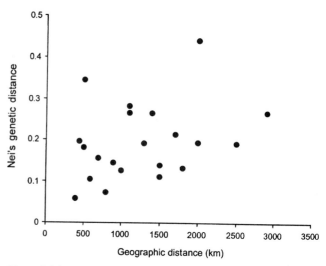

Figure 2.14 Scatter plot of the geographic versus the genetic distances between grapevine gene pools of 164 European cultivars. This suggests the relative minor role played by introduced cultivars from Greek or Phoenician sources. (From Sefc *et al.*, 2003, reproduced by permission)

'Gouais blanc' (syn. 'Heunisch weiss') (Bowers *et al.*, 1999); 'Pinot noir' is likely derived from a 'Traminer' × 'Schwarzriesling' (syn. 'Müllerrebe') crossing (Regner *et al.*, 2000); 'Silvaner,' a cross between 'Traminer' and 'Österreichisch weiß' (Sefc *et al.*, 1998); and 'Syrah,' a progeny of two southern French cultivars, 'Dureza' and 'Mondeuse blanche' (Bowers *et al.*, 2000).

Although it may be impossible to establish parentage, as the actual parents may no longer exist, it may suggest potential lineage sources. Recombination and mutation can complicate simple interpretation of DNA banding patterns. However, DNA fingerprinting can clearly indicate which parentages are not possible. For example, DNA techniques have shown that 'Müller-Thurgau' is neither the result of a self-crossing of 'Riesling,' nor the crossing of 'Riesling' and 'Silvaner' (Büscher *et al.*, 1994). Current evidence indicates that 'Müller-Thurgau' is the progeny of a crossing between 'Riesling' and 'Madeleine Royale' (Dettweiler *et al.*, 2000).

DNA fingerprinting has also been applied to DNA isolated from seed remains found at archeological sites (Manen *et al.*, 2003). Amplification has generated useful microsatellite markers from both waterlogged and charred seeds, dating back to the fifth century B.C. Although the number of markers is few, the data are not inconsistent with their involvement in the origin of currently grown cultivars. Such data are much more informative than seed morphology, which may suggest, but does not permit, the unambiguous differentiation between domesticated and wild vines, let alone cultivar groupings.

DNA fingerprinting has also proven particularly useful in establishing synonymies among cultivars. For this, it is the technique *par excellence*. DNA techniques have also been used to understand the interconnection between many of multiple variants of 'Muscat' cultivars (Fig. 2.15).

One of the potential weaknesses of current DNA fingerprinting techniques involves the use of markers of unknown genetic significance. For example, all the DNA is sampled for markers, whereas only about 4% of the DNA codes for functional genes in grapevines (Lodhi and Reisch, 1995). The additional DNA consists primarily of variously repeated segments of unknown function. Thus, relationships based on DNA banding patterns are more than likely based on differences that have no significance to vine evolution. Ideally, evolutionary relationships should be based on DNA sequences that have been of selective value, rather than properties that have probably diverged randomly.

Until little more than a century ago, deliberate grapevine breeding was limited (or unrecorded). Some of the earliest examples of suspected crosses occurred between indigenous *Vitis* spp. and varieties of *Vitis vinifera* planted by colonists in New England. 'Concord' and 'Ives' are thought to be chance crossings between *V. labrusca* and *V. vinifera*. Another likely product of chance crossing is 'Delaware' – this time among three species (*V. vinifera*, *V. labrusca*, and *V. aestivalis*). In contrast, 'Dutchess' is believed to be derived from the intentional crossing of *V. labrusca* with *V. vinifera*. Other *V. labrusca* cultivars, such as 'Catawba' and 'Isabella' are thought to be straight selections from local *V. labrusca* strains. Aside from *V. labrusca*, the involvement of native North American species in the development of indigenous North American cultivars is limited. Exceptions include 'Noah' and 'Clinton' (*V. labrusca* × *V. riparia*), 'Herbemont' and 'Lenoir' (*V. aestivalis* × *V. cinerea* × *V. vinifera*), and 'Cynthiana' and 'Norton' (possibly *V. vinifera* × *V. aestivalis*). These early cultivars are termed American hybrids, to distinguish them from hybrids developed in France between *V. vinifera* and one or more of *V. rupestris*, *V. riparia* and *V. aestivalis* var. *lincecumii*. The latter are variously termed French-American hybrids, French hybrids, or direct producers. The last designation comes from their tendency to grow ungrafted in phylloxera-infested soils. Some French-American hybrids are of complex parentage, based on subsequent backcrossing to one or more *V. vinifera* cultivars.

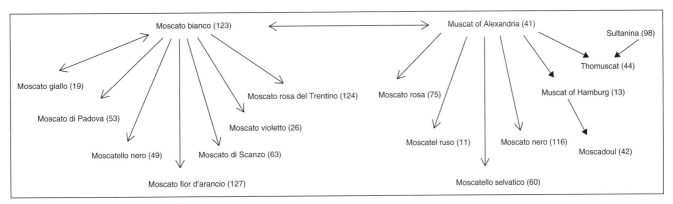

Figure 2.15 Putative genetic relationships (parent–offspring) among 16 of the 20 identified Muscat cultivars. Double arrows indicate uncertain direction; single arrows indicate probably direction of origin; thick arrows indicate certain origin. (From Crespan and Milani, 2001, reproduced by permission)

Backcrossing was used to enhance the expression of *vinifera*-like winemaking qualities.

The original aim of most French breeders was to develop cultivars containing the wine-producing attributes of *V. vinifera* and the phylloxera resistance of American species. It was hoped that this would avoid the expense and problems associated with grafting existing cultivars onto American rootstocks. Grafting was the only effective control against the devastation caused by phylloxera (*Daktulosphaira vitifoliae*). It also was hoped that breeding would incorporate resistance to pathogens, such as powdery and downy mildew. More by accident than design, several of the hybrids proved more productive than their European counterparts. They became so popular that by 1955 about one-third of French vineyards were planted with French-American hybrids. This success, and their different aromatic character, came to be viewed as a threat to the established reputation of viticultural regions, such as Bordeaux and Burgundy. Their actions resulted in laws designed to restrict the use of French-American hybrids and prohibiting new plantings. Subsequently, similar legislation has been passed in most European countries.

Although largely rejected in the land of their origin, French-American hybrids have found broad acceptance in much of northeastern North America. In addition to growing well in many areas of the United States and Canada, French-American hybrids have helped the production of regionally distinctive wines. French-American hybrids are also extensively grown in several other parts of the world, notably Brazil and Japan.

In the coastal plains of the southeastern United States, commercial viticulture is based primarily on selections of *Vitis rotundifolia* var. *rotundifolia*. 'Scuppernong' is the most widely known cultivar, but has the disadvantage of being unisexual. Newer cultivars such as 'Noble,' 'Magnolia,' and 'Carlos' are bisexual, thus avoiding the necessity of interplanting male vines to achieve adequate fruit set. Another complication is the tendency of berries to separate (shatter) from the cluster as they ripen. Newer cultivars such as 'Fry' and 'Pride' show less tendency to shatter. In addition, ethepon (2-chloroethyl phosphonic acid) has been applied to induce more uniform fruit ripening. The resistance of muscadine grapes to most indigenous diseases and pests in the southern United States has permitted a local wine industry to develop where the commercial cultivation of *V. vinifera* cultivars is difficult to impossible. Although Pierce's disease has limited the cultivation of most nonmuscadine cultivars in the region, the varieties 'Herbemont,' 'Lenoir,' and 'Conquistador' are exceptions. The first two are thought to be natural *V. aestivalis* × *V. cinerea* × *V. vinifera* hybrids, whereas 'Conquistador' is a complex cross involving several local *Vitis* spp. and *V. vinifera*.

Although grafting prevented the demise of grape growing and winemaking in Europe, early rootstock cultivars created their own problems. Most of the initial rootstock varieties were direct selections from *V. riparia* or *V. rupestris*. As most were poorly adapted to the high-calcium soils frequently found in Europe, the incidence of lime-induced chlorosis in the shoot (scion) increased markedly. Nevertheless, both species rooted easily, grafted well, and were relatively phylloxera-tolerant. *Vitis riparia* rootstocks also provided some cold hardiness, resistance to *coulure* (unusually poor fruit set), restricted vigor on deep rich soils, and favored early fruit maturity. In contrast, *V. rupestris* selections showed acceptable resistance to lime-induced chlorosis and tended to root deeply. Although the latter provided some drought tolerance on deep soils, these rootstocks were unsuitable on shallow soils.

The problems associated with early rootstock selections, such as 'Gloire de Montpellier' and 'St. George,' led to the development of new hybrid rootstocks, incorporating properties from several *Vitis* species (Fig. 2.16).

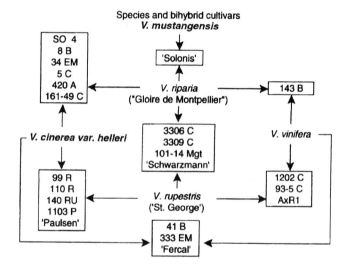

Figure 2.16 Origin of some important rootstock varieties: ( ) = selection from a single species; ☐ = interspecies crossings. The Champin varieties 'Dog Ridge' and 'Salt Creek' (Ramsey) are considered to be either natural *V. mustangensis* × *V. rupestris* hybrids, or strains of *V.* × *champinii*. 'Solonis' is considered by some authorities to be a selection of *V. acerifolia* (*V. longii*), rather than a *V. riparia* × *V. mustangensis* hybrid. On this basis, some multiple-species hybrids noted above are bihybrids. (Modified from Pongrácz, 1983, reproduced by permission)

For example, *V. cinerea* var. *helleri* (*V. berlandieri*) can donate resistance to lime-induced chlorosis; *V. × champinii* can supply tolerance to root-knot nematodes; and *V. vulpina* (*V. cordifolia*) can provide drought tolerance under shallow soil conditions. Alone, however, they have major drawbacks as rootstocks. Both *V. cinerea* var. *helleri* (*V. berlandieri*) and *V. vulpina* root with difficulty; *V. mustangensis* (*V. candicans*) has only moderate resistance to phylloxera; and both *V. mustangensis* and *V. vulpina* are susceptible to lime-induced chlorosis. Important cultural properties of some of the more widely planted rootstocks are given in Tables 4.6 and 4.7.

Grapevine Improvement

Standard Breeding Techniques

The focus of grape breeding has changed little since the work of early breeders such as F. Baco, A. Seibel, and B. Seyve (Neagu, 1968). The goals involve improving the agronomic properties of rootstocks and enhancing the winemaking properties of fruit-bearing (scion) cultivars. The major changes involve the availability of analytical techniques and a better understanding of genetics. Such knowledge can dramatically increase breeding effectiveness.

Of breeding programs, those developing new rootstocks are potentially the simplest. Improvement often involves the addition of properties, such as lime or drought tolerance, to established rootstock varieties. When one or a few genes are involved, incorporation may require only the crossing of a desirable variety with a source of the new trait, followed by selection of offspring for the desired trait(s) (Fig. 2.17A). Backcross breeding (Fig. 2.17B) can be particularly useful when integrating single dominant traits into a cultivar already possessing many desirable properties. In grapevines, though, the repeated backcrossing required to reestablish expression of desirable characteristics may diminish seed viability and vine vigor. This phenomenon is termed **inbreeding depression**. This occasionally can be countered if more than one cultivar can serve as a recurrent parent, or if embryo rescue can be employed.

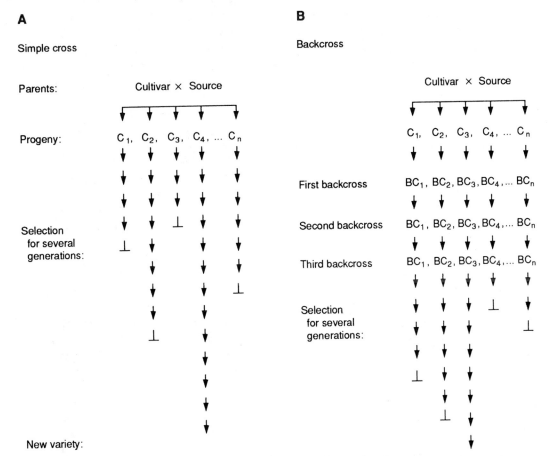

A

Simple cross

B

Backcross

Figure 2.17 Genetic strategies for improving rootstocks. (**A**) Simple-cross breeding program with vegetative propagation following the initial cross. (**B**) Backcross breeding program involving a cross followed by several backcrosses (BC) to a common (parental variety) to reestablish the desirable properties of the parental variety.

Embryo rescue involves the isolation and cultivation of seedling embryos on culture media before they abort (Spiegel-Roy *et al.*, 1985; Agüero *et al.*, 1995).

Ideally, evolutionary relationships should be based on DNA sequences that have been of selective value, rather than properties that have probably diverged randomly.

Developing new scion (fruit-bearing) varieties tends to be more complex, due to the greater number and genetic complexity of the properties associated with commercial acceptability. Genetic complexity greatly complicates the selection process. It blurs the distinction between environmentally and genetically controlled variation. These problems may be partially diminished if the genetic basis of complex traits is known. For example, selection for cold hardiness would be facilitated if the nature and relative importance of the regulatory genetic loci were known. Individual aspects of this complex property involve diverse features, such as control of cellular osmotic potential (timing and degree of starch-to-sugar conversion), the unsaturated fatty-acid content of cell membranes, and the population of ice-nucleation bacteria on leaf surfaces. Chemical indicators and other readily detected features are especially useful in the early selection of suitable offspring. The early and efficient elimination of undesirable progeny assure that only potentially useful offspring are retained for propagation and further study. This liberates vineyard space for a more in-depth investigation of traits such as winemaking quality. For example, the presence of methyl anthranilate and other volatile esters has been used in the early elimination of progeny possessing a *labrusca* fragrance (Fuleki, 1982). Improved knowledge of color stability in red wines would be helpful in the early selection of offspring possessing better color retention. Even selection for disease resistance could be aided by simple chemical analysis. The production of phytoalexins, or resistance to fungal toxins may be useful in the early screening of seedlings for disease resistance. Regrettably, laboratory trials are not always good indicators of field characteristics. In addition, high **combining ability** in the donor would be helpful. Combining ability refers to a strain's relative success rate in breeding programs. Regrettably, potential combining ability is impossible to assess in advance. For example, it was a great surprise when it was discovered that one of the parents of several important cultivars, such as 'Chardonnay' and 'Gamay noir,' was 'Heunisch weiss' ('Gouais blanc'). The latter is widely considered to possess few winemaking qualities.

Adding to the complexities of grape breeding is the time required for the expression of most traits. For example, vines begin to bear sufficient fruit for testing and selection of berry characteristics when 3 or more

years-old. Thus, increasing the population of a promising offspring to a degree whereby a detailed assessment of features such as winemaking potential can occur can take more than 25 years.

The induction of precocious flowering in cuttings and seedlings can speed selection, especially for properties assessable with small fruit samples (Mullins and Rajasekaran, 1981). For example, flowering can be induced within 4 weeks of seed germination by the application of synthetic cytokinins, such as benzyladenine or 6-(benzylamino)-9-(2-tetrahydropyranyl-9H), to tendrils (Fig. 2.18). Precocious flowering and fruit set can also shorten dramatically the interval between crosses. Generation cycling can be further condensed by exposing seed to peroxide and gibberellin prior to chilling (Ellis *et al.*, 1983). The treatment cuts the normally prolonged cold treatment required for seed germination. In theory, the combination of precocious flowering and shortened seed dormancy could reduce the generation time from 3 to 4 years to about 8 months. However,

Figure 2.18 Induction of precocious flowering and fruiting in grapevine seedlings: (**A**) Inflorescence formation by a tendril of a 4-week-old seedling of 'Cabernet Sauvignon.' The seeding was treated daily for 7 days with 6-(benzylamino)-9-(2-tetrahydropyranyl) 9-H-purine (PBA). (**B**) Flower formation by four tendrils of a 6-week-old seedling of 'Muscat Hamburg' after treatment with PBA. (**C**) Inflorescence formation in a 6-month-old seedling of 'Muscat Hamburg' when sprayed with PBA and chlormequat when 3 months old. (**D**) Two bunches of ripe grape produced by an 8-month-old seedling of 'Muscat Hamburg'–bunches developed from tendrils of a lateral shoot. (From Mullins, 1982, reproduced by permission)

practical problems have limited their joint use. Inducing dormant cutting to flower is another means of accelerating the breeding process. It permits crossing to occur year-round under greenhouse conditions.

Although clear indicators of gene expression are highly desirable, few genetic traits of agronomic importance are amenable to early detection. Assessing properties such as winemaking quality can easily extend beyond the career of an individual breeder. The long 'gestation' period for new varieties also results from conservatism in the regulation and marketing of wines. In Europe, the more prestigious appellations are restricted to wines made from existing *vinifera* cultivars. This automatically places legal and marketing disadvantages on wines produced from other cultivars, new or old. This situation is even more severe for interspecies crosses, despite many possessing *vinifera*-like aromas (Becker, 1985). Correspondingly, the incorporation of disease and pest resistance into existing *V. vinifera* cultivars is almost nonexistent. Although species other than *V. vinifera* are the primary sources of disease and pest resistance, untapped resistance may still exist within existing *V. vinifera* cultivars. For example, a crossing of two 'Riesling' clones, 'Arnsburger,' possesses resistance to *Botrytis* not apparent in the parents (Becker and Konrad, 1990).

Despite North American *Vitis* species remaining the primary source for new traits used in grapevine breeding, Asian species, such as *V. amurensis* and *V. armata*, possess several desirable traits. For example, *V. amurensis* (and some strains of *V. riparia*) are potential sources of mildew and *Botrytis* resistance (along with *V. armata* in the latter case). In addition, *V. amurensis* (and northern strains of *V. riparia*) are valuable sources of cold-hardiness, early maturity, and resistance to *coulure*. For subtropical and tropical regions, sources of disease and environmental stress resistance include *V. aestivalis* and *V. shuttleworthii* from the southeastern United States and Mexico, *V. caribaea* from Central America (Jimenez and Ingalls, 1990), and *V. davidii* and *V. pseudoreticulata* from southern China (Fengqin *et al.*, 1990). The use of these genetic resources will be assisted by the development of genetic linkage maps, such as those developed for *Vitis rupestris* and *V. arizonica* (Doucleff *et al.*, 2004).

Although most interspecific hybridization has involved crossings within the subgenus *Vitis*, success in producing some partially fertile crossings between *V. vinifera* × *V. rotundifolia* is encouraging. Crossing is limited both by the inability of the *Vitis* pollen to penetrate the style of muscadine grapes (Lu and Lamikanra, 1996), and the aneuploidy caused by unbalanced chromosome pairing during meiosis (Viljoen and Spies, 1995). Backcrossing to one of the parental species (usually the pollen source)

has been used in several crops to restore fertility to interspecies hybrids. Although this procedure is not simple, genes from *V. rotundifolia* are the principal source of resistance to Pierce's disease. *Vitis rotundifolia* also possesses remarkable resistance to most grapevine viruses, downy and powdery mildew, anthracnose, black rot, and phylloxera, as well as tolerance to root-knot and dagger nematodes, and heat exposure. Advances in the vegetative propagation of *V. vinifera* × *V. rotundifolia* hybrids may permit the commercial availability of these lines (Torregrossa and Bouquet, 1995). Similar breeding programs could also be used to introduce *V. vinifera*-like winemaking properties into muscadine cultivars.

Although most interspecies hybrids possess aromas distinct from those shown by *V. vinifera* cultivars, their aroma is not necessarily less enjoyable. Distinctive interspecific hybrid and non-*vinifera* aromas can form the basis of regional wine styles. It was the regional distinctiveness of certain European wines that originally provided them with much of their appeal. Regrettably, the current wine scene is so tradition-bound that it seems deathly opposed to change. Such resistance to new flavors seems to have petrified what is acceptable. This can only be to the long-term detriment of wine sales.

The conservatism just noted also limits the benefits of most grape breeding programs. Traditional breeding demands that the progeny (even of a self-cross) be given a new name. Thus, a new cultivar loses the marketing advantage of the names of its parents. Most wine consumers, possibly because of the negativism of wine critics, are wary of new cultivars. They are generally considered as inferior to established cultivars. This may also be due to new cultivars being relegated to regions where premium varieties do not grow profitably, or where increased yield or reduced production costs offset the loss of the varietal name recognition. The increasing popularity of "organically grown" wines may enhance the acceptance and cultivation of new varieties. New cultivars often require less pesticide and fertilizer use than older cultivars.

Genetic Engineering

Standard breeding techniques have been successful, but involve considerable time, effort, and expense. In contrast, genetic engineering introduces selected genes, often without disrupting varietal characteristics. Consequently, the central attributes of the cultivar remain unchanged.

Genetic engineering involves one of several techniques that isolate, amplify and insert desirable genes into an unrelated organism (Vivier and Pretorius, 2000; Pretorius and Høj, 2005). Viticultural examples include the incorporation of the GNA gene from *Galanthus nivialis* (snowdrop) and a protein-coat gene for the grapevine fanleaf virus (GFLV) into several rootstock varieties (Mauro *et al.*, 1995). They were inserted to enhance resistance to fanleaf degeneration, by either reducing feeding of viral-transmitting nematodes, or preventing infection by GFLV, respectively. Alternately, the addition of a superoxide dismutase gene from *Arabidopsis thaliana* has been incorporated into scion varieties to improve cold resistance.

Protocols for genetic transformation, using *Agrobacterium* as the gene carrier, are described in Agüero *et al.* (2006) and Bouquet *et al.* (2006). Despite limited success, recent discoveries in genetics, such as the widespread occurrence of pseudogenes and retrotransposons, suggest that progress may be fraught with unsuspected difficulties. In addition, the majority of important grapevine traits appear to be under complex (multigene) control. If these are not closely linked on a single chromosome, successful transfer and expression will be especially difficult. Success will also depend on governmental and consumer acceptance of genetically modified (GM) crops. Surprisingly, those who would benefit the most from genetically engineered crops seem to be those most opposed to their development and use – organic producers. The incorporation of disease and pest resistance would greatly reduce the need for pesticide and fertilizer use (synthetic or 'natural').

Even if GM vines are never cultivated commercially, data obtained by the investigation should provide vital information about grapevine genetics (Thomas *et al.*, 2003). Basic information about issues such as the nature disease resistance is essential if efforts toward reducing the environmental impact of disease control are to be optimally successful. It should also facilitate the streamlining of traditional breeding procedures.

Another issue in genetic engineering that still needs to be resolved is the difficulty in inducing grape cells to develop into whole plants in tissue culture. Success typically is genotype, tissue, and culture medium specific. Genetic engineering requires isolated cells grown in tissue culture. For transformed cells to be of value, they must be induced to form and differentiate into whole plants. Nevertheless, progress has been made in both understanding some of the origins of poor plantlet yield from grape cells (Maes *et al.*, 1997), and in enhancing the proportion of embryonic tissue that develops into vines (Reustle *et al.*, 1995). Examples of stages in the process are shown in Plate 2.3.

Another problem, associated with propagating vines from tissue-cultured cells is progeny phenotypic variability. This may, however, be only an expression of vine juvenility, as variability often decreases with vine age or repeated propagation (Mullins, 1990). It apparently can be minimized by using buds from the fortieth node on, and grafting onto a desirable rootstock (Grenan,

1994). For this, shoots must be very long, as found on vines grown in a greenhouse.

Clonal Selection

Although genetic engineering possesses great potential, clonal selection continues to be the primary means by which cultivar characteristics can be modified, without significantly changing varietal characteristics (or name). However, improvement is limited to the genetic variation that already exists, or can be generated within a cultivar. Variation consists primarily of mutations that have accumulated and been propagated vegetatively since the cultivar's origin. The most obvious mutations are those that affect grape pigmentation or vine morphology. Nevertheless, most mutations have more subtle effects, such as affecting enzymatic properties. These could influence attributes from the aromatic character to climatic sensitivity. Most mutations arise from random changes in the nucleotide sequence of genes. Because most genes regulate the structure of enzymatic proteins, mutations result in the production of isozymes (functional and/or structural variants of an enzyme). Other mutations can result from phenomena such as gene inversions, translocations, or the insertion of retrotransposons. Once a mutation has occurred, it will be reproduced essentially unmodified in cuttings. Because mutations tend to accumulate slowly, older cultivars have a greater tendency to possess mutations than newer cultivars. Each mutant strain of a cultivar can be considered a distinct clone (a cultivar with a distinctive phenotype). Old cultivars such as 'Pinot noir' consist of a wide variety of more or less differentiable clones.

Clonal selection normally refers to a series of procedures designed to isolate and provide premium stock to grape growers. Selection usually goes through several steps, where cuttings are multiplied and repeatedly assessed for their viti- and vinicultural traits, and cleared of systemic pathogens (Fig. 2.19).

Because clonal selection depends on the degree of clonal variability, it is particularly useful with older cultivars. Monoclonal isolates (derived from a single parental plant) may possess few to many mutations, depending on its age and genetic stability. Polyclonal isolates (derived from a collection of related individuals) possess variation derived from mutations in the individual lines as well as genetic differences among the lines.

Fruit-color differences have often been recognized at the varietal name level, for example the gray and white variants of 'Pinot noir' – 'Pinot gris' and 'Pinot blanc.' Other variants have been designated as biotypes, as with many of the 'Sangiovese' clones (Calò *et al.*, 1995). In still other instances, clones have been grouped relative to their possession of similar traits. In 'Pinot noir,'

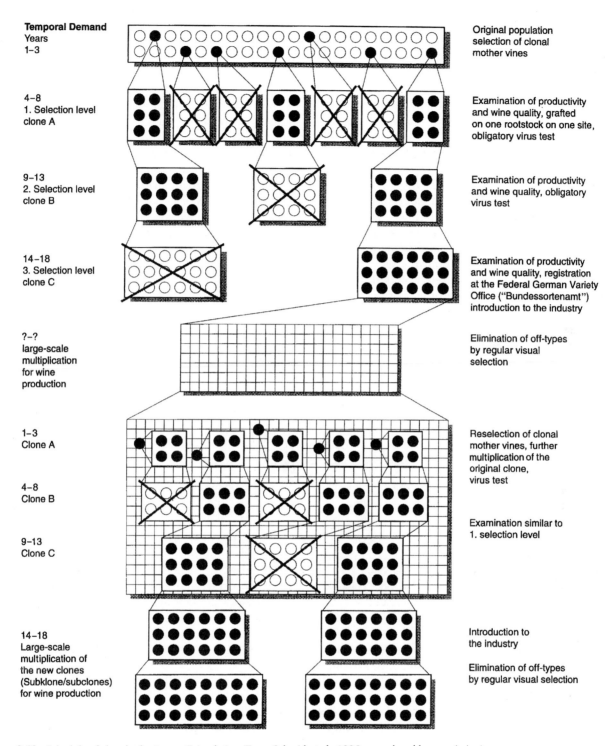

Temporal Demand
Years
1–3

Original population
selection of clonal
mother vines

4–8
1. Selection level
clone A

Examination of productivity
and wine quality, grafted
on one rootstock on one site,
obligatory virus test

9–13
2. Selection level
clone B

Examination of productivity
and wine quality, obligatory
virus test

14–18
3. Selection level
clone C

Examination of productivity
and wine quality, registration
at the Federal German Variety
Office ("Bundessortenamt")
introduction to the industry

?–?
large-scale
multiplication
for wine
production

Elimination of off-types
by regular visual
selection

1–3
Clone A

Reselection of clonal
mother vines, further
multiplication of the
original clone,
virus test

4–8
Clone B

9–13
Clone C

Examination similar to
1. selection level

14–18
Large-scale
multiplication of
the new clones
(Subklone/subclones)
for wine production

Introduction to
the industry

Elimination of off-types
by regular visual selection

Figure 2.19 Schedule of clonal selection at Geisenheim. (From Schmid *et al.*, 1995, reproduced by permission)

these may be grouped into the *Pinot fin* grouping – trailing, low-yielding vines, with small tight clusters; *Pinot droit* – higher-yielding vines with upright shoots; *Pinot fructifer* – high-yielding strains; and Mariafeld-type strains – loose clustered, moderate-yielding vines (Wolpert, 1995). Although clones may differ significantly

in physiologic or morphologic expression, their corresponding nucleotide differences in gene structure are only beginning to yield their secrets.

An integral part of all modern clonal selection procedures is the elimination of systemic pathogens. Because viruses and other systemic pathogenic agents generally invade most tissues, selection alone is unable to eliminate these agents. In this regard, thermotherapy has been particularly valuable in eliminating several systemically inhabiting pathogens. One version involves exposing dormant cuttings to high temperatures (~38°C) for several weeks. Alternately, the excision and propagation of plants from small portions of bud tissue can eliminate viruses that do not invade apical meristematic tissues. Nevertheless, isolating clones free of all known systemic agents (bacterial, phytoplasmal, viral, and viroidal) is difficult to impossible. In addition, reinfection often recurs where the agent is well established in the region. In addition, some systemic agents are not known to provoke recognizable disease symptoms in some cultivars. Thus, their presence may go unnoticed and the vine remains a reservoir of infection for sensitive cultivars.

For rootstock varieties, clonal selection is often limited to certifying them as disease-free. Their recent origin means that insufficient time has elapsed for the accumulation of somatic mutations.

Genetic stability and phenotypic consistency are desirable traits in any cultivar. Correspondingly, elimination of phenotypic variability is usually one of the principal goals. Regrettably, detection of such instability is often difficult to assess in practice. Expression is often dependent on specific environmental conditions. These conditions may not occur in the sites where clonal selection is conducted. Furthermore, genetic uniformity is not itself always desirable, potentially leading to limited aromatic complexity. Where desirable, phenotypic variability is preferably achieved through the selective planting of several clones, rather than through uncontrolled phenotypic plasticity of a particular clone.

Most instances of clonal instability are due to chimeric mutations – genetic modifications that have occurred in apical meristematic cells (Franks *et al.*, 2002). As a result, the vine produces tissue layers that differ genetically in one or more traits. For example, periclinal chimeras (Fig. 2.20) possess buds in which the **outer tunica** (L1) differs genetically from the **inner tunica** and **corpus** (L2) layers. Cells of the outer tunica divide only perpendicularly (at right-angles) to the plant surface and give rise to epidermal tissues. Thus, the epidermis may express properties distinct from other tissues in the vine.

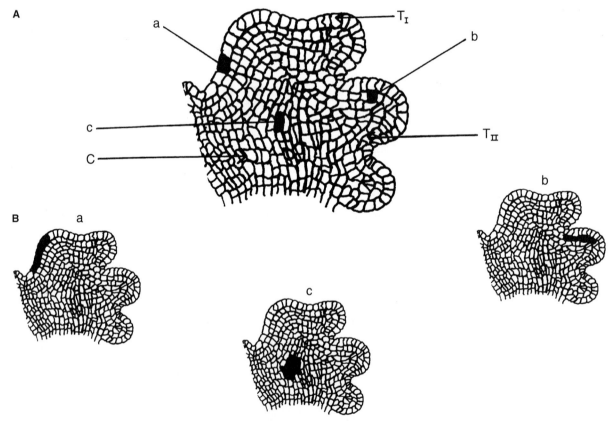

Figure 2.20 Chimeric structures and origins. Apical meristems of grapevine bud are composed of three tissues: two outer layers of tunica, layers T-I and T-II, enclosing the inner corpus, C. A mutation can arise (**A**) in a single cell of any of the layers (a, b, c), which can develop as single layers, or a core of mutated tissue (**B**) to produce a stable periclinal (sectoral) chimera, an unstable mericlinal chimera, and a stable reversion to the original type. Swellings are primordial leaves.

Such properties cannot be transmitted sexually (via seed) because only cells of the inner tunica generate gametes. Genetic properties possessed by the epidermis are propagated only vegetatively. Although chimeras may be a source of clonal instability, most periclinal chimeras are quite stable. Examples are the color chimeras of 'Pinot noir' – 'Pinot blanc' and 'Pinot gris'[2] (Hocquigny *et al.*, 2004); 'Cabernet Sauvignon' – 'Malian' and 'Shalistin' (Walker *et al.*, 2006); and the hairy leaf phenotype of 'Pinot Meunier' (Boss and Thomas, 2002). Plate 2.4 illustrates an example of an unstable chimera.

In addition to concerns about systemic pathogens and phenotypic instability, improved crop yield and grape quality are central to all clonal selection procedures. Improved yield has often been associated with increased vegetative vigor, both of which can result in poorer grape quality. This undesirable correlation can often be avoided by appropriate adjustment in viticultural practice, such as a more open canopy, basal-leaf removal, or increased planting density. In addition, increased yield is not necessarily associated with reduced quality (e.g., sugar content) (Fig. 2.21). Furthermore, in a study of 'Riesling' clones, the one that showed the highest yield and °Brix values showed the weakest growth (Schöfflinger and Stellmach, 1996). Such clones have the economic and environmental benefits of reducing fertilizer input while maximizing yield and quality.

In clonal selection, quality is always of significance. Unfortunately, there are no generally accepted objective measures of quality. Typically, microvinification tests are conducted over several years from vines grown at different locations. The resultant wines need to be aged and assessed by a panel of experienced (and consistent) wine judges (McCarthy and Ewart, 1988). These demands make the establishment of significant differences long, complex, and costly. This becomes particularly critical when choosing between clones that may differ in yield by as much as 60%. Correspondingly, the use of regionally significant, but imprecise, measures of grape maturity, such as °Brix or pH are often substituted. Correspondingly, additional measures of potential quality, such as glycosyl glucose (G-G) content, and improvements in our understanding of varietal aroma compounds, are of particular interest. For red wines, spectrophotometric measures of the total phenol content, wine color density and hue (Somers, 1998) may provide additional, objective criteria of grape quality. The use of objective, readily quantifiable, measures of grape and wine quality would greatly facilitate clonal selection and ease the comparison of data from different researchers. One caveat, however, is that quality as

perceived by wine judges and winemakers often differs considerably from that of wine consumers.

Other features frequently integrated into clonal selection involve factors such as a unique varietal aroma, for example, 'Chardonnay' clones 77 and 809 with a muscat-like nuance (Boidron, 1995); differences in winemaking potential, for example, 'Gamay' 222 vs. 'Gamay' 509 in the production of the *nouveau* vs. *cru* Beaujolais (Boidron, 1995); selection of *Pinot fin* clones for red burgundy vs. *Pinot droit* clones for champagne production (Bernard and Leguay, 1985); choice of growth habit – erect for mechanical harvesting; preferences in berry shape and size to affect flavor and color potential (Watson *et al.*, 1988); or factors such as berry-cluster morphology or other features that may affect resistance to pathogens or physiologic disorders.

Although clonal selection has many benefits, it is essential that varietal variation not be eroded in the process. To avoid this, less desirable clones are often retained as a library of diversity that may become useful in the future.

An import issue in clonal selection relates to how the data are used. Because the properties of a clone may be significantly modified by environmental conditions, desirable features may not always express themselves locally or seasonally. For example, the mild muscat character of some 'Chardonnay' clones may vary from year to year, depending on fruit maturity (Versini *et al.*, 1992). Consequently, vineyard owners need to make their own comparative assessments of apparently desirable clones. Growers must also be aware that clonal traits may require adjustment in viticultural practice to achieve the benefits expected. For example, traits often vary with the rootstock used and training system employed.

It is generally recommended that more than one clone be planted. Boidron (1995) suggests at least two clones in small plots, and up to five or six clones in large vineyards. This helps to avoid problems that may arise due to the limited genetic variation in a single clone, and the resultant

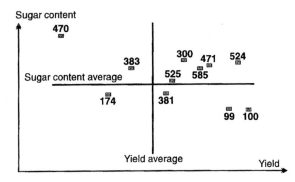

Figure 2.21 Results of 10 years of comparative experiments on the behavior of clones of Syrah. Each clone position for yield (abscissa) and for sugar content (ordinate) is indicated in comparison with the average of all clones. (From Boidron, 1995, reproduced by permission)

---

[2] Data from Walker *et al.* (2006) indicate that 'Pinot gris' and 'Pinot blanc' cultivars are composed of genetically distinct groups of clones.

increase in environmental sensitivity. Environmental flexibility is maximized by planting several clones.

When soil-borne viral problems are a significance factor, it is important that the land be virus-free (fumigated or laid fallow for 6 or more years). Otherwise, nematode-transmitted viruses may soon infect the newly planted, disease-free vines, resulting in a potentially rapid drop in vine productivity. The use of nematode-resistant rootstocks, where available and acceptable, may be a more economical solution than leaving valuable vineyard land fallow for extended periods.

Somaclonal Selection and Mutation

The elimination of viral infection, in association with clonal selection, has made significant improvements in the planting material available to grape growers. However, further significant advances may depend on the incorporation of new genetic variation (mutation or genetic engineering), or on the means by which the expression of existing variation can be enhanced (Kuksova *et al.*, 1997). Variation can be induced by exposing meristematic tissue or cells in tissue culture to mutagenic chemicals or radiation (Fig. 2.22). In addition, somaclonal selection can enhance the expression and isolation of clonal variation.

Somaclonal selection involves the selective growth enhancement of particular lines of cells. For example, lines possessing tolerance for salinity or fungal toxins (Soulie *et al.*, 1993) may be isolated by exposing cells to these toxic agents added to the tissue culture medium. However, the cellular tolerance selected in tissue culture occasionally is not expressed as tissue or whole-plant tolerance (Lebrun *et al.*, 1985). Selective toxic agents are often used to isolate transgenic cell lines in tissue culture (Mauro *et al.*, 1995; Kikkert *et al.*, 1996).

Liquid tissue culture conditions may also be used to disrupt the normal distinction between the tunica and corpus layers of apical meristems. As a result, cells of the outer tunica may relocate and act as inner tunica or corpus cells, and vice versa, in regenerated plantlets. Consequently, traits found in different layers of a periclinal chimera may be expressed throughout the vine. It also permits traits located only in epidermal cells to be transmitted by sexual reproduction. In addition, traits located in separate layers of a chimera may be isolated and used to propagate nonchimeric clones by somatic embryogenesis. For example, the distinct L_1 layer of a chimeric strain of 'Chardonnay 96' has been achieved in pure form (Bertsch *et al.*, 2005).

Grapevine Cultivars

With the number of named grapevine cultivars approaching 15,000 (many of which are synonyms), a comprehensive system of cultivar classification would be useful. Regrettably no such system exists. In some countries, there have been attempts to rationalize local cultivars into related groups (Fig. 2.23). These ecogeographic associations have been primarily on ampelographic attributes. Because of the localized distribution of each group, they may have been derived from one or a few related individuals. This view has received support from microsatellite DNA (SSR) analysis (Aradhya *et al.*, 2003). French cultivars appear to be relatively distinct and related to wild accessions from the Pyrenees or Tunisia. As well, Armenian, Georgian and Turkish varieties appear to be isolated genetically (Vouillamoz *et al.*, 2006). Nevertheless, the same study found evidence that some Western European cultivars ('Chasselas,' 'Nebbiolo,' 'Pinot noir,' and 'Syrah'), or their ancestors, appear to cluster genetically with Georgian cultivars.

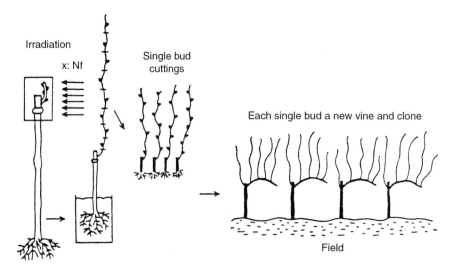

Figure 2.22 Method for the isolation of somatic mutation following irradiation. (From Becker, 1986, reproduced by permission)

Currently, definitive answers on the origin of most cultivars are still unavailable.

Former methods of assessing relatedness involved comparing aroma profiles, for example Fig. 2.24. Numerical taxonomic procedures were also used to reduce the

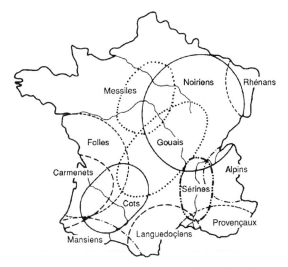

Figure 2.23 Zones of origin or extension of the principal groups of French grape varieties. Example of some cultivars considered to belong to these ecogroups are *Alpins* – 'Corbeau' and 'Durif'; *Carmenets* – 'Cabernet Franc,' 'Cabernet Sauvignon,' 'Merlot,' and 'Petit Verdot'; *Cots* – 'Malbec' and 'Tannat'; *Folles* – 'Folle blanche,' 'Jurançon,' and 'Gros Plant;' *Gouais* – 'Aligoté' and 'Muscadelle;' Languedociens – 'Cinsaut' and 'Piquepouls;' *Mansiens* – 'Mansenc'; Messiles – 'Chenin blanc' and 'Sauvignon blanc;' *Noiriens* – 'Chardonnay,' 'Gamay,' and 'Pinot noir;' *Provençaux* – 'Clairette' and 'Colombeau;' *Rhénans* – 'Savagins;' *Sérines* – 'Marsanne,' 'Syrah,' and 'Viognier.' (Modified from Bisson, 1989, reproduced by permission)

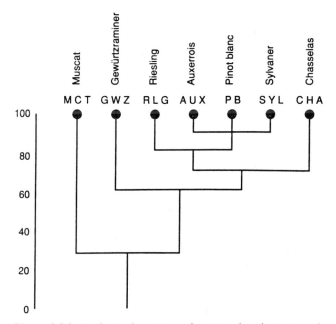

Figure 2.24 Similarity description of varieties based on aromatic profile. (From Lefort, 1980, reproduced by permission)

subjectivity of traditional classification techniques (Fanizza, 1980). However, as noted above, DNA techniques have supplied more objective and evolutionarily significant data than all older techniques combined.

Nevertheless, for pragmatic reasons ampelographic procedures (Galet, 1979) remain the typical means by which growers identify cultivars – despite the plasticity of the traits employed. Ampelography uses vegetative characteristics, such as leaf shape and shoot-tip pubescence (hairiness). The development of a computer-aided digitizing system for determining ampelographic measurements (AmpeloCAD) should simplify grapevine identification for the nonampelographer (Alessandri *et al.*, 1996).

On a broader scale, cultivars are grouped according to their specific or interspecific origin. Most commercial varieties are pure *V. vinifera* cultivars. French-American hybrids constitute the next largest group. They were derived from crosses between *V. vinifera* and one or more of the following: *V. riparia*, *V. rupestris*, and *V. aestivalis*. Early American cultivars are either selections from indigenous grapevines, or are hybrids between them and *V. vinifera*. Interspecific cultivars refer to modern crosses between *V. vinifera* and species such as *V. amurensis*, *V. riparia*, *V. armata*, and *V. rotundifolia*. The origin of such cultivars is usually clearly identified, for example *V. amurensis* × *V. vinifera*, or *V. armata* × *V. vinifera* hybrids.

Vitis vinifera Cultivars

Because of the large number of named cultivars, only a few of the better-known varieties are discussed here. Although often world famous, these cultivars seldom constitute the principal varieties grown, even in their country of origin. Their productivity is usually lower, and their cultivation is typically more demanding. Their reputation usually comes from their yielding varietally distinctive, balanced wines with long-aging potential. In favorable locations, their excellent winemaking properties compensate for reduced yield and increased production costs.

Many of the varieties listed below are French cultivars. This is undoubtedly a geographic or historic accident, with many other equally worthy cultivars being little recognized outside their homelands. The cool climate of the best vineyards in France has favored the retention of subtle fragrances, complete fermentation, and long aging. The former position of France as a major political and cultural power, and its proximity to rich connoisseur-conscious countries, encouraged the selection and development of the best local cultivars and wines. The long and frequent contacts between England and France, and the global expansion of British colonial influence, fostered the dispersal of French cultivars throughout much of the English-speaking world. Regrettably, fine cultivars from southern and eastern Europe did not receive the same opportunities afforded French, and some German, cultivars.

Examples of several regionally important, but less well-known, aromatically distinctive varieties are 'Arinto' (white, w) and 'Ramisco' (red, r) from Portugal; 'Corvina' (r), 'Dolcetto' (r), 'Negro Amaro' (r), 'Fiano' (w), 'Garganega' (w), and 'Torbato' (w) from Italy; 'Rhoditis' (w or rosé) from Greece; 'Furmint' (w) from Hungary; and 'Malvasia' (w), 'Parellada' (w), and 'Graciano' (r) from Spain. These could contribute significantly to increasing the diversity of quality wines available worldwide.

RED CULTIVARS

'Barbera' is the most widely cultivated variety in Piedmont and is very important in the rest of Italy. It is the third most cultivated variety in Italy, after 'Sangiovese' and 'Trebbiano Toscano.' Outside Italy, it is primarily planted in California and Argentina. It is moderately high yielding, producing fruit that is intensely colored, high in acidity, but moderate in tannin content. The clusters possess long green stalks, making mechanical and manual picking easy. Cultivation is also uncomplicated, due to its adaptability to different soils. 'Barbera' can make a varietally distinctive, fruity wine, but is commonly blended with other cultivars to add acidity and 'fruit' to wines high in pH. With regard to disease susceptibility, 'Barbera' is particularly sensitive to grapevine leafroll. Some clones are severely affected by bunch rot.

'Cabernet Sauvignon' is undoubtedly the best-known red cultivar. This is due to both its association with one of Europe's best-known red wines (bordeaux) and its production of equally superb wines in many parts of the world. Under optimal conditions, it produces a fragrant wine possessing a black-currant aroma (described as violet in France). Under less favorable conditions, it generates a bell-pepper aroma. The berries are small, acidic, seedy, and possess a darkly pigmented, tough skin. The cultivar is frequently cane-pruned to accentuate production and provide better sun exposure for its upright shoots. The easy separation of the small round berries from the cluster facilitates mechanical harvesting. The cultivar is highly susceptible to several fungal diseases, notably Eutypa and Esca wood decays, powdery mildew, and phomopsis. DNA fingerprinting techniques indicate that 'Cabernet Sauvignon' is the result of a crossing of 'Cabernet Franc' and 'Sauvignon blanc' (Bowers and Meredith, 1997). In Bordeaux, and increasingly in other regions, wines made from 'Cabernet Sauvignon' are blended with wines produced from other, related cultivars, such as 'Cabernet Franc' and 'Merlot.' The latter moderate the tannin content and accelerate maturation.

'Merlot' has the advantage of growing in cooler, more moist soils than 'Cabernet Sauvignon,' but is more susceptible to *coulure*. The tendency of 'Merlot' to mature more quickly has made it a popular substitute for 'Cabernet Sauvignon.' 'Ruby Cabernet' is a Davis cross between 'Carignan' and 'Cabernet Sauvignon.' It possesses a Cabernet aroma, but grows better in hot climates than 'Cabernet Sauvignon.'

'Dolcetto' is an Italian cultivar grown almost exclusively in Piedmont. Nevertheless, it still ranks eighth in overall hectarage in Italy. It has rarely been tried outside its homeland. In growth, 'Dolcetto' possesses comparatively weak vigor, producing small to medium-sized clusters, with small rounded berries. It produces a wine usually light, but bright in color, medium-bodied, and with a mild distinctive aroma.

'Gamay noir à jus blanc' is the primary (white-juiced) Gamay cultivar. Its reputation has risen in association with the increased popularity of Beaujolais wines. Half of the hectarage of this cultivar is in Beaujolais. It is little cultivated outside of France. Crushed and fermented by standard procedures, 'Gamay' produces a light red wine with few distinctive attributes. This may partially result from its generally high productivity. When processed by carbonic maceration, it yields a distinctly fruity wine. Most of these features come from the grape fermentative process that precedes alcoholic fermentation (see Chapter 9). 'Gamay' produces medium-size fruit with tough skins. It is sensitive to most fungal grapevine diseases. The 'Gamay Beaujolais' and 'Napa Gamay' grown in California are not directly related to 'Gamay noir.' They are considered to be clones of 'Pinot noir' and 'Valdiguié,' respectively.

'Garnacha' ('Grenache,' 'Cannonau') is a widely planted cultivar in Spain, southern France, southern Italy, Sardinia, Sicily, California and Australia. It exists in several phenotypic color variants (white and gray), in addition to the standard red clones. The vine has an upright growth habit, suitable to head training and spur pruning. It is well adapted to hot, dry conditions, but tends to be excessively productive with irrigation. Clusters are broad and compact, varying from pink to red, depending on the crop load. The cultivar is not well adapted to mechanical harvesting. By itself, it is usually used to make rosé or fortified wines. It typically is blended in with other varieties to speed maturation. Although sensitive to powdery mildew, bunch rot and *coulure*, 'Garnacha' is relatively resistant to downy mildew.

'Graciano' is little cultivated outside of northern Spain, notably Rioja and Navarra, possibly because of its relatively low yield. Nevertheless, it is comparatively resistant to most fungal diseases and is drought resistant. The fruit stalks are often woody, making the cultivar ill-suited to mechanical harvesting. The fruit is one of the most aromatic of Spanish varieties and possesses good acidity. It is an integral component of many of the best Rioja wines.

'Nebbiolo' is generally acknowledged as producing one of the most highly regarded red wines in northwestern Italy. With traditional vinification, it produces a wine high in tannin and acid, requiring many years to mellow. The color has a tendency to oxidize rapidly. Common varietal descriptors include tar, violets, and truffles. 'Nebbiolo' yields well only when cane-pruned (due to low basal fertility) and adapts well to a wide range of soil pHs and types. The variety has not been planted extensively outside northern Italy. It is particularly susceptible to powdery mildew, but is relatively resistant to bunch rot. The variety's weak skin and juicy fruit make it ill-suited to mechanical harvesting.

'Pinot noir' is the famous red grape of Burgundy. It appears to be one of the most environmentally sensitive of varieties, and consists of a large number of distinctive clones. Fruit-color mutants have given rise to 'Pinot gris' and 'Pinot blanc.' 'Meunier' is another named 'Pinot noir' variant. Usually, the more prostrate, lower-yielding clones produce more flavorful wines. The upright, higher-yielding clones are more suited to the production of rosé and sparkling wines. 'Pinot noir' produces an aromatically distinctive wine under optimal conditions; otherwise, it produces nondistinctive wines. Various authors have proposed terms for its aroma, such as beets, peppermint, or cherries, but none seems appropriate. The cultivar produces small clusters of small- to medium-size fruit with large seeds. If the clusters are compact, it is particularly sensitive to bunch rot. Crossed with 'Cinsaut,' it has produced one of the most distinctive of South African cultivars, 'Pinotage.' The Californian cultivar 'Gamay Beaujolais' is a clone of 'Pinot noir' (Bowers *et al.*, 1993).

'Sangiovese' is probably an ancient cultivar that now comprises an extensive diversity of distinctive clones. It is grown extensively throughout central Italy. It is most well known for the light- to full-bodied wines from Chianti and produces many of the finest red wines in Italy. 'Sangiovese' is also grown under local synonyms, such as 'Brunello' and 'Prugnolo.' 'Sangiovese' is relatively vigorous, but variable in yield. Its clusters are small- to medium-sized, possessing oval berries. The fruit is compatible with mechanical harvesting. Under optimal conditions, it yields a wine possessing an aroma reminiscent of cherries, violets, and licorice. 'Sangiovese' has achieved little of the international recognition it deserves (a fate typical of most Italian grape cultivars). 'Sangiovese' is particularly sensitive to both bunch rot and powdery mildew. It appears to be the progeny of two ancient Italian cultivars, 'Ciliegiolo' and 'Calabrese di Montenuova' (Vouillamoz *et al.*, 2007).

'Syrah' is the most renowned French cultivar from the Rhône Valley. Lower yielding strains produce a deep red, tannic wine with long aging potential. 'Syrah' is little grown outside France, except in Australia, where it is grown extensively under the name 'Shiraz.' 'Syrah' is a vigorous grower with a spreading growth habit. The fruit clusters are elongated with small round or oval berries. It yields deep-colored, flavorful wines with aspects reminiscent of violets, raspberries, and currants, and possesses a peppery finish. 'Syrah' is particularly prone to drought, bunch rot, and infestation by grape berry moths. Most vines labeled 'Petit Sirah' in California are identical to Rhône cultivar 'Durif' (Meredith *et al.*, 1999).

'Tempranillo' ('Ull de Llebre') is probably the finest Spanish red-grape variety. Under favorable conditions, it yields a fine, subtle wine that ages well. In addition, it can produce delicate *nouveau*-style wines. It is the most important red cultivar in Rioja (occupying about 33,000ha) and is grown extensively through much of Spain. Outside Spain, it is primarily grown in Argentina. In California, it usually goes under the name 'Valdepeñas.' 'Tempranillo' generates an aroma distinguished by a complex, berry-jam fragrance, with nuances of citrus and incense. 'Tempranillo' produces mid-size, thick-skinned fruits that are subject to both powdery and downy mildews. The vine is comparatively vigorous and produces upright shoots.

'Touriga National' is one of the preeminent Portuguese grape varieties. It is grown predominantly in the Upper Douro for the production of port, but is also cultivated in other regions to produce red table wines. The wine is deep in color and richly flavored. The vine is fairly vigorous, with a trailing growth habit. Its clusters are of small- to medium-size, containing small berries.

'Zinfandel' is extensively grown in California. It appears to be synonymous with (or a clone of) the Italian variety 'Primitivo' (Bowers *et al.*, 1993) or its Croatian equivalent 'Crljenak kastelanski' (Fanizza *et al.*, 2005). 'Zinfandel' is used to produce a wide range of wines, from ports to light blush wines. In rosé versions, it shows a raspberry fragrance, whereas full-bodied red wines possess rich berry flavors. Some of the difficulties with 'Zinfandel' are the uneven manners with which the fruit ripens, and its tendency to produce a second crop later in the season. Both properties complicate harvesting fruit of uniform maturity.

WHITE CULTIVARS

'Chardonnay' is undoubtedly the most widely grown white French cultivar. This stems from both its appealing fruit fragrance and its ability to do well in most wine-producing regions. In addition to producing fine table wines, it also yields one of the finest sparkling wines (champagne). The fruit is comparatively small, round, and forms compact clusters. Under optimal conditions, the wine develops aspects reminiscent of various fruits, including apple, peach, and melon. The vine and fruit are predisposed to powdery mildew and bunch rot.

'Chenin blanc' comes from the central Loire Valley of France. There it yields fine sweet and dry table wines, as well as sparkling wines. 'Chenin blanc' is also grown extensively in Australia, California, and South Africa (as 'Steen'). Fine examples of its wine often exhibit a delicate fragrance loosely resembling guava fruit or camellia blossoms. The fruit is tough-skinned and of medium size. It is especially susceptible to both downy and powdery mildews, bunch rot and grape berry moths.

'Ehrenfelser' is one of the best German varieties, being derived from a 'Riesling' × 'Silvaner' cross. It has many of the characteristics of its 'Riesling' parent, such as its flavor, acidity, disease resistance, and cold-hardiness. In addition, it has the advantages of ripening somewhat earlier. 'Ehrenfelser' is still largely cultivated in Germany, but is being grown with considerable success in Canada.

'Müller-Thurgau' is possibly the most well known modern *V. vinifera* cultivar, constituting nearly 30% of German hectarage. It was developed by H. Müller-Thurgau, supposedly from a crossing between 'Riesling' and 'Silvaner' in 1882. Nevertheless, microsatellite analysis has indicated that 'Müller-Thurgau' is presumably a crossing between 'Riesling' and 'Chasselas de Courtillier'(Sefc *et al.*, 1997) or 'Madeleine Royale' (Dettweiler *et al.*, 2000). The first commercial plantings of Müller-Thurgau began in 1903. It is now extensively grown in most cool regions of Europe and formerly in New Zealand. Its mild acidity and subtle fruity fragrance are ideal for producing light wines. 'Müller-Thurgau' is a high-yielding cultivar that often produces lateral (side) clusters of mid-size fruit. It grows best in rich porous soils. The fruit is subject to both powdery and downy mildews and bunch rot.

'Muscat blanc' is one of many related Muscat varieties grown extensively throughout the world. Their aroma is so marked and distinctive that it is described in terms of the cultivar name, muscaty. Because of the intense flavor, slight bitterness (due to a high level of flavonoid extraction during maceration), and tendency to oxidize, Muscat grapes have most commonly been used in the production of dessert wines. Muscat grapes are also characterized by the presence of high levels of soluble proteins. Consequently, special precautions must be taken to avoid haze formation. The new Muscat cultivar 'Symphony' is less bitter and its lower susceptibility to oxidation gives it better aging ability. 'Muscat blanc' ('Moscato bianco' in Italy) is the primary variety used in the flourishing sparkling wine industry in Asti. Other named Muscat varieties include the 'Orange Muscat' ('Muscato Fiori d'Arinico' in Italy), 'Muscat

of Alexandria,' 'Muscat Ottonel,' and the darkly pigmented 'Black Muscat' ('Muscat Hamburg').

'Parellada' is a variety distinctive to the Catalonian region of Spain. It produces an aroma that is apple- to citrus-like in character, occasionally also showing hints of licorice or cinnamon.

'Pinot gris' and 'Pinot blanc' (respectively 'Ruländer' and 'Weissburgunder' in Germany, and 'Pinot grigio' and 'Pinot bianco' in Italy) are named color mutants of 'Pinot noir.' Both are cultivated throughout the cool climatic regions of Europe for the production of dry, botrytized and sparkling wines. Neither has gained much popularity outside Europe. 'Pinot gris' can vary in color from blue to white, depending on the microclimatic conditions of the cluster. It typically yields subtly fragrant wines with aspects of passion fruit, whereas 'Pinot blanc' is more fruity, with suggestions of hard cheese.

'Riesling' ('White Riesling' or 'Johannisberg Riesling') is without doubt Germany's most highly esteemed grape variety. Outside Germany, it largest plantings are in California and Australia. It can produce fresh, aromatic, well-aged wines, which can vary from dry to sweet. Its floral aroma, commonly reminiscent of roses, has made it popular throughout central Europe and much of the world. This renown is reflected in the number of cultivars whose names have incorporated the word Riesling (i.e., 'Hunter Riesling,' 'Goldriesling,' 'Frankenriesling,' and 'Wälschriesling'), none of which bear genetic or aromatic relationship to 'Riesling.' 'Riesling' produces clusters of small- to medium-size berries that are particularly sensitive to powdery mildew and bunch rot. Although relatively cold-hardy, the fruit matures slowly. The yield tends to be moderate.

'Sauvignon blanc' is one of the primary white varieties in Bordeaux, and the main white cultivar in the upper Loire Valley. It has become popular in California and New Zealand in recent years. It also is grown in northern Italy and eastern Europe. Often, its aroma shows elements of green peppers, as well as a herbaceous aspect, especially in cooler climates. Better clones possess a subtly floral character. Its modest clusters produce small berries that are sensitive to powdery mildew and black rot, but possess partial resistance to bunch rot and downy mildew.

'Sémillon' is widely grown in Bordeaux and is most well known for its use in producing Sauternes wines. In Bordeaux, it is commonly blended with 'Sauvignon blanc.' Significant plantings also occur in Australia and Argentina. 'Sémillon' produces small clusters of medium-size fruit notably susceptible to bunch rot. The vine is also especially sensitive to fanleaf degeneration. When fully mature, and without the intervention of noble rot, 'Sémillon' yields a dry wine said to contain nuances of fig or melon that develop primarily on aging.

'Traminer' is a distinctively aromatic cultivar grown throughout the cooler regions of Europe and much of the world. Its clones can be grouped into three color classes: white, pink and reddish. The intensity of fragrance tends to correlate with color depth. The pink appears to be a mutant of the white and the red is reported to have subsequently arisen in the Rheinpfalz between 1750 and 1870 (Bourke, 2004). DNA sequencing techniques indicate that all clones are near identical (Imazio *et al.*, 2002). Despite the color difference, the cultivar is processed as a white grape. 'Traminer' is used to generate both dry and sweet styles, depending on regional preferences. Intensely fragrant, reddish clones ('Gewürztraminer') generally possess an aroma resembling that of lichi fruit. White, mildly aromatic clones of the variety are called 'Savagnin' in southeastern France, 'Albariño' in Spain, and 'Alvarinho' in Portugal. All forms produce modest clusters of small fruit with tough skins. The variety is prone to powdery mildew and bunch rot and often expresses *coulure*.

'Viognier' has been a variety largely restricted to the Condrieu appellation of the Rhône Valley. Comparatively recently, it has caught the fancy of several North American and Australian producers. The variety has a tendency toward poor fertility and, correspondingly, low yield. Thus, it needs to be cane-pruned. The vine generally requires excellent drainage. 'Viognier' produces small round berries possessing a muscaty fragrance. Correspondingly, the wine matures quickly. The variety is also reported to show peach or apricot aspects.

'Viura' is the main white variety in Rioja. It produces few clusters, but they are of great size. In cool regions, it produces a fresh wine possessing a subtle floral aroma with aspects of lemon. After prolonged aging in wood, it develops a golden color and rich butterscotch or banana fragrance that characterizes the traditionally aged white wines of Rioja.

Interspecies Hybrids

AMERICAN HYBRIDS

Although decreasing in significance throughout North America, early selections from native American grapevines or accidental interspecies hybrids are still of considerable importance. American hybrids constitute the major plantings in eastern North America, and they are grown commercially in South America, Asia, and Eastern Europe.

Of the American hybrid cultivars, the most important are based on *V. labrusca*. They possess a wide range of flavors. Some, such as 'Niagara,' are characterized by the presence of a foxy aspect. Others, however, are characterized more by a strawberry fragrance ('Ives'),

the grapy aspect of methyl anthranilate ('Concord'), or a strong floral aroma ('Catawba'). The high acidity and low sugar content of American hybrid grapes have made chaptalization (the addition of sugar to the juice) necessary for table-wine production.

Various methods have been used to diminish what is often viewed as the overabundant flavor of most American cultivars. Long aging results in a dissipation of *labrusca* flavors. This is generally non-feasible, as most *V. labrusca* wines are consumed young. The presence of high levels of carbon dioxide, as in sparkling wines, tends to mask most *labrusca* fragrances. Processing the grapes via carbonic maceration is another means of reducing (masking) the intensity of *labrusca* flavors. However, the most generally accepted mechanism is early picking and cold fermentation. These limit the development and extraction, respectively, of *labrusca* flavors, while still producing a wine with fruitiness.

The varieties 'Norton' and 'Cynthiana' (possibly clonal variants; Reisch *et al.*, 1993) are predominantly of *Vitis aestivalis* origin. These cultivars are primarily planted in Arkansas and Missouri. Locally derived from wild vines, and possibly introgressed to *V. vinifera*, they possess resistance to the local indigenous diseases and pests that make cultivation of *vinifera* varieties difficult.

The other major group of American cultivars are those derived from *V. rotundifolia*. Although 'Scuppernong' is the most well known, it is rarely grown presently. New self-fertile varieties possessing different aromatic properties are the principal muscadine cultivars grown in commercial vineyards (AAES Special Report 203). The excellent resistance of these cultivars to indigenous diseases, especially Pierce's disease, has allowed them to flourish in the southeastern coastal United States. Similar to the *V. labrusca* cultivars, the low sugar content of the fruit usually requires juice chaptalization before vinification. The pulpy texture, tough skin, differential fruit maturation, and separation of fruit from the pedicel on maturation complicate their use in winemaking.

Most muscadine cultivars have a distinctive and marked fragrance, containing aspects of orange blossoms and roses. Some fertile crossings with *V. vinifera* show *vinifera*-like flavors, combined with the fruiting characteristics and disease resistance of their muscadine parent.

FRENCH-AMERICAN HYBRIDS (DIRECT PRODUCERS)

French-American hybrids were developed to avoid the necessity, complexity, and expense of grafting *V. vinifera* cultivars to phylloxera-resistant rootstocks. The easier cultivation, reduced sensitivity to several leaf pathogens, and higher yield made them popular with many grape growers in France. Although the tendency of base buds to grow and bear fruit increased yield, it also tended to result in overcropping, that exacerbated the serious grape overproduction in France, a problem that still plagues much of Europe. This factor, combined with the nontraditional fragrance of these hybrids, led to a general ban on new plantings. Their use in appellation control (AC) wines was also prohibited. Restrictions against new plantings of French-American hybrids subsequently spread throughout the European Economic Community (EEC). The remaining, temporary, exception is the cultivation of 'Baco blanc' for armagnac production.

In North America, with the exception of most of the southern, gulf, and western coastal states, French-American hybrids formed the basis of the expanding wine industry in the early 1960s. They are also grown extensively in some South American and Asian countries. In Europe, as well as other areas, French-American hybrids are often used in breeding programs as a source of resistance to several foliar, stem, and fruit pathogens. Many French-American hybrids suffer from sensitivity to soil-borne virus diseases, such as those of the tomato ringspot group (Alleweldt, 1993). Where this is a problem, grafting to resistant rootstocks is required.

Unlike American hybrids, few French-American hybrids possess *Vitis labrusca* parentage. *Vitis rupestris*, *V. riparia*, and *V. aestivalis* var. *lincecumii* were the principal species used. Although some French-American hybrids are simple crosses between an American *Vitis* species and a *V. vinifera* parent, most are derived from complex crosses between an American species and several *V. vinifera* cultivars. This is clearly evident from their genetic diversity (Pollefeys and Bousquet, 2003).

A brief description of some of the better American and French-American hybrids is provided below.

'Baco noir' is a 'Folle blanche' × *V. riparia* hybrid. Its acidity, flavor, and pigmentation yield a wine with considerable aging potential. It develops a fruity aroma associated with aspects of herbs. It is sensitive to bunch rot and several soil-borne viruses. Poor cane maturity is often a problem in cold climatic regions due to vine vigor.

'de Chaunac' is a Seibel crossing of unknown parentage. Once widely planted in eastern North America, due to its cold-hardiness and high yield, its tendency to produce wine of neutral character, and its susceptibility to several soil-borne viruses has resulted in its loss of favor.

'Chambourcin' is a Joannes Seyve hybrid of unknown parentage. Its popularity increased markedly in the Loire Valley during the 1960s and 1970s. It is considered one of the best of French-American hybrids. 'Chambourcin' has also done remarkably well in Australia, possibly due to the long growing season. This permits full grape maturity and the development of a rich, wonderfully complex flavor. 'Chambourcin' has also found a following in eastern North America. The variety possesses good resistance to both downy and powdery mildews.

'Maréchal Foch' is a Kuhlmann hybrid derived from crossing a *V. riparia* × *V. rupestris* selection with 'Goldriesling' ('Riesling' × 'Courtiller musqué'). It yields deeply colored, berry-scented, early-maturing wines. The variety's characteristics of winter-hardiness, productiveness, early maturity, and resistance to downy mildew have also given it appeal in eastern North America.

'Delaware' is one of the finest, early ripening, light-red American hybrids. It is generally thought to be a *V. labrusca* × *V. aestivalis* × *V. vinifera* hybrid. Nevertheless, its susceptibility to phylloxera and various fungal pathogens, tendency to crack, and the need for well-drained soil has limited its widespread cultivation. It was once extensively used in sparkling-wine production.

'Dutchess' is another highly rated, older, late-ripening, white American hybrid. This *Vitis labrusca* × *V. vinifera* hybrid has a mild, fruity aroma with little *labrusca* flavor. As with 'Delaware,' difficulty in growing the cultivar negates most of its enologic qualities.

'Magnolia' is one of the more popular, new, muscadine cultivars. It produces bisexual flowers and is self-fertile. It yields sweet, bronze-colored fruit.

'Noble' is a dark-red muscadine cultivar. Its deep-red color and bisexual habit have made it popular in the southeastern United States.

'Seyval blanc' is a Seyve-Villard hybrid of complex *V. vinifera*, *V. rupestris*, and *V. aestivalis* var. *lincecumii* parentage. The variety yields a mildly fruity white wine with a pomade fragrance and bitterish finish. Although susceptible to bunch rot, it is relatively winter-hardy, tolerant of many soil types, and a consistent producer.

'Vidal blanc' is possibly the best of the white French-American hybrids. This Vidal hybrid of complex ancestry has both excellent winemaking and viticultural properties. Under optimal conditions 'Vidal blanc' yields a wine of Riesling-like character. Its tough skin and late maturity assists in its being used in the production of icewines of excellent quality. The variety is relatively cold-hardy, but less so than 'Seyval blanc.'

Although most interspecies hybridization ceased by the 1920s in France, it has continued unabated in Germany. Varieties such as 'Orion,' 'Phoenix,' and 'Regent' show winemaking qualities equal or superior to several currently used *V. vinifera* cultivars. Breeding has also continued in eastern North America. 'Veeblanc' is one of the newer white cultivars of complex parentage developed in Ontario. It generates a mildly fruity wine of good quality. New York also has an active breeding program. One of their most commercially successful introductions to date is 'Cayuga White.' It is well adapted to a range of wine styles; possesses a fruity fragrance resembling apples, citrus, and tropical fruit; and produces a wine without bitterness and a rich mouth-feel. When harvested early, it produces an excellent base for sparkling wines. Although very productive, 'Cayuga White' seldom requires cluster thinning. Finally, 'Cayuga White' is resistant to most common fungal diseases and, thus, needs little fungicidal protection. Another popular new introduction from Cornell is 'Traminette.'

Suggested Readings

Taxonomy and Evolution of *Vitis*

Levadoux, L. (1956) Les populations sauvages et cultivées de *Vitis vinifera* L. *Ann. Amélior. Plantes, Sér. B* **1**, 59–118.

Levadoux, L., Boubals, D., and Rives, M. (1962) Le genre *Vitis* et ses espèces. *Ann. Amélior. Plantes* **12**, 19–44.

Núñez, D. R., and Walker, M. J. (1989) A review of paleobotanical findings of early *Vitis* in the Mediterranean and of the origins of cultivated grape-vines, with special reference to prehistoric exploitation in the western Mediterranean. *Rev. Paleobot. Palynol.* **61**, 205–237.

Olmo, H. P. (1995) The domestication of the grapevine *Vitis vinifera* L. in the Near East. In: *The Origins and Ancient History of Wine* (P. E. McGovern *et al.*, eds.), pp. 31–44. Gordon and Breach, Luxembourg.

Shetty, B. V. (1959) Cytotaxonomical studies in Vitaceae. *Bibliogr. Genet.* **18**, 167–272.

Zapriagaeva, V. I. (1964) Grapevine – *Vitis* L. In: *Wild Growing Fruits in Tadzhikistgan* (Russian, with English summary), pp. 542–559. Nauka, Moscow.

Zohary, D., and Hopf, M. (1993) *Domestication of Plants in the Old World*, 2nd edn. Oxford University Press (Clarendon), Oxford.

Rootstock

Catlin, T. (1991) *Alternative Rootstock Update.* ASEV Technical Projects Committee and UC Cooperative Extension, University of California Agriculture Publications, Oakland, CA.

May, P. (1994) *Using Grapevine Rootstocks – The Australian Perspective.* Winetitles, Adelaide, Australia.

Morton, L. T., and Jackson, L. E. (1988) Myth of the universal rootstock: The fads and facts of rootstock selection. In: *Proceedings of the 2nd International Symposium of Viticulture and Oenology* (R. E. Smart *et al.*, eds.), pp. 25–29. New Zealand Society of Viticulture and Oenology, Auckland, New Zealand.

Pongrácz, D. P. (1983) *Rootstocks for Grape-vines.* David Philip, Cape Town, South Africa.

Wolpert, J. A., Walker, M. A., and Weber, E. (eds.) (1992) *Proceedings of the Rootstock Seminar: A Worldwide Perspective.* American Society of Enology and Viticulture, Davis, CA.

Grape Breeding and Selection

Becker, H. (1988) Breeding resistant varieties and vine improvement for cool climate viticulture. In: *Proceedings of the 2nd International Symposium of Viticulture and Oenology* (R. E. Smart *et al.*, eds.), pp. 63–64. New Zealand Society of Viticulture and Oenology, Auckland, New Zealand.

Boidron, R. (1995) Clonal selection in France. Methods, organization, and use. In: *Proceedings of the International Symposium on Clonal Selection* (J. M. Rantz, ed.), pp. 1–7. American Society of Enology and Viticulture, Davis, CA.

Bouquet, A., and Boursiquot, J.-M. (eds.) (2000) 7th International Symposium on Grapevine Genetics and Breeding. *Acta Horticulturae* **528**.

Hajdu, E., and Borbas, E. (eds.) (2003) 8th International Conference on Grape Genetics and Breeding. ISHS *Acta Horticulturae* **603**.

Martinelli, L., and Gribaudo, I. (2001) Somatic embryogenesis in grapevine. In: *Molecular Biology and Biotechnology of the Grapevine* (K. Roubelakis-Angelakis, ed.), pp. 327–351. Kluwer Adacemic Publ., Dordrecht, Netherlands.

Martinelli, L., and Mandolino, G. (2001) Transgenic grapes (*Vitis* species) In: *Biotechnology in Agriculture and Forestry* (Y. P. S. Bajaj, ed.), pp. 325–338. Springer-Verlag, Berlin.

Meredith, C. P., and Reiche, B. I. (1996) The new tools of grapevine genetics. In: *Proceedings of the 4th International Symposium for Cool Climate Viticulture and Enology* (T. Henick-Kling, *et al.*, eds.), pp. VIII-12–18. New York State Agricultural Experimental Station, Geneva, NY.

Roubelakis-Angelakis, K. A. ed. (2001) *Molecular Biology and Biotechnology of Grapevine.* Kluwer Academic Publishers, Amsterdam.

Schöffling, H., and Stellmach, G. (1996) Clone selection of grape vine varieties in Germany. *Fruit Var. J.* **50**, 235–247.

Vivier, M. A., and Pretorius, I. S. (2000) Genetic improvement of grapevine: Tailoring grape varieties for the third millennium – a review. *S. Afr. J. Enol. Vitic.* **21** (Special Issue), 5–26.

Grape Cultivars

Alleweldt, G. (1989) *The Genetic Resources of Vitis: Genetic and Geographic Origin of Grape Cultivars, Their Prime Names and Synonyms.* Federal Research Centre for Grape Breeding, Geilweilerhof, Germany.

Bailey, L. H. (1989) *Sketch of the Evolution of Our Native Fruits.* Macmillan, New York.

Bettega, L. (2003) *Wine Grape Varieties in California.* Division of Agriculture and Natural Resources, University of California, Pbu. # 3419.

Blouin, M. S. (2003) DNA-based methods for pedigree reconstruction and kinship analysis in natural populations. *Trends Ecol Evol.* **18**, 503–511.

Galet, P. (1979) *A Practical Ampelography – Grapevine Identification* (L. T. Morton, ed. and trans.). Cornell University Press, Ithaca, NY.

Galet, P. (1988, 1990) *Cépages et Vignobles de France*, Vol. 1 (*Les Vignes Américaines*), Vol. 2 (*L'Ampélographie Française*), Vol. 3 (*Les Cépages de Cuves*) 2nd edn. Galet, Montpellier, France.

Hillebrand, W., Lott, H., and Pfaff, F. (2002) *Taschenbuch der Rebsorten.* Fachverlag Fraund, Weisbaden.

Kerridge, G., and Antcliff, A. (1996) *Wine Grape – Varieties of Australia.* CSIRO, Collingwood, Australia.

Olien, W. C. (1990) The muscadine grape: Botany, viticulture, history, and current industry. *HortScience* **25**, 732–739.

Peñin, J., Cervera, A., Cabello, F., Diez, R., and Peñin, P. (1997) *Cepas del Mundo.* Ediciones Pl y Erre, Madrid.

Reisch, B. J., Pool, R. M., Peterson, D. V., Martins, M-H., and Henick-Kling, T. (1993) *Wine and Juice Grape Varieties for Cool Climates.* Cornell Cooperative Extension Publication, Information Bulletin No. 233. Cornell University, Ithaca, NY.

References

AAES Special Report 203. http://www.uark.edu/depts/agripub/Publications/specialreports/203-text.pdf.

Agüero, C. B., Meredith, C. P., and Dandekar, A. M. (2006) Genetic transformation of *Vitis vinifera* L. cvs Thompson Seedless and Chardonnay with the pear PGIP and GFP encoding genes. *Vitis* **45**, 1–8.

Agüero, C., Riquelme, C., and Tizio, R. (1995) Embryo rescue from seedless grapevines (*Vitis vinifera* L.) treated with growth retardants. *Vitis* **34**, 73–76.

Alessandri, S., Vignozzi, N., and Vignini, A. M. (1996) *AmpeloCADs* (Ampelographic Computer-Aided Digitizing System): An integrated system to digitize, file, and process biometrical data from *Vitis* spp. leaves. *Am. J. Enol. Vitic.* **47**, 257–267.

Alleweldt, G. (1993) Disease resistant varieties. In: *Proceedings of the 8th Australian Wine Industry Technical Conference* (C. S. Stockley *et al.*, eds.), pp. 116–119. Winetitles, Adelaide, Australia.

Ambrosi, H., Dettweiler, E., Rühl, E., Schmid, J., and Schumann, F. (1994) *Farbatlas Rebsorten. 300 Sorten und ihre Weine.* Ulmer Verlag, Stuttgart.

Ammerman, A. C., and Cavalli-Sporza, L. L. (1971) Measuring the rate of spread of early farming in Europe. *Man* **6**, 674–688.

Anzani, R., Failla, O., Scienza, A., and Campostrini, F. (1990) Wild grapevine (*Vitis vinifera* var. *silvestris*) in Italy: Distribution, characteristics and germplasm preservation – 1989 report. In: *Proceeding of the 5th International Symposium on Grape Breeding*, pp. 97–113. (Special issue of *Vitis*) St Martin, Pfalz, Germany.

Aradhya, M. K., Dangle, G., Prins, B. H., Boursiquot, J.-M., Walker, M. A., Meredith, C. P., and Simon, C. J. (2003) Genetic structure and differentiation in cultivated grape, *Vitis vinifera* L. *Genet. Res. Camb.* **81**, 179–192.

Arnold, C., Gillet, F., and Gobat, J. M. (1998) Situation de la vigne sauvage *Vitis vinifera* spp. *silvestris* en Europe. *Vitis* **37**, 159–170.

Arroyo-García, R., Lefort, F., de Andrés, M. T., Ibáñez, J., Borrego, J., Jouve, N., Cabello, F., and Martínez-Zapater, J. M. (2002) Chloroplast microsatellite polymorphisms in *Vitis* species. *Genome* **45**, 1142–1149.

Arroyo-García, R., Ruiz-García, L., Bolling, L., Ocete, R., López, M. A., Arnold, C., Ergul, A., Söylemezoğlu, G., Uzun, H. I., Cabello, F., Ibáñez, J., Aradhya, M. K., Atanassov, A., Atanassov, I., Balint, S., Cenis, J. L., Costantini, L., Goris-Lavets, S., Grando, M. S., Klein, B. Y., McGovern, P. E., Merdinoglu, D., Pejic, I., Pelsy, F., Primikirios, N., Risovannaya, V., Roubelakis-Angelakis, K. A., Snoussi, H., Sotiri, P., Tamhankar, S., This, P., Troshin, L., Malpica, J. M., Lefort, F., and Martinez-Zapater, J. M. (2006) Multiple origins of cultivated grapevine (*Vitis vinifera* L. ssp. *sativa*) based on chloroplast DNA polymorphisms. *Mol. Ecol.* **15**, 3707–3714.

Bailey, L. H. (1933) The species of grapes peculiar to North America. *Gentes Herbarum* **3**, 150–244.

Baker, H. G. (1972) Human influences on plant evolution. *Econ. Bot.* **26**, 32–43.

Becker, H. (1985) White grape varieties for cool climate. In: *International Symposium on Cool Climate Viticulture and Enology* (B. A. Heatherbell, P. B. Lombard, F. W. Bodyfelt and S. F. Price, eds.), OSU Agric. Exp. Stn. Tech. Publ. No. 7628, pp. 46–62. Oregon State University, Corvallis, OR.

Becker, H. (1986) Induction of somatic mutations in clones of grape cultivars. *Acta Hortic.* **180**, 121–128.

Becker, H., and Konrad, H. (1990) Breeding of Botrytis tolerant *V. vinifera* and interspecific wine varieties. In: *Proceeding of the 5th International Symposium on Grape Breeding*, p. 302 (Special Issue of *Vitis*). St Martin, Pfalz, Germany.

Bernard, R., and Leguay, M. (1985) Clonal variability of Pinot Noir in Burgundy and its potential adaptation under other cooler climate. In: *International Symposium on Cool Climate Viticulture and Enology* (B. A. Heatherbell *et al.*, eds.), *OSU Agric. Exp. Stn. Tech. Publ.* No. 7628, pp. 63–74. Oregon State University, Corvallis, OR.

Bertsch, C., Kieffer, F., Maillot, P., Farine, S., Butterlin, G., Merdinoglu, D., and Walter, B. (2005) Genetic chimerism of *Vitis vinifera* cv. Chardonnay 96 is maintained through organogenesis but not somatic embryogenesis. *BMC Plant Biol.* **5**, 20.

Bisson, J. (1989) Les Messiles, groupe ampélographique du bassin de la Loire. *Connaiss. Vigne Vin* **23**, 175–191.

Boidron, R. (1995) Clonal selection in France. Methods, organization, and use. In: *Proceedings of the International Symposium on Clonal Selection* (J. M. Rantz, ed.), pp. 1–7. American Society of Enology and Viticulture Davis, CA.

Boss, P. K., and Thomas, M. R. (2002) Association of dwarfism and floral induction with a grape 'green revolution' mutation. *Nature* **416**, 163–170.

Bouquet, A., Torregrosa, L., Iocco, P., and Thomas, M. R. (2006) Grapevine (*Vitis vinifera* L.). *Methods Mol. Biol.* **344**, 273–286.

Bourke, C. (2004) Is Traminer Gewurz, or is it Roter or Rose, and if Bianco, what about Albarino? Goodness only knows! *Aust. NZ Grapegrower Winemaker* **488**, 19–22, 24.

Bourquin, J.-C., Sonko, A., Otten, L., and Walter, B. (1993) Restriction fragment length polymorphism and molecular taxonomy in *Vitis vinifera* L. *Theor. Appl. Genet.* **87**, 431–438.

Boursiquot, J. M. (1990) Évolution de l'encépagement du vignoble français au cours des trente dernières années. *Prog. Agric. Vitic.* **107**, 15–20.

Bowers, J. E., and Meredith, C. P. (1996) Genetic similarities among wine grape cultivars revealed by restriction fragment-length polymorphism (RFLP) analysis. *J. Am. Soc. Hort. Sci.* **121**, 620–624.

Bowers, J. E., and Meredith, C. P. (1997) The parentage of a classic wine grape, Cabernet Sauvignon. *Nature Genetics* **16**, 84–87.

Bowers, J. E., Bandman, E. B., and Meredith, C. P. (1993) DNA fingerprint characterization of some wine grape cultivars. *Am. J. Enol. Vitic.* **44**, 266–274.

Bowers, J., Boursiquot, J.-M., This, P., Chu, K., Johansson, H., and Meredith, C. (1999) Historical genetics: The parentage of Chardonnay, Gamay, and other wine grapes of Northeastern France. *Science* **285**, 1562–1565.

Bowers, J. E., Siret, R., Meredith, C. P., This, P., and Boursiquot, J. M. (2000) A single pair of parents proposed for a group of grapevine varieties in Northeastern France. *Acta Hort.* **528**, 129–132.

Briggs, D., and Walters, S. M. (1986) *Plant Variation and Evolution.* Cambridge University Press, Cambridge.

Bronner, A., and Oliveira, J. (1990) Creation and study of the Pinot noir variety lineage. In: *Proceeding of the 5th International Symposium on Grape Breeding*, p. 69. (Special Issue of *Vitis*) St Martin, Pfalz, Germany.

Buck, C. D. (1949) *A Dictionary of Selected Synonyms in the Principal Indo-European Languages.* University of Chicago Press, Chicago.

Büscher, N., Zyprian, E., Bachmann, O., and Blaich, R. (1994) On the origin of the grapevine variety Müller-Thurgau as investigated by the inheritance of random amplified polymorphic DNA (RAPD) *Vitis* **33**, 15–17.

Caetano-Anolles, G., and Gresshoff, P. M. (1997) *DNA Markers – Protocols, Applications and Overviews.* Wiley, New York.

Calò, A., Costacurta, A., Paludetti, G., Crespan, M., Giust, M., Egger, E., Grasselli, A., Storchi, P., Borsa, D., and di Stefano, R. (1995) Characterization of biotypes of Sangiovese as a basis for clonal selection. In: *Proceedings of the International Symposium on Clonal Selection* (J. M. Rantz, ed.), pp. 99–104. American Society of Enology and Viticulture, Davis, CA.

Chicki, L., Nichols, R. A., Barbujani, G., and Beaumont, M. A. (2002) Y genetic data support the Neolithic demic diffusion model. *Proc. Natl Acad. Sci.* **99**, 11008–11013.

Crespan, M., and Milani, N. (2001) The Muscats: A molecular analysis of synonyms, homonyms and genetic relationships within a large family of grapevine cultivars. *Vitis* **40**, 23–30.

de Blij, H. J. (1983) *Wine, a Geographic Appreciation.* Rowman and Allanheld, Totowa, NJ.

de Wet, J. M. J., and Harlan, J. R. (1975) Weeds and domesticates: Evolution in the man-made habitat. *Econ. Bot.* **29**, 99–109.

Dettweiler, E., Jung, A., Zyprian, E., and Töpfer, R. (2000) Grapevine cultivar Müller-Thurgau and its true to type descent. *Vitis* **39**, 63–65.

Dietsch, M.-F. (1996) Gathered fruits and cultivated plants at Bercy (Paris), a Neolithic village in a fluvial context. *Veg. Hist. Archaeobot.* **5**, 89–97.

Donini, P., Elias, M. L., Bougourd, S. M., and Koebner, R. M. D. (1997) AFLP fingerprinting reveals pattern differences between template DNA extracted from different plant organs. *Genome* **40**, 521–526.

Dorf, E. (1960) Climatic change of the past and present. *Am. Sci.* **48**, 341–364.

Doucleff, M., Jin, Y., Gao, F., Riaz, S., Krivanek, A. F., and Walker, M. A. (2004) A genetic linkage map of grape, utilizing *Vitis rupestris* and *Vitis arizonica. Theor. Appl. Genet.* **109**, 1178–1187.

Ellis, R. H., Hong, T. D., and Roberts, E. H. (1983) A note on the development of a practical procedure for promoting the germination of dormant seed of grape (*Vitis* spp.) *Vitis* **22**, 211–219.

Fanizza, G. (1980) Multivariate analysis to estimate the genetic diversity of wine grapes (*Vitis vinifera*) for cross breeding in southern Italy. In: *Proceedings of the 3rd International Symposium on Grape Breeding*, pp. 105–110. University of California, Davis, CA.

Fanizza, G., Lamaj, F., Resta, P., Ricciardi, L., and Savino, V. (2005) Grapevine cvs Primitivo, Zinfandel and Crljenak kastelanski: Molecular analysis by AFLP. *Vitis* **44**, 147–148.

Fengqin, Z., Fangmei, L., and Dabin, G. (1990) Studies on germplasm resources of wild grape species (*Vitis* spp.) in China. In: *Proceeding of the 5th International Symposium on Grape Breeding*, pp. 50–57. (Special Issue of *Vitis*) St Martin, Pfalz, Germany.

Fernandez, L., Romieu, C., Moing, A., Bouquet, A., Maucourt, M., Thomas, M. R., and Torregrosa, L. (2006) The grapevine fleshless berry mutation. A unique genotype to investigate differences between fleshy and nonfleshy fruit. *Plant Physiol.* **140**, 537–547.

Forbes, R. J. (1965) *Studies in Ancient Technology*, 2nd edn, Vol. 3. E. J. Brill, Leiden, The Netherlands.

Franks, T., Botta, R., and Thomas, M. R. (2002) Chimerism in grapevines: implications for cultivar identity, ancestry and genetic improvement. *Theor. Appl. Genet.* **104**, 192–199.

Fuleki, T. (1982) The vineland grape flavor index – a new objective method for the accelerated screening of grape seedlings on the basis of flavor character. *Vitis* **21**, 111–120.

Galet, P. (1979) *A Practical Ampelography – Grapevine Identification* (L. T. Morton, ed. and trans.) Cornell University Press, Ithaca, NY.

Galet, P. (1988) *Cépages et Vignobles de France*, Vol. 1 (*Les Vignes Américaines*) 2nd edn. Dehan, Montpellier, France.

Gamkrelidze, T. V., and Ivanov, V. V. (1990) The early history of Indo-European languages. *Sci. Am.* **262** (3), 110–116.

Goto-Yamamoto, N., Mouri, H., Azumi, M., and Edwards, K. J. (2006) Development of grape microsatellite markers and microsatellite analysis including oriental cultivars. *Am. J. Enol. Vitic.* **57**, 105–108.

Gramotenko, P. M., and Troshin, L. P. (1988) Improvement in the classification of *Vitis vinifera* L. In: *Prospects of Grape Genetics and Breeding for Immunity* (in Russian), pp. 45–52. Naukova Dumka, Kiev.

Grassi, F., Labra, M., Imazio, S., Spada, A., Sgorbati, S., Scienza, A., and Sala, F. (2003) Evidence of a secondary grapevine domestication centre detected by SSR analysis. *Theor. Appl. Genet.* **107**, 1315–1320.

Grenan, S. (1994) Multiplication *in vitro* et caractéristiques juvéniles de la vigne. *Bull. O.I.V.* **67**, 5–14.

Haas, H. U., Budahn, H., and Alleweldt, G. (1994) *In situ* hybridization in *Vitis vinifera* L. *Vitis* **33**, 251–252.

Hartmann, H. T., Kester, D. E., and Davies, F. T. (1990) *Plant Propagation. Principles and Practices.* Prentice Hall, Englewood Cliffs, NJ.

Hocquigny, S., Pelsy, F., Dumas, V., Kindt, S., Heloir, M.-C., and Merdinoglu, D. (2004) Diversification within grapevine cultivars goes through chimeric states. *Genome*, **47**, 579–589.

Hopf, M. (1983) Jericho plant remains. In: *Excavations at Jericho* (K. M. Kenyon and T. A. Holland, eds.), Vol. 5, pp. 576–621. British School of Archaeology in Jerusalem, London.

Imazio, S., Labra, M., Grassi, F., Winfield, M., Bardini, M., and Scienza, A. (2002) Molecular tools for clone identification: the case of the grapevine cultivar Traminer. *Plant Breed.* **121**, 531–535.

Jimenez, A. L. G., and Ingalls, A. (1990) *Vitis caribaea* as a source of resistance to Pierce's disease in breeding grapes for the tropics. In: *Proceeding of the 5th International Symposium on Grape Breeding*, pp. 262–270. (Special Issue of *Vitis*) St Martin, Pfalz, Germany.

Jones, G., and Legge, A. (1987) The grape (*Vitis vinifera* L.) in the Neolithic of Britain. *Antiquity* **61**, 452–455.

Jongmans, W. (ed.) (1939) *Fossilium Catalogus. II. Plantae. Pars 24: Rhamnales I: Vitaceae.* W. Junk, Verlag für Natruwissenschaften, The Hague, The Netherlands.

Kikkert, J. R., Hébért-Soulé, D., Wallace, P. G., Striem, M. J., and Reisch, B. I. (1996) Transgenic plantlets of 'Chancellor' grapevine (*Vitis* sp.) from biolistic transformation of embryonic cell suspensions. *Plant Cell Rep.* **15**, 311–316.

Kuksova, V. B., Piven, N. M., and Gleba, Y. Y. (1997) Somaclonal variation and *in vitro* induced mutagenesis in grapevine. *Plant Cell Tiss. Org. Cult.* **49**, 17–27.

Lebrun, L., Rajasekaran, K., and Mullins, M. G. (1985) Selection *in vitro* for NaCl-tolerance in *Vitis rupestris* Scheele. *Ann. Bot.* **56**, 733–739.

Lefort, P.-L. (1980) Biometrical analysis of must aromagrams: Application to grape breeding. *Proc. 3rd Int. Symp. Grape Breeding, June 15–18, 1980*, pp. 120–129. University of California, Davis, CA.

Levadoux, L. (1946) Étude de la fleur et de la sexualité chez la vigne. *Ann. Éc. Natl Agric. Montpellier* **27**, 1–90.

Levadoux, L. (1956) Les populations sauvages et cultivées de *Vitis vinifera* L. *Ann. Amélior. Plant. Sér. B* **1**, 59–118.

Lodhi, M. A., and Reisch, B. I. (1995) Nuclear DNA content of *Vitis* species, cultivars and other genera of the Vitaceae. *Theor. Appl. Genet.* **90**, 11–16.

Longbottom, M. L., Dry, P. R., and Sedgley, M. (2004) A research note on the occurrence of 'star' flowers in grapevines: observations during the 2003–2004 growing season. *Aust J. Grape Wine Res.* **10**, 199–202.

López-Pérez, A. J., Carreño, J., Martínez-Cutillas, A., and Dabauza, M. (2005) High embryogenic ability and plant regeneration of table grapevine cultivars (*Vitis vinifera* L.) induced by activated charcoal. *Vitis* **44**, 79–85.

Lu, J., and Lamikanra, D. (1996) Barriers to intersubgeneric crosses between *Muscadinia* and *Euvitis. HortScience* **31**, 269–271.

Maes, O., Coutos-Thevenot, P., Jouenne, T., Boulay, M., and Guern, J. (1997) Influence of extracellular proteins, proteases and protease inhibitors on grapevine somatic embryogenesis. *Plant Cell Tissue Organ Cult.* **50**, 97–105.

Maletić, E., Pejić, I., Kontić, J. K., Piljac, J., Dangl, G. S., Vokurka, A., Lacombe, T., Mirošević, N., and Meredith, C. P. (2004) Zinfandel, Dobričić, and Plavac mali: the genetic relationship among three cultivars of the Dalmatian Coast of Croatia. *Am. J. Enol. Vitic.* **55**, 174–180.

Manen, J.-F., Bouby, L., Dalnoki, O., Marinval, P., Turgay, M., and Schlumbaum, A. (2003) Microsatellites from archaeological *Vitis vinifera* seeds allow a tentative assignment of the geographical origin of ancient cultivars. *J. Archaeol. Sci.* **30**, 721–729.

Martinez, M., and Mantilla, J. L. G. (1993) Behaviour of *Vitis vinifera* L. *cv.* Albariño plants, produced by propagation *in vitro*, when using single bud cuttings. *J. Intl Sci. Vigne Vin.* **27**, 159–177.

Mauro, M. C., Toutain, S., Walter, B., Pinck, L., Otten, L., Coutos-Thevenot, P., Deloire, A., and Barbier, P. (1995) High efficiency regeneration of grapevine plants transformed with the GFLV coat protein gene. *Plant Sci.* **112**, 97–106.

McCarthy, M. G., and Ewart, A. J. W. (1988) Clonal evaluation for quality winegrape production. In: *Proceedings of the 2nd International Symposium for Cool Climate Viticulture and Oenology* (R. E. Smart et al., eds.), pp. 34–36. New Zealand Society for Viticulture and Oenology, Auckland, New Zealand.

McGovern, P. E., Glusker, D. L., Exner, L. J., and Voigt, M. M. (1996) Neolithic resinated wine. *Nature* **381**, 480–481.

Meredith, C. P., Bowers, J. E., Riaz, S., Handley, V., Bandman, E. B., and Dangl, G. S. (1999) The identity and parentage of the variety known in California as Petit Sirah. *Am. J. Enol. Vitic.* **50**, 236–242.

Miller, N. F. (1991) The Near East. In: *Progress in Old World Paleoethnobotany* (W. Van Zeist et al., eds.), pp. 133–160. Balkema, Rotterdam, The Netherlands.

Moore, M. O. (1991) Classification and systematics of eastern North American *Vitis* L., Vitaceae, north of Mexico. *SIDA* **14**, 339–367.

Mullins, M. G. (1982) Growth regulators and the genetic improvement of grapevines. In: *Grape Wine Centennial Symposium Proceedings 1981*, pp. 143–147. University of California, Davis, CA.

Mullins, M. G. (1990) Applications of tissue culture to the genetic improvement of grapevines. In: *Proceedings of the 5th International Symposium on Grape Breeding*, pp. 399–407. (Special Issue of *Vitis*) St Martin, Pfalz, Germany.

Mullins, M. G., and Rajasekaran, K. (1981) Fruiting cuttings: Revised method for producing test plants of grapevine cultivars. *Am. J. Enol. Vitic.* **32**, 35–40.

Neagu, M. M. (1968) Génétique et amélioration de la vigne. Rapport général. *Bull. O.I.V.* **41**, 1301–1337.

Negrul, A. M. (1938) Evolution of cultivated forms of grapes. *C. R. (Doklady) Acad. Sci. U.R.S.S.* **18**, 585–588.

Núñez, D. R., and Walker, M. J. (1989) A review of paleobotanical findings of early *Vitis* in the Mediterranean and of the origins of cultivated grape-vines, with special reference to prehistoric exploitation in the western Mediterranean. *Rev. Paleobot. Palynol.* **61**, 205–237.

Okamoto, G., Fujii, Y., Hirano, K., Tai, A., and Kobayashi, A. (1995) Pollen tube growth inhibitors from Pione grape pistils. *Am. J. Enol. Vitic.* **46**, 17–21.

Olmo, H. P. (1976) Grapes. In: *Evolution of Crop Plants* (N. W. Simmonds, ed.), pp. 294–298. Longman, London.

Patel, G. I., and Olmo, H. P. (1955) Cytogenetics of *Vitis*. I. The hybrid *V. vinifera* × *V. rotundifolia*. *Am. J. Bot.* **42**, 141–159.

Planchais, N. (1972–1973) Apports de l'analyse pollinique à la connaissance de l'extension de la vigne au quaternaire. *Naturalia Monspeliensia, Sér. Bot.* **23–24**, 211–223.

Planchais, N., and Vergara, P. (1984) Analyse pollinique de sédiments lagunaires et côtiers en Languedoc, en Roussillon et dans la province de Castgellon (Espagne); Bioclimatologie. *Bull. Soc. Bot. Fr.* **131**, *Actual. Bot.* (2/ 3/ 4), 97–105.

Pollefeys, P., and Bousquet, J. (2003) Molecular genetic diversity of the French-American grapevine hybrids cultivated in North America. *Genome* **46**, 1037–1048.

Pongrácz, D. P. (1983) *Rootstocks for Grape-vines*. David Philip, Cape Town, South Africa.

Pratt, C. (1988) Grapevine structure and growth stages. In: *Compendium of Grape Diseases* (R. C. Pearson and A. C. Goheen, eds.), pp. 3–7. APS Press, St Paul, Minnesota.

Pretorius, I. S., and Høj, P. B. (2005) Grape and wine biotechnology: Challenges, opportunities and potential benefits. *Aust. J. Grape Wine Res.* **11**, 83–108.

Rausing, G. (1990) *Vitis* pips in Neolithic Sweden. *Antiquity* **64**, 117–122.

Regner, F., Stadlbauer, A., Eisenheld, C., and Kaserer, H. (2000) Genetic relationships among Pinots and related cultivars. *Am. J. Enol. Vitic.* **51**, 7–14.

Reisch, B. I., Goodman, R. N., Martens, M.-H., and Weeden, N. F. (1993) The relationship between Norton and Cynthiana, red wine cultivars derived from *Vitis aestivalis*. *Am. J. Enol. Vitic.* **44**, 441–444.

Renfrew, C. (1989) The origins of Indo-European languages. *Sci. Am.* **261** (4), 106–116.

Renfrew, J. M. (1973) *Palaeoethnobotany: The Prehistoric Food Plants of the Near East and Europe*. Columbia University Press, New York.

Renfrew, J. M. (1995) Palaeoethnobotanical find of *Vitis* from Greece. In: *The Origins and Ancient History of Wine* (P. E. McGovern et al., eds.), pp. 255–268. Gordon and Breach, Luxembourg.

Reustle, G., Harst, M., and Alleweldt, G. (1995) Plant regeneration of grapevine (*Vitis* sp.) protoplast isolated from embryogenic tissue. *Plant Cell Rep.* **15**, 238–241.

Rives, M. (1975) Les origines de la vigne. *Recherche* **53**, 120–129.

Rogers, D. J., and Rogers, C. F. (1978) Systematics of North American grape species. *Am. J. Enol. Vitic.* **29**, 73–78.

Schmid, J., Ries, R., and Rühl, E. H. (1995) Aims and achievements of clonal selection at Geisenheim. In: *Proceedings of the International Symposium on Clonal Selection* (J. M. Rantz, ed.), pp. 70–73. American Society of Enology and Viticulture, Davis, CA.

Schöfflinger, H., and Stellmach, G. (1996) Clone selection of grape vine varieties in Germany. *Fruit Var. J.* **50**, 235–247.

Schumann, F. (1997) Rebsorten und Weinarten im mittelalterlichen Deutschland. In: *Weinwirtschaft im Mittelalter*, Band 9 (C. Schrenk and H. Weckbach, eds.). Stadtarchiv Heilbronn.

Scienza, A., Villa, P., Tedesco, G., Parini, L., Ettori, C., Magenes, S., and Gianazza, E. (1994) A chemotaxonomic investigation on *Vitis vinifera* L. II. Comparison among ssp. *sativa* traditional cultivars and wild biotypes of spp. *silvestris* from various Italian regions. *Vitis* **33**, 217–224.

Sefc, K. M., Regner, F., Glössl, J., Steinkellner, H. (1998) Genotyping of grapevine and rootstock cultivars using microsatellite markers. *Vitis* **37**, 15–20.

Sefc, K. M., Steinkellner, H., Wagner, H. W., Glössl, J., and Regner, F. (1997) Application of microsatellite markers to parentage studies in grapevine. *Vitis* **36**, 179–184.

Sefc, K. M., Steinkellner, H., Lefort, F., Botta, R., da Câmara Machado, A., Borrego, J., Maletić, J., and Glössl, J. (2003) Evaluation of the genetic contribution of local wild vines to European grapevine cultivars. *Am. J. Enol. Vitic.* **54**, 15–21.

Shetty, B. V. (1959) Cytotaxonomical studies in Vitaceae. *Bibliogr. Genet.* **18**, 167–272.

Snoussi, H., Slimane, M. H. B., Ruiz-Garcia, L., Martinez-Zapater, J. M., and Arroyo-Garcia, R. (2004) Genetic relationship among cultivated and wild grapevine accessions from Tunisia. *Genome* **47**, 1211–1219.

Somers, C. (1998) *The Wine Spectrum*. Winetitles, Adelaide, Australia.

Soulie, O., Roustan, J.-P., and Fallot, J. (1993) Early *in vitro* selection of eupypine-tolerant plantlets. Application to screening of *Vitis vinifera* cv. Ugni blanc somaclones. *Vitis* **32**, 243–244.

Spiegel-Roy, P., Sahar, N., Baron, I., and Lavi, U. (1985) In: *vitro* culture and plant formation from grape cultivars with abortive ovules and seeds. *J. Am. Soc. Hort. Sci.* **110**, 109–112.

Srinivasan, C., and Mullins, M. G. (1981) Physiology of flowering in the grapevine – a review. *Am. J. Enol. Vitic.* **32**, 47–59.

Stanislawski, D. (1975) Dionysus westward: Early religion and the economic geography of wine. *Geogr. Rev.* **65**, 427–444.

Stevenson, A. C. (1985) Studies in the vegetational history of S.W. Spain. II. Palynological investigations at Laguna de los Madres, Spain. *J. Biogeogr.* **12**, 293–314.

Stummer, A. (1911) Zur Urgeschichte der Rebe und des Weinbaues. *Mitt. der Anthropol. Ges. Wien.* **41**, 283–296.

This, P., Lacombe, T., and Thomas, M. R. (2006) Historical origins and genetic diversity of wine grapes. *Trends Genet.* **22**, 511–519.

Thomas, M. R., Cain, P., and Scott, N. S. (1994) DNA typing of grapevines: A universal methodology and database for describing cultivars and evaluating genetic relatedness. *Plant Mol. Biol.* **25**, 939–949.

Thomas, M., and Dry, I., Barker, C., Donald, T., Adam-Blondon, A.-F., Pauquet, J., and Bouquet, A. (2003) Use of molecular techniques for the transfer of powdery mildew resistance from a wild American grapevine to elite winegrape cultivars. *Aust. NZ Grapegrower Winemaker* **473a**, 97–99.

Torregrosa, L., and Bouquet, A. (1995) *In vitro* propagation of *Vitis* × *Muscadinia* hybrids by microcuttings or axillary budding. *Vitis* **34**, 237–238.

Tsertsvadze, N. V. (1986) Classification of cultured grape *Vitis vinifera* L. in Georgia. (in Russian) *Tech. Progress Vitic. Georgia, Tbilisi*, pp. 229–240.

Turner, S. D., and Brown, A. G. (2004) *Vitis* pollen dispersal in and from organic vineyards I. Pollen trap and soil pollen data. *Rev. Paleobot. Palynol.* **129**, 117–132.

Vaquer, J., Geddes, D., Barbaza, M., and Erroux, J. (1985) Mesolithic plant exploitation at the Balma Abeurador (France) *Oxford J. Archaeol.* **5**, 1–18.

Versini, G., Dalla Serra, A., Falcetti, M., and Sferlazzo, G. (1992) Rôle du clone, du millésiime et de l'époque de la récolte sur le potentiel aromatique du raisin de Chardonnay. *Rev. Oenologues* **18**, 19–23.

Versini, G., Rapp, A., Volkmann, C., and Scienza, A. (1990) Flavour compounds of clones from different grape varieties. In: *Proceeding of the 5th International Symposium on Grape Breeding*, pp. 513–524. (Special Issue of *Vitis*) St Martin, Pfalz, Germany.

Viljoen, T. A., and Spies, J. J. (1995) Cytogenetical studies of three *Vitis* species. *Vitis* **34**, 221–224.

Vivier, M. A., and Pretorius, I. S. (2000) Genetic improvement of grapevine: Tailoring grape varieties for the third millennium – A review. *S. Afr. J. Enol. Vitic.* **21** (Special Issue), 5–26.

von Babo, A. F., and Mach, E. (1923) *Handbuch Weinbaues und der Kellerwirtschaft*, 4th edn, Vol. 1, Part 1. Parey, Berlin.

Vouillamoz, J. F., and Grando, M. S. (2006) Genealogy of wine grape cultivars: 'Pinot' is related to 'Syrah.' *Heredity* **97**, 102–110.

Vouillamoz, J. F., McGovern, P. E., Ergul, A., Söylemezoğlu, G., Tevzadze, G., Meredith, C. P., and Grando, M. S. (2006) Genetic characterization and relationships of traditional grape cultivars from Transcaucasia and Anatolia. *Plant Genet. Res.* **4**, 144–158.

Vouillamoz, J. F., Monaxo, A., Constantini, L., Stefanini, M., Scienza, A., and Grandon, M. S. (2007) The parentage of 'Sangiovese', the most important Italian wine grape. *Vitis* **46**, 19–22.

Walker, A. R., Lee, E., and Robinson, S. P. (2006) Two new grape cultivars, bud sports of Cabernet Sauvignon bearing pale-coloured berries, as the result of deletion of two regulatory genes of the berry colour locus. *Plant Mol. Biol.* **62**, 623–635.

Wang, X., Shi, X., Hao, B., Ge, B., and Luo, J. (2005) Duplication and DNA segmental loss in the rice genome: implications for diploidization. *New Phytol.* **165**, 937–946.

Watson, B., Lombard, P., Price, S., McDaniel, M., and Heatherbell, D. (1988) Evaluation of Pinot noir clones in Oregon. In: *Proceeding of the 2nd International Symposium for Cool Climate Viticulture and Oenology* (R. E. Smart *et al.*, eds.), pp. 276–278. New Zealand Society for Viticulture and Oenology, Auckland, New Zealand.

Wolpert, J. A. (1995) An overview of Pinot noir closes tested at UC-Davis. *Vineyard Winery Manage.* **21**, 18–21.

Zapriagaeva, V. I. (1964) Grapevine – *Vitis* L. In: *Wild Growing Fruits in Tadzhikistgan* (Russian, with English summary), pp. 542–559. Nauka, Moscow.

Zohary, D. E. (1995) The domestication of the grapevine *Vitis vinifera* L. in the Near East. In: *The Origins and Ancient History of Wine* (M. P. E. McGovern, S. J. Fleming, and S. H. Katz, eds.), pp. 23–30. Gordon Beach Publ., Luxembourg.

Zukovskij, P. M. (1950) *Cultivated Plants and their Wild Relatives*. State Publication House Soviet Science, Moscow. Abridged edition (1962) (P. S. Hudson, trans.) Commonwealth Agriculture Bureau, Wallingford, UK.

3

Grapevine Structure and Function

Vegetative Structure and Function

The uniqueness of some aspects of plant structure is obvious, even to the casual observer. However, many fundamental features become apparent only when studied microscopically. Unlike animal cells, plant cells are enclosed in rigid cell walls. Nevertheless, each cell initially possesses direct cytoplasmic connections with adjacent cells, through thin channels called plasmodesmata. Thus, embryonic plant tissue resembles one huge cell, divided into multiple interconnected compartments, each possessing cytoplasm and a single nucleus. As the cells differentiate, many die and the plant begins to resemble longitudinal, semi-independent cones of tissue, connected primarily by specialized conductive (vascular) tissue.

Most vascular cells elongate longitudinally, permitting the rapid movement of water and nutrients between superimposed sections of root and shoot tissue. The conduction of water and nutrients between adjacent cells (apoplasmic) is limited, and direct translocation between tissues on opposite sides of shoots and roots is nonexistent.

The vascular system consists of two structurally and functionally distinct components. The main water and mineral conducting elements, the xylem tracheae and tracheids, become functional on the disintegration of their cytoplasmic contents. The empty cell walls act as passive conduits. The primary cells translocating organic nutrients are the sieve tube elements of the phloem. They become functional only after their nuclei disintegrate.

Plants also show a distinctive growth habit. Growth in length typically occurs behind special embryonic (meristematic) cells located in the shoot and root tips. Growth in breadth is initially limited to that resulting from the enlargement of cells produced in the shoot or root tip. Further growth in diameter occurs when a circular band of cells, the vascular cambium, becomes active and produces cells both laterally (to the sides) and radially (around the circumference). In addition, plants show distinctive growth patterns that generate leaves, and their evolutionary derivatives, flower parts. In these latter plant organs, sites of growth (plate meristems) occur dispersed throughout the young leaf or flower part. Nonuniform rates and patterns of growth of the plate meristems generate the characteristic shape of the respective plant part.

The Root System

The root system possesses several cell types, tissues, and regions, each with a particular function. It also consists of a series of root types (fine lateral, leader, and large permanent roots), each with their own functions and relative life spans. Thus, the root system functions as a community of distinct but interconnected members, each contributing to success of the whole.

The root tip performs many of the most significant root functions, such as water and inorganic nutrient uptake, synthesis of growth regulators, and the elongation that pushes the root tip into regions untapped of their readily available water and nutrient supplies. Not surprisingly, this region is the most physiologically active, and has a highest risk of early mortality (Anderson *et al.*, 2003). In addition, the root tip is the site for the initiation of mycorrhizal associations, while secretions from the root cap soften the soil in advance of penetration. Root cap secretions also promote the development of a unique rhizosphere microbial flora on and around the root tip. Older mature portions of the root system transport water and inorganic nutrients upward to the shoot system, and organic nutrients downward to growing portions of the root. The outer secondary tissues of mature roots restrict water and nutrient loss from the root into the soil, and help to protect the root from parasitic and mechanical injury. Permanent parts of the root system anchor the vine and act as an important nutrient storage organ during the winter (Yang *et al.*, 1980). Reserves generally decline during the winter and spring months, associated with root maintenance, and shoot and leaf production, respectively. Reserve accumulation after shoot growth slows later in the year. A maximum is usually reached by late autumn. This probably explains why roots produced before bloom have the shortest life span (due to the lowest carbohydrate reserve) (Anderson *et al.*, 2003). Because of the significant differences in the structure and function of young and old roots, they are discussed separately.

THE YOUNG ROOT

Structurally and functionally, the young root can be divided into several zones. The most apical is the root tip, containing the root cap and apical meristem (the metabolic 'hot spot' of the root system). The latter consists of cells that remain embryonic as well as those that differentiate into the primary tissues of the root. Primary tissues are defined as those that develop from apically located meristematic tissues, whereas secondary tissues develop from laterally positioned meristematic tissues (cambia). The embryonic cells are concentrated in the center of the meristematic zone, called the quiescent region. In addition, this region is important as a major site for cytokinin and gibberellin synthesis. These growth regulators are translocated upward to the shoot, where they help maintain a balanced shoot/root ratio. They may also influence inflorescence (flower cluster) initiation and fruit development.

Surrounding the quiescent zone are rapidly dividing meristematic cells and their derivatives in early stages of differentiation. Cells at the apex develop into root-cap cells. These are short-lived cells that produce mucilaginous polysaccharides that ease root penetration into the soil. The cap also appears to cushion embryonic cells of the quiescent zone from physical damage during soil penetration. Laterally, cells differentiate into the epidermis, the hypodermis, and cortex, the last consisting of several layers of largely undifferentiated parenchyma cells. Behind the apical meristem, vascular tissues begin to differentiate and elongate. Cell enlargement occurs primarily along the axis of root growth and forces the extension of the root tip into the soil. This is primarily limited to a short (~2 mm) region called the elongation zone. Except for some enlargement in cell diameter,

most lateral root expansion results from the production of new (secondary) tissue behind the root apex.

Behind the elongation zone is the region in which most cells complete their differentiation (the differentiation zone). The most apical region is commonly called the root-hair zone, due to the fine extensions (root hairs) that grow out of epidermal cells. The formation, length, and period of activity of root hairs depend on many factors. For example, alkaline conditions suppress root-hair development, as does formation of a symbiotic association with mycorrhizal fungi. Root hairs are usually thought to significantly increase root–soil contact and, thereby, water and mineral uptake. However, due to their small size, direct evidence for this function is lacking. Root hairs also release organic nutrients into the soil, promoting the development of a unique microbial flora on and around the root. The flora helps to protect the root from soil-borne pathogens, and favors solubilization of inorganic soil nutrients. Root hairs are generally short-lived, and their collapse provides additional nutrients for microbial activity.

The first of the vascular tissues to differentiate is the phloem. Early phloem development facilitates the translocation of organic nutrients to the dividing and differentiating cells of the root tip. The xylem, involved in long-distance transport of water and inorganic nutrients, develops further back in the differentiation zone. This sequence probably evolved to avoid drawing water and inorganic nutrients away from the root tip. These compounds, locally absorbed, are required for the cell growth that typifies the region.

Directly adjacent to the primary xylem, and forming the innermost cortical layer, is the endodermis. The endodermis deposits a band of wax and lignin around its radial walls, called the Casparian strip. This prevents the diffusion of water and nutrients between the cortex and vascular cylinder. Thus, the movement of water and nutrients between these two regions must pass through the cytoplasm of the endodermal cells. This gives the root metabolic control over the transport of material into and out of the vascular tissues. Further back in the differentiation zone, the pericycle develops just inside of the endodermis. It generates a protective cork layer if the root continues to grow and differentiate. The pericycle also initiates lateral root growth.

Although still considered the major site for water and mineral uptake, the root-hair zone is no longer considered the only site so involved. Behind the root-hair zone, the epidermis eventually dies and the underlying hypodermis becomes encased in waxy suberin. Suberization occurs rapidly during the summer and may advance to include the root tip under dry conditions (Pratt, 1974). The endodermis also thickens and becomes nonuniformly suberized behind the root tip. These changes

reduce but do not prevent water uptake. The influence of suberization on mineral uptake depends on the ion involved. Water and potassium uptake by heavily suberized roots approximates 20 and 4%, respectively, of that absorbed by young unsuberized roots (Queen, 1968). In maturing sections of feeder roots, the cortical tissues collapse and the root becomes encased in cork, especially in dry soil. However, as only the inner wall surfaces of cork cells are suberized, and may be penetrated by plasmodesmata, diffusion of water and solutes through pores in the region of the wall between adjacent cells is possible (Atkinson, 1980).

These anatomical changes are reflected in a rapid decline in nitrate uptake and root respiration. Nevertheless, the observed drop in nitrogen uptake is more precipitous than would be expected solely based on morphological changes. Volder *et al.* (2005) found that nitrate uptake declined by about 50% within two days of emergence, and had fallen to about 10–20% within a few days. It subsequently remained relatively constant for a month or more. Respiration followed a similar trend. The source of the rapid decline in nitrogen uptake may be due to the rapid depletion of nitrate in the immediate vicinity of the growing tip. The marked decline in root functionality was also associated with a marked drop in root nitrogen concentration, presumably due to translocation to other parts of the vine. Nutrient and water uptake in many plants is largely regulated by the timing and duration of new root production.

Root life span depends on multiple factors. These include soil depth at which the roots are found, timing of emergence, and root diameter. For example, fine lateral roots had a life span of little more than 50 days, whereas deeper and coarser roots survived considerably longer (Anderson *et al.*, 2003). McLean *et al.* (1992) obtained similar results, finding at 60–80% of fine roots last no more than one season. Pruning treatment and grape yield also influence root longevity. For example, higher grape yields were correlated with reduced root viability, whereas roots of minimally pruned vines tended to remain metabolically active for a longer period than those of balanced pruned vines (Comas *et al.*, 2000).

Regardless of site, water uptake is predominantly induced by passive forces. In the spring, absorption and movement may be driven by the conversion of stored carbohydrates into soluble, osmotically active sugars. The negative osmotic potential produced generates what is termed root pressure. Root pressure produces the bleeding that occurs at cut ends of spurs and canes in the spring. Once the leaves have expanded sufficiently, transpiration creates the negative vascular pressures that maintain the upward flow of water and inorganic nutrients throughout the growing season. In contrast, mineral uptake and unloading into the xylem involve the expenditure of metabolic energy. Metabolic energy is also required for loading growth regulators synthesized in the root tip. The subsequent movement of inorganic nutrients and growth regulators appears to be passive, being simply transported along with water drawn up into the xylem. Only infrequently are organic molecules absorbed from the soil and translocated in the conducting elements of the root system. Examples include systemic fungicides and herbicides, some fungal toxins, and growth regulators produced by soil microorganisms.

Although essential to plant growth, the regulation of mineral uptake, and nutrient loading into the xylem sap are still incompletely understood. Mapfumo *et al.* (1994) found large lignified xylem vessels possessing protoplasm up to 22.5 cm back from the root tip. In several crops, such cells are involved in potassium accumulation prior to transport (McCully, 1994). Whether this is equally true for grapevines is unknown. Nevertheless, there is a marked remobilization of stored Ca, K, Mg, Zn, and Fe. Their movement out of the root in the spring is dependent on the flow rate in the xylem. In contrast, the movement of NH_4, NO_3, P, Cu, Na, and Cl appears to take place independently of xylem flow rate (Campbell and Strother, 1996a, 1996b).

MYCORRHIZAL ASSOCIATION

An important factor influencing mineral uptake in the majority of plants is the formation of a symbiotic association between the root tip and a mycorrhizal fungus.

Grapevines are among the more than 80% of plant species that develop mycorrhizal associations. Plants with weedy tendencies are an interesting exception. Such plants depend on their ability to germinate and rapidly produce a fine root-hair system to establish a foothold in disturbed sites. In contrast, most plants grow more slowly and frequently depend on mycorrhizal associations to exploit soil water and nutrient reserves. Of the three major groups of mycorrhizal fungi, only vesicular–arbuscular fungi invade grapevine roots (Schubert *et al.*, 1990).

Glomus species are the primary vesicular–arbuscular fungi associated with grapevines, although *Acaulospora*, *Gigaspora*, *Scutellospora* and *Sclerocystis* are occasionally involved. These fungi produce chlamydospores or sporocarps (containing many chlamydospores), from which infective hyphae invade the epidermis of host roots. From the epidermis, the fungus penetrates and colonizes the cortex. Here, the fungus produces large swollen vesicles and highly branched arbuscules. The fungi do not markedly alter root morphology, as other mycorrhizal associations tend to do (Fig. 3.1). Nonetheless, they do reduce root-hair development, increase lateral root production, and promote dichotomous root branching. Infection may result in the normally white root tip becoming yellowish. Although it reduces root elongation and soil exploration, mycorrhizal infection produces a more economical root system for nutrient acquisition (Schellenbaum *et al.*, 1991).

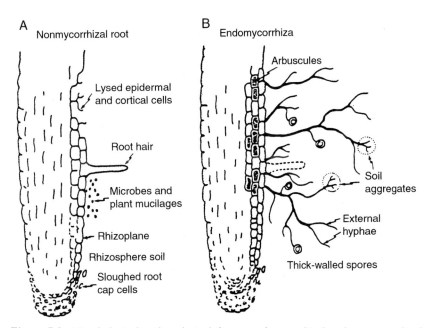

Figure 3.1 Morphological and ecological features of mycorrhizal and nonmycorrhizal roots. Mycorrhizal roots develop an equivalent of the rhizosphere called the mycorrhizosphere. (From Linderman, 1988, reproduced by permission)

New mycorrhizal associations begin near the root tip, and advance apically as the root grows and produces new cortical tissue. Mycorrhizal activity declines and ceases rearward, as the root matures and the cortex is sloughed off. Mycorrhizal roots are strong sinks for plant carbohydrates, which may partially explain why they tend to live longer than nonmycorrhizal roots.

As the fungus establishes itself in the cortex, wefts of hyphae grow outward and ramify extensively into the surrounding soil. The hyphal extensions absorb and translocate minerals and water back to the root. Mycorrhizal fungi are more effective at mineral absorption than the root itself. The high surface area to volume ratio of fungal hyphae, and their small diameter, permit their penetration into more minute soil pores than root hairs. In addition, the hyphae produce hydroxyamates (peptides) that combine with and facilitate nutrient uptake. The hyphae also secrete oxalate, citrate, and malate that mobilize various mineral elements. The effectiveness of mycorrhizal hyphae in mineral uptake is especially notable with poorly soluble inorganic nutrients, such as phosphorus, zinc, and copper. Phosphorus is one of the most immobile of essential soil nutrients, and its diffusion in dry soils may be reduced to 1–10% of that in moist soils.

In the cortex, sugars from the root provide the carbohydrates required for fungal growth. These are 'exchanged' for inorganic nutrients, notably phosphorus and ammonia, but also nickel, sulfur, manganese, boron, iron, zinc, copper, calcium and potassium. Mycorrhizal associations also appear to protect the vine from the toxic effects of soil contamination with lead and cadmium (Karagiannidis and Nikolaou, 2000). The benefits of mycorrhizal association are particularly noticeable in soils low in readily available nutrients and under dryland farming conditions. Under limiting growth conditions, the shoot conducts addition carbohydrate to the roots, which encourages not only mycorrhizal development but also fungal growth. In contrast, where access to inorganic nutrients is fully adequate, less carbohydrate is allocated to the roots and mycorrhizal growth is limited. Under drought conditions, mycorrhizal fungi augment water uptake and transport to the host, enhance stomatal conductance and transpiration, and accelerate recovery from stress (see Allen *et al.*, 2003).

Both root and mycorrhizal exudates appear to influence soil texture and the soil flora. These beneficial effects partially result from the selective favoring of nitrogen-fixing and ethylene-synthesizing bacteria around the roots (Meyer and Linderman, 1986). These changes usually increase root resistance, or at least tolerance to fungal and nematode infections. For example, inoculation of roots with *Glomus mosseae* reduced

the incidence of grapevine 'replant disease' in affected nursery soil (Waschkies *et al.*, 1994). Mycorrhizal associations also tend to reduce vine sensitivity to salinity and mineral toxicities. By affecting the level of growth regulators, such as cytokinins, gibberellins, and ethylene, mycorrhizal fungi further influence vine growth. An additional indirect effect may be the promotion of chelator production, such as catechols by soil bacteria. These help keep minerals, such as iron, available for plant growth. Finally, mycorrhizae contribute significantly to the formation of a stable soil crumb structure.

In most vineyard soils, mycorrhizal associations arise spontaneously from the naturally occurring soil inoculum. Artificial inoculation of cuttings in a nursery is complicated and often ineffective, due to extensive rootlet mortality following transplantation in the vineyard (Conner and Thomas, 1981). Artificial inoculation tends to be more effective when apical tissues are micropropagated, or when vines are planted in fumigated soils devoid of an indigenous inoculum. Although spontaneous infection is poorly understood, it is known to be influenced by soil type and soil cover, retarded by fungicidal applications, facilitated by the rhizosphere flora, and modified by the indigenous mycorrhizal population. Soils low in available phosphorus favor mycorrhizal association (Karagiannidis *et al.*, 1997), whereas soils rich in nutrients limit its development. Differences in the incidence of colonization also occur among rootstocks (usually higher in cultivars imparting greater vigor to the scion). Nonetheless, these differences are small in comparison to factors such as crop load (higher yield associated with reduced numbers of roots with arbuscules) (Schreiner, 2003).

Although mycorrhizal association is usually beneficial, there are situations where its formation may be detrimental. If phosphorus is seriously deficient, colonization by mycorrhizal fungi may not overcome the deficiency (Ryan *et al.*, 2005). Under such conditions, metabolism of plant carbohydrates by the mycorrhizal fungus may reduce crop yield due to its 'parasitic' action. In addition, under some circumstances, mycorrhizae may increase the availability of, and thereby toxicity of, trace elements such as aluminum. Mycorrhizal fungi can also occasionally increase the incidence of lime-induced chlorosis (Biricolti *et al.*, 1992).

SECONDARY TISSUE DEVELOPMENT

As noted above, the epidermis and root cortex often become infected by mycorrhizal fungi. These tissues also may succumb to limited microbial and nematode attack. The latter probably generates the brown and collapsed regions commonly observed along otherwise healthy young roots. Regardless of health, the cortex and outer tissues soon die, especially if the central

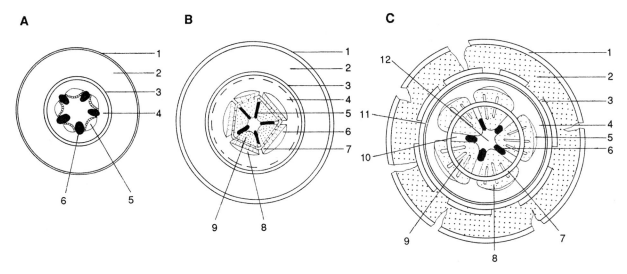

Figure 3.2 Diagrammatic representation of the secondary growth in grapevine roots: 1, epidermis; 2, cortex; 3, endodermis; 4, pericycle; 5, primary phloem; 6, primary xylem; 7, vascular cambium; 8, secondary phloem; 9, secondary xylem; 10, medullary rays; 11, periderm; 12, pith. (From Swanepoel and de Villiers, 1988, reproduced by permission)

region of the root commences secondary (lateral) growth (Fig. 3.2). Secondary growth includes the production of a cork layer (periderm) next to the endodermis. Its formation restricts and eventually cuts off water and nutrient flow to the cortex and epidermis. A lateral meristem (vascular cambium) develops between the initial (primary) xylem and phloem cells, and produces new (secondary) vascular tissues – xylem and phloem to the inside and outside of the cambium, respectively. Periodically, ray cells are generated from the vascular cambium. These act to facilitate the movement of water and nutrients between adjacent regions of phloem and xylem. As the root enlarges in diameter, new cork layers may develop in the secondary phloem, cutting off the older cork layers and nonfunctional phloem.

ROOT-SYSTEM DEVELOPMENT

In exploiting the soil, the root system employs both extension growth and branching. Extension growth entails the rapid growth of thick leader roots into unoccupied soil. If the site is favorable, many thin, highly branched, lateral roots develop from the leader root. Most of the lateral (side) roots are short-lived and are replaced by new laterals. Thus, few laterals survive and add to the permanent structural framework of the root system. Large roots are retained much longer than most young roots, but are themselves subject to being aborted periodically (McKenry, 1984).

The largest roots generally develop about 0.3–0.35 m below ground level. Their number and distribution tend to stabilize a few years after planting (Fig. 3.3). From this basic framework, smaller permanent roots spread horizontally outward and vertically downward

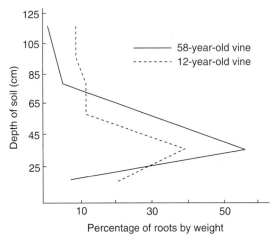

Figure 3.3 Influence of vine age on root-system distribution. (From Branas and Vergnes, 1957, with permission)

into the surrounding soil. These, in turn, produce the short-lived lateral (feeder) roots of the vine. Tentative evidence suggests that a ratio of >3.5 for fine roots (<2 mm) to larger roots (>2 mm) indicates an effective root system (Archer and Hunter, 2005). That permits the vine to quickly take advantage of favorable conditions to fully mature the fruit.

Root-system expansion is dependent on both environmental and genetic factors. Spread may be limited by layers or regions of compacted soil, high water tables, or saline, mineral, and acidic zones. Thus, it is imperative that the soil be properly assessed and corrective measures taken before planting. Tillage, mulching, and irrigation favor the optimal positioning of major root development, but cannot offset undesirable soil conditions. Increasing planting density tends to decrease root mass, but increases root density (roots per soil volume). Differences in the angle and depth of penetration vary with the genetic potentialities of the rootstock and rootstock–shoot interaction. For example, *V. rupestris* rootstocks are generally thought to sink their roots at a steeper angle and penetrate more deeply than *V. riparia* rootstocks. The latter produce a shallower, more spreading root system (Bouard, 1980). However, an extensive review of literature by Smart *et al.* (2006) suggests that local soil conditions are more significant than genetic factors in affecting root distribution.

Grapevines are characterized by their extensive colonization of the soil, principally laterally, but at low rooting densities (Zapriagaeva, 1964). Grapevine roots commonly occupy only about 0.05% of the available soil volume (McKenry, 1984). Most root growth develops within a zone 4–8 m around the trunk. Depth of penetration varies widely, but is largely confined to the top 60 cm of the soil. Nevertheless, roots up to 1 cm in diameter can penetrate to a depth of more than 6 m. The tendency for deep penetration may reflect evolutionary pressures related to competition with the roots of trees on which they tended to climb. The finest roots, which do most of the absorption, generally occur within the uppermost 0.1–0.6 m of the soil. This portion of the root system may be located lower, depending on the depth of tilling, or the presence of a permanent ground cover. Conversely, mulches or clean, no-till conditions favor root growth nearer the soil surface.

Unlike the situation in many other woody perennials, resumption of root growth in the spring lags significantly behind shoot growth. New root production initially commences slowly after bud break, and only accelerates significantly just before anthesis. Peak production may coincide with anthesis or fruit set (Fig. 3.4). A second, autumnal root-growth period may occur in warm climatic regions, where the postharvest growth period may be long (van Zyl and van Huyssteen, 1987). In temperate climates, root growth tends to be unimodal. Growth commences slowly (at temperatures >6 °C) following bud break, occasionally reaching a peak only in late summer (Mohr, 1996). In contrast, Comas *et al.* (2005) found that root production reached a peak near midseason (between bloom and *véraison*). Root production closely followed canopy production, but was also influenced by yearly fluctuations in climatic conditions. Heavy pruning and delayed canopy development also affected root formation. Signs of root mortality appear most prominently in association with fruit development and ripening. Root production is most pronounced in years following high yields, when starch levels in the canes are low.

The effect of soil moisture on root production appears to be primarily indirect, by affecting soil aeration and mechanical resistance. Some cultivars appear to be particularly sensitive to anaerobiosis, with oxygen tensions as low as 2% being toxic (Iwasaki, 1972). Waterlogging also enhances salt toxicity (West and Taylor, 1984). Furthermore, soil moisture content affects the rate at which soil warms in the spring, slowing the resumption of root growth, and the availability and movement of nutrients in the soil. Even relatively short periods of waterlogging (2 days) can significantly retard root growth (McLachlan *et al.*, 1993). Root growth and distribution are also markedly influenced by soil cultivation, compaction, and irrigation.

The Shoot System

The shoot system of grapevines possesses an unusually complex developmental pattern. This complexity provides the vine with a remarkable ability to adjust its development throughout much of the growing season. Three or more successive sets of shoots may develop in a single year. In addition, dormant buds from previous seasons may become active. In recognition of this complexity, shoots may be designated according to their origin, age, position, or length. The buds from which

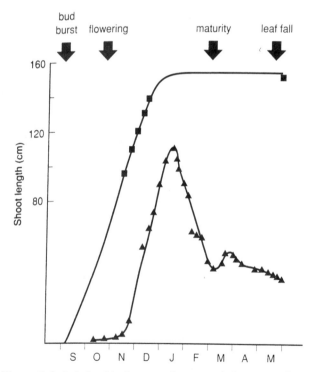

Figure 3.4 Relationship between the state of shoot growth (■) and root growth (▲) for the variety 'Shiraz' grown in the Southern Hemisphere. (After Richards, 1983; from Freeman and Smart, 1976, reproduced by permission)

shoots arise are identified as to their position, germination sequence, and fertility.

BUDS

All buds (immature shoot systems) produced by grapevines may be classed as **axillary buds**. That is, they are formed in the axils of foliar leaves, or their modifications (bracts). Axils are defined as the positions along shoots where leaves develop. As such, axils are a part of a circular region of the stem called the **node** (see Fig. 2.1). Most structures that develop from shoots – leaves, buds, tendrils, and inflorescences – develop at nodes. The first two nodes of a shoot are very close together, and subtend bracts (modified leaves). Only the subsequent nodes that generate photosynthetic leaves are usually considered when counting nodes on a shoot. These are frequently referred to as **count nodes**.

Most stem elongation occurs in the region separating adjacent nodes, the **internodes**. Buds occurring in the axils of bracts (leaves modified for bud protection) are typically termed **accessory buds**. When located at the base of mature canes, however, they are more commonly referred to as **base buds**. Buds of any kind that remain dormant for one or more seasons may also be designated **latent buds**. Depending on their location, latent buds may give rise to **water sprouts** or **suckers** (see Fig. 4.11), based on whether the latent bud originated above or below ground level, respectively.

Each shoot node potentially can develop an axillary-bud complex, consisting of four buds. These include the **lateral** (prompt or true axillary) **bud**, positioned to the dorsal side of the shoot, and a **compound** (latent) **bud**, which is positioned more ventrally (Fig. 3.5; Plate 3.1). The compound bud typically possesses three buds in differing states of development, referred to as the primary, secondary, and tertiary buds.

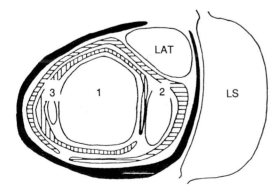

Figure 3.5 Transverse section through an axillary bud complex. LS, leaf scar; LAT, lateral bud; primary (1), secondary (2), and tertiary (3) buds of the compound bud, bracts of the lateral (solid) and primary (hatched) structures are also shown. (From Pratt, 1959, reproduced by permission)

Depending on genetic and environmental factors, lateral buds may differentiate into shoots during the year they are produced. More commonly, they may remain dormant, becoming latent buds. Shoots developing from lateral buds are termed **lateral** (summer) **shoots**.

There is considerable dissension among grape growers on whether to retain, trim or remove lateral shoots. Initially lateral shoots act as a drain on vine photosynthate. However, later they can act as net exporters – when possessing two or more fully expanded leaves (Hale and Weaver, 1962). Lateral shoots (especially those that terminate growth in mid-season) can benefit fruit ripening (Candolfi-Vasconcelos and Koblet, 1990), but may reduce fruit set earlier in the season (Vasconcelos and Castagnoli, 2000). These divergent effects may be explained by the relative maturity of the leaves at different stages in fruit production. In vigorous vines, lateral shoots can lead to dense canopies with excessive shading, reduced fruit quality, and enhanced disease susceptibility. In addition, fruit-bearing lateral shoots may produce what is called a 'second' crop. Lateral buds, formed in the leaf axils of lateral shoots, also possess the potential to develop during the current year. If they do, they may generate a second series of lateral shoots.

The **primary bud** of a compound bud develops in the axil of the bract (prophyll) produced by the lateral bud. **Secondary** and **tertiary buds** develop in the axils of the bracts produced by the primary and secondary buds, respectively. All three buds typically remain inactive during the growing season in which they form, unless stimulated by severe summer pruning. Subsequently, the compound bud develops endogenous (self-imposed) dormancy and is now usually referred to as a **dormant bud** ('eye').

During the development of dormancy, the moisture content of buds declines from about 80% to 50%, starch grains become prominent, signs of mitosis are absent, catalase activity declines, and respiration falls to its lowest level (Pouget, 1963). The mechanism of dormancy induction is still unknown, but is associated with an increase in the content of *cis*-abscisic acid (Koussa *et al.*, 1994), and the production of a particular set of glycoproteins (Salzman *et al.*, 1996). The subsequent release from dormancy is usually thought to require exposure to cold temperatures during the late fall and winter months. The mechanism of bud break is unknown, but may be associated with an increase in hydrogen peroxide production and the development of oxidative stress associated with low levels of catalase. This is consistent with the action of hydrogen cyanamide in promoting early bud break (Lavee and May, 1997). Hydrogen cyanamide depresses transcription of the catalase gene (Or *et al.*, 2002). Hydrogen cyanamide application is also associated with activation of genes that produce a dormancy-breaking protein kinase (GDBRPK), and genes producing pyruvate decarboxylase and alcohol dehydrogenase in buds (Or et al., 2000). In tropical and subtropical regions, severe pruning usually can induce premature bud break. However, this treatment is more effective and uniform if combined with hydrogen cyanamide treatment.

Assuming that the primary buds are not destroyed by freezing temperatures, insect damage, pathogenic influences or physiological disturbances, they generate the **primary shoots** (major shoot systems) of each year's growth. The secondary and tertiary buds become active only if the primary and secondary buds, respectively, die.

The primary bud is the most developed in the compound bud. It usually possesses several primordial (embryonic) leaves, inflorescences, and lateral buds before becoming dormant at the end of the season (Morrison, 1991). The secondary bud occasionally is fertile, but the tertiary bud is typically infertile (does not bear inflorescences). The degree of inflorescence differentiation, prior to the onset of dormancy, is a function of the genetic characteristics of the cultivar, vigor of the rootstock, bud location, and the surrounding environment during bud formation.

Buds that bear nascent (primordial) inflorescences are termed **fruit** (fertile) **buds**, whereas those that do not are called **leaf** (sterile) **buds**. In most V. vinifera cultivars, fruit buds form with increasing frequency up from the basal leaf, often reaching a peak between the fourth and tenth nodes. However, this property varies considerably among cultivars (Fig. 3.6). Some cultivars, such as 'Nebbiolo,' do not form fruit buds at the base of the

shoot. Such cultivars are not spur-pruned, as it would drastically limit fruit production. Bud fruitfulness also influences pruning procedures. For example, the production of small fruit clusters by 'Pinot noir' usually requires the retention of many buds on long canes.

Shoots generally produce two (one to four) flower clusters per shoot. Flower clusters usually develop opposite the third and fourth, fourth and fifth, or fifth and sixth leaves on a shoot. Although inflorescence number and position are predominantly scion characteristics, they may be modified by hormonal signals coming from the rootstock (see Richards, 1983).

SHOOTS AND SHOOT GROWTH

Components of the shoot system obtain their names from the type and position of the bud from which they are derived, as well as their age, position, and relative length (Fig. 4.11). Once the outer photosynthetic subepidermal tissues degenerate and turn brown, the shoots are termed **canes**. If a cane is pruned to possess only a few buds, it is designated a **spur**. Canes retained for 2 years or more, and supporting fruiting wood (spurs or canes), are called **arms**. When arms are positioned horizontally, they are referred to as **cordons**. The **trunk** is the major permanent upright structure of the vine. When it is old and thick, trunks need no support. Nevertheless, most vines require trellising to support the arms, canes, and growing shoots of the vine. All stem tissue 2 or more years old is designated **old wood**.

Continued shoot growth is controlled primarily by environmental factors. Because there are no terminal buds, growth could theoretically continue as long as climatic conditions permit. This **indeterminant** growth

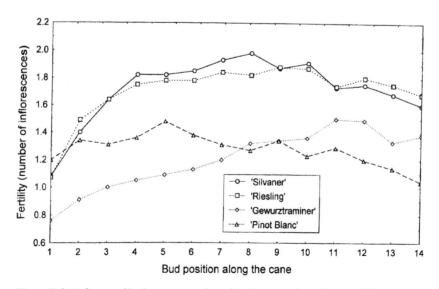

Figure 3.6 Influence of bud position on shoot fertility. (Data from Huglin, 1986)

habit is much more characteristic of tropical than temperate woody plants. Growth is favored by warm conditions, especially warm nights. Low-light conditions promote shoot elongation, but are detrimental to inflorescence induction. Genetic factors, probably acting through hormone production, influence the characteristic growth patterns of various cultivars. These tendencies can be modified by pruning and other cultural practices. For example, minimal pruning induces more, but shorter and thinner shoots.

Bud break and shoot growth have generally been thought to begin when the mean daily temperature reaches 10 °C or above, following the loss of endogenous dormancy. However, bud break in some varieties may commence at temperatures as low as 0.4 °C (average 3.5 °C) and leaf production may begin at 5 °C (average 7 °C) (Moncur *et al.*, 1989). The rate of bud break and shoot growth increases rapidly above the minimum temperature (Fig. 3.7). Once initiated, growth rapidly reaches a maximum, after which growth progressively slows and may cease. Further shoot growth during the summer generally originates from the activation of lateral buds.

The slow initiation of growth in the spring and the potential for lateral shoot growth throughout the season probably reflect the ancestral growth habit of the vine. Slow bud activation in the spring would have delayed leaf production until the supporting tree had completed most of its foliage production. Thus, the vine could position its leaves, relative to the foliage of

the host, in locations optimally suited for vine photosynthesis. In addition, the potential to generate several series of lateral shoots allows the vine to position new leaves in favorable sites where and when necessary.

Older portions of the vine provide the support and translocation needs of the growing shoots, leaves, and fruit. The woody parts of the vine constitute the majority of its structure. Mature wood also acts as a significant storage organ, thereby helping to cushion the effects of unfavorable growth conditions on fruit production. These reserves also tend to increase average vine productivity. Berry sugar content and pH increase slightly, relative to the portion of old wood (Koblet *et al.*, 1994). In addition, increased berry aroma and fruit flavor have been correlated with vine age (Heymann and Noble, 1987), or with the proportion of old wood (Reynolds *et al.*, 1994).

Shoot growth early in the season depends primarily on previously stored nutrients in the woody parts of the vine (May, 1987). Nitrogen mobilization initiates in the canes and progresses downward into older parts of the vine, finally reaching the roots (Conradie, 1991b). Significant mobilization and translocation of nutrient reserves to the developing fruit also may occur following *véraison* (the onset of color change in the fruit). The vine stores organic nutrients, primarily starch and arginine, as well as inorganic nutrients such as potassium, phosphorus, zinc, and iron. The movement of nutrients, such as nitrogen, into woody parts of the grapevine probably occurs throughout the growing season, but is most marked following fruit ripening (Conradie, 1990, 1991a). It appears that some of the first nutrients mobilized in the spring are those last stored in the fall.

Although shoot growth can continue into the fall, this is usually undesirable. Not only is shoot maturation delayed and bud survival reduced, but it also draws nutrients away from the ripening fruit. Typically, new shoot growth should ideally cease with the onset of *véraison*. Various procedures may be used to promote the termination of new vegetative growth. These may include inducing limited water stress, trimming off the shoot tips, or use of devigorating rootstocks. In hot Mediterranean climates, shoot growth may terminate shortly after flowering, when little more than a meter long. In contrast, vigorous shoots may reach ≥4 m in length.

TISSUE DEVELOPMENT

Shoot growth develops from an apical meristem consisting of **tunica** cells covering an inner collection of **corpus** cells (Fig. 3.8). The outer tunica layer produces the epidermal tissues of the stem, leaves, tendrils, flowers, and fruit, as well as initiating the development of

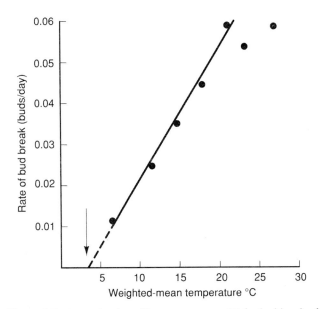

Figure 3.7 Determination of base temperature (↓) for bud break of 'Pinot noir' from the effect of temperature on the rate of bud break. (From Moncur *et al.*, 1989, reproduced by permission)

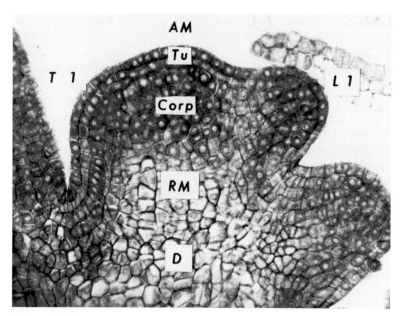

Figure 3.8 Median longitudinal section through a dormant shoot apex of *Vitis labrusca*. AM, Apical meristem; Corp, corpus; D, diaphragm; L, leaf primordium; RM, rib meristem; T, tendril initiation; Tu, two-layered tunica. (From Pratt, 1959, reproduced by permission)

leaf and flower primordia. The corpus generates the inner tissues of the shoot and associated organs. The shoot apex, unlike its root equivalent, has a highly complex morphology. Buds, leaves, tendrils, and flower clusters all begin their development within a few millimeters of the apex (Fig. 3.9). Subsequent cell division, differentiation, and enlargement or elongation produces the mature structures of each organ.

Outgrowths of the vascular tissue into the leaves and other subtended organs, called traces, are equally complex. They translocate nutrients to and from the developing structures.

As in the root, phloem is the first vascular tissue to differentiate. The need for organic nutrients by the rapidly dividing apical tissues undoubtedly explains this early differentiation. Cell elongation, which primarily entails water uptake, occurs later. Xylem differentiation occurs further back. As tissue elongation ruptures the connections between the initially formed phloem cells, additional sieve tubes differentiate to replace those destroyed.

The outer tissues of the young stem consist of a layer of epidermis and several layers of cortical cells (Fig. 3.10). The epidermis is initially photosynthetic and bears stomata, hair cells and pearl glands, especially along the veins. Most cortical cells also contain chloroplasts and are photosynthetic. Collections of cells with especially thickened side walls (collenchyma) differentiate opposite the vascular bundles. These regions generate the ridges (ribs) of the young shoot.

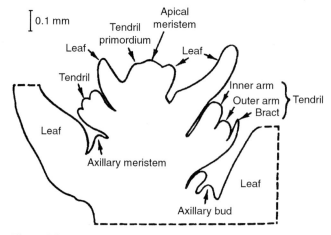

Figure 3.9 Longitudinal section of the shoot tip of 'Concord' grape. (From Pratt, 1974, reproduced by permission)

Under the cortical tissues lie several vascular bundles, consisting of phloem on the outside and xylem on the inside. Interior to this region are relatively undifferentiated parenchyma cells. These become increasingly large toward the center. They constitute the pith.

When cell elongation ceases, a layer of cells between the phloem and xylem differentiates into a lateral meristem, the **vascular cambium**. It generates the secondary vascular and ray tissues of the maturing stem (Plate 3.2). The secondary phloem, similar to the primary, contains translocating cells (sieve tubes), companion cells, fibers

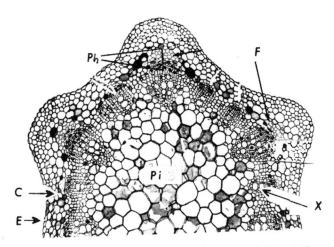

Figure 3.10 Transverse section of a seedling stem of *Vitis vinifera*. E, epidermis; C, cortex; F, collenchyma fibers; Ph, phloem; X, xylem; Pi, pith. (Modified from Esau, 1948)

and storage parenchyma. The secondary xylem consists of large-diameter xylem vessels, tracheids, as well as structural fibers. Ray cells elongate horizontally along the stem radius and transport nutrients between the xylem and phloem. In the phloem, ray tissue expands radially to form V-shaped segments (Plate 3.2). These cells are extensively involved in the storage of starch, along with xylem parenchyma. As secondary phloem and xylem accumulate, the vascular bundles take on the appearance of elongated lenses in cross-section. The innermost stem tissue, the pith, consists predominantly of thin-walled parenchyma cells. In species belonging to the section *Vitis*, the pith soon disintegrates in all but the nodal region, where it develops into the woody **diaphragm** (Fig. 2.3A).

Environmental conditions during growth can significantly affect xylem development. For example, leaf shading (Schultz and Matthews, 1993) and water stress (Lovisolo and Schubert, 1998) slow growth, possibly as a consequence of reduced water conductance and disrupted hormonal movement. Positioning shoots downward also reduces vessel diameter, while increasing their density (Schubert *et al.*, 1999). This influence has occasionally been used to reduce vine vigor. The effects of shoot orientation on water conductance appear to be related to increased auxin levels (Lovisolo *et al.*, 2002). This promotes vessel division but limits expansion. Smaller diameter vessels increase resistance to water flow, but reduce the likelihood of embolism (the formation of gas pockets in vessels that break direct water flow).

During shoot maturation, a layer of parenchyma cells in the phloem differentiates into a cork cambium (phellogen) (Plate 3.3). The cork (periderm) produced prevents nutrient and water supplies from reaching the outer tissues (mostly cortex and epidermis). These

tissues subsequently die, begin to slough off, and create the brown appearance of maturing shoots.

In most regions, with the probable exception of the tropics, the end walls (sieve plates) of the sieve tubes become plugged with callose in the autumn, terminating translocation of organic nutrients. This occurs as the leaves die and are shed.

When buds become active in the spring, phloem cells progressively regain their ability to translocate. Activation progresses longitudinally up and down the cane from each bud, and outward from the vascular cambium (Aloni and Peterson, 1991). The release of auxins from buds appears to be involved in stimulating callose breakdown (Aloni *et al.*, 1991). Although cambial reactivation is also associated with bud activation, it shows marked apical dominance. Thus, activation starts in the uppermost buds and progresses downward, and laterally around the canes and trunk until the enlarging discontinuous patches meet (Esau, 1948). The activated cambium produces new xylem cells to the interior and phloem cells to the exterior (Plate 3.4). Unlike either the phloem or cambium, mature xylem vessels contain no living material and are potentially functional whenever the temperature permits water to exist in a liquid state.

Phloem cells in the trunk may remain viable for up to 4 years, but most sieve tubes are functionally active only during the year in which they form (Aloni and Peterson, 1991). In the xylem, vessel inactivation begins about 2–3 years after formation, but is complete only after 6–7 years. Inactivation involves tyloses that grow into the vessel cavity from surrounding parenchyma cells.

In members of the subgenus *Vitis*, new cork cambia develop at infrequent intervals in nontranslocatory regions of the secondary phloem. The outer tissues (largely old phloem, and initially remnants of the cortex and epidermis) die and turn brown. The tissues subsequently split and are eventually shed. In contrast, in the *Muscadinia*, the cork cambium that forms under the epidermis persists for several years. Thus, the canes of the *Muscadinia* do not form shedding bark like *Vitis* species (Fig. 2.3D).

Tendrils

Tendrils are considered to be modified flower clusters, which are themselves viewed as modified shoots. Not surprisingly, tendrils bear obvious morphological similarities to shoots. Unlike vegetative shoots, tendril growth is **determinant**, that is, its growth is strictly limited. Tendril development also passes through three developmental and functional phases. Initially, tendrils develop water-secreting openings called **hydathodes** at their tips. Subsequently, the hydathodes degenerate and pressure-sensitive cells develop along the tendril. On contact with solid objects, the pressure-sensitive cells activate the elongation and growth of cells on the opposite side of the tendril. This induces twining of the tendril around the object touched. With the development of collenchyma cells in the cortex and xylem, and lignification of the ray cells, the tendril becomes woody and rigid at maturity.

With the exception of *Vitis labrusca* and its cultivars, tendril production develops in a discontinuous manner in *Vitis*. In other words, tendrils are produced opposite the first two of every three leaves, distal to the first two to three basal leaves. In *V. labrusca*, tendrils are produced opposite most leaves. On bearing shoots, flower clusters replace the tendrils in the lower two or more locations in most *V. vinifera* cultivars, and in the basal three to four tendril locations in *V. labrusca* cultivars.

Leaves

Leaves initially develop as localized outgrowths in the inner tunica of growing shoots. Corpus cells differentiate to connect the developing vascular system of the stem (traces) with that of the expanding leaf. The first two leaf-like structures of a shoot develop into bracts. The internodes between the bracts are very short. Subsequent internodes elongate normally, and the leaf primordia subtended expand into mature leaves typical of the cultivar.

Maturing leaves consist of a broad photosynthetically active blade, the supportive and conductive petiole, and two basal semicircular stipules. The last soon die and dehisce, leaving only the petiole and blade. Unlike the growth of most plant structures, leaf growth is not induced by apical or lateral meristems. Leaf growth entails the action of many, variously positioned, **plate meristems** (Fig. 3.11). These generate the flat lobed appearance of the leaf.

The leaf blade consists of an upper and lower epidermis, a single palisade layer, about three layers of spongy mesophyll, and a few large (and several small) veins (Fig. 3.12). The upper epidermis is covered by a waxy cuticle consisting of an inner layer of cutin and several outer waxy plates on the surface. The cuticle retards water and solute loss, helps limit the adherence of pathogens, minimizes mechanical abrasion of the leaf, and slows the diffusion of chemicals into the leaf. The lower epidermis shows a less well developed cuticle, but contains more stomata and commonly possesses leaf hairs and pearl glands. The latter derive their name from the small bead-like secretions they produce. Water-secreting hydathodes commonly develop at the pointed tips (teeth) of the blade.

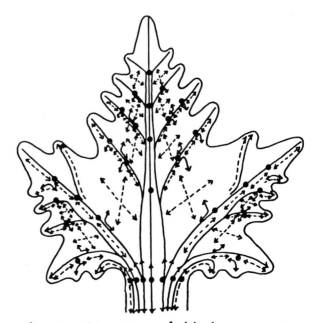

←--→ : Direction of blade expansion. ←--→ : Direction of development of vessels. ● : Area of slow down or cessation of growth.

Figure 3.11 Illustration of a young grape leaf ('Pinot noir' C.3309) showing the directions of blade expansion and xylem differentiation. Blade expansion occurs by intercalary growth along and between the major veins. Vessel differentiation occurs away from areas where growth has slowed or ceased. (From Fournioux, 1972, reproduced by permission)

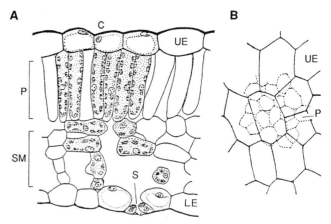

Figure 3.12 Transverse (**A**) and epidermal view (**B**) of a grape leaf. C, Cuticle; LE, lower epidermis; P, palisade layer; S, stomatal apparatus with guard cells; SM, spongy mesophyll; UE, upper epidermis. (After Mounts, 1932, reproduced by permission)

The palisade layer consists of cells directly below and elongated at right-angles to the upper epidermis. When the leaf is young, the cells are tightly packed, but intercellular spaces develop as the leaf matures.

In leaves that develop in bright sunlight, the palisade cells are shorter but thicker than those formed in the shade. Cells of the palisade layer are the primary photosynthetic cells of the plant. Directly below the palisade layer are five to six layers of photosynthetic spongy mesophyll. The number of mesophyll layers decreases when the leaf develops under brightly lit conditions. The cells of the mesophyll are extensively lobed and generate large intracellular spaces in the lower half of the leaf. The large surface area generated, along with the stomata, facilitate the diffusion of water and gases between the inner leaf cells and the surrounding air. Without efficient evapotranspiration, the leaf would rapidly overheat in full sun, suppressing photosynthesis. Effective gas circulation is equally important for the rapid exchange of CO_2 and O_2. Carbon dioxide is an essential ingredient in photosynthesis, and oxygen (one of its by-products) inhibits the crucial action of ribulose bisphosphate carboxylase (RuBPCase).

The branching vascular network of the leaf consists of a few large veins containing several vascular bundles, and many smaller veins, containing a single bundle. Each vascular bundle consists of xylem in its upper portion and phloem on the lower side. The vascular tissues are surrounded by a set of thickened cells, the bundle sheath. The latter often extends upward and downward to the upper and lower epidermis. Xylem translocates water and minerals such as calcium, manganese and zinc, whereas the phloem transports organic compounds and minerals such as potassium, phosphorus, sulfur, magnesium, boron, iron and copper. The organic compounds include sugars, amino acids, and growth regulators.

Because of the divisions imposed by the bundle sheath extensions covering the veins, the leaf is divided into many airtight compartments with little lateral gas interchange. Under drought, low atmospheric humidity, or saline conditions, each compartment may show distinct differences (patchiness) in their rate of transpiration and photosynthesis (Düring and Loveys, 1996). Leaves of this kind are termed **heterobaric**.

In the autumn, abscission layers form at the base of the petiole and leaf blade. The connections between the shoot and the leaf cease when a periderm forms between the stem and the abscission layer.

PHOTOSYNTHESIS AND OTHER LIGHT-ACTIVATED PROCESSES

The production of organic compounds from carbon dioxide and water, using sunlight, is the quintessential attribute of plant life. While sugar is the major organic by-product of photosynthesis, it also generates the intermediates for the synthesis of fatty acids, amino acids, and nitrogen bases. From these, the plant can

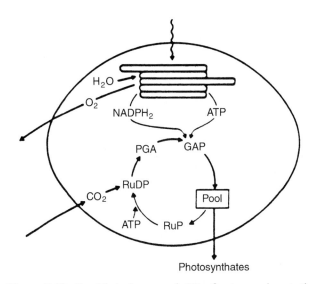

Figure 3.13 Simplified diagram of CO_2 fixation and assimilation via the Calvin cycle. GAP, glyceraldehyde-3-phosphate; PGA, 3-phosphoglyceric acid; RuBP, ribulose bisphosphate; RuP, ribulose phosphate. (From Larcher, 1991)

derive all its organic needs. The process is summarized in Fig. 3.13.

Chlorophyll molecules, in a complex called photosystem II (PS II), captures light energy. This is used to split water molecules into its component elements. Oxygen atoms combine to form molecular oxygen that escapes into the surrounding air, whereas the hydrogen ions dissolve in the lumen (central fluid) of the chloroplast thylakoid. Energized electrons, removed from hydrogen atoms, are passed along an electron-transport chain embedded in the thylakoid membrane. Associated with the electron transport, the hydrogen ions are pumped across the membrane into the central fluid (stroma) of the chloroplast. Because the thylakoid membrane is highly impermeable to hydrogen ions, an electrochemical differential develops across the membrane. Hydrogen ions move back into the thylakoid through ATPase complexes embedded in the thylakoid membrane. Connection of this reentry with phosphorylation generates adenosine triphosphate (ATP), from adenosine diphosphate (ADP) and inorganic phosphate. The partially de-energized electrons are passed along a transport chain to chlorophylls in photosystem I (PS I), also situated in the thylakoid membrane. These electrons are further energized by photon energy absorbed PS I. The electrons are subsequently passed to a second electron transport chain, finally being transferred to nicotinamide adenine dinucleotide phosphate ($NADP^+$). This results in the production of reducing power (NADPH). Both NADPH and ATP are required in the Calvin cycle for the synthesis of sugars during the 'dark' CO_2-fixing reactions of photosynthesis.

The initial by-product of CO_2 fixation, with RuBP (ribulose bisphosphate), is extremely unstable and splits almost instantaneously into two molecules of PGA (3-phosphoglyceric acid). As sufficient carbon dioxide is incorporated, intermediates of the Calvin cycle can be directed toward the synthesis of sucrose and other organic compounds (Fig. 3.13). Quantitatively, sucrose is the most important by-product of photosynthesis, as well as being the major organic compound translocated in the phloem. Sucrose may be stored temporarily, as starch in the leaf, but is usually translocated out of the leaf to other parts of the plant. It may be stored subsequently as starch, polymerized into structural components of the cell wall, metabolized into any of the other organic components, or respired as an energy source for cellular metabolism.

Because photosynthesis is fundamental to plant function, providing an optimal environment for its occurrence is one of the most essential aspects of viticulture, especially in canopy management. A favorable light environment also influences other photo-activated processes. These include such vital aspects as inflorescence initiation, fruit ripening, and cane maturation.

For photosynthesis, radiation in the blue and red regions of the visible spectrum is particularly important. For most other light-activated processes, it is the balance between the red and far-red portions of the electromagnetic spectrum that is crucial. These differences relate to both the energy requirements and the pigments involved.

In photosynthesis, the important pigments are chlorophylls and carotenoids, whereas in most other light-activated processes, phytochrome is involved. Both chlorophylls *a* and *b* absorb optimally in the red and blue regions of the light spectrum; carotenoids absorb significantly only in the blue. The splitting of water, the major light-activated process in photosynthesis, is energy-intensive. Nonetheless, it becomes maximally effective at about one-third full sunlight intensity ($700-800\,\mu E/m^2/s$). This apparently is due to rate-limiting factors imposed by the pace of the dark (carbon-fixing) reactions in photosynthesis. In the center of a dense vine canopy, the intensity of sunlight may fall to $15-30\,\mu E/m^2/s$.

In contrast, phytochrome-activated processes require very little energy. For these, interconversion between the two physiological states of phytochrome depends on the relative energies available in the red and far-red portions of the spectrum. These regions correspond to the absorption peaks of the two states of phytochrome (P_r and P_{fr}). Red light converts P_r to physiologically active P_{fr}, whereas far-red radiation transforms it back to the physiologically inactive P_r state. In sunlight, the natural red/far-red balance of 1.1–1.2 generates

a 60 : 40 balance between P_{fr} and P_r. However, when light passes through the leaf canopy, the strong red absorbency of chlorophyll shifts the red/far-red balance to about 0.1 (toward the far-red). This probably means that some processes active in sunlit tissue are inactive in shaded tissue. The precise levels and actions of phytochrome in grapevine tissues are poorly understood.

Because of the negative influence of shade on photosynthesis and phytochrome-induced phenomena, most pruning and training systems are designed to optimize light exposure. The effects of shading on spectral intensity and the red/far-red balance are illustrated in Fig. 3.14A. In contrast, cloud cover produces little spectral modification in the visible and far-red spectra, although markedly reducing light intensity (Fig. 3.14B). Cloud cover, however, significantly reduces the intensity of the solar infrared (heat) radiation received directly from the sun. Figure 3.15 illustrates the daily cyclical effects of shade on the light conditions of different canopy levels. Because photosynthesis is usually maximal at light intensities equivalent to that provided by a lightly overcast sky, cloud cover affects most severely the photosynthesis of shaded leaves.

Canopy shading affects both leaf structure and physiology. In response to shading, pigment content increases (carotenoids and chlorophylls a and b) and the compensation point decreases (the light intensity at which the rates of the photosynthesis and respiration are equal) to about $80-30\,\mu E/cm^2/s$. Nevertheless, the lower respiratory rate of cooler, shaded leaves, enhanced pigmentation, and the reduced compensation point do not fully compensate for the poorer spectral quality and reduced intensity of shade light. As a result, shade leaves generally do not photosynthesize sufficiently to export sucrose or contribute significantly to vine growth. This was dramatically shown by Williams *et al.* (1987), where removal of shade leaves (30% of the foliage) did not delay berry growth or maturity. This situation might change in windy environments. The movement of the exterior canopy could dramatically increase the exposure of shaded leaves to sunflecking (periodic exposure to sunlight through the canopy). Although sunflecking is highly variable, average rates of 0.6 s per 2 s interval could significantly improve net photosynthesis (Kriedemann *et al.*, 1973). Sunflecking also appears to delay premature leaf senescence and dehiscence.

When leaves are young and rapidly unfolding, they act as a carbohydrate sink, rather than as a net source of photosynthate. Leaves begin to export photosynthates (sugars and other organic by-products) only when they reach about 30% of their full size. At this time they develop their mature green coloration. Nonetheless, leaves continue to import carbohydrates

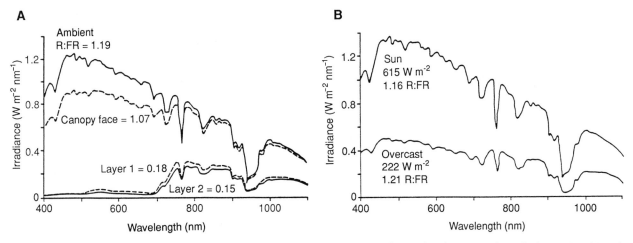

Figure 3.14 Spectral distribution of ambient sunlight radiation at the canopy surface and under one- and two-leaf canopies (**A**), and between sunny and overcast days (**B**). (From Smart, 1987, reproduced by permission)

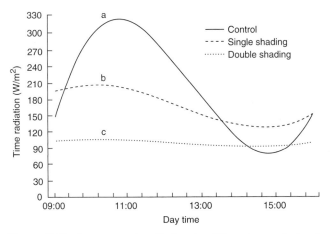

Figure 3.15 Average total radiation in differently manipulated 'Sauvignon blanc' canopies. (From Marais *et al.*, 1996, reproduced by permission)

until they have reached 50–75% of their mature size (Koblet, 1969). Maximal photosynthesis and sugar export occur about 40 days after unfolding, when the leaves are fully expanded (Fig. 3.16). Old leaves continue to photosynthesize, but their contribution to vine growth declines markedly (Poni *et al.*, 1994). The rate of this photosynthetic decline varies considerably, depending on shade, nitrogen deficiency, and the activity of new leaves. Leaves cease to be net producers when they lose their typical dark-green coloration. Leaves formed early in the season may photosynthesize at up to twice the rate of those produced later.

The angle of the sun, relative to the leaf blade, also influences photosynthetic rate. Nevertheless, even leaf blades aligned parallel to the sun's rays may photosynthesize at rates up to 50% that of perpendicularly positioned leaves (Kriedemann *et al.*, 1973). This may reflect the importance of diffuse sky light and the cooler leaf temperature to photosynthesis.

The effect of temperature on photosynthesis varies slightly throughout the growing season. In the summer, optimal fixation tends to occur at between 25 and 30 °C, whereas in the autumn, the optimum may fall to between 20 and 25 °C (Stoev and Slavtcheva, 1982). Photosynthesis is more sensitive to (suppressed by) temperatures below (<15 °C) than to temperatures above the optimum (>40 °C).

Fruit load and canopy size affect photosynthesis, presumably through feedback regulation. Within limits, the rate of photosynthesis can adjust to demand. For example, fruit removal results in a reduction of the photosynthetic rate of adjacent leaves (Downton *et al.*, 1987). Conversely, photosynthetic rate increases as a consequence of basal leaf removal. Improved photosynthetic efficiency appears to be associated with wider stomatal openings and increased gas exchange. Greater sugar demand could also activate export, thus reducing the concentration of Calvin-cycle intermediates that could act as feedback inhibitors.

Water conditions also markedly influence the rate of photosynthesis. In particular, water stress increases the production of abscisic acid. Because abscisic acid can reduce stomatal opening, transpiration and gas exchange declines, leaf temperature rises, and the rate of photosynthesis falls. Because water status affects the relative incorporation of stable isotopes of carbon (the $^{13}C/^{12}C$ ratio) (Brugnoli and Farquhar, 2000), this feature has been used to assess the cumulative effects of water status on vine photosynthesis (Gaudillère *et al.*, 2002). Plate 3.5 illustrates a technique for assessing whole plant gas exchange.

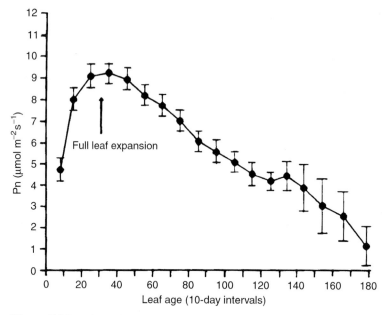

Figure 3.16 Relationship between leaf age, given as days after unfolding, and net photosynthesis (Pn), calculated on a seasonal basis, of 'Sangiovese' vines. Vertical bars represent 2 SEM. (From Poni *et al.*, 1994, reproduced by permission)

The direction of carbohydrate export from leaves varies with their position along the shoot, and the time of year. Export from young maturing leaves generally is apical, toward new growth. Subsequently, photosynthate from the upper maturing leaves is directed basally, toward the fruit (Fig. 3.33). By the end of berry ripening, basal leaves are rarely involved in export. This function is taken over by the upper leaves. Translocation typically is restricted to the side of the cane and trunk on which the leaves are produced.

Phytochrome is associated with a wide range of photoperiodic and photomorphogenic responses in plants. Typically, grapevines are relatively insensitive to photoperiod, but may show several other phytochrome-induced responses. For example, bud dormancy in *Vitis labrusca* and *V. riparia* is induced by short-day photoperiods (Fennell and Hoover, 1991; Wake and Fennell, 2000). Several important enzymes in plants are known to be partially activated by changes in phytochrome balance. These include enzymes involved in the synthesis and metabolism of malic acid (malic enzyme), the production of phenols and anthocyanins (phenylalanine ammonia lyase and dihydroflavonol reductase (Gollop *et al.*, 2002), sucrose hydrolysis (invertase), and possibly nitrate reduction and phosphate accumulation (see Kliewer and Smart, 1989). Phytochrome is also suspected of being implicated in the regulation of berry growth (Smart *et al.*, 1988). Nevertheless,

the actual importance of the shift in the P_{fr}/P_r ratio in shaded grapevine leaves is unclear. Also, the significance of sunflecking (the intermittent penetration of sunlight into the leaf canopy) on phytochrome balance is unknown.

A third light-induced phenomenon in grapevines is activated by ultraviolet radiation, namely the toughening of the cuticle. The strong absorption of ultraviolet radiation by the epidermis results in its almost complete absence in shade light. Consequently, the cuticle of shaded leaves and fruit is softer than their sun-exposed counterparts. This factor, combined with the slower rate of drying and higher humidity in the understorey, may explain the greater sensitivity of shaded tissue to fungal infection. It is suspected, but unconfirmed, that either or both of the blue/UV-A or UV-B photoreceptors (such as cryptochrome) are involved. Exposure to short-wave ultraviolet radiation (UV-C, 100–290 nm) appears to activate several metabolic pathways. Of particular importance is general disease resistance. This includes the synthesis of phytoalexins, chitinases and pathogenesis-related (PR) proteins (Bonomelli *et al.*, 2004).

TRANSPIRATION AND STOMATAL FUNCTION

To optimize its photosynthetic potential, the leaf must act as an efficient light trap. Because of the importance of diffuse radiation coming from the sky, most leaves provide a broad surface for light impact. Even on clear days, up to 30% of the radiation impacting a leaf may come from diffuse skylight. However, the large surface area of the leaf blade also makes it an effective heat trap. Heating can be reduced by evaporative cooling, but, if unrestricted, this would place an unacceptably high water demand on the root system. The cuticular coating of the leaf limits water loss, but also retards gas exchange essential for photosynthesis. These conflicting demands result in a series of complex compromises.

When the water supply is adequate, transpirational cooling minimizes leaf heating (to about 1–2 °C above ambient) during direct exposure to sunlight (Millar, 1972). This, however, places considerable demands on the water supply. Even under mild transpiration conditions, water loss can occur at rates up to $10\,mg/cm^2/h$. Except for the small proportion of water (<1%) used in photosynthesis, and other metabolic reactions, most of the water absorbed by the root system is lost via transpiration. Water stress usually does not limit transpiration, or photosynthesis, until leaf water potentials (ψ) fall below −13 to −15 bar.

To regulate the conflicting demands of limiting transpiration and enabling gas exchange, leaves depend primarily on stomatal control. Stomatal closure results in both a rise in leaf temperature and O_2 content, as well as a reduction in CO_2 level. These influences suppress net carbon fixation by increasing photorespiration and limiting photosynthesis. The effects of water deficit on stomatal function and photosynthesis often linger long after the return of turgor. The slow return to normal leaf function may result from the accumulation of abscisic acid produced during water deficit, or from disruption of the chloroplast photosynthetic apparatus. *Vitis labrusca* cultivars appear to be less sensitive to such disruptions than *V. vinifera*. When water deficit develops slowly, increased root growth may offset the effects of reduced soil-water availability (Hofäcker, 1976). The root systems of some varieties also appear to make osmotic adjustments to drought conditions (Düring, 1984), thereby enhancing their ability to extract water from the soil.

The stoma, or more correctly the stomatal apparatus, consists of two guard cells and associated accessory cells. The lower epidermis of grapevine leaves possess about $10\text{–}15 \times 10^3$ stomata/cm^2 (Kriedemann, 1977). Although stomata constitute only about 1% of the lower leaf surface, they permit transpiration at a rate equivalent to about 25% of the total leaf surface. This paradoxical finding is explained by the spongy mesophyll functioning as the actual surface over which transpiration occurs. The stomata act primarily as openings through which the evaporated water escape into the surrounding atmosphere. Mesophyll surfaces also promote the efficient exchange of carbon dioxide and oxygen with the atmosphere.

The control of stomatal opening and closing is complex, as befits its central role in leaf function. Water deficit is the principal factor controlling stomatal function, but it action is indirect. A decrease in water potential (Ψ) under dry conditions may result in xylem vessel cavitation, slowing water flow. In addition, excessive transpiration can activate genes leading to the synthesis of abscisic acid (ABA) in leaves (Soar *et al.*, 2006). Both can induce stomatal closure. This results in an increase in the CO_2 concentration within the leaf that further activates closing. Split-root experiments have shown that when water availability in the soil is low, water-stressed root systems induce the production of abscisic acid and release it in the xylem sap (Düring *et al.*, 1996). On reaching the leaves, stomatal closure is activated. High leaf temperatures (partially a result of reduced transpiration from reduced stomatal opening) further promotes stomatal closure, possibly by activating abscisic acid synthesis in the leaf. High temperatures also suppress photosynthesis and spurs respiration, further increasing the leaf CO_2 concentration. In contrast, in the absence of a water deficit, light exposure tends to activate stomatal opening by inducing malic acid synthesis from starch in guard cells, as well as by reducing leaf CO_2 content via activating photosynthesis.

Stomatal opening and closing involve the active uptake and release, respectively, of potassium ions by the guard cells. Malic acid synthesis may provide the hydrogen ions needed for K^+ ion exchange in the guard cells. As a result of the K^+ influx or efflux, water moves into or out of the guard cells, respectively. Water uptake provides the turgor pressure that forces the guard cells to elongate and curve in shape, causing stomatal opening. Conversely, water loss from the guard cells induces closing.

Reproductive Structure and Development

Inflorescence (Flower Cluster)

INDUCTION

As noted, only some buds produce inflorescences. At inception, fertile and sterile buds are identical. The precise factors that activate inflorescence inception are unclear, but many of the prerequisite conditions are known.

A young vine typically does not begin to bear flowers and fruit until the second or third seasons. Juvenility is not based simply on vine size, because cytokinin application can induce flowering in young seedlings (Srinivasan and Mullins, 1981). Thus, flowering appears to be controlled by physiological rather than chronological age.

The induction of inflorescence differentiation generally coincides with blooming and the slowing of vegetative growth. Induction is particularly sensitive to water stress, and its occurrence during blooming can severely reduce bud fruitfulness. Induction usually occurs about 2 weeks before morphological signs of differentiation become evident. A period of several weeks may separate the initiation of the first and second inflorescence primordia (*anlagen*) (Fig. 3.17). This period tends to correspond to growth stages 13 to 18 of the Eichhorn and Lorenz (E–L) system (Fig. 4.2).

Environmental conditions around the bud and closely associated leaves can markedly affect induction. This partially reflects changes in the hormonal or nutrient status of the bud. Cool conditions favor gibberellin synthesis, which promotes vegetative growth and limits nutrient accumulation. Relatively high gibberellin contents often occur in apical internodes. In contrast, warm conditions (25–35 °C) promote cytokinin synthesis, which favors inflorescence differentiation. Optimal nitrogen, potassium, and phosphorus supplies also promote the synthesis and translocation of cytokinins from the roots. The action of the growth-retardant chlormequat (CCC) in favoring fruit-bud initiation may be due to inhibition of gibberellin synthesis and stimulation of cytokinin production. Exogenously applied gibberellin can shift inflorescence development into tendril formation. In contrast, auxin application improves inflorescence induction, as well as the number of flowers formed per cluster.

The timing and deposition of carbohydrates in the shoots correlate well with the period and node location of fruit-bud development. This period corresponds to the interval between rapid shoot growth and the acceleration of fruit development. Correspondingly, factors such as high vigor, defoliation, or untimely drought, which disrupt carbohydrate accumulation, can interfere with inflorescence initiation (Bennett *et al.*, 2005). Drought is also known to suppress the upward transport of cytokinins in the xylem.

Sun exposure is generally more important than day length in inducing fruit-bud development in *vinifera* cultivars. By contrast, long days appear to improve inflorescence induction in *V. labrusca*. The level of sun exposure required for maximal inflorescence induction varies markedly from cultivar to cultivar (Sánchez and Dokoozlian, 2005). Depending on the cultivar, the frequency and size of fruit buds increase progressively from the shoot base, often reaching a maximum about the tenth node (Fig. 3.6). Fruitfulness tends to decline beyond that point. Although lateral shoots seldom

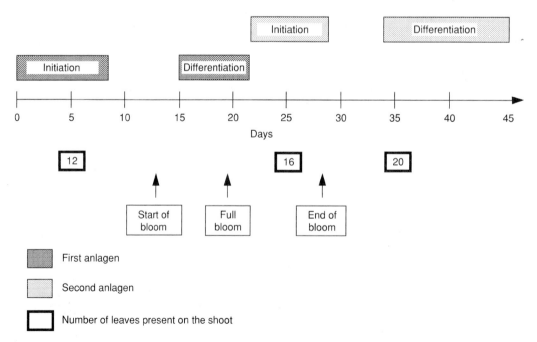

Figure 3.17 Diagrammatic representation of events associated with the initiation and primary development of the two inflorescence primordia of 'Chenin blanc' fruit buds. (From Swanepoel and Archer, 1988, reproduced by permission)

produce inflorescences, woody laterals can favor the fruitfulness of compound buds associated with the laterals (Christensen and Smith, 1989).

As indicated, fruitfulness is primarily associated with the primary bud. The fruitfulness of secondary buds is generally important only if the primary bud dies. The formation of inflorescence primordia in base buds is seldom of importance because they remain inactive. However, base-bud activation in French-American hybrid cultivars can lead to overcropping, poor wood maturation, and weakening of the vine. In contrast, little is known about the conditions affecting the flowering of lateral shoots. The production of a second crop on lateral shoots can seriously complicate attempts to regulate crop yield and quality (nonuniform fruit maturity).

By midsummer, when branch initials have developed, the *anlagen* (primordia) become dormant. Differentiation recommences in the spring when the buds begin to swell and individual flowers form.

The most fruitful buds generally develop in leaf axils from the fourth to the twelfth node. Nevertheless, flower buds may develop beginning with the first leaf (**count**) node along a cane. Many European cultivars are fruitful from the base, whereas varieties of western and central Asiatic origin often are barren at the base. Varieties that produce sterile base buds are unsuitable for spur pruning.

INFLORESCENCE MORPHOLOGY AND DEVELOPMENT

The grapevine inflorescence is a complex, highly modified branch system containing reduced shoots and flowers (Fig. 3.18). The complete branch system is called the **rachis**. It is composed of a basal stem (the **peduncle**), two main branches (the **inner** and **outer** (lateral) **arms**), and various subbranches terminating in pedicels. **Pedicels** give rise to individual flowers. Flowers commonly occur in groups of threes, called a **dichasium**, but may occur singly or in clusters of modified dichasia. Individual flowers are considered to be truncated shoots, with leaves modified for reproductive functions.

The branching of the inflorescence into arms occurs very early in differentiation (Fig. 3.19), similar to tendrils. The outer arm develops into the lower and smaller branch of the inflorescence. Occasionally it develops into a tendril or fails to develop at all. The inner arm develops into the major branch of the flower cluster. In buds that develop early, usually those between the fourth and twelfth nodes, branching is complete by the end of the growing season. Such buds also may contain up to 6–10 leaf primordia by fall. The rudiments of the flower primordia may be present, with nascent flower clusters already showing the shape typical of the cultivar. Less mature buds generally possess

atypically shaped, smaller clusters. Flower development may continue during warm winter spells.

Unlike most perennial plants, grapevine inflorescences develop opposite a leaf, usually in a position that otherwise would have produced a tendril. The vine may produce up to four flower clusters per shoot, though two is more common. They often develop at the third and fourth, fourth and fifth, or fifth and sixth nodes, depending on the variety.

FLOWER DEVELOPMENT

By the end of the growing season, flowers may have developed to the point of producing receptacles – the swollen base of the pedicel from which the flower parts originate. As buds swell in the spring, and cytokinins flow upward in the sap, cells in the flower primordia begin to divide. Floral differentiation progresses centripetally from the receptacle rim, as the shoot and rachis elongate.

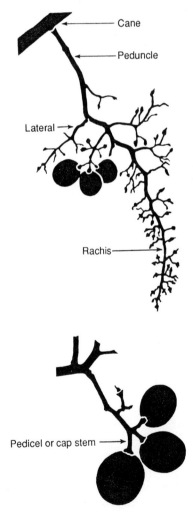

Figure 3.18 Structure of the grape cluster and its attachment to the cane. (From Flaherty *et al.*, 1981, reproduced by permission)

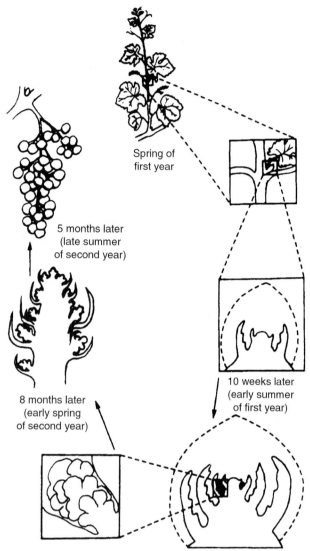

Spring of
first year

5 months later
(late summer
of second year)

10 weeks later
(early summer
of first year)

8 months later
(early spring
of second year)

Figure 3.19 Development of grape inflorescence. Buds develop in leaf axils in the spring and early summer. At midsummer, the buds illustrated have eight leaf primordia with fruiting primordia opposite leaf primordia six and seven. Early the following spring, the club-like fruiting primordia differentiate into individual florets with bracts, and subsequently berries develop from about one-third of the flowers. (From Buttrose, 1974, reproduced by permission)

Shortly after the leaves begin to unfold, sepal differentiation becomes evident. These leafy structures do not develop significantly and are barely visible in mature flowers. About a week later, **petals** begin to differentiate. They arch upward, past the sepals, and fuse into a unified enclosing structure, the **calyptra**. Petal fusion occurs through the action of special interlocking cells that form on the edges of the petals (Fig. 3.20A). Within about 3 weeks, stamens have formed, their elongating filaments pushing the pollen-bearing anthers upward. Each of the five **anthers** contains two elongated

pollen sacs, attached to the tip of the filament at a central juncture. Each anther produces thousands of pollen grains, containing a generative and a pollen-tube nucleus. Adjacent to the filaments are small **nectaries**. These produce a mild floral scent. About 1 week after stamen genesis, two carpels form, each producing two ovules. The two carpels fuse to form the developing **pistil**, composed of a basal swollen ovary, a short style, and a slightly flared stigma (Fig. 3.20D). The timing of egg development in the **ovary** closely parallels that of pollen in the anthers.

As the pollen matures, the base of the petals separate from the receptacle (Fig. 3.20B) and curve outward and upward. The freed calyptra dries and falls, or is blown off. Shedding of the calyptra (Plate 3.6) occurs most frequently in the early morning. The act of dehiscence often triggers rupture of the pollen sacs (anthesis). Rupture occurs along a line of weakness adjacent to where they adhere to the filaments. The region contains a layer of thickened cells, which can rip the epidermis of the pollen sac on drying. The violent release of pollen sheds pollen onto the stigma. This feature, plus the short style of most varieties, favors self-fertilization.

TIMING AND DURATION OF FLOWERING

Flowering normally occurs within 8 weeks of bud break. The precise timing varies with weather conditions and cultivar characteristics. In warm temperate zones, flowering often begins when the mean daily temperature reaches 20° C. In cooler climates, increasing day-length may be an important stimulus. Where flowering is staggered over several weeks, cyanamide treatment may be used to improve synchronization. This can favor more uniform maturation and fruit quality in cultivars tending to show asynchronous flowering, such as 'Merlot' and 'Zinfandel.'

Pollen release, as measured in the surrounding air, has been recommended as a means of assessing probable fruit yield (Besselat and Cour, 1990). Presumably, the warm sunny conditions favoring aerial pollen dispersal correlate with the conditions that favor self-pollination.

Flowering usually begins on the uppermost shoots, similar to the sequence of bud break. For individual clusters, however, blooming commences from the base of the inflorescence. Under favorably warm, sunny, conditions, individual flower clusters may complete blooming within a few days. Because of timing differences throughout the vine, blooming usually lasts about 7–10 days. Under cold rainy conditions, flowering may extend over several weeks. Under such conditions, the calyptra may not be lost. Although this does not prevent pollination, such conditions usually reduce

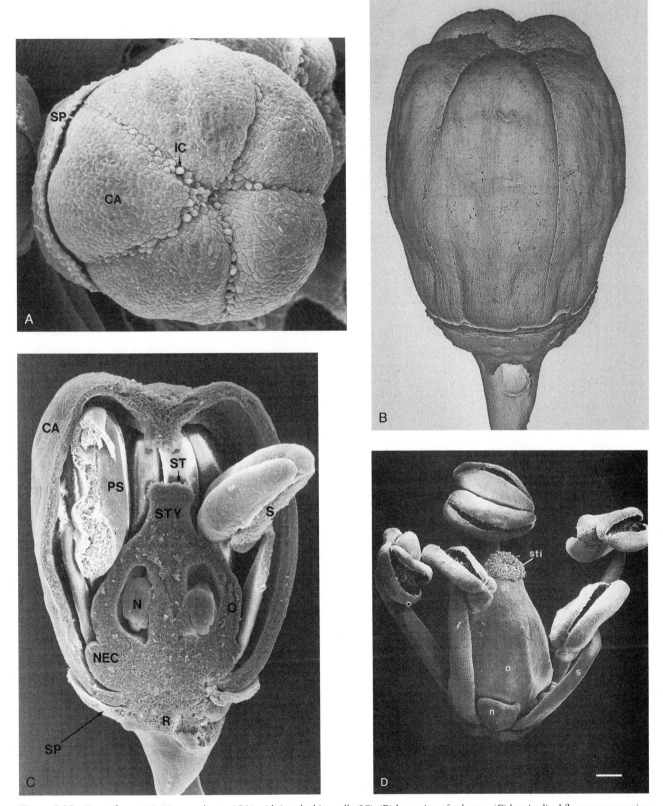

Figure 3.20 Grape flower. (**A**) Young calyptra (CA) with interlocking cells (IC); (**B**) loosening of calyptra; (**C**) longitudinal flower cross-section with nucellus (N), nectary (NEC), ovary (O), anthers (PS), receptacle (R), stamens (S), sepals (SP), stigma (ST), and style (STY); (**D**) flower structure at anthesis. (A–C from Swanepoel and Archer, 1988, reproduced by permission; D from Hardie *et al.*, 1996, reproduced by permission)

the proportion of flowers fertilized. It also tends to lead to asynchronous fruit set. Whether induced by genetic or environmental factors, poor synchrony produces an undesirable range of fruit maturity at harvest.

POLLINATION AND FERTILIZATION

Self-pollination appears to be the rule for most grapevine cultivars. Under vineyard conditions, wind and insect pollination appear to be of little significance. Even in areas where grapes are the dominant agricultural crop, pollen levels in the air are generally low during flowering – in contrast to the relatively high levels associated with cross-pollinated wild grapevines (see Stevenson, 1985). Yields of about 1.4×10^4 pollen grains/m²/day, sedimenting out of the air during July, have been recorded in Montpellier, France (Cour *et al.*, 1972–1973). In Portugal, peak pollen counts were recorded at 24 pollen grains per m³ air (Cunha *et al.*, 2003). Despite the apparent insignificance of airborne pollen to successful pollination, they were able to predict wine production from pollen counts in dryer climatic regions of Portugal.

The minimal importance of insect pollination to vine fertilization may accrue from the nearly simultaneous blooming of innumerable vines over wide areas. Nevertheless, syrphid flies, long-horned and tumbling flower beetles, as well as bees occasionally visit grape flowers. The principal attractant appears to be the scent produced by nectaries. Sesqiterpenes and monoterpenes appear to be the primary aromatic compounds produced (Buchbauer *et al.*, 1995). Nectaries are located between the stamens and pistil (Fig. 3.20C) and are particularly prominent in male flowers on wild *V. vinifera* vines. Nectaries are modified for scent, rather than nectar production as the name might suggest. Visiting insects feed on pollen, not nectar. Pollen fertility has no influence on flower attractiveness, but the presence or absence of anthers does affect the duration of visits to female flowers (Branties, 1978).

Shortly after landing on the stigma, the pollen begins to swell. The sugary solution produced by the stigma is required both for pollen growth and to prevent osmotic lysis of the germ tube. The stigmatic fluid also occurs in the intercellular spaces of the style. This may explain why rain does not significantly inhibit or delay pollen germination, or delay the penetration of the **germ tube** into the **style**. However, pollen germination and germ-tube growth are markedly affected by temperature (Fig. 3.21), even though viability is less affected. In contrast, **ovules** show obvious signs of degeneration within about a week at temperatures below 10°C. Cool temperatures, just before flowering at favorably warm temperatures, can delay pollen germinability and germ-tube growth. Similar conditions can equally reduce

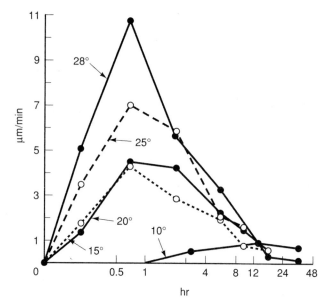

Figure 3.21 Rate of pollen germ-tube growth in relation to temperature. (From Staudt, 1982, reproduced by permission)

fertility by disrupting aspects of ovule development (Ebadi *et al.*, 1995).

As the pollen tube penetrates the style, the germinative nucleus divides into two sperm nuclei, if this has not occurred previously. On reaching the opening of the ovule (micropyle), one sperm nucleus fuses with the egg nucleus, whereas the other fuses with the two polar nuclei. The fertilization of the egg nucleus initiates embryo development, whereas fusion with polar nuclei induces endosperm differentiation. Fertilization also inaugurates a series of events that transforms the ovules into seeds and the ovary wall into the skin and flesh of the berry. Fertilization is usually complete within 2–3 days of pollination.

In certain varieties, abnormalities result in the absence of viable seed following pollination. Even in fully fertile cultivars, only 20–30% of the flowers successfully develop fruit. In **parthenocarpic** cultivars, ovules fail to develop in the flower. Although pollination stimulates sufficient auxin production to prevent abscission of the fruit (shatter), it is inadequate to permit normal berry enlargement. 'Black Corinth,' the primary commercial source of dried currants, is the most important parthenocarpic variety. Although parthenocarpic varieties produce no seed, other seedless varieties such as 'Thompson seedless' contain seeds. As these usually abort a few weeks after fertilization, the seeds are empty, small and soft. Because of partial seed development, greater auxin production induces medium-size fruit development. This situation is called **stenospermocarpy**. If abortion occurs even later, as in the cultivar 'Chaouch,' normal-size fruit develop containing hard empty seeds. In contrast to the well-known examples of parthenocarpy and stenospermocarpy, the development of fruit and viable seeds in the absence of fertilization (apomixis) is unconfirmed in grapevines.

FLOWER TYPE AND GENETIC CONTROL

Most cultivated grapevines produce predominantly perfect flowers. That is, they contain both functional male and female parts. The flowers are also self-fertilizing. Nevertheless, cultivars often produce a series of morphologically different flowers. Male (staminate) flowers have erect stamens and a reduced pistil, or ovaries that abort before forming a mature embryo (Fig. 2.2). Female (pistillate) flowers have a well-developed functional pistil, but produce reflexed stamens possessing sterile pollen. Sterile pollen is often characterized by one or more of the following properties: disruption of chromosome separation during nuclear division, inner-wall abnormalities, absence of surface furrows, and failure to produce germ pores under suitable conditions (Fig. 2.11). Such a wide variation suggests that sexual expression in grapevines is under complex genetic control. Studies on the genetics of sexuality in *Vitis vinifera* support this view.

Because both male and female parts usually form in the unisexual flowers of wild vines (Fig. 2.2), the ancestral state of the genus is considered to have been bisexual (**hermaphroditic**). This situation still exists in the more primitive Vitaceae, for example *Cissus*. Early in the evolution of *Vitis* there must have been strong selective pressure for unisexuality. It would have had the advantage of imposing cross-fertilization and, thus, maximizing genetic diversity. Conversely, self-fertilization was probably inadvertently selected during cultivation. It would have resulted in pollination (and fruit production) being less susceptible to environmental disruption. Because of the agricultural advantages associated with self-fertilization, mutations reinstating both male and female functionality may have occurred repeatedly and independently during domestication.

Sexual determination in grapevines, unlike many organisms, is not associated with morphologically distinct chromosomes (Negi and Olmo, 1971). In *Vitis* spp., sexuality is primarily under the control of a single gene (Dalbó *et al.*, 2000). Nevertheless, there is evidence that an epistatic (modifier) gene (Carbonneau, 1983), as well as environmental factors can affect sexual expression. The potential of some male strains of *V. vinifera* f. *sylvestris* to periodically develop functional hermaphroditic flowers is probably environmentally induced (Negi and Ough, 1971). In addition, female flowers have occasionally been observed on male vines of *V. acerifolia* (*V. longii*),*V. riparia*, and *V. rupestris* (Negi and Olmo, 1971).

The primary sex-determinant locus in grapevines was designated *Su* (suppressor) by Negi and Olmo (1971). It was considered to exist in three allelic states – Su^+, Su^m and Su^F (Table 3.1). The ancestral form, Su^+, permitted the development of bisexual (hermaphroditic) flowers. It presumably mutated millions of years ago to alternate forms, Su^F and Su^m. Wild vines exist as either $Su^F Su^m$ (male) or $Su^m Su^m$ (female), as suggested by the relatively equal number of male and female vines. Su^F is dominant over both Su^+ and Su^m. It suppresses the development of functional pistils, but permits the development of functional anthers. In its turn, Su^+ is dominant over Su^m, permitting the development of fully functional hermaphroditic flowers. Su^m, when homozygous, permits functional pistil development, but results in recurved anthers and the production of sterile pollen.

Reversion from the unisexual to functional bisexual state during domestication probably involved the action of back-mutations. Most significant would have been the mutation of the Su^F or Su^m allele to Su^+, producing bisexual ($Su^+ Su^m$) flowers. Predominantly unisexual female cultivars still exist, such as 'Picolit.' Its fertility is dependent on the presence of other hermaphroditic varieties.

An indication of the possible biochemical effects of the various *Su* alleles has been provided by the artificial kinin SD8339. Application to immature inflorescences of a male strain can induce the development of functional pistils (Negi and Olmo, 1966). Confirmation of the following proposal awaits further studies.

Su^F – marked degradation of cytokinins, suppressing the development of functional pistils
Su^+ – limited accumulation of cytokinins, permitting development of both functional pistils and anthers
Su^m – accumulation of cytokinins to the point that anther (or functional pollen) development is suppressed

An alternate proposal has been suggested by Carbonneau (1983). He proposes that the principal gene exists in three states (termed *M* – male, *H* – hermaphroditic, and *F* – female). These have a dominance hierarchy similar to those proposed by Negi and Olmo (1971). However, the *F* allele is proposed to occur in three states – F_m, F_h, and F'_m. Their differentiation being based on how their expression is affected by the epistatic gene (*E*). For example, the $F'_m F'_m$ genotype may be male, hermaphrodite, and female, depending on its association with *EE*, *Ee*, and *ee*, respectively. Negi and Olmo (1971) also suggest that other genetic and/ or environmental factors are probably involved in sex expression – there often being considerable variation in

Table 3.1 Presumed sexual genotypes (and phenotype) in the Vitaceae as proposed by Negi and Olmo (1971)

Form of *Vitis*	Genotype (phenotype)		
Proposed ancestral state	$Su^+ Su^+$ (bisexual)		
Vitis vinifera f. *sylvestris*	$Su^F Su^F$ (male – rare),	$Su^F Su^m$ (male),	$Su^m Su^m$ (female)
Vitis vinifera f. *sativa*	$Su^+ Su^+$ (bisexual),	$Su^+ Su^m$ (bisexual),	$Su^m Su^m$ (female – rare)

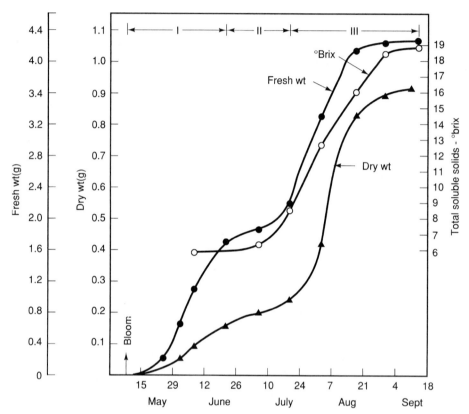

Figure 3.22 Relationship of growth stages I, II, and III and the accumulation of total soluble solids in 'Tokay' berries. (From Winkler *et al.*, 1974, reproduced by permission)

sexual expression among vines, and on an annual basis (see Fig. 2.2).

Berry Growth and Development

Following fertilization, the ovary begins its growth and development into the berry. Typically, only a portion of the fruit matures, with a variable number aborting. Fruit set is affected by genetic factors, climatic influences, and viticultural practices (Collins and Dry, 2006). In addition, increased flower production per inflorescence tends to reduce the proportion that mature (Vasconcelos and Castagnoli, 2000). The latter may be endogenous compensation, favoring fruit ripening by reducing intracluster competition. The carbohydrate

supply typically comes exclusively from photosynthetically active leaves (Candolfi-Vasconcellos *et al.*, 1994). Mobilization of reserves from old wood usually is of minor significance. Leaves closest to the cluster (either on the primary shoot or adjacent lateral shoots) supply most of the photosynthate.

Grapes typically exhibit a slightly double-sigmoid growth curve (Fig. 3.22). In stage I, the berry shows rapid cell division, associated with cell enlargement and endosperm development. The initial phase typically lasts from 6 weeks to 2 months. Stage II is a transitional period in which growth slows, the embryo develops and the seed coats harden. Stage II is the most variable in duration (1–6 weeks), and largely establishes a cultivar's early or late maturing character. At the end of stage II, the berry begins to lose its green color. This turning point, called **véraison**, signifies the beginning of a fundamental physiological shift that culminates in berry maturation (stage III). Stage III is associated with seed maturation, and the final enlargement of the berry. Cell enlargement occurs primarily at night and is not associated with the daytime volume fluctuations that tend to characterize stage I. Ripening is associated with tissue softening, a decrease in acidity, the accumulation of sugars, the synthesis of anthocyanins (in red-skinned varieties), and the acquisition of aroma compounds. This phase generally lasts about 5–8 weeks. Although the differences between stages I and III are clear, the lag component of stage II may be minimal or even undetectable (Staudt *et al.*, 1986). Overripening, that occurs if harvesting is delayed, is occasionally termed stage IV.

Genetic regulation of berry development is only beginning to be understood. Particularly intriguing are results from studies on *fbl* (fruitless berry) mutants (Fernandez *et al.*, 2006) (Plate 3.7). Its action shows that fertility, as well as seed, epidermis and vascular development are physiologically independent of mesocarp development. The *flb* mutant principally impairs cell division and differentiation of the inner mesocarp. Because these cells are the largest in the berry (most vacuolated), fruit weight and size are drastically reduced. Accumulation of malic acid is also significantly reduced. Although total sugar accumulation is reduced, the dynamics of accumulation in the fruit is relatively similar to normal fruit. Softening and color changes typically associated with *véraison* are unaffected.

BERRY STRUCTURE

As with other aspects of vine morphology, the terms applied to berry structure can vary from authority to authority. According to standard botanical use, the ovary wall develops into the fruit wall or **pericarp**. The pericarp is in turn subdivided into an outer **exocarp**, the middle **mesocarp**, and the inner **endocarp**. Various authors have applied these terms differently to the **skin** and various fleshy portions of the berry (**pulp**), depending on their respective views on berry development. One representation is given in Fig. 3.23. Because of disagreement concerning the correct use of technical anatomical terms, the common terms "skin" and "flesh" will be used primarily in this book.

The grape skin consists of two anatomically distinguishable regions, the outer epidermis and an inner hypodermis. The latter consists of several cell layers (Fig. 3.24). The internal fleshy tissues (mesophyll) of the berry also may be subdivided into regions – the outer portion (outer wall) consisting of the tissues

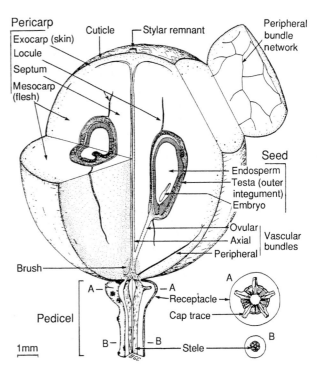

Figure 3.23 Diagrammatic representation of a grape berry. (After Coombe, 1987a, reproduced by permission)

between the hypodermis and the peripheral vascular strands, and the inner section (inner wall) being delimited by the peripheral and axial vascular strands. The basal portion of the axial and seed vascular bundles, and associated parenchyma, are designated the brush. The layer of cells that corresponds to the boundary between the flesh and the seed locules is occasionally termed the inner epidermis. The septum is the central region where the two carpels of the pistil join.

Cell division in the flesh occurs most rapidly about 1 week after fertilization. It subsequently slows and may stop within 3 weeks. Cell division ceases earliest close to the seeds and last next to the skin. Cell enlargement may occur at any stage of berry development, but is most marked after cell division ceases.

The epidermis consists of a single layer of flattened, disk-shaped cells with irregularly undulating edges. These nonphotosynthetic cells possess vacuoles containing large oil droplets. Depending on the variety, they may develop thickened or lignified walls. The epidermis does not differentiate hair cells and produces few stomata. The stomata are sunken within raised (peristomatal) regions on the fruit surface (Fig. 3.25). The stomata soon become nonfunctional. Subsequently, transpiration occurs primarily through the cuticle. Heat dissipation results primarily from wind-facilitated convection from the berry surface.

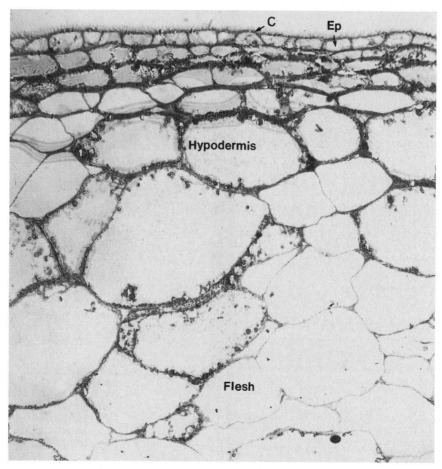

Figure 3.24 Cross-section through the epidermis, hypodermis, and outermost flesh cells of 'Chardonnay' fruit (×400): C, cuticle; Ep, epidermis. (Photograph courtesy of J. C. Audran)

The surface of the peristomatal region accumulates abnormally high concentrations of silicon and calcium (Blanke *et al.*, 1999). As the fruit mature, the region becomes considerably suberized and accumulates polyphenolics. Microfissures may develop next to the stomata during periods of rapid fruit enlargement (see Fig. 4.42A). Berry diameter can fluctuate up to 6% during the pre-*véraison* period due to transpriational water loss during the day (Greenspan *et al.*, 1996).

As with most other aerial plant surfaces, a relatively thick layer of cuticle and epicuticular wax develops over the **epidermis** (Fig. 3.26A). The cuticle is present almost in its entirety before anthesis. It forms as a series of compressed ridges a few weeks prior to anthesis (Fig. 3.26B). During subsequent growth, the ridges spread apart and eventually form a relatively flat layer over the epidermis (Fig. 3.26C). Exposure to intense sunlight and high temperatures appear to destroy its crystalline structure, and is correlated with susceptibility to grape sunburn (Greer *et al.*, 2006). Initially, the wax platelets are small and upright, occurring both between and on cuticular ridges. The number, size, and structural complexity of the platelets increase during berry growth (Fig. 3.26D). Changes in platelet structure during berry maturation may correspond to changes in the chemical nature of these wax deposits. The cuticular covering is not necessarily uniform, as evidenced by the occurrence of **microfissures** in the epidermis (see Fig. 4.42A) and **micropores** in the cuticle (see Fig. 4.42B). The occurrences of such features, as well as cuticular thickness and wax plate structure, are affected by both hereditary and environmental factors. Thus, their influence on disease resistance is complex and may vary from season to season.

The **hypodermis** consists of a variable number of tightly packed layers of mesophyll cells (Fig. 3.24). The cells are flattened, with especially thickened corner walls. The hypodermis commonly contains 10 layers of cells, but this can vary from one to 17, depending on the cultivar. When young, the cells are photosynthetic. After *véraison*, the plastids lose their chlorophyll and starch contents, and begin to accumulate oil droplets (Fig. 3.27). These modified plastids appear to be the site of terpenoid and norisoprenoid synthesis

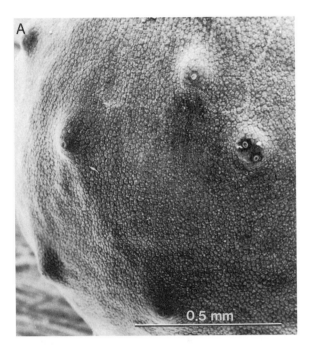
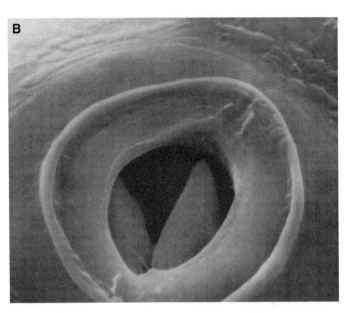

Figure 3.25 Scanning electron micrograph of (**A**) the epidermal surface of berries at *véraison* showing raised regions with one or two stomata and (**B**) open stomata (×2300). (A from Pucheu-Plante and Mércier, 1983, reproduced by permission; B from Blanke and Leyhe, 1987, reproduced by permission)

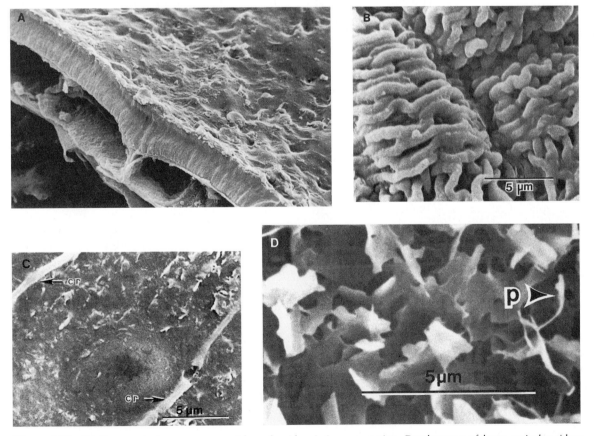

Figure 3.26 (**A**) Berry cuticular structure, cuticle and epidermis in cross-section. Development of berry cuticular ridges (cr), with epicuticular waxes removed, (**B**) 2 weeks before anthesis, and (**C**) 11 days after bloom. (**D**) Appearance of epicuticular wax plates on mature berry, 13 weeks after anthesis. (A from Blaich *et al.*, 1984, reproduced by permission; B–D from Rosenquist and Morrison, 1988, reproduced by permission)

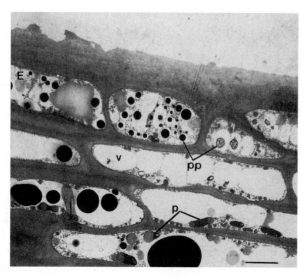

Figure 3.27 Cross-section of the skin of grape berry near maturity. Plastids (p) show little starch, but prominent lipid-like globules. Vacuoles (v) show polyphenolic deposits. Those of the reflective droplet type (pp) occur in two forms. Those in the cell to the right are at an earlier ontogenetic stage than those in the cells to the left. (From Hardie *et al.*, 1996, reproduced by permission)

and storage. They are suspected to be by-products of carotenoid degradation. Most hypodermal cell vacuoles also accumulate flavonoid phenolics (Fig 3.27) – notably anthocyanins in the outermost layers of red grape varieties. In addition, hypodermal cells may contain vacuolar collections of **raphides** (needle-like crystals of calcium oxalate monohydrate). These become less abundant during maturation. Calcium oxalate also occurs as star-shaped crystalline formations in cells of the flesh, called **druses**. The cell wall thickens, due to hydration, as connections between the cells loosen during maturation. In 'slip-skin' varieties, the cells of the flesh adjacent to the hypodermis become thinner and lose much of their pectinaceous cell-wall material. This produces a zone of weakness, permitting the skin to separate readily from the flesh.

Most cells of the **flesh** are round to ovoid, containing large vacuoles that may store phenolics. Although the outer mesocarp cells generally contain plastids, they seldom contribute significantly to photosynthesis. Pericarp tissue constitutes about 65% of the berry volume.

Cell division in the flesh begins several days before anthesis, and continues for about another 3 weeks. Subsequent growth results from cell enlargement. Fruit size appears to be predominantly regulated by cell enlargement, as the number of pericarp cells is little influenced by the external environment. The most rapid period of pericarp enlargement occurs shortly after anthesis. Although associated with rapid cell division, this period is also when water uptake is most rapid. This may explain the high sensitivity of fruit set to

water stress during this period (Nagarajah, 1989). After *véraison*, growth is exclusively by enlargement. During this period, the cell vacuole greatly expands in diameter. These vacuoles become the primary site for sugar accumulation. The vacuoles also are sites for the deposition of phenolics and most grape acids. The cells remain viable late into maturity.

The vascular tissue of the fruit develops directly from that of the ovary. It consists primarily of a series of **peripheral bundles** that ramify throughout the outer circumference of the berry, and **axial bundles** that extend directly up through the **septum**. The **locules** of the fruit correspond to the ovule-containing cavities of the ovary and are almost undetectable in mature fruit. The locular space is filled either by the seeds or by growth of the septum.

Fruit abscission may occur either at the base of the pedicel or where the fruit joins the receptacle. Separation at the pedicel base (**shatter**) results from the localized formation of thin-walled parenchymatous cells. This commonly occurs if the ovules are unfertilized or the seeds abort early. Some varieties, such as 'Muscat Ottonel,' 'Grenache,' and 'Gewürztraminer,' are abnormally susceptible to fruit abscission shortly after fertilization – a physiological disorder called inflorescence necrosis (*coulure*). In contrast, dehiscence that follows fruit ripening in muscadine cultivars results from an abscission layer that forms in the vascular tissue at the apex of the pedicel, next to the receptacle.

SEED MORPHOLOGY

As noted, seed development is associated with the synthesis of growth regulators vital to fruit growth. Therefore, fruit size is normally a partial function of the number of seeds formed.

Seeds constitute only a small fraction of the fresh-weight of the fruit. Equally, the embryo makes up only a modest proportion of the seed volume (Fig. 3.28). The embryo consists of two seedling leaves (cotyledons), a nascent shoot (epicotyl), and an embryonic root (radicle) at the tip of the hypocotyl. The embryo is surrounded by a nutritive endosperm, which constitutes the bulk of the seed. The endosperm is enclosed in a pair of seed coats (integuments), of which only the outer integument (testa) develops significantly. The testa consists of a hard inner section, a middle parenchymatous layer, and an outer papery component or epidermis.

CHEMICAL CHANGES DURING BERRY MATURATION

Due to the major importance of both the quantitative and qualitative aspects of berry development, the subject has received extensive study. However, as most research has been directed primarily at improving grape yield

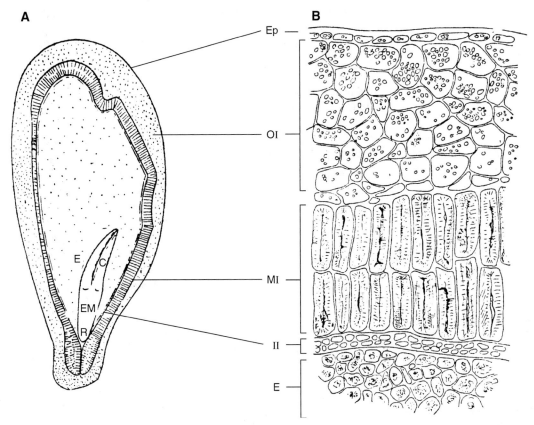

Figure 3.28 Diagrammatic representation of the (**A**) grape-seed and (**B**) seed-coat structures. C, cotyledons; E, endosperm; EM, embyro; Ep, epidermis; II, inner integument; MI, middle integument; OI, outer integument; R, radicle. (A from Levadoux, 1951, reproduced by permission; B modified by Levadoux, 1951 from Ravaz, 1915, reproduced by permission)

and quality, no unified theory of berry development has emerged. Several investigators have been pursuing this goal. A general theory of berry development might permit grape growers to predict the probable consequences of various viticultural options. In this regard, the following discussion will incorporate, where possible, physiological explanations of berry development. Because of the chemical nature of certain topics, some readers may wish to refer to Chapter 6 for clarification.

Following chemical changes is often complicated due to the various ways in which data may be presented (fresh vs. dry weight, or per berry). The potential for translocation, chemical modification, and polymerization further complicates data interpretation. Finally, the enologic significance of these changes is often dependent on extraction during vinification, a factor not necessarily or directly correlated with fruit composition. This is especially true when dealing with phenolic and flavor compounds.

As noted, a major shift in metabolism occurs simultaneously with berry color change (*véraison*). That these changes are controlled by plant-growth regulators is beyond doubt. Regrettably, the specific actions of

individual growth regulators remain unclear. Unlike animal hormones, plant growth regulators have many and differing actions. Their effects can depend on the tissue involved, its physiological state, and the relative and absolute concentration of other growth regulators. This plasticity has often made the prediction of growth-regulator effects excruciatingly difficult.

During stage I of berry growth, auxin, cytokinin, and gibberellin contents tend to increase. They undoubtedly stimulate cell division and enlargement up to *véraison*. Their further role in cell enlargement is unclear, because their concentration subsequently declines. The drop in auxin and gibberellin content coincides with an increase in the concentration of abscisic acid. The decline in auxin content also correlates with a fall in fruit acidity, the accumulation of sugars, and a reduction in shoot growth (Fig. 3.29). The importance of this change is suggested by the delay in ripening provoked by the application of auxins to grapes (Davies *et al.*, 1997). Subsequently, the level of gibberellins may rise. Gibberellin content is highest in the seeds (Pérez *et al.*, 2000).

Ethylene has traditionally been considered to play an insignificant role in grape ripening. However, data from

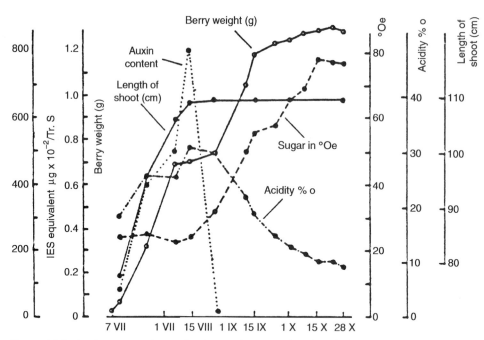

Figure 3.29 Auxin content of grape berries in relation to shoot length, berry weight, sugar content, and acidity of the cultivar 'Riesling'. (From Alleweldt and Hifny, 1972)

El-Kereamy *et al.* (2003) suggest a role in anthocyanin synthesis. In addition, Hilt and Blessis (2003) have demonstrated its role in the abscission of young fruit, presumably by its action on the synthesis of abscisic acid. In this regard, its action is dependent on the phenologic stage of development. Ethylene concentration tends to rise slightly until *véraison*, but returns to a low level by maturity. Recently, Symons *et al.* (2006) have proposed a role for a newly discovered group of growth regulators, brassinosteroids, in berry ripening.

Because of their influence on berry development, growth regulators have been used in controlling fruit development and berry spacing in clusters (Plate 3.8). Synthetic growth regulators have been more effective than their natural counterparts because they are less affected by feedback control mechanisms. Artificial auxins such as benzothiazole-2-oxyacetic acid can delay ripening up to several weeks when applied to immature fruit. Conversely, the artificial growth retardant methyl-2-(ureidooxy) propionate markedly hastens ripening when applied before *véraison* (Hawker *et al.*, 1981). Ripening also can be shortened marginally by applying growth regulators, such as ethylene or abscisic acid after *véraison*.

During development, structural changes in the skin and vascular tissues influence the types and amount of substances transported to the berry. The degeneration of stomatal function, disruption of xylem function, and development of a thicker waxy coating can combine to reduce transpirational water loss (Fig. 3.30).

This also reduces gas exchange between the berry and the atmosphere, as well as increases the likelihood of fruit overheating – and marked day–night temperature fluctuations.

As berries near maturity, water uptake declines (Rogiers *et al.*, 2001). This can result in a tenfold reduction in hydraulic conductance to berries from *véraison* to ripeness (Tyerman *et al.*, 2004). This has normally been interpreted as resulting from xylem vessels rupture in the peripheral vascular tissues of the berry (Creasy *et al.*, 1993). This interpretation has been challenged (Chatelet *et al.*, 2005). Although tracheids do not stretch during berry growth, and some rupture, others remain functionally intact, while others develop. Data from Bondada *et al.* (2005) suggest that the decrease in xylem flow results from a reduction in the hydrostatic gradient of the berry apoplast. This would explain why most of the water supplied to post-*véraison* berries comes from the phloem (Fig. 3.31) and the disappearance of diurnal fluctuations in berry diameter (Greenspan *et al.*, 1994). It would also explain why minerals such as calcium, manganese and zinc, transported primarily in the xylem, become restricted post-*véraison* (Rogiers *et al.*, 2006b). In contrast, potassium, which is transported equally in the phloem and xylem, continues to accumulate throughout ripening (Hrazdina *et al.*, 1984). The increasing importance of phloem transport for water as well as organic and inorganic nutrients during ripening has been equally found in other fruits, for example apples and tomatoes. Evidence is contradictory as to whether the xylem acts

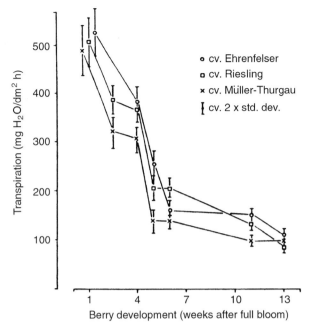

Figure 3.30 Transpiration of berry bunches per fruit surface area for three grape varieties. Measurements were performed at 20°C, 52% r.h., 335 vpm CO_2, and 800 µmol/m^2/s (about 56 kLux). (From Blanke and Levhe, 1987, reproduced by permission)

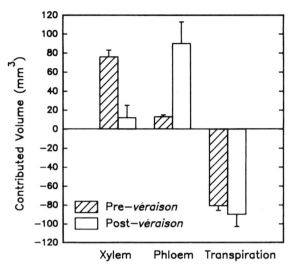

Figure 3.31 Pooled data for daily water flow for each component of the water budget of well-irrigated 'Cabernet Sauvignon' berries. (From Greenspan *et al.*, 1994, reproduced by permission)

as a conduit for the movement of water out of the berry (Lang and Thorpe, 1989; Greenspan *et al.*, 1996). At maturity, the phloem ceases to function, and a periderm forms at the base of the berry. The abscission layer so formed prevents further translocation between the vine and the fruit. Its formation facilitates separation of the fruit from the pedicel at harvest, and probably explains shrinkage of the fruit at this stage (Rogiers *et al.*, 2006a). 'Shiraz,' is particularly sensitive to shriveling at this stage.

Sugars Quantitatively, sucrose is the most significant organic compound translocated into the fruit. Not surprisingly, it is also the primary organic compound transported in the phloem. Early in berry development, much of the carbohydrate used in growth is photosynthesized by the berry itself. As the berry enlarges and approaches *véraison*, the surrounding leaves take over the role of major carbohydrate supplier. The trunk and arms of the vine may also occasionally act as important carbohydrate sources. It has been estimated that up to 40% of the carbohydrate accumulated in fruit can be supplied from the permanent, woody parts of the vine (Kliewer and Antcliff, 1970).

Sugar accumulation is particularly marked following *véraison*. This coincides with a pronounced decline in glycolysis. As primary shoot growth usually slows dramatically at this time, a major redirection of vine photosynthate toward the fruit is possible. Root growth also declines following *véraison*, further reducing its

drain on photosynthate. Nevertheless, an adequate, coordinated explanation of these events has yet to be proposed (see Coombe, 1992).

Part of the accumulation may relate to reduced sugar need, associated with a switch to respiring malic acid. Malic acid accumulates in considerable amounts early in fruit development, and it is a major intermediary in the TCA (tricarboxylic acid) cycle. A minor factor contributing to the increase in sugar content may also be the biosynthesis of glucose from malic acid. Gluconeogenesis may also partially explain the reduction in other TCA intermediates, such as citric and oxaloacetic acids during maturation. The increase in abscisic acid content, associated with the beginning of *véraison*, may activate this metabolic shift (Palejwala *et al.*, 1985). Nevertheless, the major factors in sugar accumulation probably relate to redirection of phloem sap from leaves to the fruit (Dreier *et al.*, 2000), and evaporative concentration.

On reaching the berry, most of the sucrose carried in the phloem is hydrolyzed to fructose and glucose. Although invertase generates equal amounts of glucose and fructose from sucrose, the levels of the two sugars are seldom equal in the fruit. In young berries, the proportion of glucose is generally higher. During ripening, the glucose/fructose ratio often falls, with fructose accumulation (or retention) being slightly higher than that of glucose. By maturation, if not before, fructose is marginally more prevalent than glucose (Kliewer, 1967). The reasons for this disequilibrium are unknown, but may originate from differential rates of glucose and fructose metabolism, or selective synthesis of fructose from malic acid.

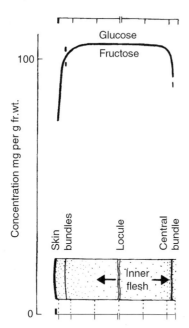

Figure 3.32 Smoothed graph showing glucose and fructose concentrations into the center of a grape – vertical lines in the berry cross-section indicate the position of vascular bundles. (From Coombe, 1990, reproduced by permission)

As with other cellular constituents, sugar accumulation varies with the cultivar, maturity and prevailing environmental conditions. Depending on these factors, the sugar content may vary from 12 to 28% at harvest. Further increases during over-ripening appear to reflect concentration, due to water loss, rather than additional sugar uptake. For winemaking, optimal sugar contents usually range from 21 to 25%.

Depending on the species or cultivar, some of the sucrose translocated to the fruit is not metabolized. This is most marked in muscadine cultivars, in which 10–20% of the sugar content may remain as sucrose. In some French-American hybrids, sucrose constitutes about 2% of the sugar content. In most *vinifera* cultivars, the sucrose content averages about 0.4% (Holbach *et al.*, 1998).

Small amounts of other sugars are found in mature berries, such as raffinose, stachyose, melibiose, maltose, galactose, arabinose, and xylose. Their presence is considered insignificant because they occur in small amounts, are not metabolized by most wine yeasts, and do not impact the wine's perceived sweetness.

Sugar storage occurs primarily in cell vacuoles. Glucose and fructose accumulation occurs predominantly in the flesh, with lesser amounts being deposited in the skin (Fig. 3.32). This can vary with the cultivar and stage of ripening. In contrast, the small amount of sucrose that occurs in the fruit of *vinifera* cultivars is restricted to the axial vascular bundles, and the skin adjacent to the peripheral vascular strands.

During ripening, much of the sugar comes from leaves located on the same side of the shoot as the fruit cluster and directly above the cluster. Carbohydrate supplies come increasingly from the shoot tip as it develops (Fig. 3.33), from associated lateral shoots, and from old wood. Estimates of the foliage cover (leaf area/fruit weight) required to fully ripen grape clusters vary considerably, from 6.2 cm^2/g (Smart, 1982) to 10 cm^2/g (Jackson, 1986). The latter appears to be more common. The variation may reflect varietal and environmental influences on fruit load and leaf photosynthetic efficiency.

Acids   Next to sugar accumulation, reduction in acidity is quantitatively the most marked chemical modification during ripening. Tartaric and malic acids account for about 70–90% of the berry acid content. The remainder consists of variable amounts of other organic acids (e.g., citric and succinic acids), phenolic acids (e.g., quinic and shikimic acids), amino acids, and fatty acids.

Although structurally similar, tartaric and malic acids are synthesized and metabolized differently. Tartaric acid is derived via a complex transformation from vitamin C (ascorbic acid). This appears to involve L-idonic acid, as a rate limiting step (DeBolt *et al.*, 2006). In contrast, malic acid is an important intermediate in the TCA cycle. As such, it can be variously synthesized from sugars (via glycolysis and the TCA cycle), or via carbon dioxide fixation from phosphoenolpyruvate (PEP). Malic acid also can be readily respired, or decarboxylated to PEP via oxaloacetate in the gluconeogenesis of sugars. Not surprisingly, the malic acid content of berries changes more rapidly and strikingly than that of tartaric acid.

The degree and nature of acid conversions can vary widely, depending on the cultivar and environmental factors. Nevertheless, several major trends are usually observed. After an initial, intense synthesis of tartaric acid in the ovary, both tartaric and malic acids increase slowly in content up to *véraison*. Subsequently, the amount of tartaric acid tends to stabilize, whereas that of malic acid declines. It is hypothesized that the initial accumulation of malic acid acts as a nutrient reserve to replace glucose as the major respired substrate following *véraison*. This would explain the frequent rapid drop in malic acid content during the later stages of ripening. However, it leaves unexplained the slow decline in malic acid content characteristic of vines grown in cool climates.

This pattern differs in some significant regards in the cultivars of *V. rotundifolia*. In muscadine cultivars, acidity, donated primarily by tartaric acid, declines from a maximum at berry set. Succinic acid, insignificant

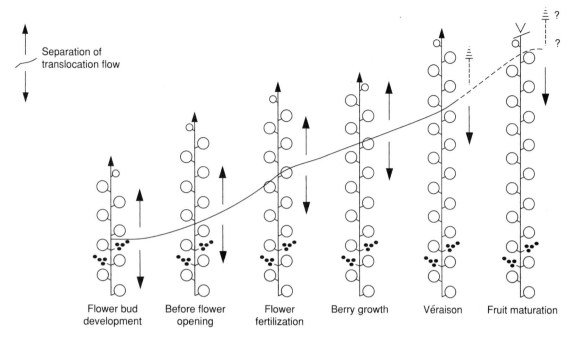

Figure 3.33 Development in the translocation of photosynthate from leaves during shoot growth. (After Koblet, 1969, reproduced by permission)

in *V. vinifera* cultivars, is a major component in very young muscadine berries (Lamikanra *et al.*, 1995). Its concentration declines rapidly thereafter. Malic acid content tends to increase up to *véraison*, declining thereafter.

It has long been known that grapes grown in hot climates often metabolize all their malic acid before harvest. Conversely, grapes grown in cool climates may retain most of their malic acid into maturity. Although exposure to high temperatures activates enzymes that catabolize malic acid, this alone appears insufficient to explain the effect of temperature on berry malic acid content. Reduced synthesis, and possibly heightened gluconeogenesis, may also play a role in the drop in malic acid content.

The drop in acid concentration during ripening is particularly marked when assessed on fresh-berry weight. This results from berry enlargement (water uptake) markedly exceeding the accumulation of tartaric acid and the metabolism of malic acid.

Shoot vigor is another factor influencing fruit acidity. Vigor (as indicated by an increased leaf surface/fruit ratio) is correlated with reduced grape acidity and higher pH (Jackson, 1986). This connection has usually been ascribed to leaf and fruit shading associated with increased vigor. Nevertheless, experiments where shading was prevented showed a direct effect of vigor on reduced fruit quality (lower °Brix, higher pH and lower acidity) (Jackson, 1986).

In addition to environmental influences, hereditary factors can significantly affect berry acid content. Some varieties such as 'Zinfandel,' 'Cabernet Franc,' 'Chenin blanc,' 'Syrah,' and 'Pinot noir' are proportionally high in malic acid, whereas others, such as 'Riesling,' 'Sémillon,' 'Merlot,' 'Grenache,' and 'Palomino' are higher in tartaric acid content (Kliewer *et al.*, 1967).

Tartaric acid tends to accumulate primarily in the skin, with lower amounts relatively evenly distributed throughout the flesh (Fig. 3.34). This difference may slowly diminish during ripening. In contrast, malic acid deposition is high in the epidermis, low in hypodermis, and increases again to a maximum near the berry locule. As ripening advances, the differences across the flesh become less marked. By maturity, the malic acid concentration in the skin, although low, may surpass that in the flesh. This may result from malate metabolism initiating around the axial vascular bundles and progressing outward.

The free versus salt state of tartaric acid generally changes throughout maturation. Initially, most of the tartaric acid exists as free acid. During ripening, progressively more tartaric acid combines with cations, predominantly K^+. Salt formation with Ca^{2+} ions usually results in the deposition of calcium tartrate crystals in the vacuoles of skin cells (Ruffner, 1982). In contrast, malic acid generally remains as a free acid. The combination of the changing acid concentrations and their salt states results in the lowest titratable acidity occurring

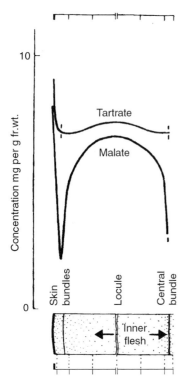

Figure 3.34 Smoothed graph showing tartaric and malic acid concentrations into the center of a grape – vertical lines in the berry cross-section indicate the position of vascular bundles. (From Coombe, 1990, reproduced by permission)

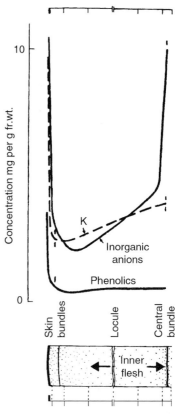

Figure 3.35 Smoothed graph showing potassium, inorganic ions and phenolic concentration into the center of a grape – vertical lines in the berry cross-section indicate the position of vascular bundles. (From Coombe, 1990, reproduced by permission)

next to the skin. The highest titratable acidity normally exists next to the seeds, in the center of the berry.

The factors regulating the synthesis of tartaric and malic acids are poorly understood. The effect of temperature on malic acid degradation has already been mentioned. Nutritional factors can considerably influence berry acidity and acid content.

Increased potassium availability results in slight increases in the levels of both acids. The amplification is probably driven by the need to produce additional acids to permit potassium uptake. Transport of K^+ into cells requires the simultaneous export of H^+. Although potassium uptake may enhance acid synthesis, it also results in a rise in pH, due to its formation of acid salts. This effect may not be simple, however, because the proportion of free tartaric acid in skin cells may rise during maturation, even though up to 40% of potassium accumulation occurs in the skin. This apparent anomaly may result from a differential deposition of potassium and acids in vacuoles of the skin (Iland and Coombe, 1988).

Nitrogen fertilization enhances malic acid synthesis, but may induce a decline in tartaric acid content. The increase in malic acid synthesis may involve cytoplasmic pH stability. Nitrate reduction to ammonia

tends to raise cytoplasmic pH by releasing OH^+ ions. The synthesis of malic acid could neutralize this effect. However, because of the low level of malic acid ionization in grape cells, and its metabolism via respiration during ripening, the early synthesis of malic acid does not permanently prevent a rise in pH during ripening.

Although both malic and tartaric acids appear to be involved in maintaining a favorable ionic and osmotic balance in berry cells, they tend to be stored in cell vacuoles to avoid excessive cytoplasmic acidity. Here, the distinctive transport characteristics of the vacuolar membrane isolate the acids from the cytoplasm.

Potassium and Other Minerals   As noted, potassium uptake during maturation affects both acid synthesis and ionization. The factors controlling the accumulation of potassium in berry tissues are poorly understood. It is transported in both the xylem and phloem, and can be redistributed from older to younger tissues. Its uptake increases after *véraison*, especially in the skin (Fig. 3.35). The skin, constituting about 10% of the berry weight, may contain up to 30–40% of the potassium content. The high correlation between potassium and sugar accumulation in the skin and flesh, respectively, fits the view that potassium acts as an osmoticum in skin cells, as sugar does in the flesh. It also correlates with berry softening, associated with apoplast acidification. H^+ ions export compensates for the import of K^+ ions. Because potassium accumulation in the vacuole affects its permeability, potassium uptake may affect the release of malic acid, favoring its metabolism in the cytoplasm. Potassium is also a well-known enzyme activator. However, the reasons for the nonuniform accumulation of potassium in skin cells (Storey, 1987) remain unclear.

Rogiers *et al.* (2006b) has investigated the deposition pattern of other minerals in the fruit. They found that the pulp and skin were the principal sinks for boron. In contrast, seeds were the major deposition sites for calcium, manganese, phosphorus, sulfur and zinc.

Phenolics In red grapes, flavonoid phenolics constitute the third most significant group of organic compounds. They not only donate the color to red wines, but also provide most of their characteristic taste and aging properties. White grapes have lower total phenolic contents and do not synthesize anthocyanins. Aside from seed phenolics, both red and white grapes contain most of their phenolics in the skin (Fig. 3.35). Because seed phenolics are slowly extracted, the primary source of grape phenolics in most wines comes from the skin. They are primarily located in phenolic granules that accumulate in epidermal and hypodermal vacuoles (Fig. 3.27). Those granules that adhere to the vacuole membrane or cell wall are not easily extracted. Hydroxycinnamic acid esters of tartaric acid form the predominant phenolic components of the flesh.

Pigmentation in most red cultivars is restricted to the epidermis and outer hypodermal layers of the skin (Walker *et al.*, 2006). In white grapes, color comes from the presence of carotenoids, xanthophylls, and flavonols such as quercetin. Carotenoids accumulate predominantly in plastids, whereas flavonoid pigments are deposited in cell vacuoles. Similar pigments occur in red grapes, but the predominant color comes from the production of anthocyanins.

Activation of anthocyanin synthesis is under the control of adjacent regulator *MYB* genes (*VvMYBA1* and *VvMYBA2*). Either can regulate anthocyanin synthesis through production of a regulatory protein. Two other versions of the gene appear not to be involved in anthocyanin synthesis, at least in fruit (Walker *et al.*, 2007). The role of *VvMYBA1* is illustrated in 'Pinot blanc,' where skin cells possess two copies of *VvMYBAa*, the nonfunctional allele. In contrast, 'Pinot noir' is heterozygous, possessing a copy each of *VvMYBAc* (the dominant, functional allele) and *VvMYBAa* (Yakushiji *et al.*,

2006). Location of the retrotransposon *Gret1* in the promotor region of *VvMYBA1* is frequently, but not exclusively, associated with white-fruited phenotypes (This *et al.*, 2007). Inactive *VvMYBA2* alleles have a substitution mutations and a deletion (Walker *et al.*, 2007). White varieties possess both color loci genes in inactive forms. What is amazing is that of 55-white berries cultivars, all appear to bear similar mutated *MYB* genes. As well, many famous red cultivars are heterozygous for the color locus (Walker *et al.*, 2007), for example 'Cabernet Franc,' 'Cabernet Sauvignon,' 'Dolcetto,' 'Gamay,' 'Merlot,' 'Nebbiolo,' 'Pinot Noir,' Sangiovese,' 'Shiraz,' and 'Zinfandel.' This suggests a much greater genetic relationship between domesticated grape cultivars than has hitherto been suspected.

Several genes are involved in anthocyanin synthesis. Of these, the *Ufgt* gene is critical in the synthesis of UDP glucose–flavonoid 3–o–glucosyl transferase (UFGT). It directs anthocyanin glycosidation. The other six flavonoid genes involved in anthocyanin (and other flavonoid) syntheses are expressed in all tissues, while only *Ufgt* is expressed in skin cells. *Ufgt* is functional but inactive in white varieties (Boss *et al.*, 1996a). The pathway of synthesis of anthocyanins is represented in Fig. 3.36.

Expression of flavonoid synthesis genes, and those associated with anthocyanin synthesis, are enhanced by the presence of abscisic acid (Jeong *et al.*, 2004). This could explain why pigment formation is suppressed at high temperatures. Heat reduces abscisic acid content (Yamane *et al.*, 2006). *Ufgt* is also activated by the application of 2-chloro-ethylphosphonic acid, an ethylene-releasing compound (El-Kereamy *et al.*, 2003). Activation is also associated with an upsurge in the presence of chalcone synthase and flavanone 3-hydroxylase. The color enhancing effect of spraying dilute ethanol (or methanol) on grapes may result from their activation of ethylene production (Chervin *et al.*, 2001). These data suggest that grapes may actually be a climacteric fruit.

Only the epidermis and first hypodermal layer of the skin of red varieties is normally darkly pigmented. The next two hypodermal layers may contain smaller amounts of anthocyanins, and subsequent layers tend to be sporadically and weakly pigmented. Pigmentation seldom occurs deeper than the sixth hypodermal layer. An exception are *teinturier* varieties. These possess weakly pigmented flesh (Hrazdina and Moskowitz, 1982). Otherwise, anthocyanins only occur in the flesh of overripe fruit. As skin cells senescence after ripening, their pigmentation may diffuse into the flesh.

Anthocyanins are transported and stored in cell vacuoles after synthesis on the endoplasmic reticulum and being glycosylated. The anthocyanin content in the outer hypodermal layer(s) soon approaches saturation.

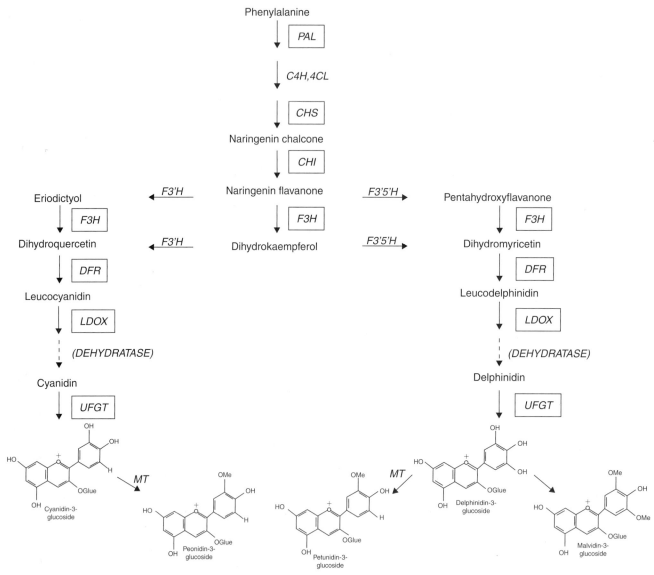

Figure 3.36 Simplified schematic representation of the anthocyanin biosynthetic pathway in grapes: PAL, phenylalanine ammonia lyase; CHS, chalcone synthase; CHI, chalcone isomerase; F3H, flavanone-3-hydroxylase; F3'H, flavonoid 3'hydroxylase; F3'5'H, flavonoid 3'5'-hydroxylase; DFR, dihydroflavonol 4-reductase; LDOX, leucoanthocyanidin dioxygenase; UFGT, UDP glucose-flavonoid 3-O-glucosyl transferase; C4H, cinnamate 4-hydroxylase; 4CL, 4-coumarate CoA ligase; MT, methyltransferase. (From Boss *et al.*, 1996b, reproduced by permission)

They subsequently combine in self-association or co-pigment complexes. The decline in anthocyanin content that occurs in some cultivars near or after maturity is probably caused by β-glycosidases and peroxidases. These also occur in skin vacuoles (Calderón *et al.*, 1992). These may become active as the grapes mature.

In addition to anthocyanins and tannins, red-skinned cultivars contain flavonols, various derivatives of benzoic and cinnamic acids and aldehydes, and hydroxycinnamic acid esters of tartaric acid. They occur in small amounts in the grape skin and at much lower concentrations in the flesh. The predominant flavonol in *V. vinifera* grapes tends to be kaempferol, whereas in

V. labrusca cultivars, quercetin appears to predominate. Most flavonols are glycosidically linked to glucose, rhamnose, or glucuronic acid. The predominant hydroxycinnamic acid ester is caffeoyl tartrate, with smaller amounts of coumaroyl tartrate and feruloyl tartrate. As with most other chemical constituents, the specific concentrations vary widely from season to season, and from cultivar to cultivar.

Flavonoid synthesis begins with the condensation of erythrose, derived from the pentose phosphate pathway (PPP), and phosphoenolpyruvate, derived from glycolysis. The phenylalanine so generated is converted to 4-coumaroyl-CoA. Its binding with acetate (from malonyl

CoA) to form the B ring and A rings of flavonoids, respectively (Hrazdina *et al.*, 1984). Subsequent modification gives rise to flavonoid phenolics, including anthocyanins (Figs 3.36, 3.37). Their condensation generates tannins (Boss *et al.*, 1996b; Dixon *et al.*, 2005). Induction of flavonoid synthesis appears to be under the control of another *MYB* gene, in this case *VvMYBPA1* (Bogs *et al.*, 2007).

The synthesis of phenolics begins shortly after berry development commences. Some anthocyanins are synthesized early, but most production involves other flavonoid or nonflavonoid phenolics. Anthocyanin synthesis becomes pronounced only after *véraison*. The timing and degree of anthocyanin synthesis depends on a variety of factors, such as temperature, light exposure, water status, sugar accumulation, and genetic factors. After reaching a maximum (usually at grape maturity), anthocyanin concentration tends to decline slightly. Tannin content tends to mirror changes in anthocyanin content, especially those more readily extracted from the skin. However, a decline in solubility, associated with increasing polymer size, contributes to a drop in fruit astringency.

Although the anthocyanin content of red grapes increases during ripening, the proportion of the various anthocyanins, and their acylated derivatives can change markedly (González-SanJosé *et al.*, 1990). As the final distribution can significantly influence the hue, intensity, and color stability of a wine, these changes have great enologic significance. Methylation of the oxidation-sensitive *o*-diphenol sites improves anthocyanin stability. Thus, the monophenolic anthocyanins, peonin and malvin, are the most stable. Susceptibility to oxidation also is decreased by bonding with sugars and acyl groups (Robinson *et al.*, 1966). Independent factors also can affect color development, for example, the influence of mild water stress in enhancing flavonol synthesis. Flavonols may be important in the formation of anthocyanin copigments.

The precise conditions that initiate anthocyanin synthesis during *véraison* are unknown. One hypothesis suggests that sugar accumulation provides the substrate needed for synthesis. The marked correlation between sugar accumulation in the skin and anthocyanin synthesis is consistent with this view. The accumulation of sugars, beyond the immediate needs of a tissue, often favors the synthesis of secondary metabolites, such as anthocyanins and various flavorants. Nevertheless, sugar accumulation may act indirectly, through its effect on the osmotic potential of skin cells. In culture, high osmotic potentials can induce anthocyanin synthesis and their methylation in grape cells (Do and Cormier, 1991). However, as abscisic acid content is associated with sugar accumulation, it may be abscisic acid that is

the initial trigger for anthocyanin synthesis (Ban *et al.*, 2003). This could explain why abscisic acid application enhances anthocyanin accumulation (Peppi *et al.*, 2005). Alternately, anthocyanin synthesis may be activated by changes in the potassium and calcium contents of skin cells. Potassium accumulates in the skin following *véraison*, and its cellular distribution is as nonuniform as initial anthocyanin synthesis. Conversely, the decline in calcium content is inversely correlated with anthocyanin synthesis.

During berry development there are corresponding changes in the concentration and composition of skin tannins (Downey *et al.*, 2003). The degree of polymerization increases, as does the proportion of epigallocatechin subunits and the inclusion of anthocyanin monomers in procyanidin polymers (Kennedy *et al.*, 2001). The principal subunits are (−)-epicatechin and (−)-epigallocatechin, with the terminal units being primarily (+)-catechin. The epicatechins are biosynthesized from anthocyanidins, whereas catechin is derived from leucocyanidins (flavan-3,4-diols) (Fig. 3.37). The average number of subunits in the polymers is about 25 in 'Shiraz.' Peak concentration occurs about *véraison*.

Synthesis of nonflavonoid phenolics tends to decline and may cease following *véraison*. As a result, the concentrations of hydroxycinnamic acid esters and catechins usually decline, probably due to dilution during berry enlargement. Nevertheless, some low-molecular-weight phenolic compounds (i.e., benzoic and cinnamic acid derivatives) may increase during ripening (Fernández de Simón *et al.*, 1992).

In white grapes, the phenolic changes during ripening primarily involve hydroxycinnamic tartrates and catechins. Although the dynamics of hydroxycinnamic tartrate metabolism are unclear, their levels may fall strikingly during ripening (Lee and Jaworski, 1989). As the degree and nature of the decline are influenced by genetic and environmental factors, their role in oxidative browning can change markedly from variety to variety, and from year to year. Catechin levels may rise following *véraison*, but decline again to low levels by maturity. Catechin monomers tend not to polymerize into large tannins in white cultivars.

The predominant phenolics in seeds are flavan-3-ols (catechin, epicatechin, and their procyanidin polymers). Those in the seed wall are more polymerized and contain a higher proportion of epicatechin gallate than those in inner parts of the seed (Geny *et al.*, 2003). Extension subunits are principally epicatechin and epicatechin gallate, with the terminal subunits being about equally epicatechin, epicatechin gallate and catechin (Downey *et al.*, 2003). In 'Shiraz' the average number of subunits in procyanidin polymers is about 5. Although they constitute the major source of phenols

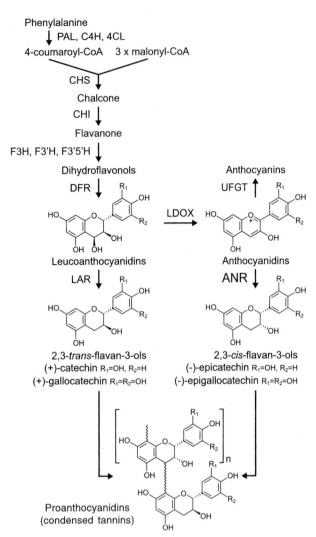

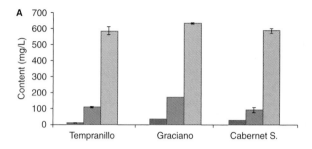

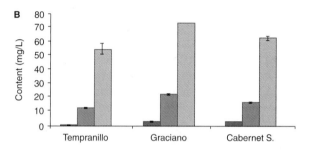

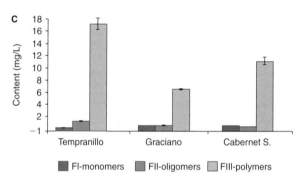

Figure 3.37 Schematic representation of the biosynthetic pathway of proanthocyanidin and anthocyanin. Enzyme abbreviations: PAL, phenylalanine ammonia-lyase; C4H, cinnamate 4-hydroxylase; 4CL, 4-coumarate:CoA-ligase; CHS, chalcone synthase; CHI, chalcone isomerase; F3H, flavanone 3-hydroxylase; F3'H, flavonoid 3'-hydroxylase; F3'5'H, flavonoid 3',5'-hydroxylase; DFR, dihydroflavonol 4-reductase; LDOX, leucoanthocyanidin dioxygenase; UFGT, UDP-glucose:flavonoid 3-O-glucosyl transferase; LAR, leucoanthocyanidin reductase; ANR, anthocyanidin reductase. R_1, R_2 = H or OH. (From Fujita *et al.*, 2005, reproduced by permission)

Figure 3.38 Flavan-3-ol content of the monomeric, oligomeric, and polymeric fractions of (**A**) wine, (**B**) seeds, and (**C**) skins from 'Tempranillo,' 'Graciano,' and 'Cabernet Sauvignon.' Bars represent standard deviation. (From Monagras *et al.*, 2003, reproduced by permission. Copyright 2003 American Chemical Society)

in grapes (about 60% vs. 20% each in the skins and stalks), they are seldom extracted in significant amounts. The lipid coating of seeds retards tannin extraction during wine production, at least until the alcohol content of the ferment facilitates their release. The proportion of monomeric, oligomeric and polymeric flavonols in seeds, skins and wine from three cultivars is illustrated in Fig. 3.38.

After a pronounced increase early in berry development, the content of individual flavanols declines as they polymerize into procyanidins and condensed tannins.

Because this is more pronounced in vines with low vigor (Peña-Neira *et al.*, 2004), it may partially explain why wines made from such vines tend to be less bitter and astringent than equivalent wines made from grapes harvested from vigorous vines. Reduced astringency may also result from the association of pectins with procyanidins (Kennedy *et al.*, 2001).

A group of phenolic compounds that has attracted considerable attention recently are the stilbene phytoalexins. Phytoalexins are compounds synthesized predominantly in response to localized stress, such as infection by pathogens. The major phytoalexin in grapes is resveratrol (and its glycoside piceid). Smaller amounts of several related compounds, derived from resveratrol (pterostilbene and viniferins), also may occur. Phytoalexin synthesis generally declines after *véraison*,

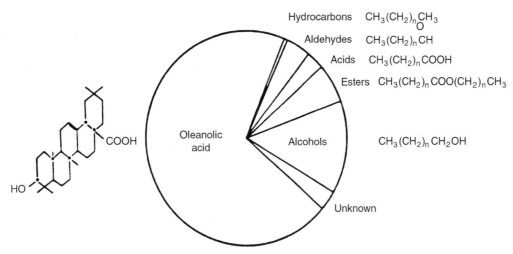

Hydrocarbons $CH_3(CH_2)_n CH_3$

Aldehydes $CH_3(CH_2)_n \overset{O}{C}H$

Acids $CH_3(CH_2)_n COOH$

Esters $CH_3(CH_2)_n COO(CH_2)_n CH_3$

$CH_3(CH_2)_n CH_2OH$

Oleanolic acid

Alcohols

Unknown

Figure 3.39 Cuticular wax content of 'Sultana' grapes. (From Grncarevic and Radler, 1971, reproduced by permission)

coinciding with the synthesis of anthocyanins in red grape varieties. Because the synthesis of anthocyanins and stilbenes have the same precursor, it is hypothesized that the synthesis of anthocyanins may provoke a decline in the potential of berries to synthesize resveratrol (Jeandet *et al.*, 1995).

Pectins One of the more obvious changes during berry ripening is its softening. Softening facilitates juice release from the flesh and the extraction of phenolic and flavor components from the skin. Softening results from a loosening of the bonds holding plant cells together. It is associated with an increase in water-soluble pectins and a decrease in wall-bound pectins (Silacci and Morrison, 1990). These changes may result from enzymatic degradation or a decline in the calcium content of cell-wall pectins. The latter is associated with acidification of the apoplast. Calcium is important in maintaining the solid pectin matrix that characterizes most plant cell walls. Calcium uptake essentially ceases following *véraison*. The precise role of cellulases and hemicellulases, in addition to that of pectinases, in softening is unknown.

Lipids The lipid content of grapes consists of cuticular and epicuticular waxes, cutin fatty acids, membrane phospho- and glyco-lipids, and seed oils. Seed oils are important as an energy source during seed germination, but they are seldom found in juice or wine. Their presence in significant amounts could generate a rancid odor. Because the lipid concentration of the berry changes little during growth and maturation, synthesis appears to be equivalent to berry growth.

The most abundant fraction of grape lipids are the membrane phospholipids, followed by neutral lipids. Glycolipids and sterols (primarily β-sitosterol) constitute

the least common groups (Le Fur *et al.*, 1994). The predominant fatty components of grape lipids are the long-chain fatty acids linoleic, linolenic, and palmitic acids (Roufet *et al.*, 1987).

Most of the cuticular layer is deposited in folds by anthesis and expand during berry growth (Fig. 3.26). In contrast, epicuticular wax plates are deposited throughout berry growth, commencing about anthesis (Rosenquist and Morrison, 1988). Most of the epicuticular coating consists of hard wax, composed primarily of oleanolic acid (Fig. 3.39). The softer, underlying cuticle contains the fatty acid polymer called cutin. It consists primarily of C_{16} and C_{18} fatty acids, such as palmitic, linoleic, oleic, and stearic acids and their derivatives. Cutin is embedded in a complex mixture of relatively nonpolar waxy compounds.

About 30–40% of the long-chain fatty acids of the berry are located in the skin. They could act as important sources of unsaturated fatty acids for the synthesis of yeast cell membranes during wine fermentation. Although the relative concentration of fatty acids changes little during maturation, the concentration of the individual components may alter considerably. The most significant change involves a decline in linolenic acid content during ripening (Roufet *et al.*, 1987). This could reduce the involvement of its oxidation products in the generation of herbaceous odors in wine.

Another important class of lipids are the carotenoids. In red varieties, the concentration of both major (β-carotene and lutein) and minor (5,6-epoxylutein and neoxanthin) carotenoids falls markedly after *véraison* (Razungles *et al.*, 1988). The decline appears to be the result of dilution during berry growth in stage III, following cessation of synthesis. Because of their insolubility, most carotenoids remain with the skins and pulp following crushing. Nevertheless, enzymatic and acidic hydrolysis during maceration may release water-soluble carotenoid derivatives. These can include important aromatic compounds such as damascenone and β-ionone. For example, the increasing concentration of norisoprenoids in ripening 'Muscat of Alexandra' is correlated with a decrease in carotenoid content (Razungles *et al.*, 1993).

Nitrogen-Containing Compounds Although soluble proteins may induce haze formation in wine, comparatively little is known about the conditions affecting their production during berry maturation. The concentration of soluble proteins in grapes can increase markedly during maturation, reaching levels from 200 to 800 mg/liter. In addition to yearly variation, cultivars differ considerably in protein content (Tyson *et al.*, 1982). One of the most abundant proteins in mature grapes is a thaumatin-like protein (VVTL1). Its production coincides with sugar accumulation and berry softening (Tattersall *et al.*, 1997). Although its precise role is unknown, it is structurally related to proteins having antifungal properties. Another major group of soluble proteins in maturing grapes are the chitinases.

The most common peptide in ripe grapes is glutathione, a tripeptide consisting of glutamine, cysteine, and glycine. Its accumulation begins with the onset of *véraison* and closely follows the rise in berry sugar content (Adams and Liyanage, 1993). It functions as an important antioxidant, by maintaining ascorbic acid in its reduced form. As such, glutathione is a significant, but short-lived, component in grape juice. It rapidly reacts with caftaric acid and related phenolics as they oxidize, forming *S*-glutathionyl complexes.

The free-amino acid content of grapes rises at the end of ripening, the primary amino acids being proline and arginine. The latter is the principal form of nitrogen transported in the phloem. The relative concentrations of these two amino acids can vary up to 20-fold, depending on the variety and fruit maturity (Sponholz, 1991, Stines *et al.*, 2000). Their ratio may shift during maturation from a preponderance of arginine to that of proline (Polo *et al.*, 1983) or vice versa (Bath *et al.*, 1991). In many plants, proline acts as a cytoplasmic osmoticum (Aspinall and Paleg, 1981), similar to potassium. Whether proline accumulation has the same role in grapes is unclear.

Amino acids usually constitute the principal group of soluble organic nitrogen compounds released during crushing and pressing. Nevertheless, up to 80% of berry nitrogen content may remain with the pomace (nonsoluble components of the fruit).

Changes during ripening also occur in the level of inorganic nitrogen, primarily ammonia. Ammonia may

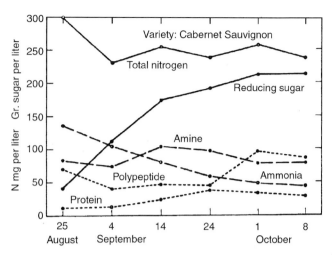

Figure 3.40 Changes in the nitrogen fraction of grapes during maturation. (Data from Peynaud and Maurié, 1953, from Amerine *et al.*, 1980, reproduced by permission)

show either a decline (Solari *et al.*, 1988) or an increase (Gu *et al.*, 1991). Although a decline might appear to negatively affect fermentation, it probably is more than compensated by the simultaneous increase in the amino acid content. Amino acids are incorporated by yeasts at up to 5–10 times the rate of ammonia. Their incorporation also alleviates the requirement of diverting metabolic intermediates for amino acid synthesis. However, as individual amino acids are accumulated by yeast cells at different rates, the specific amino acid content of the juice can significantly affect its fermentability. Some of the changes in nitrogen grape content are summarized in Fig. 3.40.

Aromatic Compounds Most research on fruit ripening has focused primarily on the major constituents of grapes, notably sugars and acids. Investigations of aromatic constituents have lagged behind due to lack of equipment with sufficient resolving power. Thus, long-held views about the synthesis and location of aromatic compounds in grapes remained unsubstantiated. The development and refinement of gas chromatography and other procedures now permit these beliefs to be tested. Of particular importance is the possibility that harvesting may be timed to coincide with the accumulation of important aroma compounds in the fruit.

Much of the research on grape aroma has involved Muscat cultivars, whose aroma is based primarily on monoterpenes. In general, monoterpenes accumulate during ripening. Along with synthesis, however, they may be subsequently converted to nonvolatile forms. These may exist as glycosides, oxidized derivatives, or polymers of the parent compound (Wilson *et al.*, 1986). Thus, mature grapes often generate less aroma than the level of their aromatic compounds would suggest.

Similar trends have been also noted with norisoprenoids and nonflavonoid phenols (Strauss, Gooley *et al.*, 1987).

The formation of glycosides increases the solubility of the aglycone (nonsugar) component of the complex, thus facilitating accumulation in vacuoles. Similar complexes may accumulate at even higher concentrations in the leaves (Skouroumounis and Winterhalter, 1994). These, however, appear not to be translocated to maturing fruit. At least, this is the interpretation from approach-grafting studies that show that the accumulation of aromatics in the fruit is not affected by the genotype of the shoot supporting the grape cluster (Gholami *et al.*, 1995).

Specific details of the accumulation and inactivation of terpene aroma compounds are shown for 'Muscat of Alexandria' (Fig. 3.41). The very young berry possesses a high concentration of geraniol. This rapidly declines, with a corresponding rise in its glycosidically bound form. After *véraison*, concentration of the major terpene linalool rises. Nevertheless, the proportion of the glycosidically bound form also increases. After reaching a maximum at maturity, both the free and bound fractions tend to decline.

Although most terpenes accumulate in the skin, notably geraniol and nerol, not all do so. For example, free linalool and diendiol I occur more frequently in the flesh than the skin. In addition, the proportion of free and glycosidically bound terpene may differ between the skin and the flesh.

In contrast, norisoprenoids, such as TDN (1,1,6-trimethyl-1,2-dihydronaphthalene), vitispirane, and damascenone, are localized predominantly in the flesh. They occur primarily in glycosidically bound forms, with only trace amounts of damascenone occurring as free volatile molecules. As with monoterpenes, their concentrations increase during ripening (Strauss, Wilson *et al.*, 1987).

In addition to monoterpenes and norisoprenoids, other aromatic compounds are present in nonvolatile, conjugated (glycosidically bound) forms. One group consists of low-molecular-weight aromatic phenols. Examples in 'Riesling' include vanillin, propiovanillone, methyl vanillate, zingerone, and coniferyl alcohol. Even phenolic alcohols, such as 2-phenylethanol and benzyl alcohol, may become glycosylated during ripening. Conjugated aromatic phenols have also been isolated from 'Sauvignon blanc,' 'Chardonnay,' and 'Muscat of Alexandria' (Strauss, Gooley *et al.*, 1987).

Another group of nonvolatile precursors of varietal significance are *S*-cysteine conjugates. They are metabolized by yeasts during fermentation, liberating their volatile thiols moieties. These play significant roles in the aroma of many varieties in the Cabernet family, and to a lesser extent in other *vinifera* cultivars. The S-cysteine conjugates are largely restricted to the skin (Peyrot des Gachons *et al.*, 2002).

Unlike the compounds mentioned above, others remain in a free, volatile form through maturity. For example, methyl anthranilate increases in concentration throughout the later stages of maturation (Robinson *et al.*, 1949). In other instances, the concentrations of impact compounds decline during maturation, for example, methoxypyrazines (Fig. 3.42). In this instance, it is a favorable change. Methoxypyrazines can easily dominate a wine's flavor, suppressing the potential expression of a rich fragrance. Fermentation conditions do little to diminish methoxypyrazine content (Roujou de Boubée *et al.*, 2002).

In several varieties, maturity affects the production of fusel alcohols and esters during fermentation. This has normally been interpreted as a result of their differing sugar contents. However, as Fig. 7.26 shows, other changes during ripening may affect the formation of esters during fermentation.

Studies comparable to those on white grapes have not been conducted with red grapes. In some varieties, however, aroma development has been correlated with ripeness and sun exposure. In 'Cabernet Sauvignon,' the total volatile component of the fruit was found to be about equally distributed between the skin and flesh (Bayonove *et al.*, 1974). Because the skin constitutes only about 5 to 12% of the fruit mass, its aromatic concentration in the skin must be considerably higher than that in the flesh.

CULTURAL AND CLIMATIC INFLUENCES ON BERRY MATURATION

Any factor affecting grapevine growth and health directly or indirectly influences the ability of the vine to nourish and ripen its fruit. Consequently, most viticultural practices are directed at regulating these factors to achieve the maximum yield, consistent with grape quality and long-term vine health and productivity. Although macro- and mesoclimatic factors directly affect berry maturation, they are usually beyond the control of the grape grower.

Yield Because of the obvious importance of yield to commercial success, much research has focused on increasing fruit production. However, yield increases can reduce the ability of the vine to mature the fruit, or its potential to produce subsequent crops. This has led to a debate over the appropriate balance between grape yield, fruit quality, and long-term vine health.

In France, yield is viewed so directly connected with grape and wine quality that theoretical maximum crop yields have been set for Appellation Control regions. This is partially justified on the moderately negative

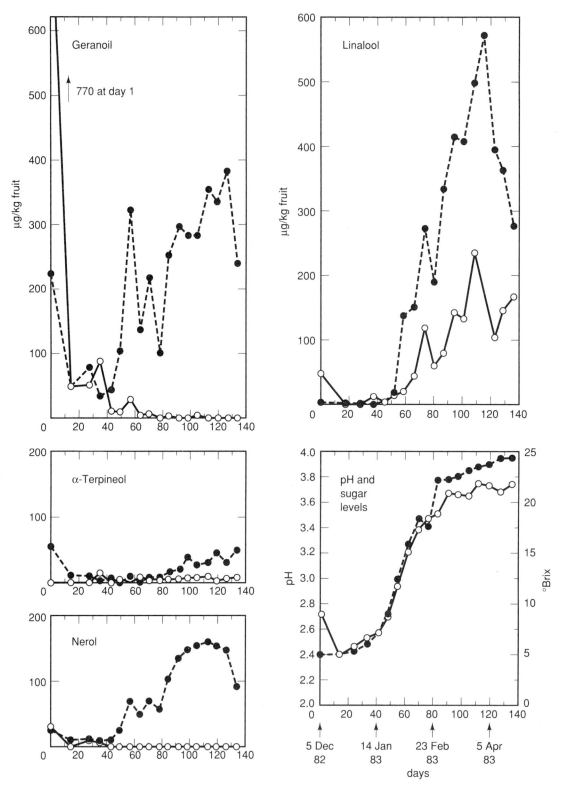

Figure 3.41 Changes in juice concentration of free (solid lines) and glycosidically bound (dashed lines) monoterpenes in developing 'Muscat of Alexandria' grapes. (Reprinted by permission from Wilson *et al.*, 1984. Copyright 1984. American Chemical Society)

correlation between increased yield and sugar accumulation (Fig. 3.43, and see Fig. 4.4). Nevertheless, focusing attention simply on yield deflects attention from other factors of equal or greater importance, such as improved nutrition and light exposure (Reynolds *et al.*, 1996). The elimination of viral infection often improves the ability of vines to produce and ripen fruit. The yield quadrupling in German vineyards in the twentieth century (Fig. 3.44) is a dramatic, but not isolated, example of yield increases without a comparable change in ripeness (as measured in °Oechsle). Clonal selection has also identified lines able to produce more and better-quality fruit. Improved fertilization, irrigation, weed and pest control, and appropriate canopy management can often further improve the capacity of a vine to produce fully ripened fruit (see Management of Vine Growth, Chapter 4).

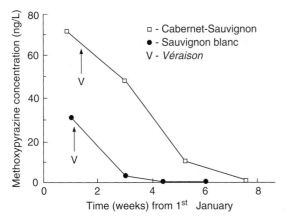

Figure 3.42 Typical concentration of methoxypyrazines of 'Cabernet Sauvignon' grape juice and 'Sauvignon blanc' wine compared to their aroma threshold. (From Allen *et al.*, 1996, reproduced by permission)

Although serious overcropping (>110–150 hl/ha) reduces wine quality, the specific cause–effect relationships are poorly understood. Factors frequently involved include delayed ripening, higher acidity, enhanced potassium content, reduced anthocyanin synthesis, diminished sugar accumulation, and limited flavor development. Achieving the ideal yield is one of the most perplexing demands of grape growing. It depends not only on the grape variety and soil characteristics, but also on the type of wine desired and the specifics of the prevailing climate.

Although overcropping is commonly thought to reduce grape and wine quality, and negatively affect subsequent yield and vine health, low yield does not necessarily improve quality. Undercropping can prolong shoot growth and leaf production, increase shading, depress fruit acidity, and undesirably influence berry nitrogen and inorganic nutrient contents. Reduced yield also tends to induce the vine to produce larger berries. The reduced skin/flesh ratio can negatively affect attributes derived primarily from the skin. This may partially explain that while reduced yield increased the intensity of "good" wine aroma, it also enhanced taste intensity (Fig. 3.45). The enhanced taste intensity partially offset the benefits of increased aroma. An overly intense aroma can compromise the desirable subtlety of a wine, resulting in diminished appreciation. In a recent study by Chapman *et al.* (2004), reduced yield was associated with a slight increase in the bell pepper odor of 2-methoxy-3-isobutylpyrazine, bitterness, and astringency in 'Cabernet Sauvignon' wines. In contrast, high yield generated wine moderately higher in red/black berry aromas, jammy flavors and fruitiness. The effects were less marked than those found for 'Zinfandel' (Fig. 3.45). Excessive yield reduction also can jeopardize the financial viability of a vineyard. Of particular

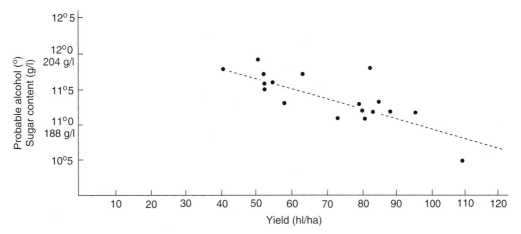

Figure 3.43 Relation between crop load and sugar accumulation in the variety 'Gewürztraminer.' (From Huglin and Balthazard, 1976, reproduced by permission)

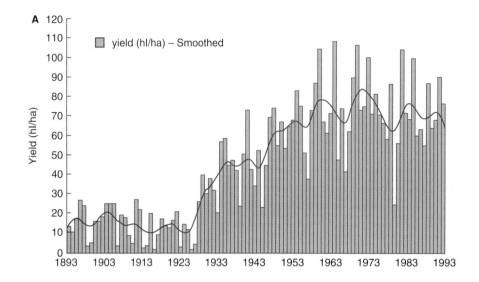

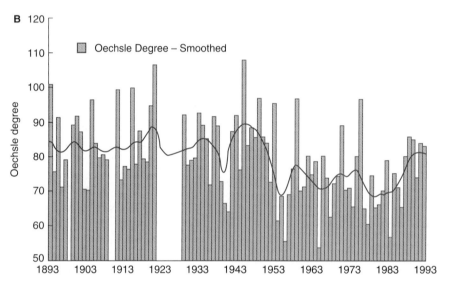

Figure 3.44 Fruit production (**A**) and soluble solids (**B**) from 'Riesling' at Johannesberg from 1885 to 1991 and 1890 to 1991, respectively. (From Hoppmann and Hüster, 1993, reproduced by permission)

interest was the observation that the method of achieving yield variation was important (Chapman *et al.*, 2004). Winter pruning, which had less effect on yield manipulation, had a greater effect on sensory characteristics than did the more marked yield variations generated by cluster thinning.

It has generally been believed that wines produced from vines bearing light to intermediate crops are preferred (Cordner and Ough, 1978; Gallander, 1983; Ough and Nagaoka, 1984). Nevertheless, the yield/quality ratio is not necessarily constant within the mid-yield range (Sinton *et al.*, 1978).

An improved understanding of the yield/quality relationship will come with a better grasp of the factors

that lead to grape and wine quality (see Criteria for Harvest Timing, Chapter 4). In addition, it is difficult to separate the indirect effects of vigorous growth, such as excessive canopy shading, from its more direct effects on yield and grape maturity. Canopy shading can influence fruit quality even when the grapes themselves are not shaded (Schneider *et al.*, 1990).

An excellent example of research in this area is provided by Cortell *et al.* (2005). The study was conducted in a commercial vineyard planted with vines of the same age, clone, and rootstock. Georeferencing permitted sampling and wine preparation from sites showing differing levels of vigor. Fruit from vines showing low vigor possessed higher levels of proanthocyanidins in the skin,

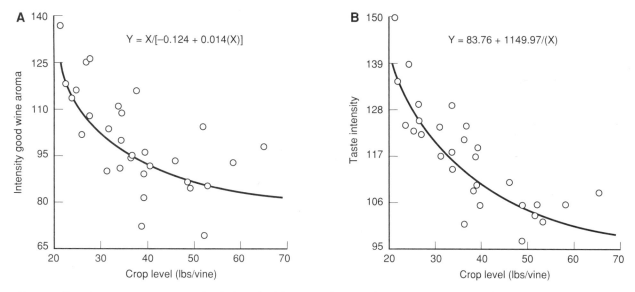

Figure 3.45 Relationship between crop level of 'Zinfandel' vines and (**A**) good wine aroma and (**B**) taste intensity. (From Sinton *et al.*, 1978, reproduced by permission)

as well as an increased proportion of epigallocatechin, proanthocyanin size and pigmented polymers. These attributes were also reflected in the wines produced.

It is also essential that studies on yield/quality relationships investigate more than just its sensory effects on young wines. The effects of yield on wine-aging potential is also required.

Another commonly held belief is that "stressing" vines increases their ability to produce fine-quality wine. This view arose from the reduced vigor/improved grape quality association of some low-nutrient-status, renowned, European vineyards. Nevertheless, balancing the leaf area/fruit ratio can improve the microclimate within and around the vine, increase photosynthetic efficiency, and promote fruit maturation and flavor development (see Leaf Area/Fruit Ratio, Chapter 4). Appropriate balancing of the vegetative and reproductive functions of the vine also appears to affect fruit acidity and pH favorably.

Sunlight As the energy source for photosynthesis, sunlight is without doubt the single most important climatic factor affecting berry development. Because sunlight contains both visible and infrared (heat) radiation, and much of the absorbed radiation is released as heat, it is often difficult to separate light from temperature effects. Because of the early cessation of berry photosynthesis, most of the direct effects of light on berry maturation are either thermal or phytochrome-induced.

Because of the selective absorption of light by chlorophyll, there is a marked increase in the proportion of far-red light within vine canopies. This shifts the proportion of physiologically active phytochrome (P_{fr}) from 60% in sunlight, to below 20% in the shade (Smith and Holmes, 1977). Evidence suggests that this may delay the initiation of anthocyanin synthesis, decrease sugar accumulation, and increase the ammonia and nitrate content in the fruit (Smart *et al.*, 1988). The red/far-red balance of sunlight, and occasionally ultraviolet radiation, are known to influence flavonoid biosynthesis in several plants (Hahlbrock, 1981). Whether sunflecks can offset the effects of shading is unknown, but it is theoretically possible. P_r is more efficiently converted to P_{fr} by red light than is the reverse reaction (on exposure to far-red radiation). However, because some plants also show high-intensity blue and far-red light responses, the effects of leaf shading may be more complex than the red/far-red balance alone might suggest.

Berry pigmentation, and its change following *véraison*, can influence the P_r/P_{fr} ratio in maturing fruit (Blanke, 1990). Whether this significantly influences fruit ripening is unknown, but it could affect the activity of light-activated enzymes such as phosphoenolpyruvate carboxylase (PEPC), a critical enzyme in the metabolism of malic acid (Lakso and Kliewer, 1975).

Fruit exposure is generally essential for flavonol synthesis, and may promote anthocyanin synthesis, inducing deep coloration in red cultivars (Rojas-Lara and Morrison, 1989). Nevertheless, this finding is not universal, seemingly being varietally dependent. For example, Weaver and McCune (1960) found little difference in pigmentation between berries covered to prevent direct-sunlight exposure and sun-exposed fruit.

Downey et al. (2004) have confirmed this with 'Shiraz' (Plate 3.9). High intensity light may actually decrease coloration in varieties such as 'Pinot noir' (Dokoozlian, 1990). Berry phenol content tends to follow the trends set by anthocyanin synthesis in response to light exposure (Morrison and Noble, 1990), but again not consistently (Price et al., 1995). Some of these differences may relate to macroclimatic differences, where increased exposure (and temperature) favors anthocyanin and phenolic synthesis in cool climates, but does not favor synthesis (or enhances degradation) in hot climates.

The influence of light exposure on the specific anthocyanin composition of grapes is unknown. Because anthocyanins differ in their susceptibility to oxidation, changes in composition could be more significant to wine-color stability than total anthocyanin accumulation.

Sun-exposed fruit may show higher titratable acidity and concentrations of tartaric acid than shaded fruit (Smith et al., 1988). The increase in acidity may or may not be reflected in a drop in pH and potassium accumulation. Shading also tends to increase magnesium and calcium accumulation (Smart et al., 1988). The lower malic acid content occasionally observed in sun-exposed fruit may result from the associated heating that increases the rate of malic enzyme activity. The lower sugar concentration in fruit associated with leaf shading is thought to result from dilution associated with increased berry volume (Rojas-Lara and Morrison, 1989). Sun exposure also can hasten the processes of ripening.

Berry size generally increases with fruit or leaf shading, but it is little affected when only the fruit is shaded (Morrison, 1988). These differences may result from reduced transpiration under shade conditions.

There are several reports of sun exposure affecting grape aroma. Usually the level of grassy or herbaceous odors in 'Sauvignon blanc' (Arnold and Bledsoe, 1990) and 'Sémillon' (Pszczolkowski et al., 1985) are reduced by sun exposure, whereas other fruit flavors may be unaffected or augmented. Sun exposure enhances the synthesis of carotenoids in young grapes, notably the UV-B portion (Steel and Keller, 2000). Following véraison, however, the carotenoid level falls, probably as a partial consequence of conversion into aromatic norisoprenoids, such as β-damascenone and vitispirane (Marais, Van Wyk et al., 1992; Baumes et al., 2002). The monoterpene content of several cultivars has also been shown to increase relative to sun exposure (Smith et al., 1988; Reynolds and Wardle, 1989). The higher levels of monoterpenes generated in sun-exposed 'Sémillon' and 'Sauvignon blanc' grapes may help to mask the vegetative odors produced by methoxypyrazines (Reynolds and Wardle, 1991). The concentrations

of methoxypyrazines in shaded 'Cabernet Sauvignon' grapes may be up to 2.5 times that of grapes grown in the sun (Allen, 1993). Similar results have also been obtained for 'Sauvignon blanc' (Marais et al., 1996). These effects can also be affected by the degree and timing of leaf removal around clusters. For 'Sauvignon blanc,' the effect was most marked when leaves around the cluster were removed several weeks before véraison (Arnold and Bledsoe, 1990).

In addition to direct light-activated influences, sun exposure indirectly influences fruit maturation through associated heating effects. As berries mature, their stomata cease to open, and the cooling induced by transpirational water loss is reduced. Although the grape skin is more water-permeable than those of many other fleshy fruits (Nobel, 1975), and therefore could show considerable transpiration-induced cooling, sun exposure can generate temperatures up to 15 °C above ambient (Smart and Sinclair, 1976). The level of heating depends primarily on the light intensity, angle of incidence, and wind velocity. Densely packed clusters of dark fruit show greater heating than loose clusters of light-colored fruit. Shaded fruit is often cooler than the surrounding air due to the reflection of heat by the canopy (see Fig. 5.22).

Temperature Temperature is known to markedly influence enzyme function and cell-membrane permeability. However, how these apply directly to the effects of temperature on fruit development is unclear. The known influences of temperature on fruit development are summarized in Fig. 3.46.

Although the effect of temperature on malate respiration is well known, its precise mode of action has yet to be elucidated. Although temperature enhances the action of malic enzymes, notably above 30 °C (Lakso and Kliewer, 1978), this influence appears inadequate to explain the marked affect of temperature on malate respiration. Whether warm temperatures induce the leakage of malic acid from cell vacuoles appears not to have been studied. In contrast, tartaric acid content is little affected by temperatures below 30 °C.

Temperature has a pronounced effect on anthocyanin synthesis (Kliewer, 1977b) and subsequent color stability in wine. In several varieties, synthesis is favored by warm daytime temperatures and cool nights (20–25 °C and 10–15 °C, respectively). However, in varieties such as 'Kyoho' optimal anthocyanin synthesis may occur at temperatures as low as 15 °C. Temperatures above 35 °C often suppress or inhibit anthocyanin synthesis (Kliewer, 1977a; Spayd et al., 2002). The synthesis of other phenolic compounds may be augmented at warm temperatures (Herrick and Nagel, 1985). In white cultivars, such as 'Riesling,' this can be undesirable. It may lead to wines possessing more bitterness than desired.

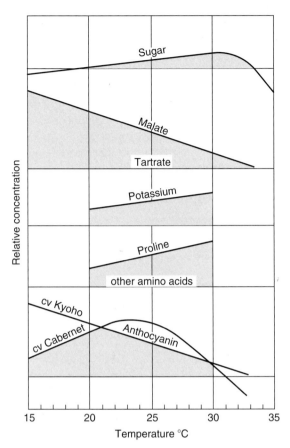

Figure 3.46 Summary of some of the effects of temperature on the concentration of chemical compounds in grapes. (Data from Buttrose *et al.*, 1971; Hale, 1981; Hale and Buttrose, 1973, 1974; Kliewer, 1971, 1972; Kliewer and Lider, 1970; Kobayashi *et al.*, 1965a, 1965b; Lavee, 1977; Radler, 1965; Tomana *et al.*, 1979a, 1979b; graphed by Coombe, 1987b, reproduced by permission)

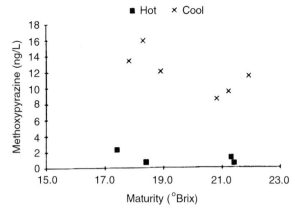

Figure 3.47 Influence of hot and cool climates on the content of methoxypyrazine in 'Sauvignon blanc' grapes. (From Allen and Lacey, 1993, reproduced by permission)

The high temperatures that can develop in sun-exposed fruit can lead to color loss in some varieties. This is probably more important in varieties that terminate anthocyanin synthesis early during berry development. Whether temperature influences the proportional composition of grape anthocyanins during maturation has been little investigated (González-SanJosé *et al.*, 1990).

Increasing temperatures generally has a beneficial influence on sugar accumulation, except above 32–35 °C. The upper limit may result from disruption of photosynthesis at these temperatures (Kriedemann, 1968). High temperatures also temporarily increase the rate of respiration. In overmature fruit, high temperatures can increase sugar concentration by enhancing water loss, and, thereby, increase the proportional sugar content.

Warm conditions can increase the amino acid content of developing fruit. Most of the change is expressed in the accumulation of proline, arginine, or other amino acid amides (Sponholz, 1991). Potassium accumulation usually increases with warmer temperatures (Hale, 1981). Whether this is driven by increased transpiration, sugar accumulation, or some other factor is unclear. Not surprisingly, the increase in potassium content is equally associated with a rise in pH, especially in association with warm night temperatures (Kliewer, 1973).

The effect of temperature on the accumulation of aromatic compounds has drawn little attention. The common view is that cool conditions (in warm climates) enhances flavor development. This may arise from reduced growth and slower respiration, which, in turn, redirects nutrients toward secondary metabolic pathways. These generate most aromatic compounds. For example, the production of aromatic compounds in 'Pinot noir' is reported to be higher in grapes grown in long cool seasons than in short hot seasons (Watson *et al.*, 1991). However, if conditions are too cool or short, the expression of berry aromas and fruit flavors may be reduced, and the development of less desirable vegetable odors enhanced (Heymann and Noble, 1987). Gladstones (1992) has suggested that this may arise from the increased production of unsaturated fatty acids, and their subsequent metabolism to methoxypyrazines, as well as various leaf aldehydes and leaf alcohols. The methoxypyrazine content is typically higher at maturity in grapes that mature in shade or have been grown under cool conditions (Fig. 3.47). Cool climatic conditions also reduce carotenoid production. This can limit the excessive production and accumulation of 1,1,6-trimethyl-1,2-dihydronaphthalene (TDN) typically found in 'Riesling' grapes (Marais, Versini *et al.*, 1992) grown in hot climates. Cooler conditions are also generally favorable to varieties such as 'Riesling,' as the production of important monoterpene flavorants is enhanced (Ewart, 1987).

High temperatures during or shortly after pollination negatively affect fertility and fruit development.

Reduced fertility may arise from direct effects on inflorescence suppression, or indirectly from increasing vine water stress (Matthews and Anderson, 1988). Heat stress early in fruit development appears to irreversibly restrict berry enlargement (Hale and Buttrose, 1974), leading to smaller fruit production, as well as delayed berry maturation.

Inorganic Nutrients   Low soil nitrogen content appears to favor anthocyanin synthesis, whereas high nitrogen levels suppress production. These effects are probably the consequences of the influence of nitrogen on vegetative growth and canopy shading, and their impact on fruit maturation and sugar accumulation. Increasing nitrogen availability also increases the accumulation of free amino acids, notably arginine.

Phosphorus and sulfur deficiencies are known to increase anthocyanin content in several plants. In grapevines, phosphorus deficiency induces interveinal reddening in leaves, but how it may affect berry anthocyanin synthesis is obscure. High potassium levels can reduce berry pH and, thereby, lower fruit color and color stability in red wines.

Water   It is often believed that mild water deficit, following *véraison*, is beneficial to grape quality. This may result from suppressing further vegetative growth, leading to a more balanced leaf area/fruit ratio. It could also enhance anthocyanin synthesis and generate berries of smaller size. A direct effect on accelerated ripening also appears possible (Freeman and Kliewer, 1983). In contrast, water deficit early in the season can lead to reduced berry set and limit berry enlargement. Thus, it appears that water deficit should be limited, by irrigation if necessary, up to *véraison*. In excess, however, irrigation can more than double fruit yield, increase berry size, and delay maturity – all potentially detrimental to fruit quality.

The effect of irrigation on sugar content is usually minor (<10%). Although typically resulting in a decline, irrigation can increase grape sugar content under certain conditions. Commonly, irrigation results in an initial increase in grape acidity, presumably by permitting sufficient transpirational cooling to reduce malic acid respiration. However, there is seldom an accompanying drop in pH. This anomaly may partially be explained by the dilution associated with fruit enlargement. When harvest is sufficiently delayed, the high acidity of the fruit may dissipate, presumably from the slower rate of malate respiration being compensated by the longer maturation period.

The poor wine pigmentation often attributed to excessive irrigation may result from reduced anthocyanin synthesis. However, poor wine color may also arise from dilution associated with increased berry volume. Poor coloration in red cultivars is one of the prime reasons why irrigation is generally ill-advised after *véraison*.

Little is known about the effects of irrigation on aroma development. It has been noted, however, that irrigation diminishes the rate of monoterpene accumulation and the development of juice aroma in 'Riesling' berries (McCarthy and Coombe, 1985).

Suggested Readings

General References

Huglin, P. (1986) *Biologie et Écologie de la Vigne*. Payot Lausanne, Paris.

May, P. (2004) *Flowering and Fruitset of Grapevines*. Winetitles, Adelaide, Australia.

Mullins, M. G., Bouquet, A., and Williams, L. E. (1992) *Biology of the Grapevine*. Cambridge University Press, Cambridge, UK.

Pongrácz, D. P. (1978) *Practical Viticulture*. David Philip, Cape Town, South Africa.

Weaver, R. J. (1976) *Grape Growing*. Wiley, New York.

Williams, L. E. (1996) Grape. In: *Photoassimilate Distribution in Plants and Crops* (E. Zamsli and A. A. Schaffer, eds.), pp. 851–881. Marcel Dekker, New York.

Williams, L. E., Dokoozlian, N. K., and Wample, R. (1994) Grape. In: *Handbook of Environmental Physiology of Fruit Crops*. Vol. 1 (*Temperate Crops*) (B. Schaffer and P. C. Andersen, eds.), pp. 85–113. CRC Press, Boca Raton, FL.

Winkler, A. J., Cook, J. A., Kliewer, W. M., and Lider, L. A. (1974) *General Viticulture*. University of California Press, Berkeley, CA.

Root System

Linderman, R. G. (1988) Mycorrhizal interactions with the rhizosphere microflora: The mycorrhizosphere effect. *Phytopathology* 78, 366–371.

Menge, J. A., Raski, D. J., Lider, L. A., Johnson, E. L. V., Jones, N. O., Kissler, J. J., and Hemstreet, C. L. (1983) Interaction between mycorrhizal fungi, soil fumigation, and growth of grapes in California. *Am. J. Enol. Vitic.* 34, 117–121.

Pratt, C. (1974) Vegetative anatomy of cultivated grapes, a review. *Am. J. Enol. Vitic.* 25, 131–148.

Richards, D. (1983) The grape root system. *Hortic. Rev.* 5, 127–168.

Van Zyl, J. L. (ed.) (1988) *The Grapevine Root and its Environment*. Department of Agricultural Water Supply Technical Communication No. 215, Pretoria, South Africa.

Shoot System

Gerrath, J. M., and Posluszny, U. (1988) Morphological and anatomical development in the Vitaceae, I. Vegetative development in *Vitis riparia*. *Can. J. Bot.* 66, 209–224.

Koblet, W., and Perret, P. (1982) The role of old vine wood on yield and quality of grapes. In: *Grape and Wine Centennial Symposium Proceedings*, pp. 164–169. University of California, Davis.

May, P. (1987) The grapevine as a perennial plastic and productive plant. In: *Proceedings of the 6th Australian Wine Industry*

Technical Conference (T. Lee, ed.), pp. 40–49. Australian Industrial Publishers, Adelaide, Australia.

Morrison, J. C. (1991) Bud development in *Vitis vinifera. Bot. Gaz.* **152**, 304–315.

Pratt, C. (1974) Vegetative anatomy of cultivated grapes, a review. *Am. J. Enol. Vitic.* **25**, 131–148.

Photosynthesis and Transpiration

Düring, H. (1988) CO_2 assimilation and photorespiration of grapevine leaves: Responses to light and drought. *Vitis* **27**, 199–208.

Düring, H., Dry, P. R., and Loveys, B. R. (1996) Root signals affect water use efficiency and shoot growth. *Acta Hortic.* **427**, 1–13.

Intrieri, C., Zerbi, G., Marchiol, L., Poni, S., and Caiado, T. (1995) Physiological response of grapevine leaves to light flecks. *Sci. Hortic.* **61**, 47–60.

Koblet, W. (1984) Influence of light and temperature on vine performance in cool climates and applications to vineyard management. In: *International Symposium on Cool Climate Viticulture and Enology* (D. A. Heatherbell *et al.*, eds.), OSU Agricultural Experimental Station Technical Publication No. 7628, pp. 139–157. Oregon State University, Corvallis, OR.

Kriedemann, P. E. (1977) Vineleaf photosynthesis. In: *International Symposium Quality Vintage*, pp. 67–87. Oenol. Viticult. Res. Inst., Stellenbosch, South Africa.

Smart, R. E. (1983) Water relations of grapevines. In: *Water Deficits and Plant Growth*, Vol. 7 (*Additional Woody Crop Plants*) (T. T. Kozlowski, ed.), pp. 137–196. Academic Press, New York.

Reproductive System

Branties, N. B. M. (1978) Pollinator attraction of *Vitis vinifera* subsp. *silvestris. Vitis* **17**, 229–233.

Carbonneau, A. (1983a) Stérilités mâle et femelle dans le genre *Vitis*, I. Modélisation de leur hérédité. *Agronomie* **3**, 635–644.

Carbonneau, A. (1983b) Stérilités mâle et femelle dans le genre *Vitis*, II. Conséquences en génétique et sélection. *Agronomie* **3**, 645–649.

Gerrath, J. M., and Posluszny, U. (1988) Morphological and anatomical development in the Vitaceae, II. Floral development in *Vitis riparia. Can. J. Bot.* **66**, 1334–1351.

Hardie, W. J., O'Brien, T. P., and Jaundzems, V. G. (1996) Cell biology of grape secondary metabolism – a viticultural perspective. In: *Proceedings of the 9th Australian Wine Industry Technical Conference* (C. S. Stockley *et al.*, eds.), pp. 78–82. Winetitles, Adelaide, Australia.

Pratt, C. (1971) Reproductive anatomy of cultivated grapes, a review. *Am. J. Enol. Vitic.* **21**, 92–109.

Ristic, R., and Iland, P. K. (2005) Relationships between seed and berry development of *Vitis vinifera* L. cv Shiraz: Developmental changes in seed morphology and phenolic composition. *Aust. J. Grape Wine Res.* **11**, 43–58.

Srinivasan, C., and Mullins, M. G. (1981) Physiology of flowering in the grapevine – a review. *Am. J. Enol. Vitic.* **32**, 47–63.

Swanepoel, J. J., and Archer, E. (1988) The ontogeny and development of *Vitis vinifera* L. cv. Chenin blanc inflorescence in relation to phenological stages. *Vitis* **27**, 133–141.

Berry Maturation

Champagnol, F. (1986) L'acidité des moûts et des vins. Partie 2. Facteurs physiologiques et agronomiques de variation. *Prog. Agric. Vitic.* **103**, 361–374.

Coombe, B. G. (1992) Research on development and ripening of the grape berry. *Am. J. Enol. Vitic.* **43**, 101–110.

Coombe, B. G., and Iland, P. F. (1987) Grape berry development. In: *Proceedings of the 6th Australian Wine Industry Technical Conference* (T. Lee, ed.), pp. 50–54. Australian Industrial Publishers, Adelaide, Australia.

Darriet, P., Boidron, J.-N., and Dubourdieu, D. (1988) L'hydrolysis des hétérosides terpéniques du Muscat à petits grains par les enzymes périplasmiques de *Saccharomyces cerevisiae. Connaiss. Vigne Vin* **22**, 189–195.

Hardie, W. J., O'Brien, T. P., and Jaudzems, V. G. 1996. Morphology, anatomy and development of the pericarp after anthesis in grape, *Vitis vinifera* L. *Aust. J. Grape Wine Res.* **2**, 97–142.

Matthews, M. A., and Shackel, K. A. (2005) Growth and water transport in fleshy fruit. In: *Vascular Transport in Plants* (N. M. Holbrook and M. A. Zwieniecki, eds.), pp. 181–197. Academic Press, London.

May, P. (2004) *Flowering and Fruitset in Grapevines*. Winetitles, Adelaide, Australia.

Ruffner, H. P. (1982a) Metabolism of tartaric and malic acids in *Vitis*: A review. Part A. *Vitis* **21**, 247–259.

Ruffner, H. P. (1982b) Metabolism of tartaric and malic acids in *Vitis*: A review. Part B. *Vitis* **21**, 346–358.

Silacci, M. W., and Morrison, J. C. (1990) Changes in pectin content of Cabernet Sauvignon grape berries during maturation. *Am. J. Enol. Vitic.* **41**, 111–115.

Williams, P. J., Strauss, C. R., Aryan, A. P., and Wilson, B. (1987) Grape flavour – a review of some pre- and postharvest influences. In: *Proceedings of the 6th Australian Wine Industry Technical Conference* (T. Lee, ed.), pp. 111–116. Australian Industrial Publishers, Adelaide, Australia.

Williams, P. J., Strauss, C. R., and Wilson, B. (1988) Developments in flavour research on premium varieties. In: *Proceedings of the 2nd International Symposium for Cool Climate Viticulture and Oenology* (R. E. Smart *et al.*, eds.), pp. 331–334. New Zealand Society of Viticulture and Oenology, Auckland, New Zealand.

Factors Affecting Berry Maturation

Clingeleffer, P. R. (1985) Use of plant growth regulating chemicals in viticulture. In: *Chemicals in the Vineyard*, pp. 95–100. Australian Society of Viticulture Oenology, Mildura, Australia.

Coombe, B. G. (1987) Influence of temperature on composition and quality of grapes. *Acta Hortic.* **206**, 23–35.

Hepner, Y., and Bravdo, B. (1985) Effect of crop level and drip irrigation scheduling on the potassium status of Cabernet Sauvignon and Carignane vines and its influence on must and wine composition and quality. *Am. J. Enol. Vitic.* **36**, 140–147.

Jackson, D. I. (1988) Factors affecting soluble solids, acid, pH, and color in grapes. *Am. J. Enol. Vitic.* **37**, 179–183.

Lavee, S. (1987) Usefulness of growth regulators for controlling vine growth and improving grape quality in intensive vineyards. *Acta Hortic.* **206**, 89–108.

Morrison, J. C., and Noble, A. C. (1990) The effects of leaf and cluster shading on the composition of Cabernet Sauvignon grapes and on fruit and wine sensory properties. *Am. J. Enol. Vitic.* **41**, 193–200.

Smart, R. E. (1987) Influence of light on composition and quality of grapes. *Acta Hortic.* **206**, 37–47.

Smart, R. E., and Robinson, M. (1991) *Sunlight into Wine: A Handbook for Winegrape Canopy Management*. Winetitles, Adelaide, Australia.

References

Adams, D. O., and Liyanage, C. (1993) Glutathione increases in grape berries at the onset of ripening. *Am. J. Enol. Vitic.* **44**, 333–338.

Allen, M. F., Swenson, W., Querejeta, J. I., Egerton-Warburton, L. M., and Treseder, K. K. (2003) Ecology of mycorrhizae: a conceptual framework for complex interactions among plants and fungi. *Annu. Rev. Phytopathol.* **41**, 271–303.

Allen, M. F. (1993) Effect of fruit exposure on methoxypyrazine concentration in Cabernet Sauvignon grapes. In: *Proceedings of the 8th Australian Wine Industry Technical Conference* (C. S. Stockley *et al.*, eds.). Winetitles, Adelaide, Australia.

Allen, M. F., and Lacey, M. J. (1993) Methoxypyrazine grape flavor: Influence of climate, cultivar and viticulture. *Wein Wiss.* **48**, 211–213.

Allen, M. F., Lacey, M. J., and Boyd, S. J. (1996) Methoxypyrazines of grapes and wines – differences of origin and behaviour. In: *Proceedings of the 9th Australian Wine Industry Technical Conference* (C. S. Stockley *et al.*, eds.) pp. 83–86. Winetitles, Adelaide, Australia.

Alleweldt, G., and Hifny, H. A. A. (1972) Zur Stiellähme der Reben, II. Kausalanalytische Untersuchungen. *Vitis* **11**, 10–28.

Aloni, R., and Peterson, C. A. (1991) Seasonal changes in callose levels and fluorescein translocation in the phloem of *Vitis vinifera*. *IAWA Bull.* **12**, 223–234.

Aloni, R., Raviv, A., and Peterson, C. A. (1991) The role of auxin in the removal of dormancy callose and resumption of phloem activity in *Vitis vinifera*. *Can. J. Bot.* **69**, 1825–1832.

Amerine, M. A., Berg, H. W., Kunkee, R. E., Ough, C. S., Singleton, V. L., and Webb, A. D. (1980) *The Technology of Wine Making*. Avi Publ., Westport, CT.

Anderson, L. J., Comas, L. H., Lakso, A. N., and Eissenstat, D. M. (2003) Multiple risk factors in root survivorship: a 4-year study in Concord grape. *New Phytologist* **158**, 489–501.

Anonymous (1979) *The Wine Industry in the Federal Republic of Germany*. Evaluation and Information Service for Food, Agriculture and Forestry, Bonn, Germany.

Archer, E., and Hunter, J. (2005) Vine roots play an important role in determining wine quality. *Wynboer* (March), p. 4 (http://www.wynboer.co.za/recentarticles/200503vineroots.php3).

Arnold, R. A., and Bledsoe, A. M. (1990) The effect of various leaf removal treatments on the aroma and flavor of Sauvignon blanc wine. *Am. J. Enol. Vitic.* **41**, 74–76.

Aspinall, D., and Paleg, L. G. (1981) Proline accumulation: Physiological aspects. In: *The Physiology and Biochemistry of Drought Stress in Plants* (L. G. Paleg and D. Aspinall, eds.), pp. 205–241. Academic Press, New York.

Atkinson, D. (1980) The distribution and effectiveness of the roots of tree crops. *Hortic. Rev.* **2**, 424–490.

Ban, T., Ishimaru, M., Kobayashi, S., Shiozaki, S., Goto-Yamamoto, N., and Horiuchi, S. (2003) Abscisic acid and 2,4-dichlorohenoxyacetic acid affect the expression of anthocyanin biosynthetic pathway genes in 'Kyoho' grape berries. *J. Hort. Sci. Biotechnol.* **78**, 586–589.

Bath, G. E., Bell, C. J., and Lloyd, H. L. (1991) Arginine as an indicator of the nitrogen status of wine grapes. In: *Proceedings of the International Symposium Nitrogen in Grapes and Wine* (J. M. Rantz, ed.), pp. 202–205. American Society of Enology and Viticulture, Davis, CA.

Baumes, R., Wirth, J., Bureau, S., Gunata, Y., and Razungles, A. (2002) Biogeneration of C13-norisoprenoid compounds: experiments supportive for an apo-carotenoid pathway in grapevines. *Analytica Chimica Acta* **458**, 3–14.

Bayonove, C., Cordonnier, R., and Ratier, R. (1974) Localisation de l'arome dans la baie de raisin: variétés Muscat d'Alexandrie et Cabernet-Sauvignon. *C. R. Séances Acad. Agric. Fr.* **6**, 1321–1328.

Bennett, J., Jarvis, P., Creasy, G. L., and Trough, M. C. (2005) Influence of defoliation on overwintering carbohydrate reserves, return bloom, and yield of mature Chardonnay grapevines. *Am. J. Enol. Vitic.* **56**, 386–393.

Besselat, B., and Cour, P. (1990) La prévison de la production viticole à l'aide de la technique de dosage pollinique de l'atmosphère. *Bull. O.I.V.* **63**, 721–740.

Biricolti, S., Ferrini, F., Molli, R., Rinaldelli, E., and Vignozzi, N. (1992) Influenza esercitata da tre specie di funghi micorrizici V.A. sull'accrescimento di barbatelle di 'Kober 5 BB' allevate in substrati calcarei. In: *Proceedings of the 4th International Symposium on Grapevine Physiology*, pp. 503–505. University of Torino, San Michele, Italy.

Biricolti, S., Ferrini, F., Rinadelli, E., Tamantini, I., and Vignossi, N. (1996) VAM fungi ans soil lime content influence rootstock growth and nutrient content. *Am. J. Enol. Vitic.* **48**, 93–99.

Blaich, R., Stein, U., and Wind, R. (1984) Perforation in der Cuticula von Weinbeeren als morphologischer Faktor der Botrytisresistenz. *Vitis* **23**, 242–256.

Blanke, M. M. (1990) Carbon economy of the grape inflorescence, 4. Light transmission into grape berries. *Wein Wiss.* **45**, 21–23.

Blanke, M. M. (1991) Kohlenstoff-Haushalt der Infloreszenz der Rebe 7. Oberflächenanalyse der Spaltöffnungen der Weinbeere. *Wein Wiss.* **46**, 8–10.

Blanke, M. M., and Levhe, A. (1987) Stomatal activity of the grape berry cv. Riesling, Müller Thurgau and Ehrenfelser. *J. Plant Physiol.* **127**, 451–460.

Blanke, M., Pring, R. J., and Baker, E. A. (1999) Structure and elemental composition of grape berry stomata. *J. Plant Physiol.* **134**, 477–481.

Bogs, J., Jaffé, F. W., Takos, A. M., Walker, A. R., and Robinson, S. P. (2007) The grapevine transcription factor VvMYBPA1 regulates proanthocyanidin synthesis during fruit development. *Plant Physiol.* **143**, 1347–1361.

Bondada, B. R., Matthews, M. A., and Shackel, K. A. (2005) Functional xylem in the post-véraison grape berry. *J. Exp. Bot.* **56**, 2949–2957.

Bonomelli, A., Mercier, L., Franchel, J., Baillieul, F., Benizri, E., and Mauro, M. C. (2004) Response of grapevine defenses to UV-C exposure. *Am. J. Enol. Vitic.* **55**, 51–59.

Boss, P. K., Davies, C., and Robinson, S. P. (1996a) Expression of anthocyanin biosynthesis pathway genes in red and white grapes. *Plant Molec. Biol.* **32**, 565–569.

Boss, P. K., Davies, C., and Robinson, S. P. (1996b) Analysis of the expression of anthocyanin pathway genes in developing *Vitis vinifera* L. cv Shiraz grape berries and the implications for pathway regulation. *Plant Physiol.* **111**, 1059–1066.

Bouard, J. (1980) Tissues et organes de la vigne. In: *Sciences et Techniques de la Vigne*, Vol. 1 (*Biologie de la Vigne, Sols de Vignobles*) (J. Ribéreau-Gayon, and E. Peynaud, eds.), pp. 3–130. Dunod, Paris.

Branas, J., and Vergnes, A. (1957) Morphologie du système radiculaire de la vigne. *Prog. Agric. Vitic.* **74**, 1–47.

Branties, N. B. M. (1978) Pollinator attraction of *Vitis vinifera* subsp. *silvestris*. *Vitis* **17**, 229–233.

Brugnoli, E., and Farquhar, G. D. (2000) Photosynthetic fractionation of carbon isotopes. In: *Photosynthesis: Physiology and Metabolism* (C. Leegood, T. D. Sharkey, and S. von Caemmerer, eds.), pp. 399–434. Kluwer Academic Publishers, Dordrecht.

Buchbauer, G., Jirovetz, L., Wasicky, M., and Nikiforov, A. (1995) Aroma von Rotweinblüten: Korrelation sensorischer Daten mit Headspace-Inhaltsstoffen. *Z. Lebems-Unter. -Forsch.* **200**, 443–446.

Buttrose, M. S. (1974) Climatic factors and fruitfulness in grapevines. *Hortic. Abstr.* **44**, 319–326.

Buttrose, M. S., Hale, C. R., and Kliewer, W. M. (1971) Effect of temperature on the composition of Cabernet Sauvignon berries. *Am. J. Enol. Vitic.* **22**, 71–75.

Calderón, A. A., Garcia-Florenciano, E., Muñoz, R., and Ros Barceló, A. (1992) Gamay grapevine peroxidase, Its role in vacuolar anthocyani(di)n degradation. *Vitis* **31**, 139–147.

Campbell, J. A., and Strother, S. (1996a) Seasonal variation in pH, carbohydrate and nitrogen of xylem exudate of *Vitis vinifera*. *Aust. J. Plant Physiol.* **23**, 115–118.

Campbell, J. A., and Strother, S. (1996b) Xylem exudate concentrations of cofactor nutrients in grapevine are correlated with exudation rate. *J. Plant Nutr.* **19**, 867–879.

Candolfi-Vasconcelos, M. C., and Koblet, W. (1990) Yield, fruit quality, bud fertility and starch reserves of the wood as a function of leaf removal in *Vitis vinifera* – evidence of compensation and stress recovering. *Vitis* **29**, 199–221.

Candolfi-Vasconcelos, M.C., Candolfi, M. P., and Koblet, W. (1994) Retranslocation of carbon reserves from the woody storage tissues into the fruit as a response to defoliation stress during the ripening period in *Vitis vinifera* L. *Planta* **192**, 567–573.

Carbonneau, A. (1983) Stérilités mâle et femelle dans le genre *Vitis*, I. Modélisation de leur hérédité. *Agronomie* **3**, 635–644.

Chapman, D. M., Matthews, M. A., and Guinard, J.-X. (2004) Sensory attributes of Cabernet Sauvignon wines made from vines with different crop yields. *Am. J. Enol. Vitic.* **55**, 325–334.

Chatelet, D. S., Matthews, M. A., and Shackel, K. (2005) Structural integrity of xylem in the post véraison grape berry. *Am. J. Enol. Vitic.* **56**, 302A.

Chervin, C., Elkereamy, A., Roustan, J.-P., Faragher, J. D., Latche, A., Pech, J.-C., and Bouzayen, M. (2001) An ethanol spray at véraison enhances colour in red wines. *J. Grape Wine Res.* **7**, 144–145.

Christensen, L. P., and Smith, R. J. (1989) Effects of persistent woody laterals on bud performance of Thompson Seedless fruiting canes. *Am. J. Enol. Vitic.* **40**, 27–30.

Collins, C., and Dry, P. (2006) Manipulating fruitset in grapevines. *Aust. NZ Grapegrower Winemaker* **509a**, 38–40.

Comas, L. H., Eissenstat, D. M., and Lakso, A. N. (2000) Assessing root death and root system dynamics in a study of grape canopy pruning. *New Phytol.* **147**, 171–178.

Comas, L. H., Anderson, L. J., Dunst, R. M., Lakso, A. N., Eissenstat, D. M. (2005) Canopy and environmental control of root dynamics in a long-term study of Concord grape. *New Phytol.* **167**, 829–840.

Conner, A. J., and Thomas, M. B. (1981) Re-establishing plantlets from tissue culture: A review. *Proc. Intl Plant Prop. Soc.* **31**, 342–357.

Conradie, W. J. (1990) Distribution and translocation of nitrogen absorbed during late spring by two-year-old grapevines grown in sand culture. *Am. J. Enol. Vitic.* **41**, 241–250.

Conradie, W. J. (1991a) Distribution and translocation of nitrogen absorbed during early summer by two-year-old grapevines grown in sand culture. *Am. J. Enol. Vitic.* **42**, 180–190.

Conradie, W. J. (1991b) Translocation and storage by grapevines as affected by time of application. In: *Proceedings of the International Symposium Nitrogen in Grapes and Wine* (J. M. Rantz, ed.), pp. 32–42. American Society of Enology and Viticulture, Davis, CA.

Coombe, B. G. (1987a) Distribution of solutes within the developing grape berry in relation to its morphology. *Am. J. Enol. Vitic.* **38**, 120–128.

Coombe, B. G. (1987b) Influence of temperature on composition and quality of grapes. *Acta Hortic.* **206**, 23–35.

Coombe, B. G. (1990) Research on berry composition and ripening. In: *Proceedings of the 7th Australian Wine Industry Technical Conference* (P. J. Williams *et al.*, eds.), pp. 150–152. Winetitles, Adelaide, Australia.

Coombe, B. G. (1992) Research on development and ripening of the grape berry. *Am. J. Enol. Vitic.* **43**, 101–110.

Cordner, C. W., and Ough, C. S. (1978) Prediction of panel preference for Zinfandel wine from analytical data: Using difference in crop level to affect must, wine, and headspace composition. *Am. J. Enol. Vitic.* **29**, 254–257.

Cortell, J. M., Halbleib, M., Gallagher, A. V., Righetti, T. L., and Kennedy, J. A. (2005) Influence of vine vigor on grape (*Vitis vinifera* L. cv. Pinot noir) and wine proanthocyanidins. *J. Agric Food Chem.* **53**, 5798–5808.

Cour, P., Duzer, D., and Planchais, N. (1972–1973) Analyses polliniques de l'atmosphère de Montpellier: Document correspondant à la phénologie de la floraison de la vigne, en 1972. *Naturalis Nonspeliensia, Sér. Bot.* **23–24**, 225–229.

Creasy, G. L., Price, S. F., and Lombard, P. B. (1993) Evidence for xylem discontinuity in Pinot noir and Merlot grapes: Dye uptake and mineral composition during berry maturation. *Am. J. Enol. Vitic.* **44**, 187–192.

Cunha, M., Abreu, I., Pinto, P., and de Castro, R. (2003) Airborne pollen samples for early-season estimates of wine production in a Mediterranean climate area of Northern Portugal. *Am. J. Enol. Vitic.* **54**, 189–194.

Dalbó, M. A., Ye, G. N., Weeden, N. F., Steinkellner, H., Sefc, K. M., and Reisch, B. I. (2000) A gene controlling sex in grapevines placed on a molecular marker-based genetic map. *Genome* **43**, 333–340.

Davies, C., Boss, P. K., and Robinson, S. P. (1997) Treatment of grape berries, a nonclimacteric fruit with a synthetic auxin, retards ripening and alters the expression of developmentally regulated genes. *Plant. Physiol.* **115**, 1155–1161.

DeBolt, S., Cook, D. R., and Ford, C. M. (2006) L-Tartaric acid synthesis from vitamin C in higher plants. *Proc. Natl Acad. Sci.* **103**, 5608–5613.

Dixon, R. A., Xie, D.-Y., and Sharma, S. B. (2005) Proanthocyanidins – a final frontier in flavonoid research? *New Phytologist* **165**, 9–28.

Do, C. B., and Cormier, F. (1991) Accumulation of peonidin-3-glucoside enhanced by osmotic stress in grape (*Vitis vinifera* L.) cell suspension. *Plant Cell Organ Cult.* **24**, 49–54.

Dokoozlian, N. K. (1990) Light quantity and light quality within *Vitis vinifera* L. grapevine canopies and their relative influence on betty growth and composition. PhD Thesis, University of California, Davis.

Downey, M. O., Harvey, J. S., and Robinson, S. P. (2003) Analysis of tannins in seeds and skins of Shiraz grapes throughout berry development. *Aust. J. Grape Wine Res.* **9**, 15–27.

Downey, M. O., Harvey, J. S., and Robinson, S. P. (2004) The effect of bunch shading on berry development and flavonoid accumulation in Shiraz grapes. *Aust J. Grape Wine Res.* **10**, 55–73.

Downton, W. J. S., Grant, W. J. R., and Loveys, B. R. (1987) Diurnal changes in the photosynthesis of field-grown grape vines. *New Phytol.* **105**, 71–80.

Dreier, L. P., Stoll, G. S., and Ruffner, H. P. (2000) Berry ripening and evapotranspiration in *Vitis vinifera* L. *Am. J. Enol. Vitic.* **51**, 340–346.

Düring, H. (1984) Evidence for osmotic adjustment to drought in grapevines (*Vitis vinifera* L.). *Vitis* **23**, 1–10.

Düring, H., Dry, P. R., and Loveys, B. R. (1996) Root signals affect water use efficiency and shoot growth. *Acta Hortic.* **427**, 1–13.

Düring, H., and Loveys, B. R. (1996) Stomatal patchiness of field-grown Sultana leaves, diurnal changes and light effects. *Vitis* 35, 7–10.

Ebadi, A., May, P., Sedgley, M., and Coombe, B. G. (1995) Effect of low temperature near flowering time on ovule development and pollen tube growth in the grapevine (*Vitis vinifera* L.), cvs Chardonnay and Shiraz. *Aust. J. Grape Wine Res.* 1, 11–18.

El-Kereamy, A., Chervin, C., Roustan, J. P., Cheynier, V., Souquet, J.-M., Moutounet, M., Raynal, J., Ford, C., Latché, A., Pech, J.-C., and Bouzayen, M. (2003) Exogenous ethylene stimulates the long-term expression of genes related to anthocyanin biosynthesis in grape berries. *Physiol. Plantarum* 199, 175–182.

Esau, K. (1948) Phloem structure in the grapevine, and its seasonal changes. *Hilgardia* 18, 217–296.

Ewart, A. J. W. (1987) Influence of vineyard size and grape maturity on juice and wine quality of *Vitis vinifera* cv. Riesling. In: *Proceedings of the 6th Australian Wine Industry Conference* (T. Lee, ed.), pp. 71–74. Australian Industrial Publishers, Adelaide, Australia.

Fennell, A., and Hoover, E. (1991) Photoperiod influences growth, bud dormancy and cold acclimation in *Vitis labruscana* and *V. riparia. J. Am. Soc. Hortic. Sci.* 116, 270–273.

Fernandez, L., Romieu, C., Moing, A., Bouquet, A., Maucourt, M., Thomas, M. R., and Torregrosa, L. (2006) The grapevine fleshless berry mutation. A unique genotype to investigate differences between fleshy and nonfleshy fruit. *Plant Physiol.* 140, 537–547.

Fernández de Simón, B., Hernández, T., Estrella, I., and Gómez-Cordovés, C. (1992) Variation in phenol content in grapes during ripening: Low-molecular-weight phenols. *Z. Lebensm. Unters. Forsch.* 194, 351–354.

Flaherty, D. L., Jensen, F. L., Kasimatis, A. N., Kido, H., and Moller, W. J. (1981) *Grape Pest Management*. Publication 4105. Cooperative Extension, University of California, Oakland.

Fournioux, J.-C. (1972) Distribution et différenciation des tissus conducteurs primaires dans les organes aeriens de *Vitis vinifera* L. PhD Thesis, University of Dijon, France.

Freeman, B. M., and Kliewer, W. M. (1983) Effect of irrigation, crop level and potassium fertilization on Carignane vines, II. Grape and wine quality. *Am. J. Enol. Vitic.* 34, 197–207.

Freeman, B. M., and Smart, R. E. (1976) A root observation laboratory for studies with grapevines. *Am. J. Enol. Vitic.* 27, 36–39.

Fujita, A., Soma, N., Goto-Yamamoto, N., Shindo, H., Kakuta, T., Koizumi, T., and Hashizume, K. (2005) Anthocyanindin reductase gene expression and accumulation of flavan-3-ols in grape berry. *Am. J. Enol. Vitic.* 56, 336–342.

Gallander, J. F. (1983) Effect of grape maturity on the composition and quality of Ohio Vidal Blanc wines. *Am. J. Enol. Vitic.* 34, 139–141.

Gaudillère, J.-P., Van Leeuwen, C., and Ollat, N. (2002) Carbon isotope composition of sugars in grapevine, an integrated indicator of vineyard water status. *J. Expt. Bot.* 53, 757–763.

Geny, L., Saucier, C., Bracco, S., Daviaud, F., Glories, Y. (2003) Composition and cellular localization of tannins in grape seeds during maturation. *J. Agric. Food Chem.* 51, 8051–8054.

Gholami, M., Hayasaka, Y., Coombe, B. G., Jackson, J. F., Robinsoni, S. P., and Williams, P. J. (1995) Biosynthesis of flavour compounds in Muscat Gordo Blanco grape berries. *Aust. J. Grape Wine Res.* 1, 19–24.

Gladstones, J. (1992) *Viticulture and Environment*. Winetitles, Adelaide, Australia.

Gollop, R., Even, S., Colova-Tsolova, V., and Perl, A. (2002) Expression of the grape dihydroflavonol reductase gene and analysis of its promoter region. *J. Expt. Bot.* 53, 1397–1409.

González-SanJosé, M. L., Barron, L. J. R., and Diez, C. (1990) Evolution of anthocyanins during maturation of Tempranillo

grape variety (*Vitis vinifera*) using polynomial regression models. *J. Sci. Food Agric.* 51, 337–343.

Greenspan, M. D., Shackel, K. A., and Matthews, M. A. (1994) Developmental changes in the diurnal water budget of the grape berry exposed to water deficits. *Plant Cell Environ.* 17, 811–820.

Greenspan, M. D., Schultz, H. R., and Matthews, M. A. (1996) Field evaluation of water transport in grape berries during water deficits. *Physiol. Plantarum* 97, 55–62.

Greer, D. H., Rogers, S. Y., and Steel, C. C. (2006) Susceptibility of Chardonnay grapes to sunburn. *Vitis* 45, 147–148.

Grncarevic, M., and Radler, F. (1971) A review of the surface lipids of grapes and their importance in the drying process. *Am. J. Enol. Vitic.* 22, 80–86.

Gu, S., Lombard, P. B., and Price, S. F. (1991) Inflorescence necrosis induced by ammonium incubation in clusters of Pinot noir grapes. In: *Proceedings of the International Symposium Nitrogen Grapes Wine* (J. M. Rantz, ed.), pp. 259–261. American Society for Enology Viticulture, Davis, CA.

Hahlbrock, K. (1981) Flavonoids. In: *The Biochemistry of Plants* (P. K. Strumpf and E. E. Conn, eds.), Vol. 7, pp. 425–456. Academic Press, New York.

Hale, C. R. (1981) Interaction between temperature and potassium in grape acids. In: CSIRO Division of Horticulture Research Report for 1979–81, pp. 87–88. Glen Osmond, Australia.

Hale, C. R., and Buttrose, M. S. (1973) Effect of temperature on anthocyanin content of Cabernet Sauvignon berries. In: *CSIRO Division of Horticulture Report for 1971–73*, pp. 98–99. Glen Osmond, Australia.

Hale, C. R., and Buttrose, M. S. (1974) Effect of temperature on ontogeny of berries of *Vitis vinifera* L. cv. Cabernet Sauvignon. *J. Am. Soc. Hortic. Sci.* 99, 390–394.

Hale, C. R., and Weaver, R. J. (1962) The effect of development stage on direction of translation of photosynthate in *Vitis vinifera. Hilgardia* 33, 89–131.

Hardie, W. J., O'Brien, T. P., and Jaudzems, V. G. 1996. Morphology, anatomy and development of the pericarp after anthesis in grape, *Vitis vinifera* L. *Aust. J. Grape Wine Res.* 2, 97–142.

Hawker, J. S., Hale, C. R., and Kerridge, G. H. (1981) Advancing the time of ripeness of grapes by the application of methyl 2-(ureidooxy)propionate (a growth retardant). *Vitis* 20, 302–310.

Herrick, I. W., and Nagel, C. W. (1985) The caffeoyl tartrate content of white Riesling wines from California, Washington, and Alsace. *Am. J. Enol. Vitic.* 36, 95–97.

Heymann, H., and Noble, A. C. (1987) Descriptive analysis of Pinot noir wines from Carneros, Napa and Sonoma. *Am. J. Enol. Vitic.* 38, 41–44.

Hilt, C., and Blessis, R. (2003) Abscission of grapevine fruitlets in relation to ethylene biosynthesis. *Vitis* 42, 1–3.

Hofäcker, W. (1976) Untersuchungen über den Einfluß wechselnder Bodenwasserversorgung auf die Photosyntheseintensität und den Diffusionswiderstand bei Rebblättern. *Vitis* 15, 171–182.

Holbach, B., Marx, R., and Steinmetz, E. (1998) Natürliche Saccharosegehalte in Weinberren. *Wein Wiss.* 53, 37–39.

Hoppmann, D., and Hüster, H. (1993) Trends in the development in must quality of 'White Riesling' as dependent on climatic conditions. *Wein Wiss.* 48, 76–80.

Hrazdina, G., and Moskowitz, A. H. (1982) Subcellular status of anthocyanins in grape skins. In: Grape and Wine Centennial Symposium Proceedings 1980, pp. 245–253. University of California, Davis.

Hrazdina, G., Parsons, G. F., and Mattick, L. R. (1984) Physiological and biochemical events during development and maturation of grape berries. *Am. J. Enol. Vitic.* 35, 220–227.

Huglin, P. (1986) *Biologie et Écologie de la Vigne*. Payot Lausanne, Paris.

Huglin, P., and Balthazard, J. (1976) Données relatives a l'influence du rendement sur le taux de sucre des raisins. *Connaiss. Vigne Vin* 10, 175–191.

Iland, P. G., and Coombe, B. G. (1988) Malate, tartrate, potassium, and sodium in flesh and skin of Shiraz grapes during ripening: Concentration and compartmentation. *Am. J. Enol. Vitic.* 39, 71–76.

Iwasaki, K. (1972) Effects of soil aeration on vine growth and fruit development of grapes. *Mem. Coll. Agric., Ehime Univ.* (*Ehime Daigaku Nogakubu Kiyo*) 16, 4–26 (in Japanese).

Jackson, D. I. (1986) Factors affecting soluble solids, acid pH, and color in grapes. *Am. J. Enol. Vitic.* 37, 179–183.

Jeandet, P., Sbaghi, M., Bessis, R., and Meunier, P. (1995) The potential relationship of stilbene (resveratrol) synthesis of anthocyanin content in grape berry skins. *Vitis* 34, 91–94.

Jeong, S. T., Goto-Yamamoto, N., Kobayashi, S., and Esaka, M. (2004) Effects of plant hormones and shading on the accumulation of anthocyanins and the expression of anthocyanin biosynthetic genes in grape berry skins. *Plant Sci.* 167, 247–252.

Karagiannidis, N. and Nikolaou, N. (2000) Influence of arbuscular mycorrhizae on heavy metal (PB and CD) uptake, growth, and chemical composition of *Vitis vinifera* L. (cv. Razaki) *Am. J. Enol. Vitic.* 51, 269–275.

Karagiannidis, N., Velemis, D., and Stavropoulos, N. (1997) Root colonization and spore population by VA-mycorrhizal fungi in four grapevine rootstocks. *Vitis* 36, 57–60.

Kennedy, J. A., Hayasaka, Y., Vidal, S., Waters, E. J., and Jones, G. P. (2001) Composition of grape skin proanthocyanidins at different stages of berry development. *J. Agric. Food Chem.* 49, 5348–5355.

Kliewer, W. M. (1967) The glucose-fructose ratio of *Vitis vinifera* grapes. *Am. J. Enol. Vitic.* 18, 33–41.

Kliewer, W. M. (1971) Effect of day temperature and light intensity on concentration of malic and tartaric acids in *Vitis vinifera* L. grapes. *J. Am. Sci. Hortic. Sci.* 96, 372–377.

Kliewer, W. M. (1973) Berry composition of *Vitis vinifera* cultivars as influenced by photo- and nycto-temperatures during maturation. *J. Am. Soc. Hortic. Sci.* 98, 153–159.

Kliewer, W. M. (1977a) Influence of temperature, solar radiation and nitrogen on coloration and composition of Emperor grapes. *Am. J. Enol. Vitic.* 28, 96–103.

Kliewer, W. M. (1977b) Grape coloration as influenced by temperature. In: *International Symposium on Quality in the Vintage*, pp. 89–106. Oenology and Viticultural Research Institute, Stellenbosch, South Africa.

Kliewer, W. M., and Antcliff, A. J. (1970) Influence of defoliation, leaf darkening, and cluster shading on the growth and composition of Sultana grapes. *Am. J. Enol. Vitic.* 21, 26–36.

Kliewer, W. M., Howorth, L., and Omori, M. (1967) Concentrations of tartaric acid and malic acids and their salts in *Vitis vinifera* grapes. *Am. J. Enol. Vitic.* 18, 42–54.

Kliewer, W. M., and Lider, L. A. (1970) Effect of day temperature and light intensity on growth and composition of *Vitis vinifera* L. fruits. *J. Am. Soc. Hortic. Sci.* 95, 766–769.

Kliewer, W. M., and Smart, R. (1989) Canopy manipulation for optimizing wine microclimate, crop yield and composition of grapes. In: *Manipulation of Fruiting. Proceedings of the 47th Easter School Agricultural Science Symposium*, pp. 275–291. Butterworths, London.

Kobayashi, A., Yukinaga, H., and Itano, T. (1965a) Studies on the thermal conditions of grapes, III. Effects of night temperature at the ripening stage on the fruit maturity and quality of Delaware grapes. *J. Jpn Soc. Hortic. Sci.* 34, 26–32.

Kobayashi, A., Yukinaga, H., and Matsunaga, E. (1965b) Studies on the thermal conditions of grapes. V. Berry growth, yield, quality of Muscat of Alexandria as affected by night temperature. *J. Jpn Soc. Hortic. Sci.* 34, 152–158.

Koblet, W. (1969) Wanderung von Assimilaten in Rebtrieben und Einfluss der Blattfläche auf Ertrag und Qualität der Trauben. *Wein Wiss.* 24, 277–319.

Koblet, W., Candolfi-Vasconcelos, M. C., Zweifel, W., and Howell, G. W. (1994) Influence of leaf removal, rootstock, training system on yield and fruit composition of Pinot noir grapevines. *Am. J. Enol. Vitic.* 45, 181–187.

Koussa, T., Broquedis, M., and Bouard, J. (1994) Importance de l'acide abscissique dans le dévéllopement des bourgeons latents de vigne (*Vitis vinifera* L., var. Merlot) et plus particulièrement dans la phase de levée de dormance. *Vitis* 33, 62–67.

Kriedemann, P. E. (1968) Photosynthesis in vine leaves as a function of light intensity, temperature and leaf age. *Vitis* 7, 213–220.

Kriedemann, P. E. (1977) Vineleaf photosynthesis. In: *International Symposium on Quality in the Vintage*, pp. 67–87. Oenology and Viticultural Research Institute, Stellenbosch, South Africa.

Kriedemann, P. E., Kliewer, W. M., and Harris, J. M. (1970) Leaf age and photosynthesis in *Vitis vinifera*. L. *Vitis* 9, 97–104.

Kriedemann, P. E., Törökfalvy, E., and Smart, R. E. (1973) Natural occurrence and photosynthetic utilization of sunflecks by grapevine leaves. *Photosynthetica* 7, 18–27.

Lakso, A. N., and Kliewer, W. M. (1975) Physical properties of phosphoenolpyruvate carboxylase and malic enzyme in grape berries. *Am. J. Enol. Vitic.* 26, 75–78.

Lakso, A. N., and Kliewer, W. M. (1978) The influence of temperature on malic acid metabolism in grape berries. II. Temperature responses of net dark CO^2 fixation and malic acid pools. *Am. J. Enol. Vitic.* 29, 145–149.

Lamikanra, O., Inyang, I. D., and Leong, S. (1995) Distribution and effect of grape maturity on organic acid content of red muscadine grapes. *J. Agric. Food Chem.* 43, 3026–3028.

Lang, A., and Thorpe, M. R. (1989) Xylem, phloem and transpirational flow in a grape: Application of a technique for measuring the volume of attached fruits to high resolution using Archimedes' principle. *J. Exp. Bot.* 40, 1069–1078.

Larcher, W. (1991) *Physiological Plant Physiology*, 2nd edn. Springer-Verlag, Berlin.

Lavee, S. (1977) The response of vine growth and bunch development to elevated winter, spring and summer day temperature. In: *International Symposium on Quality in the Vintage*, pp. 209–226. Oenology and Viticultural Research Institute, Stellenbosch, South Africa.

Lavee, S., and May, P. (1997) Dormancy of grapevine buds – facts and speculation. *Aust. J. Grape Wine Res.* 3, 31–46.

Lee, C. Y., and Jaworski, A. (1989) Major phenolic compounds in ripening white grapes. *Am. J. Enol. Vitic.* 40, 43–46.

Le Fur, Y., Hory, C., Bard, M.-H., and Olsson, A. (1994) Evolution of phytosterols in Chardonnay grape berry skins during last stages of ripening. *Vitis* 33, 127–131.

Levadoux, L. (1951) La sélection et l'hybridation chez la vigne. *Ann. Éc. Natl. Agric. Montpellier* 28, 9–195.

Linderman, R. G. (1988) Mycorrhizal interactions with the rhizosphere microflora: The mycorrhizosphere effect. *Phytopathology* 78, 366–371.

Lovisolo, C., and Schubert, A. (1998) Effects of water stress on vessel size and xylem hydraulic conductivity in *Vitis vinifera* L. *J. Expt. Bot.* 49, 693–700.

Lovisolo, C., Schubert, A., and Sorce, C. (2002) Are xylem radial development and hydraulic conductivity in downwardly-growing grapevine shoots influence by perturbed auxin metabolism? *New Phytol.* 156, 65–74.

McLachlan, G., Sinclair, P. J., and Dick, J. K. (1993) Effects of water-logging on grapevines. In: *Proceedings of the 8th Australian Wine Industry Technical Conference* (C. S. Stockley *et al.*, eds.), pp. 209. Winetitles, Adelaide, Australia.

Mapfumo, E., Aspinall, T., and Hancock, T. W. (1994) Vessel-diameter distribution in roots of grapevines (*Vitis vinifera* L. cv. Shiraz). *Plant Soil* **159**, 49–56.

Marais, J., Hunter, J. J., Haasbroeck, P. D., and Augustyn, O. P. H. (1996) Effect of canopy microclimate on the composition of Sauvignon blanc grapes. In: *Proceedings of the 9th Australian Wine Industry Technical Conference* (C. S. Stockley *et al.*, eds.), pp. 72–77. Winetitles, Adelaide, Australia.

Marais, J., van Wyk, C. J., and Rapp, A. (1992) Effect of sunlight and shade on norisoprenoid levels in maturing Weisser Riesling and Chenin blanc grapes and Weisser Riesling wines. *S. Afr. J. Enol. Vitic.* **13**, 23–32.

Marais, J., Versini, G., van Wyk, C. J., and Rapp, A. (1992) Effect of region on free and bound monoterpene and C_{13}–norisoprenoid concentrations in Weisser Riesling wines. *S. Afr. J. Enol. Vitic.* **13**, 71–77.

Matthews, M. A., and Anderson, M. M. (1988) Fruit ripening in *Vitis vinifera* L.: Responses to seasonal water deficits. *Am. J. Enol. Vitic.* **39**, 313–320.

May, P. (1987) The grapevine as a perennial plastic and productive plant. In: *Proceedings of the 6th Australian Wine Industry Technical Conference* (T. Lee, ed.), pp. 40–49. Australian Industrial Publishers, Adelaide, Australia.

McCarthy, M. G., and Coombe, B. G. (1985) Water status and winegrape quality. *Acta Hortic.* **171**, 447–456.

McColl, C. R. (1986) Cyanamide advances the maturity of table grapes in central Australia. *Aust. J. Exp. Agric.* **26**, 505–509.

McCully, M. E. (1994) Accumulation of high levels of potassium in the developing xylem elements of roots of soybean and some other dicotyledons. *Protoplasma* **183**, 116–125.

McKenry, M. V. (1984) Grape root phenology, relative to control of parasitic nematodes. *Am. J. Enol. Vitic.* **35**, 206–211.

McLean, M., Howell, G. S., and Smucker, A. J. M. (1992) A minirhizotron system for *in situ* root observation studies of Seyval grapevines. *Am. J. Enol. Vitic.* **43**, 87–89.

Meyer, J. R., and Linderman, R. G. (1986) Selective influences on populations of rhizosphere or rhizoplane bacteria and actinomycetes by mycorrhizas formed by *Glomus fasciculatum*. *Soil Biol. Biochem.* **18**, 191–196.

Millar, A. A. (1972) Thermal regime of grapevines. *Am. J. Enol. Vitic.* **23**, 173–176.

Mohr, H. D. (1996) Periodicity of root tip growth of vines in the Moselle valley. *Wein-Wiss.* **51**, 83–90.

Monagas, M., Gómez-Cordovés, C., Bartolomé, B., Laureano, O., and Ricardo da Silva, J. M. (2003) Monomeric, oligomeric, and polymeric flavan-3-ol composition of wines and grapes from *Vitis vinifera* L. cv. Graciano, Tempranillo, and Cabernet Sauvignon. *J. Agric. Food Chem.* **51**, 6475–6481.

Moncur, M. W., Rattigan, K., Mackenzie, D. H., and McIntyre, G. N. (1989) Base temperatures for budbreak and leaf appearance of grapevines. *Am. J. Enol. Vitic.* **40**, 21–26.

Morrison, J. C. (1988) The effects of shading on the composition of Cabernet Sauvignon grape berries. In: *Proceedings of the 2nd International Symposium Cool Climate Viticulture and Oenology* (R. E. Smart *et al.*, eds.), pp. 144–146. New Zealand Society of Viticulture and Oenology, Auckland, New Zealand.

Morrison, J. C. (1991) Bud development in *Vitis vinifera*. *Bot. Gaz.* **152**, 304–315.

Morrison, J. C., and Noble, A. C. (1990) The effects of leaf and cluster shading on the composition of Cabernet Sauvignon grapes and on fruit and wine sensory properties. *Am. J. Enol. Vitic.* **41**, 193–200.

Mounts, B. T. (1932) Development of foliage leaves. *University Iowa Studies Nat. Hist.* **14**, 1–19.

Nagarajah, S. (1989) Physiological responses of grapevines to water stress. *Acta Hortic.* **240**, 249–256.

Negi, S. S., and Olmo, H. P. (1966) Sex conversion in a male *Vitis vinifera* L. by a kinin. *Science* **152**, 1624.

Negi, S. S., and Olmo, H. P. (1971) Conversion and determination of sex in *Vitis vinifera* L. (*sylvestris*) *Vitis* **9**, 265–279.

Nobel, P. S. (1975) Effective thickness and resistance of the air boundary layer adjacent to spherical plant parts. *J. Exp. Bot.* **26**, 120–130.

Or, E., Vilozny, I., Eyal, Y., and Ogrodovitch, A. (2000) The transduction of the signal for grape bud dormancy breaking induced by hydrogen cyanamide may involve the SNF-like protein kinase GDBRPK. *Plant Mol. Biol.* **43**, 483–489.

Or, E., Vilozny, I., Fennell, A., Eyal, Y., and Ogrodovitch, A. (2002) Dormancy in grape buds: isolation and characterization of catalase cDNA and analysis of its expression following chemical induction of bud dormancy release. *Plant Sci.* **162**, 121–130.

Ough, C. S., and Nagaoka, R. (1984) Effect of cluster thinning and vineyard yield on grape and wine composition and wine quality of Cabernet Sauvignon. *Am. J. Enol. Vitic.* **35**, 30–34.

Palejwala, V. A., Parikh, H. R., and Modi, V. V. (1985) The role of abscisic acid in the ripening of grapes. *Physiol. Plant* **65**, 498–502.

Peña, J. P., and Tarara, J. (2004) A portable whole canopy gas exchange system for several mature field-grown grapevines. *Vitis* **43**, 7–14.

Peña-Neira, A., Dueñas, M., Duarte, A., Hernandez, T., Estrella, I., and Loyola, E. (2004) Effects of ripening stages and of plant vegetative vigor on the phenolic composition of grapes (*Vitis vinifera* L.) Cv. Cabernet Sauvignon in the Maipo Valley (Chile) *Vitis* **43**, 51–57.

Peppi, M. C., Fidelibus, M., Dokoozlian, N., and Katayama, D. (2005) Abscisic acid concentration and timing for color improvement of table grapes. *Am. J. Enol. Vitic.* **56**, 296A.

Pérez, F. J., Viani, C., and Retamales, J. (2000) Bioactive gibberellins in seeded and seedless grapes: Identification and changes in content during berry development. *Am. J. Enol. Vitic.* **51**, 315–318.

Peynaud, E., and Maurié, A. (1953) Sur l'évolution de L'azote dans les différentes parties du raisin au cours de la maturation. *Ann. Technol. Agric.* **1**, 15–25.

Peyrot des Gachons, C., Tominaga, T., and Dubourdieu, D. (2002) Localization of S-cysteine conjugates in the berry: effect of skin contact on aromatic potential of *Vitis vinifera* L. cv. Sauvignon blanc must. *Am. J. Enol. Vitic.* **53**, 144–146.

Polo, M. C., Herraiz, M., and Cabezudo, M. D. (1983) A study of nitrogen fertilization and fruit maturity as an approach for obtaining the analytical profiles of wines and wine grapes. *Instrum. Anal. Foods.* **2**, 357–374.

Poni, S., Intrieri, C., and Silverstroni, O. (1994) Interactions of leaf age, fruiting, and exogenous cytokinins in Sangiovese grapevines under non-irrigated conditions, I. Gas exchange. *Am. J. Enol. Vitic.* **45**, 71–78.

Possner, D. R. E., and Kliewer, W. M. (1985) The localisation of acids, sugars, potassium and calcium in developing grape berries. *Vitis* **24**, 229–240.

Pouget, R. (1963) Recherches physiologiques sur le repos végétatif de la vigne (*Vitis vinifera* L.): La dormance et le mécanism de sa disparition. *Ann. Amél. Plantes* **13**, no hors séries 1.

Pratt, C. (1959) Radiation damage in shoot apices of Concord grape. *Am. J. Bot.* **46**, 102–109.

Pratt, C. (1971) Reproductive anatomy of cultivated grapes, a review. *Am. J. Enol. Vitic.* **21**, 92–109.

Pratt, C. (1974) Vegetative anatomy of cultivated grapes, a review. *Am. J. Enol. Vitic.* **25**, 131–148.

Price, S. F., Breen, P. J., Valladao, M., and Watson B. T. (1995) Cluster sun exposure and quercetin in Pinot noir grapes and wine. *Am. J. Enol. Vitic.* **46**, 187–194.

Pszczolkowski, P., Morales, A., and Cava, S. (1985) Composicion quimica y calidad de mostos y vinos obtenidos de racimos diferentemente asoleados. *Ciencia e Investigacion Agraria* **12**, 181–188.

Pucheu-Plante, B., and Mércier, M. (1983) Étude ultrastructurale de l'interrelation hôte-parasite entre le raisin et le champignon *Botrytis cinerea*: Exemple de la pourriture noble en Sauternais. *Can. J. Bot.* **61**, 1785–1797.

Queen, W. H. (1968) Radial movement of water and ^{32}P through suberized and unsuberized roots of grape. *Diss. Abstr. Sect. B.* **29**, 72–73.

Radler, F. (1965) The effect of temperature on the ripening of 'Sultana' grapes. *Am. J. Enol. Vitic.* **16**, 38–41.

Ravaz, L. (1915) Les grains verts des producteurs-directs. *Ann. Éc. Natl. Agric. Montpellier* **14**, 200–211.

Razungles, A., Bayonove, C. L., Cordonnier, R. E., and Sapis, J. C. (1988) Grape carotenoids: Changes during the maturation period and localization in mature berries. *Am. J. Enol. Vitic.* **39**, 44–48.

Razungles, A., Gunata, Z., Pinatel, S., Baumes, R., and Bayonove, C. (1993) Étude quantitative de composés terpéniques, norisoprénoïds et de leurs précurseurs dans diverses variétés de raisins. *Sci. Aliments* **13**, 59–72.

Reynolds, A. G., and Wardle, D. A. (1989) Influence of fruit microclimate on monoterpene levels of Gewürztraminer. *Am. J. Enol. Vitic.* **40**, 149–154.

Reynolds, A. G., and Wardle, D. (1991) Effects of fruit exposure on fruit composition (abstract). *Am. J. Enol. Vitic.* **42**, 89.

Reynolds, A. G., Wardle, D. A., and Dever, M. (1994) Shoot density effects on Riesling grapevines: Interactions with cordon age. *Am. J. Enol. Vitic.* **45**, 435–443.

Reynolds, A. G., Yerle, S., Watson, B. T., Price, S. F., and Wardle, D. A. (1996) Fruit environment and crop level effects on Pinot noir, III. Composition and descriptive analysis of Oregon and British Columbia wines. *Am. J. Enol. Vitic.* **47**, 329–339.

Richards, D. (1983) The grape root system. *Hortic. Rev.* **5**, 127–168.

Robinson, W. B., Shaulis, N., and Pederson, C. S. (1949) Ripening studies of grapes grown in 1948 for juice manufacture. *Fruit Prod. J. Am. Food Manuf.* **29**, 36–37, 54, 62.

Robinson, W. B., Weirs, L. D., Bertino, J. J., and Mattick, L. R. (1966) The relation of anthocyanin composition to color stability of New York State wines. *Am. J. Enol. Vitic.* **17**, 178–184.

Rogiers, S. Y., Greer, D. H., Hatfield, J. M., Orchard, B. A., and Keller, M. (2006a) Solute transport into Shiraz berries during development and late-ripening shrinkage. *Am. J. Enol. Vitic.* **57**, 73–80.

Rogiers, S. Y., Greer, D. H., Hatfield, J. M., Orchard, B. A., and Keller, M. (2006b) Mineral sinks within ripening grape berries (*Vitis vinifera* L.). *Vitis* **45**, 115–123.

Rogiers, S. Y., Smith, J. A., White, R., Keller, M., Holzapfel, B. P., and Virgona, J. M. (2001) Vascular function in berries of *Vitis vinifera* (L) cv. Shiraz. *J. Grape Wine Res.* **7**, 47–51.

Rojas-Lara, B. A., and Morrison, J. C. (1989) Differential effects of shading fruit or foliage on the development and composition of grape berries. *Vitis* **28**, 199–208.

Rosenquist, J. K., and Morrison, J. C. (1988) The development of the cuticle and epicuticular wax of the grape berry. *Vitis* **27**, 63–70.

Roufet, M., Bayonove, C. L., and Cordonnier, R. E. (1987) Étude de la composition lipidique du raisin, *Vitis vinifera*: Evolution au cours de la maturation et localisation dans la baie. *Vitis* **26**, 85–97.

Roujou de Boubée, D., Cumsille, A. M., Pons, M., and Dubourdieu, D. (2002) Location of 2-methoxy-3-isobutylpyrazine in Cabernet Sauvignon grape bunches and its extractability during vinification. *Am. J. Enol. Vitic.* **53**, 1–5.

Ruffner, H. P. (1982) Metabolism of tartaric and malic acids in *Vitis*: A review, Part A. *Vitis* **21**, 247–259.

Ryan, M. H., van Herwaarden, A. F., Angus, J. F., and Kirkegaard, J. A. (2005) Reduced growth of autumn-sown wheat in a low-P soil is associated with high colonisation by arbuscular mycorrhizal fungi. *Plant Soil* **270**, 275–286.

Salzman, R. A., Bressan, R. A., Hasegawa, P. M., Ashworth, E. N., and Bordelon, B. P. (1996) Programmed accumulation of LEA-like proteins during desiccation and cold acclimation of overwintering grape buds. *Plant Cell Environ.* **19**, 713–720.

Sánchez, L. A., and Dokoozlian, N. K. (2005) Bud microclimate and fruitfulness in *Vitis vinifera* L. *Am. J. Enol. Vitic.* **56**, 319–329.

Schellenbaum, L., Berra, G., Ravolanirina, F., Tisserant, B., Gianinazzi, S., and Fitter, A. H. (1991) Influence of endomycorrhizal infection on root morphology in a micropropagated woody plant species (*Vitis vinifera* L.) *Ann. Bot.* **68**, 135–141.

Schneider, A., Mannini, F., Gerbi, V., and Zeppa, G. (1990) Effect of vine vigour of *Vitis vinifera* cv. Nebbiolo clones on wine acidity and quality. In: *Proceedings of the 5th International Symposium Grape Breeding*, pp. 525–531. (Special Issue of *Vitis*.) St Martin, Pfalz, Germany.

Schreiner, R. P. (2003) Mycorrhizal colonization of grapevine rootstocks under field conditions. *Am. J. Enol. Vitic.* **54**, 143–149.

Schubert, A. (1985) Les mycorhizes à vesicules et arbuscules chez la vigne. *Connaiss. Vigne Vin* **19**, 207–214.

Schubert, A., Lovisolo, C., and Peterlunger, E. (1999) Shoot orientation affects vessel size, shoot hydraulic conductivity and shoot growth rate in *Vitis vinifera* L. *Plant Cell Environ.* **22**, 197–204.

Schubert, A., Mazzitelli, M., Ariusso, O., and Eynard, I. (1990) Effects of vesicular-arbuscular mycorrhizal fungi on micropropagated grapevines: Influence of endophyte strain, P fertilization and growth medium. *Vitis* **29**, 5–13.

Schultz, H. R., and Matthews, M. A. (1993) Xylem development and hydraulic conductance in sun and shade shoots of grapevine (*Vitis vinifera*), evidence that low light uncouples water transport capacity from leaf area. *Planta* **190**, 393–406.

Silacci, M. W., and Morrison, J. C. (1990) Changes in pectin content of Cabernet Sauvignon grape berries during maturation. *Am. J. Enol. Vitic.* **41**, 111–115.

Sinton, T. H., Ough, C. S., Kissler, J. J., and Kasimatis, A. N. (1978) Grape juice indicators for prediction of potential wine quality. I. Relationship between crop level, juice and wine composition, and wine sensory ratings and scores. *Am. J. Enol. Vitic.* **29**, 267–271.

Skouroumounis, G. K., and Winterhalter, P. (1994) Glycosidically bound norisoprenoids from *Vitis vinifera* cv. Riesling leaves. *J. Agric. Food Chem.* **42**, 1068–1072.

Smart, D. R., Schwass, E., Lasko, A., and Morano, L. (2006) Grapevine rooting patterns: A comprehensive analysis and a review. *Am. J. Enol. Vitic.* **57**, 89–104.

Smart, R. E. (1982) Vine manipulation to improve wine grape quality. In: *Grape and Wine Centennial Symposium Proceedings 1980*, pp. 362–375. University of California, Davis.

Smart, R. E. (1987) Influence of light on composition and quality of grapes. *Acta Hortic.* **206**, 37–47.

Smart, R. E., and Sinclair, T. R. (1976) Solar heating of grape berries and other spherical fruits. *Agric. Meteorol.* **17**, 241–259.

Smart, R. E., Smith, S. M., and Winchester, R. V. (1988) Light quality and quantity effects on fruit ripening for Cabernet Sauvignon. *Am. J. Enol. Vitic.* **39**, 250–258.

Smith, H., and Holmes, M. G. (1977) The function of phytochrome in the natural environment. III. Measurement and calculation of phytochrome photoequilibria. *Photochem. Photobiol.* **25**, 547–550.

Smith, S., Codrington, I. C., Robertson, M., and Smart, R. (1988) Viticultural and oenological implications of leaf removal for New Zealand vineyards. In: *Proceedings of the 2nd International Symposium for Cool Climate Viticulture and Oenology* (R. E. Smart et al., eds.), pp. 127–133. New Zealand Society of Viticulture and Oenology, Auckland, New Zealand.

Soar, C. J., Speirs, J., Maffei, S. M., Penrose, A. B., McCarthy, M. G., and Loveys, B. R. (2006) Grape vine varieties Shiraz and Grenache differ in their stomatal response to VPD: apparent links with ABA physiology and gene expression in leaf tissue. *Aust. J. Grape Wine Res.* **12**, 2–12.

Solari, C., Silvestroni, O., Giudici, P., and Intrieri, C. (1988) Influence of topping on juice composition of Sangiovese grapevines (V. vinifera L.) In: *Proceedings of the 2nd International Symposium for Cool Climate Viticulture Oenology* (R. E. Smart et al., eds.), pp. 147–151. New Zealand Society of Viticulture and Oenology, Auckland, New Zealand.

Spayd, S. E., Tarara, J. M., Mee, D. L., and Ferguson, J. C. (2002) Separation of sunlight and temperature effects on the composition of Vitis vinifera cv Merlot berries. *Am. J. Enol. Vitic.* **53**, 171–182.

Sponholz, W. R. (1991) Nitrogen compounds in grapes, must, and wine. In: *Proceedings of the International Symposium Nitrogen in Grapes and Wine* (J. M. Rantz, ed.), pp. 67–77. American Society for Enology and Viticulture, Davis, CA.

Srinivasan, C., and Mullins, M. G. (1981) Induction of precocious flowering in grapevine seedlings by growth regulators. *Agronomie* **1**, 1–5.

Staudt, G. (1982) Pollenkeimung und Pollenschlauchwachstum in vivo bei Vitis und die Abhähgigkeit von der Temperatur. *Vitis* **21**, 205–216.

Staudt, G., Schneider, W., and Leidel, J. (1986) Phases of berry growth in Vitis vinifera. *Ann. Bot.* **58**, 789–800.

Steel, C. C., and Keller, M. (2000) Environmental effects on lipids: atmosphere and temperature. *Biochem. Soc. Trans.* **28**, 883–885.

Stevenson, A. C. (1985) Studies in the vegetational history of S.W. Spain, II. Palynological investigations at Laguna de las Madres, S.W. Spain. *J. Biogeogr.* **12**, 293–314.

Stines, A. P., Grubb, J., Gockowiak, H., Henschke, P. A., Hoj, P. B., and van Heeswick, R. (2000) Proline and arginine accumulation in developing berries of Vitis vinifera L. in Australian vineyards: Influence of vine cultivar, berry maturity and tissue type. *Aust. J. Grape Wine Res.* **6**, 150–158.

Stoev, K., and Slavtcheva, T. (1982) La photosynthèse nette chez la vigne (V. vinifera) et les facteurs écologiques. *Connaiss. Vigne Vin* **16**, 171–185.

Storey, R. (1987) Potassium localization in the grape berry. Pericarp by energy-dispersive X-ray microanalysis. *Am. J. Enol. Vitic.* **38**, 301–309.

Strauss, C. R., Gooley, P. R., Wilson, B., and Williams, P. J. (1987) Application of droplet countercurrent chromatography to the analysis of conjugated forms of terpenoids, phenols, and other constituents of grape juice. *J. Agric. Food Chem.* **35**, 519–524.

Strauss, C. R., Wilson, B., Anderson, R., and Williams, P. J. (1987) Development of precursors of C[13] nor-isoprenoid flavorants in Riesling grapes. *Am. J. Enol. Vitic.* **38**, 23–27.

Swanepoel, J. J., and Archer, E. (1988) The ontogeny and development of Vitis vinifera L. cv. Chenin blanc inflorescence in relation to phenological stages. *Vitis* **27**, 133–141.

Swanepoel, J. J., and de Villiers, C. E. (1988) The anatomy of Vitis roots and certain abnormalities. In: *The Grapevine Root and its Environment* (J. L. Van Zyl, ed.), pp. 138–146. Department of Agricultural Water Supply Technical Communication No. 215. Pretoria, South Africa.

Symons, G., Davies, C., Shavurkov, Y., Dry, I. B., Reid, J., and Thomas, M. R. (2006) Grapes on steroids. Brassinosteroids are involved in grape berry ripening. *Plant Physiol.* **140**, 150–158.

Tattersall, D. B., van Heeswijck, R., and Høj, P. B. (1997) Identification and characterization of a fruit-specific, thaumatin-like protein that assimilates at very high levels in conjunction with the onset of sugar accumulation and berry softening in grapes. *Plant Physiol.* **114**, 759–769.

This, P., Lacombe, T., Cadle-Davidson, M., Owens, C. L. (2007) Wine grape (Vitis vinifera L.) color associates with allelic variation in the domestication gene VvmybA1. *Theor Appl Genet.* **114**, 723–730.

Tomana, T., Utsunomiya, N., and Kataoka, I. (1979a) The effect of environmental temperatures on fruit ripening on the tree, II. The effect of temperatures around whole vines and clusters on the colouration of 'Kyoho' grapes. *J. Jpn Soc. Hortic. Sci.* **48**, 261–266.

Tomana, T., Utsunomiya, N., and Kataoka, I. (1979b) The effect of environmental temperatures on fruit ripening on the tree – the effect of temperatures around whole vines and clusters on the ripening of 'Delaware' grapes. *Stud. Inst. Hortic. Kyoto Univ.* (*Engeigaku Kenkyu Shuroku*) **9**, 1–5.

Tyerman, S. D., Tilbrook, J., Pardo, C., Kotula, L., and Sullivan, W. (2004) Direct measurement of hydraulic properties in developing berries of Vitis vinifera L. cv. Shiraz and Chardonnay. *Aust J. Grape Wine Res.* **10**, 170–181.

Tyson, P. J., Luis, E. S., and Lee, T. H. (1982) Soluble protein levels in grapes and wine. In: *Grape and Wine Centennial Symposium Proceedings 1980*, pp. 287–290. University of California, Davis.

van Zyl, J. L., and van Huyssteen, L. (1987) Root pruning. *Decid. Fruit Growth* **37** (1), 20–25.

Vasconcelos, M. C., and Castagnoli, S. (2000) Leaf canopy structure and vine performance. *Am. J. Enol. Vitic.* **51**, 390–396.

Volder, A., Smart, D. R., Bloom, A. J., and Eissenstat, D. M. (2005) Rapid decline in nitrate uptake and respiration with age in fine lateral roots of grape: implications for root efficiency and competitive effectiveness. *New Phytologist* **165**, 493–502.

Wake, C. M. F., and Fennell, A. (2000) Morphological, physiological and dormancy responses of three Vitis genotypes to short photoperiod. *Physiol. Plantarum* **109**, 203–210.

Walker, A. R., Lee, E., Bogs, J., McDavid, D. A. J., Thomas, M. R., and Robinson, S. P. (2007) White grapes arose through the mutation of two similar and adjacent regulatory genes. *Plant J.* **49**, 772–785.

Walker, A. R., Lee, E., and Robinson, S. P. (2006) Two new grape cultivars, bud sports of Cabernet Sauvignon bearing pale-coloured berries, as the result of deletion of two regulatory genes of the berry colour locus. *Plant Mol. Biol.* **62**, 623–635.

Waschkies, C., Schropp, A., and Marschner, H. (1994) Relations between grapevine replant disease and root colonization of grapevine (Vitis sp.) by fluorescent pseudomonads and endomycorrhizal fungi. *Plant Soil* **162**, 219–227.

Watson, B. T., McDaniel, M., Miranda-Lopez, R., Michaels, N., Price, S., and Yorgey, B. (1991) Fruit maturation and flavor development in Pinot noir (abstract). *Am. J. Enol. Vitic.* **42**, 88–89.

Weaver, R. J., and McCune, S. B. (1960) Influence of light on color development in Vitis vinifera grapes. *Am. J. Enol. Vitic.* **11**, 179–184.

West, D. W., and Taylor, J. A. (1984) Response of six grape cultivars to the combined effects of high salinity and rootzone waterlogging. *J. Am. Soc. Hortic. Sci.* **109**, 844–851.

Williams, L. E., Biscay, P. J., and Smith, R. J. (1987) Effect of interior canopy defoliation on berry composition and potassium distribution in Thompson Seedless grapevines. *Am. J. Enol. Vitic.* **38**, 287–292.

Wilson, B., Strauss, C. R., and Williams, P. J. (1984) Changes in free and glycosidically bound monoterpenes in developing Muscat grapes. *J. Agric. Food Chem.* **32**, 919–924.

Wilson, B., Strauss, C. R., and Williams, P. J. (1986) The distribution of free and glycosidically-bound monoterpenes among skin, juice, and pulp fractions of some white grape varieties. *Am. J. Enol. Vitic.* **37**, 107–114.

Winkler, A. J., Cook, J. A., Kliewer, W. M., and Lider, L. A. (1974) *General Viticulture*. University of California Press, Berkeley, CA.

Yakushiji, H., Kobayashi, S., Goto-Yamamoto, N., Tae Jeong, S., Sueta, T., Mitani, N., and Azuma, A. (2006) A skin color mutation of grapevine, from black-skinned Pinot noir to white-skinned Pinot blanc, is caused by deletion of the functional *VvmybA1* allele. *Biosci. Biotechnol. Biochem.* **70**, 1506–1508.

Yamane, T., Jeong, S. T., Goto-Yamamoto, N., Koshita, Y., and Kobayashi, S. (2006) Effects of temperature on anthocyanin biosynthesis in grape berry skins. *Am. J. Enol. Vitic.* **57**, 54–59.

Yang, Y., Hori, I., and Ogata, R. (1980) Studies on retranslocation of accumulated assimilates in Delaware grape vines, II and III. *Tokyo J. Agric. Res.* **31**, 109–129.

Zapriagaeva, V. I. (1964) Grapevine – *Vitis* L. In: *Wild Growing Fruits in Tadzhikistgan* (Russian, with English summary), pp. 542–559. Nauka, Moscow.

Vineyard Practice

In Chapter 3, details were given concerning the major physiological processes of grapevines. In this chapter vineyard practice, and how it impacts grape yield and quality, are the central focus. Because vineyard practice is so closely allied to the yearly cycle of the vine, a brief description of the growth cycle and its association with vineyard activities is given below.

Vine Cycle and Vineyard Activity

The end of one growth cycle and preparation for another coincides in temperate regions with the onset of winter dormancy. Winter dormancy provides grape growers with the opportunity to do many of the less urgent vineyard activities for which there was insufficient time during the growing season. In addition, pruning is conducted more conveniently when the vine is dormant, and the absence of foliage permits easier wood selection and cane tying.

With the return of warmer weather, both the metabolic activity of the vine, and the pace of vineyard endeavors quicken. Usually, the first sign of renewed activity is the "bleeding" of sap from the cut ends of canes or spurs. The sap contains organic constituents such as sugars and amino acids, as well as growth regulators, inorganic nutrients, and trace amounts of other organic compounds. This is reflected in a marked drop in carbohydrate reserves in the roots and trunk (Bennett *et al.*, 2005). Between bud burst and 50% bloom, starch and sugar reserves of the roots may decline by 42% and 72%, respectively. Soluble sugars in the trunk can decline by 92% during the same period. These levels usually return to normal by leaf fall. When the temperature rises above a critical value, which depends on the variety, buds begin to burst (Fig. 4.1). Activation progresses downward from the tips of canes and spurs, as the phloem again becomes functional. The cambium subsequently resumes its meristematic activity and produces new xylem and phloem to the interior and exterior, respectively.

Typically, only the primary bud in the dormant bud becomes active in the spring. The secondary and tertiary buds remain inactive unless the primary bud is killed or severely damaged. Once initiated, shoot growth rapidly reaches its maximum as the climate continues to warm. The development and enlargement of the primordial leaves, tendrils, and inflorescence clusters recommence (Fig. 4.2). As growth continues, the shoot differentiates new leaves and tendrils. Simultaneously, new buds arise in the leaf axils. Those that form early may give rise to lateral shoots.

Root growth lags behind shoot growth, often coinciding with flowering (five- to eight-leaf stage). Peak root development commonly occurs between the end of flowering and the initiation of fruit coloration (*véraison*). Subsequently, root development declines, with possibly a second growth spurt in the autumn.

Flower development in the spring progresses from the outermost ring of flower parts (the sepals) inward to the pistil. By the time the anthers mature and split open (anthesis), the cap of fused petals (calyptra) separates from the ovary base and is shed. In the process, self-pollination typically occurs as liberated pollen falls onto the stigma. If followed by fertilization, embryo and berry development commence.

Coincident with flowering, inflorescence induction begins in the nascent primary buds of the leaf axils. These typically will become dormant by mid-season and complete their development only in the subsequent season.

Several weeks after bloom, many small berries dehisce (shatter). This is a normal process in all varieties. Nascent fruit sheds from clusters in which fertilization did not occur, or the embryos aborted. At least one seed is normally required for berry development to continue.

Once the shoots have developed sufficiently, they typically are tied to a support system. Shoots derived from latent buds on old wood (water sprouts and suckers) are usually removed early in the season. Spraying for disease and pest control usually commences early. Spraying or cultivation for weed control often occurs before bud burst. Irrigation, where required and permissible, is usually limited after *véraison*. This restricts

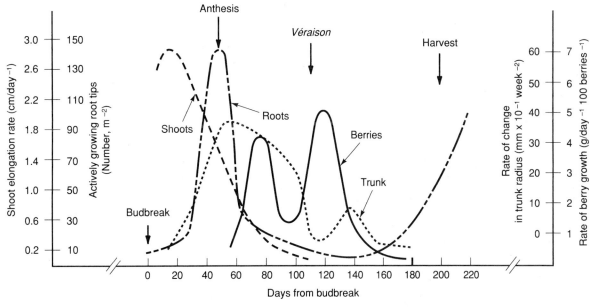

Figure 4.1 Growth rate of various organs of 'Colombar' grapevines grown in South Africa throughout the season. (After van Zyl, 1984, from Williams and Matthews, 1990, reproduced by permission)

Grapevine growth stages – The modified E-L system

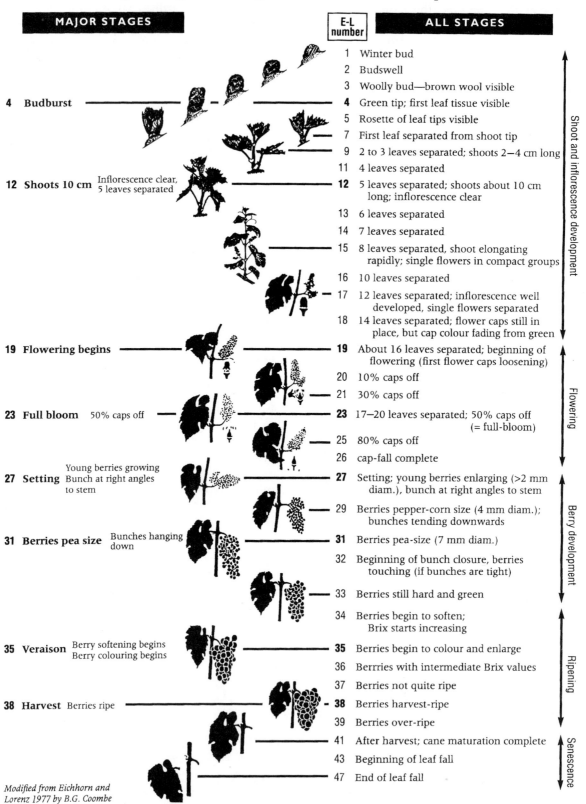

MAJOR STAGES

| E-L number | **ALL STAGES** |

4 Budburst

12 Shoots 10 cm Inflorescence clear, 5 leaves separated

19 Flowering begins

23 Full bloom 50% caps off

27 Setting Young berries growing Bunch at right angles to stem

31 Berries pea size Bunches hanging down

35 Veraison Berry softening begins Berry colouring begins

38 Harvest Berries ripe

1   Winter bud
2   Budswell
3   Woolly bud—brown wool visible
4   Green tip; first leaf tissue visible
5   Rosette of leaf tips visible
7   First leaf separated from shoot tip
9   2 to 3 leaves separated; shoots 2–4 cm long
11   4 leaves separated
12   5 leaves separated; shoots about 10 cm long; inflorescence clear
13   6 leaves separated
14   7 leaves separated
15   8 leaves separated, shoot elongating rapidly; single flowers in compact groups
16   10 leaves separated
17   12 leaves separated; inflorescence well developed, single flowers separated
18   14 leaves separated; flower caps still in place, but cap colour fading from green
19   About 16 leaves separated; beginning of flowering (first flower caps loosening)
20   10% caps off
21   30% caps off
23   17–20 leaves separated; 50% caps off (= full-bloom)
25   80% caps off
26   cap-fall complete
27   Setting; young berries enlarging (>2 mm diam.), bunch at right angles to stem
29   Berries pepper-corn size (4 mm diam.); bunches tending downwards
31   Berries pea-size (7 mm diam.)
32   Beginning of bunch closure, berries touching (if bunches are tight)
33   Berries still hard and green
34   Berries begin to soften; Brix starts increasing
35   Berries begin to colour and enlarge
36   Berrries with intermediate Brix values
37   Berries not quite ripe
38   Berries harvest-ripe
39   Berries over-ripe
41   After harvest; cane maturation complete
43   Beginning of leaf fall
47   End of leaf fall

Shoot and inflorescence development

Flowering

Berry development

Ripening

Senescence

Modified from Eichhorn and Lorenz 1977 by B.G. Coombe

Figure 4.2 Phenological (growth) stages of the grapevine – the modified E–L system. (From Coombe, 1995, reproduced by permission)

continued vegetative growth that often negatively affects fruit maturation. Irrigation may be reinitiated after harvest to avoid stressing the vine and favor optimal cane maturation.

When berry development has reached *véraison*, primary shoot growth has usually ceased. Shoot topping may be practiced if primary or lateral shoot growth continues, or to facilitate machinery movement in dense plantings. Fruit cluster thinning may be performed if potential fruit yield appears excessive. Flower-cluster thinning early in the season can achieve a similar yield reduction. Basal leaf removal may be employed to improve cane and fruit exposure to light and air, as well as to open access to disease- and pest-control chemicals.

As the fruit approaches maturity, vineyard activity shifts toward preparation for harvest. Fruit samples are taken throughout the vineyard for chemical analysis. Based on the results, environmental conditions, and the desires of the winemaker, a tentative harvest date is set. In cool climates, measures are put in place for frost protection.

Once harvest is complete, vineyard activity is directed toward preparing the vines for winter. In cold climates, this may vary from mounding soil up around the shoot–rootstock union, to removing the whole shoot system from its support system for burial.

Management of Vine Growth

Vineyard practice is primarily directed toward obtaining the maximum yield of desired quality. One of the major means of achieving this goal involves training and pruning. Because of this, they have been extensively analyzed for the last century and studied empirically for millennia. The result is a bewildering array of systems. Although a full discussion of this diversity is beyond the scope of the book, the following is an overview of the sources of this heterogeneity. For descriptions of local training and pruning systems, the reader is directed to regional governmental publications, universities and research stations.

However, before discussing vine management, it is advisable to first define several commonly used terms. **Training** refers to the development of a permanent vine structure and the location of renewal wood. It is intended to position shoot growth in favorable locations. **Renewal spurs** often consist of short cane segments, retained to generate canes (**bearing wood**), from which **bearing shoots** may originate in subsequent years (see Fig. 4.11). Alternately, spurs are used to reposition bearing wood closer to the head or cordon (**replacement spurs**). Training usually is associated with a support (**trellis**). Training

ideally takes into consideration factors such as the prevailing climate, harvesting practices, and the fruiting characteristics of the cultivar. **Canopy management** is generally viewed as positioning and maintaining bearing (growing) shoots and their fruit in a microclimate optimal for grape quality, inflorescence initiation, and cane maturation. **Pruning** may involve the selective removal of canes, shoots, wood, and leaves, or the severing of roots to obtain the goals of training and canopy management. However, "pruning" most commonly refers to the removal of unnecessary shoot growth at the end of each year's growth. **Thinning** comprises the removal of whole or parts of flower and fruit clusters to improve the berry microclimate and leaf area/fruit balance. Finally, **vigor** refers to the rate and extent of vegetative growth, whereas **capacity** denotes to the amount of growth and the vine's ability to mature fruit.

Yield/Quality Ratio

Central to all vine management is the yield/quality ratio. This is largely a function of the photosynthetic surface area (cm^2) to fruit mass (g) borne by the vine. It provides a measure of the vine's capacity to ripen a crop. The leaf area/fruit (LA/F) ratio is related to the more familiar concept of yield per hectare. The LA/F ratio focuses directly on the fundamental relationship between the supply and demand for energy and organic carbon.

Regrettably, application of the LA/F ratio is limited by the absence of a rapid and simple means of obtaining the requisite data (Tregoat *et al.*, 2001). Suggestions by Grantz and Williams (1993), Mabrouk and Carbonneau (1996) and Lopes and Pinto (2005) may provide efficient means of assessing leaf area. Multispectral remote sensing (Johnson *et al.*, 2003) has potential as it can be readily correlated with field data. In addition, Costanza *et al.* (2004) have shown a strong correlation between shoot length and leaf area. Complications may include the effects of nutrient and water status, temperature and light conditions, disease pressures, and pruning and training systems on photosynthetic efficiency and carbohydrate shunting throughout the vine. These factors may explain the diversity in LA/F ratios calculated to be necessary to achieve full ripening (roughly between 7 and 14 cm^2/g). Another source of variability involves cultivar response (Fig. 4.3). For most effective use, the values derived should incorporate data only from leaves significantly contributing to vine growth (eliminating young, old, and heavily shaded leaves). When data from different canopy systems are compared, differences in the proportion of exterior to interior leaf canopy need to be taken into account.

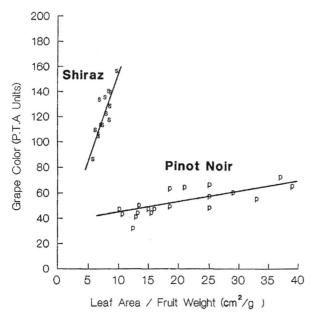

Figure 4.3 The relationship between grape color and LA/F for 'Shiraz' and 'Pinot noir' vines. (From Iland *et al.*, 1993, reproduced by permission)

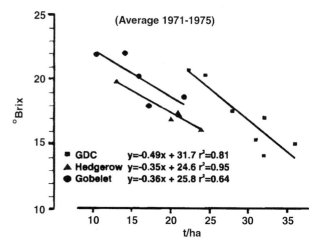

Figure 4.4 Relationship between yield and soluble solids (3-year average) using three distinct training systems. (From Intrieri and Poni, 1995, reproduced by permission)

A simplistic view would suggest that an increased leaf area should directly correlate into an increased ability to produce additional fully ripened fruit. However, the grapevine is an uncommonly complex and adaptive plant. The fruit-bearing capacity of a year's growth is largely defined in the previous year. Thus, most of the flower clusters for the next year are initiated within a 4-week period, bracketing the blooming of the current season's flowers. Thus, as typical with other perennial crops, conditions in the previous year place outer limits on the current season's crop. This is particularly marked in cool-climatic regions, where seasonal variations are often pronounced. Nutrient availability and growth may also be markedly influenced by vine health (and ability to store nutrients) during the previous year.

Another problem with simple yield/quality associations is the potential for the vine to produce several shoot sets per year. Both the primary and, to a lesser extent lateral shoots, are potential sources of flowers and fruit. Thus, some cultivars may sequentially produce two sets of fruit, whose inceptions may be separated by 4–6 weeks. This not only retards ripening of the developing fruit, but also leads to an additional source of poor fruit quality – a wide range of maturity at harvest. Induction of a second (or third) crop can be caused by overly zealous pruning, especially on rich moist soils. In contrast, heavy pruning of vines grown on comparatively nutrient-poor dry soils can direct photosynthetic capacity to fully ripen a restricted fruit load. Figure 4.4 illustrates how different training

systems affect the yield/°Brix ratio. There can also be marked variation within a cultivar (Fig. 4.5), site, or vintage (Plan *et al.*, 1976). Sensory quality may actually decrease with reduced yield, by enhancing the presence of undesirable flavorants (Fig. 4.6) and diminishing varietal aroma (Chapman *et al.*, 2004a). Thus, a universal relationship between vine yield and grape (wine) quality does not exist.

The view of a universal yield/quality ratio is especially in error when pruning techniques, developed for dry low-nutrient hillside sites, are applied to vines grown on rich lowland moist sites. Severe pruning only spurs excess vegetative growth, rather than limiting it. As a result, nutrients are directed to shoot growth, and the activation of dormant buds, while the existing fruit becomes shaded, disease prone, and matures slowly, incompletely, and nonuniformly. Furthermore, cool climatic conditions provide less opportunity for the vine to compensate for temporary poor conditions, in contrast to the extended growing season and higher light intensities that typify warmer Mediterranean climates (see Howell, 2001).

The important question is not whether there is some ideal yield/hectare ratio, but what is the optimal canopy size and placement to adequately nourish the crop to its optimal state of ripeness. Insufficient leaf surface can be as detrimental to quality as is excessive cover.

A centuries-old view holds that when vines are grown on relatively nutrient-poor soils, and experience moderate water limitation, the potential for vigorous growth is restrained and fruit ripens optimally. Although generally accepted as valid, adequate scientific validation has been difficult to obtain. However, a recent study from Oregon has compared data from two commercial vineyards with vines of the same clone, rootstock, age and

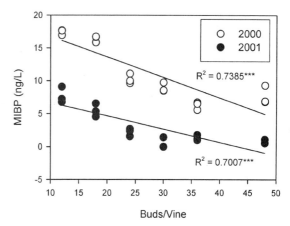

Figure 4.6 Concentration of 2-methoxy-3-isobutylpyrazine (bell pepper odor) of 'Cabernet Sauvignon' wines in relation to pruning treatment. (From Chapman *et al.*, 2004b, reproduced by permission)

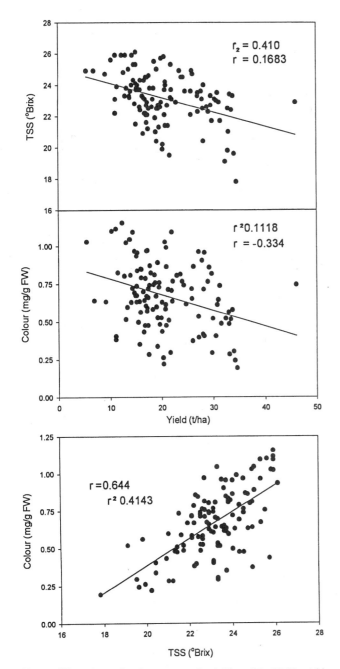

Figure 4.5 Relationship between total soluble solids (TTS), yield, and color in 'Shiraz' showing the correlation coefficient (r) and the fitted line. (From Holzapfel *et al.*, 1999, reproduced by permission)

vineyard management practices (Cortell *et al.*, 2005). In regions of the vineyards having reduced vine vigor, proanthocyanidin, epigallocatechin and pigmented polymer contents were significantly higher than from vines in blocks showing higher vigor. The differences in vine vigor could be correlated to differences in soil depth and water holding capacity at the two sites. The results also indicate that, for 'Pinot noir,' the relative contribution of skin to seed proanthocyanidins increases with a

reduction in vine vigor (Fig. 4.7). Although enhancing potential wine color, this was not necessarily correlated with improved wine flavor (Chapman *et al.*, 2004a). In addition, data from Chapman *et al.* (2004a, 2005) indicate that the manner of yield reduction (pruning vs. water deficit) may be more significant than yield reduction itself. This indicates that more needs to be done to understand how vineyard practice affects not only soluble solids and color, but also the concentration of varietal flavorants and wine flavor.

One of the techniques historically used to reduce yield has been dense planting. Intervine competition tends to promote the production of many fine roots. This, in turn, may enhance nutrient extraction from the soil. Some roots may also grow deep to reach subsurface supplies of water. Severe pruning delays bud break and suppresses fruiting potential (by removing fruit buds). In the New World, grapevines have often been planted on rich loamy soil, with ample supplies of moisture. Wide vine spacing, appropriate for standard farm equipment, further encouraged vigorous growth. Under such conditions, it is essential that the increased growth potential of the vine be directed, not pruned away, causing dormant bud activation and excessive vegetative growth. The modern view is that when an appropriate functional LA/F ratio is developed, both fruit yield and wine quality can be simultaneously enhanced.

The development of training systems, based partially on achieving an optimal LA/F ratio, is one of the major successes of modern viticulture. The first successful application of the concept of balancing photosynthetic capacity to yield was the Geneva Double Curtain system. This has been followed by others more suitable to *V. vinifera* cultivars, notably Vertical Shoot Positioning (VSP), the Scott Henry trellis, the Smart–Dyson trellis, the Lyre, and Ruakura Twin Two Tier.

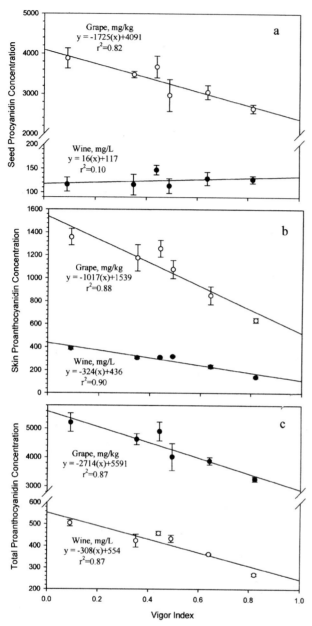

Figure 4.7 Concentration of proanthocyanidins in seed (a), skin (b), and skin plus seed (c) in grapes at harvest and in the corresponding wine, error bar indicating ±SEM ($n = 3$). (From Cortell *et al.*, 2005, reproduced by permission. Copyright 2005 American Chemical Society)

These training systems, in addition to favoring an optimal LA/F ratio, provide better fruit exposure to light and air. These features promote good berry coloration, flavor development, fruit health, and flower-cluster initiation. Optimal LA/F ratios tend to fall in the range of 10 cm^2/g for several important cultivars (about 5–6 mature main leaves per medium-size grape cluster) (Jackson, 1986). Values lower than 6–10 cm^2/g

(0.6–1.0 m^2/kg) are generally insufficient to fully ripen the fruit of a wide range of cultivars (Kliewer and Dokoozlian, 2000). Higher values often indicate excessive shading, reduced anthocyanin content, and delayed ripening, without adequate °Brix values. However, as shown in Fig. 4.3, the relationship between leaf area and grape color can vary markedly between cultivars. Within limits, higher LA/F ratios favor earlier and more intense fruit coloration. Such features have usually been associated with enhanced wine flavor (Iland *et al.*, 1993).

Physiological Effects of Pruning

Pruning is often used for separate but related reasons. It permits the grape grower to establish a particular training system and manipulate vine yield. Pruning permits the selection of bearing wood (spurs and canes), and thereby influences the location and development of the canopy. This in turn can affect grape yield, health, and maturation, as well as pruning and harvesting costs.

The vine, in common with most other perennial fruit crops, shows inflorescence initiation a season in advance of development. However, unlike most orchard crops, inflorescence development occurs opposite leaves on vegetative shoots. In apples, pears, and peaches, for example, the fruit develops on specialized flower spurs (short determinant shoots). The property of inflorescence initiation, a year in advance of flower production, allows the grape grower to limit fruiting capacity by bud removal long before flowering. Winter pruning is the primary means of restricting grapevine yield. However, as a perennial crop, the vine stores considerable energy reserves in its woody parts. Thus, pruning removes nutrients that limit the ability of the vine to initiate rapid growth in the spring. Vines with little mature wood are less able to fully ripen their crop in poor years than are vines possessing large cordons or trunks. Some of these influences may also result from the balance between the primary source of growth regulators, such as cytokinins and gibberellins (the roots), and auxins (produced by the smaller shoot system). When performed properly, pruning can limit shoot vigor, permitting the vine to fully ripen its fruit.

It is now clear that both excessive pruning and excessive overcropping suppress vine capacity (Fig. 4.8). The reduced capacity associated with overcropping has long been known. It has been part of the rationale for removing 85–90% of the yearly shoot growth at the end of the season. However, severe pruning can lead to both reduced yield and grape quality. Severe pruning also delays bud break and leaf production, in addition to activating dormant buds – potentially desirable in the case of severe frost or winter damage. Normally, delayed activation results in leaf production continuing

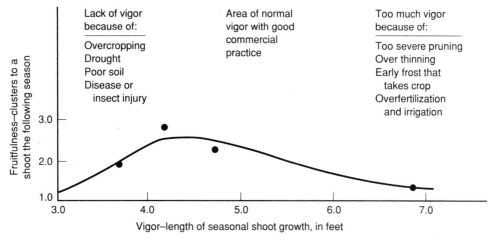

Figure 4.8 The relation of vigor (length) of shoot growth to fruitfulness of buds of 'Muscat of Alexandria' and 'Alicante Bouschet' at Davis, California. (From Winkler *et al.*, 1974, reproduced by permission)

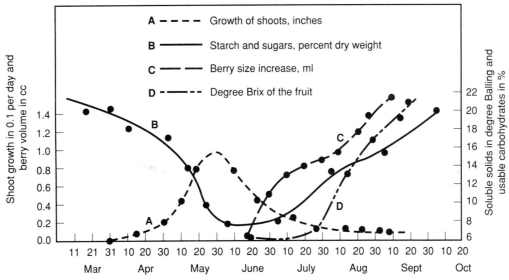

Figure 4.9 The annual cycle of the vine, showing shoot and fruit development as well as seasonal carbohydrate changes in berries and at the bases of 1-year-old wood. (From Winkler *et al.*, 1974, reproduced by permission)

into the fruiting period. This can limit early photosynthate translocation to ripening fruit. In contrast, the absence of pruning permits the rapid development of the leaf canopy and the early initiation of fruit development. However, excessive production can delay berry maturation. Yields can be 2.5 times that of normally pruned vines. This can suppress yield in subsequent years by reducing inflorescence initiation and diminishing nutrient reserves in the old wood. If left unpruned, however, the vine eventually self-regulates itself, and good- to excellent-quality fruit is borne (see Minimal Pruning). Because both inadequate and excessive pruning can result in insufficient or excessive vigor, respectively, compromised fruit quality, and diminished fruitfulness, it is crucial that pruning be matched to long-term vine capacity. The general relationship of shoot to berry growth, and the concentration of carbohydrate in the shoot and fruit are illustrated in Fig. 4.9.

Winter pruning is the most practical but least discriminatory type of pruning. Capacity can be only imprecisely predicted. In contrast, **fruit thinning** is the most precise, but the most labor intensive. This results from the difficulty of selectively removing fruit from clusters partially obscured by foliage. The preferable compromise, where feasible, is judicious winter pruning, followed by remedial flower-cluster thinning in the spring if necessary. This restricts fruit production, while delaying final adjustment until a fairly accurate estimate of grape yield is possible. It also has the advantage of early leaf production, maximizing carbohydrate production over the year, while minimizing competition between foliage and fruit during berry ripening.

Additional augmentation in vine capacity and fruit quality may be obtained with canopy management, by improving the microclimate in and around the grape cluster. Positioning the basal region of shoots in a favorable light environment improves inflorescence induction and helps maintain fruitfulness from year to year.

Clonal selection, the elimination of debilitating viruses, improved disease and pest control, irrigation, and fertilization have significantly amplified potential vine vigor. One of the challenges of modern viticulture is to channel this potential into enhanced fruit and wine quality, and away from undesirable, excessive, vegetative growth. Interest has been shown in minimal pruning, based on the potential of the vine to self-regulate fruit production. However, as the applicability of this technique to most cultivars and climates is unestablished, most of the following discussion deals with the fundamentals of pruning.

In general, the principles of pruning enunciated by Winkler *et al.* (1974) still remain valid:

1. Pruning reduces vine capacity by removing both buds and stored nutrients in canes. Thus, pruning should be kept to the minimum necessary to permit the vine to fully ripen its fruit.

2. Excessive or inadequate pruning depresses vine capacity for several years. To avoid this, pruning should attempt to match bud removal to vine capacity.

3. Capacity partially depends on the ability of the vine to generate a leaf canopy rapidly. This usually requires light pruning, followed by cluster thinning to balance crop production to the existing canopy.

4. Increased crop load and shoot number depress shoot elongation and leaf production during fruit development and, up to a point, favor full ripening. Moderately vigorous shoot growth is most consistent with vine fruitfulness.

5. In establishing a training system, the retention of one main vigorous shoot enhances growth and suppresses early fruit production. Both speed development of the permanent vine structure. With established vines, balanced pruning augments yield potential.

6. Cane thickness is a good indicator of bearing capacity. Thus, to balance growth throughout the vine, the level of bud retention should reflect the diameter of the cane. Alternately, if bud number should remain constant, the canes or spurs retained should be relatively uniform in thickness.

7. Optimal capacity refers to the maximal fruit load that the vine can ripen fully within the normal growing season. Reduced fruit load has no effect on the rate of ripening. In contrast, overcropping delays maturation, increases fruit shatter, and decreases berry quality. Capacity is a function of the current environmental conditions, those of the past few years, and the genetic potential of the cultivar.

Pruning Options

When considering pruning options, it is necessary to evaluate features other than just the principles noted above. The prevailing climate, genetic characteristics of the rootstock and scion (shoot section), soil fertility, and training system can significantly influence the consequences of pruning. Several of these factors are discussed below.

The prevailing climate imposes constraints on where and how grapevines can be grown. In cool continental climates, severe winter temperatures may kill most buds. Equally, late frosts can destroy newly emerged shoots and inflorescences. Where such losses occur frequently, more buds need to be retained than theoretically ideal, to compensate for potential bud kill. Subsequent disbudding, or flower-cluster thinning, can adjust the potential yield downward if required. Alternately, varieties may be pruned in late winter or very early spring to permit pruning level to reflect actual bud viability. Another possibility, with spur-pruned wines, is to leave a few canes as insurance (Kirk, 1999). Because buds at the base of canes usually break late, they may survive and replace emerging shoots killed by late frosts. Also, if no frost damage occurs, the canes can be easily removed before significantly affecting growth from the spurs.

As deep fertile soil enhances vigor, light pruning is often employed to permit sufficient early shoot growth to restrict late vegetative growth during fruit ripening. Vines on poorer soils usually benefit from more extensive pruning. This channels nutrient reserves into the buds retained, assuring sufficient vigor for canopy development, fruit ripening, and inflorescence initiation.

Varietal fruiting characteristics, such as the number of bunches per shoot, number of flowers per cluster, berry weight, and bud position along the cane, can influence the extent and type of pruning. Some cultivars, such as 'Sultana' and 'Nebbiolo,' commonly produce sterile buds near the base of the cane. As spur pruning would leave few fruit buds, these varieties are cane-pruned. Varieties showing strong apical dominance also may fail to bud out at the base or middle portion of the vine (Fig. 4.10). Arching and inclining the cane downward can reduce apical dominance in varieties where this is a problem, and balance bud break along the cane. Alternately, spur pruning limits the expression of apical dominance. Varieties tending to produce small clusters are left with more buds than those bearing large clusters, to correlate yield with varietal fruiting characteristics.

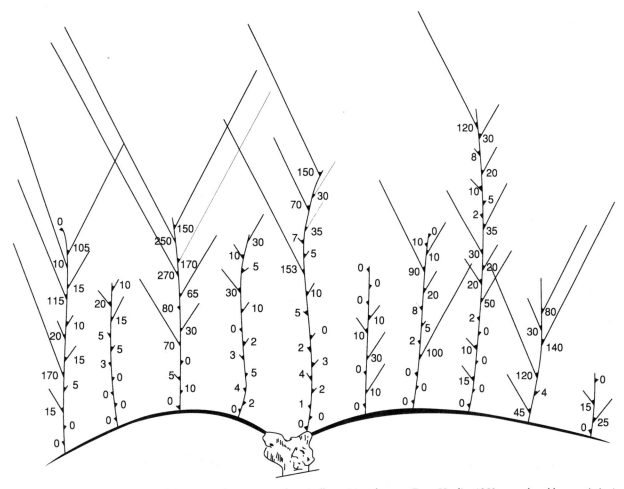

Figure 4.10 Apical dominance of shoot growth on unpruned vertically positioned canes. (From Huglin, 1958, reproduced by permission)

Many of the more renowned cultivars produce small clusters, for example 'Riesling,' 'Pinot noir,' 'Chardonnay,' 'Cabernet Sauvignon,' and 'Sauvignon blanc.' Large-clustered varieties include 'Chenin blanc,' 'Grenache,' and 'Carignan.' The susceptibility of varieties, such as 'Gewürztraminer,' to inflorescence necrosis (*coulure*) can influence how many buds should be retained. In addition, the ideal number of buds retained depends on the tendency of the variety to produce fertile buds. Examples of cultivars having a high percentage of fertile buds are 'Muscat of Alexandria' and 'Rubired.'

As noted later, the type of harvesting (manual vs. mechanical) can alter the pruning and training system chosen. Spur pruning is generally more adapted to mechanical harvesting than is cane pruning. It is also easier for inexperienced pruners to learn than cane pruning.

Not least among grower concerns is the cost of pruning. Although the most economical is winter pruning, it is the most difficult to employ skillfully, because it is impossible to predict bud viability precisely. Cluster and fruit thinning permit the most precise regulation of yield, but are the most expensive. Economic pragmatism generally favors winter pruning as the method of choice.

Pruning Level and Timing

Widely spaced vines of average vigor and pruning degree may generate up to 25 canes, possessing about 30 buds each. Thus, before pruning, the vine may possess upward of 750 buds. Even without pruning, only a small number of the buds (about 100–150) would burst the following spring. Despite this, they would generate a fruit load considerably in excess of the wine's ability to mature. Thus, pruning is normally required. However, as noted previously, excessive removal can undesirably impact fruit maturity and subsequent fruitfulness. Thus, determining the appropriate number of buds to retain is one of the critical yearly tasks of the grape grower. The job is made all the more complex because actual fruitfulness is known in the spring, after winter pruning is complete.

This has led to several attempts to improve assessment of yield potential. One method involves direct observation of sectioned buds. Although reliable, it is not easily used. Another procedure is balanced pruning. It estimates the vine's capacity based on the weight of cuttings removed during pruning. From this, an optimal number of buds is established. Depending on previous experience, additional buds are retained to account for bud mortality during the winter. Only buds separated by obvious internodes are counted because base (noncount) buds usually remain dormant.

The number of buds retained is indicated in increments (weight-of-prunings) above a minimum, established empirically for each variety (see Table 4.1). Pruning recommendations are usually given in parentheses, for example (30 + 10). The first value indicates the number of buds (count nodes) retained for the first pound of prunings, and the second value suggests the number of buds to retain for each additional pound of prunings. Values for *vinifera* cultivars under Australian conditions are in the range of 30–40 buds/kg of prunings (Smart and Robinson, 1991). Vines producing less than a minimum weight-of-prunings are pruned to less than the specified number of buds.

Alternate measures of pruning level are presented in terms of buds retained per vine or per meter row. For example, severe, moderate, and light pruning are noted as <20, 20–70, and >75 nodes per meter row (3 m spacing) in Australia (Tassie and Freeman, 1992).

In **balanced pruning**, the current year's growth is used as a measure of vine fruitfulness and capacity.

The technique has been used primarily with *labrusca* varieties in eastern North America. The application of the technique to other varieties has been somewhat less successful. When used, pruning recommendations must account for cultivar fertility and prevailing climatic conditions. For example, varieties with low fertility need more buds to improve yield potential, whereas bud retention must be adjusted downward in regions with short growing seasons to permit the full ripening of the crop. In addition, the tendency of buds to produce more than the standard two clusters per bud (up to 4) and the production of flower clusters from shoots derived from noncount (base) buds has limited the usefulness of balanced pruning with most French-American hybrids (Morris *et al.*, 1984). Preliminary evidence indicates that flower cluster and shoot thinning may be a more effective means of limiting overcropping (Morris *et al.*, 2004), though results can vary from cultivar to cultivar.

Balanced pruning has found its greatest application in research, where objective values are required (values of 0.3–0.6 kg/m prunings are frequently considered optimal). It has found less use in normal vineyard practice. Once pruners obtain sufficient experience, they can adjust bud number to the capacity of the vine without the need for quantification. Furthermore, the convenience and economics of modern mechanical pruning may offset the advantages of individually tailoring pruning to each vine (see Tassie and Freeman, 1992).

Pruning is often done in midwinter, but may be conducted from leaf fall until after bud break. Pruning

Table 4.1 Suggested pruning severity for balanced pruning of mature vines of several American (*Vitis labrusca*) and French-American varieties in New York State

Grape variety	Number of nodes to retain per pound (0.45 kg) of cane prunings		
	First pound (0.45 kg)		Each extra pound (0.45 kg)
American cultivars			
'Concord'	30	+	10
'Fredonia'	40	+	10
'Niagara,' 'Delaware,' 'Catawba'	25	+	10
'Ives', 'Elvira,' 'Dutchess'	20	+	10
French-American hybrid Cultivars[a]			
Small-clustered varieties ('Maréchal Foch,' 'Leon Millot,' etc.)	20	+	10
Medium-clustered varieties ('Aurore,' 'Cascade,' 'Chelois,' etc.)	10	+	10
Large-clustered varieties ('Seyval,' 'de Chaunac,' 'Chancellor,' etc.)	20	+	10

[a]All require suckering of the trunk, head, and cordon during the spring and early summer.

Source: After Jordan *et al.*, 1981, reproduced by permission

should begin only after the phloem has sealed itself with callose for the winter. This avoids nutrient loss and the activation of partially dormant buds. Pruning during the winter has many advantages, the most important of which may be that most other vineyard activities are at a minimum. Also, bud counting and cane selection are more easily performed with bare vines. Equally, prunings can be more easily disposed of, either by chopping for soil incorporation or by burning.

For tender varieties, pruning is normally performed late, when the danger of killing frost has passed. The proportion of dead buds can then be assessed directly by cutting open a sampling of buds. Dead buds show blackened primary and possibly secondary buds (Plate 5.1). An alternative procedure involves **double pruning**. Most of the pruning and cleanup occur during the winter, with the final removal delayed until bud break. The delay in bud burst induced by this technique often has been used for frost protection. It can retard bud break by one to several weeks in precocious varieties. Late pruning also has been used to delay bud break in regions experiencing serious early season storms, such as the Margaret River region of Western Australia. Because the delay induced by late pruning often persists through berry maturity, harvesting may be equally retarded. Thus, the advantages of delayed bud break must be weighed against the potential disadvantages of deferred ripening. For some cool-climate cultivars, delayed maturation can be desirable when they are grown in warm regions. It can produce fruit higher in acidity, enriched in color, and augmented flavor (Dry, 1987). Regrettably, it can also reduce yield. Late pruning can also be used to reduce the incidence of Eutypa dieback.

Pruning and thinning in late spring and early summer involve adjustments that can improve vine microclimate, minimize wind damage, as well as limit fruit production. These activities may involve disbudding, pinching, suckering, topping, and thinning of flower or fruit clusters.

Disbudding is commonly used in training young vines to a desired mature form. It is also employed to remove unwanted base or latent buds that may have become active. Early removal economizes nutrient reserves and favors strong shoot growth.

Once growth has commenced, summer pruning may involve the partial or complete removal of shoots. **Suckering** usually entails the manual removal of all suckers and water sprouts (Fig. 4.11). Occasionally, the application of naphthalene acetic acid (NAA), mixed with latex paint or asphalt, can control suckers for 2–3 years (Pearn and Bath, 1992). Application to the trunk is best done after bud break. Repeat activation of suckers and water sprouts throughout the growing season is often a sign of overpruning and insufficient energy being directed to fruit production.

Although **suckers** are undesirable (they divert photosynthate away from ripening fruit, disrupt vine training, and complicate herbicide application), some **water sprouts** can be of value. If favorably positioned, water sprouts can act as replacement spurs in repositioning the growing points of the vine (Fig. 4.11). As the vine grows, the location of renewal wood tends to move outward. Replacement spurs reestablish, and thus maintain, the shape and training system of the vine. Depending on their position, water sprouts also may supply nutrients for fruit development.

Shoot thinning conducted at flowering may be used to correct imperfections in the positioning of shoots or to remove weak or nonbearing shoots. If the vine is in balance, and winter or late frosts did not cause excessive bud or shoot kill, the need for shoot thinning is often negligible to nonexistent.

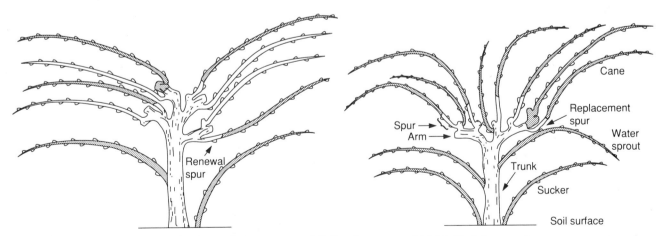

Figure 4.11 Illustration showing a head-trained vine spur-pruned (*left*) and cane-pruned (*right*). Shaded areas refer to water sprouts, suckers and portions of canes removed at pruning. (From Weaver, R J. *Grape Growing*, Copyright © 1976, John Wiley & Sons Inc. Reprinted by permission of John Wiley & Sons Inc.)

Partial shoot removal, called **trimming**, can vary widely in degree and timing. Pinching refers to the removal of the uppermost few centimeters of shoot growth. In varieties with poor fruit set, pinching to coincide with flowering may improve fruit set (Collins and Dry, 2006). More extensive trimming is called tipping, topping, or hedging, depending on the length of shoot removed. In general, where trimming is needed, it is kept to a minimum. In addition, if extensive removal is required, it is less physiologically disruptive if performed periodically and moderately, rather than all at once.

Pinching is usually conducted in early season. When conducted during flowering, fruit set may be enhanced. The procedure reduces inflorescence necrosis in varieties predisposed to the disorder, presumably by reducing carbohydrate competition between the developing leaves and embryonic fruit. Pinching can also be used to maintain shoots in an upright position. The activation of lateral shoot growth may provide desirable fruit shading in hot sunny climates.

Tipping (topping) is performed later and may be repeated periodically throughout the growing season. It usually removes only the shoot tip and associated young leaves, leaving at least 15 or more mature leaves per shoot. Depending on timing, it can reduce competition between developing flowers, or fruit and young leaves for photosynthate. Tipping redirects carbohydrate movement from growing leaves (Quinlan and Weaver, 1970) to the inflorescence or developing fruit. Tipping also can improve canopy microclimate by removing draping leaf cover. In windy environments, tipping produces shorter, more sturdy, upright shoots. These are less susceptible to wind damage. Trimming tends to lower berry potassium and pH values. However, tipping can reduce cane and pruning weights, and fewer shoots and grape clusters in the succeeding year (Vasconcelos and Castagnoli, 2000).

Hedging (trimming of the vine to produce vertical canopies resembling a hedge) removes entangling vegetation, facilitating the movement of machinery through the vineyard. This is often necessary where vines are densely planted (rows ≤2 m wide). Like tipping, it reduces carbohydrate competition between new, expanding leaves and the fruit, but also lowers vine capacity. Hedging increases the number of shoots, but reduces their relative length, increasing light and atmospheric exposure of the leaves and fruit.

Trimming shoots to less than 15 leaves is generally undesirable. If this occurs during or before fruit set, it can stimulate undesirable lateral bud activation. The extent to which this occurs is partially dependent both on the variety and the form of training (upright vs. downward shoot growth) (Poni and Intrieri, 1996). In contrast, late trimming (after *véraison*) seldom activates lateral growth. However, a deficiency in photosynthetic potential may persist, delaying both fruit and cane maturation, as well as decreasing cold hardiness. Physiological compensation by the remaining leaves, through delayed leaf senescence and higher photosynthetic rates, is usually inadequate. The degree of expression of these problems depends on the severity of trimming.

Trimming can variously affect fruit composition, depending on its timing, severity and vine vigor. Many of the effects resemble those of basal leaf removal discussed below. The potassium content of the fruit increases more slowly and peaks at a lower value (Solari *et al.*, 1988). The rise in berry pH and the decline in malic acid content associated with ripening may be less marked. Total soluble solids may be little influenced, or reduced, in trimmed vines. Anthocyanin synthesis may be adversely affected in varieties such as 'de Chaunac' (Reynolds and Wardle, 1988). This tendency presumably is not found in Bordeaux, where hedging is common. Improvements in amino acid and ammonia nitrogen levels associated with light trimming shortly after fruit set (Solari *et al.*, 1988) may be one of the more desirable features induced by trimming.

Considerable interest has been shown in single or repeat leaf removal around fruit clusters (**basal leaf removal**). Leaf removal selectively improves air and light exposure around clusters. It also eases the effective application of protective chemicals to the fruit. The reduced incidence of bunch rot was one of the first benefits detected with basal leaf removal. This has been attributed to the increased evaporation potential, wind speed, higher temperature, and improved light exposure in and around the fruit (Thomas *et al.*, 1988).

The desirability of basal leaf removal depends largely on timing and the cost/benefit ratio relative to implementation. If the vine canopy is open and the fruit adequately exposed to light and air, leaf removal is unnecessary and removes photosynthetic potential. It could expose the fruit to sunburn in hot environments, delay maturity, and reduce (modify) wine quality (Greer and La Borde, 2006).

An effect generally associated with basal leaf removal is a drop in titratable acidity (Fig. 4.12). This is associated with a reduction in potassium uptake and enhanced malic acid degradation. Tartaric and citric acid levels are seldom affected. Anthocyanin levels generally are significantly enhanced, whereas the total phenol content may rise or remain unchanged. Levels of grassy, herbaceous, or vegetable odors decline, whereas fruity aromas may rise or remain unaffected. Level of terpenes (Table 4.2), norisoprenoids (Marais *et al.*, 1992a), and the precursor of several fruity-smelling, volatile thiols (Murat and Dumeau, 2005) rises with increased fruit exposure in some cultivars. Basal leaf removal may increase the concentration of glycosyl-glucose in

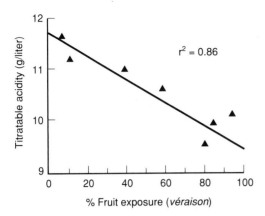

Figure 4.12 Relationship between fruit exposure at *véraison* and titratable acidity at harvest of the cultivar 'Sauvignon blanc'. (From Smith *et al.*, 1988, reproduced by permission)

Table 4.2 Aroma profile analysis and protein content of 'Sauvignon blanc' fruit with and without basal leaf removal

	Control	Basal leaf removal
Free aroma constituents (μg/liter)		
Geraniol	1.6	9.5
trans-2-Octen-1-al	0.3	3.1
1-Octen-3-ol	2.4	6.6
trans-2-Octen-1-ol	6.5	11.7
trans-2-Penten-1-al	4.6	12.7
α-Terpineol	1.9	4.1
Potential volatiles (μg/liter)		
Linalool	23	49
trans-2-Hexen-1-ol	29	61
cis-3-Hexen-1-ol	5.1	6.3
trans-2-Octen-1-ol	321	830
2-Phenylethanol	17.9	50
β-Ionone	26	66
Protein (mg/liter)[a]		
Molecular weight > 66,000	32	33
Molecular weight < 20,000	62	81

[a]Based on bovine serum albumin as standard.

Source: From Smith *et al.*, 1988, reproduced by permission

'Riesling' grapes (Zoecklein *et al.*, 1998), an indicator of flavor potential. Although usually beneficial, excessive production of some norisoprenoids, for example, 1,1,6-trimethyl-1,2-dihydronaphthalene (TDN), and their precursors can generate undesirable kerosene-like fragrances in 'Riesling' wines (Marais *et al.*, 1992b). Berry sugar content is seldom affected significantly.

Most studies have not noted a change in cold-hardiness or fruitfulness as a result of basal leaf removal (see Howell *et al.*, 1994). This probably results because leaf removal typically occurs after inflorescence initiation has taken place and involves the more mature buds near the base of the shoot.

The effects of basal leaf removal depend considerably on timing. Its benefits generally show most clearly when trimming occurs after blooming and before *véraison*. Earlier removal (during bloom) tends to increase inflorescence necrosis and reduces inflorescence initiation (Candolfi-Vasconcelos and Koblet, 1990). Basal leaves are typically the primary exporters of carbohydrate to flowers and fruit set (import of carbohydrate reserves from perennial parts of the vine has usually ceased by this time). Later removal, when more leaves have matured, does not significantly affect fruit set or the initial stages of fruit development (Vasconcelos and Castagnoli, 2000). However, removal after *véraison* may be detrimental by reducing sugar accumulation by the fruit (Iacono *et al.*, 1995), or show few benefits (Hunter and Le Roux, 1992).

The impact of basal leaf removal is also influenced by the number of leaves removed. Removal usually involves the leaves positioned immediately above and below, and opposite the fruit cluster. The vine compensates for these leaves by increasing the number of leaves produced on lateral shoots, delaying leaf senescence, or enhancing the photosynthetic efficiency of the remaining leaves (Candolfi-Vasconcelos and Koblet, 1991).

By adjusting carbohydrate supply during flowering by basal leaf removal (or tipping and lateral shoot removal), the grower can increase or decrease fruit set, as well as influence fruit ripening. These canopy management activities affect the average age and photosynthetic productivity of the canopy. Improved fruit set will be desirable in cultivars showing poor fruit set (e.g., Gewürztraminer), whereas reduced fruit set can be beneficial in cultivars susceptible to bunch rot.

In practice, the use of basal leaf removal has been limited, presumably due to its manual and labor-intensive nature. This may change with the introduction of efficient mechanical leaf-removers. One model involves suction that draws leaves toward a set of cutting blades.

Early defoliation (Poni *et al.*, 2006) is another leaf removal procedure. It is primarily aimed at yield reduction, and improves fruit maturity of a smaller crop. It is being studied as an alternate method to the complexities often associated with manual cluster thinning.

Another method of adjusting fruit yield involves **flower- or fruit-cluster (bunch) thinning**. Its principal purpose is to prevent overcropping. Flower-cluster thinning has an effect similar to delayed winter pruning, in permitting a more precise adjustment of fruit load to vine capacity. Fruit-cluster thinning can be even more precise, but is more difficult, owing to the more advanced canopy development. Fruit-cluster thinning is especially useful with French-American hybrids, in which overcropping is a potential due to base-bud activation. As fruit set tends to suppress the growth of lateral shoots, delaying crop reduction (as results with fruit-cluster thinning) achieves the desired level of fruit production without promoting undesirable bud activation. Bunch thinning may also be considered if leaf damage from pathogen or pest attack is severe. By reducing the carbohydrate sink, the remaining grapes will have a better chance of ripening fully, and the vine regains its strength for subsequent growth.

Occasionally, flower-cluster thinning favors the development of more, but smaller, berries in the remaining clusters. This can be valuable in reducing bunch rot in varieties that form compact clusters, for example 'Seyval blanc' and 'Vignoles' (Reynolds et al., 1986). However, within the usual range of vine capacity, fruit-cluster thinning may show marginal or insignificant improvement in fruit chemistry (Keller et al., 2005), and no detectable effect on wine quality (McDonald, 2005). Thus, the yield losses incurred by cluster thinning, and its expense, must be weighed against the potential benefits it may provide. It may be of value only with cultivars or clones that overproduce regularly (Bavaresco et al., 1991).

Bearing Wood Selection

Although balancing yield to capacity is important, it is equally important to choose the best canes for fruit bearing. The proper choice of bearing wood is especially significant in cool climates. Fully matured healthy canes produce buds that are the least susceptible to winter injury. Canes that develop in well-lit regions of the vine also tend to be the most fruitful. The browning of the bark is the most visible sign of cane maturity. Typical internode distances are a varietally useful indicator of fruitfulness. The presence of short, mature lateral shoots on a cane often signifies good bearing wood. Canes of moderate thickness (about 1 cm) are generally preferred because their buds tend to be the most fruitful. Canes of similar diameter tend to be of equal vigor and, thus, maintain balanced growth throughout the vine. Other indicators of healthy, fruitful canes are the presence of round (vs. flattened) buds, brown coloration to the tip, and hardness to the touch.

Where canes of different diameters are retained, the number of buds (nodes) per cane should reflect the capacity of the individual cane. Thus, weaker canes are pruned shorter than thicker canes.

Equally important to balanced growth is an appropriate spacing of the bearing wood. Retained canes should permit the optimal positioning of the coming season's bearing shoots for photosynthesis and fruit production. Thus, canes are normally selected that originate from similar positions on the vine. If the vine is cordon-trained, canes originating on the upper side of the cordon are generally preferred.

Pruning Procedures

Manual pruning requires a knowledge of how varietal traits and the prevailing climate influence vine capacity. An assessment of the health of each vine and the appropriate location, size, and maturity of individual canes are essential. This requires considerable skill in a labor force, which is becoming increasingly harder to find and retain. With the savings obtained using machine harvesting, pruning has become one of the major costs of vineyard maintenance. These factors have led to increased interest in **mechanical pruning**, which can frequently result in a cost reduction of 40–45% compared to pneumatic manual pruning (Bath, 1993). Even greater savings are possible, compared to pruning without the use of power secateurs. Pruning quality, as measured by variation in spur length, can be the equivalent of manual pruning (Intrieri and Poni, 1995). Variation can come from irregularities in the terrain, the angle of cut, anomalies in cordon shape, and human error. Nevertheless, remedial manual pruning is often required every few years, to avoid congestion from the retention of old wood.

Mechanical pruning is used most successfully on cordon-trained, spur-pruned vines. The cutting planes can be adjusted easily to remove all growth except that in a designated zone around the cordon (Plate 4.1). The cutting planes also can be adjusted for skirting or hedging before harvest to ease mechanical harvesting. Pivot cutting arms, associated with a feeder device, minimize damage to the trunk or trellis posts. If subsequent manual pruning is required, the mechanical component is called **prepruning**. Mechanical plus prepruning can reduce manual pruning time by up to 80%. The applicability of mechanical pruning is often cultivar dependent. For example, 'Malbec,' 'Riesling,' and 'Sémillon' may respond well, whereas others such as 'Shiraz' and 'Sangiovese' generally do not.

With some cultivars, mechanical pruning has the additional advantage of generating more uniform vines. These tend to produce smaller but more numerous grape clusters. The yield may initially be higher than with manual pruning, due to the retention of a higher bud number. Nonetheless, subsequent self-adjustment results in average yield being unaffected.

A distinctly novel approach to pruning has been suggested by work conducted in Australia (Clingeleffer, 1984). The experiments have shown that some cultivars can regulate their own growth, and yield good-quality fruit, without regular pruning. The results have spawned the **minimal pruning system** now fairly common in some parts of Australia. It usually involves only light mechanical pruning – along the sides or at the base of the vine during the winter, and possibly some summer trimming (skirting) to prevent shoot trailing on the ground.

Initially, vines may overcrop, but this diminishes as vines reduce the number of buds that mature and become active in succeeding years. Spontaneous abscission of most immature shoots in the autumn essentially eliminates the need for winter pruning.

Minimal pruning had become sufficiently popular by 1987 that more than 1500 ha of commercial vineyards in Australia were already using it (Clingeleffer and Possingham, 1987). Initially developed for 'Sultana' grown in hot, irrigated vineyards, minimal pruning has been used with considerable success with several premium grape varieties. Some of Australia's most prestigious vineyards are minimally pruned. It has been successful in both the hot and cooler viticultural regions in Australia (Clingeleffer and Possingham, 1987).

In viticulture, pruning normally refers to the removal of aerial parts of the vine. In other perennial crops, such as fruit trees, pruning frequently entails root trimming. Although much less common in viticulture, **root pruning** has been investigated as a technique to restrict excessive shoot vigor (Dry *et al.*, 1998). Clear cultivation in the vineyard, by repeatedly severing surface feeder roots, may have the same effect. Cutting large-diameter roots of grapevines to a depth of 60 cm can promote shoot growth under conditions in which growth is retarded by soil compaction. Under such conditions, root pruning is recommended in adjacent rows in alternate years (van Zyl and van Huyssteen, 1987). Susceptibility of the roots to pathogenic attack does not seem enhanced by pruning. New lateral-root development may occur up to 5 cm back from the pruning wound.

Training Options and Systems

Training involves the development and maintenance of the woody structure of the vine in a particular form. The form attempts to achieve optimal fruit quality and yield, consistent with prolonged vine health, and economic vineyard operation. Because grapevines have remarkable regenerative powers, established vines can buffer the effects of a change in training system for several years. Thus, studies on training systems need to be conducted over many years to assess the actual, long-term effects. In addition, factors such as varietal fruiting habits, prevailing climate, disease prevalence, desired fruit quality, type of pruning and harvesting, and grape pricing can all influence the choice of a training system. Many of these factors are similar to those that affect pruning choices discussed previously. Regrettably, tradition and the intransigency of Appellation Control legislation have slowed the acceptance of newer, more efficient systems.

Because many training systems are only of local interest, and new and improved ways of training are being developed, the discussion focuses primarily on the components that distinguish training systems. Illustrative diagrams outlining the establishment of several common training systems are given in Pongrácz (1978). Details on regional training systems are generally available from provincial or state viticulture research stations.

There is no universally accepted system of classifying training systems. Nonetheless, most systems can be characterized by the origin and length of the bearing wood – head vs. cordon and canes vs. spurs, respectively. These groupings can be further subdivided by the height and position of the bearing wood, the placement of the bearing shoots, and the number of trunks retained. Ancient systems of training on trees do not conveniently fit into this classification and resemble natural grapevine growth.

BEARING WOOD ORIGIN

One of the most distinguishing features of a training system relates to the origin of the bearing wood. On this basis, most training systems can be classed as either head-trained or cordon-trained (Fig. 4.13). **Head training** positions the canes or spurs that generate the fruit-bearing shoots radially on a swollen apex, or as several radially positioned short arms at the trunk apex. In contrast, **cordon training** positions the bearing wood equidistantly along angled portion(s) of the trunk (cordons). In most cordon-trained systems, the cordon is developed along a horizontal plane, usually parallel to the row. Occasionally, however, the cordon is inclined at an intermediate angle or is positioned vertically.

Historically, head training was much more common. Its simplicity made it easy and inexpensive to develop. Also, maintenance was fairly simple. The bearing shoots of one season often provided the bearing wood for the next year's crop. As the trunk thickened, the vine became self-supporting, avoiding the expense and complexity of a trellis and shoot positioning. The lack of support wires also permitted cross-cultivation, desirable when weed control was done manually. Head training is particularly suitable with soils of low fertility and where water availability is limited.

The major drawback of head training is its tendency for shoot crowding – especially where vines have medium to high vigor. Shoot crowding leads not only to undesirable fruit, leaf, and cane shading, but also to higher canopy humidity. The latter favors disease development, whereas other aspects lead to poor fruit maturation, as well as reduced photosynthesis, bud fruitfulness, and cane maturation. However, avoiding these problems with severe pruning creates other problems, notably curtailed yield and delayed development of shoot growth. Severe pruning also can accentuate vigorous vegetative growth, to the detriment of fruit development and maturation. Because limited bud retention increases the

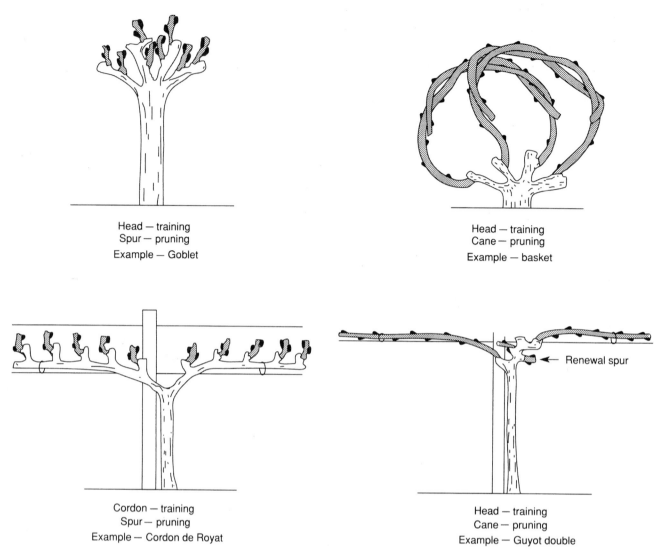

Head — training
Spur — pruning
Example — Goblet

Head — training
Cane — pruning
Example — basket

Cordon — training
Spur — pruning
Example — Cordon de Royat

Head — training
Cane — pruning
Example — Guyot double

← Renewal spur

Figure 4.13 Diagrams illustrating head- and cordon-training systems. Unshaded areas represent old wood, shaded areas bearing wood, and black areas buds. (After Weaver, R. J. *Grape Growing*, Copyright © 1976, John Wiley & Sons Inc. Reprinted by permission of John Wiley & Sons Inc.)

potential damage caused by early frosts, head training has rarely been used in cool regions.

The disadvantages of head training, except for vines of low vigor, are augmented with spur pruning. Spur pruning increases the tendency to produce compact canopies around the head. Although it provides useful fruit shading in hot, dry climates, it favors disease development and poor maturation in cooler, moist climates. These effects may be partially offset by using fewer but longer spurs. However, this makes maintaining vine shape more difficult. Moreover, head-trained, spur-pruned systems are unsuited to mechanical harvesting. They most commonly have been used in dry Mediterranean-type climates, even though they may lead to higher evapotranspiration rates than other training systems (van Zyl and van Huyssteen, 1980). This may result from increased evaporation from soil not

protected by foliage from sun and wind. The *Goblet* training system, used extensively in southern Europe (Fig. 4.13), is an example of a head-trained, spur-pruned system. It is suitable only for cultivars with medium-size fruit clusters that produce fruitful buds at the base of the cane.

In contrast, head training associated with cane pruning has been more popular in cooler, moister climates. In addition to providing a more favorable canopy microclimate, cane pruning permits more buds to be retained by dispersing the fruit-bearing shoots away from the head. This improves yield with cultivars bearing small fruit clusters, such as 'Riesling,' 'Chardonnay,' and 'Pinot noir.' Larger-clustered varieties such as 'Chenin blanc,' 'Carignan,' and 'Grenache,' or those with very fruitful buds, such as 'Muscat of Alexandra' and 'Rubired,' tend to overproduce with cane pruning. The Mosel Arch

(Fig. 4.15A) used in Germany is a novel and effective approach by which cane bending supplies both shoot support and diminished apical dominance along the cane. Shoot dispersion, possible with cane pruning, requires the use of a trellis and wires along each row to support the crop. It provides both better light exposure and enhanced yield. With this system, canes may eventually arch or roll downward. In cooler climates, this may be desirable to provide direct fruit exposure to the sun. In hot, dry climates, this is usually undesirable, as it may result in fruit burn. Under such conditions, additional foliage-support (catch) wires may be used to raise the foliage and provide some necessary shading, or simply to support the shoots to prevent excessive arching.

In cane pruning, it is necessary to select renewal spurs (Fig. 4.11). They are required because the canes of one season usually cannot be used as bearing wood the next season. Their use would quickly result in relocating the bearing wood away from the head. The same tendency occurs with spur pruning, but at a much slower rate. The conscious selection of renewal shoots may be reduced if water sprouts develop from the head. Common examples of head-trained, cane-pruned systems are the Guyot in France and the Kniffin in eastern North America.

In contrast to the central location of the bearing wood in head training, cordon training positions the bearing wood along the upper portion of an elongated trunk. Most systems possess either one (unilateral) or two (bilateral) horizontally positioned cordons. Occasionally, the trunk is divided into two horizontal trunks, directed at right-angles (laterally) to the vine row. These subsequently branch into two or more cordons running in opposite directions, parallel to the row. This is quadrilateral cordon training. Vertical (upright) cordon systems are uncommon as apical dominance and shading combine to promote growth at the top. These features complicate maintaining a balanced cordon system. Vertical cordons are also poorly adapted to mechanized pruning and harvesting. Several obliquely angled cordon-trained systems are popular in northern Italy.

Horizontal cordons experience little apical dominance because of the uniform height of the buds. However, the horizontal positioning of the cordon places considerable stress on its junction with the trunk. This requires the use of one or more support wires to carry the weight of the shoot system and crop. Because bearing wood is selected to be uniformly spaced along the cordon, the shoots generally develop a canopy microclimate favorable for optimal fruit ripening. Higher yields can result from improved net photosynthesis, production of more fruitful buds, and increased nutrient reserves located in the enlarged woody vine structure.

Location of the fruit in a common zone along the row makes cordon training well suited to mechanical harvesting. It also provides relatively homogeneous growing conditions, which favors uniform maturation. Furthermore, positioning the bearing wood in a narrow region above or below the cordon makes the vine more amenable to mechanical pruning.

The disadvantages of cordon training include its higher costs, involving the use of a strong trellis and support wires, the time and expense required in its establishment, and the greater skill demanded in selecting and positioning the arms that bear the spurs or canes. In cool climates, the more synchronous bud break can increase the damage caused by late spring frosts. Because cordon-trained vines tend to be more vigorous, they are commonly spur-pruned to minimize overcropping. This feature limits the combined use of spur pruning and cordon training to varieties that bear fruit buds down to the base of the cane.

Cane pruning is usually combined with cordon training only for strong vines capable of maturing heavy fruit crops. Because of the large permanent woody structure of the trunk and cordon(s), cordon training is often inappropriate for French-American hybrids. The woody component increases the number of base (non-count) buds that may become active and accentuates the tendency to overcrop.

Because of the many advantages of cordon training, most newer training systems, such as the Geneva Double Curtain, the Ruakura Twin Two Tier, Lyre and Scott Henry employ it. Several older systems are also examples of cordon training, for example the Cordon de Royat in France, the Hudson River Umbrella in eastern North America, the Dragon in China, and pergolas in Italy.

BEARING WOOD LENGTH

As noted, the choice of spur vs. cane pruning often depends on the training system used. Conversely, the training system may be chosen based on the advantages provided by spur or cane pruning.

Cane pruning is especially appropriate for cultivars producing small clusters that need the retention of extra buds. However, for the development of a desirable canopy microclimate, the variety also must possess relatively long internodes to minimize shoot overcrowding. In addition, cane pruning enhances vine capacity by retaining more apically positioned buds. These are generally more fruitful than basally positioned buds (see Fig. 3.6). This is particularly important for cultivars that produce sterile base buds. Cane pruning not only allows precise shoot positioning along the row, but it also facilitates the development of wide-topped trellises, extending both along and perpendicular to the row. Cane-pruned vines tend to develop their canopy sooner in the season. This is particularly valuable with varieties susceptible to inflorescence necrosis, such as

'Gewürztraminer' and 'Muscat of Alexandra,' or with vigorous vines.

The disadvantages of cane pruning include the expense involved in trellising and tying the shoots. Because the crop develops from only a few canes, particular care must be taken in choosing them. Pruning must be done manually by skilled workers. In addition, damage or death of even one cane can seriously reduce vine yield. A further complicating factor is the removal of the current year's bearing wood. Bearing wood for the next season's crop typically comes from shoots that develop from renewal spurs. The long length of the bearing wood can result in uneven shoot development owing to apical dominance. This can lead to nonuniform canopy development and asynchronous fruit ripening. Arching or positioning the canes obliquely downward can often minimize apical dominance, but it places the bearing shoots and fruit in diverse environments. Finally, converting a vine from spur to cane pruning often results in temporary overcropping.

Spur pruning tends to show properties that are the inverse of cane pruning. Because of its greater simplicity and uniformity, spur pruning requires less skill. Because spur pruning tends to restrict fruit production to predetermined locations, it is particularly amenable to mechanical harvesting. If the spurs are located equidistant from the ground, the resulting absence of apical dominance favors uniform bud break.

The tendency of spur pruning to limit productivity can be either beneficial or detrimental, depending on the vigor and capacity of the vine. Up to a point, decreased productivity may be compensated for by leaving more buds. Berry size is generally reduced with spur pruning. This has the advantage of increasing the surface area/volume ratio, thereby tending to enhance wine flavor and color potential. Spurs are usually left with two nodes, but occasionally may be reduced to one node for bountiful varieties such as 'Muscat Gordo.' For other varieties a combination of two- and four- to six-node spurs ('finger and thumb' pruning) has proven appropriate.

Restricting yield may be a desirable feature in cool climates or under poor-nutrient conditions, but it is less important in warm climates on deep rich soils. Spur pruning is inappropriate for varieties that bear sterile buds at the base of the bearing wood, or those susceptible to inflorescence necrosis (*coulure*). The delay in leaf production associated with spur pruning can result in increased competition between expanding leaves and the young developing fruit. Delayed canopy production may also explain the higher concentration of methoxypyrazines (delayed degradation) when used with 'Cabernet Sauvignon.' Finally, without removal of malpositioned spurs, shoot crowding is likely.

Commonly accepted norms for most *V. vinifera* cultivars are about 15 shoots per meter (positioned more or less equidistantly along the cordon), with each shoot bearing not more than 12–18 count nodes. These values can vary depending on features such as row spacing, vine density, soil fertility, water availability, likelihood of winter kill, rootstock and scion cultivar, and the quality requirements of the grower. Yield to pruning weight ratio of 6–7:1 are generally desirable with many cultivars. For some such as 'Pinot noir,' this ratio may be high.

SHOOT POSITIONING

As with locating of the vine's woody structure, shoot placement can be used to promote vertical, horizontal, or inclined growth (Fig. 4.14). In addition, shoots may be prevented from, permitted to, or encouraged to arch and grow downward (Fig. 4.15).

The training of shoots vertically upward (by raising catch wires) is generally favored because of its suitability to mechanical harvesting and pruning. Mechanical harvesting is facilitated by the central location of the fruit on the vine. Mechanical pruning and hedging are equally aided by the largely unobstructed access to the

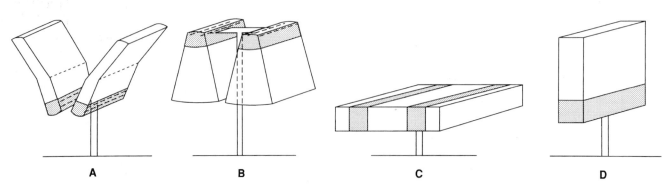

Figure 4.14 Training systems showing different canopy placements with the fruit-bearing zone indicated by hatching: (**A**) inclined upright (Lyre), (**B**) trailing (Geneva Double Curtain), (**C**) horizontal (Lincoln), and (**D**) vertical upright (Guyot).

bearing shoots. In addition, vertically trained vines generally are more vigorous and have higher capacities. Ready accessibility to most vineyard practices also makes vertical positioning more preferred than horizontally or inclined canopies. The latter systems can be useful where protection of the fruit from direct exposure to sunlight is desirable.

In contrast, pendulous or trailing growth commonly induces vine devigoration and reduced crop yield. This seems to result from narrowing of the xylem vessels, reducing sap flow (Fig. 4.16). This influence is systemic and not just limited to the bent regions. The trailing growth that results requires little shoot tying, facilitating mechanical harvesting. Nevertheless, vines need to be trained high, if not skirted, to avoid shoots trailing on the ground.

With low to moderately vigorous vines, arching or direct trailing can expose the basal portion of the shoot

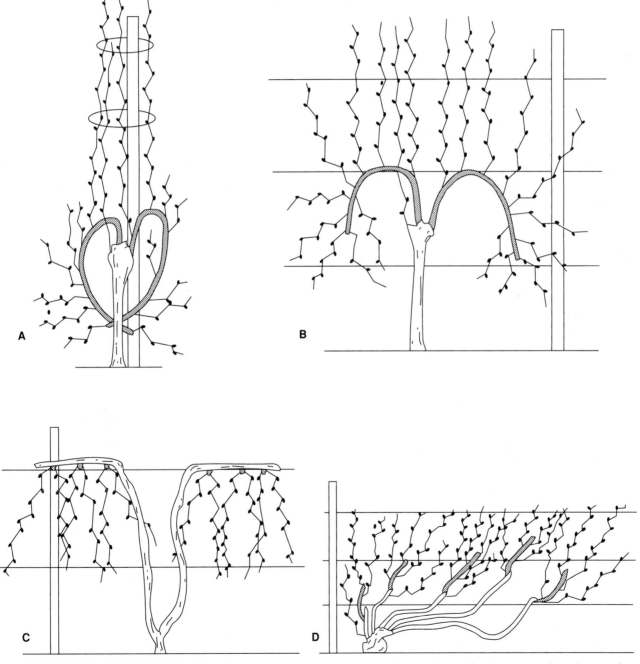

Figure 4.15 Diagrams of vines showing different shoot positioning: (**A**) upright and horizontal (Mosel Arch), (**B**) upright and procumbent (Umbrella Kniffin), (**C**) procumbent (Hudson River Umbrella), and (**D**) upright (Chablis). Old wood is unshaded and bearing wood is shaded.

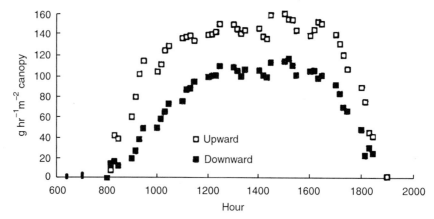

Figure 4.16 Sap flow rate per unit canopy area measured during the day in shoots of 'Nebbiolo' trained upward and downward. (From Schubert *et al.*, 1996, reproduced by permission)

to sun and wind. This favors cane maturation, bud fruitfulness, fruit maturation, and vine health. However, sunburning the fruit is a potential problem in hot sunny climates. The tendency of most *vinifera* cultivars to grow upright initially complicates the formation of trailing vertical canopies, but is the natural tendency with most *labrusca* cultivars.

Upright vertical canopies can be more expensive to maintain and often produce heavy fruit shading if vine growth is vigorous. Positioning long shoots in an upright position usually requires tying the shoots to several wires. This is costly both in terms of materials and complicates pruning (because of shoot entanglement in support wires). The latter can be minimized by trimming to short, stout shoots. Another potential disadvantage of upright canopies is the location of the fruit and renewal zones at the base of the canopy. However, with adequate canopy division, sufficient light reaches the base to permit adequate fruit ripening, inflorescence initiation, and cane maturation. Under certain conditions, basal positioning of the fruit zone can prove advantageous. It can provide protection against hail, sunburn, strong winds, and bird damage. Fruit ripening may also occur earlier, even though flowering is delayed (Kliewer *et al.*, 1989).

CANOPY DIVISION

Improving canopy microclimate is one of the newer concepts in training design. The first of these systems was the Geneva Double Curtain (GDC) (Shaulis *et al.*, 1966). It has subsequently been modified to improve its applicability to mechanical harvesting and pruning. Newer examples are the Ruakura Twin Two Tier (RT2T) (Smart *et al.*, 1990c), the Lyre (Carbonneau and Casteran, 1987), and the Scott Henry (Smart and Robinson, 1991). Such training systems have developed largely out of fundamental studies on vine microclimate.

By dividing the canopy into separate components, exposure of the fruit and vegetation to the sun is enhanced, fluctuation in berry temperature augmented, humidity decreased, and the transpiration rate increased. In the center of dense canopies, light exposure can fall to less than 1% of above-canopy levels, and wind movement can decline by more than 90%. Reduced transpiration and evaporation rates correlate with higher humidity and decreased convection, all of which favor fungal infection. Divided-canopy systems increase the leaf surface area exposed directly to the sun, similar to dense plantings. However, they are much more economic in terms of planting costs (especially with grafted vines).

In many regions, increased fruit exposure to sunlight is desired. In hot sunny climates, however, this can be a disadvantage. Shading may be essential to avoid sunburn. In addition, sun exposure may undesirably augment some flavor compounds, such as 1,1,6-trimethyl-1, 2-dihydronaphthalene (TDN) (Marais, 1996). Under most circumstances, though, enhanced sun exposure favors fruit maturity and quality. For example, the methoxypyrazine content in 'Cabernet Sauvignon' and 'Sauvignon blanc' grapes is reduced by increased sun exposure. This is most pronounced at moderately high radiation levels (Marais *et al.*, 2001). Sun exposure also tends to promote wine flavor. This is frequently associated with enhanced production of monoterpenes, norisoprenoids, and anthocyanins.

In areas of high rainfall and low evaporation, a large exposed leaf canopy increases transpiration. The occasional development of limited water deficiency can induce termination of shoot growth and favor fruit ripening. Divided canopies also can increase yield without quality loss as more buds can be retained without producing excessive within-canopy shading. In addition, the retention of more buds can restrict excessive vine vigor, a problem in many fertile vineyards. High shoot number limits shoot growth, decreases internode length, leaf number, leaf size, and lateral shoot activation – features likely to create a favorable canopy microclimate. Care must be taken, however, not to overstress the vine. Excessive yield, generated by retaining too many buds, can reduce fruit quality and shortened vine life.

Although divided canopies are often efficient and valuable means of restricting vine vigor, this is inappropriate in low-nutrient soils. Also, large exposed canopies could cause excessive water deficit in areas where rainfall or irrigation is limited.

The complex trellis required for divided-canopy systems is expensive, but is usually offset by increased yield and improved fruit quality. Furthermore, canopy division can be a more cost-effective means of achieving many of the benefits of increased vine density. Planting vines, especially grafted cuttings, is often the major expense in establishing a vineyard.

On the other hand, undivided canopies are usually simpler and less expensive to develop and maintain. They are also more common and have the sanction of centuries of use. Prior to the use of mechanized cultivation, irrigation, fertilization, effective disease control, and cultivation on rich soils, excessive vigor was rarely a problem. Correspondingly, favorable grape exposure could be obtained with undivided canopies using hedging, severe pruning, and high-density planting. However, high-quality fruit came at the price of low yield.

Where the production of quality wine at a premium price can be justified, old techniques can be commercially viable. In most viticultural regions, market forces require the use of techniques that optimize production and minimize costs. In these situations, measures that direct vine energy toward enhanced grape yield and quality are crucial. Where new vineyards are being planted, or old vineyards are being replanted, the use of divided-canopy training systems is typically a prudent move. In existing vineyards, leaf and lateral-shoot removal or hedging are more immediate and less costly means of achieving the same goals.

CANOPY HEIGHT

Trunk height is one of the more obvious features of a training system. Training is considered to be low if the principal arm is less than 0.6 m (2 ft) above ground level, standard if between 0.6 and 1.2 m (2–4 ft), and high if above 1.2 m (4 ft). Arbors often have trunks 2–2.5 m (6–7 ft) high.

Low trunks have been used most commonly in hot dry climates. Combined with spur pruning and head training, short bushy vines are produced (Plate 4.2). This form tends to equalize daily leaf and fruit light exposure over the vine, thereby minimizing drought stress and fruit sunburn. If the vines are densely planted, soil shading may reduce water loss from the soil and limit soil heating. Once the soil is heated, the vine is close to heat radiated from the soil. Closeness to the soil can, however, minimize day–night temperature fluctuation, and be an important heat source for grape maturation and frost protection in the fall. Low trunks (and the associated small vine size) can occasionally be used with benefit in cool climatic regions to minimize the portion of the vine damaged by freezing, as well as to limit vine vigor on poor soils. The location of the renewal portion of the vine close to the ground can potentially provide additional protection from a snow cover. By limiting the need for vine trellising, stout short trunks lower vineyard operating expenses.

Although low training is occasionally used in cool regions, for example in Chablis (see Fig. 10.15), high training systems are more common. By positioning the buds and shoots away from the ground, they are partially protected from cold air that accumulates at ground level. Another advantage of high training is the greater access that direct sunlight has to the ground in the spring. This can speed soil warming and encourage early vine growth.

Medium to high trunks are required for training systems with trailing shoots. High trunks reduce or eliminate the need for skirting long vegetative growth. Of greater significance, is the improved exposure of the leaves and fruit to both direct and diffuse sunlight. For this to be maximized, row width should be similar to canopy height and the rows should run north–south (Smart, 1973). The maximal ratio of exposed canopy surface area to land area in practice is about 2.2 (Smart and Robinson, 1991). High trellising also facilitates herbicide application (by positioning buds and shoots away from the ground). Furthermore, trunks of moderate height (1–1.4 m) make most manual and mechanized vineyard practices easier, by locating the canopy about breast height. Finally, high trunks possess a large woody structure that permits improved nutrient storage. More perennial structure promotes earlier bud burst, affects composition and °Brix, as well as increasing yield (Howell *et al.*, 1991; Koblet *et al.*, 1994). However, by raising the canopy, greater stress is placed on the trunk. Thus, stronger trellising and wiring become necessary.

TRUNK NUMBER

In most situations, the vine possesses a single trunk. This may be subsequently divided at the apex to form two or more cordons. In some cold climates, such as the northeastern United States, two or more trunks may be established.

Dual and multiple trunks have several advantages in cold regions. They provide "insurance" against cold

Table 4.3 Effect of plant spacing on the yield of 3-year-old 'Pinot noir' vines

Plant spacing (m)	Vine density (vine/ha)	Leaf area (m/vine)	Leaf area (cm²/g grape)	Yield (kg/vine)	Yield (kg/ha)	Wine color (520 nm)
1.0×0.5	20,000	1.3	22.03	0.58	11.64	0.875
1.0×1.0	10,000	2.7	26.27	1.03	10.33	0.677
2.0×1.0	5000	4.0	28.25	1.43	7.15	0.555
2.0×2.0	2500	4.0	15.41	2.60	6.54	0.472
3.0×1.5	2222	4.5	18.01	2.50	5.51	0.419
3.0×3.0	1111	6.3	15.36	4.12	4.57	0.438

Source: Data from Archer and Strauss, 1985, Archer, 1987, and Archer *et al.*, 1988

injury or various trunk infections. If one trunk dies, the other(s) sustain vine growth. Also, by dividing the energy of the vine between several trunks, each trunk grows more slowly and remains more flexible. This is important if the vine needs to be removed from its trellis and buried as a defense against frigid winter weather.

PLANTING DENSITY AND ROW SPACING

Although not directly an aspect of training, the effect of planting density on row spacing often significantly influences the choice of training system. Both planting density and row spacing have marked effects on the economics of vineyard operation, as well as influencing vine growth (Hunter, 1998). Examples of different planting densities are illustrated in Plates 4.3 and 4.4.

In practice, it is often difficult to separate the direct effects of planting density, such as competition for water, from indirect influences on canopy microclimate, or from the impact of unrelated factors such as the training system and soil structure. The situation is further complicated by vineyards having the same average planting density, but different between- and within-row vine spacing. Thus, vines having similar *average* soil volumes may experience markedly different degrees of root and shoot crowding. These factors help explain the diversity of opinion on the relative merits of various planting densities.

Planting densities commonly used in Europe have changed considerably since the 1850s. Before the phylloxera epidemic, planting densities occasionally reached 30,000 to 50,000 vines/ha (Freese, 1986; Champagnol, 1993). Values for narrow-row plantings in Europe tend to vary between 4000 and 5000 vines/ha, occasionally rising to above 10,000 vines/ha. In California and Australia, common figures for wide-row plantings range from about 1100 to 1600 vines/ha (2700–4000 vines/acre).

Vineyards planted at higher vine densities often, but not consistently, show desirable features, such as improved grape yield and wine color (Table 4.3). The lower productivity of individual vines is usually more

than compensated by the higher photosynthetic efficiency of the greater number of vines. Improved grape quality is usually explained in terms of limited vegetative vigor (a lower level of bud activation and restricted shoot elongation) and the improved canopy microclimate that results in better flavor and wine color. These benefits are similar to those of canopy division – vine devigoration, enhanced light exposure, and improved air flow in and around grape clusters. What canopy division lacks is the "prestige" of traditional use in Europe.

One of the advantages of high-density planting is a foreshortening of the time taken for a vineyard to come to full production. This suggests that some aspect of intervine competition is involved. Increased vine (and bud) numbers per hectare also may provide some protection against yield loss due to winterkill. Although little noted in winery promotional literature, increased yield can be a significant factor encouraging its use.

The suppression of vegetative vigor associated with dense planting may result from root competition. Shoot and root growth are strongly interrelated. Dense planting promotes deeper root penetration, but restricts lateral extension (Table 4.4). The proportions of fine, medium, and large roots generally are unaffected by planting density. Although root production per vine is reduced, high-density planting increases overall root density (Archer and Strauss, 1989) and, thus, total soil volume used (Hunter *et al.*, 1996). This can be of considerable value where vines are planted on hillsides, and in less fertile deep soils under dry conditions. However, high-density planting may result in excessive water deficit in shallow soils. Moderate water deficit between berry set and *véraison* may enhance grape quality by initiating early cessation of vegetative growth. In contrast, low-density conditions promote a more extensive, but generally shallower root system. Low-density plantings are preferable on deep, fertile soils with an adequate water supply, but with sufficient water deficit to restrain excessive vegetative growth (Archer, 1987).

The major disadvantage of dense vine planting is a marked increase in vineyard establishment cost.

Table 4.4 The effect of plant spacing on the root pattern of 3-year-old vines of 'Pinot noir' on '99 R' Rootstock

Parameter	Plant spacing (m)					
	3×3	3×1.5	2×2	2×1	1×1	1×0.5
Primary roots (m)	2.21 (37%)	1.76 (38%)	1.67 (35%)	1.63 (39%)	1.09 (37%)	0.89 (36%)
Secondary roots (m)	2.99 (50%)	2.31 (49%)	2.58 (53%)	1.95 (47%)	1.38 (47%)	1.12 (46%)
Tertiary roots (m)	0.77 (13%)	0.61(13%)	0.59 (12%)	0.56 (14%)	0.46 (16%)	0.46 (18%)
Total root length (m)	5.96×10^3	4.68×10^3	4.84×10^3	4.13×10^3	2.93×10^3	2.45×10^3
Root density (m/m³)	1.10×10^3	1.73×10^3	2.02×10^3	3.44×10^3	4.89×10^3	8.21×10^3
Angle of penetration	15.3°	22.6°	30.9°	41.1°	58.6°	77.5°

Source: From Archer and Strauss, 1985, reproduced by permission

The expense of planting grafted vines, even at low density, can exceed the cost of all other aspects of vineyard development. Thus, the expense of planting at high density may offset the potential benefits of increased yield and quality. In addition, improved grape quality is not always observed (Eisenbarth, 1992). As already noted, the yield/quality equation is not simple or direct (Chapman *et al.*, 2004a, 2004b; Chapman *et al.*, 2005). Because increased planting density usually involves the use of narrow rows, additional expense may be incurred in the purchase of special narrow-wheel-base equipment, specifically when shifting from low- to high-density planting. Close planting complicates soil cultivation and may increase the need for herbicide use. The requirement for more severe and precise pruning can further add to the maintenance costs of high-density vineyards. Increased expenditures also can result from treating more vines per hectare with protective chemicals. Finally, deep rich soils may counteract the devigoration produced by vine competition.

A significant feature favoring the retention of wide-row planting, which typify most New World vineyards, is its adaptation to existing vineyard machinery. With new training systems, widely spaced vines can achieve or surpass the yield and quality of traditional, narrow-row, dense plantings. These features are achieved at lower planting costs, both initially and during replanting. Large vines also appear to live longer than the smaller vines characteristic of densely planted vineyards. Finally, the more extensive root system of large vines may limit the development of water deficit during dry spells.

The major disadvantage of wide-row spacing is its potential for shoot crowding and poor canopy microclimate. In addition, doubling vine row spacing may double the amount of sunlight impacting the ground rather than the vine canopy (Pool, 2000). As noted later, most of these disadvantages can be limited or offset by various vineyard procedures, and the greater vine capacity directed to economic advantage.

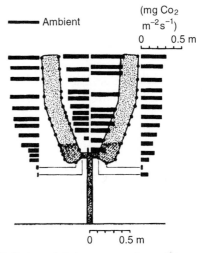

Figure 4.17 Estimated daily average values of photosynthetic rate for leaves at the canopy exterior based on PAR measurements. Note that leaves at the bottom of the canopy show low rates due to shading. (From Smart, 1985, reproduced with permission)

ROW ORIENTATION

Many factors can influence row orientation (Intrieri *et al.*, 1996), but the slope of the land is often the major determinant. To minimize soil erosion, it is advisable to position rows perpendicular to the slope. Thus, slope can significantly influence the normally preferred north–south alignment.

Where possible, row orientation is positioned to achieve maximal canopy exposure to direct sunlight. For most vertical canopy-training systems, this is obtained by a north–south orientation. This is especially true at higher latitudes and with most narrow-canopy systems. The north–south alignment exposes the canopy's largest surface area (the sides) to direct sunlight during the mid-morning and midafternoon hours (Fig. 4.17), when light intensity is optimal for photosynthesis. Maximal light intensity (at noon) falls on the canopy top. Although the intensity is often greater than surface leaves can utilize, second- and possibly third-layer leaves may still be able to photosynthesize effectively at midday.

Nevertheless, angling rows toward the southwest can improve photosynthesis by increasing exposure to the early morning sun. Conversely, a southeast angling could improve fruit heat accumulation in the autumn. Conditions such as the prevailing wind direction also can influence optimal row orientation.

Canopy Management and Training System Development

In the discussion of training systems above, several divided-canopy systems were mentioned. These developed out of fundamental studies on vine microclimate. The findings have been condensed by Smart *et al.* (1990b) into a series of canopy-management principles:

1. The rapid development of a large canopy surface area/volume ratio increases photosynthetic efficiency, as well as fruit set and ripening. Tall thin vertical canopies aligned along a north–south axis permit maximal sun exposure.

2. To avoid both excessive interrow shading and energy loss by insufficient canopy development, the ratio of canopy height to interrow width between canopies should approximate unity.

3. Shading in the renewal or fruit-bearing zone of the canopy should be minimized. Shading has several undesirable influences on fruit maturation and health. These include augmented potassium levels, increased pH and herbaceous character, retention of malic acid, enhanced susceptibility to powdery mildew and bunch rot, and reduced sugar, tartaric acid, monoterpene, anthocyanin, and tannin levels. Shading also reduces inflorescence initiation, favors primary bud necrosis, suppresses fruit set, and slows berry growth and ripening. There is no precise indication of the level of shading at which undesirable influences begin.

4. Excessive and prolonged shoot growth, causing a drain on the carbohydrate available for fruit maturation and vine storage, should be restrained by trimming or devigoration procedures (see later). There should be no vegetative growing point activity between *véraison* and harvest.

5. The location of different parts of the vine in distinct regions not only favors uniform growing conditions and even fruit maturation, but also facilitates mechanized pruning and harvesting practices.

These principles have also been combined into a training system ideotype (Table 4.5) and a vineyard score sheet (Smart and Robinson, 1991). These assess features such as the termination of shoot growth after *véraison* (an indicator of moderate vegetative vigor and slight water and nutrient deficit) and exposure of the fruit to the sun. Vineyard scoring has proven valuable, in association with wine assessment, in quantifying vineyard practices that are the most significant in defining grape quality.

CHOICE OF TRAINING SYSTEM

Training systems obviously should be chosen on the basis of a serious assessment of the limitations imposed by the climate. In most established regions, there are usually a number of alternative systems that have been used for years, or are under active evaluation. Several modern training systems, devised to solve particular problems, are noted below. However, in new viticultural regions, without clearly recognizable equivalents, designing a training system to match local conditions may be worthwhile. While a novel and experimental approach, it offers the possibility of developing a system specifically designed to address local conditions. It could involve selecting choices based on the principles noted above. This is probably preferable to attempting to adjust existing systems to situations distant from those for which they were originally designed.

Selected Training Systems

Smart and Robinson (1991) provide an exhaustive discussion of training systems designed to improve canopy management, as well as yield. What follows below is a brief discussion of some of these systems. Regrettably, the literature provides few experimental comparisons of these or other systems. The work of Wolf *et al.* (2003) is a welcome exception.

VERTICAL SHOOT POSITIONING (VSP)

Vertical shoot positioning (VPS) (Plate 4.5) refers to a group of training systems popular and widely used in Europe and elsewhere. They possess undivided canopies that resemble hedgerows. They are particularly useful with vines of low to medium vigor; have vines arranged in narrowly spaced rows (1.5–2 m apart); and are particularly valuable in regions prone to fungal disease. Where the vines are cane pruned, four canes are usually retained. Pairs of canes are directed in opposite directions along two support wires, typically positioned about 0.2 m apart. If the vines are spur pruned, two cordons are directed in opposite directions along a single support wire. The shoots are trained upwards using two foliage wires. The shoots are trimmed at the top and frequently along the sides. The result is a hedge some 0.4–0.6 m wide and about 1 m high.

VSP systems position the fruit in a common zone about 1–1.2 m above the ground. This eases most vineyard activities, such as mechanical harvesting and

Table 4.5 Canopy characteristics promoting improved grape yield and quality

Character assessed	Optimal value	Justification of optimal value
Canopy characters		
Row orientation	North–south	Promotes radiation interception (Smart, 1973, although Champagnol (1984) argues that hourly interception should be integrated with other environmental conditions (i.e., temperature) that affect photosynthesis to evaluate optimal row orientation for a site; wind effects can also be important (Weiss and Allen, 1976a, 1976b)
Ratio of canopy height to alley width	~1:1	High values lead to shading at canopy bases, and low values lead to inefficiency of radiation interception (Smart *et al.*, 1990b)
Foliage wall inclination	Vertical or nearly so	Underside of inclined canopies is shaded (Smart and Smith, 1988)
Renewal/fruiting area location	Near canopy top	A well-exposed renewal/fruiting area promotes yield and, generally, wine quality, although phenols may be increased above desirable levels
Canopy surface area	~21,000 m²/ha	Lower values generally indicate incomplete sunlight interception; higher values are associated with excessive cross-row shading
Ratio of leaf area to surface area	<1.5	An indication of low canopy density is especially useful for vertical canopy walls (Smart, 1982; Smart *et al.*, 1985)
Shoot spacing	~15 shoots/m	Lower values are associated with incomplete sunlight interception, higher values with shade; optimal values is for vertical shoot orientation and varies with vigor (Smart, 1988)
Canopy width	300–400 mm	Canopies should be as thin as possible; values quoted are minimum likely width, but actual value will depend on petiole and lamina lengths and orientation
Shoot and fruit characters		
Short length	10–15 nodes, ~600–900 mm length	These values are normally attained by shoot trimming; short shoots have insufficient leaf area to ripen fruit, and long shoots contribute to canopy shade and cause elevated must and wine pH
Lateral development	Limited, say, less than 5–10 lateral nodes total per shoot	Excessive lateral growth is associated with high vigor (Smart *et al.*, 1985, 1990b; Smart and Smith, 1988; Smart, 1988)
Ratio of leaf area to fruit mass	~10 cm²/g (range 6–15 cm²/g)	Smaller values cause inadequate ripening, and higher values lead to increased pH (Shaulis and Smart, 1974; Peterson and Smart, 1975; Smart, 1982; Koblet, 1987); a value around 10 is optimal
Ratio of yield to canopy surface area	1–1.5 kg fruit/m² canopy surface	This is the amount of exposed canopy surface area required to ripen grapes (Shaulis and Smart, 1974); values of 2.0 kg/m² have been found to be associated with ripening delays in New Zealand, but higher values may be possible in warmer and more sunny climates
Ratio of yield to total cane mass	6–10	Low values are associated with low yields and excessive shoot vigor; higher values are associated with ripening delays and quality reduction
Growing tip presence after *véraison*	Nil	Absence of growing tip encourages fruit ripening since actively growing shoot tips are an important alternate sink to the cluster (Koblet, 1987)
Cane mass (in winter)	20–40 g	Values indicate desirable vigor level: leaf area is related to cane mass, with 50–100 cm² leaf area/g cane mass, but values will vary with variety and shoot length (Smart and Smith, 1988; Smart *et al.*, 1990a)
Internode length	60–80 mm	Values indicate desirable vigor level (Smart *et al.*, 1990a) but will vary with variety
Ratio of total cane mass to canopy length	0.3–0.6 kg/m	Lower values indicate canopy is too sparse, and higher values indicate shading; values will vary with variety and shoot length (Shaulis and Smart, 1974; Shaulis, 1982; Smart, 1988)
Microclimate characters		
Proportion of canopy gaps	20–40%	Higher values lead to sunlight loss, and lower values can be associated with shading (Smart and Smith, 1988; Smart, 1988)
Leaf layer number	1–1.5	Higher values are associated with shading and lower values with incomplete sunlight interception (Smart, 1988)
Proportion of exterior fruit	50–100%	Interior fruit has composition defects
Proportion of exterior leaves	80–100%	Shaded leaves cause yield and fruit composition defects

Source: From Smart *et al.*, 1990b, reproduced by permission

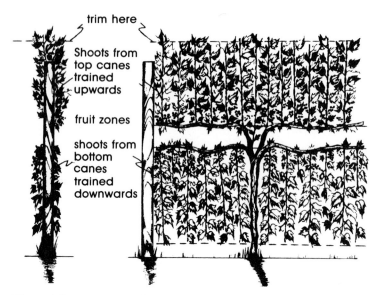

Figure 4.18 An example of a cane-pruned Scott Henry system. (From Smart and Robinson, 1991, reproduced by permission)

selective fruit spraying. As long as the vines are not overly vigorous, fruit shading is usually not a serious problem. Where shading is likely to lower fruit quality, basal leaf removal is facilitated by the fruit zone being at chest height. Improved light and air exposure, and an efficient canopy generally favor good to excellent fruit yield and quality.

SCOTT HENRY AND SMART-DYSON SYSTEMS

The Scott Henry is a specific variant of a VSP system. It may be either cane- or spur-pruned. With cane pruning, four canes are retained, two each attached to an upper and lower wire. These are positioned in opposite directions along the fruit wire (Fig. 4.18). It differs from standard VSP systems by directing the shoots from the upper canes upward, while directing those from the lower canes downward. With spur pruning, cordons from adjacent vines are trained alternately to the upper and lower support wires. Upward positioned spurs are retained on the higher cordons (the direction in which the shoots will be directed), whereas downward located spurs are retained on the lower cordons. Several foliage wires hold the shoots in position.

The Scott Henry system generates a vertically divided canopy with the fruit-bearing zones of the upward and downward directed canopies adjacent to each other. This increases by about 60% the effective canopy surface area (and reduces its density) compared with simple hedge row or VSP systems. These benefits are obtained by directing half of the shoots downward. The division also improves light exposure and air circulation around the crop. The conditions are sufficiently similar in both canopies to ripen essentially simultaneously. The

devigoration imposed on the lower canopy is of especial value for vines of medium vigor. As a consequence, photosynthate otherwise consumed in unnecessary vegetative growth can be directed into enhanced production of high-quality fruit. Improved sugar content and reduced acidity are combined with increased yield.

The Scott Henry Trellis, as with other VSP systems, is most effective when associated with vine rows between 1.5 and 2 m apart. It is well adapted to use with conventional mechanical harvesters. Its conversion from other systems is comparatively simple.

The Smart–Dyson Trellis is a modification of the Scott Henry Trellis (Smart, 1994b). It involves using a single cordon to generate both the upward- and downward-facing canopies. These are derived from upward- and downward-facing spurs, respectively. The timing and manner of shoot positioning are essentially the same as the Scott Henry. The primary advantage of the Smart–Dyson modification is its adaptation to mechanical pruning – all the cutting can be conducted in the same plane. Training costs are also somewhat less.

For spur-pruned, cordon-trained vineyards desiring an inexpensive retrofit to a divided-canopy system, Smart (1994a) suggests the **Ballerina** modification of the Smart–Dyson. Because the cordons do not possess buds originating on the underside, some of the shoots are trained upward, whereas others are trained outward and downward.

GENEVA DOUBLE CURTAIN (GDC)

The Geneva Double Curtain (GDC) was the first training system based primarily on microclimate analysis (Shaulis *et al.*, 1966) (Fig. 4.14B). It is a tall (1.5–1.8 m), bilateral, cordon system pruned to short spurs (four to six buds). These are directed downward. The cordons diverge laterally and then bend, to be held about 1.2 m apart by parallel wires running along the row. Alternately, two short lateral cordons are pruned to four long canes. The latter are supported on wires. The bearing shoots of upright-growing *V. vinifera* cultivars must be positioned downward about flowering time with the aid of movable catch wires. The system works well for vines of medium to high vigor, and with rows about 3–3.6 m apart. This provides interrow canopy spacings of about 2.4 m. Some further shoot positioning during the season may be necessary to keep the two canopies separate and to minimize shading.

Initially developed for *V. labrusca* varieties, such as 'Concord,' the GDC has been used with French-American hybrids and *V. vinifera* cultivars in several parts of the world. Consistent with its divided canopy and increased bud retention, fruit quality is often excellent and yield is enhanced. Although the GDC system demands more in terms of skill and materials, the higher yield (excellent sun exposure for inflorescence induction) usually more than offsets the higher establishment costs. The use of hinged side supports on the trellis can easily permit the canopies to be pulled toward the post, facilitating mechanical pruning and harvesting (Smith, 1991).

Because the GDC positions the fruit-bearing or renewal zone at the apex of the canopy, it is especially valuable where maximal direct-sun exposure is desired, as in regions with considerable cloud cover during the summer. In some regions, however, this can result in increased fruit sunburn, hail injury, and bird damage. Locating the arms on the upper portion of the cordon or less shoot arching may increase protection of the fruit by the leaves. Because trailing shoots need little support, GDC has the lowest wiring costs of any divided-canopy system.

LYRE OR U SYSTEM

The Lyre system is a divided-canopy system appropriate for vines of medium vigor (Carbonneau, 1985). The system consists of a short trunk branching into bilateral cordons that diverge laterally, and then bend along parallel cordon wires positioned about 0.7 m apart (Plate 4.6). The bearing wood consists of equidistantly positioned spurs. The shoots are trained to two inclined trellises, supported by fixed and movable catch wires. Rows are placed about 3–3.6 m apart, with about 2.4 m separating vines within the row. The Lyre training system has been described as an inverted GDC. Trimming excessive growth may be needed to keep the canopies separate and minimize basal shading.

The inclined canopies are ideally suited for maximizing direct sun exposure in the early morning and late evening, when photosynthetic efficiency is at a maximum. However, this advantage comes at the cost of extensive shading at the exterior base of the canopy by the overhanging inclined vegetation. In addition, its establishment costs are higher than standard wide-row systems, including an increase in the number of foliage wires required.

The Lyre system disperses capacity over a larger canopy. It generally provides increased yield at equal or better quality than that produced using more traditional, dense plantings, such as with the double Guyot (Carbonneau and Casteran, 1987; Carbonneau, 2004). In Bordeaux, the value of the Lyre system is particularly evident on less-favored sites and during poorer vintage years. These advantages may arise from the increased canopy size and the beneficial microclimate generated. Also, it has been noted that Lyre-trained vines are less susceptible to winter injury and inflorescence necrosis (*coulure*). The system is amenable to both mechanical harvesting and pruning, with adjustment of existing equipment.

When vines are converted from vertical training to the Lyre system, there is no concomitant increase in root volume (Hunter and Volschenk, 2001). The result is an increase in cordon length to root volume ratio. Total cane growth increases although individual canes are shorter.

RUAKURA TWIN TWO TIER *(RT2T)*

The Ruakura Twin Two Tier (RT2T) system was specifically developed for high-fertility conditions. It differs from the previous systems by dividing the canopy both vertically and laterally (Fig. 4.19). Each cordon bends along and is supported by wires running parallel to the row (Plate 4.7). Spur pruning facilitates both equal and uniform distribution of the canopies along the row. The RT2T system is compatible with mechanical pruning. The vertical canopy division is achieved by training alternate vines high and low, to higher and lower cordon wires, respectively. This is necessary to avoid gravitrophic effects on growth in the individual vines, where buds positioned higher on a vine tend to grow more vigorously than those nearer the ground. Rows are placed 3.6 m apart. As the between-row canopies are positioned 1.8 m apart, the same as the within-row canopies, all canopies are equally separated. Also, as the combined height of the two vertical canopies (tiers) is equivalent to the width between the canopies, the ratio of canopy height to interrow canopy separation is unity. Individual vines are planted about 2 m apart in the rows.

To limit shading of the lower tier by the upper tier, a gap of about 15 cm (6 in) is maintained between the canopies by trimming. An alternative technique places the two cordons of each tier about 15 cm apart, with

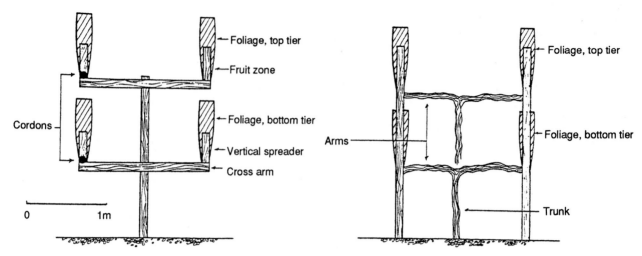

Figure 4.19 Two versions of support systems for the Ruakura Twin Two Tier training system. (From Smart *et al.*, 1990c, reproduced by permission)

the upper canopy trained upward and the lower trailing downward. By positioning the fruit-bearing regions of both tiers under approximately the same environmental conditions, chemical differences between the fruit of both tiers are minimized.

The advantages of RT2T training involve a high leaf surface area to canopy volume ratio and extensive cordon development. The former favors the creation of limited water deficit that helps restrict shoot growth following blooming. As a result, most of the photosynthate is available for fruit development or storage. The formation of 4 m of cordon per meter of row provides many well-spaced shoots per vine. These further act to limit vine vigor by restricting internode elongation and leaf enlargement, thereby lessening canopy shading. Increased shoot numbers also enhance vine productivity, and the improved canopy microclimate maintains or improves grape quality. Because of the strong vigor control provided by RT2T training, it is particularly useful for vigorous vines grown on deep, rich soils with an ample water supply.

Although RT2T systems are more complex and expensive to establish than more traditional systems, wide row spacing limits planting costs. Also, the narrow vertical canopies ease mechanical pruning and harvesting, and increase the effectiveness of protective chemical application.

TATURA TRELLIS

Several training systems have been developed from canopy management principles designed for tree fruit crops. An example is the modified Tatura Trellis (van den Ende, 1984). It possesses a 2.8 m high, V-shaped trellis, arranged with support wires to hold six tiered, horizontally arranged cordons on both inclined planes. Each vine is divided near the base into two inclined

trunks. Each trunk gives rise to six short cordons, three on each side that run parallel to the row, or the vines are trained alternately high and low to limit gravitrophic effects while still providing six cordon tiers. The vines are spur-pruned. A third placement system consists of using the bilateral trunks directly as inclined cordons (van den Ende *et al.*, 1986). The vines are then pruned with alternate regions of the trellis used for fruit and replacement shoot development.

In the Tatura Trellis system, the vines are densely planted at one vine per meter along the row, with the rows spaced 4.5 to 6 m apart. Because of root competition between the closely spaced vines, and the large number of shoots developed, excessive vigor is restricted, shading is limited, and fruit productivity is increased. The Tatura Trellis favors the early development of the fruiting potential of the vine.

The most serious limitation of the Tatura Trellis is its tendency to concentrate fruit production in the upper part of the trellis. In addition, the vine must be trained, pruned, and harvested manually, as the trellising is unsuited to mechanical harvesting and pruning.

MINIMAL PRUNING (MPCT)

Problems with the tendency to overprune, combined with a desire to reduce production costs, led to the development of the minimal pruning system (Plate 4.8). Without pruning, many cultivars regulate their own growth and yield good-quality fruit. Although vines may overcrop or undercrop in the first few years, especially young vines, this typically ceases by the fifth year. Spontaneous dehiscence of immature shoots largely eliminates the need for pruning old growth.

Minimally pruned vines produce more but smaller shoots; possess fewer, more closely positioned nodes; and develop smaller, paler leaves. Nevertheless, net photosynthesis and carbon gain are significantly higher than with other systems, for example VSP (Weyand and Schultz, 2006). The more open canopy also reduces the incidence of fungal diseases, notably bunch rot. Most cultivars maintain their shape and vigor when minimally pruned. Vines not already cordon-trained are so developed on a high (1.4–1.8 m) single wire. Summer trimming is limited to a light skirting along the sides and bottom, as deemed necessary to facilitate machinery movement.

Crop yield is either sustained or considerably enhanced, depending on the variety, clone, and rootstock employed. The fruit is carried on an increased number of bunches, each containing fewer and smaller berries. Commonly, the fruit is borne uniformly over the outer portion of the vine, in well-exposed locations. Fruit maturity is generally delayed about one to several weeks. Grape soluble solids may be slightly reduced, pH decreased, and acidity increased (McCarthy and Cirami, 1990). Fruit color in red cultivars is generally diminished slightly, but this may be offset by mechanical thinning – passing a mechanical harvester through the vineyard about a month after flowering. This involves removal of the lower shoots and fruit (Kidd, 1987). The effect is to reduce yield by about 30%, increase soluble solids, improve acidity, elevate the proportion of ionized anthocyanins, and enhance the color density of the resultant wine (Clingeleffer, 1993; Petrie and Clingeleffer, 2006). Mechanical thinning is commercially used for premium quality wine production. Data from trials in eastern North America are given in Fendinger *et al.* (1996).

The enhanced yield of minimally pruned vines may result from nutrients stored in the shoots (not lost as a result of pruning). The ready availability of nutrients may also explain the rapid completion of canopy development following bud burst. This helps limit leaf–fruit competition during berry development. An additional advantage of minimal pruning comes from easier fruit removal during mechanical harvest. Finally, fewer leaves are removed along with the fruit because of the more flexible canopy.

An unexpected benefit of minimal pruning has been a reduction in the incidence of diseases such as Eutypa dieback. The elimination of most pruning wounds, especially those on wood 2 years old or older, decreases incidence of the disease. However, the retention of canes infected with *Phomopsis viticola* may lead to an increased incidence of Phomopsis cane and leaf spot, a rare disease in conventionally pruned vineyards (Pool *et al.*, 1988).

Minimal pruning appears to be best suited to situations in which the vines are moderately vigorous and grown in dry climates. The system appears to be less suitable to vines grown on poor soils, arid conditions

without the option of irrigation, or in cool, wet climates. In cool regions, ripening may be critically delayed and vine self-regulation less pronounced. When used in appropriate climates, the fruit produce well-balanced wine, although occasionally lighter in color than those derived from vines trained and pruned traditionally.

Outside of Australia, minimal pruning has found particular favor with *V. labrusca* growers in the eastern USA. Vines initiate growth earlier, producing canopies of similar size, but significantly in advance of heavily pruned vines (Comas *et al.*, 2005). They also produce more fruit. The more efficient use of light appears to be the source of increased fruit productivity. Root development tends to be more shallow, initiate earlier, and is more extensive (by up to 24%) than in vines balance-pruned. Success has also been found in Germany with 'Riesling' and 'Pinot noir' (Schultz *et al.*, 2000).

Although different in several aspects, all the training systems mentioned are designed to direct the benefits of improved plant health and nutrition toward enhanced fruit yield and quality. These goals are achieved by reducing individual shoot vigor and increasing sun and air exposure. The simpler divided-canopy systems are more suitable for vines of lower capacity (~0.5 kg/m prunings), whereas the more complex systems are more appropriate for vines of high capacity (~1.5 kg/m prunings). Although losing their advantages under situations of marked water and disease stress, poor drainage, or salt buildup, divided-canopy systems provide long-term economic benefits in several situations. Whether the yield and quality improvements justify their additional costs will depend on the profit derived when compared with more traditional procedures.

Ancient Roman Example

Although much has been learned about training systems in the past few decades, it is somewhat humbling to realize how many of our present "discoveries" were known to the ancients. What follows is a short digression into Roman viticulture.

The only precise descriptions of ancient viticultural practice come from the writing of Roman authors, such as Columella, Pliny the Elder, Varro, and Cato. They often give detailed instructions on how to plant vineyards, providing distances between vines, advice on training systems and pruning, suggested yields, as well as specifics on how to get the best out of one's slaves. Until the late 1960s, no vineyards had been discovered that could give confirmation of the suggestions provided by these Roman authors.

In 1968, Jashemski began excavating a site in Pompeii. Her work uncovered the largest known vineyard in the ancient world (Jashemski, 1968, 1973). The eruption of

Vesuvius, in late August 79 A.D., buried the city under lapilli and volcanic ash, forming a time-capsule of Roman life. The site of the vineyard (*Foro Boario*) was first investigated in 1755, but mistakenly identified. Only some two hundred years later did Jashemski recognize the actual function of the site. Her study has revealed how closely vineyard layout followed the directions noted by Roman authors.

The vineyard is comparatively small and could have been worked by a single family. The site also contains a press room, fermentation area, as well as a wine shop and portico, where its produce was sold, probably to patrons of the nearby amphitheater. The site also included two *triclinia* (stone lounges), associated with relaxed outdoor dining in the Roman manner.

The vineyard covered about 1.4 acres (0.56 ha) and was planted with about 4000 vines (~7200 vines/ha) (Fig. 4.20). The vines were planted about 4 Roman feet apart (~1.2 m). Although Columella suggested rows 5 ft apart for hand cultivation (at least 7 ft for cultivation with oxen and a plow), Pliny the Elder proposed 4 ft spacing on rich volcanic soils like Pompeii. Each vine was supported by a stake averaging 2.5–5.5 cm in diameter. The vines were probably pergola trained. This system was highly recommended by ancient authors for hot, dry climates such as Pompeii. It is still used in some regions of Italy. This, and other training systems used by the Romans, are illustrated in Jashemski (1973). Most vines also showed several depressions around the vine, presumably designed to catch rain.

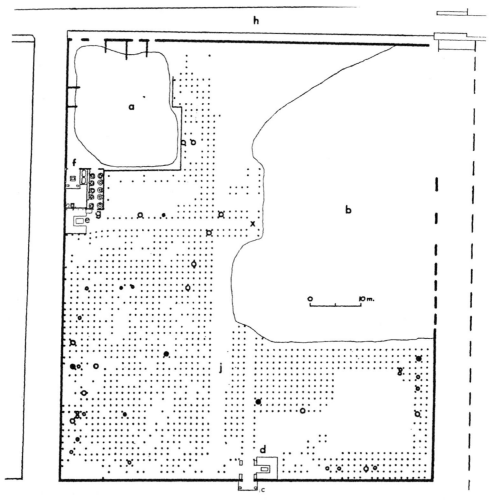

Figure 4.20 Vineyard at Pompeii. (a,b) unexcavated areas; (c) south entrance; (d,e) masonry triclinia; (f) room with wine press; (g) shed with ten dolia embedded in ground; (h) Via dell'Abbondanza; (i) Sarno gate; (j) unexcavated backfill; (k) path along north wall; (m) wine shop and portico; (x) intersection of paths. Dots indicate grapevine roots; small circles, small tree roots 10 cm or less in longest diameter; large circles, medium tree roots; large black circles, large tree roots 30 cm or more in longest diameter. (From Jashemski, 1973, reproduced by permission)

The vineyard was divided into quadrants by two paths between the vines (producing sections approximating those suggested by Columella). Not only did these facilitate transporting material in and out of the vineyard, but their slope acted as sites for water drainage during downpours. Because of the presence of large diameter roots along the paths, and the thicker stakes in this region, it is suspected that the paths also functioned as vine arbors. A third path occurred along the northern edge of the vineyard. All these features are found in recommendation by ancient authors.

Another fascinating feature illustrated by the vineyard is the implantation of trees. Not only was there a row of trees around the edges of the vineyard (usually between the second and third rows from the wall), but also a few trees randomly interspersed, and widely spaced trees between the first and second rows along the central paths. The trees were often fruit trees, such as fig, pear, plum, cherry or apple, as well as popular or willow (used as a source of stakes or tying the vines).

Although there is no evidence of the method or frequency of cultivation, ancient authors agree as to its importance. Frequent hoeing was strongly recommended to remove weeds and grass, their remains being left on the ground as a mulch. Columella specifically encouraged root-pruning every three years in established vineyards. Other "modern" approaches recommended by ancient authors include basal leaf removal, monoculture, and the removal of unripe grapes from clusters before fermentation.

If the ancient proprietor in Pompeii followed accepted practice, one can assume that each vine was pruned to two buds left on each of four spurs. This should have resulted in at least sixteen clusters per vine, generating about 16 tons of fruit. This equates to about 29 tons/ha, or some 160 hl of wine per ha. This estimate seems reasonable in terms of the volume of wine that could have been produced in the ten *dolia* (large earthenware containers imbedded into the floor of the *cella vinaria*). Each *dolium* could contain up to about 10 hl (the equivalent of 40 amphoras). Although this would be considered overcropping in modern terms, it is consistent with yields noted by Varro and Cato. It also would coincide with the recommendation of Columella that, on level land (not optimal for grape quality), quantity be aimed for.

In 1996, Piero Mastroberardino was permitted to replant several vineyard regions in Pompeii. The vines chosen were those presently grown in Campania and which most resemble those shown in frescos found in Pompeii. The principal varieties are two red cultivars, 'Sciascinoso' and 'Piedirosso.' They have been planted in a manner similar to that of the *Foro Boario* vineyard described above, and following instructions given by Pliny the Elder. Despite the ancient viticultural methods,

the wine produced from the vineyards (called Villa dei Misteri) is made to modern specifications.

Control of Vine Vigor (Devigoration)

As noted previously, the rich soils of many New World vineyards stimulate vegetative growth, to the potential detriment of fruit ripening. This feature may be accentuated by the use of vigorous rootstock, irrigation, fertilization, weed control, and the elimination of viral infections. The problem is not that the vines grow too well, but that too much of the potential has gone into extended vegetative growth – an excess pruned away at the end of the season. The intention of vigor control is to limit vegetative growth and redirect the improved capacity into increased yield and improved fruit ripening.

An old technique of restricting vine vigor is hedging. However, the effect expresses itself slowly and risks inducing excessive loss in capacity in vigorous vines. Another procedure used extensively in Europe has been high-density planting. Adjusting the type and breadth of a groundcover is another measure used to induce vine devigoration by limiting root growth. Root pruning, as noted previously, is another alternative. However, permanent vigor restriction can be achieved when scions are grafted onto devigorating rootstocks, such as '3309 Couderc,' '420 A,' '101-14 Mgt,' and 'Gloire de Montpellier.' In contrast, rootstock cultivars such as '99 Richter' and '140 Ruggeri' accentuate vine vigor (see Table 4.6).

Additional measures employed to restrain vigor are restricting nitrogen fertilization and irrigation. Limiting fertilization, notably nitrogen, minimizes vegetative growth, as does limited water deficit. For example, shoot growth can terminate more than 1 month early under conditions of water deficit (Matthews *et al.*, 1987). Water deficit also tends to have a more pronounced effect on reducing leaf growth on lateral shoots than on primary shoots (Williams and Matthews, 1990). The enhanced transpiration typical of divided-canopy systems can produce mild water deficit, which can arrest continued vegetative growth. Where applicable, promoting a trailing growth habit can also retard shoot elongation and restrain lateral shoot initiation.

Although soil type can indirectly affect vine vigor, choosing soil type is an option only when selecting a vineyard site. For example, stony to sandy soils provide sufficient, but restricted access to water and nutrients that can limit vegetative vigor.

Another alternative devigoration technique involves the application of growth regulators such as ethephon and paclobutrazol. Although effective, they may have undesirable secondary effects. For example, ethephon reduces

Table 4.6 Important cultural characteristics, other than resistance to phylloxera, of commercially cultivated rootstocks

			Propagation by			
Rootstock[a]	Vigor of grafted vine	Vegetative cycle	Cutting (rooting)	Bench grafting	Field grafting	Affinity with *V. vinifera*
'Rupestris du Lot'	xxxx	Long	xxx	xxx	xxx	xxxx
'Riparia Gloire'	xx	Short	xxx	xxx	xxx	xx
'99 Richter'	xxxx	Medium	xxxx	xxxx	xxxx	xxxx
'110 Richter'	xxx	Very long	xxx	xxx	xxx	xxxx
'140 Ruggeri'	xxxx	Very long	xxx	xxx	xxx	xxxx
'1103 Paulsen'	xxx	Long	xxx	xxx	xxx	xxxx
'SO 4'	xx	Medium	xx	xx	xx	xxx
'5 BB Teleki'	xx	Medium	xx	xx	xx	x
'420 A Mgt'	xx	Long	xx	xx	xxx	xx
'44–53 Malègue'	xxx	Medium	xxxx	xxxx	xxxx	xxxx
'3309 Couderc'	xx	Medium	xxx	xx	xxx	xx
'101–14 Mgt'	xx	Short	xxx	xx	xx	xx
'196–17 Castel'	xxx	Medium	xxx	xxx	xxx	xxx
'41 B Mgt'	xx	Short	x	xx	xxx	xxx
'333 EM'	x	Medium	x	xx	xxx	xxx
'Salt Creek'	xxxx	Very long	x	xx	xx	x

[a]Rootstocks that have proved insufficiently resistant to phylloxera, and for this reason abandoned nearly everywhere (e.g., '1202 C,' 'ARG,' and '1613 C'), are not included.

Source: From Pongrácz (1983), reproduced by permission. Summarized from data of Branas (1974), Boubals (1954, 1980), Cosmo *et al.* (1958), Galet (1971, 1979), Mottard *et al.* (1963), Pàstena (1972), Pongrácz (1978), and Ribérau-Gayon and Peynaud (1971)

the photosynthetic rate of sprayed leaves (Shoseyov, 1983).

Ideally, devigoration should be obtained by directing the potential for excessive vegetative growth into additional fruit production and improved grape quality. For example, the increased fruit yield associated with higher bud retention in divided-canopy and minimal-pruning systems limits vegetative growth. As long as a favorable canopy microclimate is developed, the increased fruit load has a good chance of maturing fully without adversely affecting subsequent fruitfulness and vine life span. However, the use of mechanical pruning in such techniques must be carefully watched. There is the possibility that weaker vines may be permitted to repeatedly overproduce. This can lead to extensive reserve loss, leading to vine death (Miller *et al.*, 1993).

Rootstock

The initial rationale for grafting grapevines was to check the destruction being caused by phylloxera in European vineyards. Although still the principal reason for grafting, rootstocks can also limit the damage caused by other soil factors. In addition, rootstock choice can modify scion attributes. Thus, rootstock selection offers the grower an opportunity to change varietal traits, without genetically modifying the scion.

Limiting the potential of rootstock use has been the difficulty in predicting how the two components will interact. Interaction results from the mutual translocation of nutrients and growth regulators between the scion and rootstock. A clear example of this interaction is the influence of particular rootstocks on scion vigor. More subtle examples include the induced susceptibility of *V. vinifera* scions to phylloxera leaf galling, when grafted onto rootstocks susceptible to leaf galling (Wapshere and Helm, 1987), and the reduction in sensitivity of some rootstocks to lime-induced chlorosis when grafted to particular *V. vinifera* scions (Pouget, 1987). In the latter instance, citric acid translocated to the roots from the scion enhances the formation of ferric citrate. This facilitates the transport of iron up to the leaves.

Because climatic and soil conditions can modify the expression of both rootstock and scion traits, their interaction can vary from year to year, and from location to location. Thus, although general trends can be noted in Tables 4.6 and 4.7 (see also Howell, 1987; Ludvigsen, 1999; Anonymous, 2003), the applicability of particular rootstocks with specific cultivars must be assessed empirically, and ideally in consultation with local growers and viticultural specialists.

In selecting a rootstock, ranking desired properties is necessary as each rootstock has its benefits and deficits. Selection cannot be taken lightly. Once a rootstock has

Table 4.7 Important cultural characteristics of commercially cultivated rootstocks

Rootstock	Adaptation to					Tolerance to		
	Humidity ("wet feet")	Dry shallow clay	Deep silt or dense loam	Deep, dry, sandy soil	Nematode resistance	Drought	Active lime (%)	Salt[a,b]
'Rupestris du Lot'	x	xx	xxx	x	xx	xx	14	0.7 g/kg
'Riparia Gloire'	xxx	x	xx	xx	xx	x	6	N/A
'99 Richter'	x	xx	xxxx	xx	xxx	xx	17	N/A
'110 Richter'	xxx	xxxx	xxx	xxx	xx	xxxx	17	N/A
'140 Ruggeri'	xx	xxx	xxx	xxxx	xxx	xxxx	20	N/A
'1103 Paulsen'	xxx	xxx	xxx	xxx	xx	xxx	17	0.6 g/kg
'SO 4'	xxx	x	xx	x	xxxx	x	17	0.4 g/kg
'5 BB Teleki'	xxx	xx	xx	x	xxx	x	20	N/A
'420 Mgt'	xx	xxx	xx	xx	xx	xx	20	N/A
'44–53 Malègue'	xxx	xx	xxx	xx	xxx	xx	10	N/A
'3309 Couderc'	xxx	xx	xx	xx	x	x	11	0.4 g/kg
'101–14 Mgt'	xxx	xx	xx	x	xx	x	9	N/A
'196–17 Castel'	xx	x	xx	xxx	x	xxx	6	N/A
'41 B Mgt'	x	x	x	x	x	xxx	40	Nil
'333 EM'	x	x	x	x	x	xx	40	Nil
'Salt Creek'	xx	x	xxx	xxxx	xxx	xx	?	N/A

[a]Approximate levels of tolerance are as follows: American species, 1.5 g/kg absolute maximum; *V. vinifera*, 3 g/kg absolute maximum.
[b]N/A, not available.

Source: From Pongrácz (1983), reproduced by permission. Summarized from data of Branas (1974), Cosmo *et al.* (1958), Galet (1971), Mottard *et al.* (1963), Pàstena (1972), Pongrácz (1978), and Ribéreau-Gayon and Peynaud (1971)

been chosen, it remains a permanent component of the vineyard until replanting.

The most basic criterion for acceptability is compatibility (affinity) between the rootstock and scion. Compatibility refers to the formation and stability of the graft union. Early and complete fusion of the adjoining cambial tissues is critical to effective translocation of water and nutrients between the rootstock and scion. Areas that do not join shortly after grafting never fuse. Such gaps leave weak points that provide sites for invasion by various pests and disease-causing agents. Recommendations on cultivar compatibility are given in Furkaliev (1999).

For many rootstock varieties, data are available on basic properties (Table 4.6). Delas (1992) gives current recommendations for France, relative to drought, waterlogging, nematodes, soil depth, lime sensitivity, vigor, and so on. Views from other countries are also supplied by regional specialist in Wolpert *et al.* (1992). In some regions, desirable rootstock combinations have already been identified for local cultivars. However, for new scion–rootstock combinations, or in new viticultural regions, existing data can only provide a guide as to what appears appropriate for local field trials.

Because of the large number of rootstock–scion combinations, it is important to predict unsuitable combinations. Although the parentage of a rootstock gives clues to compatibility, accurate prediction remains illusive. Whether compatibility can be determined using electrophoretic similarity between scion and rootstock phosphatases, as suggested by Masa (1989) is contentious. Were it, or some other rapid measurement, found to be universally applicable, rapid screening could identify incompatible unions. Regrettably, we are not at that state.

Although incompatibility often originates from an unexplained physiological disparity between the scion and rootstock, poor union between otherwise compatible cultivars may be caused by pathogens. Grafting healthy scions to virally infected rootstocks (i.e., GLRaV-2, RSPaV, and fleck viruses) can result in a poor union, often expressed as a swelling at the graft site or xylem disruption (Golino, 1993). The presence of several fungi has also been correlated with graft failure, notably *Phaeoacremonium parasiticum* (often associated with Petri disease) and *Botryosphaeria* spp. (causal agents of several Diplodia diseases).

In the initial rootstock trials conducted in the late 1800s, the most successful were selections from *V. riparia*, *V. rupestris*, crosses between these two, or *V. cinerea* var. *helleri* (*V. berlandieri*) hybrids. Their progeny still constitute the bulk of rootstock cultivars (Howell, 1987). Subsequent breeding has incorporated traits from species such as *V. vinifera*, *V. mustangensis* (*V. candicans*), and *V. rotundifolia*.

In regions where phylloxera (*Daktulosphaira vitifoliae*) occurs, grafting *V. vinifera* cultivars to resistant rootstock is generally essential. Even where phylloxera is not present, serious consideration should be given to using phylloxera-resistant rootstock. The past history of accidental phylloxera introduction suggests that the eventual infestation of all compatible sites is only a matter of time.

The occurrence of several *D. vitifoliae* biotypes, along with the presence of differential tissue sensitivity in different grapevine species, indicates that phylloxera resistance is complex (Wapshere and Helm, 1987). Phylloxera biotypes often are distinguished on the basis of differential rates of multiplication on particular (tester) rootstocks. For example, biotype B phylloxera multiplies twice as rapidly as biotype A on 'A×R#1' ('Ganzin 1') (Granett *et al.*, 1987). When 'A×R#1' became the predominant rootstock in much of northern California, the selection of a biotype capable of multiplying on this rootstock was probably inevitable. Although most commercial rootstock cultivars possess some phylloxera resistance, derived from *V. riparia*, *V. rupestris*, or *V. cinerea* var. *helleri* (*V. berlandieri*), additional potential sources of resistance are *V. rotundifolia*, *V. mustangensis* (*V. candicans*), *V. cinerea*, and *V. vulpina* (*V. cordifolia*).

Although phylloxera resistance is the prime reason for most rootstock grafting, nematode resistance is more significant in some regions. Grapevine roots may be attacked by several pathogenic nematodes, but the most important are the root-knot (*Meloidogyne* spp.) and dagger (*Xiphinema* spp.) nematodes. Dagger nematodes are also transmitters of fanleaf degeneration. Because *V. rotundifolia* is particularly resistant to fanleaf degeneration and nematode damage, it has been used in breeding several new rootstock cultivars resistant to these maladies, notably 'VR O39-16' and 'VR O43-43' (Walker *et al.*, 1994). Regrettably, lime-susceptibility and sensitivity to phylloxera limit the use of 'VR 043-43' to noncalcareous soils, and regions devoid of phylloxera. Another valuable source of nematode resistance is *V. vulpina*. Some of its resistance genes have been incorporated into varieties such as 'Salt Creek' ('Ramsey'), 'Freedom,' and possibly '1613 C.'

The importance of local soil factors for grapevine growth can significantly influence rootstock choice. For example, tolerance to high levels of active lime ($CaCO_3$) is essential in many European sites. In such regions, the use of varieties such as 'Fercal' or '41 B' are preferred. In contrast, low sensitivity to aluminum is crucial in some acidic Australian and South African soils. Because of the importance of soil factors, most commercial rootstocks have been studied to determine their tolerance to such factors (Table 4.7). Where conditions vary considerably

within a single vineyard, the use of several rootstocks may be required.

Drought tolerance can be another factor crucial in selecting a rootstock, especially in arid regions where irrigation is limited, unavailable, or not permitted. As with most traits, drought tolerance is based on complex physiological, developmental, and anatomical properties. Differences in root depth, distribution, and density appear to be partially involved (Fig. 4.21). Several drought-tolerant varieties reduce stomatal conductance in the scion. In addition, '110 Richter' may induce the production of fewer and smaller stomata (Scienza and Boselli, 1981; Düzenlı and Ergenoğlu, 1991). However, some rootstock varieties do not affect scion transpiration efficiency (Virgona *et al.*, 2003). The merits of using a drought-tolerant rootstock may be enhanced under conditions that restrict root growth, such as high-density plantings. Although *V. cinerea* var. *helleri* is one of the most drought-tolerant grapevine species, the expression of this trait varies considerably in *V. cinerea* var. *helleri*-based rootstocks. *Vitis vulpina*-based rootstocks are often particularly useful on shallow soils in drought situations.

In regions having short growing seasons, early fruit ripening and cane maturation are essential. Most rootstocks that favor early maturity have *V. riparia* in their parentage. Where yearly variation in cold severity is marked, random grafting of vines to more than one rootstock may provide some protection against climatic vicissitudes (Hubáčková and Hubáček, 1984).

Another vital factor influencing rootstock selection is its effect on grapevine yield. Although the rapid establishment of a vineyard is aided by vigorous vegetative growth, this property may be undesirable in the long term. Thus, rootstocks may be chosen specifically to induce devigoration. This property of '3309 C' has made it particularly popular under dryland viticultural conditions. Devigorating rootstocks usually limit yield and may improve fruit quality, as well as repressing physiological disorders such as inflorescence and bunch-stem necrosis. Fruit yield generally shows a weak negative correlation with quality, as measured by sugar content (see Fig. 3.43). The specific yield vs. quality influence of any rootstock varies considerably (Fig. 4.22) and is influenced by conditions such as vineyard layout, canopy management, and irrigation (Whiting, 1988; Foott *et al.*, 1989).

Some of the effects of rootstocks on vine vigor and fruit quality may accrue from differential nutrient uptake. For example, the preferential accumulation of potassium can antagonize the uptake of other cations. For rootstocks, such as 'SO 4' and '44–53 M,' this can lead to magnesium deficiency with cultivars such as 'Cabernet Sauvignon' and 'Grenache' (Boulay, 1982). Limited zinc

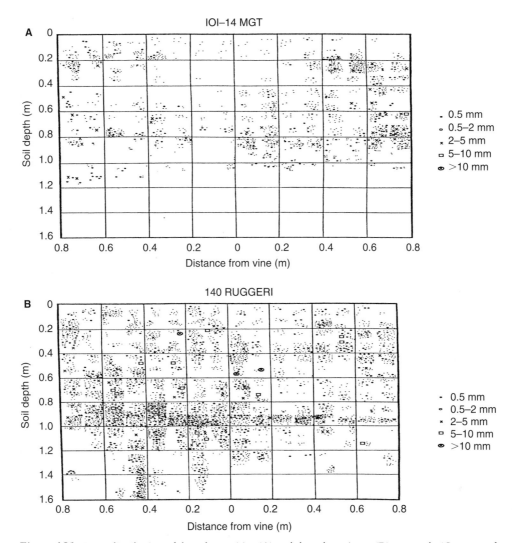

Figure 4.21 Root distribution of drought-sensitive (**A**) and drought-resistant (**B**) rootstock. (Courtesy of E. Archer, University of Stellenbosch, SA)

uptake by 'Rupestris St George' may be a source of the poor fruit set occasionally associated with its use (Skinner *et al.*, 1988). Because the root system supplies most of the nitrogen required in the early part of the growing season (Conradie, 1988), variation in rootstock nitrogen uptake and storage may influence scion fruitfulness. The importance of rootstock selection in limiting lime-induced chlorosis has already been mentioned. Although most of these effects are undoubtedly under the direct genetic control of the rootstock, some variation may result indirectly from differential colonization of the roots by mycorrhizae.

Rootstock selection can significantly alter scion fruit composition. By affecting berry size, rootstocks can influence the skin/flesh ratio and, thereby, wine attributes. Additional indirect effects on fruit composition may result from increased vegetative growth, augmenting leaf–fruit competition, or shading. For example, the use of

'Ramsey' for root-knot nematode control and drought tolerance in Australia has inadvertently increased problems due to excessive vine vigor. Nevertheless, other rootstock effects are probably direct, via differential mineral uptake from the soil (Fig. 4.34). The rootstock's impact on potassium uptake (Ruhl, 1989), and its accumulation in the vine (Failla *et al.*, 1990) can be especially influential. Potassium distribution affects not only growth, but also juice pH and potential wine quality. Rootstock modifications of fruit amino acid content have been correlated with the rate of juice fermentation (Huang and Ough, 1989).

Few studies have investigated the significance of rootstocks on grape aroma. As indicative of the complexity of the relationship, some studies have shown a decrease in monoterpene content associated with rootstocks promoting high yield (McCarthy and Nicholas, 1989), whereas others have found little influence

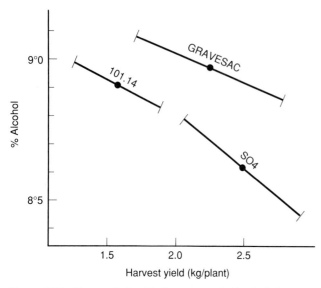

Figure 4.22 Linear relationship between probable alcohol content (ordinate) and yield (abscissa) of 'Cabernet Sauvignon' on three different rootstocks. (From Pouget, 1987, reproduced by permission)

associated with changes in yield of up to 250% (Whiting and Noon, 1993).

Although rootstock grafting can be valuable, if not essential, in most viticultural regions, it is expensive. In addition, the cost of special rootstocks may be higher, owing to limited demand or difficulty in propagation. These features can also make them difficult to obtain. Nevertheless, the long-term benefits of using the most suitable rootstocks usually outweigh the additional expenditure associated with their procurement.

One of the problems associated with rootstock use have been errors in identification. Many of the cultivars are morphologically similar, making amphelographic recognition difficult. This should become less frequent due to the introduction of polymerase chain reaction (PCR)-based fingerprinting (Guerra and Meredith, 1995). The technique can ascertain whether existing stocks are correctly identified.

An unintentional consequence of the use of grafting has probably been the unsuspected propagation and worldwide distribution of grapevine viruses and viroids (Szychowski *et al.*, 1988). Graft unions also may act as sites for the invasion of several pests and pathogens. Nevertheless, the grafting procedure itself can apparently induce, at least temporarily, resistance to infection and transmission to tomato ringspot virus (Stobbs *et al.*, 1988).

In the popular press, there is continual musing over the possible (usually undesirable) influence of grafting on wine quality. In most situations, the question is purely academic. Grape culture in many regions of the world would be commercially nonviable without grafting. As in other aspects of grape production, choice of

a rootstock can either enhance or diminish the quality of the grapes and the wine produced.

Vine Propagation and Grafting

Grapevines typically are propagated by vegetative means to retain their unique genetic constitution. Sexual reproduction, by rearranging grapevine traits, usually disrupts desirable gene combinations. Thus, seed propagation is limited to breeding cultivars where new genetic arrangements are desired.

Several techniques may be used to vegetatively propagate grapevines. The method of choice depends primarily on pragmatic matters, such as the number of plants required, the rapidity of multiplication, when propagation is conducted, whether grafting is involved, and, if so, the thickness of the trunk. Regardless of the method, some degree of callus formation occurs.

Callus tissue consists of undifferentiated cells that develop in response to physical damage. Callus develops most prominently in and around meristematic tissues. In grafting, the callus establishes the union between adjacent vascular and cortical tissues of the rootstock and scion (Fig. 4.23). For the union to persist, it is crucial that the thin cambial layer of both rootstock and scion be aligned adjacent to one another. Callus formation also is associated with, but not directly involved in, the formation of roots from cuttings. New (adventitious) roots typically develop from or near cambial cells between the vascular bundles of the cane. Most roots emanate from a region adjacent to the basal node of the cutting. Finally, callus cells developed in tissue culture may differentiate into shoot and root meristems, from which whole plants can develop (Plate 2.3).

Callus tissue is particularly metabolically active, and its formation is favored by warm conditions and ample oxygen. Because of the predominantly undifferentiated state of the callus, its thin-walled cells are very sensitive to drying and sun exposure. Correspondingly, the union is often covered by grafting tape to prevent drying and light exposure until protective layers have formed over the graft site.

Multiplication Procedures

The simplest means of vegetative propagation is **layering**. The technique involves bending a cane down to the ground and mounding soil over the section. Once rooted sufficiently, connection to the parent plant can be severed. Layering has the advantage of continuously supplying water and nutrients throughout root formation and development. This is particularly useful with difficult-to-root vines, such as muscadine cultivars.

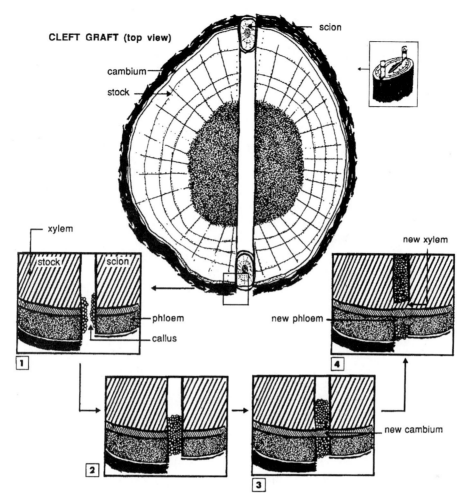

Figure 4.23 Development of a cleft graft union. (From Nicholas *et al.*, 1992, reproduced by permission)

Otherwise, layering is seldom used. Other techniques are often as effective, and do not interfere with viticultural practices, such as cultivation and weed control.

The most common means of grapevine propagation involves cane **cuttings**. These are typically taken from prunings collected during the winter. The best cane sections are usually those 8–13 mm in diameter, uniformly brown, and possessing internode lengths typical of the variety. These features indicate that cane development occurred under favorable conditions and is well matured. In addition, the sections need to be sufficiently long to supply the new root and shoot system with ample nutrients, until the plant becomes self-sufficient. Cane length also depends on the water retention properties of the soil into which the vine is to be planted, and the availability of irrigation water subsequent to planting. Appropriate cane wood is usually cut into sections about 35–45 cm in length. Soaking in a disinfectant, such as 8-hydroxyquinoline sulfate, guards against infection by pathogens such as *Botrytis cinerea*. If deemed necessary, submersion in hot (50–55°C) water can inactivate pests such as

nematodes and several contaminant or systemic fungal and bacterial pathogens. The canes are kept cool and moist until adventitious root production commences.

A lower perpendicular cut is made just below a node, and an upper 45° diagonal cut is made about 20–25 mm above the uppermost bud. The diagonal cut facilitates rapid identification of the apical location. This is important because the original apical–basal orientation of the cutting must be retained when rooted. Cane polarity restricts root initiation to the basal region of the cutting. In addition, the diagonal section provides some physical protection for the apical bud. Protecting the terminal bud is especially important in rootstock varieties. In this case, all buds except the terminal one are removed before rooting to limit subsequent rootstock suckering. The apical bud in this case is the sole source of auxin for the activation of root development.

If the cuttings are to be rooted in the vineyard, it is important to leave about 10 cm of the cane above ground. This minimizes the likelihood of scion rooting. If scion rooting were to occur, its roots might outgrow those of the rootstock, only to succumb to the conditions for which grafting was conducted.

The canes of most *V. vinifera* varieties root easily. The same is also the case for most rootstock cultivars that are selections of *V. rupestris* and *V. riparia*, hybrids between them, or hybrids with *V. vinifera*. Most rootstocks containing *V. cinerea* var. *cinerea*, *V. cinerea* var. *helleri*, *V. mustangensis*, *V. vulpina*, or *V. rotundifolia* heritage are to varying degrees difficult to root (see Table. 4.6). Because the latter contain many useful properties, researchers have spent considerable effort attempting to enhance rooting success. Generally, the most effective activators include soaking in water for 24 hours, dipping in a solution of about 2000 ppm indolebutyric acid (IBA), applying bottom heat (25–30°C) to the rooting bed, and periodic misting to maintain high humidity. Additional factors of potential value have involved spraying parental vines with chlormequat (CCC) in the spring before cane selection (Fabbri *et al.*, 1986), and aquaculturing after callus formation (Williams and Antcliff, 1984).

Rooting success is improved when cuttings are produced and planted directly after cane harvesting. If rooting cannot be initiated immediately, the canes are best kept refrigerated (1–5°C), moist, and mold free. Upright storage of the canes in moist sand or sawdust is common. During this period, a basal callus forms, from which the roots will develop after planting.

An alternative method of cane propagation involves **green cuttings**. These are single-node pieces cut from growing shoots. Rooting occurs under mist propagation in the greenhouse. This permits rapid multiplication. It is particularly useful when the source of desirable canes is limited. The procedure is also valuable for cultivars that do not root well from cane wood, notably those from *Vitis cinerea* var. *helleri* (*V. berlandieri*) and *V. rotundifolia*. The technique is more complex and demanding, both in equipment and protection after rooting. In addition, their tender nature requires considerable caution in hardening the rooted plants to withstand vineyard conditions.

Although cuttings are the most common means of grapevine propagation, it may be inadequate for the rapid multiplication of speciality stock. **Micropropagation** from axillary buds is the simplest tissue culture method. In some instances, as with *Vitis × Muscadinia* crosses, it may be the only convenient method (Torregrosa and Bouquet, 1996). If the financial return is adequate, vines can be multiplied even more quickly using **shoot–apex fragmentation** (Barlass and Skene, 1978) or **somatic embryogenesis** (Reustle *et al.*, 1995; Zhu *et al.*, 1997).

More complex and demanding than other reproduction techniques, tissue culture is the only means of mass propagating a cultivar. Regrettably, micropropagation

is complicated by the need to adjust the procedures for different varieties or tissues (Martinelli *et al.*, 1996). However, because strict hygiene is required, infection of disease-free stock is avoided.

Grafting

Where conditions obviate the need for grafting, self-rooted scion cuttings can be directly planted in the vineyard. However, in most viticultural areas, profitable grape culture depends on grafting the scion to a suitable rootstock. This typically involves inserting one-bud scion sections at the apex of a rootstock cutting. When done indoors, as in a nursery or greenhouse, it is called **bench grafting**. When grafting occurs at or shortly following rootstock planting in the vineyard, it is termed **field grafting**. The other major use of grafting is converting (**topworking**, **grafting over**) existing vines to another fruiting variety. When the scion piece consists of a cane segment, the process is termed **grafting** to distinguish it from the use of only a small side piece from a cane, designated **budding**.

Bench grafting has the advantage of being more amenable to mechanized mass production. It also can be performed over a longer period, as it commonly uses dormant cuttings. To facilitate proper cambial alignment when using grafting machines, it is necessary to presort the rootstock and scion pieces by size. After making the cuts and joining the two sections (Fig. 4.24), the grafted cutting is placed in a callusing room under moist warm conditions. This favors rapid callus development and graft union. Grafting machines permit junctions of sufficient strength that grafting tape is not needed while the union forms. If the grafted rootstock has already been rooted, the vine is ready for planting shortly after the

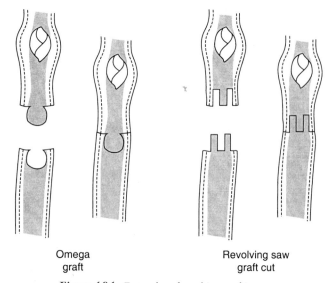

Figure 4.24 Examples of machine grafting.

union has formed, and the exposed callus hardened off and coated with wax. If a difficult-to-root dormant rootstock is grafted, the base may be treated with IBA and placed in a heated rooting bed, while the upper graft union is kept cool. This favors root development before the scion bud bursts, placing water demands on the young root system. With easily rooted rootstock, canes usually root sufficiently rapidly to supply the needs of the developing scion without special treatment.

Occasionally, actively growing shoots are grafted directly onto growing rootstocks in a process called **green grafting**. Graft union is usually rapid and highly successful. It can avoid some of the incompatibility problems that occasionally plague grafting distantly related *Vitis* species (Bouquet and Hevin, 1978). In most cases, though, the higher labor costs and more demanding environmental controls usually do not warrant its use. Nevertheless, modern developments may reduce the expense of green grafting (Alleweldt *et al.*, 1991; Collard, 1991).

Where labor and timing are appropriate, field grafting is the least expensive means of grafting new vines. Field grafting is also the only means of converting existing vines to another variety. Field grafting preferably occurs shortly after growth has commenced in the spring. By this time, the cambium in the rootstock has become active, and graft union develops quickly. This permits prompt growth of the scion. The rootstocks are planted leaving about 8–13 cm projecting above the ground. This both limits scion rooting and places the root system sufficiently deep to minimize damage during manual weeding. Grafting unrooted rootstock in the field is not recommended because the success rate is poor.

Commonly used manual techniques for grafting include whip grafting and chip budding. **Whip grafting** bonds scion and rootstock canes of equivalent diameter (Fig. 4.25). Two cuts are made about 5 mm above and below a scion bud. The upper cut is shallowly angled and directed away from the bud to identify the polarity of the scion piece. The lower cut is long and steep (15–25°), usually 2.5 times longer than the diameter of the cane. A "tongue" is produced in the lower cut by making an upward slice, and gently pressing outward away from the bud with the pruning knife. A set of cuts matching those in the lower end of the scion piece is made in the rootstock to receive the scion. After connection and alignment, the union is secured with grafting tape, raffia, or other appropriate material. Plastic grafting tape is popular because it is both quickly and easily applied, will not cause girdling, and helps limit drying of the graft union. This obviates the need for mounding and the subsequent removal of moist soil from over the union site.

Whip grafting provides an extensive area over which the union can establish itself. Its main disadvantage is the skill and time required in performing the procedure.

In addition, the large wound produced creates a large potential invasion site for a complex of wood decay fungi. Over many years, these could weaken the trunk and cause progressive yield decline.

Chip budding provides less union surface than whip grafting, but the smaller size of graft piece demands less contact area. Chip budding is often preferred because it requires less skill in preparing matching cuts. Also, because the scion source does not need to be identical in diameter to the rootstock, time is saved by avoiding the requirement of matching scion and rootstock pieces.

In chip budding, two oblique downward cuts are made above and below the scion bud (Fig. 4.26). The upper cut is more acute and meets the lower incision, making a wedge-shaped chip about 12 mm long and 3 mm deep at the base. A matching section is cut out about 8–13 cm above ground level on the rootstock. The chip is held in position with grafting tape or equivalent material.

For either grafting procedure, it is imperative that each set of cuts and the insertion of the scion piece be performed rapidly to avoid drying. Drying of the cut surfaces dramatically reduces the chances of successful union.

Various techniques are used in topworking existing vines to another fruit-bearing cultivar. For trunks less than 2 cm in diameter, whip grafting is commonly used, whereas for trunks between 2 and 4 cm in diameter, side-whip grafting is often preferred. Trunks more than

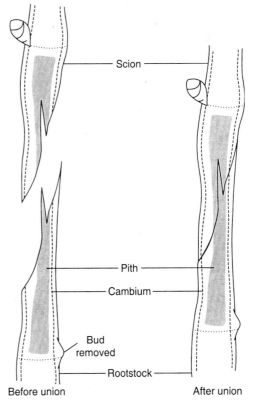

Figure 4.25 Whip-graft union.

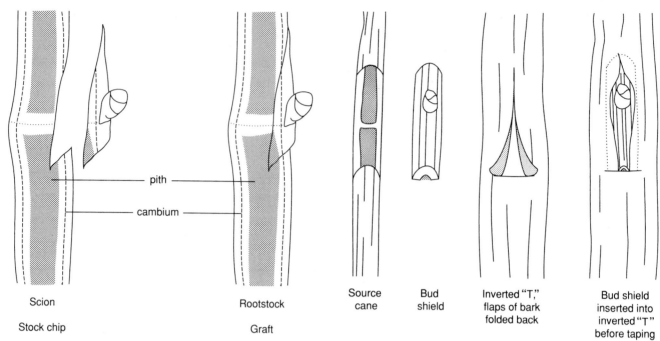

Figure 4.26 Chip bud grafting.

Figure 4.27 Inverted-T graft.

4 cm may be **notch-**, **wedge-**, **cleft-** or **bark-grafted** (Alley, 1975). In all size classes, chip budding can be used, whereas T-budding is largely limited to trunks more than 4 cm in diameter. Budding techniques are often preferred to the use of larger scion pieces because they require less skill and can be as successful (Steinhauer *et al.*, 1980). Conversion high on the trunk allows most of the existing trunk to be retained, thus speeding the vine's return to full productivity.

T-budding derives its name from the shape of the two cuts produced in the vine being converted. After making the cuts, the bark is pulled back to form two flaps. This generates a gap into which the scion piece (**bud shield**) is slid. The bud shield is produced by making a shallow downward cut behind the bud on the source vine. The slice begins and ends about 2 cm above and below the bud. A second oblique incision below the bud liberates the bud shield. To facilitate a better union, Alley and Koyama (1981) recommend that T-cuts in the trunk be inverted. In the inverted-T graft (Fig. 4.27), the rounded upper end of the bud shield is pushed upward and the bark flaps cover the bud shield. In the original version of the technique, failure of the top of the shield to join well with the trunk created a critical zone of weakness. Because winds could easily split the growing shoot from the trunk, shoots had to be tied to a support shortly after emergence. This occurrence is much less likely with inverted T-budding.

Depending on the trunk diameter, several bud shields may be grafted around the trunk. The shoots so derived are necessary to provide sufficient photosynthate to the vine's trunk and root system. Each bud shield is grafted at the same height to conserve grafting tape and speed grafting.

T-budding has the advantage of requiring the least skill of any vine conversion technique. In addition, demands on cold storage space are minimal because bud shields rather than cuttings can be stored (Gargiulo, 1983). It has the disadvantage that the period during which it can be performed most effectively is short. The grape grower must wait until the trunk cambium has become active after bud break in the spring. Only then can the bark be easily separated from the wood to permit insertion of a bud shield. Postponing T-budding much beyond this point delays bud break and may result in poor shoot maturation by autumn.

In contrast, chip budding can be performed earlier, thereby allowing the grape grower greater flexibility in timing budding-over. With chip budding, it is necessary to use a section of the trunk with a curvature similar to that of the chip. Otherwise, the cambial alignment may be inadequate and the union fail. Depending on trunk thickness, two or more buds are grafted per vine. Because of the hardness of the wood, cutting out slots for chip budding is more difficult than for T-budding. Nevertheless, the longer period over which chip budding can be performed may make it preferable. The success rate of chip budding is usually equivalent to T-budding (Alley and Koyama, 1980).

Although vine conversion is usually performed in the spring, this has the disadvantage that the season's crop is lost. It is necessary to remove the existing top to permit the newly grafted scion pieces to develop into the new top. To offset this loss, chip budding may be performed in early autumn, when the buds have matured but weather conditions are still favorable ($\geq 15\,°C$) for graft union (Nicholson, 1990). Full union is usually complete by leaf fall, but the bud remains dormant until spring. Angling the upper and lower cuts away from the bud produces a bud chip that slides into a matching slot made in the host vine. Although the technique is more complex, the interlocking of the chip in the trunk assures a firm connection with the vine. Grafting tape protects the graft site from drying. In the spring, the tape is cut to permit the bud to sprout. An encircling incision above the grafted buds stimulates early bud break while restraining growth of the existing top. This technique allows the vine to bear a crop while the grafted scion establishes itself. At the end of the season, the existing top is removed and the grafted cultivar trained as desired.

The older techniques of cleft, notch, and bark grafting are still used, but less frequently. Not only do the older techniques require more skill in cutting and aligning the scion and trunk cambia, but they also take longer and require grafting compound to protect to graft while the union develops. Readers desiring details on these and other grafting techniques are directed to standard references (e.g., Winkler *et al.*, 1974; Alley, 1975; Weaver, 1976; Alley and Koyama, 1980, 1981).

Soil Preparation

Before planting rooted cuttings in a vineyard, the soil should be analyzed and prepared to receive the vines. The degree of preparation depends on the soil texture, degree of compaction, previous use, drainage conditions, nutrient deficiencies or toxicities, pH, irrigation needs, and prevailing diseases and pests. If the land is virgin, noxious perennial weeds and rodents should be eliminated and obstacles to efficient cultivation removed. Where the soil has already been under cultivation, providing sufficient drainage and soil loosening for excellent root development are the primary concerns. The effects of soil characteristics on root distribution are illustrated in Plates 4.9, 4.10, 4.11, and 4.12.

Although certain soil conditions are known to be unfavorable to root growth (e.g., acidic, saline, sodic, waterlogged, low nutrient conditions), it is impossible to provide universal recommendations. What is "ideal" will depend on a multiple of factors, notably the genetics of the scion and rootstock (if grafted) and the climate. Local recommendations should be obtained from regional authorities.

Inadequate drainage is most effectively improved by laying drainage tile. Winkler *et al.* (1974) recommend

draining soil to a depth of about 1.5 m in cool climates and to 2 m in warm to hot climates. Narrow ditches are a substitute, but can complicate vineyard mechanization. In addition, ditches remove valuable vineyard land from production. Drainage efficiency may be further improved by breaking hardpans or other impediments to water percolation.

Deep ripping (0.3–1 m), used to break hardpans, can also be used to loosen deep soil layers. This is especially useful in heavy, nonirrigated soil where greater soil access can minimize water deficit under drought conditions (van Huyssteen, 1988a). Homogeneity of soil loosening is also important in favoring effective soil use by vines (Saayman, 1982; van Huyssteen, 1990). Nevertheless, ripping can incorporate nutrient poor, deep soil horizons into the top soil, and enhance erosion on slopes. Ripping the soil when wet can generate columns of compacted soil between the rows, which complicate rather than aid drainage.

In sites possessing considerable heterogeneity, earth moving, leveling, and mixing should be seriously considered. This can minimize, if not eliminate, serious local variations in soil acidity and nutrient or water availability. Where the soil is deficient in poorly mobile nutrients, this is the optimum time to incorporate nutrients such as potassium and zinc. Vineyards exhibiting a wide diversity of conditions usually yield fruit of unequal uniformity and wines of lower quality (Long, 1987; Bramley and Hamilton, 2004; Cortell *et al.*, 2005).

Where nematodes are a problem, it is often beneficial to fumigate the soil even when nematode-resistant rootstocks are used. Fumigation reduces the level of infestation and enhances the effectiveness of resistant or tolerant rootstocks in maintaining healthy vines.

Where surface (furrow) irrigation is desired, the land must be flat or possess only a slight slope. Thus, land leveling may be required if this irrigation method is planned.

Vineyard Planting and Establishment

Various planting procedures can be used, but mechanized planting is favored because of its time and cost savings. Where bare-rooted cuttings are planted, it is critical to protect the plants from drying. Because roots are trimmed to the size of the planting hole, they should be sufficiently large to permit retention of most of the root system. Direct planting of rooted cuttings from tubes or pots maximizes root retention. If sufficiently acclimated to field conditions, potted vines suffer minimal transplantation shock. This approach also gives the grape grower more flexibility in scheduling planting.

Where permanent stakes are not already in position, it is advisable to angle the planting hole away from the stake location and parallel to the row. This minimizes subsequent root damage from the use of posthole diggers and cultivators. Often soil is hilled around the exposed portion of vine until shoot development is well established. Alternately, planting may take place on mounds of earth covered by meter-wide sheets of black plastic. The two techniques promote root development and minimize the manual weeding normally required during the first year.

Proper hole preparation is important to assure adequate root development. Poor root development can lead to restricted vine growth not only in the first few years, but may also later when fruit production puts increased demand on a confined root system. Vertical penetration promotes greater access to water in dry spells and provides better use of mobile nutrients such as nitrogen. Horizontal proliferation of the root system is necessary to reach nutrients poorly mobile in soil such as phosphate.

Louw and van Huyssteen (1992) recommend square holes, finished with a garden fork, to produce uneven sides and loosen the bottom. Although beneficial, the expense of these measures must be weighed against the economy of more automated planting systems. Only soil of the same type and texture should be used to fill the hole. Lighter textured filling soil can induce roots to remain within the planting hole, rather than grow into the surrounding soil. This not only produces a pot-bound effect, but can increase the likelihood of water logging, poor soil aeration, and soil pathogen problems. In addition, roots tend to penetrate soil pores of a diameter equal or greater than the diameter of growing root tips. The smearing of moist soil, as with an auger, can seal off most soil cavities. As a result, roots may grow in a circle along the smeared sides of the planting hole. Thus, the moisture condition of the soil must be ideal to use an auger. With use of an automatic planting machine or water lance (a jet of water that creates a hole), care should be taken to assure that air pockets do not remain below the roots. This is frequently assessed by a gentle push downward on the shoot.

Transplanted vines are watered at least once after planting, and again if drought conditions develop. Irrigation is avoided after midsummer to restrict continued vegetative growth and favor cane maturation. Cane maturation is required to permit the vine to withstand early frosts and winter cold.

During the first growing season, the vine is permitted to grow largely at will, to facilitate the establishment of an effective root system. Most vineyard activities are limited to weed, disease, and pest control to protect the young succulent tissues. Frequent localized fertilizer application (primarily nitrogen) promotes early and vigorous shoot and root development. Topping is conducted only if watering is insufficient to prevent severe water deficit. Pruning occurs after growth has ceased and the leaves have fallen. For the majority of training systems, only one strong well-positioned cane is retained, often pruned to four buds.

It has become popular to enclose young vines in a commercially prepared housing. The enclosure provides the graft region with additional protection, guards against wind, sand, and rodent damage, as well as accidental exposure to herbicides used in weed control. The guards also direct the upward growth of the shoot that will become the trunk. Tubes about 9 cm in diameter are preferred in hot climates to avoid heat-induced damage by the encasement. Despite reduced photosynthesis, vine growth is enhanced due to the improved moisture conditions provided. The degree of improved growth is influenced both by the relative benefit provided by reduced water deficit, and the absence of restrictions to root growth in the soil. In cool climate regions, growth tubes should be removed in late season, to avoid delayed cane hardening in the autumn.

What appears not to have been studied is how these protective tubes affect long-term health of the vine. The sheltering provided is likely to increase the proportion of parenchyma tissue and reduce the woody structure in the stem and root system (as has been found in other perennial plants). This could make the vine more susceptible to inclement conditions, and may favor the development of early trunk diseases, such a Petri disease. In cooler regions, there have also been incidences of severe powdery mildew infection of young vines.

In the spring of the second season, the strongest shoot is retained and tied to the trellising stake to form the future vine trunk. Subsequent pruning varies depending on the training systems desired.

Irrigation

Grape growing possesses one of the longest historical records of irrigation of any crop. Records of vineyard irrigation go as far back as 2900 B.C. (Younger, 1966). Mesopotamian agriculture also provides one of the earliest examples of improper irrigation use, leading to salination and loss of soil productivity. Excessive irrigation, besides being wasteful of a precious resource, leads to nutrient leaching from the soil and increases the potential for soil acidification.

In Europe, irrigation is prohibited in most Appellation Control areas. This may be the source of the myth that irrigation is *ipso facto* inimical to grape quality. If used excessively, irrigation can have undesirable effects on cane maturation and fruit ripening (large berries, reduced sugar and anthocyanin concentrations, compact clusters, and increased disease incidence). Used wisely, it not only permits grape culture in arid and semiarid regions, but also can facilitate the production of premium quality grapes. In contrast, drought stress can produce small, unflavorful berries, reduced yield, premature growth termination (diminished production of photosynthates), shoot tip death, and reduction in root development. The effects of water deficit may also affect growth in subsequent years (Petrie *et al.*, 2004). However, moderate water deficit has several benefits, in addition to early termination of shoot growth. For example, Chapman *et al.* (2005) found that wine produced from minimally irrigated 'Cabernet Sauvignon' had more marked red/blackberry aroma, jammy and fruit flavors, whereas standard irrigation yielded wines with higher vegetal, bell pepper and black pepper attributes. Anthocyanin and proanthocyanidin polymer concentrations also tended to increase. Although reducing yield, yield alone seems not be the cause. Where pruning was used to reduce yield, the effect on wine quality was almost the reverse (Chapman *et al.*, 2004b).

In dry regions, Regulated Deficit Irrigation (RDI) has been used to control vine vigor and favor optimal grape ripening (Hardie and Martin, 1990; McCarthy, 1998). For example, berry weight was sufficiently sensitive to water deficit that it could irreversibly limit fruit size after berry set (McCarthy, 1997). Whether this is directly due to water shortage, or associated reduction in nutrient availability, is uncertain. Limited water availability, notably between fruit set and *véraison*, curtails continued (undesired) vegetative growth. Moderate water deficit may be maintained after *véraison*, if required, to constrain further shoot growth, while supplying enough water for post harvest root growth. During this period, limited shoot growth tends to neither adversely affect fruit development (Dry *et al.*, 2001), nor root growth (van Zyl, 1984). RDI can improve fruit coloration, enhance ripening, and reduce disease incidence. However, enhanced water deficit can limit assimilable fruit nitrogen, resulting in sluggish fermentation and a reduction in wine quality. Limited RDI seems most applicable to vigorous vines on soils of relatively uniform texture, without salinity problems, and a ready supply of groundwater. Experience with an automatic irrigation scheduling system is desirable before assessing the applicability of RDI. Although useful for several cultivars, it may be ill-advised for those whose varietal character is based on monoterpenes (e.g., 'Gewürztraminer,' 'Muscat,' and 'Riesling'). There are indications that the concentration of monoterpenes is reduced relative to the duration of water deficit (Reynolds and Wardle, 1996).

An alternate irrigation technique is termed Partial Rootzone Drying (PRD). With it, problems associated with the timing and intensity of RDI are much reduced. It involves the use of two drip irrigation lines, positioned on opposite sides of the row. Irrigation events occur alternately, first on one side then the other. Cycles may vary from 3 to 5 days, to 10 to 14 day intervals, depending on the severity of water stress (temperature, relative humidity, wind) and nature of the soil. The system appears to work best with sandy soil (those with high infiltration rates), and where most of the water is derived via irrigation. The result is a significant reduction in water use (up to 50%), particularly valuable where irrigation water is a scarce and expensive resource, and increasing soil salinity is a concern. In addition, PRD has many of the benefits of RDI on fruit quality (higher acidity, lower pH, and increased anthocyanin and glycosylglucose content), but without the yield reduction that can accrue from whole-vine water deficit (Dry *et al.*, 1996). By exposing part of the root system to water deficit, root synthesis of abscisic acid (ABA) is markedly increased, while cytokinin concentration decreases (Stoll *et al.*, 2000; Antolín *et al.*, 2006). Translocation of abscisic acid throughout the vine, via the vascular system, leads to a generalized reduction of stomatal opening and transpiration. The resultant drop in gas exchange limits photosynthesis, diminishing vegetative growth. Partial stomatal closure also tends to promote more efficient water use, or, at least change water use (Collins *et al.*, 2005). Root growth also tends to be enhanced.

Because the effects of PRD are short term (Fig. 4.28), it is essential that the portion of the root system exposed to drying be rotated. Because sufficient water is supplied to meet actual vine needs, the vine does not actually experience water deficit. However, the shortage of water on one side triggers that portion of the root system to signal the whole vine to adjust water usage. Under appropriate conditions, grape quality is enhanced, while water use is diminished.

Although RDI or PRD have distinct benefits, the precise conditions under which they are most advantageous have yet to be clearly defined (de Souza *et al.*, 2004; Collins *et al.*, 2005). These differences in opinion may result from our incomplete understanding of the degree and importance of soil variability to root function. In addition, it is very difficult to fully control influences under field conditions (see Richards *et al.*, 2005). Because the root system and its microclimate are literally out-of-sight, knowledge and appreciation of these issues are still underappreciated and studied with difficulty. This lamentable situation applies to all aspects of viticulture concerning the soil.

Soil acts not only as an anchorage and supply of inorganic nutrients, but also as the source of water. Of the water present in soil, only a variable fraction is available for plant use. This portion depends on the textural

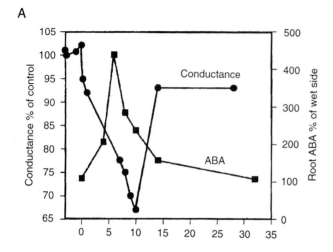

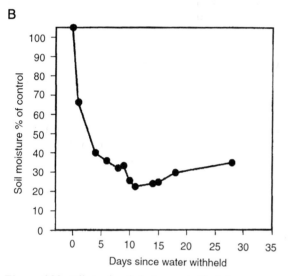

Figure 4.28 Effect of soil drying on half of the root system on root abscisic acid (ABA) content and leaf stomatal conductance of twin-rooted 'Chardonnay' vines (**A**); the decline in soil moisture of the dried half is illustrated in (**B**). (From Loveys *et al.*, 1998, reproduced by permission)

properties of the soil (percentage of stones, sand, silt, and clay) and its organic components (humus, plant, and animal remains).

When soil becomes saturated after a rain or irrigation, all its cavities are filled with water. Within several hours, water contained in the larger soil voids percolates out of the root zone. Because this **gravitational water** is lost so rapidly, it seldom plays a significant role in plant growth. The amount of water that remains is termed the **field capacity**. This component is held by forces sufficient to counteract the action of gravity. These forces include those involved in the adsorption and hygroscopic bonding of water to soil particles, as well as the capillary action of small soil pores and fissures. Some of the water is held weakly and can be

readily absorbed by roots. As this readily available portion is absorbed, the roots increasingly must extract water held more strongly in minute pores or on the surface of soil particles. If dry conditions prevail, the roots extract all the available water and the **permanent wilting percentage** is reached. At this point, plants can no longer extract water from the soil. Consequently, **available water** refers to the difference between the soil's field capacity and its permanent wilting percentage.

Different instruments measure various aspects of soil/ water relations, but none directly assesses water availability. Those typically used in vineyard applications are borehole techniques that provide indicators of local soil water content. **Tensiometers** measure soil water potential, an indicator of the force required to extract water. Accurate to about −0.1 MPa, tensiometer readings are of value only when the grower wishes to maintain the soil at or near field capacity. Various forms of **resistance blocks** measure the electrical resistance of the soil, and are accurate to water potentials between −0.7 and −1.5 MPa. **Neutron probes** do not directly measure water content, but instead estimate its presence by the slowing of fast neutrons by the soil's hydrogen content. Neutron probes are valuable because they function over the full range of soil water contents, can estimate water content at any depth, and averages the measurement within a spherical column about 20–30 cm around the probe. The instrument is particularly useful in soils low in organic content, where essentially all the hydrogen is associated with water molecules, and in the absence of high chloride contents (Hanson and Dickey, 1993). Some recent models have been adjusted and can account for such errors. Although more accurate than other techniques, its measurements near the surface tend to be unreliable, due to inclusion of the air-ground interface. In addition, the averaging of data over its detection zone can mask sharp changes in the actual soil-water profile.

A limitation to all these measuring devices is their ability to assess water content only in the region immediately around the probe. Thus, they do not provide an accurate map of spatial and temporal variability in moisture conditions throughout the vineyard. Such variation can significantly affect vine growth and irrigation efficacy. Airborne imaging is also limited in that it can assess moisture content to a depth of only 5 cm, and on bare ground.

For mapping soil moisture content in a vineyard, land-based **ground penetrating radar** (GPR) appears especially applicable (Huisman *et al.*, 2003; Hubbard and Rubin, 2004; Lunt *et al.*, 2005). GPR units project high-frequency electromagnetic pulses into the soil. The reflection travel time taken for the wave to move through the soil is used to estimate the dielectric constant of the soil. These data can be calibrated to soil's water content by comparison with data from other instruments, such as the neutron probe. GPR units can assess water content both near the surface and deeper in the soil. By generating both ground waves (shallow) and reflected waves (deep), measurements taken at short distances throughout the vineyards can provide a map of soil water content. Although soil moisture content fluctuates throughout the year, patterns are generally consistent, depending largely on constants such as soil texture and topography. The data can generate a map locating problem sites that need modifying, for example waterlogged areas (requiring drainage or, at least, less irrigation), or shallow or sandy regions (that may require additional irrigation). Nevertheless, since GPR does not provide continuing data, it cannot be used to time irrigation.

The primary function of water measurement is to assess the effectiveness of irrigation in maintaining an adequate supply of available moisture to the grapevine. When water flows into soil, it raises the immediate wetted area to its field capacity, before it moves vertically or laterally into the surrounding soil. Thus, once the zone effectively used by the root system, and the soil's field capacity have been determined, a neutron probe can chart water loss in the region. In addition, the amount of water required to reestablish the soil to field capacity can be estimated.

The proportion and amount of available water vary widely with soil type. Sandy soils have the lowest water retentive properties, but have the highest proportion (upward of 70%) in the available form. They also tend to permit deeper root penetration that enhances water access. Clay soils often possess the highest field capacities, but only about 35% may be held weakly enough to permit easy removal by plants (Milne, 1988). There are also increased risks of the presence of hardpans that limit deep root penetration. Silt soils often retain slightly less water than clay soils, but more of the aqueous component is available (up to 200 mm/m). Water availability problems are accentuated in young vines that have as yet not developed an extensive or deep root system.

In addition, root distribution and, thereby, water extraction is uneven throughout the soil. To account for this, critical measurements now include a weighting that reflects the root distribution profile in the soil (Stevens *et al.*, 1995). This is called the **root-weighted soil matric potential** (τ_{sRW}). Regrettably, because of its cost, the assessment is used only in research studies.

Another important aspect of soil structure and composition relates to the ease with which the available water can be absorbed by roots. Up to 90% of the available water in sandy soils can be readily removed at tensions equal to or less than −0.2 MPa (2 bar) (Hagan, 1955). This corresponds to the typical osmotic tension (water

potential) of cell cytoplasm. In contrast, only about 30–50% of the available water in clay soils can be removed at such low tensions. Below a soil moisture tension of −0.2 MPa, transpiration from leaf surfaces provides the force needed to extract water and transport it up the vine. Under periods of water deficit, leaf and root water potentials can decline by an additional −0.4 MPa (Düring, 1984). This facilitates water uptake, while retarding transpirational water loss.

Another significant aspect of soil–vine water relations concerns the rate at which the soil water potential declines. The more rapidly the soil water potential falls, the sooner vines are likely to experience water deficit. This means that during drought conditions, water stress tends to develop both earlier and more suddenly on sandy soil than on silt or clay soils. Grapevines cannot extract significant amounts of water from soil below −1.5 MPa. Leaf stomata generally close when the water potential, developed by transpiration, falls below −1.3 MPa (Kriedemann and Smart, 1971).

Despite the usefulness of soil-water measurements, they still do not indicate the amount of water biologically available to the roots. In addition, they cannot indicate vine water demands – these can and typically vary throughout the day, as well as the season. Leaf wilting, the standard sign of high water deficit in most flowering plants, is not readily apparent in mature grapevine leaves. When it does occur, it is typically restricted to young leaves and the shoot apex. An earlier indicator of water deficit is a decrease in the angle subtended by the petiole and the plane of the leaf blade (Smart, 1974). If leaf wilt occurs, it develops quickly and systemically. This situation typically occurs only on shallow or sandy soils. On deep or silty–clayey soils, water deficit tends to build up slowly, and osmotic adjustments and stomatal closure limit leaf wilting. Although wilting is seldom expressed in grapevines, other signs of water deficit can develop early and under conditions of mild water deficit. Thus, wilting occurs too late to be useful in directing water use strategy.

Other factors influencing the development and severity of water deficits are the temperature (affecting the rate of transpiration) and light exposure (increasing temperature and favoring photosynthesis that increases transpiration). As a result, the type of training system can affect water demand. More open and divided training systems, like the Lyre and GDC, increases light and air penetration. Thus, they tend to have higher water demands than smaller, single canopy systems. Minimally pruned vines also tend to have proportionally greater water demands in early season than other training systems, due to earlier canopy development. In addition, varieties differ in tolerance to water deficit. For example, 'Grenache' is relatively insensitive (anisohydric), in contrast to 'Shiraz,' which is sensitive (isohydric). Sensitivity may also be influenced by the rootstock, not only because of its own attributes, but those it may donate to the scion. For example the reduction in the number and size of stomata by '110 Richter,' noted previously.

Water deficit develops not only when roots experience difficulty in extracting water, but also when air voids form in xylem vessels, breaking the water column (cavitation) (Schultz and Matthews, 1993). This forces water to move sideways into adjacent vessels, increasing friction and retarding flow. Under prolonged water deficit, vessel formation is disrupted (Lovisolo and Schubert, 1998).

The sensitivity of different tissues and physiological processes to water deficit varies widely in grapevines. Because the shoot tip is particularly sensitive, suppression of shoot elongation is one of the earliest signs of water deficit. As shoot growth slows, the yellow-green color of young leaves and shoot tips changes to the gray-green of mature leaves. The root system reacts markedly, but differently, to water deficit. Initially, growth and fine-root production are stimulated by mild water deficit (van Zyl, 1988), whereas severe drought restricts root growth, especially at the soil surface. Both roots and leaves adjust osmotically by increasing their solute concentration (Düring, 1984). Roots also increase the production of abscisic acid. Its transportation to the shoot induces stomatal closure in the leaves and limits photosynthesis. However, because transpiration is suppressed more than photosynthesis, photosynthetic efficiency relative to water loss increases (Düring et al., 1996). The relative significance of reduced water uptake, impaired transport, and diminished transpiration to the effects of water shortage are unresolved.

Stomata also open later and close earlier in the day as soil water potential falls. This results in a rise in leaf temperature and suppression of photosynthesis. Cultivars showing drought tolerance may make better use of their water supply. For example, 'Riesling' shows earlier stomatal closure than do drought-sensitive varieties, such as 'Shiraz.' The responsiveness of leaves to water deficit allows them to minimize water loss, while optimizing carbon dioxide uptake for photosynthesis (Düring, 1990).

Water deficit can reduce fruit set, berry size, and inflorescence initiation and development. The degree to which these occur depends both on the timing and duration of the deficit (Smart and Coombe, 1983). The most sensitive period is between flowering and fruit set. Stress throughout this stage will reduce successful pollination and fertilization, resulting in marked flower and nascent fruit abscission. Much of the increase in berry size that occurs shortly after anthesis results from cell division (Harris et al., 1968). Correspondingly, water stress can permanently limit berry enlargement. Limited water

deficit later on has been used to limit berry size and cluster compactness and thereby diminish the incidence of bunch rot. Limited water deficit can also enhance proportional anthocyanin content. Although some of this is due to the increased skin/pulp ratio, part of it is independent of size influences (Roby et al., 2004). It is suspected that this is due to a differential growth sensitivity of the inner pulp vs. the skin. °Brix, and anthocyanin contents increase with berry size, but not proportional to the berry enlargement, resulting in a concentration effect. Several other researches have obtained data suggesting that other factors, and not just surface area to volume ratio, are involved in improved wine quality with smaller fruit size (Chapman et al., 2005; Walker et al., 2005). However, the potential benefit of improved fruit quality by moderate water deficit has to be balanced against suppression of leaf formation and carbon fixation. The latter are essential for the crop to mature.

Subsequent to *véraison*, fruit expansion is less influenced by water deficit (Matthews and Anderson, 1989). Mild stress following *véraison* can be beneficial in hastening fruit ripening, enhancing sugar and anthocyanin contents, and diminishing excessive acidity. The effects on pH tend to be more variable, probably because pH depends not only on the concentration of the various organic acids, but also on the accumulation and location of potassium in the berry. However, marked water deficit clearly has adverse effects on fruit composition and the latter stages of maturation, and can lead to shoot tip death, basal leaf drop, and the curling and spotting of young leaves.

By suppressing shoot growth, mild water deficit can limit competition between vegetative growth and fruit development and, thus, favor the formation of a more open and desirable canopy microclimate. Moderate water deficit also favors periderm formation in shoots. Although photosynthesis is not markedly affected by moderate water deficit, the transport and accumulation of sugars in berries and old wood are favored (Schneider, 1989).

After harvest, both drought and overwatering should be avoided to favor leaf function into the autumn, restrict late vegetative growth, promote cane maturation and encourage an autumn surge in root growth. Frost penetration during the winter will also be deeper in dry soils, leading potentially to more root damage.

Although problems associated with water deficit are more common, the saturation of the soil with water also can have undesirable consequences. If it occurs late in the growing season, it can induce skin cracking and favor bunch rot. Protracted periods of soil saturation suppress root growth. Even maintaining the soil surface at or near field capacity by irrigation limits the growth of surface roots (van Zyl, 1988).

Timing and Need for Irrigation

In areas supplied with adequate rainfall and at appropriate times, irrigation is unnecessary or rarely required. Even in semiarid regions, if rooting is deep, irrigation may be unessential. Irrigation water in these situations may be useful primarily in frost control. However, in arid and most semiarid regions, irrigation provides the grape grower with a supplemental means of influencing fruit yield and quality. With RDI or PRD (noted above), berry size may be reduced, fruit color intensified, and flavor potential enhanced (McCarthy *et al.*, 1996). Limited water deficit can also result in an increase in the proportion of pigment-tannin complexes (Kennedy *et al.*, 2002). This might explain the increased anthocyanin extraction associated with mild water deficit, although little increase in anthocyanin accumulation was noted. The increase in skin/flesh ratio can also raise juice pH (most of the potassium coming from skin). The main obstacle to applying the potential of water deficit regulation is the lack of a simple, inexpensive means of assessing vine water deficit and soil-water availability in a timely fashion and throughout the vineyard.

An indicator of grapevine water deficit is the difference between the ambient and canopy temperatures, and the rapidity of its development. The difference between ambient and leaf temperature influences stomatal opening and, therefore, the development of low water potentials in the vine. As water potential falls, the stomata close, the cooling produced by transpiration diminishes, and leaf temperature rises. The rapidity with which a temperature differential develops during the day indicates the degree to which the roots are experiencing difficulty in extracting water from the soil. Thus, the dynamics of the temperature differential is a biological indicator of how soil, atmosphere, and canopy conditions impact vine water demand. Although less sensitive to water stress than shoot elongation, the ambient canopy-temperature differential can be more easily and frequently measured. Hand-held infrared thermometers have made leaf-temperature measurement relatively simple. Regrettably, proper interpretation of the data is far less simple (Stockle and Dugas, 1992). More complex and instrumentally demanding indicators, based on estimates of evapotranspiration potential and canopy temperature are being assessed. Evapotranspiration refers to the water lost both by leaf transpiration and by evaporation from the soil. Other devices, such as neutron probes and tensiometers indirectly predict water deficit by assessing water availability in the soil.

Because most water deficit indicators are influenced by current and past vineyard conditions, proper interpretation of the results needs to include several independent factors. The influence of intermittent sunny periods on these indicators is clear. Less obvious are factors such as the lingering effect of wind exposure. Windy conditions can affect stomatal opening and, thereby, the canopy temperature several days after the winds have ceased (Kobriger *et al.*, 1984). Also, soil texture and depth can markedly affect the speed of water-deficit development. Moreover, canopy size and structure greatly affect water demand.

A new technique – **sap-flow sensors** – may provide the effective assessment of water use and need (Eastham and Gray, 1998). This is especially so with drip irrigation, in which the standard measures of assessing soil water availability have little applicability. Sap-flow sensors use temperature sensors to measure the rate at which heat applied to the trunk is dissipated by water flowing through the xylem vessels. The rate of heat lost can be converted into the rate of water flow. Not only does the technique indicate the current rate of transpiration of the vine, but it can also be used to assess water use on an hourly, daily, weekly, or other-time-frame basis. Thus, the technique has the potential to maximize efficient water use and application, as well as optimize control over vine and berry development. Trunk and berry diameter measurements can perform the same function (Ton and Kopyt, 2004; Kopyt and Ton, 2007), but its vineyard application is limited due to the instrument's fragility.

Currently, the standard indicator of vine water deficiency is **water potential**. This may be determined as the predawn **stem water potential** (stem Ψ) (Choné *et al.*, 2001b), or midday **leaf water potential** (leaf Ψ) (Padgett-Johnson *et al.*, 2003). These are, respectively, a measure of the relative ability of the vine to extract water from the soil (average water deficit), or current water stress. There is disagreement as to which is the better indicator of irrigation need. Predawn stem Ψ involves the removal of a healthy mature leaf before sunrise and placing it in a sealed chamber with the petiole extending. The leaf is placed in a plastic container and kept out of light for about 1 hour before the measurement is taken. As gas pressure is applied to the chamber, fluid drawn into the petiole when the leaf was cut is forced back into the petiole. When fluid begins to appear at the cut end of the leaf petiole, the pressure indicates the stem Ψ. Midday leaf Ψ involves removing a healthy, sunlit, fully expanded leaf, within one-half hour of solar noon (about 1 p.m.). The interval between cutting, insertion into the chamber, and covering from light exposure should be as short as possible.

Research from Williams at University of California, Davis suggests that application of water is not required unless stem Ψ or leaf Ψ fall below -0.75 MPa (-7.5 bar) or -1.0 MPa (-10 bar), respectively. Vines without water deficit frequently possess stem Ψ in the range of -0.2 MPa. It seems most important to avoid anything more than slight water stress during the first phase of berry development (pea size), and limit it during the later portion of fruit maturation.

Although water potential is a valuable research tool in studying vine water use, it is not a convenient tool in directing water use application. Simpler techniques are needed for the grape grower. **Infrared thermography** (Guisard and Tesic, 2006) has the potential of becoming such a tool, if its use can be simplified (with intelligent software), making it more intuitive and user friendly.

Precision irrigation has been plagued by imprecise knowledge on the vertical and horizontal diversity of soil conditions, technical problems in measuring soil-water availability, as well as a precise understanding of vine needs. Advances in understand vine physiology and modeling of soil water distribution may provide better solutions in the future (Fuentes *et al.*, 2004). Regrettably, where water comes from a single common source (as in most irrigation districts), irrigation may occur more often on a preset timed-allotment basis, rather than on need. Thus, there may be little opportunity to capitalize on technical improvement and increased understanding.

Water Quality and Salinity

Water's polar nature means that it typically contains a variety of dissolved salts, ions, and suspended particles. Unless absorbed by the plant, precipitated in insoluble forms, or leached from the soil, the salts and ions can accumulate and lead to salination. The significance of dissolved salts in irrigation water depends not only on their concentration and chemical nature, but also on soil texture, depth and drainage, annual precipitation, and irrigation method. Typically, salt toxicity develops only under arid to semiarid conditions. Under such conditions, upward (capillary) water flow, induced by evaporation from the soil surface, can result in the concentration of salts in the upper soil horizon.

The most common toxic salts found in water are borates and chlorides. Of these, grapevines are particularly sensitive to chlorides. The initial symptoms of chlorine toxicity include leaf chlorosis. It often begins along the margins, which turn necrotic, and progresses inward ("leaf burn"). Physiological effects, such as delayed fruit maturation, smaller berry size, and reduced sugar accumulation, are probably the consequence of disrupted photosynthesis. Even where salt uptake does not induce leaf burn, or result in detectable changes in grape phenolic composition, significant negative sensory changes may be detectable in the wine (Walker *et al.*, 2003). Effects appear to be directly related to the duration of exposure and degree of salinity (Shani and Ben-Gal, 2005).

Salt tolerance varies considerably among scion and rootstock cultivars. Under conditions where precipitation or drainage is inadequate to leach the salts, or the water quality is poor (saline), grafting to relatively salt-tolerant rootstocks is advisable, and may be necessary. Of the commercial rootstock cultivars tested by Downton (1977), 'Rupestris du Lot,' 'Schwarzmann,' '99R,' and '34EM' effectively exclude both chlorine and sodium. There is evidence, though, that the salt tolerance of some rootstock varieties declines over time, for example 'Ramsey' and '1103 Paulsen' (Walker *et al.*, 2003). Rootstock salt tolerance may also be affected by scion characteristics. Information on rootstock characteristics for sodium and chlorine exclusion may be found in McCarthy (1997).

Sodium occasionally can reach toxic levels, but sodium accumulation is generally more significant through its disruption of soil structure and permeability. Sodium also can result in the displacement and subsequent loss of calcium and magnesium from the soil. The exchange of divalent cations for monovalent sodium weakens the association between clay particles that help generate soil-aggregate formation. With aggregate structure lost, clay particles can flow downward with the water, eventually plugging soil capillaries. Over time, this can lead to the establishment of a **claypan**. Drying at the surface generates a rigid top layer that is both difficult to cultivate and relatively impermeable to water infiltration. Sodium accumulation also raises soil pH, releasing caustic carbonate and bicarbonate ions into the soil solution. Such conditions can result in sodic (alkali) soils. Salt accumulation decreases water availability by decreasing soil water potential and increasing the force required by roots to extract water. Sodic soils are also sticky when wet, and nearly water impermeable when dry.

Because of the many variables affecting salt buildup in the soil, and its effects, it is difficult to make panoptic statements about water quality. Nevertheless, water possessing electrical conductivity (EC_e) values below 0.75 mmhos/cm,[1] a ratio of sodium to calcium and magnesium content (SAR) below 8, and slightly acid to alkaline pH (6.5–8.5), generally do not create problems. Low chloride (<100 ppm) and boron ($\leq$1 ppm) levels are also desirable. Water with higher levels of these indicators can be safely used in some circumstances, where natural conditions or increased irrigation leach them out of the root zone. Adding powered gypsum (calcium sulfate) to the water also can counteract high SAR values. The relatively soluble calcium in gypsum can displace sodium on clay particles, permitting the sodium to

---

[1] An EC_e value of 1 mmho/cm is produced by about 640 ppm of salt.

be leached away. Calcium can also limit sodium toxicity in grape vines. This is most efficiently achieved by adding liquid calcium to drip irrigation water (Rodgers, 1999). The property of divalent calcium ions (Ca^{2+}) to bind clay particles is particularly useful with hard-setting clays. It can promote water infiltration and ease machine access to land after rains or irrigation. A precise method of surveying the extent and location of salinity problems in a vineyard uses an electromagnetic induction meter (Evans, 1998). It can measure soil conductivity at several depths.

In areas characterized by shallow saline water tables, poor irrigation management can raise the level of the saline water into the root zone. Subsequent capillary action may continue to add salts to the upper soil horizons. Waterlogging and the associated poor aeration of the soil can enhance the uptake and damage caused by chloride and sodium ions (West and Taylor, 1984).

Types of Irrigation

Where the availability of irrigation water is limited, the use of drought-tolerant rootstock and scion varieties may be crucial to vineyard success. However, where irrigation is permissible, necessary, and the water of adequate quality, factors such as soil texture, depth, and slope, heat, wind, cost, and tillage practice are critical factors in choosing an appropriate system. Additional factors affecting choice may be its use for frost and heat protection, and the potential benefits of simultaneous fertilizer and pesticide application.

Of the factors influencing irrigation decisions, water pricing and availability are typically beyond the control of the grower. Where irrigation water is in ample supply and has a low cost, systems requiring low initial costs, such as furrow irrigation, may be viable options. However, where water is costly or is in limited supply, the use of systems such as drip irrigation is much more efficient and cost-effective. Water quality also greatly affects system feasibility. For example, use of saline water is less damaging when used with either broad-bottom furrows or drip irrigation. Broad-bottom furrows disperse the water over a large surface area, and thereby delay the accumulation of high salt concentrations. Natural rainfall or additional irrigation may be sufficient to prevent serious salt buildup. With drip irrigation, the slow but frequent addition of water tends to move salts to the edge of the wetted area and away from the region of root concentration.

Furrow irrigation is an ancient but effective means of irrigation. Where water supply is abundant and inexpensive, inefficient water use may not be a critical factor. This situation is becoming increasingly rare. Plant water use may be as low as 30%, but is commonly around 60–70%. Where the water is marginally saline, additional irrigation or rainfall is required to flush out salt accumulations. In such situations, it is essential to have adequate drainage to avoid raising the water table to the root zone.

Typically, furrow irrigation involves several evenly spaced, shallow, V-shaped trenches, or a few wide, flat-bottomed furrows between each row. Where the soil is sandy and penetration rapid, the furrows are kept relatively short. This avoids erosion resulting from the rapid filling needed to achieve even water penetration. Broad, flat-bottomed furrows are often used to offset the minimal lateral movement of water in sandy soils and favor uniform irrigation. Broad furrows are also preferred because of their large surface area where water salinity is a problem. Silt and loam soils, with their considerable lateral water flow and moderate infiltration rates can effectively use long, narrow furrows. Clay soil, with their very slow infiltration rates (and correspondingly slow filling rates) are seldom suitable to furrow irrigation. Furrow irrigation in block form (Fig. 4.29) has often been used in California.

In furrow irrigation, sufficient water is added to moisten the effective rooting depth to field capacity. Typically this is about 1–1.5 m. Additional water may be added periodically to flush out salt accumulations.

Furrow irrigation is most easily used where the ground is flat, has been leveled, or possesses no more than a minimal slope. Otherwise, a series of checks along the furrows are required to divide the channel into self-contained segments, each showing an acceptable level drop. Furrow irrigation is not feasible on hilly terrain because of water runoff and erosion.

In fine-textured soils, furrow irrigation has the tendency to cause clay-particle dispersion and hardening of the soil surface. This increasingly causes long retention times for water infiltration and enhanced evaporative water loss. Where there is insufficient earthworm activity to keep the soil porous, periodic cultivation may be required to maintain adequate water permeability.

Because of the large area of wet soil, root growth is promoted through much of the upper soil volume. Where the soil alone acts as the primary source of inorganic nutrients, this is desirable. However, it can be wasteful when chemical fertilizer is applied. The large soil volume requires more fertilizer than is required with drip irrigation. Thus, furrow irrigation often results in greater nutrient loss by leaching, volatilization, or uptake by weeds and microbes in comparison with drip irrigation. Furrow irrigation, by moistening most of the soil often accentuates weed problems.

Sprinkler irrigation has distinct advantages on sloping terrain, where runoff and erosion are potential problems with other systems. Sprinklers also have advantages in

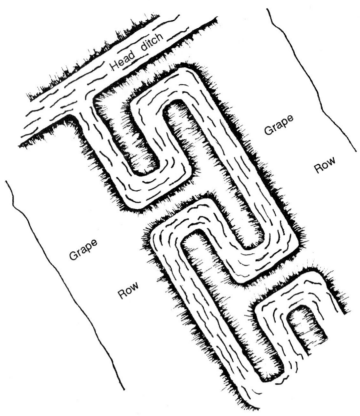

Figure 4.29 Diagram of a block furrow irrigation system. (From Bishop *et al.*, 1967, reproduced by permission)

both highly and poorly porous soil. In porous soils, there is little lateral movement of water and its non-uniform distribution can be a problem. The widespread dispersal of water that is possible with sprinklers can assure uniform soil moistening. With heavy soils, possessing low infiltration rates, judicious choice of nozzle permits water to be applied slowly and as fine droplets. Fine droplets tend not to disrupt soil-aggregate structure and, therefore, do not exacerbate existing permeability problems. Because of uniform water application, sprinkler irrigation is especially valuable in leaching saline soils and minimizing wind erosion. In addition, sprinkler systems can be used for the foliar application of micronutrients and pesticides. Although the initial installation costs of a fixed sprinkler system are high, subsequent labor costs are low. Generally between 25 and 35 sprinklers are used per hectare.

Despite these advantages, sprinklers owe much of their popularity to their use in frost control. Sprinkler irrigation has also been investigated as a means of heat control. However, its use in cooling requires that the water quality be high; otherwise, the vegetation may become coated with toxic levels of salts, notably borates and chlorides.

The major drawbacks of sprinkler irrigation are the high costs of installation and operation. Its property of uniform soil moistening also can lead to increased evaporative water loss from the soil and increased weed growth. Where fairly saline water must be used, irrigation should occur at night or on overcast days to avoid toxic salt buildup on the foliage and fruit. Sprinkler irrigation also tends to increase salt accumulation in leaf tissue, in comparison with equivalent root irrigation (Stevens *et al.*, 1996). By prolonging foliage wetting and enhancing pesticide removal, sprinkler irrigation potentially favors disease development. Although the strong drying associated with arid climates typically counteracts the development of most disease problems, washing off insecticides may increase pesticide use.

Movable sprinkler systems, such as the **wheel line** and **center pivot** systems, frequently used with annual crops, are rarely used in vineyards. Their movement requires gaps in trellised vineyards. Alternately, irrigation equipment may be moved down rows. Nevertheless, the periodic and often heavy application of water can lead to soil compaction, disease, and erosion problems.

For most new installations, especially where water costs are high or availability low, **drip (trickle) irrigation**

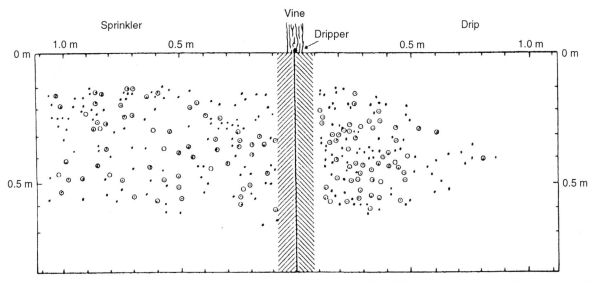

Figure 4.30 Comparison of root distribution in shallow soil, comparing sprinkler (*left*) and drip (*right*) irrigation. Diameter of roots (mm): ·, 1; ⊙, 1.5–4; ○, 5–7. (From Safran *et al.*, 1975, by permission)

is preferred. Water is supplied under low pressure and released through special emitters that generate a slow trickle. Emitters are spaced to produce a relatively uniform zone of irrigation along the length of each row. Consequently, the emitters themselves do not need to be close to vine trunks. The number and placement of emitters are primarily determined by the soil texture and how this influences the lateral water flow (Fig. 5.2). Water is supplied to the emitters through an array of surface or buried plastic pipes that run through the vineyard. Because the root system quickly becomes focused within the moist zone created by drip irrigation (Fig. 4.30), the system can be used even in established vineyards. Root-zone concentration permits the efficient application of fertilizers and nematicides to grapevines.

Root concentration limits the amount of water that must be supplied and permits application to be more specifically related to vine need. Not only can this improve efficient water use, but it also permits irrigation to regulate vegetative growth and modifying fruit quality. The slow rate of water release is especially useful in soils possessing slow infiltration properties, minimizing runoff. A slow trickle also means that little water percolates out of the root zone. Efficient water use can be further enhanced by pulse application and burial of the system. For example, pulse application every 2 hours minimizes percolative loss, whereas burial limits surface moistening. Both reduce surface evaporative loss and decrease salt accumulation at the soil surface. In addition, new types of buried drippers retard water release and disperse the water over a broader surface area. This should favor the development of an expanded root zone.

Drip irrigation is as effective on steep slopes as on rolling or flat surfaces. It is uninfluenced by wind conditions, which can restrict the timing of sprinkler irrigation. The localization of water application facilitates the control of all but drought-tolerant weeds. Furthermore, drip irrigation permits the better use of shallow soils, or those with saline water tables close to the soil surface.

Other advantages of drip irrigation include limited energy consumption (owing to low-wattage pumps), avoidance of salt accumulations on leaves, and improved efficiency of nutrient uptake (and thereby reduced fertilizer costs and nitrate contamination of groundwater). Moreover, it does not offset the benefits of arid environments in limiting most disease development.

Although possessing many advantages, drip irrigation is not without its problems and limitations. Primary among these is the tendency of emitters to plug. Plugging caused by particulate matter in the water is usually avoided by the use of filtration. Growth of slime-producing microbes in the system can usually be controlled by the continuous or periodic addition of chlorine to the water line (1 ppm or 10–20 ppm, respectively). Chlorination and flushing are particularly important if the water line is also used for fertilizer application. These nutrients could promote the abundant growth of algae and bacteria in the tubing and around the emitters. In addition, careful formulation is essential to avoid precipitation in the tubing. Corrosion-resistant emitters are also required due to the caustic properties of some fertilizers. Plugging from the deposition of calcium carbonate (lime) on emitters may be minimized by the inclusion of a homopolymer of maleic anhydride (Meyer *et al.*, 1991). The deposition of iron salts from well-water can be a serious if uncommon problem. Water with ≥ 0.4 ppm iron is typically oxygenated in holding ponds to promote precipitation before use. Finally, obstruction by root growth around the emitters in buried systems can be deterred by the incorporation of minute amounts of herbicides and the use of acid fertilizers.

Water application by drip irrigation must be frequent because the root system is concentrated in a comparatively small region of the soil. This is especially important in sandy soils, in which the wetted zone is narrow, due to the limited lateral movement of water (see Fig. 5.2). The addition of fertilizer to the water supply is often required due to the restriction of the soil volume used by the root system and the tendency for leaching. Although fertilization with irrigation is an added expense, it further enhances the ability of irrigation to regulate vine growth and fruit ripening.

Fertilization

In the previous section, the potential use of irrigation water in applying fertilizers was introduced. Although it is possible with any system, fertilization via drip irrigation has a unique potential. Bravdo and Hepner (1987) have stressed the ability of combined fertilization and irrigation (**fertigation**) to regulate vine growth and grape quality. It provides the opportunity, under field conditions, to achieve some of the control possible with hydroponics. Fertigation is most applicable on sandy soil with minimal nutrient retention properties. The appropriate use of fertigation to regulate vine growth requires the knowledge of both the factors that affect water and nutrient availability and their effects on the various stages of vine growth. This is particularly true in reference to potassium, where excess fertilization can disrupt magnesium uptake. When employed, fertilizer should be added near the end of the irrigation period to avoid flushing the fertilizer out of the root zone. In the previous section, the influences of water availability were discussed. In this section, nutrient availability and its effects on grapevine growth are discussed.

Based on relative need, inorganic nutrients are grouped into macro- and micro-nutrient classes. Macronutrients include the three elements typically found in most commercial fertilizers – nitrogen (N), phosphorus (P), and potassium (K) – as well as calcium (Ca), magnesium (Mg), and sulfur (S). There is no need to discuss the other major elements required by living cells – carbon (C), hydrogen (H), and oxygen (O) – because they come from the atmosphere and water. They are also required in much higher amounts than any of the mineral elements. Micronutrients are required only in trace amounts. These include boron (B), chlorine (Cl), copper (Cu), iron (Fe),

manganese (Mn), molybdenum (Mo), and zinc (Zn). Most are involved as catalysts in pigments, enzymes, and vitamins, or in their activation.

Factors Affecting Nutrient Supply and Acquisition

Although nutrient availability is primarily dependent on the mineral and organic makeup of the soil, nutrient uptake is dependent on the physiological characteristics of the scion and rootstock. In comparison with other crops, grapevines have relatively limited nutrient demands (Olson and Kurtz, 1982). This, combined with the considerable accumulation of nutrients in the woody parts of the plant, makes assessing the response of grapevines to fertilizer application particularly exacting.

Although the soil acts as a nutrient reservoir, most of it is in an unavailable form. Most assimilable nutrients occur dissolved in the soil solution (10^{-6} to 10^{-3} M). This constitutes less than 0.2% of that present in the soil. Nearly all nutrients (about 98%) are bound in unavailable forms in the humus and mineral fractions of the soil. These become available only slowly, as the humus decomposes and the mineral fraction weathers. Most of the remaining 2% is bound by weak electrostatic forces to colloids in the soil (humus and clay particles), or as chelates with organic compounds. Soil colloids possess an immense surface – 600–800 m²/g for montmorillonite clays and upward of 700 m²/g for humus (Brady, 1974). These nutrients become available as a result of shifts in the equilibria between sorbed and dissolved forms (Scheidegger and Sparks, 1996), as well as through ion exchange. Their presence significantly enhances nutrient availability and acquisition. In contrast, sandy soils contain little colloidal material. Because sandy soils retain few nutrients, and what is presence is easily lost by leaching, they require higher or more frequent fertilizer application.

Both organic and inorganic soil colloids possess a net negative charge. Thus, they effectively retain extractable positively charged ions (**cations**). How readily these dissolve into the soil solution depends on their valence charge, their tendency to become hydrated, the soil pH, and the presence of other ions. Most negatively charged ions (**anions**) exist organically bound in the humus. As free ions, they do not sorb well onto soil particles. Thus, anions such as nitrates and sulfates are comparatively mobile and readily leached out of soil (Fig. 4.31). Nevertheless, this property allows them to be applied effectively on the soil surface, where rainfall or irrigation water can move them down into the root zone. In contrast, the low solubility of phosphate salts and their rapid combination with aluminum, iron, and calcium ions restrict phosphate movement in soil. Thus, although

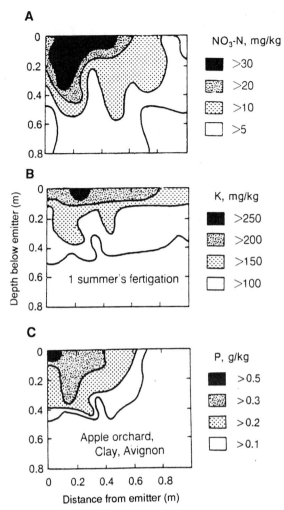

Figure 4.31 Measured pattern of nitrogen, potassium, and phosphorus in soil after one summer's application of soluble fertilizer from a drip emitter. (Redrawn from Guennelon *et al.*, 1979, in Elrick and Clothier, 1990, reproduced by permission)

The sorptive binding of most mineral elements to soil colloids has many advantages. It limits nutrient loss from the root zone and retains them in a readily dissolved form. Sorption also helps keep the nutrient concentration in the soil solution low (10^{-6} to 10^{-3} M). Consequently, the water potential of the soil remains high, easing vine access to water. It also minimizes the development of toxic nutrient levels in the soil solution. Finally, the dynamic equilibria between sorbed and free forms help maintain a relatively stable supply of nutrient cations.

Roots may gain access to nutrients by several means. Usually, nutrients are directly assimilated from dissolved ions in the soil solution. This shifts the equilibrium between free and sorbed forms, replenishing the supply of available nutrients. In addition, the release of H^+ ions, carbon dioxide, and organic acids from roots makes H^+ ions available for cation exchange with sorbed cations. This further releases cations into a state accessible to roots. Hydrogen ions are also involved in converting insoluble ferric ions (Fe^{3+}) into the more soluble ferrous (Fe^{2+}) state. A somewhat similar release of negatively charged nutrients occurs in alkaline soils by anion exchange, but is largely restricted to phosphate salts. Organic phosphates are released through the action of extracellular phosphatases liberated by microbes and roots. Finally, the release of chelating and reducing compounds by roots and microbes helps to keep metallic ions, such as iron and zinc, in readily available forms. Most metallic cations tend to be in limited supply in neutral and alkaline soils, due to the formation of insoluble oxides, sulfides, silicates, and carbonates.

In addition to the uptake of nutrients for growth, cations such as potassium may be incorporated to maintain the electrical and osmotic balance of the cytoplasm. This may be required to counter the negative charges associated with the uptake of the major nutrient anions, NO_3^- and PO_4^{3-}.

In spite of the processes releasing soil nutrients, plant demand often outstrips the ability of the soil to replenish its nutrient supply adjacent to feeder roots. Consequently, root extension into new regions is usually vital to maintaining an adequate nutrient supply. This is especially important for nutrients, such as phosphates, zinc, and copper, which do not migrate significantly in soil.

Although uptake increases the liberation of sorbed ions into the soil solution, the addition of fertilizer can reverse the process, resulting in the precipitation of soluble nutrients. This, combined with deep rooting and nutrient storage in the vine (Conradie, 1988), helps to explain why grapevines often respond slowly and marginally to fertilizer application.

phosphate is usually in adequate supply, direct deposition in furrows within the root zone is required if phosphorus is deficient. Most nutrient cations show limited movement in soils.

The tendency of mineral cations to sorb to soil colloids decreases in the order Ca^{2+}, Mg^{2+}, NH_3^+, K^+; whereas anions decrease in the order PO_4^{3-}, SO_4^{3-}, NO_3^-, Cl^-. Heavy metal ions such as Zn^{2+} are sorbed, but only in trace amounts. The relative tendencies of these ions to adhere to soil colloids influences how they affect each other's sorption–desorption equilibria. This, in turn, affects their retention and plant availability. For example, liming soil provides active calcium, which replaces hydrogen and other cations, whereas the liberal application of potassium fertilizer liberates calcium, and other ions. In addition, solutes in irrigation water both add and displace nutrients in the soil.

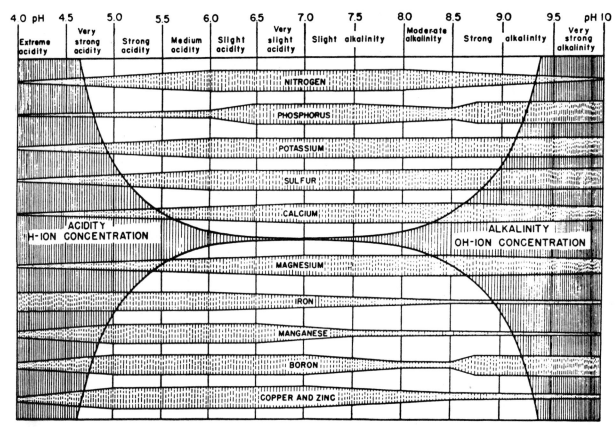

Figure 4.32 General relationship between soil pH (and associated factors) and availability of plant nutrients. The width of each band representing an element indicates the potential availability of the nutrient at a particular pH value. The width of the heavy cross-hatched area between the curved lines is proportional to the relative excess of hydrogen ions (*left*) or hydroxyl ions (*right*) at the corresponding pH. (From Truog, 1946, reproduced by permission)

Of soil factors, pH probably has the greatest influence on soil nutrient availability (Fig. 4.32). Depending on the chemical nature of the parental rock, the degree of weathering, and the organic content of the soil, most soils are buffered within a narrow pH range. In calcareous (lime) soils, the primary buffering salts are $CaCO_3$ and $Ca(HCO_3)_2$. Because of this buffering, the availability of Fe^{2+}, Mn^{2+}, Zn^{2+}, Cu^+, Cu^{2+}, and PO_4^{3-} is restricted. Through prolonged leaching, most soils in high-rainfall areas are acidic. Such soils are often deficient in available Ca^{2+}, Mg^{2+}, K^+, PO_4^{3-}, and MoO_4^{2-}. In contrast, the low rainfall of arid regions tends to generate saline or sodic soils, especially in the subsoil where leachates often accumulate.

Soil pH also influences the activity of microbes, especially bacteria, most of which are unable to grow under acidic conditions. The consequent slowing of microbial activity retards the oxidation of NH_4^+ to NO_3^+. Soil temperature also influences microbial activity, and thus the oxidation of ammonia to nitrate. Cool temperatures also slow the liberation of organically bound phosphates into the soil solution. Thus, if fertilizer is applied in the spring, nitrate and inorganic phosphate should be used because soil microbial activity is insufficient to effectively oxidize ammonia and liberate organically bound phosphates.

If vineyard soils are acidic, they are often treated with crushed limestone to raise their pH, preferably before planting. Alternately, in a procedure called **soil slotting**, surface and subsurface soil horizons are mixed with lime and gypsum. Slots 80 cm deep, 15 cm wide, and 1 m apart can significantly increase root penetration in acidic soils (Kirchhof *et al.*, 1990). The amount required depends on the neutralizing power of the lime source, as well as the soil pH, organic content, and texture (Fig. 4.33). Acid-tolerant rootstocks such as '140 Ruggeri,' '110 Richter,' or 'Gravesac' can minimize the detrimental effects of acidic soils, notably toxicity resulting from the increased solubility of aluminum, copper, and manganese.

Lime-tolerant rootstocks are extensively used on calcareous soils. In addition, sulfur or gypsum (calcium sulfate) may be added to alkaline and sodic soils. Elemental sulfur is microbially oxidized to sulfuric acid in the soil, neutralizing hydroxides, whereas calcium sulfate both neutralizes hydroxides and permits sodium leaching (by displacing it from soil colloids and carbonates).

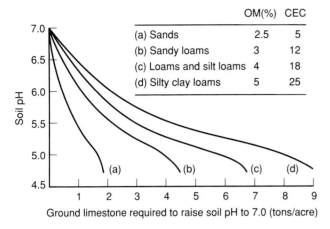

	OM(%)	CEC
(a) Sands	2.5	5
(b) Sandy loams	3	12
(c) Loams and silt loams	4	18
(d) Silty clay loams	5	25

Figure 4.33 Relationship between soil texture and amount of limestone required to raise the soil pH to 7.0. CEC, cation exchange capacity; OM, Organic matter. (From Peech, 1961, reproduced by permission)

An alternative or additional procedure is to rip the soil in mid-row and mound the soil in which the vines are planted. This is particularly useful where the water table is high (Cass *et al.*, 2004).

Rootstock choice can also differentially affect nutrient uptake. Although rootstocks differ little in the accumulation of nitrogen, phosphorus, and zinc, they vary widely in the absorption of potassium, calcium, magnesium, and chlorine (Fig. 4.34A). Such variation probably arises from the selective activity of transport systems in root cell membranes. Differences in nutrient concentration in grapevines also may develop because of differential accumulation by the scion (Fig. 4.34B).

Although nutrient uptake may be initially passive, all nutrients must pass into cytoplasm on, or before, reaching the endodermis. Direct access to the vascular system is prevented by the Casparian strip, which makes the endodermal cell wall impermeable to water-soluble substances. Thus, transport into the vascular tissues is under the metabolic control of one or more transport systems in the endodermis. Unloading into the xylem and phloem also is likely to be under metabolic, and thus genetic, control. Accumulation within different grapevine tissues is undoubtedly under genetic control.

Some rootstock-derived differences in mineral uptake may arise indirectly from the association of roots with particular mycorrhizal fungi. This is especially likely for phosphate, for which mycorrhizal association favors uptake from soils low in phosphates. Nevertheless, the leaf phosphorus content is correlated more to the specific demand of the scion than to phosphorus availability, or the degree of root mycorrhizal colonization (Karagiannidis *et al.*, 1997). In addition to phosphorus, mycorrhizae also improve zinc and copper uptake.

Assessment of Nutrient Need

Deficiency and toxicity symptoms are often sufficiently distinctive to be diagnostic. However, detrimental effects can occur before diagnostic symptoms appear. Thus, the need for more sensitive indicators of nutrient deficiency has long been known. The primary method of nutrient analysis is based on sampling plant tissue. Most standard soil analyses are unreliable in assessing nutrient stress because of marked differences between availability and vine uptake. Grapevine requirements also vary throughout the year. Nevertheless, soil nutrient and pH analyses can be useful in predicting major deficiency and toxicity problems.

Technological developments are facilitating the assessment of soil nutrient status. **Electroultrafiltration** (EUF), for example, significantly improves determining nutrient availability (Schepers and Saint-Fort, 1988). The technique measures the desorption of nitrogen, phosphorus, potassium, calcium, magnesium, manganese, and zinc, as well as the presence of phytotoxic levels of aluminum in soil. Although EUF is not widely used in viticulture, EUF values have shown high correlation with grapevine-nutrient uptake in some soils (Eifert *et al.*, 1982, 1985).

Possibly even more significant are advancements in *on-the-go* measurements of soil nutrient status (Adamchuk *et al.*, 2004). Although only a few sensors are currently available, this is an active area of research. Hopefully in the near future, assessment of the spatial distribution of nutrients throughout a vineyard may be possible. Not only would they permit high-resolution mapping of nutrients in a timely and efficient manner, but also improve adjusting application to need and diminish its environmental impact.

Multiple nitrate and potassium assessments can be obtained with the sampler mounted on the back of a tractor (Adamchuk *et al.*, 2003). Small soil samples collected as the tractor passes down the rows are brought in contact with sensitive membranes associated with ion-selective electrodes. Measurements are made in seconds. Electrode surfaces are quickly and automatically rinsed and ready for a new sample. Data can be georeferenced using a global positioning system (GPS). Alternatively, data could be used immediately to adjust fertilizer application.

As with other soil measurements, nutrient assessments only indicate potential availability. For determining actual uptake and accumulation, grapevine samples must be analyzed. These results can then be correlated with the soil nutrient availability. Standard analysis involves in-laboratory assessment of leaf petioles (Cook, 1966). Leaf petioles act as sensitive indicators of most elemental nutrients, and are easily collected in large,

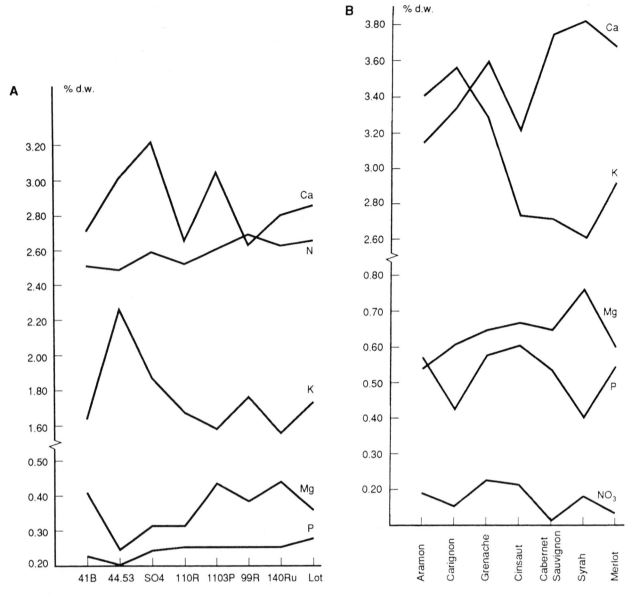

Figure 4.34 Level of mineral nutrients (% dry weight) in the leaves of 'Grenache' as function of eight rootstocks (**A**) and of the peptides of eight scion varieties on 'SO4' rootstock (**B**). Values of P and MG are double in both (A) and (B) and for NO_3 in (B). Data in (A) and (B) come from different sites and were obtained under different growth conditions. (From Loué and Boulay, 1984, reproduced by permission)

statistically valid numbers. In California, one sampling at full bloom from leaves opposite clusters is considered sufficient. In France, both leaf and petiole analyses are taken at the end of flowering and at *véraison*. In either method, consistent timing of measurement is essential as the relative nutrient content of the vine declines during the season (Williams, 1987).

Except for quick nitrogen analyses, samples are dried for 48 h at 70 °C, and ground to 20-mesh fineness. For nitrogen, fresh leaves can be assessed for nitrate content by cutting the petiole and leaf base lengthwise. A drop of indicator solution (1 g diphenylamine/100 ml

H_2SO_4) is added to the basal 2 cm of the cut surface. Deficiency is indicated when less than 25% of a 20-leaf sample shows bluing. When more than 75% of a 20-leaf sample shows a positive reaction, nitrogen availability may be in excess. Other nitrogen and nutrient assessments are usually conducted in the laboratory on dried samples by colorimetry and atomic absorption spectrometry.

Another component influencing the level of fertilization, or any other component in vineyard practice, is the relative financial return. For example, improved fruit yield and quality are maximal when fertilizer

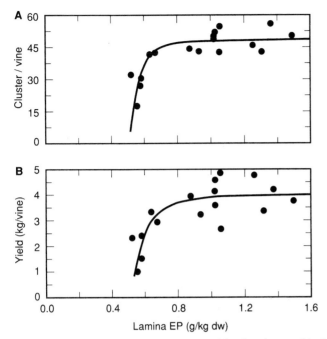

Figure 4.35 Relationship between extractable phosphorus of leaf blades (EP) and cluster number (**A**) and yield (**B**) in the second season after phosphorus application. (From Skinner *et al.*, 1988, reproduced by permission)

Table 4.8 Approximate range of nutrients needed by grapevines[a]

Nutrient	Deficiency	Adequate	Excess
Nitrogen (NO_3^-)	<50 ppm	600–1200 ppm	>2000 ppm
Phosphorus	<0.15%	0.15–0.2%	>0.3–0.6%
Potassium	<1%	1.2–2.5%	>3%
Magnesium	<0.3%	0.5–0.8%	>1%
Zinc	<15 ppm	25–150 ppm	>450 ppm
Boron	<25 ppm	25–60 ppm	>300 ppm (leaves)
Chlorine	<0.05%	0.05–0.15	>0.5%
Iron	<50 ppm	100–200 ppm	>300 ppm
Magnesium	<20 ppm	30–200 ppm	>500 ppm
Copper	<4 ppm	5–30 ppm	>40 ppm

[a]Based on petiole analysis of leaves, opposite clusters in full bloom except where noted.

Source: Data from Christensen *et al.*, 1978, Cook and Wheeler, 1978, and Fregoni, 1985

addition is low (root uptake high and runoff minimal) and fertilizer need is high. Frequently, however, additional application results in minimal or no amelioration. Optimal fertilizer addition is roughly equivalent to vine need, minus the amount already available in the soil (Löhnertz, 1991). Supply in excess of need seldom enhances vine growth or fruit quality, and can be detrimental.

Nutrient Requirements

Although soil nutrient status can be assessed fairly accurately, the interpretation of the results relative to nutrient availability is more difficult. Nutrient availability, uptake, and accumulation often vary considerably throughout the growing season (Fregoni, 1985) and can differ significantly among cultivars (Christensen, 1984). Thus, critical values need to be established for each cultivar and nutrient.

Below a certain value, a deficiency becomes progressively more severe. Within broad midrange concentrations, nutrient supply is adequate, and plant response is typically unmodified by increasing nutrient presence (Fig. 4.35). However, plants may show "luxury" storage beyond need. Various tissues frequently respond differentially in this regard. For example, fruit nitrogen levels commonly rise with increasing nitrogen fertilization, whereas leaf and petiole values may remain only

slightly modified. Eventually, excessive nutrient accumulation becomes increasingly toxic.

Adding to the complexity of determining nutrient need is the extensive storage and mobilization of nutrients in the woody parts of the vine. This may explain the occasional delay in vine response to fertilization. Under deficiency conditions, the response to fertilization is usually rapid, but may decline in subsequent years (Skinner and Matthews, 1990). Nutrient balance in the soil and vine tissues also can be important. For example, phosphorus can limit the uptake of potassium (Conradie and Saayman, 1989), whereas potassium antagonizes the adsorption of calcium and magnesium (Scienza *et al.*, 1986). Conversely, the addition of calcium in lime, used to increase soil pH, may interfere with potassium uptake. In addition, phosphorus deficiency can induce magnesium deficiency by limiting its translocation from the roots (Skinner and Matthews, 1990).

Because of these factors, only general ranges for most nutrient requirements are presently possible for grapevines (see Table 4.8). Deficient, adequate, and toxic levels for several micronutrients have not been determined for grapevines. Table 4.9 summarizes the known deficiency and toxicity symptoms in grapevines.

NITROGEN

Nitrogen is an essential constituent of amino acids and nucleotides and, therefore of proteins and nucleic acids. These are found in their highest concentrations in actively growing roots and photosynthesizing leaves. In the latter, a single photosynthetic enzyme, RuBP carboxylase, may constitute up to 50% of the protein content of

Table 4.9 Essential nutrient deficiency/toxicity symptoms

Nutrient	General deficiency symptoms	Additional deficiency systems	Toxicity symptoms due to excess
Nitrogen (N)	Shoot growth reduced	Diffuse yellow-green chlorosis, young shoots pinkish, leaves prematurely reddening in the fall	Deposition of white amino acid salts at leaf edges, severe burning in cases of extreme nitrogen excess
Phosphorus (P)	Shoot growth reduced, difficult to detect	Petioles and veins red in adult leaves, interveinal regions dark green, blades down turning; small fruit and poor yield, *véraison* delayed	
Potassium (K)	Shoot growth reduced, easily mistaken for Mg deficiency or moisture stress	Chlorosis/bronzing, then browning and necrosis, progress inward from the leaf edges, first seen in the central region of the primary shoot, sun-exposed basal leaves may become violet to dark brown ('black leaf'), internodes shortened; fruit small, of delayed or uneven maturity and of poor yield	
Magnesium (Mg)	Little effect unless severe, can be induced by excessive application of potassium or ammonium	Midveinal discoloration with leaf margins initially green, later becoming necrotic, petiolar region often remains green, shows initially in mid-season discoloration starts in the basal leaves and progresses upward, white cultivars become pale creamy colored while red cultivars develop a red/violet border next to the necrosis	
Calcium (Ca)	May be confused easily with potassium deficiency	Narrow yellow border along leaf edges that progresses inward, small, brown knobs may form on the internodal bark, shoots may dry, starting from the tip	
Iron (Fe)	Growth terminated with ends appearing burnt, especially frequent in calcareous soils	Terminal leaves are the first to show severe chlorosis (except along the veins), turning ivory and finally brown and necrotic, initial chlorosis is partially reversible on application of a foliar iron spray; fruit yield much reduced	
Boron (B)	Terminal shoot regions may be shortened, twisted and die back early, bushy appearance from activated lateral growth	First symptoms occur as dark bulges on the apical tendrils, misshapen leaves with patchy interveinal chlorosis or browning, beginning in terminal leaves of the primary shoot, brown speckling on leaf margin in late season internodes selective may swell, associated with necrosis of the pith; flower clusters may dry, fruit shatter excessively, or be small and seedless	Leaves may be severely distorted, leaf necrosis begins in the serrated outer margins of older leaves, progressing inward interveinally, activation of lateral shoot growth
Molybdenum (Mo)	Growth reduced with central region burnt	Leaves develop a brownish discoloration on edges of blades that progress to base of the leaf blade, unaffected zones very distinct; marked reduction in fruit yield	
Zinc (Zn)	Appears short of foliage, easily confused with Mn deficiency	Interveinal chlorotic mottling, with narrow green band along cleared veins, first seen in terminal leaves, angle with the petiolar sinus enlarged, frequent occurrence of shortened shoots and 'little-leaf' syndrome; straggly clusters with many small and 'shot' berries	
Manganese (Mn)	Shoot dieback at tip, deficiency on alkaline soils, toxicity on acid soils	Spotty interveinal yellowing surrounded by small green veins (herringbone pattern), first seen in the basal leaves but progresses up the shoot symptoms especially marked on sun-exposed leaves; yield reduction only in cases of extreme deficiency	Partially associated with Ca and Mg deficiency, where older leaves develop coalescing yellow to brown spots after flowering, red cultivars show bright red spots

chloroplasts. Nucleotides, in addition to their involvement in nucleic acid synthesis, function as essential electron carriers in cellular metabolism. Nitrogen is also a constituent of chlorophyll and several growth regulators. Although nitrogen is required in larger quantities than any other inorganic soil nutrient, grapevine needs are considerably less than those of most other agricultural crops. Yearly use is estimated to vary from approximately 40 to 70 kg N/ha (Champagnol, 1978; Löhnertz, 1991).

Although constituting about 78% of the Earth's atmosphere, nitrogen is unavailable to plants in the gaseous dinitrogen (N_2) form. Under natural conditions, nitrogen must first be reduced to ammonia (NH_4^+) by one of several nitrogen-fixing bacteria. Ammonia can be absorbed directly by grapevines, but nitrogen is usually assimilated as nitrate (NO_3^-). This occurs after ammonia has been oxidized to nitrate by soil-inhabiting nitrifying bacteria. Because of its negative charge, nitrate is more mobile in soil than positively charged ammonia. Ammonia commonly adheres to soil colloids and remains relatively immobile. Thus, nitrogen supplied as ammonia (or urea, which breaks down to ammonia) should be applied early enough for its microbial conversion to nitrate to coincide with the primary nutrient-uptake period of grapevine growth.

Until the use of inorganic nitrogen fertilizers, vineyard nitrogen supply was dependent primarily on three factors – the activity of free-living nitrogen-fixing bacteria in the soil, nitrogen fixed by endosymbiotic bacteria in the nodules of legumes, and the addition of manure. Unlike other soil nutrients, nitrogen is not a component of the soil mineral makeup. Its availability is particularly dependent on the effect of seasonal factors. These include features such as soil moisture, aeration and temperature, and on how these affect the activity of soil microorganisms and cover crops. In addition, as nitrate is poorly sorbed by soil colloids, nitrate tends to be leached out if not rapidly assimilated by plants and microbes. Nitrogen tends to be the nutrient most frequently deficient in vineyard soils.

The lower cost of urea and ammonia salts, combined with ammonia's ready sorption to soil particles, generally makes it the preferred form of nitrogen fertilizer. However, the tendency of ammonia to promote vegetative growth may predispose grapevines to increased disease susceptibility (Rabe, 1990). In contrast, most organic fertilizers have the potential advantage of prolonging nitrogen release. In addition, because liberation of organically bound nitrogen is dependent on microbial decomposition, its release may correspond more closely to grapevine need and uptake. Where inorganic forms of nitrogen are used, application is normally timed to have it available when the need is the greatest – during bud break and fruit set. Fall application can supply adequate nitrogen uptake for spring growth, but is not associated with an equivalent nitrogen accumulation in the fruit (Goldspink and Pierce, 1993).

Nitrogen status in grapevines commonly has been assessed by petiole analysis using the diphenylamine-sulfuric acid test noted previously. The test is particularly useful because visible deficiency symptoms are slow to develop and may not express themselves until deficiency is severe. Several other procedures, such as arginine analysis, have been used to assess grapevine nitrogen status (Kliewer, 1991). However, arginine is a general stress indicator in many plants (Rabe, 1990). Arginine appears to accumulate as a means of detoxifying ammonia. Ammonia accumulates when growth is retarded under conditions such as elevated temperatures, salinity, pollution, and disease. Although activation of glutamate dehydrogenase is closely related to stress-related functions in some organisms (Syntichaki et al., 1996), there is no direct metabolic connection between glutamine and arginine. Under most conditions, ammonia is assimilated by glutamine synthetase. Most of the nitrogen translocated in the phloem is in the form of glutamine or glutamate.

An additional means of assessing nitrogen need involves determining if soil nitrate availability is below the known vine requirements (Löhnertz, 1991). Because soil nitrate availability varies throughout the year, it is important to take measurements when the vine is actively assimilating nitrogen. For early growth, mobilization of stored reserves may supply much of the nitrogen required (Löhnertz et al., 1989). Subsequently, most of the nitrogen comes directly from the soil. Nitrogen uptake occurs slowly until root growth commences in the spring, peaks about flowering, and declines following *véraison* (Fig. 4.36; Araujo and Williams, 1988; Löhnertz, 1991). Thus, uptake is most efficient and effective when peak availability coincides with flowering, and later around *véraison*. During *véraison*, adequate nitrogen supply increases the average number of berries per bunch reaching maturity. In warm climatic regions, renewed root production in the autumn can provide an alternate opportunity to apply nitrogen. The nitrogen stored by the still photosynthesizing vine can provide much of the nitrogen demand the following spring (Conradie, 1992). Appropriate timing and degree of fertilization are the components maximizing the efficiency of nitrogen uptake while minimizing groundwater nitrate pollution.

Chlorotic or light-green foliage and reduced shoot growth are the primary symptoms of nitrogen deficiency. Chlorosis is first observed in basal leaves as nitrogen is recycled (translocated) from older to younger tissues. In severe cases, apical shoot regions, as well as cluster

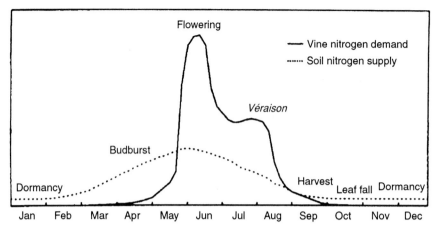

Figure 4.36 General patterns of nitrogen supply and demand in a typical European vineyard throughout the year. The supply curve is higher in wet and warm soils and lower in dry and cool soils. (From Keller, 1997, reproduced by permission)

stems and petioles, may take on a pink to red discoloration. A decrease in fruit yield is one of the first consequences of marginal nitrogen deficiency. During vinification, nitrogen deficiency in the fruit can result in stuck fermentation. Deficiency problems are enhanced when vines are lightly pruned, or a groundcover is grown on nitrogen-poor soils.

Moderate nitrogen deficiency (like mild water deficit) has been correlated with improved fruit quality (Choné et al., 2001a). In general, stage I berry growth is favored by restricted nitrogen availability (Rodriguez-Lovelle and Gaudillière, 2002). Modest nitrogen deficiency more affects vegetative growth, limiting carbohydrate competition between the fruit and the shoot. However, subsequently, adequate nitrogen availability is essential for grape maturation and to sustain vinification. Timing nitrogen availability to need can be most effectively achieved by fertigation or foliar spraying.

Although nitrogen deficiency is a problem in some areas, excessive application can induce toxicity. This is most noticeable in older leaves, in which the edges accumulate salt-like deposits, become water soaked, and finally necrotic. At excessive but nontoxic concentrations, undesirable activation of shoot growth can result in both poor fruit quality and reduced bud fruitfulness (due to shading). In addition, application beyond need is profligate, contributes to nitrate pollution, and aggravates molybdenum and phosphorus nutrient deficiencies (by increasing the acidification of marginally acidic soils).

A marked tendency in several European countries is the reduction or elimination of nitrogen fertilization. This is combined with the use of groundcovers to supply and store nitrogen, plus mowing or tilling to restrict water and nitrogen competition at critical periods during vine growth. This trend to low-input (green) viticulture is spurred by the equal desire to reduce "chemical" applications and depend on biological control of pests and diseases. Judicious selection of an appropriate groundcover can enhance the numbers of predators and parasites of vine pests. Unfortunately, it may also provide overwintering sites of grapevine pests or act as hosts of pathogenic nematodes. The concept of green viticulture is laudable, but possibly overly idealistic for mainstream grape production (Keller, 1997).

Nitrogen fertilization enhances vegetative growth, but usually does not induce luxury accumulation in the foliage. Nevertheless, fruit may show enhanced storage with increased nitrogen availability. Nitrogen excess is most noticeable in the increase in free amino acid content, notably proline and arginine, and soluble proteins. Although enhanced nitrogen content facilitates rapid juice fermentation, the accompanying increase in pH may be undesirable. In addition, an increase in arginine level may augment the production of ethyl carbamate (a suspected carcinogen) (Ough, 1991). Fruit maturation also may be delayed because of competition with enhanced vegetative growth under high-nitrogen conditions, leading to reduced fruit quality.

High nitrogen levels are suspected to be involved, at least partially, in the development of several grapevine physiological disorders. These include a "false potassium deficiency" (Christensen et al., 1990), as well as inflorescence and bunch-stem necroses. Excessive nitrogen availability also tends to enhance susceptibility to several fungal infections, notably bunch rot. This may result from suppressed phytoalexin synthesis (Bavaresco and Eibach, 1987) and/or increased canopy density.

PHOSPHORUS

Phosphorus is an important component of cell-membrane lipids, nucleic acids, energy carriers such as ATP, and some proteins. It also is required for sugar metabolism. Phosphorus accumulates primarily in meristematic regions, seeds and fruit. It is translocated principally in the phloem.

Phosphorus deficiency is rare in grapevines. This probably results from the limited phosphorus requirements of the vine, remobilization of phosphorus throughout the vine, and the ample presence of inorganic and organic phosphates in most vineyard soils. Nevertheless, phosphorus deficiency has been recognized in grapevines grown on acidic soils and weathered, low-phosphorus, hillside sites.

Deficiency symptoms include the formation of dark-green leaves, downturning of leaf edges, reduced shoot growth, purple discoloration in the main veins of older leaves, as well as interveinal discoloration, premature fruit ripening, and reduced yield. More subtle consequences of phosphorus deficiency involve the reduced initiation, development, and maintenance of flower primordia (Skinner and Matthews, 1989). Fertilization can alleviate these problems, increase grape color and sugar content, and augment free monoterpene accumulation in some cultivars (Bravdo and Hepner, 1987). When required, phosphate is frequently applied as superphosphate. Subsurface (15–30 cm) incorporation is required because the element diffuses slowly in soil.

Because phosphorus deficiency in grapevines is uncommon, fertilization should be used only if deficiency has been clearly demonstrated. This avoids potential interference with potassium, manganese, magnesium, and iron acquisition. Disruption may result directly from metabolic effects on the vine, or indirectly from modified cation exchange in the soil. For example, the addition of alkaline rock phosphate ($3Ca_3(PO_4)_2.CaF_2$) can retard manganese solubilization, whereas the calcium released by the more soluble superphosphate ($Ca(H_2PO_4)_2$ and $CaHPO_4$) can liberate manganese, making it more readily available for root uptake (Eifert *et al.*, 1982).

POTASSIUM

Potassium is the only macronutrient that is not a structural component of cellular macromolecules. Its presence is required for cellular osmotic and ionic balance, electrochemical processes, neutralization of organic acids, regulation of stomatal function, cell division, enzyme activation, protein synthesis, and synthesis and translocation of sugars.

Potassium is absorbed as free K^+ ions dissolved in the soil solution and translocated in the phloem. Because mineral weathering only slowly replenishes the potassium supply, soil depth is important to potassium fertility. Nonuniform redistribution of fertile topsoil, resulting from land leveling or erosion on slopes, can expose less productive shallow subsoil. This can produce patchy areas of potassium deficiency in vineyards. Sandy soils in high rainfall regions, and vineyards characterized by high calcium or magnesium contents also tend to be potassium-deficient. In addition, potassium deficiency has occasionally been observed in virgin soil, where it is called "pasture burn" (*Wasenbrand*) (P. Perret, 1992, personal communication).

Grapevines may express one or possibly two foliar forms of potassium deficiency. "Leaf scorch" occurs initially in the middle of primary shoots and apically on lateral shoots. The symptoms begin as a loss of green color (chlorosis) or bronzing along the edges of leaves that progresses inward. Margins subsequently dry and roll. In contrast, "black leaf" begins with interveinal necrosis on the upper surface of sun-exposed leaves, subsequently spreading to cover the leaf surface. Vines low in potassium are also more drought-prone and less cold-tolerant.

When adequate, potassium favors grape quality by enhancing fruit coloration and sufficient acidity. However, excess potassium may raise juice pH undesirably, and can potentially suppress magnesium uptake by the roots. In cases of potassium excess, the juice or wine often needs pH adjustment, usually by the addition of tartaric acid. In the vineyard, reduced potassium uptake may be achieved by the selective use of rootstock cultivars, canopy management, and irrigation procedures such as partial rootzone drying (Mpelasoka *et al.*, 2003). Potassium deficiency may be induced if phosphate availability is excessive.

Because potassium diffuses poorly in soil, potassium fertilizers are best added deeply to position the element within the root zone. Potassium sulfate is generally preferred to potassium chloride, to avoid increasing the soil's chloride content.

CALCIUM

Calcium is a vital constituent of plant-cell walls, where it reacts with pectins, making them relatively water insoluble and rigid. Calcium also plays important roles in regulating cell-membrane permeability, ion and hormone transport, and enzyme function. Along with potassium, calcium helps detoxify organic acids in cell vacuoles by inducing acid precipitation. Calcium is primarily translocated throughout the vine in the xylem.

Calcium deficiency typically occurs only on strongly acidic quartz gravel. It is expressed as a narrow zone of necrosis along the edge of leaves, which may progress toward the petiole. Minute, brown, slightly raised regions may develop in the bark of shoots. Clusters may show dieback from the tip. When in excess, as in calcareous soils, it can cause lime-induced chlorosis with sensitive rootstocks. As gypsum (calcium sulfate), calcium may be added to improve water permeability in sodic soils, whereas as dolomite or slaked or burnt lime, it is used to raise the pH of acidic soils.

MAGNESIUM

Magnesium is a vital cofactor in the absorption of light energy by chlorophyll. It stabilizes ribosome, nucleic acid, and cell-membrane structure, and is involved in the activation of phosphate-transfer enzymes. Magnesium is principally translocated in the phloem.

Deficiencies are frequently experienced in sandy soils in high-rainfall regions, poorly drained sites, and high-pH soils. This results from the relative ease with which magnesium is leached from the soil. Symptoms first begin to develop in basal leaves, due to magnesium's trans-location to growing points under deficient conditions. Interveinal regions develop a straw-yellow chlorotic discoloration, whereas bordering regions remain green. Early in the season, symptoms may appear as small brownish spots next to the leaf margins. It occurs more frequently in vines grafted to certain rootstocks, such as 'SO 4.' Magnesium deficiency may also be involved in the physiological disorder termed bunch-stem necrosis.

SULFUR

Sulfur is an integral component of the amino acids cysteine and methionine. The cross-linking of their sul-fur atoms is often crucial in the formation of functional protein structure. Sulfur is also an integral component of the vitamins thiamine and biotin, and is present in coenzyme A. Sulfur is principally translocated in the phloem.

Although an essential nutrient, sulfur appears in ample supply in most vineyard soils. It is often applied to vines, not as a nutrient but as a fungicide. Sulfur also may be supplied unintentionally, as sulfate with potassium or magnesium fertilizers, or with calcium in gypsum. Sulfur is not known to occur naturally at toxic levels in soil.

ZINC

Zinc plays an important role as a cofactor in several enzymes, such as carbonic anhydrase, and in the syn-thesis of the indoleacetic acid (IAA). Zinc can be trans-located in both the phloem and xylem.

Although zinc is required in small amounts, the sup-pression of solubility in alkaline soils can lead to the development of deficiency symptoms. Zinc deficiency can develop in sandy soils where low levels of inorganic or humus colloidal material limit zinc availability. High levels of phosphate may precipitate the metal as zinc phosphate and restrict its availability to plants. Land leveling also can lead to patchy zones of zinc deficiency. Rootstocks such as 'Salt Creek' and 'Dog Ridge' are particularly susceptible to zinc deficiency.

Zinc deficiency can generate a series of distinctive symptoms. It produces a mottled interveinal chloro-sis, with an irregular green border along clear veins. Modified leaf development produces atypically wide angles between the main leaf veins. Leaves at the apex of primary shoots, and along lateral shoots, are much reduced in size and often asymmetrically shaped. The latter symptom is called "little leaf." Zinc deficiency also can lead to poor fruit set and clusters containing small, green, immature, "shot" berries. The symptoms may resemble fanleaf degeneration.

For spur-pruned vines, the application of zinc sulfate to freshly cut spur ends can be beneficial. Alternately, leaves may be sprayed with a foliar fertilizer containing zinc. Application just before anthesis has the advantage of reducing the formation of "shot" berries.

Although less common, zinc toxicity can disrupt root growth, and is occasionally associated with the use of contaminated compost. Compost also may be a source of toxic levels of other heavy metal, such as lead (Pb), cad-mium (Cd), and copper (Cu) (Perret and Weissenbach, 1991). Because of the amounts of compost required to adequately supplement vine nutrition (usually in excess of 10^3 kg/ha), tainting with heavy metals could lead to significant uptake and wine contamination.

MANGANESE

Manganese functions in the activation of, or as a cofac-tor in, several enzymes, for example the oxygen-releasing component of the photosystem (PS) II complex in the chloroplast. Manganese is directly involved in the syn-thesis of fatty acids, in the neutralization of toxic oxygen radicals, and in the reduction of nitrates to ammonia. Manganese is principally translocated in the xylem.

As with most bivalent cations, the availability of man-ganese to plants is reduced in alkaline and humic sandy soils. Deficiency symptoms produce a chlorosis, associ-ated with a green border along the veins, similar to that found in zinc deficiency. However, manganese deficiency neither modifies leaf-vein angles nor induces "little leaf." In addition, chlorosis develops early in the season and commences with the basal leaves. The leaves may take on a geranium-like appearance. On poorly aerated acidic soil, manganese may reach toxic levels due to increased solubility of its reduced (Mn^{2+}) oxides.

IRON

Iron plays a role in chloroplast development, is essen-tial in the synthesis of chlorophyll, acts as a cofactor in redox reactions in electron transport, and is a constituent in enzymes such as catalase and peroxidase.

As with other nutrients closely associated with photo-synthesis, deficiency of iron results in chlorosis. Chlorosis usually begins apically and early in the season. Yellowing often is marked, with only the fine veins remaining green. Severely affected leaves may wither and fall. Fruit yield and berry size are often severely reduced.

Lime-induced chlorosis is most frequently observed on calcareous soils. Although the high pH of calcareous soils may limit iron solubility, this, in itself, seems not to induce iron-deficiency chlorosis. Leaves showing chlorosis may have iron contents as high as healthy green leaves (Mengel *et al.*, 1984; Bavaresco, 1997). Thus, chlorosis may result from other factors associated with the high lime content of soil, notably high ash alkalinity (Bavaresco *et al.*, 1995). This could disrupt cellular incorporation or use of iron as a result of bicarbonate accumulation in leaf tissue. Chlorosis has also been attributed to the tendency of calcareous soils to compact easily, especially those with 5–10% fine particulate (>2 mm) calcium carbonate. This could aggravate conditions that limit root growth, such as anaerobiosis following heavy rains, increased ethylene synthesis in the soil (Perret and Koblet, 1984), and enhanced risk of heavy-metal toxicity. Overcropping has also been correlated with increased incidence of chlorosis in the subsequent year (Murisier and Aerny, 1994). Other factors implicated in the complex relationship between iron and lime-induced chlorosis include rootstock cultivar, symbiosis with mycorrhizal fungi, and the presence of endophytic bacteria (Bavaresco and Fogher, 1996).

Young roots are the primary sites of iron acquisition. In a study by Bavaresco *et al.* (1991), iron uptake and resistance to chlorosis were closely correlated with root diameter and root-hair development. Resistance has also been closely associated with acidification of the rhizosphere, production of Fe^{3+} reductase, increased membrane redox potential, and augmented synthesis of organic acids (Brancadoro *et al.*, 1995). Acidification increases the solubility of the ferric (Fe^{3+}) ion, and favors its reduction to the more soluble ferrous (Fe^{2+}) state, the form in which iron is translocated into the root. Nevertheless, the mechanism of resistance or tolerance varies with the species involved (Bavaresco *et al.*, 1995). Once in the plant, ferrous ion is reoxidized, chelated primarily with citrate, and moved in the xylem stream to the shoot system. In the leaf, it may be stored in a protein complex called **phytoferritin**. Redistribution may occur in the phloem.

Most North American grapevines are sensitive to lime-induced chlorosis on calcareous soils. Thus, choice of rootstock, such as 'Fercal' or '41 B,' is important when the grafted vine is planted in calcareous soils. Alternatively, the foliage may be sprayed with iron chelate or, if the soil is not too calcareous or alkaline, the pH raised with sulfur.

BORON

Boron is required in nucleic acid synthesis, in the maintenance of cell membrane integrity, and in calcium use. It is variably translocated in the xylem or phloem. Although boron is required only in trace amounts, the range separating deficiency and toxicity is narrow. Deficiency symptoms develop as blotchy yellow chlorotic regions on terminal leaves. Chlorosis soon spreads to the leaf margin. Apical regions of the shoot may develop slightly swollen dark-green bulges. Early dieback may result in the development of stunted lateral shoots. Boron deficiency also induces poor fruit set (due to retarded germ-tube elongation) and the development of many "shot" berries. Deficiency symptoms tend to develop on sandy soils in high-rainfall areas, on soils irrigated with water low in boron, or on strongly acidic soils.

Application of borax to counteract deficiency can lead to toxicity if it is not evenly distributed. Toxicity symptoms begin with the development of brown to black specks on the tips of leaf serrations. The necrotic regions spread and become continuous. The inhibition of growth may result in leaves wrinkling and puckering along the margin.

COPPER

Copper acts primarily as a cofactor in oxidative reactions involved in respiration and the synthesis of proteins, carbohydrates, and chlorophyll. It is variably translocated in the phloem or xylem. Copper-deficiency symptoms are only rarely observed in grapevines, possibly because of the use of copper in fungicides such as Bordeaux mixture. When deficiency occurs, the leaves are dwarfed and pale green, shoots develop short internodes, cane bark has a rough texture, and root development is poor.

Toxicity produces leaf chlorosis similar to that of lime-induced chlorosis. Toxicity most commonly occurs in soils where copper has accumulated following prolonged use of copper-containing fungicides such as Bordeaux mixture (Scholl and Enkelmann, 1984). Copper buildup in the soil is also implicated in some vineyard "replant diseases."

MOLYBDENUM

Molybdenum is required for nitrate reduction as well as in the synthesis of proteins and chlorophyll. Only rarely has molybdenum been found to be deficient in vineyard soils. When molybdenum is deficient, necrosis develops and spreads rapidly from the leaf margins inward. The demarcation zone between healthy and necrotic tissue is pronounced, and the unaffected areas appear normal. Affected leaves often remain attached to the shoot and fall with difficulty.

CHLORINE

Chlorine is involved in both osmotic and ionic balance, as well as the splitting of water during photosynthesis. Although required in fairly large amounts, chlorine is not known to be limiting in vineyard soils. More commonly, it is associated with toxicity under saline conditions. It produces a progressive, well-defined necrosis that moves from the edges of the leaf toward the midvein and petiole. Because physiological disruption occurs at levels well below those causing visible toxicity, the use of chlorine-tolerant rootstocks on saline soils is advisable, for example 'Salt Creek' and 'Dog Ridge' (Table 4.7).

Organic Fertilizers

The addition of manure to vineyards was once standard practice. With specialization and concentration in agricultural industries, ready access to animal manure became more difficult. At the same time, production costs for inorganic fertilizers fell. These features added to the convenience of commercial fertilizer use.

Several trends have encouraged some growers to return to organic fertilization. Increased energy (and therefore fertilizer) costs, concerns about groundwater pollution, and the belief that inorganic fertilizers encourage the destruction of the soil's humus content have led to a reduction in commercial fertilizer use. Regrettably, this issue is far more complex than commonly believed (Kirchmann and Ryan, 2004).

Nitrogen availability is typically the most limiting factor in the decomposition (mineralization) of organic material. Because most manures have a relatively high carbon to nitrogen ratio, the initial effect of manure addition can be reduction in nitrogen availability. This results because microbes are more numerous and effective at incorporating nitrogen than vine roots. In addition, fresh manure may contain sufficient urea or uric acid to be toxic ('burning roots') before it is metabolized to ammonia. Tillage, associated with manure incorporation into the soil, increases oxygen availability and rapid nitrogen consumption by soil microbes.

In most instances, the growing market for organically grown grapes and wine has promoted a "back to nature" philosophy. In addition, many governments are enacting laws designed to limit nitrate water pollution from all sources. This has spurred interest in means by which vineyard production may be maintained with limited or no use of inorganic fertilizers. To date, this has yet to be established as a long-term viable option, or under what conditions.

Animal manure, if well matured (composted), is a source of slowly released nutrients and humic material.

Manure may also increase the activity of nitrogen-fixing bacteria, such as *Azotobacter*. In contrast, its activity is repressed by the application of nitrogen fertilizer. Additional benefits of manure application include improved soil aeration (Baumberger, 1988), and a shift in soil pH toward neutrality (Fardossi *et al.*, 1990). However, fresh manure, high in liquid components, must be applied sparingly. Even during warm weather, microbial activity may be insufficient to rapidly bind nutrients and limit runoff.

Alternative organic fertilizers include compost, green manures, chopped vine prunings, and winery waste. As with liquid manure, winery wastes must be added with caution. Otherwise, nutrient loss may pollute groundwater. The direct use of winery or distillery waste is not recommended on acid soils. Manganese has been noted to occasionally build up to toxic levels (Boubals, 1984). Pomace can also be a source of viral contamination, increasing the incidence of nematode-transmitted viral infections.

Humic colloids play a vital role in the maintenance of soil fertility. As humus decomposes, sorbed and bound nutrients are released into the soil solution. The microbial activity associated with humus formation also binds nutrients otherwise lost by leaching.

Although microbial activity can positively affect soil structure, there is no clear evidence of improved nutrient availability (Kirchmann *et al.*, 2004). The effect of microbial activity on soil structure is a consequence of humification. Humic substances consist primarily of refractory compounds, derived principally from microbial changes to the structure of lignins, other plant phenolics, and complex polysaccharides. These form colloidal particles that, with clay colloids, form soil aggregates. Cations such as Ca^{2+} and Fe^{3+} help stabilize aggregates, forming the desirable crumb structure of fertile soils. Humus also provides much of the surface on which most soil microbes grow. They, in turn, function as the main food source for soil invertebrates. Of these, the most visible are earthworms. Through their action, earthworms increase soil porosity, thereby facilitating soil aeration, water penetration and retention, as well as distributing nutrients throughout the upper layers of the soil. Smaller invertebrates, such as collembolae and other insect larvae, act as the principal herbivores, chewing dead plant remains and exposing their remnants to more effective microbial action. The mucilage microbes produce further aids in the formation and stability of a friable soil structure. Soil-aggregate structure is important not only for water infiltration, but also for minimizing soil erosion.

Manure and compost is most easily incorporated at the inception of vineyard establishment. This is especially advantageous if the organic content of the soil is low. Alternatively, organic amendments may be disced in

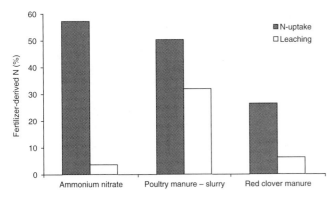

Figure 4.37 Nitrogen in harvested crops and leaching during periods of 2 (red clover manure) and 3 years (NH_4NO_3 and poultry manure), in percent of N applied with the respective N source during the initial year. (From Dahlin *et al.*, 2005, reproduced by permission)

periodically to improve soil structure. Nevertheless, earthworm activity can often incorporate material within a short period, if added periodically in small amounts (such as a top dressing). This is best done in the early spring or autumn. Additional information on composts and other means of enhancing soil organic content can be found in Cass and McGrath (2005) and McGourty and Reganold (2005).

Despite the benefits of composts and manures, there are several disadvantages. Synchronization of nutrient release to grapevine need is difficult. Mineralization and nutrient release is affected not only by a wide variety of soil and atmospheric conditions (rainfall, temperature, pH, moisture content, and texture), but also by the chemical composition of the organic material (Dahlin *et al.*, 2005). Inadequate release can lead to nutrient deficiencies, whereas untimely liberation can result in loss by leaching (Kirchmann and Thorvaldsson, 2000). In contrast, application of inorganic fertilizers can usually be timed to vine need and root uptake, resulting in more efficient and effective nutrient use (Fig. 4.37). Leaching is particular likely during cool, wet periods.

For organic fertilizers to be effective, the timing of nutrient release from the particular source must be known. Thereby, application can be arranged for nutrient release to coincide with the production of fine feeder roots. These roots have a comparatively short life span and absorb nutrients primarily in the first few days following emergence (Volder *et al.*, 2005). Their production typically occurs in early summer, when shoot growth has started and the flowers are open (Comas *et al.*, 2005).

ANIMAL MANURES

Although animal manures have advantages, they possess several inconveniences beyond local unavailability. Variability in composition (Table 4.10) is a major concern, because it greatly complicates calculation of

appropriate application rates. Important factors influencing nutritional composition are the animal source, the nature and amount of litter (bedding) incorporated, and the handling and storage before application. Additional potential difficulties include toxicity resulting from copper supplementation of pig feed, development of zinc deficiency from the use of poultry manure, induction of nitrogen deficiency from the decomposition of large amounts of incorporated straw, and nitrogen loss from ammonia volatilization during urine breakdown.

Because animals use only a portion of the organic and nutrient content of their feed, much nutritive value remains in the manure. About half of the nitrogen, most of the phosphates, and nearly 40% of the potassium found in the original feed remain in the feces. Fresh manure contains both readily and slowly available nitrogen. The readily available form (mostly ammonia) is rapidly lost within a few weeks by volatilization, if not bound to soil colloids, or taken up by microbes and plant roots. The slowly available, organically bound nitrogen is more stable (Klausner, 1995). Most of these decompose within one year, whereas the more resistant may take several years to mineralize. Thus, repeat application over many years is essential for development of relatively uniform nutrient availability. Brady (1974) estimated that roughly 1000 kg (~1 ton) of manure supplies about 2.5 kg nitrogen, 0.5 kg phosphorus, and 2.5 kg potassium. Most of the plant remains are either hemicellulose, lignins, or lignin–protein complexes. Additional organic materials consist of the cell remains of bacteria. They can constitute up to 50% of the manure's dry mass.

GREEN MANURES

The use of green manures is another long-established procedure for improving soil structure and nutrient content. The procedure involves growing a crop and plowing it under while still green. If plowing occurs while the crop is still succulent, its nitrogen content is typically adequate to promote rapid decomposition, without limiting nitrogen availability to the vine.

Green manures commonly are planted in midsummer to early fall, to limit or avoid inducing early water and nutrient stress in the vines. This is especially important under nonirrigated arid conditions or on poor soils. However, some competition may be desired to limit vine vigor under conditions of high rainfall or on rich soils (Caspari *et al.*, 1996). Green manures, also grown as a groundcover, are also useful in limiting water, fertilizer, and soil loss on slopes. For example, groundcovers can dramatically restrict water runoff during downpours – to 15% versus 80% (Rod, 1977). Although loss is less marked when measured over the whole season, groundcovers may limit water runoff

Table 4.10 Average values for moisture and nutrient content of farm animal manures[a]

Source	Portion	Percent	Moisture Content	Nitrogen (%)	Phosphorus (%)	Potassium (%)
Horse	Manure	80	75	0.55 (0.50–0.60)	0.33 (0.25–0.35)	0.40 (0.30–0.50)
	Urine	20	90	1.35 (1.20–1.50)	Trace	1.25 (1.00–1.50)
	Mixture	—	78	0.7	0.25	0.55
Cattle	Manure	70	85	0.40 (0.30–0.45)	0.20 (0.15–0.25)	0.10 (0.05–0.15)
	Urine	30	92	1.00 (0.80–1.20)	Trace	1.35 (1.30–1.40)
	Mixture	—	86	0.6	0.15	0.45
Swine	Manure	60	80	0.55 (0.50–0.60)	0.50 (0.45–0.60)	0.40 (0.35–0.50)
	Urine	40	97	0.40 (0.30–0.50)	0.10 (0.07–0.15)	0.45 (0.20–0.70)
	Mixture	—	87	0.5	0.35	0.4
Sheep	Manure	67	60	0.75 (0.70–0.80)	0.50 (0.45–0.60)	0.45 (0.30–0.60)
	Urine	33	85	1.35 (1.30–1.40)	0.05 (0.02–0.08)	2.10 (2.00–2.25)
	Mixture	—	68	0.95	0.35	1
Poultry	Mixture	—	55	1.00 (0.55–1.40)	0.80 (0.35–1.00)	0.40 (0.25–0.50)

[a] Ranges in parentheses.

Source: Data from van Slyke, 1932

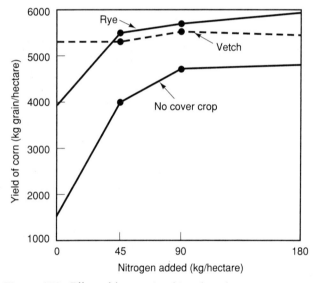

Figure 4.38 Effect of legume (vetch) and nonlegume (rye) cover crops on the 7-year-average yield of corn receiving three levels of nitrogen fertilizer on a sandy loam soil. Cover crops were grown each year. (Data from Adams *et al.*, 1970)

to 1.5% versus 19% on slopes more than 40° (Rod, 1977). In addition, groundcovers can facilitate weed control, bind nutrients otherwise lost by leaching, and keep dust down. If green-manure crops are left until the spring, they can trap snow and limit frost penetration into the ground.

Several leguminous and cereal crops have been used as green manures. Legumes have the advantage of forming symbiotic associations with *Rhizobium*. These root-nodule forming bacteria can easily add the nitrogen that is lost with harvested grapes (Fig. 4.38). The degree of nitrogen incorporation by rhizobial bacteria depends on the strain of *Rhizobium*, the host legume, soil moisture conditions, and the nitrate content of the soil – the higher the nitrate level, the lower the degree of nitrogen fixation. Production levels can vary from 10 to more than 200 kg/ha/year in Europe (Keller, 1997).

High seed costs, the need for *Rhizobium* inoculation, and higher water demands can limit legume feasibility. These disadvantages occasionally may be avoided with self-seeding legumes, such as burr clover. Cereal crops require less water and can often penetrate and loosen compacted soil. Consequently, legumes and grasses are frequently used together (Winkler *et al.*, 1974). This has the additional advantage of providing a range of diverse habitats for beneficial insects that can be of value in biological pest control. Nevertheless, the habitat provided must be carefully studied to avoid favoring rodent populations. It has been found that the positioning of nesting or resting sites for owls and hawks, in or adjacent to vineyards, helps to control rodent pests and can minimize bird damage.

COMPOSTING

Except for green manures, most organic fertilizers are best if initially composted. For example, composted material becomes less variable in chemical composition, and their nutrient contents are more slowly liberated. Compost can also have additional benefits, such as acting as a mulch (reducing water evaporation from the soil) and helping to control annual weeds. Co-composting with winery wastes (stalks, seeds, and pomace), winery or vineyard waste water, or sewage sludge helps convert these wastes into a product that can be safely added to vineyard soils. Experimentation is required to establish an effective formula for efficient and effective composting. For example, high water content complicates composting. Occasionally, contamination of sewage sludge with heavy metals can make its use unacceptable (Perret and Weissenbach, 1991).

Composting typically involves mounding organic waste into large piles (about 2 m high). These have sufficient size to insulate the interior, permitting the temperature to rise to 55–70 °C. The heat, generated by microbial activity, selects thermophilic bacteria that quickly decompose the organic material. The heat also tends to pasteurize the waste, killing pathogenic organisms and weed seeds. Because composting is most efficient when microaerobic, the mounds need to be turned frequently (typically four times over a period of several weeks) to incorporate sufficient oxygen, as well as incorporate material on the exterior for composting.

Once bacterial mineralization brings the carbon/nitrogen ratio close to 20 : 1, microbial activity slows, the temperature falls, and further composting comes under the action of fungi. The latter are particular important in humification of much of the remaining organic material. This partially involves the condensation of phenolics with ammonia. Subsequent maturation of the compost involves degradation or volatilization of potential phytotoxins, such as ethylene oxide, ammonia, and low-molecular-weight fatty acids (notably acetic, propionic and butyric acids). After some 12–24 weeks (depending on the temperature, moisture and nutrient content, and turning frequency), the compost is ready for use as an organic amendment or mulch.

Disease, Pest, and Weed Control

Changes similar to those affecting vineyard fertilization are affecting the practice of disease, pest, and weed control. At the extreme edge of this change are those dedicated to organic viticulture (Jenkins, 1991). Occasionally, "natural" pesticides, such as phosphoric acid (Foli-R-Fos®), canola oil (Synertrol®) (Magarey *et al.*, 1993), and potassium silicate (Reynolds *et al.*, 1996), can be as effective as standard pesticides. This is not a consistent finding, though, with more natural pesticides usually being required (Schilder *et al.*, 2002). Little if any research has been conducted on the effects of pesticides (natural or man-made) on the aromatic composition of wine. One study by Oliva *et al.* (1999) showed an effect, but most influences were below threshold values, with the exceptions of ethyl acetate and isoamyl acetate.

Although minimizing environmental damage is a laudable goal, natural pesticides are no more guaranteed to be safer than their synthetic equivalents. For example,

Stylet-Oil reduced grapevine photosynthesis and the accumulation of soluble solids (Finger *et al.*, 2002). The effects were volume- and frequency-dependent. In an extensive study of natural vs. synthetic pesticides, the percentage potentially carcinogenic was the same[2] (about 50%) (Gold *et al.*, 1992; Ames and Gold, 1997). In addition, accepted pesticides used in organic viticulture may disrupt the action of natural disease control agents. For example, sulfur can increase the mortality of *Anagrus* spp., a biocontrol of leafhoppers (Martinson *et al.*, 2001), and suppress tydeid mite (*Orthotydeus lambi*) populations. The latter can reduce the incidence of powdery mildew (English-Loeb *et al.*, 1999).

Natural pesticides should require the same exhaustive assessment for efficacy, safety, and residue accumulation as current agents now require. For example, the increased use of pyrethroids in California has led to their accumulation in stream sediments to toxic levels for bottom dwellers (Weston *et al.*, 2004). In addition, various oils used in organic pest control can affect the taste and odor of grapes and wine (Redl and Bauer, 1990). Although health concerns about the risks of pesticide use have induced several governments to contemplate pesticide deregistration, little thought has gone into the human health risks of increased mycotoxin contamination. This could have unconscionable consequences, as have other well-intentioned but ill-advised government reactions. Nevertheless, the inability of current pesticides and herbicides to provide adequate crop protection, and the need to reduce production costs, have wisely spurred research into alternative control measures.

One of the main concepts in providing better and less expensive pest and disease control is called **integrated pest management** (IPM). The term *management* reflects a shift in attitude – that limiting damage to an economically acceptable level (its economic threshold) is more feasible and prudent than attempting what is often impossible, *control*. Integrated pest management combines the expertise of specialists in cognate fields, such as plant pathology, economic entomology, plant nutrition, weed control, soil science, microclimatology, statistics, and computer science. Coordinated programs usually reduce pesticide use while improving effectiveness. IPM programs tend to be more pragmatic and less philosophically based than organic approaches. It often includes factors such as environmental modification and assessment, biological control, and better synchronization of pesticide rotation and application. Timing pesticide application to coincide with vulnerable periods in

---

[2] Despite these results, their concentration in food is so low that none constitute a significant cancer health risk.

the life cycle of a pathogen usually reduces the number of applications, while increasing effectiveness. It also promotes pesticide selection and application, appropriate to the situation. For example, use of a simpler (less specific) protective agent when disease stress is low, and application of a curative agent (more selective and often systemic) when disease incidence is high. This approach is much less disruptive of indigenous control agents. Advances in disease forecasting, combined with monitoring pest or disease incidence, permit more accurate prediction of their potential economic damage. Because most models have been developed with data collected over a fairly large area, their predictive value may be limited on a local scale. Only with extensive use is it possible to adjust prediction to specific vineyards or vineyard sites. Nevertheless, risk assessment (the cost/benefit ratio of pesticide use) is being increasingly used to regulate if, and when, pesticide application is justified. For example, spider mites are often considered to be a serious grapevine pest. However, little significant effect on yield and quality may develop even with heavy infection by European red mites (*Panonychus ulmi*) (Candolfi *et al.*, 1993). Vine capacity may be less affected than appearance might suggest, due to compensation by the root and/or shoot system to foliage loss (Hunter and Visser, 1988; Candolfi-Vasconcelos and Koblet, 1990; Fournioux, 1997). Compensation is more evident early in the season; less so later.

Timing applications has been greatly facilitated by developments in computer hardware and software, miniaturization of solar-powered weather stations (Plate 4.13) (Hill and Kassemeyer, 1997), and GIS technology. These systems have permitted the development of expert systems that can predict the incidence of disease and pest outbreaks. These stations also can provide data to adjust irrigation schedules. Computer programs are now available for the major diseases in several viticultural regions.

Although the economic and ecological benefits of IPM are obvious, its successful implementation is far from simple. A major factor limiting its wider application involves its considerable developmental costs. It takes years to develop an effective program, requiring the dedication of specialists in diverse fields. Without their various skills, predicting the consequences of a program is impossible. For example, the most efficient fungicides against a particular plant pathogen may be toxic to predatory mites of an equivalently significant pest. As well, reduction in pesticide use may lead to secondary pest outbreaks. For example, adoption of an IPM program, largely dependent on biological control agents for the western grape leafhopper in California, coincided with an infestation with variegated leafhoppers. Even unrelated changes in viticultural practice can affect IPM efficiency. For example, avoiding problems associated with cultivation by the use of herbicides can favor the germination of the resting stages (e.g., sclerotia) of fungal pathogens. In addition, improved nitrogen fertilization may result in a suppression of disease resistance.

IPM systems must also be sufficiently flexible to accommodate variability in disease/pest severity, associated with regional and yearly climatic fluctuations. Finally, predicting the economic benefits of IPM programs is fraught with enigmas. In most instances, the financial losses *actually* ascribable to specific pest and disease agents are unknown. In some instances, reduced yield may permit the optimal ripening of the remaining fruit. Thus, calculating the cost/benefit ratio of pesticide use often tends to be no more than a crude guesstimate. Most pesticide application is driven more by fear than a rational understanding of the consequences. Because of these factors, IPM implementation tends to be regional, and only directed at the most economically significant of disease or pest agents.

Another valuable avenue of research has focused on pesticide combinations. This can avoid adverse interactions and minimize the number of treatments required (Marois *et al.*, 1987). Improvements in nozzle and sprayer design now provide better and more uniform chemical spread, reducing runoff, and achieving better pest or pathogen contact. Assuring that a greater percentage of the target receives a toxic dose should delay the development of resistant strains. Sublethal doses selectively favor the survival (reproduction) of pathogens and pests that inherently possess some resistance to the control agent.

The most efficient nozzles available are those that give a small negative charge to droplets as they are released (electrostatic sprayers). Plant surfaces (positively charged) attract the droplets, facilitating their uniform deposition on plant surfaces, and limiting their removal by rain. Restricted variation in droplet size (optimum between 25 and 100 μm), and the reduced volume of water needed minimizes runoff and drift from vaporization of very small droplets. A greater efficiency of application may be achieved by spraying at night, when conditions are calm and evaporation from droplets is retarded.

Integrated pest management is placing emphasis on multiple means of control, rather than relying on single techniques (Basler *et al.*, 1991). Experience has indicated that dependence on chemical control is ultimately unsatisfactory. The same will probably be true if overconfidence is placed in biological or genetic control, and other forms of control are abandoned. Because most pathogenic agents multiply exceedingly rapidly, they are disappointingly adept at developing resistance to single selective pressures, whether anthropogenic or natural.

Control of Pathogens

CHEMICAL METHODS

Faced with increasing pesticide resistance, it has been imperative that techniques be developed to retard loss of pesticide effectiveness (Staub, 1991). For fungal pathogens, the use of relatively nonselective fungicides has value. Their broad-spectrum toxic action provides general protection against a vast range of pathogenic fungi and bacteria. This reserves the curative action of selective fungicides for situations in which their precise action is required. Disease forecasting, the prediction of disease outbreaks based on meteorological data, is particularly useful in this regard. It reduces the need for frequent (prophylactic) pesticide application, increases the effectiveness of what is applied, and can retard the rate at which resistance accumulates in pest or pathogen populations. In the absence of disease forecasting, timing of pesticide application is most effective when based on the phenologic (growth) stage of the vine. The precision permitted by the Modified E–L System (see Fig. 4.2) is well adapted to the timing of sprays.

Development of resistance to nonspecific (contact) pesticides tends to be slow or nonexistent. Single mutations rarely (ever?) provide adequate protection against the nonselective, extensive disruption of membrane structure and enzyme function induced by contact pesticides. When the use of nonselective fungicides began to enter the market in the late 1940s, the destructive potential of several pests of the past was largely forgotten. Their potential destructiveness reappeared when their use was replaced by selective pesticides in the early 1960s (Mur and Branas, 1991). For example, the increased incidence of omnivorous leafrollers in California can be partially attributed to reduced use of nonselective pesticides against grape leafhoppers (Flaherty *et al.*, 1982). This has, in the case of organic viticulture, encouraged the use of nonselective pesticides, such as oils (e.g., Stylet-Oil), plant poisons (e.g., pyrethroids), or bacterial toxins (e.g., Bt toxin).

Selective fungicides and insecticides often have the advantage of producing less damage against harmless (and potentially beneficial) insects and fungi. Selective pesticides also tend to be active at lower concentrations and require fewer applications to be effective. In addition, most selective pesticides are incorporated by the plant and may be translocated systemically throughout the vine. As a result, they can kill pathogens and pests within host tissues, something contact pesticides rarely do. Regrettably, their highly specific toxic action often permits negation by single mutations in the pathogen. Initially, resistant populations tend to be less pathogenic, due to resistance being accompanied with reduced biological fitness. However, after prolonged selective pressure, subsequent mutations often arise, restoring biological fitness. At this point, removal of the pesticide has minimal effect on reducing the population of resistant strains. Thus, for long-term effectiveness, selective (systemic) chemicals should be restricted to use only in situations where their precise in-tissue toxicity is required (Delp, 1980; Northover, 1987).

Another approach extending the long-term effectiveness of selective pesticides involves rotational application. By switching from a pesticide in one chemical group to another, the advantage of resistance genes against individual chemicals may be counteracted. Frequently, the initial resistance genes are detrimental to the growth or reproduction of the pest or pathogen. Thus, in the absence of constant pesticide pressure, selection acts against the retention of resistance genes. Continual pesticide pressure, however, tends to eventually select mutations without metabolic liabilities. Unfortunately, success using rotation is not guaranteed (Baroffio *et al.*, 2003).

A similar but alternative approach involves the combination of nonselective and selective pesticides. Nonselective fungicides reduce the size of the pest or pathogen population – thereby reducing the likelihood of resistant gene selection. This approach has been accredited with delaying the emergence of resistance genes in some pathogen populations (Delp, 1980). Survey of resistance development in local pathogen populations is an important component in assessing the continuing efficacy and value of any pesticide.

Another approach in delaying the stabilization of pesticide resistance is to jointly apply a combination of two or more selective pesticides. Theoretically, this reduces the chance of resistance developing to any one of the pesticides – for example, if the probability of resistance development in a fungal pathogen against three fungicides were equal and 1×10^{-8}, the combined likelihood of simultaneous resistance to all three fungicides would be 1×10^{-24} (effectively nil).

In both approaches, it is important that each of the pesticides belong to a different class of chemical compound – possessing different modes of toxic action (Table 4.11) (for a compete listing, see www.pesticideinfo.org or www.alanwood.net/pesticides/class_fungicides.html). If the chemicals have similar modes of toxicity, the selective pressures they produce are likely to be equivalent. Thus, resistance to one member of a class usually provides resistance against other members of the same class – a phenomenon called **cross-resistance**.

Another means of retarding pesticide resistance is to increase the effectiveness of pesticide application. This can involve the use of better sprayers, thereby providing more uniform crop cover. Sites where pesticide coverage is inadequate to kill the pathogen may inadvertently provide conditions favoring the subsequent

Table 4.11 Examples of fungicides used in controlling grape diseases, arranged according to chemical family

Chemical group	Activity group[a]	Common and trade name(s)	Structure (Principal example)
Anilinopyrimidine	I	pyrimethanil (Scala) other examples: cyprodinil (Vangard, Chorus)	
Benzimidazole	A	carbendazim (Delsene) other examples: benomyl (Benlate) triopnanate methyl (Topsin-M)	
Benzothiadiazole	*	acibenzolar-S-methyl (Actigard)	
Dicarboximide	B	iprodione (Rovral) other examples: vinclozolin (Ronilan) chlozoline (Serinal) procymidone (Sumisclex)	
Dithiocarbamate	Y	maneb (Maneb) other examples: zineb (Zineb) ferbam (Carbamate) maneb + Zn (Mancozeb, Dithane)	
DMI-Pyrimidine	C	fenarimol (Rubigen)	
DMI-Triazole	C	myclobutanil (Rally, Nova) other examples: difenoconazole (Score) hexacibazike (Anvil) tebuconazole (Elite) triadimefon (Bayleton)	
Hydroxyanilide	J	fenhexamid (Elevate, Teldor)	
Phenylamide	D	mefenoxam (Metalaxyl, Ridomil) other examples: furalaxyl (Fongarid)	
Phosphonate	Y	fosetyl-aluminum (Aliette) other examples: phosphorous acid	
Phenoxy quinoline	M	quinoxyfen (Legend)	
Phthalimide	Y	captan (Captan, Merpan) other examples: chlorothalonil (Bravo)	
Strobilurin	K	trifloxystrobin (Flint) other examples: azoxystrobin (Abound, Amistar) kresoxim-methyl (Stroby) pyraclostrobin (Cabrio)	

[a] Grouping of fungicides by their similar mode of action.

* Unclassified fungicide/plant activator.

development of resistance. The use of spreaders in a pesticide formulation can facilitate the uniform dispersion of the chemical over plant surfaces. Stickers can be beneficial as well, by minimizing dilution during rainy periods. Regrettably, stickers limit the subsequent spread of the pesticide to regions that develop after application. Dusts, by virtue of their generally smaller mass, diffuse more uniformly over plant surfaces, and into dense canopies, than do droplets of a wettable pesticide. Conversely, dusts tend to settle and adhere more poorly to plant surfaces, with more potential for wastage due to wind dispersion.

Finally, application of other control measures (e.g., hygiene, environmental modification) tends to retard selection for pesticide resistance. Because mutation is largely a random event, and therefore partially dependent on population size, the smaller the pest population, the less likely (more slowly) resistant mutations will occur.

None of these techniques will prevent the eventual development of resistance, but they can significantly delay its occurrence. The extent of the delay will depend on factors such as the widespread use of protective measures, the number of growth cycles per year, the existing genetic variability of the pest or pathogen population, the population size, and the facility with which the pest or pathogen is dispersed geographically.

Complicating the application of some of these protective procedures are archaic regulations, not only in the viticultural region of origin, but also in the importing country (e.g., restrictions on pesticide residues). Because these restrictions can vary from year to year, the topic must be addressed on a local basis, with the appropriate governmental agencies.

Recently, a new and novel class of pest control chemicals have entered the market. They have their effect by inducing systemic acquired resistance in the host. Their appearance is encouraging, as they take a distinctly new, and "natural" approach to disease and pest control. Examples are jasmonic acid against Pacific spider mites and phylloxera (Omer *et al.*, 2000), β-aminobutrytic acid against downy mildew (Cohen, 2002), acidbenzolar-*S*-methyl against powdery mildew (Campbell and Latorre, 2004), and benzothiadiazol against bunch rot (Iriti *et al.*, 2004).

BIOLOGICAL CONTROL

Although pesticides will remain one of the primary means of controlling pests and disease-causing agents into the foreseeable future, increasing emphasis is being placed on biological control. Although some predators of insect pests have developed pesticide resistance (Englert and Maixner, 1988), most remain sensitive to insecticides. Thus, to use indigenous insect- or mite-control agents,

pesticide applications must be delayed, minimized, or avoided.

In addition to restricting insecticide use, it is often necessary to maintain a broad diversity of plants in the vicinity of the vineyard. This frequently provides the range of alternate hosts necessary for maintaining or enhancing the population of pest-control agents. Nevertheless, this does not consistently improve biological control.

Occasionally, the release of competitive species can provide adequate control for serious pests. By establishing themselves on the host, the competitors restrict the colonizing potential of the pest species. An example of this phenomenon, called **competitive exclusion**, is the action of the Willamette spider mite (*Eotetranychus willamettei*) against the Pacific spider mite (*Tetranychus pacificus*) (Karban *et al.*, 1997).

Although insects are susceptible to many bacterial, fungal, and viral pathogens, few have shown promise in becoming practical control agents. Exceptions include the granulosis virus against the western grape leaf skeletonizer (*Harrisina brillians*) (Stern and Federici, 1990) and *Beauveria bassiana* (GHA strain) against the western flower thrips. The major success story has been with the Bt toxin, produced by *Bacillus thuringiensis*. It is active against several lepidopteran pests, such as the omnivorous leafroller (*Platynota stultana*) and the grape leaffolder (*Desmia funeralis*). Bt toxin produces "holes" in the intestinal membrane of insects, permitting the entrance of intestinal bacteria. These can induce septicemia in the host's hemolymph (Broderick *et al.*, 2006). In some regions, where *B. thuringiensis* toxin has been used extensively in insect control, signs of resistance to Bt toxin have appeared. This may make the incorporation of the gene for toxin production into crop plants of little long-term value. The same might occur with fungal resistance, supplied by endochitinase genes from *Trichoderma harzianum* (Lorito *et al.*, 2001), genetically engineered into grapevines.

Pheromone application is another biological technique employed in insect management. These are species-specific, airborne hormones used by insects to locate receptive members of the opposite sex. Pheromones have been used effectively to attract insects to pesticide traps, or to cause mating disorientation when applied throughout the vineyard. The release of a large number of artificially reared, sterile individuals during the mating season also can disrupt successful mating.

Last, but definitely not least, is the use of resistant rootstocks. This strategy was first used to control the phylloxera infestation in Europe in the late 1800s. Grafting to resistant or tolerant rootstocks can also be used to limit the damage caused by several nematodes, viruses, and environmentally induced toxicities.

The biological control of fungal pathogens is less developed than for arthropod pests. This partially reflects the growth of fungal pathogens within plants, away from exposure to parasites. Nevertheless, inoculation of leaf surfaces with epiphytic microbes may prevent the germination and subsequent penetration of fungal pathogens. The phylloplane flora may inhibit pathogens by competing for organic nutrients required for germination. This has been developed as a commercial control with *Trichoderma harzianum* (Trichodex®). This and related products appear to be especially effective against wood-invading fungi, such as those inducing Eutypa, Esca and Petri ('black goo') diseases (Hunt, 1999). Occasionally, however, *Botrytis* control is inadequate and requires rotation with conventional fungicides. Another commercial biofungicide active against *B. cinerea* is Greygold®. It is a mixture of the filamentous fungus *Trichoderma hamatum*, the yeast *Rhodotorula glutinis*, and the bacterium *Bacillus megaterium*. With powdery mildew, where the mycelium remains predominantly on the plant surface, the mycoparasite *Ampelomyces quisqualis* (AQ10®) has proven commercially effective (Falk *et al.*, 1995).

One of the most challenging aspects of biological control can be the special requirements for storing and applying biological control agents. Because spores and biological toxins often have relatively short half-lives, they need to be kept under dry, cool conditions. In addition, they are often quickly inactivated by solar UV radiation. Thus, they are best applied in the evening, to extend their effective period. These limitations have retarded grower acceptance (Hofstein, 1996).

Plowing under of a green manure is a long-established technique favoring the elimination of fungal pathogens that survive for short periods in the soil. This is thought to enhance competition and destroy sites in which the pathogen can survive saprophytically. A somewhat similar approach is being employed with foliar pathogens. However, in this instance, a solution of organic nutrients is sprayed on the leaves to promote the growth of the phylloplane flora. Their activity may both directly and indirectly suppress the germination or growth of leaf and fruit pathogens (Sackenheim *et al.*, 1994).

One of the more intriguing examples of biological control involves **cross-protection**. Cross-protection is a phenomenon in which virally infected cells are immune from further infection by the same or related viruses. The selective incorporation of particular viral genes into plants can induce a similar phenomenon (Beachy *et al.*, 1990). Thus, the incorporation of genes coding for coat or dispersal proteins may produce transgenic cultivars resistant to the source virus (Krastanova *et al.*, 1995).

ENVIRONMENTAL MODIFICATION

Modifying the microclimate around plants has long been known to potentially minimize disease and pest incidence. By improving the light and air exposure around grapevines, canopy management can increase the toughness and thickness of the epidermis and its cuticular covering. An open-canopy structure also facilitates the rapid drying of fruit and foliage surfaces. This reduces the time available for fungal penetration and may reduce or inhibit spore production. Furthermore, an exposed canopy enables more efficient application of chemical or biological pesticides.

Exposure of fruit to the drying action of sun and air can be further enhanced by applying gibberellic acid. This would be especially useful in tight-clustered varieties. Gibberellin induces cluster stem elongation and opens the fruit cluster (Plate 3.8). Reduction in berry compactness also favors the production and retention of a typical epicuticular wax coating on the fruit (Fig. 4.39). However, the effect of gibberellic acid on grape composition is complex, and appears to be cultivar-specific (Teszlák *et al.*, 2005). For example, it can affect grape coloration, phenolic content, as well as mineral content. In addition, it apparently increases the formation of o-aminoacetophenone (Christoph *et al.*, 1998; Pour Nikfardjam *et al.*, 2005b), a compound frequently associated with the development of an "untypical aged flavor" in wine. By suppressing the action of IAA oxidase, gibberellic acid can initiate an increase in auxin content, that appears to favor o-aminoacetophenone synthesis (Pour Nikfardjam *et al.*, 2005a). Fruit exposure can be enhanced still further by basal leaf removal. This technique has been so successful in reducing dependence on fungicidal sprays that several wineries have written it into their grower contracts (Stapleton *et al.*, 1990). Basal leaf detachment also removes most of the first-generation nymphs of the grape leafhopper (*Erythroneura elegantula*) (Stapleton *et al.*, 1990). This can improve biological control by the leafhopper parasitic wasp (*Anagrus epos*). It provides additional time for the wasp population to buildup and spread to grapevines from overwintering sites on wild blackberries.

Balanced plant nutrition generally favors disease and pest resistance, by promoting the development of the inherent anatomical and physiological resistance of the vine. Nutrient excess or deficiency may have the reverse effect. For example, high-nitrogen levels suppress the synthesis of a major group of grape antifungal compounds, the phytoalexins (Bavaresco and Eibach, 1987).

Irrigation can decrease the consequences of nematode damage by providing an ample water supply to the damaged root system. Irrigation also may favor healthy

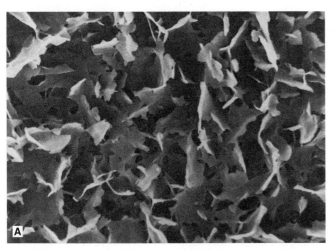
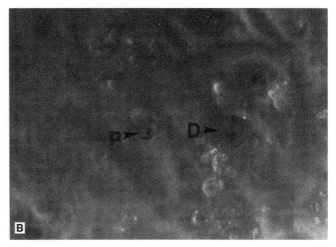

Figure 4.39 Scanning EM of the epicuticular wax of mature berries: (**A**) typical platelet structure of noncontact areas; (**B**) shallow depression (D) and pore (P) typical of contact area between adjacent berries. (From Marois *et al.*, 1986, reproduced by permission)

vine growth by reducing water deficit. However, excessive irrigation can favor disease development by promoting luxurious canopy development and increasing berry-cluster compactness. This is particularly important for cultivars such as 'Zinfandel' and 'Chenin blanc' that normally produce compact clusters.

Weed control generally reduces disease incidence. This may result from the removal of alternate hosts, on which pests and disease-causing agents may survive and propagate. For example, dandelions and plantain often are carriers of tomato and tobacco ringspot viruses, and Bermuda grass is a reservoir for sharpshooter leafhoppers – the primary vectors of Pierce's disease.

Soil tillage occasionally can be beneficial in disease control. For example, the burial of the survival stages of *Botrytis cinerea* promotes their destruction in soil. In addition, the emergence of adult grape root borers (*Vitacea polistiformis*) is restricted by burial of the pupae during soil cultivation (All *et al.*, 1985).

Although environmental modification can limit the severity of some pathogens, it can itself enhance other problems. For example, soil acidification used in the control of Texas root rot (*Phymatotrichum omnivorum*) has increased the incidence of phosphorus deficiency in Arizona vineyards (Dutt *et al.*, 1986). The elimination of weeds as carriers of pests and disease-causing agents may inadvertently limit the effectiveness of some forms of biological control. It can remove survival sites for predators and parasites of grapevine pests. The carrier of one pest may be the reservoir of predators for another.

GENETIC CONTROL

Improved disease resistance is one of the major aims of grapevine breeding. It was first seriously investigated as a means of avoiding the expense of grafting in phylloxera

control. Subsequently, breeding has focused primarily on developing rootstocks possessing improved drought, salt, lime, virus, and nematode resistance. Work has also progressed, but more slowly, in developing new scions with improved pest and disease resistance. In the premium wine market, consumer (or critic) resistance to new varieties has limited acceptance. However, at the lower end of the market, new varieties often have a distinct advantage because of their reduced production costs. Resistant varieties may be even more acceptable for "organic" viticulture, where synthetic pesticides are forbidden. Here, avoidance of synthetic (but not "natural") pesticide use is probably more critical than varietal origin.

In most European countries, there are legal restrictions against the use of interspecies crosses in the production of Appellation Control wines. This is to their disadvantage, because the best sources of disease and pest resistance come from species other than *V. vinifera*. Within *V. vinifera*, the few remaining wild vines growing in Europe and southwestern Asia probably contain much of what remains of its former genetic diversity (Avramov *et al.*, 1980). Although hidden sources of resistance still reside within existing cultivars of *V. vinifera* (Becker and Konrad, 1990), they are probably limited.

Because of the restrictions imposed by European countries on interspecies crosses, and the market dominance of a relatively few cultivars, the greatest likelihood of successful genetic improvement in disease resistance will depend on genetic engineering. Nevertheless, the insertion of specific genes for resistance is in the experimental stage. In some cultivars, the activity of genes regulating chitinase, β-glucanase, and basic peroxidase isoenzyme production is correlated with resistance to several important fungal diseases. Their isolation and insertion into important cultivars could theoretically improve their insensitivity to several pathogens. Nonetheless, long-term disease resistance is likely to require the incorporation of many genes, affecting both pre- and postinfection host responses. Studies on the biochemical basis of disease resistance will probably increase the number of candidate genes that might be incorporated. Although not involving resistance genes *per se*, the incorporation of genes coding for one of several viral genes, noted earlier, can induce immunity (cross-protection) against the source virus.

In most instances, complete resistance (immunity) to infection is preferable. However, even slowing the rate of disease spread (tolerance) may provide adequate protection in most years. In some cases, tolerance to one pest can favor infection by another (e.g., tolerance to *Xiphinema index* favors transmission of the grapevine fanleaf virus). In addition, tolerance to a systemic pathogen (by masking its presence) favors the likelihood of its spread by grafting or other forms of mechanical transmission.

ERADICATION AND SANITATION

In most situations, the eradication of established pathogens is impossible, due to their survival on alternate hosts, such as weeds or native plants. Eradication can be useful, however, in the elimination of systemic pathogens. Seed propagation is often used for this purpose, but this is inapplicable with grapevines – it would disrupt the combination of traits that makes each cultivar unique. Thus, except where disease-free individuals can be found, the elimination of systemic pathogens involves **thermotherapy**, **meristem culture**, or the combination of both.

Thermotherapy typically involves placing vines at 35–38 °C for 2–3 months (or dormant canes at higher temperatures for shorter periods). It is effective against several viruses, such as the grapevine fanleaf, tomato ringspot, and fleck viruses, as well as the leafroll agent (typically associated with closteroviruses). For most other viruses, culturing small meristematic fragments of shoot apex may permit the isolation and propagation of uninfected vines. Meristem culture has been used in the elimination of the virus-like agents of stem pitting, corky bark, leafroll, the viroid of yellow speckle, and the bacterium *Agrobacterium vitis*. The last also may be eliminated by a short, hot-water treatment (Bazzi *et al.*, 1991). Hot-water treatments, typically between 45 and 55 °C for from 10 min to 2.5 h (depending on the purpose), can eliminate other pathogens, such as the bacteria *Xylella fastidiosa*, *Xanthomonas ampelina*, the phytoplasma associated with *flavescence dorée*, the fungus *Phytophthora cinnamoni*, and nematodes, such as *Xiphinema index* and *Melidoyne* spp. The procedures and recommended precautions for hot-water treatment are given in Hamilton (1997) and Waite *et al.* (2001). Cleansing vines infected by several systemic pathogens may require the combined use of several procedures.

The use of thermotherapy and meristem culture often cause grower concern. They may modify the morphological and physiological traits of the treated vines. Expression of juvenile traits, such as spiral phyllotaxy, reduction in tendril production, more jagged and pubescent leaves, stem coloration by anthocyanins, and reduced fertility are fairly typical. They usually disappear, however, as the vines mature or are propagated repeatedly (Mullins, 1990; Grenan, 1994).

Once freed of systemic pathogens, vines usually remain disease-free, if grafted to disease-free rootstock and planted on soil free of the pathogen. Although most serious grapevine viruses are not transmitted by leaf-feeding insects, the use of resistant rootstock is advisable where the soil is infested with nematodes. Alternatives are leaving the land fallow for up to 10 years or fumigating the soil.

Sanitation may not eliminate disease or pest problems, but it usually reduces their severity, by destroying resting stages or by removing survival sites.

QUARANTINE

Most, if not all, wine-producing countries possess laws regulating the importation of grapevines. Some of the best examples illustrating the need for quarantine laws are those involving the transmission of grapevine pathogens. Two of the major grapevine diseases in Europe (downy and powdery mildew) were imported unknowingly from North America in the nineteenth century. The phylloxera root louse was also accidentally introduced from North America, probably on rooted cuttings. Several viral and virus-like agents are now widespread in all major wine-growing regions. They are thought to have been spread, unsuspectedly, through the importation of asymptomatic but infected rootstock cultivars. Examples are the agents causing leafroll, corky bark, and stem pitting.

Thankfully, some other potentially devastating diseases have as yet to become widespread. For example, Pierce's disease is still largely confined to southeastern North America and Central America. However, it is now causing severe problems in parts of California, due to spread by the glassy-winged sharpshooter. Even phylloxera has not spread throughout all regions in countries where it now occurs. Thus, limiting grapevine movement within regions of a single country can still have a significant impact on the spread of pathogens with low potential for self-dispersal.

Because it may be difficult to detect the presence of some pests and disease-causing agents, only dormant canes are usually permitted entrance in most jurisdictions.

This impedes the introduction of root and foliar pathogens, but fungal spores, insect eggs, and other minute dispersal agents may go undetected. Consequently, imported canes are typically quarantined for several years, until they are determined to be free, or freed, of known pathogens. Most pests and fungal and bacterial pathogens express their presence during this period. However, the presence of latent viruses and viroids usually requires grafting or mechanical transmission to **indicator plants**. The detection of systemic pathogens, through their transmission to sensitive (indicator) plants, is called **indexing**. Less time-consuming analytic techniques, such as ELISA (enzyme-linked immunosorbent assay) (Clark and Adams, 1977), cDNA (Koenig *et al.*, 1988), or PCR probes (Rowhani *et al.*, 1995) improve and speed the detection of systemic pathogens. It can also permit the earlier release of imported cultivars.

Unfortunately, quarantine is not only expensive but also can seriously delay importer access to his stock. A possible solution is the development of encapsulated somatic embryos. Being produced under sterile conditions from healthy tissue, they could be shipped without the potential for incidental pathogen importation associated with whole cutting. On arrival, cultivation and multiplication could occur almost immediately under sterile conditions.

Consequences of Pathogenesis for Fruit Quality

The negative influence of pests and diseases is obvious in symptoms such as blighting, distortion, shriveling, decay, and tissue destruction. More subtle effects involve vine vigor, berry size, and fruit ripening. Sequelae such as reduced root growth, poor grafting success, decreased photosynthesis, or increased incidence of bird damage on weak vines (Schroth *et al.*, 1988) can be easily overlooked. In some instances, detection of infection is impeded by minimal symptoms and the absence of uninfected individuals. For example, the almost universal prevalence of viroids in grapevine cultivars was not recognized for years because uninfected vines for comparison were rare.

Most pest and disease research is concerned with understanding the pathogenic state and how it can be managed. However, in making practical decisions on disease control, it is important to know the effects of disease not only on vine health and yield, but also on grape and winemaking quality.

All pests and disease agents disrupt vine physiology and, therefore, can influence fruit yield and quality to some degree. However, agents that attack berries directly have the greatest impact on fruit quality. These include three of the major fungal grapevine pathogens, *Botrytis cinerea*, *Plasmopara viticola*, and *Uncinula necator*. Grapevine viruses and viroids, being systemic, can both directly and indirectly affect berry characteristics. For example, leaf roll-associated viruses reduce grape °Brix and increase titratable acidity. Insect pests can cause fruit discoloration and malformation, as well as create lesions favoring invasion by microbes.

Of fruit-infecting fungi, the effects of *B. cinerea* have been the most extensively studied. Under special environmental conditions, infection produces a "noble" rot, often yielding superb wines. Normally, however, the fungus produces an ignoble (bunch) rot. The early invasion of infected fruit by acetic acid bacteria probably explains the high levels of fixed and volatile acidity in the fruit. Under moist conditions, additional secondary invaders such as *Penicillium* and *Aspergillus* may contribute additional off-flavors. More serious are those fungi producing mycotoxins, notably isofumigaclavine, festuclavine, roquefortine by *Penicillium* spp. (Moller *et al.*, 1997); ochratoxin A from species of *Aspergillus*, notably *A. carbonarius* and *A. ochraceus* (O'Brien & Dietrich, 2005); and trichothecines from *Trichothecium roseum* (Schwenk *et al.*, 1989). *Mucor* species produce compounds inhibitory to malolactic fermentation (San Romáo and Silva Alemão, 1986), but their toxicity to mammals, if any, is unknown. *B. cinerea* does not produce mycotoxins (Krogh and Carlton, 1982), but it can synthesize several antimicrobial compounds, active against yeasts and other fungi (Blakeman, 1980). In addition to increased fixed and volatile acidity, bunch rot produces reduced nitrogen and sugar levels that create difficulties during juice fermentation. Large accumulations of β-glucans create clarification problems. The polyphenol oxidase, laccase, oxidizes anthocyanins and generates phenol flavors that generally make infected red grapes unacceptable for winemaking. *Botrytis* infection can also increase problems associated with protein haze (Girbau *et al.*, 2004). This appears to result from the increased presence of disease protective, thaumatin-like (PR) proteins.

Little information is available on the direct consequences of other pathogens on grape and wine quality. One exception is the higher pH and phenol contents of wine produced from fruit infected by powdery mildew. They may donate a bitterish attribute (Ough and Berg, 1979), or other flavor modifications (Fig. 4.40). The flavor consequences of infection are augmented with increased skin contact prior to or during fermentation. This may involve the conversion of several ketones to 3-octanone and (Z)-5-octen-3-one (Darriet *et al.*, 2002). Off-flavors may be detected in wine made from grapes with as little as 3% infection. Yield, °Brix, and anthocyanin synthesis in red grapes are reduced as a consequence of infection. Browning is common, but can

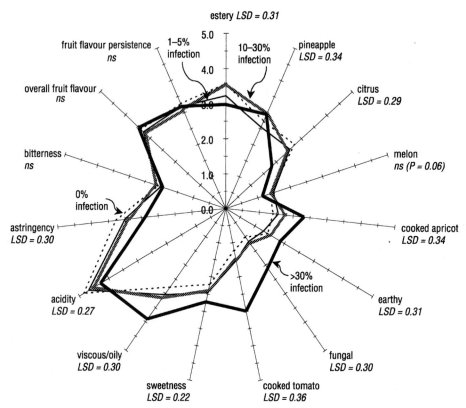

Figure 4.40 Effect of the level of powdery mildew (0, 1–5%, 10–30%, 30%) on the sensory attributes for the 'Chardonnay' wine. LSD: Least significant difference (P = 0.05), ns: not significant. (From Stummer *et al.*, 2005, reproduced with permission)

be partially offset by the bleaching action of sulfur dioxide. Infected white grapes also possess higher concentrations of haze-producing proteins. Fermentation may also take up to twice as long to complete (Ewart *et al.*, 1993). In *flavescence dorée*, the production of a dense, bitter pulp makes wine production from diseased fruit virtually impossible.

Leafroll has been the most investigated of virus or virus-like infections relative to grape and winemaking quality. Potassium transport is affected and berry titratable acidity decreases. This typically generates wines of higher pH and poor color. Sugar accumulation in the berries is usually decreased, due to the suppression of transport from the leaves. Ripening is often delayed.

The physiological disorder, bunch-stem necrosis (*dessèchement de la rafle*, *Stiellähme*), causes grape shriveling and fruit fall around and after *véraison*. Wines produced from vines so affected are often imbalanced, high in acidity, and low in ethanol, as well as several higher alcohols and esters (Ureta *et al.*, 1982). It is probably an inherited trait, as some cultivars (i.e., 'Silvaner' and the Pinot family) are particularly resistant to the disorder.

The effects of disease on aroma have seldom been reported. Exceptions are the reduction in the varietal character of grapes infected by *B. cinerea*, or their modification by the ajinashika virus (Yamakawa and Moriya, 1983).

Although not a direct consequence of pathogenesis, the application of protective chemicals may indirectly affect wine quality. For example, the copper in Bordeaux mixture can compromise the quality of wines from 'Sauvignon blanc' and related cultivars. It can reduce the concentration of important varietal aroma compounds, such as 4-mercapto-4-methylpentan-2-one. This effect can be reduced by prolonged "skin" contact (Hatzidimitriou *et al.*, 1996), or avoiding the use of Bordeaux mixture (Darriet *et al.*, 2001). Fungicides, herbicides, and other pesticides may also have phytotoxic effects on the vine. These can vary from direct visible damage to more subtle changes, such as reduced sugar accumulation in berries (Hatzidimitriou *et al.*, 1996). Even more indirect may be influences on the soil flora and fauna. For example, long-term use of Bordeaux mixture (in the control of mildew) has resulted in a substantial accumulation of copper in many French vineyards. This apparently results due to the reaction between copper and iron, forming oxyhydroxides that bind to the soil's clay fraction. This seemingly delays microbial activity, augmenting soil organic content (Parat *et al.*, 2002).

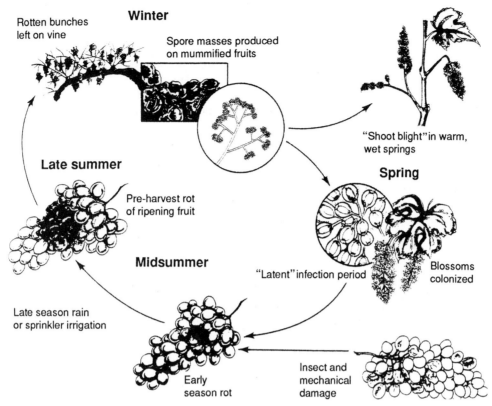

Figure 4.41   Disease cycle of Botrytis bunch rot. (From Flaherty *et al*., 1982, reproduced by permission)

Examples of Grapevine Diseases and Pests

Grapevines can be attacked by a wide diversity of biologic agents. There is insufficient space in this book to deal with all these maladies. Thus, only a few important and/or representative examples of the major categories of grapevine disorders are given here. Detailed discussions of grapevine maladies for specific various countries can be found in specialized works such as Pearson and Goheen (1988) and Flaherty *et al*. (1992) (North America), Galet (1991) and Larcher *et al*. (1985) (Europe), and Coombe and Dry (1992) and Nicholas *et al*. (1994) (Australia).

FUNGAL PATHOGENS

With few exceptions, fungal pathogens grow vegetatively as long, thin, branched, microscopic filaments, individually called **hyphae,** and collectively termed **mycelia.** Most fungi produce cell-wall ingrowths along the hyphae termed **septa.** The ingrowths are usually incomplete and leave a central opening through which nutrients, cytoplasm, and cell organelles may pass. Thus, fungi possess the potential to adjust the number and proportion of nuclei within the organism. This gives fungi a degree of genetic flexibility unknown in other organisms. The filamentous growth habit also provides them with the ability to physically puncture plant-cell walls. This property, combined with their degradative powers and prodigious spore production, helps explain why fungi are the predominant disease-causing agents of plants.

Botrytis Bunch Rot   Several hyphomycetes can generate what is called bunch rot, either alone or together. However, the principal causal agent is *Botrytis cinerea.* The pathogen appears to exist as two coinhabiting subpopulations – *transposa* and *vacuma* (Giraud *et al.*, 1999). Both infect grapes, may occur sympatrically, but tend to possess different frequencies of fungicide resistance. Unlike many grape pathogens, and most other *Botrytis* species (Staats *et al.*, 2005), *B. cinerea* is a nonspecialized pathogen, infecting a wide diversity of plants and plant tissues. As a consequence, spores may arise from a diversity of host species, within and around vineyards. Nevertheless, most early infections probably develop from spores produced on overwintered mycelia in the vineyard (Fig. 4.41). In addition, black, multicellular resting structures called **sclerotia** may generate spores. These are usually conidia (asexual spores). Occasionally, ascospores are produced from multicellular fructifications called apothecia.

Initial infections usually develop on aborted and senescing flower parts. When the remnants of flowers are trapped within growing fruit clusters, they may initiate fruit infections later in the season. Another source of fruit infection comes from latent infections that occur in the spring. These form when hyphae invade the vessels of young green berries (Pezet and Pont, 1986). The fungus subsequently becomes inactive, until the level of acidity and other antifungal compounds in the fruit declines during ripening. Latent infections likely act as the primary source of disease under dry autumn conditions. For this reason, it is important to spray vines to limit early infections in regions plagued by gray mold. Spraying at flowering, 80% cap fall, and bunch closure often provides considerable protection. Nevertheless, under protracted rainy spells, bunch rot can rapidly develop from *de novo* infections. It often involves secondary invaders, such as *Penicillium*, *Aspergillus*, *Cladosporium*, *Rhizopus*, and *Acetobacter* spp. Several insects, such as the European grape berry moth, light-brown apple moth, and fruit flies can aggravate disease incidence, by transporting and infecting fruit with *B. cinerea* conidia (Mondy, Charrier *et al.*, 1998). Infections of leaves, shoots, and other vine parts occur, but are primarily important as sites for winter survival.

During the growing season, both physiological and anatomical changes can increase the susceptibility of fruit to fungal attack. Microfissures develop around stomata (Fig. 4.42A), and micropores may form in the cuticle (Fig. 4.42B). Both provide sites for fungal penetration and the release of plant nutrients that aid spore germination. The weathering of cuticular waxy plates also favors infection by facilitating spore adherence. The loss of wax is most noticeable where berries press and rub against one another (Marois *et al.*, 1986). The cutin content can decrease to less than 40% of its preanthesis level by *véraison* (Comménil *et al.*, 1997). Rapid berry enlargement, especially during heavy rains, can induce splitting and the release of juice, favoring infection.

Many factors affect bunch rot susceptibility. Skin toughness and open fruit clusters reduce bunch rot incidence, whereas heavy rains, protracted periods of high humidity, and shallow vine-rooting increase susceptibility. Shallow rooting exposes the vine to waterlogging, which can favor rapid water uptake and berry splitting. Berry splitting also can result from the osmotic uptake of water through the skin under rainy and cloudy conditions (Lang and Thorpe, 1989). In addition, protracted moist periods provide conditions that favor spore germination and production.

After germination, spores produce one or more germ tubes that grow out through the spore wall. Fruit penetration occurs shortly thereafter, often through microfissures in the epidermis. Subsequent ramification progresses more or less parallel to the berry surface through the hypodermal tissues. Infection may fail to spread to the mesocarp (Glidewell *et al.*, 1997).

Depending on the temperature and humidity, new spores are produced within several days or weeks. Spores are borne on elongated, branched filaments that erupt either through the stomata or epidermis. The white to gray color of the young spores gives rise to the common name for most *Botrytis* diseases, gray mold. On maturity, the spores turn brown, and often are so

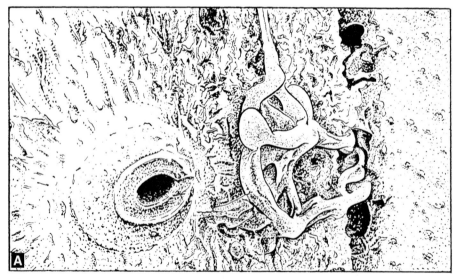

Figure 4.42 (A) Drawing of *Botrytis cinerea* penetrating a berry peristomatal microfissure (from an scanning electron micrograph by Bessis, 1972, in Ribéreau-Gayon *et al.*, 1980); (B) scanning electron micrograph of a section through a cuticular micropore (from Blaich *et al.*, 1984, reproduced by permission)

densely packed as to give the infected tissues a felt-like appearance. Early in infection, white grapes may take on a purplish coloration. All infected fruit eventually turn brown, presumably due to the phenol oxidizing action of laccase.

Effective control often requires both fungicidal sprays and environmental modification. Some fungicides remain localized on the surface and act protectively; others are incorporated into plant tissues and possess both protective and curative properties. With the development of resistance to fungicides (dicarboximides) or their deregulation (benzimidazoles), chemicals such as iprodione (Rovral®), cyprodinil (Vangard®), pyrimethanil (Scala®), and fenhexamid (Elevate®), and purified paraffinic oil (Stylet oil) are becoming the primary substitutes. Neither sulfur- nor copper-based fungicides are effective. Effective fungicide application is enhanced by leaf removal around the clusters. Benzothiadiazole, a new class of disease control agents, possesses the novel property of inducing systemic acquired resistance. It has been found to activate phenol synthesis, including anthocyanin and resveratrol in grapes. It enhances grape resistance to *Botrytis cinerea* (Iriti *et al.*, 2004).

Plowing under infested plant remains, along with a green manure, helps to reduce vineyard survival of the fungus. By itself, though, it is ineffective. The use of less vigorous rootstocks, canopy management, and basal leaf removal can help generate a more open canopy and speed drying of vine surfaces. The application of gibberellic acid also can favor drying by opening tight fruit clusters, but probably is acceptable only with table or raisin grapes. This results from gibberellic acid increasing the incidence of the untypical aged flavor in wine.

Because of the significance of *Botrytis* infection to many crops, biological control is under investigation in many parts of the world. Many of these studies are still in their early stages, but suggest alternative control strategies. These vary from the application of compost and manure extracts, to adding suppressive viral (dsRNA mycovirus), bacterial, yeast (*Pichia membranifaciens*), or filamentous fungal (*Pythium radiosum*) agents. To date, most have shown inconsistent field results. The most commercially successful of these preparations has been Trichdex®, a formulation of *Trichoderma harzianum* spores (O'Neill *et al.*, 1996). Because the action of biocontrol agents may be suppressed by fungicides, they usually are not used in combination. Because bacteria are seldom affected by fungicides, the action of suppressive *Bacillus* spp. is unaffected by joint or subsequent application of some fungicides.

Powdery Mildew (Oidium) Powdery mildew is induced by a specialized pathogen, *Uncinula necator*. It specifically attacks members of the *Vitaceae* (Halleen and Holz, 2001). Distinctive hyphal extensions, called haustoria, grow into living epidermal cells (Fig. 4.43). Nevertheless, adjacent palisade cells are physiologically disrupted and soon become necrotic. Most of the fungus remains external to the vine.

Fungal overwintering often depends on dormant hyphae that survive on the inner bud scales (prophylls) of grapevines. Sporulation may commence within the bud. In cool climates, survival also may involve microscopic, round, reddish-black resting structures called **cleistothecia**. Mature cleistothecia, washed from diseased tissue, may lodge in bark crevices. In this position they are ideally located to initiate infections in the spring (Gubler and Ypema, 1996). After rains, overwintering cleistothecia swell, rupture, and eject ascospores. The ascospores may then wash or be blown onto young tissues and initiate early infection following bud burst (Pearson and Gadoury, 1987). For reasons that are still unclear, cleistothecia rarely form in several European countries. Thus, the principal means of survival in several regions depends on dormant hyphae. These survive on infected tissues (notably canes and bud scales). They reactivate in the spring, produce spores, and initiate infection upon bud break. If sufficiently infected, stunted shoots ("flag shoots") and leaves are covered with a cottony covering of powdery mildew.

In regions where cleistothecia participate in the infection cycle, it is essential to initiate spraying earlier than when infection develops only from overwintered hyphae. Application of lime-sulfur or flowable sulfur, prior to bud break, can effectively limit cleistothecial-based infection (Gadoury *et al.*, 1994).

Infection can result in leaf and fruit distortion, by killing surface tissues before they reach maturity. Severe infection leads to leaf and fruit drop, as well as death of the shoot tip. Spore production produces a white powdery appearance on the infected surfaces. Fungal growth and sporulation are optimal between about 20 and 30°C. At above 32°C, fungal metabolism essentially stops. Later in the season, the production of cleistothecia can give infected tissues a distinctive red- to black-speckled appearance. Fruit is most sensitive to infection shortly after initiation, rapidly developing immunity thereafter in many cultivars (Gadoury *et al.*, 2003). This appears to involve hyphal inability to effectively penetrate the mature epidermis. Immunity may be complete within as little as four weeks (Ficke *et al.*, 2003). This may be due to enhanced synthesis of stress and pathogenesis-related (PR) proteins, such as osmotin and traumatin-like proteins (Monteiro *et al.*, 2003). Leaves more than 8-weeks old are also seldom infected.

Early control is not only important for fruit protection, but also minimizes foliage damage and its consequences on growth the following year. Although the rachis remains

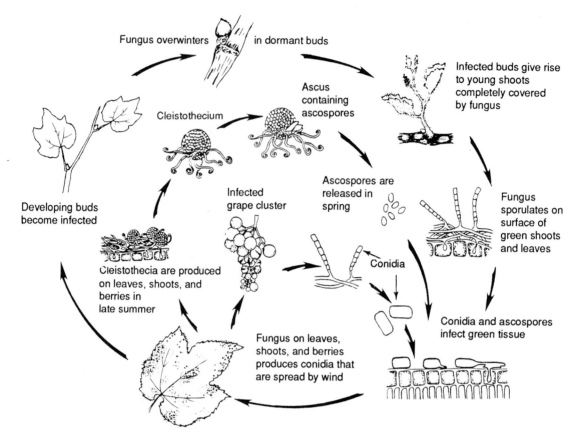

Figure 4.43 Disease cycle of powdery mildew. (Drawing by R. Sticht (Kohlage), from Pearson and Goheen, 1988, reproduced by permission)

susceptible for a prolonged period, this does not appear to affect fruit development or enhance fruit infection. Thus, fungicide application is most effective at and shortly after flowering and fruit set. Where leaf infection is typically serious, spraying should commence earlier and be extended possibly up to *véraison*. Early termination of treatment has the added advantage of reducing the residual sulfur level in crushed juice, where elemental sulfur can disrupt yeast growth. It also avoids favoring yeast production of hydrogen sulfide during fermentation (Thomas *et al.*, 1993).

Disease management is based primarily on hygiene and fungicidal sprays. Because spore dispersal during the season is limited, removal of overwintering fungal tissue on leaves, stems and fruit limits early disease onset. Delaying early onset can often postpone development of severe disease development until after harvest. The development of an open canopy also benefits in reducing disease incidence and improves spray contact with vine surfaces. During the growing season, wettable or sulfur dusts have commonly been used. Although not incorporated into plant tissues, sulfur acts both as a preventive and curative agent. This double benefit results from the majority of the fungal tissue residing exterior to the

plant and, thereby, being directly exposed to the fungicide. Unfortunately, sulfur effectiveness is temperature-dependent – being much less active below 20 °C, and unacceptably phytotoxic above 30 °C. In addition, sulfur use has the potential for increasing spider mite populations by suppressing predators. Demethylation-inhibiting (DMI) fungicides (such as Bayleton®, Rally®, and Rubigan®), or strobilurin fungicides (such as Abound®, Flint®, and Sovran®) are more effective and less phytotoxic than sulfur, but more costly. The potential for rapid resistance to these highly specific agents requires that they be used sparingly, and never more than twice in sequence. Agents such as silicon, bicarbonates (e.g., baking soda), oils (e.g., canola oil), cinnamic aldehyde, and phosphate fertilizer also may be effective. Most of these are available in commercial form. Finally, *Ampelomyces quisqualis* (AQ[10]) can be applied as a biological control agent (Falk *et al.*, 1995).

Early prevention is essential because infection can spread extensively throughout the crop before disease severity becomes evident. In addition, yield loss is often directly related to the timing of the initial infection; thus, the importance of lime-sulfur application in killing overwintering cleistothecia in early spring. Even in

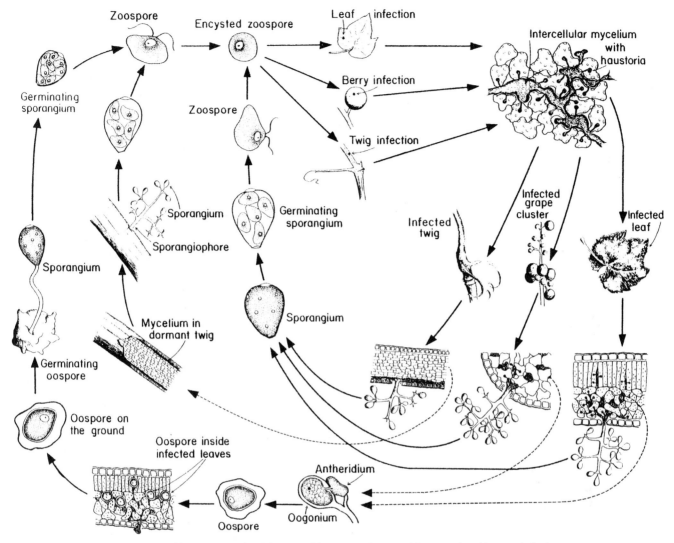

Figure 4.44 Disease cycle of downy mildew. (From Agrios, 1997, reproduced by permission)

regions where cleistothecia are not vital in overwintering, the removal and destruction of infected cane wood may be of importance.

The development and widespread use of disease-risk models in several major wine-producing regions have effectively reduced fungicide application, as well as provided better or equal disease control.

Downy Mildew (Peronospora) Downy mildew is induced by *Plasmopara viticola* (Fig. 4.44), a fungus unrelated to the fungus causing powdery mildew. The term "mildew" refers to the cottony white growth that develops on infected tissue under moist conditions.

The spores, called **sporangia**, germinate to produce several flagellated zoospores. Zoospores possess a short motile stage, during which they swim toward the stomata. After adhering to the plant surface, the spore begins to penetrate the host. Sporangial production usually occurs at night, and spores remain viable for only a few hours after sunrise. Consequently, downy mildew is a serious problem only under conditions in which rainfall is prevalent throughout much of the growing season.

Similar to powdery mildew, downy mildew attacks all green parts of the vine and produces haustoria. However, *Plasmopara viticola* hyphae do not remain exterior to the plant; they ramify extensively throughout host tissues. Under moist conditions, sporulation develops rapidly. On leaves, spore-bearing hyphae erupt preferentially through stomata on the lower surface. Leaf invasion is the primary source for spores inducing fruit infections. Infected shoot tips become white with spore production and show a distinct 'S'-shaped distortion. The shoot subsequently turns brown and dies.

The fruit is most vulnerable to infection when young, but all parts of the fruit cluster remain susceptible until maturity. As with leaf infection, severe development of the disease can result in premature abscission.

During the summer, the fungus produces a resting stage called an **oospore**. Oospores may remain dormant for several years, until conditions favorable for germination result in spore production. Typically, this occurs in the spring. In mild climates, both oospores and dormant mycelia (in infected leaves) may initiate spring infections.

No effective biological control measures are known against *P. viticola*. The production of an open canopy has only a minor beneficial effect on disease incidence. Consequently, chemicals remain the only effective treatment for this pathogen under conditions favorable to disease development.

Bordeaux mixture and several other nonsystemic fungicides are toxic to *P. viticola*, but they are only preventive in action. Newer systemic fungicides, such as fosetyl aluminum and phenylamides (e.g., Metalaxyl®), are especially effective due to their incorporation into plant

tissues. The uptake of the fungicide by plant tissues also reduces their removal by rain.

Black Rot of Grapes   Unlike the diseases noted so far, black rot is of economic significance only in eastern North America and selected regions in Europe and South America. Most indigenous *Vitis* spp. show considerable resistance to attack, having evolved in the presence of the pathogen for millions of years. In contrast, cultivars of *V. vinifera* are very susceptible. Depending on weather conditions, and the initial inoculum, the disease can cause crop losses of up to 80%.

Three subspecies of the causal fungus, *Guignardia bidwellii* (*Phyllosticta ampelicida*), are recognized. One affects only species of the *Vitis* subgenus. A second infects both subgenera (*Vitis* and *Muscadinia*). The third attacks only species of the related genus, *Parthenocissus*.

Infection can occur on new growth at any time during the growing season (Fig. 4.45), but does not develop on mature leaves or fruit following *véraison*. On leaves, lesions develop as small creamy circular spots. As these enlarge (up to 10 mm), they darken to a tan color, and

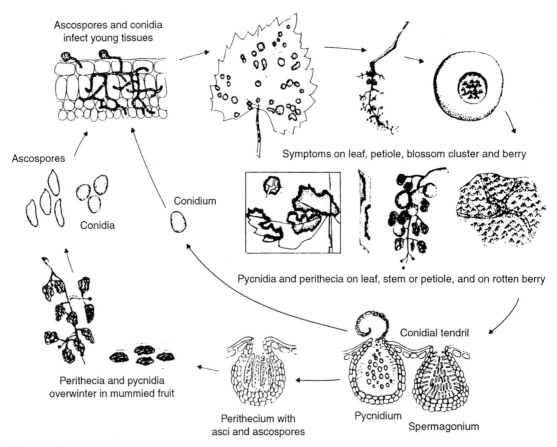

Figure 4.45   Disease cycle of black rot of grapes caused by *Guignardia bidwellii*. (From Agrios, 1997, reproduced by permission)

finally turn reddish brown. The spots are surrounded by a band of dark-brown tissue. Characteristic small black roundish raised structures (spore-bearing pycnidia) develop in older parts of the lesion. Black elongated lesions usually develop on petioles and the fruit stalk. These may encircle the organ, killing any tissue distal to the lesion. On young shoots, similar but larger black cankers develop. These produce pycnidia during the growing season. Lesions develop surprisingly quickly on the fruit. Whole berries may be converted to blue-black, shriveled "mummies" within a few days.

Many cycles of localized infection may occur during a single season. During the fall and winter months, structures superficially similar to pycnidia develop in mummies on the ground. These fruiting bodies, called pseudothecia, develop as a result of sexual reproduction. In the spring, the ascospores mature and are ejected into the air following even light rainfalls. They initiate the new round of infections.

On wild vines, the fungus probably survives most effectively on fruit mummies. However, in commercial vineyards, most mummies are collected and destroyed. Thus, survival is most probably associated with infected portions of canes not removed during pruning. Spore production from overwintered tissue can continue well into midsummer.

Control is based primarily on the use of protective, contact fungicides, such as maneb and ferbam. The systemic fungicide tridimefon (Bayleton®) is also useful due to its curative property. Nevertheless, sanitation (destruction of mummified fruit and infected cane wood) is particularly important in reducing the inoculum load in the spring. This can significantly delay and reduce the economic significance of infection.

Eutypa Dieback   Eutypa dieback is a serious disease inducing a slow but insidious attack on the woody components of the vine. The pathogen, *Eutypa lata* (*E. armeniacae*) preferentially invades wounds of the perennial shoot system (Fig. 4.46). It begins by infecting the xylem exposed by graft conversion, or pruning wounds created in the maintenance or conversion of training systems. It normally does not invade cane wounds, produced during annual shoot pruning.

Growth of the fungus in the xylem is slow, and the enlarging, elongated, lens-shaped canker generated can remain hidden for years by overlying bark. Disease symptoms often appear months or years after infection, and are typically detected as a result of the development of distinctive leaf and shoot symptoms. These are usually restricted to the shoots directly associated with and above the lesion. Symptoms are most apparent in the

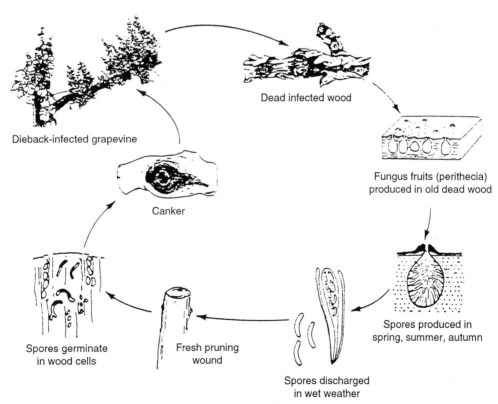

Figure 4.46   Disease cycle of Eutypa dieback. (From Flaherty *et al.*, 1982, reproduced by permission)

spring, when the stunted shoots are about 25–50 cm long. The young leaves of affected shoots are usually upturned, small, distorted, chlorotic, and possess a tattered margin. Shoot internodes are markedly dwarfed and fruit yield significantly reduced. These effects are caused by mycotoxins produced by the fungus. Eventually, the portion of the vine associated with the lesion dies, resulting in "dead arm."

The canker usually can be observed only by removing the outer bark. Well-established cankers contain rows of flask-shaped **perithecia**, structures in which spores are produced. Cutting tangentially along the wood through the canker exposes the perithecia as round objects appearing to contain a jelly-like material (translucent when wet, white and sheet-like when dry, or mat black when empty after spore discharge). The infected wood typically possesses a V-shaped appearance and has a light-grayish to dark-brown color.

Spores in new perithecia generally mature in late winter or early spring. Subsequent spore release follows periods of sufficient rainfall (>1 mm), often being most marked in very early spring. Sprinkler irrigation may be an important source of water for spore discharge in regions with insufficient rain.

Eutypa dieback is particularly difficult to control. Except in areas where grapevines are the major woody plant, sanitation has relatively little effect in reducing the number of spores found in the air. This results from the very extensive host range of *E. lata*. This includes some 80 woody species commonly indigenous to grape-growing regions. Spores also can be effectively dispersed by wind more than 100 km. Thus, a local source of infection is not required for the development of a serious disease outbreak. Finally, normal (surface) fungicidal spraying is ineffective in preventing spore germination within the xylem.

Management techniques include the early removal of infected wood to about 5–10 cm below any area of staining. This is required as the fungus may occur considerably in advance of a stain reaction. The wound produced must be treated with an antifungal preparation. Destruction (preferably burning) of the infected wood is recommended. Diseased tissue can remain a source of spore production long after removal from the vine. Pruning wounds on 2-year or older wood should be treated with a creamy suspension of Flusilozolr®, Carbendazim®, or 20% boric acid. Application is performed as soon as possible after pruning, especially when pruning occurs in the fall or winter months. This protects the site from infection by airborne fungal spores. The fungicide must soak into the exposed wood to provide adequate protection. Alternately, a suspension of *Trichoderma harzianum* may be used. Where possible, pruning ideally should occur when the xylem is active and spore production low. The formation of lignin and suberin in reaction to wounding can markedly reduce disease incidence (Munkvold and Marois, 1990). Thus, very early-autumn or late-spring pruning can be useful in control.

ESCA, BLACK MEASLES, PETRI AND BLACK FOOT DISEASES

These terms refer to various expressions of a disease complex caused principally, but not exclusively, by *Phaeomoniella chlamydospora*. Esca and Black Measles are terms for its expression in mature vines in Europe and North America, respectively. Petri disease, slow decline and black goo are names given to its expression in young grafted vines. Drought stress is an important factor inducing symptomatic expression, which otherwise remains largely or completely hidden. Grapevine "replant disease" may partially be another expression of this disease complex, although assorted bacteria and nematodes have also been implicated.

Much confusion about the origin of the disease has been caused by the frequent association of symptoms with the presence of other fungi and uncertainties as to their correct identification. Fungi not infrequently isolated from diseased and dying vines are *Phaeoacremonium aleophilum*, several other hyphomycetes, as well as several white rot fungi (notably *Fomitiporia punctata* and *F. mediterreanea*). The latter are thought to be particularly important in the Esca expression in Europe.

The disease has been present for centuries, but only recently has its serious consequence become particularly apparent. One of the factors awakening people to its significance was the extensive replanting necessitated in California by an outbreak of phylloxera in vines grafted to 'A×R#1'. The replanting provided conditions where the appearance of slow vine decline was accentuated by their number. Once identified, other trunk syndromes were recognized as alternative expressions of possibly the same disease. Another probable cause for an increase in disease expression, especially in mature vines, is the abandonment of sodium arsenate several weeks after pruning. Depending on when pruning is done, wounds may remain susceptible to infection by *P. chlamydospora* for weeks or months.

Esca expression is intermittent, and may reveal itself either in chronic or acute forms. Symptoms of chronic disease are characterized by a progressive foliage deterioration, whereas the acute phase results in a sudden death of all or parts of the vine. In the chronic form, leaf symptoms can begin at any time. They start in basal leaves and move apically. Symptoms appear as yellow to red patches. These develop necrotic centers as the patches coalesce ("tiger strips"). They eventually become irregular brown zones of necrosis between the veins and margins of the leaf, leading eventually to leaf dehiscence. Fruit symptoms vary with the region and cultivar. In France and northern Italy, berries are often unaffected, except for not filling or maturing properly. In southern Europe and California, affected berries may develop brown violet patches (the black measles syndrome). Foliar and fruit symptoms may occur simultaneously or separately. Vines showing chronic symptoms one year may appear healthy and symptomless in subsequent years.

Internal signs of infection appear as pale brown discolored sections of the xylem, surrounded by a dark ring, especially in the trunk or cordon. This lesion may occur alone, or in combination with lesions produced by white rot fungi, or Eutypa dieback (see Creaser *et al.*, 2002). Frequently, infection occurs in the absence of leaf or fruit symptoms.

Expression of the disease in young vines (Petri disease) most frequently results in, or from, partial graft failure associated with infection by *P. chlamydospora*. However, infection does not itself necessarily induce disease development (Edwards and Pascoe, 2003). Stress conditions, such as water deficit, or overcropping before the vine is established, appear to favor symptom expression. The vine may show a slow decline within the next few years. If decline occurs later, it would normally be considered an example of Esca. On investigation, the graft site possesses lesions (streaks or dots) that often exude a thick black ooze ("black goo"). Microscopically, xylem vessels may show tyloses that have grown in from infected parenchyma cells. The xylem exudate involves phenolic substances synthesized by parenchyma cells in response to infection. External symptoms include poor bud break, stunted shoot growth, mild foliar chlorosis associated with necrotic edges, wilting, and dieback. This presumably results from toxins produced by the fungus. These prevent full development of the callus. The resultant poor xylem connections severely restrict water and nutrient transport under hot, dry conditions. This probably explains why extensive watering and nutrient application may revive diseased plants. Although most prevalent and severe in grafted vines, own-rooted vines may also develop the disease.

Although the origin of infection in young vines is still unclear, the prevailing view is that it originates principally from symptomless infected vines, from which the rootstock or scion wood was obtained. In contrast, Esca may develop from infection of pruning wounds, though infection from scion mother plants is a distinct possibility.

Another disease that develops under similar conditions is *Cylindrocarpon* black foot disease. It is an additional cause of grapevine decline in new plantations. In this case, though, the fungus causes a root and butt rot.

Subterranean symptoms include few feeder roots, low root biomass, and necrotic root lesions, while the vegetation is chlorotic and stunted. No effective control exists, although interesting results with endomycorrhizal infection are promising (Petit and Gubler, 2005). Inoculation of roots with the mycorrhizal fungus *Glomus intraradices* reduced both the number of root lesions, as well as disease severity, while increasing root dry weight.

BACTERIAL PATHOGENS

Bacteria are an ancient group of microorganisms, existing predominantly as colonies of independent cells. Most possess a rigid cell wall that generally gives them spherical to rod shapes. They may or may not be motile, depending on the species and strain. They are typically restricted to entering plants either via natural openings (e.g., stomata and lenticels) or through wounds. Nevertheless, root pathogens can produce cell wall-degrading enzymes that permit direct entrance into young roots. Alternately, bacteria may enter the host through grafting wounds or are transmitted via leafhoppers. Overwintering occurs as dormant cells in soil, or in plant parts or vectors. Grapevines may be a host for many fastidious, endophytic bacteria (Bell *et al.*, 1995), as well as rhizosphere bacteria. Some of these are growth-promoting, such as *Burkholderia* (Compant *et al.*, 2005). Their presence may reduce disease severity to some pathogens. Only a few endophytic bacteria induce disease development.

Crown Gall The primary agent of crown gall in grapevines is *Agrobacterium vitis* (Fig. 4.47). It is unique in being found only in association with grapevines. *A. tumifaciens* is a more ubiquitous agent of crown gall. It can infect grapevines, but is rarely isolated from diseased vines. Exterior to the vine, *A. vitis* is isolated only from the rhizosphere around young grapevine roots. It can survive for at least 2 years in dead and dying roots, and possibly longer (Burr *et al.*, 1995).

Agrobacterium vitis can cause lesions on young roots, possibly through its production of polygalacturonidases. These may also assist the bacterium in invading the xylem of healthy roots. Alternately, wounds produced by the feeding of nematodes, such as *Meloidogyne hapla*, appear to facilitate penetration (Süle *et al.*, 1995). From the roots, the bacterium can be translocated into the trunk. Movement of the bacterium into new shoot tissue appears to occur in early spring (with sap rise). Subsequently, the bacterial population drops precipitously, rising again only in the fall (Bauer *et al.*, 1994). Thus, although the bacterium grows systemically throughout the vine, its distribution is far from uniform and is most frequently isolated from the roots.

Alternative sources of infection are mechanical wounds produced during grafting or cultivation around

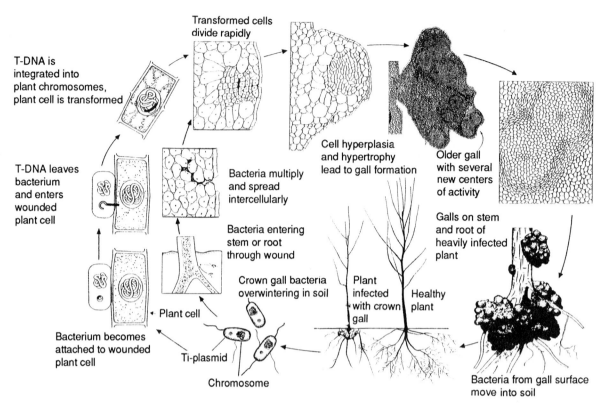

Figure 4.47   General disease cycle of crown gall. (From Agrios, 1997, reproduced by permission)

the trunk base. Because of the bacterium's host specificity and its ability to reside nonsymptomatically in the xylem, rootstock importation has probably been the principal origin of the now worldwide distribution of *Agrobacterium vitis*.

Despite the more frequent occurrence of the bacterium in the root system, its serious pathogenicity occurs in the trunk. The bacterium can induce uncoordinated cell division in the phloem and xylem, generating gall formation. Individual galls are commonly self-limiting and may subsequently rot and separate from the stem. New galls often originate next to old galls. In young vines, the galls can girdle the trunk, killing the vine. In large vines, the consequences of infection often depend on the number, size, and distribution of the galls. Heavy galling significantly disrupts vascular flow, reducing vine vigor and fruit yield (Schroth *et al.*, 1988). Galls are also invasion sites for other grapevine pathogens, such as *Pseudomonas syringae* f. *syringae* and *Armillariella mellea*.

Minor wounds act as the stimulus inducing gall formation. Frost damage appears to be the most important source of such wounds. Correspondingly, gall initiation is most marked in the spring in cold climatic regions. Galls develop most commonly near the trunk base (crown) – hence the name "crown gall." Graft conversion of existing vines also can be a significant activator of gall formation, resulting in graft failure.

Wounding indirectly stimulates gall induction, by activating the division of associated parenchyma cells. This in turn promotes bacterial attachment and subsequent transfer of a tumorigenic (T_i) plasmid from the bacterium to dedifferentiating parenchyma cells. These plasmid-infected cells subsequently transform into tumorous cells. Upon transformation, they begin to overproduce auxins and cytokinins, multiplying to form galls. The plasmid also codes for the production of a unique group of amino compounds, the opines (nopaline, octapine or vitopine). The opines so produced serve as specific nutrients for *A. vitis*.

No fully effective means of crown gall control currently exists. The best option, where possible, is to plant disease-free stock in virgin soil. In already-infested vineyard soils, the best choices are either to graft to one of the crown-gall-resistant rootstocks, or to use biological control. Rootstocks such as 'Gloire de Monpellier,' 'C3309,' or '101-14 MGT' do not prevent infection, but are highly resistant to transformation. Biological control involves prior infection with either the HLB-2 strain of *Agrobacterium radiobacter* (Pu and Goodman, 1993) or one of the nonvirulent (T_i plasmid-deficient) *A. vitis* strains (Zäuner *et al.*, 2006). In addition, the inoculation of disease-free rootstocks with *Pseudomonas aureofaciens* and *P. fluorescens* can markedly reduce the incidence and severity of *A. vitis* infection (Khmel *et al.*,

1998). Where crown gall is serious, as in northeastern North America, multiple trunking is often a valuable precaution. It does not prevent crown gall, but as a management practice, it diminishes the damage caused by the disease. It is unlikely that all trunks of a vine will succumb simultaneously. The disease is typically more serious on *V. vinifera* cultivars than on *V. labrusca* and French-American hybrids.

Freeing scions and rootstocks of the bacterium can be achieved with hot-water treatment, but not consistently (Burr *et al.*, 1996). Thus, the more complex procedure of apical shoot micropropagation is the most successful means of eliminating *A. vitis* infection. The use of pasteurized soil and equipment, when propagating disease-free clones, should prevent the spread of *A. vitis* during bench grafting.

Pierce's Disease Pierce's disease is produced by another xylem-inhabiting bacterium, *Xylella fastidiosa*. It infects and causes a range of diseases in several important tree fruit crops, in addition to grapes. There appears to be considerable genetic diversity between strains infecting these crops (Hendson *et al.*, 2001). In addition, the xylem of several common annual agricultural crops, vines, and vineyard weeds may possess *X. fastidiosa* as a nonpathogenic endosymbiant (Wistrom and Purcell, 2005). From these, xylem-feeding insects can ingest and transmit the bacterium to grapevines or other susceptible plants (Redak *et al.*, 2004). Because of bacterial host specificity and differential cultivar sensitivity, transmission does not necessarily result in disease. The primary insect transmitters are sharpshooter leafhoppers (Cicadellidae) and spittle bugs (Cercopidae).

The geographical distribution of the disease is limited by the presence of suitable vectors and the bacterium. Except for several isolated cases in Europe (Boubals, 1989; Berisha *et al.*, 1998), Pierce's disease has primarily been isolated to subtropical and tropical coastal regions of the southeastern United States, Mexico and Central America. This limitation suggests that winter temperatures are a significant factor limiting disease spread. In areas where the pathogen is endemic, indigenous species of *Vitis* are resistant or relatively tolerant to infection. Disease severity is often a reflection of the relative incidence of the pathogen (Ruel and Walker, 2006). Although its occurrence can severely restrict the commercial cultivation of *V. vinifera*, there is considerable variability in cultivar susceptibility. Of popular varieties, 'Chardonnay' and 'Pinot noir' are particularly sensitive, 'Cabernet Sauvignon' and 'Sauvignon blanc' being less sensitive, and 'Riesling' and 'Zinfandel' showing moderate resistance.

Until recently, Pierce's disease was of limited significance in California – presumably due to the absence of an effective vector. This situation has changed significantly since the appearance of the glassy-winged sharpshooter (*Homalodisca coagulata*). It is an indigenous insect within the historical range of the disease in the southeastern USA and northeastern Mexico.

In sensitive vines, invasion often results in bacterial colonization, associated with adherence of the bacterium to interior surfaces of xylem vessels. This is accompanied with the accumulation of xanthan gums in the vessel lumina. Subsequently, tyloses grow in the vessels, further occluding vessels (Stevenson *et al.*, 2004). All these factors disrupt water flow and place the vine under potential water stress. Correspondingly, symptoms tend to be more severe in hot or arid climates. However, drought alone does not induce symptomatic expression (Thorne *et al.*, 2006). The bacterium also produces several phytotoxins that could disrupt cellular function. Physiological disruptions produced by these toxins are probably critical to full symptom development.

Leaf symptoms (scorch) develop as a progressive inward browning and desiccation of the blade. In advance of the necrosing region, concentric areas of discoloration commonly develop. They are yellow in white cultivars and red-purple in red cultivars. The blade eventually may drop, leaving the petiole still attached to the shoot (Stevenson *et al.*, 2005). It produces a symptom called "matchstick." Late in the season, "green islands" may remain on canes, surrounded by brown mature bark. These islands are associated with regions where periderm differentiation is absent.

In severely affected vines, bud break is delayed, and shoot growth is slow and stunted. The first four to six leaves are dwarfed, and the main veins are bordered by dark-green bands. Subsequent leaves generally are more typical in size and appearance.

Infected *V. vinifera* cultivars may survive for 1 to 5 years, depending on the age of the vine when infected, the variety, and local conditions. Young vines are particularly susceptible and frequently succumb within two years. Until the late 1990s, with the spread of the glassy-winged sharpshooter, infection in southern California developed in mid- to late season. Frequently, the bacterium did not survive the winter, and recovery was common (Hill and Purcell, 1995). This is no longer the situation with the new, more effective vector.

In warm climates, where the pathogen and vector are common, the only effective control is growing resistant or tolerant cultivars. Where the disease is established, but localized, vineyard planting should ideally avoid areas where reservoirs of the pathogen and vector are common, notably river banks populated with vines and shrubs such as blackberry and elderberry. Because transfer from indigenous plants is less efficient, replanting riparian environments with native plants can reduce the likelihood of transmission. Transfer is predominantly from symptomless carriers, rather than from vine to vine. Thus, planting a 7-m-wide conifer or hardwood belt around vineyards in susceptible sites has been investigated as a transmission buffer. Insecticidal control of vectors has generally been unsuccessful in halting disease spread, but can be effective in dramatically reducing local vector populations. For example, imidacloprid and thiamethoxam are particularly active against sharpshooters. These insecticides appear to have little effect on sharpshooter egg-parasitoids. In California, release of egg-parasitoid wasps (*Gonatocerus* spp.) has markedly reduced the population of glassy-winged sharpshooters. Other biological control agents of potential interest include mycoparasites, such as *Hirsutella* sp. and *Beauveria bassiana,* as well as various natural predators.

Yellows Diseases   Several grapevine yellows diseases have been identified in various parts of the world. These are all believed to be induced by one or several forms of phytoplasmas (Prince *et al.,* 1993). Because the bacteria are difficult to recognize microscopically and disease symptoms are insufficiently diagnostic, confirmation of infection can be determined only using molecular techniques (e.g., PCR). Although probably members of the genus *Acholeplasma* or *Candidatus,* they are usually classified by their 16S ribosomal RNA fragment. For example, *flavescence dorée* is induced by a member of the elm yellows disease group (16SrV); grapevine yellows in the eastern United States and northern Italy is associated with a member of the western X disease group (16SrIII); *Vergilbungskrankheit* (Germany), *bois noir* (France), and southern European (Mediterranean) grapevine yellows may develop following infection by members of the stolbur subgroup of aster yellows (16SrI$_G$); and Australian grapevine yellows is incited by an Australian subgroup of aster yellows (16SrI$_J$). Most of these phytoplasmas appear to be transmitted by different insect vectors: *Vergilbungskrankheit* and *bois noir* by the planthopper *Hyalesthes obsoletus;* Australian grapevine yellows probably by the leafhopper *Orosius argentatus;* and *flavescence dorée* and the eastern American forms by the leafhopper *Scaphoideus titanus* (*littoralis*). In the vine, phytoplasmas multiply exclusively in cells of the host phloem. Thus, the vector acquires and accumulates the bacterium in its salivary glands during feeding. Subsequently, the phytoplasma reproduces in the vector, and accumulates in the salivary glands. This favors transmission via the fluid ejected by the proboscis just prior to feeding.

Because *S. titanus* is endemic to eastern North America, it was thought that *flavescence dorée* may have been introduced into Europe, along with *S. titanus,* in the late 1940s (Maixner *et al.,* 1993). However, this appears unlikely because the phytoplasmas causing yellows diseases in France and eastern North America are different. Some forms of the disease are also transmissible at a low frequency by grafting.

Grapevine yellows diseases are generally characterized by the following set of symptoms. Newly infected vines show delayed bud burst and shoot growth in the spring. Internodes are shortened and cane development may show a zigzag pattern. Leaf blades may become partially necrotic and roll downward, more or less overlapping one another, and become brittle. The foliage often turns yellow in white cultivars, and red in red cultivars. Alternately, angular colored spots may develop on leaf blades. In sensitive cultivars, the most distinctive symptom is the drooping posture the vine develops in the summer. This results from poor development and lignification of xylem vessels and phloem fibers. As a consequence, shoots do not turn brown or develop only patchy brown areas. The shoots usually die during the winter. Shoot tips may die back and develop black pustules. The fruit tends to shrivel and develop a dense, fibrous, bitter pulp. Symptoms typically begin to develop in the year following infection – the "crisis" year. Symptoms may be more pronounced in association with the simultaneous occurrence of virus infection.

Two distinct expressions of *flavescence dorée* occur in Europe. In the *Nieluccio* type, the disease becomes progressively more severe each year until the vine dies. In the *Baco 22A* type, the vine recovers after symptoms develop in the crisis year. If they are reinfected within a few years of a previous infection, recovered vines show only a localized, rather than systemic reaction. There is also considerable variation in the sensitivity of different cultivars to the various forms of grapevine yellows. For example, 'Pinot noir' is particularly susceptible to *flavescence dorée,* but is little affected by *bois noir.* Other forms of yellows diseases may or may not show remission. In European regions, where both *flavescence dorée* and *bois noir* occur, symptoms of *flavescence dorée* usually appear much earlier in the season than those of *bois noir* (Angelini *et al.,* 2006).

No effective disease control is known for regions where both vectors and pathogens are established. The only effective option is growing varieties that show recovery (*Baco 22A* expression), or are relatively insensitive. Insecticide spraying delays but does not stop pathogen spread. Care should be taken in vine propagation as the pathogen may be spread by grafting. Unfortunately, the risk of graft spread is most serious during the infection year, when the vine is symptomless. Recovered vines apparently are not infectious and can be used safely as stock for propagation. Dormant scion wood can be cured of infection by hot-water treatment.

Eggs of the leafhopper vector are also killed by hot-water treatment (Caudwell *et al.*, 1997).

VIRUSES, VIRUS-LIKE, AND VIROID PATHOGENS

Viruses and related pathogens are submicroscopic, noncellular, self-replicating, infective agents. Those that attack grapevines possess only RNA as their genetic material. Differentiation between viruses and viroids is based primarily on the presence or absence, respectively, of a protein coat enveloping the nucleic acid. Viroids also possess a much smaller RNA genome than the majority of plant viruses. Virus-like diseases are those resembling the transmission characteristics of viruses, but for which consistent association with a pathogenic agent has yet to be established. Infection is usually systemic, affecting all tissues, with the possible exception of the apical meristem, or pollen and seed. The absence of apical-meristem infection permits the elimination of viruses by apical tissue culture (micropropagation). This is particularly important in grapevines, where vegetative propagation can lead to the spread and perpetuation of systemic infections. Because these agents are translocated systemically in grafted vines, the worldwide occurrence of most viruses and viroids may be the consequence of rootstock grafting to control phylloxera and other root problems. Although grapevine viruses and viroid infections are graft transmissible, some are also spread by nematode, aphid, mealybug, and fungal vectors (Walter, 1991). Some also can be mechanically transmitted via pruning equipment.

In addition to micropropagation, cultivars may be cured of some viral infections by heat treating dormant stem cuttings or young vines. Thermotherapy is ineffective in the elimination of viroids (Duran-Vila *et al.*, 1988).

The production and propagation of virus- and viroid-free nursery stock are ongoing projects in most wine-producing countries. This is based on the belief that cured clones grow better and produce better grapes than infected vines. Although commonly valid, clones free of all known systemic pathogens do not consistently perform better than their infected counterparts (Woodham *et al.*, 1984). Nevertheless, the elimination of all systemic infections is desirable because symtomless carriers can be a source of agents that induce debilitating disease or limit grafting success in susceptible cultivars.

Detection of viral and viroid infection has historically been based on indexing, which is the transmission of the agent to an "indicator" plant that produces distinctive disease symptoms after infection. Identification with serological techniques, especially with enzyme-enhanced serology (ELISA), cDNA, or PCR probes, can confirm indexing results and may eventually replace the long and expensive indexing procedure. Because disease-free vines, micropropagated in culture vessels remain pathogen free, their use may reduce, if not eliminate, the need for quarantining and indexing imported vine cuttings.

The major viruses infecting grapevines fall into one of three main groups – the nepoviruses, the closteroviruses, and the vitiviruses (formerly classified under the trichoviruses). Nepoviruses are polyhedral (spherical), single-stranded RNA (ssRNA) viruses, with a genome divided unequally into two linear segments. They include the fanleaf, tomato ringspot, tobacco ringspot and peach rosette mosaic viruses. All are nematode transmitted and cause several forms of grapevine decline. Closteroviruses are flexuous, filamentous ssRNA viruses with a single linear genome. A few are known to be occasionally transmissible by mealybugs or scale insects. They are the probable cause of most instances of leafroll. Vitiviruses are also flexuous, filamentous ssRNA viruses. The main examples are GVA and GVB (grapevine viruses A and B). They are most frequently associated with instances of stem pitting (Kober stem grooving) and corky bark, respectively. Both viruses can be transmitted by mealybugs. Examples of other viral groups causing or associated with grapevine diseases are trichoviruses (grapevine berry inner necrosis virus), luteoviruses (grapevine Ajinashika virus), grapevine fleck virus (possibly belonging to the tymovirus group), a foveavirus inducing Rupestris stem pitting, and a capillovirus-like virus occasionally associated with rugose wood. Most, if not all, grapevine viruses are graft-transmissible.

The nature of many presumably viral and viroid diseases in grapevines is unclear. Investigation is confounded by many vines being symptomless carriers, the slow development of symptoms, the apparent inconsistent association with recognizable pathogen(s), and the complex etiology of symptom expression. In some instances, several viroids or viruses, or both may be required for expression, or may modify disease expression. For example, vein-banding disease only develops when vines are jointly infected with both the fanleaf virus (GFLV) and a grapevine yellow speckle viroid (GYSVd-1) (Szychowski *et al.*, 1995).

Fanleaf Degeneration Fanleaf degeneration is the viral grapevine disease with the longest known historical record, being identifiable from herbarium specimens more than 200 years old. Because of its long history in Europe and the absence of infection in free-living North American grapevines, grapevine fanleaf virus (GFLV) is assumed to be of European or Near Eastern origin. Its current distribution is believed to be due to grafting and the spread of European cultivars worldwide. Natural spread in vineyards is slow because of the limited movement in the soil of its major nematode vectors, *Xiphinema index* and *X. italiae* (about 1.5 m/year).

Because grapevine roots do not form natural grafts, vine-to-vine transfer does not occur. Although transferable to other plants, the disease is limited to grapevines under field conditions. This may result from the limited host range of its nematode vectors.

The impact of infection varies widely, depending on cultivar tolerance and environmental conditions. Tolerant cultivars are little affected by infection, whereas susceptible varieties show progressive decline. Yield losses may be up to 80%, with the fruit of poor quality. Infected vines have a shortened life span, increased sensitivity to environmental stress, and reduced grafting and rooting potential. Three distinctive syndromes have been recognized, based on particular strain–cultivar combinations. These are malformation, yellow mosaic, and vein-banding.

Fan-shaped leaf malformation is the most distinctive, but not only, foliage expression. Chlorotic speckling and a leathery texture commonly accompany leaf distortion. Shoots may be misshapen, showing fasciation (stem flattening), a zigzag pattern at the nodes, atypically variable internode lengths, double nodes, and other aberrations. Fruit set is poor and bunches are reduced in size. The yellow mosaic (chromatic) syndrome develops as a strikingly bright-yellow mottling of the leaves, tendrils, shoots, and inflorescences in the early spring. Discoloration can vary from isolated chlorotic spots to uniform yellowing. The third expression, veinbanding, develops as a speckled yellowing on mature leaves, bordering the main veins in mid- to late-summer. In both discoloration syndromes, leaf shape is normal, but fruit set is poor, with many shot berries ("hen and chickens" appearance). Fanleaf degeneration also shows a characteristic intracellular development of trabeculae. These appear as strands of cell-wall material spanning the lumen of xylem vessels.

Where both vector and virus are established, planting vines grafted on rootstock resistant to the virus is usually required (tolerance to the vector can still permit infection). Fumigation of the soil with nematicides can, to varying degrees, reduce but not eradicate nematode populations. Allowing the land to lie fallow can be useful in reducing nematode populations, but the strategy often requires 6–10 years. The long requisite fallow period probably results from nematode survival on undislodged roots. These occasionally remain viable for up to 6 years following vine uprooting.

Leafroll   Leafroll is a widespread virus-like disease associated with one or more of up to eight or nine grapevine leafroll-associated viruses (GLRaVs) (Boscia *et al.*, 1995). GLRaV-1 and GLRaV-3 are the most widespread and economically significant. They are all members of the closterovirus group. Two vitiviruses (GVA and GVB)

have also been associated with leafroll, although they are more commonly associated with Kober stem grooving and corky bark, respectively. Symptomatic expression varies considerably, but it does not generally lead to vine degeneration. Many scion and rootstock cultivars are symptomless carriers of the infectious agent(s). As with fanleaf degeneration, the pathogens probably originated in Europe or the Near East; feral North American grapevines are not infected.

The spread of the causal agents depends primarily on graft transmission. Nevertheless, insect vectors may occasionally be involved. For example, GLRaV-1, -3, -5 can be transmitted by mealybugs or soft scale insects (Sforza *et al.*, 2003). Several species of mealybugs have been reported to transfer grapevine viruses A (GVA) and B (GVB). Nevertheless, healthy and infected vines often coexist side by side without transmission. That the mealybug, *Planococcus ficus*, has been found on the roots of grapevines (Walton and Pringle, 2004) is of concern since most control measures have been aimed at aboveground insects. It is a key pest in South Africa, the Mediterranean, Argentina, and now occurs in some regions of California.

The disease complex derives its name from a marked down-rolling of the basal leaves that occurs late in the season in some varieties. The interveinal areas of leaves also may turn pale yellow or deep red, depending on the cultivar, whereas the main veins remain distinctly green. Infected vines can occasionally be detected by their retention of leaves much longer than adjacent healthy vines. In addition, leaf blades may fall, leaving petioles still attached to the cane. Whole vines, as well as individual shoots and leaves are dwarfed in comparison to healthy plants. Fruit production may be depressed by up to 40%, and berries may show delayed ripening, reduced sugar levels, and altered pigmentation.

Control is dependent on the destruction of infected vines and replacement by disease-free stock. Disease-free nursery stock may be generated by thermotherapy or by micropropagation. Because the identity of all causal agents still remains unestablished, confirmation of elimination of the infectious agent(s) is performed by grafting to sensitive cultivars (indexing).

Yellow Speckle   Yellow speckle is a widespread but relatively minor viroid disease of grapevines. Other viroids occur in grapevines, but their economic significance and relationship to recognized grapevine diseases are unclear. Symptoms of infection by grapevine yellow speckle viroids (GYSVd1 and GYSVd2) are often short-lived and develop only under special climatic conditions. Foliar symptoms generally develop at the end of the summer and consist of leaf spotting. When sufficiently marked, the scattered chlorotic spots may resemble the vein-banding symptom of fanleaf degeneration. Studies suggest that shoot growth is slightly curtailed and grape acidity is reduced (Wolpert *et al.*, 1996).

Control is dependent on planting viroid-free vines. Elimination of the causal agent can be achieved by micropropagation. Thermotherapy is ineffective (Barlass and Skene, 1987).

NEMATODE PATHOGENS

Nematodes are a large group of microscopic, unsegmented, roundworms that live predominantly as saprobes in soil. However, some are parasitic on plants, fungi, and animals. Those attacking grapevines are restricted to feeding on the root system. They derive their nutrition from extracting the cytoplasmic fluids from root cells. This is accomplished with a spear-like stylet that punctures the host cells.

Feeding may be restricted to the surface of roots or, following burrowing, may occur in the root cortex. In addition to the direct damage caused by feeding and the resultant root disruption, nematodes may transmit viruses and facilitate infection by other pathogens. Active dispersal by nematodes in soil is both slow and limited, with most long-distance movement being through the action of wind and water, or by the translocation of infested plants.

Reproduction occurs via egg production, with or without the interaction of males. The prolonged survival of eggs in a dormant state often markedly reduces the effectiveness of fallowing in nematode control. Soil fumigation, especially in combination with fallow, can dramatically reduce nematode populations in shallow sandy soils. It is of limited value, however, in deep or clayey soils, where fumigant penetration is restricted. The most widely used nematicide, methyl bromide, is scheduled for deregulation in 2005 in the United States (although special exemptions are scheduled up to 2009). Methyl iodide appears to be an equally effective soil fumigant (Ohr *et al.*, 1996), and is not a stratosphere ozone depleter. Other potential alternatives are sodium methyldithiocarb (Vapam®), chloropicrin + iodomethane, propargyl bromide, and possibly sodium and potassium azide. Whether they will be registered for grapevines is currently unknown. Another potential substitute is DiTera®, a selective nematicide that reportedly does not kill beneficial mycorrhizal fungi or saprophytic nematodes. It contains a formulation of the hyphomycete *Myrothecium*.

Generally, the most effective means of limiting nematode damage entails the use of nematode-resistant or -tolerant rootstocks. Although no current rootstock cultivar is resistant to all grapevine-attacking nematodes, some are resistant or tolerant to one or more of the important pathogenic genera. Therefore, determination

of both the actual and potential nematode pests of a site is required in rootstock selection (Walker, 2001). Where groundcovers are used in vineyards infested with pathogenic nematodes, nematode-resistant crops such as Cahaba white vetch (*Vicia sativa*), barley, and Blando bromegrass is advisable. Mulching of underrow cover crops, such as biofumigant *Brassica* spp., has also shown promise in reducing nematode numbers (McLeod and Steel, 1999). Conversely, nematode-favorable hosts, such as peas, lupins, clovers, and most vetch crops should be avoided.

The detection and identification of nematode problems usually require microscopic examination of the root system. Aboveground symptoms are insufficient diagnostic. Many nematode species can attack grapevine roots, but few induce significant damage. The most serious are the root-knot (*Meloidogyne*) and dagger (*Xiphinema*) nematodes. Other nematodes occasionally found feeding on grapevine roots include species of *Pratylenchus* (lesion nematodes), *Tylenchulus* (citrus nematodes), and *Criconemella* (ring nematodes).

Root-Knot Nematodes Root-knot nematodes (*Meloidogyne* spp.) are sedentary endoparasites that penetrate young feeder roots. After penetration, adjacent cells are stimulated to divide and increase in size, producing gall-like swellings (Fig. 4.48). Where multiple infections occur, knot formation may give the root a chain-of-beads appearance. Reproduction commonly occurs during the spring and fall root-growth periods (de Klerk and Loubser, 1988). Each female may produce up to 1500 eggs. Second-stage juveniles have been shown to travel up to 30cm in 3 days through sandy loam (Flaherty *et al.*, 1992). The most important species are *M. incognita*, *M. javanica*, *M. arenaria*, and *M. hapla*.

Infection results in a decline in vigor and yield, as well as increased susceptibility to water and nutrient stress. Symptomatic severity can be limited by increasing the water supply and pruning to prevent overcropping. Damage is typically most marked in young developing vines planted in highly infested, light-textured soils. Older vines with deep root systems seem to be less affected by root-knot nematodes.

Dagger Nematodes In contrast to root-knot nematodes, dagger nematodes (*Xiphinema* spp.) are migratory and feed on epidermal cells near the root tip. Concentrated feeding may initiate root bending, followed by lesion darkening. Extensive attack causes root death and induces the production of tufts of lateral roots. In addition to destroying roots, *X. index* can act as a vector of the grapevine fanleaf virus. The combined action of both nematode and viral infections can quickly make a vineyard commercially unproductive.

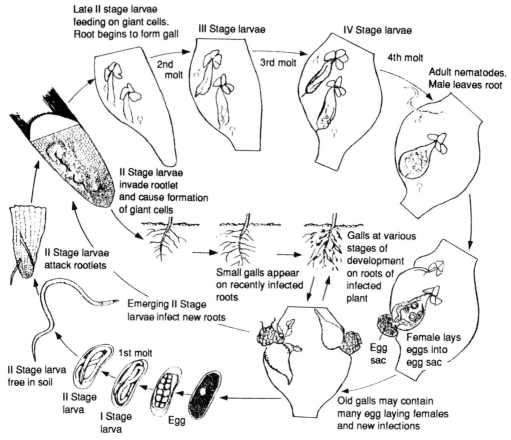

Figure 4.48 Disease cycle of root-knot nematodes. (From Agrios, 1997, reproduced by permission)

Other species of *Xiphinema* may transmit a range of other, but less serious, viruses to grapevines (see Walter, 1991).

INSECT AND MITE PESTS

Several insects and mites cause extensive damage to grapevines. Although most species primarily infest only one part of the grapevine, all parts are attacked by one or more species. Most control measures have been based on the use of synthetic pesticides. Because of problems associated with pesticide use, greater emphasis is being placed on cultural and biological controls.

As a group, insects and mites are distinguished by a hard exoskeleton. This is shed several times during growth. Mites and many insects pass through several immature, adult-like stages before becoming reproductive, whereas other insects pass through several worm-like (larval) and a pupal stage before reaching adulthood.

Insects are distinguished by possessing three main body parts (head, thorax, and abdomen), sensory antennae, three pairs of legs attached to the central thorax, and a segmented abdomen. In contrast, mites possess a fused cephalothorax, broadly attached to an unsegmented abdomen, lack antennae, and possess four pairs of legs,

except in the initial immature stage. Two pairs of legs are attached to the cephalothorax and the other two pairs are located on the abdomen. Both groups reproduce primarily by egg production.

Although some insects have evolved intimate relationships with grapevines, notably phylloxera, others infest a wide range of hosts. As with fungal pathogens, infestation does not necessarily correlate with severity. Severity depends more of the specific genetic properties on the cultivar–pest association, the macro- and microclimate, soil conditions, and resistance to applied pesticides.

Phylloxera Of all grapevine pests, phylloxera (*Daktulosphaira vitifoliae*) has had the greatest impact on viticulture and wine production. A relatively minor endemic pest of *Vitis* species in eastern North America, phylloxera devastated *V. vinifera* vineyards when it was inadvertently introduced into Europe about 1860. The subsequent destruction of more than 2 million hectares of vineyards both demonstrated the need for quarantine laws and illustrated the possible effectiveness of biological control. Grafting sensitive fruiting varieties (scions) to resistant rootstocks was so successful that the danger still posed by the pest often has been forgotten. Common

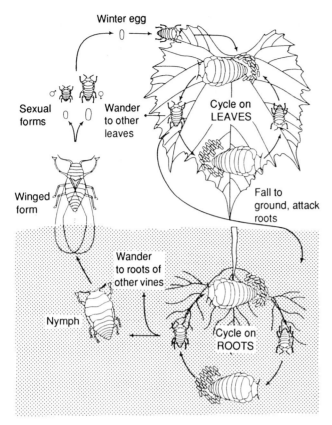

Figure 4.49 Life cycle of grape phylloxera. (From Fergusson-Kolmes and Dennehy, 1991, reproduced by permission)

examples are 'SO 4,' '5BB' and '5C.' The explosive spread of a phylloxera biotype capable of rapidly growing on 'A×R#1' (formerly the dominant rootstock used in California) in the late 1980s caused considerable hardship. It required replanting of some 5000 ha in Napa and Sonoma between 1988 and 1995. Biotypes, other than the two main types (A and B), have also been identified in California (De Benedictis *et al.*, 1996).

The need to replant vineyards on a massive scale has apparently permitted several relatively minor vascular pathogens of older grapevines to make their presence noticeable. Both *Cylindrocarpon* blackfoot disease and *Phaeomoniella chlamydospora* grapevine decline (Petri disease) can induce significant losses in young vineyards (Scheck *et al.*, 1998).

The life cycle of the aphid-like phylloxera is illustrated in Fig. 4.49. For the sexual stage to be expressed, the insect must pass through both leaf- and root-galling phases. However, most species of *Vitis* are resistant to either the leaf- or root-galling phase (Table 4.12). Thus, the absence of species susceptible to the initial (**fundatrix**) leaf-galling phase (Fig. 4.50) probably explains the absence of the sexual cycle in California, and most other viticultural areas where phylloxera is not indigenous.

Occasionally, the leaves of *V. vinifera* are attacked and galled by phylloxera nymphs that develop on the leaves of American rootstocks (Remund and Boller, 1994).

Phylloxera biotypes, differentially pathogenic on grapevine cultivars, appear to be common (King and Rilling, 1991), despite the rarity of the sexual stage in most infested areas. Differences in pathogenicity may be expressed in the ability of the insect to feed, stimulate gall formation, or reproduce rapidly. The pest usually goes through several asexually reproduced generations per year. Long-distance spread without human involvement is slow.

Susceptible hosts respond to phylloxera feeding by increasing the supply of nutrients to the damaged region and by forming extensive galls. Phylloxera feeding on young roots provokes distinctive hook-shaped bends and swellings, called **nodosities** (Plate 4.14). These soon succumb to secondary infection by fungi, such as *Fusarium*, *Pythium*, and *Cephalosporium* spp. (Granett *et al.*, 1998). The rapid destruction of fine roots may explain both vine decline and the anomaly that phylloxera populations and vine damage are often poorly correlated (decaying roots do not support phylloxera growth). Older roots produce semispherical swellings called **tuberosities**. These give the root a roughened, warty appearance. The development of tuberosities is the more serious expression of infestation, because it seriously disrupts water and nutrient flow. The insect population increases primarily on tuberosities, because nodosities quickly rot and do not support dense populations of feeding insects (Williams and Granett, 1988). In contrast, resistant vines show limited production of nodosities and healing following attack. Tolerant vines show galling, but are little affected by infestation and support only limited insect reproduction (Boubals, 1966).

The aboveground symptoms of root infestation are relatively indistinct and are primarily expressed as a decline in vine vigor. The first clear indications of infection usually appear as stunted growth and premature leaf yellowing. With sensitive cultivars, these effects annually become more marked, until the vine dies.

Phylloxera infestation can be confirmed only by root observation. The distinctive lemon-yellow eggs (Plate 4.15) and clusters of yellowish-green nymphs and adults on young roots are diagnostic. Tuberosities isolated from dying vines usually possess few phylloxera. Nevertheless, the presence of the tyroglyphid mite (*Rhizoglyphus elongatus*) appears to be an indicator of the past presence of phylloxera. The mite lives on the decaying cortical tissues of tuberosities.

Many environmental conditions can affect the severity of vine attack. It is well known that phylloxera infestation is much less significant in sandy soils. It has been suggested that this results from the higher level

Table 4.12   Susceptibility of *Vitis* species to root and leaf expressions of phylloxera

Host reaction	Resistant (bearing none to few galls on roots)	Tolerant (bearing many galls on roots)
***Vitis* species already exposed to phylloxera** (eastern North America)		
Resistant (bearing none to few galls on leaves)	*V. rotundifolia* *V. berlandieri* *V. candicans* *V. cinerea* *V. cordifolia* *V. rubra*	*V. aestivalis* *V. girdiana* *V. labrusca* *V. lincecumii* *V. monticola*
Tolerant (bearing many galls on leaves)	*V. riparia* (= *V. vulpina*) *V. rupestris*	None
***Vitis* species not previously exposed to phylloxera** (western North America, Asia, Europe, and Middle East)		
Resistant (bearing none to few galls on leaves)	*V. coignetiae*	*V. arizonica* *V. californica* *V. davidii* *V. romanetti* *V. ficifolia* (= *V. flexuosa*) *V. vinifera* (incl. *V. sylvestris*)
Susceptible (bearing many galls on leaves)	*V. betulifolia* *V. reticulata*	*V. amurensis* *V. piazeskii*

Source: From Wapshere and Helm, 1987, reproduced by permission

Figure 4.50 Vine leaf infested by phylloxera; insert, gall in cross-section showing phylloxera. (From *The Gardener's Chronicle*, reproduced in Barron, 1900)

of silicon, either in the soil solution or in vine roots (Ermolaev, 1990). High soil temperatures (above 32 °C) are unfavorable to phylloxera and limit its damage (Foott, 1987). Irrigation and fertilization can occasionally diminish the significance of phylloxera damage, whereas drought increases its severity (Flaherty *et al.*, 1982).

Quarantine is still useful in limiting phylloxera dispersal into uninfested areas. Even where the pest is present quarantine may prevent the importation and distribution of new biotypes.

In nurseries, vines can be disinfected by placing the washed root systems into hot water (52–54 °C) for about 5 min. Nursery soil can be disinfected by pasteurization or fumigation. If the vineyard soil is already infested, the major control measure remains grafting sensitive scion varieties onto resistant rootstocks. Typically, rootstocks with some *V. vinifera* parentage should be avoided (Granett *et al.*, 1996). Nevertheless, '1202C' and 'O39-16' seem to be exceptions to this rule, at least in California. Whether rootstocks of pure North American *Vitis* spp. will continue to remain resistant to phylloxera under vineyard monoculture is unknown. Reports of decline have been noted in Germany. Hopefully, advances in categorizing phylloxera strains via DNA fingerprinting will assist understanding the occasionally conflicting worldwide data on rootstock resistance or tolerance to phylloxera.

Although chemical agents such as aldicarb (Temik®) (Loubser *et al.*, 1992) have shown some promise, sodium tetrathiocarbonate (Enzone®) appears to be of little value (Weber *et al.*, 1996). Thus, for the foreseeable future, control of phylloxera will continue to depend on quarantine and grafting to resistant rootstock.

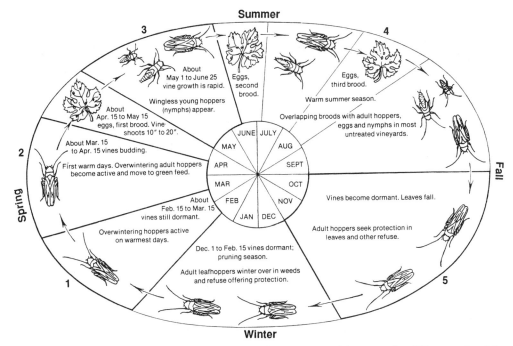

Figure 4.51 Life cycle of grape leafhoppers in California. (From Flaherty *et al.*, 1982, reproduced by permission)

Leafhoppers and Sharpshooters Several leafhoppers are destructive pests on grapevines as well as vectors of bacterial pathogens. The species of significance tend to vary from region to region. In many grape-growing regions of North America, the western and eastern grape leafhoppers (*Erythroneura elegantula* and *E. comes*, respectively) are the most important species; in southern California, the variegated leafhopper (*E. variabilis*) and the nonindigenous glassy-winged sharpshooter (*Homalodisca coagulata*) are the major pest species; whereas the potato leafhopper (*Empoasca fabae*) is particularly significant in the southeastern United States. About 20 species of sharpshooter leafhoppers are vectors of *Xylella fastidiosa*, the causal agent of Pierce's disease. In Europe, *Empoasca* leafhoppers tend to be the most damaging group, especially *E. vitis*. *Scaphoideus littoralis* (synonym, *S. titanus*) is the primary vector of *flavescence dorée*. In South Africa, *Acia lineatifrons* is the prominent pest species. Most leafhoppers have fairly similar life cycles and are controlled by similar techniques.

Depending on the species and prevailing conditions, leafhoppers may go through one to three generations per year. Those species passing through several generations per year are generally the most serious, as their numbers can increase dramatically throughout the season.

Leafhopper eggs typically are laid under the epidermal tissues of leaves. After hatching, the nymphs usually pass through five molts before reaching the adult stage (Fig. 4.51). All stages feed on the cytoplasmic fluid of leaf and fruit tissue. Feeding results in the formation of white spots, which, on heavily infested leaves, leads to a marked loss in color. Growth of one or more hyphomycetes on escaped sap and insect honey dew can produce a sooty appearance on plant surfaces. Pronounced damage is usually caused only when infestations reach >10–15 leafhoppers/leaf. Severe infestation can lead to leaf necrosis, premature defoliation, delayed berry ripening, and reduced fruit quality. Some varieties, notably late-maturing cultivars, tend to sustain greater damage than early-maturing varieties. Such cultivars not only endure leafhopper infestation for a longer period, but also may suffer from the migration of leafhoppers from early-maturing varieties. Most leafhopper species affecting grapevines do not infest them exclusively. Thus, in the fall and early spring, they often survive and multiply on other plants. Overwintering occurs as adults under leaves, weeds, or debris in and adjacent to vineyards.

Control measures have been increasingly based on enhancing the population of indigenous parasites and predators. In California, the wasp *Anagrus epos* is an effective parasite on the eggs of the western grape leafhopper (less so on the variegated leafhopper). Its short life cycle permits up to 10 generations per year. By July, the parasite population may reach levels sufficient to destroy 90–95% of the leafhopper eggs. The numbers of *A. epos* can be augmented by selective habitat diversification, such as planting prune trees upwind from vineyards (Murphy *et al.*, 1996). The windbreak produced by the trees further enhances the concentration of the parasites. Another parasitic wasp, *Aphelopsis cosemi*, attacks nymphs, resulting in sterilization of the adult. Several predatory insects, such as lacewings and ladybugs, as well as the general predatory mite, *Anystis agilis*, attack leafhoppers. The release of commercially reared green lacewings, *Chrysoperla* spp., can be both effective and economically feasible if the timing is correct. Release should coincide with egg hatching and at population of about 15–25 leafhoppers per leaf (Daane *et al.*, 1993). Of cultural practices, basal leaf removal is particularly useful in removing most first-generation leafhoppers. These occur most frequently on basal leaves.

At present, members of the genus *Gonatocerus* (Pilkington *et al.*, 2005) are the principal parasitoid wasps feeding on the eggs of the glassy-winged sharpshooter. They can infest 10–50% of the eggs during the first generation of sharpshooters, and up to 90–100% on the second, late-summer population. *Anagrus epos* may also prove effective. One of the problems associated with this, or any other, control measure is that its major significance comes from being an effective vector for *Xylella fastidiosa*, the agent of Pierce's disease. As such, short exposure to only a single insect may result in effective transmission of *X. fastidiosa*.

An alternate biocontrol mechanism of potential value is the use of *Alcaligenes*. It is a symbiotic bacterium that limits the multiplication of *X. fastidiosa* in the insect vector (Bextine *et al.*, 2004). Symbiotic control is also under investigation for control of the bacterium causing *flavescence dorée* in *Scaphoideus titanus* (Marzarati *et al.*, 2006).

Chemical control of leafhoppers has shifted from synthetic pesticides, such as organophosphates, to 'softer' nicotine-based compounds (e.g., imidacloprid). When its use is required, it has the advantage of causing minimal disruption of beneficial biocontrol agents.

Tortricid Moths   Grapevines are attacked by a wide diversity of tortricid moths, the species varying from region to region and country to country. In southern regions of Europe, *Lobesia botrana* is the major pest species, whereas in more northern regions *Eupoecilia ambiguella* is the significant species. In much of eastern and central North America, the important tortricid is the grape berry moth, *Endopiza viteana* (Fig. 4.52A). In California, the omnivorous leafroller, *Platynota stultana* (Fig. 4.52B) and the orange tortrix, *Argyrotaenia citrana*, are the notable forms in warmer and cooler regions, respectively. In Australia, the light-brown apple moth, *Epiphyas postvittana*, is the major tortricid. It has recently been found in California. The species is now naturalized throughout much of New Zealand.

Because of their taxonomic affinity, all tortricid moths possess relatively similar life cycles. Adult females lay egg

clusters on or close to flowers and grape bunches. Those less specialized to grapevines lay eggs on leaves. The eggs hatch into pale-colored larvae that feed predominantly on flowers and developing fruit, or on leaves, depending on the species. Following several molts, the larvae form a web-like cocoon in which they metamorphose into pupae. After a variable period, the adult moths emerge and mate, initiating the next generation. Adults are small, relatively inconspicuous, brown moths generally possessing a bell-shaped wing profile and projecting snout-like mouth parts. Depending on the species and prevailing climatic conditions, two to four generations develop per year.

Although tortricid moths possess many properties in common, significant differences occur. For example, European berry moths form cocoons under rough bark and in the crevices and cracks on trellis posts. American berry moths form folds in leaves in which they spin their cocoon. These eventually fall to the ground. Although tortricid moths affecting grapes in California and Australia form cocoons in leaf folds, the insects do not hibernate as pupae, as do other American and European berry moths. The Californian tortricid moths survive as larvae in web nests formed in mummified grape clusters, or on other vine and vineyard debris. These hibernating characteristics influence the type and success of sanitation used in reducing their overwintering success.

Second- and third-generation larvae are the most damaging, as their numbers typically increase throughout the season. In addition to direct-feeding damage, larvae produce wounds subsequently infected by bunch rot fungi, yeasts, and bacteria. Their action further attracts infestations by fruit flies (*Drosophila* spp.).

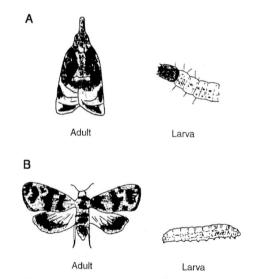

Figure 4.52   Diagrams of the adult and larval stages of (**A**) the grape berry moth (*Endopiza viteana*), and (**B**) the omnivorous leafroller (*Platynota stultana*)

Not only can tortricid larvae produce wounds facilitating bunch rot, but several species are carriers of *Botrytis*. The details of transmission are most well established for the European berry moth (*Lobesia botrana*), but the light-brown apple moth has the same potential in Australia (Bailey *et al.*, 1997). Adult females of *L. botrana* preferentially select *Botrytis*-infected berries on which to lay their eggs, while the larvae selectively feed on infected berries (Mondy, Pracros *et al.*, 1998). In addition, females raised on *Botrytis* infected grapes produce more eggs (Mondy, Charrier *et al.*, 1998).

As with other insect pests, increased emphasis is being placed on control by endemic pests and parasites. The diminished use of pesticides and the establishment of habitats for sustaining populations of indigenous parasites and predators are essential to successful biological control (Sengonca and Leisse, 1989). In addition, synthetic pesticides may be replaced by commercial preparations of *Bacillus thuringiensis*, or by pheromone applications to disrupt mating success. The effectiveness of pheromones in disorientating male *Lobesia botrana* is affected by the height of applicators and the development of the leaf canopy (Sauer and Karg, 1998). Grape leaves "fix" pheromones and slowly release them into the atmosphere (Schmitz *et al.*, 1997). Release of artificially reared *Trichogramma minutum* and *T. embryophagum* (Plate 4.16), egg parasites of several insects, including grape berry moths, has been successful in controlling tortricid pests. *Trichogramma carverae* has been investigated as a commercially viable control for the light-brown apple moth (*Epiphyas postvittana*). Green lacewings (*Crysoperla* spp.) and many spiders are also

active predators in summer. Egg and pupae parasites kill the insect before damage can be done, whereas larval predators and parasitoids limit population buildup by preventing reproduction. Where insects overwinter on the ground, row cultivation to bury hibernating pupae or larvae can be potentially valuable. It often takes several years for tortricid populations to build up to critical levels (Flaherty *et al.*, 1992). For species not specialized to grapevines, such as *Lobesia botrana*, the use of grass vs. broadleaf cover crops (which the pest prefers) can diminish their incidence on vines.

Mites　Several types of mites inflict damage on grapevines. Of these, the most generally significant are spider mites.

Spider mites are most serious under dusty conditions, especially when vines are water stressed. The most significant species typically differs from region to region. The European red spider mite (*Panonychus ulmi*) and the two-spotted spider mite (*Tetranychus urticae*) tend to be the most important species in much of Europe, whereas the yellow vine spider mite (*Eotetranychus carpini*) is the primary species in Mediterranean France and Italy. In eastern North America, the European red spider mite is the principal damaging species, whereas in California, the Pacific spider mite (*Tetranychus pacificus*) is the most destructive form (Fig. 4.53). The Willamette spider mite (*Eotetranychus willamettei*) is commonly found in California, but is less damaging than the Pacific spider mite. It has been released in vineyards as a biological control agent against infestation by the Pacific spider mite (Karban *et al.*, 1997). It can also serve as a host

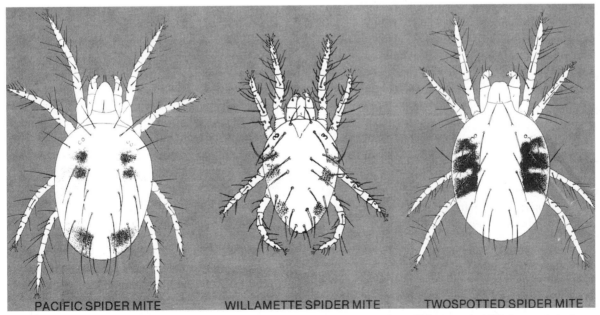

PACIFIC SPIDER MITE　　　WILLAMETTE SPIDER MITE　　　TWOSPOTTED SPIDER MITE

Figure 4.53　Diagrams of three types of spider mites occurring on grapevines. (From Flaherty *et al.*, 1982, by permission)

for enhancing the population of predators of the Pacific spider mite. In Chili, *Oligonychus vitis* is the most injurious species found.

Spider mites typically overwinter as females under rough bark on trunks and cordons. They begin to emerge early in the spring. If present in high numbers, mites can kill the margins of growing leaves, permanently stunting leaf growth.

The initial (larval) stage of spider mites resembles the adult, except in size and the possession of only three pairs of legs. After feeding, the larvae molt and pass through the eight-legged protonymph and deutonymph states, to become either a male or female adult. Under favorable conditions, spider mites can pass through their life cycle in about 10 days. This can lead to explosive population increases.

Spider mites typically feed on the undersurfaces of leaves by injecting their mouth parts into epidermal cells. Initial damage results in fine yellow spots on the leaf. With extensive feeding, the foliage turns yellow in white varieties and bronze in red varieties. Web formation is more or less pronounced, usually occurring in the angles formed by leaf veins. If attack is heavy, leaves usually drop prematurely. Although spider mites seldom attack the fruit, foliage damage may result in delayed ripening, or in severe cases, fruit shriveling and dehiscence.

Effective management of damage can often be achieved by favoring conditions that diminish vine susceptibility and enhance natural predation. Grass groundcovers, where water and fertilization are ample, diminishes vine susceptibility by limiting dust production. Sprinkler irrigation discourages spider mite development, without affecting its predators. Planting vegetation that maintains high levels of spider mite predators, and the avoidance of pesticides known to be toxic to their predators, promote effective biological control (James and Rayner, 1995). In some instances, application of natural (i.e., Canola) oils can effectively suppress the destructive phytophagous mite without damaging beneficial predatory mites (Kiss *et al.*, 1996). In Europe, effective predator phytoseiid mites include *Typhlodromus pyri* and several *Amblyseius* spp. (Duso, 1989); in California, the primary predators are *Metaseiulus occidentalis* and *T. caudiglans*. In Australia, *Typhlodromus doreenae* and *Amblyseius victoriensis* are highly effective against the distantly related erineum (eriophytid) mites (Fig. 4.54), such as *Colomerus vitis*. In California, *Metaseiulus occidentalis* has been observed actively feeding on the same eriophytid species. For several predatory mites, sheltered habitats and pollen food sources are important in maintaining high predator populations in vineyards, whereas for *M. occidentalis*, tydeid mites act as important alternate hosts when spider mite populations are low. The minute pirate bug (*Orius vicinus*) is an equally

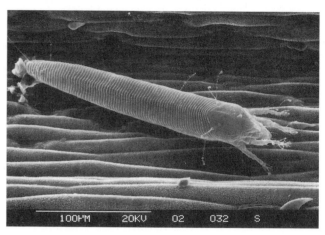

Figure 4.54 Electron micrograph of an eriophytid mite (typically less than 0.05 mm long). They possess only two pairs of relative short legs. (From Frost, 1996, reproduced by permission)

important predator (Plate 4.17), but regrettably of both spider mites and their mite predators.

Several eriophytid mites can also induce significant damage when conditions are favorable to their reproduction. The most common is the rust mite (*Calepitrimerus vitis*) and the erineum mite (*Colomeris vitis*). Both are minute (0.2 mm) and so pale colored as to be hardly visible even with a 10× hand lens. When abundant early in the season, rust mites can cause severe leaf deformation, causing one form of restricted spring growth (RSG) (Bernard *et al.*, 2001). Later in the season, they can induce leaf bronzing. In contrast, the erineum mite occurs in three distinct strains, each characterized by the damage it causes. The erineum strain causes gall-like deformation of the leaf, associated with excessive leaf hair growth on the undersurface of the concave puckerings. The bud-mite strain limits its damage to the buds, resulting in a varied pattern of damage to leaf and shoot growth. A leaf-curling strain affects growth in the summer.

Adults tend to overwinter under the outer bud scales, or occasionally in crevices of the bark. Damage is often limited due to the action of the western predatory mite (*Metaseiulus occidentalis*). Outbreaks may result from reduced use of sulfur applied to control powdery mildew, or the application of strays that unintentionally disrupt their natural predators (Bernard *et al.*, 2001).

MAMMALIAN AND BIRD DAMAGE

Mammals such as deer and rodents (notably gophers, voles, and rabbits) can cause considerable vine and crop damage. However, control of these locally important pests is complicated because of regulations on the use of poisons, traps, and hunting. Management of bird damage can be even more intractable. Bird damage can cause greater economic loss than grape splitting and fungal diseases combined (Duke, 1993). Control measures include scarecrows, noise and distress-call generators, electric fences, chemical repellants, and vine netting (Sinclair and Hathaway, 2005). Of these, the most effective is netting. It can reduce bird damage by up to 99%. In areas where birds are a persistent pest problem, netting can be cost-effective, but coverage must be complete (Fuller-Perrine, 1993). An electronic deterrent system (Muehleback and Bracher, 1998) employs radar to time distress calls, predator sounds, or other noises to bird arrival. The types and sequence of sounds should be varied at frequent intervals to avoid bird habituation, and to suit the bird species (Berge *et al.*, 2007). Activation of visual deterrents, such as flashing lights and hawk replicas, to bird arrival enhances effectiveness. Where feasible, encouraging the nesting of predatory birds near vineyards can be an effective and a long-term deterrent. Of the repellant sprays, methyl anthranilate (0.75%) often remains effective for several weeks. Methyl anthranilate is also nontoxic, being a natural component of some grapes, and employed as a grape flavoring in confectionary and fruit juices (Sinclair *et al.*, 1993). Although locally effective, deterrents may only move the pest problem from one vineyard to another. Long-term solutions need regional management.

PHYSIOLOGICAL DISORDERS

Grapes are susceptible to several physiological disorders of ill-defined etiology. This has made differentiation difficult and has led to a wide range of terms whose exact equivalents may be unclear. Nevertheless, three groups of phenomena appear to be fairly distinct. These include the death of the primary bud in compound buds (primary bud-axis necrosis); abnormal flower drop and aborted berry development shortly following fruit-set (inflorescence necrosis, shelling, early bunch-stem necrosis, or *coulure*); and premature fruit shriveling and drop following *véraison* (bunch-stem necrosis, shanking, waterberry, *dessèchement de la rafle*, or *Stiellähme*).

Primary bud(-axis) necrosis can cause serious yield loss in several grape varieties, such as 'Shiraz.' Rootstock, pruning method, harvesting technique, and irrigation strategy may have a significant influence on the expression primary bud necrosis (Collins and Rawnsley, 2004). In general, conditions that favor excessive shoot vigor are associated with the development of the disorder. Deterioration of the primary bud becomes evident some 1–3 months subsequent to flowering, and occurs principally in basal buds. The buds may show a normal exterior appearance or exhibit a "split-bud" look (sunken in the center). Accurate assessment of its significance on vine yield requires selective bud dissection throughout the vineyard. Although death of the primary bud

promotes secondary bud development, they are rarely fruitful.

Inflorescence necrosis (*coulure*) is associated with a wide range of conditions, such as cold wet weather during flowering and high vine vigor. The flowers and rachis show increasing necrosis and young fruit abscission. It may result from the premature formation of an abscission layer at the base of the pedicel, or unexplained abortion of berry development. Ammonia and ethylene accumulation have been implicated in the induction of this disorder (Gu *et al.*, 1991; Bessis and Fournioux, 1992). Keller and Koblet (1995) have linked both inflorescence and basal-stem necrosis with stress induced by poor light conditions, and the associated carbon starvation. Nitrogen deficiency can also induce inflorescence necrosis (Plate 4.18). Flower abscission has also been correlated with a lack of carbohydrate reserves in buds in sensitive cultivars such as 'Gewürztraminer' (Lebon *et al.*, 2004), especially during the critical period when meiosis is occurring.

The association of NH^{4+} with both inflorescence necrosis and primary bud necrosis may result from protein degradation induced by carbon starvation. At high concentrations ammonia can be toxic to plant cells. Disrupted nitrogen metabolism may also explain the observed correlation between a related phenomenon, called *millerandage* (unequal development of berries in a cluster). It has been associated with heightened levels of the growth regulator, abscisic acid (Brouquedis *et al.*, 1995), as well as sustained levels of polyamines (Colin *et al.*, 2002). The polyamine content of millerandage berries at maturity was similar to that of normal berries at fruit setting.

Bunch-stem necrosis (BSN) is associated with vine vigor and heavy or frequent rains. It starts after the onset of ripening, as expanding soft water-soaked regions turn into dark sunken necrotic spots. These develop on the rachis, its branches, or berry pedicels, usually starting around the stomata. The fruit fail to ripen properly, develop little flavor, become flaccid, and separate from the cluster. The necrosis is associated with suppression of xylem development just distal to each branch of the peduncle. Vessel constriction and the presence of only narrow vessels could restrict sap flow. There is good correlation between varieties susceptible to bunch-stem necrosis and the production of a xylem "bottleneck" at the base of the fruit (Düring and Lang, 1993). This could curtail fruit development and result in the yield loss associated with the malady. Disrupted xylem connections could also explain the reduced calcium uptake by the affected fruit–calcium translocation being predominantly in the xylem. The disorder has also been associated with magnesium or calcium deficiency. In some cases, its incidence has been reduced by spraying the fruit, beginning at *véraison*, with one or more applications of magnesium sulfate (Bubl, 1987). This is occasionally supplemented with calcium chloride. Both calcium and especially magnesium ions activate glutamine synthetase, involved in the assimilation (and detoxification) of NH^{4+} (Roubelakis-Angelakis and Kliewer, 1983). Nevertheless, results from Holzapfel and Coombe (1998) suggest that magnesium, alone or with calcium, do not affect the incidence of bunch-stem necrosis. In addition, development of bunch-stem necrosis has been associated with ammonia accumulation and high nitrogen fertilization (Christensen *et al.*, 1991). In contrast, Capps and Wolf (2000) have obtained data correlating low tissue nitrogen with bunch-stem necrosis. These conflicting findings may denote that bunch-stem necrosis may refer to several syndromes, each caused by different environmental conditions, or is of considerable etiological complexity.

AIR POLLUTION

The economic damage caused by air pollution has been little studied in grapevines. Of known air pollutants, ozone and hydrogen fluoride appear to be the most important in provoking visible injury. Sulfur dioxide can produce injury, but at atmospheric contents much higher than typically found in vineyards. Even in the form of acid rain, sulfur dioxide is not known to affect grapevines severely (Weinstein, 1984).

The magnitude of damage caused by air pollution has been difficult to assess and predict due to a marked differential in sensitivity at various growth stages. In addition, the duration, concentration, and environmental conditions of exposure significantly influence sensitivity. Furthermore, it is generally considered that vineyard influences, such as overcropping and other stresses (water, nutrient, weeds, and disease) are more significant in determining the degree of damage than pollutant concentration.

Ozone   Ozone is the most injurious of common air pollutants. Grapevines were also one of the first plants found to be sensitive to the pollutant. Ozone is produced in the upper atmosphere when oxygen is exposed to short-wave ultraviolet radiation, and during lightning. However, most of the ozone in the lower atmosphere comes indirectly from automobile exhaust. Nitrogen dioxide (NO_2), released in automobile emissions, is photochemically split into nitric oxide and singlet oxygen (O). The oxygen radical reacts with molecular oxygen (O_2) to form ozone (O_3).

$$NO_2 \underset{\text{light}}{\rightleftharpoons} NO + O$$
$$O + O_2 \rightarrow O_3$$
$$NO + \text{hydrocarbons} \rightarrow PAN$$

Nitrogen dioxide would reform by a reversal of the reaction were it not for the associated release of hydrocarbons in automobile exhaust. The hydrocarbons react with nitric oxide, forming peroxyacetyl nitrate (PAN). This reaction limits the reformation of nitrogen dioxide and results in the accumulation of ozone. Ozone is the most significant air pollutant affecting grapevines in North America and severely limits grape yields in some areas.

Ozone primarily diffuses into grapevine leaves through the stomata. Ozone, or a by-product, reacts with membrane constituents disrupting cell function. The palisade cells appear to be the most sensitive, often collapsing after exposure. Their collapse and death generate small brown lesions on the upper leaf surface. These coalesce to produce the interveinal spotting called oxidant stipple. A severe reaction produces a yellowing or bronzing of the leaf and premature leaf fall. Basal leaves, and mature portions of new leaves, are particularly susceptible to ozone injury. Damage may result in reduced yield in the current year and, by depressing inflorescence induction, in the subsequent year. The severity of damage is markedly affected by cultivar sensitivity and prevailing climatic conditions. Sensitivity to ozone damage may be reduced by maintaining relatively high nitrogen levels, avoiding water stress, planting cover crops, and spraying with antioxidant compounds such as ethylene diurea.

One of the significant indirect effects of ozone damage is root mortality. Ozone reduces photosynthetic ability and the damage caused results in a shift in carbohydrate allocation to repair. This, in turn, reduces the allocation to the root system, resulting in mycorrhizal starvation and fine root death (Anderson, 2003).

Hydrogen Fluoride   Hydrogen fluoride is an atmospheric contaminant derived from emissions released in several industrial processes, such as aluminum and steel smelting, ceramics production, and the fabrication of phosphorus fertilizer. Although the leaves do not accumulate fluoride in large amounts, grapevines are still one of the more sensitive plants to this pollutant.

Symptoms begin with the development of a gray–green discoloration at the margins of younger leaves. Subsequently, the affected region expands and turns brown, often being separated from healthy tissue by a dark-red, brown, or purple band, and a thin chlorotic transition zone. Young foliage is more severely affected than are older leaves.

Vines may often be protected from airborne fluoride damage by the application of calcium salts. Thus, the presence of slaked lime ($Ca(OH)_2$) in Bordeaux mixture may unintentionally provide protection against hydrogen fluoride injury.

Chemical Spray Phytotoxicity Herbicide drift onto grapevine vegetation can cause a wide variety of leaf malformations and injury, depending on the herbicide involved. Several fungicides may also produce phytotoxic effects, notably sulfur (applied at above 30 °C) and Bordeaux mixture. In addition, some pesticides induce leaf damage under specific environmental conditions, or if applied improperly, notably endosulfan, phosalone, and propargite. Details are given in Pearson *et al.* (1988).

WEED CONTROL

Weed control is probably older than viticulture. The most ancient advance in weed control, manual hoeing, is still occasionally used in commercial viticulture, notably on slopes where mechanical tillage is impractical. Until recently, tillage was the principal method of weed control. Increasing energy and labor costs, combined with the development of effective herbicides during the 1950s, resulted in a decline in tillage. Environmental concerns have again shifted interest toward other methods of weed control, notably mulches, groundcovers, and biological control. Because each technique has different effects on root distribution (Fig. 4.55), no system of weed control is ideal for all situations. In addition,

different systems can variously affect water and nutrient supply, disease control, vine growth, and fruit quality. Finally, the choice of one system over another may depend on economic concerns, the relative benefits and disadvantages of clean cultivation, government regulations, and restrictions imposed by organic-grower associations.

Tillage Tillage to a depth of 15–20 cm is widely used in weed control. In addition, tillage is used to break up compacted soil, incorporate organic and inorganic fertilizers, prepare the soil for sowing cover crops, and bury diseased and infested plant remains for pest and disease control. However, awareness of its disadvantages has combined with other factors to curtail its use. Disruption of soil aggregate structure is among its major drawbacks. This can lead to "puddling" under heavy rain and the progressive formation of a hardpan under the tilled layer. The latter results from the transport, and subsequent accumulation, of clay particles deeper in the soil. Hardpans delay water infiltration, increase soil erosion, reduce water conservation, limit root penetration and tend to restrict root development to near the soil surface. Soil cultivation destroys infiltration and aeration channels produced by cracks

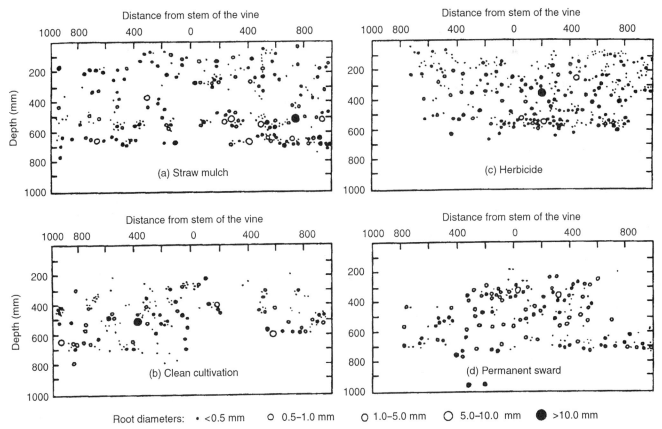

Figure 4.55 Root distribution under four soil-management systems. (From van Zyl and van Huyssteen, 1984, reproduced by permission)

in the soil and drastically reduces earthworm activity. This effect is most marked when the soil is cultivated when wet. In addition, by facilitating oxygen infiltration, tillage increases the rate of soil microbial action. This favors mineralization of organic material, disrupting soil crumb structure and elevating soil nutrient loss. Finally, soil compaction by heavy equipment can limit root growth between rows in shallow soils (van Huyssteen, 1988b).

Herbicides   No-till cultivation became popular in many parts of the world when herbicides became readily available. No-till cultivation avoids most of the disadvantages of cultivation, while achieving its benefits.

Herbicides fall into one of a number of functional categories. Some, such as simazine, primarily kill seedlings after germination, but do not affect existing weeds. They are typically nonselective and chiefly useful in controlling annual weeds. Herbicides such as diquat and paraquat destroy plant vegetation on contact, whereas others, such as aminotriazole and glyphosate begin their destructive action after being translocated throughout the plant. To limit vine damage, herbicides are usually applied before bud break and with special rigs designed to direct application only to the base of the vines. In addition, the use of controlled droplet applicator spray heads has become more common. They markedly reduce the amount of chemical needed. Nevertheless, limited herbicide use is recommended during the first few years of vineyard establishment, as well as mid- to high-training of the vines.

In cool climatic regions, bindweed (*Convolvulus* spp.) and quack grass (*Agropyrons repens*) tend to be the most widely distributed noxious weeds. In warm regions, Bermuda grass (*Cynodon dactylon*) is often the most common noxious weed.

Although some herbicides, such as paraquat, decompose slowly in soil, most degrade readily. Thus, they neither accumulate in the soil, disrupt fermentation, nor contaminate the finished wine. Nevertheless, increasing opposition to herbicide use is a widespread phenomenon. For example, increase in the incidence of some pest problems, such as omnivorous leafroller, have been attributed to reliance on herbicide use (Flaherty *et al.*, 1982). In addition, some herbicides reduce the population of desirable predatory mites, such as *Rhodecarellus silesiacus* (Jörger, 1990). Finally, the activity of a wide range of microbes and soil invertebrates (notably earthworms) decrease in association with herbicide use (Encheva and Rankov, 1990; el Fantroussi *et al.*, 1999; Wechert *et al.*, 2005). This is generally viewed as negative, despite our inability to accurately measure the soil flora (Kirk *et al.*, 2004). It is estimated that only about 1% of the soil's bacterial population, and much of its

fungal flora, cannot be cultured or enumerated. Thus, the actual extent of herbicide influence on the soil fauna is largely speculation. Nevertheless, some of the observed reduction in the soil flora and fauna, associated with herbicide use, probably results from a decrease in the soil's organic content, resulting from clean cultivation. The soil's organic content provides the nutrient base for the flora and fauna.

Mulches   Straw mulches have long been used for weed control and water conservation in agriculture (Walpole *et al.*, 1993). Alternate materials have included bark compost, leaf and twig compost, and sewage sludge (treated to destroy pathogens and weed seeds). Municipal solid-waste compost has seldom been used due to potential contamination with heavy metals (Pinamonti, 1988). Nevertheless, organic mulches have seen limited use in viticulture. This situation remains despite the potential advantages of cooler soil-surface temperatures, enhanced root activity in the upper soil horizon, reduced likelihood of erosion, enhanced invertebrate and microbial activity (improving soil structure), reduced salt-encrustation (reduced upward water flow and surface evaporation), and a reduction in some pathogen problems (Hoitink *et al.*, 2002). Mulches can also improve vine establishment by promoting root development (Mundy and Agnew, 2004), enhance yield (especially in low-rainfall regions) and reduce irrigation water use in dry climates (Buckerfield and Webester, 1999).

Disadvantages to organic mulches include their being a potential overwintering site for pests and pathogens, slower initiation of vine growth in the spring (due to cool moist soils), enhancing the risk of frost damage (due to slower upward movement of heat in the soil), acting as a site for rodent nests, and aggravating poor soil-drainage problems. The latter can be offset by planting vines on raised mounds (see Cass *et al.*, 2004).

In contrast, plastic mulches have had more success in penetrating viticultural practice, probably due to lower cost. Polyethylene sheets can also be engineered to be wavelength-selective. Thus, they can be designed to exclude photosynthetically active radiation (retard weed growth), but allow penetration of far-red radiation (permitting soil warming). As such, plastic mulches have proven particularly useful in establishing vineyards, where they maintain higher moisture levels near the soil surface and promote faster root development. Enhanced surface rooting does not impede deep root development (van der Westhuizen, 1980). Plastic mulches also enhance vine vigor, fruit yield, and eliminate the potential damage caused by hoeing or herbicide application around young vines (Stevenson *et al.*, 1986).

Plastic mulches may consist of either impermeable or porous woven sheets of polyethylene. Porous sheeting has the benefit of better air and water permeability, but it increases evaporative water loss. Black plastic has been most commonly used, but colored plastic may provide better heat and photosynthetic light reflection up into the canopy. The reflection of light into the vine canopy is also one of the principal rationales for using crustacean shells as a mulch (Creasy *et al.*, 2006). The effect of a surface reflective stone layer on vineyard climate and grape ripening is well known. Jute matting has the advantage of biodegradeability, but difficulties with application and cost do not recommend its use.

The cost/benefit ratio of mulch use will vary considerably from site to site (climate, need for water preservation, costs of alternative weed control measures, etc.). Thus, small grower trials over several years are required to establish the relative benefits of mulch use.

Cover crops Planting cover crops is another but more complex means of weed control. Depending on the relative benefits of a groundcover, they may be restricted to between-row strips or form a complete groundcover. Although cover crops usually contain a particular selection of grasses and legumes, natural vegetation may be used. Where natural vegetation is used, application of a dilute herbicide solution can restrict excessive and undesirable seasonal growth (Summers, 1985). This is particularly useful early in the season, at about a month after bloom. It limits competition for water during this critical period of vine and berry growth. Dilute herbicide use also restricts seed production and, thereby, limits self-seeding under the vines. Another option is mowing or mulching the cover crop, sometimes in alternate rows, to restrict nitrogen demand and limit seed production (W. Koblet, 1992, personal communication). Mowing is less effective in reducing water competition. Nitrogen fertilization may be necessary to compensate for competition between a perennial grass cover and the vines. This is unnecessary with most legume crops, such as Lana woollypod vetch (*Vicia villosa* spp. *dasycarpa*) (Ingels *et al.*, 2005).

Seeding a cover crop usually occurs during the fall or winter months, depending on the periodicity of rainfall and the desirability of a winter groundcover. Cover crops usually possess a mixture of one or more grasses (rye, oats, barley) and legumes (vetch, bur clover, or subterranean clover). Rye, for example, has the advantage of providing supplemental weed control due to its allelopathic effects. This can help reduce garden weevil populations by decreasing the taprooted weeds on which weevils overwinter (Hibbert and Horne, 2001). In Mediterranean climates, native perennial grasses may be of particular value because they grow primarily during the winter months (when the vines are dormant). Thus, they compete little with the vine during the summer. Regrettably, seed costs are considerably higher for native species than for common annuals. Establishment also takes longer, but once it is complete, perennial grasses are self-sustaining.

Strain selection can be as important as species selection. Water and nutrient use vary considerably among strains. Water and nutrient competition is of particular concern on shallow soils under nonirrigated, dryland conditions (Lombard *et al.*, 1988). Competition may be regulated either by mowing, plowing under, or the application of a throwdown herbicide such as Roundup® or Touchdown®, usually just before vine flowering. Where cover crops are also used to regulate vine vigor, limiting the cover growth may be delayed to have the desired result. Seeded cover crops usually require one mowing per year to prevent self-seeding.

In addition to weed control, groundcovers can promote the development of a desirable microbial and invertebrate population in the soil and limit soil erosion. The latter is particularly valuable on steep slopes, but can also be useful on level ground. Vegetation breaks the force of water droplets that can destroy soil-aggregate structure. Cover crop roots and their associated mycorrhizal fungi improve soil structure by binding soil particles, thus limiting sheet erosion. Furthermore, as roots decay, water infiltration is improved and organic material is added to the soil. This is especially so for legumes which can incorporate organic nitrogen into the soil. Poorly mobile nutrients, such as potassium, are translocated down into the soil by the roots (Saayman, 1981), as well as by the burrowing action of soil invertebrates. In addition to limiting soil erosion, cover crops can improve water conservation by reducing water runoff. Finally, cover crops facilitate machinery access to vineyards when the soil is wet, as well as reducing the incidence of some pests.

Groundcovers of diverse composition can provide a variety of habitats and pollen sources for parasites and predators of vineyard pests, notably ladybugs, green lacewings, spiders, and a myriad of parasitic wasps. Mowing is done in alternate rows to maintain a continuing habitat for desirable insects. Species heterogeneity in the groundcover also enhances the earthworm population. In arid regions, the ground vegetation may restrict dust production, and thus assist minimizing mite damage. Nonetheless, cover crops may equally be potential carriers of grapevine pest and disease-causing agents. For example, creeping red fescue is a host for the larvae of the black vine weevil; common and purple vetch is a host of root-knot and ring nematodes (Flaherty *et al.*, 1992); grasses can support populations of sharpshooter leafhoppers; and legumes can be hosts of lightbrown apple moths. Dieback or mowing of the cover crop can precipitate movement of the pest to vines.

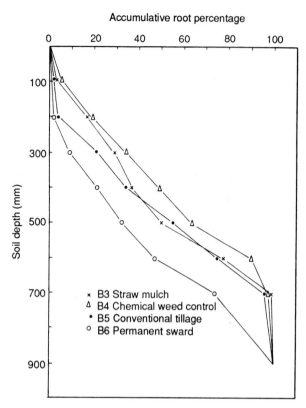

Figure 4.56 Root distribution with depth under different soil tillage practices in a dry land vineyard. (From van Huyssteen, 1988a, reproduced by permission)

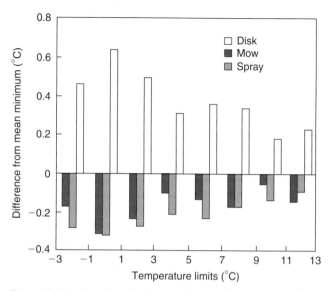

Figure 4.57 The effect of various weed control treatments on deviation of the daily mean minimum temperature versus various daily mean minimum temperature categories. (From Donaldson *et al.*, 1993, reproduced by permission)

Because cover crops usually suppress root growth near the soil surface (Fig. 4.56), the applicability of cover crop use often depends on soil depth, water availability, and desired vine vigor. In such situations, tilling every second row, adding fertilizer such as with farmyard manure, or periodically irrigating are possible solutions. In some locations, the possibility of groundcovers increasing the likelihood of frost occurrence (by lowering the rate of heat radiation from the ground) may be important (Fig. 4.57). To minimize this occurrence, herbicides such as Roundup® or Stomp® may be applied along a relatively narrow strip down the length of each row, to leave the soil bare directly underneath the vines. During the growing season, however, groundcovers appear to have little significance on the interrow temperature of vineyards (Lombard *et al.*, 1988).

Biological Control   One of the newest techniques in weed control involves the action of weed pests and diseases. Although less investigated than with insect control, biological control may become an adjunct in weed control.

Harvesting

The timing of harvest is probably the single most important viticultural decision taken each season. The properties of the harvested fruit set limits on the potential quality of the wine produced. Certain winemaking practices can ameliorate deficiencies in grape quality, but they cannot fully offset inherent defects. Timing is most critical when all the fruit is harvested concurrently. Only rarely is it economically feasible to selectively and repeatedly harvest a vineyard for fruit of a particular quality.

Where the grape grower is also the winemaker, and premium quality a priority, there is little difficulty in justifying the time and effort involved in precisely assessing fruit quality. However, when the grape grower and winemaker are different, adequate compensation for the development and harvesting of fruit at its optimal quality is required. The practice of basing grape payment on variety, weight, and sugar content is inadequate. Greater recognition and remuneration for practices enhancing grape quality could improve wine quality beyond the considerable, present-day standards.

Criteria for Harvest Timing

The problem facing the grape grower in choosing the optimal harvest time is knowing how to most appropriately assess grape quality (Reynolds, 1996). Objective criteria require chemical or physicochemical measurements. Unfortunately, the chemical basis of wine quality

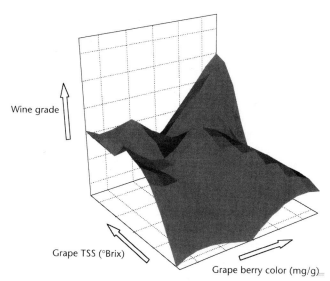

Wine grade

Grape TSS (°Brix)

Grape berry color (mg/g)

Figure 4.58 Schematic representation of a relationship between grape total soluble solids and berry color with the quality grading of the resultant wine. (From Gishen *et al.*, 2002, reproduced with permission)

is still ill-defined and varies with the style desired. For example, grapes of intermediate maturity may produce fruitier, but less complex, wines than fully mature grapes (Gallander, 1983). Historically, harvest date depended (assuming favorable weather conditions) on subjective visual, textural and flavor clues to fruit ripeness. Except for the skins, the juice from most grape varieties, with the major exception of Muscat cultivars, is essentially odorless, regardless of ripeness (Murat and Dumeau, 2005). In addition, the varietal flavor attributes of some cultivars (derived from precursor compounds) develop only during fermentation or aging, and are poorly detectable even in mature grapes. Examples are the important varietal thiols such as 3-mercaptohexanol and 4-mercapto-4-methylpentan-2-one (Tominaga *et al.*, 1998). This does not deny, though, that for specific sites and with experience, a strong correlation between grape favor and eventual wine character cannot be developed. Centuries of anecdotal accounts support this view. Nevertheless, subjective criteria are inadequate for large scale commercial operations, using a variety of cultivars, originating from a diversity of vineyard sites.

The shift from subjective to objective criteria for selecting harvest date began with the discovery of the central roles of grape sugar and acid contents to wine-making, and the development of convenient means for assessing their concentrations. They have now become the standard indicators of fruit maturity and harvest timing. For red grapes, color intensity is another standard (but more difficult to quantify easily). The complexity of trying to correlate wine quality with chemical measurements is illustrated in Fig. 4.58. This relationship is

likely to vary from cultivar-to-cultivar, from place-to-place, and from year-to-year. In addition, what is desirable for one wine style will almost certainly be different for another. Advances in near infrared spectroscopy (NIRS) may soon permit rapid (real-time), inexpensive and accurate winery assessment of sugar, acid and color indicators (Gishen *et al.*, 2005).

The sugar/acid ratio has usually been the preferred maturity indicator in temperate climates. Desirable changes in both factors occur more or less concurrently, making the ratio a good index of grape quality. In many parts of the world, the sugar/acid ratio is as important as yield limitations in assessing the price given growers for their crop. In contrast, in cool climates, where insufficient sugar content is a primary concern, reaching the desired level has often been the principal harvest indicator. In hot climates, adequate levels of soluble solids are typical, but avoiding an excessive rise in pH (drop in acidity) is crucial. Thus, harvesting may be timed to avoid pH values greater than 3.3 for white wines and 3.5 for red wines.

In addition to the importance of sugar to fermentation, grape soluble solids have been correlated for generations with berry-flavor development. Although this belief is supported by data on the synthesis of anthocyanins and vitispirane precursors, the association is far from simple (Dimitriadis and Williams, 1984; Roggero *et al.*, 1986). For example, the concentration of β-ionone increases during maturation in 'Pinotage,' but decreases with 'Cabernet Sauvignon' (Waldner and Marais, 2002). In addition, sugar and pH are inadequate predictors of grape-flavor potential in New York (Henick-Kling, personal communication). In another cool climatic region (British Columbia), the concentration of varietal flavors, based on monoterpenes, did not readily correlate with values of soluble solids, acidity, or pH (Reynolds and Wardle, 1996). Finally, the assessed quality of 'Gewürztraminer' wines was not correlated with the accumulation of monoterpenes in South Africa (Marais, 1996). Correspondingly, no simple or fully adequate means of assessing grape flavor content is currently (or may ever be) available. Unfortunately, there is a tendency among some winemakers to directly associate increased sugar content with more flavorful wines. This is clearly incorrect. It only assures that the wine (if fermented to dryness) will have a higher alcohol content. Although alcohol content affects the volatility of wine aromatics, whether this has a desirable sensory effect is far from clear.

Because of the importance of anthocyanins and tannins to red wine quality, their presence is often viewed as an indicator of grape quality. In addition, average berry size is often used as an index of the potential color and flavor of red wines (Singleton, 1972; Somers

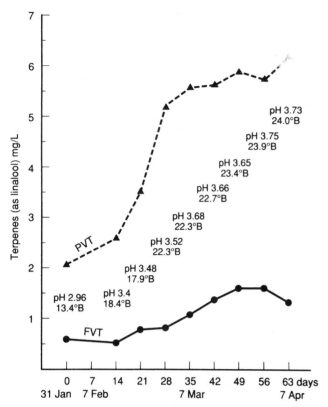

Figure 4.59 Change in free volatile terpenes (FVT) and potential volatile terpenes (PVT) after *véraison* for 'Muscat of Alexandria' grapes. °Brix (°B) and pH are shown at each sampling. (From Dimitriadis and Williams, 1984, reproduced by permission)

and Pocock, 1986). Presumably this results from the inverse relationship between berry volume and surface area, and the localization of anthocyanin pigments in the skin. Nonetheless, the correlation between grape phenolic measurements and subsequent wine quality is weak (Roggero *et al.*, 1986; Somers and Pocock, 1986). This probably arises from difficulties in both rapidly and accurately measuring their presence in small grape samples, but especially from predicting the influence of vinification practices on their uptake, retention and physicochemical state in wine.

Greater success has been achieved in correlating grape flavor content with wine quality in cultivars dependent on terpenes for much of the varietal aroma. In some varieties, the volatile monoterpene content of grapes continues to increase for several weeks, after appropriate sugar and acid levels have been reached (Fig. 4.59). This may also be associated with an increasing proportion being converted to nonvolatile forms. Although free terpenes are important in the fragrance of some young wines, high levels of nonvolatile terpenes have been correlated with subsequent flavor and quality development.

With several rosé wines, fruit flavor has been associated with the presence of 3-mercaptohexan-1-ol, 3-mercaptohexyl acetate and phenethyl acetate. The latter is a fermentation by-product, but the first two (volatile thiols) are derived from a grape precursor during fermentation. Although only a small and variable proportion of the precursor (*S*-3-hexan-1-ol-L-cysteine) is metabolized to aromatic thiols, its presence is correlated with the wine's eventual flavor. Thus, precursor concentration may serve as a useful index to direct viticultural practice and harvest timing (Murat and Dumeau, 2005). Regrettably, for most cultivars, there are no simple measures of grape varietal flavor – this property developing only during or subsequent to fermentation.

Where flavor development is not concurrent with optimal sugar/acid balance, the grape grower is placed in a dilemma. The solution requires a knowledge of the sensory significance of particular flavor constituents, as well as of the acceptability and applicability of sugar and acid amelioration procedures. For example, where alcohol content is an important legal measure of quality, and chaptalization is illegal, harvesting at an appropriate °Brix may be more important than grape flavor content. However, where flavor is the primary quality indicator, harvesting when flavor content is optimal, and adjusting the sugar and acid content after crushing, may be preferable.

Another potential indicator for harvest timing involves measuring the grape glycosyl-glucose (G-G) content. Because many grape flavorants are weakly bound to glucose, assessment of the G-G content (Williams, 1996; Francis *et al.*, 1999) may provide a direct measure of berry flavor potential (Iland *et al.*, 1996; Williams *et al.*, 1996). The G-G content is more general than the volatile terpene content. Free and potential volatile terpene (FVT and PVT) contents are relevant only to varieties whose aroma is largely dependent on terpenes. It should be noted, however, that many grape glycosides are not associated with aromatic compounds. For example, the G-G measurement must be adjusted in red grapes to account for the glucose glycosidically bound to anthocyanins.

In warm regions, the sugar content of red grapes is closely correlated with the G-G content (Francis *et al.*, 1998), possibly due to the close association between anthocyanin synthesis (and its glucose component) with sugar accumulation. However, when the glucose component associated with anthocyanins is removed, the red-free G-G content shows a marked increase only during the latter stages of ripening (Fig. 4.60).

Only additional research will establish if there is a direct connection between the G-G content and perceived wine aromatic quality. In the interim, research is progressing on finding faster, more convenient ways of

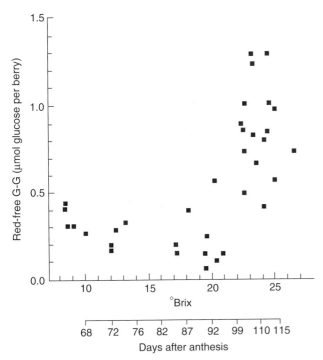

Figure 4.60 The content of red-free G-G (as μmol glucose/berry) against juice total soluble solids (°Brix) and days after flowering of grape berries cv. Shiraz, sampled at intervals during ripening. (From Coombe and McCarthy, 1997, reproduced by permission)

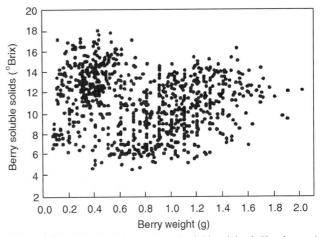

Figure 4.61 Variation in berry size and soluble solids of 'Chardonnay.' Fruit was taken from two bunches from each of 15 shoots randomly selected from within the vineyard. (From Trought, 1996, reproduced by permission)

determining the G-G content, and adapting it for on-site use (Gishen and Dambergs, 1998).

An objective measure of grape-flavor potential would be of tremendous value to grape grower, winemaker, and researcher alike. It would give the researcher a quantitative measure of grape maturity, against which the effects of modified viticultural practice could be measured. For the winemaker, it would offer an objective criterion by which to harvest grapes, depending on the style desired. For the grape grower, it would provide an objective measure of grape quality, and correspondingly, monetary value.

Although assessment of aroma constituents is likely to receive greater attention in the future, such as with the use of electronic noses, perceived wine quality is not always directly associated with grape volatile aroma content (Whiting and Noon, 1993). The prediction of wine quality based on grape chemistry is still in its infancy.

In addition to assessing properties correlated with quality, grape growers must also take into consideration factors that may lower fruit quality. These are usually difficult to predict because they depend on local climatic conditions. The detrimental effects of early frosts and protracted rainy periods on grape quality are well known, but forecasting their imminent occurrence is still regrettably imprecise.

Sampling

Despite progress in ferreting out chemical indicators of grape maturity, fundamental problems still exist. Part of the problem relates to difficulties in applying them in the vineyard. Fruit quality is as dependent on grape uniformity as grape chemistry. Extensive variation in maturity can negate maturity indicators derived from inappropriate sampling. Various studies, such as those comparing fruit size and soluble solids (Fig. 4.61) and data from yield monitors mounted on grape pickers (Plate 4.19) show that vineyard variability is often more pronounced than previously thought. Such fluctuation can arise from a host of causes, including differences in vine age and health, as well as variations in soil structure, texture, nutrition, moisture content, and microclimate. Additional diversity may arise from protracted flowering and the location of clusters on, and within, the vine canopy.

Advancements in measuring vineyard variation have led to the newest buzz phrase in viticulture – **precision viticulture** (PV). It often involves a combination of onsite analyses (soil, disease, yield, grape composition), combined with spectrophotometric analyses via remote sensing. Its aim is to determine the sources of vineyard variation so that it can be reduced as much as possible. In the future, minimizing ripening heterogeneity will become one of the principal tasks of the grape grower (Smith, 2004).

Many methods of fruit sampling have been investigated to determine the optimal combination of adequacy and ease of execution. Simple application is important because vineyards may be checked almost daily for weeks before harvest. Although different sampling procedures are used worldwide, that proposed by Amerine and Roessler (1958) has been a standard for much of the wine industry. It entails the collection of about 100–200 berries from many grape clusters, selected at random throughout the vineyard. Berries from clusters at row ends, or from obviously aberrant vines, are avoided. Sampling usually begins 2–3 weeks before the grapes are likely to reach maturity. Chemical analysis is performed on the juice extracted from the berry sample. The basic methods of must analysis are given in Ough and Amerine (1988) and Zoecklein et al. (1995).

Recently, procedures associated with PV (zonal vineyard management) indicate that considerable improvements in yield and maturity prediction are possible (Proffitt and Malcolm, 2005). Developments in remote sensing are making mapping of vineyard variability, down to individual vines, both feasible and increasingly affordable. If considerable vineyard variability exists, then it is important that random fruit selection be adjusted to reflect this disparity. For example, if excessive and low vigor were to, respectively, represent 25% and 10% of the vineyard area, then 25% of the fruit should be randomly selected from vigorous parcels, 10% from low vigor sites, and the rest from medium sites. Adjusted selection more accurately represents fruit yield and quality at harvest. It can also be the basis for selective timing of harvest for distinct vineyard sites.

Improvements in timing can provide one of the dreams of winemakers – improved fruit uniformity at the vineyard door. Improved uniformity ideally should permit wines of more predictable character, tuned to the desires of the consumer. Nevertheless, there is still the conundrum of the interrelationship between grape chemical analysis and subjective wine quality. Thus, some of the potential benefits of better fruit uniformity at the winery still await a better understanding of the chemical basis of wine quality. In addition, the geographic distribution of factors affecting chemical indicators of fruit quality, such as °Brix, pH, total acidity, anthocyanin and phenolic content, may not correlate with more easily measured features, such as yield or vine vigor (Bramley, 2005).

Harvest Mechanisms

Until the late 1960s, essentially all grapes were harvested manually. Subsequently, market forces as well as labor shortages combined to make mechanical harvesting progressively more cost-effective. In some regions, mechanical harvesters have been estimated to reduce harvesting costs by about 75% (Bath, 1993). Mechanical harvesters also permit rapid collection at the optimum time (determined by grape maturity, weather conditions, and winery desires). Market forces

may permit the retention of manual harvesting in some regions, but most hectarage in both the Old and New Worlds is now mechanically harvested.

Many different types of containers have been employed in collecting and transporting grapes to the winery. In Europe, traditional wicker and wooden containers reflect cultural traditions and the terrain of the region. In the New World, plastic or aluminum boxes are commonly used. From these, grapes are transferred to containers of various sizes for transport to the winery. Mechanically harvested grapes are often directly conveyed to transport containers. Any system is adequate, if bruising and breaking of the fruit are minimized, and if rapid movement to the winery permits processing shortly after harvesting.

MANUAL HARVESTING

Manually harvesting grapes still has several advantages over mechanical harvesting. This is especially true for thin-skinned cultivars that break open easily, for example, 'Sémillon.' Hand-harvesting also permits the rejection of immature, raisined, or diseased fruit, as well as the selection of grapes at particular states of maturity. Except for special wines, such as those produced from the outer arm of grape clusters in Amarone production, or noble-rotted grapes for botrytized wines, this option is seldom employed.

For the advantages of manual harvesting to be realized, the clusters must be collected and placed in containers with all due care to minimize breakage. The grapes also must be transported quickly to the winery for rapid processing. This minimizes grape heating and the growth of undesirable microorganisms on the berries or released juice.

Although hand-harvesting is beneficial, or required under special circumstances, such as steeply sloped vineyards, the disadvantages of manual harvesting often outweigh its advantages. In addition to labor costs and inadequate availability, manual harvesting is slower, stops during inclement weather, and seldom occurs 24 hours a day. Nevertheless, a California producer now manufactures a boom containing banks of fluorescent lights that can be attached to a tractor. It can illuminate up to four rows for manual picking at night.

Relative to haze-forming proteins, there is little difference between manually and mechanically harvested grapes. However, if mechanical harvested fruit is transported long distances, the extraction of protein from the skins and flesh can increase clarification problems (Pocock et al., 1998). These are primarily pathogenesis-related proteins, notably thaumatin-like proteins and chitinases. Long delays between harvesting and crushing can also result in excessive oxidative browning and microbial contamination of the juice before fermentation.

Figure 4.62 Types of mechanical harvesters based on the mechanism of head functioning. (**A**) Slapper head with two paired banks of beaters; (**B**) impactor head for use with T-trellised wines; (**C**) pulsator head with two rails that shake the vine trunk. (From Hamilton and Coombe, 1992, reproduced by permission)

Under most circumstances, though, mechanical harvesters are the most economic and logical method of harvesting. Conveniences such as four-wheel drive, automated leveling, expanded size options, and improved performance continue to extend their use into a wider range of vineyard situations.

MECHANICAL HARVESTERS

All mechanical harvesters ostensibly use the same means to remove fruit. Force applied to one or more parts of the vine induces rapid and abrupt swinging that detaches the berries or fruit clusters. Harvesters are usually classified according to the mechanism by which they apply the force. Most harvesters fall into one of two main categories (Fig. 4.62). Those that apply force directly to the bearing shoot are called pivotal striker, cane shaker, striker, or impactor machines. Those that supply the force directly to the vine trunk are termed either pulsator, trunk shaker, or shaker machines. Some harvesters combine both actions and may be referred to as pivotal pulsators. A third category, the slapper-type, directs force to the support wire bearing the shoots.

Most machines are designed to harvest vines trained as single vertical canopies. Special modifications are required to harvest vines trained on T-trellises, such as

the Geneva Double Curtain, and major changes in design are required for training systems such as the Tendone or pergola (Cargnello and Piccoli, 1978).

Once the berries or clusters have been shaken free, they are collected in a series of plates resembling enlarged fish scales. These open and close around the vine trunk or trellis supports. The plates are made of nylon or polyethylene, and are arranged to slope outward. Thus, the fruit roll toward belts or buckets on each side of the vine and are conveyed for collection in one of a series of bins or gondolas. Where skin contact is not desired, the fruit may be crushed immediately and only the juice transported to the winery.

Pivotal Striker Harvesters Striker (impactor) harvesters possess a double bank of upright flexible rods arranged parallel to, and on each side of the vine (Fig. 4.62A). Formerly, the rods were solid fiberglass tubes with bent ends. Current preference is for curved rods in the shape of a bow (Fig. 4.63). The bend in the bow can be increased to enlarge the surface area of the rod impacting the vine. This reduces the amount of berry rupture (juicing) and minimizes contamination with material-other-than-grapes (MOG). The banks of rods oscillate back and forth, striking the vine canopy and

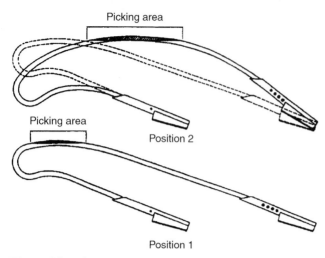

Figure 4.63 The bow picking rod, showing the bowed shape of the rod for "soft pick" (Position 2) and the straight rod profile for "penetration pick" (Position 1) operation. (From Burke, 1996, reprinted by permission)

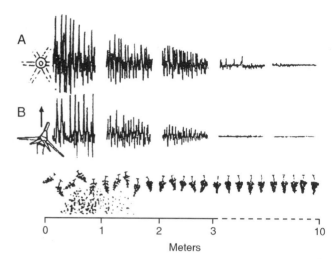

Figure 4.64 Vine cordon oscillation on the horizontal (**A**) and vertical (**B**) plane, and cluster movement induced by a horizontal impactor. (From Intrieri and Poni, 1995, reproduced with permission)

shaking the fruit loose. For vines possessing less foliage, models having both the front and back banks (quad arrangement) oscillate together. Under heavy foliage, oscillating alternately tends to be more effective. With bow rods, the alternation can be adjusted so that when one set of rods reaches their maximum curvature, the other is at its straightest. Further adjustments to achieve the desired level of fruit quality and removal include the rod stroke (extent of side-to-side movement) and oscillation speed (rpm).

Striker machines are generally more suitable for cane-pruned vines, when cordons are young, or with other systems where the fruit is borne away from the permanent vine structure. Strikers are not as efficient at dislodging fruit close to the vine head. Increasing the striking velocity to dislodge the fruit accentuates vine damage and increases fruit contamination with leaves, leaf fragments, and other vine parts (MOG). However, striker machines are easier on the trellis structure and accommodate themselves more readily to vine rows of imperfect alignment.

Trunk-shaker Harvesters The trunk-shaker machine typically possesses two parallel, oscillating rails that impart vibration to the cordon or upper trunk a few centimeters below the fruit zone (Fig. 4.62C). This shakes the vine back and forth several centimeters.

Shaker (pulsator) machines are most effective in removing fruit in close proximity to cordons or the trunk (spur-pruned). Insufficient force tends to spread along flexible canes to efficiently remove fruit on cane-pruned vines. However, the machines do have the advantage of dislodging fewer leaves and other vine material. The major drawback is the limitation in vine training required

for optimal function. In addition, the stakes must be flexible and in perfect alignment to withstand the force applied during shaking. The cordons must be firmly attached to support wires and these in turn securely affixed to stakes. Magnets are typically used in the harvester to remove nails, staples, or other metallic parts shaken loose from the trellis.

Striker–Shaker Combination Harvesters Because both striker and shaker processes have limitations, machines combining both principles have been introduced. They possess a pair of oscillating rails and several short pulsating rods, and operate at lower speeds (50–75 rpm) than other machines. They tend to produce less vine and trellis damage, may be used with a wide range of training systems, and generate lower levels of MOG contamination.

Horizontal Impactor The horizontal impactor (Fig. 4.62B) is designed for use with training systems employing wide-topped (T-) trellises, such as the Geneva Double Curtain. For use of the harvester, canopy wires must by attached to movable crossarm supports, or the canopy wire must be held loosely in a slot on rigid crossarms. This is necessary because the jarring action to dislodge the fruit comes from a vertically rotating wheel whose spokes strike and raise the canopy or cordon wire from below (Fig. 4.64). Because of the extra stress placed on the trellis system, stronger wire than usual is required.

Robotic Harvesters Although not currently used in commercial production, robot harvesters are being investigated (Kondo, 1995). The prototype uses a video camera to measure light frequency, and from that, determine fruit position. Secateurs then selectively remove fruit clusters. Is this a preview of vineyard harvesting in the future?

FACTORS AFFECTING HARVESTER EFFICIENCY

In addition to the training system, the growth and fruiting habits of the cultivar can significantly affect harvester efficiency. The two major fruiting characteristics influencing varietal suitability for machine harvesting are the ease with which the grapes separate from the vine and berry fragility. Features such as ready separation as whole bunches, easy fragmentation at rachis divisions, or dehiscence with facility at the pedicel are ideal attributes for mechanical harvesting. Ease of detachment releases the stress applied by the harvester and minimizes the chance of fruit rupture. Because wine grapes are predominantly juicy, firm attachment and fibrous rachis and cluster structure can lead to extensive berry rupture and juice loss, as, for example with 'Emerald Riesling' and 'Zinfandel.' Soft-skinned varieties, such as 'Sémillon' and 'Muscat Canelli,' also lose much of their juice if mechanically harvested (Table 4.13). This problem is alleviated if prolonged skin contact is not desired and crushing occurs simultaneously with fruit harvesting.

The vegetative characteristics of the vine most affecting harvester function are cane flexibility, plus foliage density and size. Dense canopies can interfere with force transmission and fruit dislodgement, as well as plug and disrupt conveyor-belt operation. The detrimental effects of excessive MOG contamination can be partially offset by fans blowing over the collector belts. However, small leaf fragments tend to become wet with juice and stick to, or are propelled into, the fruit. Brittle wood can produce spur breakage and conveyor plugging. Preharvest thinning often diminishes the severity of undesirable growth characteristics on harvester function. Additional problems can develop when harvesting old vineyards and trellises. Bits of wood from posts, trellis wire, staples, bolts and other metal can be dislodged. These can result in significant repair costs and downtime caused by damaging the crushers.

Other factors influencing harvester efficiency are the time of harvesting, vineyard slope, and soil condition. As grapes and other vine parts are typically more turgid at night, berry and cluster separation usually requires less force. Night harvesting also has the advantage of removing fruit at a cool temperature in warm climates. However, poor visibility, even using headlights, can make the adequate observation of harvesting performance difficult. Slopes of more than about 7% require harvesters that can adjust their wheels independently to level the catching frame of the harvester. Soil grading also may be required to facilitate steering and to remove

Table 4.13 Adaptability of several grape varieties to mechanical harvesting

Variety	Ease of harvest	Amount of juicing
White varieties		
'Chardonnay'	Hard	Medium
'Chenin blanc'	Medium	Medium
'Emerald Riesling'	Very hard	Very hard
'Flora'	Easy	Light
'French Colombard'	Medium	Medium
'Gewürztraminer'	Easy/medium	Light
'Gray Riesling'	Easy/medium	Light
'Malvasia bianca'	Medium/hard	Medium
'Muscat Canelli'	Hard	Heavy
'Palomino'	Easy/medium	Medium
'Pedro Ximénez'	Easy/medium	Heavy
'Pinot blanc'	Medium	Medium
'Riesling'	Medium	Medium
'Sauvignon blanc'	Medium	Medium
'Sémillon'	Medium/hard	Heavy
'Silvaner'	Medium	Light
'Trebbiano'	Medium	Heavy
Red varieties		
'Aleatico'	Hard	Heavy
'Barbera'	Medium/hard	Medium
'Cabernet Sauvignon'	Easy/medium	Light/medium
'Carignan'	Medium/hard	Medium/heavy
'Gamay'	Medium	Medium
'Grenache'	Hard	Medium/heavy
'Petite Sirah'	Medium	Medium/heavy
'Pinot noir'	Medium	Medium
'Rubired'	Easy	Medium
'Tempranillo'	Medium/hard	Medium/heavy
'Zinfandel'	Hard	Medium/heavy

Source: Data from Christensen *et al.*, 1973

ridges that can interfere with keeping the catching frame close to the ground.

RELATIVE MERITS OF MECHANICAL HARVESTING

With improved harvester design and increasing awareness of the importance of vine training and uniform trellising, fruit of equivalent quality often can be obtained by either manual or mechanical means (Table 4.14). Wines produced from grapes harvested by different means can occasionally be distinguished, but no clear sensory preference has developed either in North America or in Europe (see Clary *et al.*, 1990). Expected increases in phenolic contents, due to fruit rupture, often do not materialize. This absence may be due to rapid oxidation, and subsequent precipitation of the phenolics extracted. Nevertheless, mechanical harvesting has been correlated with increased problems with protein instability in wine, especially if coinciding with delays or extended transport to the winery (Pocock *et al.*, 1998).

Table 4.14   Effect of harvesting method on crop and juice characteristics[a]

	Treatment		
Property	Striker harvester	Shaker harvester	Manual harvester
Yield (kg/ha)	12,509	12,475	12,800
Stem content (kg/ha)	155	288	524
Ground loss (kg/ha)	317	249	338
Juice loss (%)	5.7	8.0	0
MOG[b] (%)	1.3	0.7	0.5
°Brix	22.2	22.2	22.6
Total acidity (g/100 ml)	0.75	0.76	0.74
pH	3.26	3.28	3.27
Malic acid (g/liter)	4.552	4.409	4.137

[a] The slightly higher °Brix and lower total and malic acidity of the hand-harvested grapes is likely explained by the absence of second crop fruit.

[b] MOG, material other than grapes.

Source: Data from Clary *et al.*, 1990

With cultivars suitable for mechanical harvesting, the choice of method often depends on factors other than the direct effects on fruit quality. Features such as the potential for night and rapid harvesting, cost and availability of manual labor, juice loss, and vineyard size become deciding factors. For example, where grapes grown in hot climates must be transported long distances to the winery, harvesting during the cool of the night can result in grapes arriving in a healthier condition. Field crushing and storage under cool anaerobic conditions, immediately following harvest, is another solution that preserves the juice quality of fruit picked at a distance from the winery. For high-priced, low-yielding cultivars, juice losses between 5 and 10% may negate the economic benefits of mechanical harvesting. Also, for French-American hybrids that may produce an extensive second crop, the detrimental effects of harvesting immature grapes on wine quality may outweigh the cost benefits of mechanical harvesting. Whether features usually associated with mechanical harvesting, such as lower stem content, marginally higher MOG contents (Clary *et al.*, 1990), increased potential for juice oxidation, and increased fungal growth on the juice-soaked canes of the vine (Bugaret, 1988) are of practical significance is uncertain. If they are, the significance probably varies considerably, depending on the cultivar, harvesting conditions, and wine style desired. Nevertheless, leaf defoliation, if pronounced, can have a negative influence by reducing the vine's ability to produce and store carbohydrate reserves for the subsequent season. This could decrease vine productivity in the succeeding year, especially if the current year's crop were high. Because

of the perennial nature of the vine, such influences can be accumulative (Holzapfel *et al.*, 2006).

Measurement of Vineyard Variability

The major theme in this chapter has been means by which the grape grower can achieve maximum yield relative to optimum fruit quality and long-term vine health. The measure of success has usually been in terms of average yield or wine production per hectare. This, however, masks important variation in yield and quality within the vineyard, and their causes.

One of the most significant viticultural developments during the past decade has been the introduction of **precision viticulture** (PV). Its philosophy could be said to be based on the old adage:

"You can't manage what you can't measure."

The tools of precision viticulture include global positioning[3] (GPS), mechanical harvesters (connected to continuous yield monitors), airborne remote sensing (using multispectral imaging) (Hall *et al.*, 2002), improved and automated soil sensing (Adamchuk *et al.*, 2004), and telemetric vineyard monitoring. These can now be instantly integrated due to the increasing availability of computing power. These changes are bringing detailed vineyard assessment within the financial reach of an increasing number of grape growers. Resolution with remote sensing devices on low flying aircraft (150 m to 3 km) is currently adequate to assess individual vines, and distinguish between vines and other vineyard features. Multispectral sensors can simultaneously record data from at least four distinct wavelengths (infrared, red, green and blue). Sensors that detect up to 18 wavelengths are available. On-the-go grape assessment systems are also under development that could make spatial fruit assessment during harvesting as easily precise.

These techniques can quantify the spatial distribution of intra-vineyard variation to a degree heretofore unknown (Plate 4.19). Although measures of grape quality often differ more among sites and individual vines than between vintages, these variations are often considerably less than differences in yield (Bramley and Lamb, 2003). Yield among vines can easily differ by a factor of 8 to 10 (Bramley and Hamilton, 2004). Objective assessment of such nonuniformity has spurred a desire to reduce its occurrence, either by selective management of distinct areas, or by selective harvesting. Although some

[3] High-accuracy GPS receivers allow measurement not only of longitude and latitude but also altitude. This permits the calculation of slope, aspect, and other pertinent topographic parameters.

attributes may average out on blending, features such as flavor do not. Rarely do under- or over-mature grapes possess 'ideal' flavor attributes. PV techniques should also provide grape growers with more precise tools for estimating yield and quality.

Bramley (2005) gives an example of precision viticulture, what he terms **zonal vineyard management**. It employs remote sensing to map variation in vine growth throughout the vineyard. The spatial variation maps generated are correlated with on-site assessments of grape quality and health, soil, and topographic conditions. The data so derived can be used to direct customized fertilization, irrigation, pruning, salinity adjustment, and pest management. Subsequent assessments can be used to assess the degree to which management adjustments have succeeded in enhancing vineyard uniformity. By reducing the sources of intra-vineyard heterogeneity, average grape quality can be increased. In addition, savings or improved efficiency of fertilizer, water, pesticide or other applications are possible. Proffitt and Malcolm (2005) found that improved fruit uniformity was achieved by decreasing irrigation in vigorous regions, and augmented it in areas with impoverished growth. Not only can this enhance synchronization of berry development, but also reduce costly procedures such as basal-leaf removal. Alternatively, grapes from distinct parcels of a vineyard may be harvested at different times, or separated at harvest. Overall wine quality is partially dependent on grape uniformity.

Several features are prerequisites for converting zonal vineyard management from an option to standard practice. Under most circumstances, spatial variation in yield and vigor are relatively constant from year to year. Although readily observable features, such as yield and vine vigor tend to correlate well with important quality features, such as °Brix and phenolic content, other chemical attributes may not. Although critical vineyard factors, such as nutrition, irrigation and pest management can be selectively adjusted in an existing vineyard; other attributes, such as slope, soil depth, or drainage can be most effectively modified only at inception or replanting. In addition, although underlying factors such as drainage and soil depth remain constant, the influence of these factors can vary, depending on yearly variations in temperature, frost prevalence, or rainfall (Lamb, 2000). Finally, the economic feasibility of zonal vineyard management will depend on several independent factors, including the extent and severity of vineyard variation, the costs of implementing remedial action, the commercial value of the variety, and the demonstration that the improved and more uniform grape quality lead to superior wine.

Finally, knowledge of actual vineyard variation allows more precise assessment of the effect of such variables on vine growth and wine-producing quality of the fruit. Without this knowledge, researchers are forced to use randomized plot designs to statistically account for hidden variables. Although better than nothing, such procedures can lead to statistically significant results that are invalid, as the plots may not accurately represent actual variability. Precision viticulture offers grape growers and researchers new opportunities to accurately assess the significance of vineyard variables on grape growth and wine quality (Panten and Bramley, 2006).

Suggested Readings

General References

Coombe, B., and Dry, P. (eds.) (2004) *Viticulture*, Vol. 1 (*Resources*), 2nd edn. Winetitles, Adelaide, Australia.

Coombe, B., and Dry, P. (eds.) (1992) *Viticulture*, Vol. 2 (*Practices*). Winetitles, Adelaide, Australia.

Hidalgo, L. (1999) *Tratado de Viticultura General*, 2nd edn. Mundi-Prensa, Madrid.

Jackson, D., and Schuster, D. (2001) *The Production of Grapes and Wines in Cool Climates*, 3rd edn. Dunmore Press, Auckland, New Zealand.

Morton, L. T. (1985) *Wine Growing in Eastern America*. Cornell University Press, Ithaca, NY.

Mullins, M. G., Bouquet, A., and Williams, L. E. (1992) *Biology of the Grapevine*. Cambridge University Press, Cambridge, UK.

Weaver, R. J. (1976) *Grape Growing*. Wiley, New York.

Winkler, A. J., Cook, J. A., Kliewer, W. M., and Lider, L. A. (1974) *General Viticulture*. University of California Press, Berkeley.

Growth–Yield Relationship

Howell, G. S. (2001) Sustainable grape productivity and the growth-yield relationship: A review. *Am. J. Enol. Vitic.* 52, 165–174.

May, P. (1987) The grapevine as a perennial plastic and productive plant. In: *Proceedings of the 6th Australian Wine Industry Conference* (T. Lee, ed.), pp. 40–49. Australian Industrial Publishers, Adelaide.

Pruning

Bailey, L. H. (1919) *The Pruning-Manual*. Macmillan, New York.

Jordan, T. D., Pool, R. M., Zabadal, T. J., and Tomkins, J. P. (1981) *Cultural Practices for Commercial Vineyards*. Miscellaneous Bulletin. 111, Cornell Cooperative Extension Publications, Cornell University, Ithaca, NY.

Possingham, J. V. (1994) New concepts in pruning grapevines. *Hortic. Rev.* 16, 235–254.

Training and Canopy Management

Intrieri, C., and Filippetti, I. (2000) Innovations and outlook in grapevine training systems and mechanization in North-Central Italy. In: *Proceedings of the ASEV 50th Anniversary Annual Meeting*, Seattle, Washington June 19–23, 2000, pp. 170–184. American Society for Enology and Viticulture, Davis, CA.

Jackson, D. I., and Lombard, P. B. (1993) Environmental and management practices affecting grape composition and wine quality. A review. *Am. J. Enol. Vitic.* **44**, 409–430.

Kliewer, W. M. (ed.) (1987) *Symposium on Grapevine Canopy and Vigor Management*. Special Issue of *Acta Hortic.* **206**, 1–168.

Morris, J. (1996) A total vineyard mechanization system and its impact on quality and yield. In: *Proceedings of the 4th International Symposium Cool Climate Viticulture and Enology* (T. Henick-Kling *et al.*, eds.), pp. IV-6–10. New York State Agricultural Experimental Station, Geneva, NY.

Morris, J. R. (2000) Past, present, and future of vineyard mechanization. In: *Proceedings of the ASEV 50th Anniversary Annual Meeting*, Seattle, Washington June 19–23, 2000, pp. 155–164. American Society for Enology and Viticulture, Davis, CA.

Smart, R. E. (1996) The modern difficulty of choosing a trellis for your situation. *Vineyard Winery Manage.* **22**(1), 42–44, 46, 70–71.

Smart, R. E., and Robinson, M. (1991) *Sunlight into Wine. A Handbook for Winegrape Canopy Management*. Winetitles, Adelaide, Australia.

Vasconcelos, M. C. and Castagnoli, S. (2000) Leaf canopy structure and vine performance. *Am. Enol. Vitic.* **51**, 390–396.

Rootstock

Cowham, S. (2003) *A Growers' Guide to Choosing Rootstocks in South Australia*. Phylloxera and Grape Industry Board of South Australia. Adelaide. http://www.phylloxera.com.au/resources/index.asp?view=Phylloxera%20resources

May, P. (1994) *Using Grapevine Rootstocks – The Australian Perspective*. Winetitles, Adelaide, Australia.

Morton, L. T., and Jackson, L. E. (1988) Myth of the universal rootstock. The fads and facts of rootstock selection. In: *Proceedings of the 2nd International Symposium Cool Climate Viticulture and Oenology* (R. E. Smart *et al.*, eds.), pp. 25–29. New Zealand Society of Viticulture and Oenology, Auckland, New Zealand.

Pongrácz, D. P. (1983) *Rootstocks for Grape-vines*. David Philip, Cape Town.

Southey, J. M., and Archer, E. (1988) The effect of rootstock cultivar on grapevine root distribution and density. In: *The Grapevine Root and its Environment* (J. L. van Zyl, comp.), pp. 57–73. Technical Communication No. 215, Department of Agricultural Water Supply, Pretoria, South Africa.

Wolpert, J. A., Walker, M. A., and Weber, E. (eds.) (1992) *Rootstock Seminar: A Worldwide Perspective*. American Society of Enology Viticulture, Davis, CA.

Propagation and Grafting

Hathaway, S. (Editor) *A Grower's Guide to Top-Working Grapevines* Phylloxera and Grape Industry Board of South Australia. http://www.phylloxera.com.au/Viticulture/Topworking/Phy%20Top%20Woring%Guide/20wh

Loyola-Vargas, V. M., and Felipe Vazquez-Flota, F. eds. (2005) *Plant Cell Culture Protocols*, 2nd edn. Humana Press, Totowa, NJ.

Martinelli, L., and Gribaudo, I. (2001) Somatic embryogenesis in grapevine. In: *Molecular Biology and Biotechnology of the Grapevine* (K. Roubelakis-Angelakis, ed.), pp. 327–351. Kluwer Adacemic Publ., Dordrecht, Netherlands.

Steinhauer, R. E., Lopez, R., and Pickering, W. (1980) Comparison of four methods of variety conversion in established vineyards. *Am. J. Enol. Vitic.* **31**, 261–264.

Irrigation and Water Relations

Anonymous (2005) *Introduction to Irrigation Management*. Evaluating your Pressurised Systems. NSW Department of Primary Industries. http://www.agric.nsw.gov.au/waterwise.

Anonymous (2006) Land, air and water resources (LAWT). University of California Cooperative Extension, San Joaquin County. http://cesanjoaquin.ucdavis.edu/Custom_Program/Publications_Available_for_Download.htm?$=733

Behboudian, M. H., and Singh, Z. (2001) Water relations and irrigation scheduling in grapevine. *Hortic. Rev.* **27**, 189–225.

Dry, P. R., and Loveys, B. R. (1998) Factors influencing grapevine vigour and the potential for control with partial rootzone drying. *Aust. J. Grape Wine Res.* **4**, 140–148.

Goodwin, I. (1995) *Irrigation of Vineyards*. Winetitles, Adelaide, Australia.

Keller, M. (2005) Deficit irrigation and vine mineral nutrition. *Am. J. Enol. Vitic.* **56**, 267–283.

Kramer, P. J., and Boyer, J. S. (1995) *Water Relations of Plants and Soils*. Academic Press, San Diego, CA.

Lakso, A. N., and Pool, R. M. (2000) Drought stress effects on vine growth, function, ripening and implications for wine quality. pp 86–90. In: *29th Annual NY Wine Industry Workshop*. New York State Agric. Exper. Sta., Geneva, NY.

van Zyl, J. L. (1988) Response of grapevine roots to soil water regimes and irrigation systems. In: *The Grapevine Root and Its Environment* (J. L. van Zyl, comp.), Technical Communication No. 215, pp. 30–43. Department of Agricultural Water Supply, Pretoria, South Africa.

Williams, L. E., and Matthews, M. A. (1990) Grapevines. In: *Irrigation of Agricultural Crops*. (B. A. Stewart, and D. R. Nielsen, eds.), pp. 1019–1055. Agronomy Series Monogr. No. 30. American Society of Agronomy, Madison, Wisconsin.

Nutrition

Bell, S.-J., and Henschke, P. A. (2005) Implication of nitrogen nutrition for grapes, fermentation and wine. *Aust. J. Grape Wine Res.* **11**, 242–295.

Christensen, P., and Smart, D. (eds) (2005) *Proceeding of the Soil Environment and Vine Mineral Nutrition Symposium*. American Society for Enology and Viticulture, Davis, CA.

MacRae, R. J., and Mehuys, G. R. (1985) The effects of green manuring on the physical properties of temperate-area soils. *Adv. Soil Sci.* **3**, 71–94.

Mpelasoka, B. S., Schachtman, D. P., Treeby, M. T., and Thomas, M. R. (2003) A review of potassium nutrition in grapevines with special emphasis on berry accumulation. *Aust J. Grape Wine Res.* **9**, 154–168.

Rantz, J. M. (ed.) (1991) In: *Proceedings of the International Symposium Nitrogen in Grapes and Wine*. American Society of Enology Viticulture, Davis, CA.

Wolf, B., and Snyder, G. H. (2004) *Sustainable Soils – the Place for Organic Matter in Sustaining Spoils and their Productivity*. Food Products Press, New York.

Disease, Pest and Weed Control

Anonymous (1993) *Cover Crops: A Practical Tool for Vineyard Management*. Technical Projects Committee ASEV, American Society of Enology Viticulture, Davis, CA.

Durrant, W. E., and Dong, X. (2004) Systemic acquired resistance. *Annu. Rev. Phytopathol.* **42**, 185–209.

Flaherty, D. L., Christensen, L. P., Lanini, W. T., Marois, J. J., Phillips, P. A., and Wilson, L. T. (1992) *Grape Pest Management*, 2nd edn. Publication No. 3343, Division of Agricultural and Natural Resources, University of California, Oakland.

Gubler, W. D., Baumgartner, G. T., Browne, G. T., Eskalen, A., Latham, S. R., Petit, E., and Bayramian, L. A. (2004) Four diseases of grapevines in California and their control. *Aust. Plant Pathol.* **32**, 47–52.

Ingels, C. A., Bugg, R. L., McGourty, G. T., and Christensen, L. P. (1998) *Cover Cropping for Vineyards: A Grower's Handbook*. University of California Division of Agricultural and Natural Resources, Davis, CA.

Martelli, G. P. (2000) Major graft-transmissible diseases of grapevines: Nature, diagnosis, and sanitation. In: *Proceedings of the ASEV 50th Anniversary Annual Meeting*, Seattle, Washington, June 19–23, 2000, pp. 231–236. American Society for Enology and Viticulture, Davis, CA.

Milholland, R. D. (1991) Muscadine grapes: Some important diseases and their control. *Plant Dis.* **75**, 113–117.

Pearson, R. C. and Goheen, A. C. (eds) (1988) *Compendium of Grape Diseases*. APS Press, St Paul, MI.

Penfold, C. (2004) Mulching and its impact on weed control, vine nutrition, yield and quality. *Aust. NZ Grapegrower Winemaker* **485**, 32–38.

Roehrich, R., and Boller, E. (1991) Tortricids in vineyards. In: *Tortricid Pests. Their Biology, Natural Enemies and Control* (L. P. S. van der Geest, and H. H. Evenhuis, eds.), pp. 507–514. Elsevier, Amsterdam.

Walter, B. ed. (1997) *Sanitary Selection of the Grapevine*. INRA Éditions, Versailles, France.

Weinstein, L. H. (1984) Effects of air pollution on grapevines. *Vitis* **23**, 274–303.

Wilson, M. (1997) Biocontrol of aerial plant diseases in agriculture and horticulture: Current approaches and future prospects. *J. Indust. Microbiol. Biotechnol.* **19**, 188–191.

Harvesting

Carnacini, A., Amati, A., Capella, P., Casalini, A., Galassi, S., and Riponi, C. (1985) Influence of harvesting techniques, grape crushing and wine treatments on the volatile components of white wines. *Vitis* **24**, 257–267.

Christensen, L. P., Kasimatis, A. N., Kissler, J. J., Jensen, F., and Luvisi, D. (1973) *Mechanical Harvesting of Grapes for the Winery*. Agricultural Extension, University of California, Oakland.

Clary, C. D., Steinhauer, R. E., Frisinger, J. E., and Peffer, T. E. (1990) Evaluation of machine- vs. hand-harvested Chardonnay. *Am. J. Enol. Vitic.* **41**, 176–181.

du Plessis, C. S. (1984) Optimum maturity and quality parameters in grapes: A review. *S. Afr. J. Enol. Vitic.* **5**, 35–42.

Gishen, M., Iland, P. G., Dambergs, R. G., Esler, M. B., Francis, I. L., Kambouris, A., Johnstone, R. S., and Høj, P. B. (2002) Objective measures of grape and wine quality. In: *11th Australian Wine Industry Technical Conference*. Oct. 7–11, 2001, Adelaide, South Australia. (R. J. Blair, P. J. Williams and P. B. Høj, eds.), pp. 188–194. Winetitles, Adelaide, Australia.

Morris, J. R., Main, G. L., and Oswald, O. L. (2004) Flower cluster and shoot thinning for crop control in French-American hybrid grapes. *Am. J. Enol. Vitic* **55**, 423–426.

Morris, J. R., Sims, C. A., Bourque, J.E., and Oakes, J. L. (1984) Influence of training system, pruning severity, and spur length on yield and quality of six French-American hybrid grape cultivars. *Am. J. Enol. Vitic.* **35**, 23–27.

Reynolds, A. G. (1996) Assessing and understanding grape maturity. In: *25th Annual Wine Industry Workshop* (T. Henick-Kling, ed.) pp. 75–88. New York State Agricultural Experimental Station, Cornell University, Geneva, NY.

Precision Viticulture

Bramley, R. (ed.) (2004) *Proceedings from the Workshop on Precision Viticulture*. 12th Australian Wine Industry Technical Conference and Trade Exhibition, Melbourne, Victoria, July 25–28, 2004. http://awitc.com.au/workshops/Workshop_30B_Proceedings.pdf

Bramley, R. G. V., and Lamb, D. W. (2003) Making sense of vineyard variability in Australia. In: *Precision Viticulture* (Ortega, R. and A Esser, eds.), pp. 35–54. Proceedings of an International Symposium IX Congreso Latinoamericano de Viticultura y Enologia, Santiago, Chile.

Hall, A., Lamb, D. W., Holzapfel, B., and Louis, J. (2002) Optically remote sensing applications in viticulture – a review. *J. Grape Wine Res.* **8**, 36–47.

Krstic, M., Moulds, G., Panagiotopoulos, B., and West, S. (2003) *Growing Quality Grapes to Winery Specifications*. Winetitles, Adelaide, Australia.

Lamb, D. W., Hall, A., Louis, J., and Frazier, P. (2004) Remote sensing for vineyard management. Does (pixel) size really matter? *Aust. NZ Grapegrower Winemaker* **485a**, 139–142.

References

Adamchuk, V. I., Hummel, J. W., Morgan, M. T., and Upadhyaya, S. K. (2004) On-the-go soil sensors for precision agriculture. *Comput. Electron. Agric.* **44**, 71–91.

Adamchuk, V. I., Lund, E., Dobermann, A., and Morgan, M. T. (2003) On-the-go mapping of soil properties using ion-selective electrodes. In: *Precision Agriculture*. (J. Stafford and A. Werner, eds.) Wageningen Adacemic Publishers, Wageningen, The Netherlands. pp. 27–33.

Adams, W. E., Morris, H. D., and Dawson, R. N. (1970) Effects of cropping systems and nitrogen levels on corn (*Zea mays*) yields in the southern Piedmont region. *Agron. J.* **62**, 655–659.

Agrios, G. N. (1997) *Plant Pathology*, 4th edn. Academic Press, New York.

All, J. N., Dutcher, J. D., Saunders, M. C., and Brady, U. E. (1985) Prevention strategies for grape root borer (Lepidoptera, Sesiidae) infestations in Concord grape vineyards. *J. Econ. Entomol.* **78**, 666–670.

Allaway, W. H. (1968) Agronomic controls over the environment cycling of trace elements. *Adv. Agron.* **20**, 235–274.

Alleweldt, G., Reustle, G., and Binzel, A. (1991) Die maschinelle Grünveredlung – Entwicklung eines Verfahrens für die Praxis. *Dtsch. Weinbau* **46**, 976–978.

Alley, C. J. (1975) Grapevine propagation. VII. The wedge graft – a modified notch graft. *Am. J. Enol. Vitic.* **26**, 105–108.

Alley, C. J., and Koyama, A. T. (1980) Grapevine propagation, XVI. Chip-budding and T-budding at high level. *Am. J. Enol. Vitic.* **31**, 60–63.

Alley, C. J., and Koyama, A.T. (1981) Grapevine propagation, XIX. Comparison of inverted with standard T-budding. *Am. J. Enol. Vitic.* **32**, 20–34.

Amerine, J. A., and Roessler, E. B. (1958) Field testing of grape maturity. *Hilgardia* **28**, 93–114.

Ames, B. N., and Gold, L. S. (1997) Environmental pollution, pesticides, and the prevention of cancer: misconceptions. *FASEB* **11**, 1041–1052.

Anderson, C. P. (2003) Source-sink balance and carbon allocation below ground in plants exposed to ozone. *New Phytol.* **157**, 213–228.

Angelini, E., Filippin, L., Michielini, C., Bellotto, D., and Borgo, M. (2006) High occurrence of *Flavence dorée* phytoplasma early in the season on grapevines infected with grapevine yellows. *Vitis* **45**, 151–152.

Anonymous. (2003) *A Grower's Guide to Choosing Rootstocks in South Australia.* Phylloxera and Grape Industry Board of South Australia.

Antolín, M. C., Ayari, M., and Sánchez-Díaz, M. (2006) Effects of partial rootzone drying on yield, ripening and berry ABA in potted Tempranillo grapevines with split roots. *Aust. J. Grape Wine Res.* **12**, 13–20.

Araujo, F. J., and Williams, L. E. (1988) Dry matter and nitrogen partitioning and root growth of young field-grown Thompson seedless grapevines. *Vitis* **27**, 21–32.

Archer, E. (1987) Effect of plant spacing on root distribution and some qualitative parameters of vines. In: *Proceedings of the 6th Australian Wine Industry Conference* (T. Lee, ed.), pp. 55–58. Australian Industrial Publishers, Adelaide.

Archer, E., and Strauss, H.C. (1989) The effect of plant spacing on the water status of soil and grapevines. *S. Afr. J. Enol. Vitic.* **10**, 49–58.

Archer, E., and Strauss, H.C. (1990) The effect of vine spacing on some physiological aspects of *Vitis vinifera* L. (cv. Pinot noir) *S. Afr. J. Enol. Vitic.* **11**, 76–87.

Archer, E., and Strauss, H. C. (1985) Effect of plant density on root distribution of three-year-old grafted 99 Richter grapevines. *S. Afr. J. Enol. Vitic.* **6**, 25–30.

Archer, E., Swanepoel, J. J., and Strauss, H. C. (1988) Effect of plant spacing and trellising systems on grapevine root distribution. In: *The Grapevine Root and Its Environment* (J. L. van Zyl, comp.), Technical Communication No. 215, pp. 74–87. Department of Agricultural Water Supply, Pretoria, South Africa.

Avramov, L., Pemovski, D., Lovic, R., Males, P., Ulicevic, M., and Jurcevic, A. (1980) Germ plasm of *Vitis vinifera* in Yugoslavia. In: *Proceedings of the 3rd International Symposium Grape Breeding*, pp. 197–203. University California, Davis.

Bailey, P. T., Ferguson, K. L., McMahon, R., and Wicks, T. J. (1997) Transmission of *Botrytis cinerea* by light brown apple moth larvae on grapes. *Aust. J. Grape Wine Res.* **3**, 90–94.

Baldini, E. (1982) Italian experience of double curtain training systems with special reference to mechanization. In: *Grape and Wine Centennial Symposium Proceedings 1980* (A. D. Webb, ed.), pp. 195–200. University California, Davis.

Barlass, M., and Skene, K. G. M. (1978) *In vitro* propagation of grapevine (*Vitis vinifera* L.) from fragmented shoot apices. *Vitis* **17**, 335–340.

Barlass, M., and Skene, K. G. M. (1987) Tissue culture and disease control. In: *Proceedings of the 6th Australian Wine Industry Technical Conference* (T. Lee, ed.), pp. 191–193. Australian Industrial Publ., Adelaide, Australia.

Baroffio, C. A., Siegfried, W., and Hilber, U.W. (2003) Long-term monitoring for resistance of *Botryotinia fuckeliana* to anilinopyrimidine, phenylpyrrole, and hydroxyanilide fungicides in Switzerland. *Phytopathology* **87**, 662–666.

Barron, A. F. (1900) *Vines and Vine Culture*, 4th edn. Journal of Horticulture, London.

Basler, P., Boller, E. F., and Koblet, W. (1991) Integrated viticulture in eastern Switzerland. *Practic. Winery Vineyard*, **May/June**, 22–25.

Bath, G.I. (1993) Vineyard mechanization. In: *Proceedings of the 8th Australian Wine Industry Technical Conference* (C. S. Stockley *et al.*, eds.), p. 192. Winetitles, Adelaide, Australia.

Bauer, C., Schulz, T. F., Lorenz, D., Eichhorn, K. W., and Plapp. R. (1994) Population dynamics of *Agrobacterium vitis* in two grapevine varieties during the vegetation period. *Vitis* **33**, 25–29.

Baumberger, I. (1988) Regenwürmer – Schützenwerte Nützlinge im Boden. *Weinwirtschaft Anbau* **124**(6), 19–21.

Bavaresco, L. (1997) Relationship between chlorosis occurrence and mineral composition of grapevine leaves and berries. *Communication Soil Sci. Plant Anal.* **28**, 13–21.

Bavaresco, L., and Eibach, R. (1987) Investigations on the influence of N fertilizer on resistance to powdery mildew (*Oidium tuckeri*), downy mildew (*Plasmopara viticola*) and on phytoalexin synthesis in different grape varieties. *Vitis* **26**, 192–200.

Bavaresco, L., and Fogher, C. (1996) Lime-induced chlorosis of grapevine as affected by rootstock and root infection with arbuscular mycorrhiza and *Pseudomonas fluorescens*. *Vitis* **35**, 119–123.

Bavaresco, L., Fregoni, M., and Perino, A. (1994) Physiological aspects of lime-induced chlorosis in some *Vitis* species. I. Pot trial on calcareous soil. *Vitis* **33**, 123–126.

Bavaresco, L., Fregoni, M., and Perino, A. (1995) Physiological aspects of lime-induced chlorosis in some *Vitis* species, II. Genotype response to stress conditions. *Vitis* **34**, 232–234.

Bavaresco, L., and Lovisolo, C. (2000) Effect of grafting on grapevine chlorosis and hydraulic conductivity. *Vitis* **39**, 89–92.

Bavaresco, L., Zamboni, M., and Corazzina, E. (1991) Comportamento produttivo di alcune combinazioni d'innesto di Rondinella e Corvino nel Bardolino. *Rev. Vitic. Enol.* **44**, 3–20.

Bazzi, C., Stefani, E., Gozzi, R., Burr, T. J., Moore, C. L., and Anaclerio, F. (1991) Hot-water treatment of dormant grape cuttings. Its effects on *Agrobacterium tumifaciens* and on grafting and growth of vine. *Vitis* **30**, 177–187.

Beachy, R. N., Loesch-Fries, S., and Tumer, N. E. (1990) Coat proteinmediated resistance against virus infection. *Annu. Rev. Phytopathol.* **28**, 451–474.

Becker, H., and Konrad, H. (1990) Breeding of Botrytis tolerant *V. vinifera* and interspecific wine varieties. In: *Proceedings of the 5th International Symposium Grape Breeding*, p. 302. (Special Issue of *Vitis*) St Martin, Pfalz, Germany.

Bell, C. R., Dickie, G. A., Harvey, W. L. G., and Chan, J. E. Y. F. (1995) Endophytic bacteria in grapevine. *Can. J. Microbiol.* **41**, 46–53.

Bennett, J., Jarvis, P., Creasy, G. L., and Trough, M. C. (2005) Influence of defoliation on overwintering carbohydrate reserves, return bloom, and yield of mature Chardonnay grapevines. *Am. J. Enol. Vitic.* **56**, 386–393.

Berge, A., Delwiche, M., Gorenzel, W. P., and Salmon, T. (2007) Bird control in vineyards using alarm and distress calls. *Am. J. Enol. Vitic.* **58**, 135–143.

Berisha, B., Chen, Y. D., Zhang, G. Y., Xu, B. Y., and Chen, T. A. (1998) Isolation of Pierce's disease bacteria from grapevines in Europe. *Eur. J. Plant Pathol.* **104**, 427–433.

Bernard, M., Horne, P. A., and Hoffmann, A. A. (2001) Preventing restricted spring growth (RSG) in grapevines by successful rust mite control – spray application, timing and eliminating sprays harmful to rust mite predators are critical. *Aust. NZ Grapegrower Winemaker* **452**, 16–17, 19–22.

Bessis, M. R. (1972) Étude en microscopie electronique à balayage des rapports entre l'hôte et le parasite dans le cas de la pourriture grise. *C. R. Acad. Sci. Paris, Sér. D.* **274**, 2991–2994.

Bessis, R., and Fournioux, J. C. (1992) Zone d'abscission en coulure de la vigne. *Vitis* **31**, 9–21.

Bextine, B., Lauzon, C., Potter, S., Lampe, D., and Miller, T. A. (2004) Delivery of a genetically marked *Alcaligenes* sp. to the glassy-winged sharpshooter for use in a paratransgenic control strategy. *Curr. Microbiol.* **48**, 327–331.

Bishop, A. A., Jensen, M. E., and Hall, W. A. (1967) Surface irrigation systems. In: *Irrigation of Agricultural Lands* (R. M. Hagan *et al.*, eds.), pp. 865–884. Agronomy Monograph **11**, American Society of Agronomy, Madison, WI.

Blaich, R., Stein, U., and Wind, R. (1984) Perforation in der Cuticula von Weinbeeren als morphologischer Faktor der Botrytisresistenz. *Vitis* **23**, 242–256.

Blakeman, J. P. (1980) Behaviour of conidia on aerial plant surfaces. In: *The Biology of Botrytis* (J. R. Coley-Smith *et al.*, eds.), pp. 115–151. Academic Press, New York.

Boscia, D., Greif, C., Gugerli, P., Martelli, G. P., Walter, B., and Gonsalves, D. (1995) The nomenclature of grapevine leafroll-associated putative closteroviruses. *Vitis* **34**, 171–175.

Boubals, D. (1954) Les németodes parasites de la vigne. *Prog. Agric. Vitic.* **71**, 141–173.

Boubals, D. (1966) Étude de la distribution et des causes de la résistance au *Phylloxera radicicole* chez les Vitacées. *Ann. Amelior. Plant.* **16**, 145–184.

Boubals, D. (1980) Conduite pour établer une vigne dans un milieu infecté par l'anguillule ou nématode des racines (*Meloidogynae* sp.) *Prog. Agric. Vitic.* **3**, 99.

Boubals, D. (1984) Note on accidents by massive soil applications of grape byproducts. *Prog. Agric. Vitic.* **101**, 152–155.

Boubals, D. (1989) La maladie de Pierce arrive dans les vignobles d'Europe. *Bull. O.I.V.* **62**, 309–314.

Boulay, H. (1982) Absorption différenciée des cépages et des porte-greffes en Languedoc. *Prog. Agric. Vitic.* **99**, 431–434.

Bouquet, A., and Hevin, M. (1978) Green-grafting between Muscadine grapes (*Vitis rotundifolia* Michx.) and bunch grapes (*Euvitis* spp.) as a tool for physiological and pathological investigations. *Vitis* **17**, 134–138.

Brady, N. C. (1974) *The Nature and Properties of Soils*, 8th edn. Macmillan, New York.

Bramley, R. G. V. (2005) Understanding variability in winegrape production systems. 2. Within vineyard variation in quality over several vintages. *Aust. J. Grape Wine Res.* **11**, 33–42.

Bramley, R. G. V., and Hamilton, R. P. (2004) Understanding variability in winegrape production systems. 1. Within vineyard variation in yield over several vintages. *Aust. J. Grape Wine Res.* **10**, 32–45.

Bramley, R. G. V., and Lamb, D. W. (2003) Making sense of vineyard variability in Australia. In: *Precision Viticulture* (R. Ortega and A. Esser, eds.), pp. 35–54. Proceedings of an International Symposium IX Congreso Latinoamericano de Viticultura y Enologia, Santiago, Chile.

Bramley, R., and Proffitt, T. (1999) Managing variability in viticultural production. *Aust. Grapegrower Winemaker* **427**, 11–12, 15–16.

Branas, J. (1974) *Viticulture*. J. Branas, Montpellier, France.

Brancadoro, L., Rabotti, G., Scienza, A., and Zocchi, G. (1995) Mechanisms of Fe-efficiency in roots of *Vitis* spp. in response to iron deficiency stress. *Plant Soil* **171**, 229–234.

Bravdo, B., and Hepner, Y. (1987) Irrigation management and fertigation to optimize grape composition and vine performance. *Acta Hortic.* **206**, 49–67.

Broderick, N. A., Raffa, K. F., and Handelsman, J. (2006) Midgut bacteria required for *Bacillus thuringiensis* insecticidal activity. *PNAS* **103**, 15196–15199.

Broquedis, M., Lespy-Labaylette, P., and Bouard, J. (1995) Role des polyamines dans la coulure et le millerandage. *Act. Colloque C.I.V.B. Bordeaux*, 23–26.

Bubl, W. (1987) Control of stem necrosis with magnesium and micronutrient fertilizers during the period 1983 to 1985. *Mitt. Klosterneuburg* **37**, 126–129.

Buckerfield, J., and Webester, K. (1999) Compost as mulch for vineyards. *Aust. Grapegrower Winemaker* **426a**, 112–114, 117–118.

Bugaret, Y. (1988) L'influence des traitements anti-mildiou et de la récolte méchanique sur l'état sanitaire des bois. *Phytoma* **399**, 42–44.

Burke, D. (1996) Bow picking rod innovation explained. *Aust. Grapegrower Winemaker* **386**, 72–74.

Burr, T. J., Reid, C. L., Splittstoesser, D. F., and Yoskimura, M. (1996) Effect of heat treatments on grape bud mortality and survival of *Agrobacterium vitis in vitro* and in dormant grape cuttings. *Am. J. Enol. Vitic.* **47**, 119–123.

Burr, T. J., Reid, C. L., Yoshimura, M., Monol, E. A., and Bazzi, C. (1995) Survival and tumorigenicity of *Agrobacterium vitis* in living and decaying grape roots and canes in soil. *Plant Dis.* **79**, 677–682.

Campbell, P. A., and Latorre, B. A. (2004) Suppression of grapevine powdery mildew (*Uncinula necator*) by acibenzolar-S-methyl. *Vitis* **43**, 209–210.

Candolfi, M. P., Wermelinger, B., and Boller, E. F. (1993) Influence of the European red mite (*Panonychus ulmi* KOCH) on yield, fruit quality and plant vigour of three *Vitis vinifera* varieties. *Wein Wiss* **48**, 161–164.

Candolfi-Vasconcelos, M. C. (2000) *Phylloxera-Resistant Rootstock for Grapevines*. Northwest Berry & Grape Information Network. http://berrygrape.oregonstate.edu/fruitgrowing/grapes/phyrtsk.htm

Candolfi-Vasconcelos, M. C., and Koblet, W. (1990) Yield, fruit quality, bud fertility and starch reserves of the wood as a function of leaf removal in *Vitis vinifera* – evidence of compensation and stress recovering. *Vitis* **29**, 199–221.

Candolfi-Vasconcelos, M. C., and Koblet, W. (1991) Influence of partial defoliation on gas exchange parameters and chlorophyll content of field-grown grapevines – mechanisms and limitations of the compensation capacity. *Vitis* **30**, 129–141.

Capps, E. R., and Wolf, T. K. (2000) Reduction of bunch stem necrosis of Cabernet Sauvignon by increased tissue nitrogen concentration. *Am. J. Enol. Vitic.* **51**, 319–328.

Carbonneau, A. (1985) Trellising and canopy management for cool climate viticulture. In: *International Symposium Cool Climate Viticulture and Enology* (D. A. Heatherbell *et al.*, eds.), pp. 158–183. OSU Agriculture Experimental Station Technical Publication No. 7628, Oregon State University, Corvallis, OR.

Carbonneau, A. (2004) The lyre trellis for viticulture. Bulletin 4242. http://agspsrv34.agric.wa.gov.au/agency/ pubns/bulletin/bull4242/

Carbonneau, A., and Casteran, P. (1987) Optimization of vine performance by the lyre training systems. In: *Proceedings of the 6th Australian Wine Industry Technical Conference* (T. Lee, ed.), pp. 194–204. Australian Industrial Publishers, Adelaide, Australia.

Cargnello, G., and Piccoli, P. (1978) Vendemmiatrice per vigneti a pergola e a tendone. *Inf. Agrario* **35**, 2813–2814.

Caspari, H. W., Neal, S., Taylor, A., and Trought, M.C.T. (1996) Use of cover crops and deficit irrigation to reduce vegetative vigour of 'Sauvignon Blanc' grapevines in a humid climate. In: *Proceedings of the 4th International Symposium Cool Climate Vitic. Enology* (T. Henick-Kling *et al.*, eds.), pp. II-63–66. New York State Agricultural Experimental Station, Geneva, NY.

Cass, A., Lanyon, D., and Hansen, D. (2004) Mounding and mulching to overcome soil restrictions. *Aust. NZ Grapegrower Winemaker* **485a**, 27–30.

Cass, A., and McGrath, M. C. (2005) Compost benefits and quality for vineyard soils. In: *Proceeding of the Soil Environment and Vine Mineral Nutrition Symposium* (P. Christensen and D. Smart, eds.), pp. 135–143. American Society for Enology and Viticulture, Davis, CA.

Caudwell, A. (1990) Epidemiology and characterization of *flavescence dorée* (FD) and other grapevine yellows. *Agronomie* **10**, 655–663.

Caudwell, A., Larrue, J., Boudon-Padieu, E., and McLean, G. D. (1997) Flavescence dorée elimination from dormant wood of grapevines by hot-water treatment. *Aust. J. Grape Wine Res.* 3, 21–25.

Champagnol, F. (1978) Fertilisation optimale de la vigne. *Prog. Agric. Vitic.* 95, 423–440.

Champagnol, F. (1984) *Elements de Physiologie de la Vigne et du Viticulture Générale.* F. Champagnol, Montpellier, France.

Champagnol, F. (1993) Incidences sur la physiologie de la vigne, de la disposition du feuillage et des opérations en vert de printemps. *Prog. Agric. Vitic.* 110, 295–301.

Chapman, D. M., Roby, G., Ebeler, S. E., Guinard, J.-X., and Matthews, M. A. (2005) Sensory attributes of Cabernet Sauvignon wine made from vines with different water status. *Aust. J. Grape Wine Res.* 11, 339–347.

Chapman, D. M., Thorngate, J. H., Matthews, M. A., and Guinard, J.-X. (2004a) Sensory attributes of Cabernet Sauvignon wines made from vines with different crop yields. *Am. J. Enol. Vitic.* 55, 325–334.

Chapman, D. M., Thorngate, J. H., Matthews, M. A., Guinard, J.-X., and Ebeler, S. E. (2004b) Yield effects on 2-methoxy-3-isobutylpyrazine concentration in Cabernet Sauvignon using a solid phase microextraction gas chromatography/mass spectrometry method. *J. Agric. Food Chem.* 52, 5431–5435.

Choné, X., van Leeuwen, C., Chéry, P., and Ribéreau-Gayon, P. (2001a) Terroir influence on water status and nitrogen status of non-irrigated Cabernet Sauvignon (*Vitis vinifera*) vegetative development, must and wine composition. (Example of a Medoc top estate vineyard) *S. Afr. J. Enol. Vitic.* 22, 8–15.

Choné, X., van Leeuwen, C., Dubourdieu, D., and Gaudillère, J. P. (2001b) Stem water potential is a sensitive indicator of grapevine water status. *Ann. Bot.* 87, 477–483.

Christensen, L. P. (1984) Nutrient level comparisons of leaf petioles and blades in twenty-six grape cultivars over three years (1979 through 1981). *Am. J. Enol. Vitic.* 35, 124–133.

Christensen, L. P., Boggero, J., and Adams, D.O. (1991) The relationship of nitrogen and other nutritional elements to the bunch stem necrosis disorder waterberry. In: *Proceedings of the International Symposium Nitrogen Grapes Wine* (J. M. Rantz, ed.), pp. 108–109. American Society of Enology Viticulture, Davis, CA.

Christensen, L. P., Boggero, J., and Bianchi, M. (1990) Comparative leaf tissue analysis of potassium deficiency and a disorder resembling potassium deficiency in Thompson Seedless grapevines. *Am. J. Enol. Vitic.* 41, 77–83.

Christensen, L. P., Kasimatis, A. N., and Jensen, F. L. (1978) *Grapevine Nutrition and Fertilization in the San Joaquin Valley.* University of California, Division of Agriculture and Natural Resources. Publ. No. 4087.

Christensen, L. P., Kasimatis, A. N., Kissler, J. J., Jensen, F., and Luvisi, D. (1973) Mechanical Harvesting of Grapes for the Winery. Agricultural Extension Publ. No. 2365, University of California, Berkeley.

Christoph, N., Bauer-Christoph, C., Geßner, M., Köhler, H. J., Simat, T. J., and Hoenicke, K. (1998) Bildung von 2 :Aminoacetophenon und Formylaminoacetophenon im Wein durch Einwirkung von schwefliger Säure auf Indol-3–essigsäure. *Wein-Wissenschaft* 53, 79–86.

Clark, M. F., and Adams, A. N. (1977) Characteristics of the microplate method of enzyme-linked immunosorbent assay (ELISA) *J. Gen. Virol.* 34, 475–483.

Clary, C. D., Steinhauer, R. E., Frisinger, J. E., and Peffer, T. E. (1990) Evaluation of machine- vs. hand-harvested Chardonnay. *Am. J. Enol. Vitic.* 41, 176–181.

Clingeleffer, P. R. (1984) Production and growth of minimal pruned Sultana vines. *Vitis* 23, 42–54.

Clingeleffer, P. R. (1993) Development of management systems for low cost, high quality wine production and vigour control in cool climate Australian vineyards. *Aust. Grapegrower Winemaker* 358, 43–48.

Clingeleffer, P. R., and Possingham, J. V. (1987) The role of minimal pruning of cordon trained vines (MPCT) in canopy management and its adoption in Australian viticulture. *Aust. Grapegrower Winemaker*, 280, 7–11.

Cohen, Y. R. (2002) β-Aminobutrytic acid-induced resistance against plant pathogens. *Plant Dis.* 86, 448–457.

Colin, L., Cholet, C., and Geny, L. (2002) Relationship between endogenous polyamines, cellular structure and arrested growth of grape berries. *J. Grape Wine Res.* 8, 101–108.

Collard, B. (1991) Greffe en vert. C'est au point. *Vigne* 14, 26–27.

Collins, C., and Dry, P. (2006) Manipulating fruitset in grapevines. *Aust. NZ Grapegrower Winemaker* 509a, 38–40.

Collins, C., and Rawnsley, B. (2004) National survey reveals primary bud necrosis is widespread. *Aust. NZ Grapegrower Winemaker* 485a, 46–49.

Collins, M., Fuentes, S., and Barlow, S. (2005) Water-use of grapevines to PRD irrigation at two water levels. A case study in North-Eastern Victoria. *Aust. NZ Grapegrower Winemaker* 502, 41–45.

Comas, L. H., Anderson, L. J., Dunst, R. M., Lakso, A. N., Eissenstat, D. M. (2005) Canopy and environmental control of root dynamics in a long-term study of Concord grape. *New Phytol.* 167, 829–840.

Comménil, P., Brunet, L., and Audran, J.-C. (1997) The development of the grape berry cuticle in relation to susceptibility to bunch rot disease. *J. Expt. Bot.* 48, 1599–1607.

Compant, S., Reiter, B., Sessitsch, A., Nowak, J., Clément, C., and Ait Barka, E. (2005) Endophytic colonization of *Vitis vinifera* L. by plant growth-promoting bacterius *Burkholderia* sp. strain PsJN. *Appl. Environ. Microbiol.* 71, 1685–1693.

Conner, A. J., and Thomas, M. B. (1981) Re-establishing plantlets from tissue culture: A review. *Proc. Intl Plant Prop. Soc.* 31, 342–357.

Conradie, W. J. (1988) Effect of soil acidity on grapevine root growth and the role of roots as a source of nutrient reserves. In: *The Grapevine Root and its Environment* (J. L. van Zyl, comp.), pp. 16–29. Technical Communication No. 215. Department of Agricultural Water Supply, Pretoria, South Africa.

Conradie, W. J. (1992) Partitioning of nitrogen in grapevines during autumn and the utilization of nitrogen reserves during the following growing season. *S. Afr. J. Enol. Vitic.* 13, 45–51.

Conradie, W. J., and Saayman, D. (1989) Effects of long-term nitrogen, phosphorus, and potassium fertilization on Chenin Blanc vines I. Nutrient demand and vine performance. *Am. J. Enol. Vitic.* 40, 85–90.

Cook, J. A. (1966) Grape nutrition. In: *Nutrition of Fruit Crops Temperate, Sub-tropical, Tropical* (N. F. Childers, ed.), pp. 777–812. Somerset Press, Somerville, NJ.

Cook, J. A., and Wheeler, D. W. (1978) Use of tissue analysis in viticulture. In: *Soil and Plant-Tissue Testing in California* (H. M. Reisenaure, ed.), pp. 14–16. *Calif. Div. Agric. Sci. Bull.* No. 1879.

Coombe, B. G. (1995) Adoption of a system for identifying grapevine growth stages. *Aust. J. Grape Wine Res.* 1, 100–110.

Coombe, B., and Dry, P. (eds.) (1992) *Viticulture*, Vol. 2. Winetitles, Adelaide, Australia.

Coombe, B. G., and McCarthy, M. G. (1997) Identification and naming of the inception of aroma development in ripening grape berries. *Aust. J. Grape Wine Res.* 3, 18–20.

Cortell, J. M., Halbleib, M., Gallagher, A. V., Righetti, T. J., and Kennedy, J. A. (2005) Influence of vine vigor on grape (*Vitis vinifera* L. cv. Pinot noir) and wine proanthocyanidins. *J. Agric. Food Chem.* 53, 5798–5808.

Cosmo, I., Comuzzi, A., and Polniselli, M. (1958) *Portinesti della Vite*. Agricole, Bologna, Italy.

Costanza, P., Tisseyre, B., Hunter, J. J., and Deloire, A. (2004) Shoot development and non-destructive determination of grapevine (*Vitis vinifera* L.) leaf area. *S. Afr. J. Enol. Vitic* 25, 43–47.

Creaser, M., Wicks, T., Edwards, J., and Pascoe, I. (2002) Identification of grapevine trunk diseases. http://www.phylloxera.com.au/vine%20health/pdfs/Trunkdiseases.pdf.

Creasy, G. L., Crawford, M., Ibbotson, L., Gladstone, P., Kavanagh, J., Sutherland, A. (2006) Mussel shell mulch alters Pinot noir grapevine development and wine qualities. *Aust. NZ Grapegrower Winemaker* 509a, 12, 14, 16–18.

Daane, W. M., Yokata, Glenn Y., Rasmussen, Y. D., Zheng, Y., and Hagen, K. S. (1993) Effectiveness of leafhopper control varies with lacewing release methods. *Cal. Agric.* 47(6), 19–23.

Dahlin, S., Kirchmann, H., Katterer, T., Gunnarsson, S., and Bergstrom, L. (2005) Possibilities for improving nitrogen use from organic materials in agricultural cropping systems. *Ambio.* 34, 288–295.

Darriet, P., Bouchilloux, P., Poupot, C., Bugaret, Y., Clerjeau, M., Sauris, P., Medina, B., and Dubourdieu, D. (2001) Effects of copper fungicide spraying on volatile thiols of the varietal aroma of Sauvignon blanc, Cabernet Sauvignon and Merlot wines. *Vitis* 40, 93–99.

Darriet, P., Pons, M., Henry, R., Dumont, O., Findeling, V., Cartolaro, P., Calonnec, A., and Dubourdieu, D. (2002) Impact odorants contributing to the fungus type aroma from grape berries contaminated by powdery mildew (*Uncinula necator*); incidence of enzymatic activities of the yeast *Saccharomyces cerevisiae*. *J. Agric. Food Chem.* 50, 3277–3282.

De Benedictis, J. A., Granett, J., and Taormino, S. P. (1996) Differences in host utilization by California strains of grape phylloxera. *Am. J. Enol. Vitic.* 47, 373–379.

de Klerk, C. A., and Loubser, J. V. (1988) Relationship between grapevine roots and soil-borne pests. In: *The Grapevine Root and Its Environment* (J. L. van Zyl, comp.), Technical Communication No. 215, pp. 88–105. Department of Agricultural Water Supply, Pretoria, South Africa.

Delas, J. J. (1992) Criteria used for rootstock selection in France. In: *Proceedings of the Rootstock Seminar: A Worldwide Perspective* (J. A. Wolpert *et al.*, eds.), pp. 1–14. American Society of Enology Viticulture, Davis, CA.

Delp, C. J. (1980) Coping with resistance to plant disease. *Plant Dis.* 64, 652–657.

de Souza, C. R., Maroco, J. P., Chaves, M. M., dos Santos, T., Rodriguez, A. S., Lopes, C., and Pereira, J. S. (2004) Effects of partial root drying on the physiology and production of grapevines composition. In: *International Symposium on Irrigation and Water Relations in Grapevine and Fruit Trees* (R. C. Vallone, ed.). *Acta Hortic.* 646, 121–126.

Dimitriadis, E., and Williams, P. J. (1984) The development and use of a rapid analytical technique for estimation of free and potentially volatile monoterpene flavorants of grapes. *Am. J. Enol. Vitic.* 35, 66–71.

Donaldson, D. R., Snyder, R. L., Elmore, C., and Gallagher, S. (1993) Weed control influences vineyard minimum temperatures. *Am. J. Enol. Vitic.* 44, 431–434.

Downton, W. J. S. (1977) Photosynthesis in salt-stressed grapevines. *Aust. J. Plant Physiol.* 4, 183–192.

Dry, P. R. (1987) How to grow cool climate grapes in hot regions. *Aust. Grapegrower Winemaker* 283, 25–26.

Dry, P. R., Loveys, B. R., Botting, D. G., and Düring, H. (1996) Effects of partial root-zone drying on grapevine vigour, yield, composition of fruit and use of water. In: *Proceedings of the 9th Australian Wine Industry Technical Conference* (C. S. Stockley *et al.*, eds.), pp. 128–131. Winetitles, Adelaide, Australia.

Dry, P., Loveys, B. R., Johnstone, A., and Sadler, L. (1998) Grapevine response to root pruning. *Aust. Grapegrower Winemaker* 414a, 73–74, 76–78.

Dry, P. R., Loveys, B. R., McCarthy, M. G., and Stoll, M. (2001) Strategic irrigation management in Australian vineyards. *J. Int. Sci. Vigne Vin* 35, 129–139.

Duke, G. F. (1993) The evaluation of bird damage to grape crops in southern Victoria. In: *Proceedings of the 8th Australian Wine Industry Technical Conference* (C. S. Stockley *et al.*, eds.), p. 197. Winetitles, Adelaide, Australia.

Duran-Vila, N., Juárez, J., and Arregui, J. M. (1988) Production of viroid-free grapevines by shoot tip culture. *Am. J. Enol. Vitic.* 39, 217–220.

Düring, H. (1984) Evidence for osmotic adjustment to drought in grapevines (*Vitis vinifera* L.). *Vitis* 23, 1–10.

Düring, H. (1990) Stomatal adaptation of grapevine leaves to water stress. In: *Proceedings of the 5th International Symposium Grape Breeding*, pp. 366–370. (Special Issue of *Vitis*) St Martin, Pfalz, Germany.

Düring, H., and Lang, A. (1993) Xylem development and function in the grape peduncle, Relations to bunch stem necrosis. *Vitis* 32, 15–22.

Düring, H., Dry, P. R., and Loveys, B. R. (1996) Root signals affect water use efficiency and shoot growth. *Acta Hortic.* 427, 1–13.

Duso, C. (1989) Role of the predatory mites *Amblyseius aberrans* (Oud.) *Typhlodromus pyri* Scheuten and *Amblyseius andersoni* (Chant) (Acari, Phytoseiidae) in vineyards, I. The effects of single and mixed phytoseiid population releases on spider mite densities (Acari, Tetranychidae). *J. Appl. Entomol.* 107, 474–492.

Dutt, E. C., Olsen, M. W., and Stroehlein, J. L. (1986) Fight root rot in the border wine belt. *Wines Vines* 67(3), 40–41.

Düzenli, S., and Ergenoğlu, F. (1991) Studies on the density of stomata of some *Vitis vinifera* L. varieties grafted on different rootstocks trained up various trellis systems. *Doğa-Tr. J. Agric. For.* 15, 308–317.

Eastham, J., and Gray, S. A. (1998) A preliminary evaluation of the suitability of sap flow sensors for use in scheduling vineyard irrigation. *Am. J. Enol. Vitic.* 49, 171–176.

Edwards, J., and Pascoe, I. (2003) Incidence of *Phaeoconiella chlamydospore* infection in symptomless young vines. *Aust. NZ Grapegrower Winemaker* 473a, 90–92.

Eichhorn, K. W., and Lorenz, D. H. (1977) Phänologische Entwicklongsstadien der Rebe. *Nachrichtenbl. Dtsch. Pflanzenschutzdienstes* (*Braunschweig*) 29, 119–120.

Eifert, J., Varnai, M., and Szöke, L. (1982) Application of the EUF procedure in grape production. *Plant Soil* 64, 105–113.

Eifert, J., Varnai, M., and Szöke, L. (1985) EUF-nutrient contents required for optimal nutrition of grapes. *Plant Soil* 83, 183–189.

Eisenbarth, H. J. (1992) Der Einfluss unterschiedlicher Belastungen auf die Ertragsleistung der Rebe. *Dtsch. Weinbau* 47, 18–22.

el Fantroussi, S., Verschuere, L., Verstrawte, W., Top, E. M. (1999) Effect of phenylurea hervicides on soil microbial communities estimated by analysis of 16S rRNA gene fingerprints and community-level physiological profiles. *Appl. Environ. Microbiol.* 65, 982–988.

Elrick, D. E., and Clothier, B. E. (1990) Solute transport and leaching. In: *Irrigation of Agronomic Crops* (B. A. Stewart and D. R. Nielsen, eds.), pp. 93–126. Agronomy Monograph 30, American Society of Agronomy, Crop Science Society of America, and Soil Science Society of America Publication, Madison, WI.

Encheva, K., and Rankov, V. (1990) Effect of prolonged usage of some herbicides in vine plantation on the biological activity of the soil (in Bulgarian). *Soil Sci. Agrochem.* (*Prchvozn. Agrokhim.*) 25, 66–73.

Englert, W. D., and Maixner, M. (1988) Biologische Spinnmibenbekämpfung im Weinbau durch Schonung der Raubmilbe *Typhlodromus pyri*. Schonung und Förderung von Nützlingen.

Schriftenr. Bundesminist. Ernährung, Landwirtschaft Forsten, Reihe A: Angew. Wiss. **365**, 300–306.

English-Loeb, G., Norton, A. P., Gadoury, D. M., Seem, R. C., and Wilcox, W.F. (1999) Control of powdery mildew in wild and cultivated grapes by a tydeid mite. *Biol. Cont.* **14**, 97–103.

Ermolaev, A. A. (1990) Resistance of grape to phylloxera on sandy soils (in Russian). *Agrokhimiya* **2**, 141–151.

Evans, T. (1998) Mapping vineyard salinity using electromagnetic surveys. *Aust. Grapegrower Winemaker* **415**, 20–21.

Ewart, A. J. W., Walker, S., and Botting, D.G. (1993) The effect of powdery mildew on wine quality. In: *Proceedings of the 8th Australian Wine Industry Technical Conference* (C. S. Stockley *et al.*, eds.), p. 201. Winetitles, Adelaide, Australia.

Fabbri, A., Lambardi, M., and Sani, P. (1986) Treatments with CCC and GA3 on stock plants and rootings of cuttings of the grape rootstock 140 Ruggeri. *Am. J. Enol. Vitic.* **37**, 220–223.

Failla, O., Scienza, A., Stringari, G., and Falcetti, M. (1990) Potassium partitioning between leaves and clusters, Role of rootstock. In: *Proceedings of the 5th International Symposium Grape Breeding*, pp. 187–196. (Special Issue of *Vitis*) St Martin, Pfalz, Germany.

Falk, S. P., Gadoury, D. M., Pearson, R. C., and Seem, R. C. (1995) Partial control of grape powdery mildew by the mycoparasite *Ampelomyces quisqualis*. *Plant Dis.* **79**, 483–490.

Fardossi, A., Barna, J., Hepp, E., Mayer, C., and Wendelin, S. (1990) Einfluß von organischer Substanz auf die Nährstoffaufnahme durch die Weinrebe im Gefäßversuch. *Mitt. Klosterneuburg* **40**, 60–67.

Fendinger, A. G., Pool, R. M., Dunst, R. M., and Smith, R. L. (1996) Effect of mechanical thinning of minimally pruned 'Concord' grapevines on fruit composition. In: *Proceedings of the 4th International Symposium Cool Climate Viticulture Enology* (T. Henick-Kling *et al.*, eds.), pp. IV-13–17. New York State Agricultural Experimental Station, Geneva, NY.

Fergusson-Kolmes, L., and Dennehy, T. J. (1991) Anything new under the sun? Not phylloxera biotypes. *Wines Vines* **72**(6), 51–56.

Ficke, A., Gadoury, D. M., Seem, R. C., and Dry, I. B. (2003) Effects on ontogenic resistance upon establishment and growth of *Uncinula necator* on grape berries. *Phytopathology* **93**, 556–563.

Finger, S. A., Wolf, T. K., and Baudoin, A. B. (2002) Effects of horticultural oils on the photosynthesis, fruit maturity, and crop yield of winegrapes. *Am. J. Enol. Vitic.* **53**, 116–124.

Flaherty, D. L., Christensen, L. P., Lanini, W. T., Marois, J. J., Phillips, P. A., and Wilson, L. T. (1992) *Grape Pest Management*, 2nd ed. Publication No. 3343. Division of Agriculture and Natural Resources, University of California, Oakland.

Flaherty, D. L., Jensen, F. L., Kasimatis, A. N., Kido, H., and Moller, W. J. (1982) *Grape Pest Management*. Publication No. 4105. Cooperative Extension, University of California, Oakland.

Foott, J. H. (1987) A comparison of three methods of pruning Gewürztraminer. *Cal. Agric.* **41**(1), 9–12.

Foott, J. H., Ough, C. S., and Wolpert, J. A. (1989) Rootstock effects on wine grapes. *Cal. Agric.* **43**(4), 27–29.

Forneck, A., Kleinmann, S., Blaich, R., and Anvari, S. F. (2002) Histochemistry and anatomy of phylloxera (*Daktulospaira vitifoliae*) nodosities on young roots of grapevine (*Vitis* spp). *Vitis* **41**, 93–98.

Fournioux, J. C. (1997) Influences foliaires sur le développement végétativ de la vigne. *J. Int. Sci. Vigne Vin* **31**, 165–183.

Francis, L., Armstrong, H., Cynkar, W., Kwiatkowski, M., Iland, P., and Williams, P. (1998) A national vineyard fruit composition survey – evaluating the G-G assay. *Aust. Grapegrower Winemaker* **414a**, 51–53, 55–58.

Francis, I. L., Kassara, S., Noble, A C., and Williams, P. J. (1999) The contribution of glycoside precursors to Cabernet Sauvignon and Merlot aroma: sensory and compositional studies. In: *Chemistry of Wine Flavour* (A. L. Waterhouse and S. E. Ebeler, eds.), pp. 13–30. Amer. Chem. Soc. Washington, DC.

Freese, P. (1986) Here's a close look at vine spacing. *Wines Vines* **67**(4), 28–30.

Fregoni, M. (1985) Exigences d'éléments nutritifs en viticulture. *Bull. O.I.V.* **58**, 416–434.

Frost, B. (1996) Eriophytid mites and grape production in Australian vineyards. *Aust. Grapegrower Winemaker* **393**, 25–27, 29.

Fuentes, S., Camus, C., Rogers, G., and Conroy, J. (2004) Precision irrigation in grapevines under RDI and PRD. Wetting Pattern Analysis (WPA©), a novel software tool to visualise real time soil wetting patterns. *Aust. NZ Grapegrower Winemaker* **485a**, 120–122, 125.

Fuller-Perrine, L. D., and Tobin, M. E. (1993) A method for applying and removing bird-exclusion netting in commercial vineyards. *Wildlife Soc. Bull.* **21**, 47–51.

Furkaliev, D. G. (1999) Hybrids between *Vitis* species – the development of modern rootstocks. *Aust Grapegrower Winemaker* **426a**, 18–22, 24–25.

Gadoury, D. M., Pearson, R. C., Riegel, D. G., Seem, R. C., Becker, C. M., and Pscheidt, J. W. (1994) Reduction of powdery mildew and other diseases by over-the-trellis applications of lime sulfur to dormant grapevines. *Plant Dis.* **78**, 83–87.

Gadoury, D. M., Seem, R. C., Ficke, A., and Wilcox, W.F. (2003) Ontogenic resistance to powdery mildew in grape berries. *Phytopathology* **93**, 547–555.

Galet, P. (1971) *Précis d'Ampélographie*. Dehan, Montpellier.

Galet, P. (1979) *A Practical Ampelography—Grapevine Identification* (L. T. Morton, trans. and ed.). Cornell University Press, Ithaca, NY.

Galet, P. (1991) *Précis de Pathologie Viticole*. Dehan, Montpellier.

Gallander, J. F. (1983) Effect of grape maturity on the composition and quality of Ohio Vidal Blanc wines. *Am. J. Enol. Vitic.* **34**, 139–141.

Gargiulo, A. A. (1983) Woody T-budding of grapevines – storage of bud shields instead of cuttings. *Am. J. Enol. Vitic.* **34**, 95–97.

Giraud, T., Fortini, D., Levis, C., Lamarque, C., Leroux, P., LoBouglio, K., and Brygoo, Y. (1999) Two sibling species of the *Botrytis cinerea* complex, *transposa* and *vacuma*, are found in sympatry on numerous host plants. *Phytopathology* **89**, 967–973.

Girbau, T., Stummer, B. E., Pocock, K. F., Baldock, G. A., Scott, E. S., and Waters, E. J. (2004) The effect of *Uncinula necator* (powdery mildew) and *Botrytis cinerea* infection of grapes on the levels of haze-forming pathogenesis-related proteins in grape juice and wine. *Aust J. Grape Wine Res.* **10**, 125–133.

Gishen, M., and Dambergs, B. (1998) Some preliminary trials in the application of scanning near infrared spectroscopy (NIRS) for determining the compositional quality of grape wine and spirits. *Aust. Grapegrower Winemaker* **414a**, 43–45, 47.

Gishen, M., Dambergs, R. G., and Coxxolino, D. (2005) Grape and wine analysis – enhancing the power of spectroscopy with chemometrics. A review of some application in the Australian wine industry. *Aust. J. Grape Wine Res.* **11**, 296–305.

Gishen, M., Iland, P. G., Dambergs, R. G., Esler, M. B., Francis, I. L., Kambouris, A., Johnstone, R. S., and Høj, P. B. (2002) Objective measures of grape and wine quality. In: *11th Australian Wine Industry Technical Conference*, October 7–11, 2001, Adelaide, South Australia (R. J. Blair, P. J. Williams and P. B. Høj, eds.), pp. 188–194. Winetitles, Adelaide, Australia.

Glidewell, S. M., Williamson, B., Goodman, B. A., Chudek, J. A., and Hunter, G. (1997) A NMR microscopic study of grape (*Vitis vinifera*). *Protoplasma* **198**, 27–35.

Gold, L. S., Slone, T. H., Stern, B. R., Manley, N. B., and Ames, B. N. (1992) Rodent carcinogens: setting priorities. *Science* **258**, 261–265.

Goldspink, B. H., and Pierce, C. A. (1993) Post-harvest nitrogen applications, are they beneficial? In: *Proceedings of the 8th Australian Wine Industry Technical Conference* (C.S. Stockley *et al.*, eds.), p. 202. Winetitles, Adelaide, Australia.

Golino, D. A. (1993) Potential interactions between rootstocks and grapevine latent viruses. *Am. J. Enol. Vitic.* **44**, 148–152.

Granett, J., Goheen, A. C., Lider, L. A., and White, J. J. (1987) Evaluation of grape rootstocks for resistance to Type A and Type B grape phylloxera. *Am. J. Enol. Vitic.* **38**, 298–300.

Granett, J., Omer, A. D., Pessereau, P., and Walker, M. A. (1998) Fungal infections of grapevine roots in phylloxera-infested vineyards. *Vitis* **37**, 39–42.

Granett, J., Walker, A., de Benedictis, J., Fong, G., Lin, H., and Weber, E. (1996) California grape phylloxera more variable than expected. *Cal. Agric.* **50**(4), 9–13.

Grantz, D. A., and Williams, N. E. (1993) An empirical protocol for indirect measurement of leaf area index in grape (*Vitis vinifera* L.). *HortScience* **28**, 777–779.

Greer, D. H., and La Borde, D. (2006) Sunburn of grapes affects wine quality. *Aust. NZ Grapegrower Winemaker* **506**, 21–23.

Grenan, S. (1994) Multiplication *in vitro* et caractéristiques juvéniles de la vigne. *Bull. O.I.V.* **67**, 5–14.

Gu, S., Lombard, P. B., and Price, S. F. (1991) Inflorescence necrosis induced by ammonium incubation in clusters of Pinot noir grapes. In: *Proceedings of the International Symposium on Nitrogen in Grapes and Wine* (J. M. Rantz, ed.), pp. 259–261. American Society of Enology Viticulture, Davis, CA.

Gubler, W. D., and Ypema, H. L. (1996) Occurrence of resistance in *Uncinula necator* to triadimefon, myclobutanil, and fenarimol in California grapevines. *Plant Dis.* **80**, 902–909.

Guennelon, R., Habib, R., and Cockborn, A. M. (1979) Aspects particuliers concernant la disponibilité de N, P et K en irrigation localisée fertilisante sur arbres fruitiers. In: *Séminaires sur l'Irrigation Localisée I*, pp. 21–34. L'Institut d'Agronomie de l'Université de Bologne, Italy.

Guerra, B., and Meredith, C. P. (1995) Comparison of *Vitis berlandieri* x *Vitis riparia* rootstock cultivars by restriction fragment length polymorphism analysis. *Vitis* **34**, 109–112.

Guisard, Y., and Tesic, D. (2006) Infrared thermography to evaluate stomatal conductance in grapevines: an overview of the method. *Aust. NZ Grapegrower Winemaker* **509**, 29–32.

Hagan, R. M. (1955) Factors affecting soil moisture-plant growth relations. In: *Report of the 14th International Horticultural Congress*, pp. 82–102. The Hague, The Netherlands.

Hall, A., Lamb, D. W., Holzapfel, B., and Louis, J. (2002) Optically remote sensing applications in viticulture – a review. *J. Grape Wine Res.* **8**, 36–47.

Halleen, F., and Holz, G. (2001) An overview of the biology, epidemiology and control of *Uncinula necator* (powdery mildew) on grapevine with reference to South Africa. *S. Afr. J. Enol. Vitic.* **22**, 111–121.

Hamilton, R. (1997) Hot water treatment of grapevine propagation material. *Aust. Grapegrower Winemaker* **400**, 21–22.

Hamilton, R. P., and Coombe, B. G. (1992) Harvesting of winegrapes. In: *Viticulture*, Vol. 2 (*Practices*) (B. G. Coombe and P. R. Dry, eds.), pp. 302–327. Winetitles, Adelaide, Australia.

Hanson, B. R., and Dickey, G. L. (1993) Field practices affect neutron moisture meter accuracy. *Calif. Agric.* **47**(6), 29–31.

Hardie, W. J., and Martin, S. R. (1990) A strategy for vine growth regulation by soil water management. In: *Proceedings of the 7th Australian Wine Industry Technical Conference* (P. J. Williams et al., eds.), pp. 51–57. Winetitles, Adelaide, Australia.

Harris, J. M., Kriedemann, P. E., and Possingham, J. V. (1968) Anatomical aspects of grape berry development. *Vitis* **7**, 106–119.

Hatzidimitriou, E., Bouchilloux, P., Darriet, P., Bugaret, Y., Clerjeau, M., Poupot, C., Medina, B., and Dubourdieu, D. (1996) Incidence d'une protection viticole anticryptgamique utilisant une formulation cuprique sur le niveau de maturité des raisins et l'arôme variétal des vins de Sauvignon. Bilan de trois années d'expérimentation. *J. Intl Sci. Vigne Vin.* **30**, 133–150.

Hendson, M., Purcell, A. H., Chen, D., Smart, C., Guilhabert, M., and Kirkpatrick, B. (2001) Genetic diversity of Pierce's disease strains and other pathotypes of *Xylella fastidiosa*. *Appl. Environ. Microbiol.* **67**, 895–903.

Hibbert, D., and Horne, P. (2001) IPM: how inter-row cover crops can encourage insect population in vineyards. *Aust. NZ Grapegrower Winemaker* **452**, 76.

Hill, B. L., and Purcell, A. H. (1995) Multiplication and movement of *Xylella fastidiosa* within grapevine and four other plants. *Phytopathology* **85**, 1368–1372.

Hill, G. K., and Kassemeyer, H.-H. (Organizers) (1997) Proceedings of the 2nd International Workshop on Grapevine Downy Powdery Mildew Modeling. *Wein Wiss.* **3–4**, 115–231.

Hofstein, R., Daoust, R. A., and Aeschlimann, J. P. (1996) Constraints to the development of biofungicides: The example AQ10, a new product for controlling powdery mildews. *Entomophaga* **41**, 455–460.

Hoitink, H. A. J., Wang, P., and Changa, C. M. (2002) Role of organic matter in plant health and soil quality. In: 11th Aust. Wine Ind. Tech. Conf. Oct. 7–11, 2001, Adelaide, South Australia. (R. J. Blair, P. J. Williams and P. B. Høj, eds.), pp. 57–60. Winetitles, Adelaide, Australia.

Holzapfel, B. P., and Coombe, B. G. (1998) Interaction of perfused chemicals as inducers and reducers of bunchstem necrosis in grapevine bunches and the effects on the bunchstem concentrations of ammonium ion and abscisic acid. *Aust. J. Grape Wine Res.* **4**, 59–66

Holzapfel, B. P., Rogiers, S., Degaris, K., and Small, G. (1999) Ripening grapes to specification: effect of yield on colour development of Shiraz grapes in the Riverina. *Aust. Grapegrower Winemaker* **428**, 24, 26–28.

Holzapfel, B. P., Smith, J. P., Mandel, R. M., and Keller, M. (2006) Manipulating the postharvest period and its impact on vine productivity of Semillon grapes. *Am. J. Enol. Vitic.* **57**, 148–157.

Howell, G. S. (1987) *Vitis* rootstocks. In: *Rootstocks for Fruit Crops* (R. C. Rom and R. F. Carlson, eds.), pp. 451–472. John Wiley, New York.

Howell, G. S. (2001) Sustainable grape productivity and the growth-yield relationship: A review. *Am. J. Enol. Vitic.* **52**, 165–174.

Howell, G. S., Candolfi-Vasconcelos, M. C., and Koblet, W. (1994) Response of Pinot noir grapevine growth, yield, and fruit composition to defoliation the previous growing season. *Am. J. Enol. Vitic.* **45**, 188–191.

Howell, G. S., Miller, D. P., Edson, C. E., and Striegler, R. K. (1991) Influence of training system and pruning severity on yield, vine size, and fruit composition of Vignoles grapevines. *Am. J. Enol. Vitic.* **42**, 191–198.

Huang, Z., and Ough, C. S. (1989) Effect of vineyard locations, varieties, and rootstocks on the juice amino acid composition of several cultivars. *Am. J. Enol. Vitic.* **40**, 135–139.

Hubáčková, M., and Hubáček, V. (1984) Frost resistance of grapevine buds on different rootstocks (in Russian). *Vinohrad* **22**, 55–56.

Hubbard, S., and Rubin, Y. (2004) The quest for better wine using geophysics. *Geotimes* **49**(8), 30–34.

Huglin, P. (1958) Recherches sur les bourgeons de la vigne. Initiation florale et développement végétatif. Doctoral Thesis, University of Strasbourg, France.

Huisman, S., Hubbard, S., Redman, D., and Annan, P. (2003) Monitoring soil water content with ground-penetrating radar: A review. *Vadose Zone J.* **2**, 476–489.

Hunt, J. S. (1999) Wood-inhabiting fungi: protective management in the vineyard. *Aust. Grapegrower Winemaker* **426a**, 125–126.

Hunter, J. J. (1998) Plant spacing implications for grafted grapevine, II. Soil water, plant water relations, canopy physiology, vegetative and reproductive characteristics, grape composition, wine quality and labour requirements. *S. Afr. J. Enol. Vitic.* **19**, 35–51.

Hunter, J. J., and Le Roux, D. J. (1992) The effect of partial defoliation on development and distribution of roots of *Vitis vinifera* L. cv. Cabernet Sauvignon grafted onto rootstock 99 Richter. *Am. J. Enol. Vitic.* **43**, 71–78.

Hunter, J. J., and Visser, J. H. (1988) The effect of partial defoliation, leaf position and developmental stage of the vine on the photosynthetic activity of *Vitis vinifera* L. cv Cabernet Sauvignon. *S. Afr. J. Enol. Vitic.* **9**(2), 9–15.

Hunter, J. J., and Volschenk, C. G. (2001) Effect of altered canopy: Root volume ratio on grapevine growth compensation. *S. Afr. J. Enol. Vitic.* **22**, 27–31.

Hunter, J. J., Volschenk, C. G., Fouché, G. W., Le Roux, D. J., and Burger, E. (1996) Performance of *Vitis vinifera* L. cv. Pinot noir/99Richter as affected by plant spacing. In: *Proceedings of the 4th International Symposium Cool Climate on Viticulture and Enology* (T. Henick-Kling *et al.*, eds.), pp. I-40–45. New York State Agricultural Experimental Station, Geneva, NY.

Iacono, F., Bertamini, M., Scienza, A., and Coombe, B.G. (1995) Differential effects of canopy manipulation and shading of *Vitis vinifera* L. cv. Cabernet Sauvignon. Leaf gas exchange, photosynthetic electron transport rate and sugar accumulation in berries. *Vitis* **34**, 201–206.

Iland, P. G., Gawel, R., Coombe, B. G. and Henschke, P. M. (1993) Viticultural parameters for sustaining wine style. In: *Proceedings of the 8th Australian Wine Industry Technical Conference* (C. S. Stockley *et al.*, eds.), pp. 167–169. Winetitles, Adelaide, Australia.

Iland, P. G., Gawel, R., McCarthy, M. G., Botting, D. G., Giddings, J., Coombe, B. G., and Williams, P. J. (1996) The glycosyl-glucose assay – its application to assessing grape composition. In: *Proceedings of the 9th Australian Wine Industry Technical Conference* (C. S. Stockley *et al.*, eds.), pp. 98–100. Winetitles, Adelaide, Australia.

Ingels, C. A., Scow, K. M., Whisson, D. A., and Drenovsky, R. E. (2005) Effects of cover crops on grapevines, yield, juice composition, soil microbial ecology and gopher activity. *Am. J. Enol. Vitic* **56**, 19–29.

Intrieri, C., and Poni, S. (1995) Integrated evolution of trellis training systems and machines to improve grape quality and vintage quality of mechanized Italian vineyards. *Am. J. Enol. Vitic.* **46**, 116–127.

Intrieri, C., Silvestroni, O., Rebucci, B., Poni, S., and Filippetti, I. (1996) The effects of row orientation on growth, yield, quality, and dry matter partitioning in Chardonnay wines trained to simple curtain and spur-pruned cordon. In: *Proceedings of the 4th International Symposium on Cool Climate Viticulture and Enology* (T. Henick-Kling *et al.*, eds.), pp. I-10–15. New York State Agricultural Experimental Station, Geneva, NY.

Iriti, M., Rossoni, M., Borgo, M., and Faoro, F. (2004) Benzothiadiazole enhances resveratrol and anthocyanin biosynthesis in grapevine, meanwhile improving resistance to *Botrytis cinerea*. *J. Agric. Food Chem.* **52**, 4406–4413.

Jackson, D. I. (1986) Factors affecting soluble solids, acid, pH, and color in grapes. *Am. J. Enol. Vitic.* **37**, 179–183.

James, D. G., and Rayner, M. (1995) Toxicity of viticultural pesticides to the predatory mites *Amblyseius victoriensis* and *Typhlodromus doreenae*. *Plant Protect. Quart.* **10**, 99–102.

Jashemski, W. F. (1968) Excavation in the Foro Boario at Pompeii: A preliminary report. *Am. J. Archaeol.* **72**, 69–73.

Jashemski, W. F. (1973) The discovery of a large vineyard at Pompeii: University of Maryland Excavations, 1970. *Am. J. Archaeol.* **77**, 27–41.

Jenkins, A. (1991) Review of production techniques for organic vineyards. *Aust. Grapegrower Winemaker* **328**, 133, 135–138, 140–141.

Johnson, L. F., Roczen, D. E., Youkhana, S. K., Nemani, R. R., and Bosch, D.F. (2003) Mapping vineyard leaf area with multispectral satellite imagery. *Computers Electronics Agric.* **38**, 33–44.

Jordan, T. D., Pool, R. M., Zabadal, T. J., and Tomkins, J. P. (1981) *Cultural Practices for Commercial Vineyards*. Miscellaneous Bulletin 111, New York State College of Agriculture, Cornell University, Ithaca, NY.

Jörger, V. (1990) Ökasystem Weinberg aus der Sisht des Bodenlebens, Teil I. Grundlagen der Untersuchungen und Auswirkungen der Herbizide. *Wein Wiss.* **45**, 146–155.

Kalantidis, K. (2004) Grafting the way to the systemic silencing signal in plants. *PloS Biol.* **2**, 1059–1061.

Karagiannidis, N., Velemis, D., and Stavropoulos, N. (1997) Root colonization and spore population by VA-mycorrhizal fungi in four grapevine rootstocks. *Vitis* **36**, 57–60.

Karban, R., English-Loeb, G., and Hougen-Eitzman, D. (1997) Mite vaccinations for sustainable management of spider mites in vineyards. *Ecol. Applic.* **7**, 183–193.

Keller, M. (1997) Can soil management replace nitrogen fertilisation? A European perspective. *Aust. Grapegrower Winemaker* **408**, 23–24, 26–28.

Keller, M., and Koblet, W. (1995) Stress-induced development of inflorescence necrosis and bunch-stem necrosis in *Vitis vinifera* L. in response to environmental and nutritional effects. *Vitis* **34**, 145–150.

Keller, M., Kummer, M., and Vasconcelos, M. C. (2001) Soil nitrogen utilisation for growth and gas exchange by grapevines in response to nitrogen supply and rootstock. *J. Grape Wine Res.* **7**, 2–11.

Keller, M., Mills, L. J., Wample, R. L., and Spayd, S. E. (2005) Cluster thinning effects on three deficit-irrigated *Vitis vinifera* cultivars. *Am. J. Enol. Vitic.* **56**, 91–103.

Kennedy, J. A., Matthews, M. A., and Waterhouse, A. L. (2002) Effect of maturity and vine water status on grape skin and wine flavonoids. *Am. J. Enol. Vitic.* **53**, 268–274.

Kerk, J. L., Beaudette, L. A., Hart, M., Moutoglis, P., Klironomos, J. N., Lee, H., and Trevors, J. T. (2004) Methods of studying soil microbial diversity. *J. Microbiol. Meth.* **58**, 169–188.

Khmel, I. A., Sorokina, T. A., Lemanova, N. B., Lipasova, V. A., Metlitski, O. Z., Burdeinaya, T. V., and Chernin, L. S. (1998) Biological control of crown gall in grapevine and raspberry by two *Pseudomonas* spp. with a wide spectrum of antagonistic activity. *Biocont. Sci. Technol.* **8**, 45–57.

Kidd, C.H. (1987) Canopy management. In: *Proceedings of the 6th Australian Wine Industry Conference* (T. Lee, ed.), pp. 212–213. Australian Industrial Publishers, Adelaide, Australia.

King, P. D., and Rilling, G. (1991) Further evidence of phylloxera biotypes, variations in the tolerance of mature grapevine roots related to the geographical origin of the insect. *Vitis* **30**, 233–244.

Kirchhof, G., Blackwell, J., and Smart, R. E. (1990) Growth of vineyard roots into segmentally ameliorated acid subsoils. *Plant Soil* **134**, 121–126.

Kirchmann, H., Haberhauer, G., Kandeler, E., Sissitsch, A., and Gerzabek, H. H. (2004) Effects of level and quality of organic matter input on soil carbon storage and microbial activity in soil – synthesis of a long-term experiment. *Global Biogeochem.* **18**, GB4011.

Kirchmann, H., and Ryan, M. H. (2004) Nutrients in organic farming – are there advantages from the exclusive use of organic manures and untreated minerals? In: 4th International Crop Science Congress. *New Directions for a Diverse Planet*, Sept. 26 to Oct. 1, Brisbane, Australia, pp. 32.

Kirchmann, H., and Thorvaldsson, G. (2000) Challenging targets for future agriculture. *Eur. J. Agron.* **12**, 145–161.

Kirk, J. L., Beaudette, L. A., Hart, M., Moutoglis, P., Klironomos, J. K., Lee, H., and Trevors, J. T. (2004) Methods of studying soil microbial diversity. *J. Microbiol. Meth.* **58**, 169–188.

Kirk, J. T. O. (1999) Performance of 'frost insurance' canes at Clonakilla during the 1998/1999 season. *Aust. Grapegrower Winemaker* **247**, 61–62.

Kiss, J., Szendrey, L., Schlösser, E. and Kotlár, I. (1996) Application of natural oil in IPM of grapevine with special regard to predatory mites. *J. Environ. Sci. Health.* **B31**, 421–425.

Klausner, S. (1995) Managing animal manures. In: *Organic Grape and Wine Symposium.* 21–22 March, 1995. New York State Agricultural Experimental Station, Geneva, NY. http://www.nyases. cornell.edu/hort/faculty/pool/ organicvitwkshp/%208Klausner.pdf

Kliewer, W. M. (1991) Methods for determining the nitrogen status of vineyards. In: *Proceedings of the International Symposium on Nitrogen in Grapes and Wine* (J. M. Rantz, ed.), pp. 133–147. American Society of Enology Viticulture, Davis, CA.

Kliewer, W. M., Bowen, P., and Benz, M. (1989) Influence of shoot orientation on growth and yield development in Cabernet Sauvignon. *Am. J. Enol. Vitic.* **40**, 259–264.

Kliewer, W. M., and Dokoozlian, N. K. (2000) Leaf area/crop weight ratios of grapevines: influence on fruit composition and wine quality. In: *Proceedings of the ASEV 50th Anniversary Meeting,* Seattle, Washington, June 19–23, 2000, pp. 285–295. American Society for Enology and Viticulture, Davis, CA.

Koblet, W. (1987) Effectiveness of shoot topping and leaf removal as a means of improving quality. *Acta Hortic.* **206**, 141–156.

Koblet, W., Candolfi-Vasconcelos, M.C., Zweifel, W., and Howell, G. S. (1994) Influence of leaf removal, rootstock, training system on yield and fruit composition of Pinot noir grapevines. *Am. J. Enol. Vitic.* **45**, 181–187.

Kobriger, J. M., Kliewer, W. M., and Lagier, S. T. (1984) Effects of wind on water relations of several grapevine cultivars. *Am. J. Enol. Vitic.* **35**, 164–169.

Koenig, R., An, D., and Burgermeister, W. (1988) The use of filter hybridization techniques for the identification, differentiation and classification of plant viruses. *J. Virol. Methods* **19**, 57–68.

Kondo, N. (1995) Harvesting robot based on physical properties of grapevine. *Jpn Agr. Res. Q.* **29**, 171–178.

Kopyt, M., and Ton, Y. (2007) Trunk and berry size monitoring: applications for irrigation. *Aust. NZ Grapegrower Winemaker* **517**, 33–39.

Kotzé, W. A. G. (1973) The influence of aluminum on plant growth. *Decid. Fruit Growth* **23**, 20–22.

Kramer, P. J. (1983) *Water Relations in Plants.* Academic Press, New York, NY.

Krastanova, S., Perrin, M., Barbier, P., Demangeat, G., Cornuet, P., Bardonnet, N., Otten, L., Pinck, L., and Walter, B. (1995) Transformation of grapevine rootstocks with the coat protein gene of grapevine fanleaf nepovirus. *Plant Cell Reports* **14**, 550–554.

Kriedemann, P. E., and Smart, R. E. (1971) Effects of irradiance, temperature, and leaf water potential on photosynthesis of vine leaves. *Photosynthetica* **5**, 6–15.

Krogh, P., and Carlton, W. W. (1982) Nontoxicity of *Botrytis cinerea* strains used in wine production. In: *Grape and Wine Centennial Symposium Proceedings 1980* (A. D. Webb, ed.), pp. 182–183. University of California, Davis.

Kurl, W. R., and Mowbray, G. H. (1984) Grapes. In: *Handbook of Plant Cell Culture* (W. R. Sharp *et al.*, eds.), Vol. 2, pp. 396–434. Macmillan, New York.

Lamb, D. W. (2000) The use of qualitative airborne multispectral imaging for managing agricultural crops – a case study in southeastern Australia. *Aust. J. Expt. Agric.* **40**, 727–738.

Lang, A., and Thorpe, M. R. (1989) Xylem, phloem and transpirational flow in a grape, application of a technique for measuring the volume of attached fruits to high resolution using Archimedes' principle. *J. Exp. Bot.* **40**, 1069–1078.

Larcher, W., Häckel, H., and Sakai, A. (1985) *Handbuch der Pflanzenkrankheiten*, Vol. 7. Parey, Berlin.

Lebon, G., Duchêne, E., Brun, O., Magné, C., and Clément, C. (2004) Flower abscission and inflorescence carbohydrates in sensitive and non-sensitive cultivars of grapevine. *Sex. Plant Reprod.* **17**, 71–79.

Löhnertz, O. (1991) Soil nitrogen and uptake of nitrogen in grapevines. In: *Proceedings of the International Symposium on Nitrogen in Grapes and Wine* (J. A. Rantz, ed.), pp. 1–11. American Society of Enology Viticulture, Davis, CA.

Löhnertz, O., Schaller, K., and Mengel, K. (1989) Nährstaffdynamik in Reben, III. Mitteiling, Stickstoffkonzwntration und Verlauf der Aufnahme in der Vegetation. *Wein Wiss.* **44**, 192–204.

Lombard, P., Price, S., Wilson, W., and Watson, B. (1988) Grass cover crops in vineyards. In: *Proceedings of the 2nd International Symposium for Cool Climate Viticulture and Oenology* (R. E. Smart *et al.*, eds.), pp. 152–155. New Zealand Society of Viticulture and Oenology, Auckland, New Zealand.

Long, Z. R. (1987) Manipulation of grape flavour in the vineyard, California, North Coast region. In: *Proceedings of the 6th Australian Wine Industry Conference* (T. Lee, ed.), pp. 82–88. Australian Industrial Publishers, Adelaide, Australia.

Lopes, C., and Pinto, P. A. (2005) Easy and accurate estimation of grapevine leaf area with simple mathematical models. *Vitis* **44**, 55–61.

Lorito, M., Woo, S. L., Ferandez, I. G., Colucci, G., Harman, G. E., Pintor-Toro, J. A., Filippone, E., Muccifora, S., Lawrence, C. B., Zoina, A., Tuzun, S., and Scala, F. (2001) Genes from mycoparasitic fungi as a source for improving plant resistance to fungal pathogens. *Proc. Natl Acad. Sci.* **95**, 7860–7865.

Loubser, J. T., van Aarde, I. M. F., and Höppner, G. F. J. (1992) Assessing the control potential of aldicarb against grapevine phylloxera. *S. Afr. J. Enol. Vitic.* **13**, 84–86.

Loué, A., and Boulay, N. (1984) Effets des cépages et des portegreffes sur les diagnostics de nutrition minérale sur la vigne. In: *6th Colloqium Internationale des Nutrition en Plantes*, pp. 357–364. Martin-Prevel, Montpellier, France.

Louw, P. J. E., and van Huyssteen, L. (1992) The effect of planting holes on the root distribution of grapevines. *Aust. Grapegrower Winemaker* **340**, 7–15.

Loveys, B., Stoll, M., Dry, P., and McCarthy, M. (1998) Partial rootzone drying stimulates stress responses in grapevine to improve water use efficiency while maintaining crop yield and quality. *Aust. Grapegrower Winemaker* **414a**, 108–110, 113.

Lovisolo, C., and Schubert, A. (1998) Effects of water stress on vessel size and xylem hydraulic conductivity in *Vitis vinifera* L. *J. Expt. Bot.* **49**, 693–700.

Ludvigsen, K. (1999) Rootstocks – a complex and difficult choice. *Aust. Grapegrower Winemaker* **426a**, 36–39.

Lunt, I., Hubbard, S., and Rubin, Y. (2005) Soil moisture content estimation using ground-penetrating radar reflection data. *J. Hydrol.* **307**, 254–269.

Mabrouk, N., and Carbonneau, A. (1996) Une méthode simple de détermination de la surface folaire de la vigne (*Vitis vinifera* L.) *Prog. Agric. Vitic.* **113**, 392–398.

Magarey, P.A., Biggins, L.T., Wachtel, M.F., and Wicks, T.J. (1993) Soft solutions for foliage diseases in Australian vineyards. In: *Proceedings of the 8th Australian Wine Industry Technical Conference* (C.S. Stockley *et al.*, eds), pp. 207–208. Winetitles, Adelaide, Australia.

Maixner, M., Pearson, R. C., Boudon-Padieu, E., and Caudwell, A. (1993) *Scaphoideus titanus*, a possible vector of grapevine yellows in New York. *Plant Dis.* **77**, 408–413.

Marais, J. (1996) Fruit environment and prefermentation practices for manipulation of monoterpene, norisoprenoid and pyraxine flavorants. In: *Proceedings of the 4th International Symposium on Cool Climate Viticulture and Enology* (T. Henick-Kling et al., eds.), pp. V-40–48. New York State Agricultural Experimental Station, Geneva, NY.

Marais, J., Calitz, F., and Haasbroek, P. D. (2001) Relationship between microclimatic data, aroma component concentrations and wine quality parameters in the prediction of Sauvignon blanc wine quality. *S. Afr. J. Enol. Vitic.* 22, 22–26.

Marais, J., van Wyk, C. J., and Rapp, A. (1992a) Effect of sunlight and shade on norisoprenoid levels in maturing Weisser Riesling and Chenin blanc grapes and Weisser Riesling wines. *S. Afr. J. Enol. Vitic.* 13, 23–32.

Marais, J., van Wyk, C. J., and Rapp, A. (1992b) Effect of storage time, temperature and region on the levels of 1,1,6-trimethyl-1, 2-dihydronaphthalene and other volatiles, and on quality of Weisser Riesling wines. *S. Afr. J. Enol. Vitic.* 13, 33–44.

Marois, J. J., Bledsoe, A. M., Bostock, R. M., and Gubler, W. D. (1987) Effects of spray adjuvants on development of *Botrytis cinerea* on *Vitis vinifera* berries. *Phytopathology* 77, 1148–1152.

Marois, J. J., Nelson, J. K., Morrison, J. C., Lile, L. S., and Bledsoe, A. M. (1986) The influence of berry contact within grape clusters on the development of *Botrytis cinerea* and epicuticular wax. *Am. J. Enol. Vitic.* 37, 293–295.

Martinelli, L., Poletti, V., Bragagna, P., and Poznanski, E. (1996) A study on organogenic potential in the *Vitis* genus. *Vitis* 35, 159–161.

Martinson, T., Williams, L., III, and English-Loeb, G. (2001) Compatibility of chemical disease and insect management practices used in New York vineyards with biological control by *Anagrus* spp. (Hymenoptera: Mymaridae), parasitoids of *Erythroneura* leafhoppers. *Biol. Control* 22, 227–234.

Marzorati, M., Alma, A., Sacchi, L., Pajoro, M., Palermo, S., Brusetti, L., Raddadi, N., Balloi, A., Tedeschi, R., Clementi, E., Corona, S., Quaglino, F., Bianco, P. A., Beninati, T., Bandi, C., and Daffonchio, D. (2006) A novel *Bacteroidetes* symbiont localized in *Scaphoideus titanus*, the insect vector of Flavescence dorée in *Vitis vinifera*. *Appl. Environ. Microbiol.* 72, 1467–1475.

Masa, A. (1989) Affinité biochimique entre le greffon du cultivar Albariño (*Vitis vinifera* L.) et différents porte-greffes. *Connaiss. Vigne Vin* 23, 207–213.

Matthews, M. A., and Anderson, M. M. (1989) Reproductive development in grape (*Vitis vinifera* L), responses to seasonal water deficits. *Am. J. Enol. Vitic.* 40, 52–59.

Matthews, M. A., Anderson, M. M., and Schultz, H. R. (1987) Phenological and growth responses to early and late season water deficiency in Cabernet Franc. *Vitis* 26, 147–160.

McCarthy, M. G. (1997) The effect of transient water deficit on berry development of cv. Shiraz (*Vitis vinifera* L.). *Aust. J. Grape Wine Res.* 3, 102–108.

McCarthy, M. G. (1998) Irrigation management to improve winegrape quality – nearly 10 years on. *Aust. Grapegrower Winemaker* 414a, 65–66, 68, 70–71.

McCarthy, M. G., and Cirami, R. M. (1990) Minimal pruning effects on the performance of selections of four *Vitis vinifera* cultivars. *Vitis* 29, 85–96.

McCarthy, M. G., and Neidner, K. (1997) Summer drought and vine rootstocks – an update. *Aust. Grapegrower Winemaker* 402a, 68–70.

McCarthy, M. G., and Nicholas, P. R. (1989) Terpenes – a new measure of grape quality? *Aust. Grapegrower Winemaker* 297, 10.

McCarthy, M. G., Iland, P. G., Coombe, B. G., and Williams, P. J. (1996) Manipulation of the concentration of glycosyl-glucose in Shiraz grapes with irrigation management. In: *Proceedings of the 9th Australian Wine Industry Technical Conference* (C. S. Stockley et al., eds.), pp. 101–104. Winetitles, Adelaide, Australia.

McDonald, C. (2005) Bunch thinning: when is the best time, and is it worthwhile? *Aust. NZ Grapegrower Winemaker.* 497a, 41–43, 45–46.

McGourty, G. T., and Reganold, J. P. (2005) Managing vineyards soil organic matter with cover crops. In: *Proceeding of the Soil Environment and Vine Mineral Nutrition Symposium* (P. Christensen and D. Smart, eds.), pp. 145–151. American Society for Enology and Viticulture, Davis, CA.

McLeod, R. W., and Steel, C. C. (1999) Effects of brassica-leaf green manures and crops on activity and reproduction of *Meliodogyne javanica*. *Nematology* 1, 613–624.

Mengel, K., Bubl, W., and Scherer, H. W. (1984) Iron distribution in vine leaves with HCO^{3-} induced chlorosis. *J. Plant Nutr.* 7, 715–724.

Meyer, J. L., Snyder, M. J., Valenzuela, L. H., Harris, A., and Strohman, R. (1991) Liquid polymers keep drip irrigation lines from clogging. *Cal. Agric.* 45(1), 24–25.

Miller, D. P., Howell, G. S., and Striegler, R. K. (1993) Reproductive and vegetative response of mature grapevines subjected to differential cropping stresses. *Am. J. Enol. Vitic.* 44, 435–440.

Milne, D. (1988) Vine responses to soil water. In: *Proceedings of the 2nd International Symposium Cool Climate Viticulture and Oenology* (R. E. Smart et al., eds.), pp. 57–58. New Zealand Society of Viticulture and Oenology, Auckland, New Zealand.

Moller, T., Akerstroud, K., and Massoud, T. (1997) Toxin-producing species of *Penicillium* and the development of mycotoxins in must and homemade wine. *Nat. Toxins* 5, 86–89.

Mondy, N., Charrier, B., Fermaud, M., Pracros, P., and Coriocostet, M. F. (1998) Mutualism between a phytopathogenic fungus (*Botrytis cinerea*) and a vineyard pest (*Lobesia botrana*). Positive effects on insect development and oviposition behaviour. *C. R. Acad Sci. Serie III, Life Sci.* 321, 665–671.

Mondy, N., Pracros, P., Fermaud, M., and Coriostet, M. F. (1998) Olfactory and gustatory behaviour by larvae of *Lobesia botrana* in response to *Botrytis cinerea*. *Entomol. Experiment. Appl.* 88, 1–7.

Monteiro, S. M., Barakat, M., Piçarra-Pereira, M. A., Teixeira, A. R., and Ferreira, R. B. (2003) Osmotin and thaumatin from grape: a putative general defense mechanism against pathogenic fungi. *Phytopathology* 93, 1505–1512.

Morris, J. R., Main, G. L., and Oswald, O. L. (2004) Flower cluster and shoot thinning for crop control in French-American hybrid grapes. *Am. J. Enol. Vitic* 55, 423–426.

Morris, J. R., Sims, C. A., Bourque, J. E., and Oakes, J. L. (1984) Influence of training system, pruning severity, and spur length on yield and quality of six French-American hybrid grape cultivars. *Am. J. Enol. Vitic.* 35, 23–27.

Mottard, G., Nespoulus, J., and Marcout, P. (1963) Les port-greffes de la vigne. *Bulletin des Informations d'Agriculture, Paris*, 182.

Mpelasoka, B. S., Schachtman, D. P., Treeby, M. T., and Thomas, M. R. (2003) A review of potassium nutrition in grapevines with special emphasis on berry accumulation. *Aust J. Grape Wine Res.* 9, 154–168.

Muehleback, J., and Bracher, P. (1998) Using electronics to keep birds away from grapes. *Aust. Grapegrower Winemaker* 417, 65–67.

Mullins, M. G. (1990) Applications of tissue culture to the genetic improvement of grapevines. In: *Proceedings of the 5th International Symposium Grape Breeding*, pp. 399–407. (Special Issue of *Vitis*) St Martin, Pfalz, Germany.

Mundy, D., and Agnew, R. (2004) Investigation of the effects of mulch on growth and root development of young grapevines in Marlborough. *Aust. NZ Grapegrower Winemaker* 486, 91–92.

Munkvold, G., and Marois, J. J. (1990) Relationship between xylem wound response in grapevines and susceptibility to *Eutypa lata*. (abstract) *Phytopathology* **80**, 973.

Mur, G., and Branas, J. (1991) La maladie de vieux bois: Apoplexie et eutypiose. *Prog. Agric. Vitic.* **5**, 108–114.

Murat, M.-L. (2005) Recent findings on rosé wine aromas. Part 1: identifying aromas studying the aromatic potential of grapes and juice. *Aust. NZ Grapegrower Winemaker* **497a**, 64–65, 69, 71, 73–74, 76.

Murat, M.-L., and Dumeau, F. (2005) Recent findings on rosé wine aromas. Part II: optimising winemaking techniques. *Aust. NZ Grapegrower Winemaker* **499**, 49–52, 54–55.

Murisier, F., and Aerny, J. (1994) Influence du niveau de rendement de la vigne sur les réserves de la plante et sur la chlorose. Rôle du porte-greffe. *Rev. Suisse Vitic. Arboricult. Horticult.* **26**, 281–287.

Murphy, B. C., Rosenheim, J. A., and Granett, J. (1996) Habitat diversification for improving biological control: Abundance of *Anagrus epos* (Hymenoptera, Mymaridae) in grape vineyards. *Environ. Entomol.* **25**, 495–504.

Nicholas, P. R., Chapman, A. P., and Cirami, R. M. (1992) Grapevine propagation. In: *Viticulture*, Vol. 2 (*Practices*) (B. G. Coombe and P. R. Dry, eds.), pp. 1–22. Winetitles, Adelaide, Australia.

Nicholas, P., Magarey, P., and Wachtel, M. (1994) *Diseases and Pests*. Winetitles, Adelaide, Australia.

Nicholson, C. (1990) 'Birebent' graft is hailed as a breakthrough in viticulture. *Wines Vines* **71**(9), 16–18.

Northover, J. (1987) Infection sites and fungicidal prevention of *Botrytis cinerea* bunch rot of grapes in Ontario. *Can. J. Plant Path.* **9**, 129–136.

O'Brien, E., and Dietrich, D. R. (2005) Ochratoxin A: the continuing enigma. *Crit. Rev. Toxicol.* **35**, 33–60.

O'Conner, B. P. (ed.) (1987) *New Zealand Agrichemical Manual*, 2nd edn. Agpress/Novasearch, Wellington/Manawatu, New Zealand.

Ohr, H. D., Sims, J. J., Grech, N. M., Becker, J. O., and McGiffen, M. E., Jr. (1996) Methyl iodide, an ozone-safe alternative to methyl bromide as a soil fumigant. *Plant Dis.* **80**, 731–735.

Oliva, J., Navarro, S., Barba, G., Navarro, G., and Şalinas, M. R. (1999) Effect of pesticide residues on the aromatic composition of red wines. *J. Agric. Food Chem.* **47**, 2830–2836.

Olson, R. A., and Kurtz, L. T. (1982) Crop nitrogen requirements, utilization, and fertilization. In: *Nitrogen in Agricultural Soils* (F. J. Stevenson, ed.), Vol. 22, pp. 567–604. American Society of Agronomy, Crop Science Society of America, and Soil Science Society of America, Madison, WI.

Omer, A. D., Thaler, J. S., Granett, J., and Karban, R. (2000) Jasmonic acid induced resistance in grapevines to a root and leaf feeder. *J. Econ. Entomol.* **93**, 840–845.

O'Neill, T. M., Elad, Y., Shtienberg, D., and Cohen, A. (1996) Control of grapevine grey mould with *Trichoderma harzianum* T39. *Biocont. Sci. Technol.* **6**, 139–146.

Ough, C. S. (1991) Influence of nitrogen compounds in grapes on ethyl carbamate formation in wines. In: *Proceedings of the International Symposium on Nitrogen in Grapes and Wine* (J. M. Rantz, ed.), pp. 165–171. American Society of Enology and Viticulture, Davis, CA.

Ough, C. S., and Amerine, M. A. (1988) *Methods for Analysis of Musts and Wines*. Wiley, New York.

Ough, C. S., and Berg, H. W. (1979) Powdery mildew sensory effect on wine. *Am. J. Enol. Vitic.* **30**, 321.

Padgett-Johnson, M., Williams, L. E., and Walker, M. A. (2003) Vine water relations, gas exchange, and vegetative growth of seventeen *Vitis* species grown under irrigated and nonirrigated conditions in California. *J. Am. Soc. Hortic. Sci.* **128**, 269–276.

Panten, K., and Bramley, R. (2006) A new approach to viticultural experimentation. *Aust. NZ Grapegrower Winemaker* **509a**, 7–8, 10–11.

Parat, C., Chaussod, R., Lévèque, J., Dousset, S., and Andreux, F. (2002) The relationship between copper accumulated in vineyard calcareous soils and soil organic matter and iron. *Eur. J. Soil Sci.* **53**, 663–669.

Pàstena, B. (1972) *Trattato di Viticultura Italiana*. Palermo, Italy.

Pearn, P., and Bath, G. I. (1992) Control of grapevine suckers – a review. *Aust. Grapegrower Winemaker* **346**, 45–49.

Pearson, R. C., and Gadoury, D. M. (1987) Cleistothecia, the source of primary inoculum for grape powdery mildew in New York. *Phytopathology* **77**, 1509–1608.

Pearson, R. C., and Goheen, A. C. (eds.) (1988) *Compendium of Grape Diseases*. APS Press, St Paul, MI.

Pearson, R. C., Pool, R. M., and Jubb, G. L., Jr. (1988) Pesticide toxicity. In: *Compendium of Grape Diseases* (R. C. Pearson and A. C. Goheen, eds.), pp. 69–71. APS Press, St Paul, MI.

Peech, M. (1961) Lime requirements *vs* soil pH curves for soils of New York State. Mimeograph. Department of Agronomy, Cornell University, Ithaca, NY.

Perret, P., and Koblet, W. (1984) Soil compaction induced iron-chlorosis in grape vineyards, presumed involvement of exogenous soil ethylene. *J. Plant Nutr.* **7**, 533–539.

Perret, P., and Weissenbach, P. (1991) Schwermetalleintrag in den Rebberg aus organischen Düngern. *Schweiz. Z. Obst- Weinbau* **127**, 124–130.

Peterson, J., and Smart, R. (1975) Foliage removal effects on Shiraz grapevines. *Am. J. Enol. Vitic.* **26**, 119–124.

Petit, E., and Gubler, W. D. (2005) The endomycorrhizal fungus *Glomus intraradices*: A potential biocontrol of *Cylindrocarpon* black foot disease of grapevine. *Am. J. Enol. Vitic.* **56**, 300A.

Petrie, P. R., and Clingeleffer, P. R. (2006) Crop thinning (hand *versus* mechanical) grape maturity and anthocyanin concentration: outcomes from irrigated Cabernet sauvignon (*Vitis vinifera* L.) in a warm climate. *Aust. J. Grape Wine Res.* **12**, 21–29.

Petrie, P. R., Cooley, N. M., and Clingeleffer, P. R. (2004) The effect of post-véraison water deficit on yield components and maturation of irrigated Shiraz (*Vitis vinifera* L.) in the current and following season. *Aust. J. Grape Wine Res.* **10**, 203–215.

Pezet, R., and Pont, V. (1986) Infection florale et latence de *Botrytis cinerea* dans les grappes de *Vitis vinifera* (var. Gamay) *Rev. Suisse Vitic. Arboric. Hortic.* **18**, 317–322.

Pickering, G., Lin, J., Riesen, R., Reynolds, A., Brindle, I., and Soleas, G. (2004) Influence of *Harmonia axyridis* on the sensory properties of white and red wine. *Am. J. Enol. Vitic.* **55**, 153–159.

Pilkington, L. J., Irwin, N. A., Boyd, E. A., Hoddle, M. S., Tripitsyn, S. V., Carey, B. G., Jones, W. A., and Morgon, D. J. W. (2005) Introduced parasitic wasps could control glassy-winged sharpshooter. *Calif. Agric.* **59**, 223–228.

Pinamonti, F. (1998) Compost mulch effects on soil fertility, nutritional status and performance of grapevine. *Nutr. Cycl. Agroecosyst.* **51**, 239–248.

Plan, C., Anizan, C., Galzy, P., and Nigond, J. (1976) Remarques sur la relation degré alcoolique-rendement chez la vigne. *Vitis* **15**, 236–242.

Pocock, K. F., Hayasaka, Y., Peng, Z., Williams, P. J., and Waters, E. J. (1998) The effect of mechanical harvesting and long-distance transport on the concentration of haze-forming proteins in grape juice. *Aust. J. Grape Wine Res.* **4**, 23–29.

Pongrácz, D. P. (1978) *Practical Viticulture*. David Philip, Cape Town, South Africa.

Pongrácz, D. P. (1983) *Rootstocks for Grape-vines*. David Philip, Cape Town, South Africa.

Poni, S., and Intrieri, C. (1996) Physiology of grape leaf ageing as related to improved canopy management and grape quality. In: *Proceedings of the 9th Australian Wine Industry Technical Conference* (C. S. Stockley *et al.*, eds.), pp. 113–122. Winetitles, Adelaide, Australia.

Poni, S., Casalini, L., Bernizzoni, F., Civardi, S., and Intrieri, C. (2006) Effects of early defoliation on shoot photosynthesis, yield components, and grape composition. *Am. J. Enol. Vitic.* **57**, 397–407.

Pool, R. (2000) Training systems for New York vineyards. http://www.nysaes.cornell.edu/hort/faculty/ pool/train/trainandrootstocks.html

Pool, R. M., Kasimatis, A. N., and Christensen, L. P. (1988) Effects of cultural practices on disease. In: *Compendium of Grape Diseases* (R. C. Pearson and A. C. Goheen, eds.), pp. 72–73. APS Press, St Paul, MI.

Pouget, R. (1987) Usefulness of rootstocks for controlling vine vigour and improving wine quality. *Acta Hortic.* **206**, 109–118.

Pour Nikfardjam, M. S., Gaál, K., and Teszlák, P. (2005a) Influence of grapevine flower treatment with gibberellic acid (GA₃) on indole-3-acetic acids (IAA) contents of white wine. *Mitt. Klosterneuburg* **55**, 114–117.

Pour Nikfardjam, M. S., Gaál, K., Teszlák, P., Kreck, P., and Dietrich, H. (2005b) Influence of grapevine flower treatment with gibberellic acid (GA₃) on *o*-aminoacetophenone (AAP) content and sensory properties of white wine. *Mitt. Klosterneuburg* **55**, 184–190.

Prince, J. P., Davis, R. E., Wolf, T. K., Lee, I.-M., Mogen, B.D., Dally, E. L., Bertaccini, A., Credi, R., and Barba, M. (1993) Molecular detection of diverse mycoplasmalike organisms (MLOs) associated with grapevine yellows and their classification with aster yellows, X-disease, and elm yellows MLOs. *Phytopathology* **83**, 1130–1137.

Proffitt, T., and Malcolm, A. (2005) Zonal vineyard management through airborne remote sensing. *Aust. NZ Grapegrower Winemaker* **502**, 22–24, 25–26.

Pu, X., and Goodman, R. N. (1993) Tumor formation by *Agrobacterium tumifaciens* is suppressed by *Agrobacterium radiobacter* HLB-2 on grape plants. *Am. J. Enol. Vitic.* **44**, 249–254.

Quinlan, J. Q., and Weaver, R. J. (1970) Modification of the pattern of the photosynthate movement within and between shoots of *Vitis vinifera* L. *Plant Physiol.* **46**, 527–530.

Rabe, E. (1990) Stress physiology, the functional significance of the accumulation of nitrogen-containing compounds. *J. Hortic. Sci.* **65**, 231–243.

Redak, R. A., Purcell, A. H., Lopes, J. R. S., Blua, M. J., Mizell, R. F. III, and Andersen, P. C. (2004) The biology of xylem fluid-feeding insect vectors of *Xylella fastidiosa* and their relation to disease epidemiology. *Annu. Rev. Entomol.* **49**, 243–270.

Redl, H., and Bauer, K. (1990) Prüfung alternativer Mittel auf deren Wirkung gegenüber *Plasmopara viticola* und deren Einfluß auf das Ertragsgeschehen unter österreichischen Weinbaubedingungen. *Mitt. Klosterneuburg, Rebe Wein Obstbau Früchteverwertung* **40**, 134–138.

Remund, U., and Boller, E. (1994) Die Reblaus – wieder aktuell? *Obst- u Weinbau* **10**, 242–244.

Reustle, G., Harst, M., and Alleweldt, G. (1995) Plant regeneration of grapevine (*Vitis* sp.) protoplasts isolated from embryogenic tissue. *Plant Cell Rep.* **15**, 238–241.

Reynolds, A. G. (1996) Assessing and understanding grape maturity. In: *Twenty-fifth Annual Wine Industry Workshop* (T. Henick-Kling, ed.), pp. 75–88. New York State Agricultural Experimental Station, Cornell University, Geneva, NY.

Reynolds, A. G., Pool, R. M., and Mattick, L. R. (1986) Effect of shoot density and crop control on growth, yield, fruit composition, and wine quality of 'Seyval Blanc' grapes. *J. Am. Soc. Hortic. Sci.* **111**, 55–63.

Reynolds, A. G., Veto, L. J., Sholberg, P. L., Wardle, D. A., and Haag, P. (1996) Use of potassium silicate for the control of powdery mildew [*Uncinula necator* (Schwein) Burrill] in *Vitis vinifera* L. cultivar Bacchus. *Am. J. Enol. Vitic.* **47**, 421–428.

Reynolds, A. G., and Wardle, D. A. (1988) Canopy microclimate of Gewürztraminer and monoterpene levels. In: *Proceedings of the 2nd International Symposium Cool Climate Viticulture and Oenology* (R. E. Smart *et al.*, eds.), pp. 116–122. New Zealand Society of Viticulture and Oenology, Auckland, New Zealand.

Reynolds, A. G., and Wardle, D. A. (1996) Impact of viticultural practices on grape monoterpenes, and their relationship to wine sensory response. In: *Proceedings of the 4th International Symposium Cool Climate Viticulture Enology* (T. Henick-Kling *et al.*, eds.), pp. V-1–17. New York State Agricultural Experimental Station, Geneva, NY.

Ribéreau-Gayon, J., and Peynaud, E. (1971) *Traité d'Ampélologie. Sciences et Techniques du Vin*, Vols 1–2. Dunod, Paris.

Ribéreau-Gayon, J., Ribéreau-Gayon, P., and Seguin, G. (1980) *Botrytis cinerea* in enology. In: *The Biology of Botrytis* (J. R. Coley-Smith *et al.*, eds.), pp. 251–274. Academic Press, London.

Richards, A., Pudney, S., and McCarthy, M. (2005) Growing grape to a specification with minimal water inputs. *Aust. NZ Grapegrower Winemaker* **493**, 28–30.

Roby, G., Harbertson, J. F., Adams, D. A., and Matthews, M. A. (2004) Berry size and vine water deficits as factors in winegrape composition: Anthocyanins and tannins. *Aust J. Grape Wine Res.* **10**, 100–107.

Rod, P. (1977) Observations sur le ruissellement dans le vignoble vaudois. *Bull. Assoc. Romande Protect. Eau Air* **82**, 41–45.

Rodgers, G. (1999) Liquid calcium supplied through trickle drastically reduces sodium concentration in leaves. *Aust. Grapegrower Winemaker* **426a**, 129.

Rodriguez-Lovelle, B., and Gaudillère, J.-P. (2002) Carbon and nitrogen partitioning in either fruiting or non-fruiting grapevines: Effects of nitrogen limitation before and after véraison. *J. Grape Wine Res.* **8**, 86–94.

Roggero, J. P., Coen, S., and Ragonnet, B. (1986) High performance of liquid chromatography survey on changes in pigment content in ripening grapes of Syrah. An approach to anthocyanin metabolism. *Am. J. Enol. Vitic.* **37**, 77–83.

Rogiers, S. Y., Smith, J. A., White, R., Keller, M., Holzapfel, B. P., and Virgona, J. M. (2001) Vascular function in berries of *Vitis vinifera* (L) cv. Shiraz. *J. Grape Wine Res.* **7**, 47–51.

Roubelakis-Angelakis, K. A., and Kliewer, W. M. (1983) Ammonia assimilation in *Vitis vinifera* L., II. Leaf and root glutamine synthetase. *Vitis* **22**, 299–305.

Rowhani, A., Maningas, M. A., Lile, L. S., Daubert, S. D., and Golino, D. A. (1995) Development of a detection system for viruses of woody plants based on PCR analysis of immobilized virions. *Phytopathology* **85**, 347–352.

Ruel, J. J., and Walker, M. A. (2006) Resistance to Pierce's disease in *Muscadinia rotundifolia* and other native grape species. *Am. J. Enol. Vitic.* **57**, 158–165.

Ruhl, E. H. (1989) Uptake and distribution of potassium by grapevine rootstocks and its implication for grape juice pH of scion varieties. *Aust. J. Expt. Agric.* **29**, 707–712.

Saayman, D. (1981) Wingerdvoeding. In: *Wingerdbou in Suid-Afrika* (J. Burger and J. Deist, eds.), pp. 343–383. Oenology and Viticulture Research Institute, Stellenbosch, South Africa.

Saayman, D. (1982) Soil preparation studies, II. The effect of depth and method of soil preparation and organic material on the performance of *Vitis vinifera* (var. Colombar) on Clovelly/Hutton soil. *S. Afr. J. Enol. Vitic.* **3**, 61–74.

Sackenheim, R., Weltzien, H. C., and Kast, W. K. (1994) Effects of microflora composition in the phyllosphere on biological regulation of grapevine fungal diseases. *Vitis* **33**, 235–240.

Safran, R. M., Bravdo, B., and Bernstein, Z. (1975) L'irrigation de la vigne par goutte à goutte. *Bull. O.I.V.* **531**, 405–429.

San Romão, M. V., and Silva Alemáo, M. F. (1986) Premières observations sur les activités enzymatiques developpées dans le moût de raisins par certains champignons filamenteux. *Connaiss. Vigne Vin* **20**, 39–52.

Sauer, A. E., and Karg, G. (1998) Variables affecting pheromone concentration in vineyards treated for mating disruption of grape vine moth *Lobesia botrana*. *J. Chem. Ecol.* **24**, 289–302.

Scheck, H. J., Vasquez, S. J., Fogle, D., and Gubler, W. D. (1998) Young grapevine decline in California. *Practic. Winery Vineyard.* **May/June**, 32–38.

Scheidegger, A. M., and Sparks, D. L. (1996) A critical assessment of sorption and desorption mechanisms at the soil mineral/water interface. *Soil Sci.* **161**, 813–831.

Schepers, J. S., and Saint-Fort, R. (1988) Comparison of the potentially mineralized nitrogen using electroultrafiltration and four other procedures. In: *Proceedings of the 3rd International EUF Symposium*, pp. 441–450. Suedzucker AG, Mannheim, Germany.

Schilder, A. M. C., Gillett, J. M., Sysak, R. W., and Wise, J. C. (2002) Evaluation of environmentally friendly products for control of fungal diseases of grapes. In: *Tenth International Conference on Cultivation Technique and Phytopathological Problems of Organic Fruit-Growing and Viticulture Proceedings*. 4–7 February, 2002 (Fördergemeinschaft Ökologisher Obstbau e.e., ed.), pp. 163–167. Weinberg, Germany.

Schmitz, V., Charlier, L., Roehrich, R., and Stockel, J. (1997) Étude de mécanisme de la confusion sexuelle pour l'eudémis de la vigne *Lobesia botrana* Den & Schiff. (Lep., Tortricidae), IV. Quel est le rôle de la fixation de la phéromone par le feuillage? *J. Appl. Entomol.* **121**, 41–46.

Schneider, C. (1989) Introduction à l'écophysiologie viticole. Application aux systèmes de conduite. *Bull. O.I.V.* **62**, 498–515.

Scholl, W., and Enkelmann, R. (1984) Zum Kupfergehalt von Weinbergsböden. *Landwirtsch. Forschung* **37**, 286–297.

Schroth, M. N., McCain, A. H., Foott, J. H., and Huisman, O. C. (1988) Reduction in yield and vigor of grapevine caused by crown gall disease. *Plant Dis.* **72**, 241–245.

Schubert, A., Lovisolo, C., and Pererlunger, E. (1996) Sap flow and hydraulic conductivity of grapevine shoots trained to the upward or downward position. In: *Proceedings of the 4th International Symposium on Cool Climate Viticulture and Enology* (T. Henick-Kling *et al.*, eds.), pp. II-39–42. New York State Agricultural Experimental Station, Geneva, NY.

Schultz, H. R., Kraml, S., Werwitzke, U., Zimmer, T., and Schmid, J. (2000) Adaptation and utilization of minimal pruning systems. In: *Proceedings of the ASEV 50th Anniversary Meeting*, Seattle, Washington, June 19–23, 2000, pp. 185–190. American Society for Enology and Viticulture, Davis, CA.

Schultz, H. R., and Matthews, M. (1993) Resistance to water transport in shoots of *Vitis vinifera* L. *Plant Phys.* **88**, 718–724.

Schwenk, S., Altmayer, B., and Eichhorm, K. W. (1989) Untersuchungen zur Bedeutung toxischer Stoffwechselprodukte des Pilzes *Trichothecium roseum* Link ex Fr. für den Weinbau. *Z. Lebensm. Unters. Forsch.* **188**, 527–530.

Scienza, A., and Boselli, M. (1981) Fréquence et caractéristiques biométriques des stomates de certains porte-greffes de vigne. *Vitis* **20**, 281–292.

Scienza, A., Failla, O., and Romano, F. (1986) Untersuchungen zur sortenspezifischen Mineralstoffaufnahme bei Reben. *Vitis* **25**, 160–168.

Sengonca, C., and Leisse, N. (1989) Enhancement of the egg parasite *Trichogramma semblidis*, AURIV., Hym., Trichogrammatidae, for control of both grapevine moth species in the Ahr valley (abstract). *J. Appl. Entomol.* **107**, 41–45.

Sforza, R., Boudon-Padieu, E., and Greif, C. (2003) New mealybug species vectoring grapevine leafroll-associated viruses-1 and −3 (GLRaV-1 and −3). *Europ. J. Plant Pathol.* **109**, 975–981.

Shani, U., and Ben-Gal, A. (2005) Long-term response of grapevines to salinity: osmotic effects and ion toxicity. *Am. J. Enol. Vitic.* **56**, 148–154.

Shaulis, N. (1982) Responses of grapevines and grapes to spacing of and within canopies. In: *Grape and Wine Centennial Symposium Proceedings* (A. D. Webb, ed.), pp. 353–361. University of California, Davis.

Shaulis, N., Amberg, H., and Crowe, D. (1966) Response of Concord grapes to light, exposure and Geneva Double Curtain training. *Proc. Am. Soc. Hortic. Sci.* **89**, 268–280.

Shaulis, N., and Smart, R. (1974) Grapevine canopies, management, microclimate and yield responses. In: *Proceedings of the 19th International Horticulture Congress*, pp. 254–265. Warsaw, Poland.

Shoseyov, O. (1983) Out of season grape production of one year old cuttings (in Hebrew, with English summary) M.Sc. Thesis. Faculty of Agriculture, Hebrew University Jerusalem.

Sinclair, R. G., Campbell, K. N., and James, E. J. (1993) Current bird control research. In: *Proceedings of the 8th Australian Wine Industry Technical Conference* (C. S. Stockley *et al.*, eds.), p. 212. Winetitles, Adelaide, Australia.

Sinclair, R. and Hathaway, S. (2005) *A Grower's Guide to Managing Birds*. Phylloxera and Grape Industry Board of South Australia.

Singleton, V. L. (1972) Effects on red wine quality of removing juice before fermentation to stimulate variation in berry size. *Am. J. Enol. Vitic.* **23**, 106–113.

Skinner, P. W., Cook, J. A., and Matthews, M. A. (1988) Responses of grapevine cvs. Chenin blanc and Chardonnay to phosphorus fertilizer applications under phosphorus-limited soil conditions. *Vitis* **27**, 95–109.

Skinner, P. W., and Matthews, M. A. (1989) Reproductive development in grape (*Vitis vinifera*) under phosphorus-limited conditions. *Sci. Hortic.* **38**, 49–60.

Skinner, P. W., and Matthews, M. A. (1990) A novel interaction of magnesium translocation with the supply of phosphorus to roots of grapevine, *Vitis vinifera*. *Plant Cell Environ.* **13**, 821–926.

Smart, R. E. (1973) Sunlight interception by vineyards. *Am. J. Enol. Vitic.* **24**, 141–147.

Smart, R. E. (1974) Aspects of water relations of the grapevine (*Vitis vinifera*). *Am. J. Enol. Vitic.* **25**, 84–91.

Smart, R. E. (1982) Vine manipulation to improve wine grape quality. In: *Grape and Wine Centennial Symposium Proceedings* (A. D. Webb, ed.), pp. 362–375. University of California, Davis.

Smart, R. E. (1985) Principles of grapevine canopy microclimate manipulation with implications for yield and quality. A review. *Am. J. Enol. Vitic.* **36**, 230–239.

Smart, R. E. (1988) Shoot spacing and canopy light microclimate. *Am. J. Enol. Vitic.* **39**, 325–333.

Smart, R. E. (1994a) Dancing girls in California vineyards – Introducing the Smart-Dyson Ballerina. *Practic. Winery Vineyard* **Nov/Dec**, 46–48.

Smart, R. E. (1994b) Introducing the Smart-Dyson trellis. *Aust. Grapegrower Winemaker* **365**, 27–28.

Smart, R. E., and Coombe, B. G. (1983) Water relations of grapevines. In: *Water Deficits and Plant Growth* (T. T. Kozlowski, ed.), Vol. 7, pp. 137–195. Academic Press, NY.

Smart, R. E., Dick, J., and Gravett, I. (1990a) Shoot devigoration by natural means. In: *Proceedings of the 7th Australian Wine Industry Technical Conference* (P. J. Williams *et al.*, eds.), pp. 58–65. Winetitles, Adelaide, Australia.

Smart, R. E., Dick, J. K., Gravett, I. M., and Fisher, B. M. (1990b) Canopy management to improve grape yield and wine quality – principles and practices. *S. Afr. J. Enol. Vitic.* **11**, 3–18.

Smart, R. E., Dick, J. K., and Smith, S. M. (1990c) A trellis for vigorous vineyards. *Wines Vines* **71**(6), 32–36.

Smart, R. E., and Robinson, M. (1991) *Sunlight into Wine. A Handbook for Winegrape Canopy Management*. Winetitles, Adelaide, Australia.

Smart, R. E., Robinson, J. B., Due, G. R., and Brien, C. J. (1985) Canopy microclimate modification for the cultivar Shiraz, II. Effects on must and wine composition. *Vitis* **24**, 119–128.

Smart, R. E., and Robinson, M. (1991) *Sunlight into Wine. A Handbook for Winegrape Canopy Management.* Winetitles, Adelaide, Australia.

Smart, R. E., and Smith, S. M. (1988) Canopy management, identifying the problems and practical solutions. In: *Proceedings of the 2nd International Symposium for Cool Climate Viticulture and Oenology* (R. E. Smart *et al.*, eds.), pp. 109–115. New Zealand Society of Viticulture and Oenology, Auckland, New Zealand.

Smith, F. (2004) Never mind the quality, feel the weight. *Aust. N Z Grapegrower Winemaker* **488**, 25–27.

Smith, S. (1991) Retrofitting a GDC. In: *Sunlight into Wine* (R. Smart and M. Robinson, eds.), p. 55. Winetitles, Adelaide, Australia.

Smith, S., Codrington, I. C., Robertson, M., and Smart, R. (1988) Viticultural and oenological implications of leaf removal for New Zealand vineyards. In: *Proceedings of the 2nd International Symposium for Cool Climate Viticulture and Oenology* (R. E. Smart *et al.*, eds.), pp. 127–133. New Zealand Society of Viticulture and Oenology, Auckland, New Zealand.

Solari, C., Silvestroni, O., Giudici, P., and Intrieri, C. (1988) Influence of topping on juice composition of Sangiovese grapevines (V. *vinifera* L.). In: *Proceedings of the 2nd International Symposium for Cool Climate Viticulture and Oenology* (R. E. Smart *et al.*, eds.), pp. 147–151. New Zealand Society of Viticulture and Oenology, Auckland, New Zealand.

Somers, T. C., and Pocock, K. F. (1986) Phenolic harvest criteria for red vinification. *Aust. Grapegrower Winemaker* **256**, 24–30.

Staats, M., van Baarlen, P., van Kan, J. A. L. (2005) Molecular phylogeny of the plant pathogenic genus *Botrytis* and the evolution of host specificity. *Mol. Biol. Evol.* **22**, 333–346.

Stapleton, J. J., Barnett, W. W., Marois, J. J., and Gubler, W. D. (1990) Leaf removal for pest management in wine grapes. *Cal. Agric.* **44**(4), 15–17.

Staub, T. (1991) Fungicide resistance, practical experience with antiresistance strategies and the role of integrated use. *Annu. Rev. Phytopathol.* **29**, 421–442.

Steinhauer, R. E., Lopez, R., and Pickering, W. (1980) Comparison of four methods of variety conversion in established vineyards. *Am. J. Enol. Vitic.* **31**, 261–264.

Stern, W. M., and Federici, B. A. (1990) Granulosis virus, Biological control of western grapeleaf skeletonizer. *Cal. Agric.* **44**(3), 21–22.

Stevens, R. M., Harvey, G., and Aspinall, D. (1995) Grapevine growth of shoots and fruit linearly correlate with water stress indices based on root-weighted soil matric potential. *Aust. J. Grape Wine Res.* **1**, 58–66.

Stevens, R. M., Harvey, G., and Davies, G. (1996) Separating the effects of foliar and root salt uptake on growth and mineral composition of four grapevine cultivars on their own roots and on Ramsey rootstock. *J. Am. Soc. Hortic. Sci.* **121**, 569–575.

Stevenson, D. S., Neisen, G. H., and Cornelsen, A. (1986) The effect of woven plastic mulch, herbicides, grass sod, and nitrogen on 'Foch' grapes under irrigation. *HortScience* **21**, 439–441.

Stevenson, J. F., Matthews, M. A., Greve, L. C., Labavitch, J. M., and Rost, T. L. (2004) Grapevine susceptibility to Pierce's disease II: Progression of anatomical symptoms. *Am. J. Enol. Vitic.* **55**, 238–245.

Stevenson, J. F., Matthews, M. A., and Rost, T. L. (2005) The developmental anatomy of Pierce's disease symptoms in grapevines: green islands and matchsticks. *Plant Dis.* **89**, 543–548.

Stobbs, L. W., Potter, J. W., Killins, R., and van Schagen, J. G. (1988) Influence of grapevine understock in infection of de Chaunac scion by tomato ringspot virus. *Can. J. Plant Pathol.* **10**, 228–231.

Stockle, C. O., and Dugas, W. A. (1992) Evaluating canopy temperature-based indices for irrigation scheduling. *Irrig. Sci.* **13**, 31–38.

Stoll, M., Loveys, B., and Dry, P. (2000) Whole plant integration and agriculture exploitation. Hormonal changes induced by partial rootzone drying of irrigated grapevine. *J. Exp. Botany* **51**, 1627–1634.

Stummer, B. E., Francis, I. L., Zanker, T., Lattey, K. A., and Scott, E. S. (2005) Effects of powdery mildew on the sensory properties and composition of Chardonnay juice and wine when grape sugar ripeness in standardised. *Aust. J. Grape Wine Res.* **11**, 66–76.

Süle, S., Lehoczky, J., Jenser, G., Nagy, P., and Burr, T. J. (1995) Infection of grapevine roots by *Agrobacterium vitis* and *Meloidogyne hapla*. *J. Phytopathol.* **143**, 169–171.

Summers, P. (1985) Managing cover. It requires care just like any other crop. *Calif. Grape Grower* (**June**), 26–27.

Syntichaki, K. M., Loulakakis, K. A., and Roubelakis-Angelakis, K. A. (1996) The amino-acid sequence similarity of plant glutamate dehydrogenase to the extremophilic archaeal enzyme conforms to its stress-related function. *Gene* **168**, 87–92.

Szychowski, J. A., Goheen, A. C., and Semancik, J. S. (1988) Mechanical transmission and rootstock reservoirs as factors in the widespread distribution of viroids in grapevines. *Am. J. Enol. Vitic.* **39**, 213–216.

Szychowski, J. A., McKenry, M. V., Walter, M. A., Wolpert, J. A., Credi, R., and Semancik, J. S. (1995) The vein-banding disease syndrome: A synergistic reaction between grapevine viroids and fanleaf virus. *Vitis* **34**, 229–232.

Tassie, E., and Freeman, B. M. (1992) Pruning. In: *Viticulture*, Vol. 2 (*Practices*) (B. G. Coombe and P. R. Dry, eds.), pp. 66–84. Winetitles, Adelaide, Australia.

Teszlák, P., Gaál, K., and Pour Nikfardjam, M. S. (2005) Influence of grapevine flower treatment with gibberellic acid (GA_3) on phenol content of *Vitis vinifera* L. wine. *Anal. Chim. Acta* **543**, 275–281.

Thomas, C. S., Gubler, W. D., Silacci, M. W., and Miller, R. (1993) Changes in elemental sulfur residues on Pinot noir and Cabernet Sauvignon grape berries during the growing season. *Am. J. Enol. Vitic.* **44**, 205–210.

Thomas, C. S., Marois, J. J., and English, J. T. (1988) The effects of wind speed, temperature, and relative humidity on development of aerial mycelium and conidia of *Botrytis cinerea* on grape. *Phytopathology* **78**, 260–265.

Thorne, E. T., Stevenson, J. F., Rost, T. L., Labavitch, J. M., and Matthews, M. A. (2006) Pierce's disease symptoms: comparison with symptoms of water deficit and the impact of water deficits. *Am. J. Enol. Vitic.* **57**, 1–11.

Tominaga, T., Peyrot des Gachons, C., and Dubourdieu, D. (1998) A new type of flavor precursor in *Vitis vinifera* L. cv. Sauvignon blanc: *S*-cysteine conjugates. *J. Agric. Food Chem.* **46**, 5215–5219.

Ton, Y., and Kopyt, M. (2004) Phytomonitoring in realization of irrigation strategies for wine grapes. *Acta Horticulturae* **652**, 167–173.

Torregrosa, L., and Bouquet, A. (1996) Adventitious bud formation and shoot development from *in vitro* leaves of *Vitis × Muscadinia* hybrids. *Plant Cell, Tiss. Org. Cult.* **45**, 245–252.

Tregoat, O., Ollat, N., Grenier, G., and van Leeuwen, C. (2001) Étude comparative de la précision et de la rapidité de mise en oeuvre de différented méthodes d'estimation de la surface foliaire de la vigne. *J. Int. Sci. Vigne Vin* **35**, 31–39.

Trought, M. C. (1996) The New Zealand terroir: sources of variation in fruit composition in New Zealand vineyards. In: *Proceedings of the 4th International Symposium on Cool Climate Viticulture and Enology* (Henick-Kling *et al.*, eds.), pp. I-23–27. New York State Agricultural Experimental Station, Geneva, NY.

Truog, E. (1946) Soil reaction influence on availability of plant nutrients. *Proc. Soil Sci. Soc. Am.* **11**, 305–308.

Ureta, F., Boidron, J. N., and Bouard, J. (1982) Influence of dessèche-ment de la rafle on wine quality. In: *Grape and Wine Centennial Symposium Proceedings* (A. D. Webb, ed.), pp. 284–286. University of California, Davis.

van den Ende, B. (1984) The Tatura trellis. A system of growing grapevines for early and high production. *Am. J. Enol. Vitic.* 35, 82–87.

van den Ende, B., Jerie, P. H., and Chalmers, D. J. (1986) Training Sultana vines on the Tatura trellis for early and high production. *Am. J. Enol. Vitic.* 37, 304–305.

van der Westhuizen, J. H. (1980) The effect of black plastic mulch on growth, production and root development of Chenin blanc vines under dryland conditions. *S. Afr. J. Enol. Vitic.* 1, 1–6.

van Huyssteen, L. (1988a) Grapevine root growth in response to soil tillage and root pruning practices. In: *The Grapevine Root and its Environment* (J. L. van Zyl, comp.), Technical Communication No. 215, pp. 44–56. Department of Agricultural Water Supply, Pretoria, South Africa.

van Huyssteen, L. (1988b) Soil preparation and grapevine root dis-tribution – a qualitative and quantitative assessment. In: *The Grapevine Root and Its Environment* (J. L. van Zyl, comp.), pp. 1–15. Technical Communication No. 215, Department of Agricultural Water Supply, Pretoria, South Africa.

van Huyssteen, L. (1990) The effect of soil management and fertiliza-tion on grape composition and wine quality with special reference to South African conditions. In: *Proceeding of the 7th Australian Wine Industry Technical Conference* (P. J. Williams *et al.*, eds.), pp. 16–25. Winetitles, Adelaide, Australia.

van Slyke, L. L. (1932) *Fertilizers and Crop Production.* Orange Judd, New York.

van Zyl, J. L. (1984) Response of Colombar grapevines to irriga-tion as regards quality aspects and growth. *S. Afr. J. Enol. Vitic.* 5, 19–28.

van Zyl, J. L. (1988) Response of grapevine roots to soil water regimes and irrigation systems. In: *The Grapevine Root and Its Environment* (J. L. van Zyl, comp.), Technical Communication No. 215, pp. 30–43. Department of Agricultural Water Supply, Pretoria, South Africa.

van Zyl, J. L., and van Huyssteen, L. (1980) Comparative studies on wine grapes on different trellising systems, I. Consumptive water use. *S. Afr. J. Enol. Vitic.* 1, 7–14.

van Zyl, J. L., and van Huyssteen, L. (1984) Soil water manage-ment towards optimum grape yield and quality under conditions of limited or no irrigation. In: *Proceedings of the 5th Australian Wine Industry Technical Conference* (T. H. Lee and T. C. Somers, eds.), pp. 25–66. Australian Wine Research Institute, Adelaide, Australia.

van Zyl, J. L., and van Huyssteen, L. (1987) Root pruning. *Decid. Fruit Grow.* 37(1), 20–25.

Vasconcelos, M. C., and Castagnoli, S. (2000) Leaf canopy structure and vine performance. *Am. J. Enol. Vitic.* 51, 390–396.

Virgona, J. M., Smith, J. P., and Holzapfel, B. P. (2003) Scions influ-ence apparent transpiration efficiency of *Vitis vinifera* (cv. Shiraz) rather than rootstocks. *Aust J. Grape Wine Res.* 9, 183–185.

Volder, A., Smart, D. R., Bloom, A. J., and Eissenstat, D. M. (2005) Rapid decline in nitrate uptake and respiration with age in fine lateral roots of grape: implications for root efficiency and com-petitive effectiveness. *New Phytologist* 165, 493–502.

Waite, H., Crocker, J., Fletcher, G., Wright, P., and deLaine, A. (2001) Hot water treatment in commercial nursery practice – an overview. *Aust. Grapegrower Winemaker* 449a, 39–443.

Waldner, M., and Marais, J. (2002) Impact aroma components in South African red wines: A preliminary study. *Wynboer* Dec. 5pp (http://www.wynboer.co.za/recentarticles/1202impact.php3).

Walker, G. (2001) Lesion nematode pathogenicity and a new nema-ticide and fungicide for potential use to Australian viticulture. *Aust. NZ Grapegrower Winemaker* 452, 58, 60.

Walker, M. A., Wolpert, J. A., and Weber, E. (1994) Field screening of grape rootstock selection for resistance to fanleaf degeneration. *Plant Dis.* 78, 134–136.

Walker, R. R., Blackmore, D. H., Clingeleffer, P. R., Godden, P., Francis, L., Valente, P., and Robinson, E. (2003) Salinity effects on vines and wines. *Bull. O.I.V.* 76, 200–227.

Walker, R. R., Blackmore, D. H., Clingeleffer, P. R., Kerridge, G. H., Rühl, E. H., and Nicholas, P. R. (2005) Shiraz berry size in rela-tion to seed number and implications for juice and wine composi-tion. *Aust. J. Grape Wine Res.* 11, 2–8.

Walpole, M., Whiting, J., and Code, G. (1993) An evaluation of mulches as substitutes for herbicides during vineyard estab-lishment. In: *Proceedings of the 8th Australian Wine Industry Technical Conference* (C. S. Stockley *et al.*, eds.), pp. 100–104. Winetitles, Adelaide, Australia.

Walter, B. (1991) Sélection de la vigne, Le dépistage des maladies de la vigne transmissibles par les bois et plants. *Bull. O.I.V.* 64, 691–701.

Walton, V. M., and Pringle, K. L. (2004) A survey of mealybugs and associated natural enemies in vineyards in the Western Cape Province, South Africa. *S. Afr. J. Enol. Vitic.* 25, 23–25.

Wapshere, A. J., and Helm, K. F. (1987) Phylloxera and *Vitis*, an experimentally testable coevolutionary hypothesis. *Am. J. Enol. Vitic.* 38, 216–222.

Weaver, R. J. (1976) *Grape Growing.* John Wiley, New York.

Weber, E., de Benedictis, J., Smith, R. and Granett, J. (1996) Enzone does little to improve health of phylloxera-infested vineyards. *Cal. Agric.* 50(4), 19–23.

Wechert, M., Hutton, R., Rouse, E., and Lamont, R. (2005) Herbicide use affects vineyard soil microbes. *Aust. NZ Grapegrower Winemaker* 497, 39–40.

Weinstein, L. H. (1984) Effects of air pollution on grapevines. *Vitis* 23, 274–303.

Weiss, A., and Allen, L. H., Jr. (1976a) Air-flow patterns in vineyard rows. *Agric. Meteorol.* 16, 329–342.

Weiss, A., and Allen, L. H., Jr. (1976b) Vertical and horizontal air flow above rows of a vineyard. *Agric. Meteorol.* 16, 433–452.

West, D. W., and Taylor, J. A. (1984) Response of six grape cultivars to the effects of high salinity and rootzone water logging. *J. Am. Soc. Hort. Sci.* 109, 844–851.

Weston, D. P., You, J. C., and Lydy, M. J. (2004) Distribution and toxicity of sediment-associated pesticides in agriculture-dominated water bodies of California's Central Valley. *Environ. Sci. Technol.* 38, 2752 – 2759.

Weyand, K. M., and Schultz, H. R. (2006) Light interception, gass exchange and carbon balance of different canopy zones of mini-mally and cane-pruned field-grown Riesling grapevines. *Vitis* 45, 105–114.

Whiting, J. R. (1988) Influences of rootstocks on yield, juice com-position and growth of Chardonnay. *Proceedings of the 2nd International Symposium Cool Climate Viticulture Oenology,* Jan. 11–15, 1988 (R. E. Smart *et al.*, eds.), pp. 48–50. New Zealand Society of Viticulture and Oenology, Auckland, New Zealand.

Whiting, J. R., and Noon, D. (1990) Wine quality assessment of Pinot noir selections. In: *Proceeding of the 7th Australian Wine Industry Technical Conference* (P. J. Williams *et al.*, eds.), p. 261. Winetitles, Adelaide, Australia.

Whiting, J. R., and Noon, D. (1993) Influence of rootstocks on muscat flavour in Brown Muscat juice and wine. In: *Proceeding of the 8th Australian Wine Industry Technical Conference* (C. S. Stockley *et al.*, eds.), p. 217. Winetitles, Adelaide, Australia.

Williams, L. E. (1987) Growth of 'Thompson Seedless' grapevine, II. Nitrogen distribution. *J. Am. Soc. Hort. Sci.* **112**, 330–333.

Williams, L. E., and Matthews, M. A. (1990) Grapevines. In: *Irrigation of Agricultural Crops* (B. A. Stewart and D. R. Nielsen, eds.), pp. 1019–1055. Agronomy Series Monograph No. 30. American Society of Agronomy, Madison, WI.

Williams, P. J. (1996) Grape and wine quality and varietal flavor. In: *Proceedings of the 9th Australian Wine Industry Technical Conference* (C. S. Stockle *et al.*, eds.), pp. 90–92. Winetitles, Adelaide, Australia.

Williams, P. J., Francis, I. L., and Black, S. (1996) Changes in concentration of juice and must glycosides, including flavor precursors. In: *Proceedings of the 4th International Symposium on Cool Climate Viticulture and Enology* (T. Henick-Kling *et al.*, eds.), pp. VI-5–10. New York State Agricultural Experimental Station, Geneva, NY.

Williams, P. L., and Antcliff, A. J. (1984) Successful propagation of *Vitis berlandieri* and *Vitis cinerea* from hardwood cuttings. *Am. J. Enol. Vitic.* **35**, 75–76.

Williams, R.N., and Granett, J. (1988) Phylloxera. In: *Compendium of Grape Diseases* (R. C. Pearson and A. C. Goheen, eds.), p. 63. APS Press, St. Paul, MI.

Winkler, A. J., Cook, J. A., Kliewer, W. M., and Lider, L. A. (1974) *General Viticulture*. University of California Press, Berkeley.

Wistrom, C., and Purcell, A. H. (2005). The fate of *Xylella fastidiosa* in vineyard weeds and other alternate hosts in California. *Plant Dis.* **89**: 994–999.

Wolf, T. K., Dry, P. R., Iland, P. G., Botting, D., Dick, J., Kennedy, U., and Ristic, R. (2003) Response of Shiraz grapevines to five different training systems in the Barossa Valley, Australia. *Aust J. Grape Wine Res.* **9**, 82–96.

Wolpert, J. A., Walker, M. A., and Weber, E. (eds.) (1992) *Rootstock Seminar: A Worldwide Perspective.* American Society of Enology and Viticulture, Davis, CA.

Wolpert, J. A., Szychowski, J. A., and Semancik, J. S. (1996) Effect of viroids on growth, yield, and maturity indices of Cabernet Sauvignon. *Am. J. Enol. Vitic.* **47**, 21–24.

Woodham, R. C., Emmett, T. N., and Fletcher, G. G. (1984) Effects of thermotherapy and virus status on yield, annual growth and grape composition of Sultana. *Vitis* **23**, 268–273.

Yamakawa, Y., and Moriya, M. (1983) Ripening changes in some constituents of virus-free Cabernet franc grape berries. *J. Jpn Soc. Hortic. Sci.* **52**, 16–21.

Younger, W. (1966) Gods, Men and Wine. George Rainbird, London.

Zäuner, S., Creasap, J. E., Burr, T. J., and Ullrich, C. I. (2006) Inhibition of crown gall induction by *Agrobacterium vitis* strain F2/5 in grapevine and *Ricinus*. *Vitis* **45**, 131–139.

Zhu, Y. M., Hoshino, Y., Nakano, M., Takahashi, E., and Mii, M. 1997. Highly efficient system of plant regeneration from protoplasts of grapevine (*Vitis vinifera* L) through somatic embryogenesis by using embryogenic callus culture and activated charcoal. *Plant Sci.* **123**, 151–157.

Zoecklein, B., Fugelsang, K. C., Gump, B. H., and Nury, F. S. (1995) *Wine Analysis and Production.* Chapman and Hall, New York.

Zoecklein, B., Wolf, T. K., Marcy, J. E., and Jasinski, Y. (1998) Effect of fruit zone leaf thinning on total glycosides and selected aglycone concentrations of Riesling (*Vitis vinifera*) grapes. *Am. J. Enol. Vitic.* **49**, 35–43.

5

Site Selection and Climate

The view that the vine needs to "suffer" to produce fine quality fruit is long established in wine folklore. If interpreted as restrained grapevine vigor, open-canopy development, and fruit yield consistent with capacity, the concept of vine suffering has more than just an element of truth. Several aspects of this have already been discussed in Chapter 4. Regions where local conditions have tended to produce these results have come to be noted for their better-quality wines. In addition, grape growers and winemakers have come to recognize vineyard practices that enhance these natural tendencies. It also was realized that sites may possess undesirable properties that viticultural practice could not offset. These beneficial and detrimental aspects of soil, topography, microclimate, and macroclimate now form the basis for choosing favorable viticultural sites. This knowledge allows grape growers not only to produce better-quality grapes in traditional wine-producing regions, but also rationally expand production into new viticultural areas.

Probably the first feature recognized as favoring finer grape production was limited soil fertility. Soils with just adequate nutrient levels restrict vegetative growth and permit a higher proportion of photosynthate to be directed toward fruit maturation. This is particularly important because flavor formation tends to develop near the end of ripening. In addition, many low-nutrient soils are highly porous. This feature improves drainage, resulting in periods of mild water deficit. It also favors rapid warming of, and heat radiation from, the soil. This, in turn, can improve the micro-climate around the vine and delay or minimize frost severity. Excellent drainage also promotes early-spring growth and limits fruit cracking following heavy rains. Finally, vines grown on well-drained soil develop fewer micro- and macro-fissures in the berry skin, which favor infection by fungi and bacteria.

Another feature recognized early on as benefitting grape quality was medium to low rainfall. Dry conditions enhance the inherent resistance of vines to several pathogens. In Europe, most of the southern regions receive the majority of their precipitation during the winter. Thus, sufficient moisture is available for early growth. During the summer, however, the vines may be exposed to varying degrees of water deficit. It is known that avoidance of water stress is most important in the spring and early summer, up to the beginning of ripening (véraison). Subsequently, restricted water availability tends to improve fruit quality and advance ripening. With limited late vegetative growth, more nutrients are directed toward berry ripening. Because grapevines tend to root deeply, they may avoid serious water stress in deep soil, even during periods of drought. The ability of the grapevine to root deeply probably has helped limit the development of nutrient deficiencies in impoverished soils through subsoil nutrient sources. Grapevines are one of the few crops that do well on relatively poor soils.

Possibly during the move of grape culture into central Europe during Roman times, it became apparent that growing cultivars near the northern limit of fruit ripening was beneficial to quality. Wines from cooler mountainous sites in Italy were already renowned to classical Romans for their superior quality. Cool conditions are now known to retain fruit acidity, which improves the microbial and color stability of wines. Temperate conditions also appear to favor the development and retention of grape aroma compounds. In addition, clement weather has value in relation to winemaking and storage (in the absence of refrigeration), independent of its effects on viticulture. However, cultivating grapevines at the higher latitude (altitude) limit of the growing range enhances the risks of crop failure, due to the shorter growing season. It also increases the

likelihood of frost damage. This is probably how the benefits of sites on hillside slopes, or in proximity to large bodies of water were discovered.

Soil Influences

Of climatic influences, soil type appears to be the least significant factor affecting grape and wine quality (Rankine et al., 1971; Wahl, 1988), or to be poorly correlated with wine characteristics (Morlat et al., 1983). Soil influences tend to be expressed indirectly through features such as heat retention, water holding capacity, and nutritional status. For example, soil color and textural composition affect heat absorption by the soil and, thereby, fruit ripening and frost protection. Thus, when discussing soil and its effects on grapevine growth, it is important to distinguish among the various physicochemical properties of soil – its texture, aggregate structure, nutrient availability, organic content, effective depth, pH, drainage, and water availability. The uniformity of soil conditions is likely to be more important than any of these properties alone. Soil variability is a major source of asynchronous berry development, and lower wine quality.

Geologic Origin

The geologic origin of the parental material of the soil has little direct influence on grape quality. Fine wines are produced from grapes grown on soils derived from all three basic rock types – igneous (derived from molten magma, e.g., granite), sedimentary (originating from consolidated sediments, e.g., shale, chalk, and limestone), or metamorphic (arising from transformed sedimentary rock, e.g., slate, quartzite, and schist). Examples of famed wine regions where the soils are primarily derived from a single rock type are Champagne and Chablis (chalk), Jerez (limestone), and Porto and Mosel (schist). However, equally famous regions have soils derived from a mixture of rock types, namely the Rheingau, Bordeaux, and Beaujolais (Wallace, 1972; Seguin, 1986). Some cultivars are reported to do better on soils composed of specific rock types (Fregoni, 1977; Seguin, 1986), but the evidence is predominantly circumstantial. Experimental evidence for these claims is lacking.

Texture

Soil texture refers to the size and proportion of its mineral component. Internationally, four standard size categories are recognized – coarse sand, fine sand, silt, and clay. Chapman's (1965) recognition of a larger number of categories, including gravels, pebbles and cobbles, is particularly relevant when dealing with several important vineyard regions of the world. Nevertheless, most agricultural soils are classified only by their relative contents of sand, silt, and clay. Heavy soils have a high proportion of clay, whereas light soils have a high proportion of sand.

Particles larger than clay and silt consist of unmodified parental rock material. In contrast, clay particles are chemically and structurally transformed minerals, bearing little resemblance to the parental material. Clay consists primarily of a microscopic complex of plates adhering to one another. The plates are often a sheet of alumina sandwiched between two sheets of silica. Silt particles are partially weathered rock material, possessing properties transitional between sand and clay.

Clay, with its large surface area to volume (SA/V) ratio, plate-like structure, and negative charge, has a major influence on the physical and chemical attributes of soil. Clay particles are so minute that they have colloidal properties. Thus, they are gelatinous and slippery when wet (the plates slide relative to one another), but hard and cohesive when dry. After wetting, most clay particles expand like a sponge. As water infiltration forces the plates apart, soil pore diameter decreases. This can markedly reduce water percolation into soils high in clay content. The large SA/V ratio, combined with the negative charge, permits clay plates to attract, retain, and exchange large quantities of positively charged ions (i.e., Ca^{2+}, Mg^{2+}, and H^+), as well as water. Both bivalent ions and water help bind clay plates together. This is critical to the formation and maintenance of the aggregate structure of good agricultural soils. The large SA/V ratio also allows a clayey soil to absorb large quantities of water. However, the bonding is so strong that much of the water is unavailable to plants. In contrast, soils with a coarse texture allow most of the water to percolate through the soil. What remains is held weakly and can be readily extracted by plant roots.

Because important features such as aeration, water availability and nutrient availability are markedly influenced by soil texture, this property significantly affects grapevine growth and fruit maturation. For example, anecdotal reports suggest that phylloxera infestation is minimal in sandy soils, possibly by severely restricting insect movement in the soil. Nevertheless, there are comparatively few reports that have directly studied the effects of soil texture on vine growth (Nagarajah, 1987).

An important property based on the textural character of the soil is heat retention. In fine-textured soils, much of the heat absorbed during sun exposure is transferred to water as it evaporates. This energy is subsequently lost as the water evaporates. In contrast, stony soils tend to retain most of the heat they absorb within its structural components. This heat is subsequently radiated back into the air during the night. The heat so derived can significantly reduce the likelihood of frost damage and accelerate fruit ripening in the autumn (Verbrugghe *et al.*, 1991). Soil compaction can also moderate the temperature in vine rows, and potentially reduce frost damage on cool nights (Bridley *et al.*, 1965).

Structure

Structure often refers to the association of soil particles into complex aggregates. Aggregate formation starts with the binding of mineral (clay) and organic (humus) colloids by bivalent ions, water, microbial filamentous growths, and plant, microbial, and invertebrate mucilages. These agglomerates bind with particles of sand and silt, as well as organic residues, to form a variety of aggregates of differing size and stability. These are subsequently rearranged or modified by the burrowing action of the soil fauna, root growth, and frost action.

Soils high in aggregate structure are friable, well aerated, and easily penetrated by roots; have high water holding capacities; and are considered to be agriculturally superior. Heavy clay soils are more porous, but their small diameter makes root penetration difficult and results in poorly aerated conditions when wet. As a consequence, roots remain at or near the surface, exposing vines to severe water stress under drought conditions. Lighter soils are well drained and aerated, but the large pores retain relatively little water. Nonetheless, vines on light soils may experience less severe water deficit under drought conditions, if the soil is sufficiently deep to permit root access to groundwater. Soil depth also may offset the poor nutrient status of many light soils. The negative effects of the small and large pores of heavy and light soils, respectively, may be counteracted by humus. Humus modulates pore size, facilitating the upward and lateral movement of water, increases water absorbency, and retains water at tensions that permit roots ready access to the water.

Although soil structure affects aeration, as well as mineral and water availability, these may be modified by vineyard practices such as tillage. Consequently, they are not a constant feature of a site, and their significance to grape and wine quality is difficult to assess accurately. Under zero tillage, the number of pores and pore area are significantly higher than under cultivation. Conventional tillage results in greater total porosity, but this consists primarily of a few, large, irregularly shaped cavities (Pagliai *et al.*, 1984). Generally, root development is better under zero tillage (Soyer *et al.*, 1984). Under no-till conditions, most root development occurs in the upper portion of the soil, whereas

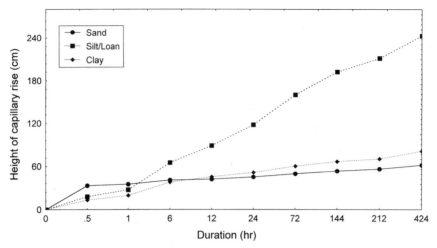

Figure 5.1 Upward movement of moisture by capillary action through soils of various textures and structures. The pores of silt or loam are more favorable to movement than are those in clay. The rate of movement often is more important than the distance raised. (Data from Bear, 1953)

conventional cultivation limits root growth to deeper portions of the soil. Under grass cover, root distribution is relatively uniform in the top one meter of the soil. Cultivated vineyards show lower levels of organic material (Pagliai *et al.*, 1984). This may result from enhanced aeration and solar heating. Both stimulate the microbial mineralization of the soil's organic content.

Drainage and Water Availability

As mentioned, both soil texture and structure have effects on water infiltration. Both properties also affect water availability. Once water has moved into the soil, water may be bound to colloidal materials by electrostatic forces, adhere to pore surfaces by cohesive forces, or percolate through the soil under the action of gravity. Depending on the clay content of the soil, and its tendency to swell (diminishing pore diameter), the rate of infiltration may slow. Capillary flow is dependent on pore diameter. Because of the variable pore diameter, and its discontinuity in the soil, water rarely rises more than 1.5 m above the water table. Only water retained by cohesive forces or sorbed to soil colloids remains available for root uptake.

Capillary water is important because cohesive forces permit both upward (Fig. 5.1) and lateral (Fig. 5.2) movement. Often it is the rate, not the distance traversed, that is the critical factor influencing the relative importance of capillary flow. For example, the soil surface can dry to the wilting point even when the water table is less than a meter below the surface. As indicated in Fig. 5.1, the capillary flow of water has its greatest significance in silty soil. In sandy soils, capillary movement is rapid but

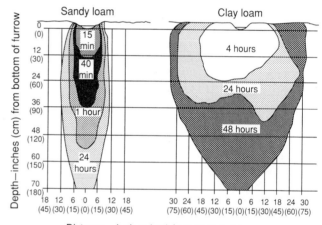

Figure 5.2 Comparative rates and direction of irrigation water movement into and through sandy and clay loams. (From Coony and Pehrson, 1955, reproduced by permission)

limited, whereas in clayey soils it is marked but too slow to be of viticultural significance. In most situations, the rate of capillary flow is less than the rate at which plant roots absorb water, requiring continued root growth into unexplored soil regions. Additional details of the relationship between soil texture and hydraulic properties can be obtained from Saxton *et al.* (1986, 1999).

Water that fills large soil pores (free water) rapidly percolates through the soil and is lost to the plant, unless the root system penetrates deeply. Hygroscopically bound water is largely unavailable for root extraction. During drought, however, hygroscopic water may evaporate and move upward in the soil. If it condenses as the soil cools at night, it may become available to the roots. This may be of importance in sandy soils in which high porosity permits greater air circulation.

In Bordeaux, the ranking of *cru classé* estates has been correlated with the presence of deep, coarse-textured soils located on elevations close to rivulets or drainage channels (Seguin, 1986). These features promote rapid drainage and are thought to permit deep root penetration. Free water can percolate through the soil to a depth of 20 m within 24 h. Thus, grapevines are less likely to suffer from waterlogging in heavy rains, or water deficit during drought. Although some cultivars are relatively tolerant of waterlogged soils, high soil-moisture content accentuates cracking of the berry skin and susceptibility to bunch rot (Seguin and Compagnon, 1970).

Even when the soil is shallow, there may be factors that diminish the probability of water deficit under drought conditions, or of waterlogging during rainy conditions. For example, the compact limestone that underlies the shallow soils in St Émilion permits the effective upward flow of water from the water table. It is estimated that 70% of the water uptake by vines in 1985 came from this source (Seguin, 1986).

In general, a good vineyard soil has been characterized by the following water infiltration and retention traits (Cass and Maschmedt, 1998):

>500 mm infiltration rate per day
>150 mm total available water in the root zone (water extracted by roots < -0.15 MPa)
>75 mm of readily available water in the root zone (water extracted by roots < -0.2 MPa)
>15% air-filled pore space
<1 MPa penetration resistance at field capacity (or 3 MPa at wilting point)
<1 d soil saturation per irrigation cycle or rainfall occurrence

Where drainage is poor, waterlogging can be a recurring problem. Not only does it retard vine growth, favor the development of chlorosis in lime soils, and encourage attack by several root pathogens, but it also causes problems with the movement of machinery in the vineyard. Some vine problems are caused by the combined effects of reduced oxygen availability in the soil and increased concentrations of carbon dioxide and ethylene. Under prolonged waterlogging, toxic amounts of hydrogen sulfide may also accumulate due to the anaerobic metabolism of soil bacteria. In arid regions, poor drainage significantly enhances salt buildup in the root zone. This results from insufficient leaching of salts precipitated by the evaporation of water drawn up by capillary action.

Open ditches may provide adequate drainage in regions with shallow grades, especially when covered with grass to minimize erosion. However, in most situations where waterlogging is a problem, the laying of drainage tiles or pipes is required. It is desirable to have unrestricted drainage to a depth of at least 2–3 m in most situations. Where a shallow hardpan is the source of the poor drainage, deep ripping to break the pan may be the best solution. For details on drainage systems, consult Webber and Jones (1992).

Soil Depth

Soil depth, in addition to soil texture and structure, can influence water availability. Shallow hardpans reduce the usable soil depth, and enhance the tendency of soil to waterlog in heavy rains, and fall below the permanent wilting percentage under drought conditions. Limiting root growth to surface layers also can influence nutrient access. For example, potassium and available phosphorus tend to predominate near the surface, especially in clay soils, whereas magnesium and calcium more commonly characterize the lower horizons. Soils vary in the accumulation of nutrients through their soil horizons.

Effective soil depth is best achieved before planting. Breaking up hardpans by soil ripping is a standard technique in several countries. An alternate procedure is mounding topsoil in regions where vines are planted. It is particularly useful in situations where high water tables are unavoidable and under saline conditions. Planting a permanent groundcover or mulching helps to minimize erosion from the mounds. Where root penetration is limited by high acidity in one or more soil horizons, soil slotting can significantly increase root soil-exploration.

Effective soil depth may decrease as a consequence of various viticultural techniques. For example, cultivation promotes microbial metabolism and degradation of organic material. This weakens the crumb structure of the soil, leading to the release and downward movement of clay particles. In addition, salinization as a result of improper irrigation can disrupt aggregate structure, releasing clay particles. If clay particles flow downward, they tend to plug soil capillaries. Over time, this can result in the formation of a claypan.

Soil Fauna and Flora

The detrimental effects of pathogenic soil microorganisms on grapevines are well known, as are the beneficial influences of mycorrhizal fungi. Far less appreciated is the activity of the tens of thousands of other members of the soil fauna and flora. These include innumerable species of bacteria, fungi, algae, protozoans, nematodes, springtales (Collembolae) and other insect larvae, mites, and earthworms. Bacteria occur in numbers in excess of 10^8 cells/g soil. It is variously estimated that from 30 to 80% of soil bacteria have yet to be cultured or identified due to their complex nutritional requirements or symbiotic relationships. Most of the fauna are the soil equivalent of terrestrial herbivores. Figure 5.3 illustrates the relative size and shape of some of the biota in small soil aggregates. Most of their effects are known only in general and from investigations unrelated to viticulture.

Among the major beneficial effects of the soil fauna and flora is the generation of the aggregate structure of the soil. This is the intermediate end-result of their metabolic activities. Bacteria are especially active in releasing polysaccharides that bind the cells to soil particles and, consequently, soil particles to each other. Additional polysaccharides are secreted during the feeding activities of earthworms and other organisms. Fungi help to hold soil particles together with their long filamentous growths.

Although algae and a few bacteria are net producers of organic material in soil, most of the nutrients on which the soil biota survive come from green plants. These are derived primarily from leaves and from the death of feeder roots. The initial decomposers are bacteria and fungi. These are, in turn, grazed by the fauna, notably protozoa, nematodes, and mites, or consumed along with soil during the feeding of earthworms or various insect larvae. Their feeding releases inorganic nutrients bound in the microbial flora, which promotes an additional round of microbial decomposition on material defecated by the fauna. The grinding action of most of the fauna destroys the morphological and cellular structure of the plant remains. This especially helps to expose plant cell-wall constituents to further decomposition. If conditions are favorable (warm and moist), most organic material (with the exception of woody tissues) are rapidly mineralized. However, in cooler or drier conditions, mineralization is only partial. What remains tends to be a collection of highly complex, oxidized phenolic material. It forms the bulk of what is called humus. Humus, along with polysaccharides released by the soil fauna and flora, constitutes the organic component of the aggregate structure of the best agricultural soils.

Another significant contribution of the soil microbiota, notably several genera of bacteria, is in the interconversion of various forms of nitrogen. Ammonia,

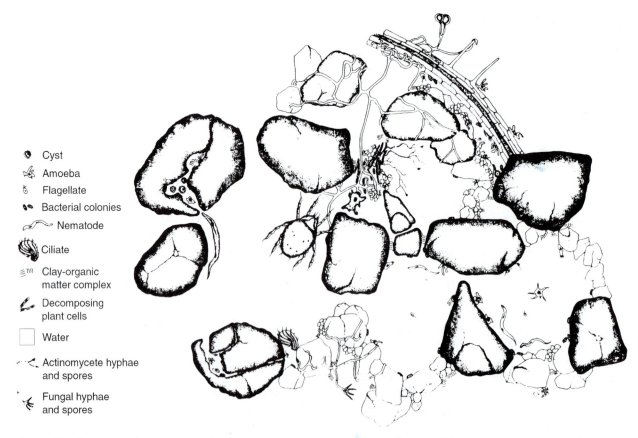

Cyst
Amoeba
Flagellate
Bacterial colonies
Nematode
Ciliate
Clay-organic matter complex
Decomposing plant cells
Water
Actinomycete hyphae and spores
Fungal hyphae and spores

Figure 5.3 Illustration of several groups of soil organisms in approximately an 1 cm² section of an aggregate in the surface horizon of a grassland soil. (From Paul and Clarke, 1989, reproduced with permission of S. Rose and T. Eliott)

released during decomposition or added as fertilizer, is converted to nitrate by nitrifying bacteria. Other bacteria, under anaerobic conditions (in anoxigenic centers of soil aggregates), perform the reverse reaction (ammonification). In addition, anaerobic bacteria can release nitrogen gas from nitrate in a process termed denitrification. Under low-nitrogen conditions, other groups of bacteria fix nitrogen gas, releasing nitrates to the soil. Nitrogen fixation is particularly well known, relative to the action of rhizobial bacteria in the root nodules of legumes. It also occurs in free-living nitrogen fixing soil bacteria and cyanobacteria.

Acids released by bacteria and fungal metabolism are also important in the extraction (and eventual solubilization) of inorganic nutrients from the mineral content of the soil. Although slow, metabolism does assist the slow weathering of the parental rock material from which soil is derived. By aiding solubilization, soil microbes participate in the transformation of sand and silt to clay.

Nutrient Content and pH

Nutrient availability in soil is influenced by many, often interrelated factors. These include the parent material, particle size, humus content, pH, water content, aeration, temperature, root-surface area, and mycorrhizal development. Nevertheless, the mineral content of soil is primarily derived from the parental rock substrata. Consequently, it has been thought that the superiority of certain vineyard sites might be due to the nutrient status of its soil substrata. Differences in nitrogen accumulation from different soils have been associated with wine quality (Ough and Nagaoka, 1984), but this has no connection with the soil's mineral base. In contrast, the mineral content of rock substrata varies widely. Fregoni (1977) interpreted differences in wine quality as resulting from mineral differences in their associated vineyard soils. Although interesting, few researchers would readily accept such correlations as being causally related without supporting experimental evidence. In Bordeaux, prestigious vineyard sites have been noted as possessing higher humus and available nutrient content than less highly ranked sites (Seguin, 1986). Occasionally, such data have been used to explain the historical ranking of the sites. However, the nutrient status of the sites may be equally, and possibly be better, explained as a result of the appropriate maintenance and a long history of manure addition. The latter were possible partially as a consequence of the financial benefits gained from the higher ranking.

Soil pH is one of the most well known factors affecting mineral solubility and thereby availability (see Fig. 4.32). Nevertheless, actual absorption by the vine is primarily regulated by root physiology (cultivar

genotype) and mycorrhizal association. Thus, grapevine mineral content does not necessarily or directly reflect the mineral content of the soil. In addition, excellent wines may be derived from grapes grown on acidic, neutral, and alkaline soils. Thus, except where deficiency or toxicity is involved, there seems little justification for assuming that wine quality is dependent on either a specific soil pH or mineral composition.

Color

Soil color is influenced by the moisture content, mineral composition, and organic content. For example, soils high in calcium tend to be white, those high in iron are reddish, and those high in humus are dark brown to black. Soil needs only about 5% organic material to appear black when wet. Soil color is also a reflection of its age, and the temperature and moisture characteristics of the climate. Thus, cooler regions tend to have topsoils grayish to black, due to the accumulation of humus. In moist, warm regions, the soils tend to be more yellowish-brown to red, depending on the hydration of ferric oxide and extensive weathering of the parental mineral content. Rapid mineralization of the organic material in warm moist regions means that insufficient humus accumulates to influence soil color. Arid soils tend to be light in color (little staining from the low-organic content) and primarily show the color of its parental mineral content. In some soils, manganese oxidation may significantly stain the inorganic component of the soil.

Following rain, water temporarily darkens the soil's color by increasing light absorption. In addition, moisture can have long-term effects on soil color. For example, by enhancing anaerobic conditions, waterlogging results in iron oxides being primarily in the ferrous state. These can give the soil a subtle bluish-gray tint. A mottled rusty or streaked appearance may indicate variably or improperly drained soils.

Color influences the rate of soil warming in the spring and cooling in the fall. Dark soils, irrespective of moisture content, absorb more heat than do the more reflective light-color soils (Oke, 1987). Soils of higher moisture content, being darker, absorb more solar radiation (Fig. 5.4), but warm more slowly than drier soils. This apparent anomaly arises from the high specific heat of water, which absorbs large amounts of energy during warming. Consequently, the surfaces of sandy and coarse soils both warm and cool more rapidly than do clay soils of the same color. Rapid cooling can significantly warm the air and fruit close to the ground during the night. Reflective groundcovers can slow the rise in soil temperature during the spring, but moderate its decline in the autumn. Analogous variation in daily soil temperature

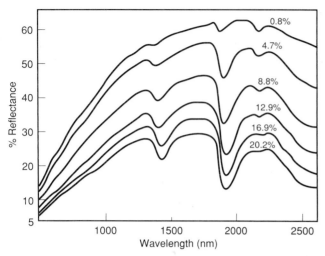

Figure 5.4 Reflectivity of a loam as a function of wavelength and water content. (From S. A. Bowers and R. J. Hanks (1965) Reflection of radiant energy from soils. *Soil Sci.* 100, 130–138. © Williams and Wilkins)

occurs under groundcovers and straw mulches (Whiting *et al.*, 1993). In contrast, plastic mulches often enhance early soil warming in vineyards (Ballif and Dutil, 1975).

The microclimatic effects of soil color and moisture content on temperature are most significant during the spring and fall. In the summer, temperature differences caused by soil surface characteristics and shading generally have little effect on vine growth and fruit maturity (Wagner and Simon, 1984). However, warm soils may enhance microbial nitrification, enhance potassium uptake, and depress magnesium and iron absorption by the vine.

Occasionally, red varieties have been selectively grown in dark soils and white varieties in light-colored soils. In marginally cool climates, this could provide the greater heat needs required for full color development in red cultivars. Nevertheless, excellent results can be obtained where the cultivation of white and red varieties on light and dark soils is reversed (Seguin, 1971).

Soil color can also influence vine growth directly by reflecting photosynthetically active radiation (PAR) up into the canopy. It can significantly influence grape yield as well as sugar, anthocyanin, polyphenol, and free amino acid contents (Robin *et al.*, 1996).

Organic Content

The organic content of soil improves water retention and permeability, as well as enhancing its aggregate structure and nutrient availability (see Chapter 4). However, roots rarely absorb organic compounds from soil. Exceptions are systemic pesticides and highly volatile compounds, such as ethylene (released by many soil microorganisms).

There is no evidence supporting the common contention that soil directly influences the aromatic character of wine. The "earthy," "barnyardy," and "flinty" qualities of certain regional wines undoubtedly arise during wine production and maturation, and do not originate from the soil. Some "terroir" attributes are actually due to improper barrel hygiene and the development of "Brett" off-odors, often characterized as being "barnyardy." Thankfully, the aromatic compounds produced by manure and the earthy odors generated by actinomycetes are not absorbed and translocated to ripening grapes. Thus, the soil organic content typically affects vine growth and potential wine quality only indirectly.

In situations in which the organic content is low (sandy soils) or has been reduced by cultivation, the most common means of increasing the humus content is with compost, such as manure or a groundcover. Where available, farm manure is an excellent means of enriching the soil's organic content, if well seasoned. Straw, used as a mulch, is much less effective, and its incorporation into the soil is slow. Earthworms eventually incorporate the straw into the soil, along with the production of macropores that aid the infiltration of water. However, because of the high carbon-to-nitrogen ratio of straw, its incorporation (and microbial decomposition) may generate temporary nitrogen deficiency. Thus, nitrogen needs to be added to compensate for the consumption of nitrogen by decay microorganism, until humifaction of the straw is complete.

Topographic Influences

Similar to the data on soil attributes, much of the information on the influences of slope on grape quality is circumstantial. Nevertheless, the effects tend to become more evident with increasing latitude and altitude. The beneficial influences of sunward-angled sites on microclimate include enhanced exposure to solar photosynthetic and heat radiation, earlier soil warming, diminished frost severity, and improved drainage. For the grapevine, photosynthetic potential may be increased, fruit ripening advanced, berry color and sugar–acid balance improved, and the growing season extended. Microclimatic disadvantages include increased potential for soil erosion, nutrient loss, water stress, and early loss of snow cover. The potential for bark splitting during the winter is increased, and cold acclimation may be lost prematurely. As the slope increases, the performance of vineyard activities becomes progressively more difficult – eventually making mechanization impossible. The net benefit of a sloped site depends on its inclination (vertical deviation), aspect (compass orientation), latitude, and soil type, as well as cultivar and viticultural choices.

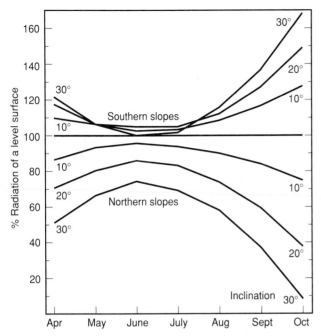

Figure 5.5 Reception of direct sunlight relative to position and inclination of slope (48°15′N) in the upper Rhine Valley. (From Becker, 1985b, reproduced by permission)

Solar Exposure

When the beneficial influences of a sloped vineyard are sufficient, they may offset the difficulties associated with performing vineyard activities. Of particular significance is the improved solar exposure created by a sun-facing aspect. The benefits of an inclined location become progressively important with increased vineyard latitude or altitude. Not surprisingly, Germany – the most northerly major wine-producing region in Europe – is renowned for its steep, south-facing vineyards.

The primary advantage created by a favorable slope orientation and inclination relates to the reduced angle of incidence at which solar radiation impacts the vineyard. This increases both solar exposure and heating.

At the highest latitudes for commercial viticulture (~50°), the optimal inclination for light exposure of a sun-facing slope is about 50°. Although slopes this steep are too difficult to work, solar exposure is only slightly less at a slope of 30° (Pope and Lloyd, 1974). This is generally considered the upper limit for manual vineyard work. Machines seldom work well at inclinations much above 6° (a slope of 10.5%). Sun exposure on east- and west-facing slopes is little affected by inclination, except those above 50°. Polar-facing slopes have a distinctly negative effect on solar input. Figure 5.5 illustrates the influence of slope inclination and aspect on seasonal influences of solar input. These influences are most marked when the altitude of the sun (position

above the horizon) is lowest (winter), and least noticeable when the solar altitude is highest (summer).

Another factor influencing the significance of slope on light incidence is the frequency of cloudiness. Cloud cover, by dispersing solar radiation across the sky, eliminates the solar benefits of equatorial-facing slopes. Cloud cover also eliminates the reflection of heat and photosynthetic radiation off water surfaces (Muñez *et al.*, 1972). Occasionally this can be beneficial in the spring. It would diminish premature cold deacclimation and bud break, and the associated increase in the likelihood of frost damage. In contrast, sunny weather in the fall is desirable as it favors maximal heat accumulation during ripening and harvest.

For increasing solar exposure, the best slopes are those directed toward the equator. In practice, however, the ideal aspect may be influenced by local factors. If fog commonly develops during cool autumn mornings, the preferred aspect may be southwest. The scattering of light by fog eliminates the radiation advantage of an equatorial-facing slope. In the late afternoon, when skies are more commonly clear, a southwest aspect provides optimal solar exposure in the Northern Hemisphere. Such situations are not uncommon along the Mosel and Rhine river valleys in Germany and the Neusiedler See in Austria.

Another important property of sunward-facing slopes is radiation reflected from water and soil surfaces (albedo). This is particularly significant at low sun altitudes. At high solar elevations, the albedo off water is low (2–3% for a smooth surface and 7–8% for a rough surface). However, at low sun elevations (<10°), reflected solar radiation can reach more than 50% of that received on a slope (Büttner and Sutter, 1935). Consequently, light reflected from water bodies is especially significant during the spring and fall. This has particular importance for sloped vineyards in high latitudes – the steeper the slope, the greater the potential interception of reflected radiation. Radiation reflected off the Main River in Germany (49°48′N) can constitute 39% of total radiation received on south-facing vineyards in early spring (Volk, 1934). This level of additional exposure could advance early initiation of growth in the spring and enhance photosynthesis and fruit ripening in the autumn. The potential significance of reflected light is indicated in a series of experiments conducted by Robin *et al.* (1996).

The reflection of solar radiation from water does not directly augment heating. Most of the infrared radiation is absorbed by the water, even at low sun altitudes. Nonetheless, heating may result indirectly from the absorption of the additional visible radiation received.

Although augmenting solar exposure is generally beneficial at high latitudes and altitudes, the opposite may be true at low latitudes and altitudes. Here, diminished sun exposure may favor a cooler microclimate leading to retention of more acidity and grape flavor. Correspondingly, east-facing slopes may be preferable. An eastern aspect exposes vines to the cooler morning sun and provides increased shading from the hot afternoon sun.

Additional influences associated with slope are increased drainage, both of water in the soil and of cold air flow away from the vines. The improved water drainage can be an advantage if the region has ample and seasonally well-distributed rainfall. Increased drainage can be a disadvantage in arid conditions, because it increases the likelihood of water deficit. Slopes also promote erosion and nutrient loss. Correspondingly, soils at the top of a slope are typically comparatively thin and nutrient poor, compared to soils further down the slope and in the adjacent flat land. Cold air flow, as noted later in this chapter, can significantly prevent the development of either late-spring or early-autumn frost conditions.

Wind Direction

The prevailing wind direction can significantly influence the features provided by a slope. Heat accumulation achieved on sunward slopes can be lost if winds greater than 7 km/h blow down vineyard rows. Crosswinds require twice the velocity to produce the same effect (Brandtner, 1974). Updrafts through vineyards also may diminish the heat-accumulation potential of sunward-facing slopes (Geiger et al., 2003).

At high latitudes, vine rows commonly are planted directly up steep slopes to facilitate cultivation. Offsetting row orientation to minimize the negative influences of the prevailing wind direction is generally impractical on steep slopes. In contrast, terracing vineyards on slopes can permit the alignment of rows relative to the prevailing winds. Regrettably, terracing eliminates many of the advantages of steeply sloped sites. Terracing may also increase soil erosion problems (Luft et al., 1983).

In humid climates, positioning vine rows 90° to the prevailing winds can increase the drying of plant surfaces by enhancing wind turbulence. This can enhance the action of fungicides in controlling fungal diseases. If the vineyard faces sunward, the enhanced solar radiation further speeds the drying action of the wind. In dry environments, rows aligned parallel to the prevailing winds may reduce foliage wind drag and potentially reduce evapotranspiration (Hicks, 1973), whereas a perpendicular alignment may lead to increased water stress, due to the stomata remaining open longer during the day (Freeman et al., 1982). Thus, the most appropriate row alignment will depend on the climatic limitations it is designed to alleviate. The presence of natural or artificial shelterbelts, modifying the velocity, turbulence, and flow of the wind may further influence optimal row alignment and slope orientation.

Frost and Winter Protection

A considerable advantage of sun-facing slopes in cool climates comes from the additional number of frost-free days provided. Part of this microclimatic benefit results from improved heat accumulation, but the flow of cold air away from the vines may be of equal or greater importance.

Under cool, clear atmospheric conditions, heat radiation from the soil and grapevines can be considerable. Without wind turbulence, an inversion layer can form, resulting in temperatures falling near or below the freezing point under cold conditions. Sloped sites often experience some protection from this phenomenon. Cold air can flow downward and away from the vines. The movement of cold air into low-lying areas can extend the frost-free season on slopes by several days or weeks (Fig. 5.6). Depending on the height of the slope, maximal protection may be achieved either at the top, or more commonly, in the mid-region of the slope. The degree of protection often depends on wind barriers, such as tree shelters, or topographical features. These can facilitate the movement of air between and away from the vines.

The flow of cool air away from grapevines on even slight slopes can also be important at temperatures above freezing. Chilling to temperatures below 10°C has been reported to permanently disrupt fruit maturation in some varieties (Becker, 1985a).

Where sloped vineyard sites are associated with lakes and rivers, the water can further modify vineyard microclimate. By acting both as a heat source and sink, water can buffer major temperature fluctuations. Large lakes and oceans generate even more marked modulation of the climate.

In regions frequently experiencing severe winter conditions, east- or west-facing slopes may be preferred. For example, in the Finger Lakes region of New York, south-facing slopes promote the early loss of an insulating snow cover. Sharratt et al. (1992) and others have clearly demonstrated the insulating properties of snow cover. South-facing slopes also may increase the likelihood of bark splitting due to sudden fluctuations in temperature due to rapid changes in sun exposure, and the "black-body" effect of dark-colored bark.

Altitude

The annual temperature (isotherm) tends to decrease by about 0.5°C/100 m elevation in altitude. Thus, altitude

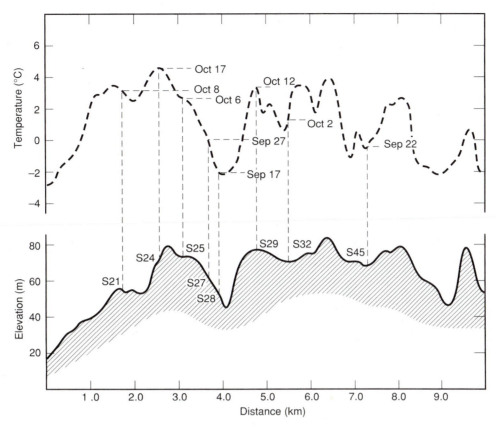

Figure 5.6 Relationship between the estimated average frost date, topography, and temperature on a clear night: S+number represents particular site locations. (From Bootsma, 1976, reproduced by permission)

can significantly affect grape maturation and the length of the growing season. Typically, lower altitudes are preferable at high latitudes, and higher altitudes more desirable at lower latitudes. The intensity of light and especially ultraviolet radiation increases with altitude. These influences can, however, be markedly modified by the local climate, notably the percentage of cloud cover.

Few direct investigations of the effects of altitude on grape and wine quality have been conducted. Thus, the data presented by Scrinzi *et al.* (1996) are of particular value. They studied the effects of altitude and soil conditions on the characteristics of 'Sauvignon blanc' wines in the alpine region of Trentino, Italy. From their investigation, they were able to make specific recommendation on vineyard location, relative to the flavor characteristics found in the wine.

Drainage

Because of erosion, the soils on slopes tend to be coarsely textured. This provides better drainage and permits soil surfaces to dry more quickly. Thus, less heat is expended in the vaporization of soil moisture from

the surface, and sun-facing slopes warm more readily. For example, it takes at least twice as much heat to raise the temperature of a moist soil compared to its dry equivalent. On the negative side, slopes show greater tendencies to be nutrient deficient (due to leaching) and require the periodic addition of topsoil. Enhanced drainage may also increase the potential for groundwater pollution from nutrients added as fertilizer.

Atmospheric Influences

Historically, cultivars appropriate to a particular region were selected or discovered empirically. How well they grew and the quality of the fruit indicated suitability. Improvements in measuring the physical parameters of a region have led to attempts to classify regions relative to compatible grape varieties. The most well known is the system devised by Amerine and Winkler (1944). They used heat-summation units. The units, called **degree-days**, are calculated for months having average temperatures above 10 °C (50 °F). For those months, a sum is calculated by multiplying the number of days by

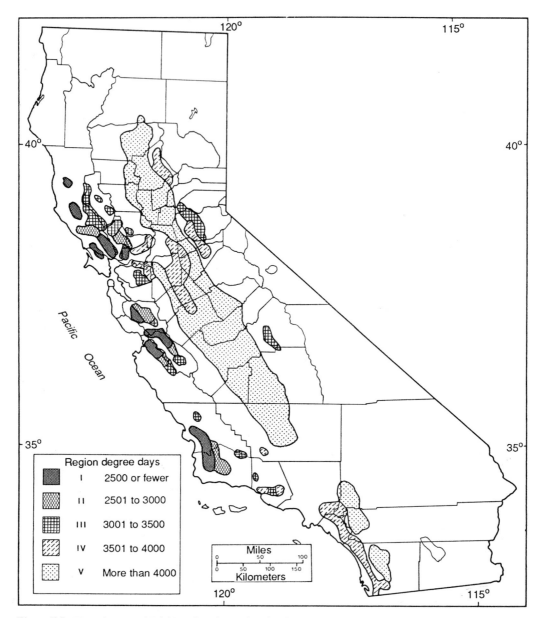

Region degree days

	I	2500 or fewer
	II	2501 to 3000
	III	3001 to 3500
	IV	3501 to 4000
	V	More than 4000

Figure 5.7 Wine districts of California based on Fahrenheit heat-summation units. (From de Blij, 1983, reproduced by permission)

their respective average temperatures (minus 10). For example the corresponding degree-day values for days with temperatures of 15 and 25 °C, respectively, would be 5 [1 × (15–10)] and 15 [(1 × (25–10)]. The 10 °C cutoff point was chosen because few grape varieties initiate significant growth below this temperature. With this system they divided California into five climatic regions (Fig. 5.7). A comparison of the Winkler and Amerine viticultural climatic regions, based on Celsius and Fahrenheit degree-day ranges, is given in Table 5.1.

Since its introduction, the degree-day formula has been used widely in many countries. However, it has not met with universal success, even in California, due to the

impact of additional climatic factors (McIntyre *et al.*, 1987). In an attempt to find a more generally applicable measure, modifications to the degree-day formula have been investigated. These have included the incorporation of factors such as indicators of humidity and water stress, or modifications to the temperature formula. One of the modified formulas is the **latitude-temperature index** proposed by Jackson and Cherry (1988). It is calculated as the product of the mean temperature (°C) of the warmest month, multiplied by 60 (minus the latitude). Another system, proposed by Bentryn (1988), plots sites in relation to the mean lowest and highest monthly temperatures, or the mean temperature and daily relative

Table 5.1 Comparison of viticultural climatic regions based on equivalent ranges of Celsius and Fahrenheit degree-days

Region	Celsius degree-days	Fahrenheit degree-days
I	≤1390	≤2500
II	1391–1670	2501–3000
III	1671–1940	3001–3500
IV	1941–2220	3501–4000
V	≥2220	≥4000

humidity of the warmest month. The latter is useful in predicting the potential severity of several fungal diseases. Tonietto and Carbonneau (2004) have presented a new worldwide climatic classification. It incorporates factors such as a cool night index (CI), a dryness index (DI), and Huglin's heliothermal index (HI).

Although useful, these formulae give only a rough indication of varietal suitability. For example, regions affected by continental influences may have more than twice the average day–night temperature variation as an equivalent maritime region. In addition, differences in north–south slope orientation become increasingly important with increasing latitude. In an attempt to obtain more sensitive indicators of local climatic conditions, some researchers have recommended use of floristic maps to determine cultivar–site compatibility (Becker, 1985b). Because of the sensitivity of some native plants to climatic conditions, they can be accurate indicators of the long-term macro- and microclimatic conditions of a site.

In long-established vineyard regions, there is little need for climatic indicators. Cultivar suitability is already known. However, for new viticultural regions, the prediction of cultivar–site suitability can avoid expensive errors and the need to replant. For some *Vitis vinifera* cultivars, there are generally accepted, empirical data indicating minimum and preferred climatic conditions. For example, Becker (1985b) noted that cool-adapted cultivars typically require more than 1000 (Celsius) degree-days, a 180-day frost-free period, an average coldest monthly temperature not less than $-1\,°C$, temperatures below $-20\,°C$ occurring less than once every 20 years, and annual precipitation greater than 400–500 mm. Phenologic stages, such as bud burst, flowering, and fruit maturation, are also well established for several major cultivars (Galet, 1979; McIntyre *et al.*, 1982). Regrettably, similar data are not available for the majority of grapevine cultivars.

Members of the genus *Vitis* grow over a wide range of climates, from the continental extremes of northern Canada, Russia, and China to the humid subtropical climates of Central America and northern South America. Nevertheless, most individual species are limited to a much narrower range of latitudes and environmental extremes. For *Vitis vinifera*, adaptation includes the latitude range between 35 and 50°N, including Mediterranean, maritime and moderate continental climates. These are characterized by wet winters and hot dry summers, relatively mild winters and dry cool summers, and cool winters and warm summers with comparatively uniform annual precipitation, respectively.

Although these indicators give clues as to site appropriateness, the only precise measure of suitability is planting. The increasing availability of chemical indicators of wine quality will hopefully facilitate more rapid and objective measures of site–cultivar compatibility (Somers, 1998; Marais *et al.*, 1999).

Temperature

Temperature and grapevine growth are markedly affected by site latitude, through its influence on the periodicity and intensity of solar radiation. Nevertheless, grapevine growth is equally controlled by the annual temperature cycle. For example, cold acclimation is predominately influenced by cooling autumn temperatures, not the shortening photoperiod. In addition, bud activation in the spring does not require a specific cold treatment, as in most temperate-zone plants. Bud activation responds progressively to temperatures above a cultivar-specific minimum temperature (Moncur *et al.*, 1989). Above this temperature, bud break and other phenological responses become increasingly rapid, up to an optimum temperature (Fig. 5.8). This type of response suggests that temperature control is relatively nonspecific, and functions through its effects on the shape or flexibility of specific regulator proteins and cell-membrane lipids. This interpretation is strengthened by the nonspecific enhancement of cellular respiration during bud break by a diverse range of treatments (Shulman *et al.*, 1983). Up to a maximum value, every $10\,°C$ increase in temperature doubles the reaction rate of most biochemical processes. As cellular reactions have dissimilar temperature-response curves, overall vine response will depend on the combined effects of temperature on the individual reactions involved.

The slow activation of bud growth in the spring may reflect the ancestral trailing-climbing habit of the vine. For several cultivars, the average minimal temperature for bud break and leaf production are 3.5 and $7.1\,°C$, respectively (Moncur *et al.*, 1989). Delay in grapevine bud break permits potential shrub and tree supports to partially produce their foliage, before the vine commences its growth. Thus, the vine can position its leaves optimally for light exposure, relative to the foliage of the support plant. During the remainder of the growing season, lateral bud activation appears to be regulated

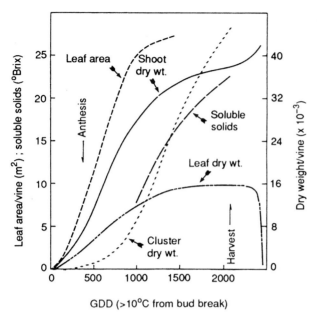

Figure 5.8 Increment of dry weight, leaf area, and soluble solids of 'Thompson Seedless' as a function of degree-days (GDD) exceeding 10°C. Average bud break in the San Joaquin valley was March 9. (Data from Williams, 1987; graphed in Williams and Matthews, 1990, reproduced by permission)

by growth substances released from apical meristematic regions. This plasticity permits the vine to respond to favorable environmental conditions, or foliage loss, throughout much of the growing season.

Temperature has marked and critical effects on both the duration and effectiveness of flowering and fruit set. Flowering typically does not occur until the average temperature reaches 20°C (18°C in cool regions). Low temperatures slow anthesis, as well as pollen release, germination, and pollen-tube growth. For example, pollen germination is low at 15°C, but high at 30–35°C. Style penetration and fertilization may take 5–7 days at 15°C, but only a few hours at 30°C (Staudt, 1982). If fertilization is delayed significantly, ovules abort. Cold temperatures slowly reduce pollen viability.

Although pollen germination and tube growth are favored by warm temperatures, optimal fertilization and subsequent fruit set become progressively poorer as temperatures rise above 20°C. Ovule fertility, seed number per berry, and berry weight are greater at lower than at high temperatures. Even soil temperature can affect vine fertility. For example, cool soil temperatures tend to suppress bud break, but enhance the number of berries produced per cluster (Kliewer, 1975). Fruit set seems proportional to the leaf area generated.

Temperature conditions affect the photosynthetic rate, but they are not known to dramatically influence the overall vine growth under normal conditions.

In fact, leaves partially adjust to seasonal temperature fluctuations by changing their optimal photosynthetic temperature. In midsummer, the optimal range generally varies between 25 and 32°C, but in the autumn it may decline to between 22 and 25°C (Stoev and Slavtcheva, 1982).

For centuries, it has been known that temperature markedly affects berry ripening and quality. This is the basis of the degree-day formula of site selection. Temperature also differently affects specific reactions that occur during berry maturation. Consequently, the fruit composition and potential wine quality can be significantly influenced.

Higher temperatures generally result in increased sugar levels, but reduced malic acidity. Because taste, color, stability and aging potential are all influenced by grape sugar and acid contents, temperature conditions throughout the season have a prominent effect in delimiting grape quality at harvest. Based on the sugar and malic acid levels, the optimal temperature range for grape maturation lies between 20 and 25°C; for anthocyanin synthesis, slightly cooler temperatures may be preferable (Kliewer and Torres, 1972). Daytime temperatures appear to be more important than nighttime temperatures for pigment formation. Although moderate temperature conditions often favor fruit coloration, cool temperatures may limit the commercial cultivation of red cultivars due to poor coloration.

Surprisingly little research has investigated the generally held view that cool temperatures favor varietal aroma development in grapes. Cool temperatures do increase the frequency of vegetable odors in 'Cabernet Sauvignon,' whereas berry aroma formation was favored under warmer conditions (Heymann and Noble, 1987). With 'Pinot noir,' several aroma compounds appear to be produced in higher amounts during long, cool seasons, compared to early, hot seasons (Watson *et al.*, 1988).

CHILLING AND FROST INJURY

Although cool sites may be beneficial in warm to hot climatic regions, prolonged exposure to temperatures below 10°C can induce irreversible physiological damage, retarding ripening (Becker, 1985a). This type of damage, called **chilling injury**, is usually reversible when of short duration. Injury apparently results from the excessive gelling of the semifluid cell membrane. This increases cell permeability, resulting in electrolyte loss and the disruption of respiratory, photosynthetic, and other cellular functions (George and Lyons, 1979). Several plant enzymes may also be irreversibly denatured by cool temperatures, for example chloroplast ATPase and RuBP carboxylase. Sensitive species are generally characterized by higher proportions of saturated fatty acids in their cell membranes. Saturated fatty acids maintain appropriate membrane fluidity at high temperatures, but make the membrane overly rigid at cool temperatures. In some plants, exposure to cool temperatures increases the proportion of linolenic acid (an unsaturated fatty acid) in cellular membranes.

Cool temperatures also increase the potential for dew formation, by raising atmospheric relative humidity. Dew on vine surfaces can increase the frequency and severity of several fungal infections.

Further cooling may result in ice crystal formation on and in plant tissues. As heat is lost from the tissues, water in the larger xylem vessels begins to freeze. Crystallization spreads to water in intercellular spaces and the cell walls of adjacent tissues. Water diffuses (or sublimes) out of the cytoplasm to replace that removed due to ice crystal formation. The resulting dehydration induces cytoplasmic shrinkage. Pulling of the membrane away from the cell wall increases protoplasmic concentration and osmolarity. This initiates protein and nucleic acid denaturation. If the tissues are not sufficiently cold acclimated, these changes can create irreversible damage, causing cell death (Guy, 2003). The degree of dehydration that results depends largely on the osmotic potential of the cell – the lower the osmotic potential, the less likely the chance of damaging dehydration. In addition, low osmotic potential depresses the freezing point of the cytoplasm, so that supercooling occurs rather than crystallization. If the temperature continues to drop, freezing eventually occurs, inducing irreparable cytoplasmic damage. This is thought to result from puncturing of cellular membranes when ice forms intracellularly. However, on a molecular level, the sequence of events leading to cell death is unclear. Further damage may result from tissue deformation, caused by differential expansion of the phloem and xylem (Meiering *et al.*, 1980).

The degree of disruption often depends as much on the rate of cooling and subsequent thawing as on the minimum temperature reached. If the temperature falls rapidly, extracellular water crystallization may occur more rapidly than water can diffuse out of the cytoplasm. As a result, the cytoplasm cannot supercool sufficiently rapidly to avoid intracellular crystallization. Rapid thawing can be as damaging to living cytoplasm as rapid temperature drops.

The tissues most sensitive to winter and spring cold damage are the primary buds (Plates 5.1 and 5.2), followed by the phloem and cambium of the bearing wood. Xylem tissue is the least vulnerable, but can be damaged, typically starting at the pith and moving outward. Pool (2000) and Goffinet (2004) present beautifully illustrated articles on the anatomical damage caused by freezing in the winter and spring, respectively.

If damage to the wood tissues is not too severe, there may be recovery. This involves undamaged cells in the region dedifferentiating into embryonic cells. They form a callus tissue, from which new cambial, and subsequently phloem and xylem cells develop. The rapidity of recovery depends on the extent to which nutrient flow is disrupted in the shoot. If severe, bud growth and development are poor. Thus, inadequate hormonal stimuli from the shoot apex retard development of new vascular tissue. Full recovery, if it occurs, can take months to years.

During the autumn and winter months, vine tissues progressively develop a degree of frost tolerance. The rate of acclimation appears to be time- and temperature-dependent (shorter periods at cold temperatures being equivalent to longer exposure at less frigid temperatures).

Xylem vessels, being nonliving, can withstand considerable ice crystal formation. Nevertheless, as winter approaches, water flow ceases and xylem vessels evacuate. This increases their relative insensitivity to cold damage. In the phloem, callose builds up at the cell plates, inhibiting nutrient and water flow. In buds, pectinaceous material appears to isolate the bud from shoot vascular tissues (Jones *et al.*, 2000). Anatomical features (Goffinet, 2004) are also thought to severely restrict the migration of water crystallization into buds from the shoot. Cells in the bud supercool, further diminishing the likelihood of intracellular water crystallization. This has been associated with increased levels of soluble sugars (Wample and Bary, 1992), oligosaccharides of the raffinose group (Hamman *et al.*, 1996), concentrations of proline (Aït Barka and Audran, 1997), and partial tissue dehydration (Wolpert and Howell, 1985). All these features decrease the osmotic potential and freezing point of the sap and, correspondingly, reduce the chance of ice-crystal formation in the cytoplasm. In addition, there is the synthesis of late embryogenesis abundant (LEA)-like proteins in bud tissue (Salzman *et al.*, 1996). In other plants, dehydrins (a subgroup of LEA proteins) are known to protect several enzymes and cellular membranes from freezing damage, providing a degree of frost tolerance. Finally, many organisms are known to produce extracellular antifreeze proteins (Griffith and Antikainen, 1996). Acclimation is especially marked in the vascular tissue and dormant buds. These tissues eventually can withstand temperatures down to $-20\,^{\circ}C$ or below.

Cold acclimation is speculated to have evolved from responses to high salinity and drought (Guy, 2003). The molecular changes involved are similar to those that have developed in response to osmotically related stress reactions.

Frost tolerance varies markedly among *Vitis* species (Table 5.2) as well as *V. vinifera* cultivars (Fig. 5.9).

Table 5.2 Cultivar classification and bud survival after minimum temperatures of $-20°$ to $-30°C$ on December 24, 1980

Cultivar classification	Bud survival, 1981 (%)		
	Primary	Secondary	Tertiary
Vitis labrusca	54.7	64.8	74.1
French-American hybrids	31.3	54.0	63.5
Vitis vinifera	11.3	14.7	23.8

Source: From Pool and Howard, 1985, reproduced by permission

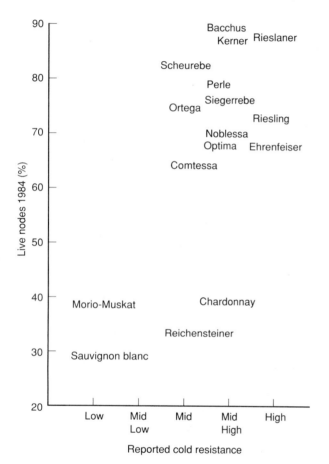

Figure 5.9 Effect of cold temperatures ($-24°C$) on bud survival of various cultivars compared with reported hardiness. (From Pool and Howard, 1985, reproduced by permission)

For example, 'Riesling,' 'Gewürztraminer,' and 'Pinot noir' are relatively winter-hardy, being able to withstand short exposures to temperatures down to $-26°C$ (when fully acclimated). Varieties such as 'Cabernet Sauvignon,' 'Sémillon,' and 'Chenin blanc' possess moderate hardiness, suffering damage at -17 to $-23°C$. 'Grenache' is considered tender and may suffer severe damage at $-14°C$. *Vitis labrusca* and French-American hybrid cultivars are relatively cold-hardy, and some *V.*

amurensis hybrids can survive prolonged exposure to temperatures well below $-30°C$.

Winter-hardiness is a complex factor and inadequately represented by the minimum temperature the cultivar can survive. Cold-hardiness is influenced by many genetic, viticultural, and environmental conditions. In addition, various parts of the vine show differential sensitivity to frost and winter damage. In the dormant bud, the most differentiated (primary) bud is the most sensitive, whereas the least differentiated (tertiary bud) is the most cold-hardy. Buds and the root system are generally less winter-hardy than old wood. For roots, this may be associated with their lower content of soluble sugars that accumulate during winter.

Although cold-hardiness is primarily a physiological property, anatomical features influence frost sensitivity. For example, hardy rootstocks generally have less bark tissue, containing relatively small phloem and ray cells, and possess woody tissues with narrow xylem vessels. Winter-hardy rootstocks appear to increase the resistance of the scion to cold damage. This may result from factors such as restrained scion vigor, modified contents of growth regulators, and earlier limitation of water availability.

In general, small vines have been considered more cold tolerant than large vines, except where vine size is restricted by stress factors such as disease, overcropping, and lime-induced chlorosis. Although this view has been challenged (Striegler and Howell, 1991), there is little doubt that late-season growth prolongs cellular activity and delays cold acclimation. This may develop because of growth-regulator or nutrient influences. Reduced carbohydrate accumulation could limit the supercooling of cellular fluids. During cold acclimation, starch is hydrolyzed to oligosaccharides (Fig. 5.10) and simple sugars. Fluctuations in the levels of glucose, fructose, raffinose and stachyose are particularly associated with cold hardiness in 'Chardonnay' and 'Riesling' vines (Hamman *et al.*, 1996). This change decreases the osmotic potential of the cytoplasm, reducing the freezing point.

The ability of surviving buds to replace winter-killed tissues is often critical to commercial success of cultivars in cool climates. The tendency of base buds of French-American hybrids to burst and develop flowering shoots often compensates for periodic severe winter damage. Even some *V. vinifera* cultivars, such as 'Chardonnay' and 'Riesling,' can withstand severe (80–90%) bud kill and produce a substantial crop from the activation of remaining buds (Pool and Howard, 1985). Correspondingly, adequate bud retention on well-matured canes is typically essential for producing profitable yields in cold viticultural regions.

The rates of cold acclimation and deacclimation also can be decisive to varietal success. With several cultivars,

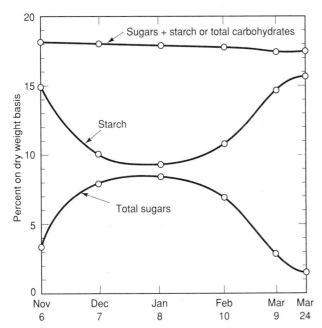

Figure 5.10 Seasonal interconversion of carbohydrates in cane wood, from starch to sugars and back, during fall cold acclimation and winter deacclimation, respectively. Total carbohydrate content remains almost constant. (From Winkler, 1934, reproduced by permission)

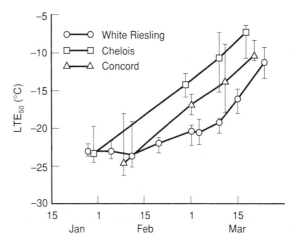

Figure 5.11 Deep supercooling (LTE_{50}) of dormant vines and deacclimation the primary buds of *V. vinifera* ('Riesling'), *V. labrusca* ('Concord'), and French-American hybrid ('Chelois') cultivars in 1983. Vertical bars indicate the limits of LTE_{10} and LTE_{90}. (From Andrews *et al.*, 1984, reproduced by permission)

such as 'Concord' (*V. labrusca*), 'Chelois' (French-American hybrid), and 'Riesling' (*V. vinifera*), winter-hardiness is inversely proportional to the rate of deacclimation during the winter (Fig. 5.11). As cold acclimation is influenced, albeit slightly, by the root-stock, an appropriate rootstock choice might augment commercial viability in particular sites (Miller *et al.*, 1988). Because rapid temperature changes are often particularly destructive, the frequency of rapid temperature fluctuations can be more limiting than the lowest yearly temperature might suggest.

MINIMIZING FROST AND WINTER DAMAGE

In an ideal world, vineyards would be situated in locations not subject to damaging frosts. However, avoiding frost-prone regions is often not an acceptable option. Thus, assessment of the likelihood of frost and winter damage is required if it is an unavoidable feature of the site. If the vineyard is distant from a regional meteorological station, on-site data, collected over several seasons, may show whether there is a strong or valuable correlation with regional averages. Because the cost-effectiveness of preventive measures varies considerably, the type and likely severity of damage are the best indicators of the appropriate means of minimizing cold temperature effects.

For frost protection, choices vary depending on whether it is **advective** or **radiative**. Advective frost results from the inflow of cold air, whereas radiative frost develops as heat is lost to the sky on cool still nights.

Where possible, selecting sites with slopes descending toward adjacent open valleys, lakes, and rivers minimizes the risk of advective frost. The slope allows cold air to flow down and away from the vineyard. The optimal site is often midway up the slope, associated with what is called the **thermal belt**, away from cold air collecting in the valley floor and air cooling along the ridge (see Geiger *et al.*, 2003). Forests or shelterbelts along the ridge can also deflect cold air around and away from vineyards. In addition, air circulation patterns created by the temperature differential between land and water can diminish the development of temperature inversions and radiative-frost development (Fig. 5.12). In the absence of, or in addition to, location along a slope, positioning fruiting wood high on tall trunks places the buds above the coldest air. The lowest temperatures usually occur at or just above ground level (Fig. 5.13). An alternate approach, where early fall frosts are a prime concern, vines on slopes may be trained close to the ground, as is typical in Champagne and Chablis. In this particular instance, heat radiated from the soil can combine with the flow of cold air to minimize the drop in air temperature and retard frost development.

Where practical, water sprinklers are particularly useful in reducing the incidence and severity of radiative frost. Permanently raised micro-sprayers are preferable. The heat of fusion, released as water freezes, can protect underlying tissue from freezing. It is critical that sprinkling continue until the sun or air temperature melts the ice coating. Otherwise, heat removed from the tissues, as the coating of ice melts, can cause frost injury. Water

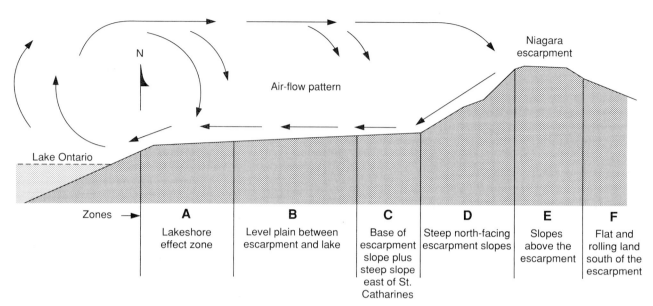

Figure 5.12 Influence of Lake Ontario on air circulation during frosty conditions over the Niagara Peninsula up to the Niagara escarpment. (From Wiebe and Anderson, 1976, reproduced by permission)

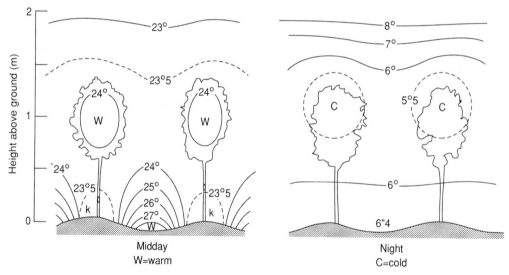

Figure 5.13 Temperatures at midday (*left*) and night (*right*) in a Rheinpfalz vineyard on September 17, 1933. (After Sonntag, 1934, reproduced by permission)

dispersion should also be as uniform and constant as possible. Otherwise, parts of the plant may freeze as heat is lost to the surrounding air or ice. Sprinkling is usually timed to begin when temperatures fall to about 1 °C. Application rates of between 2.5 and 3.5 mm/h are often adequate.

Keeping the soil immediately below vines free of vegetation can also be of value. Bare soil both absorbs more heat during the day and releases more heat during the night than does a grassy sward (Fig. 5.14). This can be sufficient to prevent frost damage in regions

susceptible to late spring frosts. This is especially effective if the soil surface is stony. Rock has a thermal conductivity 3 times that of wet sand, and 25 times that of dry sand.

Another frost-protective technique involves mixing air using wind machines (Shaw, 2002). These possess rotating blades positioned 4–7 m above the ground (Frazer *et al.*, 2006; Plate 5.3). The blades are angled downward at about 6° from vertical. Their rotation draws warmer air from above and blows it downward and outward about 100 m. Slow rotation of the head 360° every few

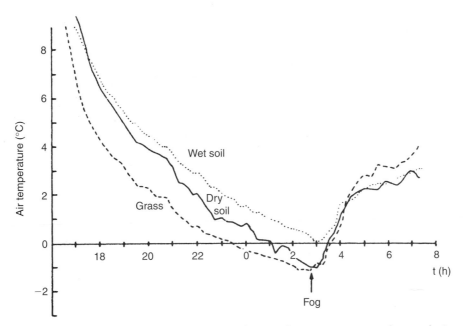

Figure 5.14 Illustration of the effect of soil covering and moisture content and atmospheric conditions (fog) on air temperatures measured 5 mm above grass (*dashed line*), dry soil (*solid line*) and wet soil (*dotted line*). (From Leuning and Cremer, 1988, reproduced with permission)

minutes directs air mixing in a wide circle. Where the temperature gradation of the inversion layer is marked and the land is flat, cold air at ground level can be effectively combined with warmer air above the vine canopy. Winds of about 3.5 km/h (2 miles/h) are often adequate to prevent frost damage. Wind machines may also be used in winter to protect against thermal inversion on cold, still, winter nights. Temperature differentials may reach 5–10 °C between air near the ground (0.6 m) and air at 20 m (Frazer *et al.*, 2006). Although effective, where urban housing developments approach (encroach) viticultural areas, home owners frequently complain about the droning noise generated by wind machines. Smudge pots, an older technique for increasing air turbulence (as well as air warming), are rarely used today due to air-pollution concerns.

Another option in the frost protection arsenal involves inoculating vines with ice-nucleation-deficient bacteria. By occupying the same ecological niche as ice-nucleating bacteria, they displace problem bacteria such as *Pseudomonas syringae*, *P. fluorescens* and *Erwinia herbicola*. Most strains of these bacteria possess a membranous lipoglycoprotein that facilitates ice nucleation at temperatures significantly above that of pure water. Frost injury has been correlated directly with epiphytic bacterial populations under frost conditions (Fig. 5.15). In their absence, ice crystallization may not commence until temperatures reach −8 to −11 °C. Without competition, the population of ice-nucleating bacteria on leaf surfaces generally increases in the autumn. Therefore, displacing them with ice-nucleation-deficient strains

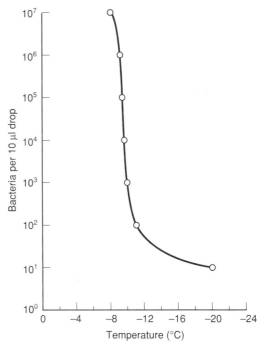

Figure 5.15 Effect of cell concentration on the nucleating activity of *Pseudomaonas fluorescens* F-12. Data points represent the temperature at which 90% of the test crops were frozen. (From Maki and Willoughby, 1978; *J. Appl. Microbiol.*, reproduced by permission of the American Meteorological Society)

can reduce the possibility of frost damage. Although the elimination ice-nucleating bacteria might also be achieved by selective antibiotic use, or copper-based sprays, biological control through the application of competitive ice-nucleation-deficient strains is preferable (Lindemann and Suslow, 1987). A commercially available preparation of *P. fluorescens* A506, a nutritionally versatile ice-nucleation-deficient bacterium, is marketed under the name BlightBan® A506.

A potential alternative is spraying with a linear polymer, such as polyglycerol (Supercool Z-1000). It effectively suppresses the action of ice nucleating proteins of bacteria (Wowk and Fahy, 2002). Other experimental cryoprotectants function by maintaining the supercooled status of fluids in dormant buds, especially in the late winter, when vines commence cold deacclimation (Himelrick *et al.*, 1991; Dami *et al.*, 1996). Another cryoprotectant, sodium alginate, appears to work by delaying bud break until the weather is frost-free (Dami *et al.*, 2000). Still other cryoprotectants may protect vine tissues from frost damage after bud break. Commercial cryoprotectants, such as Antistress® (an acrylic copolymer), form a gas-permeable membrane when sprayed on plant surfaces. The membrane is reported to protect plant surfaces from frost (as well as reducing evaporative water loss during summer droughts). An additional chemical means of reducing cold damage involves the application of growth retardants. Application about flowering has been noted to enhance cold-hardiness during the subsequent winter (Shamtsyan *et al.*, 1989).

Where late-spring frosts are frequent, delayed pruning is often recommended. By deferring pruning, the degree of pruning can reflect the extent of winter injury. This may be assessed by opening a representative sampling of buds and looking for indications of bud

viability (see Pool, 2000). An alternate approach is double-pruning. In this procedure, more buds are left per vine than normally considered appropriate. After bud break, when the degree of damage is known, the unnecessary buds are removed.

Where trunk winterkill is a major problem, retaining multiple trunks is often practiced. Double trunking is common in northeastern North America, and occasionally vines with up to five trunks of differing ages are developed. It is unlikely that all trunks will suffer equal damage in the same year. Additional trunks also limit the damage caused by crown gall or Eutypa dieback, both of which tend to be considerable problems in cold vineyard regions.

A potentially useful technique, especially for grafted vines, is placing bails of straw, closed-cell polyethylene sheeting (Bordelon, 1966), or geotextile fabric (Khanizadeh *et al.*, 2005) around the vines in late fall. Some of their beneficial effects result from favoring the collection of isolating layers of snow. These techniques are effective and much simpler than laying down and burying the whole vine for the winter. This is frequently done in frigid regions of China, Russia, and the north central United States. The difficulty and labor involved in this procedure, as well as the potential damage caused to the trunk(s), limit its use. Nevertheless, some protective layerings (if not covered with snow) may need to be removed on sunny days. Otherwise, heat accumulation could induce undesirably early bud deacclimation.

In less rigorous locations, positioning empty tubes around the trunk (Fig. 5.16) may prove useful. Earth

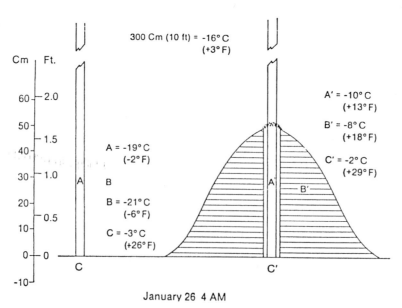

Figure 5.16 Temperatures at various locations on January 26, 1983, associated with exposed or hilled tissues. The vine trunk is contained in 15 cm plastic tubes. (From Pool and Howard, 1985, reproduced by permission)

pushed up around the tube provides insulation for the enclosed trunk. Although the graft union is colder than when buried, it is considerably warmer than unhilled graft regions.

Cultural conditions also affect winter-hardiness. In general, those features that favor balanced growth favor cold-hardiness. Reduced nitrogen availability, by limiting vegetative growth, might promote winter-hardiness. However, data from Wample *et al.* (1991) suggest its influence is modest at best. Standard levels of nitrogen fertilization have not been observed to have negative effects (Wample *et al.*, 1993). Phosphorus alone, or its proportion relative to nitrogen (Mikhailuk *et al.*, 1978), or calcium application (Eifert *et al.*, 1986), have all been reported to significantly increase winter bud survival. Limiting irrigation, by restricting water uptake, can facilitate xylem evacuation – a component of cold acclimation. Deacclimation is partially associated with the refilling of vessels with water. Groundcovers, which tend to collect snow, favor insulating the trunk and root from rapid temperature fluctuations and severe cold. Windbreaks also can favorably influence snow cover and its distribution.

Solar Radiation

As noted earlier, temperature plays a major role in regulating the yearly growth cycle of the vine. Nevertheless, it is solar radiation that generates the annual climatic cycle, and provides the energy for photosynthesis. The degree to which soil and vines receive light and heat radiation from the sun is largely a function of the angle at which the solar rays impact the Earth's surface.

The solar angle varies with the time of day, the season, the latitude, and the inclination and orientation of the site. To comprehend these factors, it is beneficial to understand certain aspects of the Earth's rotation and movement around the sun, and some principles of atmospheric and radiation science.

Solar radiation provides the energy that heats the Earth. Because its distribution is nonuniform, both geographically and temporally, it generates the forces that produce the Earth's weather patterns. The major fluctuations in solar energy input are caused by the cyclic nature of the Earth's two principal motions, rotation and revolution. Rotation refers to Earth's spin around its inclined axis. This creates our earthly 24 h day–night cycle, and the arc through which the sun seems to pass during the day. Because the sun's position in the sky is constantly changing, the angle at which solar radiation impacts the Earth is continually changing. This creates the varying level of energy reaching any point on the Earth's surface throughout the day.

The other motion, revolution, is the movement of the Earth on its orbit around the sun. This produces the cyclical changes in the Earth's inclination toward the sun, and the changing seasons away from the equator (Fig. 5.17). During the summer solstice, the Northern Hemisphere is maximally exposed to the sun's energy and is experiencing its longest day of the year, whereas the Southern Hemisphere is experiencing its shortest day. As the Earth continues its revolution around the sun, the days in the Northern Hemisphere progressively become shorter as the Earth's axis tilts away from the sun, and the sun appears to rise and set further south. During the winter solstice, the Northern

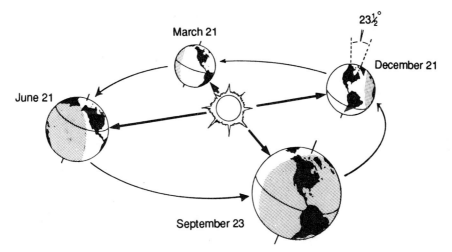

Figure 5.17 Path of Earth around the sun. The orbital characteristics include the tilt of the Earth's axis (23.5°), the direction of the axis, and the elipticity of the Earth's orbit (exaggerated here). The Northern Hemisphere summer is shown at left. (From Pisias and Imbrie, 1986–1987, reproduced by permission)

Hemisphere is tilted at an angle of 23.5° away from the sun, whereas the Southern Hemisphere is tilted at 23.5° toward the sun.

The Earth's revolution around the sun has several important effects on plant and animal life. Changing day length produces a yearly photoperiodic cycle that is used by many plants to regulate yearly growth cycles. However, most *V. vinifera* cultivars are relatively insensitive to photoperiod, the yearly cycle being influenced primarily by annual temperature fluctuations.

In addition to the obvious effect on day length, the revolution of the Earth around the sun significantly affects the intensity and spectral quality of sunlight. The intensity of solar radiation reaching the Earth decreases as the sun's altitude becomes lower. This principally results from dispersion of solar energy over an increasingly large surface area, especially at high latitudes. At lower latitudes (~20°), there is little change in the intensity of noon radiation received from spring through fall (~6%). At high latitudes (50°), there is a 28% variation in radiation intensity during the same period. Corresponding winter–summer differences in energy input at 50° latitude are about 27 and 68%, respectively.

Seasonal variation in solar input at different latitudes is considerably affected by the azimuth. The azimuth refers to the arching path through which the sun appears to move across the sky. As the summer passes and the Earth's axis tilts away from the sun, the sun appears to rise closer to the equator and its arc through the sky is progressively lower. This increases the atmospheric depth through which the incoming radiation must pass, both decreasing its intensity and modifying its spectral quality. These changes result from increased absorption and reflection as the radiation interacts with atmospheric gases and particles. This is similar to, but less marked than, the effect clouds (water droplets) have on the warmth of sunlight. Part of the benefit of sun-facing slopes at high latitudes is to offset some of the negative effects of low sun altitudes on solar diffusion over a large surface.

Various wavelengths of solar radiation are affected differently by passage through the atmosphere – the longer the path, the greater the effect. These influences are particularly significant in the visible portion of the spectrum (Henderson, 1977). Scattering enriches diffuse radiation (skylight) in blue wavelengths. Although reducing the photosynthetic value of low-angle direct sunlight, scattering increases the relative importance of skylight. Atmospheric dispersion and absorption of short-wave infrared solar radiation diminish the warming effect of sunlight at low sun angles.

The effects of atmospheric passage on the relative strengths of visible and short-wave infrared radiation

Table 5.3 Contribution of clear-sky diffuse solar radiation to total solar radiation at different wavelengths at a sun altitude of 60°

Wavelength (nm)	Spectral region	Fraction arriving as diffuse	Fraction arriving in direct beam
300	Ultraviolet	0.72	0.28
400	Blue	0.33	0.67
500	Green	0.19	0.81
600	Yellow	0.13	0.87
700	Red	0.09	0.91
1000	Far red	0.05	0.95
2000	Infrared	0.02	0.98

Source: Data from Schulze, 1970, modified by Miller, 1981, reproduced by permission

in direct and diffuse light are given in Table 5.3. These effects produce not only the blue color of the sky on sunny days, but also the diffuse radiation that comes from the sky. Because of the efficiency with which blue light (400–500 nm) is absorbed by chlorophyll, diffuse sunlight is often as photosynthetically active as direct sunlight. Diffuse radiation also penetrates more deeply into leaf canopies, because of its hemispherical origin. Because the two states of phytochrome differ in blue-light absorption, this might have important physiological consequences. The nondirectional origin of diffuse light under cloud cover also neutralizes the benefits normally gained from sun-facing slopes. Even under sunny conditions, skylight contains little solar infrared radiation. Thus, diffuse radiation provides little solar heating – the coolness of tree shade in the summer is well known. Most solar heating comes from direct beam sunlight.

PHYSIOLOGICAL EFFECTS

Light has multiple effects on grapevine growth. The most significant is activating photosynthesis, the energy source for vine growth. Most of the radiant energy comes from beam radiation or skylight, with as little as 3.5% coming from light reflected from the soil or adjacent vines (Smart, 1973). The proportion of light received directly by the horizontal (top) versus vertical (sides) of the vine changes (Fig. 5.18) depending on the shape and size of the vine and the angle of the leaves to the sun. However, the influence of vine shape and size on the shading of internal leaf layers is of greater significance.

When light directly impinges on a leaf, about 85–90% of the photosynthetically active radiation (PAR) is absorbed (Fig. 5.19). Of the visible region (400–700 nm), the photosynthetically active portions (blue

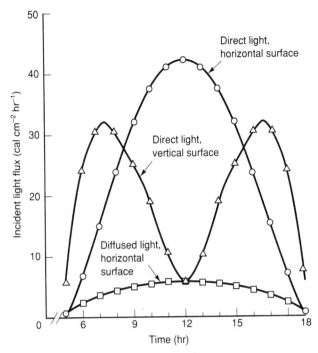

Figure 5.18 Flux densities of direct sunlight on horizontal (○) and vertical surfaces (△), and of diffused sky light on a horizontal surface (□), December 22, 35°S. (From Smart, 1973, reproduced by permission)

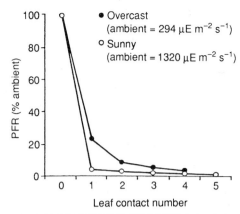

Figure 5.19 Attenuation of photosynthetically active radiation (PFR) by 'Merlot' canopies on sunny and overcast days in Bordeaux, expressed as a percentage of above-canopy ambient. (From Smart, 1985, reproduced by permission)

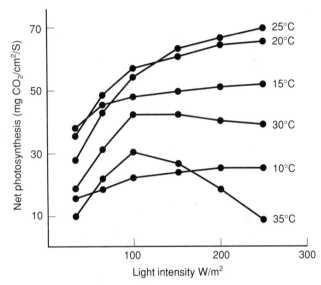

Figure 5.20 Influence of temperature on the photosynthetic capacity of grapevine leaves. (After Törökfalvy and Kriedemann, 1977, reproduced by permission)

and red) are efficiently absorbed. In contrast, radiation in the green spectrum is either reflected or transmitted, giving leaves their green color. The efficiency of light absorption by leaves greatly reduces both the intensity and quality of the radiation transmitted to inner canopy foliage. Second layer leaves under sunny conditions usually receive less than 10% of the exterior canopy radiation, and a third layer may receive only about 1–2% of the surface radiation (Fig. 5.19). In addition, the selective removal of PAR further reduces its photosynthetic value. The combined effect of the quantitative and qualitative changes in solar radiation produced by shading limits the effective photosynthetic canopy to the exterior and first inner leaf layers of the vine (Smart, 1985). About 80–90% of the photosynthate is produced by the exterior canopy. Inner leaves eventually turn yellow and drop prematurely.

The light intensity of direct sunlight can reach about 1000 W/m² (~10,000 footcandles). Grapevine leaves can use little more than about one-quarter of that value (Kriedemann and Smart, 1971). Because full intensity sunlight cannot be used photosynthetically, most of the excess absorbed radiation is liberated as heat. If the leaves are fully exposed and the roots have an ample supply of water for rapid transpiration, the increase in leaf temperature may be held to about 5°C. If the

leaves are water deficient, or there is little air movement, heating may be sufficient to suppress photosynthesis. For example, the rate of photosynthesis at 35°C may be only 15% of that at 25°C (Fig. 5.20). The actual reduction depends on the variety, the water supply, the light intensity, and the variety's adaptation to high light intensities. Nevertheless, surface leaves often do not photosynthesize maximally because of heating in full sunlight. Second-layer leaves are less likely to experience photosynthetic limitation because of solar heating, but they are exposed to a light regime close to the compensation point. Diffuse sky light or radiation reflected from the soil and adjacent vines may augment

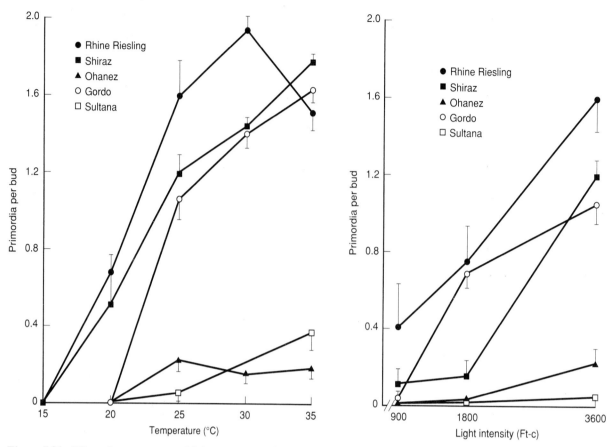

Figure 5.21 Effect of temperature and light intensity on the average number of bunch primordia per bud for the basal 12 buds of five grape varieties. (From Buttrose, 1970, reproduced by permission)

the light transmitted through the foliage. Even temporary breaks in the canopy cover (sunflecks) can significantly improve shade leaf photosynthesis. Nevertheless, most internal leaves contribute little to the photosynthate supply of the vine.

In addition to the direct and indirect effects of shading on photosynthesis, shading affects cane maturation, inflorescence initiation, grape maturation, and the aromatic attributes of the fruit. It has long been known that sun exposure improves cane maturation. Whether this is due to the intensity or spectral quality of the sunlight is unknown. It is also possible that sunlight may enhance cane maturation through an influence on heating and drying of the shoot surface.

Flower-cluster initiation begins in early summer, and development stops only in the fall. Sunlight affects both inflorescence initiation and fruitfulness. Sun-exposed buds generally bear more flowers per cluster than buds developed under shaded conditions. As with other properties, there is considerable variation in the response of particular cultivars to light-induced inflorescence initiation (Fig. 5.21).

In most regions, it is common to position shoots to increase fruit exposure to the sun. This long-standing practice favors fruit coloration and maturation. In contrast, training may be designed to protect the fruit from direct sun exposure in hot arid environments. Sun exposure often raises the temperature of dark-colored berries 5 °C above that of the surrounding air (Fig. 5.22), and berry temperatures can reach 15 °C above ambient (Smart and Sinclair, 1976). Sun exposure also influences grape composition. Although the response is cultivar-specific, sun-exposed berries generally are lower in pH, as well as in potassium and malic acid content. The berries also are generally smaller. Shaded grapes show the opposite tendencies. Sun-exposed berries typically show properties associated with higher quality, namely higher °Brix, anthocyanin, and tartaric acid levels, as well as better pH and malic acid levels. Shading is associated with greater vegetative character of some cultivars (Arnold and Bledsoe, 1990). Whether this is due to intensity, spectral quality, or thermal effects on fruit or canopy shading is unclear. Although sun exposure is usually beneficial, excessive

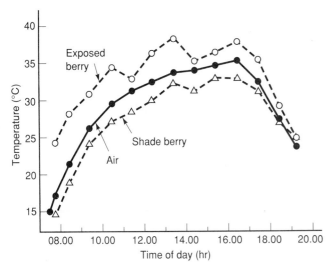

Figure 5.22 Temperature of exposed and shaded 'Carignan' berries, relative to the air temperature at cluster height on February 10, 1972. (After Millar, 1972, reproduced by permission)

exposure can produce sunburn or baking, which generate atypical wine fragrances.

Wind

Unless strong winds are a characteristic feature of a region, wind seldom is considered in vineyard selection or row alignment. In regions such as the Valtellina area of Lombardy or the Margaret River area of Western Australia, strong winds can severely limit grape cultivation. Severe winds can cause both lingering physiological and extensive physical damage to the vine. The benefits of shelterbelts (rows of trees) or windbreaks (any nonliving fence or structure) on reduced soil erosion and evapotranspiration have made them a common landscape feature in the central plains of North America and the southern Rhône Valley of France. Nevertheless, shelterbelts are not a typical feature of vineyard regions. This may reflect not only the cost of planting or construction, but also a failure to realize their considerable benefits. Instead, cultural practices such as hedging (Switzerland) and cane intertwining (Lombardy) have been more common. Where severe winds occur primarily in the early spring, late pruning can provide up to a 2-week delay in bud break (Hamilton, 1988).

With appropriate design, shelterbelts reduce both wind velocity and turbulence. If the shelterbelt is porous, the downwind influence can extend up to 25 times the height of the barrier (Fig. 5.23). Dense shelterbelts are less effective in reducing wind speed and usually increase wind turbulence.

Strong winds not only produce physical damage, but they also can reduce shoot length, leaf size, and stomatal

density; induce smaller and fewer clusters per vine; retard ripening; and generate lower soluble solids (Dry, 1993; Bettiga *et al.*, 1996). Several of these effects may result from stomatal closure, caused by an increased rate of water loss. In the leaves, carbon dioxide availability subsequently declines and the oxygen content rises, both of which reduce photosynthesis and vine growth. In addition to the immediate effects of wind exposure, physiological influences may linger long after the wind speed has diminished (Freeman *et al.*, 1982). Because the benefits of a shelterbelt increase with wind velocity, the value of shelterbelts is proportional to the average and peak wind speeds.

Occasionally, concern has been expressed about shading from shelterbelts and its effect on photosynthesis. Generally, this is not a significant factor as only the vines adjacent to the shelterbelt are likely to be affected. In addition, with a north–south alignment of the shelterbelt, the areas shaded in the morning and evening receive additional radiation reflected from the shelterbelt in the afternoon and morning, respectively. It also has been suggested that reduced evapotranspiration in the shaded area may permit prolonged photosynthetic activity due to reduced water stress (Marshall, 1967).

Because of reduced wind velocity and air mixing, temperature inversions in sheltered vineyards may be more marked. Consequently, shelterbelts may increase the frequency and severity of spring and fall frosts. This probability is reduced if the vineyard topography permits the free flow of cool air out and away from the vineyard. In addition, the potential detrimental effect on frost occurrence may be offset by wind turbulence produced between opposing stands of trees (see Geiger *et al.*, 2003).

Another potential problem associated with shelterbelt use is higher atmospheric humidity. Although this could favor disease development, the condition seldom arises as most environments benefitting from wind protection are generally dry.

The floras that develop in association with the shelterbelt can both be a refuge for grapevine pests, as well as the predators and parasites of the same pests. The relative merits of a shelterbelt, from the perspective of pest control, must be assessed on a regional basis. Alternative, abiotic, portable windbreaks can be constructed of woven plastic cloth. This avoids any shading effects and competition for nutrients and water; has almost no effect on valuable land use; and does not provide a haven for pests or pathogens.

Aligning vine rows in relation to the prevailing wind can significantly modify the effects of shelterbelts and windbreaks. Positioning rows at an oblique angle to the wind turns the outermost rows into a low shelterbelt. For example, vineyard rows aligned at angles other

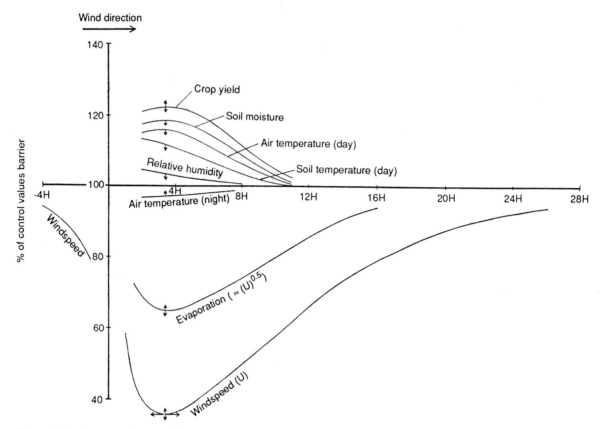

Figure 5.23 Diagram of barrier effects on microclimate. Arrows indicate the directions in which the values of different factors vary relative to unsheltered areas. H refers to barrier height. (From Marshall, 1967, reproduced by permission)

than the normally preferred north–south orientation may retain heat by limiting the wind flow down the rows (Brandtner, 1974). Thus, in windy climates, slopes facing other than due south (in northern latitudes) may be preferable because of the protection provided from northerly winds.

In contrast, there may be advantages in planting the rows parallel to the prevailing wind under dry conditions (Hamilton, 1985). Winds moving directly down the rows produce less foliage drag and minimize water loss (Hicks, 1973). It is estimated that this could produce up to a 10–20% saving in water demand. Wind moving across rows creates turbulence and greater foliage movement than moving unimpeded down rows. Turbulence may increase relative to the density of the grapevine canopy.

Water

Most of the water absorbed by roots is used as a coolant. The evaporation of water from the leaves helps maintain normal physiology by minimizing overheating. When water loss continues to exceed replacement from the roots, the stomata close. As the tissues overheat,

most metabolic activity slows. Physiological activity returns to normal only some time after the water deficit has been overcome. If the deficit persists, continued water loss through the cuticle results in cell plasmolysis and death. Mature vines seldom show marked wilting due to an extensive root system that can reach water meters below the soil surface. Nevertheless, physiological disruption occurs long before wilting. Reduced shoot growth is one of the most sensitive signs of grapevine water stress.

Where irrigation is permitted and economically feasible, the damage caused by naturally occurring water shortages can be prevented. Inexpensive and reliable sources of irrigation water are essential for commercial viticulture in most semiarid and arid climates. Water sprinkling systems may limit heat stress, but are usually unnecessary or ineffectual.

Problems with excessive rainfall, fog, or high humidity are often more serious, or at least more difficult to control. Good soil porosity and drainage diminish problems associated with heavy rainfall. Basal leaf removal and divided-canopy training systems, with their larger surface areas, can further increase water evaporation. Angling vineyard rows to increase wind turbulence around vines can further enhance the surface drying of the foliage and fruit. Nevertheless, the enhancement of fungal diseases under foggy, humid conditions means that they are likely to be controlled only with fungicides. Planting new, more resistant cultivars may be impractical because of Appellation Control restrictions (in Europe) or marketing concerns. Generally, areas of high humidity are avoided because of the severity of disease losses in such regions.

Water also has a major influence on global and regional vineyard distribution, due to the moderating effect of rivers, lakes, and oceans. This impact is seen in the proximity of most vineyard regions to the cool edges of warm continents or warm edges of cool continents (de Castella, 1912).

Suggested Readings

Soil

Brady, N. C., and Weil, R. R. (2001) *The Nature and Properties of Soils*, 13th edn. Prentice Hall, New York, NY.

Hancock, J. M., and Price, M. (1990) Real chalk balances the water supply. *J. Wine Res.* **1**, 45–60.

Kramer, P. J., and Boyer, J. S. (1995) Water Relations of Plants and Soils. San Diego, CA.

Morlat, R., Remoue, M., and Pinet, P. (1984) Influence de la densité de plantation et du mode d'entretien du sol sur l'enracinement d'un peuplement de vigne plante en sol favorable. *Agronomie* **4**, 485–491.

Paul, E. A., and Clark, F. E. (1996) *Soil Microbiology and Biochemistry*, 2nd edn. Academic Press, San Diego, CA.

Schreier, P. (1983) Correlation of wine composition with cultivar and site. In: *Grape and Wine Centenary Symposium Proceedings*, pp. 63–95. Roseworthy Agricultural College, Australia.

Seguin, G. (1986) Terroirs and pedology of wine growing. *Experientia* **42**, 861–873.

White, R. E. (2003) *Soils for Fine Wines*. Oxford University Press, New York, NY.

Climate

Becker, N. (1985) Site selection for viticulture in cooler climates using local climatic information. In: *Proceedings of the International Symposium on Cool Climate Viticulture and Enology* (D. A. Heatherbell *et al.*, eds.), pp. 20–34. Agriculture Experimental Station Technical Publication No. 7628. Oregon State University, Corvallis, OR.

Geiger, R., Aron, R. H., and Todhunter, P. (2003) *The Climate Near the Ground*, 6th edn. Rowman & Littlefield, Lanham, MD.

Gladstone, J. (1992) *Viticulture and Environment*. Winetitles, Adelaide, Australia.

Jones, H. G. (1992) *Plants and Microclimate. A Quantitative Approach to Environmental Plant Physiology*, 2nd edn. Cambridge University Press, Cambridge, UK.

Koblet, W. (1985) Influence of light and temperature on vine performance in cool climates and applications to vineyard management. In: *Proceedings of the International Symposium on Cool Climate Viticulture and Enology* (D. A. Heatherbell *et al.*,

eds.), pp. 139–157. Agriculture Experimental Station Technical Publication No. 7628. Oregon State University, Corvallis, OR.

Smart, R. E. (1985) Some aspects of climate, canopy microclimate, vine physiology, and wine quality. In: *Proceedings of the International Symposium on Cool Climate Viticulture and Enology* (D. A. Heatherbell *et al.*, eds.), pp. 1–19. Agriculture Experimental Station Technical Publication No. 7628. Oregon State University, Corvallis.

Climate Indices

Bentryn, G. (1988) World climate patterns and viticulture. In: *Proceedings of the 2nd International Symposium for Cool Climate Viticulture and Oenology* (R. E. Smart, *et al.*, eds.), pp. 9–12. New Zealand Society of Viticulture and Oenology, Auckland, New Zealand.

Jackson, D. I., and Cherry, N. J. (1988) Prediction of a district's grape-ripening capacity using a latitude-temperature index (LTI) *Am. J. Enol. Vitic.* **39**, 19–28.

Kenny, G. J., and Harrison, P. A. (1992) The effects of climate variability and change on grape suitability in Europe. *J. Wine Res.* **3**, 163–183.

McIntyre, G. N., Kliewer, W. M., and Lider, L. A. (1987) Some limitations of the degree day system as used in viticulture in California. *Am. J. Enol. Vitic.* **38**, 128–132.

Cold-Hardiness and Frost Protection

Evans, R. G. (2000) The art of protecting grapevines from low temperature injury. In: *Proceedings of the ASEV 50th Anniversary Annual Meeting*, Seattle, Washington, June 19–23, 2000, pp. 60–72. American Society for Enology and Viticulture, Davis, CA.

Goffinet, M. S. (2000) The anatomy of low-temperature injury of grapevines. In: *Proceedings of the ASEV 50th Anniversary Annual Meeting*, Seattle, Washington, June 19–23, 2000, pp. 94–100. American Society for Enology and Viticulture, Davis, CA.

Howell, G. S. (2000) Grapevine cold-hardiness: Mechanisms of cold acclimation, mid-winter hardiness maintenance, and spring deacclimation. In: *Proceedings of the ASEV 50th Anniversary Annual Meeting*, Seattle, Washington, June 19–23, 2000, pp. 35–48. American Society for Enology and Viticulture, Davis, CA.

Lee, R. E., Warren, C. J., and Gusta, L. V. (1995) *Biological Ice Nucleation and Its Applications*. APS Press, St Paul, MN.

Poghossian, K. S. (1985) Biological principles of grape cultivation in high-stemmed training systems under the continental climate conditions of southern USSR. In: *Proceedings of the International Symposium on Cool Climate Viticulture and Enology* (D. A. Heatherbell *et al.*, eds.), pp. 217–226. Agriculture Experimental Station Technical Publication No. 7628, Oregon State University, Corvallis.

Wample, R. L., and Wolf, T. K. (1996) Practical considerations that impact vine cold-hardiness. In: *Proceedings of the 4th International Symposium on Cool Climate Viticulture and Enology* (T. Henick-Kling *et al.*, eds.), pp. II-23–38. New York State Agricultural Experimental Station, Geneva, NY.

Wample, R. L., Hartley, S., and Mills, L. (2000) Dynamics of grapevine cold-hardiness. In: *Proceedings of the ASEV 50th Anniversary Annual Meeting*, Seattle, Washington, June 19–23, 2000, pp. 81–93. American Society for Enology and Viticulture, Davis, CA.

Wilson, S. (2001) *Frost Management in Cool Climates*. Grape and Wine Research and Development Corporation. Tasmania. http://www.gwrdc.com.au/downloads/ResearchTopics/UT%2099-1.pdf

Wisniewski, M., Wolf, T. K., and Fuchigami, L. (1996) Biochemical and biophysical mechanisms of cold hardiness in woody plants. In: *Proceedings of the 4th International Symposium on Cool Climate Viticulture and Enology* (T. Henick-Kling *et al.*, eds.), pp. II-14–22. New York State Agricultural Experimental Station, Geneva, NY.

Wind Protection

Hamilton, R. P. (1988) Wind effects on grapevines. In: *Proceedings of the 2nd International Symposium for Cool Climate Viticulture and Oenology* (R. E. Smart *et al.*, eds.), pp. 65–68. New Zealand Society of Viticulture and Oenology, Auckland, New Zealand.

Kobriger, J. M., Kliewer, W. M., and Lagier, S. T. (1984) Effects of wind on water relations of several grapevine cultivars. *Am. J. Enol. Vitic.* 35, 164–169.

Ludvigsen, K. (1989) Windbreaks – some considerations. *Aust. Grapegrower Winemaker* 302, 20–22.

van Eimern, J., Karschon, R., Razumova, L. A., and Robertson, G. W. (1964) Windbreaks and Shelterbelts. World Meteorological Organization Technical Note No. 59, Geneva, Switzerland.

References

Aït Barka, E., and Audran, J. C. (1997) Response of champenoise grapevine to low temperatures: Changes of shoot and bud proline concentrations in response to low temperatures and correlations with freezing tolerance. *J. Hortic. Sci.* 72, 577–582.

Amerine, M. A., and Winkler, A. J. (1944) Composition and quality of musts and wines of California grapes. *Hilgardia* 15, 493–675.

Andrews, P. K., Sandidge, C. R., and Toyama, T. K. (1984) Deep supercooling of dormant and deacclimating *Vitis* buds. *Am. J. Enol. Vitic.* 35, 175–177.

Arnold, R. A., and Bledsoe, A. M. (1990) The effect of various leaf removal treatments on the aroma and flavor of Sauvignon blanc wine. *Am. J. Enol. Vitic.* 41, 74–76.

Ballif, J. L., and Dutil, P. (1975) The warming of chalk soils by plastic films. Thermic measurements and evaluations. *Ann. Agron.* 26, 159–167.

Bear, F. E. (1953) *Soils and Fertilizers*, 4th edn. John Wiley, New York.

Becker, N. (1985a) Panel – Session I. In: *Proceedings of the International Symposium on Cool Climate Viticulture and Enology* (D. A. Heatherbell *et al.*, eds.), pp. 35–42. Agriculture Experimental Station Technical Publication No. 7628, Oregon State University, Corvallis.

Becker, N. (1985b) Site selection for viticulture in cooler climates using local climatic information. In: *Proceedings of the International Symposium on Cool Climate Viticulture and Enology* (D. A. Heatherbell *et al.*, eds.), pp. 20–34. Agriculture Experimental Station Technical Publication No. 7628, Oregon State University, Corvallis.

Bentryn, G. (1988) World climate pattern and viticulture. In: *Proceedings of the 2nd International Symposium for Cool Climate Viticulture and Oenology* (R. Smart *et al.*, eds.), pp. 9–12. New Zealand Society of Viticulture and Oenology, Auckland, New Zealand.

Bettiga, L. J., Dokoozlian, N. K., and Williams, L. E. (1996) Windbreaks improve the growth and yield of Chardonnay grapevines grown in a cool climate. In: *Proceedings of the 4th International Symposium on Cool Climate Viticulture and Enology* (T. Henick-Kling, *et al.*, eds.), pp. II-43–46. New York State Agricultural Experimental Station, Geneva, NY.

Bootsma, A. (1976) Estimating minimum temperature and climatological freeze risk in hilly terrain. *Agric. Meteorol.* 16, 425–443.

Bordelon, B. (1996) Winter protection of cold-tender grapevines with insulating materials. In: *Proceedings of the 4th International Symposium on Cool Climate Viticulture and Enology* (T. Henick-Kling, *et al.*, eds.), pp. I-32–35. New York State Agricultural Experimental Station, Geneva, NY.

Bowers, S. A., and Hanks, R. J. (1965) Reflection of radiant energy from soils. *Soil Sci.* 100, 130–138.

Brady, N. C. (1974) *Nature and Properties of Soil*, 8th edn. Macmillan, New York.

Brandtner, E. (1974) Die Bewertung geländeklimatischer Verhältnisse in Weinbaulagen. Deutscher Wetterdienst, Zentralamt, Offenbach, Germany.

Bridley, S. F., Taylor, R. J., and Webber, T. J. (1965) The effects of irrigation and rolling on nocturnal air temperatures in vineyards. *Agric. Meteorol.* 24, 373–383.

Büttner, K., and Sutter, E. (1935) Die Abkühlungsgröße in den Dünen etc. *Strahlentherapie* 54, 156–173.

Buttrose, M. S. (1970) Fruitfulness in grape-vines, the response of different cultivars to light, temperature and daylength. *Vitis* 9, 121–125.

Cass, A., and Maschmedt, D. (1998) Vineyard soils: Recognizing structural problems. *Aust. Grapegrower Winemaker* 412, 23–26.

Chapman, H. D. (1965) Chemical factors of the soil as they affect microorganisms. In: *Ecology of Soil-Borne Plant Pathogens.* (K. F. Baker and W. C. Snyder, eds.), pp. 120–141. John Murray, London.

Coony, J. J., and Pehrson, J. E. (1955) Avocado Irrigation. Leaflet No. 50. California Agriculture Extension Station, Oakland.

Dami, I., Hamman, R., Stushnoff, C., and Wolf, T. K. (2000) Use of oils and alginate to delay budbreak of grapevines. In: *Proceedings of the American Society of Enology and Viticulture 50th Anniversary Annual Meeting*, Seattle, Washington, June 19–23, 2000, pp. 73–76. American Society for Enology and Viticulture, Davis, CA.

Dami, I., Stushnoff, C., and Hamman, R. A. Jr. (1996) Overcoming spring frost injury and delaying bud break in *Vitis vinifera* vv. Chardonnay grapevines. In: *Proceedings of the 4th International Symposium on Cool Climate Viticulture and Enology* (T. Henick-Kling *et al.*, eds.), pp. II-67–70. New York State Agricultural Experimental Station, Geneva, NY.

de Blij, H. J. (1983) *Wine: A Geographic Appreciation.* Rowman and Allanheld, Totowa, NJ.

de Castella, F. (1912) The influence of geographical situation on Australian viticulture. *Vict. Geog. J.* 29, 38–48.

Dry, P. R. (1993) Exposure to wind affects grapevine performance. *Aust. Grapegrower Winemaker* 352, 73–75.

Eifert, J., Szöke, L., Varnai, M., and Nagy, E. (1986) The effect of liming on the frost resistance of grape buds. *Szölötermesztés Boraszat.* 8, 34–36. (1987. *Vitis Abstr.* 26, 99).

Frazer, H., Slingerland, K., and Ker, K. (2006) Wind machines for protecting grapes and tender fruit crops from cold injury. Ontario Ministry of Agriculture, Food and Rural Affairs. http://omafra.gov.on.ca/english/engineer/ facts.windmach_info.htm.

Freeman, B. M., Kliewer, W. M., and Stern, P. (1982) Influence of windbreaks and climatic region on diurnal fluctuation of leaf water potential, stomatal conductance and leaf temperature of grapevines. *Am. J. Enol. Vitic.* 31, 233–236.

Fregoni, M. (1977) Effects of soil and water on the quality of the harvest. In: *International Symposium on Quality in the Vintage*, pp. 151–168. Oenology and Viticulture Research Institute, Stellenbosch, South Africa.

Galet, P. (1979) *A Practical Ampelography – Grapevine Identification* (L. T. Morton, trans. and ed.). Cornell University Press, Ithaca, NY.

Geiger, R. (1966) *The Climate Near the Ground.* Harvard University Press, Cambridge, MA.

Geiger, R., Aron, R. H., and Todhunter, P. (2003) *The Climate Near the Ground,* 6th edn. Rowman & Littlefield, Lanham, MD.

George, M. F., and Lyons, J. M. (1979) Non-freezing cold temperatures as a plant stress. In: *Modification of the Aerial Environment of Plants* (B. J. Barfield and J. F. Gerber, eds.), pp. 85–96. Monograph of the American Society of Agricultural Engineering No. 2, St Joseph, MO.

Goffinet, C. M. (2004) Anatomy of grapevine winter injury and recovery. http://www.nysaes.cornell.edu/hort/faculty/goffinet/AnatomyWinterInjury.pdf.

Griffith, M., and Antikainen, M. (1996) Extracellular ice formation in frost–tolerant plants. *Adv. Low Temp. Biol.* **3**: 107–139.

Guy, C. (2003) Freezing tolerance of plants: current understanding and selected emerging concepts. *Can. J. Bot.* **81**, 1216–1223.

Hamilton, R. P. (1985) *Severe Wine Grape Losses in the Margaret River Region of Western Australia.* Research Report, Western Australian Department of Agriculture, South Perth, Australia.

Hamilton, R. P. (1988) Wind effects on grapevines. In: *Proceedings of the 2nd International Symposium for Cool Climate Viticulture and Oenology* (R. E. Smart *et al.*, eds.), pp. 65–68. New Zealand Society of Viticulture and Oenology, Auckland, New Zealand.

Hamman, R. A., Jr., Dami, I.-E., Walsh, T. M., and Stushnoff, C. (1996) Seasonal carbohydrate changes and cold hardiness of Chardonnay and Riesling grapevines. *Am. J. Enol. Vitic.* **47**, 31–36.

Henderson, S. T. (1977) *Daylight and Its Spectrum.* John Wiley, New York.

Heymann, H., and Noble, A. C. (1987) Descriptive analysis of Pinot noir wines from Carneros, Napa and Sonoma. *Am. J. Enol. Vitic.* **38**, 41–44.

Hicks, B. B. (1973) Eddy fluxes over a vineyard. *Agric. Meteorol.* **12**, 203–215.

Himelrick, D. G., Pool, R. M., and McInnis, P. J. (1991) Cryoprotectants influence freezing resistance of grapevine bud and leaf tissue. *HortScience* **26**, 406–407.

Jackson, D. I., and Cherry, N. J. (1988) Prediction of a district's grape-ripening capacity using a latitude-temperature index (LTI) *Am. J. Enol. Vitic.* **39**, 19–28.

Jones, K. S., McKersie, B.D., and Paroschy, J. (2000) Prevention of ice propagation by permeability barriers in bud axes of *Vitis vinifera. Can. J. Bot.* **78**, 3–9.

Khanizadeh, S., Rekika, D., Lavasseur, A., Groleau, Y., Richer, C., and Fisher, H. (2005) The effects of different cultural and environmental factors on vine growth, winter hardiness and performance in three locations. *Small Fruits Rev.* **4**, 3–28.

Kliewer, W. M. (1975) Effect of root temperature on budbreak, shoot growth and fruit set of Cabernet Sauvignon grapevines. *Am. J. Enol. Vitic.* **26**, 82–84.

Kliewer, W. M., and Torres, R. E. (1972) Effect of controlled day and night temperatures on grape coloration. *Am. J. Enol. Vitic.* **23**, 71–76.

Kriedemann, P. E., and Smart, R. E. (1971) Effects of irradiance, temperature and leaf water potential on photosynthesis of vine leaves. *Photosynthetica* **5**, 6–15.

Leuning, R., and Cremer, K. W. (1988) Leaf temperatures during radiation frosts. Part I. Observations. *Agric. Forest Meterol.* **42**, 121–133.

Lindemann, J., and Suslow, T. V. (1987) Competition between ice nucleation-active wild type and ice nucleation-deficient deletion mutant strains of *Pseudomonas syringae* and *P. fluorescens* Biovar l and biological control of frost injury on strawberry blossoms. *Phytopathology* **77**, 882–886.

Long, Z. R. (1987) Manipulation of grape flavour in the vineyard, California, North Coast region. In: *Proceedings of the 6th Australian Wine Industrial Conference* (T. Lee, ed.), pp. 82–88. Australian Industrial Publishers, Adelaide, Australia.

Luft, G., Morgenschweis, G., and Vogelbacher, A. (1983) Influence of large-scale changes of relief on runoff characteristics and their consequences for flood-control design. In: *Scientific Procedures Applied to the Planning, Design and Management of Water Resources Systems* (E. Plate and N. Buras, eds.). International Association of Hydrological Sciences. (IAHS) Publ. No. 147. IAHS Press, Wallingford, UK.

Maki, L. R., and Willoughby, K. J. (1978) Bacteria as biogenic sources of freezing nuclei. *J. Appl. Meteorol.* **17**, 1049–1053.

Marais, J., Hunter, J. J., and Haasbroek, P. D. (1999) Effect of canopy microclimate, season and region on Sauvignon blanc grape composition and wine quality. *S. Afr. J. Enol. Vitic.* **20**, 19–30.

Marshall, J. K. (1967) The effect of shelter on the productivity of grasslands and field crops. *Field Crop Abstr.* **20**, 1–14.

McIntyre, G. N., Kliewer, W. M., and Lider, L. A. (1987) Some limitations of the degree day system as used in viticulture in California. *Am. J. Enol. Vitic.* **38**, 128–132.

McIntyre, G. N., Lider, L. A., and Ferrari, N. L. (1982) The chronological classification of grapevine phenology. *Am. J. Enol. Vitic.* **33**, 80–85.

Meiering, A. G., Paroschy, J. H., Peterson, R. L., Hostetter, G., and Neff, A. (1980) Mechanical freezing injury in grapevine trunks. *Am. J. Enol. Vitic.* **31**, 81–89.

Mikhailuk, I., Kykharski, M. and Mikhalake, I. (1978) *High Stemmed Grape Culture* (green book) (in Russian), Kisinev, Moldavia. (Data reported from Moze and Ponomorev, pg. 158).

Millar, A. A. (1972) Thermal regime of grapevines. *Am. J. Enol. Vitic.* **23**, 173–176.

Miller, D. H. (1981) *Energy at the Surface of the Earth: An Introduction to the Energetics of Ecosystems.* Academic Press, New York.

Miller, D. P., Howell, G. S., and Striegler, R. K. (1988) Cane and bud hardiness of own-rooted White Riesling and scions of White Riesling and Chardonnay grafted to selected rootstocks. *Am. J. Enol. Vitic.* **39**, 60–66.

Moncur, M. W., Rattigan, K., Mackenzie, D. H., and McIntyre, G. N. (1989) Base temperatures for budbreak and leaf appearance of grapevines. *Am. J. Enol. Vitic.* **40**, 21–26.

Morlat, R., Asselin, C., Pages, P., Leon, H., Robichet, J., Remoue, M., Salette, J., and Caille, M. (1983) Caractérisation integrée de quelques terroirs du val de Loire influence sur les qualité des vins. *Connaiss. Vigne Vin* **17**, 219–246.

Muñez, M., Davies, J. A., and Robinson, P. J. (1972) Surface albedo of tower site in Lake Ontario. *Bound. Lay. Meterol.* (B) **1**, 108–114.

Nagarajah, S. (1987) Effects of soil texture on the rooting patterns of Thompson seedless vines on own roots and on Ramsey rootstock in irrigated vineyards. *Am. J. Enol. Vitic.* **38**, 54–58.

Oke, T. R. (1987) *Boundary Layer Climates,* 2nd edn. Routledge, London.

Ough, C. S., and Nagaóka, R. (1984) Effect of cluster thinning and vineyard yields on grape and wine composition and wine quality of Cabernet Sauvignon. *Am. J. Enol. Vitic.* **35**, 30–34.

Pagliai, M., La Marca, M., Lucamante, G., and Genovese, L. (1984) Effects of zero and conventional tillage on the length and irregularity of elongated pores in a clay loam soil under viticulture. *Soil Tillage Res.* **4**, 433–444.

Paul, E. A., and Clark, F. E. (1989) *Soil Microbiology and Biochemistry*. Academic Press, San Diego, CA.

Pisias, N. G., and Imbrie, J. (1986–87) Orbital geometry, CO_2, and Pleistocene climate. *Oceanus* 29(4), 43–49.

Pool, R. M. (2000) Assessing and responding to winter cold injury to grapevine buds. http://www.nysae.cornell.edu/hort/faculty/pool/budcoldinjury/Assessingbudcoldinjury.html

Pool, R. M., and Howard, G. E. (1985) Managing vineyards to survive low temperatures with some potential varieties for hardiness. In: *Proceedings of the International Symposium on Cool Climate Viticulture and Enology* (D. A. Heatherbell *et al.*, eds.), pp. 184–197. Agriculture Experimental Station Technical Publication No. 7628. Oregon State University, Corvallis.

Pope, D. F., and Lloyd, P. S. (1974) Hemispherical photography, topography and plant distribution. In: *Light as an Ecological Factor* (G. C. Evans *et al.*, eds.), Vol. 2, pp. 385–408. Blackwell, Oxford.

Rankine, B. C., Fornachon, J. C. M., Boehm, E. W., and Cellier, K. M. (1971) Influence of grape variety, climate and soil on grape composition and on the composition and quality of table wines. *Vitis* 10, 33–50.

Robin, J. R., Sauvage, F. X., Boulet, J. C., Suard, B., and Flanzy, C. (1996) Reinforcement of the radiative and thermic stresses of the grapevine in field conditions using a reflective soil cover. Repercussions on the must composition and on the wine quality. In: *Proceedings of the 4th International Symposium on Cool Climate Viticulture and Enology* (T. Henick-Kling *et al.*, eds.), pp. II-99–104. New York State Agricultural Experimental Station, Geneva, NY.

Salzman, R. A., Bressan, R. A., Hasegawa, P. M., Ashworth, E. N., and Bordelon, B. P. (1996) Programmed accumulation of LEA-like proteins during desiccation and cold acclimation of overwintering grape buds. *Plant Cell Environ.* 19, 713–720.

Saxton, K.E. (1999) Soil texture triangle, hydraulic properties calculator. http://www.bsyse.wsu.edu/~saxton/grphtext/soilwatr.htm? 129,153.

Saxton, K. E., Rawls, W. J., Romberger, J. S., and Papendick, R. I. (1986) Estimating generalized soil-water characteristics from texture. *Soil Sci. Soc. Am. J.* 50, 1031–1036.

Schulze, R. (1970) *Strahlenklima der Erde*. Steinkopf, Darmstadt, Germany.

Scrinzi, M., Bertamini, M., and Ponchia, G. (1996) Environmental effects affecting organoleptic characteristics of Sauvignon blanc wine in Alpine viticulture of Trentino (NE Italy). In: *Proceedings of the 4th International Symposium on Cool Climate Viticulture and Enology* (T. Henick-Kling *et al.*, eds.), pp. V-80–84. New York State Agricultural Experimental Station, Geneva, NY.

Seguin, G. (1971) Influence des facteurs naturels sur les caractéres des vins. In: *Sciences et techniques de la vigne* (J. Ribéreau-Gayon and E. Peynaud, eds.), Vol. 1, pp. 671–725. Dunod, Paris.

Seguin, G. (1986) Terroirs and pedology of wine growing. *J. Biol. Chem.* 42, 861–873.

Seguin, G., and Compagnon, J. (1970) Une cause du développement de la pourriture grise sur les sols gravelo-sableux du vignoble bordelais. *Connaiss. Vigne Vin* 4, 203–214.

Shamtsyan, S. M., Tsertsvadze, T. A., and Rapava, L. P. (1989) The effect of growth retardants on frost resistance of grapevines (in Russian). *Soobshch. Akad. Nauk, Tbilisi* 133, 381–384.

Sharratt, B. S., Baker, D. G., Wall, D. B., Skaggs, R. H., and Ruschy, D. L. (1992) Snow depth required for near steady-state soil temperatures. *Agric. Forest Meteorol.* 57, 243–252.

Shaw, A. B. (2002) A climatic assessment of the Niagara Peninsula viticultural area of Ontario for the application of wine machines. *J. Wine Res.* 13, 143–164.

Shulman, Y., Nir, G., Fanberstein, L., and Lavee, S. (1983) The effect of cyanamide on the release from dormancy of grapevine buds. *Sci. Hortic.* 19, 97–104.

Smart, R. E. (1973) Sunlight interception by vineyards. *Am. J. Enol. Vitic.* 24, 141–147.

Smart, R. E. (1985) Principles of grapevine canopy microclimate manipulation with implications for yield and quality, a review. *Am. J. Enol. Vitic.* 36, 230–239.

Smart, R. E., and Sinclair, T. R. (1976) Solar heating of grape berries and other spherical fruits. *Agric. Meteriol.* 17, 241–259.

Sonntag, K. (1934) *Bericht über die Arbeiten des Kalmit-Observatoriums*. Dtsch. Met. Jahrb. f. Bayern, Anhang D., Offenbach, Germany.

Soyer, J. P., Delas, J., Molot, C., and Andral, P. (1984) Techniques d'entretien du sol en vignoble bordelais. *Prog. Agric. Vitic.* 101, 315–320.

Staudt, G. (1982) Pollenkeimung und Pollenschlauchwachstum *in vivo* bei *Vitis* und die Abhähgigkeit von der Temperatur. *Vitis* 21, 205–216.

Stoev, K., and Slavtcheva, T. (1982) La photosynthèse nette chez la vigne (*V. vinifera* L.) et les facteurs écologiques. *Connaiss. Vigne Vin* 16, 171–185.

Striegler, R. K., and Howell, G. S. (1991) The influence of rootstock on the cold hardiness of Seyval grapevines. I. Primary and secondary effects on growth, canopy development, yield, fruit quality and cold hardiness. *Vitis* 30, 1–10.

Tonietto, J., and Carbonneau, A. (2004) A multicriteria climatic classification system for grape-growing regions worldwide. *Agric. Forest Meteorol.* 124, 81–97.

Törökfalvy, E., and Kriedemann, P. (1977) Unpublished data, vineleaf photosynthesis. In: *International Symposium on Quality in the Vintage*, pp. 67–87. Oenology and Viticultural Research Institute, Stellenbosch, South Africa.

Trough, M. (2004) Assessing frost risk on new vineyard sites. *Aust. NZ Grapegrower Winemaker* 486, 44–46.

Verbrugghe, M., Guyot, G., Hanocq, J. F., and Ripoche, D. (1991) Influence de différents types de sol de la basse Vallée du Rhône sur les températures de surface de raisins et de feuilles de *Vitis vinifera*. *Rev. Fr. Oenol.* 128, 14–20.

Volk, O. H. (1934) Ein neuer für botanische Zwecke geeigneter Lichtmesser. *Ber. Dtsch. Bot. Ges.* 52, 195–202.

Wagner, R., and Simon, H. (1984) Studies on the microclimate of neighbouring vineyards under the same climactical conditions. *Bull. O.I.V.* 57, 573–583.

Wahl, K. (1988) Climate and soil effects on grapevine and wine. The situation on the northern borders of viticulture – the example of Franconia. In: *Proceeding of the 2nd International Symposium for Cool Climate Viticulture and Oenology* (R. E. Smart *et al.*, eds.), pp. 1–5. New Zealand Society of Viticulture and Oenology, Auckland, New Zealand.

Wallace, P. (1972) Geology of wine. In: *Proceeding of the 24th International Geology Congress*, Montreal, Canada. Section 6, *Stratigraphy and Sedimentology* (J. E. Gill, ed.), pp. 359–365. International Geological Congress.

Wample, R. L., and Bary, A. (1992) Harvest date as a factor in carbohydrate storage and cold hardiness of Cabernet Sauvignon grapevines. *J. Am. Soc. Hortic. Sci.* 117, 32–36.

Wample, R. L., Spayd, S. E., Evans, R. G., and Stevens, R. G. (1991) Nitrogen fertilization and factors influencing grapevine cold hardiness. In: *Proceedings of the International Symposium on Nitrogen in Grapes and Wine*, Seattle (M. Rantz, ed.), pp. 120–125. American Society of Enology and Viticulture, Davis, CA.

Wample, R. L., Spayd, S. E., Evans, R. E., and Stevens, R. G. (1993) Nitrogen fertilization of White Riesling grapes in Washington, nitrogen seasonal effects on bud cold hardiness and carbohydrate reserves. *Am. J. Enol. Vitic.* **44**, 159–167.

Watson, B., Lombard, P., Price, S., McDaniel, M., and Heatherbell, D. (1988) Evaluation of Pinot noir clones in Oregon. In: *Proceedings of the 2nd International Symposium for Cool Climate Viticulture and Oenology* (R. E. Smart *et al.*, eds.), pp. 276–278. New Zealand Society of Viticulture and Oenology, Auckland, New Zealand.

Webber, R. T. J. and Jones, L. D. (1992) Drainage and soil salinity. In: *Viticulture* Vol. 2 (*Practices*) (B. G. Coombe and P. R. Dry, eds.), pp. 129–147. Winetitles, Adelaide, Australia.

Whiting, J. W., Goodwin, I., and Walpole, M. (1993) Influence of mulching on soil temperature changes. In: *Proceedings of the 8th Australian Wine Industry Technical Conference* (C.S. Stockley *et al.*, eds), pp. 217–218. Winetitles, Adelaide, Australia.

Wiebe, J., and Anderson, E. T. (1976) *Site Selection of Grapes in the Niagara Peninsula.* Horticultural Research Institute of Ontario, Vineland Station, Canada.

Williams, L. E. (1987) Growth of 'Thompson Seedless' grapevines, I. Leaf area development and dry weight distribution. *J. Am. Soc. Hortic. Sci.* **112**, 325–330.

Williams, L. E., and Matthews, M. A. (1990) Grapevines. In: *Irrigation of Agricultural Crops* (B. A. Stewart and D. R. Nielsen, eds.), pp. 1019–1055. Agronomy Series Monograph No. 30, American Society of Agronomy, Madison, WI.

Winkler, A. L. (1934) Pruning *vinifera* grapevines. *Calif. Agric. Ext. Ser. Circ.* **89**, 1–68.

Wolpert, J. A., and Howell, G. S. (1985) Cold acclimation of Concord grapevines. II. Natural acclimation pattern and tissue moisture decline in canes and primary buds of bearing vines. *Am. J. Enol. Vitic.* **36**, 189–194.

Wowk, B., and Fahy, G. M. (2002) Inhibition of bacterial ice nucleation by polyglycerol polymers. *Cryobiology* **44**, 14–23.

6

Chemical Constituents of Grapes and Wine

Introduction

Our understanding of the chemical nature of grapes and wine has advanced profoundly since the late 1960s. Although much remains to be discovered, the basic picture of what makes wine distinctive is beginning to emerge. Even the mysteries of the benefits of aging and barrel maturation are yielding their secrets. This knowledge is beginning to guide vineyard and winery practice toward the production of more consistent and better-quality wine. Plant breeders are also using this information to streamline the development of new grape varieties.

Although remarkable progress has been made, there are limitations on its application. Some compounds cannot be detected by certain techniques, and some procedures produce chemical artifacts. These problems are minimized through the use of improved instrumentation and more than one analytical procedure. A more serious limitation lies in its sensory significance. Perception is separated by many neural steps from sensation in the mouth or nose (see Chapter 11). In addition, compounds may interact in complicated ways to influence sensory perception. Therefore, it is often difficult to predict how chemical composition will affect wine attributes. For example, only rarely can a particular varietal aroma or wine bouquet be ascribed to one or a few volatile compounds. Distinctive fragrances usually arise from the combined influences of many aromatic compounds, not a single varietally unique substance.

The rapid increase in the number of compounds found in wine has been spawned by developments in techniques such as gas chromatography (GC), thin-layer chromatography (TLC), high-performance liquid chromatography (HPLC), droplet countercurrent chromatography (DCCC), multilayer coil countercurrent chromatography (MLCCC), infrared spectroscopy, solid-phase microextraction (SPME), and nuclear magnetic resonance (NMR) spectroscopy (Hayasaka *et al.*, 2005). Especially valuable has been the combination of gas chromatography with mass spectrometry (Hayasaka *et al.*, 2005). More than 700 aromatic compounds have been isolated and identified from various wines. Alone, more than 160 esters have been distinguished. Most of these compounds occur at concentrations between 10^{-4} and 10^{-9} g/liter. At these levels, most are below the limit of human detection. Therefore, most compounds isolated from wine, individually, play no role in the sensory characteristics of wine. For example, Grosch (2000) estimates that <5% of the volatile compounds identified in foods contribute to their aroma. In combination, though, some may have sensory significance.

Although these techniques are very useful, most are time-consuming, expensive, and not readily available. Thus, they are of limited value under commercial winery conditions. This has led to much interest in chemometrics and visible-near-infrared spectroscopic measurement (Cozzolino *et al.*, 2006). They may provide affordable real-time fermentation assessment in the winery.

The vast majority of chemicals found in wine are the metabolic by-products of yeast activity during fermentation. By comparison, the number of aromatic compounds derived from grapes is comparatively small. Nevertheless, these often constitute the compounds that make varietal wines distinctive.

Other than alcohol, wines generally contain about 0.8–1.2 g aromatic compounds/liter. This constitutes about 1% of the wine's ethanol content. The most common aromatic compounds are fusel alcohols, volatile acids, and fatty acid esters. Of these, fusel alcohols often constitute 50% of all volatile substances other than ethanol. Although present in much smaller concentrations, carbonyls, phenols, lactones, terpenes, acetals, hydrocarbons, and sulfur and nitrogen compounds are more

important to the varietal and unique sensory features of wine fragrance.

The taste and mouth-feel sensations of a wine are due primarily to the few compounds that occur individually at concentrations above 0.1 g/liter. These include water, alcohol (ethanol), fixed acids (primarily tartaric and malic or lactic acids), sugars (glucose and fructose), and glycerol. Tannins are important sapid substances in red wines, but they occur in moderate amounts in white wines without maturation in oak cooperage.

Overview of Chemical Functional Groups

Wine consists of two primary ingredients, water and ethanol. However, the basic flavor of wine depends on an additional 20 or more compounds. The subtle differences that distinguish one varietal wine from another depend on an even larger number of compounds. Because of the bewildering array of compounds found in grapes and wine, a brief overview is given before discussing each group in detail. Tables 6.1A and 6.1B illustrate the principal groups of compounds found in wine.

Organic chemistry deals principally with compounds containing carbon and hydrogen atoms, although other elements are often involved. Carbon atoms may bond covalently to one another or to other elements to form straight chains, branched chains, or ring structures. When only hydrogen and carbon atoms are involved, the molecule is classified as a **hydrocarbon**. If some carbons in a hydrocarbon chain are joined by double covalent bonds, the compounds are said to be **unsaturated**. In contrast, **saturated** hydrocarbons contain only single covalent bonds between the carbon atoms. Linear-chain compounds are called **aliphatic**, whereas those forming at least one circular linkage of six carbon atoms (a benzene ring) are called **aromatic**. This term comes from the fact that the first identified benzene derivatives were fragrant. However, not all odorous compounds are chemically aromatic, nor are all chemically aromatic compounds odorous. Because "aromatic" is commonly used to refer to the olfactory property of volatile compounds, the term aromatic will be restricted in this book to the fragrant, not the chemical, nature of the compound.

Most organic molecules contain more than just carbon and hydrogen atoms. The additional elements, along with double covalent bonds, give organic molecules most of their chemical (reactive) characteristics. The region of a molecule that generally gives a compound its primary reactive property is called the functional group. Frequently, compounds contain more than one functional group. This complicates classification and the expression of chemical properties. Substances with more than a single type of functional group show,

Table 6.1A Some important functional and chemical groups in grapes and wine[a]

Compound class	General structure	Functional group	Example
Alcohols	R—OH	—OH	Ethanol
Carbonyls Aldehydes	R—C(=O)—H	—C(=O)—H	Acetaldehyde
Ketones	R_1—C(=O)—R_2	—C(=O)—	Diacetyl
Carboxylic acids	R—C(=O)—OH	—C(=O)—OH	Acetic acid
Esters	R_1—C(=O)—O—R_2	—C(=O)—O—	Ethyl acetate
Amides	R—C(=O)—NH_2	—C(=O)—NH_2	Acetamide
Amines	R—NH_2	—NH_2	Histamine
α-Amino acids	H_2N—CH(R)—C(=O)—OH	H_2N—CH(R)—C(=O)—OH	Alanine
Acetals	R_1—O—CH(R)—O—R_2	—O—CH(R)—O—	Acetal
Terpenes	(CH₃)(R_3)C=C(R_1)(R_2)	(CH₃)(—)C=C(—)(—)	Linalool
Thiols	R—SH	—SH	Ethanethiol
Thioesters	R_2—S(=O)—OR_1	—S(=O)—O—	Methyl thiolacetate

[a] The letter R usually designates the rest of the molecule. Complex molecules may have more than one type of R group, designated by different subscripts.

to varying degrees, the properties of the different functional groups they possess. Not surprisingly, compounds may be classified (and named) depending on the intent of the author. For example, methyl anthranilate may be classified as a benzoic acid derivative (and therefore a phenyl derivative), as an ester (because of its ester group), or as a nitrogen compound (because of the amine group).

methyl anthranilate

Most atoms in organic molecules are connected by covalent bonds, in which electrons are shared between the associated atoms. This sharing is equal between two or more carbon atoms, or between carbon and hydrogen atoms. Because no electrical charge develops around such bonds, organic molecules tend to be nonpolar and hydrophobic (poorly water soluble). Water solubility is largely dependent on the electrical attraction between charged groups on the molecule and adjacent water molecules. For most organic compounds, solubility is dependent on the presence of atoms such as oxygen and nitrogen, which often form polar covalent bonds with hydrogen. Because the electrons are unequally shared between the atoms, weakly charged (polar) areas form that can associate with the charged (polar) regions of water molecules. In some compounds, an element completely removes one or more electrons from another, forming an ionic bond. In water, the components of an ionic molecule tend to dissociate, forming free, soluble, charged components

Table 6.1B Some important functional and chemical groups in grapes and wine[a]

Compound class	General structure	Example
Phenolics		Vanillin
Lactones		3-Methyl-γ-lactone
Pyrazines		2-Methoxy-3-isobutylpyrazine
Pyridines		2-Acetyltetrahydropyridine
C_{13} Norisoprenoids		Vitaspirane
Thiolanes		2-Methylthiolane-3-ol
Thiazoles		5-(2-Hydroxyethyl-4-methythiazole)

(probable precursor)

[a] The letter R usually designates the rest of the molecule. Complex molecules may have more than one type of R group, designated by different subscripts.

called ions. The presence of ionic bonds in an organic molecule often greatly enhances its water solubility.

The symbol **R** is used to represent hydrogen, hydrocarbon, or hydrocarbon-derivative groups attached to a carbon chain. These are either alkyls, if based on an aliphatic carbon chain, or phenyls, if based on a benzene derivative.

$$-OH \quad -O-R \quad -C=O \quad -\overset{\overset{\displaystyle OH}{|}}{C}=O$$

hydroxyl   alkoxy   carbonyl   carboxyl

In grapes and wine, most organic functional groups are based on the double covalent bond (especially benzene derivatives), or on bonds with oxygen, nitrogen,

and sulfur. The most common functional groups are those based on carbon–oxygen bonds. A common example is the **hydroxyl** group that provides the properties of an alcohol to organic compounds. The substitution of the hydrogen in a hydroxyl group by a carbon yields the **alkoxy** group that characterizes ethers. Another common oxygen-based functional group is the **carbonyl** group. Depending on its location in the group, it characterizes either aldehydes (terminal position) or ketones (internal position). Association with a hydroxyl group on the same carbon generates the **carboxyl** group, giving organic molecules the properties of an acid.

When an organic alcohol (hydroxyl) group reacts with an acid (carboxyl) group, it produces the **ester** linkage. The interaction of the hydroxyl group of alcohols with the carbonyl group of an aldehyde generates the **acetal** group. Oxygen can also be involved in the formation of carbon ring structures. An example is the **cyclic ester** group called lactones.

$$
\underset{\text{ester}}{-\overset{\displaystyle\overset{O}{\|}}{C}-O-C} \qquad\qquad \underset{\text{acetal}}{-O-\overset{\displaystyle\overset{H}{|}}{\underset{\displaystyle R}{C}}-O-}
$$

An important group of aromatic compounds in wines are terpenes. Although not possessing a particular functional group, they are constructed from repeating groups of five carbon atoms in a distinctive arrangement called the **isoprene** unit.

$$
\underset{\text{isoprene}}{\diagup\!\!\diagdown\!\!\diagup\!\!\diagdown}
$$

Additional functional groups are based on carbon–nitrogen linkages. The most important of these is the **amine** group. Alone, it forms compounds called amines. If it is associated with a carbonyl group on the same carbon, it generates the **amide** grouping. With a carboxyl group on an adjacent carbon atom, the amine group forms the **amino** grouping that characterizes α-amino acids. Nitrogen also may be part of a carbon-ring structure. Important examples are pyrazines and pyridines.

$$
\underset{\text{amine}}{C-\overset{\displaystyle\overset{|}{}}{N}-} \qquad \underset{\text{amide}}{-\overset{\displaystyle\overset{H_2N}{|}}{C}=O} \qquad \underset{\text{amino acid}}{-\overset{\displaystyle\overset{H_2N}{|}}{C}-\overset{\displaystyle\overset{O}{\|}}{C}-OH}
$$

Sulfur-based functional groups generate important aromatic compounds in wine. They are often structurally analogous to functional groups based on oxygen. Examples are the thiol group of mercaptans, thioethers, and thioesters. Sulfur, alone or along with nitrogen, also may be involved in the formation of ring structures. Examples are thiolanes and thiazoles, respectively.

$$
\underset{\text{thiol, sulfhydryl}}{-SH} \qquad \underset{\text{thioether}}{-S-} \qquad \underset{\text{thioester}}{-\overset{\displaystyle\overset{O}{\|}}{C}-S-}
$$

Chemical Constituents

Water

The water content of grapes and wine is seldom discussed. Its presence is taken for granted. Nevertheless, as the predominant chemical constituent of grapes and wine, water plays a critical role in establishing the basic characteristics of wine. For example, only compounds at least slightly soluble in water play a significant role in wine. Water also governs the basic flow characteristics of wine. Even the occurrence of tears in a glass of wine is partially dependent on the properties of water. In addition, the high specific heat of water slows the warming of wine in a glass. Water is also an essential component in many of the chemical reactions involved in grape growth, juice fermentation, and wine aging.

Sugars

Sugars are a category of carbohydrates distinguishable by the presence of several hydroxyl groups and an aldehyde or ketone group. Simple sugars may bond together to form polymers such as pectins, gums, starches, hemicelluloses, and cellulose, or with other secondary metabolites such as lactones and anthocyanidins, to form glycosides. The formation of glycosides is probably essential to their accumulation in cellular vacuoles. Glycosidation increases solubility. Only some of the simpler sugars taste sweet.

$$
\underset{\text{aldehyde}}{-\overset{\displaystyle\overset{H}{|}}{C}=O} \qquad\qquad \underset{\text{ketone}}{-\overset{\displaystyle\overset{O}{\|}}{C}-}
$$

The principal sugars in grapes are glucose and fructose. They often occur in roughly equal proportions at maturity, whereas overmature grapes often have a higher proportion of fructose. Sugars other than glucose and fructose occur, but in relatively insignificant amounts. Sucrose is rarely found in *Vitis vinifera* grapes, but it may constitute up to 10% of the sugar content in non-*V. vinifera* cultivars. Sucrose, whether natural or added, is enzymatically split into glucose and fructose during fermentation.

Grape sugar content varies depending on the species, variety, maturity, and health of the fruit. Cultivars of *V. vinifera* generally reach a sugar concentration of 20% or more at maturity. Other winemaking species, such as *V. labrusca* and *V. rotundifolia*, seldom reach this level. Sugar commonly needs to be added to the juice from these species to develop the 10–12% alcohol content typical of most wines.

In North America, sugar content (**total soluble solids**) is measured in °Brix. °Brix is a good indicator of berry sugar content at levels above 18, when sugars become the predominant soluble solids in grapes (Crippen and Morrison, 1986). °Brix is compared with other specific gravity measurements used in Europe, notably Oechsle and Baumé, in Appendix 6.1.

Grape sugar content is critical to yeast growth and metabolism. *Saccharomyces cerevisiae*, the primary wine yeast, derives most of its metabolic energy from glucose and fructose. Because *S. cerevisiae* has limited abilities to ferment other substances, it is important that most grape nutrients be in the form of glucose and fructose. Unfermented sugars are collectively termed **residual sugars**. In dry wines, the residual sugar content consists primarily of pentose sugars, such as arabinose, rhamnose, and xylose, and small amounts of unfermented glucose and fructose. These levels may increase slightly during maturation in oak cooperage, via the breakdown of glycosides in the wood. In addition, the amounts of galactose and arabinose increase relative to the degree of fruit botrytization. Estimates given by Dittrich and Barth (1992) are 0.17, 0.36, 0.75, 1.17, and 1.95 for kabinett, spätkese, auslese, beerenauslese, and trochenbeerenauslese wines, respectively. Sugars also may be synthesized and released by yeast cells. Trehalose, a common fungal sugar, is found in botrytized wines.

The residual sugar content of dry wine is generally less than 1.5 g/liter. At this concentration, the perception of sweetness is undetectable on the palate. The nutritive value of these sugars is also usually insufficient to constitute a threat to the microbial stability of bottled wine. At higher concentrations, though, residual sugars increasingly pose a microbial hazard. This is particularly marked in sweet wines, with low acid and alcohol levels. Special procedures are required to prevent undesirable yeast and bacterial growth in wine containing high sugar contents. Pentoses are also a potential source for furfural synthesis in baked wines.

When the residual sugar content rises above 0.2%, individuals with acute sensory thresholds begin to detect sweetness. Generally, however, the sugar content must be more than 1% to possess distinct sweetness for the majority of people. Perceptible sweetness is markedly influenced by other wine constituents, notably ethanol, acids and tannins, as well as by individual sensitivity. Conversely, detectable sweetness has a mitigating effect on the perception of sourness and bitterness. Very sweet table wines, such as sauternes, trockenbeerenausleses, and icewines, may reach residual sugar contents well over 10%.

Although residual sugars are of obvious importance to the sweetness of wine, fermentable sugars in grapes are absolutely essential for fermentation. The single most significant by-product of fermentation is ethanol. In addition, sugars may be metabolized to higher alcohols, fatty acid esters, and aldehydes. These give different wines much of their aromatic character. It is the abundance of fermentable sugars in grapes, in contrast to other fruits, that probably made wine the first, fruit-based, alcoholic beverage discovered by humans.

Slow structural transformations in sugars may be involved in the darkening of dry white wine during aging. This is especially so in the production of brown melanoidin pigments found in sweet sherries, madeiras, and similar fortified wines (Rivero-Pérez *et al.*, 2002). These may develop by a complex series of browning or Maillard reactions. Maillard reactions occur between reducing sugars (primarily glucose and fructose in wine) and various amines (notably amino acids and proteins). The amine acts as a catalyst inducing the dehydration of reducing sugars. The imine product undergoes isomerization to form aminoketones. These typically breakdown spontaneously to release the amine and a mixture of deoxysomes. These highly reactive compounds can undergo several changes, notably cyclization to form flavorants such as furfuraldehydes. Subsequent condensation or polymerization with other nucleophilic compounds leads to the formation of brown pigments. Metal ions, notably iron, can catalyze Maillard reactions, as well as the later formation of colored complexes. Maillard reactions occur most rapidly at elevated temperatures, but still occur, albeit slowly, at cellar temperatures.

Caramelization is not commonly a cause of coloration in table wines, except when aged in toasted oak barrels. The toasting process induces the thermal degradation of sugars in the staves. The condensation products generated can be dissolved into wine during subsequent maturation. Direct caramelization of grape sugars can also occur when must is boiled to produce color for addition to some fortified wines and brandies. In grape must, the presence of organic bases (e.g., amines and amino acids) facilitates sugar degradation which leads to the formation of a caramel flavor (e.g., hydroxymethylfurfural) and generation of a golden color. The exact nature of the by-products depends on the sugars present, the pH, the temperature, and the nature of any amine catalysts (see Rizzi, 1997).

Sugar concentration also can increase the volatility of aromatic compounds (Sorrentino *et al.*, 1986). The relevance of sugars to the fragrance of fortified and sweet table wines is unknown.

Pectins, Gums, and Related Polysaccharides

Pectins, gums, and related substances typically are mucilaginous polymers of sugar acids that hold plant cells together. They commonly occur as complex branched chains. Pectins are linear polymers of galacturonic acid, often possessing multiple esterified methyl groups, and complexed to varying degrees with rhamnogalacturonan and chains consisting of arabinans and arabinoglactan. Gums are polymeric mixtures of arabinose, galactose, xylose, and fructose. Being partially water-soluble, they are extracted into the juice during crushing and pressing. Extraction is favored when whole or crushed grapes are heated to hasten anthocyanin liberation. During fermentation, the polysaccharides form complex colloids in the presence of alcohol and tend to precipitate. Consequently, grape pectins, gums, and glucosans seldom cause wine clouding or filtration problems, except with pulpy cultivars. However, pectins can cause considerable difficulty during the pressing of slip-skin grapes, notably *V. labrusca* cultivars. The addition of pectinase following crushing significantly reduces the pectin content and their effect on pressing or filtration.

In contrast to the polysaccharides of healthy grapes, β-glucans produced by *Botrytis*-infected grapes can cause serious problems. They hinder juice and wine clarification by inhibiting the precipitation of other colloidal materials, such as tannins and proteins. In addition, β-glucans form a fibrous mat on filters, plugging the pores. In contrast, grape polysaccharides produce nonplugging, spherical colloids (Ribéreau-Gayon *et al.*, 1980). Because *Botrytis* glucans are localized under the grape skin, gentle harvesting and pressing minimize their extraction.

The polysaccharide level in finished wine is generally low. The significance of polysaccharides to the sensory properties of wine has not been adequately studied. Only mannoproteins, a complex of mannans and protein, released by yeast cells during autolysis, are known to affect wine quality (Charpentier, 2000). Those higher in protein content tend to be the most significant, affecting not only tartrate and protein stability, but also may influence taste and flavor.

Alcohols

Alcohols are organic compounds containing one or more hydroxyl groups (−OH). Simple alcohols contain a single hydroxyl group, whereas diols and polyols contain two or more hydroxyl groups, respectively. Phenols are six-carbon-ring compounds containing one or more hydroxyl groups on the phenyl ring. The chemistry of phenols is sufficiently distinct that they are treated separately.

ETHANOL

Ethanol is undisputably the most important alcohol in wine. Although small quantities are produced in grape cells during carbonic maceration, the primary source of ethanol in wine is yeast fermentation. Ethanol is the principal organic by-product of fermentation.

Under standard fermentation conditions, ethanol can accumulate at up to about 14–15%. Higher levels can be reached by the sequential addition of sugar

ethanol (an alcohol)

glycerol (a triol)

phenol (a phenol)

during fermentation. Generally, however, ethanol concentrations in wine above 14–15% are the result of fortification. The prime factors controlling ethanol production are sugar content, fermentation temperature, and yeast strain.

Alcohol content is variously indicated in terms of percent by volume (vol %), percent by weight (wt %), gram per 100 ml, specific gravity, or proof. These expressions are compared in Appendix 6.2. In the book, percent by volume is used.

In addition to its significant physiological and psychological effects on humans, ethanol is crucial to the stability, aging, and sensory properties of wine. During fermentation, the increasing alcohol content progressively limits the growth of microorganisms. The relative alcohol insensitivity of *S. cerevisiae* assures that it typically dominates fermentation. Microbes that might produce off-odors are generally suppressed. The inhibitory action of ethanol, combined with the acidity of the wine, permits wine to remain stable for years in the absence of air.

Ethanol acts as an important cosolvent, along with water, in extracting constituents from grapes. In this regard, ethanol is particularly important in solubilizing non-polar aromatics. This action also affects their volatility. At the concentrations found in table wines, ethanol forms a monodispersed aqueous solution. In contrast, in fortified wines and brandies, ethanol molecules cluster to reduce hydrophobic hydration (D'Angelo *et al.*, 1994).

By affecting the metabolic activity of yeasts, ethanol also influences the types and amounts of aromatic compounds produced. Furthermore, ethanol acts as an essential reactant in the formation of several important volatile compounds and adds its own distinctive odor.

Ethanol has multiple effects on taste and mouth-feel. It directly enhances sweetness through its own sweet taste. It indirectly modifies the perception of acidity, making acidic wines appear less sour and more balanced. At high concentrations, alcohol produces a burning sensation and may contribute to the feeling of weight or body, especially in dry wines. Ethanol can also increase the intensity of bitterness, while decreasing the astringency of tannins (Lea and Arnold, 1978).

In addition to functioning as a solvent in pigment and tannin extraction from grapes, ethanol also helps to dissolve volatile compounds produced during fermentation and those formed during maturation in wood cooperage. The dissolving action of the alcohol probably reduces the escape of aromatic compounds with carbon dioxide during fermentation. However, alcohol concentrations below 7% favor the release of many aromatic compounds. This could significantly affect the aromatic distinctiveness of a wine (Guth, 1998), and expression of its finish (Williams and Rosser, 1981).

Ethanol plays several roles in the aging of wine. Along with other alcohols, ethanol slowly reacts with organic acids to produce esters. The ethanol concentration also influences the stability of esters. In addition, ethanol reacts slowly with aldehydes to produce acetals.

METHANOL

Methanol is not a major constituent in wines. Within the usual range found in wine (0.1–0.2 g/liter), methanol has no direct sensory effect. Of the over 160 esters found in wine, few are associated with methanol.

The concern occasionally about methanol relates to its metabolism to formaldehyde and formic acid. Both metabolites are toxic to the central nervous system. One of the first targets of formaldehyde toxicity is the optic nerve, causing blindness. Methanol never accumulates to toxic levels under legitimate winemaking procedures.

The limited amount of methanol found in wine is primarily generated from the enzymatic breakdown of pectins. After degradation, methyl groups associated with pectin are released as methanol. Thus, the methanol content of fermented beverages is primarily a function of the pectin content of the substrate. Unlike most fruits, grapes are low in pectin. As a result, wine generally has the lowest methanol content of any fermented beverage. Pectolytic enzymes, added to juice or wine to aid clarification, can inadvertently increase the methanol content. Adding distilled spirits to a wine also may slightly increase the methanol content. The concentration of ethanol and other flavor compounds achieved with distillation augments the methanol content of the distillate.

HIGHER (FUSEL) ALCOHOLS

Alcohols with more than two carbon atoms are commonly called higher or fusel alcohols. They may be present in healthy grapes, but seldom occur in significant amounts. Hexanols are the major exception to this generalization. They also donate a herbaceous odor in certain wines. Other potentially significant higher alcohols from grapes that survive fermentation are 2-ethyl-1-hexanol, benzyl alcohol, 2-phenylethanol, 3-octanol, and 1-octen-3-ol. However, most higher alcohols found in wine are the by-products of yeast fermentation. Their synthesis closely parallels that of ethanol production (Fig. 6.1). They commonly account for about 50% of the aromatic constituents of wine, excluding ethanol.

Quantitatively, the most important higher alcohols are the straight-chain alcohols – 1-propanol, 2-methyl-1-propanol (isobutyl alcohol), 2-methyl-1-butanol, and

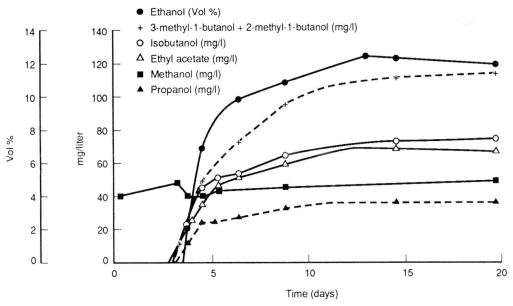

Figure 6.1 Production of various alcohols and ethyl acetate during fermentation. (From Rapp and Mandery, 1986, reproduced by permission)

3-methyl-1-butanol (isoamyl alcohol). 2-Phenylethanol (phenethyl alcohol) is the most important phenol-derived higher alcohol.

Most straight-chain higher alcohols have a strong pungent odor. At low concentrations (~0.3 g/liter or less), they generally add an aspect of complexity to the bouquet. At higher levels, they increasingly overpower the fragrance. In distilled beverages, such as brandies and whiskeys, fusel alcohols give these beverages some of their distinctive aromatic character. Only in *porto* is a distinctive fusel character expected and appreciated. This property comes from the brandy added during port production.

The formation of higher alcohols during fermentation is markedly influenced by winery practices (see Sponholz, 1988). Synthesis is favored by the presence of oxygen, high fermentation temperatures, and the presence of suspended material in the fermenting juice. Chaptalization and pressure-tank fermentation also tend to enhance synthesis of higher alcohols. Conversely, prefermentative clarification, the presence of sulfur dioxide, and low fermentation temperatures suppress production. Yeasts vary considerably in their propensity to produce higher alcohols.

Higher alcohols may originate from grape-derived aldehydes, by the reductive denitrification of amino acids, or via synthesis from sugars. The relative importance of those sources appears to vary with the specific fusel alcohol. Amino acid deamination is especially important in the generation of longer-chain higher alcohols (Chen, 1978).

Additional higher alcohols may come from the metabolic activity of spoilage yeasts and bacteria. Table wines showing a fusel off-odor usually have been infected at some stage by spoilage organisms. Occasionally, pleasant smelling higher alcohols may be produced by microorganisms. An example is the mushroom alcohol (1-octen-3-ol), synthesized during noble rotting by *B. cinerea* (Rapp and Güntert, 1986). It is produced by the enzymatic breakdown of linoleic acid.

Higher alcohols also play an indirect role in the development of an aged wine bouquet. By reacting with organic acids, they add to the number of esters found in wine. During fermentation, the production of esters occurs rapidly under the control of yeast enzymes. Certain esterification reactions continue during aging, but at a much slower, nonenzymatic pace (Rapp and Güntert, 1986).

OTHER ALCOHOLS

In addition to the alcohols already mentioned, there are important terpene- and phenol-derived alcohols. Because they show stronger terpenic and phenolic characteristics than alcoholic properties, respectively, they are discussed later under their predominant chemical associations.

DIOLS, POLYOLS, AND SUGAR ALCOHOLS

The most prominent diol in wine is 2,3-butanediol (2,3-butylene glycol). It has little odor and possesses a mildly bittersweet taste. It appears to have little sensory significance in wine.

By far the most prominent wine polyol is glycerol. In dry wine, glycerol is commonly the most abundant compound, after water and ethanol. It is often higher in red (~10 mg/liter) than white (7 mg/liter) wines. Correspondingly, it was assumed to be of sensory importance, notably in viscosity. Nevertheless, glycerol rarely reaches a concentration where it begins to significantly affect perceived viscosity ($\geq$26 g/liter) (Noble and Bursick, 1984). Glycerol has a slight sweet taste. Nevertheless, it is unlikely to be noticeable in a sweet wine. It might play a minor role in dry wines, in which the concentration of glycerol often surpasses its sweetness sensory threshold (>5 g/liter). Glycerol is an important nutrient source for the growth of flor yeasts in sherry production. Certain spoilage bacteria also metabolize glycerol.

Variety, maturity, and health all affect the amount of glycerol present in grapes. Infection by *B. cinerea* produces juice with the highest glycerol content. During fermentation, yeast strain, temperature, sulfur dioxide, and pH level can influence the synthesis of glycerol.

Sugar alcohols, such as alditol, arabitol, erythritol, mannitol, *myo*-inositol, and sorbitol, are commonly found in small amounts in wine. Higher concentrations usually are the result of fungal infection in the vineyard or bacterial growth in the wine. Sugar alcohols can be oxidized by some acetic acid bacteria to the respective sugars. The sensory and enologic significance of this conversion is unknown. Combined, polyols and sugar alcohols may have a slight effect on the sensation of body.

Acids

Acids are characterized by the ionization and release of hydrogen ions (H^+) in water. With organic compounds, this property is primarily associated with the carboxyl group. The carboxyl group dissociates into a negatively charged carboxyl radical and a free, positively charged hydrogen ion. Inorganic acids, such as carbonic acid, dissociate into a negatively charged ion and one or more positively charged hydrogen ions. The degree of ionization in wine depends primarily on the cation content (notably potassium), the pH, and the ionizing characteristics of the particular acid.

For the majority of table wines, a range of between 5.5 and 8.5 mg/liter total acidity is desirable. White wines are typically preferred at the higher end of the scale, whereas red wines are preferred at the lower end. A pH range of between 3.1 and 3.4 is suitable for most white wines, and between 3.3 and 3.6 for most red wines.

$$R - COOH \rightleftharpoons R - COO^- + H^+$$

organic   carboxyl   hydrogen
acid   radical   ion

$$H_2CO_3 \rightleftharpoons HCO_3^- + H^+ \rightleftharpoons CO_3^{2-} + 2H^+$$

carbonic   bicarbonate   carbonate
acid   ion   ion

The principal inorganic acids in wine are carbonic and sulfurous acids. Both also occur as dissolved gases, namely CO_2 and SO_2, respectively. Because they are more important in wine as gases and do not noticeably affect wine pH or perceptible acidity, discussion of these acids is provided later in this chapter (Dissolved gases).

Acidity in wine is customarily divided into two categories, volatile and fixed. **Volatile acidity** refers to acids that can be readily removed by steam distillation, whereas **fixed acidity** includes those that are poorly volatile. **Total acidity** is the combination of both categories. Total acidity may be expressed in terms of tartaric, malic, citric, lactic, sulfuric, or acetic acid equivalents (Appendix 6.3). In this text, total acidity is expressed in tartaric acid equivalents.

Acetic acid is the principal volatile acid, but other carboxylic acids such as formic, butyric, and propionic acids also may be involved. Related acids, possessing longer hydrocarbon (aliphatic) chains ($\geq C_{12}$), are commonly termed fatty acids. Their presence in wine is primarily derived from the fatty acid metabolism of yeasts and bacteria. All volatile carboxylic acids have marked odors – acetic acid is vinegary, propionic acid is characterized as fatty, butyric acid resembles rancid butter, and C_6 to C_{10} carboxylic acids possess a goaty odor. Although these acids commonly exist in wine, they typically occur at detectable levels only in microbially spoiled wines. Because acetic acid is the major volatile acid, volatile acidity is usually measured in terms of its presence alone. Acetic acid occurs in wine primarily as a by-product of yeast and bacterial metabolism, but is also formed during the chemical hydrolysis of hemicelluloses during wine maturation in oak cooperage.

Fixed acidity refers to all organic acids not included under the volatile category. Quantitatively, they control the pH of wine. In grapes, two dicarboxylic acids (tartaric and malic) often constitute more than 90% of the fixed acidity. The same two acids dominate the acid composition of wines. Depending on the climatic conditions and ripeness of the grapes, fixed acidity can vary from less than 2 g/liter to over 5 g/liter. If the wine undergoes malolactic fermentation, malic acid is replaced by the smoother tasting monocarboxylic lactic acid. Fermentation has little effect on total acidity, but it does increase their chemical diversity. The increased complexity may play a minor role in the development of an aged bouquet.

Other fixed acids in wine include di- and tri-carboxylic acids, such as those of the TCA (tricarboxylic acid) cycle (e.g., citric, isocitric, fumaric, and α-ketoglutaric acids). Those present in wine are primarily derived from yeast metabolism (see Fig. 7.15). Their precursors may come from the metabolic breakdown of sugars, amino acids, or fatty acids. Most of these acids are found in minor amounts and are generally not known to be of sensory significance. Exceptions may include succinate, α-ketoglutarate and pyruvate. Succinate may donate a salty, bitterish aspect; α-ketoglutarate can bind sulfur dioxide, reducing its free, active concentration in wine; and pyruvic acid can act in stabilizing wine color. Occasionally, citric acid may be added to acidify high pH wines. Tartaric acid may be added for the same purpose.

Sugar acids, such as gluconic, glucuronic, and galacturonic acids, are associated with grape infection by *B. cinerea*. Gluconic acid is so characteristic of the disease that its presence has been used as an indicator of the degree of infection. Nevertheless, these acids appear to be produced primarily by acetic acid bacteria that grow concurrently with *Botrytis* on or in the fruit (Sponholz and Dittrich, 1984). As a consequence, other measures of *Botrytis* infection are used or are under investigation. Examples are the presence of laccase (Roudet *et al.*, 1992), or particular antigens (Dewey *et al.*, 2005). Other than as a diagnostic tool, gluconic acid produced by *Botrytis* appears to be of no enologic significance. However, its intramolecular esterification to gluconolactone may bind sulfur dioxide (Barbe *et al.*, 2002).

None of the sugar acids affects taste or odor. However, galacturonic acid may be involved in the browning of white wines (Jayaraman and van Buren, 1972). Its oxidation to brown pigments is catalyzed by copper and iron ions. In addition to being synthesized by bacteria, galacturonic acid is liberated through the enzymatic breakdown of pectins.

Wine phenolic acids may be derived from grapes or yeast cells (synthesized via the shikimic acid pathway), or be extracted from oak cooperage. They may contribute a slight bitterness to wine. Amino acids are usually excluded from a discussion of organic acids, because of the dominating influence of the amino group on the molecule. Correspondingly, they are discussed later under nitrogen compounds. Similarly, vitamins and growth regulators with acidic groups are discussed under their main chemical headings (see Macromolecules and Growth Factors).

As a group, acids are almost as important to the characteristics of wines as alcohols. Acids not only produce a refreshing taste (or sourness, if in excess), but also modify the perception of other taste and mouthfeel sensations. This is especially noticeable in a reduction in perceived sweetness. In addition, the release of acids from cell vacuoles on crushing is probably instrumental in the initiation of acid hydrolysis of nonvolatile precursors in the fruit (Winterhalter *et al.*, 1990). Several important aroma compounds, such as monoterpenes, phenolics, C_{13} norisoprenoids, benzyl alcohol, and 2-phenylethanol, may occur in the fruit as acid-labile nonvolatile glycosides (Strauss, Gooley *et al.*, 1987), usually as disaccharide conjugates. They act as a reserve of aroma compounds that may be subsequently liberated during fermentation or aging. The sugar moieties are primarily glucose, rhamnose, and arabinose.

The role of acids in maintaining a low pH is crucial to the color stability of red wines. As the pH rises, anthocyanins lose their red color and turn bluish. Acidity also affects ionization of phenolic compounds. The ionized (phenolate) state is more readily oxidized than its nonionized form. Accordingly, wines of high pH ($\geqslant 3.9$) are very susceptible to oxidization and loss of their young color (Singleton, 1987).

Acids are involved in the precipitation of pectins and proteins that otherwise could cloud a finished wine. Conversely, acids can solubilize copper and iron, which can induce haziness (*casse*).

The low pH produced by wine acids has a beneficial antimicrobial effect. Most bacteria do not grow at low pH values. Low pH values also enhance the antimicrobial properties of fatty acids. Fatty acids are more toxic in the undissociated (nonionized) state found under acidic conditions (Doores, 1983). However, by inhibiting yeast metabolism, carboxylic acids such as decanoic acid can favor the premature termination of fermentation (sticking).

At or just below the threshold level, fatty acids contribute to the complexity of a wine's bouquet. Above their threshold, though, they begin to have a negative influence. The mild odors of lactic and succinic acids are generally considered inoffensive, as is the butter-like smell of sorbic acid.

During fermentation and aging, acids are involved in reactions leading to the formation of esters. These are often important to the fresh-fruity fragrance of wines. Acids are also important during wine aging and possibly the development of a desirable bottle bouquet. Low pH also facilitates the hydrolysis of disaccharides, such as trehalose, and various polysaccharides. This can slowly add fermentable sugars to wine.

ACETIC ACID

Small amounts of acetic acid are produced by yeasts during fermentation. At normal levels in wine (<300 mg/liter), acetic acid can be a desirable flavorant, adding to the complexity of taste and odor. It is more important, though, in the production of acetate esters that can give wine a fruity character. If it is greater than 300 mg/liter, however, acetic acid progressively gives wine a sour taste and taints its fragrance. High levels of acetic acid are usually associated with contamination of grapes, juice, or wine with acetic acid bacteria.

MALIC ACID

Malic acid may constitute about half the total acidity of grapes and wine. Its concentration in the fruit tends to decrease as grapes mature, especially during hot periods at the end of the season. This can lead to the production of wine with a flat taste, and susceptible to microbial spoilage. Conversely, under cool conditions, the malic acid level may remain high, and give the resultant wine a sour taste. Therefore, malic acid content is one of the prime indicators used in determining harvest date.

LACTIC ACID

A small amount of lactic acid is produced by yeast cells during fermentation. However, when lactic acid occurs as a major constituent in wine, it comes from the metabolic activity of bacteria. The bacteria most commonly involved are lactic acid bacteria. They produce an enzyme that decarboxylates malic acid directly to lactic acid. The process, called malolactic fermentation, is commonly encouraged in red and in some white wines. The major benefit of malolactic fermentation is the conversion of the harsher-tasting, dicarboxylic, malic acid to the smoother-tasting, monocarboxylic, lactic acid. The predominance of L-lactic acid, one of the two stereoisomers of the acid, is usually an indicator of malolactic fermentation. In contrast, yeasts and some bacteria synthesize equal quantities of both (D and L) forms of lactic acid during their metabolism of malic acid.

SUCCINIC ACID

Succinic acid is one of the more common by-products of yeast metabolism. It is resistant to microbial attack under anaerobic conditions and is particularly stable in wine. However, the bitter-salty taste of succinic acid limits its use for wine acidification. It is the second most significant organic acid in 'Noble' muscadine wines (*V. rotundifolia*) (Lamikanra, 1997). In most wines, though, its presence occurs as a consequence of yeast fermentation, rather than being a by-product of grape metabolism.

TARTARIC ACID

Tartaric acid is the other major grape acid, along with malic acid. Unlike malic acid, the concentration of tartaric acid does not decline markedly during grape ripening. In addition, tartaric acid is metabolized by few microorganisms. Thus, it is usually the preferred acid added to increase the acidity of high pH wines. Regrettably, this carries the risk of increasing bitartrate instability.

Tartaric acid is synthesized in many plants, but accumulates in significant quantities only in a few genera, most significantly, members of the Vitaceae. It commonly collects as a potassium salt in leaves and grapes. Tartaric acid is so characteristic of *V. vinifera* that its presence in Near Eastern neolithic vessels has been taken as evidence of wine production (McGovern and Michel, 1995). Although characteristic of grapes, the fruit of a few other plants accumulate tartaric acid in significant amounts, exceptions are tamarind (*Tamarindus*) and hawthorn (*Crataegus*). Some yeast species may also synthesize small amounts of tartaric acid.

As wines age, dissolved tartrates crystallize and tend to precipitate. Because chilling speeds the process, wines often are cooled near the end of maturation to enhance early tartrate precipitation and avoid crystal deposition in the bottle. Nevertheless, crystals may continue to form after bottling. This partially occurs due to the conversion of the natural (L form) of tartaric acid to the D isomer. The calcium salt of both isomers is about one-eighth as soluble as the L-tartrate salt alone. Therefore, most wines form a salt deposit when aged sufficiently long.

Phenols and Related Phenol (Phenyl) Derivatives

Phenols are a large and complex group of compounds of primary importance to the characteristics and quality of red wine. They are also significant in white wines, but occur at much lower concentrations. Phenols and related compounds can affect the appearance, taste, mouth-feel, fragrance, and antimicrobial properties of wine. Although primarily of grape origin, smaller amounts may be extracted from wood cooperage. Only trace amounts are derived from yeast metabolism.

In the past few years, there have been major advances in our understanding of chemical nature of wine phenolics, notably tannins and red pigments. However, the dynamics of their production, concentration, and stability are still inadequately understood. Thus, a detailed explanation of their sensory significance remains imperfect. Many of the compounds recently discovered may be of no more significance than most of the aromatic compounds present in wine. A clear assessment of the significant of these compounds to grape and wine quality awaits a clearer picture of this incredibly complex group.

CHEMICAL GROUPS OF WINE PHENOLICS

Chemically, phenols are cyclic benzene compounds, possessing one or more hydroxyl groups associated directly with the ring structure. Although containing alcohol groups, they do not show the properties typically associated with aliphatic alcohols.

The major phenolics found in wine are either members of the diphenylpropanoids (**flavonoids**) or phenylpropanoids (**nonflavonoids**) (Table 6.2). In addition, there are related (phenyl) compounds that do not possess one or more hydroxyl groups on the phenyl ring (and are correspondingly strictly not phenols). In conformity with standard practice in the viticultural and enologic literature, they are discussed together. Table 6.3 and de Beer *et al.* (2002) present data on the relative phenolic contents of red and white wines.

Flavonoids are characterized by a C_6-C_3-C_6 skeleton, consisting of two phenolic rings (A and B) joined by a central pyran (oxygen-containing) ring (C). The numbering of the carbons is noted below in the accompanying illustration. The most common flavonoids in wine are flavonols, catechins (flavan-3-ols), and anthocyanins (red wines). Small amounts of flavan-3,4-diols (leucoanthocyanins) also occur. Flavonoids may exist free or in polymers with other flavonoids, sugars, nonflavonoids, or a combination of these. Those esterified to sugars and nonflavonoids are called glycosides and acyl derivatives, respectively. Flavonoids are largely synthesized in the endoplasmic reticulum before being translocated and stored in the central vacuole of the producing cell. Their biosynthetic pathway is outlined in Figs 3.36 and 3.37. Their presumed function in grapes (and other plants) is as a first line defense against microbial pathogens, insect pests, and herbivores.

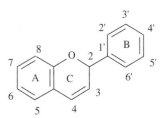

Basic flavonoid skeleton

In grapes, polymerization of catechins (flavan-3-ols) produces a class of polymers called **procyanidins** (condensed tannins). Procyanidins may be classified based on the nature of their flavonoid monomers, bonding, esterification to other compounds, or functional properties. For example, procyanidins may be subdivided into proanthocyanidins or prodelphinidins, based on their cleavage under acidic conditions, releasing either cyanidin or delphinidin, respectively. The most common structural class of procyanidins contains only a single covalent carbon bond between adjacent flavonoid subunits. Typical catechin (flavan-3-ol) subunits in grapes are (+)-catechin, (−)-epicatechin, (−)-epicatechin-gallate, and less commonly (−)-epigallocatechin (Fig. 6.2). (+)-Catechin and (−)-epicatechin differ only in

Table 6.2 Phenolic and related substances in grapes and wine

General type	General structure	Examples	Major source[a]
Nonflavonoids			
Benzoic acid	COOH	Benzoic acid	G, O
		Vanillic acid	O
		Gallic acid	G, O
		Protocatechuic acid	G, O
		Hydrolyzable tannins	G
Benzaldehyde	CHO	Benzaldehyde	G, O, Y
		Vanillin	O
		Syringaldehyde	O
Cinnamic acid	CH=CHCOOH	*p*-Coumaric acid	G, O
		Ferulic acid	G, O
		Chlorogenic acid	G
		Caffeic acid	G
Cinnamaldehyde	CH=CHCHO	Coniferaldehyde	O
		Sinapaldehyde	O
Tyrosol	CH₂CH₂OH	Tyrosol	Y
Flavonoids			
Flavonols		Quercetin	G
		Kaempferol	G
		Myricetin	G
Anthocyanins		Cyanin	G
		Delphinin	G
		Petunin	G
		Peonin	G
		Malvin	G
Flavan-3-ols		Catechin	G
		Epicatechin	G
		Gallocatechin	G
		Procyanidins	G
		Condensed tannins	G

[a] G = grape; O = oak; Y = yeast.

Source: Data from Amerine and Ough, 1980

Table 6.3 Gross phenol composition estimated in mg GAE/L for typical table wines from *Vitis vinifera* grapes

Phenol class	Source[a]	White wine		Red wine	
		Young	Aged	Young	Aged
Nonflavonoids, total		175	160–260	235	240–500
Cinnamates, derivatives	G, D	154	130	165	150
Low volatility benzene deriv.	D, M, G, E	10	15	50	60
Tyrosol	M	10	10	15	15
Volatile phenols	M, D, E	1	5	5	15
Hydrolyzable tannins, etc.	E	0	0–100	0	0–260
Macromolecular complexes					
"Protein" – "tannin"	G, D, E	10	5	5	10
Flavonoids total		30	25	1060	705
Catechins	G	25	15	200	150
Flavanols	G, D	tr	tr	50	10
Anthocyanins	G	0	0	200	20
Soluble tannins, derivatives	G, D	5	10	550	450
Other flavonoids, derivatives	G, D, E, M	?	?	60?	75?
Total phenols		215	190–290	1300	955–1215

[a] D = degradation product; E = environment, cooperage; G = grapes; M = microbes, yeast.

Source: From Singleton, 1982, reproduced by permission

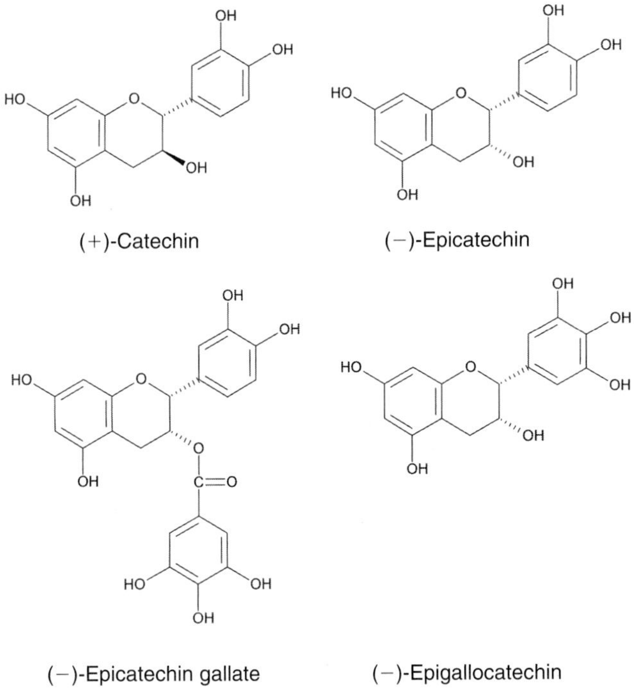

Figure 6.2 Diagrammatic representation of the structures of the principal catechins (flavan-3-ols) in grape tannins.

their stereochemistry. (+)-Catechin possesses its C_3 hydroxyl group in a plane opposite the B-ring, whereas (−)-epicatechin possesses both C_3 hydroxyl groups in the plane of the B-ring. (−)-Epigallocatechin differs from epicatechin in possessing a third hydroxyl group in its B-ring, whereas (−)-epicatechin gallate has a gallic acid esterified to C_3. For simplicity, the stereoisomerism of these compounds, indicated by (+) and (−), is not noted further in the text.

The most common covalent linkage between flavonoids occurs between the C_4 of the pyran ring of an upper flavonoid with C_8 of the A ring of a lower flavonoid (Fig. 6.3A). These form the **B type** procyanidins characteristic of grapes and wine. Procyanidin B_1 to B_4 dimers differ only in the arrangement of the initial and terminal epicatechin and catechin units. Bonding of flavonoids between C_4 and C_6 sites permits branching of the normally linear, procyanidin polymer (Fig. 6.3B). A less common association, combining bonding between C_4 and C_8, as well as between C_2 and the hydroxyl of C_7 generates **A type** procyanidins. Polymerization may also occur at other sites, depending on whether bonding occurs in association with oxidation, with acetaldehyde or glyoxylic acid (Table 6.4). Although catechin is the typical terminal unit in skin tannins, catechin, epicatechin and epicatechin-gallate may occur as terminal units in seed tannins. All catechins may occur as extension units in procyanidin polymers, with the exception of epigallocatechin. It appears to occur only in skin tannins.

The vast majority of procyanidins in wine are derived from grapes, with only trace amounts possibly being extracted from oak cooperage. Structural differences exist among skin, stem, and seed procyanidins. There is also considerable variation in the types and concentrations among cultivars (Kovać *et al.*, 1990). Seed tannins are less polymerized than skin tannins – seed tannins contain up to 28 flavanol moieties

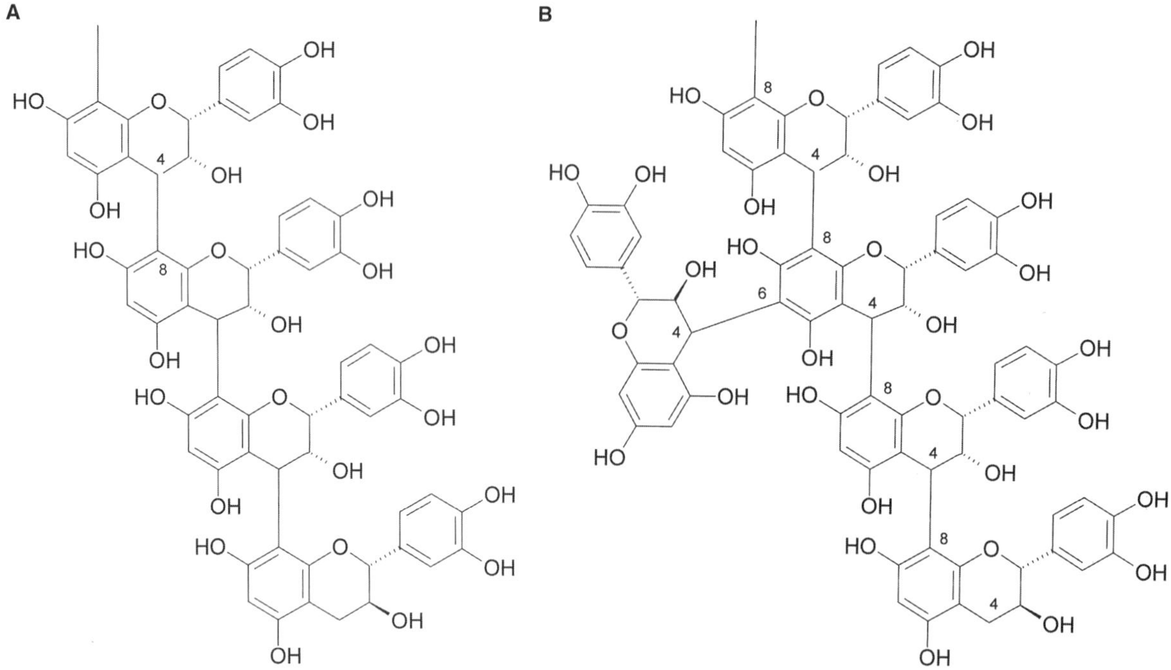

Figure 6.3 Examples of B-type condensed tannins in grapes: **A**, unbranched flavan-3-ol procyanidin; **B**, branched flavan-3-ol procyanidin.

Table 6.4 Formation of condensed tannins

Without oxidation	With oxidation	With ethyl linkage
Linkage of the C_4 with C_8 (or C_6) of the adjacent flavonoid subunit	Linkage of a $C_{3'}$, $C_{4'}$, $C_{5'}$, or $C_{6'}$ of one flavonoid unit with the C_8 (or C_6) of the adjacent flavonoid unit	Linkage of C_8 of one unit via an ethyl junction (derived from acetaldehyde or glyoxylic acid) with the C_8 (or C_6) of the adjacent unit
May involve attachment of anthocyanin units to terminal flavonoids		May involve attachment of anthocyanin units to terminal flavonoids
Slow at wine pH values	Rapid with enzymatic action	Moderate at wine pH values; major brownish pigments

(Hayasaka *et al.*, 2003), whereas skin tannins may possess up to 74 flavanol units (Labarbe *et al.*, 1999). In young wines, the majority of extracted procyanidins are dimers or timers. Red wines contain about 20 times the procyanidin content of white wines.

During wine aging, procyanidins slowly combine with monomeric flavonoids to generate polymers (tannins) of from 8 to 14 units in length. These generally possess molecular weights ranging from 2000 to more than 5000 daltons. In addition to self-polymerization, simple procyanidins and their larger polymers (condensed tannins) may condense with anthocyanins and polysaccharides. The presence of many unconjugated hydroxy-phenolic groups in flavonoid polymers is thought to give tannins their distinctive protein-binding property. As polymer size increases, they become increasing insoluble, and eventually can no longer react with, or precipitate proteins.

In grapes, flavonoids are primarily synthesized in the skins and seeds. They are produced in smaller amounts in the stems. Flavonols and anthocyanins are principally localized to the skins, whereas flavan-3-ols and their procyanidin polymers are synthesized primarily in seeds and stems (about 60 and 20% respectively), with skins producing about 15–20% (Bourzeix *et al.*, 1986; Downey *et al.*, 2003a).

In the skin, flavonols collect in cellular vacuoles of the epidermis and outer hypodermis. In this location (along with anthocyanins in red grapes), they absorb ultraviolet radiation, protecting inner tissues from the damaging effects of solar UV radiation. The principal flavonols are kaempferol, quercetin and myricetin. Flavonols are less frequently deposited in stem tissue. In wine, they may act as copigments with anthocyanins. Of grape flavonoids, flavonols occur in the lowest concentration, varying

Table 6.5 Average total phenol and anthocyanin content of the fruit of different grape varieties in the South of France

Variety	Level (mg/kg fresh weight)		Individual monoglucoside anthocyamins (%)		
	Total phenolics	Total anthocyanins	Delphinin	Petunin	Malvin and peonin
'Colobel'	10,949	9967	20	21	28
'Pinot noir'	7722	631	4	12	77
'Alicante Bouschet'	7674	4893	7	10	55
'Cabernet Sauvignon'	6124	2339	17	8	48
'Syrah'	6071	2200	11	12	45
'Tempranillo'	5954	1493	25	16	41
'Gamay noir'	5354	844	1	3	64
'Malbec'	4613	1710	7	8	54
'Chardonnay'	4126	–	–	–	–
'Grenache'	3658	1222	7	10	63
'Carignan'	3582	1638	18	14	43
'Sauvignon blanc'	2446	–	–	–	
'Villard blanc'	2280	–	–	–	–
'Cinsaut'	2154	575	6	9	51

Source: Data from Bourzeix *et al.*, 1983

from 1–10% of the total phenolic content, depending on the cultivar and growing conditions. Synthesis is primarily active at fruit set, and subsequently during ripening (Downey *et al.*, 2003b). The synthesis of flavonols (and to a lesser degree anthocyanins – depending on the cultivar) is activated by exposure to UV and blue radiation. Anthocyanin synthesis, unlike flavonol production, is tightly linked with the onset of *véraison*. Anthocyanin glycosides, when present, typically accumulate in the epidermis and outermost hypodermal cells.

Stem and seed (seed coat) flavan-3-ols occur primarily as catechin, epicatechin, epigallocatechin and the ester, epicatechin-gallate, as well as procyanidin oligomers and polymers (condensed tannins). Until oxidized, grape tannins are colorless. Seed tannins are primarily smaller than those found in the skins (mean of 10 units vs. 30 for skin tannins), possess significantly greater epigallocatechin content, and are less galloylated (esters of flavanols with gallic acid) (Souquet *et al.*, 1996; Downey *et al.*, 2003b). Stem tannins have a size distribution between those of skin and seed tannins, and consist principally of epicatechin extension units (Souquet *et al.*, 2000). Some varieties, for example 'Pinot noir,' appear to produce no condensed tannins in the skin (Thorngate, 1992). This may partially explain the poor color intensity typical of 'Pinot noir' wines, as well as why stems have often been added to the ferment. 'Pinot noir' also appears atypical in its low proportion of seed tannins, the majority of flavonoids occurring as monomers. During fermentation, skin tannins are extracted earlier than seed tannins, but this tends to change as fermentation progresses (Peyrot des Gachons and Kennedy, 2003).

Flavonoids characterize red wines more than they do white wines. In red wines, they constitute more than 85% of the phenolic content (≥ 1000 mg/liter). In white wines, flavonoids typically constitute less than 20% of the total phenolic content (≤ 50 mg/liter). The remainder consists primarily of the nonflavonoid, caffeic acid.

The degree to which flavonoids are extracted during wine production depends on many factors. Extraction is ultimately limited by the amount present in the fruit. This content varies considerably from variety to variety (Table 6.5), with climatic conditions and fruit maturity. Traditional fermentation, due to its longer maceration with the seeds and skins, extracts more phenolic compounds than carbonic maceration or thermovinification. Flavonoid extraction is also markedly influenced by the pH, sulfur dioxide content, and ethanol content of the juice, as well as the temperature and duration of fermentation. Consequently, there is no simple means of predicting wine phenolic content from grape phenolic content.

Due to the multiple factors affecting phenolic extraction, it is not surprising that the phenolic content shows greater variation than any other major wine constituent. In addition, there are greater quantitative and qualitative changes in the concentration and structure of phenolics during aging than any other wine constituent.

Nonflavonoids (possessing a C_6-C_3 skeleton) are structurally simpler, but their origin in wine is more diverse. In wines not aged in oak, the primary nonflavonoids are derivatives of hydroxycinnamic and hydroxybenzoic acids. They are stored primarily in cell vacuoles of skin and pulp, and are easily extracted on crushing. The most numerous and variable are hydroxycinnamic acid derivatives. They occur principally as esters with tartaric acid, but may also be associated with sugars, various alcohols, or other organic acids. Common examples are caftaric, coutaric, and fertaric acids – the tartaric acid esters of caffeic, *p*-coumaric, and ferulic acids, respectively. In the presence of pectin methyl esterase, the esters breakdown to their monomers. The esters also slowly hydrolyze during fermentation. The most common nonflavonoid in grapes, caftaric acid (an *o*–diphenol) is one of the primary substrates for polyphenol oxidase. It often plays an important role in oxidative browning of must. In small amounts, the oxidized derivatives of caftaric and coutaric acids may donate much of the straw yellow–gold coloration of white wines. Although equally present in red wines, the abundance of anthocyanins and procyanidins masks the presence of oxidized grape nonflavonoids.

Wines matured in oak possess elevated levels of hydroxybenzoic acid derivatives, notably ellagic acid (the dilactone formed by the association of two molecules of gallic acid). Ellagic acid comes from the hydrolytic breakdown of **ellagitannins**, polymers of hexahydrodiphenic

Gallic acid 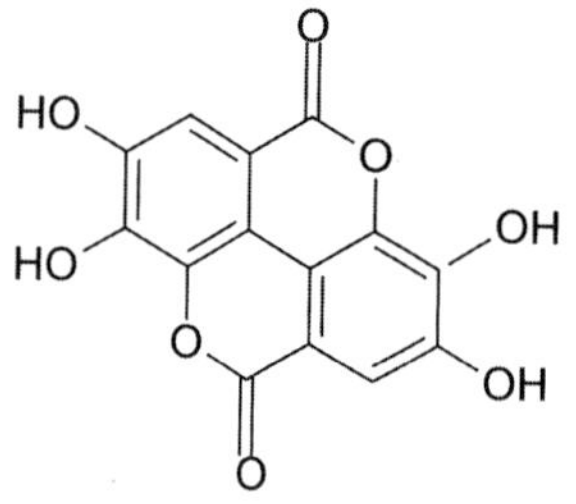   Ellagic acid

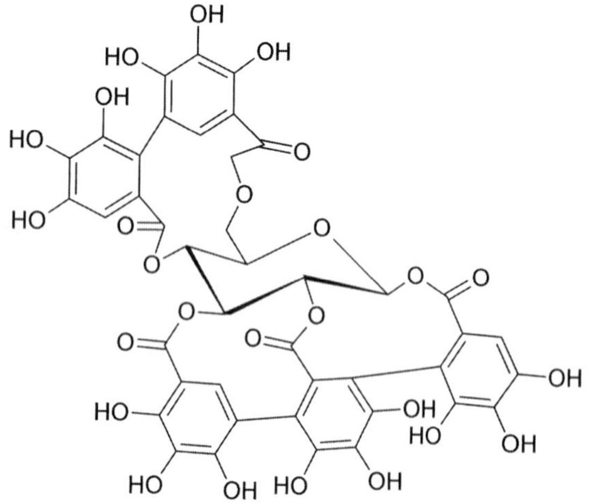

Example of an ellagitannin

Figure 6.4 Diagrammatic representation of the structures of gallic acid, ellagic acid and a hydrolyzable tannin (ellagitannin, a polymer of ellagic acid moieties esterified to glucose core).

acid, esterified to glucose (Fig. 6.4). Gallotannins are another group of hydrolyzable tannins consisting of gallic acid polymers. They make up only about 5% of the total hydrolyzable tannin content of oak.

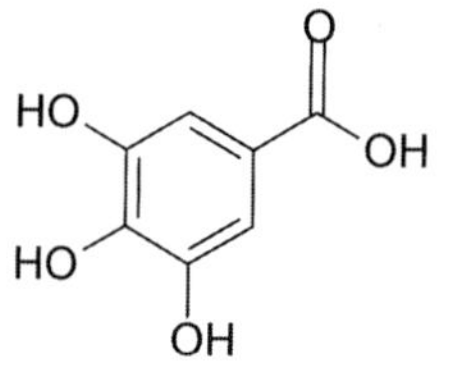

Hydrolyzable tannins represent up to 10% of the dry weight of oak heartwood, where they limit microbial decomposition. Castalagin and vescalagin are the most frequently extracted, being isomers of five gallic acid moieties bonded to a central glucose. Roburins are dimers of castalagin and vescalagin, with or without links to xylose or lyxose. On hydrolysis, the polymers yield gallic acid and its dilactone, ellagic acid. Esters of ellagic acid may enhance red wine color by forming copigments with anthocyanins. Their ready oxidation also makes them active in the consumption of oxygen in wine matured in oak cooperage. Wines aged in chestnut cooperage extract slightly different, but related, hydrolyzable tannins and nonflavonoids.

Degradation of oak lignins also liberates various volatile cinnamaldehyde and benzaldehyde derivatives, notably vanillin, sinapaldehyde, coniferaldehyde and syringaldehyde. In addition, small amounts of other nonflavonoid phenolics, such as esculin and scopoline, are usually extracted. The slow conversion of esculin to the less bitter esculetin during aging is thought to be partially responsible for a reduction in bitterness of wines aged in oak.

Both flavonoid and nonflavonoid polymers, generically termed **tannins**, derive their name from their ability to tan leather. Nonflavonoid-based, hydrolyzable tannins separate readily into their component parts under acidic conditions. The low pH weakens the bonding between the hydrogen and oxygen atoms of the associated moieties. In contrast, flavonoid-based condensed tannins are more stable under acidic conditions. They are held together by covalent bonds.

Yeast metabolism may provide additional nonflavonoid phenolics. The most prevalent is tyrosol. Tryptophol also is synthesized, but in smaller quantities.

Nonflavonoids of grape origin are initially synthesized from phenylalanine (Hrazdina *et al.*, 1984), whereas those of yeast origin are derived from acetic acid (Packter, 1980). Flavonoids are derived from the combination of derivatives synthesized from both phenylalanine (via the shikimic acid pathway) and acetic acid.

The total phenolic content of wine increases during the early stages of fermentation. The derivatives of hydroxycinnamic and hydroxybenzoic acids are the most rapidly extracted, followed by grape flavonols and anthocyanins. The slowest to dissolve are the flavan-3-ols (catechins) and their polymers (procyanidins and condensed tannins). Their extraction is initially time-dependent, reflecting the duration of contact with the seeds and skins. Subsequently, the phenolic content declines, as they oxidize and polymerize, as well as bond and precipitate with proteins and cell remnants. During fining and maturation, the phenolic content continues to decline. Aging in wood cooperage results in a temporary increase in the phenolic content.

COLOR – RED WINES

Our knowledge of the chemical complexity of red wine color continues to expand at an amazing pace. Because of the extreme complexity of the phenolic content of wine, most of the studies have been conducted in model wine solutions, with defined chemical composition. This permits individual reactions to be studied more easily, but under conditions that do not fully represent those in wine. Nevertheless, this is a necessary first step in understanding the reactions that generate both color depth and tint as well as stability. It is likely that future findings will significantly modify our current interpretation of red wine color.

The principal source of red color in wine comes from its anthocyanin content. Nevertheless, free anthocyanins are not particularly stable. Polymerization is essential to color stability. Stability develops through a complex series of mechanisms, including short-term factors, such as copigmentation, and long-term factors, such as polymerization with flavan-3-ols and procyanidins, as well as the formation of new pigments, such as pyranoanthocyanins. The latter can also condensate with tannins. Additional sources of pigmentation come from the oxidation and polymerization of both grape- and oak-derived flavonoids. An example of the latter are oaklins, coniferaldehyde-catechin condensation products.

Anthocyanins exist in grapes as glycosides – the bonding of the flavonoid component (the aglycone) with sugar. The aglycone is called an anthocyanidin. The junction occurs at C_3, and if a diglycoside, at positions C_3 and C_5. In grapes, the sugar moiety is typically glucose. The glycosidic bonding increases both the chemical stability and solubility of the anthocyanidin. Each anthocyanin may be further complexed (acylated) by a bonding of the sugar moiety, usually at C_6, with either acetic acid, coumaric acid, or caffeic acid (Table 6.6).

In young red wines, anthocyanins occur predominantly as a dynamic equilibrium among five major molecular states, one bonded to sulfur dioxide and four free-forms (Fig. 6.5). Most forms are colorless within the pH range typical of wine. Red color comes primarily from the small proportion of anthocyanins that exists in the **flavylium** state. The proportion depends on the pH and free sulfur dioxide content of the wine. Low pH increases the concentration of the flavylium state, enhancing redness. Low pH also retards the hydrolysis of the anthocyanin molecule into its sugar and aglycone moieties. As the pH rises, the color density and proportion of anthocyanins in the flavylium state rapidly decline. For example, 20–25% of anthocyanins at a pH of 3.4–3.6 (appropriate for red wines) are in the ionized flavylium state, whereas only about 10% are in the flavylium state at a pH of 4. The blue–mauve cast of high pH wines comes from a slight increase in the proportion of **quinoidal** anthocyanins. However, the most significant factor affecting color density of young red wines is not the pH, but the amount of free sulfur dioxide. Sulfur dioxide is an effective, albeit reversible, anthocyanin bleaching agent.

Anthocyanin classification is based primarily on the position of the hydroxyl and methyl groups on the B ring of the anthocyanidin (Table 6.6). On this basis, grape anthocyanins can be divided into five classes, namely cyanins, delphinins, malvins, peonins, and petunins. The proportion and amount of each class varies widely among cultivars and with growing conditions (Wenzel *et al.*, 1987). The proportion of anthocyanins markedly influences both hue and color stability. Both properties are directly affected by the hydroxylation pattern of the anthocyanidin B ring. Blueness increases with the number of free hydroxyl groups, whereas redness intensifies with the degree of methylation. Because the predominant anthocyanin in most red grapes is malvin, the reddest of anthocyanins, it donates most of the color to young red wines.

Sensitivity to oxidation in anthocyanins is influenced by the presence of *ortho*-diphenols on the B ring

Table 6.6 Anthocyanins occurring in wines

NAME	R1	R2	R3
Cyanin	OH	OH	H
Peonin	OCH$_3$	OH	H
Delphinin	OH	OH	OH
Petunin	OCH$_3$	OH	OH
Malvin	OCH$_3$	OH	OCH$_3$

DERIVATIVES

R4 – H: monoglucoside; glucose: diglucoside

R5 – acetyl, *p*-coumaroyl, caffeoyl

(Hrazdina *et al.*, 1970). The adjacent hydroxyl groups of *o*-diphenols are sensitive to enzymatic and nonenzymatic oxidation. Except for laccase, most polyphenol oxidases affect only *o*-diphenol sites. Because neither malvidin nor peonidin possess *ortho*-positioned hydroxyl groups, they are comparatively resistant to oxidation. Resistance to oxidation also is affected by conjugation with sugar and other compounds (Robinson *et al.*, 1966). Hydrated (hemiketal or carbinol)

forms of anthocyanins readily react with *o*-diquinones generated by oxidation. The reaction product is colorless and chemically unstable.

Anthocyanins may also be classified by the number of sugar molecules attached to the anthocyanidin moiety. In most grape species, both mono- and diglucosidic anthocyanins are present. Triglucosidic anthocyanins do not occur in grapes. Diglucoside anthocyanins are more stable than their monoglucosidic counterparts,

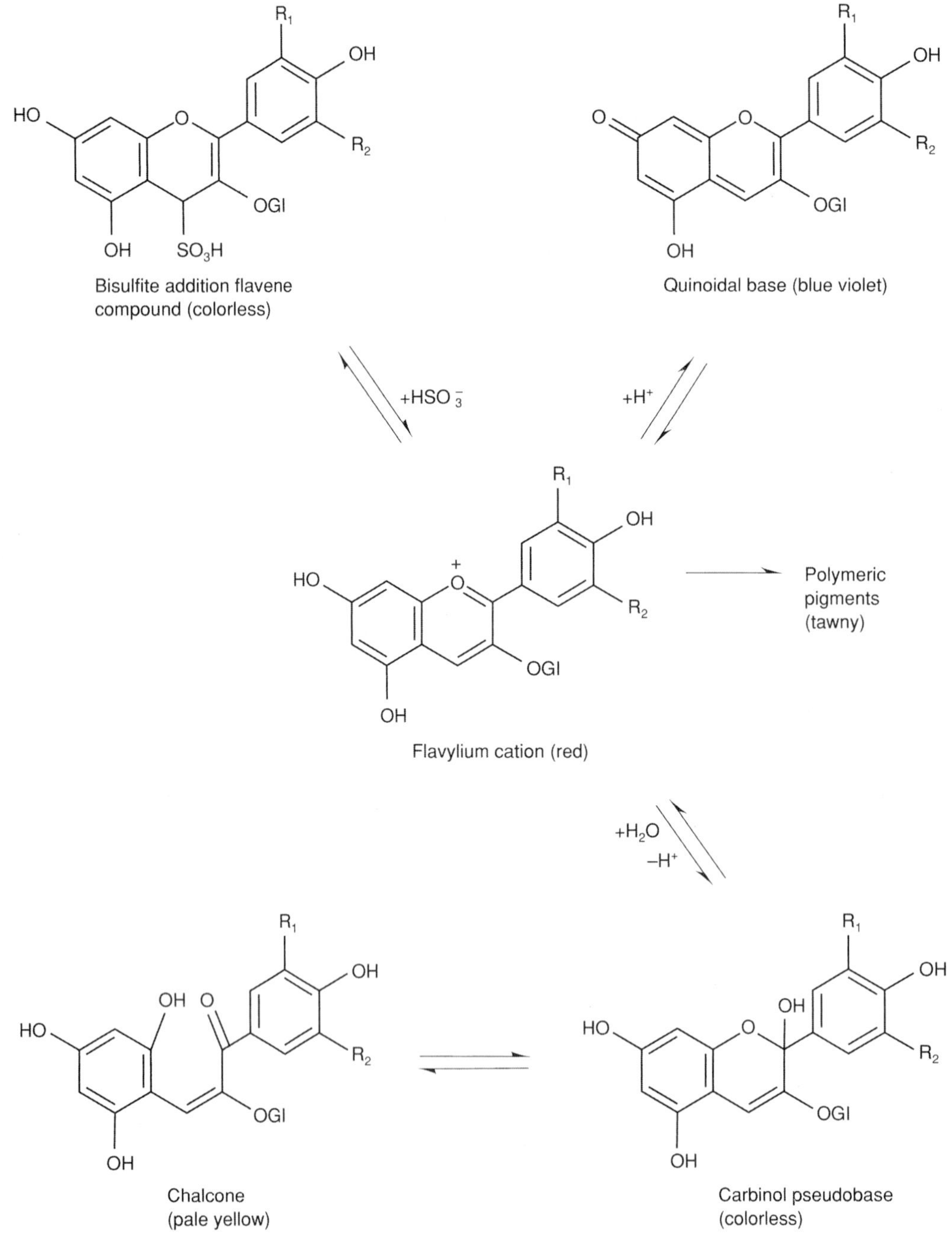

Figure 6.5 Equilibria among the various forms of anthocyanins in wine. Gl, glucose.

but are more susceptible to browning (Robinson *et al.*, 1966). Monoglucosidic anthocyanins tend to be more colored than their diglucosidic forms.

Cultivars of *Vitis vinifera* synthesize only monoglucosidic anthocyanins. They lack the dominant allele regulating the production of diglucosidic anthocyanins. Thus, first-generation interspecies crosses with *V. vinifera* produce both mono- and diglucosidic anthocyanins. Complex hybrids, involving one or more backcrosses to *V. vinifera*, produce only monoglucosidic anthocyanins, if they no longer possess the dominant allele for diglucoside synthesis (van Buren *et al.*, 1970). The presence of diglucosidic anthocyanins has been used to detect the use of hybrid grapes in Appellation Control (AC) red wines. This works because most red French-American hybrids synthesize diglucosidic anthocyanins. It would not work with 'Plantet.' The Seibel hybrid (#5455) produces only monoglucosidic anthocyanins.

In the majority of red grapes (excluding 'Pinot noir') a variable portion of the anthocyanins is acylated (bonded to acetic, caffeic or *p*-coumaric acids). Intriguingly, red-colored mutants of white grape varieties, such as red 'Chardonnay,' pink 'Sultana,' and 'Muscat Rouge' produce no acylated anthocyanins (Boss *et al.*, 1996). Muscadine grapes (*Vitis rotundifolia*) also do not produce acylated anthocyanins, nor do they produce caffeoyltartaric acid.

In the fruit of red grape varieties, anthocyanins exist primarily as loose complexes, either with themselves or with other compounds. Anthocyanin conglomerates are held together by processes called **self-association** and **copigmentation** (Somers and Vérette, 1988; Mistry *et al.*, 1991). Both lead to stacked molecular aggregates held together by hydrophobic interactions, between anthocyanins alone (self-association), or between anthocyanins and other organic compounds (copigmentation). The principal copigments in wine are flavonoids and nonflavonoids, although amino acids (notably proline and arginine), organic acids, polysaccharides and purines may be involved. Both types of complexes significantly increase color density, and may affect color tint. Self-association appears to be particularly important in grape coloration, whereas copigmentation is more important to the purplish coloration of young red wines. The stacking of anthocyanin molecules in these complexes physically limits water access to (and hydration of) red flavylium and bluish quinoidal base states to colorless carbinol (hemiketal) forms (Goto and Kondo, 1991). Both colored anthocyanin states are almost planar, making interaction between them and other anthocyanin molecules or copigments easier and more probable. At the low pH values of wine, copigmentation primarily involves anthocyanins in their flavylium state (Somers and Evans, 1986). This, in turn, shifts the free anthocyanin equilibrium toward the colored flavylium state, further enhancing coloration and tending to shift the tint toward purple. Copigmentation is estimated to contribute about 30–50% of the color of young red wines (Boulton, 2001).

Self-association complexes are sustained by both hydrophilic attractions between their glucose components and hydrophobic repulsion by water (Goto and Kondo, 1991). In the case of the stronger (more stable) copigmentation complexes, covalent bonds form between acyl groups of anthocyanins and copigments. A wide variety of compounds may act as copigments, but the principal constituents are epicatechin, procyanidins, cinnamic acids, and hydroxycinnamoyl esters. Variation in the relative proportion of these copigments from wine to wine may partially explain the different color characteristics of wines produced from the same variety. For example, although epicatechins copigment readily, catechins form more intensely colored complexes. Cinnamic acid derivatives, such as caffeic acid (Darias-Martín *et al.*, 2002) and *p*-coumaric acid (Bloomfield *et al.*, 2003), also readily participate in copigment formation. This may explain why some regions have traditionally added some white must (occasionally high in the latter nonflavonoids) to the must of red grapes prior to fermentation.

Neither self-association nor copigmentation play a significant role in white or rosé wines. White wines contain insufficient anthocyanins for copigment formation. This also applies to rosé wines. The minimum concentration required for copigmentation is about 250 mg/liter (Asen and Jurd, 1967). This is significantly more than the typical 20–50 mg/liter found in most rosés.

Various factors may lead to the disruption of anthocyanin complexes. Heating grapes or must to improve color extraction (thermovinification) also destabilizes self-association. This can lead to a serious loss in color during wine maturation, if insufficient tannins are extracted from the pomace. Tannins favor the formation of colored anthocyanin–tannin polymers. Alcohol also destabilizes the hydrogen bonding between anthocyanin aggregates (by disrupting the lattice-like interaction of water molecules). Consequently, must fermented in contact with seeds and skins for only a few days may show a pronounced loss in color as fermentation continues, despite the anthocyanin content remaining fairly constant (Fig. 6.6). Typical losses in color density vary from two- to fivefold, depending on the phenolic content, pH, and ethanol content of the wine. The loss is primarily due to a decrease in light absorbency, as anthocyanins in the stacked complexes dissociate. Contained within the complexes are anthocyanins in quinoidal states. These lose their bluish color when freed into the acidic environment of wine.

Although diminished, self-association may be the origin of some of the purple tints that characterize young

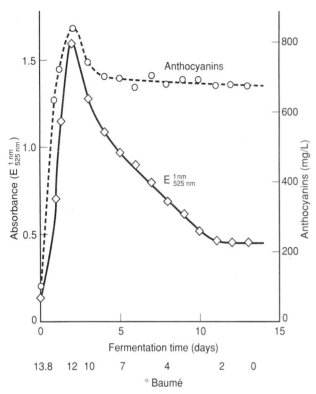

Figure 6.6 Changes in absorbency at 525 nm (color intensity) and in total anthocyanin content of the fermenting must during a traditional fermentation on skins. The must was pressed after 2 days. (From Somers, 1982, reproduced by permission)

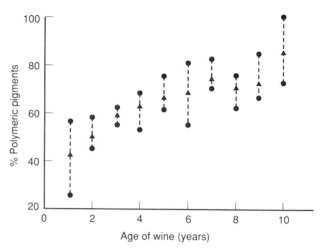

Figure 6.7 Increasing contributions of polymeric pigments to wine color density during the aging of 'Shiraz' wines: ▲, mean values; ●, extremes. (From Somers, 1982, reproduced by permission)

red wines. Cool storage temperatures also favor copigmentation and retards their disassociation. Because molecular stacking occurs in a dynamic equilibrium with free anthocyanins, red wines with low pigment and phenolic contents show enhanced dissociation and greater color loss than would be predicted from their anthocyanin content. In addition, anthocyanin molecules tend to hydrolyze to anthocyanidins, losing their acyl (acetate, caffeate, or coumarate) constituents, along with their glycosidic (glucose) component(s). Dissociated anthocyanidins are both more sensitive to irreversible oxidative color loss, as well as reversible conversion of colored flavylium forms to colorless hemiketals.

Phenolics extracted for the pomace, along with the anthocyanins, both directly and indirectly contribute to wine coloration. These compounds, mostly catechins and procyanidins, begin to polymerize with themselves, or free anthocyanins and anthocyanidins. These complexes tend to stabilize and retain the reddish color of anthocyanins. The complexes are also important to the flavor of red wines, by increasing the solubility (and retention) of flavonoid polymers (tannins). Color stability is also enhanced by polymerization with hydroxycinnamic acid derivatives, such as caffeic acid (Darias-Martín *et al.*, 2002).

By the end of fermentation, about 25% of anthocyanins may have polymerized with flavonoid or nonflavonoid phenolics. This level may rise to 40% or more within one year (Somers, 1982). Thereafter, polymerization continues at a slower pace, until the level approaches 100% after several years (Fig. 6.7). Cultivar variability in flavonoid content has been suggested as one of the reasons for differences in color stability among red wines (McCloskey and Yengoyan, 1981). The lack of appropriate flavonoids appears to be involved in the color instability of muscadine wines (Sims and Morris, 1986). In addition, the absence of acylated anthocyanins and caffeoyl tartaric acid may limit the color stability of muscadine wines.

Nonflavonoid tannins extracted from oak appear to play little role in condensation reaction with anthocyanins. Nevertheless, they appear to favor color stabilization indirectly, by protecting anthocyanins and flavanols from oxidative degradation (Guerra *et al.*, 1996).

Polymerization helps stabilize wine color by protecting the anthocyanidin molecule from oxidation or other chemical modifications, such as sulfite decoloration. In addition, a higher proportion of anthocyanin molecules are colored (flavylium and quinoidal states) when covalently bonded with tannins. For example, about 60% of polymerized anthocyanins are colored at a pH of 3.4, whereas only 20% of the equivalent amount of free anthocyanin may be colored (Fig. 6.8). This underscores the importance of both a favorable pH and polymerization to the bright color of red wines. However, polymerization and oxidation shift the absorption properties of the anthocyanin chromophore toward the yellow to yellow-brown (Glories, 1984;

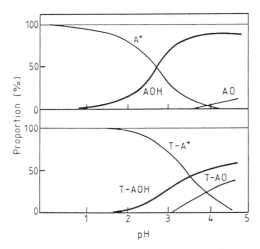

Figure 6.8 Equilibria between the different forms of free anthocyanins (A) and combined anthocyanins (T–A) extracted from wine. +, red flavylium cation; OH, colorless carbinol pseudobase; O, blue-violet quinoidal base. (From Ribéreau-Gayon and Glories, 1987, reproduced by permission)

large condensed tannins, potentially much more complex than those extracted from grapes. These large polymers are less prone to condense with anthocyanins than smaller procyanidins or catechins. Correspondingly, factors that delay anthocyanins–flavonoid polymerization increase the likelihood irreversible anthocyanin oxidation (and browning). For example, the addition of sulfur dioxide induces sulfonate formation with anthocyanins. As this involves the same carbon that binds anthocyanins and flavonoids, polymerization is prevented. This both delays and reduces eventual polymerization. Polymerization repels water, protecting the anthocyanin from nucleophilic attack. In addition, polymerization protects the anthocyanin molecule from the bleaching action sulfur dioxide.

Although anthocyanins condense with flavonoids, the reaction occurs slowly in wine. Bonding is enhanced, however, in the presence of acetaldehyde (Bakker *et al.*, 1993; Fulcrand *et al.*, 1996). At low pH values, acetaldehyde tends to exist in a reactive carbonium ion state (Fig. 6.10). In its carbonium state, acetaldehyde can react with a nucleophilic (negatively charged) C_8 of a terminal unit of a procyanidin. On dehydration, the acetaldehyde moiety can cross-link with the C_8 of an anthocyanin in its hemiketal state (Fig. 6.10). Subsequent dehydration converts the colorless hemiketal to the colored flavylium state (red) or quinoidal base (violet) form, enhancing wine color. Similar acetaldehyde-induced cross-linkages may result in the polymerization of pairs of flavan-3-ol moieties, either via adjacent nucleophilic C_8 sites (Lee *et al.*, 2004), or between the C_8 site of one flavonoid and the nucleophilic C_6 site of another (Saucier *et al.*, 1998). An alternate acetaldehyde cross-linking between tannins and anthocyanins involves the nucleophilic position of a flavan-3-ol and the C_4 of a flavylium (positively charged) anthocyanin (Asenstorfer *et al.*, 2001). Adding further complexity, Pissarra *et al.* (2004) have demonstrated that minor aldehydes, such as propionaldehyde, isovaleraldehyde, isobutyraldehyde and benzaldehyde can act like acetaldehyde. They too can form cross-linkages between anthocyanins and flavan-3-ols, or pairs of flavan-3-ols.

Although cool cellar temperatures retard ethyl-linked, aldehyde-associated, reactions between anthocyanins and catechins, this may be valuable in limiting the formation of excessively large pigment polymers. These could precipitate, causing permanent color loss.

Currently, the relative importance of skin and seed tannins vs. smaller flavan-3-ols and procyanidins in the formation of stable anthocyanin–tannin complexes remains unclear. Most studies have involved smaller flavonoids of known composition, in model wine solutions. However, the speed of pigment polymer formation, and their physical resemblance to tannins, suggest

Cheynier *et al.*, 2000). This results in the progressive color shift toward a brickish cast. Polymerization between anthocyanins and procyanidins also is valuable in limiting tannin precipitation.

Because of the sensitivity of free anthocyanins to irreversible degradation, it is desirable that polymerization occur early during wine maturation. As extracted from grapes, catechins and procyanidins are highly soluble. As such, they can form soluble polymers with anthocyanins. The reaction can generate either T-A or A-T type anthocyanin/tannin adducts. The more abundant, but colorless, hemiketal (nucleophilic) anthocyanins generate **T-A adducts**. They form as the C_8 (or C_6) of the anthocyanin moiety binds to the electrophilic C_4 of a terminal flavonoid unit of a procyanidin or small condensed tannin (Fig. 6.9). Upon dehydration, they generate colored flavylium chromophores. This enhances color expression. Such polymers may possess up to eight flavonoid subunits (Hayasaka and Kennedy, 2003). In contrast, electron sharing between C_4 of an electrophilic flavylium anthocyanin and the C_8 (or C_6) site of a nucleophilic flavan-3-ol of a procyanidin, or a catechin or epicatechin molecule, form **A-T adducts**. It initially generates a colorless (hemiketal) complex. Subsequent oxidation is thought to result in reestablishment of the colored flavylium state. Further structural rearrangement may generate yellow-orange xanthylium forms. The xanthylium structure forms by a dehydration reaction between C_5 of the anthocyanin moiety and the C_8 of the flavonoid to which it is bound. This forms a pyran ring between the two parent molecules.

Flavan-3-ols and their polymers (procyanidins) may also bind with themselves (Fig. 6.9). These generate

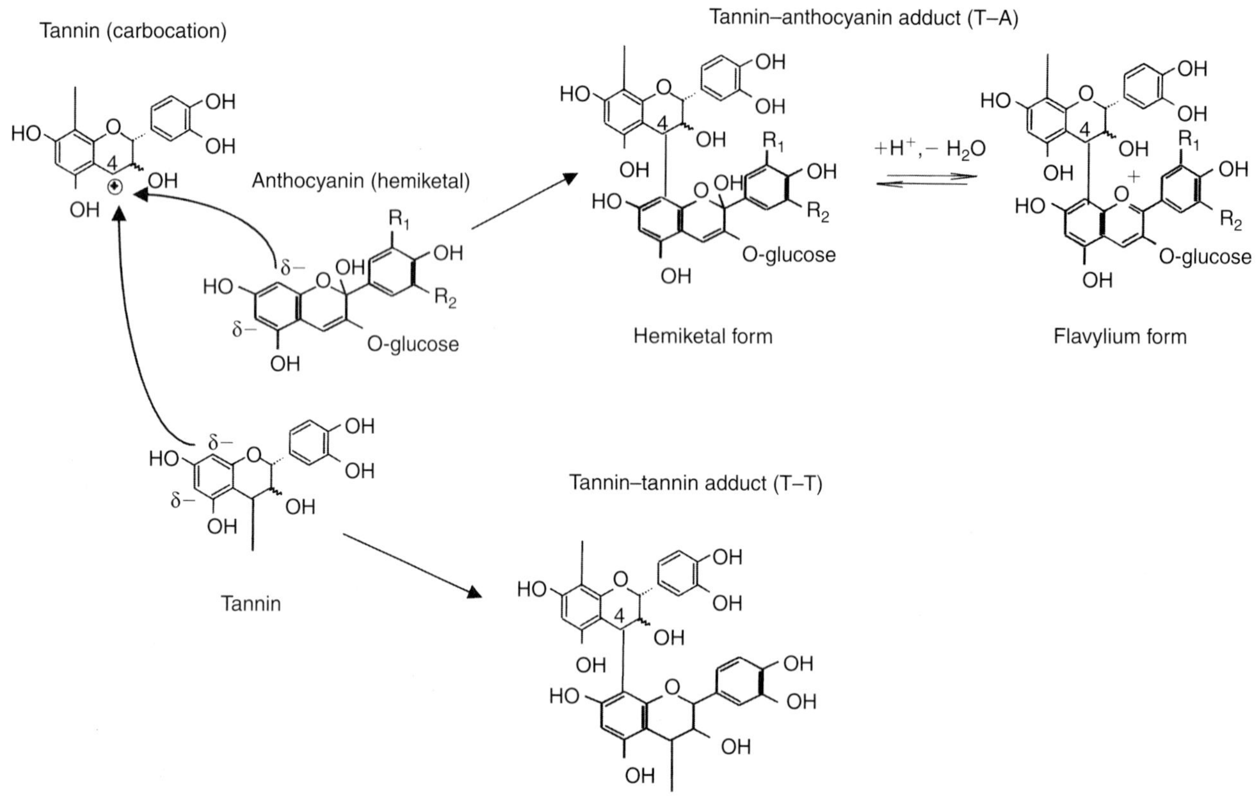

Figure 6.9   Hypothetical mechanisms leading to T–A and T–T adducts. (From Cheynier *et al.*, 2000, reproduced by permission)

that grape tannins play a major role, at least in young wine (Eglinton *et al.*, 2004). Their early formation during and just after fermentation suggests the importance of yeast-generated acetaldehyde.

In port, the augmented alcohol and acetaldehyde contents may decrease the significance of copigmentation. There is also a corresponding increase in the significance of acetaldehyde-induced polymerization.

The most reactive anthocyanin in polymer formation is malvin, the most common grape anthocyanin. The reaction occurs more rapidly than copigmentation, but requires the autooxidation of wine phenolics in the presence of oxygen. This probably explains the color enhancing and stabilizing effect of exposing young red wines to small amounts of oxygen (about 40 mg O_2/year). Traditionally, this has been associated with racking. During *o*-diphenol autooxidation, catalyzed by copper or iron ions, hydrogen peroxide is generated (Fig. 6.11). The *o*-diquinone generated can react with an *o*-diphenol, eventually regenerating an *o*-diphenol dimer. The hydrogen peroxide formed in the initial autooxidation can activate the oxidation of ethanol to acetaldehyde, again in the presence of copper or iron ions. The anthocyanin–tannin polymerization reactions, activated by acetaldehyde, are thought to promote the violet shift so characteristic of young red wines (Dallas

et al., 1996). The extent of this reaction is dependent both on the uptake of oxygen, the presence of sulfur dioxide, and the amount and types of catechins and their polymers (procyanidins).

An alternative pathway for color intensification, associated with the autooxidative generation of peroxide, may involve the oxidation of glycerol. It is the second most common alcohol in wine. The associated formation of glyceraldehyde and dihydroxyacetone could promote the formation of additional, novel, anthocyanin-based pigments (Laurie and Waterhouse, 2006).

Other mechanisms suspected to be involved in early color stabilization involve direct reaction between malvin and yeast by-products, such as acetaldehyde, pyruvic acid, and vinylphenols (e.g., cinnamic acid derivatives) (Bakker and Timberlake, 1997; Fulcrand *et al.*, 1998; Schwarz *et al.*, 2003). These compounds, called **pyranoanthocyanins** (Fig. 6.12), are cycloaddition products. They form an additional pyran ring between C_4 and the hydroxyl group on C_5 of the anthocyanin. Being highly stable and resistant to sulfur dioxide bleaching, they could significantly contribute to color stability. With the exception of some portisins, which are bluish (Mateus *et al.*, 2004), most pyranoanthocyanins possess yellow-orange colors. They probably

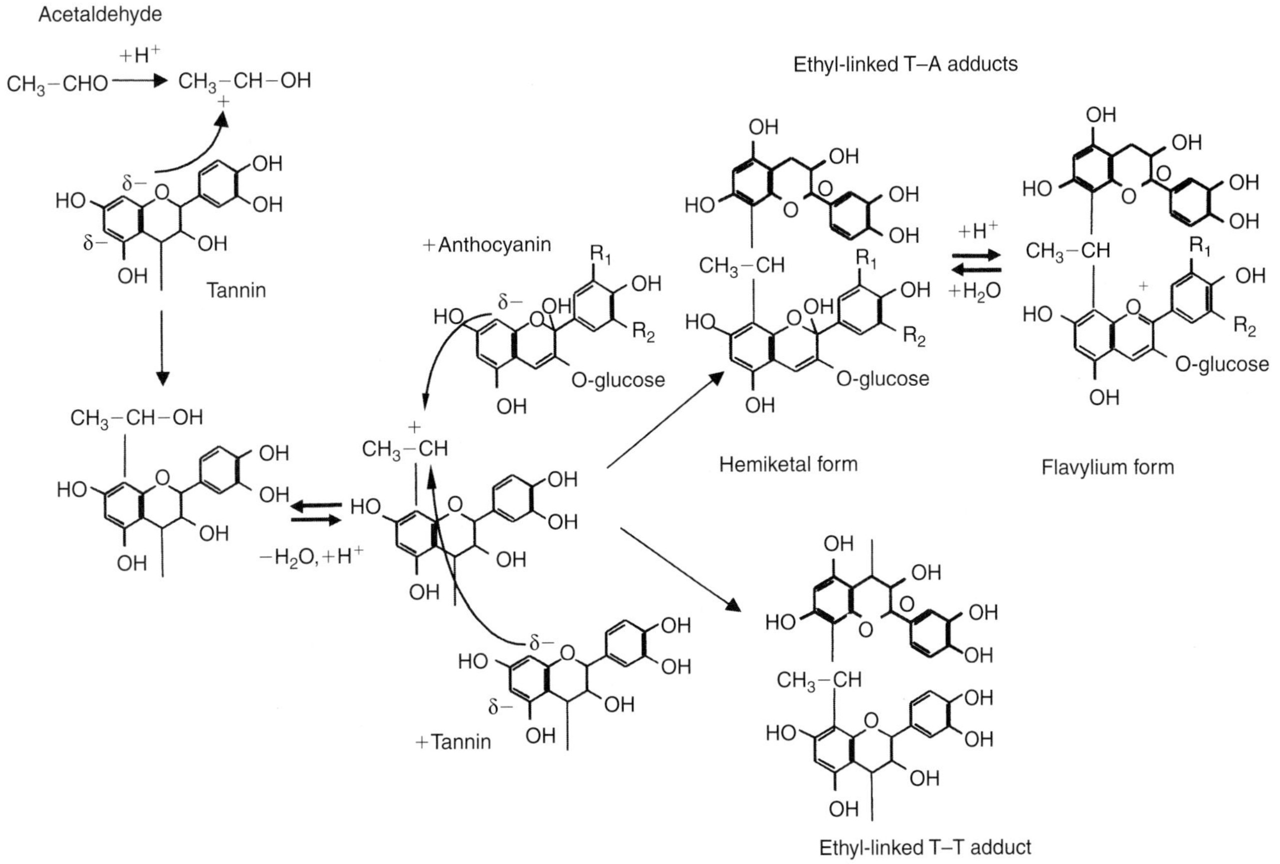

Acetaldehyde

Figure 6.10 Mechanism of acetaldehyde-induced tannin–tannin and tannin–anthocyanin additions. (From Cheynier *et al.*, 2000, reproduced by permission)

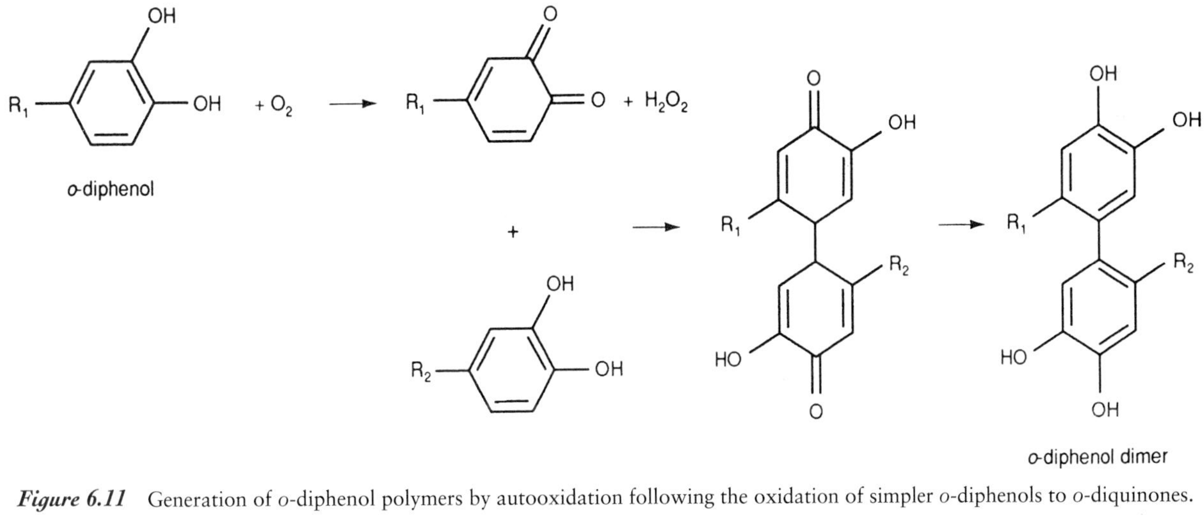

Figure 6.11 Generation of *o*-diphenol polymers by autooxidation following the oxidation of simpler *o*-diphenols to *o*-diquinones.

contribute to the tawny color shift associated with aging. In port, the principal monomeric anthocyanins remaining in aged versions are **vitisins** (Morata *et al.*, 2007) – reaction products between malvin and pyruvic acid (vitisins A), or acetaldehyde (vitisins B). Like anthocyanins, they may be acylated. They tend to

be formed early during fermentation. They can also form polymers with tannins (Atanasova *et al.*, 2002). **Pinotins** form between anthocyanins and cinnamic acid derivatives, such as caffeic acid. They tend to accumulate postfermentation. **Portisins** are derived from anthocyanin-pyruvic acid adducts and flavanols in the

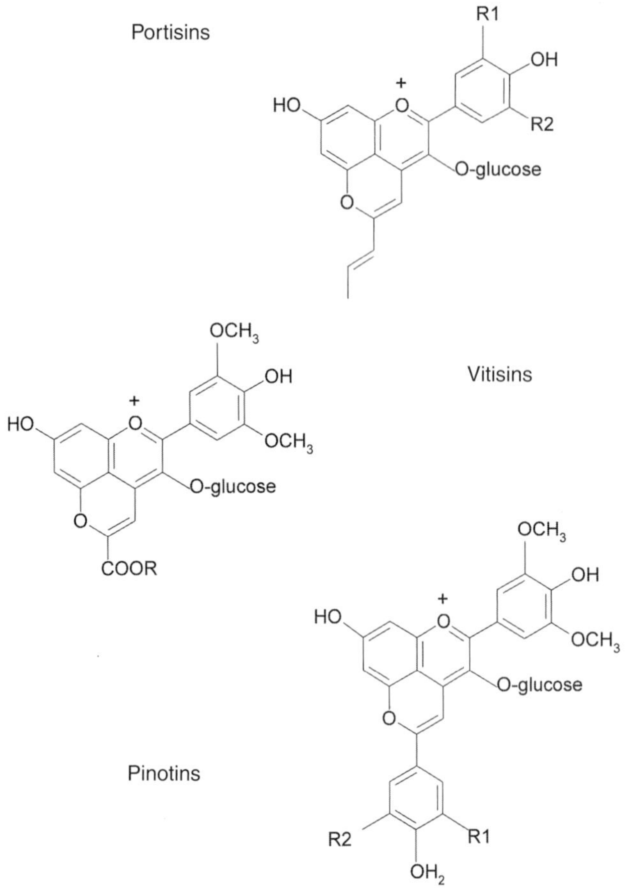

Portisins

R1
OH
HO
+
R2
O-glucose

Vitisins

OCH₃
OH
HO
+
OCH₃
O-glucose
COOR

Pinotins

OCH₃
OH
HO
+
OCH₃
O-glucose
R2
R1
OH₂

Figure 6.12 Diagrammatic illustration of three families of pyranoanthocyanins (pinotins, portisins, vitisins) found in wine.

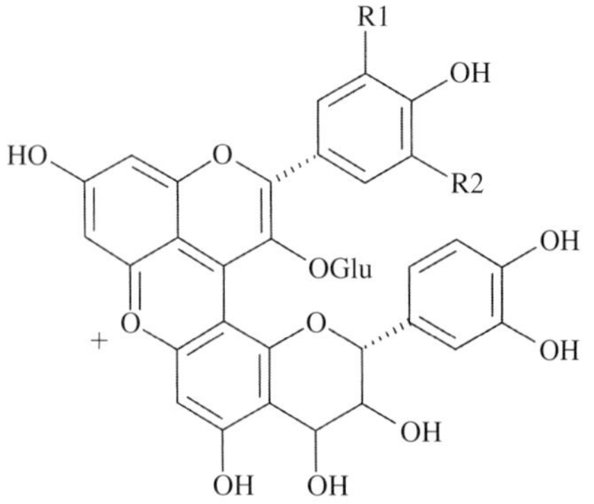

R1
OH
HO
R2
OGlu
OH
OH
+
O
OH
OH
OH
OH

Xanthylium

Table 6.7 Color and molecular weight of several wine phenols.

Nameᵃ	Color	Molecular weight
A⁺	Red	
AOH	Noncolored	500
AO	Violet	
AHSO₃	Noncolored	
P	Noncolored	600
T	Yellow	1000–2000
T-A⁺	Red	
T-AOH	Noncolored	1000–2000
T-AO	Violet	
T-AHSO₃	Noncolored	
TC	Yellow-red	2000–3000
TtC	Yellow-brown	3000–5000
TP	Yellow	5000

ᵃ A = anthocyanin; P = procyanidin; T = tannin; TC = condensed tannin; TtC = very condensed tannins; TP = tannin condensed with polysaccharides; +-flavylium; OH, Carbinol; O, quinoidal base.

Source: From Ribéreau-Gayon and Glories, 1987, reproduced by permission

presence of acetaldehyde. They form as the initial reaction products undergo cyclization and oxidation. New yellowish pigments have also isolated that are derived from anthocyanins and acetoacetic acid (He *et al.*, 2006).

Anthocyanins and flavan-3-ols may also bind with glyoxylic acid, producing orange-yellow **xanthylium** products. Glyoxylic acid is a metal ion-catalyzed oxidation product of tartaric acid. Xanthylium products may subsequently bind to flavanols to form complex tannin structures (Francia-Aricha *et al.*, 1998). Finally, glyoxylic acid may condense with anthocyanin degradation products (phloroglucinol) and catechins, forming additional yellowish pigments (Furtado *et al.*, 1993).

Although self-association, copigmentation, and various condensation products are involved in the early evolution of the wine color, during and after fermentation, long-term color stability largely reflects the subsequent formation of anthocyanin–tannin polymers. These polymeric pigments variously give yellow, yellow-red, yellow-brown, red, and violet shades, depending on their chemical nature (Table 6.7). Because most

polymers have a yellowish, orange or brownish cast, the wine progressively takes on a more brickish shade. Color density also diminishes with time. This may result from the destruction of free anthocyanins, additional structural changes that modify the hue of anthocyanin–tannin polymers, and the formation and precipitation of pigment polymers. Because anthocyanins principally bind via C_4-C_8 linkages to terminal tannin flavonoids, anthocyanin–tannin polymers generally do not enlarge as much as tannin–tannin polymers. Thus, precipitation of pigment polymers usually requires complexing with residual soluble wine proteins.

While much has been learned about chemical changes in the structure of anthocyanins, and their adjuncts with other compounds, the precise significance of these discoveries to wine color is still unclear. Until the dynamics of their formation and their relative proportions are known, much of what can now be said will remain tentative.

COLOR – WHITE WINES

In contrast to red wines, comparatively little is known about the development and chemical nature of white wine color. Of the small amount of phenolic material found in white wines, most consists of readily soluble nonflavonoids (hydroxycinnamates), such as caftaric acid (caffeoyl tartaric acid) and the related derivatives, *p*-coumaric acid and ferulic acid (Lee and Jaworski, 1987). Treatment of the juice with pectic enzymes enhances the hydrolysis of caftaric acid to its components, caffeic and tartaric acids (Singleton *et al.*, 1978). After crushing, caftaric acid and related *o*-diphenols readily oxidize in the presence of grape polyphenol oxidase and oxygen. These compounds and their oxidized products can react with the tripeptide glutathione to form colorless *S*-glutathionyl complexes. These complexes typically combine with other constituents, precipitating out of the wine. Unless further oxidized, as by laccase (derived from rotted grapes), the colorless glutathione–hydroxycinnamic acid complexes usually do not polymerize or form brown pigments (Singleton *et al.*, 1985). Consequently, protecting the juice from oxygen during crushing, or the addition of SO_2 to inactivate polyphenol oxidases may be ill advised. They limit or prevent the formation of glutathione-hydroxycinnamic acid complexes, and contribute to subsequent browning potential of the wine.

The glutathione content of grapes generally increases markedly during ripening, coinciding with *véraison* (Adams and Liyanage, 1993). Cultivars with high hydroxycinnamic acid contents, but low glutathione contents, have a heightened browning potential.

Because flavonols and other flavonoid phenols are extracted relatively slowly, they are only found in significant quantities in juice macerated with the pomace. It is believed that much of the yellow coloration in young white wine is derived from the limited extraction and oxidation of flavonols, such as quercetin and kaempferol. The flavanol content consists primarily of catechins and catechin-gallate polymers (Lee and Jaworski, 1987). Subsequent oxidative browning of white wines is closely related to the flavonoid, but not nonflavonoid content (Simpson, 1982). Some browning may result from the formation of xanthylium cation pigments from catechins and glyoxylic acid (via oxidation of tartaric acid) (George *et al.*, 2006). Nonflavonoids and lignins extracted from oak cooperage during wine maturation also may contribute to the color of white wines. For example, on oxidation, caffeic acid can polymerize forming dimers (caffeicins) and trimers (Cilliers and Singleton, 1991). The deepening yellow-gold color of older white wines probably comes from the oxidation of such phenolics and possibly galacturonic acid. The former are known to enhance absorption in the visible range (380–450 nm). However, golden shades in sweet white wines may come from the formation of melanoidin compounds by Maillard reactions, or the caramelization of sugars. These processes definitely contribute to the brownish coloration of wines exposed to heating, such as madeira, baked sherries, and *vin santo*. Whether they ever play a significant role in the color development of aged, dry, white, table wines is unknown.

TASTE AND MOUTH-FEEL

Flavonoid tannins have marked influences on taste and mouth-feel. Although anthocyanins are abundant in red wines, they contribute little directly to the taste of wine. Nevertheless, their polymerization with tannins is important to the retention (solubility) of tannins in wine (Singleton and Trousdale, 1992). Thus, the absence of anthocyanins helps to explain the lower astringency of white wines with long skin contact or fermented on the skins.

Catechins and their polymers, the procyanidins and condensed tannins, are the major sapid substances in red wine. They are the predominant source of bitter and astringent sensations. Monomeric catechins donate most of the bitter aspect of red wines. In contrast, various condensed tannins produce the diverse astringent sensations found in wines (e.g., rough, grainy, puckery, dry, dust-in-the-mouth). The gallic acid moiety commonly found in seed tannins appears to donate a coarseness and dryness not associated with skin tannins (Francis *et al.*, 2002).

Tannins bind to proteins through the formation of hydrogen bonds between hydroxyl groups and peptide links, or via hydrophobic interactions. The high proline content and open elongated structure of salivary proteins are thought to aid the interaction between these proteins and tannins. The larger condensed tannins have little influence on taste. They appear to be too massive to react well with taste receptors on the tongue or to precipitate saliva proteins.

The modest flavonoid content of white wines may still contribute to wine flavor, through the action of flavonols such as the flavanone glycoside naringin. In the cultivars 'Riesling' and 'Silvaner,' they may contribute to the bitterness characteristic of their wines (Drawert, 1970). In addition, small amounts of flavan-3-ols and flavan-3,4-diols may contribute to the wine's body and perceived quality. The classification of German wines, customarily based on fruit ripeness, has been correlated with increased flavonoid content (Dittrich *et al.*, 1974). Nevertheless, the role of flavonoids in browning and bitterness (>40 mg/liter) places an upper limit on their desirability.

Hydrolyzable tannins seldom play a significant role in the bitterness and astringency of wine (Pocock *et al.*, 1994). Although they tend to be more astringent than condensed tannins, their concentration and early degradation generally limit their sensory impact. Nevertheless, the derivatives of hydroxycinnamic acid are usually linked by ester bonds to tartaric acid. After hydrolysis, they may be liberated at concentrations sufficient to contribute to bitterness in white wines (Ong and Nagel, 1978).

Most phenolic acids, such as caftaric acid, occur at concentrations below their detection threshold in wine (Singleton and Noble, 1976). Nevertheless, combinations of phenolic acids have lower thresholds than their individual components. This property may increase with the alcoholic strength of wine. Thus, phenolic acids may contribute jointly to the bitterness and flavor of wine phenolics.

At the typical content of about 25 mg/liter, tyrosol (a yeast-generated phenolic) could contribute to bitterness in white wine. It may be more important in this regard in sparkling wines, in which its concentration increases during the second fermentation.

Some phenolics may contribute to the peppery sensation associated with certain grape cultivars. 2-Phenylethanol and methyl anthranilate appear to have this effect. Other phenol derivatives generate a pungent mouth-feel, or contribute to the varietal aroma of some cultivars, due to their volatility. In addition to direct influences on bitterness and astringency, phenols have complex indirect effects on the perception of sweetness and acidity. They also may have direct effects on the sensation of body and balance.

ODOR

Few volatile phenols or phenol derivatives come from grapes. Acetovanillone, which has a faintly vanilla-like odor, is one exception. The most important, however, is methyl anthranilate. This phenol-derived ester is an important component of the characteristic aroma of some, but not all, *V. labrusca* varieties (Fig. 6.13). Methyl anthranilate is also produced in small quantities in several *V. vinifera* cultivars, such as 'Pinot noir' (Moio and Etiévant, 1995), 'Riesling, ' and 'Silvaner' (Rapp and Versini, 1996). Another significant volatile phenol is 2-aminoacetophenone. It has been implicated in the naphthalene-like off-odor called "untypical aged (UTA) flavor" (*untypischen Alterungsnote*) (Geßner *et al.*, 1995). Another important phenolic flavorant is 2-phenylethanol. It produces the roselike fragrance often attributed to *V. rotundifolia* varieties. 2-Phenylethanol and several other volatile phenol derivatives (such as

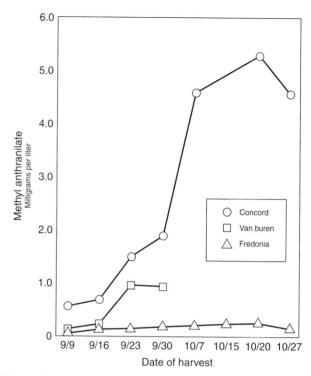

Figure 6.13 Methyl anthranilate development during ripening of several *Vitis labrusca* cultivars. (From Robinson *et al.*, 1949, reproduced by permission)

vanillin, and zingerone) occur as nonvolatile conjugates in several *V. vinifera* varieties. Their release by enzymatic or acid hydrolysis could significantly influence the sensory impact of these grape derivatives.

Although contributing to the varietal aroma of a few grape varieties, grape-derived phenolics appear to be more significant as sources of flavorants during or after fermentation. Esters of hydroxycinnamic acid derivatives, notably coumaric and ferulic acids, are particularly important in this regard. Oak is an additional source of these acids. Hydroxycinnamates can be metabolized to volatile phenols by a variety of microbes. The derivatives, vinylphenols (4-vinylguaiacol and 4-vinylphenol) and ethylphenols (4-ethylphenol and 4-ethylguaiacol) can donate spicy, pharmaceutical, clove-like odors and smoky, phenolic, animal, stable-like notes, respectively. Off-odors are frequently detected when the content of ethylphenols exceeds 400 μg/liter, or 725 μg/liter for vinylphenols. Conversion initially results in the decarboxylation of their hydroxycinnamate precursors to vinylphenols, possibly followed by reduction to ethylphenols (Chatonnet *et al.*, 1992). Although several bacteria and yeasts are capable of metabolizing hydroxycinnamates to vinylphenols, only a few yeasts, notably those belonging to the genus *Brettanomyces* (*Dekkera*), can convert substantial amounts to ethylphenols (Chatonnet *et al.*, 1992).

Red wines typically show a greater proportion of ethyl- to vinyl-phenols, and a higher absolute concentration of these compounds. The reverse is characteristic of white wines. This may be due to the more frequent maturation of red than white wines in oak cooperage, as source of volatile phenols precursors.

Eugenol, another clove-like derivative, may occur in wine. It is typically found in wine fermented or matured in oak cooperage, notably with light toasting. At usual concentrations, eugenol is sufficient only to add a general spicy note. Guaiacol may be generated as a consequence of thermal degradation of oak lignins. Nevertheless, its concentration is rarely sufficient to influence the bouquet directly. At higher concentrations, usually as a microbial degradation by-product, guaiacol and related compounds may be involved in off-odors derived from contaminated stoppers. Guaiacol has a sweet, smoky odor at threshold values (Dubois, 1983). Maltol is another characteristically sweet, caramel-selling phenolic compound extracted from new oak barrels.

Certain lactic acid bacteria generate volatile phenolics from nonphenolic compounds. An example is the synthesis of catechol from shikimic or quinic acids.

In addition to 2-phenylethanol, the other major phenolic alcohol found in wine is tyrosol. This yeast-synthesized phenolic has a mild beeswax, honey-like odor. Whether it plays a role in the honey-like bouquet of certain wines, such as botrytized wines is unknown.

Oak cooperage has already been noted as a source of several volatile and nonvolatile phenolic acids. In addition, oak is the principal source of phenolic aldehydes (Chatonnet *et al.*, 1990). These are primarily derivatives of benzaldehyde and cinnamaldehyde. Benzaldehyde is the most prominent, and possesses an almond-like odor. Benzaldehyde occurs in sufficient quantities in sherries to potentially participate in their nut-like bouquet. Benzaldehyde may also occur in wine following the oxidation of benzyl alcohol by enzymes produced in *Botrytis*-infected grapes (Goetghebeur *et al.*, 1992), or by some yeasts. Other important phenolic aldehydes are vanillin and syringaldehyde, both of which possess vanilla-like fragrances. They form during the breakdown of wood lignins.

Another source of volatile phenolic aldehydes involves the heating of must or wine. For example, fructose is rapidly converted to 5-(hydroxymethyl)-2-furaldehyde during baking, or very slowly during aging. 5-(Hydroxymethyl)-2-furaldehyde is considered to have a camomile-like odor. Furfural is commonly produced during distillation and the toasting of oak staves during barrel construction. Finally, phenolic aldehydes appear to be generated by the activation of phenylpropanoid metabolism. This is particularly prominent under the anaerobic conditions that develop during carbonic maceration (Dourtoglou *et al.*, 1994).

In addition to being a source of flavorants, phenolics can also react with wine flavorants. An important example is the removal of acetaldehyde in procyanidin polymers. It reduces the potential development of an oxidized odor in red wines. However, a more subtle, but possibly equally significant effect, is in the degradation of thiols by polyphenolics and quinones. In contrast, phenolic acids donate stability to several volatile esters and terpenes (Roussis *et al.*, 2005). Anthocyanins also provide stability to the important flavorant, 3-mercapto-hexan-1-ol, in Bordeaux rosé wines (Murat *et al.*, 2003).

OXIDANT AND ANTIOXIDANT ACTION

It may seem contradictory that wine phenolics (notably flavonoids) both participate in and limit oxidation. This apparent anomaly results from the by-product of phenolic oxidation (quinones) slowly changing structure to regenerate phenols (Fig. 6.11). These can react with additional oxygen. As a consequence, oxygen in must or wine is progressively depleted. Correspondingly, it becomes unavailable to oxidize other wine constituents. Conversely, the oxidation of phenolic to quinones generates hydrogen peroxide, a potent oxidant.

In grape must, the principal redox reactions involve the enzymatic oxidation of dihydrophenols, notably of caffeoyl tartaric acid (an ester of caffeic and tartaric acids) to its diquinone. Alternate important reactants are coumaroyl tartaric acid and flavan-3-ol (catechin) monomers. The latter produce yellow-brown pigments that, on further condensation, can cause browning. The polyphenol oxidases that catalyze these reactions are liberated from cellular vacuoles during crushing. They are activated in the presence of oxygen and acids liberated from broken vacuoles. The *o*-diquinones produced via oxidation may be reduced back to *o*-diphenols (those with adjacent hydroxyl groups) through the coupled oxidation of ascorbic acid, catechins, or procyanidins (Cheynier and Ricardo da Silva, 1991). This promotes the polymerization and early precipitation of readily oxidizable phenols during fermentation.

Grape polyphenol oxidases remain active in must for a comparatively short time. Because they remain largely bound to grape cellular constituents, enzymatic oxidation drops off dramatically after pressing. Furthermore, the quinones generated during oxidation can denature and inactivate enzymes such as polyphenol oxidases. This occurs when carbonyl groups of the quinone react with free amino or sulfhydryl groups of proteins. Tannins can also bind nonspecifically to and inactivate polyphenol oxidases. Thus, after pressing, oxidation occurs predominantly by nonenzymatic mechanisms (autooxidation). Although occurring more slowly, it affects a wider range of phenolics, not just the *o*-diphenols affected by grape polyphenol oxidase. Thus, oxidation in wine is principally nonenzymatic.

In red wines, oxidation primarily involves dihydroxyphenolics, such as catechins and cyanidins. These are converted to diquinones under the catalytic action of ferrous ions (Danilewicz, 2003). A by-product of the reaction is hydrogen peroxide (Wildenradt and Singleton, 1974), a potent oxidant. Such reactions are termed peroxidations. Ellagitannins derived from oak may also participate in peroxidation (Vivas and Glories, 1996). The peroxide generated during peroxidation oxidizes other substances, notably ethanol. Ethanol acts as the primary substrate due to its being the major oxidizable substrate in wine. The by-product is acetaldehyde. Other readily oxidizable substrates, other than phenols and ethanol, are sulfur dioxide, glycerol, and amino acids. Hydrogen peroxide is not produced during the enzymic oxidation of *o*-diphenols.

Because quinones can react with other phenols, they enhance phenol polymerization, and the generation of brown pigments. By slow structural rearrangement, the quinone–phenol dimers can generate new *o*-diphenol dimers (Fig. 6.11). This reaction can convert poorly oxidized phenolics into readily oxidizable dimeric phenolics. These can react with additional oxygen, producing more peroxide, and subsequently additional acetaldehyde. The *o*-quinone dimers so generated can again react with phenolics, producing even more complex polyphenolics. If the phenolics oxidized are catechins, the condensed tannins formed are often linked between $C_{6'}$ and C_6 or C_8 of adjacent B and A rings respectively. Nevertheless, linkages may alternately involve $C_{3'}$, $C_{4'}$, and $C_{5'}$ of the B ring (Guyot *et al.*, 1996). Thus, wine tannins are often considerably more structurally complex than grape tannins. Because of the extensive degree and structural rearrangement possible within complex polyphenols, the ability of red wine to slowly assimilate oxygen is considerable (Singleton, 1987). Singleton has estimated that red wines can interact with the oxygen found in 50–100 liters of air (the lower phenolic content of white wines permits them to consume about a tenth as much).

Tannin and anthocyanin–tannin polymerization is also directly enhanced by the production of acetaldehyde (Fig. 6.10). The latter helps stabilize the color of red wines, by protecting the anthocyanin chromophore from oxidation, or its reversible bleaching by sulfur dioxide. In addition, the reaction removes free acetaldehyde that might generate a stale, oxidized odor.

Another potentially significant source of flavonoid polymerization may result from the metal (Cu^{2+} or Fe^{2+})-induced oxidation of tartaric acid to glyoxylic acid (Fulcrand *et al.*, 1997). Glyoxylic acid, similar to acetaldehyde, can bind catechins and other phenolics. Neither of these polymerization reactions requires the direct oxidation of phenols. Figure 6.14 gives an overview of phenolic oxidation and polymerization.

The antioxidant (oxygen assimilating) action of wine phenolics is improved if oxygen absorption is slow or infrequent. Under these conditions, new oxidizable phenols regenerate sufficiently rapidly to quickly assimilate the oxygen. With the comparatively rapid removal of oxygen (approximately a week in the absence of polyphenol oxidase) less desirable oxidative reactions are limited. In addition, the consumption of hydrogen peroxide in the oxidation of ethanol to acetaldehyde restricts the oxidative degradation of important wine aromatic compounds. In addition, slight oxidation ($\sim$40 mg O_2/liter/year) can trigger the precipitation of

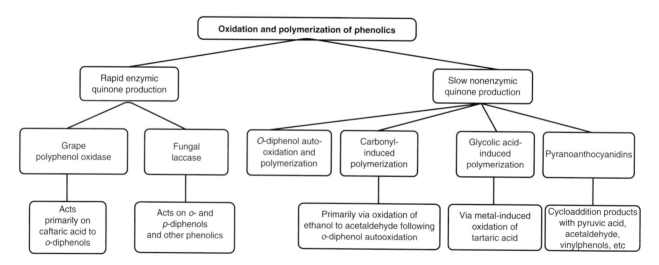

Figure 6.14   Flow chart of the various mechanism of phenol oxidation and polymerization in wines.

excess tannins (through the generation of large condensed tannins) (Ribéreau-Gayon *et al.*, 1983). This can reduce both wine bitterness and astringency.

Phenolics can also act as antioxidants through their antiradical properties – that is, by scavenging free oxygen radicals (Hagerman *et al.*, 1998). The donation of hydrogen atoms neutralizes the oxidant action of superoxide, hydroxyl and peroxyl radicals.

Because of their efficient removal of oxygen, phenols can help maintain wine at a low redox potential. This is considered central to the development of an aged bouquet during long, in-bottle maturation.

In contrast to red wines, the minimal color and limited antioxidant character of white wines makes them far more susceptible to oxidation, notably **oxidative browning**. Although the predominant phenolic in freshly crushed grape juice, caftaric acid (caffoyltartrate ester) is readily oxidized by polyphenol oxidases, browning may be limited by the combination of its *o*-quinone by-product with glutathione. This reduces the diquinone back to a colorless diphenol, 2-S-glutathionyl caftaric acid (Singleton *et al.*, 1985). The removal of quinones also results in the continued action of polyphenol oxidases – quinones do not accumulate and denature the enzyme. Following glutathione depletion, *o*-quinones may begin to accumulate, activating the oxidation and condensation of flavan-3-ols (primarily catechin and epicatechin) and other *o*-diphenols, as well as inactivating polyphenol oxidase. In white wines, flavan-3-ols are the primary browning agents (Singleton and Cilliers, 1995). If oxidation is promoted shortly after crushing, the brown polyphenolics produced tend to precipitate with the lees, leaving the wine less susceptible to in-bottle browning. Removal by binding with yeast cells increases as a function of polymer size (Lopez-Toledano *et al.*, 2004). This constitutes one of the primary advantages of hyperoxidation, or simply not preventing the exposure of juice to air during crushing.

Grape varieties have long been known to differ markedly in their sensitivity to oxidation and oxidative browning. This is at least partially explained by the differing phenolic compositions and the diverse consequences of phenolic oxidation (Yokotsuka *et al.*, 1991). Grape cultivars also differ in the presence of phenol-oxidizing enzymes.

ANTIMICROBIAL ACTION

The protective action of wine against certain gastrointestinal diseases has long been known – millennia before its mechanism was understood. Even now, the precise means by which wine, notably its phenolics, exerts their antimicrobial effect remain unclear. Even the particular phenols involved are unknown. Part of the problem arises from the diverse effects phenols have on living systems (Scalbert, 1991), and the various abilities of different phenolics to react with substances. By binding with proteins, tannins can limit enzyme action by modifying their solubility and structure. In addition, restricted movement of the enzyme's catalytic site would impair its activity. Because bacteria and fungi digest complex nutrients outside their cells, the inactivation of their digestive enzymes would be inhibitory, if not lethal. In addition, tannins can bind to phospholipids and proteins of the plasmalemma, disrupting membrane function. Finally, phenols, such as procyanidins, have strong chelating properties that can restrict microbial access to essential minerals, notably iron and zinc. Smaller flavonoids, such as quercetin, rutin, and nonflavonoids, such as caffeic and protocatechuic acids, also have antimicrobial effects.

Under normal fermentation conditions, phenols do not inhibit the growth or metabolism of yeasts and lactic acid bacteria. This insensitivity may hinge on their metabolism not being dependent on external digestive enzymes. However, phenols can complicate the initiation of the second fermentation of sparkling wine. For this reason, red wines are seldom used in the production of sparkling wine.

A special subset of antimicrobial phenolic compounds has received considerable attention recently. These are the stilbenes, notably resveratrol. They are part of the grape's response to pathogenic attack. However, it is not their antimicrobial action that has garnered the most attention. It is their antioxidant action in animal systems. Resveratrol may be involved in the health benefits derived from moderate wine consumption. This topic is discussed in depth in Chapter 12. Although resveratrol synthesis is a response to stress, such as attack by *B. cinerea*, the concentration of resveratrol in severely infected grapes may decline (Jeandet *et al.*, 1995). Several pathogenic fungi can either inhibit or destroy phytoalexins such as resveratrol.

Resveratrol

CLARIFICATION

The effectiveness of tannins in precipitating proteins, and vice versa, has often been used in wine clarification. Red wines may contain an excessive amount of tannin, which can make the wine overly astringent and generate large amounts of sediment. Protein fining agents may be added to provide the wine with a smoother mouth-feel (by precipitation of some of the tannins). Conversely, white wines may contain excessively high levels of colloidal proteins, which can lead to haziness. Enologic tannins may be added to precipitate such colloidal proteins.

Most enologic tannins are hydrolyzable tannins (Fig. 6.4), derived from either nutgalls, or oak and chestnut wood. Each of these has different chemical compositions and uses. Nutgall tannin is primarily gallotannin, yielding almost equal proportions of gallic, digallic and trigallic acids on acid hydrolysis (Salagoïty-Auguste *et al.*, 1986). It comes primarily from galls, induced by the wasp *Cynips tinctoria*, on the twigs of *Quercus infectoria* and related species. It is often used in tests for protein instability in wines. Tannins from chestnut wood are primarily gallotannins, composed of gallic acid and small amounts of digallic acid and ellagic acid. It may be added to red wines to avoid ferric casse, or combined with gelatin in the fining of white and rosé wines. In contrast, the tannins from oak wood are principally ellagic acid polymers. It is often added during brandy production to augment its color, as well as accentuate aromatic compounds derived from lignins. Most enologic tannins are not chemically purified.

A comparatively new source of enologic tannins comes from grape seeds. Depending on the variety of grape, seed maturity, and extraction procedure, these can have different properties relative to the precipitation of wine proteins, antioxidant potential, or tendency to stabilize anthocyanins.

Aldehydes and Ketones

Aldehydes are carbonyl compounds distinguished by the terminal location of the carbonyl functional group $(-C=O)$. Ketones are related compounds with the carbonyl group located on an internal carbon.

ALDEHYDES

Grapes produce few aldehydes important in the generation of varietal aromas. This may result from their reduction to alcohols during fermentation. Of those not metabolized during fermentation, the C_6 aldehydes (hexanals and hexenals) appear to be the most significant. They may be involved in the grassy to herbaceous odor associated with certain grape varieties, such as 'Grenache' and 'Sauvignon blanc', or with wines made from immature grapes. They appear to be formed during crushing by the enzymatic oxidation of grape lipids. The dienal, 2,4-hexadienal, may also be generated by the same process. However, most aldehydes found in wine are produced during fermentation, processing, or extracted from oak cooperage.

Acetaldehyde
(an aldehyde)

Diacetyl
(an α-diketone)

Phenyl acetaldehyde
(a phenolic aldehyde)

p-quinone
(a phenolic diketone)

Acetaldehyde is the major wine aldehyde. It often constitutes more than 90% of its aldehyde content. Above threshold values, it usually is considered an off-odor. Combined with other oxidized compounds, it contributes to the fragrance of sherries and other oxidized wines.

Acetaldehyde is one of the early metabolic by-products of fermentation (Fig. 7.14). It is often secreted out of the cell. As fermentation approaches completion, acetaldehyde may be transported back into yeast cells and reduced to ethanol. Thus, the acetaldehyde content usually falls to a low level by the end of fermentation (Fig. 7.16). In *fino* sherries, most of the acetaldehyde accumulated is thought to be a by-product of the respiratory metabolism of film (*flor*) yeasts.

A relatively minor, but sensorially important, source of acetaldehyde involves the coupling of the autooxidation of *o*-diphenols and ethanol. The acetaldehyde generated is important in stabilizing the color of red wines.

Other than potentially generating a temporary stale odor in newly bottled wines ("bottle sickness"), acetaldehyde seldom accumulates to detectable levels in table wines. It commonly reacts with sulfur dioxide or is consumed in the polymerization of anthocyanins and procyanidins (Fig. 6.10).

Other aldehydes, occasionally having a sensory impact on wine, are furfural and 5-(hydroxymethyl)-2-furaldehyde. Because furfural synthesis from sugars is accelerated by high temperatures, furfurals primarily occur in wine heated during processing. They add to the baked fragrance of such wines.

Phenolic aldehydes, such as cinnamaldehyde and vanillin, may accumulate in wines aged in oak. They are degradation products of lignins found in wood cooperage. Other phenolic aldehydes, such as benzaldehyde, may have diverse origins. Its bitter-almond odor is considered characteristic of certain wines, for example, those produced from 'Gamay' grapes. Benzaldehyde can also be derived from the oxidation of benzyl alcohol, used as a plasticizer in some epoxy resins; or through the metabolic action of some yeast strains or *B. cinerea*.

Although having no direct sensory effect, hydroxypropanedial (triose reductone) characteristically occurs in botrytized wines (Guillou *et al.*, 1997). It exists in a tautomeric equilibrium with 3-hydroxy-2-oxopropanal and 3-hydroxy-2-hydroxypro-2-enal. Reductones, such as hydroxypropanedial, can play a role in preserving the organoleptic qualities of a wine by fixing (diminishing the volatility of) its flavorants. Reductones also may combine with amino acids to produce browning compounds.

KETONES

Few ketones are found in grapes, but those that are present usually survive fermentation. Examples are the norisoprenoid ketones, β-damascenone, α-ionone, and β-ionone. The intense exotic flower or rose-like scent of β-damascenone, and its low odor threshold, indicate that it probably plays a contributing role in the aroma of several grape varieties, including 'Chardonnay' (Simpson and Miller, 1984) and 'Riesling' (Strauss, Wilson, Anderson *et al.*, 1987). The violet–raspberry scent generated by β-ionone, along with β-damascenone, appear to be significant in the aroma of several red grape varieties (Ferreira *et al.*, 1993).

Several important ketones may also be generated by fungal action. Significant examples are 1-octen-3-one and (Z)-1,5-octadien-3-one, produced by the grape pathogen *Uncinula necator*. These compounds are partially responsible for the mushroomy odor of grapes affected by powdery mildew (Darriet *et al.*, 2002). These compounds are metabolized to 3-octanone and (Z)-5-octen-3-one during fermentation, resulting in a reduction in the mushroomy aspect of wine produced from mildewed grapes.

Many ketones are produced during fermentation, but few appear to have sensory significance. The major exception is diacetyl (biacetyl, or 2,3-butanedione). At low concentrations (<5 mg/liter), diacetyl may donate a buttery, nutty, or toasty flavor. However, at slightly above its sensory threshold, diacetyl may begin to donate a caramel-like attribute (Rogerson *et al.*, 2001). Its ultimate sensory importance depends on its stability during aging, other volatile wine constituents, and the presence of sulfur dioxide (Bartowsky *et al.*, 2002). Diacetyl may be produced by yeasts, especially at high fermentation temperatures, but is most commonly associated with malolactic fermentation. Concentrations often fall by the end of malolactic fermentation (Bartowsky and Henschke, 2000). At concentrations significantly above threshold values, diacetyl produces a lactic off-odor. This commonly occurs in association with spoilage induced by certain strains of lactic acid bacteria. Diacetyl occurs in fairly high concentrations in sherries, along with another ketone, acetoin. Acetoin (3-hydroxy-2-butanone) has a sugary, butter-like character. Its sensory significance in table wines, in which it occurs at low concentrations, is doubtful. 2,3-Pentanedione and its related diol possess similar aromatic characteristics, varying from buttery to plastic.

Acetals

Acetals are formed when an aldehyde reacts with the hydroxyl groups of two alcohols. Acetals are typically produced during aging and distillation, potentially contributing a vegetable-like attribute. Although over 20 acetals have been isolated from wines, their concentration and volatility seem to suggest that they have little sensory impact in table wines. Acetals may play a minor role in the bouquet of sherries and similar wines, in which conditions are more favorable for their production.

$$R^1-\overset{\displaystyle O}{\underset{\displaystyle H}{C}} \; + \; 2\,R^2-OH \;\rightleftharpoons\; R^2-O-\overset{\displaystyle H}{\underset{\displaystyle R^1}{C}}-O-R^2$$

aldehyde   alcohols   acetal

Esters

Esters form as condensation products between the carboxyl group of an organic acid and the hydroxyl group of an alcohol or phenol. A prominent example is the formation of ethyl acetate from acetic acid and ethanol.

acetic acid   ethanol

ethyl acetate   + H_2O

Of all functional groups in wine, esters are the most frequently encountered. Over 160 specific esters have been identified. Because most esters are found only in trace amounts, and have either low volatility or mild odors, their importance to wine fragrance is probably negligible. However, the more common esters occur at or above their sensory thresholds. These include acetate esters, derived from acetic acid and fusel alcohols, and ethyl esters, formed between ethanol and fatty acids, or nonvolatile, fixed, organic acids. The fruity aspects are several of these esters is often important in generating the bouquet of young white wines (Marais and Pool, 1980). Mixtures of esters may not possess the same intensity or qualitative attributes as do the esters individually (van der Merwe and van Wyk, 1981). The importance of esters to the fragrance of red wines is less well understood.

CHEMICAL NATURE

Esters may be grouped into straight-chain (aliphatic) or cyclic (phenolic) categories. Most phenolic esters possess low detection thresholds. However, because of poor volatility and trace occurrence, they generally are sensorially insignificant. The primary exception is methyl anthranilate, which gives the grapy aroma to several *V. labrusca* cultivars, notably 'Concord.' *V. vinifera* cultivars may synthesize methyl anthranilate, but at a concentration up to 0.3 µg/liter (Rapp and Versini, 1996). This is well below the sensory threshold of methyl anthranilate (~300 µg/liter).

Aliphatic esters comprise the larger ester group in wine. They are subdivided into mono-, di-, or tricarboxylic acid esters (depending on the number of carboxyl groups in the acid), and hydroxy and oxo acid esters (those containing a hydroxyl or ketone group, respectively, in the acidic component). Of the subgroups, only the first is believed to be of aromatic significance.

Of the monocarboxylic acid esters, the most important are those based on ethanol and saturated fatty acids, such as hexanoic (caproic), octanoic (caprylic), and decanoic (capric) acids, and those based on acetic acid and higher (fusel) alcohols, such as isoamyl and isobutyl alcohols. The second group is often considered to give wine much of its vinous fragrance. These low-molecular-weight esters, often termed "fruit" esters, have a distinctly fruit-like fragrance. Examples of acetate esters are isoamyl acetate (3-methylbutyl acetate), which has a banana-like scent, and benzyl acetate, with an apple-like scent. Fruity ethyl esters include ethyl butyrate and ethyl hexanoate, possessing pineapple, apple-peel-like attributes, respectively. Fruit esters play an important role in the flavor of young white wines (Vernin *et al.*, 1986). With ethyl esters, the odor shifts as the length of the hydrocarbon chain of the acid increases, going from being fruity to soap-like and, finally, lard-like with C_{16} and C_{18} fatty acids. The presence of certain esters, for example, hexyl acetate and ethyl octanoate, has occasionally been considered indicators of red wine quality (Marais *et al.*, 1979).

Di- and tricarboxylic acid esters generally occur in wine at concentrations up to 1 mg/liter and above, especially ethyl lactate following malolactic fermentation. Nevertheless, because of their weak odors, they generally do not appear to be aromatically significant.

In contrast, the formation of the methanolic and ethanolic esters of succinic acid appear to contribute to the aroma of muscadine wines (Lamikanra *et al.*, 1996). Other examples found in wine are esters based on malic, tartaric, and citric acids.

Hydroxy and oxo acid esters have low volatility and, correspondingly, appear to play little sensory role. The major esters of this group are associated with lactic acid.

Ethyl and methyl esters of amino acids occur in the milligram per liter range. Their sensory significance is unknown. In contrast, the high concentration of phenolic acid esters, such as caffeoyl tartrate (caftaric acid) in 'Riesling' wines, may help to explain their typical slight bitterness (Ong and Nagel, 1978).

ORIGIN

Esters are synthesized in grapes, but seldom in amounts to be of sensory significance. The prime exceptions are the phenolic ester, methyl anthranilate, and possibly isoamyl acetate in 'Pinotage' (Marais *et al.*, 1979). The synthesis of ethyl 9-hydroxynonanoate by *B. cinerea* may contribute to the distinctive aroma of botrytized wines (Masuda *et al.*, 1984).

Most esters found in wines are produced by yeasts, after cell division has essentially ceased. Straight-chain forms are synthesized from alchololysis of the corresponding acids, which have been activated by acyl-S-CoA. However, ethyl esters of short-chain, branched, fatty acids, such as ethyl isobutanoate, are derived from the metabolism of amino acids. Subsequent synthesis and hydrolytic breakdown continue nonenzymatically, based on the chemical composition and storage conditions of the wine (Rapp and Güntert, 1986; Díaz-Maroto *et al.*, 2005).

By the end of fermentation, fruit esters are generally in excess of their equilibrium constants. As a result, many acetate esters begin to hydrolyze back to their component alcohols and acetic acid. Hydrolysis is favored at elevated temperatures and low pH values (Ramey and Ough, 1980). For wines that derive much of their fragrance from fruit esters, aging can result in a fading of the bouquet. In contrast, fusel alcohol esters are generally retained in yeast cells, rather than being released into the surrounding juice. Because their concentration in wine is commonly below their equilibrium constant at the end of fermentation, there is a slow synthesis of fusel alcohol esters. This especially applies to those based on branched, short-chain fatty acids. The esters of dicarboxylic acids also increase during aging. Nonenzymatic synthesis appears to be higher in sherries than in table wines (Shinohara *et al.*, 1979).

Ester formation during fermentation is influenced by many factors. In certain instances, the ability of the must to support ester formation declines as the grapes reach maturity (see Fig. 7.26). Esterase activity in different yeast strains is another important factor. Low fermentation temperatures (~10°C) favor the synthesis of fruit esters, such as isoamyl, isobutyl and hexyl acetates, whereas higher temperatures (15–20°C) promote the production of higher-molecular-weight esters, such as ethyl octanoate, ethyl decanoate, and phenethyl acetate (Killian and Ough, 1979). Higher temperatures also tend to limit ester accumulation by favoring hydrolysis. Both low SO_2 levels and juice clarification support ester synthesis and retention. Intercellular grape fermentation (carbonic maceration) and the absence of oxygen during yeast fermentation further augment ester formation.

Of all the esters, ethyl acetate has been the most investigated. In sound wines, the concentration of ethyl acetate generally is below 50–100 mg/liter. At low levels (<50 mg/liter), it may be suitable and add complexity to the fragrance, whereas above 150 mg/liter, it is likely to produce a nail-polish remover, sour-vinegary, off-odor (Amerine and Roessler, 1983). The development of undesirable levels of ethyl acetate is usually associated with grape, must, or wine contamination with acetic acid bacteria. The bacteria not only directly synthesize ethyl acetate, but they also produce acetic acid that can react nonenzymatically with ethanol, forming ethyl acetate. Ethyl acetate can seriously flaw the fragrance of a wine long before the acetic acid level reaches a concentration sufficient to make the wine vinegary.

Lactones and Other Oxygen Heterocycles

Lactones are a special subgroup of esters formed by internal esterification between carboxyl and hydroxyl groups of the parent molecule. The result is the formation of a cyclic ester. As with other esters, lactones exist in equilibrium with their reactants, in this case a hydroxy acid:

hydroxy acid

esterification
⇌
hydrolysis

lactone

Most lactones in wines are four-carbon esterified rings. Most also are γ-lactones; that is, the hydroxyl group involved in esterification is C_4 along the chain.

Lactones may come from grapes, be synthesized during fermentation and aging, or be extracted from oak cooperage. Lactones derived from grapes generally are not involved in the development of varietal odors. The lactone, 2-vinyl-2-methyltetrahydrofuran-5-one, is one exception. It may be involved in the distinctive aroma of 'Riesling' and 'Muscat' varieties (Schreier and Drawert, 1974). Because lactone formation is enhanced during heating, some of the raisined character of sunburned grapes may come from lactones, notably 2-pentenoic acid-γ-lactone. Sotolon (4,5-dimethyl-tetrahydro-2,3-furandione) is characteristically found in *Botrytis*-infected grapes and wine. It commonly occurs at levels above the sensory threshold (>5 ppb) in botrytized wines (Masuda *et al.*, 1984). Sotolon also may be a significant fragrance compound in sherries (Martin *et al.*, 1992). It adds a nutty, sweet, burnt odor. It also contributes to the *rancio* character of sweet fortified wines, at concentration above 600 mg/liter (Cutzach *et al.*, 1998). Another potent lactone isolated from several wines (termed "wine lactone") possesses a coconut, woody, sweet odor (Guth, 1996). The eight stereoisomers of wine lactone (3α,4,5,7α-tetrahydro-3, 6-dimethylbenzofuran-2(3*H*)-one) possess markedly different thresholds, from 0.00001 ng/liter to over 1000 ng/liter.

Most lactones in wine appear to be produced during fermentation, although some are generated in grapes under carbonic maceration (Dourtoglou *et al.*, 1994). They are apparently derived from amino or organic acids, notably glutamic and succinic acids. Solerone (4-acetyl-4-hydroxybutyric acid-γ-lactone) and pantolactone (2,4-dihydroxy-3,3-dimethylbutyric acid-γ-lactone) are examples. Although solerone has been considered to possess a bottle-aged fragrance, it apparently does not accumulate to levels above the detection threshold in either table wines or sherries (Martin *et al.*, 1991).

Oak is an additional source of lactones (Chatonnet *et al.*, 1990). The most important of these are the oak lactones, isomers of β-methyl-γ-octalactone. Small amounts may also be synthesized by yeasts. The most aromatically significant is the *cis*-isomer (Brown *et al.*, 2006). They have an oaky or coconut-like fragrance. Several γ-nonalactones also have an oaky aspect. As such, they may be involved in development of a barrel-aged fragrance.

Among other oxygen heterocyclic compounds, the spiroether vitispirane has been the most extensively investigated (Etiévant, 1991). It may be derived from several compounds, such as free or glucosidically bound 3-hydroxytheaspirane and megasigma-3,6,9-triols. Vitispirane is slowly generated during aging, reaching concentrations of 20–100 ppb. It consists of two isomers, each of which has a qualitatively different odor. The *cis*-isomer has a chrysanthemum flower-fruity odor, whereas *trans*-vitispirane has a heavier, exotic fruit scent. Other authors have considered vitispirane to have a camphoraceous or eucalyptus odor. However, its sensory significance remains in doubt. Its concentration is close to its perception threshold.

Terpenes and Oxygenated Derivatives

Terpenes are an important group of aromatic compounds characterizing the odor of many flowers, fruits, seeds, leaves, woods, and roots. As such, terpenes are often important in the fragrance of herb-flavored wines such as vermouth and fruit-flavored wines. In addition, they also characterize the aromas of several grape varieties. About 50 monoterpenic compounds have been isolated from wines.

Chemically, terpenes are grouped together because of their distinctive carbon skeleton. It consists of a basic five-carbon isoprene unit (2-methyl-1,3-butadiene). Terpenes generally are composed of two, three, four, or six isoprene units. These are called monoterpenes, sesquiterpenes, diterpenes, and triterpenes, respectively.

where they are synthesized in plastids. In grapes, terpenes may exist in three states (Fig. 6.15). Most are found as free monoterpene alcohols or oxides. In this form they are volatile and may contribute to wine fragrance. A variable proportion also exists complexed with glycosides, or occurs as di- or triols. Neither of the latter groups are aromatic.

Considerable interest has been shown in augmenting the release of free (volatile) monoterpenes from their glycosidic linkages, by the addition of enzyme preparations (Günata *et al.*, 1990). However, the preparations may favor the production of off-odors, generated by vinylphenols. These are derived from enzymatically liberated hydroxycinnamic acid derivatives (Dugelay *et al.*, 1993). Although the enzymatic liberation of terpenes has been most studied in white wines, nonspecific hydrolytic glycosidases may be useful in increasing the desirable flavor of wines made from red grapes, for example 'Shiraz' (Abbott *et al.*, 1991). The specific chemical nature of the flavor compound(s) involved is unknown. Although crushing, maceration, or fermentation generally do not modify the terpene content, some yeast strains can convert geraniol and nerol to citronellol.

The terpene content varies considerably from cultivar to cultivar. In some wines, the monoterpene content is sufficiently distinctive to permit varietal identification (Rapp and Mandery, 1986). Geographical origin appears not to modify monoterpene content significantly. The

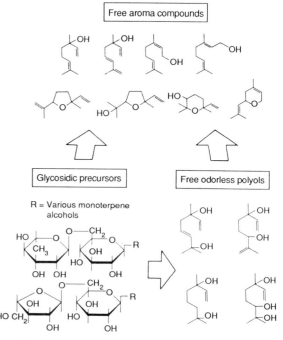

Figure 6.15 Categories of monoterpenes in grapes. Glycosidic precursors and free odorless polyols are a reserve of odorless precursors in the fruit. Only free aroma compounds make a direct contribution to fruit character. (From Williams *et al.*, 1987, reproduced by permission)

Terpenes may contain a variety of functional groups. Many important terpenes contain hydroxyl groups, making them terpene alcohols. Other terpenes are ketones. Terpene oxides are terpenes having an oxygen-containing ring structure, as well as the basic isoprenoid structure. As such, they contain a cyclic ether (C-O-C) bond.

Unlike many of the aromatic constituents of wine, terpenes are primarily derived from grapes (Strauss *et al.*, 1986). They are principally located in the skin,

varieties most easily characterized by terpene contents are members of the 'Muscat' and 'Riesling' families. Other grape cultivars may possess terpenes, but they are less dependent on them for their varietal distinctiveness (Strauss, Wilson and Williams, 1987).

Although the terpene concentration of healthy grapes is generally stable throughout fermentation, infection by *B. cinerea* can both reduce and modify grape terpene content. The loss undoubtedly plays a major role in the destruction of varietal aromas that characterizes botrytized wines (Bock *et al.*, 1988).

During aging, the types and proportions of terpenes may change considerably (Rapp and Güntert, 1986). Some increase may result from the release of free terpenes from glucosidically bound forms (Fig. 8.20). Losses also may occur due to oxidation or other transformations. Although these occur more slowly than changes in fruit esters, a marked loss of aroma can result over several years. This is particularly significant in 'Muscat' varieties, which depend on monoterpene alcohols for much of their distinctive fragrance. Most monoterpene alcohols are replaced by terpene oxides. These often have sensory thresholds up to 10 times higher than their precursors. Nevertheless, additive or synergistic effects of the various terpenes make prediction of sensory consequences difficult. Changes in terpene content also affect odor quality. For example, the muscaty, iris-like attribute of linalool is progressively replaced by the musty, pine-like scent of α-terpineol.

During aging, additional changes can modify terpene structure. Some terpenes become cyclic and form lactones. Other terpenes may transform into ketones, such as α- and β-ionone, or spiroethers, such as vitispirane.

The minimal contribution of terpenes to the fragrance of red wines means that changes in their structure have little sensory significance. The major exception involves wines made from 'Black Muscat.'

Although most terpenes have pleasant odors, some may be decidedly unpleasant. For example, the musky-smelling sesquiterpenes produced by *Penicillium roquefortii* growing in cork (Heimann *et al.*, 1983). *Streptomyces* may also synthesize earthy-smelling sesquiterpenes on cork or cooperage wood. Their presence can severely compromise the sensory quality of the wine.

Of nonaromatic terpenes, the most important is the triterpene oleanolic acid. It is a primary constituent of the waxy covering on grapes. It can act as a precursor for sterol synthesis by yeasts during fermentation.

Nitrogen-containing Compounds

Many nitrogen-containing compounds are found in grapes and wine. These include inorganic forms such as ammonia and nitrates, and diverse organic forms, including amines, amides, amino acids, pyrazines, nitrogen bases, pyrimidines, proteins, and nucleic acids. Complex organic nitrogen compounds (pyrimidines, proteins, and nucleic acids) are essential for the growth and metabolism of grape and yeast cells, but are seldom involved directly in the sensory attributes of wine. However, colloidal proteins can cause haze problems in wine.

$$NH_3 \qquad R{-}NH_2$$
ammonia   amines

$$CH_3CH_2NH_2 \qquad C_6H_5CH_2CH_2NH_2$$
ethylamine   phenethylamine

The simplest of organic nitrogen compounds in wine are the amines. They are small organic compounds associated with an ammonia group. Several simple volatile amines have been found in grapes and wine, including ethylamine, phenethylamine, methylamine, and isopentylamine. Their concentration tends to decline during fermentation as they are metabolized by yeasts. Nevertheless, yeasts may also synthesize amines, especially during the early phases of fermentation (Garde-Cerdán *et al.*, 2006). Their retention is favored at both high and low fermentation temperatures. Subsequently, the level may increase due to release during yeast autolysis. Their importance to flavor development is uncertain (Etiévant, 1991), but some may become volatile as salt forms dissociate at the more neutral pH values in the mouth. In beer, volatile amines are known to produce harsh tastes. The higher flavor of red wines probably precludes a similar effect in red wines, but white wines could be affected.

Wine also contains small amounts of nonvolatile, biogenic amines. The most well studied is histamine. Other physiologically active amines include tyramine and phenethylamine (volatile). They usually occur in wines at a few mg/liter (Lehtonen, 1996). At significantly higher concentrations, biogenic amines can induce headaches, hypertension, and allergic reactions, in sensitive individuals. These effects are enhanced in the copresence of ethanol and acetaldehyde. This may result from suppression of the action of diamine oxidase in the small intestine, and monoamine oxidase in the liver. However, biogenic amines rarely occur in wines at levels capable of inducing these effects (Radler and Fäth, 1991).

Spoilage bacteria have usually been considered the primary source of biogenic amines in wines. Nevertheless, studies, such as those of Coton *et al.* (1998), suggest that some strains of *Oenococcus oeni* may be a significant histamine producer. Nonvolatile amines are primarily by-products of amino acid metabolism. Features such as low ethanol content and long contact with lees favor the production of histamine. Biogenic amines may also be produced in the early stages of alcoholic fermentation.

Polyamines, such as putrescine and cadaverine, are usually present in wine only as a result of bacterial contamination, or the use of moldy or nutrient-deficient grapes. These compounds are produced by the vine in response to stress factors, for example fungal infection or potassium deficiency. Polyamines, partially due to their polycationic property, are thought to bind to polyanionic nucleic acids, acting as gene regulators. In grapes, they frequently occur complexed with coumaric, caffeic and ferulic acids. They appear to have no sensory relevance to wine flavor.

AMIDES

Amides are amines with a carbonyl group associated with the ammonia-associated carbon. Although a number of amides may occur in wine, none appear to significantly affect wine flavor.

Urea is a simple nitrogen compound related to amides. It consists of two ammonia groups attached to a common carbonyl. Urea is produced in wine as a by-product of arginine metabolism, and was previously added to juice to promote yeast growth. Its presence in wine used to be considered of little significance. However, if urea is incomplete metabolized to ammonia, it can react spontaneously with ethanol to generate ethyl carbamate (urethane) (Ough *et al.*, 1990). Because ethyl carbamate is a suspected human carcinogen, minimizing its production is important, especially if the wine is heated during processing (Stevens and Ough, 1993). Heating also promotes the generation of ethyl carbamate from citrulline or carbamyl phosphate (an intermediate metabolic breakdown product of arginine) (Sponholz *et al.*, 1991). This also involves a reaction with ethanol. Recently, Hasnip *et al.* (2004) have developed an equation that predicts potential ethyl carbamate concentration, based on the time and temperature of storage. Although the addition of acid urease can reduce the urea content (Kodama *et al.*, 1991), it does not affect other ethyl carbamate precursors in grapes (Tegmo-Larsson and Henick-Kling, 1990).

The timing and occurrence of aeration during fermentation can significantly influence the production and degradation of urea (Henschke and Ough, 1991).

The degradation of arginine (frequently the most common amino acid in grapes) by some lactic acid bacteria favors the potential synthesis of ethyl carbamate. In this regard, the relative inability of *Oenococcus oeni* to metabolize arginine effectively at wine pH values (Mira de Orduña *et al.*, 2001) is another reason why it is the preferred malolactic bacterium.

AMINO ACIDS

Amino acids are another class of amine derivatives. They contain a carboxyl group attached to the amine-containing carbon. Amino acids are most important as subunits in the biosynthesis of enzymes and other proteins. In addition, amino acids may act both as nitrogen and energy sources for yeast metabolism. This may indirectly generate important flavor constituents. For example, amino acids may be metabolized to organic acids, higher alcohols, aldehydes, phenols, and lactones. During brandy distillation, heating may convert amino acids into aromatic pyrazines. Amino acids are also associated with the caramelization of sugars during the heat processing of baked sherries and madeira. Although some amino acids have bitter, sweet, or sour tastes, their low concentration in finished wine means that they are unlikely to have an appreciable sensory influence.

OTHER COMPOUNDS

Pyrazines (cyclic nitrogen-containing compounds) contribute significantly to the flavor of many natural and baked foods. They also are important to the varietal aroma of several grape cultivars. 2-Methoxy-3-isobutylpyrazine plays a major role in the green-pepper defect often detectable in 'Cabernet Sauvignon' and related cultivars. It frequently occurs at concentrations above its detection threshold (1–2 ng/liter). At concentrations about 8–20 ng/liter, it may be desirable, but above this, it generates an overpowering vegetative or herbaceous aroma. It is quickly extracted from skins during crushing. Most subsequent vinification procedures have little effect on its concentration (Roujou de Boubée *et al.*, 2002). Other related methoxypyrazines occur, but generally at concentrations just at or below their detection thresholds (Allen *et al.*, 1996).

Another group of cyclic nitrogen compounds are the pyridines. Thus far, their involvement in wine flavor appears to be restricted to the production of mousy off-odors. This odor is associated with the presence of 2-acetyltetrahydropyridines (Heresztyn, 1986). The presence and significance of proteins and nucleic acids are discussed later.

Sulfur-containing Compounds

Inorganic sulfites are the principal sulfur compounds found in grape juice and wine. They come primarily from the deliberate addition of sulfur dioxide as an antimicrobial or antioxidant. Even where sulfur dioxide is not added, yeast may produce between 10–30 mg/liter of sulfite. Some strains can generate up to 100 mg/liter (Eschenbruch, 1974). Additional sources may involve sulfur or sulfur-containing fungicides, applied to grapes in diseased control. The major, organic, sulfur-containing compounds are two amino acids (cysteine and its derivative methionine), a tripeptide (glutathione[1]) containing cysteine, and proteins containing cysteine and methionine. Their concentration is considerably lower in wine than in must. Sulfur-containing vitamins (thiamine and biotin) also occur at higher concentration in must than in the corresponding wine.

The metabolism of cysteine, the generation of H_2S, and the presence of cysteinylated conjugates in grapes appear to be the principal sources of thiol compounds in wine. Methionine seems to be less involved (Moreira *et al.*, 2002). Although the degradation to thiamine is known to generate thiols, this is primarily associated with exposure to heat. Thus, except in baked wines, this route is unlikely to be of enologic significance.

As a consequence of yeast metabolism, heat treatment, light exposure, or other nonenzymatic reactions, a wide diversity of volatile sulfur-containing compounds may be produced during fermentation, maturation, and post-bottling. Although generally occurring in trace amounts, their high volatility and low sensory thresholds (often a few parts per trillion), can give them great significance. At or above recognition thresholds, hydrogen sulfide and most volatile thiol compounds produce odors that in most instances vary from unpleasant to nauseating. Thus, limiting their presence is of major concern in winemaking.

Generally, hydrogen sulfide (H_2S) is the most common volatile sulfur compound in wine. It is easily recognized by its rotten-egg odor. Nevertheless, at near-threshold levels, hydrogen sulfide generates part of the yeasty odor to newly fermented wines. It may form during fermentation, as yeasts reduce residual elemental sulfur (Schütz and Kunkee, 1977), or metabolize sulfur-containing amino acids, notably cysteine (Henschke

and Jiranek, 1991). Nitrogen or vitamin (pantothenate or pyridoxine) deficiency promotes H_2S biosynthesis.

Carbon disulfide (CS_2) and carbonyl sulfide (COS) are also formed, but generally occur at below threshold values. They initially appear as reaction products during the crushing of grapes (Eschenbruch *et al.*, 1986). Their concentration tends to decline during fermentation. Subsequent production may come from the degradation of sulfur-containing fungicides.

Volatile organosulfur compounds include a wide diversity of straight-chain and cyclic molecules. They form principally during yeast metabolism of sulfur-containing amino acids, peptides, and proteins. Figure 6.16 shows the close association between fermentation and the production of some organosulfur compounds. Reduced-sulfur off-odors may also originate during yeast autolysis. Light exposure can also activate the catalytic production of organosulfur compounds, such as those that produce a *goût de lumière* off-odor in champagne (Charpentier and Maujean, 1981). In a related process, prenyl mercaptan (3-methyl-2-butene-1-thiol) synthesis generates a sun-struck off-odor in beer.

Structurally, mercaptans are the simplest organosulfur compounds. They are hydrocarbons attached to a sulfhydryl (-SH) group. A significant member of this group is ethanethiol (ethyl mercaptan) (CH_3CH_2SH). It produces a rotten onion, burnt-rubber off-odor at threshold levels. At higher levels, it has a skunky, fecal odor. Of related thiols, 2-mercaptoethanol ($SHCH_2CH_2OH$) produces a barnyard-like (*böxer*) odor (Rapp *et al.*, 1985), while methanethiol (methyl mercaptan) (CH_3SH) generates a rotten cabbage odor. Ethanedithiol ($SHCH_2CH_2SH$) is another compound occasionally producing sulfur–rubber

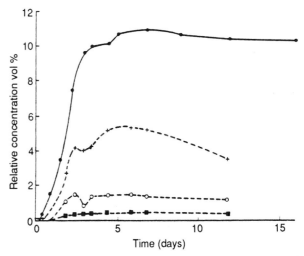

Figure 6.16 Production of organosulfur compounds during fermentation: ●, ethanol (%); +, 3-methylthio-1-propanol; ○, (3-methylthiopropyl) acetamide; ■, 2-methylthiolane-3-ol. (From Tucknott, 1977, in Rapp and Mandery, 1986, reproduced by permission)

---

[1] Glutathione acts as the principal antioxidant in chloroplasts protecting them from reactive oxygen radicals; it is also a major antioxidant in animal tissues.

off-odors in wine. Ethanedithiol is formed in the presence of hydrogen sulfide and acetaldehyde. It can also combine with other wine constituents to cause additional off-odors (Rauhut *et al.*, 1993), for example *cis*- or *trans*-3, 6-dimethyl-1,2,4,5-tetrathiane. It also possesses a rubbery smell. Two additional sulfur compounds, 2-mercaptoethyl acetate and 3-mercaptopropyl acetate, have been detected as the source of a grilled, roasted meat odor in wines made from 'Sauvignon blanc' and 'Sémillon' grapes (Lavigne *et al.*, 1998). When present, they tend to become detectable during fermentation, and may increase further in concentration during barrel aging.

Thioethers are organosulfur compounds characterized by the presence of one or more sulfur atoms bonded between two carbon atoms. The most frequent example is dimethyl sulfide (CH_3-S-CH_3). It may contribute a cooked-cabbage, shrimp-like odor at above threshold values. At low levels, it apparently has asparagus-, corn-, and molasses-like aspects. Its concentration has occasionally been found to increase during aging (Goto and Takamuro, 1987). As such, dimethyl sulfide may contribute to the complexity of an aged bouquet (Simpson, 1979) and to truffle and black olive attributes (Segurel *et al.*, 2004). The origin of dimethyl sulfide is still unclear, but may arise from the metabolism of cysteine (de Mora *et al.*, 1986). Other thioethers found in wine may include dimethyl and diethyl disulfides (CH_3-S-S-CH_3 and CH_3CH_2-S-S-CH_2CH_3). Although usually not causing serious off-odor problems themselves, they can slowly break down under reducing conditions to their respective mercaptans, methanethiol and ethanethiol. This is of particular importance if it occurs after bottling. Both mercaptans have thresholds significantly lower than their corresponding thioethers.

Thiolanes are ring structures containing ether-like sulfur bonds. An example is 2-methylthiolane-3-ol. It has a faint onion-like smell.

Thiazoles are additional cyclic compounds that contain both sulfur and nitrogen as part of the ring. The medicinal, peanut-like smell of 5-(2-hydroxyethyl)-4-methylthiazole has been detected in wine and grape distillates. It is not known whether the compound occurs at levels sufficient to directly influence the fragrance of wine. Another thiazole of potential significance is 2-acetyl-2-thiazoline. It possesses roasted hazelnut attributes and possesses a very low threshold (<5 μg/liter in water).

Thioesters are formed between a carboxyl-containing thiol and an alcohol. The most important of these may

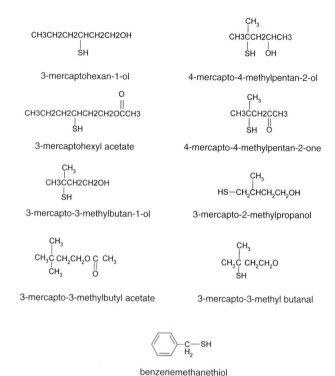

Figure 6.17 Diagrammatic illustration of several of varietally significant thiol compounds.

be ethyl 3-mercaptopropionate, one of the proposed sources of the foxy (fox-den) odor of some *V. labrusca* varieties (Kolor, 1983). The thioester *S*-methylthioacetate and *S*-ethylthioacetate may generate cheesy, onion-like, or burnt odors at above threshold values (Leppänen *et al.*, 1980). However, it is their subsequent hydrolysis to their corresponding mercaptans, methanethiol and ethanethiol, that is of primary winemaking concern.

Many other volatile sulfur compounds have been isolated from wine (see Rauhut, 1993). Although individually occurring below their threshold values, they may act synergistically to augment the detection of other reduced-sulfur odorants.

Although most thiols generate off-odors, a few contribute significantly to the varietal aroma of certain cultivars (Fig. 6.17). For example, 4-mercapto-4-methylpentan-2-ol and 3-mercaptohexan-1-ol produce odors reminiscent of citrus zest/cat urine and grapefruit, respectively. Both compounds are important in the varietal character of 'Sauvignon blanc' (Tominaga *et al.*, 1998), whereas the former is an important odorant in 'Scheurebe' (Guth, 1997). In addition, 4-mercapto-4-methylpentan-2-one plus 3-mercaptohexyl acetate (Tominaga *et al.*, 1996), and especially benzenemethanethiol (benzyl mercaptan) (Tominaga *et al.*, 2003) contribute to the varietal box-tree (smoky) aroma of

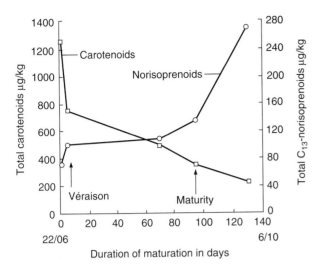

Figure 6.18 Changes in carotenoids and C_{13}-norisoprenoids during the maturation of 'Muscat of Alexandria' grapes. (From Razungles *et al.*, 1993, reproduced by permission)

some 'Sauvignon blanc' wines. In contrast, 3-mercapto-3-methylbutan-1-ol, also found in 'Sauvignon blanc' wines, donates a cooked-leek odor.

Thiols are equally important in the flavor attributes of other beverages. For example, furfuryl mercaptan is an essential component of the flavor of roasted coffee, and thiomenthone is used to supply a black currant aroma in soft drinks.

Most volatile sulfur compounds in wines appear to be derived from the microbial transformation of sulfur-containing amino acids (cysteine and methionine), or occasionally from elemental sulfur applied in disease control. However, several sulfur off-odors are generated by the breakdown of sulfur-containing organic pesticides. Bisdithiocarbamates fungicides can be the source of yeast-generated mercaptans, H_2S, CS_2, and ethylenethiourea. The acephate-containing insecticide Orthene® may be a source of methanethiol and dimethyl sulfide off-odors in some wines.

Hydrocarbons and Derivatives

Hydrocarbons are compounds composed solely of carbon and hydrogen atoms. Because of their poor solubility in water, they usually remain associated with grape cellular debris and are lost before or during clarification, or at pressing. Thus, they typically do not directly influence the sensory characteristics of wine. Nevertheless, hydrocarbon-degradation products may produce important aromatics. Examples are β-damascenone, α- and β-ionone, vitispirane, 1,1,6-trimethyl-1,2-dihydronaphthalene (TDN), and the phenolic, (*E*)-1-(2,3,6-trimethylphenyl)buta-1,3-diene. Several are derived

from the hydrolysis and degradation of megastigmane derivatives (Sefton *et al.*, 1989). Megastigmane is itself frequently generated via carotenoid degradation. The relationship between carotenoid degradation and norisoprenoid synthesis is illustrated in Fig. 6.18. The conjugated double bonds of the carbon chain linking the two opposing terminal ring structures of carotenoids are particularly susceptible to oxidative rupture (producing a variety of aromatic by-products). The content of these compounds in wine can be increased by inducing the hydrolysis of their glycosidic precursors.

Possibly the most significant aromatic hydrocarbon present in grapes, or subsequently generated in wine, is the norisoprenoid, TDN. One of the suspected precursors, probably 2,6,10,10-tetramethyl-1-oxaspiro[4,5]dec-6-ene-2,8-diol, appears to accumulate coincident with the increase in sugar content during the ripening of 'Riesling' grapes (Winterhalter, 1991). In wine, the concentration of TDN has been noted to rise from undetectable to about 40 ppb after several years (Rapp and Güntert, 1986). The sensory threshold is about 20 ppb. It has a smoky, kerosene, bottle-aged fragrance, and may be desirable at low concentrations. It may also induce a hydrocarbon off-odor in young brandy (Vidal *et al.*, 1993). Its production in grapes appears to be enhanced by warm temperatures and sun exposure (Marais *et al.*, 1992). Different clones and yeast strains also significantly affect the TDN content of wine (Sponholz and Hühn, 1997).

A cyclic hydrocarbon occasionally tainting wine is styrene. Although, styrene is synthesized by yeast cells from 2-phenylethanol, amounts sufficient to give a plastic taint usually come from storage in plastic cooperage or transport containers (Hamatschek, 1982).

Additional hydrocarbon taints may come from microbially contaminated corks. For example, methyl tetrahydronaphthalene has been implicated in one of the types of corky off-odors occasionally found in wine (Dubois and Rigaud, 1981).

Macromolecules

Macromolecules are the polymers that constitute the structural and major regulatory molecules of cells. These include carbohydrates, proteins, nucleic acids, and some lipids. The specific roles of macromolecules in the growth and reproduction of grape, yeast, and bacterial cells are beyond the scope of this book. However, without them, life as we know it would not exist. Few occur in wine in significant quantities. Due to poor solubility, macromolecules usually remain in, or precipitate with cellular debris during clarification or fermentation.

CARBOHYDRATES

The major carbohydrate polymers of plant cells are cellulose, hemicelluloses, pectins, and starch. They function primarily as structural elements in cell walls or as forms of energy storage. Cellulose is too insoluble to be extracted into wine and remains with the pomace. Hemicelluloses are poorly soluble and, if extracted, precipitate during fermentation. Pectins usually precipitate as the alcohol content rises during fermentation or are enzymatically degraded. Occasionally, though, they can cause clarification or filtration problems. With pulpy grape varieties, pectins can seriously complicate pressing. Starch, the major storage carbohydrate of plants, is not found in significant quantities in mature grapes. Grapes are atypical in using soluble sugars as their primary storage carbohydrate.

Mannans and glucans, the major carbohydrate polymers of yeast cell walls, are either insoluble or precipitate during fermentation. Related, but smaller, polysaccharides are soluble and may be released in significant amounts (>400 mg/liter). Most are mannans combined with proteins (mannoproteins). In combination with cell-wall degradation products, they participate in generating the effervescence attributes of sparkling wines.

The glucans and chitins of most fungal cell walls are too insoluble to be incorporated into wine. However, the extracellular glucans produced by *B. cinerea* can cause serious winemaking problems. High-molecular-weight forms can induce severe plugging during juice or wine filtration, whereas the low-molecular-weight forms can inhibit yeast metabolism (Ribéreau-Gayon, 1988).

Yeast cell walls may be added to fermenting juice to prevent the premature termination of fermentation. They appear to act by removing toxic carboxylic acids and as a source of vital yeast nutrients (Munoz and Ingledew, 1990).

LIPIDS

Plants possess two major groups of lipids. One group is based on fatty acids often esterified to a polyol, usually glycerol. The other is based primarily on isoprene subunits. The first group includes phospholipids, fats, oils, waxes, glycolipids, and sulfolipids; the second includes steroids. All lipids are vital to the structure and function of plant and yeast cells. However, only oils, waxes, and steroids directly influence wine quality.

Plant oils generally do not occur in wine. Their presence would probably indicate the use of excessive pressure during grape crushing, which could rupture grape seeds and release their oils. After oxidation, the oils could generate a rancid taint. Modern crushers have almost eliminated this source of wine contamination. Fruit, and especially leaves accidentally macerated with the grapes, may release small amounts of linoleic and linolenic acids into the juice (Roufet *et al.*, 1986). Grape lipoxygenases, activated during crushing, can rapidly oxidize these compounds and release aromatic C_6 aldehydes and alcohols, such as *trans*-2-hexenal and *cis*-2-hexenol (Iglesias *et al.*, 1991). These compounds produce both herbaceous odors and bitter tastes. They are occasionally termed "leaf" aldehydes and alcohols.

Both the growth and metabolic activity of yeast cells require the presence of sterols and unsaturated fatty acids. In the presence of oxygen, yeast cells synthesize their own lipid requirements. However, the anaerobic conditions that develop during fermentation severely restrict the ability of yeasts to produce some lipid constituents. Oleanolic acid (oxytriterpenic acid), a major component of grape wax, can be incorporated and used under anaerobic conditions in the synthesis of essential yeast sterols. The unsaturated-fatty-acid requirement also may be satisfied by linoleic and linolenic acids released from grape cells. Both types of lipids help maintain membrane function and enhance yeast tolerance to alcohol during and after fermentation. The extraction of lipids is significantly improved by leaving the juice in contact with the skins for several hours after crushing.

PROTEINS

During ripening, the soluble protein content of grapes increases, the degree of enrichment being cultivar-dependent. After crushing, the soluble protein content may increase by a further 50% during cold settling (Tyson *et al.*, 1982). The proteins come primarily from the pulp. The addition of bentonite reverses this trend. During fermentation, the soluble protein content may increase, decrease, or fluctuate markedly, depending on the cultivar. By the end of fermentation, many proteins have precipitated with tannins, especially in red wines. Those that remain are highly resistant to proteolysis and low pH values. They are relatively homogeneous and consist primarily of pathogenesis related (PR) proteins (Waters *et al.*, 1996), notably chitinases, thaumatin- and osmotin-like proteins.

In most wines, soluble proteins are considered undesirable, because they may induce haze formation. In sparkling wines, however, mannoproteins help to stabilize the effervescence (Maujean *et al.*, 1990). Mannoproteins released during alcoholic fermentation also enhance the growth and malolactic fermentation of lactic acid bacteria. In addition, mannoproteins can bind important flavorants, such as β-ionone, ethyl hexanoate, and octanal, as well as enhance the volatility of others, such as ethyl octanoate and ethyl decanoate (Lubbers *et al.*, 1994).

In juice, the most important group of enzymatic proteins are hydrolases, activated after their release from grape cells during crushing or pressing. Phenol oxidases (*o*-diphenol oxidases) are the most well known class. They activate the early browning of juice in the presence of oxygen. Because the enzymes tend to remain with grape cellular debris, enzyme activity is largely restricted to the period before pressing. The addition of sulfur dioxide further limits their action. The most troublesome are the phenol oxidases (laccases) produced by *Botrytis cinerea*. Not only are laccases readily soluble and relatively insensitive to SO_2 inhibition, they oxidize a wider range of phenols. These include monophenols, *o*-, *m*-, and *p*-diphenols, *o*-triphenols, anthocyanins, catechins, procyanidins and 2-*S*-glutathionyl caftaric acid (see Macheix *et al.*, 1991). The solubility of laccases means that they are not easily removed by most clarification techniques, including bentonite addition. One of the potential advantages of ultrafiltration is its removal of laccase. It may also be inactivated during pasteurization. Laccase activity can be measured using the syringaldazine test developed by Dubourdieu *et al.* (1984).

Pectinases are important in fruit softening and maceration following crushing. Tissue disintegration eases juice release and flavor liberation from the pomace. Thus, juice from white grapes may be left in contact with the skins and pulp for several hours before pressing. Commercial pectinases may be added to enhance tissue breakdown and reduce the high pectin levels that characterize some grape varieties.

Lipoxygenases are being viewed with increasing interest, due to their ability to oxidize fruit and leaf oils, notably linoleic and linolenic acids. This action could generate several aromatic C_6 and C_9 aldehydes, including hexanals, hexenals, nonenals, nonadienal, and their corresponding alcohols. Although C_6 ("leaf") aldehydes and alcohols can give wine a grassy off-odor, lipoxygenases may also generate aromatic compounds from carotenoids, such as norisoprenoids, and release oleanolic acid from grape skins.

Proteases (protein-hydrolyzing enzymes) have been detected in grape must, but their significance is unknown. By releasing amino acids, they may increase nitrogen availability to yeast cells.

NUCLEIC ACIDS

Nucleic acids are long polymers of nucleotides that function in the storage, transmission, and translation of genetic information. The molecular weight of nucleic acids is so great that they are not released in significant amounts from grape cells on crushing. Although the degradation products of nucleic acids are readily soluble and easily assimilated, yeasts can synthesize their own nucleotide requirements.

During extended yeast autolysis, such as during *sur lies* maturation, there is considerable release of nucleotides and nucleic acids from the dead and dying cells.

The release is most significant during the first few months. Several nucleotides (5'-guanosine monophosphate and 5'-inosine monophosphate) have potent flavor enhancement attributes. Whether these play a significant role in wine is currently unknown (Courtis *et al.*, 1998).

Vitamins

Vitamins encompass a series of diverse chemicals involved in the regulation of cellular activity. They are found in small quantities in grape cells, juice, and wine. The concentration of vitamins generally falls during fermentation and aging. For example, ascorbic acid (vitamin C) is oxidized rapidly following crushing; thiamine (vitamin B_1) is degraded by reaction with SO_2, exposure to heat, or absorption to bentonite; and riboflavin (vitamin B_2) is oxidized after exposure to light. The only vitamin to increase notably during fermentation is *p*-aminobenzoic acid (PABA).

Vitamins occasionally are added to juice to encourage vigorous fermentation, diminish the use of sulfur dioxide, or reduce the likelihood of sticking. Nevertheless, except in situations where the need for supplementation has been established, the addition of vitamins is generally ill-advised. They have been noticed to increase the risk of generating high levels of acetic acid (volatile acidity) during fermentation (Eglinton *et al.*, 1993).

Vitamin levels in wine are inadequate to be of major significance in human nutrition, but they usually are ample for microbial growth. Biotin (vitamin H) and nicotinic acid (niacin) contents are adequate for most yeast strains. In addition, indigenous vitamin and growth factor levels are usually adequate for lactic acid bacterial needs.

Dissolved Gases

Wines contain varying amounts of several gases. All except nitrogen can have marked effects on the sensory properties of wine. Nitrogen gas is both chemically inert and poorly soluble in wine.

CARBON DIOXIDE

Carbon dioxide in wine comes primarily from yeast metabolism. Additional small amounts may be generated by lactic acid bacteria. Minute amounts may arise from the breakdown of amino acids and phenols during aging.

Most of the carbon dioxide produced by yeast action escapes during fermentation. Nevertheless, wine usually remains supersaturated with carbon dioxide at the end of fermentation. During maturation, much of this carbon dioxide escapes, and the CO_2 concentration falls to about 2 g/liter (saturation) at bottling. At this concentration, carbon dioxide has no sensory effect. If wine is bottled while still supersaturated, bubbles are likely to form in the glass when the wine is poured. At above 5 g/liter, carbon dioxide begins to produce a prickling sensation on the tongue (Amerine and Roessler, 1983).

Refermentation is the primary source of detectable effervescence in still wines. It may be associated with off-odors and haziness. This potential fault can be turned to advantage, if fermentation is induced by a desirable yeast strain and haziness is avoided. The result is sparkling wine. The production and role of carbon dioxide to sparkling wine is discussed in Chapter 9.

OXYGEN

Before being crushed, grapes contain very low levels of oxygen. Crushing results in the rapid uptake of about 6 ml (9 mg) O_2/liter – at 20°C. The use of crushers employing minimal agitation limits oxygen uptake. Slight juice aeration is preferable as it favors complete fermentation. The oxygen allows yeasts to synthesize essential compounds, such as unsaturated fatty acids, sterols, and nicotinic acid. It also limits browning, by converting caftaric acid to a less-oxidizable, colorless complex with glutathione. It also promotes the early oxidation and precipitation of other readily oxidized phenolic compounds. Occasionally, juice from white grapes is purposely hyperoxygenated to encourage these processes.

Because oxygen is rapidly consumed in various oxidative reactions, the majority of fermentation occurs in the absence of oxygen. Oxygen uptake is limited by the generation of carbon dioxide, which rapidly blankets the fermenting juice. Once fermentation is complete, however, the wine must be protected from exposure to oxygen.

Much of the oxygen absorbed by must or wine is consumed in oxidative reactions with phenols. Oxygen consumption is comparatively rapid in red wine, often being complete within 6 days at 30°C (Singleton, 1987). Consumption goes to completion at lower temperatures, but takes longer.

The slow or periodic incorporation of small amounts of oxygen (~40 ml/liter) is thought to benefit the maturation of red wines (Ribéreau-Gayon et al., 1983). Oxygen aids color stabilization and reduces the bitterness and astringency of wine tannins. In contrast, oxygen uptake by white wine is generally detrimental. White wines are less able to bind the acetaldehyde generated, and to mask the stale oxidized odor it produces. Neither do white wines benefit from color stabilization, nor bitterness reduction associated with slight aeration.

Although red wines benefit from limited aeration during the early stages of maturation, excessive oxygen exposure produces an oxidized odor and browning. Oxygen exposure is also likely to favor the growth of spoilage organisms. Sherries begin to take on an oxidized bouquet following exposure to about 60 ml O_2/liter (Singleton et al., 1979). Table 6.8 illustrates one view on the amount of oxygen required during the maturation of different wines.

SULFUR DIOXIDE

Sulfur dioxide is a normal constituent of wine, occasionally accumulating to between 12 and 64 mg/liter as a result of yeast metabolism (Larue et al., 1985). Most yeast strains, however, produce less than 10 mg/liter. Strains isolated from sulfited must show a much stronger tendency to produce SO_2 than those from nonsulfited musts (Sussi and Romano, 1982). Strains of Saccharomyces bayanus seem particularly prone to fermentative sulfur dioxide synthesis. Major factors influencing the biosynthesis of sulfur dioxide are yeast strain, fermentation temperature, and the sulfur content of the grapes (Würdig, 1985). Nevertheless, SO_2 levels above 30 mg/liter usually result from addition during or after vinification.

Burning sulfur was used by the ancient Egyptians, Greeks and Romans as a fumigant. However, the absence of specific mention in ancient texts in association with wine suggests that its value relative to wine was largely unrecognized. The first clear reference of sulfur dioxide use in wine production comes from a report published in Rotenburg, Germany in 1487, reproduced in Kellermaisterey, 1537 (Anonymous, 1986). The publication recommends that three wood splinters, covered with powdered sulfur (mixed with viola root and incense) be burnt in upturned casks and then sealed. The treatment was recommended to avoid incidents of wine spoilage. Schumann (1997) also cites a publication by Rasch (1582) concerning the use of burning sulfur. The number of notations and formulations concerning sulfur use increase significantly from 10 recipes in Drinckwein (1585) to 30 recipes in Kellermaisterey (1705 edition). The first English report of sulfur use appears to be by Dr Beale (noted in Evelyn, 1664).

As sulfur hath some use in wine, so some do lay brimstone on a rag and by a wire let it down into the cider vessel and there fire it and when the vessel is full of smoak, the liquor speedily poured in.

About 60 mg SO_2/liter can be absorbed by wine poured into barrels after fumigation (Amerine et al., 1980). This is sufficient to act as an effective antimicrobial agent.

Although the incidental incorporation of sulfur dioxide into wine clearly has a long history, the deliberate addition of purified sulfur dioxide began only early in the twentieth century (Somers and Wescombe, 1982). It is usually added as a liquified gas (from cylinders). Alternately, it can be added in salt form – as potassium metabisulfite ($K_2S_2O_5$). Bisulfite salts rapidly ionize under

Table 6.8 Relative need for oxygen during the maturation and processing of certain types of wines

Type of wine	Oxygen demand	Typical period between production and bottling (years)	Typical aging potential (years)
Table wines			
White	None	0.3–1	1–3
Rosé	None	0.5–1	1–2
Light red	Slight	0.5–2	2–4
Deep red	Slight to moderate	2–4	5–40+
Fortified wines			
Flor sherries	Considerable	3–7	0.5–1
Oloroso sherries	Extensive	4–10	2–4
Tawny ports	Considerable	10–40+	1–2

Source: After Somers, 1983, reproduced by permission

the acidic conditions, releasing gaseous sulfur dioxide. Although undesirable in excess, sulfur dioxide has many benefits. These include its antimicrobial and antioxidant properties, as well as the potential to bleach pigments and suppress oxidized odors. The relative value of SO_2 use often depends as much on when, as on how much is added.

$$\text{(molecular)}$$
$$\Updownarrow$$
Total sulfur dioxide = Free *(bisulfite)*
$$\Updownarrow$$
$$\text{(sulfite)}$$

(unstable e.g., sulfonates with carbonyl groups or anthocyanins)
$$\Updownarrow$$
Bound
$$+$$
$$\Updownarrow$$
(stable e.g., split disulfide bonds; reactions with quinones or H_2O_2)

Sulfur dioxide exists in a variety of free and bound states in wine. Of free forms, only a small portion exists as a dissolved gas (from 7.5% at a pH of 2.9, to about 1% at a pH of 3.8). An additional small fraction exists as free sulfite ions (SO_3^{2-}) (0.004% at pH 2.9, to about 0.04% at pH 3.8). The vast majority of ionic sulfur dioxide occurs as bisulfite ions (HSO_3^{-}). Sulfur dioxide also binds reversibly or irreversibly with a diversity of wine constituents. The most common of these are hydroxysulfonates, principally with carbonyl compounds, notably acetaldehyde. Anthocyanins are another significant binding substrate in young red wines. Additional hydroxysulfonate addition products include pyruvic acid, α-ketoglutarate, sugars, sugar acids, and tannins. In must, about 40–70% of added sulfur dioxide binds with sugars, notably aldo-sugars, principally glucose in its rare aldohydro state. These sugar-sulfur dioxide complexes break down as sugar is metabolized during fermentation (releasing the sulfur dioxide). When the total sulfur dioxide content of a wine is calculated, it does not include sulfur dioxide that forms stable complexes, or has been involved in other irreversible reactions. All binding reactions can significantly reduce the active (free) sulfur dioxide concentration in the must or wine.

$$CH_3CHO \quad + \quad SO_2 \quad + \quad H_2O \quad \leftrightarrows \quad CH_3CHOH-HSO_3$$

Acetaldehyde   sulfur dioxide   water   acetaldehyde hydroxysulfonate

Because of the many interconvertible states of sulfur dioxide, determining the relationship between added SO_2 and its active (molecular) content is often difficult. In contrast to the rapid binding of sulfur dioxide with sugars in must, the equilibrium in fermenting wine is constantly changing. For example, the formation of hydroxysulfonates with acetaldehyde promotes the additional biosynthesis of acetaldehyde by yeast cells (continuing the consumption of SO_2). Only when essentially all free carbonyl compounds have combined with sulfur dioxide does the free SO_2 content show a direct and linear relationship with what is added. The final value depends on the wine's composition, as well as its temperature and pH. The latter affects both the solubility and dissociation constants of sulfur dioxide. Continuing microbial metabolism, by synthesizing or degrading carbonyl compounds, further modulates the balance. For example, the metabolism of acetaldehyde by lactic acid bacteria during malolactic fermentation releases sulfur dioxide. This can be sufficient to slow or inhibit malolactic fermentation. It can also result in color loss, notably through the bleaching of anthocyanins.

Determining the free content is essential to estimating the molecular (antimicrobial) SO_2 content.

In bottled wine, the sulfur dioxide content progressively falls. Although some may escape via the cork, this is not viewed as being significant. More important, at least in the short term, may be the formation of hydroxysulfonates (Burton *et al.*, 1963). Reaction with quinones may be another significant factor, especially in red wines (Embs and Markakis, 1965). The reduction of diquinones to their corresponding diphenols can be associated with the oxidation of sulfites to sulfates (Walker, 1975). Finally, sulfur dioxide (in its sulfite form) may be oxidized to sulfate in the presence of oxygen. The reaction does not directly involve oxygen, but is associated with hydrogen peroxide generated in the autooxidation of *o*-diphenols (Danilewicz, 2007). This may explain most of the slow disappearance of sulfur dioxide noted in bottled wines.

Because the sulfur dioxide content of bottled wine declines with time, its antioxidative activity eventually becomes inadequate to prevent oxidation. The release of carbonyls, due to the shifting equilibrium between free and bound forms, may begin to generate an oxidized odor, while the liberation of chromophoric carbonyls (initially bleached by sulfur dioxide) may result in browning. Additional oxygen ingress could directly activate the oxidative browning of phenolics, induce the oxidation of ethanol to acetaldehyde, and aggravate the oxidative loss of terpene fragrances.

As with other aspects of sulfur dioxide activity, antimicrobial activity is largely dependent on its free component. Bound forms tend to be weakly antimicrobial. The apparent toxicity of acetaldehyde hydroxysulfonate may result from the release of sulfur dioxide, following acetaldehyde degradation by microbial cells. It is for this reason that, when malolactic fermentation is desired, the addition of sulfur dioxide be delayed or kept minimal until after its completion.

It is common to consider only the molecular sulfur dioxide content when assessing antimicrobial action. Of the free states of sulfur dioxide, molecular SO_2 is the most readily absorbed by microbes. However, as cytoplasm typically has a pH of around 6.5, the molecular form quickly changes into its ionic states, bisulfite and sulfite. Binding of these ions with various cellular constituents enhances the continued uptake of sulfur dioxide.

The antimicrobial action of sulfur dioxide probably involves the splitting of disulfide bonds (see Beech and Thomas, 1985). Disulfide bonds are often essential in maintaining the functional structure of enzymes and regulatory proteins. Sulfonate formation can, thus, seriously disrupt cellular metabolism. Binding with nucleic acids and lipids may also provoke genetic and membrane dysfunction. Additional antimicrobial activity may

result from a rapid drop in the intracellular pool of ATP (due to disruption of glyceraldehyde-3-phosphate dehydrogenase), a decline in cytoplasmic pH (and an associated reduction in the electrochemical potential across the plasma membrane), interaction with $NAD/NADP^+$ and cofactors, and the cleavage of thiamine. Although thiamine destruction is potentially significant, its slow reaction rate within the pH and typical storage temperature range of wine, likely limits its importance.

Sulfur dioxide rapidly inactivates a wide range of microbes. Values of between 0.8 and 1.5 mg/liter (ppm) molecular SO_2 have generally been viewed as sufficient to inhibit the growth of most wild yeasts and bacteria (Beech *et al.*, 1979; Sudraud and Chauvet, 1985). However, Heard and Fleet (1988) question both its efficacy and need. Yeasts, such as *Kloeckera*, *Hanseniaspora*, and *Candida*, usually die out as the alcohol content rises due to the activity of *Saccharomyces* spp., regardless of sulfur dioxide addition. The addition of sulfur dioxide at crushing is probably justified only with moldy grapes, due to their high population of wild yeasts and spoilage bacteria. Moldy grapes also tend to possess high concentrations of compounds that bind sulfur dioxide, for example galacturonic acid released by pectin degradation. Addition after fermentation is essential primarily if the wine becomes contaminated with spoilage yeasts, notably *Brettanomyces* spp., *Saccharomycodes ludwigii*, and *Zygosaccharomyces bailii*. Considerably higher sulfite additions than typical are required to achieve effective control of these spoilage yeasts. All are relatively insensitive to sulfur dioxide.

The level of free sulfur dioxide required to obtain a desired amount of molecular SO_2 can be partially estimated by dividing the desired value by the percentage molecular SO_2 at the wine's pH. For example, the free sulfur dioxide concentration to achieve a value of 1.5 mg/liter molecular SO_2, at a pH of 3.2, would be 1.5 mg/liter $\div$ 0.039 = 38.5 mg/liter. The percentage molecular SO_2 at various pH values are given below:

pH 2.9 (7.5%); pH 3.0 (6.1%); pH 3.1 (4.9%); pH 3.2 (3.9%); pH 3.3 (3.1%); pH 3.4 (2.5%); pH 3.5 (2.0%); pH 3.6 (1.6%); pH 3.7 (1.3%); pH 3.8 (1.0%); pH 3.9 (0.8%)

This simple indicator for molecular SO_2 is relatively accurate for a short period after addition. However, as sulfur dioxide binds with carbonyls and phenolics, the percentage free and total SO_2 declines, until it reaches a new equilibrium between its bound and free forms. Subsequent declines result from the slow production of new oxidants (as oxygen gains access to the wine), or sulfur dioxide gas escapes from the wine.

Despite the toxicity of SO_2 to most microorganisms, several yeasts show comparative insensitivity to sulfur dioxide. The production of acetaldehyde by actively growing cells may be a major factor in providing relative insensitivity to sulfur dioxide (Pilkington and Rose, 1988). This probably reduces the effective concentration of sulfur dioxide within the cell. That many yeasts also possess proteins with atypically low cysteine content, that can be denatured by sulfur dioxide, probably provides additional protection. Furthermore, a defect in sulfate (HSO_3^-) transport in some yeasts may limit sulfur dioxide uptake, and correspondingly its toxicity. This property may have arisen as a defense against the toxicity of selenate or chromate (Schreibel *et al.*, 1997). Both metal salts share the same transport permeases as sulfate (Lachance and Pang, 1997).

Because yeasts emerging from dormancy are more sensitive to sulfur dioxide (and other stress factors) than metabolically active cells, sulfur dioxide often selectively restricts the growth of indigenous wine yeasts derived from grape skins and winery equipment. This gives advantage to inoculated strains that are either fully active upon addition, or rapidly regain full metabolic status. Nevertheless, even inactivation of 99% of a population of wild yeasts can still leave many viable cells (for example, a relatively typical population of 10^5 yeasts per ml would be reduced to 10^3 cells per ml).

Another of sulfur dioxide's beneficial effects is its action as an effective antioxidant. By suppressing the activity of oxidases, sulfite ions limit the oxidation of phenols. However, as noted below, whether this is actually beneficial is a moot point. Sulfite is considerably less effective against fungal oxidases (laccases) (Kovač, 1979). Enzymatic inactivation apparently involves reduction of copper that acts as a cofactor (Schopfer and Aerny, 1985). Sulfur dioxide can also reverse some oxidative reactions. For example, bisulfites can reduce diquinones back to their corresponding diphenols. The irreversible binding of sulfur with quinones, found in tannin pigments, bleaches their color as well as preventing their participation in additional oxidative reactions (see Taylor *et al.*, 1986). Bisulfites can also directly inhibit certain browning reactions, such as the formation of Maillard products. This is thought to involve the binding of bisulfites with the carbonyl group of sugars. Additional nonenzymatic browning reactions are inhibited by sulfite ions, such as the slow caramelization of sugars, or the conversion of dehydroascorbic acid into yellowish pigments.

A more significant antioxidant action involves the reduction of hydrogen peroxide by molecular SO_2. By rapidly reducing hydrogen peroxide to water, sulfur dioxide limits the oxidation of ethanol to acetaldehyde, as well as curtailing its reaction with ascorbic acid and *o*-diphenols. Sulfur dioxide can also directly reduce the oxidized character of wine by forming nonvolatile

hydroxysulfonate complexes with acetaldehyde, as well as bleaching brown oxidation products. These influences are of greater significance in white than red wines, due to their more subtle fragrance and paler color.

Because sulfur dioxide suppresses the reaction between caftaric acid and glutathione, the addition of sulfur dioxide at crushing can limit the oxidation of these readily oxidizable phenols, enhancing the tendency of white wines to brown during aging (Singleton *et al.*, 1985). In addition, sulfur dioxide increases the absorption of phenolics from the grape pomace or wood cooperage (sulfite addition products are more soluble in wine than the phenolics themselves). Finally, sulfur dioxide can significantly reduce the thiamine content of must (irreversible splitting thiamine into two inactive ingredients). This may require thiamine supplementation of musts low in thiamine (e.g., botrytized grapes). Correspondingly, current recommendations for sulfur dioxide addition must tend to be limited to situations where the immediate risk of microbial contamination is great.

Although opinions differ on the value of sulfur dioxide addition at crushing, or post-fermentation, there is little argument about its benefits prior to bottling, especially for white wines. Small amounts of sulfur dioxide donate a fresher odor to finished wine, possibly by forming nonvolatile sulfonates with acetaldehyde. By reacting with phenols, sulfur dioxide can reduce the nonenzymatic generation of acetaldehyde. Sulfur dioxide also bleaches brown pigments, giving the wine a paler color. However, in red wines, the bleaching action can induce a temporary loss of color. In addition, by reducing hydrogen peroxide (produced during the autooxidation of phenols), SO_2 limits acetaldehyde formation and the development of color-stabilizing anthocyanin–tannin polymers. This is not compensated for during maturation by the recoloration of anthocyanins released from their association with sulfur dioxide. Free anthocyanins are much more liable to nonreversible decoloration than are polymerized forms. In addition, delaying anthocyanin–tannin copolymerization restricts its ultimate occurrence. Catechin self-polymerization to form tannins diminishes their availability to combine with anthocyanins. Although sulfur dioxide addition may have undesirable effects on color stability, it can be beneficial in protecting varietally important volatile thiols, such as 3-mercaptohexan-1-ol, from oxidative loss (Blanchard *et al.*, 2004).

The levels of free SO_2 generally recommended for wine before bottling is in the range of 15–40 mg free/liter, depending on the amount required to achieve about 0.8 mg/liter (ppm) molecular. The concentration in red wines is usually lower.

By itself, sulfur dioxide begins to show a burnt match odor at about 100 mg/liter (Amerine and Roessler, 1983).

However, as only the molecular component is detectable, perception of sulfur dioxide is largely dependent on the wine's pH and temperature. Because the concentration of sulfur dioxide declines following bottling (Casey, 1992), a sulfur dioxide odor is rarely detectable by the time the bottle is opened.

For healthy individuals, consuming 400 mg SO_2 (free and bound)/day for several weeks appears to possess no adverse effects (Hötzel *et al.*, 1969). Regardless, the FAO/WHO Joint Expert Committee on Food Additives (1974) established an acceptable daily sulfite intake of 0.7 mg/kg body weight. Most commercial wines contain less than 100 mg/liter. The median value for American wines tested by Peterson *et al.* (2000) was 64 mg/liter total sulfur dioxide, with less than 4% possessing more than 150 mg/liter. At the average sulfur dioxide content, acceptable intake is equivalent to about a bottle of wine per day for a person of 150 lb.

Although gaseous SO_2 can precipitate asthma attacks in sensitive individuals, most wines possess insufficient molecular sulfur dioxide to induce an attack. Nevertheless, all forms of sulfur dioxide are potentially allergenic to a small portion of asthmatics. For these individuals, sulfites absorbed by blood from the digestive tract can be translocated to the lung, where it could institute an asthma attack. In addition, a few people possess a rare genetic defect (sulfituria), making them unable to inactivate sulfites to sulfates (Duran *et al.*, 1979). However, the disorder is so injurious that affected individuals seldom survive childhood. In contrast, normal individuals rapidly convert sulfites to sulfates. The kidneys subsequently eliminate sulfates in the urine. In the process, it is estimated that humans eliminate about 2.4 g sulfate/day (Institute of Food Technologists Expert Panel on Food Safety and Nutrition, 1975). Most of the sulfate found in the blood comes from sulfite liberated during the cellular metabolism of sulfur-containing amino acids. Sulfite derived from normal food metabolism is many times higher than that obtained from wine consumption. On the positive side, sulfites suppress the toxicity or mutagenicity of natural constituents in some foods and beverages (see Taylor *et al.*, 1986).

Although the sulfite content of wine is a legitimate health concern for some individuals, this has been grossly exaggerated. Although regrettable, the attention has focused research on clarifying the relative benefits and disadvantages of enologic sulfur dioxide use. Practices that promote the health and adequate acidity of grapes delivered to the winery have limited its need. In addition, understanding the benefits of limited oxygen exposure following crushing, fermentation, and maturation have indicated its preservative action is less necessary than previously thought. This is especially so with

adequate temperature control. Even the reasons for sulfur dioxide addition at bottling have been clarified and confirmed. The current trend is to add about 15–40 mg/liter SO_2 at bottling to white wines, less to red wines. The higher phenolic content of red wines partially offsets the need for sulfite addition. In addition, as noted above, sulfur dioxide can reduce the color intensity, delay the formation of stable red pigments, and can occasionally provoke the development of a phenolic haze.

Not only is sulfur dioxide the most important antimicrobial and antioxidant additive in wine, it is also the primary disinfectant of winery equipment. Nevertheless, excessive use can be detrimental. The corrosive action of SO_2 can solubilize metal ions from unprotected surfaces. Sulfur dioxide may also react with oak constituents, forming lignosulfurous acid. It has been postulated that, after decomposition, lignosulfurous acid may release hydrogen sulfide that reacts with pyrazines in the wood. This could form musty-smelling thiopyrazines (Tanner and Zanier, 1980). Subsequent extraction into the wine could donate a distinct off-odor.

Minerals

Many mineral elements are found in grapes and wine. In most situations, the mineral concentration reflects the uptake characteristics of the rootstock, accumulation by the scion, and climatic influences on the rate of transpiration. For example, grapes in hot climates typically have higher potassium contents than those grown in temperate or cool climatic regions. However, high levels of elemental sulfur may arise from fungicides applied to the vines for disease control; elevated calcium levels may occur in wines stored in unlined cement tanks; augmented chlorine and sodium contents may originate from the use of ion-exchange columns; and abnormal levels of copper and iron can result from contact with corroded winery equipment. Atypically high aluminum contents may originate from bentonite use (McKinnon *et al.*, 1992).

Elevated lead contents in wine were, in the past, correlated with vines grown near highways (Médina *et al.*, 1977), corroded lead capsules (Sneyd, 1988), or prolonged storage in lead crystal decanters (Falcone, 1991). Nevertheless, the most frequent source of lead contamination appears to originate with brass fittings and faucets occasionally used in wineries (Kaufmann, 1998). Although automotive exhaust (prior to lead-free gasoline), agricultural chemicals, and industrial pollution can increase the lead content of soils and grapes, only a minor portion of any lead on grapes is released into the juice (Stockley *et al.*, 1997). In addition, most of the lead that is incorporated is precipitated along with the lees during and after fermentation. Finally, measurement of the lead isotope ratio in wines does not support the contention that lead in metal capsules diffuses through cork (Gulson *et al.*, 1992). Despite this, wines are rarely if ever capped with tin-lined lead capsules, as in the past.

At naturally occurring levels, many minerals are important cofactors in vitamins and enzymes. However, heavy metals such as lead, mercury, cadmium, and selenium are potentially toxic. If present in the fruit, heavy metals usually precipitate during fermentation (von Hellmuth *et al.*, 1985). Thus, their occurrence in wine at above trace amounts usually indicates contamination post-fermentation. At higher than normal levels, minerals such as iron and copper also can be undesirable. They catalyze oxidative reactions, modify taste characteristics, or induce haziness (*casse*).

Copper ions can slowly associate with dissolved proteins and induce copper *casse*. In the presence of high levels of both phosphate and iron, a ferric-derived *casse* may develop. Under appropriate conditions, iron can also react with tannic acid, giving rise to blue *casse*. High calcium levels can delay tartrate precipitation and augment crystal formation in bottled wines.

Oxidative reactions may be catalyzed in the absence of molecular oxygen by copper and, to a lesser extent, by iron. An example is the oxidization of ascorbic acid to dehydroascorbic acid, and its cleavage into oxalic and threonic acids. In the presence of oxygen, iron can catalyze oxidative reactions involved in browning. In addition, iron favors the polymerization of phenolics with acetaldehyde, whereas manganese ions catalyze the synthesis acetaldehyde (Cacho *et al.*, 1995).

Although copper and iron can induce metallic/astringent tastes, this occurs only at concentrations higher than those usually found in wine (Amerine and Roessler, 1983). High sulfite contents can give wine a slightly salty-bitter taste.

Chemical Nature of Varietal Aromas

There has long been an interest in understanding the chemical origin of grape aromas. Such knowledge could benefit grape grower and winemaker alike by permitting a more precise determination of a desirable harvest date. It also would allow an assessment of how various vinicultural practices influence one of the most important determinants of wine quality, fragrance. In addition, such information could streamline the production of new grape varieties by permitting the selection of lines showing particular aromatic attributes. An objective measure of the varietal origin also would be of particular interest to those charged with enforcing Appellation Control laws. In Europe, these regulations often stipulate the varieties permitted in the production of AC wines. In some varieties, differences in aroma are sufficiently diagnostic to allow such a distinction (Rapp and Mandery, 1986). Nevertheless, changes with aging may limit its use to young wines from the same vintage.

Despite its advantages, determinating the chemical nature of a varietal aroma is fraught with difficulties. The first step usually involves the separation of grape and wine volatile components by gas chromatography. The column may be split to divert a fraction of each compound for sniffing, whereas the remainder undergoes physicochemical analysis. The study is easier if the crucial compound(s) occur in a volatile form in both grapes and wine. However, aroma compounds in grapes often occur in nonvolatile forms. They may be released only upon crushing (e.g., C_{18} fatty acids into "leaf" aldehydes and alcohols), through yeast metabolic activity (e.g., phenol into vinyl guaiacol), or during aging (e.g., linalool to α-terpineol). In addition, varietal aromas may originate from a particular combination of compounds, not from varietally specific compounds. Extraction procedures may influence the stability and, thereby, the isolation of potentially important compounds. When compounds of probable importance are isolated, both their identification and quantification are required. Only by comparing the concentration in wine with its sensory threshold can its potential significance be assessed. Because several hundred volatile compounds may occur in a wine, multivariate analysis is often used in detecting compounds that deserve more detailed investigation.

Even with the highly precise analytical tools currently available, great difficulty can be encountered in the detection of certain aromatic compounds (e.g., aldehydes bound to sulfur dioxide). However, the situation is even more demanding when the significant compounds are labile or occur in trace amounts, for example 2-methoxy-3-isobutylpyrazine. It is both highly labile and typically occurs in wine at less than 35 ppt. It has a detection threshold of about 2 ppt.

Based on their relative importance in aroma production, volatile ingredients may be classified as impact, contributing, or insignificant. **Impact** compounds are those that have a marked and distinctive effect on wine fragrance. They generally give wines their varietal distinctiveness. Although usually desirable, they may impart notoriety to wines, for example, the foxy aroma of certain *V. labrusca* varieties. In some instances, the relative importance of specific compounds to wine fragrance is clear. However, this is generally not the case.

Contributing compounds are considered to be those that add to the overall complexity of wine fragrance. For example, C_{10} and C_{12} esters of unsaturated fatty acids contribute to the fruity odor of 'Concord' fruit and wine (Schreier, 1982), but are not varietally distinctive. Ethyl esters of fatty acids as well as acetate esters of higher alcohols add significantly to the fruity odor of most young white wines (Ferreira *et al.*, 1995). Contributing compounds are also important to the development of an aged bouquet. They are equally responsible for the basic wine bouquet generated by yeast metabolism during fermentation.

The vast majority of the hundreds of aromatic compounds found in wine fall into the insignificant category. Their concentration is usually considerably below their threshold values. Thus, unless they act synergistically with other compounds, they cannot influence a wine's aromatic attributes.

Most grape varieties do not (or are not widely acknowledged to) develop distinctive varietal aromas. For most of those that do, unique impact compounds have as yet to be identified. Nevertheless, there is a growing list of cultivars in which impact compounds have been found. For example, the foxy character of some *V. labrusca* cultivars has been variously ascribed to ethyl 3-mercaptopropionate (Kolor, 1983), N-(N-hydroxy-N-methyl-γ-aminobutyryl)glycin (Boison and Tomlinson, 1988), or 2-aminoacetophenone (Acree *et al.*, 1990). The latter has also been isolated from some *V. vinifera* wines. However, this may result from its synthesis by some strains of *S. cerevisiae* (Ciolfi *et al.*, 1995), or other members of the grape epiphytic flora (Sponholz and Hühn, 1997). Other *V. labrusca* cultivars may derive some of their characteristic aroma from the strawberry odor of furaneol (2,5-dimethyl-4-hydroxy-2,3-dihydro-3-furanone) and its methoxy derivative (Schreier and Paroschy, 1981), or from methyl anthranilate and β-damascenone (Acree *et al.*, 1981).

The bell pepper attribute, often associated with 'Cabernet Sauvignon' (Boison and Tomlinson, 1990), 'Sauvignon blanc' (Lacey *et al.*, 1991), and related varieties (Allen and Lacey, 1993), is due primarily to the presence of 2-methoxy-3-isobutylpyrazine. Isopropyl and sec-butyl methoxypyrazines are also present, but at lower concentrations. The source of the desirable black-currant fragrance of some 'Cabernet Sauvignon' wines is uncertain, but may be derived from 4-mercapto-4-methyl-pentan-2-one (4MMP). At low concentrations, it may possess resemblances to black-currant or box tree. At higher concentrations, it is reported to generate a cat urine odor. In contrast, the fragrance of black-currants is attributed to their terpene composition (Marriott, 1986). Even trace amounts of methoxypyrazines have been isolated from 'Riesling' and 'Chardonnay' (Allen and Lacey, 1993). It has also been found in significant amounts in the Greek cultivar 'Xynomavro' (Kotseridis *et al.*, 1999). Desirable aromatic attributes of 'Cabernet Sauvignon' have also been correlated with the presence of β-damascenone and eugenol, and the absence of fruit-smelling esters such as ethyl butanoate and ethyl octanoate (Guth and Sies, 2002).

The spicy character of 'Gewürztraminer' wines has been associated with the production of 4-vinylguaiacol (from the conversion of ferulic acid during fermentation), in association with several terpenes (Versini, 1985). However, omission experiments (see below) indicate the critical importance of *cis*-rose oxide to the varietal aroma of 'Gewürztraminer' (Guth, 1997; Ong and Acree, 1999). The peppery aroma of 'Shiraz,' and possibly other grapes, appears to be associated with the formation of a quaiane type of sesquiterpene, termed rotundone. It occurs in trace amounts and can be detected at a few parts per billion.

The guava-like odor occasionally associated with 'Chenin blanc' and 'Colombard' wines has been attributed to the presence of the mercaptan, 4-methyl-4-pentan-2-one (du Plessis and Augustyn, 1981). Other varietally distinctive thiols are 4-mercapto-4-methyl-pentan-2-one, 3-mercaptohexyl acetate, 4-mercapto-4-methylpentan-2-ol and 3-mercaptohexan-1-ol. These are formed in wine from odorless grape precursors produced by several *V. vinifera* cultivars (Tominaga *et al.*, 2000, 2004), notably, but not exclusively, from 'Cabernet' related varieties. The first two compounds donate a box-tree (smoky) odor, whereas the last two may give citrus zest–grapefruit and passion fruit–grapefruit essences, respectively. The cysteinylated precursor of 3-mercaptohexan-1-ol has been identified as an important indicator of the aromatic potential of several rosé wines (Murat *et al.*, 2001; Murat, 2005). 4-Mercapto-4-methylpentan-2-one is also crucial to the varietal character of 'Scheurebe' (Guth, 1997) and 'Gewürztraminer' (Guth, 1998). Because thiols bind readily to metals, such as copper, limiting the use of Cu^{2+}-containing fungicides may be important in the full development of the varietal aroma in particular cultivars (Darriet *et al.*, 2001).

'Muscat' varieties are generally distinguished by the prominence of monoterpene alcohols and C_{13}-norisoprenoids in their varietal aromas. Similar monoterpene alcohols are important, but occur at lower concentrations in the aromas of cultivars similar or related to 'Riesling.' The relative and absolute concentrations of these compounds, and their respective sensory thresholds, distinguish the varieties in each group from one another. Nevertheless, thiols also contribute to the varietal fragrance of 'Muscat' cultivars (Tominaga *et al.*, 2000).

In some instances, compounds that are varietally distinctive also occur as by-products of fermentation. For example, the important impact compound of muscadine wines, 2-phenylethanol, can also be generated by yeast metabolism from grape constituents (Lamikanra *et al.*, 1996). Similarly, isoamyl acetate, a prominent flavorant of 'Pinotage' wines, may be derived from yeast action (van Wyk *et al.*, 1979). In addition, 'Colombard' wines are reported to derive most of their distinctive aroma from products formed during fermentation (Marais, 1986).

Several well-known, aromatically distinctive cultivars, for example 'Chardonnay' (Lorrain *et al.,* 2006) and 'Pinot noir' (Fang and Qian, 2005) do not appear to possess uniquely distinctive, aroma impact compounds. Nevertheless, they may possess certain aromatic compounds at higher than usual concentrations. For example, β-damascenone tends to be a prominent component in the aroma profile of 'Chardonnay' and 'Riesling' wines (Strauss, Wilson, Anderson *et al.*, 1987; Simpson and Miller, 1984), β-ionone in 'Muscat' cultivars (Etiévant *et al.*, 1983), and α-ionone and benzaldehyde in 'Pinot noir' and 'Gamay,' respectively (Dubois, 1983). Particular combinations have been considered to generate varietal characteristics. For example, Moio and Etiévant (1995) considered that four esters gave 'Pinot noir' its distinctive aroma – ethyl anthranilate, ethyl cinnamate, 2,3-dihydrocinnamate, and methyl anthranilate.

For most cultivars to date, varietally distinctive aromas appear to arise from quantitative rather than qualitative aromatic differences (Ferreira *et al.*, 1998). These are often a combination of compounds that are grape-derived, produced by yeast metabolism, or formed during maturation. In addition, the similarity of some flavorants in 'Chardonnay' to those found in oak (Sefton *et al.*, 1993), may indicate that the expression of the distinctive 'Chardonnay' aroma may be enhanced by fermentation and aging in oak. The effect of glycosidase enzymes on, and the influence of yearly variations to aromatic complexity are illustrated in Fig. 6.19.

As noted, assessing the relative significance of the hundreds of wine aromatic compounds to aroma and bouquet development is difficult. The task is partially simplified by estimating the **odor activity value** (OAV) of each compound. This is its concentration divided by its detection threshold. These and concentration ranges in different types of wines are provided by Francis and Newton (2005). Compounds with an OAV greater than 1 clearly have a greater potential to influence wine fragrance than those possessing OAVs less than unity. Regrettably, there is no direct correlation between OAV and sensory intensity or aromatic significance. In addition, several compounds with OAVs lower than one appear to interact synergistically, or may influence the odor quality of impact compounds (Janusz *et al.*, 2003). Nevertheless, determining the OAV of a wine's constituents is a major step forward in assessing potential sensory significance.

Also of particular interest in detecting odorant impact is **aroma extract dilution analysis** (AEDA) (Grosch, 2001). It involves the addition of compounds,

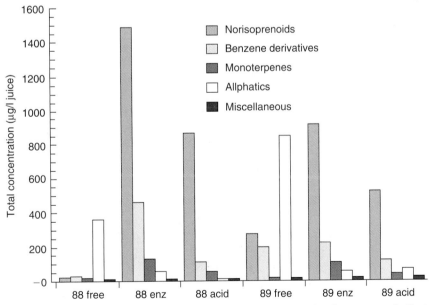

Figure 6.19 Concentrations of five categories of volatiles, observed as free compounds (free) or after release by either glycosidase enzyme (enz) or acid hydrolysis (acid) of precursor fractions from the 1988 and 1989 'Chardonnay' juices. (From Sefton *et al.*, 1993, reproduced by permission)

at concentrations above their threshold values, to aromatically neutral wines or wine-like solutions. AEDA has been used to reconstruct model wines with fragrance characteristics typical of the investigated variety. Such compounds are defined as those with a sensory impact (OAV $\geq$ 1).

The use of techniques such as AEDA may require reassessment of the presumed significance of many compounds to varietal character. These techniques have demonstrated the unequivocal importance of certain flavorants to the varietal aroma of particular cultivars. Examples include varietal aromatics distinguishing 'Grenache' wines (Ferreira *et al.*, 1998); the existence of a distinguishing caramel odor difference between 'Cabernet Sauvignon' and 'Merlot' wines – principally due to 4-hydroxy-2,5-dimethylfuran-3(2H)-one and 4-hydroxy-2(or 5)-ethyl-5 (or 2)-methylfuran-3(2H)-one

(Kotseridis *et al.*, 2000); and the probable methoxypyrazine origin of the pronounced stemmy character in 'Cabernet Sauvignon' and 'Chardonnay' wines fermented with stems (Hashizume and Samata, 1997).

In addition, AEDA has demonstrated that the characteristic flavor of 'Gewürztraminer' and 'Scheurebe' wine can be reproduced, respectively, by the 29 and 42 flavorants that occurred at above threshold values in their respective wines (Guth, 1998). Nevertheless, absence of *cis*-rose oxide and 4-mercapto-4-methyl-pentan-2-one eliminated the distinctive character of 'Gewürztraminer' and 'Scheurebe' model wines, respectively. In contrast, elimination of compounds such as acetaldehyde, β-damascenone and geraniol had little effect. Addition of 13 other aromatics, occurring at below their threshold value in the actual wine, did not affect the aroma characteristics of the model wine.

Appendix 6.1

Conversion Table for Various Hydrometer Scales Used to Measure Sugar Content of Must[a]

°Brix (Balling)	Specific gravity at 20°C	Oechsle[b]	Baumé[c]	°Brix (Balling)	Specific gravity at 20°C	Oechsle[b]	Baumé[c]
0.0	1.00000	0.0	0.00	21.2	1.08823	88	11.8
0.2	1.00078	0	0.1	21.4	1.08913	89	11.9
0.4	1.00155	1	0.2	21.6	1.09003	90	12.0
0.6	1.00233	2	0.3	21.8	1.09093	91	12.1
0.8	1.00311	3	0.45	22.0	1.09183	92	12.2
1.0	1.00389	4	0.55	22.2	1.09273	93	12.3
2.0	1.00779	8	1.1	22.4	1.09364	94	12.45
3.0	1.01172	12	1.7	22.6	1.09454	95	12.55
4.0	1.01567	15	2.2	22.8	1.09545	95	12.7
5.0	1.01965	20	2.8	23.0	1.09636	96	12.8
6.0	1.02366	24	3.3	23.2	1.09727	97	12.9
7.0	1.02770	28	3.9	23.4	1.09818	98	13.0
8.0	1.03176	32	4.4	23.6	1.09909	99	13.1
9.0	1.03586	36	5.0	23.8	1.10000	100	13.2
10.0	1.03998	40	5.6	24.0	1.10092	101	13.3
11.0	1.04413	44	6.1	24.2	1.10193	102	13.45
12.0	1.04831	48	6.7	24.4	1.10275	103	13.55
13.0	1.05252	53	7.2	24.6	1.10367	104	13.7
14.0	1.05667	57	7.8	24.8	1.10459	105	13.8
15.0	1.06104	61	8.3	25.0	1.10551	106	13.9
16.0	1.06534	65	8.9	25.2	1.10643	106	14.0
17.0	1.06968	70	9.4	25.4	1.10736	107	14.1
17.4	1.07142	71	9.7	25.6	1.10828	108	14.2
18.0	1.07404	74	10.0	25.8	1.10921	109	14.3
18.4	1.07580	76	10.2	26.0	1.11014	110	14.45
19.0	1.07844	78	10.55	26.2	1.11106	111	14.55
19.2	1.07932	79	10.65	26.4	1.11200	112	14.65
19.4	1.08021	80	10.8	26.6	1.11293	113	14.85
19.6	1.08110	81	10.9	26.8	1.11386	114	14.9
19.8	1.08198	82	11.0	27.0	1.11480	115	15.0

(Continued)

Appendix 6.1   (*Continued*)

°Brix (Balling)	Specific gravity at 20°C	Oechsle[b]	Baumé[c]	°Brix (Balling)	Specific gravity at 20°C	Oechsle[b]	Baumé[c]
20.0	1.08287	83	11.1	27.2	1.11573	116	15.1
20.2	1.08376	84	11.2	27.4	1.11667	117	15.2
20.4	1.08465	85	11.35	27.6	1.11761	118	15.3
20.6	1.08554	86	11.45	27.8	1.11855	119	15.45
20.8	1.08644	86	11.55	30.0	1.12898	129	16.57
21.0	1.08733	87	11.7				

[a] After *Methods for Analysis of Musts and Wines*, M. A. Amerine and C. S. Ough, Copyright © 1980 John Wiley & Sons, Inc. Reprinted by permission of John Wiley & Sons, Inc.

[b] The approximate sugar content on the Oechsle scale is given by dividing the Oechsle by 4 and subtracting 2.5 from the result. Thus, for a must of 80 Oechsle, 80/4 − 2.5 = 17.5. The approximate prospective percentage alcohol by volume is derived by multiplying the Oechsle by 0.125.

[c] Sugar content on the Baumé scale is approximated by the use of the equation °Brix = 1.8 × Baumé. Reducing sugar is about 2.0 less than the °Brix. The °Brix × 0.52 gives the approximate prospective alcohol production.

Appendix 6.2

Conversion Table for Various Measures of Ethanol Content at 20°C[a]

Alcohol (vol %)	Alcohol (wt %)	Alcohol (g/100 ml)	Proof	Specific gravity	Alcohol (vol %)	Alcohol (wt %)	Alcohol (g/100 ml)	Proof	Specific gravity
0.00	0.00	0.00	0.0	1.00000	11.50	9.27	9.13	23.0	0.98471
0.50	0.40	0.40	1.0	0.99925	12.00	9.679	9.52	24.0	0.98412
1.00	0.795	0.79	2.0	0.99851	12.50	10.08	9.92	25.0	0.98354
1.50	1.19	1.19	3.0	0.99777	13.00	10.487	10.31	26.0	0.98297
2.00	1.593	1.59	4.0	0.99704	13.50	10.90	10.71	27.0	0.98239
2.50	1.99	1.98	5.0	0.99633	14.00	11.317	11.11	28.0	0.98182
3.00	2.392	2.38	6.0	0.99560	14.50	11.72	11.51	29.0	0.98127
3.50	2.80	2.78	7.0	0.99490	15.00	12.138	11.90	30.0	0.98071
4.00	3.194	3.18	8.0	0.99419	15.50	12.54	12.30	31.0	0.98015
4.50	3.60	3.58	9.0	0.99360	16.00	12.961	12.69	32.0	0.98960
5.00	3.998	3.97	10.0	0.99281	16.50	13.37	13.09	33.0	0.97904
5.50	4.40	4.37	11.0	0.99214	17.00	13.786	13.49	34.0	0.97850
6.00	4.804	4.76	12.0	0.99149	17.50	14.19	13.89	35.0	0.97797
6.50	5.21	5.16	13.0	0.99084	18.00	14.612	14.28	36.0	0.97743
7.00	5.612	5.56	14.0	0.99020	18.50	15.02	14.68	37.0	0.97690
7.50	6.02	5.96	15.0	0.98956	19.00	15.440	15.08	38.0	0.97638
8.00	6.422	6.36	16.0	0.98894	19.50	15.84	15.47	39.0	0.97585
8.50	6.83	6.75	17.0	0.98832	20.00	16.269	15.87	40.0	0.97532
9.00	7.234	7.14	18.0	0.89771	20.50	16.67	16.26	41.0	0.97479
9.50	7.64	7.54	19.0	0.89711	21.00	17.100	16.66	42.0	0.97425
10.00	8.047	7.93	20.0	0.98650	21.50	17.51	17.06	43.0	0.97372
10.50	8.45	8.33	21.0	0.98590	22.00	17.993	17.46	44.0	0.97318
11.00	8.862	8.73	22.0	0.98530	22.50	18.34	17.86	45.0	0.97262

[a] Reprinted from Official Methods of Analysis, 11th ed., Appendix 6.2, table 8, 1970. Copyright 1970 by AOAC International.

Appendix 6.3

Interconversion of Acidity Units

TOTAL (TITRATABLE) ACIDITY

Conversion of Total (titratable) Acidity Expressed in Terms of One Acid to Another[a,b]

	Expressed as					
Initial units	Tartaric	Malic	Citric	Lactic	Sulfuric	Acetic
Tartaric	1.000	0.893	0.853	1.200	0.653	0.800
Malic	1.119	1.000	0.955	1.343	0.731	0.896
Citric	1.172	1.047	1.000	1.406	0.766	0.667
Lactic	0.833	0.744	0.711	1.000	0.544	0.677
Sulfuric	1.531	1.367	1.306	1.837	1.000	1.225
Acetic	1.250	1.117	1.067	1.500	0.817	1.000

[a] After *Methods for Analysis of Musts and Wines*, M. A. Amerine and C. S. Ough, Copyright © 1980 John Wiley & Sons, Inc. Reprinted by permission of John Wiley & Sons, Inc.

[b] For example, 5 g/liter (sulfuric) would be equivalent to 7.65 g/liter (tartaric) [5 × 1.531].

Suggested Readings

General Review Articles and Books

Clarke, R. J., and Bakker, J. (2004) *Wine Flavour Chemistry*. Blackwell, Oxford.

Etiévant, P. X. (1991) Wine. In: *Volatile Compounds in Foods and Beverages* (H. Maarse, ed.), pp. 483–546. Dekker, New York.

Margalit, Y. (1997) *Concepts in Wine Chemistry*. The Wine Appreciation Guild, San Francisco, CA.

Nykänen, L. (1986) Formation and occurrence of flavor compounds in wine and distilled alcoholic beverages. *Am. J. Enol. Vitic.* **37**, 84–96.

Reineccius, G. (2004) *Flavor Chemistry and Technology*, 2nd edn. CRC (Taylor and Francis), Boca Raton, FL.

Rowe, D. (ed.) (2006) *Chemistry and Technology of Flavour and Fragrance*. Blackwell, Oxford.

Usseglio-Tomasset, L. (1989) *Chimie Oenologique*. Tec & Doc–Lavoisier, Paris.

Würdig, G., and Woller, R. (1989) *Chemie des Weines*. Verlag Eugen Ulmer, Stuttgart, Germany.

Analytical Techniques

Ebeler, S. E. (2001) Analytical chemistry: unlocking the secrets of wine flavor. *Food Rev. Int.* **17**, 45–64.

Linskens, H. F., and Jackson, J. F. (eds.) (1988) *Wine Analysis*. Springer-Verlag, Berlin.

Ough, C. S., and Amerine, M. A. (1988) *Methods for Analysis of Musts and Wines*. John Wiley, New York.

Zoecklein, B., Fugelsang, K. C., Gump, B. H., and Nury, F. S. (1995) *Wine Analysis and Production*. Chapman & Hall, New York.

Acids and Amino Acids

Ough, C. S. (1988) Acids and amino acids in grapes and wine. In: *Wine Analysis* (H. F. Linskens and J. F. Jackson, eds.), pp. 92–146. Springer-Verlag, Berlin.

Spayd, S. E., and Andersen-Bagge, J. (1996) Free amino acid composition of grape juice from 12 *Vitis vinifera* cultivars in Washington. *Am. J. Enol. Vitic.* **47**, 389–402.

Volschenk, H., van Vuuren, H. J. J., and Viljoen-Bloom, M. (2006) Malic acid in wine: origin, function and metabolism during vinification. *S. Afr. J. Enol. Vitic.* **27**, 123–136.

Alcohols

Scanes, K. T., Hohmann, S., and Prior, B. A. (1998) Glycerol production by the yeast *Saccharomyces cerevisiae* and its relevance to wine: A review. *S. Afr. J. Enol. Vitic.* **19**, 17–24.

Sponholz, W. R. (1988) Alcohols derived from sugars and other sources and fullbodiedness of wines. In: *Wine Analysis* (H. F. Linskens and J. F. Jackson, eds.), pp. 147–172. Springer-Verlag, Berlin.

Carbonyls

Bartowsky, E. J., and Henschke, P. A. (2004) The 'buttery' attribute of wine – diacetyl – desirability, spoilage and beyond. *Int. J. Food Microbiol.* **96**, 235–252.

de Revel, G., Pripis-Nicolau, L., Barbe, J.-C., and Bertrand, A. (2000) The detection of α-dicarbonyl compounds in wine by formation of quinoxaline derivatives. *J. Sci. Food Agric.* **80**, 102–108.

Esters

Nykänen, L. (1986) Formation and occurrence of flavor compounds in wine and distilled alcoholic beverages. *Am. J. Enol. Vitic.* **37**, 89–96.

Ramey, D. D., and Ough, C. S. (1980) Volatile ester hydrolysis or formation during storage of model solutions and wines. *J. Agric. Food Chem.* **28**, 928–934.

Oxygen

du Toit, W. J., Marais, J., Pretorius, I. S., and du Toit, M. (2006) Oxygen in must and wine: A review. *S. Afr. J. Enol. Vitic.* **27**, 76–94.

Phenolics

Andersen, O. M., and Markham, K. R. (2006) *Flavonoids: Chemistry, Biochemistry and Applications.* CRC Press, Boca Raton, FL.

Boulton, R. B. (2001) The copigmentation of anthocyanins and its role in the color of red wine. A critical review. *Am. J. Enol. Vitic.* **52**, 67–87.

Cheynier, V., Remy, S., and Fulcrand, H. (2000) Mechanisms of anthocyanin and tannin changes during winemaking and aging. In: *Proceedings of the ASEV 50th Anniversary Annual Meeting,* Seattle, Wash. (J. M. Rantz, ed.), pp. 337–344. American Society of Enology and Viticulture, Davis, CA.

Dixon, R. A., Xie, D.-Y., and Sharma, S. B. (2005) Proanthocyanidins – a final frontier in flavonoid research? *New Phytologist* **165**, 9–28.

Ferreira, D., and Slade, D. (2002) Oligomeric proanthocyanidins: Naturally occurring O-heterocycles. *Nat. Prod. Rept.* **19**, 517–541.

Harbertson, J., Downey, M., and Spayd, S. (eds.) (2006) Papers presented at the ASEV 56 Annual Meeting Phenolic Symposium. *Am. J. Enol. Vitic.* **57**, 239–313.

Herderich, S., and Smith, P. A. (2005) Analysis of grape and wine tannins: Methods, applications and challenges. *Aust. J. Grape Wine Res.* **11**, 205–214.

Koes, R., Verweij, W., and Quattrocchio, F. (2005) Flavonoids: a colorful model for the regulation and evolution of biochemical pathways. *Trends Plant Sci.* **10**, 236–242.

Rivas-Gonzalo, J., and Santos-Buelga, C. (2001) Understanding the colour of red wines: from anthocyanins to complex pigments. In: *Polyphenols, Wine and Health* (C. Chèze, J. Vercauteren, and R. Verpoorte, eds.), pp. 99–121. Kluwer Academic Publishers, Dordrecht, Netherlands.

Singleton, V. L., and Cilliers, J. J. L. (1995) Phenolic browning: A perspective from grapes and wine research. In: *Enzymatic Browning and Its Prevention* (C. Y. Lee and J. R. Whitaker, eds.), pp. 23–48. ACS Symposium Series, No. 600. American Chemical Society, Washington, DC.

Waterhouse, A. L., and Kennedy, J. A. (eds.) (2004) *Red Wine Color. Revealing the Mysteries.* ACS Symposium Series, No. 886. American Chemical Society Publication, Washington DC.

Pyrazines

Allen, M. S., Lacey, M. J., and Boyd, S. J. (1996) Methoxypyrazines of grapes and wines – differences of origin and behaviour. In: *Proceedings of the 9th Australian Wine Industry Technical Conference* (C. S. Stockley *et al.*, eds.), pp. 83–86. Winetitles, Adelaide, Australia.

Sulfur-containing Compounds

Maujean, A. (2001) The chemistry of sulfur in musts and wines. *J. Int. Sci. Vigne Vin* **35**, 171–194.

Rauhut, D. (1993) Yeasts – production of sulfur compounds. In: *Wine Microbiology and Biotechnology* (G. H. Fleet, ed.), pp. 183–223. Gordon and Breach Science, Luxembourg.

Vermeulen, C., Gijs, L., and Collin, S. (2005) Sensorial contribution and formation pathways of thiols in food: A review. *Food Rev. Int.* **21**, 69–127.

Sulfur dioxide

Danilewicz, J. C. (2007) Interaction of sulfur dioxide, polyphenols, and oxygen in a wine model system: Central role of iron and copper. *Am. J. Enol. Vitic.* **58**, 53–60.

Romano, P., and Suzzi, G. (1993) Sulfur dioxide and wine microorganisms. In: *Wine Microbiology and Biotechnology* (G. H. Fleet, ed.), pp. 374–393. Gordon and Breach Science, Luxembourg.

Rose, A. H. (1993) Sulfur dioxide and other preservatives. *J. Wine Res.* **4**, 43–47.

Terpenes

Mateo, J. J., and Jimenez, M. (2000) Monoterpenes in grape juice and wines. *J. Chromatography A* **881**, 557–567.

Strauss, C. R., Wilson, B., Gooley, P. R., and Williams, P. J. (1986) The role of monoterpenes in grape and wine flavor – a review. In: *Biogeneration of Aroma Compounds* (T. H. Parliament and R. B. Croteau, eds.), pp. 222–242. ACS Symposium Series No. 317. American Chemical Society, Washington, DC.

Varietal Aromas

Francis, I. L., and Newton, J. L. (2005) Determining wine aroma from compositional data. *Aust. J. Grape Wine Res.* **11**, 114–126.

Grosch, W. (2001) Evaluation of the key odorants of foods by dilution experiments, aroma models and omission. *Chem. Senses* **26**, 533–545.

Leland, J. V., Scheiberle, P., Buettner, A., and Acree, T. E. (eds.) (2005) *Gas Chromatography-Olfactometry. The State of the Art.* ACS Symposium Series, No. 782. Oxford University Press, Oxford, UK.

Strauss, C. R., Wilson, B., and Williams, P. J. (1987) Flavour of non-muscat varieties. In: *Proceedings of the 6th Australian Wine Industry Technical Conference* (T. Lee, ed.), pp. 117–120. Australian Industrial Publishers, Adelaide, Australia.

Waterhouse, A. L., and Ebeler, S. E. (eds.) (1999) *Chemistry of Wine Flavor.* ACS Symposium Series No. 714, American Chemical Society Publication, Washington, DC.

Williams, P. J., Strauss, C. R., Aryan, A. P., and Wilson, B. (1987) Grape flavour – a review of some pre- and postharvest influences. In: *Proceedings of the 6th Australian Wine Industry Technical Conference* (T. Lee, ed.), pp. 111–116. Australian Industrial Publishers, Adelaide, Australia.

References

Abbott, N. A., Coombe, B. G., and Williams, P. J. (1991) The contribution of hydrolyzed flavor precursors to quality differences in Shiraz juice and wines, an investigation by sensory descriptive analysis. *Am. J. Enol. Vitic.* **42**, 167–174.

Acree, T. E., Braell, P. A., and Butts, R. M. (1981) The presence of damascenone in cultivars of *Vitis vinifera* (Linnaeus), *rotundifolia* (Michaux), and *labruscana* (Bailey). *J. Agric. Food Chem.* **29**, 688–690.

Acree, T. E., Lavin, E. H., Nishida, R., and Watanabe, S. (1990) *o*-Aminoacetophenone, the foxy smelling component of Labruscana grapes. In: *Flavour Science and Technology* (Y. Bessiere and A. F. Thomas, eds.), pp. 49–52. 6th Weurman Flavour Research Symposium, John Wiley, Chichester, UK.

Adams, D.O., and Liyanage, C. (1993) Glutathione increases in grape berries at the onset of ripening. *Am. J. Enol. Vitic.* **44**, 333–338.

Allen, M. S., and Lacey, M. J. (1993) Methoxypyrazine grape flavor: Influence of climate, cultivar and viticulture. *Wein Wiss.* **48**, 211–213.

Allen, M. S., Lacey, M. J., and Boyd, S. J. (1996) Methoxypyrazines of grapes and wines – differences of origin and behaviour. In: *Proceedings of the 9th Australian Wine Industry Technical Conference*. (S. Stockley *et al.*, eds.), pp. 83–86. Winetitles, Adelaide, Australia.

Amerine, M. A., Berg, H. W., Kunkee, R. E., Ough, C. S., Singleton, V. L., and Webb, A. D. (1980) *Technology of Wine Making*. Avi, Westport, CT.

Amerine, M. A., and Ough, C. S. (1980) *Methods for Analysis of Musts and Wines*. John Wiley, New York.

Amerine, M. A., and Roessler, E. B. (1983) *Wines, Their Sensory Evaluation*. Freeman, New York.

Anonymous. (1986) The history of wine: Sulfurous acid – used in wineries for 500 years. *German Wine Rev.* 2, 16–18.

AOAC. (1970) *Official Methods of Analysis*, 11th edn. Association of Official Analytical Chemists, Washington, DC.

Asen, S., and Jurd, L. (1967) The constitution of a crystalline blue cornflower pigment. *Phytochemistry* 6, 577–584.

Asenstorfer, R. E., Hayasaka, Y., and Jones, G. P. (2001) Isolation and structures of oligomeric wine pigments by bisulfite-mediated ion-exchange chromatography. *J. Agric. Food Chem.* 49, 5957–5963.

Atanasova, V., Fulcrand, H., Cheynier, V., and Moutounet, M. (2002) Effect of oxygenation on polyphenol changes occurring in the course of wine-making. *Anal. Chim. Acta* 458, 15–27.

Bakker, J., and Timberlake, C. F. (1997) Isolation, identification, and characterization of new color-stable anthocyanins occurring in some red wines. *J. Agric. Food Chem.* 45, 35–43.

Bakker, J., Picinelli, A., and Bridle, P. (1993) Model wine solutions: Colour and composition changes during aging. *Vitis* 32, 111–118.

Barbe, J.-C., de Revel, G., and Bertrand, A. (2002) Gluconic acid, its lactones, and SO_2 binding phenomena in musts from botrytized grapes. *J. Agric. Food Chem.* 50, 6408–6412.

Bartowsky, E. J., Francis, I. L., Vellon, J. R., and Henschke, P. A. (2002) Is buttery aroma perception in wines predictable from the diacetyl concentration? *J. Grape Wine Res.* 8, 180–185.

Bartowsky, E. J., and Henschke, P. A. (2000) Management of malolactic fermentation for the 'buttery' diacetyl flavour in wine. *Aust. Grapegrower Winemaker* 438a, 58, 60–62, 64–67.

Beech, F. W., Burroughs, L. F., Timberlake, C. F., and Whiting, G. C. (1979) Progrès recent sur l'aspect chimique et l'action antimicrobienne de l'anhydride sulfureux. *Bull. O. I. V.* 52: 1001–1022.

Beech, F. W., and Thomas, S. (1985) Action antimicrobienne de l'anhydride sulfureux. *Bull. O.I.V.* 58, 564–579.

Blanchard, L., Darriet, P., and Dubourdieu, D. (2004) Reactivity of 3-mercaptohexanol in red wine: impact of oxygen, phenolic fractions, and sulfur dioxide. *Am. J. Enol. Vitic.* 55, 115–120.

Bloomfield, D. G., Heatherbell, D. A., and Pour Nikfardjam, M. S. (2003) Effect of *p*-coumaric acid on the color in red wine. *Mitt. Klosterneuburg* 53, 195–199.

Bock, G., Benda, I., and Schreier, P. (1988) Microbial transformation of geraniol and nerol by *Botrytis cinerea*. *Appl. Microbiol. Biotechnol.* 27, 351–357.

Boison, J., and Tomlinson, R. H. (1988) An investigation of the volatile composition of *Vitis labrusca* grape must and wines, II. The identification of N-(N-hydroxy-N-methyl-γ-aminobutyryl)glycin in native North American grape varieties. *Can. J. Spectrosc.* 33, 35–38.

Boison, J. O. K., and Tomlinson, R. H. (1990) New sensitive method for the examination of the volatile flavor fraction of Cabernet Sauvignon wines. *J. Chromatogr.* 522, 315–328.

Boss, P. K., Davies, C., and Robinson, S. P. (1996) Anthocyanin composition and anthocyanin pathway gene expression in grapevine sports differing in berry skin colour. *Aust J. Grape Wine Res.* 2, 163–170.

Boulton, R. B. (2001) The copigmentation of anthocyanins and its role in the color of red wine. A critical review. *Am. J. Enol. Vitic.* 52, 67–87.

Bourzeix, M., Heredia, N., and Kovač, V. (1983) Richesse de différents cépages en composés phénoliques totaux et en anthocyanes. *Prog. Agric. Vitic.* 100, 421–428.

Bourzeix, M., and Kovač, V. (1989) Mise au point, procyanidines ou proanthocyanidols? *Bull. O.I.V.* 62, 167–175.

Bourzeix, M., Weyland, D., Heredia, N., and Desfeux, N. (1986) Etude des catéchines et des procyanidols de la grappe de raisin, du vin et d'autres dérivés de la vigne. *Bull. O.I.V.* 59, 1171–1254.

Brouillard, R., and Dangles, O. (1994) Anthocyanin molecular interactions, the first step in the formation of new pigments during wine aging? *Food Chem.* 51, 365–372.

Brown, R. C., Sefton, M. A., Taylor, D. K., and Elsey, G. M. (2006) An odour detection threshold determination of all four possible stereoisomers of oak lactone in a white and a red wine. *Aust. Grape Wine Res.* 12, 107–114.

Burton, H. S., McWeeny, D. J., and Biltcliffe, D. O. (1963) Nonenzymic browning: The role of unsaturated carbonyl compounds as intermediates and of SO_2 as an inhibitor of browning. *J. Sci. Food Agric.* 14, 911–920.

Cacho, J., Castells, J. E., Esteban, A., Laguna, B., and Sagristá, N. (1995) Iron, copper, and manganese influence on wine oxidation. *Am. J. Enol. Vitic.* 46, 380–384.

Casalone, E., Colella, C. M., Ricci, F., and Polsinelli, M. (1989) Isolation and characterization of *Saccharomyces cerevisiae* mutants resistant to sulphite. *Yeast* 5, S287–291.

Casey, J. A. (1992) Sulfur dioxide levels in bottled wine. *Aust. Grapegrower Winemaker* 348, 21–24.

Charpentier, C. (2000) Yeast autolysis and yeast macromolecules? Their contribution to wine flavor and stability. In: *Proceedings of the ASEV 50th Anniversary Annual Meeting*, Seattle, Washington June 19–23, 2000, pp. 271–277. American Society for Enology and Viticulture, Davis, CA.

Charpentier, N., and Maujean, A. (1981) Sunlight flavours in champagne wines. In: *Flavour '81. Proceedings of the 3rd Weurman Symposium* (P. Schreier, ed.), pp. 609–615. de Gruyter, Berlin.

Chatonnet, P., Boidron, J. N., and Pons, M. (1990) Élevage des vins rouges en fûts de chêne, Évolution de certains composés volatils et leur impact arômatique. *Sci. Aliments* 10, 565–587.

Chatonnet, P., Dubourdieu, D., Boidron, J.-N., and Pons, M. (1992) The origins of ethylphenols in wine. *J. Sci. Food Agric.* 60, 165–178.

Chatonnet, P., Dubourdieu, D., and Boidron, J. N. (1995) The influence of *Brettanomyces/Dekkera* sp. yeasts and lactic acid bacteria on the ethylphenol content of red wines. *Am. J. Enol. Vitic.* 46, 463–468.

Chen, E. C.-H. (1978) The relative contribution of Ehrlich and biosynthetic pathways to the formation of fusel alcohols. *J. Am. Soc. Brew. Chem.* 35, 39–43.

Cheynier, V., and Ricardo da Silva, J. M. R. (1991) Oxidation of grape procyanidins in model solutions containing *trans*-caffeoyltartaric acid and polyphenol oxidase. *J. Agric. Food Chem.* 39, 1047–1049.

Cheynier, V., Remy, S., and Fulcrand, H. (2000) Mechanisms of anthocyanin and tannin changes during winemaking and aging. In: *Proceedings of the ASEV 50th Anniversary Annual Meeting*, Seattle, WA (J. M. Rantz, ed.), pp. 337–344. American Society of Enology and Viticulture, Davis, CA.

Cilliers, J. J. L., and Singleton, V. L. (1991) Characterization of the products of nonenzymic autoxidative phenolic reactions in a caffeic acid model system. *J. Agric. Food Chem.* 39, 1298–1303.

Ciolfi, G., Garofolo, A., and Di Stefano, R. (1995) Identification of some *o*-aminophenones as secondary metabolites of *Saccharomyces cerevisiae*. *Vitis* 34, 195–196.

Coton, E., Rollan, G., Bertrand, A., and Lonvaud-Funel, A. (1998) Histamine-producing lactic acid bacteria in wines: Early detection, frequency, and distribution. *Am. J. Enol. Vitic.* 49, 199–204.

Courtis, K., Todd, B., and Zhao, J. (1998) The potential role of nucleotides in wine flavour. *Aust. Grapegrower Winemaker* **409**, 31–33.

Cozzolino, D., Parker, M., Dambergs, R. G., Herderich, M., and Gishen, M. (2006) Chemometrics and visible – near infrared spectroscopic monitoring of red wine fermentation in a pilot scale. *Biotechnol. Bioeng.* **95**, 1101–1107.

Crippen, D. D., Jr., and Morrison, J. C. (1986) The effects of sun exposure on the phenolic content of Cabernet Sauvignon berries during development. *Am. J. Enol. Vitic.* **37**, 243–247.

Cutzach, I., Chatonnet, P., and Dubourdieu, D. (1998). Rôle du sotolon dans l'arôme des vins doux naturels. Influence des conditions d'élevage et de vieillissement. *J. Int. Sci. Vigne Vin.* **32**, 223–233.

Dallas, C., Ricardo-da-Silva, J. M., and Laureano, O. (1996) Products formed in model wine solutions involving anthocyanins, procyanidin B_2, and acetaldehyde. *J. Agric. Food Chem.* **44**, 2402–2407.

D'Angelo, M., Onori, G., and Santucci, A. (1994) Self-association of monohydric alcohols in water: compressibility and infrared absorption measurements. *J. Chem. Phys.* **100**, 3107.

Danilewicz, J. C. (2003) Review of reactions mechanisms of oxygen and proposed intermediate reduction products in wine: central role of iron and copper. *Am. J. Enol. Vitic.* **54**, 73–85.

Danilewicz, J. C. (2007) Interaction of sulfur dioxide, polyphenols, and oxygen in a wine model system: Central role of iron and copper. *Am. J. Enol. Vitic.* **58**, 53–60.

Darias-Martín, J., Martín-Luis, B., Carrillo-López, M., Lamuela-Raventós, R., Díaz-Romero, C., and Boulton, R. (2002) Effect of caffeic acid on the color of red wine. *J. Agric. Food Chem.* **50**, 2062–2067.

Darriet, P., Bouchilloux, P., Poupot, C., Bugaret, Y., Clerjeau, M., Sauris, P., Medina, B., and Dubourdieu, D. (2001) Effects of copper fungicide spraying on volatile thiols of the varietal aroma of Sauvignon blanc, Cabernet Sauvignon and Merlot wines. *Vitis* **40**, 93–99.

Darriet, P., Pons, M., Henry, R., Dumont, O., Findeling, V., Cartolaro, P., Calonnec, A., and Dubourdieu, D. (2002) Impact odorants contributing to the fungus type aroma from grape berries contaminated by powdery mildew (*Uncinula necator*): Incidence of enzymatic activities of the yeast *Saccharomyces cerevisiae*. *J. Agric. Food Chem.* **50**, 3277–3282.

de Beer, D., Joubert, E., Gelderblom, W. C. A., and Manley, M. (2002) Phenolic compounds: A review of their possible role as *in vivo* antioxidants of wine. *S. Afr. J. Enol. Vitic.* **23**, 48–61.

de Freitas, V. A. P., Glories, Y., and Monique, A. (2000) Developmental changes of procyanidins in grapes of red *Vitis vinifera* varieties and their composition in respective wines. *Am. J. Enol. Vitic.* **51**, 397–403.

de Mora, S. J., Eschenbruch, R., Knowles, S. J., and Spedding, D. J. (1986) The formation of dimethyl sulfide during fermentation using a wine yeast. *Food Microbiol.* **3**, 27–32.

Dewey, F. M., Meyer, U., and Danks, C. (2005) Rapid immunoassays for stable Botrytis antigens in pre- and postsymptomatic grape berries, grape juice, and wines. *Am. J. Enol. Vitic.* **56**, 302A–303A.

Díaz-Maroto, M. C., Schneider, R., and Baumes, R. (2005) Formation pathways of ethyl esters of branched short-chain fatty acids during wine aging. *J. Agric. Food Chem.* **53**, 3503–3509.

Dittrich, H. H., and Barth, A. (1992) Galactose und Arabinose in Mosten und Weinen der Auslese-Gruppe. *Wein Wiss.* **47**, 129–131.

Dittrich, H. H., Sponholz, W. R., and Kast, W. (1974) Vergleichende Untersuchungen von Mosten und Weinen aus gesunden und aus *Botrytis*-infizierten Traubenbeeren, I. Säurestoffwechsel, Zuckerst offwechselprodukte. Leucoanthocyangehalte. *Vitis* **13**, 36–49.

Doores, S. (1983) Organic Acids. In: *Antimicrobials in Foods* (A. L. Branden and P. M. Davidson, eds.), pp. 75–99. Marcel Dekker, New York.

Dourtoglou, V. G., Yannovits, N. G., Tychopoulos, V. G., and Vamvakis, M. M. (1994) Effect of storage under CO_2 atmosphere on the volatile, amino acid and pigment constituents in the red grape (*Vitis vinifera* L. var Agiorgitiko). *J. Agric. Food Chem.* **42**, 338–344.

Downey, M. O., Harvey, J. S., and Robinson, S. P. (2003a) Analysis of tannins in seeds and skins of Shiraz grapes throughout berry development. *Aust J. Grape Wine Res.* **9**, 15–27.

Downey, M. O., Harvey, J. S., and Robinson, S. P. (2003b) Synthesis of flavonols and expression of flavonol synthase genes in the developing grape berries of Shiraz and Chardonnay (*Vitis vinifera* L.). *Aust J. Grape Wine Res.* **9**, 110–121.

Drawert, F. (1970) Causes déterinant l'amertume de certains vins blancs. *Bull. O.I.V.* **43**, 19–27.

Dubois, P. (1983) Volatile phenols in wines. In: *Flavour of Distilled Beverages* (J. R. Piggott, ed.), pp. 110–119. Ellis Horwood, Chichester, UK.

Dubois, P., and Rigaud, J. (1981) A propos de goût de bouchon. *Vignes Vins* **301**, 48–49.

Dubourdieu, D., Grassin, C., Deruche, C., and Ribéreau-Gayon, P. (1984) Mise au point d'une mesure rapide de l'activité laccase dans le moûts et dans les vins par la méthode à la syringaldazine. Application à l'appréciation de l'état sanitaire des vendages. *Connaiss. Vigne Vin* **18**, 237–252.

Dugelay, I., Gunata, Z., Sapis, J.-C., Baumes, R., and Bayonove, C. (1993) Role of cinnamoyl esterase activities from enzyme preparations on the formation of volatile phenols during winemaking. *J. Agric. Food. Chem.* **41**, 2092–2096.

du Plessis, C. S., and Augustyn, O. P. H. (1981) Initial study on the guava aroma of Chenin blanc and Colombar wines. *S. Afr. J. Enol. Vitic.* **2**, 101–103.

Duran, K., Korteland, J., Beemer, F. A., Heiden, C. V. D., de Bree, P. K., Brink, M., and Wadman, S. K. (1979) Variability of sulfituria: Combined deficiency of sulfite oxidase and xanthine oxidase. In: *Models for the Study of Inborn Errors of Metabolism* (F. A. Hommes, ed.), pp. 103–107. Elsevier, Amsterdam.

Eglinton, J. M., Buckingham, L., and Henschke, P. A. (1993) Increased volatile acidity of white wines by chemical vitamin mixtures is grape juice dependent. *Proceedings of the 8th Australian Wine Industry Technical Conference*, 25–29 Oct., 1992, Melbourne, Victoria (C. S. Stockley, P. A. Johnstone, P. A. Leske, and T. H. Lee, eds.), pp. 197–198. Winetitles, Adelaide, Australia.

Eglinton, J., Griesser, M., Henschke, P., Kwiatkowski, M., Parker, M., and Herderich, M. (2004) Yeast-mediated formation of pigmented polymers in red wine. In: *Red Wine Color. Revealing the Mysteries* (A. L. Waterhouse and J. A. Kennedy, eds.), pp. 7–21. ACS Symposium Series, No. 886. American Chemical Society, Washington, DC.

Embs, R. J., and Markakis, P. (1965) The mechanism of sulfite inhibition of browning caused by polyphenol oxidase. *J. Food Sci.* **30**, 753–758.

Eschenbruch, R. (1974) Sulfite and sulfide formation during wine making. A review. *Am. J. Enol. Vitic.* **25**, 157–161.

Eschenbruch, R., de Mora, S. J., Knowles, S. J., Leonard, W.K., Forrester, T., and Spedding, D.J. (1986) The formation of volatile sulphur compounds in unclarified grape juice. *Vitis* **25**, 53–57.

Etiévant, P. X. (1991) Wine. In: *Volatile Compounds in Foods and Beverages* (H. Maarse, ed.), pp. 483–546. Marcel Dekker, New York.

Etiévant, P. X., Issanchou, S. N., and Bayonove, C. L. (1983). The flavour of Muscat wine, the sensory contribution of some volatile compounds. *J. Sci. Food Agric.* **34**, 497–504.

Evelyn, J. (1664) Pomona, or an appendix concerning fruit trees. In: *Relation to Ciders the Making and Several Ways of Ordering It.* Supplement: Aphorism Concerning Cider by Beale (p. 24). Martyn & Allestry, London.

Falcone, F. (1991) Migration of lead into alcoholic beverages during storage in lead crystal decanters. *J. Food Protect.* **54**, 378–380.

Fang, Y., and Qian, M. (2005) Aroma compounds in Oregon Pinot Noir wine determined by aroma extract dilution analysis (AEDA) *Flavour Fragr. J.* **20**, 22–29.

FAO/WHO Joint Expert Committee on Food Additives. (1974) *Toxicological Evaluation of Certain Food Additives with a Review of General Principles and of Specifications.* 17th Report. Food and Agriculture Organization, Rome.

Ferreira, V., Fernández, P., Gracia, J. P., and Cacho, J. F. (1995) Identification of volatile constituents in wines from *Vitis vinifera* var. Vidadillo and sensory contribution of the different wine flavour fractions. *J. Sci. Food. Agric.* **69**, 299–310.

Ferreira, V., Fernández, P., Peña, C., Escudero, A., and Cacho, J. (1995) Investigation on the role played by fermentation esters in the aroma of young Spanish wines by multivariate analysis. *J. Sci. Food Agric.* **67**, 381–392.

Ferreira, V., López, R., Escudero, A., and Cacho, J. F. (1998) The aroma of Grenache red wine: Hierarchy and nature of its main odorants. *J. Sci. Food Agric.* **77**, 259–267.

Ferreira, V., Peña, C., Escudero, C. L., Fernández, P., and Cacho, J. (1993) Method for the HPLC prefractionation of wine flavour extracts. Part II – Sensory aspects. Profiling wine aroma. In: *Progress in Food Fermentation* (C. Benedito *et al.*, eds.), Vol. 2, pp. 69–74. IATA, CSIC, Valencia, Spain.

Francia-Aricha, E. M., Rivas-Gonzalo, J. C., and Santos-Buelga, C. (1998) Effect of malvidin-3-monoglucoside on the browning of monomeric and dimeric flavanols. *Z. Lebensm. Unters. Forsch.* **207**, 223–228.

Francis, I. L., and Newton, J. L. (2005) Determining wine aroma from compositional data. *Aust. J. Grape Wine Res.* **11**, 114–126.

Francis, I. L., Gawel, R., Iland, P. G., Vidal, S., Cheynier, V., Guyot, S., Kwiatkowski, M. J., and Waters, E. J. (2002) Characterising mouth-feel properties of red wine. In: *11th Australian Wine Industry Technical* Conference, Oct. 7–11, 2001, Adelaide, South Australia (R. J. Blair, P. J. Williams and P. B. Høj, eds.), pp. 123–127. Winetitles, Adelaide, Australia.

Frankel, E. N., Waterhouse, A. L., and Teissedre, P. L. 1995. Principal phenolic phytochemicals in selected California wines and their antioxidant activity in inhibiting oxidation of human low-density lipoproteins. *J. Agric. Food Chem.* **43**, 890–894.

Fulcrand, H., Benabdeljalil, C., Rigaud, J., Chenyier, V., and Moutounet, M. (1998) A new class of wine pigments generated by reaction between pyruvic acid and grape anthocyanins. *Phytochemistry* **47**, 1401–1407.

Fulcrand, H., Cameira dos Santos, P. J., Sarni-Manchado, P., Cheynier, V., and Favre-Bonvin, J. (1996) Structure of new anthocyanin-derived wine pigments. *J. Chem. Soc. Perkin Trans.* **1**, 735–739.

Fulcrand, H., Cheynier, V., Oszmianski, J., and Moutounet, M. (1997) The oxidized tartaric acid residue as a new bridge potentially competing with acetaldehyde in flavan-3-ol condensation. *Phytochemistry* **46**, 223–227.

Furtado, P., Figueiredo, P., Chaves, H., and Pinar, F. (1993) Photochemical and thermal degradation of anthocyanidins. *J. Photochem. Photobiol. A. Chem.* **75**, 113–118.

Garde-Cerdán, T., Arias-Gil, M., and Ancín-Azpilcueta, C. (2006) Formation of biogenic amines throughout spontaneous and inoculated wine alcoholic fermentations: effect of SO$_2$. *Food Cont.* [in press]

Geßner, M., Köhler, H. J., Christoph, N., Bauer-Christoph, C., Miltenberger, R., and Schmitt, A. (1995) Untypische Alterungsnote im Wein, Beschreibende Verkostung von UTA-Weinen; Beziehungen zwischen Sensorik und chemisch-physikalischen Analysenwerten. *Rebe Wein* **48**, 388–394.

George, N., Clark, A. C., Prenzier, P. D., and Scollary, G. R. (2006) Factors influencing the production and stability of xanthylium cation pigments in a model white wine system. *Aust. J. Grape Wine Res.* **12**, 57–68.

Glories, Y. (1984) La couleur des vins rouges. Mesure, origine et interpretation. *Connaiss. Vigne Vin* **18**, 253–271.

Godden, P., Francis, L., Field, J., Gishen, M., Coulter, A., Valente, P., Høj, P., and Robinson, E. (2001) Wine bottle closures: physical characteristics and effect on composition ans sensory properties of a Semillon wine. 1. Performance up to 200 months post-bottling. *J. Grape Wine Res.* **7**, 64–105.

Goetghebeur, M., Nicolas, M., Blaise, A., Galzy, P., and Brun, S. (1992) Étude sur le rôle et l'origine de la benzyl alcool oxydase responsable du goût d'amande amère des vins. *Bull. O.I.V.* **65**, 345–360.

Goto, T., and Kondo, T. (1991) Structure and molecular stacking of anthocyanins flower colour variation. *Angew. Chem. Int. Ed. Engl.* **30**, 17–33.

Goto, S., and Takamuro, R. (1987) Concentration of dimethyl sulfide in wine after different storage times. *Hakkokogaku* **65**, 53–57.

Grosch, W. (2000) Specificity of the human nose in perceiving food odorants. In: *Frontiers of Flavour Science.* Proceedings 9th Weurman Flavour Research Symposium (P. Schieberle and K.-H. Engel, eds.), pp. 213–219. Deutsche Forschungsanstalt für Lebensmittelchemie, Garching, Germany.

Grosch, W. (2001) Evaluation of the key odorants of foods by dilution experiments, aroma models and omission. *Chem. Senses* **26**, 533–545.

Guerra, C., Glories, Y., and Vivas, N. (1996) Influence of ellagitannins on flavanols/anthocyans/acetaldehyde condensation reactions. *J. Sci. Tech. Tonnellerie* **2**, 97–103.

Guillou, I., Bertrand, A., De Revel, G., and Barbe, J. C. (1997) Occurrence of hydroxypropanedial in certain musts and wines. *J. Agric. Food Chem.* **45**, 3382–3386.

Gulson, B. L., Lee, T. H., Mizon, K. J., Korsch, M. J., and Eschnauer, H. R. (1992) The application of lead isotope ratios to the determine the contribution of the tin-lead to the lead content of wine. *Am. J. Enol. Vitic.* **43**, 180–190.

Günata, Y. Z., Bayonove, C. L., Tapiero, C., and Cardonnier, R. E. (1990) Hydrolysis of grape monoterpenyl β-D-glucosides by various β-glucosidases. *J. Agric. Food Chem.* **38**, 1232–1236.

Guth, H. (1996) Wine lactone – a potent odorant identified for the first time in different wine varieties. In: *Flavour Science* (A. J. Taylor and D. S. Mottram, eds.), pp. 163–167. Royal Society of Chemistry, Cambridge, UK.

Guth, H. (1997) Identification of character impact odorants of different white wine varieties. *J. Agric. Food Chem.* **45**, 3022–3026.

Guth, H. (1998) Comparison of different white wine varieties by instrumental and analyses and sensory studies. In: *Chemistry of Wine Flavor* (L. A. Waterhouse and S. E. Ebeler, eds.), ACS Symposium Series #714. American Chemical Society, Washington, DC.

Guth, T., and Sies, A. (2002) Flavour of wines: towards an understanding by reconstruction experiments and an analysis of ethanol's effect on odour activity of key compounds. In: *Proceedings of 11th Australian Wine Industry Technical Conference* (R. J. Blair, P. J. Williams, and P. B. Høj, eds.), pp. 128–139. Australian Wine Industry Technical Conference Inc., Urrbrae, Australia.

Guyot, S., Vercauteren, J., and Cheynier, V. (1996) Structural determination of colourless and yellow dimers resulting from

(+)-catechin coupling catalysed by grape polyphenoloxidase. *Phytochemistry* **42**, 1279–1288.

Hagerman, A. E., Riedl, K. M., Jones, G. A., Sovik, K. N., Ritchard, N. T., Hartzfeld, P. W., and Riechel, T. L. (1998) High molecular weight plant polyphenolics (tannins) as biological antioxidants. *J. Agric. Food Chem.* **46**, 1887–1892.

Hamatschek, J. (1982) Aromastoffe im Wein und deren Herkunft. *Dragoco Rep.* (*Ger. Ed.*) **27**, 59–71.

Hashizume, K., and Samuta, T. (1997) Green odorants of grape cluster stem and their ability to cause a stemmy flavor. *J. Agric. Food Chem.* **45**, 1333–1337.

Haslam, E. (1981) Vegetable tannins. In: *The Biochemistry of Plants* (E. E. Conn, ed.), Vol. 7, pp. 527–556. Academic Press, New York.

Haslam, E., and Lilley, T. H. (1988) Natural astringency in foods, a molecular interpretation. *Crit. Rev. Food Sci. Nutr.* **27**, 1–40.

Hasnip, S., Caputi, A., Crews, C., and Brereton, P. (2004) Effects of storage time and temperature on the concentration of ethyl carbamate and its precursors in wine. *Food Addit. Contam.* **21**, 1155–1161.

Hayasaka, Y., and Kennedy, J. A. (2003) Mass spectrometric evidence for the formation of pigmented polymers in red wine. *Aust. J. Grape Wine Res.* **9**, 210-220.

Hayasaka, Y., Baldock, G. A., and Pollnitz, A. P. (2005) Contributions of mass spectrometry in the Australian Wine Research Institute to advances in knowledge of grape and wine constituents. *Aust. J. Grape Wine Res.* **11**, 188–204.

Hayasaka, Y., Waters, E. J., Cheynier, V., Herderich, M. J., and Vidal, S. (2003) Characterization of proanthocyanidins in grape seeds using electrospray mass spectrometry. *Rapid Commun. Mass Spectrom* **17**, 9–16.

He, J., Santos-Buelga, C., Silva, A. M., Mateus, N., and Freitas, V. D. (2006) Isolation and structural characterization of new anthocyanin derived yellow pigments in aged red wines. *J. Agric. Food Chem.* **54**, 9598–9603.

Heard, S. J., and Fleet, G. H. (1988) The effects of temperature and pH on the growth of yeast species during the fermentation of grape juice. *J. Appl. Bacteriol.* **65**, 23–28.

Heimann, W., Rapp, A., Völter, J., and Knipser, W. (1983) Beitrag zur Entstehung des Korktons in Wein. *Dtsch. Lebensm.-Rundsch.* **79**, 103–107.

Henschke, P. A., and Jiranek, V. (1991) Hydrogen sulfide formation during fermentation, effect of nitrogen composition in model grape musts. In: *Proceedings of the International Symposium on Nitrogen in Grapes and Wine* (J. M. Rantz, ed.), pp. 177–184. American Society of Enology and Viticulture, Davis, CA.

Henschke, P. A., and Ough, C. S. (1991) Urea accumulation in fermenting grape juice. *Am. J. Enol. Vitic.* **42**, 317–321.

Heresztyn, T. (1986) Formation of substituted tetrahydropyridines by species of *Brettanomyces* and *Lactobacillus* isolated from mousy wines. *Am. J. Enol. Vitic.* **37**, 127–131.

Hötzel, D., Muskat, E., Bitsch, I., Aign, W., Althoff, J.-D., and Cremer, H. D. (1969) Thiamin-Mangel und Unbedenklichkeit von Sulfit für den Menschen. *Int. Z. Vitaminforsch.* **39**, 372–383.

Hrazdina, G., Borzell, A. J., and Robinson, W. B. (1970) Studies on the stability of the anthocyanidin-3,5-diglucosides. *Am. J. Enol. Vitic.* **21**, 201–204.

Hrazdina, G., Parsons, G. F., and Mattick, L. R. (1984) Physiological and biochemical events during development and maturation of grape berries. *Am. J. Enol. Vitic.* **35**, 220–227.

Iglesias, J. L. M., Dabiila, F. H., Marino, J. I. M., De Miguel Gorrdillo, C., and Exposito, J. M. (1991) Biochemical aspects of the lipids of *Vitis vinifera* grapes (Macabeo var.) Linoleic and linolenic acids as aromatic precursors. *Nahrung* **35**, 705–710.

Institute of Food Technologists Expert Panel on Food Safety and Nutrition. (1975) Sulfites as food additives. *Food Technol.* **29**, 117–120.

Janusz, A., Capone, D. L., Puglisi, C. J., Perkins, M. V., Elsey, G. M., and Sefton, M. A. (2003) (*E*)-1-(2,3,6-Trimethylphenyl)buta-1,3-diene: A potent grape-derived odorant in wine. *J. Agric. Food Chem.* **51**, 7759–7763.

Jayaraman, A., and van Buren, J. P. (1972) Browning of galacturonic acid in a model system stimulation fruit beverages and white wine. *J. Agric. Food Chem.* **20**, 122–124.

Jeandet, P., Bessis, R., Sbaghi, M., Meunier, P., and Trollat, P. (1995) Resveratrol content of wines of different ages: Relationship with fungal disease pressure in the vineyard. *Am. J. Enol. Vitic.* **4**, 1–4.

Kaufmann, A. (1998) Lead in wine. *Food Addit. Contam.* **15**, 437–445.

Killian, E., and Ough, C. S. (1979) Fermentation esters – formation and retention as affected by fermentation temperature. *Am. J. Enol. Vitic.* **30**, 301–305.

Kodama, S., Suzuki, T., Fujinawa, S., De la Teja, P., and Yotsuzuka, F. (1991) Prevention of ethyl carbamate formation in wine by urea degradation using acid urease. In: *Proceedings of the International Symosium on Nitrogen in Grapes and Wine* (J. M. Rantz, ed.), pp. 270–273. American Society of Enology and Viticulture, Davis, CA.

Kolor, M. K. (1983) Identification of an important new flavor compound in Concord grape, ethyl 3-mercaptopropionate. *J. Agric. Food Chem.* **31**, 1125–1127.

Kotseridis, Y., Beloqui, A. A., Bayonove, C. L., Baumes, R. L., and Bertrand, A. (1999) Effects of selected viticultural and enological factors on levels of 2-methoxy-3-isobutylpyrazine in wines. *J. Int. Sci. Vigne Vin* **33**, 19–23.

Kotseridis, Y., Razungles, A., Bertrand, A., and Baumes, R. (2000) Differentiation of the aromas of Merlot and Cabernet Sauvignon wines using sensory and instrumental analysis. *J. Agric. Food Chem.* **48**, 5383–5388.

Kovać, V. (1979) Étude de l'inactivation des oxidases du raisin par des moyens chimiques. *Bull. O.I.V.* **53**, 809–815.

Kovać, V., Bourzeix, M., Heredia, N., and Ramos, T. (1990) Étude des catéchines et proanthocyanidols de raisins et vins blancs. *Rev. Fr. Oenol.* **125**, 7–15.

Labarbe, B., Cheynier, V., Brossaud, F., Souquet, J. M., and Moutounet, M. (1999) Quantitative fractionation of grape proanthocyanidins according to their degree of polymerization. *J. Agric. Food Chem.* **47**, 2719–2723.

Lacey, M. J., Allen, M. S., Harris, R. L. N., and Brown, W. V. (1991) Methoxypyrazines in Sauvignon blanc grapes and wines. *Am. J. Enol. Vitic.* **42**, 103–108.

Lachance, M.-A., and Pang, W.-M. (1997) Predacious yeasts. *Yeast* **13**, 225–232.

Lamikanra, O. (1997) Changes in organic acid composition during fermentation and aging of Noble muscadine wine. *J. Agric. Food Chem.* **45**, 935–937.

Lamikanra, O., Grimm, C. C., and Inyang, I. D. (1996) Formation and occurrence of flavor components in Noble muscadine wine. *Food Chem.* **56**, 373–376.

Larue, F., Park, M. K., and Caruana, C. (1985) Quelques observations sur les conditions de la formation d'anhydride sulfureux en vinification. *Connaiss. Vigne Vin* **19**, 241–248.

Laurie, V. F., and Waterhouse, A. L. (2006) Oxidation of glycerol in the presence of hydrogen peroxide and iron in model solutions and wine. Potential effects on wine color. *J. Agric. Food Chem.* **54**, 4668–4673.

Lavigne, V., Henry, R., and Dubourdieu, D. (1998) Identification et dosage de composés soufrés intervenant dans l'arôme grillé des vins. *Sci. Alim.* **18**, 175–191.

Lea, A. G. H., and Arnold, G. M. (1978) The phenolics of ciders, bitterness and astringency. *J. Sci. Food Agric.* **29**, 478–483.

Lee, C. Y., and Jaworski, A. W. (1987) Phenolic compounds in white grapes grown in New York. *Am. J. Enol. Vitic.* **38**, 277–281.

Lee, D. F., Swinny, E. E., and Graham P. Jones. (2004) NMR identification of ethyl-linked anthocyanin-flavanol pigments formed in model wine ferments. *Tetrahedr. L.* **48**, 1671–1674.

Lehtonen, P. (1996) Determination of amines and amino acids in wine. A review. *Am. J. Enol. Vitic.* **47**, 127–133.

Leppänen, O. A., Denslow, J., and Ronkainen, P. P. (1980) Determination of thioacetates and some other volatile sulfur compounds in alcoholic beverages. *J. Agric. Food Chem.* **28**, 359–362.

Lopez-Toledano, A., Villaño-Valencia, D., Mayen, M., Merida, J., and Median, M. (2004) Interaction of yeasts with the products resulting from the condensation reaction between (+)-catechin and acetaldehyde. *J. Agric. Food Chem.* **52**, 2376–2381.

Lorrain, B., Ballester, J., Thomas-Danguin, T., Blanquet, J., Meunier, J. M., and Le Fur, Y. (2006) Selection of potential impact odorants and sensory validation of their importance in typical Chardonnay wines. *J. Agric. Food Chem.* **54**, 3973–3981.

Lubbers, S., Voilley, A., Feuillat, M., and Charpontier, C. (1994) Influence of mannoproteins from yeast on the aroma intensity of a model wine. *Lebensm.–Wiss. u. Technol.* **27**, 108–114.

Macheix, J. J., Sapis, J. C., and Fleuriet, A. (1991) Phenolic compounds and polyphenoloxidases in relation to browning grapes and wines. *Crit. Rev. Food Sci. Nutr.* **30**, 441–486.

Marais, J. (1986) Effect of storage time and temperature of the volatile composition and quality of South African *Vitis vinifera* L. cv. Colombar wines. In: *The Shelf Life of Foods and Beverages* (G. Charalambous, ed.), pp. 169–185. Elsevier, Amsterdam.

Marais, J., and Pool, H. J. (1980) Effect of storage time and temperature on the volatile composition and quality of dry white table wines. *Vitis* **19**, 151–164.

Marais, J., van Rooyen, P. C., and du Plessis, C. S. (1979) Objective quality rating of Pinotage wine. *Vitis* **18**, 31–39.

Marais, J., van Wyk, C. J., and Rapp, A. (1992) Effect of sunlight and shade on norisoprenoid levels in maturing Weisser Riesling and Chenin blanc grapes and Weisser Riesling wines. *S. Afr. J. Enol. Vitic.* **13**, 23–32.

Marriott, R. J. (1986) Biogenesis of blackcurrant (*Ribes nigrum*) aroma. In: *Biogeneration of Aromas* (T. H. Parliament and R. Crouteau, eds.), pp. 184–192. ACS Symposium Series 317, American Chemical Society, Washington, DC.

Martin, B., Etiévant, P. X., and Le Quéré, J.-L. (1991) More clues of the occurrence and flavor impact of solerone in wine. *J. Agric. Food Chem.* **39**, 1501–1503.

Martin, B., Etiévant, P. X., Le Quéré, J. L., and Schlich, P. (1992) More clues about sensory impact of Sotolon in some flor sherry wines. *J. Agric. Food Chem.* **40**, 475–478.

Masuda, J., Okawa, E., Nishimura, K., and Yunome, H. (1984) Identification of 4,5-dimethyl-3-hydroxy-2(5H)-furanone (Sotolon) and ethyl 9-hydroxynonanoate in botrytised wine and evaluation of the roles of compounds characteristic of it. *Agric. Biol. Chem.* **48**, 2707–2710.

Mateus, N., Oliveira, J., Haettich-Motto, M., and de Freitas, V. (2004) New family of bluish pyranoanthocyanins. *J. Biomed. Biotechnol.* **5**, 299–305.

Maujean, A., Poinsaut, P., Dantan, H., Brissonnet, F., and Cossiez, E. (1990) Étude de la tenue et de la qualité de mousse des vins effervescents II. Mise au point d'une technique de mesure de la moussabilité, de la tenue et de la stabilité de la mousse des vins effervescents. *Bull. O.I.V.* **63**, 405–427.

McCloskey, L. P., and Yengoyan, L. S. (1981) Analysis of anthocyanins in *Vitis vinifera* wines and red color *versus* aging by HPLC and spectrophotometry. *Am. J. Enol. Vitic.* **32**, 257–261.

McGovern, P. E., and Michel, R. H. (1995) The analytical and archaeological challenge of detecting ancient wine: Two case studies from the Ancient Near East. In: *The Origins and Ancient History of Wine* (P. E. McGovern *et al.*, eds.), pp. 57–65. Gordon and Breach Science, Luxembourg.

McKinnon, A. J., Cattrall, R. W., and Schollary, G. R. (1992) Aluminum in wine – its measurement and identification of major sources. *Am. J. Enol. Vitic.* **43**, 166–170.

Médina, B., Guimberteau, G., and Sudraud, P. (1977) Dosage de plomb dans les vins. Une cause d'enrichissement, les capsules de surbouchage. *Connaiss. Vigne Vin* **11**, 183–193.

Mira de Orduña, R. M., Patchett, M. L., Liu, S.-Q., and Pilone, G. J. (2001) Growth and arginine metabolism of the wine lactic acid bacteria *Lactobacillus buchneri* and *Oenococcus oeni* at different pH values and arginine concentrations. *Appl. Environ. Microbiol.* **67**, 1657–1662.

Mistry, T. V., Cai, Y., Lilley, T. H., and Haslam, E. (1991) Polyphenol interactions. Part 5. Anthocyanin co-pigmentation. *J. Chem. Soc. Perkin Trans.* **2**, 1287–1296.

Moio, L., and Etiévant, P. X. (1995) Ethyl anthranilate, ethyl cinnamate, 2,3-dihydrocinnamate, and methyl anthranilate: Four important odorants identified in Pinot noir wines of Burgundy. *Am. J. Enol. Vitic.* **46**, 392–398.

Morata, A., Calderón, F., González, M. C., Gómez-Cordovés, M. C., and Suárez, J. A. (2007) Formation of the highly stable pyranoanthocyanins (vitisins A and B) in red wines by the addition of pyruvic acid and acetaldehyde. *Food Chem.* **100**, 1144–1152.

Moreira, N., Mendes, F., Pereira, O., Guedes de Pinho, P., Hogg, T., and Vasconcelos, I. (2002) Volatile sulphur compounds in wines related to yeast metabolism and nitrogen composition of grape musts. *Analytica Chimica Acta* **458**, 157–167.

Muller, C. J., Kepner, R. E., and Webb, A. D. (1973) Lactones in wines – A review. *Am. J. Enol. Vitic.* **24**, 5–8.

Munoz, E., and Ingledew, W. M. (1990) Yeast hulls in wine fermentations, a review. *J. Wine Res.* **1**, 197–209.

Murat, M.-L. (2005) Recent findings on rosé wine aromas. Part 1: identifying aromas studying the aromatic potential of grapes and juice. *Aust. NZ Grapegrower Winemaker* **497a**, 64–65, 69, 71, 73–74, 76.

Murat, M.-L., Tominaga, T., and Dubourdieu, D. (2001) Assessing the aromatic potential of Cabernet Sauvignon and Merlot musts used to produce rose wine by assaying the cysteinylated precursor of 3-mercaptohexan-1-ol. *J. Agric. Food Chem.* **49**, 5412–5417.

Murat, M. L., Tominaga, T., Saucier, C., Glories, Y., and Dubourdieu, D. (2003) Effect of anthocyanins on stability of a key-odorous compound, 3-mercaptohexan-1-ol, in Bordeaux rosé wines. *Am. J. Enol. Vitic.* **54**, 135–138.

Noble, A. C., and Bursick, G. F. (1984) The contribution of glycerol to perceived viscosity and sweetness in white wine. *Am. J. Enol. Vitic.* **35**, 110–112.

Ong, B.Y., and Nagel, C. W. (1978) High-pressure liquid chromatographic analysis of hydroxycinnamic acid tartaric acid esters and their glucose esters in *Vitis vinifera*. *J. Chromatogr.* **157**, 345–355.

Ong, P., and Acree, T. E. (1999) Similarities in the aroma chemistry of Gewürztraminer variety wines an lychee (*Litchi chinesis* Sonn.) Fruit. *J. Agric Food Chem.* **47**, 667–670.

Ough, C. S., Stevens, D., Sendovski, T., Huang, Z., and An, A. (1990) Factors contributing to urea formation in commercially fermented wines. *Am. J. Enol. Vitic.* **41**, 68–73.

Packter, N. M. (1980) Biosynthesis of acetate-derived phenols (polyketides)) In: *The Biochemistry of Plants* (P. K. Strumpf, ed.), Vol. 4, pp. 535–570. Academic Press, New York.

Peterson, G. F., Kirrane, M., Hill, N., and Agapito, A. (2000) A comprehensive survey of the total sulfur dioxide concentrations of American wines. *Am. J. Enol. Vitic.* 51, 189–191.

Peynaud, E. (1984) *Knowing and Making Wine*. John Wiley, New York.

Peyrot des Gachons, C., and Kennedy, J. A. (2003) Direct method for determining seed and skin proanthocyanidinidextaction in red wine. *J. Agric. Food Chem.* 51, 5877–5881.

Pilkington, B. J., and Rose, A. H. (1988) Reactions of *Saccharomyces cerevisiae* and *Zygosaccharomyces bailii* to sulphite. *J. Gen. Microbiol.* 134, 2823–2830.

Pissarra, J., Lourenco, S., Gonzalez–Paramas, A.-M., Mateus, N., Santos-Buelga, C., Silva, A. M. S., and de Freitas, V. (2004) Structural characterization of new malvidin 3-glucoside-catechin aryl/alkyl-linked pigments. *J. Agric. Food Chem.* 52, 5519–5526.

Pocock, K. F., Sefton, M. A., and Williams, P .J. (1994) Taste thresholds of phenolic extracts of French and American oakwood: The influence of oak phenols on wine flavor. *Am. J. Enol. Vitic.* 45, 429–434.

Radler, F., and Fäth, K. P. (1991) Histamine and other biogenic amines in wines. In: *Proceedings of the International Symposium on Nitrogen in Grapes and Wine*, Seattle, WA, June, 1991. (J. M. Rantz, ed.), pp. 185–195. American Society of Enology and Viticulture, Davis, CA.

Ramey, D. D., and Ough, C. S. (1980) Volatile ester hydrolysis or formation during storage of model solutions and wines. *J. Agric. Food Chem.* 28, 928–934.

Rapp, A., and Güntert, M. (1986) Changes in aroma substances during the storage of white wines in bottles. In: *The Shelf Life of Foods and Beverages* (G. Charalambous, ed.), pp. 141–167. Elsevier, Amsterdam

Rapp, A., and Mandery, H. (1986) Wine aroma. *Experientia* 42, 873–880.

Rapp, A., and Versini, G. (1996) Vergleichende Untersuchungen zum Gehalt von Methylanthranilat (Foxton) in Weinen von neueren pilzresistenten Rebsorten und *vinifera*–Sorten. *Vitis* 35, 215–216.

Rapp, A., Güntert, M., and Almy, J. (1985) Identification and significance of several sulfur-containing compounds in wine. *Am. J. Enol. Vitic.* 36, 219–221.

Rauhut, D. (1993) Yeasts – production of sulfur compounds. In: *Wine Microbiology and Biotechnology* (G. H. Fleet, ed.), pp. 183–223. Gordon and Breach Science, Luxembourg.

Rauhut, D., Kürbel, H., and Dittrich, H. H. (1993) Sulfur compounds and their influence on wine quality. *Wein Wiss.* 48, 214–218.

Razungles, A., Gunata, Z., Pinatel, S., Baumes, R., and Bayonove, C. (1993) Étude quantitative de composés terpéniques, norisoprénoïds et de leurs précurseurs dans diverses variétés de raisins. *Sci. Aliments* 13, 59–72.

Ribéreau-Gayon, P. (1964) Les composés phénoloques du raisin et du vin. I, II, III. *Ann. Physiol. Veg.* 6, 119–147, 211–242, 259–282.

Ribéreau-Gayon, P. (1988) *Botrytis*, Advantages and disadvantages for producing quality wines. In: *Proceedings of the 2nd International Symposium for Cool Climate Viticulture and Oenology* (R. E. Smart *et al.*, eds.), pp. 319–323. New Zealand Society of Viticulture and Oenology, Auckland, New Zealand.

Ribéreau-Gayon, P., and Glories, Y. (1987) Phenolics in grapes and wines. In: *Proceedings of the 6th Australian Wine Industry Technical Conference* (T. Lee, ed.), pp. 247–256. Australian Industrial Publishers, Adelaide, Australia.

Ribéreau-Gayon, P., Pontallier, P., and Glories, Y. (1983) Some interpretations of colour changes in young red wines during their conservation. *J. Sci. Food Agric.* 34, 505–616.

Ribéreau-Gayon, J., Ribéreau-Gayon, P., and Seguin, G. (1980) *Botrytis cinerea* in enology. In: *The Biology of Botrytis* (J. R. Coley-Smith, K. Verhoeff, and W. R. Jarvis, eds.), pp. 251–274. Academic Press, London.

Rivero-Pérez, M. D., Pérez-Magariño, S., and González-San José, M. L. (2002) Role of melanoidins in sweet wines. *Anal. Chim. Acta* 458, 169–175.

Rizzi, G. (1997) Chemical structure of colored Maillard reaction products. *Food Rev. Int.* 13, 1–28.

Robinson, W. B., Shaulis, N., and Pederson, C. S. (1949) Ripening studies of grapes grown in 1948 for juice manufacture. *Fruit Prod. J. Am. Food Manuf.* 29, 36–37, 54, 62.

Robinson, W. B., Weirs, L. D., Bertino, J. J., and Mattick, L. R. (1966) The relation of anthocyanin composition to color stability of New York State wines. *Am. J. Enol. Vitic.* 17, 178–184.

Rogerson, F. S., Castro, H., Fortunato, N., Azevedo, Z., Macedo, A., and De Freitas, V. A. (2001) Chemicals with sweet aroma descriptors found in Portuguese wines from the Douro region: 2,6,6-trimethylcyclohex-2-ene-1,4-dione and diacetyl. *J. Agric. Food Chem.* 49, 263–269.

Roudet, J., Prudet, S., and Dubos, B. (1992) Relationship between grey mould of grapes and laccase activity in the must. In: *Recent Advances in Botrytis Research: Proceedings of the 10th International Botrytis Symposium* (K. Verhoeff *et al.*, eds.), pp. 83–86. Pudoc, Wageningen, The Netherlands.

Roufet, M., Bayonove, C. L., and Cordonnier, R. E. (1986) Changes in fatty acids from grape lipidic fractions during crushing exposed to air. *Am. J. Enol. Vitic.* 37, 202–205.

Roujou de Boubée, D., Cumsille, A. M., Pons, M., and Dubourdieu, D. (2002) Location of 2-methoxy-3-isobutylpyrazine in Cabernet Sauvignon grape bunches and its extractability during vinification. *Am. J. Enol. Vitic.* 53, 1–5.

Roussis, I. G., Lambropoulos, I., and Papadopoulou, D. (2005) Inhibition of the decline of volatile esters and terpenols during oxidative storage of Muscat-white and Xinomavro-red wine by caffeic acid and N-acetyl-cysteine. *Food Chem.* 93, 485–492.

Salagoïty-Auguste, M., Tricard, C., Marsal, F., and Sudraud, P. (1986) Preliminary investigation for the differentiation of enological tannins according to botanical origin: Determination of gallic acid and its derivatives. *Am. J. Enol. Vitic.* 37, 301–303.

Saucier, C., Pianet, I., Laguerre, M., and Glories, Y. (1998) NMR and molecular modeling: application to wine aging. *J. Chim. Phys.* 95, 357–365.

Scalbert, A. (1991) Antimicrobial properties of tannins. *Phytochemistry* 30, 3875–3883.

Scheibel, T., Bell, S., and Walke, S. (1997) *S. cerevisiae* and sulfur: A unique way to deal with the environment. *FASEB J.* 11, 917–921.

Schopfer, J. F., and Aerny, J. (1985) Le rôle de l'anhydride sulfureux en vinification. *Bull. O.I.V.* 58, 515–535.

Schreier, P. (1982) Volatile constituents in different grape species. In: *Grape and Wine Centennial Symposium Proceedings* (A. D. Webb, ed.), pp. 317–321. University of California, Davis.

Schreier, P., and Drawert, F. (1974) Gaschromatographisch-massenspektometrische Untersuchung flüchtiger Inhaltsstoffe des Weines, V. Alkohole, Hydroxy-Ester, Lactone und andere polare Komponenten des Weinaromas. *Chem. Mikrobiol. Technol. Lebensm.* 3, 154–160.

Schreier, P., and Paroschy, J. H. (1981) Volatile constituents from Concord, Niagara (*Vitis labrusca*) and Elvira (*V. labrusca* x *V. riparia*) grapes. *Can. Inst. Food Sci. Technol. J.* 14, 112–118.

Schumann, F. (1997) Rebsorten und Weinarten im mittelalterlichen Deutschand. *Quel. Forsch. Geschichte Stadt Heinbronn* 9, 221–254; cites Rasch, J. (1582) *Weinbuch: Von Baw/Pfleg und Brauch des Weins ...* Munich, Germany.

Schütz, M., and Kunkee, R. E. (1977) Formation of hydrogen sulfide from elemental sulfur during fermentation by wine yeast. *Am. J. Enol. Vitic.* 28, 137–144.

Schwarz, M., Wabnitz, T. C., and Winterhalter, P. (2003) Pathway leading to the formation of anthocyanin-vinylphenol adducts and related pigments in red wines. *J. Agric. Food Chem.* **51**, 3682–3687.

Sefton, M. A., Skouroumounis, G. K., Massey-Westropp, R. A., and Williams, P. J. (1989) Norisoprenoids in *Vitis vinifera* white wine grapes and the identification of a precursor of damascenone in these fruits. *Aust. J. Chem.* **42**, 2071–2084.

Sefton, M. A., Francis, I. L., and Williams, P. J. (1993) The volatile composition of Chardonnay juices: A study by flavor precursor analysis. *Am. J. Enol. Vitic.* **44**, 359–370.

Segurel, M. A., Razungles, A. J., Riou, C., Salles, M., and Baumes, R. L. (2004) Contribution of dimethyl sulfide to the aroma of Syrah and Grenache noir wines and estimation of its potential in grapes of these varieties. *J. Agric. Food Chem.* **52**, 7084–7093.

Shinohara, T., Shimizu, J., and Shimazu, Y. (1979) Esterification rates of main organic acids in wines. *Agric. Biol. Chem.* **43**, 2351–2358.

Simpson, R. F. (1979) Aroma composition of bottle aged white wine. *Vitis* **18**, 148–154.

Simpson, F. R. (1982) Factors affecting oxidative browning of white wine. *Vitis* **21**, 233–239.

Simpson, R. F., and Miller, G. C. (1984) Aroma composition of Chardonnay wine. *Vitis* **23**, 143–158.

Sims, C. A., and Morris, J. R. (1986) Effects of acetaldehyde and tannins on the color and chemical age of red Muscadine (*Vitis rotundifolia*) wine. *Am. J. Enol. Vitic.* **37**, 163–165.

Singleton, V. L. (1982) Grape and wine phenolics background and prospects. In: *Grape and Wine Centennial Symposium Proceedings* (A. D. Webb, ed.), pp. 215–227. University of California, Davis.

Singleton, V. L. (1987) Oxygen with phenols and related reactions in must, wines and model systems, observations and practical implications. *Am. J. Enol. Vitic.* **38**, 69–77.

Singleton, V. L., and Cilliers, J. J. L. (1995) Phenolic browning: A perspective from grapes and wine research. In: *Enzymatic Browning and Its Prevention* (C. Y. Lee and J. R. Whitaker, eds.), pp. 23–48. ACS Symposium Series No. 600, American Chemical Society, Washington, DC.

Singleton, V. L., and Noble, A. C. (1976) Wine flavour and phenolic substances. In: *Phenolic, Sulfur and Nitrogen Compounds in Food Flavors* (G. Charalambous and A. Katz, eds.), pp. 47–70. ACS Symposium Series No. 26. American Chemical Society, Washington, DC.

Singleton, V. L., and Trousdale, E. (1992) Anthocyanin–tannin interactions explaining differences in polymeric phenols between white and red wines. *Am. J. Enol. Vitic.* **43**, 63–70.

Singleton, V. L., Salgues, M., Zaya, J., and Trousdale, E. (1985) Caftaric acid disappearance and conversion to products of enzymic oxidation in grape must and wine. *Am. J. Enol. Vitic.* **36**, 50–56.

Singleton, V. L., Timberlake, C.F., and Lea, A. G. H. (1978) The phenolic cinnamates of white grapes and wine. *J. Sci. Food Agric.* **29**, 403–410.

Singleton, V. L., Trousdale, E., and Zaya, J. (1979) Oxidation of wines, l. Young white wines periodically exposed to air. *Am. J. Enol. Vitic.* **30**, 49–54.

Sneyd, T. N. (1988) Tin lead capsules. *Aust. Wine Res. Inst. Techn. Rev.* **56**, 1.

Somers, T. C. (1982) Pigment phenomena – from grapes to wine. In: *Grape and Wine Centennial Symposium Proceedings* (A. D. Webb ed.), pp. 254–257. University of California, Davis.

Somers, T. C. (1983) Influence du facteur temps de conservation. *Bull. O.I.V.* **57**, 172–188.

Somers, T. C., and Evans, M. E. (1986) Evolution of red wines, I. Ambient influences on colour composition during early maturation. *Vitis* **25**, 31–39.

Somers, T. C., and Vérette, E. (1988) Phenolic composition of natural wine types. In: *Wine Analysis* (H. F. Linskens and J. F. Jackson, eds.), pp. 219–257. Springer-Verlag, Berlin.

Somers, T. C., and Wescombe, L. F. (1982) Red wine quality, the critical role of SO_2 during vinification and conservation. *Aust. Grapegrower Winemaker* **220**, 1–7.

Sorrentino, F., Voilley, A., and Richon, D. (1986) Activity coefficients of aroma compounds in model food systems. *AIChE J.* **32**, 1988–1993.

Souquet, J. M., Cheynier, V., Brossaud, F., and Moutounet, M. (1996) Polymeric proanthocyanidins from grape skins. *Phytochemistry* **43**, 509–512.

Souquet, J. M., Cheynier, V., and Moutounet, M. (2000) The proanthocyanidins of grape. *Bull. OIV* **73**, 601–609.

Sponholz, W. R. (1988) Alcohols derived from sugars and other sources and fullbodiedness of wines. In: *Wine Analysis* (H. F. Liskens and J. F. Jackson, eds.), pp. 147–172. Springer-Verlag, Berlin.

Sponholz, W. R., and Dittrich, H. H. (1984) Galacturonic, glucuronic, 2- and 5-oxo-gluconic acids in wines, sherries fruit and dessert wines. *Vitis* **23**, 214–224.

Sponholz, W. R., and Hühn, T. (1997) Aging of wine: 1,1,6-Trimethyl-1,2-dihydronaphthalene (TDN) and 2-aminoacetophenone. In: *Proceedings of the 4th International Symposium on Climate Viticulture and Enology*, Rochester (T. Henick-Kling *et al.*, eds.), pp. VI-37–57. State Agricultural Experimental Station, Geneva, NY.

Sponholz, W. R., Kürbel, H., and Dittrich, H. H. (1991) Beiträge zur Bildung von Ethylcarbamat in Wein. *Wein Wiss.* **46**, 11–17.

Stevens, D. F., and Ough, C. S. (1993) Ethyl carbamate formation, reaction of urea and citrulline with ethanol in wine under low to normal temperature conditions. *Am. J. Enol. Vitic.* **44**, 309–312.

Stockley, C. S., Smith, L. H., Guerin, P., Brückbauer, H., Johnstone, R. S., Tiller, K. G., and Lee, T. H. (1997) The relationship between vineyard soil lead concentration and the concentration of lead in grape berries. *Aust. J. Grape Wine Res.* **3**, 133–140.

Strauss, C. R., Gooley, P. R., Wilson, B., and Williams, P. J. (1987) Application of droplet countercurrent chromatography to the analysis of conjugated forms of terpenoids, phenols, and other constituents of grape juice. *J. Agric. Food Chem.* **35**, 519–524.

Strauss, C. R., Wilson, B., Anderson, R., and Williams, P. J. (1987) Development of precursors of C_{13} nor-isoprenoid flavorants in Riesling grapes. *Am. J. Enol. Vitic.* **38**, 23–27.

Strauss, C. R., Wilson, B., Gooley, P. R., and Williams, P. J. (1986) The role of monoterpenes in grape and wine flavor – A review. In: *Biogeneration of Aroma Compounds* (T. H. Parliament and R. B. Croteau, eds.), pp. 222–242. ACS Symposium Series No. 317. American Chemical Society, Washington, DC.

Strauss, C. R., Wilson, B., and Williams, P. J. (1987) Flavour of non-muscat varieties. In: *Proceedings of the 6th Australian Wine Industry Technical Conference* (T. Lee, ed.), pp. 117–120. Australian Industrial Publishers, Adelaide, Australia.

Sudraud, P., and Chauvet, S. (1985) Activité antilevure de l'anhydride sulfureux moléculaire. *Connaiss. Vigne Vin* **19**, 31–40.

Sussi, G., and Romano, P. (1982) Induced changes by SO_2 on the population of *Saccharomyces* as agents of the natural fermentation of musts. *Vini d'Italia* **24**, 138–145.

Tanner, H., and Zanier, C. (1980) Der Kork als Flaschenverschluß aus der Sicht des Chemikers. *Mitt. Geb. Lebensmittelunters. Hyg.* **71**, 62–68.

Taylor, S. L., Higley, N. A., and Bush, R. K. (1986) Sulfites in foods, uses, analytical methods, residues, fate, exposure assessment, metabolism, toxicity, and hypersensitivity. *Adv. Food Res.* **30**, 1–75.

Tegmo-Larsson, I.-M., and Henick-Kling, T. (1990) Ethyl carbamate precursors in grape juice and the efficiency of acid urease on their removal. *Am. J. Enol. Vitic.* **41**, 189–192.

Thorngate, J. (1992) Flavan-3-ols and Their Polymers in Grapes and Wines: Chemical and Sensory properties. Doctoral dissertation, University of California, Davis.

Timberlake, C. F., and Bridle, P. (1976) Interactions between anthocyanins, phenolic compounds, and acetaldehyde and their significance in red wines. *Am. J. Enol. Vitic.* **27**, 97–105.

Tominaga, T., Baltenweck-Guyot, R., Peyrot des Gachons, C. and Dubourdieu, D. (2000) Contribution of volatile thiols to the aromas of white wines made from several *Vitis vinifera* grape varieties. *Am. J. Enol. Vitic.* **51**, 178–181.

Tominaga, T., Darriet, P., and Dubourdieu, D. (1996) Identification de l'acétate de 3-mercaptohexanol, composé à forte odeur de buis, intervenant dans l'arôme des vins de Sauvignon. *Vitis* **35**, 207–210.

Tominaga, T., Furrer, A., Henry, R., and Dubourdieu, D. (1998) Identification of new volatile thiols in the aroma of *Vitis vinifera* L. var. Sauvignon blanc wines. *Flavour Fragr. J.* **13**, 159–162.

Tominaga, T., Guimbertau, G., and Dubourdieu, D. (2003) Contribution of benzenemethanethiol to smoky aroma of certain *Vitis vinifera* L. wines. *J. Agric. Food Chem.* **51**, 1373–1376.

Tominaga, T., Masneuf, I., and Dubourdieu, D. (2004) Powerful aromatic volatile thiols in wines made from several *Vitis vinifera* L. grape varieties and their releasing mechanism. In: *Nutraceutical Beverages: Chemistry, Nutrition, and Health Effects* (F. Shahidi and D. K. Weerasinghe, eds.), pp. 314–337. ACS Symp. Ser. # 871, American Chemical Society, Washington, DC.

Tucknott, O. G. (1977) The Mousy Taint in Fermented Beverages. Doctoral Thesis, University of Bristol, UK.

Tyson, P. J., Luis, E. S., and Lee, T. H. (1982) Soluble protein levels in grapes and wine. In: *Grape and Wine Centennial Symposium Proceedings* (A. D. Webb, ed.), pp. 287–290. University of California, Davis, California.

van Buren, J. P., Bertino, J. J., Einset, J., Remaily, G. W., and Robinson, W. B. (1970) A comparative study of the anthocyanin pigment composition in wines derived from hybrid grapes. *Am. J. Enol. Vitic.* **21**, 117–130.

van der Merwe, C. A. and van Wyk, C. J. (1981) The contribution of some fermentation products to the odor of dry white wines. *Am. J. Enol. Vitic.* **32**, 41–46.

van Wyk, C. J., Augustyn, O. P. H., de Wet, P., and Joubert, W. A. (1979) Isoamyl acetate – a key fermentation volatile of wines of *Vitis vinifera* cv Pinotage. *Am. J. Enol. Vitic.* **30**, 167–173.

Vernin, G., Metzger, J., Rey, C., Mezieres, G., Fraisse, D., and Lamotte, A. (1986) Arômes des cépages et vins du sud-est de la France. *Prog. Agric. Vitic.* **103**, 57–98.

Versini, G. (1985) Sull'aroma del vino Traminer aromatico o Gewürztraminer. *Vignevini* **12**, 57–65.

Vidal, J. P., Cantagrel, R., Mazerolles, G., Lurton, L., and Gaschet, J. (1993) Mise en évidence de l'influence du triméthyl-1,1,6-dihydro-1,2-naphtalène sue la qualité organoleptique des eaux-de-vie traditionelles d'origine viticole. In: *Premier Symposium International, les Eaux-de-vie Traditionnelles d'Origine Viticole* (R. Cantagrel, ed.), pp. 165–173. Lavoisier-TEC & Doc, Paris.

Vivas, N., and Glories, Y. (1996) Role of oak wood ellagitannins in the oxidation process of red wines during aging. *Am. J. Enol. Vitic.* **47**, 103–107.

von Hellmuth, K. H., Fischer, E., and Rapp, A. (1985) Über das Verhalten von Spurenelementen und Radionukliden in Traubenmost bei der Gärung und beim Weinausbau. *Dtsch. Lebensm.-Rundsch.* **81**, 171–176.

Walker, J. R. L. (1975) Enzymic browning in foods: A review. *Enz. Technol. Dig.* **4**, 89–100.

Waters, E. J., Shirley, N.J., and Williams, P. J. (1996) Nuisance proteins of wine are grape pathogenesis-related proteins. *J. Agric. Food Chem.* **44**, 3–5.

Wenzel, K., Dittrich, H. H., and Heimfarth, M. (1987) Die Zusammensetzung der Anthocyane in den Beeren verschiedener Rebsorten. *Vitis* **26**, 65–78.

Wildenradt, H. L., and Singleton, V. L. (1974) The production of aldehydes as a result of oxidation of polyphenolic compounds and its relation to wine aging. *Am. J. Enol. Vitic.* **25**, 119–126.

Williams, A. A., and Rosser, P. R. (1981) Aroma enhancing effects of ethanol. *Chem. Senses* **6**, 149–153.

Williams, P. J., Strauss, C. R., Aryan, A. P., and Wilson, B. (1987) Grape flavour – a review of some pre and postharvest influences. In: *Proceedings of the 6th Australian Wine Industry Technical Conference* (T. Lee, ed.), pp. 111–116. Australian Industrial Publishers, Adelaide, Australia.

Winterhalter, P. (1991) 1,1,6-Trimethyl-1,2-dihydronaphthalene (TDN) formation in wine, 1. Studies on the hydrolysis of 2,6,10,10-tetramethyl-1-oxaspiro[4.5]dec-6-ene-2,8-diol rationalizing the origin of TDN and related C_{13} norisoprenoids in Riesling wine. *J. Agric. Food Chem.* **39**, 1825–1829.

Winterhalter, P., Sefton, M. A., and Williams, P. J. (1990) Volatile C_{13}–norisoprenoid compounds in Riesling wine are generated from multiple precursors. *Am. J. Enol. Vitic.* **41**, 277–283.

Würdig, G. (1985) Levures produisant du SO_2. *Bull. O.I.V.* **58**, 582–589.

Yokotsuka, K., Shimizu, T., and Shimizu, T. (1991) Polyphenoloxidase from six mature grape varieties on their activities towards various phenols. *J. Ferm. Bioeng.* **71**, 156–162.

7

Fermentation

The theory and practice of winemaking have changed fundamentally since its beginnings some 7500 years ago. Advancements, once sporadic, have come at an ever-quickening pace, reflecting developments in science and technology. Improvements in glass production and the use of cork favored the development of wine styles that benefitted from long aging. Sparkling wine also became possible. The research by Pasteur into problems of the wine industry during the 1860s led to solutions to several wine "diseases." These studies also laid the foundation of our understanding of the nature of fermentation. Subsequent work has perfected winemaking skills to their present-day high standards. Further study should result in premium wines showing more consistently the quality characteristics that connoisseurs expect. In addition, distinctive features based on varietal, regional, or stylistic differences should become more discernible and controllable. Dr Richard Peterson, a highly respected winemaker in California, has commented that Mother Nature is "a nasty old lady, who must be controlled." Modern enologic and viticultural science is increasingly providing the means by which many of the vicissitudes of Mother Nature can be moderated, if not controlled.

Modern wine production makes significant demands on the winemaker's ability to make the "right" choices among multiple options available at almost every stage from harvest to bottling. In addition, there exist major philosophical differences concerning how and why wine is made. Some producers claim (at least in public) that "wine is made in the vineyard." The winemaker being only a midwife, permitting the unique attributes of the grapes to manifest themselves. Others feel that grapes are "putty" in the hands of the producer. Production procedures are selected to mold the wine (within limits) to possess the characteristics desired by the artisan cellar master, or shaped to possess the attributes thought to be desired by the target consumer. These differences are the enologic equivalent of the nature-nurture debate in human development. Superimposed on these viewpoints may be limitations imposed by traditional styles or Appellation Control regulations. Fundamentally, there are no inherently right or wrong decisions, just choices that are more or less appropriate or required under particular conditions. Because grape characteristics vary from vintage to vintage, no set production formula is possible. What is crucial is that the winemaker be fully aware of the benefits and shortcomings of the techniques available, permitting the selection of the optimal, or most judicious, procedure. In the following chapters, the advantages and disadvantages of alternative procedures are presented to facilitate making rational decisions.

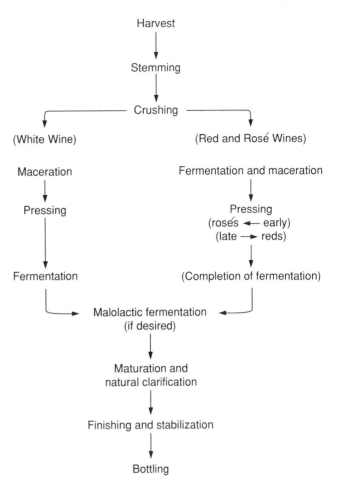

Figure 7.1 Flow diagram of winemaking.

Basic Procedures of Wine Production

Vinification formally begins when the grapes, or juice, reach the winery. The basic steps in the production of table wines are outlined in Fig. 7.1.

The first step involves removing the leaves and other extraneous material from the grapes. The fruit is then crushed (or pressed) to release the juice and begin the process of **maceration**. Maceration facilitates the extraction of nutrients, flavorants, and other constituents from the pulp, skins, and seeds. Initially, hydrolytic enzymes released from ruptured cells promote this liberation. The cytotoxic action of pectic enzymes further promotes the release of cellular constituents into the **must** (grape macerate). Enzymes released or activated by cell death may also activate the syntheses of flavor compounds and hydrolyze macromolecules to forms utilizable by yeast and bacteria.

For white wines, maceration is kept to a minimum and seldom lasts more than a few hours. The juice that runs freely from the crushed grapes (**free-run**) is usually combined with that released by pressing. The free-run and first pressings are usually combined and fermented together. Subsequent pressings are typically fermented separately.

For red wines, maceration is prolonged and occurs simultaneously with alcoholic fermentation. The alcohol generated by yeast action enhances the extraction of anthocyanin and promotes the uptake of tannins from the seeds and skins (**pomace**). The phenolic compounds solubilized give red wines their basic properties of appearance, taste, and flavor. They are also required to give red wines their aging and mellowing characteristics. In addition, ethanol augments the liberation of aromatic ingredients from the pulp and skins. After partial or complete fermentation, the **free-run** is allowed to flow away under gravity. Subsequent pressing extracts **press fractions** (most of the remaining juice). Press fractions may be incorporated with the free-run in proportions determined by the type and style of wine desired.

Rosé wines are typically made from red grapes exposed to a short prefermentative maceration. The grapes are crushed or gently broken, and the juice left on the pomace at cool temperatures until sufficient color has been extracted (generally between 12 and 24h). The free-run juice is subsequently drawn off and fermented similarly to that of a white wine. Alternately, the grapes may be pressed whole (a slow process associated with limited color extraction). The free-run juice and some of the first press-run juice are fermented without further contact with the skins. Where the coloration of the grapes is low, the grapes may be crushed and fermented with the skins until sufficient pigment has been extracted. Subsequent fermentation of the free-run juice occurs without further skin contact. When color depth is used to time pressing, consideration must be taken for both pigmentation loss during fermentation and the bleaching action of any sulfur dioxide added. Because of the short (or incidental) maceration, alcoholic fermentation often begins in earnest only after the juice is separated from the skins.

Fermentation may start spontaneously, due to endemic yeasts derived from the grapes, but more frequently from crushing equipment. Standard practice, however, is to inoculate the juice or must with a yeast strain of known characteristics. Yeasts not only produce the alcohol, but also generate the general bouquet and flavor attributes that typify wines.

After completing **alcoholic** fermentation, the wine may be treated to foster a second, **malolactic** fermentation. Malolactic fermentation is particularly valuable in cool climatic regions, where a reduction in acidity ameliorates the wine's taste characteristics. Although most red wines benefit from malolactic fermentation, fewer white wines profit from its occurrence. The milder fragrance of most white wines makes them more susceptible to potentially undesirable flavor changes induced by malolactic fermentation. The retention of acidity also adds to their fresh taste. In warm viticultural regions, malolactic fermentation is often unneeded and undesirable. Its development is usually discouraged by practices such as the addition of sulfur dioxide, early clarification, and storage under cool conditions.

Newly fermented wine is protected from, or given only minimal exposure to oxygen during maturation. This limits oxidation and microbial spoilage. During storage, excess carbon dioxide escapes, yeasty odors dissipate, and suspended material precipitates. Changes in aroma and development of an aged bouquet may begin during maturation. Exposure to air is usually restricted to that which occurs during racking, or *battonage* (during *sur lies* maturation). Such slow or limited exposure can help oxidize hydrogen sulfide and favor color stability in red wines.

After several weeks or months, the wine is racked. **Racking** separates the wine from solids that settle out during spontaneous or induced clarification. The sediment consists primarily of yeast and bacterial cells, grape cell remains, and precipitated tannins, proteins and potassium tartrate crystals. If left in contact with wine, they can lead to the production of off-odors, as well as favor microbial spoilage.

Prior to bottling, the wine may be **fined** to remove traces of dissolved proteins and other material. Otherwise, they could generate haziness, especially on exposure to heat. Fining may also be used to soften the wine's taste by removing excess tannins. Wines are commonly chilled and filtered to further enhance clarification and stability.

At bottling, wines are generally given a small dose of sulfur dioxide to limit oxidation and microbial spoilage (between 0.8 and 1.5 mg/liter free molecular SO_2). Sweet wines are usually sterile-filtered as a further protection against microbial spoilage.

Newly bottled wines are normally aged at the winery for several months to years before release. This period permits wines blended shortly before bottling to "harmonize." In addition, it allows acetaldehyde produced following bottling (as a consequence of incidental oxygen uptake) to be converted to nonaromatic compounds. Thus, the "bottle sickness" induced by acetaldehyde usually dissipates before the wine reaches the consumer.

Prefermentation Practices

Stemming and crushing are typically conducted as soon as possible after harvesting. During the harvest, some grapes are unavoidably broken and their juice released, whereas others may be bruised. Thus, oxidative browning often begins before the grapes reach the winery and crushing begins. The juice also becomes field-inoculated with the yeast and bacterial flora present on grape surfaces. If the berries are harvested during the heat of the day, undesirable microbial contamination can rapidly develop. To minimize this occurrence, grapes may be sulfited after harvest or harvested during the cool parts of the day.

Left in containers, harvested fruit quickly warm due to the endogenous metabolic activity of the grapes and the insulating influence of the volume. This can aggravate contamination by speeding microbial activity. In addition, warming may necessitate cooling to bring the temperature down to an acceptable prefermentation value.

Destemming

The present-day trend is to separate the processes of stemming and crushing physically, if not temporally. The removal of stems, leaves and grape stalks (termed MOG – material other than grapes) before crushing has several advantages. Notably, it minimizes the uptake of phenolics and lipids from vine parts. The extraction of stem phenols is of potential value only when dealing with red grape varieties low in phenol content, such as 'Pinot noir.' Stem phenols are intermediate in astringency and bitterness, relative to the less strident tastes of skin tannins and the more assertive seed tannins. The phenolics extracted from stems include catechins, flavonols (notably quercetin), and caftaric acid (Sun *et al.*, 1999).

In the past, stems were often left with the must throughout fermentation, especially in the production of red wines. Occasionally, some of the grape clusters were left uncrushed (Henderson, 1824) – thus, permitting partial carbonic maceration. The presence of stems made pressing easier, presumably by creating drainage channels along which the juice or wine could escape. Modern improvements in press design have made stem retention unnecessary, unless specifically desired. The enhanced tannin content derived from prolonged stem contact gave wines, made during poor vintage years, extra body and improved color density.

In addition to augmenting phenol extraction and facilitating pressing, maceration with the stems may increase the rate of fermentation. This effect appears to be due to the enhanced uptake of oleanolic acid (Bréchot *et al.*, 1971). It supplements the amount of long-chain unsaturated fatty acids available to yeasts. It especially helps them complete fermentation under cool cellar conditions.

Leaf removal before crushing helps limit the production of C_6 ("leaf") aldehydes and alcohols. They are produced during the enzymatic oxidation of linoleic and linolenic acids extracted from the leaf cuticle. Leaf aldehydes and alcohols can taint wine with a grassy to herbaceous odor. Nonetheless, they may contribute to the typical aroma of some wines in small amounts. High leaf content in the must may also result in the excessive uptake of quercetin. If the wine is bottled shortly after fermentation, quercetin can lead to the production of a yellowish haze in white wines (Somers and Ziemelis, 1985). When the wines are matured sufficiently, much of the quercetin precipitates before bottling. High flavonol contents can also generate bitterness in white wines.

For convenience and efficiency, the same piece of equipment often performs both stemming and crushing. Stemmers usually contain an outer perforated cylinder that permits the berries to pass through, but limits the passage of stems, stalks, and leaves (Fig. 7.2). Often there is a series of spirally arranged arms, possessing flexible paddle ends, situated on a central shaft. Shaft rotation draws grape clusters into the stemmer, forces the fruit through the perforations, and expels the stems and leaves out the end. When stemmer-crushers are working optimally, the fruit is separated from the leaves and stems with minimal breakage. Expelling the stems and leaves in as dry a state as possible avoids juice loss

Figure 7.2 Internal view of a stemmer-crusher. (Photo courtesy of the Wine Institute)

and facilitates disposal. Stems and other vine remains may be chopped for subsequent soil incorporation.

Sorting

Although stemmers may effectively remove MOG (stems and leaves), it does not remove smaller material such as trellis clips, staples, snails, or insects. Stemmers also do not remove immature (green), oxidized (brownish), raisined, or other forms of substandard berries. Where their presence is likely to detectably affect the attributes of the wine, their elimination can be critical. This has usually required manual sorting. Due to increasing labor costs, sorting is economically feasible only for premium wines. This situation may change with the development of automatic sorters. These can differentiate and selectively remove undesirable material from harvested grapes (Falconer and Hart, 2005). Rejection can be selected to function on color and/or size categories. As the fruit passes under the detector, located above the conveyor belt, the color intensity of the grapes in the green, red and infrared parts of the spectrum is recorded. Depending on instructions supplied by the operator, the computer determines whether the sample should be rejected. If so, a jet of air expels the undesired material. The sorter can be simultaneously programed to reject grapes, or MOG, based on size criteria.

Although not inexpensive, the automated machine processes grapes more rapidly and at a lower cost (based on several years of use) than manual sorting. Whether such an investment is merited will depend on the economic return derived from improved wine quality.

Crushing

Crushing typically follows stemming immediately. Some berries are unavoidably broken in the process and the juice released is highly susceptible to oxidative browning and microbial contamination. Crushing the fruit without delay permits fermentation to commence almost immediately (if desired), limits microbial spoilage, and provides better oxidation control.

Crushing is accomplished by any of a number of procedures. Those generally preferred involve pressing the fruit against a perforated wall or passing the fruit through a set of rollers. In the former, the berries are broken open, and the juice, pulp, seeds, and skins pass through openings to be collected and pumped to a retaining tank or vat. In the latter process, berries are crushed between a pair of rollers turning in opposite directions. The rollers usually have spiral ribbing or contain grooves with interconnecting profiles to draw the grapes down and through the rollers. Spacing between the rollers typically is adjustable to accommodate the variation in berry size found among different cultivars or vintages. It is important to avoid crushing the seeds. Otherwise, contamination with seed oils can eventually lead to the development of rancid odors.

Crushing also can be achieved using centrifugal force. In centrifugal crushers, the fruit is spun against the sides of the crusher. Because they tend to turn the fruit into a pulpy slurry, centrifugal crushers generally are undesirable. Clarification of the juice is difficult, and seeds may be ruptured.

Although grapes are customarily crushed prior to vinification, there are exceptions. The juice for sparkling wine production is typically obtained by pressing whole grape clusters. Special presses extract the juice with a minimum of pigment and tannin extraction. The absence of pigments and tannins is particularly important when white sparkling wines are made from red-skinned grapes. Pressing intact grape clusters is now becoming popular with some table wine producers.

Botrytized grapes are also frequently pressed, rather than crushed. The gentler separation of the juice minimizes liberation of fungal dextran (β-glucans) polymers into the juice. The latter can plug filters used in clarification. In the production of the famous botrytized wine Tokaji Eszencia, the juice is derived solely from liquid that drains away freely from heavily infected grapes. No pressure other than the weight of the fruit activates juice release.

In the production of wines employing carbonic maceration, such as *vino novello* and beaujolais, it is essential that most of the fruit remain uncrushed, at least at the beginning. Only in intact berries can an internal grape-fermentation occur. This is essential for development of the characteristic fragrance shown by these wines. After a variable period of autofermentation, berries that have not broken under their own weight are pressed to release their juice. Fermentation is completed by yeast action.

Supraextraction

An alternative to crushing is supraextraction (Defranoux *et al.*, 1989). It involves cooling the grapes to −4 °C, followed by warming to about 10 °C before pressing. Freezing causes both grape-cell rupture and skin splitting. These facilitate the escape of juice during pressing. Although increasing the extraction of sugars and phenolics, supraextraction reduces total acidity and raises the pH. The latter may result from crystallization and removal of tartaric acid.

Maceration (Skin Contact)

WHITE WINES

Maceration refers to the release of constituents from the pomace (seeds, skins and pulp) following crushing. The process is facilitated by the liberation and activation of hydrolytic enzymes from crushed cells. However, the shift to light fruity white wines in the 1970s resulted in minimizing skin contact. This trend was encouraged by the widespread adoption of mechanical harvesting. However, depending on the tendency of the grapes to rupture, some inevitable maceration occurred on way to the winery – its extent depending on the duration separating harvest and crushing/pressing, and the temperature of the grapes. Reduced maceration also diminished the uptake of heat-unstable proteins, decreasing the need for protein stabilization products.

Unfortunately, minimizing or eliminating maceration simultaneously reduces the uptake of varietal flavorants located in the skins, such as *S*-cysteine conjugates in 'Sauvignon blanc' (Peyrot des Gachons *et al.*, 2002). For wines depending on aromatics extracted from the grapes, this became increasingly important with the adoption of gentler pressing, such as provided by pneumatic presses or whole-grape pressing. To offset this deficiency, use of the first and second press-run fractions increased. This option is often easier to manipulate than maceration, due to the complexities of temperature and duration on extraction, precipitation, and degeneration of compounds during maceration. Nevertheless, the addition of press fractions augments the wine's phenolic content. Like most choices in winemaking, each decision has its pros and cons. In this instance, a potential increase in varietal flavor vs. a potential deterioration in mouth-feel (Tamborra, 1992). What the winemaker needs to know is the relative importance of flavorants derived from the grapes, relative to the ease and extent of phenolic extraction. These properties are largely cultivar dependent, but also vary with vineyard and vintage conditions.

Although prolonged maceration enhances the phenolic content of a white wine, it does not lead to the same degree of astringency typical of red wine. This anomaly appears to result from the absence of anthocyanins. Anthocyanins, which are themselves tasteless, bind with catechins and flavonoid tannins (see Chapter 6). This increases the solubility of tannins, keeping them in suspension (retaining their bitter and astringent properties). In white wines, most of the tannins precipitate during fermentation, limiting their potential to affect the wine's sensory aspects.

Grape varieties differ considerably in the amount of phenolics released during crushing or extracted during maceration (skin contact). For example, few flavonoids accumulate in the musts of 'Palomino' and 'Sauvignon blanc;' moderate amounts collect in the musts of 'Riesling,' 'Sémillon,' and 'Chardonnay;' whereas extensive extraction occurs with 'Muscat Gordo,' 'Colombard,' 'Trebbiano,' and 'Pedro Ximénez' (Somers and Pocock, 1991). An

increased phenolic extraction favors subsequent in-bottle browning. This feature may be partially offset by hyperoxidation of the must.

The major physical factors influencing extraction from the skin and pulp are temperature and duration. Extraction is often linearly related to both factors. For example, cool maceration temperatures and short duration minimize flavonoid uptake (Fig. 7.3), and thereby limit potential bitterness and astringency. Occasionally, the concentration of extracted compounds decreases with prolonged maceration, presumably due to precipitation or degradation. Extraction also varies markedly with the class of compounds involved. Although many nonflavonoids are quickly liberated into the juice, subsequent extraction of flavonoid phenolics occurs more readily than nonflavonoids (Fig. 7.3).

As with phenolic compounds, the concentration of flavorants and nutrients is markedly influenced by maceration. For example, skin contact increases the uptake of monoterpenes (Marais, 1996). The content of amino acids, fatty acids and higher alcohols may rise, whereas total acidity tends to fall (Soufleros and Bertrand, 1988; Guitart *et al.*, 1997). The decline in acidity appears to be caused by the increased release of potassium. The latter induces tartrate salt formation and precipitation. Other changes result from indirect effects on yeast metabolism. For example, increased amino acid availability has been correlated with a reduction in the production of hydrogen sulfide (Vos and Gray, 1979).

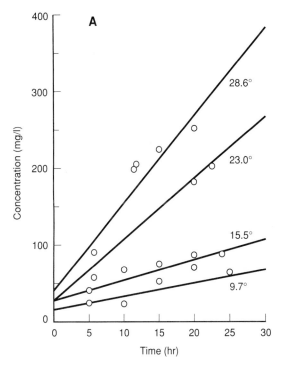
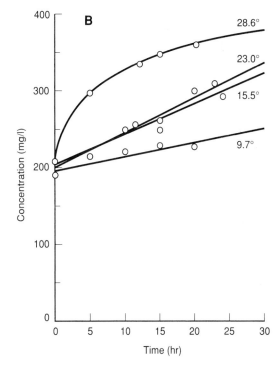

Figure 7.3 Flavonoid (**A**) and nonflavonoid (**B**) phenol content in 'Chardonnay' must during skin contact. Temperatures are in °C. (From Ramey *et al.*, 1986, reproduced by permission)

Occasionally, a short exposure (15 min) to high temperatures (70 °C) greatly increases the release of volatile compounds, such as monoterpenes (Marais, 1987, 1996). Although the concentration of most monoterpenes increases on short-term exposure to high-temperature maceration, not all follow this trend. For example, the concentration of geraniol decreases.

Generally, maceration is conducted at cool temperature. This has not only the advantage of suppressing the growth of potential spoilage organisms, before the onset of active fermentation, but also affects the subsequent synthesis of yeast flavorants during fermentation. For example, the synthesis of volatile esters may increase with a rise in maceration temperature up to 15 °C, whereas it decreases at higher temperatures. The synthesis of most alcohols (except methanol) is reduced following maceration at warmer temperatures (Fig. 7.4). Methanol content increases due to the action of grape pectinases, which release methyl groups from pectins.

The sensory influence of maceration can also be influenced by the degree of oxygen exposure. This can come from oxygen absorbed during crushing or via intentional exposure (**hyperoxidation**). Oxygen promotes the enzymatic oxidation of the primary phenolics in white must (nonflavonoid o-diphenols – notably caftaric acid). Although their polymerization can cause juice browning, the polymers usually precipitate during fermentation. This leaves the wine less sensitive to subsequent in-bottle oxidation, as well as lower in bitterness. Although potentially used from some cultivars, hyperoxidation is probably ill-advised with varieties deriving much of their varietal character from volatile thiols, notably 'Sauvignon blanc.' Oxygen can degrade volatile thiols.

Partially to facilitate early phenolic oxidation and removal, the addition of sulfur dioxide at crushing is generally avoided. In addition, sulfur dioxide can undesirably enhance the production of acetaldehyde during fermentation, enhance phenolic extraction, and retard the initiation of malolactic fermentation. Adding sulfur dioxide, at or just after crushing, is now largely limited to situations where a significant proportion of the crop is diseased, or where the interval between harvesting and crushing is protracted. Sulfur dioxide limits the action of polyphenol oxidases, and retards the multiplication of bacteria or other microbes in juice released from broken fruit.

Maceration has been observed to improve juice fermentability (Ollivier et al., 1987) and enhance yeast viability. Part of these effects is due to the release of particulate matter, lipids, and soluble nitrogen compounds into the juice. Particulate matter is well known to increase microbial growth. The solids provide surfaces for yeast and bacterial growth, the adsorption of nutrients, the binding of toxic C_{10} and C_{12} carboxylic fatty acids, and the escape of carbon dioxide. This increases must agitation and, therefore, more uniform nutrient distribution. Skin contact facilitates the extraction of long-chain (C_{16} and C_{18}) saturated and unsaturated fatty acids, such as palmitic, oleanolic, linolenic, and linoleic acids. The enhanced extraction of long-chain fatty acids also reduces the synthesis of toxic mid-chain (C_{10} and C_{12}) fatty acids (Guilloux-Benatier et al., 1998). The former lipids are important in permitting yeast cells to synthesize essential steroids and build cell membranes under anaerobic fermentation conditions. In addition, the small amounts of oxygen absorbed during crushing and during other prefermentation cellar activities probably promote the synthesis of sterols by yeast cells.

Extended skin contact also improves (more than doubles) the production of extracellular mannoproteins, formed during alcoholic fermentation. The effects of the increased mannoprotein content, and the reduced concentration of C_{10} and C_{12} fatty (carboxylic) acids, combine to facilitate malolactic fermentation by *Oenococcus oeni* (Guilloux-Benatier et al., 1998).

Minimal maceration at cool temperatures often leads to the production of young, fresh, fruity wines. Longer, warmer maceration typically produces a wine deeper in

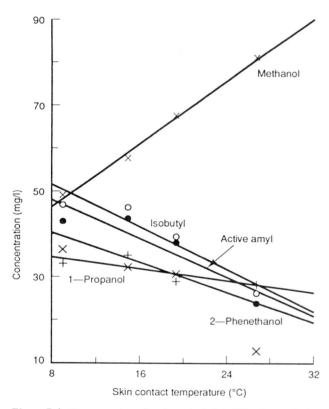

Figure 7.4 Concentration of various alcohols in 'Chardonnay' wine as a function of skin contact temperature. (From Ramey *et al.*, 1986, reproduced by permission)

color and of fuller flavor. The latter may mature more quickly and develop a more complex character than wines produced with minimal skin contact (Ramey *et al.*, 1986). Thus, varietal characteristics (Singleton *et al.*, 1980), fruit quality, equipment availability, and market response all play a role in the decision of a wine-maker to use maceration, and to what extent.

A new procedure, called cell-cracking, has been investigated as a complement to, or replacement for, maceration (Bach *et al.*, 1990). Cell-cracking involves forcing the must through narrow gaps separating steel balls positioned in a small bore. It promotes the rapid extractions of flavorants.

ROSÉ WINES

Occasionally, rosé wines are made from juice released by pressing whole red grapes. Nevertheless, the more common practice is to gently crush the grapes. This may or may not be followed by a period of maceration, lasting up to 24 hours. Maceration at ≤20 °C tends to retard microbial action. Data from Murat (2005) suggest that the top of this range might be preferable, due to improved extraction of S-3-hexan-1-ol-L-cysteine. This is an important fragrance precursor primarily located in the skin. Its metabolic conversion to the fruity smelling 3-mercaptohexan-1-ol is also favored during fermentation at about 20 °C. Short maceration also limits anthocyanin uptake, donating only the desired, slight pinkish coloration. However, because few tannins are extracted, rosé wines tend to show poor color stability (much of the color being derived from free anthocyanins, or their self-association or copigment complexes). Despite their relatively low concentrations, anthocyanins still appear to act as important antioxidants. For example, they protect 3-mercaptohexan-1-ol from oxidation (and rapid loss of this essential ingredient in the fruity fragrance of some rosé wines). Phenethyl acetate and isoamyl acetate are also significant contributors to the fruity flavor to many rosé wines (Murat, 2005).

It is important that maceration occur under anaerobic conditions. This not only limits oxidation of important volatile thiols, but equally protects anthocyanins from oxidative discoloration. Salinas *et al.* (2003) have found that adding pectolytic enzymes during maceration not only improves flavor development, but also promotes color stability. Both are features important to the shelf-life of rosé wines.

Typically, only free-run juice is used in rosé production. The press-run juice may be added to fermenting red wine or used in other wine products. Unless the grapes were comparatively immature (before full coloration), the anthocyanin content of the press-run is often too high for use in rosé production. Anthocyanin contents in the range of 20–50 mg/liter are standard for rosé wines.

In addition, tannins in the press-run juice can give a bitterness undesirable in most rosés.

Occasionally, in the production of a red wine, a portion of the juice is drawn to produce a rosé. The remaining, concentrated must is used to produce the red wine. The technique, called *saignée*, is primarily used in years, or with cultivars, where color extraction is likely to be less than desired for production of a red wine.

RED WINES

In red wine production, maceration studies have focused primarily on the extraction of pigments and tannins. Anthocyanins are the first to be extracted, their being more soluble than tannins. As fermentation becomes active, ethanol production not only enhances solubilization, but also facilitates anthocyanin escape by increasing membrane porosity. Tannin extraction is much more dependent on increasing ethanol content for its solubilization.

Both the style and consumer acceptance of wine can be dramatically altered by the duration and conditions of maceration. Thus, maceration provides one of the principal means by which winemakers can adjust the character of their wines. Short macerations (<24 h) commonly produce a rosé. For early consumption, red musts are commonly pressed after 3–5 days. This provides good coloration, but avoids extraction of harsh seed tannins, but extracts sufficient skin tannins to promote color stability. Most flavonoids reach a temporary peak in solubilization within approximately 5 days. However, there may be a second phase of extraction that begins after 15 days (Soleas *et al.*, 1998). Longer maceration periods are correlated with increased concentrations of higher-molecular-weight tannins. It may also increase the extraction of undesirable flavorants, such as methoxypyrazines (Kotseridis *et al.*, 1999). Wines for long aging have often been macerated on the seeds and skins for as long as 3 weeks. Extended maceration results in a decline in free anthocyanin content, but enhances color stability by encouraging their early polymerization with procyanidins. However, whether the prolonged skin contact, traditional in some parts of France and Italy, is required for wines to age well is a moot point. Little agreement exists among winemakers or enologists.

Although the general correlation between maceration, pumping-over, and style of wine is well known, the ease with which anthocyanins are extracted varies with the cultivar. Romero-Cascales *et al.* (2005) promote the idea of determining extractability and seed maturity indices to facilitate determining the timing (and method) of maceration. Their extractability index involves measuring anthocyanin uptake at two pH values – 3.6 and 1. Assessment of the seed maturity index is equally important, as it markedly affects color stability. It was obtained using a method described by Saint-Criq et al. (1998).

An old technique undergoing renewed interest, notably with 'Pinot noir,' is **cold maceration**. It involves a prefermentation maceration period (3–4 days) at cool (15°C) to cold (4°C) temperatures. This is somewhat equivalent to the cooling that usually occurs in small, unheated, Burgundian wine cellars in the fall. This cooling can be more effectively controlled by the judicious addition of dry ice or liquid nitrogen. The procedure has been reported to slow but facilitate the progressive extraction of phenolics, especially anthocyanins (Cuénat et al., 1996; Feuillat, 1996). Color density is enhanced, whereas flavor development becomes more complex and intense. Heatherbell et al. (1996) reported that cool maceration temperatures generated 'Pinot noir' wines with more of a peppery or bitter aspect, whereas cold temperatures tend to accentuate a sweet, blackberry aspect. Differential flavor effects have also been noted using 'Airen' and 'Macabeo' cultivars (Peinado et al., 2004). Beneficial effects have also been reported for cultivars such as 'Pinotage,' 'Sangiovese' and 'Syrah.' However, the improved color shown early in maturation is often short-lived, with color intensity becoming similar to traditional treatments on aging (Heatherbell et al., 1996).

A potential disadvantage of cold maceration is an increased risk of early spoilage by *Brettanomyces* (Renouf et al., 2006). It provides time for adaptation of the indigenous grape flora, and their subsequent growth, when the must is warmed and fermentation begins.

The addition of sulfur dioxide promotes anthocyanin extraction, especially at cooler temperatures. Sulfite addition products are more soluble in aqueous alcohol solutions than native anthocyanins. The potential disadvantage of sulfur dioxide addition is that it may delay early polymerization between anthocyanins and tannins.

Another technique for improving color and/or flavor extraction is the addition of supplementary skins, or seeds and skins, to red must during fermentation (Revilla et al., 1998). The procedure is partially based on an old Spanish technique called *double pasta*. It involves the addition of extra pomace during fermentation. Enriching the must with grape skins or seeds may enhance the varietal aroma and flavor of the wine. The process also promotes color stabilization. To avoid producing an excessively tannic wine, due to the increased uptake of catechins and dimeric procyanidins, supplementation may be limited to approximately one-third that of the pomace in the original grape must.

Except for cold maceration, little attention has been given to the extraction of aromatic compounds during skin contact. In one of the few studies on the subject, the berry aspect of 'Cabernet Sauvignon' was increased by long maceration, whereas the less desirable canned bean–asparagus aspect was diminished (Schmidt and Noble, 1983).

Dejuicing

Dejuicers are especially useful when dealing with large volumes of must. By removing most of the free-run juice, press capacity can be used more economically to extract the rest (**pressings**).

Dejuicers often consist of a columnar tank possessing a perforated basket at the end. The mass of the juice forces much of the fluid from the crushed grapes into the basket, from which the juice flows into a receiving tank (sump). Carbon dioxide pressure may be used to speed the separation. When drainage is complete, the basket is raised to permit the pomace to discharge for transport to a press.

Continuous dejuicers of simpler design consist of a sloped central cylinder containing perforations that permit the juice to escape, but retains the pomace. The crushed grapes are moved up the cylinder by rotation of a central screw. The dejuiced grapes are dumped into a hopper for loading into a press. The upward flow of the crush in the dejuicer generates the gravitational force needed to speed juice release. The main disadvantage of dejuicers is the additional clarification that may be required from the increased uptake of suspended solids.

Pressing

Presses of various designs have been used for at least 5000 years. The earliest illustrations appear in ancient Egyptian tombs. Their initial function was simply to separate the juice from the seeds and skins. Only later did technological advances permit presses to become a tool by which winemakers could influence wine attributes.

The earliest "presses" must have been inefficient and very cumbersome to use. The pomace and juice were placed in a cloth sac held by poles at each end. Stylized drawings show men twisting the poles in opposite directions. Subsequently, the "press" was stretched by pulling on one end of the polls, while another horizontally positioned worker pushed the poles apart in the center with his arms and legs (see Darby et al., 1977). By the third Dynasty (2650–2575 B.C.), presses are shown being held fixed at one end, while workers twist at the other end, shortening the press and bring the cloth taught against the crushed fruit. There seems little evidence of further advancement until Classical times. The beam-press is thought to have been developed by the Myceneans, possibly about 1600 B.C. In this type of press, a large beam, affixed at one end and weighted down at the other by a heavy object, applies pressure to grapes in a sack or other flexible porous container. The next advance involved the use of a screw to control the application of pressure by the beam on the grapes in a slotted holder. It appears to have been invented in Greece about second century B.C. A wall fresco discovered in Pompeii illustrates its use in Roman times. The next advance eliminated the need for the beam, affixing the screw to an U-shaped support, with the pressure applied to a rounded plate positioned over grapes in a slotted, wooden container (Forbes, 1957; White, 1984). Its advantages were discussed by Pliny the Elder (*Historia Naturalis*) and Vitruvius (*Architectura*). Subsequent major developments in press design did not occur again until the 1800s.

The first major modern advance in press design involved the use of hydraulic force. It replaced muscle power associated with a massive screw and lever action (Plate 7.1). Incorporation of a removable bottom permitted easier pomace discharge. Previously, presses had to be dismantled or the pomace shoveled out at the end of each press cycle. Both tasks were unpleasant, time-consuming, and labor intensive.

Increasing the drainage surface area was another momentous development. Not only did it speed juice release, but it also reduced the flow path for fluid escape. By diminishing the force required for juice extraction, higher quality juice or wine is liberated.

Placing the press on its side (horizontally) was another means of increasing surface area for liquid escape. It permitted the length (former height) of the press to be increased without difficulty. The horizontal orientation

also permitted a section of the press to be hinged, providing access for both convenient filling and emptying. By suspending the press on heavy gears, the press could both be rotated for pomace **crumbling** (tumbling), as well as be inverted for emptying (Plate 7.2). Crumbling breaks the compacted pomace produced during pressing and helps entrapped juice escape on subsequent pressing. Previously, chains or manual mixing was used to achieve crumbling. This had the disadvantages of both potentially crushing the seeds, and increasing juice clouding, due to the greater release of solids with the juice.

Another significant innovation was the development of the continuous screw press. By permitting uninterrupted operation, such equipment avoids time-consuming filling and emptying cycles. This is especially valuable when large volumes of must or wine need to be pressed in a short period. Their principal disadvantage involves an increase in suspended solids, requiring additional fining or clarification.

Because the free- and press-run fractions produced during pressing possess differing physicochemical properties, winemakers can use press design, and how they are used, to influence wine character. Free-run fractions are clearer, possess lower levels of suspended solids, phenolic contents, and flavorants principally derived from the skins. Subsequent press-run fractions contain increasing amounts of suspended solids, anthocyanins, tannins and skin flavorants. Press-run fractions also are more likely to oxidize (possess more polyphenol oxidase), possess lower

Press type	Vertical	Horizontal	Pneumatic
		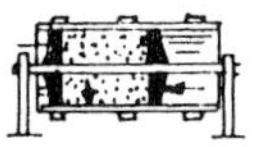	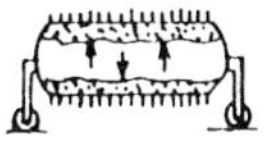
Size of the basket (cm)	113 x 90	215 x 73	215 x 73
Volume (m³)	0.9	0.9	0.9
Pressure area (m²)	1	0.42	4.95
Pressure per 1 cm² (MPa)	1.25–1.6	1.2	0.6
Pressure over the whole area (MPa)	12,500–16,000	5000	29,700
Pressure per 1 dm of pomace (MPa)	13.9–17.8	5.6	33.0
Average size of the cake (cm) at one half of the original volume	113 x 18	73 x 43	215 x 239 x 3.3
Shape of the cake			
Flowing out of the must (time)	long	short	very short
Time of one pressing (min)	100–120	100–120	50–90
Number of pressings	2	1	1
Total time of pressing (hr)	3 – 4	2	1

Figure 7.5 Comparison of various types of presses. (Fom Farkaš, 1988, reproduced by permission)

acidity (higher potassium contents), and have higher concentrations of polysaccharides, gums, and soluble proteins. Most wines are a judicious blending of both free- and the first press-run fractions. Depending on the intentions of the winemaker, and characteristics of the grapes, a portion of the second and possibly third pressing may be incorporated.

Not surprisingly, views vary considerably on the relative merits of using press-run fractions, and the various types of presses. Until more is known about the dynamics of flavor extraction during pressing, and how to predict their sensory consequences, the choice of press type will depend more on subjective or anecdotal, rather than objective information.

Brief descriptions of the major types of presses in use are given below. Figure 7.5 compares the operational characteristics of equivalent volumes of wine or juice pressed in vertical, horizontal, and pneumatic presses.

VERTICAL (BASKET) PRESSES

Vertical presses generally consist of a series of concentrically arranged slats between which the juice can escape. Pressure is applied hydraulically from above. The plate that presses the grapes is usually retracted before the sides are removed to extract or crumble the press cake. Alternately, the bottom may be lowered.

Vertical presses are principally used with small lots of fruit, or where use of horizontal or pneumatic presses may be impractical or uneconomic. In addition, it has particular advantage in the pressing of frozen grapes, as in the production of icewine. In this instance, breaking the press cake is facilitated by direct access (Plate 9.4).

A major variation of the vertical press is that used in the production of sparkling wine. It is much broader than standard versions. Although relatively slow in operation, maximizing the surface area to volume ratio increases its effectiveness in pressing uncrushed grapes.

HORIZONTAL (MOVING HEAD) PRESSES

A well-known version of the horizontal press is illustrated in Fig. 7.6. Both crushed and uncrushed grapes, as well as fermented juice, are effectively pressed in horizontal presses.

Loading occurs through an opening in the upper, raised portion of the press. Pressing is conducted by hydraulically forcing end-plate(s) inward from one or both ends. The rate at which pressure is applied can be modified to suit the needs of the grape variety and characteristics of the press fractions desired. Fluid escape occurs between slats of, or slits in, the pressing cylinder. Chains or rotation of the press breaks the pomace cake between successive pressings. Once pressing is complete, retracting the end-plate(s) and inversion permits dumping.

The primary drawback to horizontal presses is the progressive reduction in drainage surface during pressing. Thus, the force required to maintain a rapid discharge increases during pressing. This correspondingly increases the extraction of suspended solids and tannins. To maintain juice quality, the last portion of the juice may need to be sacrificed. Presence of the stems can improve this situation by creating channels for easier juice escape. Slower pressing is also another factor favoring juice quality, by providing more time for juice escape (and the release of aromatics from the skins and pulp).

PNEUMATIC (TANK, BLADDER OR MEMBRANE) PRESSES

Pneumatic presses also effectively press crushed or uncrushed grapes, as well as fermented must. The press is filled through an elongated opening in the top. Once filled and closed, the press is inverted to allow the free-run juice or wine to escape. Compressed gas, entering between a plastic bladder and solid outer cylinder wall, compresses the grape mass against perforated plates that project along the central cavity (Fig. 7.7). More recent models possess a central or side-positioned bladder that forces the must or grapes against a perforated outer cylinder. Presses with a central bladder tend to be more efficient in extracting juice at lower pressures, due to the must or grapes being pressed more uniformly against the whole surface area of the cylinder wall. Lower pressures exerted over a larger surface area liberate juice more quickly and with reduced release of suspended solids and phenolics. This is often of particular concern when extracting juice with little or minimal color extraction, as in the production of white or rosé wines. Crumbling the pomace cake between successive presses is achieved by rotating the pressing cylinder. Opening of the filling trap and inversion discharges the pomace.

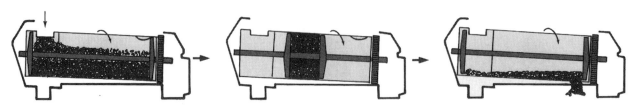

Figure 7.6 Diagram of the operation of a horizontal press. (Courtesy of CMMC, Chalonnes-sur-Loire, France)

Both small, medium, and large volume (5–22 hl) versions are commercially available. Smaller versions are of particular value when dealing with select lots of high-quality juice or wine.

Both horizontal and pneumatic presses yield high-quality pressings. The pressings are relatively low in suspended solids, and press operation neither crushes the seeds nor extracts undue amounts of tannins. A common drawback involves the time associated with their repeated filling and emptying, and fairly fixed press cycle (about 1–2 hours).

A recent development has been the design of presses extracting juice by negative pressure (vacuum), rather than positive pressure.

CONTINUOUS SCREW PRESS

Continuous-type presses have the advantage of running uninterruptedly. They avoid the time and labor costs associated with cyclical filling and emptying. Although they work best with fermented must, they can be adjusted to handle crushed, nonpulpy grapes. They do not function adequately with uncrushed grapes.

Crushed grapes, as well as fermenting or fermented must, are pumped into the press via a hopper at one end of the press (Fig. 7.8). A fixed helical screw forces the material into a pressing chamber whose perforated wall allows the juice or wine to escape. Pressed pomace accumulates at the end of the pressing cylinder, where it periodically is discharged through an exit portal.

The primary disadvantage of the continuous press is the poorer quality of the juice or wine liberated. For example, the production of fruit esters during fermentation tends to be lower in juice derived from continuous screw presses. This is particularly noticeable in older models, in which separation of different press fractions was impossible. Newer models permit such separation. The first fractions (closest to the intake) possess characteristics similar to free-run fractions. Fractions obtained

nearer the end of the pressing cylinder progressively resemble the first, second, and third pressings of conventional presses. Slower pressing can decrease the incorporation of suspended solids that diminish juice or wine quality, but also reduces one of the principal advantages of continuous-type presses, namely speed.

Pressing aids, such as cellulose or rice hulls (Plate 7.2), may be added to improve extraction. The use of such preparations can increase free-run yields by up to 5–15%. Occasionally, however, their addition may influence subsequent fragrance development. In addition, pectinases may be added to facilitate juice release – especially with slip-skin (*V. labrusca*) or other pulpy cultivars. Pectinase also eases filtering and improves clarity.

Must Clarification

White must typically is clarified before fermentation to favor the retention of a fruity character. Fruitiness may be masked by the excessive production of fusel alcohols, associated with juice containing high levels of suspended solids. The largest particles in the solids fraction seem particularly active in inducing fusel alcohol synthesis (Klingshirn *et al.*, 1987). In addition, much of the polyphenol oxidase activity is associated with particulate material. Thus, the early removal of suspended material is important to minimizing such enzyme-catalyzed

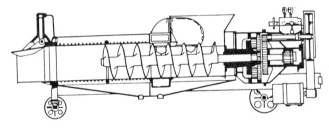

Figure 7.8 Schematic representation of a continuous press with hydraulic control. (Courtesy of Diemme)

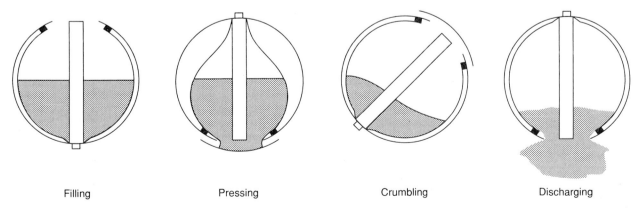

| Filling | Pressing | Crumbling | Discharging |

Figure 7.7 Diagram of the operation of a pneumatic press. Note the centrally located perforated plates for drainage and the inward-moving bladder membrane. (Courtesy of Willmes)

oxidation. High levels of suspended solids have also been reported to increase hydrogen sulfide production (Singleton *et al.*, 1975).

Although large amounts of suspended solids are generally undesirable, excessive clarification is equally inappropriate. Filtration and centrifugation can remove more than 90% of the fatty acids from must (Bertrand and Miele, 1984), as well as much of its sterol content (Delfini *et al.*, 1993). This loss can slow alcoholic fermentation, retard malolactic fermentation, diminish yeast viability, and may provoke excessive acetic acid production during alcoholic fermentation. The latter favors formation of ethyl acetate, which has a lower sensory threshold than acetic acid. Part of this relates to the removal of polyphenols, which, in combination with unsaturated fatty acids, suppress acetic acid production by yeasts (Delfini *et al.*, 1992).

Because suspended solids favor early malolactic fermentation, the retention of small amounts of suspended solids in the juice can be beneficial. Although concentrations of suspended solids between 0.1 and 0.5% seem desirable for several white grapes (Groat and Ough, 1978), the optimal value can vary with the cultivar and the wine style wanted. Within the range noted, juice fermentation usually goes to completion and is associated with the production of a suitable amount of fruit esters and higher alcohols. The precise reasons for these varied influences are poorly understood, but may involve factors such as the adsorption of toxic carboxylic acids produced during fermentation and the availability of essential nutrients. The colloidal content of the juice is also directly related to the production and release of extracellular macromolecules (notably mannoproteins) during fermentation (Guilloux-Benatier *et al.*, 1995).

White juice commonly is allowed to settle spontaneously for several hours (~12 h) before racking. Bentonite or potassium caseinate may be added to facilitate settling and subsequent protein stability. When used, bentonite is commonly added after an initial period of spontaneous settling. This minimizes production of a voluminous loose sediment and associated loss of juice. Occasionally, some of the precipitate may be left with the juice during alcoholic fermentation. This permits vital nutrients, such as sterols and unsaturated fatty acids, to remain available. Although bentonite is the most commonly used clarifying agent, its effects on wine quality are still contentious (Groat and Ough, 1978). Its removal of important fatty acids, such as palmitic, oleic, linoleic and linolenic acid, as well as squalene, stigmatanol, and β-sitosterol is difficult to predict or control. In contrast, potassium caseinate tends to remove only polyphenolics. Alternately, regulation of fatty acid removal during clarification can be regulated with flotation (Cocito and Delfini, 1997).

To speed clarification, juice may be centrifuged. By removing only suspended particles, centrifugation affects the chemical composition of the juice the least of any clarification technique. Although centrifugation equipment is expensive, minimal juice loss and speed have made it particularly popular. In contrast, an alternate technique, vacuum filtration, tends to remove solids excessively, resulting in longer fermentation times and higher volatile acidity (Ferrando *et al.*, 1998). Filtration with diatomaceous earth also may be used to promote prefermentative clarification.

Another technique that has been investigated is **flotation** (Ferrarini *et al.*, 1995). Flotation involves the implosion of microbubbles of gas (air, nitrogen, or oxygen) into the juice. As the gas rises to the surface, suspended solids adhere to the bubble surface. These are subsequently skimmed off. The procedure has several advantages over other clarification techniques, notably control over the degree of clarification desired. The technique also permits the associated hyperoxidation of the juice, if desired. In addition, it can be used as a pretreatment, to improve the efficiency of procedures such as cross-flow filtration.

Adjustments to Juice and Must

ACIDITY AND PH

Juice and must that fail to possess the desired acidity and pH may be adjusted before fermentation. **Acidification** at this time has the benefit of limiting the growth of spoilage microorganisms. In addition, postfermentative acidification is illegal in some jurisdictions. Flavor production is also generally better in musts fermented at a low pH.

In contrast, deacidification typically occurs after fermentation, when changes to wine acidity that may occur during fermentation have taken place. Deacidification can thus be based on actual, rather than predicted need. In addition, postfermentative deacidification permits the process to be delayed until spring, when other winery activities are less urgent.

No precise recommendations for optimum acidity and pH values are possible. The ideal characteristics of a wine are too dependent on the style and preferences of the winemaker and consumer. Nevertheless, the acceptable range for total acidity in most wines is generally between 5.5 and 8.5 mg/liter. White wines are generally preferred at the higher end of the scale, whereas red wines are more appreciated at the lower end. For pH, a range of between 3.1 and 3.4 is favored for white wines, and between 3.3 and 3.6 for most red wines. Somewhat lower values are usually preferred in the juice or must, because the pH often increases slightly during or after fermentation (frequently as a result of tartrate crystallization). Acidity is important not only in producing a clean fresh taste and favoring color stability, but also is crucial to proper aging and protection from microbial spoilage.

One of the oldest pH adjustment procedures is plastering. Gypsum ($CaSO_4 \cdot 2H_2O$) converts some of the potassium bitartrate to the free acid form:

$$CaSO_4 \cdot 2H_2O + 2 KH(C_4H_4O_6) \longrightarrow K_2SO_4 + H_2(C_4H_4O_6)$$
$$+ Ca(C_4H_4O_6) + 2H_2O$$

Calcium sulfate + Potassium $\longrightarrow$ Potassium + Tartaric
dihydrate   bitartrate   sulfate   acid

                                        + Calcium + Water
                                         tartrate

This form of acidification has fallen out of favor. It not only increases the wine's sulfur content, but also organic acids are readily available, relatively inexpensive, and do not similarly affect wine chemistry.

The high pH usually associated with low total acidity is corrected by direct **acidification** (addition of organic acids) (Buechsenstein and Ough, 1979). Tartaric acid is typically used because of its relative insensitivity to microbial decomposition. In addition, tartaric acid decreases the pH by inducing the precipitation of excess potassium (as a bitartrate salt). Acidification is commonly required in warm to hot viticultural regions, due to the extensive metabolism of malic acid near the end of grape maturation. Another alternative is lactic acid. It not only increases the acidic taste, but is also considered by some to enhance the perception of "body." Although citric acid may be substituted because it facilitates iron stabilization, it is susceptible to microbial degradation by lactic acid bacteria. This could lead to undesirably high concentrations of diacetyl.

Deacidification of excessively acidic juice, low in pH, may involve blending with juice of lower acidity and higher pH. Alternatively, some of the acid may be neutralized by the addition of calcium carbonate, potassium carbonate, or Acidex.

The most difficult situation occurs when juice shows both high total acidity and high pH. This situation is particularly common in cool climatic regions where grapes may possess both high malic acid and potassium contents. Nagel *et al.* (1988) suggest adding tartaric acid to adjust the malic/tartaric ratio to unity. This is followed by precipitation of the excess potassium by Acidex. Subsequent addition of tartaric acids establishes a desirable pH and acidity.

Amelioration is a means of deacidification involving dilution with water. Because it also reduces sugar content, the addition of sugar is required to readjust the °Brix upward. Although amelioration is illegal in most countries, it has the distinctive property of having little

affect on juice pH. This results from the dicarboxylic nature of tartaric acid and its low dissociation constant. The dilution of H^+, which results from the addition of water, is counterbalanced by increased dissociation of tartaric acid. Thus, acidity falls, but the pH is only slightly affected. Although reduced color, body and flavor may be consequences of amelioration, these effects may be desirable in intensely flavored varieties. Nevertheless, the greatest disadvantage of amelioration is its consumer image. The addition of sugar and, especially, water is commonly viewed as unprofessional, if not unscrupulous.

Postfermentation adjustments of acidity and pH are discussed in Chapter 8.

SUGAR CONTENT

The sugar content (**total soluble solids**) of juice is commonly measured with a hydrometer in units frequently referred to as °Brix (alternate units are Balling, Baumé, or Oechsle) (see Appendix 6.1). Because sugars constitute the major component of soluble grape solids over 18 °Brix (Crippen and Morrison, 1986), °Brix is a fairly accurate indicator of capacity of the juice to support alcohol production. More precise measurements of sugar content are available, but hydrometer determinations are usually adequate early in the winemaking process. In the field, refractometer readings of grape juice are typically used to measure °Brix.

As fermentation progresses, hydrometer readings become imprecise measures of sugar content. This results from the alcohol produced during fermentation affecting specific gravity reading, the property measured by hydrometers. Nevertheless, specific gravity can still be used to adequately indicate the termination of fermentation in dry table wines. For sweet fortified wines, correction tables are necessary for this purpose (Amerine and Ough, 1980).

Following the completion of fermentation, a precise chemical analysis of the residual sugar content of a wine is usually required (Zoecklein *et al.*, 1995). Even small amounts of residual sugars can affect the wine's microbial stability and, therefore, how the wine should be treated up to and including bottling.

When the juice sugar content is insufficient to generate the desired alcohol content, chaptalization may be used. **Chaptalization** usually involves the addition of a concentrated sugar solution. It was first advocated by Dr Chaptal in 1801 to improve the stability and character of wines produced from immature or rain-swollen grapes. The increased alcohol content, generated by the added sugar, improved both features.

Chaptalization is typically illegal in regions or countries where warm growing conditions obviate its need, but is usually permissible in areas where cool climates may prevent the full ripening of grapes.

Where permissible, chaptalization usually occurs under strict governmental regulation.

Although many factors influence the conversion of sugar to alcohol (Jones and Ough, 1985), 17 g sucrose typically yields about 10 g ethanol. For chaptalization, the sugar is first dissolved in grape juice and added near the end of the exponential phase of yeast growth (about 2–4 days after the commencement of fermentation) (Ribéreau-Gayon et al., 1987). By this time, yeast multiplication is essentially complete, and the sugar does not disrupt fermentation. The disaccharide sucrose is quickly converted into equimolar amounts of glucose and fructose. Concurrent aeration of the fermenting juice or must is often recommended.

In addition to elevating the alcohol content, chaptalization slightly augments the production of certain compounds in wine, for example, glycerol, succinic acid, and 2,3-butanediol. The synthesis of some aromatically important esters may also be increased, whereas others are decreased (Fig. 7.26). In some varietal wines, such as 'Riesling,' chaptalization can diminish the green or unripe taste derived from immature fruit (Bach and Hess, 1986). However, these influences cannot compensate for the lack of varietal character found in immature grapes, or grapes diluted by rains.

Various techniques have been investigated to improve the character of wines produced in poor vintages, without the addition of sugar. **Reverse osmosis** is one such technique (Duitschaever et al., 1991). Although first designed as an economical means of obtaining freshwater from seawater, reverse osmosis has found many applications in other industries, from sewage treatment to fruit-juice concentration. It is the latter application that has attracted the attention of enologists. In addition to offsetting problems, such as juice dilution during preharvest rains, reverse osmosis can concentrate fruit flavors in the juice. Reverse osmosis operates by forcing water out of the juice through a membrane that retains most of the sugars and flavoring components. Regrettably, membranes are not fully impermeable to varietal aromatics. Thus, important aroma components may be lost (Mietton-Peuchot et al., 2002). Small, highly volatile, water-soluble compounds, such as esters and aldehydes, are the most likely to be lost. The addition of untreated juice to concentrated juice may partially alleviate this problem. Concentration of volatiles removed with water and reintroduction into the treated juice constitute another possible solution. Development of filters with improved selective permeability may eliminate this problem. Another potential problem is augmentation in total acidity. This can be sufficiently marked as to require deacidification.

Cryoextraction is an alternative technique investigated to overcome deficiencies in sugar and flavor content (Chauvet et al., 1986). As with reverse osmosis, cryoextraction can be used with immature grapes, or berries swollen with water after rains. It also may be used to augment the sugar and flavor content of grapes in the production of sweet table wines. Cryoextraction is the technical equivalent of icewine production, except that overmature grapes are not used. Cryoextraction involves freezing the grapes, and the subsequent crushing and pressing of the partially frozen grapes. As the water in grapes cools and forms ice, dissolved substances become increasingly concentrated in the remaining liquid. Because berries of greater maturity (greater sugar content) freeze more slowly than immature grapes, preferential extraction of juice from the more mature grapes can be achieved. Although temperatures down to −15°C increase solute concentration, temperatures between −5°C and −10°C are generally sufficient to remove unwanted water. Cryoextraction appears not to produce undesirable sensory consequences.

Another technique that has been investigated is the **Entropie concentrator** (Froment, 1991). It involves juice concentration under vacuum at moderate temperatures (~20°C).

Although sugar adjustment is usually designed to increase °Brix, there is a growing market for low-alcohol wines. Reduced alcohol contents are usually generated by some form of dealcoholization, following completion of fermentation. However, a new technique offers the possibility of diminishing the capacity of juice to support ethanol production (Villettaz, 1987). The process entails the action of two enzymes, glucose oxidase and peroxidase. Glucose oxidase converts glucose to gluconic acid, a nutrient that yeasts cannot ferment. Hydrogen peroxide, produced as a by-product of glucose oxidation, is destroyed by peroxidase. The two reactions involved are:

$$2 \text{ Glucose} + 2 H_2O + O_2 \xrightarrow{\text{glucose oxidase}} 2 \text{ gluconic acid} + 2 H_2O_2$$

$$2 H_2O_2 \xrightarrow{\text{peroxidase}} 2 H_2O + O_2$$

With glucose oxidase, alcohol production can be reduced by about one-half, equivalent to the proportional concentration of glucose in the juice. Thus, ethanol production is dependent on the remaining fructose content of the juice. Because a steady supply of oxygen is required for this method of dealcoholization, the juice becomes oxidized and turns brown. Although most of these oxidized color compounds precipitate during fermentation, the wine is still left with a distinct golden color. The generally undesirable sensory aspects of this process are discussed in Pickering et al. (1999).

For smaller adjustments, choice of yeast strain may be sufficient. Modern tendencies to harvest grapes later, in the hopes of achieving higher flavor potential, have led to a general increase in wine alcohol content. This can also result in wines appearing too "hot" or unbalanced. The selection and breeding of strains that divert more fermentation by-products to glycerol can result in a reduction in alcohol content (de Barros Lopes *et al.*, 2003). However, the excess NAD^+ generated in glycerol synthesis may be used in oxidizing acetaldehyde to acetic acid. The undesirable increase in volatile acidity might be avoided by mutating aldehyde dehydrogenase. However, this would result in an undesirable increase in the accumulation of acetaldehyde. Thus, this solution, like so many others, is fraught with hidden compromises.

COLOR ENHANCEMENT – THERMOVINIFICATION (HOT PRESSING)

Several red cultivars seldom produce a dark red wine using standard vinification techniques, for example 'Pinot noir.' Standard procedures often extract only about 30% of the anthocyanin content. In such situations, various techniques have been developed to improve color extraction. Some of these have already been noted (cold maceration, pressing variations) or are discussed later (enzyme addition, various pumping-over options, choice of yeast strain, or the addition of enologic tannins). In many instances, their benefits depend on whether the grapes are deficient in cofactors for copigmentation, or result from augmented extraction of flavonoids involved in polymeric pigment formation. Another option is thermovinification.

Thermovinification is primarily used with cultivars relatively low in anthocyanin content, or with diseased grapes (contaminated with laccase). The procedure involves heating intact or crushed grapes to between 50 and 80 °C. Some versions entail exposing whole grapes to steam or boiling water (flash heating). Such treatments are typically short (~1min) and heat only the outer pigment-containing layers of the fruit to ~80 °C. This kills the skin cells, resulting in the quick release of anthocyanins during subsequent maceration (~45 °C for 6–10 h). Other versions consist of heating some or all of the pomace, or both the pomace and juice. The juice and pomace are typically heated rapidly (in a tubular heat exchanger) to between 55 and 70 °C. Heating may occur with or without continuous stirring. Maceration time varies inversely with the temperature (the higher the temperature, the shorter the duration (Wiederkehr, 1997), and the desires of the winemaker. If these treatments damage subtle varietal aromas, temperatures as low as 50 °C may be used. For especially delicate varieties, such as 'Pinot noir,' heating may be as low as 32 °C for 12 h (Cuénat *et al.*, 1991). The must is pressed immediately after heating (leading to the term "hot pressing"). The juice is cooled and fermentation commences.

Only mold-free grapes can safely be treated at temperatures below 60 °C, because laccase activity increases up to this temperature. Subsequent vinification may be conducted in the presence or absence of the seeds and skins. Each variation influences the wine attributes generated.

Thermovinification dramatically increases anthocyanin extraction (Fig. 7.9), and temperatures above 60 °C inactivate laccases. Thus, it is particularly useful with

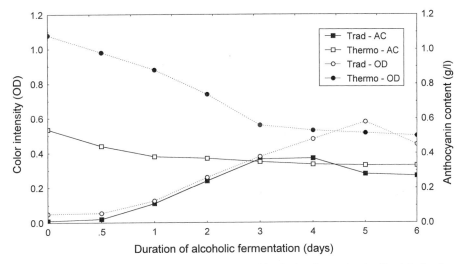

Figure 7.9 Development of color intensity (OD) and level of anthocyanins (mg/liter) during fermentation. Trad, traditional fermentation; Thermo, thermovinification; AC, anthocyanin content; OD, optical density. Fermentation ended after 5 days with a traditional fermentation and 3 days for thermovinification. (Regraphed from data in Ribéreau-Gayon *et al.*, 1976)

Botrytis-infected red grapes. Possibly because thermovinification does not also facilitate tannin extraction, the procedure is primarily used to produce wines designed for early consumption. In addition to generating a rich red color, thermovinification improves juice fermentability (both alcoholic and malolactic), produces wines low in astringency, reduces vegetal odors, and may enhance fruit flavors. Certain varietal aromas, such as those from *Vitis labrusca* cultivars, tend to be diminished and made more acceptable to consumers habituated to European wines. In addition, it may reduce the vegetative, grassy aromas that frequently mar wines produced from some red French-American hybrids. It allows the natural fruity fragrances of the grapes to express themselves. This has led to a trend in the eastern United States of blending red wines treated traditionally with those derived from thermovinified musts.

The rapid fermentation that follows thermovinification has several benefits, including the efficient use of fermentor capacity. However, the correspondingly rapid heat production increases the need for temperature control.

Occasionally, thermovinification generates undesirable bluish colors and cooked flavors. These usually can be avoided by appropriate adjustments to the technique, such as strict exclusion of oxygen and keeping the duration of heating as short as possible. Surprisingly, thermovinification provides a richer red color (less blue) with red French-American hybrid cultivars. Difficulties with clarification and filtering also may be experienced with thermovinification, due to the denaturation of grape pectinases. This can be corrected by the addition of pectinase.

ENZYME ADDITION

Advancements in industrial microbiology and chemical purification have permitted the isolation of enzymes in commercial quantities. Their use is now commonplace in many industries, including wine production. Filamentous fungi, notably *Aspergillus* and *Trichoderma* spp., are the primary sources of those authorized for wine use. They are active within the pH range and SO_2 conditions found in wine.

Enzyme preparations may be added to facilitate wine clarification, decoloration, dealcoholization, enhance flavor development, or augment anthocyanin liberation. However, because most commercial enzyme preparations are not fully purified, their effects often vary, depending on their source and conditions of use. A brief discussion of these applications is provided below. For additional details, see Lourens and Pellerin (2004) and van Rensburg and Pretorius (2000).

Most wine grapes lose their pulpy texture and become juicy as they ripen. In some situations, though, juice extraction can be improved by the addition of pectolytic enzymes immediately following crushing. This is particularly valuable with slip-skin (*Vitis labrusca*) cultivars. Pectinase preparations may also be added after pressing to improve juice clarity and filterability. Colloidal pectins can clog filters and retard the spontaneous settling of suspended particles. By partially degrading the negatively charged pectins that may surround positively charged grape solids, small particles self-attract. As their size enlarges, they precipitate, favoring the clarification of the juice. In most situations, pectinase is used only in the production of white wine.

Special pectolytic enzyme preparations are available to aid color and flavor release (Wightman *et al.*, 1997; Revilla and González-San José, 2003). These actions presumably relate to their macerating influence (inducing cell death and tissue disintegration). The benefit, if any, is often variety-specific, and largely limited to lightly colored red or slip-skin cultivars. In the first case, the principal benefit may be the release of additional color-stabilizing tannins; in the latter, degradation of pectins facilitates anthocyanin liberation. When producing red wines, the pectinase preparations must have no, or minimal, anthocyanase activity.

Although commercial pectinase preparations consist primarily of pectin lyase, they often contain additional enzymic attributes. Depending on the use, different formulations of pectinase enzymes are available. The action of pectin lyase on grape pectins (largely methylated) releases methanol, whereas preparations with polygalacturonase and pectin methyl esterase activities release less methanol (Revilla and González-San José, 1998). Additional enzymic activities (such as hemicellulases) may be incorporated to further assist color extraction and filterability.

In addition, preparations may possess glucanase activities that enhance juice or wine clarification – especially with grapes possessing significant amounts of viscous glucans (due to *Botrytis* infection). Glucanase preparations may also be used to promote earlier yeast autolysis, releasing mannoproteins and other cellular constituents.

Most preparations now possess little cinnamoylesterase (cinnamyl esterase) activities. Cinnamoylesterase breaks ester bonds between hydroxycinnamates and tartaric acid (notably caffeoyl tartrate). These can subsequently be decarboxylated to vinyl phenols during fermentation, if the yeast strain is a decarboxylase producer. At above threshold values, vinyl phenols generate a phenolic odor that is undesirable. More important, though, they can be metabolized to barnyardy-smelling ethyl phenols by many *Brettanomyces* strains.

Some pectinase preparations may also possess β-glucosidase activity. This can be a desirable attribute, if releasing glycosidically bound aromatics. This may be desired with terpenes or norisoprenoids, but undesirable for volatile phenols. Thus, pectinase preparations may affect flavor, occasionally with unanticipated effects (Lao *et al.*, 1997). Red wines are less susceptible to equivalent odor distortions. This difference partially results from the more intense flavors of red wines, and the tendency of volatile phenols to bond with tannins. Because firms preparing pectinase preparations are aware of these problems, there has been a concerted effort to minimize extraneous enzymatic activities.

Glycosidic linkages rupture naturally under acidic conditions (Mateo and Jimenez, 2000). However, acid-induced breakdown is slow, whereas heating to speed acidic hydrolysis induces flavor damage. Thus, most attention has been directed toward enzymatic hydrolysis.

Of enzyme preparations, β-glucosidases have been the most studied, even though glycosidase preparations containing α-arabinosidase, α-rhamnosidase, β-xylanosidase and β-apiosidase activities improve their effectiveness. This results from potential flavorants being frequently bound to sugars other than just glucose. These other sugars may need to be removed before glucosidase can have its effect. The effect of the enhanced flavor release must be assessed in trails. Accentuated aroma may distort the wine's traditional characteristics (e.g., 'Sauvignon blanc' and 'Chardonnay'), or sacrifice aging potential (the slow release of aromatics) for an intensely fragrant, young wine (e.g., 'Riesling').

Enzyme additions are typically supplied at the end of fermentation. Sugars in the juice inhibit the catalytic action of most of these enzymes (Canal-Llaubères, 1993). Although enzyme activity can be stopped when desired (by the addition of agents like bentonite), immobilization can provide even greater enzyme control (Caldini *et al.*, 1994).

Considerable concern was aroused several years ago when it was discovered that wines, especially those that were heated during processing, possessed a suspected carcinogen – ethyl carbamate (urethane). Ethyl carbamate can form as a reaction by-product between ethanol and urea. Urea can occur in wine as a result of arginine metabolism, addition as a nitrogen supplement, or as a nitrogen fertilizer in vineyards. Although choice of yeast strain and elimination of urea use can reduce the accumulation of ethyl carbamate, addition of urease (Ough and Trioli, 1988) can be used where past experience indicates that other control measures are insufficient.

One of the major developments in enzyme application involves immobilization on or within an inert support. Wine constituents are modified only when the wine is passed across or through the support. The technique has several distinct advantages over the addition of enzymes. It permits better control over the degree of modification, increases use efficiency, and avoids adding protein to the wine (possibly complicating protein stability). Although more costly, the advantages may outweigh the price differential.

OTHER ADJUSTMENTS

The addition of nitrogen (typically ammonium salts) and vitamins is typically unnecessary, but can improve the fermentability of botrytized and highly clarified white juice. Addition appears to be more effective when given periodically than in a single dose at the beginning of fermentation. DAP (diammonium phosphate) is occasionally added in the mistaken belief that it reduces the production of hydrogen sulfide. Yeast extract may be added to favor malolactic fermentation. If used, it is usually done at the end of fermentation.

Juice from moldy grapes may be exposed to flash heating (80–90 °C) for a few seconds to inactivate laccase, followed by rapid cooling. The treatment also denatures grape polyphenol oxidases. Although effective, the expense of the equipment, combined with the few occasions it might be used, means that it is seldom employed.

BLENDING

For white wine production, it is common to combine free-run juice with the first pressing. Occasionally, the second pressing may also be added. Other pressings usually are too tannic and difficult to clarify to be used. However, several finings and centrifugations may permit later pressings to be incorporated with the other fractions. Alternatively, late pressings and the pomace may be fermented to obtain alcohol for distillation or vinegar production.

For grape varieties such as 'Riesling,' pressings may contain two to five times the concentration of terpenes found in the free-run juice (Marais and van Wyk, 1986). Because of the importance of terpenes to the distinctive aroma of certain cultivars, the addition of pressings improves wine quality. The distribution of individual monoterpenes within the fruit varies with the cultivar, grape maturity, and the free- versus bound-state of the terpenes (Park *et al.*, 1991). Thus, the addition of pressings may affect both the quantitative and qualitative aspects of wine aroma.

Blending press fractions is also typical in red wine production. It presents the winemaker with another opportunity to adjust the flavor and color characteristics of the wine. Blending may also involve must from other vineyards or cultivars. Nevertheless, this is uncommon. Such forms of blending typically occur postfermentation. This has the advantage that the sensory attributes of the different lots of wine are known before blending begins (see Blending, Chapter 8).

DECOLORATION AND REDUCING BROWNING POTENTIAL

Although most white wines are produced from white grapes, some clones of white grapes, such as 'Gewürztraminer,' may produce small amounts of anthocyanin. In addition, white wines may be produced from red-skinned cultivars. Decoloration, if necessary, is usually conducted after fermentation, because anthocyanin levels typically decline spontaneously during fermentation. However, if experience indicates that this is inadequate, preparations with anthocyanase activity may be added before fermentation. By removing the sugar moiety from anthocyanins, their solubility declines, promoting precipitation during fermentation. Loss of the sugar moiety also makes the red pigments more susceptible to oxidative decoloration. Because anthocyanase is inactivated by sulfur dioxide, ethanol, and high temperatures, treatment normally follows juice clarification. At this point, the free sulfur dioxide content (if added) will be reduced and little ethanol will have been produced by the endemic yeast inoculum.

Another experimental method of color removal or browning prevention is the use of laccase. Laccase is a fungal polyphenol oxidase that has a wider range of substrates than grape polyphenol oxidases, and acts over a wider range of conditions. Regrettably, it can oxidize colorless glutathionyl caftaric acid complexes, enhancing subsequent browning potential. Because laccase is not a permitted food additive, immobilized laccase is being studied (Brenna and Bianchi, 1994). In this form, laccase remains attached to a reactor, through which wine is passed during treatment.

There are several means of reducing the browning potential of white wines, including gentle pressing, short maceration, phenolic removal with fining agents, acid addition to musts of high pH,[1] and the addition of sulfur dioxide before bottling. Another technique involves removing readily oxidized phenolics before fermentation. This may vary from permitting air exposure during crushing to active oxidation, such as bubbling air through the must (hyperoxidation) (Schneider, 1998).

Depending on the flavonoid content of the must, from one to three saturations of the must may be required (9–30 mg/liter) to protect the wine against subsequent oxidative browning. Without maceration, the oxygen uptake during crushing is usually inadequate to induce sufficient phenolic oxidation. Oxidized phenols usually precipitate during fermentation, leaving the wine bright and with little likelihood of in-bottle browning.

The removal is partially due to the adherence to yeast cells that precipitate post-fermentation. Hyperoxidation is usually conducted immediately after crushing and before clarification and sulfur dioxide addition. Sulfur dioxide reduces one of the primary oxidation products (caftaric acid quinone) and enhances the solubility of phenolic compounds. Hyperoxidation also reduces the bitterness and astringency of flavonoids extracted during longer skin contact. At least 2 hours after hyperoxidation clarification is required to remove the precipitated phenolics to prevent their resolubilization. Although the clarified juice appears brown, it subsequently clears during fermentation.

Whether hyperoxidation has positive, negative, or neutral effects on wine aroma depends largely on the variety employed, the potential for oxidative browning, and the type of clarification used. The effects on aroma development are highly variable and no generalities seem apparent.

ADDITION OF SULFUR DIOXIDE

Sulfur dioxide (50–100 mg/liter) may be added to the juice, depending on the health of the fruit and the maceration temperature. This practice used to be recommended on the belief that it controlled the growth and metabolism of indigenous members of the grape flora, as well as provided needed protection from must oxidation.

Recent research has thrown into question the efficacy or benefits of SO_2 addition, especially with healthy grapes, chilled and macerated at cool temperatures. Inoculation with *Saccharomyces cerevisiae* may itself be adequate to suppress other yeasts (Henick-Kling *et al.*, 1998). Although sulfur dioxide represses wild yeasts, before *Saccharomyces cerevisiae* takes over and completes fermentation (Fig. 7.10), the degree of suppression depends on the species and strain. With inoculated musts, rapid growth of the added strain speeds the development of anaerobic conditions, deletes the availability of limiting nutrients, and generates levels of ethanol that are inhibitory or toxic to most microbes at wine pH values. In spontaneous fermentations, sulfur dioxide tends to retard the onset of active fermentation, not only by suppressing *non-Saccharomyces* spp., but also most strains of *Saccharomyces* (Suzzi and Romano, 1982).

Sulfur dioxide is more effective against the bacterial flora. This action is aided by the low pH of grape juice, which increases the proportion of the most toxic form, molecular SO_2. Where the early onset of malolactic fermentation is desired, the addition of sulfur dioxide should be avoided. However, avoiding the addition of sulfur dioxide in must high in pH could favor the activity of spoilage lactic acid bacteria, notably *Pediococcus* and

---

[1] There are roughly nine times more highly oxidizable phenolate flavonoids at pH of 4.0 than at pH of 3.0 (Singleton, 1987).

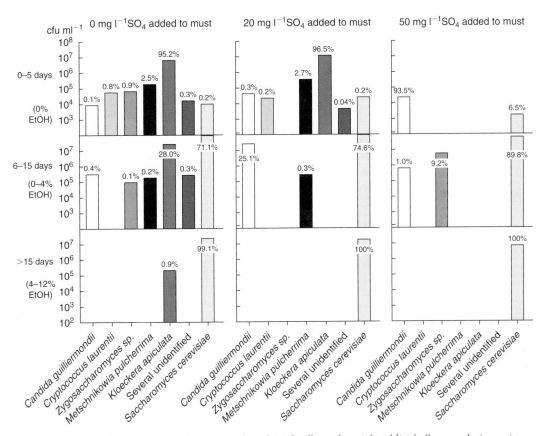

Figure 7.10 Predominance (% colonies recovered) and total cell numbers (cfu ml⁻¹) of all yeasts during uninoculated fermentations at 16 °C with three sulfite treatments. (From Henick-Kling *et al.*, 1998, reproduced by permission)

Lactobacillus spp. (Davis *et al.*, 1986a, b). Occasionally, though, sulfur dioxide can aid malolactic fermentation, by inhibiting endemic lactic acid bacteria infected with bacteriophage (Davis *et al.*, 1985).

If added, sulfur dioxide is usually supplied several hours before yeast inoculation, usually at crushing. During the settling period, the free SO_2 content declines (as it binds with sugars, carbonyls and phenolics in the must). Thus, the antimicrobial effect of sulfur dioxide is much reduced when, and if, the must is inoculated with one or more cultured yeast strains. For musts derived from moldy grapes, the dose of sulfur dioxide may need to be increased. Not only are the numbers of microbial contaminants much higher, but the presence of microbial by-products, such as glucuronic and galacturonic acids, can markedly reduce the antimicrobial influence of sulfur dioxide.

There is even more doubt about the advantages of adding sulfur dioxide to inhibit the action of grape polyphenol oxidases in white musts. Depending on the variety, between 25 and 100 mg SO_2/liter may be required to inhibit the early (enzymatic) oxidation of phenolics (White and Ough, 1973). As a result, these phenolics remain in the must, leading to an increased susceptibility to serious in-bottle browning. Regrettably, the undesirable oxidation, induced by fungal laccases (found in moldy grapes), is not controlled by commercially acceptable sulfur dioxide additions. In the past, ascorbic acid was added along with sulfur dioxide to limit early phenolic oxidation. Although effective in this regard, ascorbic acid induces even further delays in the oxidation and precipitation of readily oxidized phenolics, delaying the process until after bottling (Peng *et al.*, 1998). Ascorbic acid addition to white wine after crushing is now generally not recommended.

With red wines, sulfur dioxide can bleach anthocyanins. Although the bleaching is reversible, sulfur dioxide also binds to flavonoids, delaying the formation of stable colored complexes between anthocyanins and tannins. In addition, by binding with acetaldehyde and pyruvic acid, sulfur dioxide can delay the formation of vitisins (Morata *et al.*, 2006).

The effect of sulfur dioxide on the synthesis of flavorants by *Saccharomyces cerevisiae* and other yeast species (Herraiz *et al.*, 1990), as well as its potential to impart a metallic taste are additional points of concern. Thus, whether the antimicrobial and antioxidant effects of sulfur dioxide addition are beneficial or detrimental at this time depends on grape health and maturity, the cultivar involved, and the wine style desired.

Alcoholic Fermentation

Fermentors

Fermentors come in a wide variety of shapes, sizes, and technical designs. Most differ little in design from those used centuries ago. However, some are complex and fashioned for specific purposes. Most fermentors are straight-sided (Plates 7.3, 7.4), or have the form of slightly inverted cones. Tanks are differentiated from vats by being closed; vats have open tops. Tanks commonly double as storage cooperage, whereas vats can be used only as fermentors.

BATCH-TYPE FERMENTORS

During the fermentation of red wines, some of the carbon dioxide released by yeast metabolism becomes entrapped in the pomace. This causes the pomace to rise to the top, forming a **cap**. This severely limits contact between the pomace and most of the juice, retarding the extraction of anthocyanins and other compounds from the skins and pulp. Many of the design features of modern fermentors are intended to resolve this problem.

With vats, periodic submerging of the cap into the fermenting must (**punching down**) can be adequate. Before the 1900s, there were no simple, convenient methods of automatically achieving the benefits of manual punching down.

Tanks have almost completely replaced vats for the fermentation of all types of wine. Because they have closed tops, exposure to airborne contaminants and oxygen is largely avoided. Tanks also have the advantage of acting as potential storage cooperage throughout the year.

Since the 1950s, there has been a move away from wooden tanks (e.g., oak, chestnut, redwood) to more impervious and inert materials. Cement has been used in some regions, but stainless steel is probably the most generally preferred construction material. Fiberglass tanks are becoming more popular, because of their light weight and lower cost. Nevertheless, stainless steel has one advantage over other materials, rapid heat transfer. This property can be used to facilitate cooling of the fermenting juice. This can be achieved by flushing water over the sides of the tank, with evaporating water acting as a coolant. However, double-jacketed tanks, with coolants circulating between the inner and outer walls, provide more versatile temperature regulation. Temperature control in tanks constructed of other material usually requires the insertion of cooling coils or plates, or pumping the fermenting juice through cooling

coils. Adequate temperature control usually avoids the need for adding defoaming agents. Otherwise, these may be needed to prevent excessive froth development, and the wine loss associated with discharge through overflow valves. Common commercial defoaming agents consist of a mixture of mono- and diglucerides of oleic acid and polydimethylsiloxane.

As a construction material for fermentors, cement is difficult to surface-sterilize. Although an epoxy coating helps, it requires frequent maintenance. Coating with ceramic tiles also helps. Cement tanks have the initial advantage of being less expensive to construct than equivalent-sized, stainless-steel tanks.

Fermentors for white wine production are generally of simple design. The primary technical requirement is efficient temperature control. If not initially cool, the juice may be chilled before yeast inoculation. This is normally achieved with cooling coils, but can be more rapidly obtained by adding food-grade dry ice or liquid nitrogen. The ferment is usually maintained within a relatively narrow temperature range throughout fermentation. The desired temperature can vary considerably, depending on the desires and preferences of the winemaker. Cooler temperatures (10–15°C) tend to encourage the production and retention of fruit esters, whereas temperatures between 15 and 20°C favor the development of the varietal fragrance in certain cultivars.

Historically, fermentors for red wine production were vats of simple design. Cap formation, vigorous carbon dioxide production, and a higher phenol content often provided the fermenting juice with adequate protection from oxygen exposure. Also, if the cellar were cool and the volumes relatively low (50–100 hl), cooling during fermentation was unnecessary.

Red wines generally are fermented at, or allowed to warm to, 25–28°C. Manually punching the cap down into the fermenting must achieved several benefits. It helped slightly cool the fermenting must and partially equilibrate the temperate through the vat; facilitated the release and dispersion of potassium throughout the must (limiting an undesirable pH rise in the cap); and promoted the rapid extraction of anthocyanins. The first two features limit the potential growth of spoilage microbes in the cap, whereas the last favors optimal color extraction, while limiting excessive tannin uptake. In addition, punching down partially aerated the fermenting must (facilitating yeast growth) and limited the growth of spoilage organisms in the cap. Submerging exposes potential spoilage organisms in the cap to the toxic action of ethanol in the juice.

The shift from vats to large tanks for red must fermentation has required the development of mechanical replacements for manual punching down. This requirement has spawned an incredible array of solutions.

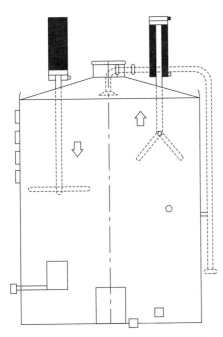

Figure 7.11 Diagram of a *pileage* fermentor showing the punching down action of a system of stainless steel plungers and flaps on the cap of fermenting red must. (Modified from Anonymous, 1983, reproduced by permission)

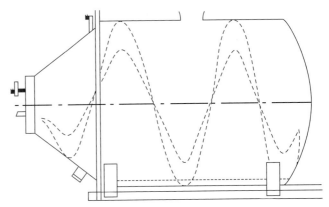

Figure 7.12 Diagram of a rotary fermentor. (Courtesy of CMMC)

One of the more novel is the *pileage* fermentor. It possesses cap punchers to simulate the action of manual punching down (Fig. 7.11). Other solutions include automatic and periodic, or continuous, pumping of the juice over the cap. This may be combined with temperature control by passing the wine through cooling coils. To limit oxygen uptake during pumping over, the headspace is often filled with inert gas (N_2 or CO_2). This is most important at the beginning and end of fermentation, when flushing of the headspace by carbon dioxide, released during fermentation, is limited.

An alternate procedure for encouraging color extraction and must cooling, combined with must aeration, is *délestage*. The procedure has several variants, but usually consists of the following. Shortly after fermentation has become sufficiently vigorous to produce a cap, the juice is drained into a holding tank. The juice may be drained and passed through a mesh to withhold seeds suspended in the juice (*déportation*). The juice, collected in the holding tank, is pumped and sprayed into a second tank, while the cap in the fermentor drains. The juice is then pumped back into the fermentor (achieving a second aeration). The procedure may be repeated several times during the course of fermentation. The procedure improves anthocyanin extraction and the early formation of polymeric pigments (Bosso *et al.*, 2001; Zoecklein *et al.*, 2004). It is also reported to be particularly valuable with incompletely ripened grapes. The latter benefit accrues by removing many of the immature ("green") seeds.

They possess a high proportion of extractable phenolics. The process also appears to result in distinct differences in the composition of fusel alcohols and various ethyl and phenyl esters (Zoecklein *et al.*, 2004).

Autofermentors are almost an automated version of *délestage* fermentors. They generally possess two superimposed fermentation chambers. The lower (main) chamber contains two traps into the smaller upper chamber. An elongated, perforated cylinder descends from one trap into the main chamber. As fermentation progresses, carbon dioxide accumulation increases the must volume. At a certain point, the increasing pressure forces this trap open and a portion of the carbon dioxide and fermenting juice gushes into the upper chamber. Here, the carbon dioxide escapes and the fermenting juice cools slightly. The weight of the juice forces the other trap to open downward. The flush of juice back in the main chamber ruptures the cap and temporarily disperses it into the fermenting must. Perforations in the cylinder prevent the pomace from escaping with the juice into the upper chamber. The cap in autofermentors is normally submerged. A simpler system for achieving a submerged cap involves a grill located below the surface of the must. Although autofermentors avoid the need of punching down, additional agitation is still required to achieve adequate mixing of the must and color extraction.

Rotary fermentors are another solution designed to improve and automate color and flavor extraction. The horizontal position of the fermentor increases the surface contact between the juice and the pomace. Rotation of the spirally shaped paddles (Fig. 7.12) gently but continuously mixes the fermenting juice with the seeds and skins. It also limits temperature stratification between the cap and fermenting juice (see Fig. 7.32). It also facilitates emptying. However, the primary benefit comes from the rapid extraction of flavor and anthocyanins. By permitting earlier pressing, the juice can be separated from the pomace before undesirable levels of bitter/astringent polyphenolics are extracted.

This provides a red wine that can be consumed early, with less barrel aging. Occasionally, the early pressing property has been used to permit the early transfer of the ferment to barrel for completion of fermentation. The disadvantage with this option is that, if the ferment is separated too early, reduced phenolic extraction can compromise the formation of stable, colored anthocyanin-procyanidin polymers.

Rotary fermentors, and other automated replacements for punching down, also can be used to reduce exposure of the fermenting must to oxygen. This avoids the likelihood of microbial spoilage, but may increase the incidence of sluggish or stuck fermentation, if used with highly clarified white juice. Although most frequently used for red wines, rotary fermentors may be used to shorten the maceration time needed to achieve flavor extraction from several white varieties, notably 'Chardonnay,' 'Gewürztraminer,' and 'Riesling.' The major disadvantage of rotary fermentors comes from the increased investment required in their purchase. Nevertheless, shorter holding periods reduce the number of fermentors needed to process the same volume of must.

Modern fermentors typically include some system to ease pomace discharge. For this purpose, sloped bottoms with trap doors are often used. Removable tank bottoms are another solution.

CONTINUOUS FERMENTATION AND RELATED PROCEDURES

Most fermentors for winemaking are of the **batch** type. In other words, separate lots (batches) are individually fermented to completion. In most industrial fermentations, continuous fermentation is the norm. In continuous fermentation, substrate is added at a relatively constant rate or at frequent intervals. Equivalent volumes of the ferment are removed to maintain a constant volume. Continuous fermentors may remain in uninterrupted operation for weeks or months. For the industrial production of single metabolic products, synthesized primarily during a particular phase of colony growth, continuous fermentation has enormous economic advantages. The technique is less compatible with wine production, especially with high-quality wines showing subtle and complex associations of hundreds of compounds produced at various times throughout fermentation.

Despite the potential advantages of continuous fermentation, it is rarely used, even to produce inexpensive wines. Because of their design and expense, continuous fermentors are most economical when used year-round. This in turn demands a steady supply of must. With the seasonal character of the grape harvest, this requires the storage of must under sterile, nonoxidizing conditions. These requirements necessitate more sophisticated storage than would be needed to store the corresponding volume of wine. Thus, technical and financial concerns generally outweigh the benefits of product uniformity and the easier alcoholic and malolactic fermentations achieved by continuous fermentors. Nevertheless, immobilization of the yeasts (Iconomou *et al.*, 1996; Verbelen *et al.*, 2006) may make the technique more accepted as a means of keeping costs down in the production of low-priced wines.

Another technique that has been investigated is the repeat use of yeasts in fermentation (Rosini, 1986). After each fermentation, the yeasts are isolated and used to inoculate successive fermentations. Removal is achieved by filtration, centrifugation, or spontaneous sedimentation. In addition to reducing yeast inoculation costs, there are further benefits to what is called **cell-recycle-batch fermentation**. The duration of fermentation is considerably diminished, and the conversion of sugars to ethanol is slightly improved. There is also a reduction in the synthesis of sulfur dioxide by yeast cells, but an increase in volatile acidity. Because yeast multiplication continues, but at progressively reduced rates, frequent monitoring for contamination by undesirable yeasts and bacteria is necessary. In addition, periodic assessment that the genetic characteristics of the yeast population have not changed is required. Both requirements may be reduced using immobilization in calcium alginate beads (Suzzi *et al.*, 1996).

Immobilization, involving the entrapment of cells, typically in alginate beads, has several potential advantages (Diviès *et al.*, 1994). Its use in sparkling wine production could significantly reduce the costs associated with disgorging (removing the yeasts prior to bottling). In addition, encapsulation appears to give cells enhanced resistance to low temperatures, and to high concentrations of ethanol and acetic acid (Krisch and Szajáni, 1997). Entrapment can also modify some of the byproducts of fermentation. For example, glycerol, propanol, and isoamyl alcohol production may be increased, whereas acetaldehyde generation is decreased.

FERMENTOR SIZE

Optimal fermentor size has more to do with the quantities of juice or must fermented than almost any other factor. When the volumes are large, immense fermentors are both needed and economically appropriate. When modest volumes are fermented, suitably small fermentors are adequate and preferable. Specially designed tanks for small must volumes are produced in a variety of materials, and are available from many manufacturers.

Small must volumes may result from limited land holdings. In addition, selective harvesting can generate small lots of special quality fruit. This is particularly striking with the higher-level Prädikat and other botrytized wines. Separate fermentation is essential to maintain the individuality of unique lots of grapes. In these situations, fermentation may be conducted in small oak cooperage.

Although **in-barrel fermentation** permits the fermentation of small lots of juice or must, it possesses a number of additional advantages, as well as disadvantages. Because cooling occurs only by passive heat radiation through the sides of the cooperage, fermentation may occur at temperatures higher than generally preferred, especially for white wines. Fruit-smelling acetate esters formed by yeasts dissipate more readily, along with the escaping carbon dioxide, at warmer temperatures (Fig. 7.29). For cultivars with distinctive aromas, this may achieve a clearer varietal expression; otherwise, it is a disadvantage. Wine fermented in small fermentors is usually left on the lees longer than in large fermentors. This favors early development of malolactic fermentation, but also increases the risk of off-odor production. This may be reduced by the use of yeast strains synthesizing little hydrogen sulfide. In addition, periodic mixing of lees and must (*battonage*) provides aeration that further decreases the development of reduced-sulfur odors. Yeast viability is also enhanced by slight aeration. This is viewed as contributing to better integration of oak flavors and tannins with the wine.

On the negative side, more effort is involved in topping, racking, cleaning, sterilizing, and maintaining small wood fermentors. There is also an increased risk of oxidation, as well as yeast and bacterial spoilage (Stuckey *et al.*, 1991). In small amounts, acetaldehyde production and the action of acetic acid bacteria, and the uptake of oak flavors, can increase wine complexity. Nevertheless, excess amounts can mar wine flavor. The reuse of barrels increases the risk of microbial contamination.

For many premium wines, fermentors commonly range in size from 50 to 100 hl. Such volumes appear to provide an appropriate balance between economics and ease of operation, and the desire to maintain wine individuality. For standard-quality wines, the economics of size favors fewer but larger fermentors. In this case, fermentors with capacities from 200 hl to more than 2000 hl (~50,000 gal) are common. Computers have facilitated the monitoring and regulation of fermentation in immense fermentors. Advancement in fiberoptic and biosensors may soon permit real-time analysis of multiple chemical parameters during fermentation.

Associated with increased fermentor volume are temperature-control problems. In large fermentors, passive heat dissipation via the surface is insufficient to prevent excessive heat buildup during fermentation. As a result, overheating and premature termination of fermentation are likely. This requires the instillation of sophisticated temperature control systems. Large volumes of must also increase the likelihood of excessive foaming and wine loss (and the need for defoamers). Sedimentation of large amounts of grape solids (and yeasts) can delay the onset of active fermentation. Without agitation, fermentation begins principally at the base, rising as fermentation becomes more active. Correspondingly, mixing of juice with yeasts, encouraged by the release of carbon dioxide, is retarded (Vlassides and Block, 2000). This may require the use of agitators, or at least automated pumping-over for red wines. Despite these potential problems, the economics of size often outweigh its disadvantages.

Fermentation

Fermentation is an energy-releasing form of metabolism in which both the substrate (initial electron donor) and by-product (final electron acceptor) are organic compounds. It differs fundamentally from respiration in not requiring the involvement of molecular oxygen. Although many fermentative pathways exist, *S. cerevisiae* possesses the most common – **alcoholic fermentation**. In it, ethanol acts as the final electron acceptor (by-product), whereas glucose is the preferred electron donor (substrate). Although *S. cerevisiae* possesses the ability to respire, it predominantly ferments, even in the presence of oxygen.

Although most organisms are able to ferment sugars, they do so only when oxygen is lacking. This partially results from the toxic action of the usual end products of fermentation, lactic acid or ethanol. In addition, fermentation is inherently an inefficient mode of energy release. For example, alcoholic fermentation converts only about 6–8% of the chemical-bond energy of glucose into readily available metabolic energy (ATP, adenosine triphosphate). Much of the energy remains bound in the terminal electron acceptor, ethanol.

The two main organisms involved in vinification, *Saccharomyces cerevisiae* and *Oenococcus oeni*, are somewhat unusual in selectively employing fermentative metabolism. *S. cerevisiae* is so adapted to fermentative metabolism that it can generate as many ATP/sec as is normally generated by respiration (Pfeiffer *et al.*, 2001). These properties are partially based on the presence of a highly efficient alcohol dehydrogenase (ADH1) (for the oxidation of acetaldehyde to ethanol), a high titer of glycolytic enzymes in the cytoplasm, and a mitochondrion that only produces respiratory enzymes in the presence of a preponderance of nonfermentable substrates (Ihmels *et al.*, 2005). In addition, both *S. cerevisiae* and *O. oeni* can withstand moderately high ethanol concentrations.

In addition to preferential alcoholic fermentation and alcohol tolerance, *Saccharomyces cerevisiae* has the properties of osmotolerance, relative insensitivity to high acidity, and acceptance of low oxygen concentrations. Thus, it is preadapted to growing in must and excluding other potential competitors in the must. Once dominating the environment, and in the absence of fermentable substrate, *S. cerevisiae* can switch to respiring the accumulated alcohol if oxygen is available.

O. oeni is less well adapted to growing in grape juice or must than *S. cerevisiae*. It typically grows slowly in juice, and most commonly in wine, after *S. cerevisiae* has completed alcoholic fermentation. In most habitats, the production of lactic acid by the bacteria lowers the pH of the substrate, excluding competitive bacteria. Lactic acid bacteria are one of the few acid-tolerant bacterial groups. However, the high acidity of grape juice and wine actually retards or inhibits the growth of most lactic acid bacteria. Thus, in this instance, the conversion of malic acid to lactic acid, a weaker acid, has the result of increasing pH, favoring their growth. Malolactic fermentation also makes excessively acidic wines more acceptable to the human palate, and may improve microbial stability by removing residual fermentable substrates.

As noted previously, wine is usually batch-fermented. Thus, nutrient availability is maximal at the beginning of fermentation, and declines progressively thereafter. By the end of fermentation, most sugars have been metabolized, leaving the wine "dry."

Batch fermentations generally show a growth pattern consisting of four phases – lag, log, stationary, and decline. Immediately following inoculation, cells need to adjust to the new environment. Because some cells do not acclimate successfully, there is an initial period in which the number of new cells produced approximates the number that die. This is called the **lag** phase.

Once adaptation is complete, most cells begin to multiply at a steady rate, until conditions become unfavorable. Because most microbes are unicellular, the growth curve approximates an exponential equation, and the phase is correspondingly called the **exponential** or **log** (logarithmic) phase. During this period, the population of viable cells rapidly increases to its maximum value.

As the nutrient content falls, toxic metabolic byproducts accumulate. Thus, after a period of rapid growth, the rate of cell division (growth) declines and approaches the rate at which cells die (or become metabolically inactive). The culture is now said to have entered the **stationary** phase. This involves considerable transcriptional modification by the nucleus (Rossignol *et al.*, 2003). As nutrient conditions continue to deteriorate, and the concentration of toxic metabolites keeps increasing, more cells die than divide. At this point, the culture enters a **decline** phase. Because most viable cells are not replaced, the colony eventually perishes, or becomes dormant.

Although similar, the population growth pattern displayed by yeast growth in grape juice shows several variations from the norm (Fig. 7.13). Typically the lag phase is short or undetectable; the exponential growth phase is relatively short (seldom amounting to more

than eight cell divisions when the must is inoculated); the stationary phase may be short and commence long before nutrients become limiting; and the decline phase is atypically long and the viable cell population can remain high for several months. As much as 40% of the sugar metabolized to alcohol may occur during the decline phase (Ribéreau-Gayon, 1985).

The brevity or apparent absence of a lag phase in yeast growth may result from the preadapted state of the cells initiating fermentation. Active dry yeast, commonly used for juice or must inoculation, comes from cultures grown exponentially in aerated media. Although these cells have functional mitochondria, capable of respiration, they also have a full complement of fermentative enzymes. Thus, little time is required for the conversion from respiration to fermentation in grape juice or must. Similarly, the indigenous yeast population on grapes requires little enzymatic adaptation to commence rapid cell growth. However, the absence of a prolonged lag period with spontaneous fermentation may be an artifact. Endemic yeast cells are commonly bathed in the juice released from broken grapes during harvesting and may pass through the lag phase before fermentation "officially" begins in the winery. In addition, yeasts growing on berry skins may exist under limited but concentrated nutrient conditions. Even the dormant yeast inocula derived from winery equipment may contain a full complement of enzymes, and be preadapted to rapidly initiate growth in grape must.

Although physiological adjustment to growth in grape juice appears minimal, a lag phase may be observed when conditions are less than optimal. Conditions such as low temperature ($\leq 10\,^{\circ}$C), and excessive protection of the juice from oxygen exposure during crushing, may disadvantage yeast cells. Active dry yeast cells are often leaky, and initially may lose vital nutrients (Kraus *et al.*, 1981). In addition, nitrogen deficiency and low juice pH can prolong the lag period. The latter probably results from the enhancement of the antimicrobial effects by any added sulfur dioxide (Ough, 1966a). High °Brix values or ethanol contents (for example, the second fermentation in sparkling wine production) also suppress yeast growth and fermentation rate (Ough, 1966a, b).

During the exponential phase, cells grow and reproduce at the maximal rate permitted by the prevailing conditions. The presence or absence of oxygen does not appear to affect the rate (Schulze *et al.*, 1996). The protein content of the cytoplasm approaches 60% (w/w) and the RNA content about 15% (w/w). Little storage carbohydrate (glycerol and trehalose) accumulates.

The early termination of exponential growth may be partially explained by the large inoculum supplied from grapes or added during inoculation. Inoculated juice or must often contains about 10^5–10^6 cells/ml at the onset

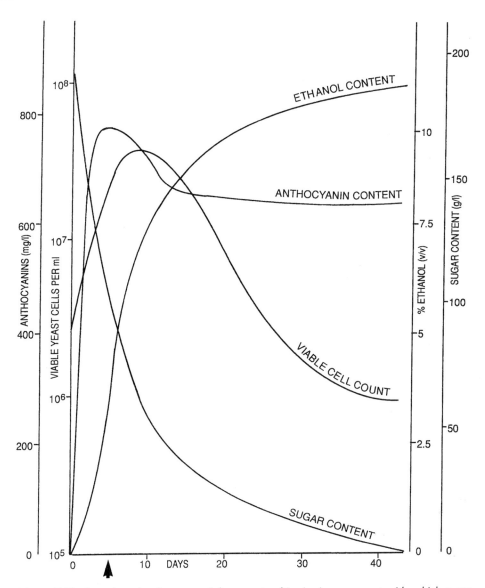

Figure 7.13 Growth cycle of yeasts and fermentation kinetics in grape must with a high sugar content. (Diagram courtesy of Herman Casteleyn)

of fermentation, which rises to little more than 10^8 viable cells/ml. Many more cell divisions occur during spontaneous fermentations, where the cell concentration is initially low, or is reduced as a result of clarification. Ethanol accumulation and sensitivity may partially explain why the viable cell counts seldom reaches more than 10^8 cells/ml. However, other factors are important because populations reaching 10^6–10^8 cells/ml still develop in juice initially fortified to 8% ethanol. These factors probably include the inability of yeasts to synthesize essential sterols and long-chain unsaturated fatty acids, and the accumulation of toxic, mid-size, carboxylic acid by-products of yeast metabolism. These disrupt cell membrane function. One factor not involved is a lack of fermentable substrate (the stationary phase commences with approximately half the sugar content unused). Nevertheless, a reduction in the supply of nitrogen may accentuate catabolic repression by glucose (Bely *et al.*, 1994), and ethanol accumulation may disrupt glucose transport into the cell. The remaining sugars are slowly metabolized during the stationary and decline phases – constituting up to 80% of the total fermentation period (Ribéreau-Gayon, 1985).

As yeast cells enter the stationary phase, there is a change in the enzyme complement, the production of several stress-related (heat-shock, Hsp) proteins (Riou *et al.*, 1997), and the accumulation of trehalose and glycerol. Trehalose stabilizes membrane fluidity (Iwahashi *et al.*, 1995) and limits protein denaturation (Hottiger *et al.*, 1994). Heat-shock proteins, produced by many organisms under stress conditions, also limit protein denaturation (Parsel *et al.*, 1994). These may play important roles in prolonging cell viability during the subsequent decline phase.

The initiation of the decline phase probably results from increasing membrane dysfunction becoming progressively disruptive to cellular function. Membrane disruption results from the combined effects of ethanol (Hallsworth, 1998) and mid-chain fatty (carboxylic) acid toxicity (Viegas *et al.*, 1998), plus a shortage in sterol precursors. The absence of oxygen may be an additional factor. Molecular oxygen is required for the synthesis of nicotinic acid, a vital component of the electron carriers NAD^+ and $NADP^+$. However, why the decline initially stabilizes at approximately 10^5–10^6 cells/ml for several weeks is unknown. Cell viability is improved by maceration on the skins before (white wines) or during fermentation (red wines). Subsequently, the remaining cells progressively die over the next several weeks to months.

Another distinction from most industrial fermentations is the nonpure status of wine fermentations. The normal procedure in most fermentations is to sterilize the nutrient medium before inoculation. Except for continuous fermentation, grape juice or must is not sterilized. In Europe, the traditional procedure has been to allow endemic yeasts to conduct fermentation. However, this is changing, with a shift toward induced fermentation with selected yeast strains (Barre and Vezinhet, 1984). Induced fermentation is standard throughout much of the world. Although sulfur dioxide often is added to limit the growth of indigenous (wild) yeasts, it is only partially effective in this regard (Martínez *et al.*, 1989; Henick-Kling *et al.*, 1998).

At one time, there was no adequate means of determining whether indigenous strains of *S. cerevisiae* were controlled by sulfur dioxide. With techniques such as mitochondrial DNA sequencing (Dubourdieu *et al.*, 1987) and gene marker analysis (Petering *et al.*, 1991), it is now possible to identify the strain(s) conducting fermentation. Although species occurring on grapes or winery equipment occasionally may dominate the fermentation of inoculated juice (Bouix *et al.*, 1981), inoculated yeasts appear to be the primary, if not the only yeasts detectable at the end of most fermentations (Figs 7.10 and 7.20).

The vinification of red must routinely occurs in the presence of high concentrations of epiphytic yeasts, regardless of yeast inoculation. White grapes, which are pressed shortly after crushing, cold settled, and quickly clarified, usually contain a diminished endemic yeast population. Nevertheless, white juice still possesses sufficient yeasts to initiate and conduct alcoholic fermentation. Although indigenous yeasts may be present and viable, they are probably not metabolically active throughout

inoculated fermentations. Admittedly, though, unequivocal evidence for this is absent.

Biochemistry of Alcoholic Fermentation

Glucose and fructose are metabolized to ethanol primarily via glycolysis (Embden–Meyerhof pathway) (Fig. 7.14). Although the primary by-product is ethanol, additional yeast metabolites generate the most common aromatic compounds found in wine. Yeast action also may influence the development of the varietal aroma by hydrolyzing nonvolatile aroma precursors, thus potentially liberating aromatic terpenes, phenols, and norisoprenoids (Laffort *et al.*, 1989). In addition, the changing physicochemical conditions produced during fermentation progressively modify yeast metabolism. This is reflected in the various phases of colony growth noted previously, related adjustments in the nutrient and energy status of the cells, and the substances released and absorbed throughout fermentation. Thus, much of the fragrance of wine can be interpreted in terms of modifications in primary and intermediary yeast cell metabolism.

ENERGY BALANCE AND THE SYNTHESIS OF METABOLIC INTERMEDIATES

During the changing phases of colony growth, yeasts have differing requirements for ATP and reducing power (in the forms of NADH or NADPH). These energy-carrying chemicals are required to activate cellular functions and maintain an acceptable ionic and redox balance in the cell. Redox balance refers primarily to the equilibrium between the oxidized and reduced forms of the two major pyridine nucleotides (NAD^+/NADH and $NADP^+$/NADPH).

As shown in Fig. 7.14, glucose and fructose are oxidized to pyruvate, primarily via glycolysis. During the process, electrons are transferred to NAD^+, reducing it to NADH. Pyruvate subsequently may be decarboxylated to acetaldehyde, which is reduced to ethanol by the transfer of electrons from NADH. In the process, NADH is reoxidized back to NAD^+.

The release of energy from glucose and fructose, and its storage in ATP and NADH, are inherently much less efficient via fermentation than by respiration. Most of the chemical energy initially associated with glucose and fructose remains bound in the end product, ethanol. Furthermore, the energy trapped in NADH is used to reduce acetaldehyde to ethanol. This process is necessary to maintain an acceptable redox balance. As most cells, yeasts contain only a limited supply of NAD^+. In the absence of oxygen, yeast cells are unable to transfer the energy stored in NADH to ADP (which is more

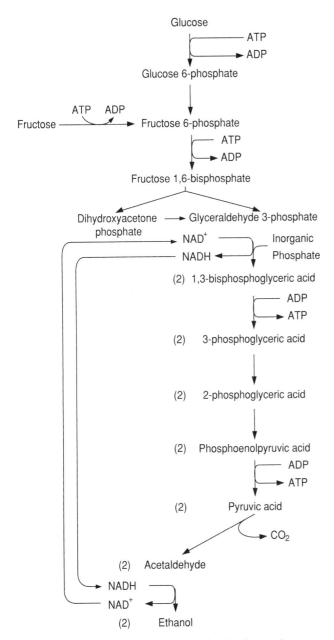

Figure 7.14 Alcoholic fermentation via the glycolytic pathway.

The low respiratory capacity of *S. cerevisiae* reflects its limited ability to produce the requisite enzymes in the presence of high sugar concentrations (glucose repression). The high proportion of glycolytic enzymes in yeast cytoplasm (about 50% of the soluble protein content) clearly demonstrates the importance of fermentation to wine yeasts (Hess *et al.*, 1969). Yeasts show high rates of glycolysis, usually about 200–300 μmol glucose/min/g cell weight (de Deken, 1966). It is estimated that about 85% of the sugars incorporated by *S. cerevisiae* are used in energy production, whereas only about 15% is involved in biosynthetic reactions. Specific values vary depending on the prevailing conditions during the fermentation, and the number of cell divisions involved in reaching a stationary population.

Although most fermentable sugars in juice are metabolized via glycolysis, some are channeled through the pentose phosphate pathway (PPP) (Fig. 7.15, *upper right*). This diversion is important to the production of pentose sugars needed for nucleic acid synthesis. The PPP also generates the NADPH required to activate particular cellular functions, such as amino acid synthesis (Gancelos and Serrano, 1989). Thus, amino acid availability in the juice can decrease the need for, and activity of the PPP. PPP intermediates not required for biosynthesis are normally directed through phosphoglycerate to pyruvate.

During alcoholic sugar fermentation the redox balance is maintained, but no NADH accumulates. However, yeasts need reducing power for growth and reproduction during the early stages of juice fermentation. Some of the required reducing power comes from the operation of the pentose phosphate pathway (notably NADPH) and the oxidation of pyruvic acid to acetic acid (yielding NADH). Additional supplies come directly from NADH generated in glycolysis. The diversion of NADH to biosynthetic functions means that acetaldehyde is not reduced to ethanol. It is released into the fermenting juice.

The changing needs of yeasts for reducing power during fermentation probably explain why compounds, such as acetaldehyde and acetic acid, are initially released into the juice during fermentation, but are subsequently reincorporated (Figs 7.16, 7.17). Early in fermentation, growth and cell division require considerable quantities of reducing power. In contrast, in the decline phase, NADH and NADPH may accumulate. This could suppress sugar fermentation by diminishing the supply of the requisite NAD$^+$ (Fig. 7.14). The reincorporation and reduction of compounds such as acetaldehyde and acetic acid help to balance the redox potential and permit continued fermentation.

The metabolic intermediates needed for cell growth and maintenance are generally synthesized from

abundant), forming ATP (adenosine triphosphate), and regenerating NAD$^+$. Under the anaerobic conditions of fermentation, the regeneration of oxidized NAD$^+$ requires the reduction of an organic molecule. In most cases, this is acetaldehyde, and the by-product is ethanol. Without the regeneration of NAD$^+$, the fermentation of sugars would quickly cease. The consequence is that alcoholic fermentation generates only about two molecules of ATP per sugar molecule, in contrast to the potential 24–34 ATPs produced via respiration. Most of the ethanol produced during fermentation escapes from the cell and accumulates in the surrounding juice.

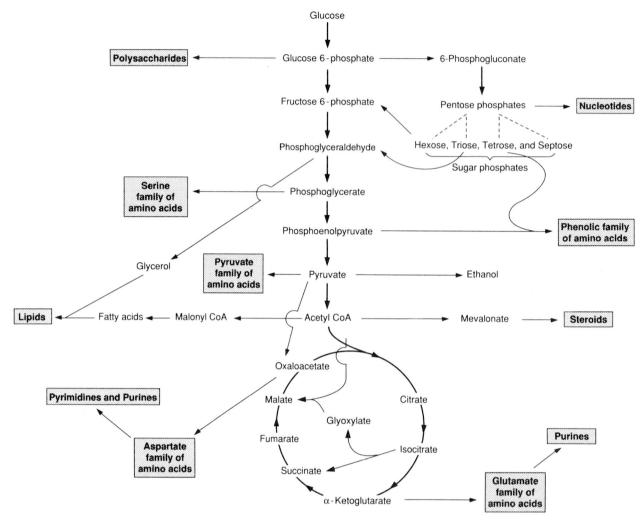

Figure 7.15 Core reactions of metabolism showing the main energy-yielding pathways (bold arrows) and the major biosynthetic products derived from central metabolism (boxes). The central pathway is the Embden–Meyerhof pathway of glycolysis, the top right shows a highly schematic pentose phosphate pathway (PPP), and the bottom is the TCA (tricarboxylic acid) cycle. Each pathway has been simplified for clarity by the omission of several intermediates. The directions of the reactions are shown as being unidirectional, although several are reversible. Energy transformations and the loss or addition of carbon dioxide are not shown. Under the anaerobic conditions of vinification, the TCA cycle does not function. However, except for the enzyme involved in the conversion of succinate to fumarate, those TCA enzymes present appear to be active only in the cytoplasm. In addition, decarboxylation of pyruvate to acetyl CoA is inactive and the glyoxylic acid pathway is suppressed (by glucose).

components of the glycolytic, PPP (pentose phosphate pathway), and TCA (tricarboxylic acid) cycles of central metabolism (Fig. 7.15). However, during vinification, most of the TCA-cycle enzymes in the mitochondrion are inactive. Isozymic versions of most of these enzymes (located in the cytoplasm) take over the function of generating the metabolic intermediates used in the biosynthesis of several amino acids and precursors of nucleotides (Fig. 7.15). The operation of the TCA cycle would produce an excess of NADH, disrupting cellular redox balance (oxidative phosphorylation is inoperative during fermentation). Disruption of redox balance is prevented because NADH produced during the oxidation of citrate to succinate (the "right-hand" side of the TCA cycle)

can be oxidized back to NAD^+ by reducing oxaloacetate to succinate (the "left-hand" side of the TCA cycle). The result is no net change in the redox balance associated with the synthesis of needed metabolic intermediates. In addition, NADH generated in glycolysis may be oxidized in the reduction of oxaloacetate to succinate, rather than in the reduction of acetaldehyde to ethanol. In both situations, an excess of succinate is generated. This probably explains why succinate is one of the major by-products of yeast fermentation. In yeasts, the PPP functions primarily in the generation of pentose sugars involved in the synthesis of nucleotides.

The replacement of TCA-cycle intermediates lost to biosynthesis probably comes from pyruvate. Pyruvate

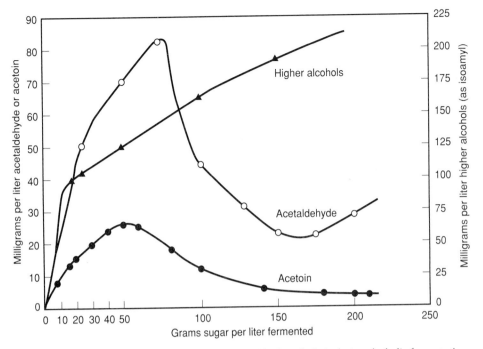

Figure 7.16 Formation of acetaldehyde, acetoin, and higher alcohols during alcoholic fermentation. The dynamics of the production of these compounds varies considerably with the yeast strain. (From Amerine and Joslyn, 1970, reproduced by permission)

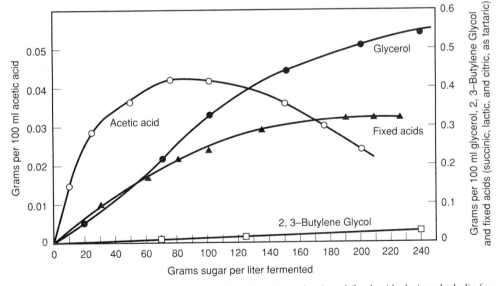

Figure 7.17 Formation of acetic acid, glycerol, 2,3-butylene glycol, and fixed acids during alcoholic fermentation. The dynamics of the production of these compounds varies with the strain. (From Amerine and Joslyn, 1970, reproduced by permission)

may be directly channeled through acetate, carboxylated to oxaloacetate, or indirectly routed via the glyoxylate pathway. The last pathway, if active, is probably active only near the end of fermentation. The glyoxylate pathway is suppressed by glucose. The involvement of biotin in the carboxylation of pyruvate to oxaloacetate probably accounts for its requirement by yeast cells.

The accumulation of another major by-product of fermentation, glycerol, also has its origin in the need to maintain a favorable redox balance, as well as its function as an osmoticum. The importance of the former function is suggested by the inability of mutants defective in glycerol synthesis to grow under anaerobic conditions (Nissen *et al.*, 2000). In addition, Roustan and Sablayrolles (2002) present evidence suggesting that glycerol synthesis, at least during the stationary phase, is associated with redox balance by eliminating excess reducing power. The reduction of dihydroxyacetone phosphate to glycerol 3-phosphate can oxidize the NADH generated in the oxidation of glyceraldehyde 3-phosphate in glycolysis (Fig. 7.18). However, the coupling of these two reactions does not generate ATP. This is in contrast to the net production of two ATP molecules during the fermentation of glucose to ethanol. The separate functions of redox balance and osmotolerance appear to be regulated by different isozymes of glycerol-3-phosphate dehydrogenase, GPD1 and GPD2 (Ansell *et al.*, 1997).

Increased glycerol production in the presence of sulfur dioxide probably comes from the need to regenerate NAD^+. The binding of sulfur dioxide to acetaldehyde inhibits its reduction to ethanol, the usual means by which NAD^+ is regenerated during alcoholic fermentation.

Throughout fermentation, yeast cells adjust physiologically to the changing conditions in the juice/must to produce adequate levels of ATP, maintain favorable redox and ionic balances, and synthesize necessary metabolic intermediates. Consequently, the concentration of yeast by-products in the juice changes continually during fermentation (Figs 7.16 and 7.17). Because several of the products are aromatic, for example, acetic acid, acetoin (primarily by conversion from diacetyl), and succinic acid, their presence can affect bouquet development. The accumulation of acetyl CoA (as a result of the inaction of the Krebs Cycle) may explain the accumulation and release of acetate esters during fermentation. The alcoholysis of acetyl CoA (and other acyl SCoA complexes) during esterification would release CoA for other metabolic functions. In addition, the formation of other aromatics,

notably higher alcohols, reflect the relative availability of amino acids and other nitrogen sources in the juice. Adequate availability also permits amino acids to be used as an energy source, or the release of organic acids, fatty acids, and/or reduced-sulfur compounds.

Although different strains of *S. cerevisiae* possess the same enzymes, their catalytic activities may vary. In addition, slight differences in regulation, or gene copy number, mean that any two yeast strains are unlikely to respond identically to the same set of environmental conditions. This variability undoubtedly accounts for many of the subtle, and not so subtle, differences between fermentations conducted by different yeast strains. For example, overexpression of cytoplasmic malate dehydrogenase increases not only the accumulation of malic acid, but also fumaric acid and citric acid (Pines *et al.*, 1997). In addition, overexpression of the enzyme glycerol 3-phosphate dehydrogenase not only results in a marked increase in glycerol production, but also augments the accumulation of acetaldehyde, pyruvate, acetate, 2,3-butanediol, succinate, and especially acetoin (Michnick *et al.*, 1997). The shift in metabolism toward glycerol synthesis is of enologic interest as it could result in "normal" alcohol contents in wines produced from grapes having high °Brix values. This is becoming more common, as delaying harvest to achieve maximum flavor development can lead to wines up to 15% ethanol.

INFLUENCE ON GRAPE CONSTITUENTS

Yeasts have their major effect on the sugar content of the juice or must. If fermentation goes to completion, only minute amounts of fermentable sugars remain (preferably ≤1g/liter). Small amounts of nonfermentable sugars, such as arabinose, rhamnose, and xylose also remain (~0.2g/liter). These small quantities of sugars have no sensory significance, and leave the wine tasting dry.

Yeasts may increase the pH by metabolizing malic acid to lactic acid. However, the proportion converted is highly variable, and can differ among strains from 3 to 45% (Rankine, 1966). Although some *Saccharomyces cerevisiae* strains can synthesize significant amounts of malic acid during fermentation (Farris *et al.*, 1989), it is a more common property of strains of *S. uvarum*. In contrast, *Schizosaccharomyces pombe* completely decarboxylates malic acid to lactic acid. It has been little used in juice deacidification because its sensory impact on wine has generally been negative. Undesirable flavors, such as hydrogen sulfide, often mask the fragrance of the wine. Delaying inoculation until after *S. cerevisiae* has been active for several days, or has completed fermentation, apparently reduces the negative impact of *Schizosaccharomyces pombe* on wine quality (Carre *et al.*, 1983). Immobilization appears to be an alternate and effective approach that does not spoil the wine's aromatic

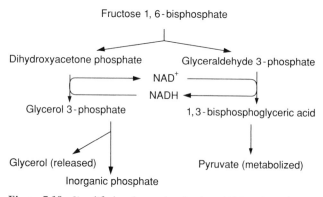

Figure 7.18 Simplified pathway showing how NADH derived from the oxidation of glyceraldehyde 3-phosphate to 1,3-bisphosphoglyceric acid is used in the reduction of dihydroxyacetone phosphate to glycerol. As a consequence, NADH is unavailable to reduce acetaldehyde to ethanol.

character (Silva *et al.*, 2003). It permits the degree of deacidification to be regulated by the winemaker.

During fermentation, the release of alcohols and other organics helps extract compounds from seeds and skins. Quantitatively, the most significant chemicals solubilized are anthocyanins and tannins. The extraction of tannins is especially dependent on the solubilizing action of ethanol. Anthocyanin extraction often reaches a maximum after 3–5 days, when the alcohol content produced during fermentation has reached about 5–7% (Somers and Pocock, 1986). As the alcohol concentration continues to rise, color intensity may begin to fall. This can result from the coprecipitation of anthocyanins with grape and yeast cells, to which they bind. Nevertheless, the primary reason for color loss is the disruption of weak anthocyanin complexes present in the juice. Freed anthocyanins may change into uncolored states in wine. Although extraction of tannic compounds occurs more slowly, tannin content often reaches higher values than that for anthocyanins. The extraction of tannins from the stems (rachis) may reach a plateau after about 7 days. Tannins from seeds are the slowest to be liberated. The accumulation of seed tannins may still be active after several weeks (Siegrist, 1985).

Ethanol also aids the solubilization of certain aromatic compounds from grape cells. Regrettably, little is known about the dynamics of the process. Conversely, ethanol decreases the solubility of other grape constituents, notably pectins and other carbohydrate polymers. The pectin content may fall by upward of 70% during fermentation.

The metabolic action of yeasts, in addition to producing many important wine volatiles, notably higher alcohols, fatty acids, and esters, also degrades some grape aromatics, notably aldehydes. This potentially could limit the expression of the herbaceous odor generated by C_6 aldehydes and alcohols produced during the grape crush. Yeast can also influence wine flavor by decarboxylating hydroxycinnamic acids to their equivalent vinylphenols. More specifically, fermentation may play a major role in the liberation of varietal aromatics, notably those bound in nonvolatile complexes with glycosides or cysteine, for example 4-mercapto-4-methylpentan-2-one (4 MMP) (Darriet *et al.*, 1993). It is an impact compound found in wines derived from 'Sauvignon blanc' and several other cultivars. Because yeast strains differ significantly in their ability to liberate 4 MMP (Howell *et al.*, 2004), choice of strain can significantly enhance or moderate the influence of this and related compounds. Yeasts are also involved in the liberation of other thiols from cysteine complexes, notably 4-mercapto-4-methylpentan-2-ol, 3-mercaptohexan-1-ol and 3-mercapto-3-methylbutan-1-ol (Tominaga *et al.*, 1998). In addition, yeasts may hydrolyze grape glycosides (Williams *et al.*, 1996), liberating aromatics such as monoterpenes and C_{13}-norisoprenoids.

Indirect effects of yeast action include modification of wine color (Eglinton *et al.*, 2004; Morata *et al.*, 2006). This may involve pigment loss, by adherence to grape cells and mannoproteins, or color stabilization by the release of carbonyls, such as pyruvic acid and acetaldehyde, and the generation of vinylphenols. Because yeast strains differ in these characteristics, strain selection can affect the color depth in red wines (Bartowsky, Markides *et al.*, 2004; Morata *et al.*, 2006).

Yeasts

CLASSIFICATION AND LIFE CYCLE

The term yeast refers to a collection of fungi that possess a particular unicellular growth habit. This is characterized by cell division that involves budding (extrusion of a daughter cell from the mother cell, Plate 7.5) or fission (division of the mother cell into one or more cells by localized ingrowths). Occasionally, they may form short chains of cells. Yeasts are also distinguished by possessing a single nucleus – in contrast to the frequently variable number of nuclei found in most filamentous fungi. In addition, the composition of yeast cell walls is unique. The major fibrous cell wall component of most fungi, chitin, is largely restricted to the bud scar (site at which daughter cells are produced). The major fibrous constituent of yeast cells is β-1,3-D-glucan. Also present are mannoproteins, covalently linked to either β-1,3-D-glucans, or smaller amounts of β-1,6-D-glucans.

Although characterized by a distinctive set of properties, yeasts are not a single, evolutionarily related group. The yeast-like growth habit has evolved independently in at least three major fungal taxa – the Zygomycota, Ascomycota, and Basidiomycota. Only yeast members of the Ascomycota (and related imperfect forms) are significant in wine production. Most imperfect yeasts are derived from ascomycete yeasts that have lost the ability to undergo sexual reproduction. Under appropriate conditions, the cells of ascomycete yeasts differentiate into asci – the structures in which haploid spores are produced through meiosis and cytoplasmic division.

In *S. cerevisiae* and related species, four haploid spores are produced as a consequence of meiosis (Fig. 7.19). After rupture of the ascal wall (originally the mother cell wall), spores typically germinate to produce new vegetative cells. Cells of opposite mating type usually fuse shortly after germination to reestablish the diploid state. Fusion to form diploid cells can occur before rupture of the ascal wall. Individual cells grow and bud about eight times before dying. Death results from disruption caused by the accumulation of circular copies

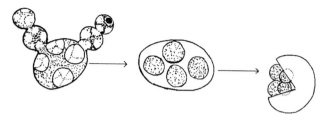

Figure 7.19 Stages of yeast development. *Left to right:* budding vegetative cells, ascospore development, spore release from ascus. A yeast cell can reproduce vegetatively about 20 times before it dies.

of rDNA in the nucleus (Sinclair and Guarente, 1997). Under appropriate conditions, diploid cells differentiate into asci and ascospores. Although wine strains of *S. cerevisiae* possess the potential for sexual reproduction, the property is rarely expressed in must or wine. Ascal development is suppressed by high concentrations of carbon dioxide and either glucose or ethanol – features characteristic of fermenting must.

If sporulation is desired, as in breeding experiments, nutrient starvation, the addition of sodium acetate, or both can promote ascospore production. The accumulation of bicarbonate in the growth medium also acts as a meiosis-promoting factor (Ohkuni *et al.*, 1998).

Until the late 1970s, yeast classification was based primarily on physiological properties, and on the few morphological traits readily observable under the light microscope. These have been supplemented with nucleotide sequence analysis and DNA–DNA hybridization. These procedures hopefully will open a new period of more stable classification based on evolutionary relationships.

In the most recent taxonomic treatment of yeasts (Kurtzman and Fell, 1998), many named species of *Saccharomyces* were reduced to synonymy. This does not refute the differences formerly used to distinguish species, but rather indicates that they are either minor genetic variants, or genetically unstable. Most of these former species are viewed as physiological races of recognized species, or occasionally of different genera. For example, *S. fermentati* and *S. rosei* are considered to be strains of *Torulaspora rosei*. However, *S. bayanus*, *S. pastorianus* (*S. carlsbergensis*) and *S. uvarum* are related to *S. cerevisiae*, but considered sufficiently distinct to justify species status (Nguyen and Gaillardin, 2005).

Saccharomyces cerevisiae and related species (*Saccharomyces sensu stricto*) are apparently evolved from an ancient chromosome doubling (autopolyploidy) (Wolfe and Shields, 1997; Wong *et al.*, 2002). This was followed by inactivation or loss of most of the duplicates (diploidization). Although this is estimated to have occurred millions of years ago, hybridization and polyploidy still occur within the genus (de Barros

Lopes *et al.*, 2002; Naumova *et al.*, 2005). For example, *S. pastorianus* is thought to be an alloploid hybrid between *S. cerevisiae* and *S. bayanus,* or another species. A listing of accepted names and synonyms of some of the more commonly found yeasts on grapes or in wine is given in Appendix 7.1. Differences between some of the physiological races of *S. cerevisiae* (formerly given species status) are noted in Appendix 7.2.

Yeast Identification Identification procedures, based on mitochondrial DNA, PCR, and other molecular technologies now permit the rapid (hours vs. days) identification of species, and even strains from small samples (Cocolin *et al.*, 2000; Martorell *et al.*, 2005). Their speed and precision are beginning to replace traditional culture techniques based on physiological properties and production of the sexual phase. Molecular techniques are particularly useful in the early detection of potential spoilage organisms. It may also permit assessment of physiological activity. Regrettably, these techniques are costly and currently available only in research centers or commercial laboratories. Alternate techniques for the rapid differentiation between fermentative and spoilage yeast involve analysis of free fatty acids using gas chromatography or pulsed field electrophoresis. Computer-based analysis of data via synoptic keys, in contrast to structured dichotomous keys (Payne, 1998) would facilitate identification using standard culture techniques. In their absence, specialized culturing procedures, such as those given by Cavazza *et al.* (1992), or the standard methods noted in Kurtzman and Fell (1998) are effective, but slow.

Standard procedures require that the organism be isolated as a single cell and grown on laboratory media. This is typically achieved by dilution from the source, containing upward of a billion cells/ml. Typically, different media are required for the isolation of different species. The isolation and sterile culture of each species or strain are essential because identification is traditionally based on the physiological capacity of the organism to grow with or without particular nutrients. Depending on how the isolation is done, data on the number of viable cells of each species and strain may be obtained. However, just because an organism was isolated does not necessarily mean that it was growing or physiologically active in the juice, must, or wine. Conversely, inability to culture an organism does not necessarily mean it was not present in a viable state. Yeast cells may survive for extended periods in a dormant (unculturable) state.

Because of the time, equipment, and experience required for the use of traditional identification techniques, these will not be discussed here. In all but large wineries, yeast identification is typically contracted out to commercial laboratories.

YEAST EVOLUTION AND GRAPE FLORA

Saccharomyces cerevisiae is undoubtedly the most important yeast species. In various forms, it may function as the wine yeast, brewer's yeast, distiller's yeast, or baker's yeast. Laboratory strains are extensively used in industry and in fundamental studies of genetics, biochemistry, and molecular biology. For all its importance, the natural habitat of *S. cerevisiae* seems only now becoming known. Current evidence indicates that its indigenous niche is the sap and bark of oak trees, and possibly adjacent soil. In these sites, it coinhabits with a sibling species, *S. paradoxus*. It is often viewed as the progenitor of *S. cerevisiae*. This view is supported by their extensive genetic and physiological similarity, equivalent natural habitats, and ability to ferment wines to dryness (Redžepović *et al.*, 2002).

Wild strains of *S. cerevisiae* and *S. paradoxus* differ primarily in being reproductively isolated, possessing genomic sequence divergence, and possessing different thermal growth profiles. They also differ in that wild strains of *S. cerevisiae* show little genetic separation across continents vs. the partial reproductive isolation between dispersed populations of *S. paradoxus* (Naumov *et al.*, 1998; Sniegowski *et al.*, 2002). Wild strains of *S. cerevisiae* are distinguishable from wine strains by the absence of chromosomal polymorphisms, being prototrophic, and being sporulation-proficient. In contrast, wine strains exhibit significant DNA sequence diversity, chromosome length polymorphisms, often possess auxotrophic or lethal mutants, express considerable heterozygosity, and show variable spore viability (see Mortimer, 2000). Currently, there is no consensus as to the evolutionary relationship between wild and domesticated strains of *S. cerevisiae*. Nevertheless, it appears that wild *S. cerevisiae* gave rise to two domesticated lines, one that developed into wine yeasts (and subsequently beer and bread yeasts). The other line gave rise to saké yeasts (Fay and Benavides, 2005). There is also support for an "out of Africa" origin of domesticated yeast (Fay and Benavides, 2005).

Although *S. cerevisiae* has occasionally been isolated from fruit flies (*Drosophila* spp.), and may be transmitted by bees and wasps, the importance of insects in the dispersal of *Saccharomyces* is unresolved (Wolf and Benda, 1965; Phaff, 1986). *S. cerevisiae* is usually absent or rare on healthy grapes. Even in long-established vineyards, the isolation of *S. cerevisiae* (in small numbers) occurs only near the end of ripening. Mortimer and Polsinelli (1999) estimate about one healthy berry per thousand carries wine yeasts. However, on surface-damaged fruit, the frequency may rise to one in four (~1 $\times 10^5$ to 1×10^6 yeast cells/berry). In Croatia, Redžepović *et al.* (2002) found that *S. paradoxus* was more frequently found on grapes than *S. cerevisiae*.

Studies on the sensory effects of *S. paradoxus* are given in Majdak *et al.* (2002).

Related *Saccharomyces* species, *S. bayanus* and *S. uvarum*, can also conduct equally effective alcoholic fermentations. Nevertheless, they are less commonly encountered and are typically employed in special wine-making situations. For example, *S. bayanus* var. *bayanus* has properties especially well adapted to the production of sparkling wines and *fino* sherries. *S. bayanus* is also well adapted to fermenting white wine from relatively neutral flavored grape varieties in warm climates. It tends to produce little volatile acidity, augments the malic acid content, and generates more aromatic alcohols and ethyl esters (Antonelli *et al.*, 1999). For cool fermentations (below 15 °C), the cryotolerant *S. uvarum* (*S. bayanus* var. *uvarum*) is of particular value. It is often involved in the production of tokaji, amarone and sauternes. In some regions, such as Alsace, it may be the predominant fermentative yeast (Demuyter *et al.*, 2004). The origin of these two species (often considered strains) is unknown, as are their natural habitats. Wild strains have occasionally been isolated from the caddis fly, some mushroom species, and hornbeam tree exudate (*Carpinus*).

There are general indications as to the effects of different species and strains, but precise prediction of their actions under specific conditions is impossible. Individual experimentation over several years is the only way to determine their effects under local conditions.

Although wine yeasts rarely occur in significant numbers on grapes, leaves or stems, other species are frequently found. These include species of *Hanseniospora* (*Kloeckera*), *Candida*, *Pichia*, *Hansenula*, *Metschnikowia*, *Sporobolomyces*, *Cryptococcus*, *Rhodotorula*, and *Aureobasidium*. Their population numbers change markedly during fruit ripening (Renouf *et al.*, 2005). These endemic yeasts may be found on the fruit pedicel, but occur more frequently on the callused terminal ends, the receptacle. They are most frequently found around stomata on the fruit, or next to cracks in the cuticle. In the last site, they routinely form small colonies (Belin, 1972). They presumably grow on nutrients seeping out of openings in the fruit, receptacle, and pedicel. Yeasts do not grow on the plates of wax that cover much of the berry surface, that produce its matte-like **bloom**. In fact, yeasts cease to grow where they come in contact with the waxy plates of the cuticle.

Commonly, these cells are in a dormant or slowly reproducing state on fruit surfaces. This unquestionably results from the predominantly dry state of the cuticle. Microbes require at least a thin coating of water to be metabolically active. This occurs only during rainy spells, or when fog or dew condenses on grape surfaces. In contrast, damaged fruit surfaces, where juice may escape, provide an ideal site for yeast (and bacterial) growth.

As noted, grapes do not appear to have been the natural (ancestral) habitat for *S. cerevisiae* (or its progenitor). It appears that winery equipment, and the winery itself, act as the major inoculum source for most spontaneous fermentations (Ciani *et al.*, 2004; Santamaría *et al.*, 2005). This is suggested by the repeated isolation of one or a few dominant strain(s) from a particular winery. Outside the winery, spread of strains seems limited and occurs only over short distances (Valero *et al.*, 2005). A possible exception occurs where pomace is spread as a vineyard fertilizer. In addition, regional strains isolated from vineyards often differ (Khan *et al.*, 2000). Thus, there may be scientific support for the view that the indigenous yeast population might contribute to a site's regional character.

The most frequently occurring yeast species on mature grapes is *Kloeckera apiculata*. In warm regions, the perfect state of *Kloeckera* (*Hanseniaspora*) tends to replace the asexual form. Other yeasts occasionally isolated from grapes include species of *Brettanomyces, Candida, Debaryomyces, Hansenula, Kluyveromyces, Metschnikowia, Nadsonia, Pichia, Saccharomycodes,* and *Torulopsis* (see Lafon-Lafourcade, 1983). Filamentous fungi, such as *Aspergillus, Botrytis, Penicillium, Plasmopara,* and *Uncinula* are rarely isolated, except from diseased or damaged fruit. Similarly, acetic acid bacteria are usually found in significant numbers only on diseased or damaged fruit.

Under most conditions, especially where must is inoculated with a cultured strain of *S. cerevisiae*, the indigenous yeast flora has generally been believed to be of little significance in winemaking. The acidic, highly osmotic conditions of grape juice restrict the growth of most yeasts, fungi, and bacteria. Also, the rapid development of anaerobic conditions and the speedy production of ethanol further limit the growth of competitive species. Nevertheless, in musts low in sugar content, the activity of yeasts less tolerant of high sugar and ethanol contents may be important (Ciani and Picciotti, 1995). Diseased grapes also have significantly modified and enhanced populations of epiphytic yeasts. These can markedly affect the outcome of fermentation, even with inoculation.

Even on healthy grapes, other yeasts may occur at concentrations as high as those of inoculated strains of *S. cerevisiae* (Fig. 7.20). Molecular evidence indicates that they continue to persist after viable cultures can no longer be isolated (Cocolin and Mills, 2003). At present, it is not established how commonly, or for how long, other yeast species remain metabolically active during induced fermentations (Millet and Lonvaud-Funel, 2000). Thus, their significance in vinification remains uncertain, and possibly underestimated. However, in some instances, they maintain or increase their numbers during fermentation (Mora and Mulet,

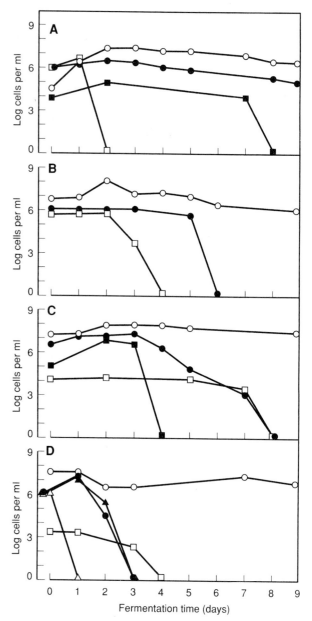

Figure 7.20 Yeast numbers during fermentation of white (A–B) and red (C–D) wines. ○, *Saccharomyces cerevisiae*; ●, *Kloeckera apiculata*; □, *Candida stellata*; ■, *C. pulcherrima*; ▲, *C. colliculosa*; △, *Hansenula anomala*. The initial population of *S. cerevisiae* comes predominantly from the inoculation conducted in the fermentations. (From Heard and Fleet, 1985, reproduced by permission)

1991; Fig. 7.20). This appears to be more common when fermentation temperatures are low (Heard and Fleet, 1988), or yeast inoculation is delayed (Petering *et al.*, 1993). Only in diseased or damaged grapes is the microbial grape flora clearly known to play an important role in vinification.

In addition to inoculation by the epiphytic grape flora, the juice and must may become inoculated from winery equipment (notably crushers, presses, and sumps) and

Plate 2.1 *Vitis rotundifolia*: (A) fruit; (B) leaf. (A, photo courtesy of Chris Evans, The University of Georgia, www.insectimages.org; B, photo courtesy of Carl Hunter. © USDA–NRCS PLANTS Database/ USDA SCS. 1991. Southern wetland flora: Field office guide to plant species. South National Technical Center, Fort Worth, TX, respectively)

Plate 2.2 Cluster of grapes of *Vitis vinifera* f. *silvestris* in southern Spain. (Photo courtesy of D. Nuñez, Universidat de Murica)

Plate 2.4 Grapes showing partial to complete chimeras in their outer pigmented tissues (Photo courtesy of A. Reynonds, Brock University)

Plate 2.3 Embryogenesis and plant regeneration. (A, B) Effect of activated charcoal on embryogenesis: callus cultured on ½MS (A) and on ½ MSAC (B). (C, D) Calli cultured on ½ MSAC, at different stages of development of somatic embryos. (E) Very soft non-embryogenic calli (Type II). (F) Different stages of germination of the somatic embryos on a medium containing IAA, GA, and AC. (G) Plant developed on half strength MA in test tubes. (H, I) Acclimatization and culture under greenhouse conditions. (From López-Pérez *et al*., 2005, reproduced by permission)

Plate 3.1 Longitudinal section of a grape bud showing the three growing points or buds. (Photo courtesy of the late R. Pool, New York State Agricultural Experimental Station, Cornell University, Geneva, NY)

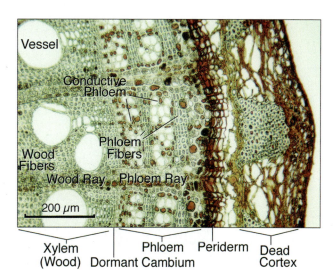

Plate 3.3 Internal anatomy of a cane: taken from a brown internode. (Photo courtesy M. Goffinet, Department of Horticultural Sciences, New York State Agricultural Experimental Station, Cornell University, Geneva, NY)

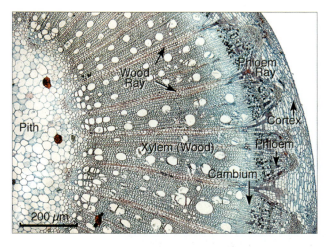

Plate 3.2 Cross-section of a young shoot: basal internode that is no longer elongating but is growing thicker. (Photo courtesy M. Goffinet, Department of Horticultural Sciences, New York State Agricultural Experimental Station, Cornell University, Geneva, NY)

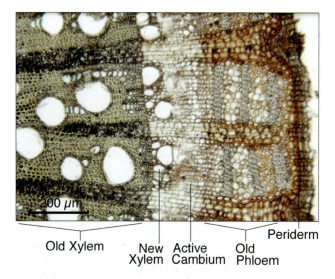

Plate 3.4 Anatomy of perennial woody stem in spring. (Photo courtesy M. Goffinet, Department of Horticultural Sciences, New York State Agricultural Experimental Station, Cornell University, Geneva, NY)

Plate 3.5 Portable whole canopy gas exchange system for several mature field-grown grapevines. (From Peña and Tarara, 2004, reproduced by permission)

Plate 3.6 Grape flower showing calyptra separated and being shed from the blossom. (Photo courtesy of D.D. Lorenz, Staatliche Lehr-und Forschungsanstalt für Landwirtshaft, Weinbau und Gartenbau, Neustadt)

Plate 3.7 Clusters of 'Ungi blanc' showing the effect of the *flb* mutant on fruit size (*top*) and internal structure before ripening (*bottom*). (Photo courtesy L. Torregrosa, UMR BEPC Campus Agro-M/INRA, Montpellier)

Plate 3.8 Effect of gibberellin (GA$_3$) application to grapevine flowers on the spacing of 'Riesling': untreated (*left*); treated (*right*). (Photo courtesy of M. Pour Nikfardjam, Staatl. Lehr- und Versuchsanstalt fuer Wein- und Obstbau, Weinsberg)

Plate 3.9 *Top:* light-proof boxes in the field at the time of application at flowering. *Bottom right:* 'Shiraz' berries 2 weeks pre-*véraison* showing differences in chlorophyll accumulation between treatments. *Bottom left:* 'Shiraz' berries 6 weeks post-*véraison*, showing no obvious difference in coloration between treatments. The shaded bunch is on the left in both pictures. (From Downey *et al.*, 2004, reproduced with permission)

Plate 4.1 Mechanical pruning of vines trained to the Geneva Double Curtain training system using the "Trimmer". (Photo courtesy of C. Intrieri, Università di Bologna)

Plate 4.2 Head-trained 100-year-old 'Zinfandel' vine shortly after bud break. (Photo courtesy of A. Reynolds, Brock University)

Plate 4.3 Wide row planting, Napa Valley. (Photo courtesy of Napa Valley Vintners)

Plate 4.4 Narrow row planting. (Photo courtesy California Wine Institute)

Plate 4.5 VSP (Vertical Shoot Positioning) training system application in Napa Valley. (Photo courtesy of Napa Valley Vintners)

Plate 4.6 Lyre training system. (Photo courtesy of Australian Grapegrower and Winemaker)

Plate 4.7 Vines trained to a RT2T trellis, shortly after bud break. (Photo courtesy of R. Smart, Smart Viticultural Services)

Plate 4.8 Minimally pruned vines, shortly after bud break. (Photo courtesy of R. Smart, Smart Viticultural Services)

Plate 4.10 Root distribution in high soil potential with poor chemical amelioration during soil preparation (acidic subsoil with low phosphates). (Photo courtesy of Dr E. Archer, University of Stellenbosch, South Africa)

Plate 4.9 Root distribution as a result of poor physical soil preparation (A). (Photo courtesy of Dr E. Archer, University of Stellenbosch, South Africa)

Plate 4.11 Well-buffered root system as a result of good physical and chemical soil preparation. (Photo courtesy of Dr E. Archer, University of Stellenbosch, South Africa)

Plate 4.12 Root depth restriction due to hardpan development through annual mechanical cultivation. (Photo courtesy of Dr E. Archer, University of Stellenbosch, South Africa)

Plate 4.14 *In vitro*-generated phylloxera nodosity on 'Cabernet Sauvignon' (three days old). (From Forneck *et al.*, 2002, reproduced by permission)

Plate 4.13 Remote weather station used in providing meteorological data for disease forecasting. (Photo courtesy of Adcom Telemetry, Inc.)

Plate 4.15 Closeup of colony of phylloxera adult female, eggs, and nymphs. (Photo by Jack Kelly Clark, from Flaherty *et al.*, 1982, reproduced by permission)

Plate 4.16 *Trichogramma* on Lightbrown apple moth eggs. (Photo courtesy of Institute for Horticultural Development, Knoxfield, Australia)

Plate 4.17 *Orius* (minute pirate bug) nymph feeds on mite. (Photo by Jack Kelly Clark, from Flaherty *et al.*, 1982, reproduced by permission).

Plate 4.18 Inflorescence necrosis in N-deficient Müller-Thurgau grapevines: (**A**), although fruit-set is low in all bunches, the proximal (basal) bunch is most severely affected; (**B**), a closeup of a proximal inflorescence. (From Keller *et al.*, 2001, reproduced with permission)

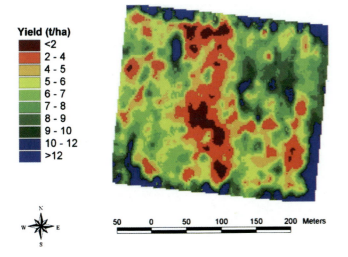

Yield (t/ha)
- <2
- 2 - 4
- 4 - 5
- 5 - 6
- 6 - 7
- 7 - 8
- 8 - 9
- 9 - 10
- 10 - 12
- >12

50   0   50   100   150   200 Meters

Plate 4.19 Variation in grape yield (1999) in a Coonawarra vineyard planted to 'Cabernet Sauvignon' in 1986. (From Bramley and Proffitt, 1999, reproduced by permission)

Plate 5.1 Illustration of winter damage. The primary bud is dead whereas the secondary bud is still alive. The tertiary bud cannot be evaluated at this level of cut. (Photo courtesy of the late Dr Bob Pool, New York State Agricultural Experimental Station, Cornell University, Geneva, NY)

Frost Injury in May

Plate 5.2 Late spring (May) frost injury to grapevine: Top left, longitudinal slice through growing healthy shoot, showing the developing vascular system inside the primary shoot as well as downward directed vascular strands into the cane; top left, healthy shoot (left) and three shoots (right) affected by frost damage; bottom left, cross-sectional slice through frost-damaged bud showing darkened cortical cells; bottom right, detail of the boxed section showing black frozen cortex cells resulting from springtime freeze. (Photo courtesy Dr. Martin Goffinet, Department of Horticultural Sciences, New York State Agricultural Experimental Station, Cornell University, Geneva, NY.)

Plate 5.3 Wind machines in Napa Valley. (Photo courtesy Napa Valley Vintners)

Plate 7.1 Old vertical wine press in Kloster Erbach, Rheingau, Germany. (Photo R. Jackson)

Plate 7.2 Horizontal press showing chains used to crumble the press cake (rice hulls were added to aid juice extraction for frozen grapes in icewine production). (Photo courtesy of E. Brian Grant, CCOVI, Brock University, St Catharines, Canada)

Plate 7.3 Winery with mid-sized stainless steel fermentation tanks. (Photo courtesy of Gary Pickering, CCOVI, Brock University, St Catharines, Canada)

Plate 7.4 Winery with large stainless steel fermentation tanks. (Photo courtesy of Gary Pickering, Brock University, St Catharines, Canada)

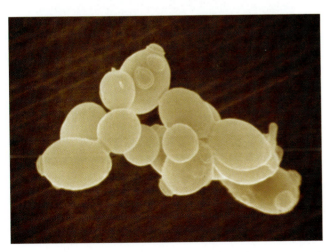

Plate 7.5 Scanning electron micrograph of *Saccharomyces cervisiae*. (Photo courtesy Lallemand)

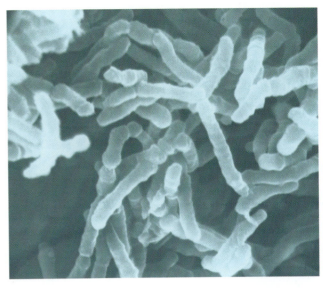

Plate 7.6 Scanning electron micrograph of *Oenococcus oeni*. (Photo courtesy Lallemand)

Plate 8.1 Roman funerary monument found at Neumagener, Germany, showing representations of wine barrels being transported by boat

Plate 8.2 Assembly of an oak wine barrel. (Photo courtesy California Wine Institute)

Plate 8.5 Bottling plant with suspended sound dampening devices. (Photo courtesy of Gary Pickering, CCOVI, Brock University, St Catharines, Canada)

Plate 8.3 Toasting of the inside of an oak barrel. (Photo courtesy California Wine Institute)

Plate 8.4 Ornately decorated oak cooperage (Stück) at Schloss Schönburg, Rheingau, Germany. (Photo R. Jackson)

Plate 9.1 Cluster of botrytized grapes showing berries in different states of noble rot. (Photo courtesy of D. Lorenz, Staatliche Lehr- und Forschungsanstalt für Landwirtschaft, Weinbau und Gartenbau, Neustadt, Germany)

Plate 9.2 'Vidal' grapes ready for picking in the production of icewine. (Photo courtesy of Inniskillin Wines Inc., Canada)

Plate 9.3 Forklift truck placing basket full of icewine grapes in preparation for crushing. (Photo courtesy of E. Brian Grant, CCOVI, Brock University, St Catharines, Ontario, Canada)

Plate 9.4 Press cake of icewine grapes from a basket press being crumbled for the second pressing. (Photo courtesy of E. Brian Grant, CCOVI, Brock University, St Catharines, Ontario, Canada)

Plate 9.5 Warehouse for the slow drying of grapes in the production of *recioto* wines. (Photo courtesy of Masi Agricola S.p.a., Italy)

Plate 9.6 Mechanical riddling machine in Champagne. (Photo courtesy of Peper-Heidsieck, France)

Plate 9.7 Growth of *flor* yeasts on the surface of wine in a sherry *butt*. (Photo courtesy of Bodegas Pedro Domecq, S.A., Spain)

Plate 10.2 Terraced vineyards of the Douro, Portugal. (Photo courtesy of Instituto do Vinho do Porto, Portugal)

Plate 9.8 *Las Copas* bodega, showing stacked rows (*criaderas*) of sherry *butts*. (Photo courtesy of Gonzalez Byass, Spain)

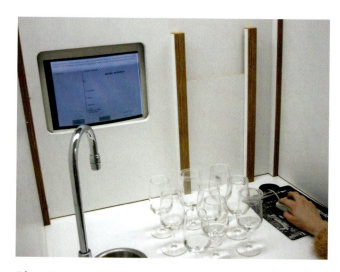

Plate 11.1 Tasting booth in a wine sensory laboratory. (Photo courtesy G. Pickering, CCOVI, Brock University, St Catharines, Canada)

Plate 10.1 Steeply sloped vineyards near Zeltingen in the Mosel, Germany. (Photo R. Jackson)

Plate 11.2 Wine dispenser divided into ambient and refrigerated sections, using nitrogen gas cylinder as the dispensing agent. (Photo courtesy of the WineKeeper®)

its environment. This is especially true in old wineries, in which the equipment and buildings are thinly covered with wine yeasts. In addition, *Hansenula anomala* and *Pichia membranaefaciens* commonly occur in wineries, with *Brettanomyces* and *Aureobasidium pullulans* often being isolated from the walls of moist cellars. The hygienic operation of present-day wineries greatly limits juice and must inoculation, or contamination from the winery and its equipment. In such situations, vineyard sources of *S. cerevisiae* could be the main source in spontaneous fermentations (Török *et al.*, 1996).

SUCCESSION DURING FERMENTATION

In spontaneous fermentation, there is a rapid and early succession of yeast species (Fig. 7.20). Initially, *Kloeckera apiculata* and *Candida stellata* often occur in the range of 10^3–10^6 cells/ml. This number increases rapidly if the grapes or must are left at warm temperatures. Although these endemic yeasts may grow at the beginning of fermentation, most strains soon pass into decline. The culturable population usually drops quickly, becoming a minor component of the yeast population. This has frequently been interpreted as a result of the increasing concentration of ethanol produced by *Saccharomyces* and/or the addition of sulfur dioxide. Recent studies indicate that other interstrain and interspecies effects may also play a role. Examples may involve the action of acetaldehyde (Cheraiti *et al.*, 2005) and a cell-contact (quorum) mediated mechanism (Nissen *et al.*, 2003). The role of quorum sensing compounds in yeast–yeast and yeast–bacterial wine interactions is just beginning to be investigated (Fleet, 2003).

During the initial stages of fermentation, non-*Saccharomyces* yeast contribute to the production of compounds such as acetic acid, glycerol, and various esters (Ciani and Maccarelli, 1998; Romano *et al.*, 2003). This can be sufficient to modify the wine's aroma (Eglinton *et al.*, 2000; Soden *et al.*, 2000). For example, some strains of *Kloeckera apiculata* produce up to 25 times the amounts of acetic acid typically produced during fermentation. *K. apiculata*, along with other members of the grape epiphytic flora, may also produce above-threshold amounts of 2-aminoacetophenone. It is associated with the naphthalene-like odor characteristic of untypical aged (UTA) off-odor (Sponholz and Hühn, 1996). In addition, *K. apiculata* may inhibit some *S. cerevisiae* strains from completing fermentation (Velázquez *et al.*, 1991), possibly due to its production of acetic acid or octanoic and decanoic acids. In addition, the indigenous grape flora may enhance amino acid availability by their proteolytic activities (Dizy and Bisson, 2000). At cool fermentation temperatures (10°C), yeasts such as *K. apiculata* may remain active (or at least viable) throughout alcoholic fermentation (Heard and Fleet, 1988; Erten, 2002).

Of more sensory significance, however, may be the action of *Candida* spp. The occasional ability of *C. stellata* to persist in fermenting juice (Fleet *et al.*, 1984), and produce high concentrations of glycerol may increase the smooth mouth-feel of wine. Jolly *et al.* (2003) report that the perceived quality of 'Chenin blanc' wine was increased when *Candida pulcherrima* was combined with *Saccharomyces cerevisiae*. *Torulaspora delbrucecki* also has the potential to positively influence the sensory properties of wine, due to its low production of acetic acid and synthesis of succinic acid. Although *Kloeckera apiculata* and *C. stellata* typically ferment only up to approximately 4 and 10% alcohol, respectively, they are able to survive much higher alcohol concentrations (Gao and Fleet, 1988).

Most other members of the indigenous grape flora either are slow growing, or are typically suppressed by the low pH, high ethanol content of the ferment, oxygen deficiency, or sulfur dioxide. Their relative inability to incorporate glucose may also play a significant role in their early demise (Nissen *et al.*, 2004). Consequently, most species of *Candida*, *Pichia*, *Cryptococcus*, and *Rhodotorula*, initially found in must, probably do not contribute significantly to fermentation. Their populations seldom rise above 10^4 cells/ml, and the species usually disappear quickly from the fermenting juice and must. However, if warm conditions prevail, and active fermentation is delayed, these yeasts may initiate severe spoilage. For example, *Pichia guilliermondii* (Barata *et al.*, 2006) can produce sufficient 4-ethyl phenol to give the wine odors variously described as burnt beans, Band-Aid, wet dog, horse sweat, and barnyard.

Because of their demonstrable effect on wine quality (Egli *et al.*, 1998), there is increasing interest among winemakers in using indigenous yeasts to give their wines distinctiveness (Fig. 7.21). Nevertheless, as with any technique, it needs to be used with discretion and tempered with experience. For example, a cool fermentation without sulfur dioxide runs the risk that non-*Saccharomyces* yeast may dominate fermentation, generating sufficient acetic acid and fusel alcohols to mask grape varietal aromas.

Most bacteria that could grow during fermentation are inhibited by *Saccharomyces cerevisiae*, with the exception of lactic acid bacteria. Thus, *S. cerevisiae* (or occasionally other *Saccharomyces* spp.) find few if any organisms capable of competing with it in grape must. Not surprisingly, *S. cerevisiae* tends to dominate and complete fermentation, even if its initial presence in the must is rare (less than 1/5000 colonies) (Holloway *et al.*, 1990). Subsequently, however, other yeasts may multiply in association with lactic acid bacteria during or following malolactic fermentation, notably species of *Pichia* (Fleet *et al.*, 1984).

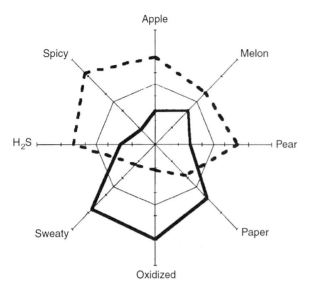

Figure 7.21 The effect of spontaneous (----) and induced (—) fermentation on the sensory characteristics of 'Riesling' wine; (——) mean score. (From Henick-Kling *et al.*, 1998, reproduced by permission)

Several investigations indicate that local populations of *S. cerevisiae* are heterogeneous, consisting of many strains, differing considerably even among regional wineries (see Mortimer, 2000). Although their numbers are usually low, they may occasionally reach 10^4–10^5 cells/ml, possibly derived from winery equipment, or as residents on damaged grapes. There may also be shifts in their dominance from year to year (Polsinelli *et al.*, 1996) or from site to site (Guillamón *et al.*, 1996). Inoculated strains typically dominate fermentation, but other strains may remain active for several days before being suppressed (Querol *et al.*, 1992). Thus, indigenous *S. cerevisiae* strains may play a significant sensory role even in induced fermentations. Shifts in population numbers may also occur throughout fermentation, when must is inoculated with several strains (Schütz and Gafner, 1993a). Because of the low incidence of sexual recombination, infrequent spore production, and poor spore viability, it is suspected that mitotic recombination and mutation are primary sources of this variability (Puig *et al.*, 2000).

MUST INOCULATION

In inoculated fermentations, *S. cerevisiae* is usually added to achieve a population of about 10^5–10^6 cells/ml. With active dry yeasts, this is equivalent to about 0.1–0.2 g/liter of must (juice). Active dry yeast often contains about 20–30×10^9 cells/g.

Before addition, the inoculum is placed in water or dilute juice. Exposure to temperatures between 38 and 40°C for 20 min is generally optimum for rehydration (Kraus *et al.*, 1981). Subsequent cooling to 25°C

is followed by a short adaptation period. During this time, cellular metabolism and membrane permeability readjusts to normal. This is important before the cells are exposed to the osmotic potential of grape must. Readaptation is estimated to involve about 2000 genes (Rossignol *et al.*, 2006). Fractional addition of further juice, before final incorporation of the inoculum with the must, avoids rapid temperature changes and promotes retention of high cell viability.

Commercially available strains of *S. cerevisiae* possess a wide range of characteristics, suitable for most winemaking situations. These include those that either enhance the release of varietal flavorants (glycosides or cysteinylated), produce an abundance of fruit esters, or are of neutral character. Strains are also available that are notable for their production of low levels of compounds such as acetic acid, hydrogen sulfide, or urea. Others may be selected because of their relative fermentation speed, ability to synthesize or degrade malic or lactic acid, augment the concentration of glycerol, ability to restart stuck fermentation, or known value in producing particular wine styles, notably carbonic maceration, late-harvest, or early- versus late-maturing reds. In addition, there are locally selected strains that are reported to produce regionally distinctive wines.

Despite all this information, the winemaker's choice is still not easy. Although the main characteristics of most commercial strains are known, most of their other properties are not, or if known, buried in research papers not readily accessed my most wineries. In addition, much information is anecdotal.

In only a few instances is intentional inoculation absolutely essential. With thermovinification or pasteurized juice, inoculation is required to achieve rapid initiation of fermentation. In addition, yeast inoculation is necessary to restart "stuck" fermentations and promote fermentation of juice containing a significant number of moldy grapes. Moldy grapes generally possess various inhibitors, such as acetic acid, that slow yeast growth and metabolism. Finally, inoculation is required to assure the initiation of the second fermentation in sparkling wine production. However, the predominant reason for using specific yeast strains is to avoid the production of undesirable flavors occasionally associated with spontaneous fermentation.

SPONTANEOUS VS. INDUCED FERMENTATION

There has been much discussion over the years concerning the relative merits of spontaneous versus induced fermentation. That various strains of *S. cerevisiae* supply distinctive sensory attributes is indisputable (Cavazza *et al.*, 1989; Grando *et al.*, 1993; Fig. 7.22). This is particularly important for aromatically neutral cultivars. Nevertheless, strain choice can equally affect the varietal

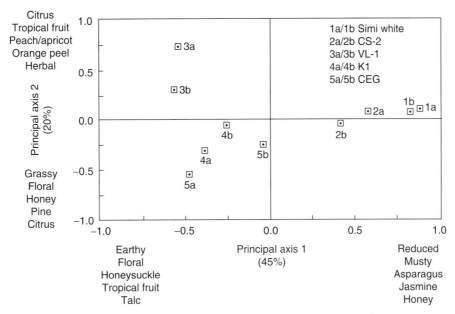

Figure 7.22 Profile of aroma of a 'Riesling' wine (after 20 months) fermented with different yeast strains. (Dumont and Dulau, 1996, reproduced by permission)

character of aromatically distinctive cultivars, by influencing the liberation of bound grape flavorants. This may be particularly significant with non-*Saccharomyces* yeasts. These appear to have greater activity in breaking glycosidic bonds (Mendes-Ferreira *et al.*, 2001). These effects can be further modified or enhanced by *sur lies* maturation. Regrettably, our knowledge of the chemical origin of their sensory effects is insufficient to permit prediction of results. Nevertheless, using established strains provides the winemaker with the greatest confidence that fermentation will be rapid and possess relatively predictable flavor and quality characteristics.

In contrast, spontaneous fermentations may accentuate yearly variations in character. It can be part of the uniqueness (mystique) associated with *terroir*. These elements are often desirable (or essential) in marketing premium wines. However, it also carries the risk of conferring off-odors or other undesirable traits. Occasionally, but not consistently, spontaneous fermentations generate higher concentrations of volatile acidity than induced fermentations. Spontaneous fermentations also tend to possess noticeable lag periods (most likely due to the low inoculum of *S. cerevisiae*) and, thus, are more susceptible to disruption by killer factors.

Those who favor spontaneous fermentation believe that the indigenous grape flora supplies a desired subtle or regional character (Mateo *et al.*, 1991), supposedly missing with induced fermentations. Large-scale wineries, where brand-name consistency is essential, cannot take risks with spontaneous fermentation. Nonetheless, even induced fermentations are not

pure-culture fermentations. The juice or must always contain a sizable population of epiphytic yeast and bacteria from the grapes, unless pasteurized or treated to thermovinification.

An alternative to either spontaneous or standard induced fermentation is inoculation with a mixture of local and commercial yeast strains (Moreno *et al.*, 1991). The combination appears to diminish individual differences, producing a more uniform and distinctive character. This may also involve the joint inoculation with species, such as *Candida stellata* (Soden *et al.*, 2000) or *Debaryomyces vanriji* (Garcia *et al.*, 2002). As noted, on-site experimentation is the only sure means of determining the value of any practice.

Another choice for winemakers searching to add a distinctive aspect to their wine is the use of cyrotolerant yeasts, primarily *S. uvarum*. *S. uvarum* is characterized not only by its ability to ferment at temperatures down to 6 °C, but also by its potential to synthesize desirable sensory characteristics. For example, cyrotolerant yeasts generally produce higher concentrations of glycerol, succinic acid, 2-phenethyl alcohol, and isoamyl and isobutyl alcohols; synthesize malic acid; and produce less acetic acid than many mesophilic *S. cerevisiae* (Castellari *et al.*, 1994; Massoutier *et al.*, 1998).

YEAST BREEDING

Genetically, *Saccharomyces cerevisiae* is the best understood eucaryote, having been used for decades as a favorite laboratory organism. It is easily cultured and has a relative small genome (~13,000 kb), possessing relatively little repetitive DNA, and few introns. The complete genome of a laboratory strain was deciphered in 1997, indicating that it possesses about 5800 protein-coding genes, located on 16 chromosomes. Furthermore, its protenome (full set of proteins) is likely to be defined in the near future. In addition, recent developments in DNA microarray technology may make it easier to understand the intricacies and interconnections of its metabolic pathways (see Cavalieri *et al.*, 2000). These advances should permit a shift from the traditional cross-and-select approach to breeding, to one based on the selective elimination, modification, or introduction of specific genes.

Although *S. cerevisiae* is admirably suited to its role as the predominant fermenter of grape must, and there exists a wide diversity of strain possessing distinctive and useful properties, improved expression or new properties would be useful. This can vary from subtle variations in aromatic synthesis (Fig. 7.23) to the incorporation of properties such as malolactic fermentation.

Genetic Modification In contrast to industrial fermentations, winemaking has to date made little use of genetically engineered microbes. This is partially explained by the complex chemical origin of wine quality, which makes delineating specific and detectable improvements difficult. In addition, other factors, such as grape variety, fruit maturity, and fermentation temperature are generally considered to be of greater importance. In addition, more is known about the negative influences of certain yeast properties than about their positive sensory effects. Nevertheless, public suspicion of genetic engineering is undoubtedly the prime reason for its minimal use.

In genetic improvement, features controlled by one or a few genes are the most easily influenced. For example,

inactivating the gene that encodes sulfite reductase limits the conversion of sulfite to H_2S. The improvement of other properties, such as flocculation, has been more difficult. Flocculation is regulated by several genes, epistatic (modifier) genes, and possibly cytoplasmic genetic factors (Teunissen and Steensma, 1995). The major locus encodes for cell-surface proteins, notably lectins and lectin receptors. These are not constitutively present on the cell surface, but appear later in colony growth. This may result from the proteins being selectively deposited where budding has occurred and at the tips of buds (Bony *et al.*, 1998). In addition, the flocculation mechanism employed by different yeast strains can differ, as in the case of top- versus bottom-fermenting brewing yeasts (Dengis and Rouxhet, 1997). Another complex genetic factor is diverting metabolic pathways. However, the consequences of such changes on the wine's sensory character are often intricate and difficult to predict (Prior *et al.*, 2000).

Many important enologic properties are under multigenic control. Alcohol tolerance and the ability to ferment steadily and cleanly at low temperatures are undoubtedly multigenic; most individual genes possessing only slight effects. Genetic improvement is possible, but probably will take considerable effort and time to achieve. Chances of improvement are significantly enhanced if the biochemical basis of such factors is understood. This knowledge pinpoints likely genes that need modifying.

Even more convoluted problems arise from our inability to predict the consequences of changing the direction of metabolic pathways. Modifying the regulation of one pathway can have important and unexpected consequences on others (Guerzoni *et al.*, 1985; de Barros Lopes *et al.*, 2003). Unforeseen metabolic disruptions are less likely when the compound concerned is the end by-product of a metabolic pathway.

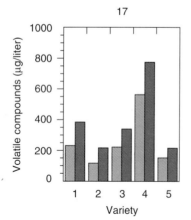 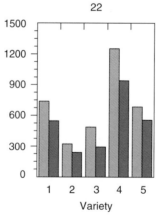 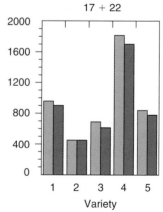

Figure 7.23 Concentration of the volatile components hexyl acetate (17), hexanol (22), and the sum (17 + 22) in wines of five cultivars (1, 'Chenin blanc'; 2, 'Sémillon'; 3, 'Muscat of Alexandria'; 4, 'Cape Riesling'; 5, 'Colombar') with the yeast strains WE 14 (light green) and WE 452 (dark green). (After Houtman and du Plessis, 1985, reproduced by permission)

Thus, terpene synthesis has been incorporated, by transfer of farnesyl diphosphate synthetase from a laboratory strain into a wine strain, without apparent undesirable side-effects (Javelot *et al.*, 1991). Use of the modified yeast gives the wine a Muscat-like attribute.

Many techniques are available to the researcher interested in improving wine yeasts. The simplest and most direct approach involves simple selection. This is much facilitated if a selective culture medium can be devised to permit only cells containing the desired trait to grow (for example antibiotic resistance linked to the desired locus). Otherwise, cells must be laboriously isolated and individually studied for the presence of the desired trait.

Initially, considerable improvement can often be obtained simply by selection. The use of adaptive evolution in strain development has had a long and very successful career in developing industrial microbes. Surprisingly, it has been little employed in the breeding of wine yeasts. However, McBryde *et al.* (2006) have illustrated its applicability with *Saccharomyces cerevisiae*. It avoids problems associated with the reticence of many winemakers and consumers to accept wine produced using generically engineered yeasts. However, for certain changes, more direct measures for modifying the genetic makeup of the yeast are required. This can involve standard procedures such as hybridization, backcrossing, and mutagenesis, or newer techniques such as transformation, somatic fusion, and genetic engineering.

Although interspecific hybridization has occurred naturally in *Saccharomyces sensu stricto* species (de Barros Lopes *et al.*, 2002), breeding wine yeasts has been difficult. One of the principal complications results from the early fusion of spores of opposite mating-type in the ascus. This precludes designed mating between haploid cells of different strains. Further complicating matters is the tendency of haploid cells to switch mating type shortly after ascospore germination, followed by fusion to form diploid cells. This precludes subsequent mating.

These problems can occasionally be avoided by tetrad analysis, the early isolation and separation of individual ascospores. Physically placing spores from desired strains next to one another often results in fusion and successful crossing. Designed mating can be made easier if the strains crossed are first made heterothallic (Bakalinsky and Snow, 1990b). Introduction of a recessive allele of the HO gene prevents switching of the mating-type gene, and correspondingly increases the probability of successful designed crosses. An alternate technique has been described by Ramírez *et al.* (1998).

Breeding wine strains is also complicated by the low frequency of sexual reproduction and spore germination (Bakalinsky and Snow, 1990a). Many strains undergo meiosis infrequently, a precondition for sexual reproduction. Of the haploid spores produced, infertility is

frequent. Spore sterility is suspected to result from the high frequency of aneuploidy (unequal numbers of similar chromosomes). It has been suggested that aneuploidy, due to its frequent occurrence, may possess unsuspected selective value under continuous culture conditions.

Even in successful matings, the typical diploid state of wine yeast can mask the presence of potentially desirable recessive alleles. In haploid organisms, the phenotype of

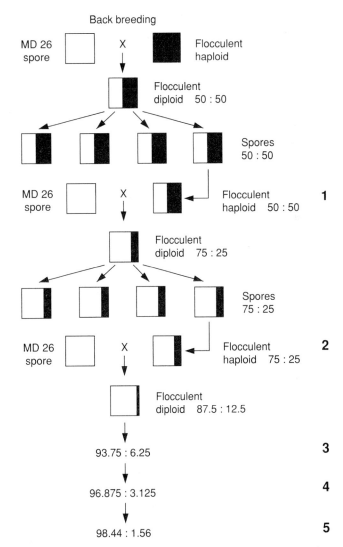

Figure 7.24 Use of backcrossing to eliminate undesired genetic information following the crossing of two yeast strains. In the example, a nonflocculent (recipient) strain is crossed with a flocculent (donor) strain to obtain the flocculant trait. Although some haploid progeny will express the flocculant property, the hybrid cells possess only 50% of the genes from the recipient (MD 26) parent. Haploid flocculant cells of the hybrid strain are backcrossed to spores of the parental MD 26 strain. This reduces the proportion of genes from the flocculant parent to 25%. Haploid flocculant progeny from this backcross are again backcrossed (2) to parental MD 26 spores. Several similar backcrossings of flocculant progeny to MD 26 (3, 4 and 5) essentially eliminate all but the desired flocculant genes derived from the flocculant strain. (From Thornton, 1985, reproduced by permission)

recessive genes is always expressed – no corresponding dominant (usually functional) allele being present. Although a limit to crossing, diploidy permits "genome renewal" by permitting the selective elimination of growth retarding alleles that slowly accumulate by mutation (Ramírez et al., 1999).

If only a single genetic trait is to be incorporated into an existing strain, **backcross breeding** is the preferred technique. Figure 7.24 illustrates a situation in which a dominant flocculant gene is transferred from a donor strain to a valuable nonflocculant strain. Hybridized cells containing the flocculant gene are backcrossed to the recipient strain. Combined with strong selection for flocculation, backcrossing rapidly eliminates undesired donor genes unintentionally incorporated in the original cross.

Where the genetic nature of the desired trait is unknown, or is under complex genetic control, crossing potentially suitable strains requires growing out thousands of progeny. These must be individually assessed for their respective characteristics. Because this is arduous, time-consuming, and expensive, breeders try to incorporate selective techniques that permit only the growth of desirable progeny.

In situations where standard procedures are not applicable (existing strains do not possess the desired trait), "unconventional" techniques must be sought. For example, **somatic fusion** may permit the incorporation of traits from yeast species with which traditional mating is impossible. Somatic fusion requires enzymatic cell-wall dissolution and subsequent mixing of the protoplasts generated. This usually occurs in the presence of polyethylene glycol. It permits the possibility of gene exchange, by promoting physical fusion of cell membranes and admixture of nuclei from both cells. Subsequent nuclear fusion offers the chance for genetic exchange.

One of the major difficulties with somatic fusion is the frequent instability of the association. Fused cells often revert to one of the original species. The incorporation of foreign genes can also interfere with expression of existing traits.

Potentially less disruptive is the incorporation of one or only a few genes from a donor organism. The procedure, called **transformation**, has the added advantage that the donor and recipient may come from distantly related organisms. In transformation, yeast protoplasts are bathed in a solution containing DNA from the donor organism. Incorporation requires the uptake of DNA containing the gene, and its insertion into the yeast chromosome or that of an endogenous plasmid. Plasmids are circular, cytoplasmic DNA segments that partially control their own replication. Although most eukaryotic cells do not possess plasmids, wine yeast frequently carry copies of a 2-μm plasmid. Although often present, it is nonessential. Insertion of a desired locus into the 2-μm plasmid may ease the gene's incorporation, expression, and maintenance in the host cell.

Using transformation, the malolactic gene from *Lactococcus lactis* has been transferred into *S. cerevisiae* along with the malate permease transport gene from *Schizosaccharomyces pombe* (Bony et al., 1997). The possession of several copies of the malolactic enzyme resulted in almost complete conversion of malic acid to lactic acid within 4 days. Recently, a commercial strain (ML01) has been released, based on a construct of genes from *S. pombe* and *Oenococcus oeni* (Husnik et al., 2007). The strain conducts simultaneous alcoholic and malolactic fermentations. This permits rapid deacidification, without the development of buttery and other flavors usually associated with bacterial malolactic fermentation. Wine stabilization procedures can commence immediately after alcoholic fermentation, without waiting, sometimes for months, for bacterial malolactic fermentation to start.

Other genes of enologic interest that have been incorporated are β-(1,4)-endoglucanase (to increase flavor by hydrolyzing glucosidically bound aromatics) and alcohol acetyltransferase (to increase fruity flavors). Both features could greatly benefit wines produced from aromatically neutral varieties.

When specific traits controlled by one or a few linked genes are desired, genetic engineering is often the most efficient method of achieving rapid success. Examples include the flocculation gene (*FLO1*) combined with the late-fermentation promotor (*HSP30*). Because the latter can be induced by a heat-shock, temperature could be used to induce flocculation "on demand." Other desirable, simply controlled factors might include polysaccharide or antibiotic genes, to degrade filter-clogging polysaccharides or prevent bacterial spoilage, respectively.

Selective deletion mutants may also be combined with overexpression of other genes. For example, a homozygous strain recessive (inactive) for the aldehyde dehydrogenase gene (*ALD6*), has been transformed with a high-copy 2 μm plasmid containing the glycerol 3-phosphate dehydrogenase gene (*GPD2*). This diverts sugars to glycerol production, without an overproduction of acetic acid (Eglinton et al., 2002). This has the potential benefit of reducing ethanol production (producing more balanced wines from fully mature grapes), increasing the fermentation rate (by enhancing osmotolerance) (Remise et al., 1999), as well as enhancing the smooth mouth-feel of the wine.

However, when more complex characteristics, based on many unlinked genes, are involved, it is a moot point whether genetic engineering has any selective advantage. In addition, genetic engineering is still complex and expensive. Part of this involves the long and complicated set of tests required before getting approval from the US Food and Drug Administration. There is also considerable controversy about the ethics of releasing genetically modified organisms into the natural environment.

A requirement for all useful strains, new and old, is genetic stability. Although a property of most yeast strains, genetic stability is not a universal characteristic. For instance, flocculant strains often lose the ability to form large clumps of cells and settle out as a powdery sediment. Genetic instability may result from either aneuploidy, mutation or epigenetic regulation.

Environmental Factors Affecting Fermentation

CARBON AND ENERGY SOURCES

The major carbon and energy sources for fermentation are glucose and fructose. Other nutrients may be used, but they are present in small amounts (amino acids), are poorly incorporated into the cell (glycerol), or can be utilized (respired) only in the presence of oxygen (acetic acid and ethanol). Sucrose can be fermented, but it is seldom present in significant amounts in grapes. It may be added, however, with techniques such as chaptalization and amelioration. Sucrose is enzymatically split into its component monosaccharides, glucose and fructose, by one of several invertases. Hydrolysis usually occurs external to the cell membrane by an invertase located between the cell wall and plasma membrane (periplasm). Most other sugars cannot be fermented by *Saccharomyces cerevisiae*. They can, however, be metabolized by several spoilage microorganisms.

Glucose is moved across the plasma membrane by a variety of transport mechanisms (Kruckeberg, 1996). Each possesses distinct regulatory and transport-kinetic properties. Some are low-affinity systems that typically work at high substrate concentrations. For example, the Hxt1p carrier is expressed only at the beginning of fermentation (Perez *et al.*, 2005). Hxt2p (an intermediate carrier) is active only during the lag period, whereas Hxt3p is functional throughout fermentation (but maximally at the onset of the stationary phase). Low-affinity systems appear to function by facilitated diffusion (without direct expenditure of metabolic energy). In contrast, high-affinity systems are energy-requiring and usually operate at low glucose concentrations. Despite this, the major high-affinity systems of wine yeasts (Hxt6p and Hxt7p) become active when the yeast colony enters the stationary phase – when considerable amounts of sugars still remain in the must.

Glucose concentration not only differentially affects the activation of sugar transport mechanisms, but also regulates expression of enzymes in the TCA and

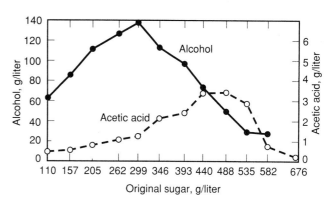

Figure 7.25 Effect of sugar concentration on alcohol and volatile acid production. (After unpublished data of C. von der Heide, from Schanderl, 1959, reproduced by permission)

glyoxylate pathways. Nevertheless, continued expression of selected chromosomal (vs. mitochondrial) genes results in some TCA enzymes being found in the cytoplasm even at high sugar concentrations. They are required for biosynthetic reactions essential for growth.

At maturity, the sugar concentration of most wine grapes ranges between 20 and 25%. At this concentration, the osmotic influence of sugar can delay the onset of fermentation. The resulting partial plasmolysis of yeast cells may be one of the causes of a lag period prior to active fermentation (Nishino *et al.*, 1985). In addition, cell viability may be reduced, cell division retarded, and sensitivity to alcohol toxicity enhanced. At sugar concentrations above 25–30%, the likelihood of fermentation terminating prematurely increases considerably. Strains of *Saccharomyces cerevisiae* differ greatly in their sensitivity to sugar concentration.

The nature of the remarkable tolerance of wine yeasts for the plasmolytic action of sugar is unclear, but appears related to increased synthesis of, or reduced permeability of, the cell membrane to glycerol (see Brewster *et al.*, 1993). These responses to increased environmental osmolarity permit glycerol to equilibrate the osmotic potential of the cytoplasm to that of the surrounding juice.

Sugar content affects the synthesis of several important aromatic compounds. High sugar concentrations increase the production of acetic acid (Fig. 7.25) and its esters. However, as indicated in Fig. 7.26, the effect of total soluble solids on esterification is not solely due to sugar content. For example, the synthesis of isoamyl and 2-phenethyl acetate rises with increasing maturity (°Brix), but decreases in juice from immature grapes, augmented with sugar to achieve the same °Brix.

Over a wide range of sugar concentrations, ethanol production is directly related to sugar content. However, above 30%, ethanol production per gram sugar begins to decline (Fig. 7.25). In some *S. cerevisiae* strains, sugar

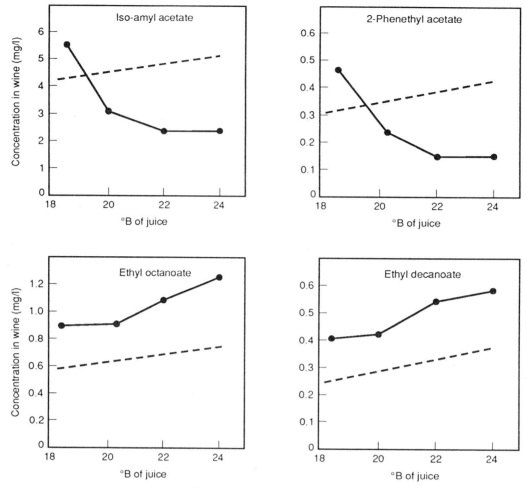

Figure 7.26 Ester concentration of wines made from grapes of varying maturity and with chaptalized must. Wines for musts of four degrees of maturity (——); average slope from wines produced from chaptalized must (– – –). (From Houtman *et al.*, 1980, reproduced by permission)

content also directly affects the synthesis of acetic acid (Henschke and Dixon, 1990).

ALCOHOLS

All alcohols are toxic to varying degrees. Because *Saccharomyces cerevisiae* shows considerable insensitivity to ethanol toxicity, much effort has been spent attempting to understand the nature of this tolerance, and why it breaks down at high concentrations. Several factors appear to be associated with ethanol tolerance. These include activation of glycerol and trehalose synthesis (Hallsworth, 1998); accumulation of Hsp104 (a general stress-related protein) and Hsp12 (Sales *et al.*, 2000); and modification of the plasma membrane (activating membrane ATPase, substitution of ergosterol for lanosterol, increasing the portion of phosphatidyl inositol vs. phosphatidyl choline (Arneborg *et al.*, 1995), and augmenting the incorporation of palmitic acid). These membrane changes decrease permeability (Mizoguchi and Hara, 1998), minimizing the loss of nutrients and cofactors

from the cell, notably magnesium and calcium. Vacuolar membrane function is also crucial for the retention of toxic substances stored in vacuoles (Kitamoto, 1989).

Although alcohol buildup eventually inhibits fermentation, it begins disrupting yeast metabolism at much lower concentrations. For example, suppression of sugar uptake can begin at about 2% ethanol (Dittrich, 1977). This property becomes increasingly marked as the fermentation temperature rises. Disruption of ammonium transport and inhibition of general amino acid permeases also occurs as alcohol content increases. These influences can be partially offset by enrichment with unsaturated fatty acids (adding yeast hulls). Although higher (fusel) alcohols are more inhibitory than ethanol, their much lower concentration substantially limits their toxic influence.

Although most strains of *S. cerevisiae* can ferment up to 13–15% ethanol, there is wide variation in this ability. Cessation of growth routinely occurs at concentrations below those that inhibit fermentation. It is generally believed that this results from disruption of the semifluid nature of the cell membrane by alcohol's effect on lowering water activity (Hallsworth, 1998). This destroys the ability of the cell to control cytoplasmic function, leading to nutrient loss and disruption of the electrochemical gradient across the membrane. The latter is vital for nutrient transport (see Cartwright *et al.*, 1989). Lowered water activity also disrupts hydrogen bonding, essential to enzyme function. High osmotic potential enhances ethanol toxicity.

Ethanol is occasionally added to fermenting must or wine, usually in the form of distilled wine spirits (unmatured brandy), to arrest yeast and other microbial activity. This property is used selectively in sherry and port production. In port, the brandy is added early during fermentation to retain about half the sugar content of the must. This leaves the wine with the aromatic attributes typical of early fermentation. For example, young port is likely to be higher in acetic acid, acetaldehyde, and acetoin content, but lower in glycerol, fixed acids, and higher alcohols (Figs 7.16, 7.17), than had it fermented to dryness. In sherry production, wine spirits are added at the end of fermentation to inhibit the growth of acetic acid bacteria (>15% ethanol) and *flor* yeasts (>18% ethanol) during solera aging.

The accumulation of alcohol during fermentation has an important dissolving action on phenolic compounds. Most of the distinctive taste of red wines depends on the extraction of flavanols by ethanol during fermentation. Ethanol aids, but is less critical to, the extraction of anthocyanins. Phenolic extraction is further enhanced in the presence of sulfur dioxide (Oszmianski *et al.*, 1988).

NITROGENOUS COMPOUNDS

Next to sugars, nitrogenous compounds are quantitatively the most important yeast nutrients. Under most circumstances, juice and must contain sufficient nitrogen for fermentation. Nitrogen contents can, however, vary considerably – values reported of juice in California range from 60 to 2400 mg/liter. Most of the nitrogen is in the form of free amino acids, notably proline and arginine. At least 150 mg/liter assimilable nitrogen (primarily free amino acids, except for proline and ammonia) is generally considered necessary for fermentation to progress to completion. Optimum levels suggested by Henschke and Jiranek (1993) are in the range of 400–500 mg/liter. Higher concentrations can be a disadvantage. They promote unnecessary cell multiplication and reduce the conversion of sugar to alcohol. In contrast, low values enhance the release of higher alcohols.

When nitrogen is required, it is usually supplied as DAP – diammonium phosphate (ammonium hydrogen phosphate). It is usually added halfway through fermentation (Bely *et al.*, 1990), when it especially benefits

synthesis of sugar transport proteins (Bely *et al.*, 1994). In addition, it favors uptake of amino acids. Their transport is depressed in the presence of ammonia. In situations where nitrogen deficiency is severe, significant amounts of hydrogen sulfide are released as a consequence of restricted amino acid synthesis. This is most marked if nitrogen becomes limiting during the exponential phase. It is countered by the availability of ammonia (or most amino acids with the notable exception of cysteine). Under such conditions, earlier addition of DAP is advisable (Jiranek and Henschke, 1995).

Although a valuable fermentation aid, addition of DAP is best minimized or avoided where possible. It has been implicated in the synthesis of ethyl carbamate (Ough, 1991), a suspected carcinogen. Excessive nitrogen may also predispose some strains to high hydrogen sulfide production. Thus, development of a rapid method to estimate assimilable nitrogen (Dukes and Butzke, 1998) now permits wineries to better assess the need for supplementation (Gardner *et al.*, 2002).

Several conditions can reduce must nitrogen content. Nitrogen deficiency in the vineyard and clarification can limit or diminish, respectively, the assimilable nitrogen content of juice. If sufficiently marked, inadequate nitrogen levels slow fermentation and is one cause of "stuck" fermentations. This may result from the irreversible inactivation of sugar transport by ammonia starvation (Lagunas, 1986). The half-life of the main glucose transport system is approximately 12 h, with complete inactivation occurring within approximately 50 h (Schulze *et al.*, 1996). This results from the cessation of protein synthesis and enzyme degradation. The lack of ammonia can also negate the allosteric activation of crucial glycolytic enzymes, such as phosphofructokinase and pyruvic kinase. This, in turn, further inhibits the uptake of glucose (Bely *et al.*, 1994).

Juice nitrogen content may also be reduced by 33–80% in grapes infected by *Botrytis cinerea* (Rapp and Reuther, 1971). In sparkling wine production, nitrogen deficiency, caused by the initial fermentation and clarification of the *cuvée* wines, is usually counteracted by the addition of ammonium salts such as diammonium phosphate.

The juice of some grape varieties is more likely to show nitrogen limitation than others, for example, 'Chardonnay' and 'Colombard.' This is especially true when the juice has been given undue preffermentative centrifugation or filtration.

Nitrogen is incorporated most rapidly during the exponential growth phase during fermentation. This correlates with the period of cell growth and division. Subsequently, there is a slow release of nitrogen-containing compounds back into the fermenting juice (Fig. 7.27).

Of inorganic nitrogen sources, ammonia is incorporated preferentially. Wine yeasts tend to overexpress ammonia (*MEP2*) transport genes (Cavalieri *et al.*, 2000). The oxidized state of ammonia permits its direct incorporation into organic compounds. Although ammonia is potentially capable of repressing the uptake of amino acids, its normal concentration in grape juice is insufficient to have this effect. *Saccharomyces cerevisiae* has several amino acid transport systems (Cartwright *et al.*, 1989). One is nonspecific and directs the uptake of all amino acids except proline. The other systems are more selective, transporting only particular groups of amino acids. These properties probably explain why certain amino acids are preferentially incorporated (phenylalanine, leucine, isoleucine, and tryptophan), whereas others are poorly assimilated (alanine, arginine, and proline) (Ough *et al.*, 1991).

The principal amino acids available in grape must are proline and arginine. Proline is not used as a nitrogen source – it requires the presence of molecular oxygen for its metabolism (Tomenchok and Brandriss, 1987). Thus, arginine is the principal amino acid used as a nitrogen source. Regrettably, arginine may be degraded to ornithine and urea. If the urea is not degraded to carbon dioxide and ammonia (the latter assimilated in amino acid synthesis), it may be excreted. This is undesirable because urea has been implicated in the production of the carcinogen, ethyl carbamate (Ough *et al.*, 1990).

Amines and peptides also may be incorporated as nitrogen sources, but protein nitrogen is unavailable. Wine yeasts are capable of neither transporting proteins into the cell nor enzymatically degrading them to amino acids outside the cell.

Nitrogen is required for the synthesis of proteins, pyrimidine nucleotides, and nucleic acids. Yeast cells generally synthesize their amino acid and nucleotide requirements from inorganic nitrogen and sugar. Consequently, most yeast strains do not need these metabolites in the must. Nevertheless, they can be assimilated from the medium when available. The assimilation of metabolites avoids the diversion of metabolic intermediates and energy to their biosynthesis.

Nitrogen content can influence the synthesis and release of aromatic compounds during fermentation. Noticeable in this regard is the reduction in fusel alcohol content in the presence of ammonia and urea. The effect can be reversed by the assimilation of certain amino acids from the juice or must. These opposing effects appear to result from the use of fusel alcohols in the biosynthesis of amino acids from ammonia and urea, and their release by deamination of amino acids assimilated from the must. The sensory impact of this equilibrium, if any, has not been established. Under nitrogen starvation, there is a dramatic increase in the production of glycerol and trehalose (Schulze *et al.*, 1996).

During, and especially after, fermentation, there is a slow release of nitrogen and other constituents into the wine, probably due to the autolysis of dead and dying yeast cells (Fig. 7.27). The release of simple nitrogen compounds may activate subsequent microbial activity. For this reason, the first racking often occurs shortly after fermentation. When malolactic fermentation is desired, however, racking is delayed until the bacterial conversion of malic to lactic acid is complete. Racking also may be delayed if lees contact with the wine is desired. When the liberation of assimilable nitrogen into the wine is desired, small cooperage is preferred. Small cooperage provides better contact between the wine and the lees (larger surface area/volume ratio).

Large molecular weight nitrogen-containing compounds may also be released, for example mannoproteins. These begin to be liberated near the end of fermentation. It is particularly marked during *sur lies* maturation. The amount that escapes depends primarily on the yeast strain and the ambient temperature.

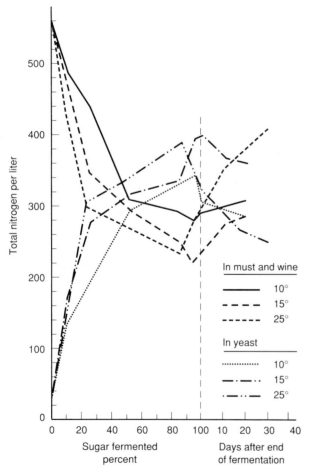

Figure 7.27 Changes in total nitrogen during and after fermentation in musts and in yeasts fermenting them. (After Nilov and Valuiko, 1958, in Amerine *et al.*, 1980, reproduced by permission)

LIPIDS

Lipids are the basic constituents of cell membranes (phospholipids and sterols), function in energy storage (oils), and act as pigments (carotenoids) as well as regulator molecules in proteins (lipoproteins) and carbohydrates (glycolipids).

Yeasts synthesize their own lipid requirements when grown aerobically, but are unable to produce long-chain unsaturated fatty acids and sterols under anaerobic conditions. This is less significant in red wine production. Adequate supplies of precursors usually can be obtained from the skins during fermentation. The anaerobic limitation of lipid synthesis, however, can result in the sluggish fermentation of highly clarified white juice. Clarification can remove more than 90% of the fatty acid content. This is particularly marked with the unsaturated fatty acids – oleic, linoleic, and linolenic acids (Bertrand and Miele, 1984). Sterols such as ergosterol and oleanolic acid are probably removed as well. Nevertheless, yeasts typically possess sufficient reserves of these vital compounds to initiate fermentation and complete several cell divisions (typically four to five when a yeast inoculum is used). However, in spontaneous fermentation, up to 16 or more cell divisions are involved in reaching the typical stationary population. Thus, the accumulated deficit in sterols and unsaturated fatty acids can aggravate ethanol-induced reduction in glucose uptake, resulting in stuck fermentation.

Wine yeasts are sensitive to the toxicity of mid-chain (C_8 and C_{10}) saturated fatty acids, such as octanoic and decanoic acid. As by-products of yeast metabolism,

they accumulate during fermentation. Because they can increase membrane fluidity, proton influx increases, acidifying the cytoplasm. Cytoplasmic acidification favors the ethanol-induced leakage of nutrients, such as amino acids (Sá Correia *et al.*, 1989), and can inhibit both high- and low-affinity sugar-transport systems (Zamora *et al.*, 1996). Because toxicity increases with a decrease in pH, injury is likely associated with the undissociated form of the acids. Toxic effects are limited or reversed by the addition of ergosterol and long-chain unsaturated fatty acids (Fig. 7.28), or the addition of various absorptive substances, such as activated charcoal, bentonite, silica gel, or yeast hulls (consisting primarily of cell-wall remnants following controlled autolysis). By removing octanoic and decanoic acids (via absorption), their potential for yeast membrane disruption is reduced. In addition, yeast hulls may act as a source of required sterols and unsaturated fatty acids (see Munoz and Ingledew, 1990).

Another possible solution to problems associated with low sterol and unsaturated fatty acid content is the use of yeasts with high sterol content. Sterol synthesis is commonly suppressed in the presence of glucose, but some strains do not show this effect. Growing yeasts under highly aerobic conditions, as in active dry yeast production, generates cells elevated in unsaturated fatty acids. They may contain up to three times the sterol content of cells grown semiaerobically (Tyagi, 1984). Musts inoculated with strains possessing enhanced sterol contents frequently ferment more sugar than strains possessing low sterol contents.

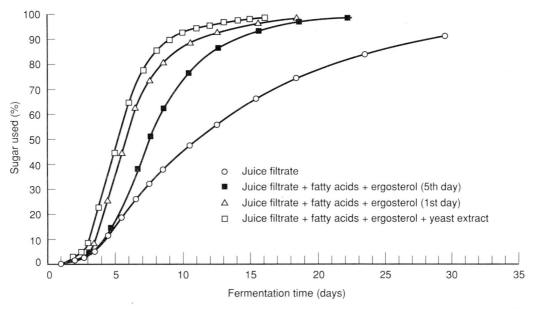

Figure 7.28 Fermentation curves of juice filtrates with the addition of yeast extract, unsaturated fatty acids, and ergosterol at inoculation and on the fifth day thereafter. (From Houtman and du Plessis, 1986, reproduced by permission)

PHENOLS

The phenolic content of the must can have various effects on the course of fermentation. For example, the phenols in red grapes, notably anthocyanins, stimulate fermentation, whereas procyanidins in white grapes can be slightly inhibitory (Cantarelli, 1989). Phenolic compounds are also a determining factor in the activation of film formation, important in *fino* production (Cantarelli, 1989). Certain phenolics, notably the esters of gallic acid, are toxic, whereas others, such as chlorogenic and isochlorogenic acids, may stimulate fermentation. The principal situation in which phenols suppress yeast metabolism is during the second fermentation in sparkling wine production. This is one of the primary reasons why few red sparkling wines are produced using standard procedures.

In addition to affecting fermentation, phenols may also be modified by yeast action. For example, ferulic and *p*-coumaric acids may be decarboxylated to aromatic vinyl phenols (4-vinyl guaiacol and 4-vinyl phenol, respectively) (Chatonnet *et al.*, 1989). Other phenolic constituents in the must may influence this conversion.

SULFUR DIOXIDE

Sulfur dioxide is synthesized by most *Saccharomyces cerevisiae* strains, but in most instances, it is bound to organic compounds in the yeast or fermenting must. Thus, the synthesis of SO_2 probably does not play a significant role in the success in *S. cerevisiae* in outcompeting other microbes during fermentation. Interestingly, the addition of sulfur dioxide favors not only the growth of strains resistant to sulfur dioxide, but also appears to select strains that produce greater amounts of sulfur dioxide.

The differential action of sulfur dioxide against the grape epiphytic flora can be used to selectively control the influence of indigenous yeasts. The increasing awareness of their presence, and their positive and negative influences on wine fragrance (Henick-Kling *et al.*, 1998), are making modulation versus inhibition an important tool in adjusting wine character.

At the concentrations typically used (usually less than 50 ppm total with healthy grapes), sulfur dioxide does not appear to affect the rate of alcoholic fermentation. However, sulfur dioxide can slow the onset of fermentation. The presence of 15–20 ppm can reduce the viability of a yeast inoculum from 10^6 to 10^4 cells/ml or less (Lehmann, 1987). Although sulfur dioxide can help restrict the growth of indigenous yeasts and bacteria, it may be unnecessary because the inoculated yeast rapidly comes to dominate fermentation (Petering *et al.*, 1993; Henick-Kling *et al.*, 1998).

Yeast resistance to sulfur dioxide is correlated with several factors. For example, the *SSU1R* gene that controls sulfite efflux is overexpressed in wine yeasts (Hauser *et al.*, 2001). This apparently results from its translocation to a position where it is under the control of the *ECM34* promotor (Pérez-Ortín *et al.*, 2002).

In addition to its antimicrobial activities, sulfur dioxide can significantly influence yeast metabolism. Sulfur dioxide readily binds with several carbonyl compounds, notably acetaldehyde, pyruvic acid, and α-ketoglutaric acid. Binding increases their biosynthesis and potential release into the fermenting must. Thus, their concentration in the finished wine often correlates with the concentration of sulfur dioxide added to the must. Sulfur dioxide also favors glycerol synthesis, whereas it tends to inhibit acetic acid production. Fixed acidity generally does not change, partially because sulfur dioxide suppresses the metabolism of both lactic acid and acetic acid bacteria.

The binding of sulfur dioxide to carbonyl compounds inadvertently increases the amount of sulfur dioxide needed to suppress the action of spoilage organisms. Bound sulfur dioxide is much less antimicrobial than molecular SO_2.

Sulfur dioxide increases the extraction of phenolic compounds, including anthocyanins from grapes. However, it can also reversibly decolorize anthocyanins. In addition, anthocyanins bound to sulfur dioxide are unable to polymerize with procyanidins. This can adversely affect long-term color stability.

Although sulfur dioxide is the best wine antimicrobial agent available, it does not control certain spoilage yeasts. Many strains of *Saccharomycodes ludwigii*, *Zygosaccharomyces bailii*, and *Brettanomyces* spp. are particularly tolerant to sulfur dioxide (Hammond and Carr, 1976; Fig. 8.9).

When present, elemental sulfur can be assimilated and used in the synthesis of sulfur-containing amino acids and coenzymes. It also may be oxidized to sulfate and sulfur dioxide, or reduced to hydrogen sulfide. The reduction of sulfur to hydrogen sulfide may be a means, albeit aromatically unpleasant, of maintaining a favorable redox balance in yeast cells under anaerobic conditions.

OXYGEN AND AERATION

The process of fermentation itself requires no oxygen. Even in the presence of oxygen, *S. cerevisiae* preferentially ferments. Nevertheless, trace amounts of oxygen can indirectly favor fermentation by permitting the biosynthesis of sterols and long-chain unsaturated fatty acids. The production and proper functioning of the yeast cell membrane require sterols (ergosterol and lanosterol), as well as C_{16} and C_{18} fatty acids. Some of the fatty acids, such as linoleic and linolenic acids, are unsaturated. Molecular oxygen is also needed for the synthesis of the vitamin nicotinic acid. In addition, anaerobic conditions favor the accumulation of toxic (C_8 and C_{10})

fatty carboxylic acids (Alexandre and Charpentier, 1995). They collect because they are not acylated in the synthesis of required long-chain fatty acids.

Juice oxidation during stemming and crushing is usually sufficient to promote adequate yeast growth during vinification. The amount of oxygen absorbed depends on the duration of skin contact (maceration). Because phenolic extraction also increases with extended skin contact, the capacity of the juice to consume oxygen rises correspondingly. By removing free oxygen, conditions shift from oxidative to reductive. Aeration beyond that which occurs coincidental to stemming and crushing (**hyperoxidation**) is variously viewed as being potentially beneficial (Cheynier *et al.*, 1991) or detrimental (Dubourdieu and Lavigne, 1990). These different views may result from strain response to oxygen, the degree and timing of oxygen uptake, and increased synthesis of higher alcohols and esters. Nevertheless, hyperoxidation does favor cell growth, promotes fermentation, and diminishes the reductive influence often associated with *sur lies* maturation (Fornairon-Bonnefond *et al.*, 2003). The initial browning commonly associated with crushing is acceptable because the colored compounds that form are lost during fermentation (attachment to yeast cells), or precipitated during clarification. It also gives white wine a degree of resistance against in-bottle oxidative browning by eliminating readily oxidizable phenols early in vinification.

During the fermentation of red wine, oxygen may be absorbed during pumping over. The resulting incorporation of about 10 mg O_2/liter (saturation) often speeds the process of fermentation. This is more marked when aeration occurs at the end of the exponential phase of yeast growth (Sablayrolles and Barre, 1986). The yeast population increases and average cell viability is enhanced. Aeration also increases the production of acetaldehyde, thus favoring color stability by assisting the early formation of anthocyanin–tannin polymers.

During the fermentation of white juice, winemakers avoid oxidation because it increases the synthesis of fusel alcohols and acetaldehyde. Increased levels of volatile acidity (acetic acid) may result from the activation of acetic acid bacteria. In addition, semiaerobic conditions can depress the synthesis of esters (Nykänen, 1986). However, the absence of oxygen can enhance the likelihood of hydrogen sulfide accumulation. Depending on the timing of aeration, oxygen uptake not only oxidizes H_2S, but also enhances its synthesis (Houtman and du Plessis, 1981). To offset H_2S accumulation, short aeration at the beginning, or a few days after the commencement of fermentation (Bertrand and Torres-Alegre, 1984), has been recommended. The timing and extent of aeration also can influence urea accumulation and, therefore, the potential for ethyl

carbamate production (Henschke and Ough, 1991). In situations where stuck fermentation has been a problem, this can occasionally be avoided by the combined addition of nitrogen and aeration at the onset of the stationary phase. Because the relative benefit of oxygen uptake varies with the yeast strain (Julien *et al.*, 2000), tests of its value need verification onsite.

The reactivation of fermentation in stuck wines usually requires aeration. *Cuvée* wines also may be aerated prior to the second fermentation in sparkling-wine production.

Following fermentation, limited slow oxygenation (~40 mg O_2/liter) may benefit the maturation of red wines. It can favor color stability and reduce the typical reduction in color associated with malolactic fermentation (Pérez-Magariño *et al.*, 2007). In contrast, most white wines are painstakingly protected from air exposure following fermentation. The exception to this practice occurs when wines are given extended contact with the lees (*sur lies* maturation). In this situation, limited oxygen uptake occurs when the wine is periodically stirred with the lees (*battonage*).

Oxygen absorption is influenced by many factors, including clarification, skin contact, phenol concentration, sulfur dioxide content, presence of polyphenol oxidases, sugar concentration, temperature, pumping over, rate of fermentation, and, of course, protection from air before, during, and after fermentation.

CARBON DIOXIDE AND PRESSURE

During fermentation, large volumes of carbon dioxide are generated, about 260 ml/g glucose. This equates to over 50 times the volume of the juice fermented. The escape of carbon dioxide is estimated to remove about 20% of the heat generated during fermentation. Some of the heat loss results from energy consumed, associated with water evaporation.

Various volatile compounds are carried off with carbon dioxide. Ethanol loss is estimated at about 1–1.5% of that produced (Williams and Boulton, 1983), but varies with sugar use and temperature. Higher alcohols and monoterpenes are lost to about the same degree (~1%). In contrast, significant losses of both ethyl and acetate esters can occur (Fig. 7.29). Depending on the grape variety, and especially the fermentation temperature, up to about 25% of these aromatically important compounds may be lost (Miller *et al.*, 1987). On average, more acetate esters are lost than ethyl esters. This loss could noticeably reduce the fruity character of the wine. Trapping these compounds for readdition to the wine is an interesting concept (Muller *et al.*, 1993; Bach, 2001).

The loss of volatiles from fermenting juice is a function of both the relative rates of synthesis and degradation, and their comparative solubilities in the

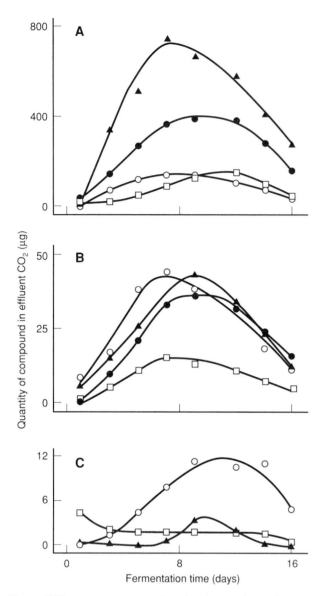

Figure 7.29 Yeast aromatics released with CO_2 during fermentation at 15 °C. (A) ▲, isoamyl acetate; ○, ethyl *n*-hexanoate; □, ethyl *n*-octanoate; ●, isoamyl alcohol; (B) □, isobutyl acetate; ○, hexyl acetate; ●, ethyl *n*-butanoate; ▲, isobutanol; (C) ○, ethyl *n*-decanoate; □, 1-hexanol; ▲, 2-phenylethanol. (From Miller *et al.*, 1987, reproduced by permission)

increasingly alcoholic juice. Evaporation is further affected by fermentor size and shape. For example, small fermentors possess higher surface area/volume ratios, and lower liquid pressures than larger fermentors. This favors volatility. In addition, although the reduction of vapor pressure at low temperatures tends to limit volatilization, the slower release of carbon dioxide could partially offset this by favoring incorporation and loss with CO_2.

The generation of carbon dioxide produces strong convection currents within the ferment. These help equilibrate the nutrient and temperature status throughout the juice. However, the presence of a floating or submerged cap can disrupt equilibration in red musts (Fig. 7.32).

In vats, and in most tank fermentors, the carbon dioxide produced during fermentation is allowed to escape into the surrounding air. When the gas is trapped, pressure in the tank rapidly rises. At pressures above 700 kPa (~7 atm), yeast growth ceases, although pressure-related effects have been reported at pressures as low as 30 kPa above ambient. Low pH and high alcohol content increase yeast sensitivity to CO_2 pressure (Kunkee and Ough, 1966). This has its most significant influence during sparkling wine production. Pressures upward of 600 kPa are often reached by the end of the second, in-bottle fermentation. Nevertheless, the fermentative ability of yeasts may not be inhibited completely until about 3000 kPa. In addition, carbon dioxide accumulation may affect metabolism by influencing the balance between carboxylation and decarboxylation reactions (Table 7.1). The effect of pressure on the synthesis of aromatic compounds during vinification has not been investigated.

Some of the consequences of high pressure on cell growth and metabolism may accrue from a decrease in water viscosity (Bett and Cappi, 1965). This could disrupt the intramolecular hydrogen bonding vital to protein structure and function. In addition, critical changes in membrane composition could disrupt cell membrane permeability. Presence of heat-shock proteins (such as Hsp104) and trehalose can limit protein denaturation (Hottiger *et al.*, 1994) and can stabilize membrane fluidity (Iwahashi *et al.*, 1995).

The pressure created by trapping the carbon dioxide produced during fermentation has occasionally been used to encourage a more constant rate of fermentation. It also has been used to induce the premature termination of fermentation, leaving the wine with a sweet finish. However, care must be used with this technique. Spoilage yeasts, such as *Torulopsis* and *Kloeckera*, are less sensitive to high pressures than is *S. cerevisiae*. The production of acetic acid by spoilage yeasts can produce a vinegary taint in the wine. Caution also needs to be taken because lactic acid bacteria (*Lactobacillus*) are little affected by the pressures that effect wine yeasts (Dittrich, 1977).

pH

The pH range normally found in juice and must has little effect on the rate of fermentation, or on the synthesis and release of aromatic compounds by yeasts. Only at abnormally low pH values (<3.0) is fermentation impeded. However, low pH may assist the uptake of some amino acids, by supplying protons used in activating transport across the cell membrane (Cartwright *et al.*, 1989).

Table 7.1   Possible effects of carbon dioxide on key enzymes of *Saccharomyces cerevisiae*

Reaction	Comment
Pyruvate → acetaldehyde → ethanol, CO_2	Reduced production of ethanol
Pyruvate (ATP → ADP) → oxaloacetate → amino acids, CO_2	Stimulation, less available pyruvate for ethanol production
Acetyl-CoA → malonyl-CoA → fatty acids, CO_2	Stimulation, less available pyruvate for ethanol production
Pyruvate → malate, CO_2	Stimulation, less available pyruvate but malate enzyme level is not high
Phosphoenolpyruvate (ADP → ATP) → oxaloacetate, CO_2	Stimulation, less available pyruvate but enzyme is repressed by glucose
6-Phosphogluconate ($NADP^+$ → NADPH) → ribulose 5-phosphate, CO_2	Reduced production of biosynthetic precursors, thus cell yield will decrease; will reduce rate of production of ethanol

Source: From Jones *et al.*, 1981, reproduced by permission

The most important effects of pH on fermentation are indirect, such as noted previously in the discussion of the antibiotic action of sulfur dioxide. Low pH also prevents many potentially competitive organisms from growing in must. In addition, pH affects the survival of some fermentation by-products in wine. The best-known effect concerns the hydrolysis of ethyl and acetate esters, in which breakdown occurs more rapidly at low pH values.

VITAMINS

Vitamins play a crucial role in the regulation of yeast metabolism as coenzymes and enzyme precursors (Table 7.2). Although vitamins are not metabolized as energy sources, their concentrations decrease markedly during fermentation (see Amerine and Joslyn, 1970). Nevertheless, yeast requirements typically are satisfied by either biosynthesis or assimilation from the juice. Certain conditions can, however, significantly reduce the concentration or availability of vitamins. Fatty acids produced during fermentation can inhibit the uptake of thiamine, oversulfiting (or long-term storage of grape juice at high sulfur dioxide concentrations) degrades thiamine, and infection of grapes or contamination of stored juice by fungi and wild yeasts can lower the vitamin content. Under such conditions, vitamin supplement may improve, or be required to reinitiate stuck fermentation.

Adequate concentrations of thiamine reduce the synthesis of carbonyl compounds that bind to sulfur dioxide, thereby diminishing the amount of SO_2 needed to control spoilage organisms. In addition to limiting carbonyl synthesis, thiamine also reduces the concentration and relative proportions of higher alcohols produced during fermentation. Although seldom a problem, deficiencies in pyridoxine and pantothenic acid can disrupt yeast metabolism, resulting in increased hydrogen sulfide synthesis.

Occasionally, vitamin addition is recommended in situations where sluggish or stuck fermentations have been a problem. Although this can be of value if there is an established deficiency, such supplements can increase the volatile acidity of some wines (Eglinton *et al.*, 1993).

INORGANIC ELEMENTS

Inorganic elements often are essential components in the active (catalytic) sites of enzymes. They also play active roles in regulating cellular metabolism and in maintaining cytoplasmic pH and ionic balance (Table 7.3). For example, magnesium is involved in the catalytic action of several key glycolytic enzymes, the activation of fermentation enzymes, and the stabilization of membrane structure. Magnesium helps adapt yeast cells to rapidly increasing alcohol concentrations (Dasari *et al.*, 1990) and limits cell damage (Walker, 1998).

Although inorganic elements are normally assumed to be in adequate supply, there is difficulty in accurately assessing yeast requirements and the available ion concentrations in grape juice and must. Not only do organic compounds, such as amino acids sequester elements, thereby reducing their effective concentration, but ions can antagonize each other's uptake.

Table 7.2   The role of vitamins in yeast metabolism

Vitamin	Active form	Metabolic role	Optimum conc. (mg/liter)
Biotin	Biotin	All carboxylation and decarboxylation reactions	0.005–0.5
Pantothenate	Coenzyme A	Keto acid oxidation reactions; fatty acid, amino acid, carbohydrate, and choline metabolism	0.2–2.0
Thiamine (B_1)	Thiamine-pyrophosphate	Fermentative decarboxylation of pyruvate; oxo acid oxidation and decarboxylation	0.1–1.0
Pyridoxine	Pyridoxal phosphate	Amino acid metabolism; deamination, decarboxylation, and racemization reactions	0.1–1.0
p-Aminobenzoic acid and folic acid	Terrahydrofolate	Transamination; ergosterol synthesis; transfer of one-carbon units	0.5–5.0
Niacin (nicotinic acid)	NAD^+, $NADP^+$	Dehydrogenation reactions	0.1–1.0
Riboflavin (B_2)	FMN, FAD	Dehydrogenation reactions and some amino acid oxidations	0.2–0.2.5

Source: From Jones *et al.*, 1981, reproduced by permission

Table 7.3   Major inorganic elements required for yeast growth and metabolism

Ion	Role	Concentration (μM)
K^+	Enhances tolerance to toxic ions; involved in control of intercellular pH; K^+ excretion is used to counterbalance uptake of essential ions, e.g., Zn^{2+}, Co^{2+}; K^+ stabilizes optimum pH for fermentation	20×10^3
Mg^{2+}	Levels regulated by divalent cation transport system; Mg^{2+} seems to buffer cell against adverse environmental effects and is involved in activating sugar uptake	5×10^3
Ca^{2+}	Actively taken up by cells during growth and incorporated into cell wall proteins; Ca^{2+} buffers cells against adverse environments; Ca^{2+} counteracts Mg^{2+} inhibition and stimulates effect of suboptimal concentrations of Mg^{2+}	1.5×10^3
Zn^{2+}	Essential for glycolysis and for synthesis of some vitamins; uptake is reduced below pH 5, and two K^+ ions are excreted for each Zn^{2+} taken up	50
Mn^{2+}	Implicated in regulating the effects of Zn^{2+}; Mn^{2+} stimulates synthesis of proteins	15
Fe^{2+}, Fe^{3+}	Present in active site of many yeast proteins	10
Na^+	Passively diffuses into cells; stimulates uptake of some sugars	0.25
Cl^-	Acts as counterion to movement of some positive ions	0.1
Mo^{2+}, Co^{2+}, B^{2+}	Stimulates growth at low concentrations	0.5

Source: From Jones *et al.*, 1981, reproduced by permission

Occasionally, as in the case of potentially toxic aluminum ions, this may be beneficial.

The abundance of potassium ions probably makes K^+ the most significant metallic cation in juice and must. A high potassium content can, however, interfere with the efficient uptake of amino acids, such as glycine. Under anaerobic conditions, potassium excretion may be necessary to maintain an acceptable ionic balance, due to the simultaneous incorporation of protons (H^+) with glycine (Cartwright *et al.*, 1989). At low pH values, abnormally high potassium contents may result in sluggish or stuck fermentation. This probably results from the joint uptake of potassium with glucose, and the subsequent decrease in pH associated with the excretion of hydrogen ions to maintain cytoplasmic ionic equilibrium (Kudo *et al.*, 1998). A high potassium concentration can also generate tartrate instability, which is associated with high juice and wine pH. High pH, in turn, can lead to microbial instability, increasing the tendency of white wines to brown, and inducing color instability in red wines.

TEMPERATURE

Temperature is one of the most influential factors affecting fermentation. Not only does temperature directly and indirectly influence yeast metabolism, but it is also one of the features over which the winemaker has the greatest control.

At the upper and lower limits, temperature can cause cell death. However, inhibitory effects are experienced well within these extremes. Relative tolerance to high temperatures appears to depend, at least partially, on production of Hsp104 (heat-shock protein 104). This limits or reverses the aggregation of essential cellular proteins (Parsel *et al.*, 1994). The disruptive influences of high temperatures are increased by growth-limiting factors, such as the presence of ethanol and certain C_8 to C_{10} carboxylic (fatty) acids. In contrast, low temperatures tend to diminish the toxic effects of ethanol. This may partially be a consequence of the higher proportion of unsaturated fatty acid residues in the plasma membrane (Rose, 1989). This property may help to explain the higher maximum viable cell count at the end of fermentations conducted at cooler temperatures (Ough, 1966a).

The growth rate of yeast cells is strongly influenced by the fermentation temperature, particularly during the exponential growth phase. For example, cell division was found to occur every 12 h at 10 °C, every 5 h at 20 °C, and every 3 h at 30 °C (Ough, 1966a). Charoenchai *et al.* (1998) report comparable data on the relative growth rates of a wide range of wine yeasts.

At temperatures above 20 °C, yeast cells experience a rapid decline in viability at the end of fermentation. At cooler temperatures, cell growth is retarded, but viability enhanced. Cool temperatures also prolong the lag phase of fermentation. For this reason, winemakers may warm white juice to 20 °C before adding the yeast inoculum. Once fermentation has commenced, the juice may be cooled to a more desirable fermentation temperature.

The rehydration temperature of active dry yeast is also critical, especially if cool fermentation temperatures are planned (Llauradó *et al.*, 2005). This can minimize the likelihood of stuck or sluggish fermentation. Alternately, or in addition, cryogenic yeast strains or species may be chosen. A rapid onset of fermentation helps limit juice oxidation and the growth of other yeasts coming from the grapes or winery environment.

In addition to affecting growth and survival, temperature has many subtle and not so subtle effects on yeast metabolism. One of the most marked is its influence on the fermentation rate. Cool temperatures dramatically slow the fermentation rate, and may cause its premature termination. Excessively high temperatures may also induce stuck fermentation, by disrupting enzyme and membrane function. Although quick onset and completion of fermentation have advantages, the preferred temperature for vinification is often less than the optimum for ethanol production or yeast growth. Because yeast strains differ in this response, the optimum temperature for vinification can vary widely.

The preference in most wine-producing regions is to conduct white wine fermentation within a range of 15–20 °C. Cooler temperatures may be used, but some traditional European regions still prefer fermentation temperatures between 20 and 25 °C. Most New World winemakers favor cool temperatures because they give fresher, more fruity wines. Fruitiness in white wine is a highly valued characteristic throughout much of the world. This partially relates to the increased synthesis of fruit esters, such as isoamyl, isobutyl, and hexyl acetates. These esters are both synthesized and retained to a greater degree at cool temperatures (Fig. 7.30A). Fatty acid ethyl esters, such as ethyl octanoate and ethyl decanoate, are produced more effectively at 15 °C, whereas 2-phenethyl acetate achieves its highest concentration at 20 °C (Fig. 7.30B). Some fatty acid esters can add a fruity note, for example caproate and caprylate ethyl esters possess apple-like aspects. A greater production of ethanol and higher alcohols may also be observed at cool fermentation temperatures. In addition, cooler fermentation temperatures reduce the release of yeast colloids, thereby facilitating clarification.

Some of the effects noted above may result from the influence of temperature on the indigenous yeast flora. Cooler temperatures reduce both the growth rate and toxicity of ethanol (Heard and Fleet, 1988). As a result, species such as *K. apiculata* can remain active for a longer period during the extended fermentation period. For example, the indigenous flora appears to contribute significantly to the highly desired fruity–flora character of 'Riesling' wines (Henick-Kling *et al.*, 1998). Such findings encourage the use of cool fermentation temperatures and reduced sulfur dioxide addition.

For fermentation at cooler temperatures, there is increased interest in the use of *Saccharomyces uvarum*. Besides cryotolerance, the species possesses several additional desirable features. For musts low in total acidity, *S. uvarum* strains typically increase the malic acid content. Thus, a flat taste can be avoided by biological acidification. *S. uvarum* may also augment the glycerol content; enhance the content of several aromatic esters and succinic acid; and generate a lower alcohol content (Massoutier *et al.*, 1998).

Red wines are typically fermented at higher temperatures than white wines. Temperatures between 24 and 27 °C are generally considered standard. Such temperatures favor not only anthocyanin but also tannin extraction. However, such temperatures are not universally preferred. For example, wines made from 'Pinotage' are reportedly better when fermented at 15 °C (du Plessis, 1983). The warmer temperatures generally preferred for red wine production probably relates more to its effect on phenol extraction than on fermentation rate. Temperature and alcohol are the major factors influencing pigment and tannin extraction from seeds and skins. Both groups of compounds dominate the characteristics of young red wines. The potentially undesirable consequences of higher fermentation temperatures, such as the

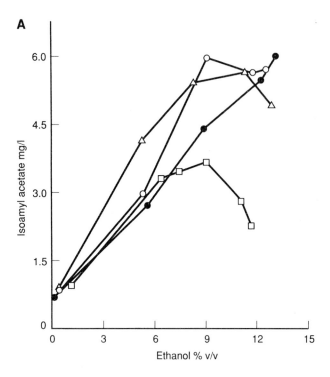

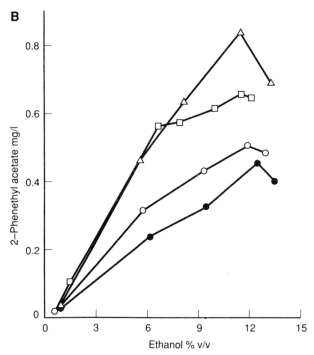

Figure 7.30 Effect of temperature and progress of fermentation on isoamyl acetate (**A**) and 2-phenethyl acetate, (**B**) content. ●, 10 °C; ○, 15 °C; △, 20 °C and □, 30 °C. (From Killian and Ough, 1979, reproduced by permission)

of glycerol at higher temperatures is often considered to give red wines a smoother mouth-feel.

Other important influences arise from factors not directly related to the effect of temperature on fermentation. For example, temperature affects the rate of ethanol loss during vinification (Williams and Boulton, 1983). Nevertheless, the volatilization of hydrophobic low-molecular-weight compounds such as esters are more marked. Consequently, their dissipation has a greater potential impact on the sensory quality of the wine than the loss of ethanol.

During fermentation, much of the chemical energy stored in grape sugars is released as heat. It is estimated that this is equivalent to about 23.5 kcal/mol glucose (see Williams, 1982). This is sufficient for juice, with a reading of 23 °Brix, to increase in temperature by about 30 °C during fermentation. If this were to occur, the yeast cells would die before completing fermentation. In practice, such temperature increases are not realized. Because the heat is liberated over several days to weeks, some of the heat is lost with escaping carbon dioxide and water vapor. Heat also radiates through the surfaces of the fermentor into the cellar environment. Nevertheless, the rise in temperature can easily reach levels critical to yeast survival, if temperature-control measures are not implemented in large fermentors.

Important in heat buildup is the initial juice or must temperature. This sets the rate at which temperature rises. Up to a point, the higher the juice temperature, the greater the initial rate of fermentation and heat release, and the sooner a lethal temperature may be reached. Thus, cool temperatures at the beginning of fermentation can limit the degree of temperature control required.

Also important to temperature control is the size and shape of the fermentor, and the presence or absence of a cap. The rate of heat lost is often directly related to the surface area/volume ratio of the fermentor. By retaining heat, the volume of juice can significantly affect the rate of fermentation – the larger the fermentor, the greater the heat retention and subsequent likelihood of over-heating. This feature is illustrated in Fig. 7.31.

The tumultuous release of carbon dioxide during fermentation may be sufficient to maintain a uniform temperature throughout the vat or tank. This is usual for fermenting white and rosé juice, in which vertical and lateral variation in temperature is seldom more than 1 °C. At cool fermentation temperatures, however, turbulence may be insufficient to equilibrate the temperature throughout the fermentor, and temperature stratification may develop.

For red wines, cap formation can disrupt the effective circulation and mixing of the must. The maximum cap-to-liquid temperature difference is often about 10 °C (Fig. 7.32). Punching down produces only transitory

production of increased amounts of acetic acid, acetaldehyde, and acetoin, and lower concentrations of some esters, are probably less noticeable against the more intense fragrance of red wines. The greater synthesis

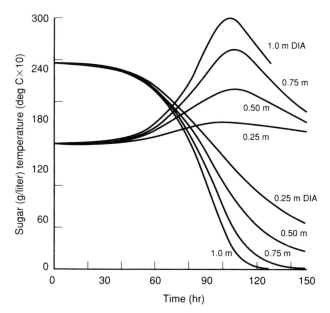

Figure 7.31 Effect of barrel diameter on fermentation rate and temperature rise during fermentation. Although the data are not presented in terms of cooperage capacity, barrels possessing maximum diameters of 0.5, 0.75 and 1.0 m could have capacities ranging from 75–150, 225–500, and 800–1200 liters, depending on barrel height and stave length. (From Boulton, 1979, reproduced by permission)

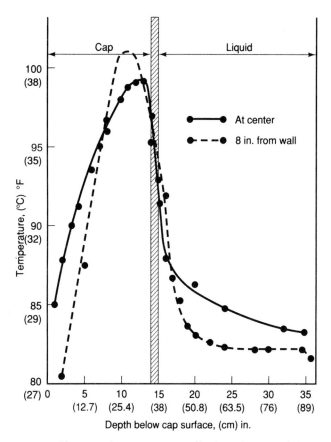

Figure 7.32 Vertical temperature profile through cap and liquid at 40 h. Crosshatching indicates that the boundary between the cap and liquid is not sharply defined. (From Guymon and Crowell, 1977, reproduced by permission)

temperature equilibration between the cap and the juice. In contrast, little temperature variation exists within the main volume of the must. Because high cap temperatures are a common feature of many traditional red wine fermentations, it follows that red wine vinification consists of two simultaneous but different phases – a liquid phase, in which the temperature is cooler and readily controlled, and a largely uncontrolled high-temperature phase in the cap (Vannobel, 1986). Because the rate of fermentation is much more rapid in the cap, the alcohol content rises quickly to above 10%. The higher temperatures found in the cap, plus the association of alcohol, probably increase the speed and efficiency of phenol extraction from the skins trapped in the cap.

Temperature regulation is achieved by a variety of techniques. Selective harvest timing has the potential to provide fruit at a desired temperature. Relatively small fermentors and vinification in cool cellars have been used for centuries to achieve a degree of temperature control. Pumping over in red wine vinification is another procedure that provides a degree of cooling. However, the maintenance of fermentation temperatures within a narrow range requires direct cooling in all but small barrels (~225 liters).

If heat transfer through the fermentor wall is sufficiently rapid, cooling the fermentor surface with water or by passing a coolant through an insulating jacket can be effective. If thermal conductance is insufficient, fermenting must can be pumped through external heat exchangers, or cooling coils may be inserted directly into the fermenting must. In special fermentors, carbon dioxide is trapped, and the pressure buildup is used to slow fermentation and, thus, heat accumulation.

PESTICIDE RESIDUES

Under most situations, no more than trace amounts of pesticide residues are found in juice or must. At such concentrations, they have little or no perceptible effect on fermentation, on the sensory qualities of the wine, or human health. Used properly, pesticides can help the fruit reach maximum quality. When used in excess or applied just before harvest, pesticides can negatively affect winemaking and potentially pose health risks.

Various factors influence the pesticide content on or in fruit. For example, heavy rains or sprinkler irrigation may wash contact pesticides off the fruit. Rain has less effect on systemic pesticides that are absorbed into plant tissues. Ultraviolet radiation in sunlight degrades some pesticides and decreases residual levels. Microbial decomposition is also possible.

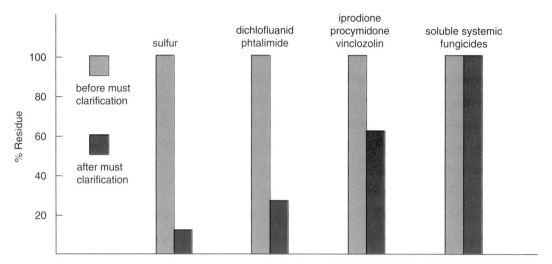

Figure 7.33 Influence of must settling on the elimination of various fungicides employed in viticulture. (After Gnaegi *et al.*, 1983, reproduced by permission)

Crushing, and especially maceration, can influence the incorporation of crop-protection chemicals into the must. Depending on the fungicide, extended maceration can either increase or decrease the amount found in the juice. Maceration generally has little effect on the content of systemic pesticides because they are already present in the juice before crushing.

Clarification, by cold settling, clarifying agents, or centrifugation, significantly reduces the concentration of contact fungicides, such as elemental sulfur, but has little effect on most systemic pesticide residues (Fig. 7.33). The persistence of pesticide residues, once dissolved, depends largely on their stability under the physicochemical conditions found in fermenting juice or wine. For example, more than 70% of dichlofluanid residues are degraded under the acidic conditions found in juice and wine (Wenzel *et al.*, 1980). In contrast, differences in racking, clarification, and filtration procedures did not markedly reduce the concentrations of fenarimol and penconazole (Navarro *et al.*, 1997). Nevertheless, for the most commonly used pesticides, degradation or precipitation reduce their levels in finished wine to trace or undetectable levels (Sala *et al.*, 1996).

Of pesticide residues, fungicides not surprisingly have the greatest effect on yeasts. Newer fungicides, such as metalaxyl (Ridomil®) and cymoxanil (Curzate®), do not appear to affect fermentation. In contrast, triadimefon (Bayleton®) can depress fermentation, presumably by disrupting sterol metabolism. The older, broad-spectrum fungicides, such as dinocap, captan, mancozeb, and maneb generally are toxic to yeasts. Fungicides such as copper sulfate and elemental sulfur seldom have a significant effect, except at abnormally high concentrations (Conner, 1983). This tolerance may be related to the relative insensitivity of wine yeasts to other sulfur compounds, notably sulfur dioxide.

Fungicides can have several direct and indirect effects on fermentation. Delaying the start of fermentation is probably the most common. As this primarily affects the lag phase, subsequent fermentation is unaffected. Increasing the inoculum size (5–10 g/hl active dry yeast) may avoid most fungicide-induced suppression of fermentation (Lemperle, 1988). The increased yeast population reduces the amount of fungicide available to react with each cell. Occasionally, the onset of fermentation occurs normally, but the rate is depressed (Gnaegi *et al.*, 1983). Such suppression may result in stuck fermentation.

Fungicides may affect the sensory qualities of wine by influencing the relative activities of various yeast biosynthetic pathways. Elemental sulfur augments the synthesis of hydrogen sulfide in some yeast strains, as may Bordeaux mixture, folpet, and zineb. Although hydrogen sulfide favors the subsequent production of mercaptans, residual copper can limit mercaptan synthesis by forming insoluble cupric sulfide with H_2S. Vinclozolin (Ronilan®) and iprodione (Rovral®) occasionally appear to affect fermentation and induce the development of off-flavors (San Romão and Coste Belchior, 1982). In addition, fungicides may occasionally react with the flavor constituents of the grape, reducing their impact on wine aroma. An example is the effect of the copper in Bordeaux mixture in reducing the content of 4-mercapto-4-methylpentan-2-one (4-MMP) in 'Sauvignon blanc' wines (Hatzidimitriou *et al.*, 1996). Copper reacts with 4-MMP during alcoholic fermentation and may also interfere with the synthesis of its precursor during grape maturation. Demonstration that newer fungicides are inactive against yeasts is often required for their registration for use on grapes.

Fungicides also may have selective effects on the endemic yeasts present during vinification. For example, captan favors the growth of *Torulopsis bacillaris* by suppressing the growth of most other yeast species (Minárik and Rágala, 1966). Several herbicides (2,4-dichlorophenoxyacetic acid [2,4-D] and simazine) and most insecticides do not disrupt yeast fermentation (Conner, 1983; Cabras *et al.*, 1995).

STUCK AND SLUGGISH FERMENTATION

Stuck fermentation refers to the premature termination of fermentation before all but trace amounts of fermentable sugars have been metabolized. Both stuck and sluggish fermentations have been a problem since time immemorial. Their occurrence in the past usually was attributed to overheating during fermentation. In the absence of adequate cooling, fruit harvested and fermented under hot conditions can readily overheat and become stuck. The resulting wines are high in residual sugar, making them particularly susceptible to microbial spoilage. Instability is increased further if the grapes are low in acidity, high in pH, or both.

The extensive use of temperature control during fermentation has essentially eliminated overheating as a significant factor in stuck fermentation. Ironically, the application of cooling during fermentation has favored the incidence of other causes of stuck wines. The desire to accentuate the fresh, fruity character of white wines has encouraged the use of cool temperatures. This can limit yeast growth and potentially favor microbial contaminants that further retard growth. The search for enhanced freshness has also induced some winemakers to use excessive juice clarification, either with bentonite, centrifugation or similar procedures. The resulting loss of sterols, unsaturated fatty acids, and nitrogenous nutrients can increase yeast sensitivity to the combined toxicity of ethanol and carboxylic acids, notably octanoic and decanoic acids and their esters. The latter enhance ethanol-induced leakage of amino acids and other nutrients (Sá Correia *et al.*, 1989). Octanoic acid (and other stress factors such as ethanol, high temperature, and nitrogen deficiency) also cause a reduction in intracellular pH, and associated enzymic and membrane dysfunction (see Viegas *et al.*, 1998). Crushing, pressing, and other prefermentative activities scrupulously conducted in the absence of oxygen heighten these effects. Molecular oxygen is required for the biosynthesis of sterols and long-chain unsaturated fatty acids essential for cell-membrane synthesis and function.

Juice from overmature or botrytized grapes generally has a very high sugar content. The osmotic effect of high sugar concentrations can partially plasmolyze yeast cells, resulting in slow or incomplete fermentation. In addition, overmature grapes may have an unusually low glucose-to-fructose ratio. This has been correlated with stuck fermentation in Switzerland (Schütz and Gafner, 1993b). It was solved by the addition of glucose to reestablish a more typical ratio. The lower concentration of available nitrogen and thiamine in botrytized must also can induce stuck fermentation. The growth of indigenous yeasts can further deplete thiamine content, thereby suppressing the activity of *S. cerevisiae* (Bataillon *et al.*, 1996). Nutrient depletion adds to the combined inhibitory effects of high sugar contents and the toxicity of ethanol and C_8 and C_{10} saturated carboxylic acids. In addition, polysaccharides synthesized by *B. cinerea* can inhibit yeast metabolism (Ribéreau-Gayon *et al.*, 1979). Occasionally, the activity of acetic acid bacteria in botrytized grapes is sufficient to further complicate yeast activity. Uptake of acetic acid produced by the bacteria can acidify yeast cytoplasm, dropping the pH from its usual 7.2 to below 6. This disrupts protein function, notably the glycolytic enzyme enolase (Pampulha and Loureiro-Dais, 1990). The presence of 10^5–10^6 acetic acid bacteria/ml can be lethal to *S. cerevisiae* (Grossman and Becker, 1984).

If the juice pH is sufficiently high, or the juice insufficiently protected by sulfur dioxide, indigenous lactobacilli may produce enough acetic acid to retard or inhibit fermentation. The best confirmed inhibitor is *Lactobacillus kunkeei* (Huang *et al.*, 1996). In this instance, eliminating sulfur dioxide addition prior to fermentation could lead to a significant increases in *L. kunkeei*-induced stuck fermentation. Sulfuring is the best-known means of controlling this relatively rare spoilage bacterium (Edwards *et al.*, 1999).

For the production of wines possessing low alcohol contents, but high residual sugar values, stuck fermentation may be purposely induced by chilling and clarification to remove the yeasts.

Another factor potentially causing stuck fermentation is the action of killer yeasts. Killer yeasts produce a protein that causes the death of cells not carrying the factor. Yeasts possessing the killer factor are infected with both a mycovirus and a satellite dsRNA. The mycovirus is considered a helper, because it regulates the replication and encapsulation of the satellite dsRNA. It is the satellite dsRNA that possesses the gene encoding the killer factor. Although most killer proteins act only on related yeast strains, forms are known that are active against other yeasts, as well as filamentous fungi and bacteria (see Magliani *et al.*, 1998). These do not appear to be important in wine fermentations.

Under optimal conditions (such as low inocula with highly sensitive strains, or continuous fermentation), killer yeasts can replace inoculated strains, even when the initial concentration of the killer strain is as low as 0.1% (Jacobs and van Vuuren, 1991). Indigenous killer *Saccharomyces cerevisiae* strains may cause sluggish or stuck fermentations, as well as donate undesirable sensory attributes. Even more serious are situations where potential spoilage yeasts, notably *K. apiculata* or *Zygosaccharomyces bailii*, possess killer factors and can inactivate inoculated strains.

This potential can largely be countered by inoculation with commercial yeast strains constructed to possess both common killer satellite dsRNAs (K1 and K2) (Boone *et al.*, 1990; Sulo *et al.*, 1992). Possession of the killer factor protects the producing cell from the effects of the toxin. Addition of sulfur dioxide may also suppress killer yeasts in the indigenous flora, if combined with inoculation with a resistant wine yeast strain. Temperature control and the addition of bentonite can also reduce the activity of killer proteins.

Killer properties have been isolated from naturally occurring *S. cerevisiae* strains in wine, as well as in other yeast genera (i.e., *Hansenula*, *Pichia*, *Torulopsis*, *Candida*, and *Kluyveromyces*). Although relatively uncommon, yeast strains resistant to killer toxins have been found that do not possess the killer factor. These are termed neutral strains. Typically, yeasts immune to a particular killer protein possess the gene that produces it.

Expression of the killer factor occurs differently in various yeasts. In S. *cerevisiae*, both K1 and K2 satellite dsRNAs are associated with a helper virus of the totiviridae group, whereas in *Kluyveromyces lactis*, linear dsDNA plasmids control the property. In some genera, chromosomal genes may be involved. In all cases, the toxic principle is associated with the production and release of a protein or glycoprotein.

With the K1 and K2 toxins, toxicity involves the attachment of the killer protein to β-1,6-D-glucans of the cell wall of sensitive strains. Subsequently, the toxin binds with a component of the cell membrane, where it forms a pore that permits the unregulated movement of ions across the membrane. This results in cell death. In contrast, the K28 toxin attaches to α-1,3-mannose of cell-wall mannoproteins. It inhibits DNA synthesis and further cell division of the affected cell. The nuclear genes that encode killer proteins (KHR and KHS) are less well known, but also result in an increase in ion permeability of sensitive cells.

Killer cells are immune to the effects of their own toxin, but may be sensitive to those produced by other strains. Eleven distinct killer (K) factors have been identified, most of which do not affect *S. cerevisiae*.

The killer proteins produced by *S. cerevisiae* act optimally at a pH above that normally found in wine, notably the K1 protein. Consequently, the K2 killer factor is of greater significance in wine production – most killer yeasts isolated from wine possess the K2 factor. Cells appear to be most sensitive to the toxin during the exponential phase of cell growth, when the growing tip of buds provide β-1,6-D-glucan receptors in close proximity to the cell membrane. The significance of killer toxins in winemaking probably depends on the juice pH, the addition of protein-binding substances (such as bentonite or yeast hulls), the ability of killer strains to ferment effectively, the presence of free ammonia nitrogen, and the degree to which the yeast population multiplies during fermentation.

The likelihood of stuck fermentation also may be reduced by limiting prefermentative clarification (restrict nutrient loss), limited aeration (~5 mg O₂/liter) at the end of exponential cell growth, the addition of ergosterol or long-chain unsaturated fatty acids (i.e., oleic, linoleic, or linolenic acids), the addition of yeast ghosts or other absorptive materials, such as bentonite, or the addition of ammonium salts periodically or about halfway through fermentation. The addition of absorptive substances, such as yeast hulls, appears to have optimal effects when applied midway in or near the end of the exponential phase of yeast growth. They have the added benefit of providing a source of sterols and long-chain unsaturated fatty acids.

The need for such treatments will be indicated by discovering the root cause(s) for past instances of stuck or sluggish fermentations. This is greatly aided by close scrutiny of the conditions and dynamics of fermentation. These details provide a basis for diagnosing the cause(s), and the commencement of early corrective measures. Bisson and Butzke (2000) have divided problem fermentations into four basic categories: slow initiation (eventually becoming normal); continuously sluggish; typical initiation, but becoming sluggish; and normal initiation but abrupt termination. Comparing sugar consumption, temperature, nutrient profiles, and records of procedures used with those from past fermentations, often provides early indications of potential problems and their possible quick resolution. Once fermentation has stopped, reinitiation is more complicated.

When stuck fermentation occurs, successful reinitiation usually requires incremental reinoculation with special yeast strains (obtained commercially), following racking off from settled lees (see Bisson and Butzke, 2000). The special strains usually possess high ethanol tolerance, as well as the ability to utilize fructose (the sugar whose proportion increases during fermentation). The addition of nutrients (if deficient), yeast hulls (to remove toxic fatty acids), must aeration, and adjustment of the fermentation temperature (if necessary) usually achieves successful refermentation.

Malolactic Fermentation

After years of intensive investigation, malolactic fermentation is finally becoming less controversial. The contention related to its respective abilities to improve, or reduce, wine quality. In addition, wines in which malolactic fermentation had its major benefits were those in which its occurrence was most difficult. In contrast, conditions in wine that favored malolactic fermentation were those where its occurrence was either unnecessary or undesirable.

The principal effects of malolactic fermentation are a rise in pH and a reduction in perceived acidity. The dicarboxylic malic acid is replaced by the monocarboxylic lactic acid, and the harsh taste of malic acid is replaced by the less aggressive sensation of lactic acid. Thus, winemakers in most cool wine-producing regions (where high acidity is usually caused by malic acid) view malolactic fermentation positively, especially for red

L-malic acid

L-lactic acid

wines. In contrast, wines produced in warm regions may be low in acidity, high in pH, or both. Malolactic fermentation can aggravate an already difficult situation, leaving the wine tasting "flat" and microbially unstable.

Although deacidification is still the paramount reason for inducing malolactic fermentation, many winemakers are increasingly viewing it as a means of adjusting wine flavor. Although more commonly promoted in red wines, it is now being applied to white wines. Malolactic fermentation is often thought to reduce the incidence of vegetal notes and accentuate fruit flavors (Laurent *et al.*, 1994). For wines marginally high in pH, tartaric acid may be added prior to the induction of malolactic fermentation.

Lactic Acid Bacteria

Lactic acid bacteria are characterized by several unique properties. Their name refers to one of these, the production of large amounts of lactic acid. Depending on the genus and species, lactic acid bacteria may ferment sugars solely to lactic acid, or to lactic acid, ethanol, and carbon dioxide. The former mechanism is termed homofermentation, whereas the latter is called heterofermentation. Bacteria capable of both types of fermentation can grow in wine. Homofermentation potentially yields two ATPs per glucose (similar to yeast fermentation), whereas heterofermentation yields but one ATP per glucose molecule.

The most beneficial member of the group is the heterofermentative species, *Oenococcus oeni* (formerly *Leuconostoc oenos*). Recently, its genome has been deciphered (Mills *et al.*, 2005). It contains 1701 ORFs (open reading frames), of which 75% have been classified as related to functional genes. This information should facilitate the study of the physiology and genetic diversity within the species.

Not only is *O. oeni* the species most frequently found in wine, but also it is the only species inducing malolactic fermentation in wines at a pH $\leqslant 3.5$. Spoilage forms are generally members of the genera *Lactobacillus* and *Pediococcus*. *Lactobacillus* contains both homo- and heterofermentative members, whereas *Pediococcus* is strictly homofermentative.

Although lactic acid bacteria are classified primarily on the basis of sugar metabolism, the extent to which sugar metabolism is important to their growth in wine is unclear. Even dry wines possess sufficient residual sugars (mostly pentoses) to support considerable bacterial growth. Occasionally, the concentration of hexose sugars (glucose and fructose) increases marginally during malolactic fermentation (Davis *et al.*, 1986a). This increase, however, is apparently unrelated to malolactic fermentation, as it can occur in its absence.

Lactic acid bacteria are further distinguished by their limited biosynthetic abilities. They require a complex series of nutrients, including B vitamins, purine and pyrimidine bases, and several amino acids. Indicative of their limited synthetic capabilities is their inability to synthesize heme proteins. As a consequence, they produce neither cytochromes nor catalase. Without cytochromes, they cannot respire. Consequently, their energy metabolism is strictly fermentative.

Most bacteria that are incapable of synthesizing heme molecules are strict anaerobes; that is, they are unable to grow in the presence of oxygen. Oxygen reacts with certain cytoplasmic components, notably flavoproteins, to produce toxic oxygen radicals (superoxide and peroxide). In aerobic organisms, superoxide dismutase and catalase rapidly inactivate these toxic radicals. Anaerobic bacteria produce neither. Lactic acid bacteria escape the fate of most anaerobic bacteria in the presence of oxygen by accumulating large quantities of Mn^{2+} ions, and producing peroxidase. Manganese detoxifies superoxide by converting it back to oxygen. The rapid action of manganese also limits the synthesis of hydrogen peroxide from superoxide. Peroxidase oxidizes any peroxide formed to water, in the presence of organic compounds. Lactic acid bacteria are the only bacterial group that is both strictly fermentative and able to grow in the presence of oxygen.

Although fermentative metabolism is an inefficient means of generating biologically useful energy, the production of large amounts of acidic wastes quickly lowers the pH of most substrates. The resulting low pH inhibits the growth of most other, potentially competitive bacteria. Lactic acid bacteria are one of the few bacterial groups capable of growing below a pH of 5. This property has preadapted lactic acid bacteria to grow in acidic environments. Although lactic acid bacteria grow under acidic conditions, growth is comparatively poor at the low pH values typical of must and wine. For example, species of *Lactobacillus* and *Pediococcus* commonly cease growing below pH 3.5. Even *Oenococcus oeni*, the primary malolactic bacterium, is inhibited by pH values below 3.0–2.9. *O. oeni* grows optimally within a pH range of 4.5–5.5. Thus, the major benefit of malolactic fermentation for the bacterium is, surprisingly, acid reduction. By metabolizing malic acid to lactic acid, the number of carboxyl groups is halved, acidity is reduced and the pH increased. The degradation of arginine, one of the major amino acids in must and wine, releases ammonia, which may also help raise the pH. Raising the pH provides conditions more suitable for bacterial growth.

The enzyme involved in malolactic fermentation by lactic acid bacteria, the **malolactic enzyme**, is unique. Unlike other conversions of malic acid to lactic acid, the enzymatic reaction directly decarboxylates L-malic acid to L-lactic acid, without free intermediates. The enzyme functions in a two-step process. First, malic acid is decarboxylated to pyruvic acid (which remains bound to the enzyme), and then pyruvic acid is reduced to lactic acid. ATP is generated through the joint export of lactic acid and hydrogen ions (protons) from the cell (Henick-Kling, 1995). The hydrogen ions that accumulate outside the cell membrane are sufficient to drive the oxidative phosphorylation of ADP to ATP. A similar energy-producing mechanism is associated with the metabolism of citric acid. The metabolism of citric acid appears to assist the fermentation of glucose (Ramos and Santos, 1996). Small amounts of reducing energy (NADH) may also result from the action of a minor alternative oxidative pathway. It directly oxidizes malic acid to pyruvic acid. Some of the pyruvic acid so generated is subsequently reduced to lactic acid.

Although the decarboxylation of malic acid to lactic acid is central to the importance of malolactic fermentation in wine production, it is not the primary energy source for lactic acid bacteria. Currently, the primary energy source for these bacteria in wine is still in question. The source appears to change through the growth phase – initially involving small amounts of sugars (exponential phase), followed by malic acid (early stationary phase), and subsequently by citric acid (late stationary phase) (Krieger *et al.*, 2000). The situation is complicated by the marked influence of pH on the ability to ferment sugars, and considerable variability between strains. *Oenococcus oeni* appears to show little ability to ferment sugars, at least below pH 3.5 (Davis *et al.*, 1986a; Firme *et al.*, 1994). However, data from Liu *et al.* (1995) suggest that sugars may be the primary carbon and energy sources for the slow-growth phase of the bacteria. Growth may be increased by the joint metabolism of several compounds, such as glucose with fructose or citrate, by improving the redox balance and increasing ATP production. The metabolism of citric acid also has enologic importance. It is the prime source of diacetyl, acetoin (Shimazu *et al.*, 1985) and acetic acid during malolactic fermentation. The fermentation of citric acid is enhanced by the presence of phenolic compounds (Rozès *et al.*, 2003). Amino acids, notably arginine, may also act as energy sources (Liu and Pilone, 1998). The activity of *O. oeni* protease on yeast and grape proteins may, thus, be important both in terms of energy metabolism and in providing essential growth factors (De Nadra *et al.*, 1997). Although, the high rate of malic acid conversion to lactic acid seems to account for much of the ATP required for maintenance during the log and stationary phases of malolactic fermentation (Henick-Kling, 1995), the energy dynamics associated with the initial growth phase under wine conditions remains controversial.

All lactic acid bacteria growing in wine assimilate acetaldehyde and other carbonyl compounds. However, the metabolism of carbonyl compounds may retard malolactic fermentation. By shifting the equilibrium between bound and free carbonyls, sulfur dioxide is liberated.

As noted for anaerobic yeast metabolism, fermentation can result in the generation of excess reduced NAD (NADH). To maintain an acceptable redox balance, the bacteria must regenerate NAD^+. How lactic acid bacteria accomplish this in wine is unclear. Some species reduce fructose to mannitol, presumably for this purpose (Salou *et al.*, 1994; Nielsen and Richelieu, 1999). This may explain the common occurrence of mannitol in wine associated with malolactic fermentation. Some strains also regenerate NAD^+ with flavoproteins and oxygen. This reaction might account for the reported improvement in malolactic fermentation in the presence of trace amounts of oxygen. However, oxygen also inactivates the pathway reducing pyruvate to ethanol – one of the means by which *O. oeni* regenerates NAD^+ for the metabolism of glucose.

In addition to important physiological differences, morphological features help to distinguish the various genera of lactic acid bacteria (Fig. 7.34). *Oenococcus* usually consists of spherical to lens-shaped cells (Plate 7.6), commonly occurring in pairs or chains, but occasionally existing singly. *Leuconostoc mesenteroides*, closely related to *Oenococcus*, occasionally is also isolated from

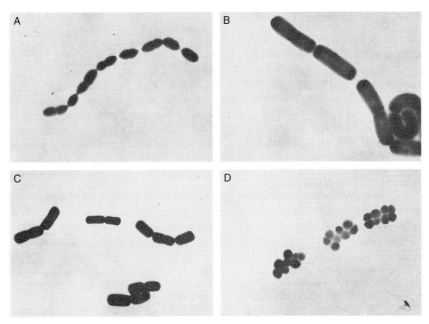

Figure 7.34 Micrographs of important members of the Lactobacillaceae found in wine: (A) *Oenococcus oeni* (×6000); (B) *Lactobacillus casei* (×8500); (C) *Lactobacillus brevis* (×5500); (D) *Pediococcus cerevisiae* (×5000). Cell shape and grouping may depend on the medium in which the bacteria grew. (From Radler, 1972, reproduced by permission of Dr Radler)

Table 7.4 Lactic acid bacteria occurring in wine

Genus	Species
Oenococcus	O. oenos
Pediococcus	P. pentosaceus, P. damnosus (P. cerevisiae), P. parvulus
Lactobacillus	L. plantarum, L. brevis, L. cellobiosis, L. buchneri, L. casei, L. hilgardii, L. trichodes, L. mesenteroides

wine. *Pediococcus* species usually occur as packets of four spherical cells. *Lactobacillus* produces long, slender, occasionally bent, rod-shaped cells, commonly occurring in chains. Some of the lactic acid bacteria that may occur in wine are listed in Table 7.4.

Effects of Malolactic Fermentation

Bacterial malolactic fermentation has three distinct, but interrelated, effects on wine quality. It reduces acidity, influences microbial stability, and affects the wine's sensory attributes.

ACIDITY

Deacidification, and a rise in pH, are the most consistent effects of malolactic fermentation. This adjustment usually does not commence until the bacterial population has reached a threshold of about 10^8 cells per ml.

A reduction in acidity increases the smoothness and drinkability of red wines, but excessive reduction generates a flat taste. The desirability of deacidification depends primarily on the initial pH and acidity of the grapes. In general, the higher the acidity and the lower the pH, the greater the benefit; conversely, the lower the acidity and the higher the pH, the greater the likelihood of undesirable consequences. In addition, the higher the relative concentration of tartaric acid in the juice, the less likely malolactic fermentation will significantly affect the acidity and pH of the wine.

As the pH of wine changes, so too does the relative proportion of the various colored and uncolored forms of anthocyanin pigments. The metabolism of carbonyl compounds (notably acetaldehyde) by lactic acid bacteria, and the accompanying release of SO_2, also may result in some pigment bleaching. In general, color loss associated with malolactic fermentation is significant only in pale-colored wines, or those with an initially high pH.

MICROBIAL STABILITY

Formerly, increased microbial stability was considered one of the prime benefits of malolactic fermentation. This view has changed considerably in recent years.

Improved microbial stability was thought to result from the metabolism of residual nutrients left after alcoholic fermentation. Metabolism of malic and citric acids left the more microbially stable tartaric and lactic acids. In addition, the complex nutrient demands of lactic acid bacteria were thought to reduce the concentration of amino acids, nitrogen bases, and vitamins. Although their levels do tend to decrease, wines having completed malolactic fermentation may continue to support the growth of *Oenococcus oeni*, or lactobacilli and pediococci (Costello *et al.*, 1983), as well as other spoilage microorganisms.

Contrary to common belief, malolactic fermentation occasionally decreases microbial stability. This can occur when the wine is marginally too high in pH. The resulting increase in pH can favor the subsequent growth of spoilage lactic acid bacteria. Spoilage organisms generally do not grow in wines at a pH below 3.5, but their ability to grow increases rapidly as the pH rises from 3.5 to 4.0 and above.

Thus, the stabilizing effect associated with malolactic fermentation may come more from the preservation practices applied after its completion, rather than as a direct consequence of malolactic fermentation. The early onset and completion of malolactic fermentation permits the application of procedures, such as sulfur dioxide addition, storage at cool temperatures, and clarification. Delayed onset, without the application of standard preservation techniques, exposes the wine to potential infection by acetic acid bacteria, spoilage lactic acid bacteria, and spoilage yeasts (notably *Brettanomyces/Dekkera* spp.). In addition, the early completion of malolactic fermentation avoids its possible occurrence post-bottling.

FLAVOR MODIFICATION

The greatest controversy concerning the relative merits or demerits of malolactic fermentation relates to flavor modification. The diversity of opinion undoubtedly reflects both the biologic and physicochemical conditions during malolactic fermentation. Considerable variability also exists among the strains and species of lactic acid bacteria (Laurent *et al.*, 1994; Krieger, 1996). In addition, the pH and temperature of wine markedly affect bacterial physiology. Thus, it is not surprising that the sensory influences of malolactic fermentation are often contested. Table 7.5 lists some of the substrates metabolized and by-products produced by lactic acid bacteria. Figure 7.35 illustrates not only the influence of different bacterial strains, but also the response variability of two sets of panelists.

Diacetyl (biacetyl, 2,3-butanedione) commonly accumulates at the end of malolactic fermentation (Rankine *et al.*, 1969). At a concentration between 1 and 4 mg/liter, diacetyl often adds a desirable complexity to the fragrance (the threshold varying with the type of wine). It is often referred to as having a buttery, nutty, or toasty character. However, at concentrations above 5–7 mg/liter, its buttery character can become pronounced and undesirable. Mild aeration dramatically increases diacetyl synthesis. Maximum accumulation correlates with the completion of malic and citric acid metabolism (Nielsen and Richelieu, 1999). Nevertheless, sensory differences among wines with and without malolactic fermentation often do not correlate with their diacetyl content (Martineau and Henick-Kling, 1995).

Other flavorants occasionally synthesized in amounts sufficient to affect a wine's sensory character include acetaldehyde, acetic acid, acetoin, 2-butanol, diethyl succinate, ethyl acetate, ethyl lactate, and 1-hexanol. Much of the acetic acid associated with malolactic fermentation appears to come from citric acid metabolism. If the strain is a significant producer of acetic acid, it can leave the wine unacceptably high in volatile acidity.

Most lactic acid bacteria produce esterases. Although this could result in important losses in the fruity character of young wines, such decreases generally are small. Conversely, nonenzymatic synthesis of some esters increases during malolactic fermentation, especially

Table 7.5 Substrates and fermentation products of lactic acid bacteria

Substrate	Products
Acids	
L-Malate	L-Lactate, CO_2, succinate, acetate
Citrate; pyruvate	Lactate, acetate, CO_2, acetoin, diacetyl
Gluconate	Lactate, acetate, CO_2
2-Oxoglutarate	4-Hydroxybutyrate, CO_2, succinate
Tartrate	Lactate, acetate, CO_2, succinate
Sorbate	2,4-Hexadien-1-ol (sorbic alcohol)
Chlorogenate	Ethylcatechol, dihydroshikimate
Sugars	
Glucose	Lactate, ethanol, acetate, CO_2
Fructose	Lactate, ethanol, acetate, CO_2, mannitol
Arabinose, xylose, or ribose	Lactate, acetate
Polyols	
Mannitol	(Probably as from glucose)
2,3-Butanediol	2-Butanol
Glycerol	1,3-Propanediol
Amino acids	
Arginine	Ornithine, CO_2, NH_4
Histidine	Histamine, CO_2
Phenylalanine	2-Phenylethylamine, CO_2
Tyrosine	Tyramine, CO_2
Ornithine	Putrescine, CO_2
Lysine	Cadaverine, CO_2
Serine	Ethanolamine, CO_2
Glutamine	Aminobutyrate, CO_2
Unknown substrates (probably sugars)	Propanol, isopropanol, isobutanol, 2-methyl-1-butanol, 3-methyl-1-butanol, ethyl acetate, acetaldehyde, *n*-hexanol, *n*-octanol, glycerol, 2,3-butanediol, erythritol, arabitol, dextran, diacetyl

Source: After Radler, 1986, reproduced by permission

ethyl acetate. Ethyl acetate also may accumulate via direct bacterial biosynthesis. Many strains of lactic acid bacteria also produce β-glucosidases. If released in sufficient quantities, and not inhibited by unfavorable temperature and ethanol conditions in the wine (Spano *et al.*, 2004), these enzymes could facilitate development of a varietal character. This would involve releasing grape aromatics from odorless glycosidic complexes (D'Incecco *et al.*, 2004).

The metabolism of arginine, which has a bitter, musty taste, could contribute to the improved taste of wines high in residual arginine. In some wines, residual arginine values can reach into the gram per liter range. The disadvantage, however, is that incomplete arginine metabolism can liberate citrulline. Its reaction with ethanol can produce ethyl carbamate. This appears more likely with other lactic acid bacteria than *Oenococcus oeni* (Mira de Orduña *et al.*, 2001). It appears to

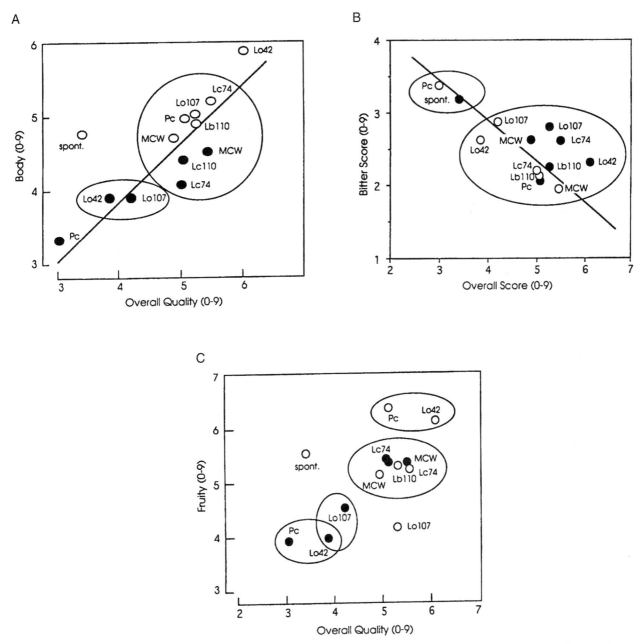

Figure 7.35 Relationship of body (**A**), bitterness (**B**), and fruitiness (**C**) to overall quality of 'Cabernet Sauvignon' wine fermented with various malolactic cultures, as evaluated by two taste panels, composed of winemakers (●) and a wine research group (○). (From Henick-Kling *et al.*, 1993, reproduced by permission)

require a high minimum pH. This occurs only at the end of malic acid decarboxylation to lactic acid.

Malolactic fermentation has traditionally been more encouraged in red than white wines. This preference may relate to the tendency of lactic acid bacteria to metabolize compounds responsible for excessively vegetative, grassy aromas, for example the herbaceous aspect of 'Cabernet Franc' wines (Gerland and Gerbaux, 1999). *Oenococcus oeni* also appears to increase the presence of C_4 to C_8 fatty acid ethyl esters and 3-methylbutyl acetate (Ugliano and Moio, 2005). Red wines, typically possessing more flavor than white wine, are less likely to be overpowered by the aromatics produced by malolactic bacteria.

Although *Oenococcus oeni* is not known to decarboxylate phenol carboxylic acids (ferulic and *p*-coumaric acids), other lactic acid bacteria can (Cavin *et al.*, 1993). For example, *Lactobacillus brevis*, *L. plantarum*, and *Pediococcus* are able to convert these acids to volatile phenols (4-ethylguaiacol and 4-ethylphenol). The production of these compounds, at amounts much above 4 mg/l, could give the wine a distinct phenolic or stable-like off-odor. At subthreshold levels they can add to a wine's aromatic complexity.

Malolactic fermentations occurring below a pH of 3.5 (typically by *Oenococcus oeni*) rarely generate undesirable off-odors. Undesirable buttery, cheesy, or milky odors are usually confined to malolactic fermentation induced by pediococci or lactobacilli at above pH 3.5.

AMINE PRODUCTION

Lactic acid bacteria, notably pediococci, produce amines through the decarboxylation of amino acids. Evidence, however, implicates *Oenococcus oeni* as the primary source of histamine production. This may occur even after the bacteria have died. The enzyme involved remains active in wine considerably longer than the producing bacteria (Coton *et al.*, 1998). Amine synthesis appears to be important only in wines of high pH. Although some biogenic amines can induce blood-vessel constriction, headaches, and other associated effects, their contents in wine appear to be insufficient to induce these physiological effects in humans (Radler and Fäth, 1991).

Origin and Growth of Lactic Acid Bacteria

The ancestral habitat of *Oenococcus oeni* is unknown. Although infrequently isolated from grape (see Bae *et al.*, 2006) or leaf surfaces (and then only in low numbers), it has no known habitat other than wine. From the relatively small variation in its genetic makeup, *O. oeni* appears to have evolved from a few individuals specialized to grow in wine (Zavaleta *et al.*, 1997). It has subsequently spread worldwide.

Species of *Pediococcus*, *Lactobacillus* and *Leuconostoc* generally occur more frequently on grapes than does *Oenococcus*. Their numbers may occur in the range of $10^3–10^4$ cells/ml shortly after crushing (Costello *et al.*, 1983). The population size depends largely on the maturity and health of the fruit – higher numbers occurring on mature or mold-infected fruit. Nevertheless, the relatively low numbers indicate that grapes are unlikely the primary habitat of any of the species.

Although malolactic fermentation may be induced by bacteria originating from grape surfaces, strains also may originate from winery equipment. Stemmers, crushers, presses, and fermentors may harbor populations of lactic acid bacteria. The relative importance of grape versus winery sources has yet to be established.

An increasing number of winemakers inoculate their wines with commercial strains when malolactic fermentation is desired. Of particular concern is avoiding indigenous strains that partially metabolize arginine to citrulline. Citrulline, or its breakdown product, carbamyl-P, can generate ethyl carbamate (a suspected carcinogen).

Unlike yeast growth during alcoholic fermentation, no consistent bacterial growth sequence develops in the must or wine during malolactic fermentation. A pattern occasionally found in spontaneous malolactic fermentations is shown in Fig. 7.36. Significant variations occur due to factors such as pH, total acidity, malic acid content, temperature, and the duration of skin contact.

Several strains of *Oenococcus oeni* may be present at the beginning of alcoholic fermentation. In spontaneous malolactic fermentation, the most common strains frequently disappear by the end of alcoholic fermentation. Malolactic fermentation is usually conducted by

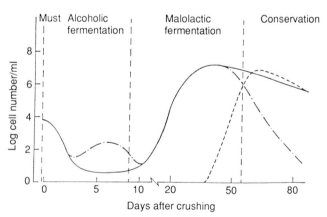

Figure 7.36 Diagram of the growth of indigenous lactic acid bacteria during the vinification of a red wine. Population development of *Oenococcus oeni* below (———) and above (——— · ———) pH 3.5, before and during malolactic fermentation; and as affected by the growth of other species during the latter stages of malolactic fermentation or wine maturation: *O. oeni* (—— · —— · ——); other species (- - -). (From Wibowo *et al.*, 1985, reproduced by permission)

one or more of the strains that are initially uncommon (Reguant and Bordons, 2003). When malolactic fermentation is induced by inoculation, it is typical for the inoculated strain to dominate deacidification or, at least, be a common member of the population inducing malolactic fermentation (Bartowsky *et al.*, 2003).

In most spontaneous fermentations, cells lyse rapidly as alcoholic fermentation begins, reducing the bacterial population from about 1×10^3 to about 1 cell/ml. Most species of lactic acid bacteria initially found die out during alcoholic fermentation. Wines with pH values higher than 3.5 may show the temporary growth of species such as *Lactobacillus plantarum*. Occasionally, when sulfiting is low, and the pH is above 3.5, *Oenococcus oeni* may induce a malolactic fermentation coincident with alcoholic fermentation.

The usual initial population decline has been variously ascribed to sulfur dioxide toxicity, acidity, the synthesis of ethanol and toxic carboxylic acids, or the increasingly nutrient-poor status of fermenting must. All these factors may be involved to some degree. At the end of alcoholic fermentation, a lag period generally ensues before the bacterial population begins to rise. This phase may be of short duration, or it may last several months. Once growth initiates, the bacterial population may rise to 10^6–10^8 cells/ml. In most wines of low pH, only *Oenococcus oeni* grows. However, there can be variations in the proportion of various *O. oeni* strains throughout malolactic fermentation. At high pH values, species of both *Lactobacillus* and, especially, *Pediococcus* may predominate. Depending on the strain or species involved, the decarboxylation of malic acid may occur simultaneously with bacterial multiplication or only after cell growth has ceased.

At the end of the exponential growth phase, the cell population enters a prolonged decline phase. The slope of the decline can be dramatically changed by cellar practices. For example, storage at above 20–25 °C, or the addition of sulfur dioxide, can result in the rapid death of *Oenococcus oeni*. If the pH is above 3.5, and other conditions are favorable, previously inactive strains of *Lactobacillus* or *Pediococcus* may begin to multiply. Their growth often produces a corresponding decline in the population of *O. oeni*. The nature of this apparently competitive antibiosis is unknown.

Factors Affecting Malolactic Fermentation The growth conditions in wine are harsh for any bacterium – cool, acidic, alcoholic, anaerobic, low in nutrients and possessing toxic fatty acids. Thus, *Oenococcus oeni* is exposed to many environmental stresses. Tolerance appears to be under the control of a master regulator (CtsR) (Grandvalet *et al.*, 2005). It controls the synthesis of a variety of stress response proteins, such

as HSPs and Clp ATP-dependent proteases. The first group facilitates maintenance of proper protein folding, while the other degrades improperly folded proteins. In addition, the bacterium produces membrane-bound ATPases, thought to maintain cytoplasmic electrolytic balance. Many of these proteins are synthesized at specific phases of colony growth. The influence of several of these stress factors on bacterial growth and physiology are outlined below.

PHYSICOCHEMICAL FACTORS

pH The initial pH of juice or wine strongly influences not only if and when malolactic fermentation occurs, but how and what species will conduct the process (Fig. 7.37). Low pH not only slows the rate, but also tends to delay its initiation. Below a pH of 3.5, *Oenococcus oeni* is the predominant species inducing malolactic fermentation, whereas above pH 3.5, *Pediococcus* and *Lactobacillus* spp. become increasingly prevalent (Costello *et al.*, 1983). Some of the inhibitory effects observed at low pH values are probably indirect, acting by increasing not only cell membrane sensitivity to ethanol, but also the concentration of molecular sulfur dioxide. pH also significantly affects the composition and degree of unsaturation of cell-membrane fatty acids, similar to the effect of ethanol (Drici-Cachon *et al.*, 1996).

The pH significantly modifies the metabolic activity of lactic acid bacteria. For example, sugar fermentation is much more effective at higher pH values. Similarly, the

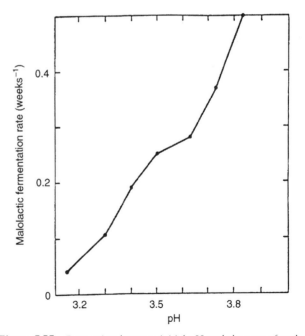

Figure 7.37 Connection between initial pH and the rate of malolactic fermentation. (From Bousbouras and Kunkee, 1971, reproduced by permission)

synthesis of acetic acid increases, whereas diacetyl production decreases in relation to increased pH (see Wibowo *et al.*, 1985). The metabolism of malic and tartaric acid is also affected, with decarboxylation of malic acid being favored at low pH, whereas the potential for tartaric acid degradation increases above pH 3.5. This results primarily from the preferential growth of *Lactobacillus brevis* and *L. plantarum* at higher pHs (Radler and Yannissis, 1972). This can lead to dramatic increases in wine pH, as both malic and tartaric acids (strong acids) are metabolized to lactic acid (a weak acid).

Temperature   The effect of temperature on malolactic fermentation has long been known. It often took place in the spring when cellars began to warm. To speed its initiation, cellars may be heated to maintain wine temperature above 20°C. Although temperature directly affects bacterial growth rate, its most significant effect is on the rate of malic acid decarboxylation. Maximal decarboxylation occurs between 20 and 25°C, whereas growth is roughly similar, within a range of 20–35°C (Ribéreau-Gayon *et al.*, 1975). Outside this range, decarboxylation slows dramatically. At temperatures below 10°C, decarboxylation essentially ceases. In addition, most strains of *Oenococcus oeni* grow very slowly or not at all below 15°C. Cool temperatures maintain cell viability, though. Thus, wines cooled after malolactic fermentation commonly retain a high viable population for months. Temperatures around 25°C favor a rapid decline in the *O. oeni* population (Lafon-Lafourcade *et al.*, 1983), but may promote the growth of pediococci and lactobacilli.

Cellar Practices   Many cellar practices can affect when, and if, malolactic fermentation occurs. Maceration on the skins commonly increases the frequency and speed of malolactic fermentation (Guilloux-Benatier *et al.*, 1989). The precise factors are unknown, but they may involve the action of phenols as electron acceptors in the oxidation of sugars during fermentation (Whiting, 1975). This would help to explain why malolactic fermentation develops more commonly in red wines (with long maceration periods) than in white wines. The higher pH of most red wines is undoubtedly involved as well.

Clarification can directly reduce the population of lactic acid bacteria by encouraging removal with yeasts and grape solids. Racking, fining, centrifugation, and other similar practices also remove nutrients or limit their uptake into wine as a consequence of yeast autolysis.

CHEMICAL FACTORS

Carbohydrates and Polyols   The chemical composition of must or wine has a profound influence on the outcome of malolactic fermentation. Carbohydrates and polyols constitute the most potentially significant group of fermentable compounds (Davis *et al.*, 1986a). Most dry wines contain between 1 and 3 g/liter residual hexoses and pentoses. There are also variable amounts of di- and trisaccharides, sugar alcohols, glycosides, glycerol, and other polyols.

There is considerable heterogeneity among strains and species of lactic acid bacteria in their use of these nutrients. Ethanol and pH also influence their ability to ferment carbohydrates. Consequently, few generalizations about carbohydrate use appear possible. The major exception may be the poor use of most polyols. Few lactobacilli metabolize glycerol, the most prevalent polyol in wine. Although uncommon, glycerol metabolism produces a bitter-tasting compound, acrolein (Meyrath and Lüthi, 1969).

Skin contact, before or during fermentation, also promotes bacterial growth and malolactic fermentation. This is associated with the release of increased amounts of mannoproteins during fermentation and with the reduced production of toxic mid-chain fatty (carboxylic) acids. A similar boost is also associated with *sur lies* maturation. Must clarification before fermentation also appears to enhance subsequent release of yeast polysaccharides. That the content of α- and β-glucosidases in lactic acid bacteria increases in the presence of yeast polysaccharides suggests that polysaccharide hydrolysis may be a carbon source for cell growth (Guilloux-Benatier *et al.*, 1993).

Occasionally, increases in the concentration of glucose and fructose have been noted following malolactic fermentation. These increases appear to be coincidental and not causally related. The sugars may arise from the nonenzymatic breakdown of various complex sugars, such as trehalose, from the hydrolysis of glycosides, or the liberation of sugars following the pyrolytic hydrolysis of hemicelluloses in oak.

Organic Acids   Although the metabolism of malic acid is the principal reason for inducing malolactic fermentation, the bacteria also degrade other acids. Of particular importance is the oxidation of citric acid. It is correlated with the synthesis of acetic acid and diacetyl (Shimazu *et al.*, 1985). Few bacteria in wine, other than *O. oeni*, appear to metabolize citric acid. Gluconic acid, characteristically found in botrytized wines, is also metabolized by lactic acid bacteria, with the exception of the pediococci. Some lactic acid bacteria, notably the lactobacilli, have been reported to degrade tartaric acid (Radler and Yannissis, 1972). This has been associated with the wine fault called *tourne*.

The relative significance of organic acid metabolism to the energy budget of lactic acid bacteria is unclear. Both malic and citric acids are fermented after the major growth phase, when the bacteria enter the stationary phase. The decarboxylation of malic acid, and the subsequent release of lactic acid from the cell, activates H$^+$ uptake via membrane-bound ATPase. This is associated with the generation of ATP. A small proportion of malic acid, metabolized directly to lactic acid, may also generate reducing energy (NADH).

Fumaric acid addition was once proposed as an inhibitor of malolactic fermentation. However, because its activity decreases dramatically at pH values above 3.5 (Pilone *et al.*, 1974), it becomes progressively ineffective under those conditions where protection is most needed.

The rise in pH associated with malic acid decarboxylation is particularly a problem when sorbic acid has been added to control spoilage yeasts in sweet wines. Metabolism of sorbic acid by lactic acid bacteria results in the formation of 2-ethoxyhexa-3,4-diene, a compound possessing a strong geranium-like odor.

The ability of lactic acid bacteria to metabolize or tolerate fatty acids is largely unknown. However, some fatty acid metabolites are toxic to lactic acid bacteria, notably decanoic and octanoic acids produced by yeasts. These are more inhibitory in their acidic forms, but more cytotoxic in their esterified forms (Guilloux-Benatier *et al.*, 1998). Adsorption onto added yeast hulls helps to diminish their toxicity.

Nitrogen-Containing Compounds Although lactic acid bacteria have complex nitrogen growth requirements, few generalizations about nitrogen composition and malolactic fermentation appear evident (Remise *et al.*, 2006). For example, the concentration of individual amino acids may increase, decrease, or remain stable during malolactic fermentation. Simple interpretation of the data is confounded by the release of bacterial proteases. These can liberate amino acids from soluble proteins. Additional amino acids are released via yeast autolysis. Only the concentration of arginine appears to change consistently, through its bioconversion to ornithine. This is associated with the generation of ATP. Arginine uptake, along with fructose, also activates stress-responsive genes that enhance survival (Bourdineaud, 2006).

Where reduction in amino acid content is evident, it is probably associated with uptake and incorporation into proteins. Further losses may result from decarboxylation to amines, as in the production of histamine from histidine and tyramine from tyrosine.

Ethanol Ethanol retards bacterial growth. It even more effectively inhibits malolactic fermentation (Fig. 7.38). Of lactic acid bacteria, *Lactobacillus* spp. are the most ethanol-tolerant. *L. trichodes* can grow in wines at up to 20% ethanol (Vaughn, 1955). A few strains of *Oenococcus oeni* can grow in culture media at up to

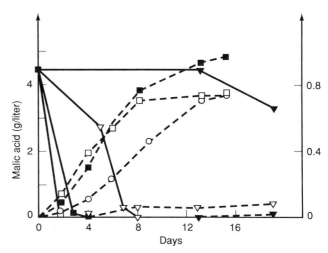

Figure 7.38 Influence of ethanol on the growth and malolactic activity of *Oenococcus oeni*. Malic acid (g/liter), ———; bacterial growth (OD$_{620nm}$), – – –; □, 0%; ■, 5%; ○, 8%; ▽, 11%; ▼, 13%. (After Guilloux-Benatier, 1987, reproduced by permission)

15% alcohol. Alcohol tolerance appears to decline, both with increasing temperature and decreasing pH values. However, at low concentrations (1.5%), ethanol appears to favor bacterial growth (King and Beelman, 1986).

The toxic mechanism of ethanol is unknown, but presumably involves changes in the semifluid nature of the cell membrane. These changes can induce enhanced passive H$^+$ ion influx and loss of cellular constituents (Graça da Silveira *et al.*, 2002). A reduction in the concentration of neutral lipids, and an increase in the proportion of glycolipids, have been correlated with high alcohol concentrations (Desens and Lonvaud-Funel, 1988). Ethanol tolerance is also correlated with an increased synthesis of certain phospholipids (phosphoethanolamine and sphingomyelin), and the incorporation of lactobacillic acid in the plasmalemma (Teixeira *et al.*, 2002).

Membrane malfunction induced by alcohol could explain growth disruption, reduced viability, and poor malolactic fermentation at high alcohol contents. An up to 80% reduction in the rate of malic acid decarboxylation has been reported with an increase in alcohol content from 11 to 13% (Lafon-Lafourcade, 1975). When malolactic fermentation is desired in wine possessing high alcoholic strength (found increasing where grapes are picked late), strains known to possess considerable ethanol tolerance are recommended.

Other Organic Compounds During malolactic fermentation, lactic acid bacteria assimilate a wide range of compounds. Occasionally, this can affect the progress of malolactic fermentation. The best-known example involves the metabolism of carbonyl hydroxysulfonates.

This can liberate sufficient SO_2 to slow or terminate malolactic fermentation.

Many lactic acid bacteria produce esterases. Despite this, the concentrations of most esters, such as 2-phenethyl acetate and ethyl hexanoate, change little during malolactic fermentation. Any reductions that occur appear to be insufficient to cause a noticeable sensory influence. Other esters, such as ethyl acetate, ethyl lactate, and diethyl succinate, may increase. Although the last two are unlikely to have a perceptible impact, due to low volatility, an increased ethyl acetate content could donate an off-odor resembling nail-polish remover (acetone).

Various phenolic acids, such as ferulic, quinic, and shikimic acids, as well as their esters, are metabolized by some lactic acid bacteria. *Lactobacillus* spp. are notable in this regard. Their metabolism can generate volatile phenolics, such as ethyl guaiacol and ethyl phenol. These generally have spicy, medicinal, creosote-like odors (Whiting, 1975). Because these fragrances do not characterize malolactic fermentations, their synthesis is probably too minimal to be perceptible.

At the concentrations generally found in wine, grape phenolics do not seriously retard malolactic fermentation, anthocyanins and gallic acid actually appearing to favor bacterial viability. Nevertheless, leaving the stems to ferment with the grapes can noticeably slow initiation (Feuillat *et al.*, 1985). Malolactic fermentation may also be retarded, as well as prolonged in oak cooperage, as compared with stainless steel (Vivas *et al.*, 1995). This apparently results from ellagitannins suppressing cell viability. In this regard, how the oak is treated during barrel assembly appears to modify this effect. For example, growth of *Oenococcus oeni* was favored when the wood was toasted, but retarded when the oak was not fired (de Revel *et al.*, 2005).

Fermentors Malolactic fermentation is typically conducted in the same fermentor as alcoholic fermentation, prior to racking and maturation. Nevertheless, there have been many anecdotal reports that wine having undergone malolactic fermentation in oak cooperage is preferable to the same wine having undergone malolactic fermentation in stainless steel. This has been confirmed by Vivas *et al.* (1995). Their results indicate that relative color intensity and stability of red wine is increased following malolactic fermentation in oak cooperage. This is correlated with increased anthocyanin–tannin polymerization. In addition, astringency is apparently reduced if malolactic fermentation occurs in oak cooperage, generating a smoother, richer wine. The researchers also commented that oak and fruit flavors were considered more balanced and harmonious. These perceptions persisted for at least 3 years after bottling.

Gases Sulfur dioxide can markedly inhibit malolactic fermentation. The effect is complex because of the differing concentrations and toxicities of the many forms of sulfur dioxide in wine. As usual, free forms are more inhibitory than bound forms (Fig. 7.39). Of the free forms, molecular SO_2 is the most antimicrobial. By affecting the relative proportions of these forms, pH significantly influences the toxicity of sulfur dioxide. Temperature also dramatically influences bacterial sensitivity to sulfur dioxide (Lafon-Lafourcade, 1981). In all cases, however, the effect is more bacteriostatic than bacteriocidal (Delfini and Morsiani, 1992). Part of the inhibitory effect of sulfur dioxide (and other factors such as ethanol and toxic fatty acids) appears to be due to disruption of ATPase activity (Carrete *et al.*, 2002).

Different species and strains vary considerably in their sensitivity to sulfur dioxide. In general, strains of *Oenococcus oeni* are particularly sensitive. Because of the greater tolerance of pediococci and lactobacilli, they may be unintentionally selected by sulfur dioxide addition, especially with high pH wines. Thus, where sulfur dioxide is added, it is important to inoculate the must or wine with a known sulfur dioxide resistant strain.

Although lactic acid bacteria are strictly fermentative, small amounts of oxygen can favor malolactic

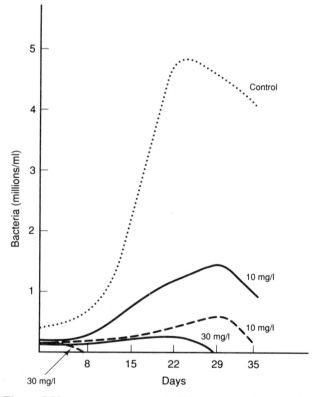

Figure 7.39 Action of two levels (10 and 30 mg/liter) of free SO_2 and acetaldehyde-bisulfite on the growth of *Oenococcus oeni*. Control, ...; acetaldehyde-bisulfite, ————; free SO_2, - - - - -. (From Lafon-Lafourcade, 1975, reproduced by permission)

fermentation. As the bacteria neither respire nor require steroids or unsaturated fatty acids for growth, the oxygen may act by maintaining a favorable redox balance through reaction with flavoproteins.

The significance of carbon dioxide is somewhat controversial. One potential mechanism for involvement may be the production of oxaloacetate, via the carboxylation of pyruvate. This could both help maintain a desirable redox balance and favor the biosynthesis of amino acids.

Pesticides Little is known about the possible effects of pesticide residues on the action of lactic acid bacteria. Vinclozolin and iprodione have been reported to depress malolactic fermentation, but increase the growth of acetic acid bacteria (San Romáo and Coste Belchior, 1982). Cymoxanil (Curzate) and dichlofluanid (Euparen) also have been reported to inhibit malolactic fermentation (Haag *et al.*, 1988). In contrast, benalaxyl, carbendazim, tridimefon, and vinclozolin were found to have no effect on lactic acid bacteria (Cabras *et al.*, 1994). Of seven fungicides and three insecticides tested by Ruediger *et al.* (2005), only the insecticide dicofol had an inhibitory effect on malolactic fermentation.

BIOLOGICAL FACTORS

Biological interactions in malolactic fermentation are as intricate as the complex interplay of physical and chemical factors already noted. Although mutually beneficial effects may occur, most interactions suppress growth.

Yeast Interactions Occasionally, alcoholic and malolactic fermentation occur together. In this instance, lactic acid bacteria occasionally exert an inhibitory effect on yeast growth, causing stuck fermentations. More commonly, though, yeasts inhibit bacterial growth (Edwards *et al.*, 1990). In this regard, there are marked sensitivity differences among *O. oeni* strains. Recently developed laboratory tests may help winemakers determine, in advance, whether, and to what degree, particular yeast and bacterial strain combinations are compatible (Arnick and Henick-Kling, 2005; Costello *et al.*, 2003). This could avoid the often prolonged delay between the completion of alcoholic fermentation and the onset of malolactic fermentation. Spoilage yeasts, such as *Pichia*, *Candida*, and *Saccharomycodes*, can also retard, if not inhibit, malolactic fermentation.

Various explanations have been offered for the inhibitory action of yeast growth. Strains of *Saccharomyces bayanus* have especially been associated with the suppression of malolactic fermentation (Nygaard and Prahl, 1966). The tendency of this species to produce SO_2, as well as its high alcohol tolerance (slow death and release of nutrients), are possible reasons. The depletion of arginine and other amino acids during the early phases

of alcoholic fermentation is another possibility. Certain yeast strains coprecipitate with bacteria, delaying malolactic fermentation. The increasing ethanol content and the accumulation of toxic carboxylic acids (notably octanoic and decanoic acid) are almost undoubtedly involved (Lafon-Lafourcade *et al.*, 1984). Proteins with antibacterial activities (e.g., lysozyme) have been isolated from some strains of *S. cerevisiae* (Dick *et al.*, 1992). Whether this is related to the selectively toxic, proteinaceous agent identified by Comitini *et al.* (2005) is unknown.

After alcoholic fermentation, yeast cells die and begin to undergo autolysis. The associated release of nutrients may explain the initiation of bacterial growth. Thus, wines left in contact with the lees for several weeks or months encourages the development of malolactic fermentation. Although nutrient release is undoubtedly involved, the maintenance of a high dissolved CO_2 concentration by the lees (Mayer, 1974), and the release of mannoproteins (Guilloux-Benatier *et al.*, 1995) appear to be involved (Fig. 7.40). Mannoproteins not only inactivate toxic fatty acids, but also provide nutrients when decomposed by bacterial β-glucosidases.

Surprisingly, malolactic fermentation may be stimulated in botrytized juice (San Romáo *et al.*, 1985). This occurs in spite of the well-known reduction in the available nitrogen content of *Botrytis*-infected grapes. Growth promotion may result from the metabolism of acetic acid (found in higher concentrations in botrytized grapes), or the removal of toxic carboxylic acids (adsorbed by *Botrytis*-synthesized polysaccharides). Regrettably, the

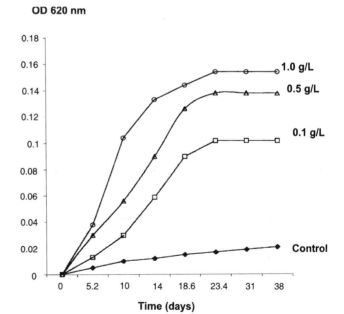

Figure 7.40 Influence of yeast mannoproteins on the growth of *Oenococcus oeni*. (From Charpentier, 2000, reproduced by permission)

increased glycerol content of the juice can favor the development of the wine fault, mannitic fermentation.

Bacterial Interactions The activity of acetic acid bacteria often favors malolactic fermentation during alcoholic fermentation. This may result indirectly from suppression of yeast growth by acetic acid (resulting in higher nutrient levels and lower concentrations of alcohol and toxic carboxylic acids), or directly by growth enhancement through the use of acetic acid as a nutrient.

Some intraspecies competition between *Oenococcus oeni* strains is suggested by changes in the relative proportion of strains throughout malolactic fermentation. More importantly, there may be antagonism between different species of lactic acid bacteria. Antagonism is rare below pH 3.5, at which typically only *O. oeni* grows. However, above pH 3.5, pediococci and lactobacilli progressively have a selective advantage over *O. oeni*. Antagonism appears most evident when a decline of *O. oeni* is mirrored by an equivalent rise in the number of lactobacilli and pediococci. The mechanism(s) of this apparent antagonism are unknown.

Viral Interactions Bacterial viruses (bacteriophages) can severely disrupt malolactic fermentation by attacking *Oenococcus oeni*. Virally infected *O. oeni* may also exist in an inactive (prophage) state (Cavin *et al.*, 1991). Because the prophage state is more unstable at higher pHs, the lytic phase may reestablish itself as the pH rises during malolactic fermentation. Under such conditions, malolactic fermentation may continue, but less predictably, under the action of *Lactobacillus* or *Pediococcus*.

O. oeni is most sensitive to phage infection during exponential growth. The virus attaches to the bacterium and injects its DNA into the cell (Fig. 7.41). In the cytoplasm, viral DNA begins to replicate and establish control over bacterial functions. After multiple copies are made of both the viral DNA and protein coat, these components self-assemble into virus particles. At this point, the cell bursts open (lyses), releasing the virus particles. These may subsequently initiate further cycles of infection and lysis.

The main defense against viral disruption of malolactic fermentation involves massive inoculation with *Oenococcus*. A large population reduces the number of cell divisions and, thereby, the duration of bacterial sensitivity to infection. The use of a mixed culture, containing several resistant strains, is also recommended. Mixed cultures minimize the likelihood that all strains will be sensitive to the phage races present. The infection of one strain might delay, but not inhibit

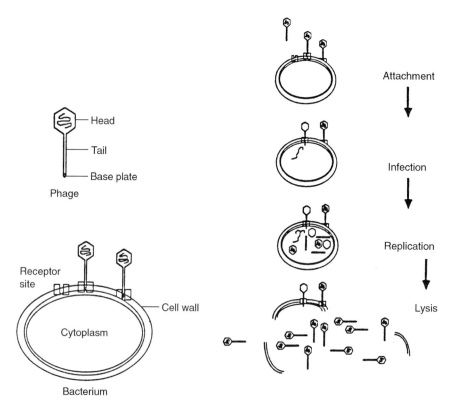

Figure 7.41 Attachment and infection cycle of phage viruses on bacteria. (From Henick-Kling, 1985, reproduced by permission)

malolactic fermentation. Besides delaying or inhibiting malolactic fermentation, phage infection can leave the wine open to the undesirable development of pediococci and lactobacilli. Specificity among phage strains means that those attacking *O. oeni* are unlikely to attack pediococci or lactobacilli, or vice versa.

Oenococcus oeni strains may possess one or more of at least three types of plasmids (Prévost *et al.*, 1995). The significance of their presence, if any, is unknown.

Control

Spontaneous malolactic fermentation has been notoriously difficult to manage. Malolactic fermentation often takes months to begin, even under conditions chosen to stimulate its development. It can also occur when least desired. The following procedures are used to either encourage or inhibit malolactic fermentation.

INOCULATION

To encourage malolactic fermentation, winemakers commonly inoculate the wine with one or more strains of *Oenococcus oeni*. This is associated with establishing conditions favorable to bacterial activity (warm temperatures, minimal or no sulfiting, and delayed racking). Their combination usually induces the rapid onset and completion of malolactic fermentation. Without inoculation, it may take weeks or months for the indigenous bacterial population to achieve a size sufficient to initiate malic acid decarboxylation.

Inoculation often involves cell populations reactivated from lyophilized or frozen concentrates. Reactivation and multiplication is usually required before addition to the fermenting must or wine. Direct addition often results in a rapid loss in viability.

Proper reactivation is critical to maintaining the malolactic ability of the cells (Hayman and Monk, 1982). Reactivation commonly takes place in nonsulfited, diluted grape juice (1:1 with water), adjusted to a pH of 3.6 or above. Yeast extract may be added at a concentration of 0.05% w/v. Reactivation generally requires at least 24 h, after which the organisms are cultured to achieve the desired population. Treating cells to various types of shock, for example 42°C for 1 hour, appears to increase resistance to subsequent inoculation (Guzzo *et al.*, 1994). This appears to involve the induction of several heat-shock proteins (Hsp). However, with several commercial preparations, the bacteria can be directly added to wine without prior reactivation or multiplication. Once malolactic fermentation has occurred in one batch of wine, it is often used to inoculate other batches.

Inoculation customarily aims at achieving a population of about 10^6–10^7 cells/ml. This usually assures the

quick onset of malolactic fermentation and the dominance of the inoculated strain. However, laboratory studies suggest that higher populations ($\geqslant 10^8$ cells/ml) may favor malolactic fermentation under highly acid conditions (Maicas *et al.*, 2000). At this population size, the cells have reached a physiological state where malic acid decarboxylation can occur without cell division. High inoculation levels also diminish the synthesis of secondary metabolites such as diacetyl and acetoin.

Although inoculation is likely to induce the rapid onset of malolactic fermentation, it is not assured. Multiple factors could result in the inoculated strain becoming inactive, or being superseded by endemic strains and species. Thus, it not possible to predict with assurance the outcome of malolactic fermentation. Various mousy, bitter, acetic flavors may develop, and viscous textures may result if spoilage lactobacilli or pediococci become dominant. Typically, though, these undesirable consequences do not occur, if the inoculated strain initiates early malolactic fermentation and the pH of the wine is below 3.6.

In situations where initiation has historically been unacceptably slow, addition of a commercial activator 24 h in advance of reinoculation is usually successful. The activator contains a combination of bacterial nutrients and components that absorb toxic fatty acids from the wine (Gindreau and Dumeau, 2005).

When simultaneous alcoholic and malolactic fermentation is desired, the reactivation medium is inoculated with both yeasts and bacteria. Occasionally, however, this has been reported to result in the production of a high concentrations of acetic acid. To avoid this, Prahl *et al.* (1988) suggested the use of *Lactobacillus plantarum*. It is insensitive to low pH, rapidly decarboxylates malic acid, does not generate acetic acid, and rapidly dies out as the alcohol content rises.

There is considerable variation in opinion as to when inoculation should occur. The Bordeaux school recommends inoculation after the completion of alcoholic fermentation, to avoid the risk of off-odor production (Lafon-Lafourcade, 1980) and the potential for stuck fermentation. Delayed inoculation avoids much of the toxicity of C_8 and C_{10} carboxylic acids (their concentration declines after alcoholic fermentation), but exposes the cells to the highest levels of ethanol.

Several researchers have not confirmed French findings concerning the development of high concentrations of acetic acid, yeast antagonism, or stuck fermentation (Beelman and Kunkee, 1985, Jussier *et al.*, 2006). They recommend concurrent alcoholic and malolactic fermentations. It permits the wine to be racked off the lees, cooled, and sulfited shortly after the completion of alcoholic fermentation. Neither do the bacteria suffer from a shortage of nutrients and exposure to the toxicity of alcohol is reduced. The sensory-significant influence of joint malolactic and alcoholic fermentation appears to be a reduction in the accumulation of diacetyl.

An alternative procedure involves inoculation midway through alcoholic fermentation. Theoretically, this should avoid the toxicity of high concentrations of ethanol and suppression by sulfur dioxide. Sulfur dioxide added during the grape crush is largely inactivated by binding to carbonyl and other compounds in the juice. However, the most intense level of yeast-induced antagonism by metabolites, such as decanoic acid, may be encountered with this timing.

Reasons for the diversity of opinion appear unclear. Possible causes may involve differences in bacterial and yeast strains, as well as grape nutrient status and vinicultural procedures. Because of this uncertainty, winemakers need to conduct their own trials to establish what works best for them.

Periodically, interest has been shown in the application of industrial approaches to malolactic fermentation. This has involved the use of both immobilized bacterial cells and enzymes. Such techniques offer the possibility of faster decarboxylation of malic acid, better control over the degree of the conversion (especially useful in wines with pH values above 3.5), and a reduction in the production of undesirable flavor compounds.

Immobilization involves coating a dense population of active cells with a gel, such as pectinate, alginate, carrageenan, or polyacrylamide (see Diviès *et al.*, 1994). The extrusion of the gel–cell mixture forms small beads, about 2–3 mm in diameter. The gel helps prevent cell division and infiltration into the wine. In many industrial uses, division of the entrapped cells is unnecessary. This is equally valid with malolactic fermentation, in which stationary phase cells actively decarboxylate malic acid. A support system (reactor) holds the beads in a rigid framework through which the wine is slowly passed. This can achieve a rapid decarboxylation of malic acid (Crapisi *et al.*, 1987). Other benefits of encapsulation include reduced bacterial sensitivity to pH, ethanol, and sulfur dioxide (Kosseva and Kennedy, 2004). Recent improvements in the useful life of the reactor may improve the procedure's economics. An alternate approach uses a cellulose sponge (given a basic charge), to which individual bacterial cells adhere (Maicas *et al.*, 2001). This avoids the development of inactive central regions, potentially found in gel–cell aggregates.

Immobilizing the malolactic enzyme into an inert permeable medium would be simpler, because it would avoid the need to maintain living cells. Enzyme immobilization also would avoid any flavor modification as only malic acid decarboxylation would be induced. However, a considerable obstacle to its successful application is the instability of the crucial enzyme cofactor NAD^+ at low pH values. This problem has been avoided by separating the enzyme reaction mixture (at a favorable pH) from the wine. A central membrane separates the reactor into two chambers, one each for the wine and enzymic solution (Formisyn *et al.*, 1997).

INHIBITION

The inhibition of malolactic fermentation involves the reverse of factors that favor its occurrence. This usually involves wine storage at or below 10 °C, early and frequent racking, early clarification, acidification of high pH musts or wines, minimal maceration, avoidance of *sur lies* maturation, and maintenance of a total sulfur dioxide content above 50 mg/liter.

Lysozyme has been approved in several countries to retard or prevent the action of lactic acid bacteria. It has the advantage of being more effective at higher pH values (where it is most needed). In addition, it has no detectable effect on wine aroma, taste (Bartowsky, Costello *et al.*, 2004), or effervescence. It is most active against growing cells. If added, it should not coincide with the addition of protein fining agents. The latter removes the enzyme along with other soluble proteins in the wine. This property also limits lysozyme effectiveness in red wines (Bartowsky, Costello *et al.*, 2004). Alternately, the bacteriocins, nisin and pediocin, appear to show promise in preventing malolactic fermentation.

Similar treatments are usually employed following malolactic fermentation to limit further metabolic activity by viable *O. oeni*, or the activity of other lactic acid bacteria. In this regard, the addition of lysozyme was found to limit the production of acetic acid and several histamines after the completion of malolactic fermentation (Gerbaux *et al.*, 1997). Because lysozyme results in cell rupture, amine generating enzymes in viable, but dormant, lactic acid bacteria are inactivated.

Depending on the likelihood of post-bottling malolactic fermentation, the wine may be given sulfur dioxide, sterile filtered into sterile bottles, or other procedures. In-bottle malolactic fermentation is undesirable. It can generate clouding, petillance (from the released carbon dioxide trapped in the bottle), and be the source of off-odors.

Appendix 7.1

Partial Synonymy of Several Important Wine Yeasts[a]

	Synonyms
Brettanomyces intermedius (Krumbholz and Tauschanoff) van der Walt and van Kerken Perfect state: *Dekkera intermedia* van der Walt	*Mycotorula intermedia* Krumbholz and Tauschanoff (1933) *Brettanomyces vini* Peynaud and Domercq (1956)
Candida stellata (Kroemer and Krumbholz) Meyer and Yarrow (1978)	*Brettanomyces italicus* Verona and Florenzano (1947) *Torulopsis bacillaris* (Kroemer and Krumbholz) Lodder (1932) *Torulopsis stellata* (Kroemer and Krumbholz) Lodder (1932) *Saccharomyces bacillaris* Kroemer and Krumbholz (1931)
Candida vini (Desmazières ex Lodder) van Uden and Buckley (1970)	*Candida mycoderma* (Reess) Lodder and Kreger-van Rij (1952) *Mycoderma cerevisiae* Desmazières (1823) ex Leberle (1909) *Mycoderma vini* Desmazières (1823) ex Lodder (1934) *Saccharomyces mycoderma* Reess (1870)
Dekkera intermedia van der Walt (1964) Imperfect state: *Brettanomyces intermedius* (Krumbholz and Tauschanoff) van der Walt and van Kerken (1971)	
Hanseniaspora uvarum (Niehaus) Shehata, Mrak and Phaff comb.nov. Imperfect state: *Kloeckera apiculata* (Reess emend. Klöcker) Janke (1928)	*Kloeckera lodderi* van Uden and Assis-Lopes (1953)
Hansenula anomala (Hansen) H. and P. Sydow (1919) Imperfect state: Candida pelliculosa Redaelli	*Saccharomyces anomalus* Hansen (1891)
Kloeckera apiculata (Reess Emend. Klöcker) Janke (1928) Perfect state: *Hanseniaspora uvarum* (Niehaus) Shehata, Mrak and Phaff	*Saccharomyces apiculatus* Reess (1870)
Metschnikowia pulcherrima Pitt and Miller (1968) Imperfect state: Candida pulcherrima (Lindner) Windisch (1940)	*Torula pulcherrima* Linder (1901) *Torulopsis pulcherrima* (Linder) Sacc. (1906) *Saccharomyces pulcherrima* (Linder) Beijerinck (1912)
Pichia fermentans Lodder (1932) Imperfect state: *Candida lambica* (Lindner and Genoud) van Uden and Buckley	*Saccharomyces dombrowskii* Sacchetti (1933) *Pichia dombrowskii* Sacchetti
Pichia membranaefaciens Hansen (1904) Imperfect state: *Candida valida* (Leberle) van Uden and Buckley (1940)	*Saccharomyces membranaefaciens* (Hansen (1888)
Saccharomyces bayanus Saccardo (1895)	*Saccharomyces abuliensis* Santa María (1978) *Saccharomyces globosus* Osterwalder (1924a) *Saccharomyces inusitatus* van der Walt (1965)
Saccharomyces cerevisiae Meyen ex Hansen (1883)	*Saccharomyces aceti* Santa María (1959) *Saccharomyces beticus* Marcilla ex Santa María (1970) *Saccharomyces capensis* van der Walt and Tscheuschner (1956) *Saccharomyces chevalieri* Guilliermond (1914) *Saccharomyces coreanus* Saito (1910) *Saccharomyces diastaticus* Andrews and Gilliland ex van der Walt (1965) *Saccharomyces ellipsoideus* Meyen ex Hansen (1883) *Saccharomyces fructuum* Lodder and Kreger-van Rij (1952) *Saccharomyces italicus* nom. nud. (Castelli 1938) *Saccharomyces norbensis* Santa María (1959) *Saccharomyces oleaceus* Santa María (1959) *Saccharomyces oleaginosus* Santa María (1959) *Saccharomyces oviformis* Osterwalder (1924) *Saccharomyces prostoserdovii* Kudriavzev (1960) *Saccharomyces steineri* Lodder and Kreger-van Rij (1952) *Saccharomyces vini* Meyer ex Kudriavzev (1960)
Saccharomyces pastorianus E. C. Hansen (1904)	*Saccharomyces carlsbergensis* E. C. Hansen (1908) *Saccharomyces monacensis* E. C. Hansen (1908)

(Continued)

Appendix 7.1   (*Continued*)

	Synonyms
Saccharomyces uvarum Beijerinck (1898b)	
Saccharomycodes ludwigii Hansen (1904)	*Saccharomyces ludwigii* Hansen (1889)
Torulaspora delbrueckii (Lindner) Lindner (1904)	*Saccharomyces delbrueckii* Lindner (1985)
Imperfect state: *Candia colliculosa* (Hartmann) Meyer and Yarrow	*Saccharomyces fermentati* (Saito) Lodder and Kreger-van Rij (1952)
	Saccharomyces rosei (Guilliermond) Lodder and Kreger-van Rij (1952)
Zygosaccharomyces bailii (Lindner) Guilliermond (1912)	*Saccharomyces acidifaciens* (Nickerson) Lodder and Kreger-van Rij (1952)
	Saccharomyces bailii Lindner (1895)
Zygosacch. bisporus (Naganishi) Lodder and Kreger-van Rij (1952)	*Saccharomyces bisporus* Naganishi (1917)
Zygosacch. florentinus Castelli ex Kudriavzev (1960)	*Saccharomyces florentinus* (Castelli ex Kudriavzev) Lodder and Kreger-van Rij (1952)
Zygosacch. rouxii (Boutroux) Yarrow (1977)	*Saccharomyces rouxii* Boutroux (1884)
Imperfect state: *Candida mogii* Vidal-Leiria	*Zygosacch. barkeri* Saccardo and Sydow (1902)

[a] Based primarily on Kurtzman and Fell (1998), Kreger-van Rij (1984), and Gouliamova and Hennebert (1998).

Appendix 7.2

Physiological Races of *Saccharomyces cerevisiae* Previously Given Species Status[a]

	Fermentation					
	Galactose	Sucrose	Maltose	Raffinose	Melibiose	Starch
S. aceti	−	−	−	−	−	−
S. capensis	−	+	−	+	−	−
S. cerevisiae	+	+	+	+	−	−
S. chevalieri	+	+	−	+	−	−
S. coreanus	+	+	−	+	+	−
S. diastaticus	+	+	+	+	−	+
S. globosus	+	−	−	−	−	−
S. heterogenicus	−	+	+	−	−	−
S. hienipiensis	−	−	+	−	+	−
S. inusitatus	−	+	+	+	+	−
S. norbensis	−	−	−	−	+	−
S. oleaceus	+	−	−	+	+	−
S. oleanginosus	+	−	+	+	+	−
S. prostoserdovii	−	−	+	−	−	−
S. steineri	+	+	+	−	−	−

[a] After van der Walt (1970), reproduced by permission.

Suggested Readings

General Texts

Blair, R. J., Francis, M. E., and Pretorius, I. S. (2005) *Advances in Wine Science*. Australian Wine Research Institute, Urrbrae, Australia.

Boulton, R. B., Singleton, V. L., Bisson, L. F., and Kunkee, R. E. (1996) *Principles and Practices of Winemaking*. Springer, New York, NY.

Delfini, C., and Formica, J. V. (2001) *Wine Microbiology: Science and Technology*. Marcel Dekker, New York, NY.

Edwards, C. G. (2006) *Illustrated Guide to Microbes and Sediments in Wine, Beer, and Juice*. WineBugs LLC, Pullman, WA.

Farkaš, J. (1988) *Technology and Biochemistry of Wine*, Vols 1 and 2. Gordon & Breach, New York.

Flanzy, C. (ed.) (1998) *Oenologie: Fondements Scientifiques et Technologiques*. Lavoisier, Paris.

Fleet, G. H. (ed.) (1993) *Wine Microbiology and Biotechnology*. Harwood Academic, New York.

Fugelsang, K. C., and Edwards, C. (2006/7) *Wine Microbiology. Practical Applications and Procedures*, 2nd edn. Springer, New York.

Jackson, D., and Schuster, D. (2001) *The Production of Grapes and Wines in Cool Climates*, 3rd edn. Dunmore Press, Auckland, New Zealand.

Ribéreau-Gayon, P., Donèche, B., Dubourdieu, D., and Lonvaud, A. (2006) *Handbook of Enology* Volume I (*The Microbiology of Wine and Vinifications*), 2nd edn. John Wiley and Sons, Chichester, UK.

Troost, R. (1988) *Technologie des Weines. Handbuch der Lebensmitteltechnologie,* 2nd edn. Ulmer, Stuttgart, Germany.

Vine, R. P., Bordelon, B., Harkness, E. M., Browning, T., and Wagner, C. (1997) *From Grape Growing to Marketplace.* Chapman & Hall, New York.

Maceration and Skin Contact

Allen, D. (2007) Prefermentative cryomaceration. *Aust. NZ Grapegrower Winemaker* **523**, 59–64.

Marais, J., and van Wyk, C. J. (1986) Effect of grape maturity and juice treatments on terpene concentrations and wine quality of *Vitis vinifera* L. cv. Weisser Riesling and Bukettraube. *S. Afr. J. Enol. Vitic.* **7**, 26–35.

Ramey, D., Bertrand, A., Ough, C. S., Singleton, V. L., and Sanders, E. (1986) Effects of skin contact temperature on Chardonnay must and wine composition. *Am. J. Enol. Vitic.* **37**, 99–106.

Sacchi, K. L., Bisson, L. F., and Adams, D. O. (2005) A review of the effect of winemaking techniques on phenolic extraction in red wines. *Am. J. Enol. Vitic.* **56**, 197–206.

Adjustments before Fermentation

Plank, P. F. H., and Zent, J. B. (1993) Use of enzymes in wine making and grape processing. In: *Beer and Wine Production – Analysis, Characterization, and Technological Advances* (B. H. Gump, ed.), pp. 181–196. ACS Symposium Series No. 536. American Chemical Society, Washington, DC.

General Fermentation Reviews

Fleet, G. H. (1990) Growth of yeasts during wine fermentations. *J. Wine Res.* **1**, 211–223.

Kunkee, R. E., and Bisson, L. F. (1993) Wine-making yeasts. In: *The Yeasts* (A. H. Rose and J. S. Harrison, eds.), 2nd edn, Vol. 5, pp. 69–127. Academic Press, London.

Moreno-Arribas, M. V., and Polo, M. C. (2005) Winemaking biochemistry and microbiology: current knowledge and future trends. *Crit. Rev. Food Sci. Nutr.* **45**, 265–286.

Yeast Biochemistry and Action

Dickinson, J. R., and Schweizer, M. (eds.) (2004) *Metabolism and Molecular Physiology of Saccharomyces cerevisiae,* 2nd edn. CRC Press, Taylor and Francis, London, UK.

Dubourdieu, D., Tominaga, T., Masneuf, I., Peyrot des Gachons, C., and Murat, M. L. (2000) The role of yeasts in grape flavor development during fermentation: the example of Sauvignon blanc. In: *Proceedings of the ASEV 50th Anniversary Annual Meeting,* Seattle, Washington, June 19–23, 2000, pp. 196–203. American Society for Enology and Viticulture, Davis, CA.

Kunkee, R. E., and Bisson, L. F. (1993) Wine-making yeasts. In: *The Yeasts* (A. H. Rose and J. S. Harrison, eds.), 2nd edn. Vol. 5, pp. 69–127. Academic Press, London.

Lambrechts, M. G., and Pretorius, I. S. (2000) Yeasts and its importance to wine aroma. *S. Afr. J. Enol. Vitic.* **21** (Special Issue), 97–129.

Swiegers, J. H., Bartowsky, E. J., Henschke, P. A., and Pretorius, I. S. (2005) Yeast and bacterial modulation of wine aroma and flavour. *Aust. J. Wine Res.* **11**, 139–173.

Walker, G. M. (1989) *Yeast Physiology and Biotechnology.* John Wiley & Sons, West Sussex, UK.

Yeast Classification and Evolution

Gouliamova, D. E., and Hennebert, G. L. (1998) Phylogenetic relationships in the *Saccharomyces cerevisiae* complex of species. *Mycotaxon* **66**, 337–353.

Kurtzman, C. P., and Fell, J. W. (eds.) (1998) *The Yeasts: A Taxonomic Study,* 4th edn. Elsevier, Amsterdam.

Liti, G., and Louis, E. J. (2005) Yeast evolution and comparative genomics. *Annu. Rev. Microbiol.* **59**, 135–153.

Mortimer, R. K. (2000) Evolution and variation of the yeast (*Saccharomyces*) genome. *Genome Res.* **10**, 403–409.

Piškur, J., Rozpedowska, E., Polakova, S., Merico, A., and Compagno, C. (2006) How did *Saccharomyces* evolve to become a good brewer? *Trends Genet.* **22**, 183–186.

Vaughan-Martini, A., and Martini, A. (1995) Facts, myths and legends on the prime industrial microorganism. *J. Ind. Microbiol.* **14**, 514–522.

Yeast Ecology

Fleet, G. H. (2003) Yeast interactions and wine flavour. *Int. J. Food Microbiol.* **86**, 11–22.

Jolly, N., Augustyn, O. P. H., and Pretorius, I. S. (2006) The role and use of non-*Saccharomyces* yeasts in wine production. *S. Afr. J. Enol. Vitic.* **27**, 15–39.

Phaff, H. J. (1986) Ecology of yeasts with actual and potential value in biotechnology. *Microb. Ecol.* **12**, 31–42.

Romano, P., Fiore, C., Paraggio, M., Caruso, M., and Capece, A. (2003) Function of yeast species and strains in wine flavour. *Int. J. Food Microbiol.* **86**, 169–180.

Yeast Genetics

Burke, D., Dawson, D., and Stearns, T. (2000) *Methods in Yeast Genetics.* Cold Spring Harbor Laboratory Press, Lab Manual edition, Cold Spring Harbor, NY.

de Winde, J. H. (2003) *Functional Genetics of Industrial Yeasts.* Springer Verlag, Berlin.

Landry, C. R., Townsend, J. P., Hartl, D. L., and Cavalieri, D. (2006) Ecological and evolutionary genomics of *Saccharomyces cerevisiae. Mol. Ecol.* **15**, 575–591.

Nielsen, J. (1998) Metabolic engineering: Techniques for analysis of targets for genetic manipulations. *Biotechnol. Bioeng.* **58**, 125–132.

Pretorius, I. S. (2000) Tailoring wine yeast for the new millennium: novel approaches to the ancient art of wine making. *Yeast* **16**, 675–727.

van Rensburg, P. (2005) Hybrid wine yeasts, offering the winemaker a unique fermentation tool. *Aust. NZ Grapegrower Winemaker* **493**, 56–58, 60.

Environmental Factors

Bauer, F. F., and Pretorius, I. S. (2000) Yeast stress response and fermentation efficiency: How to survive the making of wine – a review. *S. Afr. J. Enol. Vitic.* **21** (Special Issue), 27–51.

Bell, S.-J., and Henschke, P. A. (2005) Implications of nitrogen nutrition for grapes, fermentation and wine. *Aust. J. Grape Wine Res.* **11**, 242–295.

Birch, R. M., Ciani, M., and Walker, G. M. (2003) Magnesium, calcium and fermentative metabolism in wine yeasts. *J. Wine Res.* **14**, 3–15.

Cabras, P., and Angioni, A. (2000) Pesticide residues in grapes, wine, and their processing products. *J. Agric. Food Chem.* **48**, 967–973.

Cabras, P., Angioni, A., Garau, V. L., Pirisi, F. M., Farris, G. A., Madau, G., and Emonti, G. (1999) Pesticides in fermentative processes in wine. *J. Agric. Food Chem.* **47**, 3854–3857.

Carrasco, P., Querol, A., and del Olmo, M. (2001) Analysis of the stress resistance of commercial wine yeast strains. *Arch. Microbiol.* **175**, 450–457.

Fleet, G. H. (2003) Yeast interactions and wine flavour. *Int. J. Food Microbiol.* **86**, 11–22.

Hallsworth, J. E. (1998) Ethanol-induced water stress in yeasts. *J. Ferment. Bioeng.* **85**, 125–137.

Querol, A., Fernández-Espinar, M. T., del Olmo, M., Barro, E. (2003) Adaptive evolution of wine yeast. *Int. J. Food Microbiol.* **86**, 3–10.

Slaughter, J. C. (1989) The effects of carbon dioxide on yeasts. In: *Biotechnology Applications in Beverage Production* (C. Cantarelli and G. Lanzarini, eds.), pp. 49–64. Elsiever Applied Science, London.

van Rensburg, P., and Pretorius, I. S. (2000) Enzymes in winemaking: Harnessing natural catalysts for efficient biotransformations – a review. *S. Afr. J. Enol. Vitic.* **21** (Special Issue), 52–73.

Stuck and Sluggish Fermentation

Alexandre, H., and Charpentier, C. (1998) Biochemical aspects of stuck and sluggish fermentation in grape must. *J. Ind. Microbiol. Biotechnol.* **20**, 20–27.

Bisson, L. F. (1999) Stuck and sluggish fermentations. *Am. J. Enol. Vitic.* **50**, 107–119.

Bisson, L. F., and Butzke, C. E. (2000) Diagnosis and rectification of stuck and sluggish fermentations. *Am. J. Enol. Vitic.* **51**, 168–177.

Blateyron, L., Ortiz-Julien, A., and Sablayrolles, J. M. (2003) Stuck fermentations: Oxygen and nitrogen requirements – importance of optimising their addition. *Aust. NZ Grapegrower Winemaker* **478**, 73–79.

Magliani, W., Conti, S., Gerloni, M., Bertolotti, D., and Polonelli, L. (1997) Yeast killer systems. *Clin. Microbiol. Rev.* **10**, 369–400.

Malolactic Fermentation

Alexandre, H., Costello, P. J., Remize, F., Guzzo, J., and Guilloux-Benatier, M. (2004) *Saccharomyces cerevisiae* and *Oenococcus oeni* interactions in wine: current knowledge and perspectives. *Int. J. Food Microbiol.* **93**, 141–154.

Bartowshy, E. (2005) *Oenococcus oeni* and malolactic fermentation – moving into the molecular arena. *Aust. J. Grape Wine Res.* **11**, 174–187.

Bauer, R., and Dicks, L. M. T. (2004) Control of malolactic fermentation in wine. A review. *S. Afr. J. Enol. Vitic.* **25**, 74–88.

Davis, C. R., Wibowo, D., Eschenbruch, R., Lee, T. H., and Fleet, G. H. (1985) Practical implications of malolactic fermentation: A review. *Am. J. Enol. Vitic.* **36**, 290–301.

Davis, C. R., Wibowo, D., Fleet, G. H., and Lee, T. H. (1988) Properties of wine lactic acid bacteria: Their potential enological significance. *Am. J. Enol. Vitic.* **39**, 137–142.

Henick-Kling, T. (1993) Malolactic fermentation. In: *Wine Microbiology and Biotechnology* (G. H. Fleet, ed.), pp. 289–326. Harwood Academic, Chur, Switzerland.

Maicas, S. (2001) The use of alternative technologies to develop malolactic fermentation in wine. *Appl. Microbiol. Biotechnol.* **56**, 35–39.

Osborne, J. P., and Edwards, C. G. (2005) Bacteria important during winemaking. *Adv. Food Nutrition. Res.* **50**, 139–177.

Swiegers, J. H., Bartowsky, E. J., Henschke, P. A., and Pretorius, I. S. (2005) Yeast and bacterial modulation of wine aroma and flavour. *Aust. J. Wine Res.* **11**, 139–173.

Wibowo, D., Eschenbruch, R., Davis, C. R., Fleet, G. H., and Lee, T. H. (1985) Occurrence and growth of lactic acid bacteria in wine: A review. *Am. J. Enol. Vitic.* **36**, 302–313.

References

Alexandre, H., and Charpentier, C. (1995) Influence of fermentation medium aeration on lipid composition on fatty acids and acetic acid production by *Saccharomyces cerevisiae*. *Sci. Aliments* **15**, 579–592.

Amerine, M. A., Berg, H. W., Kunkee, R. E., Ough, C. S., Singleton, V. L., and Webb, A. D. (1980) *The Technology of Wine Making*. Avi, Westport, CT.

Amerine, M. A., and Joslyn, M. A. (1970) *Table Wines, the Technology of their Production*, 2nd edn. University of California Press, Berkeley.

Amerine, M. A., and Ough, C. S. (1980) *Methods for Analysis of Musts and Wines*. John Wiley, New York.

Anonymous (1983) Steel feet punch the cap at Buena Vista. *Wines Vines* **64**(2), 52.

Ansell, R., Granath, K., Hohmann, S., Thevelein, J. M., and Adler, L. (1997) The two isoenzymes of yeast NAD-dependent glycerol 3-phosphate dehydrogenase, *GPD1* and *GPD2* have distinct roles in osmoadaptation and redox regulation. *EMBO J.* **16**, 2179–2187.

Antonelli, A., Castellari, L., Zambonelli, C., and Carnacini, A. (1999) Yeast influence on volatile composition of wines. *J. Agric. Food Chem.* **47**, 1139–1144.

Arneborg, N., Høy, C.-E., and Jørgensen, O. B. (1995) The effect of ethanol and specific growth rate on the lipid content and composition of *Saccharomyces cerevisiae* grown anaerobically in a chemostat. *Yeast* **11**, 953–959.

Arnick, K., and Henick-Kling, T. (2005) Influence of *Saccharomyces cerevisiae* and *Oenococcus oeni* strains on successful malolactic conversion in wine. *Am. J. Enol. Vitic.* **56**, 228–237.

Bach, H.-P. (2001) Recovery of fermentation aromas. *Aust. NZ Grapegrower Winemaker* **454**, 73–78.

Bach, H. P., and Hess, K. H. (1986) Der Einfluß der Alkoholerhöhung auf Weininhaltsstoffe und Geschmack. *Weinwirsch. Tech.* **122**, 437–440.

Bach, H. P., Schneider, P., Bamberger, U., and Wintrich, K.-H. (1990) Cell-cracking und sein Einfluß auf die Qualität von Weißweinen. *Weinwirtsch. Tech.* **8**, 26–34.

Bae, S., Fleet, G. H., and Heard, G. M. (2006) Lactic acid bacteria associated with wine grapes from several Australian vineyards. *J. Appl. Microbiol.* **100**, 712–727.

Bakalinsky, A. L., and Snow, R. (1990a) The chromosomal constitution of wine strains of *Saccharomyces cerevisiae*. *Yeast* **6**, 367–382.

Bakalinsky, A. L., and Snow, R. (1990b) Conversion of wine strains of *Saccharomyces cerevisiae* to heterothallism. *Appl. Environ. Microbiol.* **56**, 849–857.

Barata, A., Nobre, A., Correia, P., Malfeito-Ferreira, M., and Loureiro, V. (2006) Growth and 4-ethylphenol production by the yeast *Pichia guilliermondii* in grape juices. *Am. J. Enol. Vitic.* **57**, 133–138.

Barre, P., and Vezinhet, F. (1984) Evolution towards fermentation with pure culture yeasts in wine making. *Microbiol. Sci.* **1**, 159–163.

Bartowsky, E. J., Costello, P. J., Villa, A., and Henschke, P. A. (2004) The chemical and sensorial effects of lysozyme addition to red and white wines over six months' cellar storage. *Aust J. Grape Wine Res.* 10, 143–150.

Bartowsky, E. J., Markides, A. J., Dillon, S. J., Dumont, A., Pretorius, I. S., Henschke, P. A., Ortiz-Julien, A., and Herderich, M. (2004) The potential of *Saccharomyces cerevisiae* wine yeast to improve red wine colour. I. *Aust. NZ Grapegrower Winemaker* 490, 83–85.

Bartowsky, E. J., McCarthy, J. M., and Henschke, P. A. (2003) Differentiation of Australian wine isolates of *Oenococcus oeni* using random amplified polymorphic DNA (RAPD) *Aust. J. Grape Wine Res.* 9, 122–126.

Bataillon, M., Rico, A., Sablayrolles, J. M., Salmon, J. M., and Barre, P. (1996) Early thiamine assimilation by yeasts under enological conditions: Impact on alcoholic fermentation kinetics. *J. Ferm. Bioeng.* 128, 1445–1447.

Beelman, R. B., and Kunkee, R. E. (1985) Inducing simultaneous malolactic-alcoholic fermentation in red table wines. In: *Malolactic Fermentation* (T. H. Lee, ed.), pp. 97–111. Australian Wine Research Institute, Urrbrae, Australia.

Belin, J. M. (1972) Recherches sur la répartition des levures à la surface de la grappe de raisin. *Vitis* 11, 135–145.

Bely, M., Sablayrolles, J. M., and Barre, P. (1990) Automatic detection of assimilable nitrogen deficiencies during alcoholic fermentation in oenological conditions. *J. Ferment. Bioeng.* 70, 246–252.

Bely, M., Salmon, J. M., and Barre, P. (1994) Assimilable nitrogen addition and hexose transport system activity during enological fermentation. *J. Inst. Brew.* 100, 279–282.

Bertrand, A., and Miele, A. (1984) Influence de la clarification du moût de raisin sur sa teneur en acides gras. *Connaiss. Vigne Vin.* 18, 293–297.

Bertrand, A., and Torres-Alegre, V. (1984) Influence of oxygen added to grape must on the synthesis of secondary products of the alcoholic fermentation. *Sci. Aliments* 4, 45–64.

Bett, K. E., and Cappi, J. B. (1965) Effect of pressure on the viscosity of water. *Nature* 207, 620–621.

Bisson, L. F., and Butzke, C. E. (2000) Diagnosis and rectification of stuck and sluggish fermentations. *Am. J. Enol. Vitic.* 51, 168–177.

Bony, M., Barre, P., and Blondin, B. (1998) Distribution of the flocculation protein, FLOP, at the cell surface during yeast growth: The availability of Flop determines the flocculation level. *Yeast* 14, 25–35.

Bony, M., Bidart, F., Camarasa, C., Ansanay, L., Dulau, L., Barre, P., and Dequin, S. (1997) Metabolic analysis of *Saccharomyces cerevisiae* strains engineered for malolactic fermentation. *FEBS Lett.* 410, 452–456.

Boone, C., Sdicu, A.-M., Wagner, J., Degré, R., Sanchez, C., and Bussey, H. (1990) Integration of the yeast K₁ killer toxin gene into the genome of marked wine yeasts and its effect on vinification. *Am. J. Enol. Vitic.* 41, 37–42.

Bosso, A., Panero, L., Guaita, M., and Marulli, C. (2001) La tecnica del delestage nelle vinificazione del Montepulciano d'Abruzzo. *Enologo* 37, 87–96.

Bouix, M., Leveau, J. Y., and Cuinier, C. (1981) Applications de l'électophorese des fractions exocellulaires de levures au contrôle de l'efficacité d'un levurage en vinification. In: *Current Developments in Yeast Research* (G. G. Stewart and I. Russel, eds.), pp. 87–92. Pergamon, New York.

Boulton, R. (1979) The heat transfer characteristics of wine fermentors. *Am. J. Enol. Vitic.* 30, 152–156.

Bourdineaud, J. P. (2006) Both arginine and fructose stimulate pH–independent resistance in the wine bacteria *Oenococcus oeni*. *Int. J. Food Microbiol.* 107, 274–280.

Bousbouras, G. E., and Kunkee, R. E. (1971) Effect of pH on malolactic fermentation in wine. *Am. J. Enol. Vitic.* 22, 121–126.

Bréchot, P., Chauvet, J., Dupuy, P., Croson, M., and Rabatu, A. (1971) Acide olénolique, facteur de croissance anaérobie de la levure de vin. *C.R. Acad. Sci. Ser. D* 272, 890–893.

Brenna, O., and Bianchi, E. (1994) Immobilised laccase for phenolic removal in must and wine. *Lett Biotechnol.* 16, 35–40.

Brewster, J. L., de Valoir, T., Dwyer, N. D., Winter, E., and Gustin, M. C. (1993) An osmosensing signal transduction pathway in yeast. *Science* 259, 1760–1763.

Buechsenstein, J., and Ough, C. S. (1979) Comparison of citric, dimalic, and fumaric acids as wine acidulants. *Am. J. Enol. Vitic.* 30, 93–97.

Cabras, P., Garau, V. L., Angioni, A., Farris, G. A., Budroni, M., and Spanedda, L. (1995) Interactions during fermentation between pesticides and oenological yeasts producing H₂S and SO₂. *Appl. Microbiol Biotechnol.* 43, 370–373.

Cabras, P., Meloni, M., Melis, M., Farris, G.A., Bufroni, M., and Satta, T. (1994) Interactions between lactic bacteria and fungicides during lactic fermentation. *J. Wine Res.* 5, 53–59.

Caldini, C., Bonomi, F., Pifferi, P. G., Lanzarini, G., and Galante, Y. M. (1994) Kinetic and immobilization studies on fungal glycosidases for aroma enhancement in wine. *Enzyme Microb. Technol.* 16, 286–291.

Canal-Llaubères, R.-M. (1993) Enzymes in winemaking. In: *Wine Microbiology and Biotechnology* (G. H. Fleet, ed.), pp. 447–506. Harwood Academic, Chur, Switzerland.

Cantarelli, C. (1989) Phenolics and yeast: Remarks concerning fermented beverages. *Yeast* 5, S53–61.

Carre, E., Lafon-Lafourcade, S., and Bertrand, A. (1983) Désacidification biologique des vins blancs secs par fermentation de l'acide malique par les levures. *Connaiss. Vigne Vin.* 17, 43–53.

Carrete, R., Vidal, M. T., Bordons, A., and Constanti, M. (2002) Inhibitory effect of sulfur dioxide and other stress compounds in wine on the ATPase activity of *Oenococcus oeni*. *FEMS Microbiol. Lett.* 211, 155–159.

Cartwright, C. P., Rose, A. H., Calderbank, J., and Keenan, M. J. (1989) Solute transport. In: *The Yeasts* (A. H. Rose and J. S. Harrison, eds.), 2nd edn. Vol. 3, pp. 5–56. Academic Press, New York.

Castellari, L., Ferruzzi, M., Magrini, A., Giudici, P., Passarelli, P., and Zambonelli, C. (1994) Unbalanced wine fermentation by cryotolerant *vs.* non-cryotolerant *Saccharomyces* strains. *Vitis* 33, 49–52.

Cavalieri, D., Townsend, J. P., and Hartl, D. L. (2000) Manifold anomalies in gene expression in a vineyard isolate of *Saccharomyces cerevisiae* revealed by DNA microarray analysis. *Proc. Natl Acad. Sci.* 97, 12369–12374.

Cavazza, A., Grando, M. S., and Zini, C. (1992) Rilevazione della flora microbica di mosti e vini. *Vignivini* 19(9), 17–20.

Cavazza, A., Versini, G., DallaSerra, A., and Romano, F. (1989) Characterization of six *Saccharomyces cerevisiae* strains on the basis of their volatile compounds production, as found in wines of different aroma profiles. *Yeast* 5, S163–167.

Cavin, J. F., Andioc, V., Etiévant, P. X., and Diviès, C. (1993) Ability of wine lactic acid bacteria to metabolize phenol carboxylic acids. *Am. J. Enol. Vitic.* 44, 76–80.

Cavin, J. F., Drici, F. Z., Prevost, H., and Diviès, C. (1991) Prophage curing in *Leuconostoc oenos* by mitomycin C induction. *Am. J. Enol. Vitic.* 42, 163–166.

Charoenchai, C., Fleet, G. H., and Henschke, P. A. (1998) Effects of temperature, pH, and sugar concentration on the growth rates and cell biomass of wine yeasts. *Am. J. Enol. Vitic.* 49, 283–288.

Charpentier, C. (2000) Yeast autolysis and yeast macromolecules? Their contribution to wine flavor and stability. In: *Proceedings of the ASEV 50th Anniversary Annual Meeting*, Seattle, Washington,

June 19–23, 2000, pp. 271–277. American Society for Enology and Viticulture, Davis, CA.

Chatonnet, P., Dubourdieu, P., and Boidron, J. N. (1989) Incidence de certains facteurs sur la décarboxylation des acides phénols par la levure. *Connaiss. Vigne Vin* **23**, 59–63.

Chauvet, S., Sudraud, P., and Jouan, T. (1986) La cryoextraction sélective des moûts. *Rev. Oenologues* **39**, 17–22.

Cheraiti, N., Guezenec, S., and Salmon, J. M. (2005) Redox interactions between *Saccharomyces cerevisiae* and *Saccharomyces uvarum* in mixed culture under enological conditions. *Appl. Environ. Microbiol.* **71**, 255–260.

Cheynier, V., Souquet, J.-M., Samson, A., and Moutounet, M. (1991) Hyperoxidation: Influence of various oxygen supply levels on oxidation kinetics of phenolic compounds and wine quality. *Vitis* **30**, 107–115.

Ciani, M., and Maccarelli, F. (1998) Oenological properties of non-*Saccharomyces* yeasts associated with wine-making. *World J. Microbiol. Biotechnol.* **14**, 199–203.

Ciani, M., Mannazzu, I., Marinangeli, P., Clementi, F., and Martini, A. (2004) Contribution of winery-resident *Saccharomyces cerevisiae* strains to spontaneous grape must fermentation. *Antonie van Leeuwenhoek* **85**, 159–164.

Ciani, M., and Picciotti, G. (1995) The growth kinetics and fermentation behaviour of some non-*Saccharomyces* yeasts associated with wine-making. *Biotechnol. Lett.* **17**, 1247–1250.

Cocito, C., and Delfini, C. (1997) Experiments for developing selective clarification techniques: Sterol and fatty acid loss from grape must related to clarification technique. *J. Wine Res.* **8**, 187–198.

Cocolin, L., Bisson, L. F., and Mills, D. A. (2000) Direct profiling of the yeast dynamics in wine fermentations. *FEMS Microbiol. Lett.* **189**, 81–87.

Cocolin, L., and Mills, D. A. (2003) Wine yeast inhibition by sulfur dioxide: a comparison of culture-dependent and independent methods. *Am. J. Enol. Vitic.* **54**, 125–130.

Comitini, F., Ferretti, R., Clementi, F., Mannazzu, I., and Ciani, M. (2005) Interactions between *Saccharomyces cerevisiae* and malolactic bacteria: preliminary characterization of a yeast proteinaceous compound(s) active against *Oenococcus oeni. J. Appl. Microbiol.* **99**, 105–111.

Conner, A. J. (1983) The comparative toxicity of vineyard pesticides to wine yeasts. *Am. J. Enol. Vitic.* **34**, 278–279.

Costello, P. J., Henschke, P. A., and Markides, A. J. (2003) Standardised methodology for testing malolactic bacteria and wine yeast compatibility. *Aust. J. Grape Wine Res.* **9**, 127–137.

Costello, P. J., Morrison, R. H., Lee, R. H., and Fleet, G. H. (1983) Numbers and species of lactic acid bacteria in wines during vinification. *Food Tech. Aust.* **35**, 14–18.

Coton, E., Rollan, G., Bertrand, A., and Lonvaud-Funel, A. (1998) Histamine-producing lactic acid bacteria in wines: Early detection, frequency, and distribution. *Am. J. Enol. Vitic.* **49**, 199–204.

Crapisi, A., Spettoli, P., Nuti, M. P., and Zamorani, A. (1987) Comparative traits of *Lactobacillus brevis, Lact. fructivorans* and *Leuconostoc oenos* immobilized cells for the control of malolactic fermentation in wine. *J. Appl. Bacteriol.* **63**, 513–521.

Crippen, D. D. Jr., and Morrison, J. C. (1986) The effects of sun exposure on the phenolic content of Cabernet Sauvignon berries during development. *Am. J. Enol. Vitic.* **37**, 243–247.

Cuénat, P., Lorenzine, F., Brégy, C.-A., and Zufferey, E. (1996) La macération préfermentaire à froid du Pinot noir, aspects technologiques et microbiologiques. *Rev. Suisse Vitic. Arboricult. Horticult.* **28**, 259–265.

Cuénat, P., Zufferey, E., Kobel, D., Bregy, C. A., and Crettenand, J. (1991) Le cuvage du Pinot noir, rôle des températures. *Rev. Suisse Vitic. Arboric. Hortic.* **23**, 267–272.

Darby, W. J., Ghalioungui, P., and Grivetti, L. (1977) *Food: The Gift of Osiris.* Vol. 2. Academic Press, London.

Darriet, P., Tominaga, T., Demole, E., and Dubourdieu, D. (1993) Mise en évidence dans le raisin de *Vitis vinifera* var. Sauvignon d'un précurseur de la 4-mercapto-4-méthylpentan-2-one. *C. R. Acad. Sci. Life Sci.* **316**, 1332–1335.

Dasari, G., Worth, M. A., Connor, M. A., and Pamment, N. B. (1990) Reasons for the apparent difference in the effects of produced and added ethanol on culture viability during rapid fermentations by *Saccharomyces cerevisiae. Biotechnol. Bioeng.* **35**, 109–122.

Davis, C. R., Wibowo, D., Eschenbruch, R., Lee, T. H., and Fleet, G. H. (1985) Practical implications of malolactic fermentation: A review. *Am. J. Enol. Vitic.* **36**, 290–301.

Davis, C. R., Wibowo, D. J., Lee, T. H., and Fleet, G. H. (1986a) Growth and metabolism of lactic acid bacteria during and after malolactic fermentation of wines at different pH. *Appl. Environ. Microbiol.* **51**, 539–545.

Davis, C. R., Wibowo, D., Lee, T. H., and Fleet, G. H. (1986b) Growth and metabolism of lactic acid bacteria during fermentation and conservation of some Australian wines. *Food Technol. Aust.* **36**, 35–40.

de Barros Lopes, M., Bellon, J. R., Shirley, N. J., and Ganter, P. F. (2002) Evidence for multiple interspecific hybridization in *Saccharomyces* sensu stricto species. *FEMS Yeast Res.* **1**, 323–331.

de Barros Lopes, M., Eglinton, J., Hensehke, P., Høj, P., and Pretorius, I. (2003) The connection between yeast and alcohol reduction in wine. *Wine Industry J.* **18**, 17–22.

de Deken, R. H. (1966) The crabtree effect: A regulatory system in yeast. *J. Gen. Microbiol.* **44**, 149–156.

Defranoux, C., Gineys, D., and Joseph, P. (1989) Le potentiel aromatique du Chardonnay. Essai d'utilisation des techniques de cryoextraction et de supraextraction (procédé Kreyer) *Rev. Oenologues* **55**, 27–29.

Delfini, C., and Morsiani, M.G. (1992) Resistance to sulfur dioxide of malolactic strains of *Leuconostoc oenos* and *Lactobacillus* sp. isolated from wines. *Sci. Aliments* **12**, 493–511.

Delfini, C., Cocito, C., Ravaglia, S., and Conterno, L. (1993) Influence of clarification and suspended grape solid materials on sterol content of free run and pressed grape musts in the presence of growing yeast cells. *Am. J. Enol. Vitic.* **44**, 452–459.

Delfini, C., Pessioni, E., Moruno, E. G. and Giunta, C. (1992) Localization of volatile acidity reducing factors in grapes. *J. Indust. Microbiol* **11**, 19–22.

Demuyter, C., Lollier, M., Legras, J.-L., and Le Jeune, C. (2004) Predominance of *Saccharomyces uvarum* during spontaneous alcoholic fermentation, for three consecutive years, in an Alsatian winery. *J. Appl. Microbiol.* **97**, 1140–1148.

De Nadra, M. C. M., Farias, M. E., Moreno-Arribas, M. V., Pueyo, E., and Polo, M. C. (1997) Proteolytic activity of *Leuconostoc oenos.* Effects on proteins and polypeptides from white wine. *FEMS Microbiol. Lett.* **150**, 135–139.

Dengis, P. B., and Rouxhet, P. G. (1997) Flocculation mechanisms of top and bottom fermenting brewing yeasts. *J. Inst. Brew.* **103**, 257–261.

de Revel, G., Bloem, A., Augustin, M., Lonvaud–Funel, A. and Bertrand, A. (2005) Interaction of *Oenococcus oeni* and oak wood compounds. *Food Microbiol.* **22**, 569–575.

Desens, C., and Lonvaud-Funel, A. (1988) Étude de la constitution lipidique des membranes de bactéries lactiques utilisées en vinification. *Connaiss. Vigne Vin.* **22**, 25–32.

Dick, K. I., Molan, P. C., and Eschenbruch, R. (1992) The isolation from *Saccharomyces cerevisiae* of two antibacterial cationic proteins that inhibit malolactic bacteria. *Vitis* **31**, 105–116.

D'Incecco, N., Bartowsky, E., Kassara, S., Lante, A., Spettoli, P., and Henschke, P. (2004) Release of glycosidically bound flavour compounds of Chardonnay by *Oenococcus oeni* during malolactic fermentation. *Food Microbiol.* 21, 257–265.

Dittrich, H. H. (1977) *Mikrobiologie des Weines, Handbuch der Getränketechnologie.* Ulmer, Stuttgart, Germany.

Diviès, C., Cachon, R., Cavin, J.-F., and Prévost, H. (1994) Immobilized cell technology in wine production. *Crit. Rev. Biotechnol.* 14, 135–153.

Dizy, M., and Bisson, L. F. (2000) Proteolytic activity of yeast strains during grape juice fermentation. *Am. J. Enol. Vitic.* 51, 155–167.

Drici-Cachon, Z., Cavin, J. F., and Diviès, C. (1996) Effect of pH and age of culture on cellular fatty acid composition of *Leuconostoc oenos. Lett. Appl. Microbiol.* 22, 331–334.

Dubourdieu, D., and Lavigne, V. (1990) Incidence de l'hyperoxidation sur la composition chimique et les qualitiés organoleptiques des vins blancs du Bordelais. *Rev. Fr. Oenol.* 30, 58–61.

Dubourdieu, D., Sokol, A., Zucca, J., Thalouarn, P., Dattee, A., and Aigle, M. (1987) Identification des souches de levures isolées de vins par l'analyse de leur AND mitochondrial. *Connaiss. Vigne Vin.* 21, 267–278.

Duitschaever, C. L., Alba, J., Buteau, C., and Allen, B. (1991) Riesling wines made from must concentrated by reverse osmosis, I. Experimental conditions and composition of musts and wines. *Am. J. Enol. Vitic.* 42, 19–25.

Dukes, B. C., and Butzke, C. E. (1998) Rapid determination of primary amino acids in grape juice using an *o*-phthaldialdehyde/N-acetyl-L-cysteine spectrophotometric assay. *Am. J. Enol. Vitic.* 49, 125–134.

Dumont, A., and Dulau, L. (1996) The role of yeasts in the formation of wine flavors. In: *Proceedings of the 4th International Symposium on Cool Climate Viticulture and Enology* (T. Henick-Kling *et al.*, eds.), pp. VI–24–28. New York State Agricultural Experimental Station, Geneva, NY.

du Plessis, C. S. (1983) Influence de la température d'élaboration et de conservation. *Bull. O.I.V.* 524, 104–115.

Edwards, C. G., Beelman, R. B., Bartley, C. E., and McConnell, A. L. (1990) Production of decanoic acid and other volatile compounds and the growth of yeast and malolactic bacteria during vinification. *Am. J. Enol. Vitic.* 41, 48–56.

Edwards, C. G., Haag, K. M., Semon, M. J., Rodriguez, A. V., and Mills, J. M. (1999) Evaluation of processing methods to control the growth of *Lactobacillus kunkeei*, a micro-organism implicated in sluggish alcoholic fermentations of grape musts. *S. Afr. J. Enol. Vitic.* 20, 11–18.

Egli, C. M., Edinger, W. D., Mitrakul, C. M., and Henick-Kling, T. (1998) Dynamics of indigenous and inoculated yeast populations and their effect on the sensory character of Riesling and Chardonnay wines. *J. Appl. Microbiol.* 85, 779–789.

Eglinton, J. M., Buckingham, L., and Henschke, P. A. (1993) Increased volatile acidity of white wines by chemical vitamin mixtures is grape juice dependent. In: *Proceedings of the 8th Australian Wine Industry Technical Conference* (C. S. Stockley *et al.*, eds.), pp. 197–198. Winetitles, Adelaide, Australia.

Eglinton, J., Griesser, M., Henschke, P., Kwiatkowski, M., Parker, M., and Herderich, M. (2004) Yeast-mediated formation of pigmented polymers in red wine. In: *Red Wine Color. Revealing the Mysteries* (A. L. Waterhouse and J. A. Kennedy, eds), pp. 7–21. ACS Symposium Series, No. 886., American Chemical Society, Washington, DC.

Eglinton, J. M., Heinrich, A. J., Pollnitz, A. P., Langridge, P., Henschke, P. A., and de Barros Lopes, M. (2002) Decreasing acetic acid accumulation by a glycerol overproducing strain of *Saccharomyces cerevisiae* by deleting the ALD6 aldehyde dehydrogenase gene. *Yeast* 19, 295–301.

Eglinton, J. M., McWilliam, S. J., Fogarty, M. W., Francis, I. L., Kwiatkowski, M. J., Hoj, P. B., and Henschke, P. A. (2000) The effect of *Saccharomyces bayanus*–mediated fermentation on the chemical composition and aroma profile of Chardonnay wine. *Aust. J. Grape Wine Res.* 6, 190–196.

Erten, H. (2002) Relations between elevated temperatures and fermentation behavious of *Kloeckera apiculata* and *Saccharomyces cerevisiae* associated with winemaking in mixed cultures. *World J. Microbiol. Biotechnol.* 18, 373–378.

Escot, S., Feuillat, M., Dulau, L., and Charpentier, C. (2001) Release of polysaccharides by yeasts and the influence of released polysaccharides on colour stability and wine astringency. *J. Grape Wine Res.* 7, 153–159.

Falconer, R. J., and Hart, A. (2005) MOG removal system makes its mark. *Aust. NZ Grapegrower Winemaker* 501, 18–22.

Farkaš, J. (1988) *Technology and Biochemistry of Wine*, Vols. 1 and 2. Gordon & Breach, New York.

Farris, G. A., Fatichenti, F., and Deiana, P. (1989) Incidence de la température et du pH sur la production d'acide malique par *Saccharomyces cerevisiae. J. Int. Sci. Vigne Vin* 23, 89–93.

Fay, J. C., and Benavides, J. A. (2005) Evidence for domesticated and wild populations of *Saccharomyces cerevisiae. PLoS Genet.* 1, 66–71.

Ferrando, M., Guell, C., and Lopez, F. (1998) Industrial wine making: Comparison of must clarification treatments. *J. Agr. Food Chem.* 46, 1523–1528.

Ferrarini, R., Celotti, E., Zironi, R., and Buiatti, S. (1995) Recent advances in the progress of flotation applied to the clarification of grape musts. *J. Wine Res.* 6, 19–33.

Feuillat, M. (1996) Vinification du Pinot noir en bourgogne par macération préfermentaire à froid. *Rev. Oenologues* 82, 29–31.

Feuillat, M., Guilloux-Benatier, M., and Gerbaux, V. (1985) Essais d'activation de la fermentation malolactique dans les vins. *Sci. Aliments* 5, 103–122.

Firme, M. P., Leitao, M. C., and San Romao, M. V. (1994) The metabolism of sugar and malic acid by *Leuconostoc oenos*, effect of malic acid, pH and aeration conditions. *J. Appl. Bacteriol.* 76, 173–181.

Fleet, G. H. (2003) Yeast interactions and wine flavour. *Int. J. Food Microbiol.* 86, 11–22.

Fleet, G. H., Lafon-Lafourcade, S., and Ribéreau-Gayon, P. (1984) Evolution of yeasts and lactic acid bacteria during fermentation and storage of Bordeaux wines. *Appl. Environ. Microbiol.* 48, 1034–1038.

Forbes, R. J. (1957) Food and drink. In: *A History of Technology.* Vol. II. (R. J. Singer, E. J. Holmyard, and A. R. Hall, eds.), pp. 103–147. Clarendon Press, Oxford.

Formisyn, P., Vaillant, H., Lantreibecq, F., and Bourgois, J. (1997) Development of an enzymatic reactor for initiating malolactic fermentation in wine. *Am. J. Enol. Vitic.* 48, 345–351.

Fornairon-Bonnefond, C., Aguera, E., Deytieux, C., Sablayrolles, J.-M., and Salmon, J.-M. (2003) Impact of oxygen addition during enological fermentation on sterol contents in yeast lees and their reactivity towards oxygen. *J. Biosci. Bioeng.* 95, 496–503.

Froment, T. (1991) Auto-enrichissement des moûts de raisin: Le concentrateur Entropie (type M.T.A.) par évaporation sous vide à très basse température (20°C). *Rev. Oenologues* 57, 7–11.

Gancelos, C., and Serrano, R. (1989) Energy-yielding metabolism. In: *The Yeasts* (A. H. Rose and J. S. Harrison, eds.), 2nd ed., Vol. 3, pp. 205–259. Academic Press, New York.

Gao, C., and Fleet, G. H. (1988) The effects of temperature and pH on the ethanol tolerance of the wine yeasts, *Saccharomyces cerevisiae, Candida stellata* and *Kloeckera apiculata. J. Appl. Bacteriol.* 65, 405–410.

Garcia, A., Carcel, C., Dulau, L., Samson, E., Aguera, E., Agosin, E., and Günata, Z. (2002) Influence of a mixed culture with *Debaryomyces vanriji* and *Saccharomyces cerevisiae* on the volatiles of a Muscat wine. *J. Food Sci.* **67**, 1138–1143.

Gardner, J. M., Poole, K., and Jiranek, V. (2002) Practical significance of relative assimilable nitrogen requirements of yeast: a preliminary study of fermentation performance and liberation of H₂S. *Aust. J. Grape Wine Res.* **8**, 175–179.

Gerbaux, V., Villa, A., Monamy, C., and Bertrand, A. (1997) Use of lysozyme to inhibit malolactic fermentation and to stabilize wine after malolactic fermentation. *Am . J. Enol. Vitic.* **48**, 49–54.

Gerland, C., and Gerbaux, V. (1999) Sélection d'une nouvelle souche de bactérie lactique: performances et impacts sur la qualité des vins. *Rev. Œnolog.* **93**, 17–19.

Gindreau, E., and Dumeau, F. (2005) Activator for malolactic fermentations: utilisation, and impact on bacterial populations and course of MLF. *Aust. NZ Grapegrower Winemaker* **502**, 84–88.

Gnaegi, F., Aerny, J., Bolay, A., and Crettenand, J. (1983) Influence des traitements viticoles antifongiques sur la vinification et la qualité du vin. *Rev. Suisse Vitic. Arboric. Hortic.* **15**, 243–250.

Graça da Silveira, M., San Romão, M. V., Loureiro-Dias, M. C., Rombouts, F. S., and Abee, T. (2002) Flow cytometric assessment of membrane integrity of ethanol-stressed *Oenococcus oeni* cells. *Appl. Environ. Microbiol.* **68**, 6087–6093.

Grando, M.S., Versini, G., Nicolini, G., and Mattivi, F. (1993) Selective use of wine yeast strains having different volatile phenols production. *Vitis* **32**, 43–50.

Grandvalet, C., Coucheney, F., Beltramo, C., and Guzzo, J. (2005) CtsR is the master regulator of stress response gene expression in *Oenococcus oeni. J. Bacteriol.* **187**, 5614–5623.

Groat, M. L., and Ough, C. S. (1978) Effect of particulate matter on fermentation rates and wine quality. *Am. J. Enol. Vitic.* **29**, 112–119.

Grossman, M. K., and Becker, R. (1984) Investigations on bacterial inhibition of wine fermentation. *Kellerwirtschaft* **10**, 272–275.

Guerzoni, M. E., Marchetti, R., and Giudici, P. (1985) Modifications de composants aromatiques des vins obtenus par fermentation avec des mutants de *Saccharomyces cerevisiae. Bull. O.I.V* **58**, 230–233.

Guillamón, J. M., Barrio, E., and Querol, A. (1996) Characterization of wine yeast strains of the *Saccharomyces* genus on the basis of molecular markers: Relationships between genetic distance and geographic or ecological origin. *Systematic Appl. Microbiol.* **19**, 122–132.

Guilloux-Benatier, M. (1987) Les souches de bactéries lactiques et les divers essais d'ensemencement de la fermentation malolactique in France. *Bull. O.I.V.* **60**, 624–642.

Guilloux-Benatier, M., Guerreau, J., and Feuillat, M. (1995) Influence of initial colloid content on yeast macromolecule production and on the metabolism of wine microorganisms. *Am. J. Enol. Vitic.* **46**, 486–492.

Guilloux-Benatier, M., Le Fur, Y., and Feuillat, M. (1989) Influence de la macération pelliculaire sur la fermentiscibilité malolactique des vins blancs de Bourgogne. *Rev. Fr. Oenol.* **29**, 29–34.

Guilloux-Benatier, M., Le Fur, Y., and Feuillat, M. (1998) Influence of fatty acids on the growth of wine microorganisms *Saccharomyces cerevisiae* and *Oenococcus oeni. J. Ind. Microbiol. Biotechnol.* **20**, 144–149.

Guilloux-Benatier, M., Son, H. S., Bouhier, S., and Feuillat, M. (1993) Activités enzymatiques: Glycosidases et peptidase chez *Leuconostoc oenos* au cours de la croissance bactérienne. Influence des macromolécules de levures. *Vitis* **32**, 51–57.

Guitart, A., Hernández-Orte, P., and Cacho, J. (1997) Effects of maceration on the amino acid content of Chardonnay musts and wines. *Vitis* **36**, 43–47.

Guymon, J. F., and Crowell, E. A. (1977) The nature and cause of cap-liquid temperature differences during wine fermentation. *Am. J. Enol. Vitic.* **28**, 74–78.

Guzzo, J., Cavin, J. F., and Divies, C. (1994) Induction of stress proteins in *Leuconostoc oenos* to perform direct inoculation of wine. *Biotechnol. Lett.* **16**, 1189–1194.

Haag, B., Krieger, S., and Hammes, W. P. (1988) Hemmung der Startezkulturen zur Einlertung des biologischen Säurenabbaus durch Spritzmittelrückstände. *Wein Wiss.* **43**, 261–278.

Hallsworth, J. E. (1998) Ethanol-induced water stress in yeasts. *J. Ferment. Bioeng.* **85**, 125–137.

Hammond, S. M., and Carr, J. C. (1976) The antimicrobial activity of SO₂ – with particular reference to fermented and nonfermented fruit juices. In: *Inhibition and Inactivation of Vegetative Microbes* (F. A. Skinner and W. B. Hugo, eds.), pp. 89–110. Academic Press, London.

Hatzidimitriou, E., Bouchilloux, P., Darriet, P., Bugaret, Y., Clerjeau, M., Poupot, C., Medina, B., and Dubourdieu, D. (1996) Incidence d'une protection viticole anticryptogamique utilisant une formulation cuprique sure le nivear de maturité des raisins et l'arôme variétal des vins de Sauvignon. Bilan de trois années d'expérimentation. *J. Int. Sci. Vigne Vin* **30**, 133–150.

Hauser, N. C., Fellenberg, K., Gil, R., Bastuck, S., Hoheisel, J. D., and Jeffries, T. W. (2001) Whole genome analysis of wine yeast strain. *Comparat. Funct. Genomics* **2**, 60–79.

Hayman, D. C., and Monk, P. R. (1982) Starter culture preparation for the induction of malolactic fermentation in wine. *Food Technol. Aust.* **34**, 14, 16–18.

Heard, G. M., and Fleet, G. H. (1985) Growth of natural yeast flora during the fermentation of inoculated wines. *Appl. Environ. Microbiol.* **50**, 727–728.

Heard, G. M., and Fleet, G. H. (1988) The effects of temperature and pH on the growth of yeast species during the fermentation of grape juice. *J. Appl. Bacteriol.* **65**, 23–28.

Heatherbell, D., Dicey, M., Goldsworthy, S., and Vanhanen, L. (1996) Effect of prefermentation cold maceration on the composition, color, and flavor of Pinot Noir wine. In: *Proceedings of the 4th International Symposium on Cool Climate Viticulture and Enology* (T. Henick-Kling *et al.*, eds.), pp. VI-10–17. New York State Agricultural Experimental Station, Geneva, NY.

Henderson, A. (1824) *The History of Ancient and Modern Wines.* Baldwin, Craddock & Joy, London.

Henick-Kling, T. (1985) Phage can block malolactic. *Wines Vines* **69** (6), 55–59.

Henick-Kling, T. (1995) Control of malo-lactic fermentation in wine: Energetics, flavour modification and methods of starter culture preparation. *J. Appl. Bacteriol.* **79**, S29–S37.

Henick-Kling, T., Acree, T., Gavitt, B. K., Kreiger, S. A., and Laurent, M. H. (1993) Sensory aspects of malolactic fermentation. In: *Proceedings of the 8th Australian Wine Industry Technical Conference* (C. S. Stockley *et al.*, eds.), pp. 148–152. Winetitles, Adelaide, Australia.

Henick-Kling, T., Edinger, W., Daniel, P., and Monk, P. (1998) Selective effects of sulfur dioxide and yeast starter culture addition on indigenous yeast populations and sensory characteristics of wine. *J. Appl. Microbiol.* **84**, 865–876.

Henschke, P. A., and Dixon, G. D. (1990) Effect of yeast strain on acetic acid accumulation during fermentation of *Botrytis* affected grape juice. In: *Proceedings of the 7th Australian Wine Industry Technical Conference* (P. J. Williams *et al.*, eds.), pp. 242–244. Australian Industrial Publishers, Adelaide, Australia.

Henschke, P. A., and Jiranek, V. (1993) Yeasts – metabolism of nitrogen compounds. In: *Wine Microbiology and Biotechnology* (G. H. Fleet, ed.), pp. 77–165. Harwood Academic, Chur, Switzerland.

Henschke, P. A., and Ough, C. S. (1991) Urea accumulation in fermenting grape juice. *Am. J. Enol. Vitic.* **42**, 317–321.

Herraiz, T., Guillermo, R., Herraiz, M., Martin-Alvarez, P. J., and Cabezudo, M. D. (1990) The influence of the yeast and type of culture on the volatile composition of wines fermented without sulfur dioxide. *Am. J. Enol. Vitic.* **41**, 313–318.

Hess, B., Boiteux, A., and Krüger, J. (1969) Cooperation of glycolytic enzymes. *Adv. Enzyme Regulation* **7**, 149–167.

Holloway, P., Subden, R. E., and Lachance, M. A. (1990) The yeasts in a Riesling must from the Niagara grape-growing region of Ontario. *Can. Inst. Food Sci. Tech. J.* **23**, 212–216.

Hottiger, T., de Virigilio, C., Hall, M., Boller, T., and Wiemken, A. (1994) The role of trehalose synthesis for the acquisition of thermotolerance in yeast, II. Physiological concentration of trehalose increases the thermal stability of protein *in vitro*. *Eur. J. Biochem.* **210**, 4744–4766.

Houtman, A. C., and du Plessis, C. S. (1981) The effect of juice clarity and several conditions promoting yeast growth on fermentation rate, the production of aroma components and wine quality. *S. Afr. J. Enol. Vitic.* **2**, 71–81.

Houtman, A. C., and du Plessis, C. S. (1985) Influence du cépage et de la souche de levure. *Bull. O.I.V.* **58**, 236–246.

Houtman, A. C., and du Plessis, C. S. (1986) Nutritional deficiencies of clarified white grape juices and their correction in relation to fermentation. *S. Afr. J. Enol. Vitic.* **7**, 39–46.

Houtman, A. C., Marais, J., and du Plessis, C. S. (1980) Factors affecting the reproducibility of fermentation of grape juice and of the aroma composition of wines, I. Grape maturity, sugar, inoculum concentration, aeration, juice turbidity and ergosterol. *Vitis* **19**, 37–84.

Howell, K. S., Swiegers, J. H., Elsey, G. M., Siebert, T. E., Bartowsky, E. J., Fleet, G. H., Pretorium, I. S., de Barros Lopes, M. A. (2004) Variation in 4-mercapto-4-methyl-pentan-2-one release by *Saccharomyces cerevisiae* commercial wine strains. *FEMS Microbiol. Lett.* **240**, 125–129.

Huang, Y.-C., Edwards, C. G., Peterson, J. C., and Haag, K. M. (1996) Relationship between sluggish fermentations and the antagonism of yeast by lactic acid bacteria. *Am. J. Enol. Vitic.* **47**, 1–10.

Husnik, J. I., Delaquis, P. J., Cliff, M. A., and van Vuuren, H. J. J. (2007) Functional analyses of the malolactic wine yeast ML01. *Am. J. Enol. Vitic.* **58**, 42–52.

Iconomou, L., Kanellaki, M., Voliotis, S., Agelopoulos, K., and Koutinas, A. A. (1996) Continuous wine making by delignified cellulosic materials supported biocatalyst – an attractive process for industrial applications. *Appl. Biochem. Biotechnol.* **60**, 303–313.

Ihmels, J., Bergmann, S., Gerami-Nejad, M., Yanai, I., McClellan, M., Berman, J., and Barkai, N. (2005) Rewiring of the yeast transcriptional network through the evolution of motif usage. *Science* **309**, 938–940.

Iwahashi, H., Obuchi, K., Fujii, S., and Komatsu, Y. (1995) The correlative evidence suggesting that trehalose stabilizes membrane structure in the yeast *Saccharomyces cerevisiae*. *Cell. Molec. Biol.* **41**, 763–769.

Jacobs, C. J., and van Vuuren, H. J. J. (1991) Effects of different killer yeasts on wine fermentations. *Am. J. Enol. Vitic.* **42**, 295–300.

Javelot, C., Girard, P., Colonna-Ceccaldi, B., and Valdescu, B. (1991) Introduction of terpene-producing ability in a wine strain of *Saccharomyces cerevisiae*. *J. Biotechnol.* **21**, 239–252.

Jiranek, V., and Henschke, P. A. (1995) Regulation of hydrogen sulfide liberation in wine-producing *Saccharomyces cerevisiae* strains by assimilable nitrogen. *Appl. Environ. Microbiol.* **61**, 161–164.

Jolly, N. P., Augustyn, O. P. H., and Pretorius, I. S. (2003) The use of *Candida pulcherrima* in combination with *Saccharomyces cerevisiae* for the production of Chenin blanc wine. *S. Afr. J. Enol. Vitic.* **24**, 63–69.

Jones, R. P., and Ough, C. S. (1985) Variations in the percent ethanol (v/v) per Brix conversions of wines from different climatic regions. *Am. J. Enol. Vitic.* **36**, 268–270.

Jones, R. P., Pamment, N., and Greenfield, P. F. (1981) Alcohol fermentation by yeasts – the effect of environmental and other variables. *Process Biochem.* **16**, 42–49.

Julien, A., Roustan, J.-L., Dulau, L., and Sablayrolles, J.-M. (2000) Comparison of nitrogen and oxygen demands of enological yeasts: technological consequences. *Am. J. Enol. Vitic.* **51**, 215–222.

Jussier, D., Morneau, A. D., and Mira de Orduña, R. (2006) Effect of simultaneous inoculation with yeast and bacteria on fermentation kinetics and key wine parameters of cool-climate Chardonnay. *Appl Environ. Microbiol.* **72**, 221–227.

Khan, W., Augustyn, O. P. H., van der Westhuizen, T. J., Lambrechts, M. G., and Pretorius, I. S. (2000) Geographic distribution and evaluation of *Saccharomyces cerevisiae* strains isolated from vineyards in the warmer, inland regions of the Western Cape in South Africa. *S. Afr. J. Enol. Vitic.* **21**, 17–31.

Killian, E., and Ough, C. S. (1979) Fermentation esters – formation and retention as affected by fermentation temperature. *Am. J. Enol. Vitic.* **30**, 301–305.

King, S. W., and Beelman, R. B. (1986) Metabolic interactions between *Saccharomyces cerevisiae* and *Leuconostoc oenos* in a model grape juice/wine system. *Am. J. Enol. Vitic.* **37**, 53–60.

Kitamoto, K. (1989) Role of yeast vacuole in sake brewing (in Japanese) *J. Brew. Soc. Japan* **84**, 367–374.

Klingshirn, L. M., Liu, J. R., and Gallander, J. F. (1987) Higher alcohol formation in wines as related to the particle size profiles of juice insoluble solids. *Am. J. Enol. Vitic.* **38**, 207–209.

Konings, W. N., Lolkema, J. S., Bolhuis, H., van Veen, H. W., Poolman, B., and Driessen, A. J. M. (1996) The role of transport processes in survival of lactic acid bacteria – energy transduction and multidrug resistance. *A. van Leewenhoek* **71**, 117–128.

Kosseva, M. R., and Kennedy, J. F. (2004) Encapsulated lactic acid bacteria for control of malolactic fermentation in wine. *Artificial Cells, Blood Substitutes, and Biotechnol.* **32**, 55–65.

Kotseridis, Y., Beloqui, A. A., Bayonove, C. L., Baumes, R. L., and Bertrand, A. (1999) Effects of selected viticultural and enological factors on levels of 2-methoxy-3-isobutylpyrazine in wines. *J. Int. Sci. Vigne Vin* **33**, 19–23.

Kraus, J. K., Scoop, R., and Chen, S. L. (1981) Effect of rehydration on dry wine yeast activity. *Am. J. Enol. Vitic.* **32**, 132–134.

Kreger-van Rij, N. J. W. (ed.) (1984) *The Yeasts: A Taxonomic Study*, 3rd edn., Elsevier, Amsterdam.

Krieger, S. (1996) Wine flavor modification by lactic acid bacteria. In: *Proceedings of the 4th International Symposium on Cool Climate Viticulture and Enology* (T. Henick-Kling et al., eds.), pp. VI-29–36. New York State Agricultural Experimental Station, Geneva, NY.

Krieger, S., Lemperle, E., and Ernst, M. (2000) Management of malolactic fermentation with regard to flavour modification in wine. In: *Proceeding of the 5th International Symposium on Cool Climate Viticulture and Oenology*. Melbourne, Australia.

Krisch, J., and Szajáni, B. (1997) Ethanol and acetic acid tolerance in free and immobilized cells of *Saccharomyces cerevisiae* and *Acetobacter aceti*. *Biotechnol Lett.* **19**, 525–528.

Kruckeberg, A. L. (1996) The hexose transporter family of *Saccharomyces cerevisiae*. *Arch. Microbiol.* **166**, 283–292.

Kudo, M., Vagnoli, P., and Bisson, L. F. (1998) Imbalance of pH and potassium concentration as a cause of stuck fermentations. *Am. J. Enol. Vitic.* **49**, 295–301.

Kunkee, R. E., and Ough, C. S. (1966) Multiplication and fermentation of *Saccharomyces cerevisiae* under carbon dioxide pressure in wine. *Appl. Microbiol.* **14**, 643–648.

Kurtzman, C. P., and Fell, J. W. (eds.) (1998) *The Yeasts: A Taxonomic Study*, 4th edn. Elsevier, Amsterdam.

Laffort, J.-F., Romat, H., and Darriet, Ph. (1989) Les levures et l'expression aromatique des vins blancs. *Rev. Oenologues* **53**, 9–12.

Lafon-Lafourcade, S. (1975) Factors of the malo-lactic fermentation of wines. In: *Lactic Acid Bacteria in Food and Beverages* (J. G. Carr *et al.*, eds.), pp. 43–53. Academic Press, London.

Lafon-Lafourcade, S. (1980) Les origines microbiologiques de l'acidité volatile des vins. *Microbiol. Ind. Aliment. Ann. Congr. Intl.* **2**, 33–48.

Lafon-Lafourcade, S. (1981) Connaissances actuelles, dans la maîtrise de la fermentation malolactique dans les moûts et les vins. In: *Actualités Oenologiques et Viticoles* (P. Ribéreau-Gayon and P. Sudraud, eds.), pp. 243–251. Dunod, Paris.

Lafon-Lafourcade, S. (1983) Wine and brandy. In: *Biotechnology*. Vol. 5 (*Food and Feed Production with Microorganisms*) (G. Reed, ed.), pp. 81–163. Verlag Chemie, Weinheim, Germany.

Lafon-Lafourcade, S., Carre, E., and Ribéreau-Gayon, P. (1983) Occurrence of lactic acid bacteria during different stages of vinification and conservation of wines. *Appl. Environ. Microbiol.* **46**, 874–880.

Lafon-Lafourcade, S., Geneix, C., and Ribéreau-Gayon, P. (1984) Inhibition of alcoholic fermentation of grape must by fatty acids produced by yeasts and their elimination by yeast ghosts. *Appl. Environ. Microbiol.* **47**, 1246–1249.

Lagunas, R. (1986) Misconceptions about the energy metabolism of *Saccharomyces cerevisiae. Yeast* **2**, 221–228.

Lao, C. L., López-Tamames, E., Lamuela-Raventós, R. M., Buxaderas, S., and De la Torre-Boronat, M. C. (1997) Pectic enzyme treatment effects on quality of white grape musts and wines. *J. Food Sci.* **62**, 1142–1144, 1149.

Laurent, M. H., Henick-Kling, T., and Acree, T. E. (1994) Changes in the aroma and odor of Chardonnay due to malolactic fermentation. *Wein Wiss.* **49**, 3–10.

Lehmann, F. L. (1987) Secondary fermentations retarded by high levels of free sulfur dioxide. *Aust. NZ Wine Ind. J.* **2**, 52–53.

Lemperle, E. (1988) Fungicide residues in musts and wines. In: *Proceedings of the 2nd International Symposium for Cool Climate Viticulture and Oenology* (R. E. Smart *et al.*, eds.), pp. 211–218. New Zealand Society for Viticulture and Oenology, Auckland, New Zealand.

Liu, S.-Q., Davis, C. R., and Brooks, J. D. (1995) Growth and metabolism of selected lactic acid bacteria in synthetic wine. *Am. J. Enol. Vitic.* **46**, 166–174.

Liu, S.-Q., and Pilone, G. J. (1998) A review: Arginine metabolism in wine lactic acid bacteria and its practical significance. *J. Appl. Microbiol.* **84**, 315–327.

Llauradó, J. M., Rozes, N., Constanti, M., and Mas, A. (2005) Study of some *Saccharomyces cerevisiae* strains for winemaking after preadaptation at low temperatures. *J. Agric. Food Chem.* **53**, 1003–1011.

Lopez-Toledano, A., Villaño-valencia, D., Mayen, M., Merida, J., and Median, M. (2004) Interaction of yeasts with the products resulting from the condensation reaction between (+)-catechin and acetaldehyde. *J. Agric. Food Chem.* **52**, 2376–2381.

Lourens, K., and Pellerin, P. (2004) Enzymes in winemaking. *Wynboer*, November, pp. 7. http://www.wynboer.co.za/recentarticles/0411enzymes.php3.

Magliani, W., Conti, S., Gerloni, M., Bertolotti, D., and Polonelli, L. (1997) Yeast killer systems. *Clin. Microbiol. Rev.* **10**, 369–400.

Maicas, S., Natividad, A., Ferrer, S., and Pardo, I. (2000) Malolactic fermentation in wine with high densities of non-proliferating *Oenococcus oeni. World J. Microbiol. Biotechnol.* **16**, 805–810.

Maicas, S., Pardo, I., and Ferrer, S. (2001) The potential of positively charged cellulose sponge for malolactic fermentation of wine, using *Oenococcus oeni. Enzyme Microb. Technol.* **28**, 415–419.

Majdak, A., Herjavec, S., Orlič, S., Redžepović, S., and Mirošević, N. (2002) Comparison of wine aroma compounds produced by *Saccharomyces paradoxus* and *Saccharomyces cerevisiae* strains. *Food Technol. Biotechnol.* **40**, 103–109.

Marais, J. (1987) Terpene concentrations and wine quality of *Vitis vinifera* L. cv. Gewürztraminer as affected by grape maturity and cellar practices. *Vitis* **26**, 231–245.

Marais, J. (1996) Fruit environment and prefermentation practices for manipulation of monoterpene, norisoprenoid and pyrazine flavorants. In: *Proceedings of the 4th International Symposium on Cool Climate Viticulture and Enology* (T. Henick-Kling *et al.*, eds.), pp. V-40–48. New York State Agricultural Experimental Station, Geneva, NY.

Marais, J., and van Wyk, C. J. (1986) Effect of grape maturity and juice treatments on terpene concentrations and wine quality of *Vitis vinifera* L. cv. Weisser Riesling and Bukettraube. *S. Afr. J. Enol. Vitic.* **7**, 26–35.

Martineau, B., and Henick-Kling, T. (1995) Performance and diacetyl production of commercial strains of malolactic bacteria in wine. *J. Appl. Bacteriol.* **78**, 526–536.

Martínez, J., Millán, C., and Ortega, J. M. (1989) Growth of natural flora during the fermentation of inoculated musts from 'Pedro Ximenez'. Grapes. *S. Afr. J. Enol. Vitic.* **10**, 31–35.

Martorell, P., Querol, A., and Fernández-Espinar, M. T. (2005) Rapid identification and enumeration of *Saccharomyces cerevisiae* cells in wine by real-time PCR. *Environ. Microbiol.* **71**, 6823–6830.

Massoutier, C., Alexandre, H., Feuillat, M., and Charpentier, C. (1998) Isolation and characterization of cryotolerant *Saccharomyces* strains. *Vitis* **37**, 55–59.

Mateo, J. J., and Jimenez, M. (2000) Monoterpenes in grape juice and wines. *J. Chromatography A* **881**, 557–567.

Mateo, J. J., Jimenez, M., Huerta, T., and Pastor, A. (1991) Contribution of different yeast isolated from musts of Monastrell grapes to the aroma of wine. *Intl J. Food Microbiol.* **14**, 153–160.

Mayer, K. (1974) Mikrobiologisch und kellertechnisch wichtige neue Erkenntnisse in bezug auf den biologischen Säureabbau. *Schweiz. Z. Obst .Weinbau.* **110**, 291–297.

McBryde, C., Gardner, J. M., de Barros Lopes, M., and Jiranek, V. (2006) Generation of novel wine yeast strains by adaptive evolution. *Am. J. Enol. Vitic.* **57**, 423–430.

Mendes-Ferreira, A., Climaco, M. C., and Mendes-Faia, A. (2001) The role of non-*Saccharomyces* species in releasing glycosidic bound fraction of grap aroma components – a preliminary study. *J. Appl. Microbiol.* **91**, 67–71.

Meyrath, J., and Lüthi, H. R. (1969) On the metabolism of hexoses and pentoses by *Leuconostoc* isolated from wines and fruit. *Lebensm. Wiss. Technol.* **2**, 22–27.

Michnick, S., Roustan, J.-L., Remize, F., Barre, P., and Dequin, S. (1997) Modulation of glycerol and ethanol yields during alcoholic fermentation in *Saccharomyces cerevisiae* strains overexpressed or disrupted for GPD1 encoding glycerol 3-phosphate dehydrogenase. *Yeast* **13**, 783–793.

Mietton-Peuchot, M., Milisic, V., and Noilet, P. (2002) Grape must concentration by using reverse osmosis. Comparison with chaptalization. *Desalination* **148**, 125–129.

Miller, G. C., Amon, J. M., and Simpson, R. F. (1987) Loss of aroma compounds in carbon dioxide effluent during white wine fermentation. *Food Technol. Aust.* **39**, 246–253.

Millet, V., and Lonvaud–Funel, A. (2000) The viable but non-culturable state of wine microorganisms during storage. *Lett. Appl. Microbiol.* **30**, 136–141.

Mills, D. A., Rawsthorne, H., Parker, C., Tamir, D., and Makarova, K. (2005) Genomic analysis of *Oenococcus oeni* PSU-1 and its relevance to winemaking. *FEMS Microbiol. Rev.* **29**, 465–475.

Minárik, E., and Rágala, P. (1966) Einfluß einiger Fungizide auf die Hefeflora bei der spotanen Mostgärung. *Mitt. Rebe Wein, Obstbau Früchteverwert. (Klosterneuburg)* **16**, 107–114.

Mira de Orduña, R. M., Patchett, M. L., Liu, S.-Q., and Pilone, G. J. (2001) Growth and arginine metabolism of the wine lactic acid bacteria *Lactobacillus buchneri* and *Oenococcus oeni* at different pH values and arginine concentrations. *Appl. Environ. Microbiol.* **67**, 1657–1662.

Mizoguchi, H. (1998) Permeability barrier of the yeast plasma membrane induced by ethanol. *J. Ferment. Bioeng.* **85**, 25–29.

Mizoguchi, H., and Hara, S. (1998) Permeability barrier of the yeast plasma membrane induced by ethanol. *J. Ferm. Bioeng.* **85**, 25–29.

Mora, J., and Mulet, A. (1991) Effects of some treatments of grape juice on the population and growth of yeast species during fermentation. *Am. J. Enol. Vitic.* **42**, 133–136.

Morata, A., Gomez–Cordoves, M. C., Calderon, F., and Suarez, J. A. (2006) Effects of pH, temperature and SO_2 on the formation of pyranoanthocyanins during red wine fermentation with two species of *Saccharomyces*. *Int. J. Food Microbiol.* **106**, 123–129.

Moreno, J. J., Millán, C., Ortega, J. M., and Medina, M. (1991) Analytical differentiation of wine fermentations using pure and mixed yeast cultures. *J. Ind. Microbiol.* **7**, 181–190.

Mortimer, R. K. (2000) Evolution and variation of the yeast (*Saccharomyces*) genome. *Genome Res.* **10**, 403–409.

Mortimer, R., and Polsinelli, M. (1999) On the origins of wine yeast. *Res. Microbiol.* **150**, 199–204.

Muller, C. J., Wahlstrom, V. L., and Fugelsang, K. C. (1993) Capture and use of volatile flavor constituents emitted during wine fermentation. In: *Beer and Wine Production – Analysis, Characterization, and Technological Advances* (B. H. Gump, ed.), pp. 219–232. ACS Symposium Series No. 536, American Chemical Society, Washington, DC.

Munoz, E., and Ingledew, W. M. (1990) Yeast hulls in wine fermentations, a review. *J. Wine Res.* **1**, 197–209.

Murat, M.-L. (2005) Recent findings on rosé wine aromas. Part I: identifying aromas studying the aromatic potential of grapes and juice. *Aust. NZ Grapegrower Winemaker* **497a**, 64–65, 69, 71, 73–74, 76.

Murat, M.-L., and Dumeau, F. (2005) Recent findings on rosé wine aromas. Part II: optimizing winemaking techniques. *Aust. NZ Grapegrower Winemaker* **499**, 49–52, 54–55.

Nagel, C. W., Weller, K., and Filiatreau, D. (1988) Adjustment of high pH-high TA musts and wines. In: *Proceedings of the 2nd International Symposium for Cool Climate Viticulture and Oenology* (R. E. Smart *et al.*, eds.), pp. 222–224. New Zealand Society for Viticulture and Oenology, Auckland, New Zealand.

Naumov, G. I., Naumova, E. S., and Sniegowski, P. D. (1998) *Saccharomyces paradoxus* and *Saccharomyces cerevisiae* are associated with exudates of North American oaks. *Can. J. Microbiol.* **44**, 1045–1050.

Naumova, E. S., Naumov, G. I., Masneuf-Pomarede, I., Aigle, M., Dubourdieu, D. (2005) Molecular genetic study of introgression betweenSaccharomyces bayanus and *S. cerevisiae*. *Yeast* **22**, 1099–1115.

Navarro, S., García, B., Navarro, G., Oliva, J., and Barba, A. (1997) Effect of wine-making practices on the concentrations of fenarimol and penconazole in rose wines. *J. Food Protect.* **60**, 1120–1124.

Nguyen, H. V., and Gaillardin, C. (2005) Evolutionary relationships between the former species *Saccharomyces uvarum* and the hybrids *Saccharomyces bayanus* and *Saccharomyces pastorianus*; reinstatement of *Saccharomyces uvarum* (Beijerinck) as a distinct species. *FEMS Yeast Res.* **5**, 471–483.

Nielsen, J. C., and Richelieu, M. (1999) Control of flavor development in wine during and after malolactic fermentation by *Oenococcus oeni*. *Appl. Environ. Microbiol.* **65**, 740–745.

Nilov, V. I., and Valuiko, G. G. (1958) Changes in nitrogen during fermentation (in Russian). *Vinodel. Vinograd. S.S.S.R.* **18**, 4–7.

Nishino, H., Miyazakim, S., and Tohjo, K. (1985) Effect of osmotic pressure on the growth rate and fermentation activity of wine yeasts. *Am. J. Enol. Vitic.* **36**, 170–174.

Nissen, T. L., Hahmann, C. W., Kielland-Brandt, M. C., Nielsen, J., and Villadsen, J. (2000) Anaerobic and aerobic batch cultivations of *Saccharomyces cerevisiae* mutants impaired in glycerol production. *Yeast* **16**, 463–474.

Nissen, P., Nielsen, D., and Arneborg, N. (2003) Viable *Saccharomyces cerevisiae* cells at high concentrations cause early growth arrest of non-*Saccharomyces* yeasts in mixed cultures by a cell–cell contact-mediated mechanism. *Yeast* **20**, 331–341.

Nissen, P., Nielsen, D., and Arneboug, N. (2004) The relative glucose uptake abilities of non-*Saccharomyces* yeasts play a role in their coexistence with *Saccharomyces cerevisiae* in mixed cultures. *Appl. Microbiol. Biotechnol.* **64**, 543–550.

Nygaard, M., and Prahl, C. (1996) Compatibility between strains of *Saccharomyces cerevisiae* and *Leuconostoc oenos* as an important factor for successful malolactic fermentation. In: *Proceedings of the 4th International Symposium on Cool Climate Viticulture and Enology* (T. Henick-Kling *et al.*, eds.), pp. VI-103–106. New York State Agricultural Experimental Station, Geneva, NY.

Nykänen, L. (1986) Formation and occurrence of flavor compounds in wine and distilled alcoholic beverages. *Am. J. Enol. Vitic.* **37**, 89–96.

Ohkuni, K., Hayashi, M., and Yamashita, I. (1998) Bicarbonate-mediated social communication stimulates meiosis and sporulation of *Saccharomyces cerevisiae*. *Yeast* **14**, 623–631.

Ollivier, C., Stonestreet, T., Larue, F., and Dubourdieu, D. (1987) Incidence de la composition colloidale des moûts blancs sur leur fermentescibilité. *Connaiss. Vigne Vin.* **21**, 59–70.

Oszmianski, J., Ramos, T., and Bourzeix, M. (1988) Fractionation of phenolic compounds in red wine. *Am. J. Enol. Vitic.* **39**, 259–262.

Ough, C. S. (1966a) Fermentation rates of grape juice, II. Effects of initial °Brix, pH, and fermentation temperature. *Am. J. Enol. Vitic.* **17**, 20–26.

Ough, C. S. (1966b) Fermentation rates of grape juice, III. Effects of initial ethyl alcohol, pH, and fermentation temperature. *Am. J. Enol. Vitic.* **17**, 74–81.

Ough, C. S. (1991) Influence of nitrogen compounds in grapes on ethyl carbamate formation in wines. In: *Proceeding of the International symposium of Nitrogen in Grapes and Wines* (J. Ratz, ed.), pp. 165–171. American Society for Enology and Viticultur, Davis, CA.

Ough, C. S., Huang, Z., An, D., and Stevers, D. (1991) Amino acid uptake by four commercial yeasts at two different temperatures of growth and fermentation: Effects on urea excretion and reabsorption. *Am. J. Enol. Vitic.* **42**, 26–40.

Ough, C. S., Stevens, D., Sendovski, T., Huang, Z., and An, A. (1990) Factors contributing to urea formation in commercially fermented wines. *Am. J. Enol. Vitic.* **41**, 68–73.

Ough, C. S., and Trioli, G. (1988) Urea removal from wine by an acid urease. *Am. J. Enol. Vitic.* **39**, 303–306.

Pampulha, M. A., and Loureiro-Dias, M. C. (1990) Activity of glycolytic enzymes of *Saccharomyces cerevisiae* in the presence of acetic acid. *Appl. Microbiol. Biotechnol.* **34**, 375–380.

Park, S. K., Morrison, J. C., Adams, D. O., and Noble, A. C. (1991) Distribution of free and glycosidically bound monoterpenes in the skin and mesocarp of Muscat of Alexandria grape during development. *J. Agric. Food Chem.* **39**, 514–518.

Parsel, D. A., Kowal, A. S., Singer, M. A., and Lindquist, S. (1994) Protein disaggregation mediated by heat-shock protein HSP104. *Nature* **372**, 475–478.

Payne, R. W. (1998) Identification of yeasts through computer-based systems. In: *Information Technology, Plant Pathology and Biodiversity* (P. D. Bridge, ed.), pp. 333–346. CAB International, Wallingford, England.

Peinado, R. A., Moreno, J., Bueno, J. E., Moreno, J. A., and Mauricio, J. C. (2004) Comparative study of aromatic compounds in two young white wines subjected to pre-fermentative cryomaceration. *Food Chem.* **84**, 585–590.

Peng, Z., Duncan, B., Pocock, K. F., and Sefton, M. A. (1998) The effect of ascorbic acid on oxidative browning of white wines and model wines. *Aust. J. Grape Wine Res.* **4**, 127–135.

Perez, M., Luyten, K., Michel, R., Riou, C., and Blondin, B. (2005) Analysis of *Saccharomyces cerevisiae* hexose carrier expression during wine fermentation: both low- and high-affinity Hxt transporters are expressed. *FEMS Yeast Res.* **5**, 351–361.

Pérez-Magariño, S., Sánchez-Iglesias, M., Ortega-Heras, M., González-Huerta, C., and González-Sanjosé, M. L. (2007) Colour stabilization of red wines by microoxygenation treatment before malolactic fermentation. *Food Chem.* **10**, 881–893.

Pérez-Ortín, J. E., García-Martínez, J., and Alberola, T. M. (2002) DNA chips for yeast biotechnology. The case of wine yeasts. *J. Biotechnol.* **98**, 227–241.

Petering, J. E., Henschke, P. A., and Langridge, P. (1991) The *Escherichia coli* β-glucuronidase gene as a marker for *Saccharomyces* yeast strain identification. *Am. J. Enol. Vitic.* **42**, 6–12.

Petering, J. E., Langridge, P., and Hensehke, P. A. (1993) Use of a marked wine yeast to determine efficiency of sulfur dioxide at low must temperature. In: *Proceedings of the 8th Australian Wine Industry Technical Conference* (C. S. Stockley *et al.*, eds.), pp. 211. Winetitles, Adelaide, Australia.

Peyrot des Gachons, C., Tominaga, T., and Dubourdieu, D. (2002) Localization of S-cysteine conjugates in the berry: effect of skin contact on aromatic potential of *Vitis vinifera* L. cv. Sauvignon blanc must. *Am. J. Enol. Vitic.* **53**, 144–146.

Pfeiffer, T., Schuster, S., and Bonhoeffer, A. (2001) Cooperation and competition in the evolution of ATP-producing pathways. *Science* **292**, 504–507.

Phaff, H. J. (1986) Ecology of yeasts with actual and potential value in biotechnology. *Microb. Ecol.* **12**, 31–42.

Pickering, G. J., Heatherbell, D. A., and Barnes, M. F. (1999) The production of reduced-alcohol wine using glucose oxidase-treated juice. Part III. Sensory. *Am. J. Enol. Vitic.* **50**, 307–316.

Pilone, G. J., Rankine, B. C., and Pilone, A. (1974) Inhibiting malolactic fermentation in Australian dry red wines by adding fumaric acid. *Am. J. Enol. Vitic.* **25**, 99–107.

Pines, O., Shemesh, S., Battat, E., and Goldberg, I. (1997) Overexpression of sytosolic malate dehydrogenase (MDH2) causes overproduction of specific organic acids in *Saccharomyces cerevisiae*. *Appl. Microbiol. Biotechnol.* **48**, 248–255.

Polsinelli, M., Romano, P., Suzzi, G., and Mortimer, R. (1996) Multiple strains of *Saccharomyces cerevisiae* on a single grape vine. *Lett. Appl. Microbiol.* **23**, 110–114.

Prahl, C., Lonvaud-Funel, A., Korsgaard, S., Morrison, E., and Joyeux, A. (1988) Étude d'un nouveau procédé de déclenchement de la fermentation malolactique. *Connaiss. Vigne Vin* **22**, 197–207.

Prévost, H., Cavin, J. F., Lamoureux, M., and Diviès, C. (1995) Plasmid and chromosome characterization of *Leuconostoc*. *Am. J. Enol. Vitic.* **46**, 43–48.

Prior, B. A., Toh, T. A., Jolly, N., Baccari, C., and Mortimer, R. K. (2000) Impact of yeast breeding for elevated glycerol production on fermentative activity and metabolite formation in Chardonnay wine. *S. Afr. J. Enol. Vitic.* **21**, 92–98.

Puig, S., Querol, A., Barrio, E., and Pérez-Ortín, J. E. (2000) Mitotic recombination and genetic changes in *Saccharomyces cerevisiae* during wine fermentation. *Appl. Environ. Microbiol.* **66**, 2057–2061.

Querol, A., Barrio, E., Huerta, T., and Ramón, D. (1992) Molecular monitoring of wine fermentations conducted by active dry yeast strains. *Appl. Environ. Microbiol.* **58**, 2948–2953.

Radler, F. (1972) Problematik des bakteriellen Säureabbaus. *Weinberg Keller* **19**, 357–370.

Radler, F. (1986) Microbial biochemistry. *Experientia* **42**, 884–893.

Radler, F., and Fäth, K. P. (1991) Histamine and other biogenic amines in wines. In: *Proceedings of the International Symposium on Nitrogen in Grapes and Wine* (J. M. Rantz, ed.), pp. 185–195. American Society for Enology and Viticulture, Davis, CA.

Radler, F., and Yannissis, C. (1972) Weinsäureabbau bei Milchsäurebakterien. *Arch. Mikrobiol.* **82**, 219–239.

Ramey, D., Bertrand, A., Ough, C. S., Singleton, V. L., and Sanders, E. (1986) Effects of skin contact temperature on Chardonnay must and wine composition. *Am. J. Enol. Vitic.* **37**, 99–106.

Ramírez, M., Peréz, F., and Regedón, J. A. (1998) A simple and reliable method for hybridization of homothallic wine strains of *Saccharomyces cerevisiae*. *Appl. Environ. Microbiol.* **64**, 5039–5041.

Ramírez, M., Regodón, J. A., Pérez, F., and Rebollo, J. E. (1999) Wine yeast fermentation vigor may be improved by elimination of recessive growth-retarding alleles. *Biotechnol. Bioengin.* **65**, 212–218.

Ramos, A., and Santos, H. (1996) Citrate and sugar cofermentation in *Leuconostoc oenos*, a ^{13}C nuclear magnetic resonance study. *Appl. Environ. Biol.* **62**, 2577–2585.

Rankine, B. C. (1966) Decomposition of L-malic acid by wine yeasts. *J. Sci. Food Agric.* **17**, 312–316.

Rankine, B. C., Fornachon, J. C. M., and Bridson, D. A. (1969) Diacetyl in Australian dry red wines and its significance in wine quality. *Vitis* **8**, 129–134.

Rapp, A., and Reuther, K. H. (1971) Der Gehalt an freien Aminosäuren in Trabenmosten von gesunden und edelfaulen Beeren verschiedener Rebsorten. *Vitis* **10**, 51–58.

Redžepovič, S., Orlič, S., Sikora, S., Majdak, A., and Pretorius, I. S. (2002) Identification and characterization of *Saccharomyces cerevisiae* and *Saccharomyces paradoxus* strains isolated from Croatian vineyards. *Lett. Appl. Microbiol.* **35**, 305–310.

Reguant, C. and Bordons, A. (2003) Typification of *Oenococcus oeni* strains by multiplex RAPD-PCR and study of population dynamics during malolactic fermentation. *J. Appl. Microbiol.* **95**, 344–353.

Remise, F., Gaudin, A., Kong, Y., Guzzo, J., Alexandre, H., Krieger, S., and Guilloux-Benatier, M. (2006) *Oenococcus oeni* preference for peptides: qualitative and quantitative analysis of nitrogen assimilation. *Arch. Microbiol.* **185**, 459–469.

Remise, F., Roustan, J. L., Sablayrolles, J. M., Barre, P., and Dequin, S. (1999) Glycerol overproduction by engineered *Saccharomyces cerevisiae* wine yeast strains leads to substantial changes in by-product formation and to a stimulation of fermentation rate in stationary phase. *Appl. Environ. Microbiol.* **65**, 143–149.

Renouf, V., Claisse, O., and Lonvaud-Funel, A. (2005) Understanding the microbial ecosystem on the grape berry surface through numeration and identification of yeast and bacteria. *Aust. J. Grape Wine Res.* **11**, 316–327.

Renouf, V., Lonvaud-Funel, A., Walling, E., and Coulon, J. (2006) Le suivi microbiologique du vin, Partie 1: de la parcelle au conditionnement: un outil pour une oenologie raisonnée. *Rev. Oenolog.* **118**, 27–31.

Revilla, I., and González-San José, M. L. (1998) Methanol release during fermentation of red grapes treated with pectolytic enzymes. *Food Chem.* **63**, 307–312.

Revilla, I., and González-San José, M. L. (2003) Addition of pectolytic enzymes: an enological practice which improves the chromaticity and stability of red wines. *Int. J. Food Sci. Technol.* **38**, 29–36.

Revilla, E., Ryan, J. M., Kovac, V., and Nemanic, J. (1998) The effect of the addition of supplementary seeds and skins during fermentation on the chemical and sensory characteristics of red wines. *Dev. Food Sci.* **40**, 583–596.

Ribéreau-Gayon, J., Peynaud, E., Ribéreau-Gayon, P., and Sudraud, P. (1975) *Traité d'Oenologie, Sciences et Techniques du Vin*, Vol. 2. Dunod, Paris.

Ribéreau-Gayon, J., Peynaud, E., Ribéreau-Gayon, P., and Sudraud, P. (1976) *Traité d'oenologie, sciences et techniques du vin*, Vol. 3. Dunod, Paris.

Ribéreau-Gayon, P. (1985) New developments in wine microbiology. *Am. J. Enol. Vitic.* **36**, 1–9.

Ribéreau-Gayon, P., Lafon-Lafourcade, S., Dubourdieu, D., Lucmaret, V., and Larue, F. (1979) Métabolism de *Saccharomyces cerevisiae* dans le môut de raisins parasités par *Botrytis cinerea*. Inhibition de la fermentation, formation d'acide acétique et de glycerol. *C. R. Acad. Sci. Paris Ser. D* **289**, 441–444.

Ribéreau-Gayon, P., Larue, F., and Chaumet, P. (1987) The effect of addition of sucrose and aeration to grape must on growth and metabolic activity of *Saccharomyces cerevisiae*. *Vitis* **26**, 208–214.

Riou, C., Nicaud, J. M., Barre, P., and Gaillardin, C. (1997) Stationary-phase gene expression in *Saccharomyces cerevisiae* during wine fermentation. *Yeast* **13**, 903–915.

Romano, P., Fiore, C., Paraggio, M., Caruso, M., and Capece, A. (2003) Function of yeast species and strains in wine flavour. *Int. J. Food Microbiol.* **86**, 169–180.

Romero-Cascales, I., Ortega-Regules, A., López-Roca, J. M., Fernández-Fernández, J. I., and Gómez-Plaza, E. (2005) Differences in anthocyanin extractability from grapes to wines according to variety. *Am. J. Enol. Vitic.* **56**, 212–219.

Rose, A. H. (1989) Influence of the environment on microbial lipid composition. In: *Microbial Lipids* (C. Ratledge and S. G. Wilkinson, eds.), Vol. 2, pp. 255–278. Academic Press, London.

Rosini, G. (1986) Wine-making by cell-recycle-batch fermentation process. *Appl. Microbiol. Biotechnol.* **24**, 140–143.

Rossignol, T., Dulau, L., Julien, A., and Blondin, B. (2003) Genome-wide monitoring of wine yeast gene expression during alcoholic fermentation. *Yeast* **20**, 1369–1385.

Rossignol, T., Postaire, O., Storai, J., and Blondin, B. (2006) Analysis of the genomic response of a wine yeast to rehydration and inoculation. *Appl. Microbiol. Biotechnol.* **71**, 699–712.

Roustan, J. L., and Sablayrolles, J. M. (2002) Impact of the addition of electron acceptors on the by-products of alcoholic fermentation. *Enzyme Microbial. Technol.* **31**, 142–152.

Rozès, N., Arola, L., and Bordons, A. (2003) Effect of phenolic compounds on the co-metabolism of citric acid and sugars by *Oenococcus oeni* from wine. *Lett. Appl. Microbiol.* **36**, 337–341.

Ruediger, G. A., Pardon, K. H., Sas, A. N., Godden, P. W., and Pollnitz, A. P. (2005) Fate of pesticides during the winemaking process in relation to malolactic fermentation. *J. Agric. Food Chem.* **53**, 3023–3026.

Sablayrolles, J. M., and Barre, P. (1986) Evaluation of oxygen requirement of alcoholic fermentations in simulated enological conditions. *Sci. Aliments* **6**, 373–383.

Sá Correia, I., Salgueiro, S. P., Viegas, C. A., and Novais, J. M. (1989) Leakage induced by ethanol, octanoic and decanoic acids in *Saccharomyces cerevisiae*. *Yeast* **5**, S124–129.

Saint-Criq, N., Vivas, N., and Glories, Y. (1998) Maturité phénolique: définition et contrôle. *Rev. Fr. Oenol.* **173**, 22–25.

Sala, C., Fort, F., Busto, O., Zamora, F., Arola, L., and Guasch, J. (1996) Fate of some common pesticides during vinification process. *J. Agric. Food Chem.* **44**, 3668–3671.

Sales, K., Brandt, W., Rumbek, E., and Lindsey, G. (2000) The LEA-like protein *HSP*12 in *Saccharomyces cerevisiae* has a plasma membrane location and protects membranes against desiccation and ethanon-induced stress. *Biochem. Biophus. Acta* **1463**, 267–278.

Salinas, M. R., Garijo, J., Pardo, F., Zalacain, A., and Alonso, G. L. (2003) Color, polyphenol, and aroma compounds in rosé wines after preferementative maceration and enzymatic treatments. *Am. J. Enol. Vitic.* **54**, 195–202.

Salou, P., Loubiere, P., and Pareilleux, A. (1994) Growth and energetics of *Leuconostoc oenos* during cometabolism of glucose with citrate or fructose. *Appl. Environ. Microbiol.* **60**, 1459–1466.

San Romáo, M. V., and Coste Belchior, A. P. (1982) Study of the influence of some antifungal products on the microbiological flora of grapes and musts (in Portuguese) *Ciénc. Téc. Vitivinic.* **1**, 101–112.

San Romáo, M. V., Coste Belchior, A. P., Silva Alemáo, M., and Gonçalves Bento, A. (1985) Observations sur le métabolisme des bacteries lactiques dans les moûts de raisins altéres. *Connaiss. Vigne Vin* **19**, 109–116.

Santamaría, P., Garijo, P., López, R., Tenorio, C., and Gutiérrez, A. R. (2005) Analysis of yeast population during spontaneous alcoholic fermentation: effect of the age of the cellar and the practice of inoculation. *Int. J. Food Microbiol.* **103**, 49–56.

Schanderl, H. (1959) *Die Mikrobiologie des Mostes und Weines*, 2nd edn. Ulmer, Stuttgart, Germany.

Scheibel, T., Bell, S., and Walke, S. (1997) *Saccharomyces cerevisiae* and sulfur: A unique way to deal with the environment. *FASEB J.* **11**, 917–921.

Schmidt, J. O., and Noble, A. C. (1983) Investigation of the effect of skin contact time on wine flavor. *Am. J. Enol. Vitic.* **34**, 135–138.

Schneider, V. (1998) Must hyperoxidation: A review. *Am. J. Enol. Vitic.* **49**, 65–73.

Schulze, U., Lidén, G., Nielsen, J., and Villadsen, J. (1996) Physiological effects of nitrogen starvation in anaerobic batch culture of *Saccharomyces cerevisiae*. *Microbiology* **142**, 2299–2310.

Schütz, M., and Gafner, J. (1993a) Analysis of yeast diversity during spontaneous and induced alcoholic fermentations. *J. Appl. Bacteriol.* **75**, 551–558.

Schütz, M., and Gafner, J. (1993b) Sluggish alcoholic fermentation in relation to alternations of the glucose-fructose ratio. *Chem. Mikrobiol. Technol. Lebensm.* **15**, 73–78.

Shimazu, Y., Uehara, M., and Watanbe, M. (1985) Transformation of citric acid to acetic acid, acetoin and diacetyl by wine making lactic acid bacteria. *Agric. Biol. Chem.* **49**, 2147–2157.

Siegrist, J. (1985) Les tanins et les anthocyanes du Pinot et les phénomènes de macération. *Rev. Oenologues* **11**(38), 11–13.

Silva, S., Ramón-Portugal, F., Andrade, P., Abreu, S., de Fatima Texeira, M., and Strehaiano, P. (2003) Malic acid consumption by dry immobilized cells of *Schizosaccharomyces pombe*. *Am. J. Enol. Vitic.* **54**, 50–55.

Sinclair, D.A., and Guarente, L. (1997) Extrachromosomal rDNA circles – a cause of aging in yeast. *Cell* **91**, 1033–1042.

Singleton, V. L. (1987) Oxygen with phenols and related reactions in musts, wines, and model systems: Observations and practical implications. *Am. J. Enol. Vitic.* **38**, 69–77.

Singleton, V. L., Sieberhagen, H. A., de Wet, P., and van Wyk, C. J. (1975) Composition and sensory qualities of wines prepared from white grapes by fermentation with and without grape solids. *Am. J. Enol. Vitic.* **26**, 62–69.

Singleton, V. L., Zaya, J., and Trousdale, E. (1980) White table wine quality and polyphenol composition as affected by must SO_2 content and pomace contact time. *Am. J. Enol. Vitic.* **31**, 14–20.

Sniegowski, P. D., Dombrowski, P. G., and Fingerman, E. (2002) *Saccharomyces cerevisiae* and *Saccharomyces paradoxus* coexist in a natural woodland site in North America and display different levels of reproductive isolation from European conspecifics. *FEMS Yeast Res.* **1**, 299–306.

Soden, A., Francis, I. L., Oakey, H., and Henschke, P. A. (2000) Effects of co-fermentation with *Candida stellata* and *Saccharomyces cerevisiae* on the aroma and composition of Chardonnay wine. *Aust. J. Grape Wine Res.* **6**, 21–30.

Soleas, G. J., Tomlinson, G., and Goldberg, D. M. (1998) Kinetics of polyphenol release into wine must during fermentation of different cultivars. *J. Wine Res.* **9**, 27–42.

Somers, T. C., and Pocock, K. F. (1986) Phenolic harvest criteria for red vinification. *Aust. Grapegrower Winemaker* **256**, 24–30.

Somers, T. C., and Pocock, K. F. (1991) Phenolic assessment of white musts: Varietal differences in free-run juices and pressings. *Vitis* **26**, 189–201.

Somers, T. C., and Ziemelis, G. (1985) Flavonol haze in white wines. *Vitis* **24**, 43–50.

Soufleros, E., and Bertrand, A. (1988) Les acides gras libres du vin, observations sur leur origine. *Connaiss. Vigne Vin* **22**, 251–260.

Spano, G., Rinaldi, A., Ugliano, M., Moio, L., Beneduce, L., and Massa, S. (2005) A β-glucosidase gene isolated from wine *Lactobacillus plantarum* is regulated by abiotic stresses. *J. Appl. Microbiol.* **98**, 855–861.

Sponholz, W. R., and Hühn, T. (1996) Aging of wine: 1,1,6-Trimethyl-1,2-dihycronaphthalene (TDN) and 2-aminoacetophenone. In: *Proceedings of the 4th International Symposium on Cool Climate Viticulture and Enology* (T. Henick-Kling *et al.*, eds.), pp. VI-37–57. New York State Agricultural Experimental Station, Geneva, NY.

Stuckey, W., Iland, P., Henschke, P. A., and Gawel, R. (1991) The effect of lees contact time on chardonnay wine composition. In: *Proceedings of the International Symposium on Nitrogen in Grapes and Wine* (J. M. Rantz, ed.), pp. 315–319. American Society for Enology and Viticulture, Davis, CA.

Sulo, P., Michačáková, S., and Reiser, V. (1992) Construction and properties of K_1 type killer wine yeast. *Biotechnol. Lett.* **14**, 55–60.

Sun, B. S., Pinto, T., Leandro, M. C., Ricardo-Da-Silva, J. M., and Spranger, M. I. (1999) Transfer of catechins and proanthocyanidins from solid parts of the grape clusters into wine. *Am. J. Enol. Vitic.* **50**, 179–184.

Suzzi, G., and Romano, P. (1982) Induced changes by SO_2 on the population of *Saccharomyces* as agents of the natural fermentation of musts. *Vini d'Italia* **24**, 138–145.

Suzzi, G., Romano, P., Vannini, L., Turbanti, L., and Domizio, P. (1996) Cell-recycle batch fermentation using immobilized cells of flocculent *Saccharomyces cerevisiae* wine strains. *World J. Microbiol. Biotechnol.* **12**, 25–27.

Tamborra, T. (1992) Influenza della macerazione sul contenuto di terpeni liberi e glucosidi nel Moscatello selvatico. *Riv. Vitic. Enol.* **45**, 35–45.

Teixeira, H., Gonçalves Rozès, N., Ramos, A., and San Romão, M. V. (2002) Lactobacillic acid accumulation in the plasma membrane of *Oenococcus oeni*: a response to ethanol stress? *Microb. Ecol.* **43**, 146–153.

Teunissen, A. W. R. H., and Steensma, H. Y. (1995) The dominant flocculation genes of *Saccharomyces cerevisiae* constitute a new subtelomeric gene family. *Yeast* **11**, 1001–1013.

Thornton, R. J. (1985) The introduction of flocculation into a homothallic wine yeast. A practical example of the modification of winemaking properties by the use of genetic techniques. *Am. J. Enol. Vitic.* **36**, 47–49.

Tomenchok, D. M., and Brandriss, M. C. (1987) Gene-enzyme relationships in the proline biosynthetic pathway of *Saccharomyces cerevisiae*. *J. Bacteriol.* **169**, 5364–5372.

Tominaga, T., Furrer, A., Henry, R., and Dubourdieu, D. (1998) Identification of new volatile thiols in the aroma of *Vitis vinifera* L. var. Sauvignon blanc wines. *Flavour Fragr. J.* **13**, 159–162.

Török, T., Mortimer, R. K., Romano, P., Suzzi, G., and Polsinelli, M. (1996) Quest for wine yeasts – an old story revisited. *J. Indust. Microbiol.* **17**, 303–313.

Tyagi, R. D. (1984) Participation of oxygen in ethanol fermentation. *Process Biochem.* **19**(4), 136–141.

Ugliano, M., and Moio, L. (2005) Changes in the concentration of yeast-derived volatile compounds of red wine during malolactic fermentation with four commercial starter cultures of *Oenococcus oeni*. *J. Agric. Food Chem.* **53**, 10134–10139.

Valero, E., Schuller, D., Cambon, B., Casal, M., and Dequin, S. (2005) Dissemination and survival of commercial wine yeast in the vineyard: a large-scale, three-years study. *FEMS Yeast Res.* **5**, 959–969.

van der Walt, J. P. (1970) The genus *Saccharomyces* (Meyer) Reess. In: *The Yeasts: A Taxonomic Study* (J. Lodder, ed.), 2nd edn, pp. 555–718. North-Holland, Amsterdam.

Vannobel, C. (1986) Réfexions sur l'évolution récente de la technologie en matière de maîtrise des températures de fermentation. *Prog. Agric. Vitic.* **21**, 488–494.

van Rensburg, P. and Pretorius, I. S. (2000) Enzymes in winemaking: Harnessing natural catalysts for efficient biotransformations. *S. Afr. J. Enol. Vitic.* **21**, 52–73.

Vaughn, R. H. (1955) Bacterial spoilage of wines with special reference to California conditions. *Adv. Food Res.* **7**, 67–109.

Velázquez, J. B., Longo, E., Sieiro, C., Cansado, J., Calo, P., and Villa, T. G. (1991) Improvement of the alcoholic fermentation of grape juice with mixed cultures of *Saccharomyces cerevisiae* wild strains, negative effect of *Kloeckera apiculata*. *World J. Microbiol. Biotechnol.* **7**, 485–489.

Verbelen, P. J., De Schutter, D. P., Delvaux, F., Verstrepen, K. J., and Delvaux, F. R. (2006) Immobilized yeast cell systems for continuous fermentation applications. *Biotechnol Lett.* **28**, 1515–1525.

Viegas, C. A., Almeida, P. F., Cavaco, M., and Sá-Correia, I. (1998) The H^+-ATPase in the plasma membrane of *Saccharomyces cerevisiae* is activated during growth latency in octanoic acid-supplemented medium accompanying the decrease in intracellular pH and cell viability. *Appl. Environ. Microbiol.* **64**, 779–783.

Villettaz, J. C. (1987) A new method for the production of low alcohol wines and better balanced wines. In: *Proceedings of the 6th Australian Wine Industry Technical Conference* (T. Lee, ed.), pp. 125–128. Australian Industrial Publishers, Adelaide, Australia.

Vivas, N., Lonvaud-Funel, A., Glories, Y., and Augustin, M. (1995) The effect of malolactic fermentation in barrels and in tanks on the composition and the quality of red wines. *J. Sci. Tech. Tonnellerie* **1**, 65–80.

Vlassides, S., and Block, D. E. (2000) Evaluation of cell concentration profiles and mixing in unagitated wine fermentors. *Am. J. Enol. Vitic.* **51**, 73–80.

Vos, P. J. A., and Gray, R. S. (1979) The origin and control of hydrogen sulfide during fermentation of grape must. *Am. J. Enol. Vitic.* **30**, 187–197.

Walker, G. M. (1998) Magnesium as a stress-protectant for industrial strains of *Saccharomyces cerevisiae*. *J. Am. Soc. Brew. Chem.* **56**, 109–113.

Wenzel, K., Dittrich, H. H., Seyffardt, H. P., and Bohnert, J. (1980) Shwefelrüchstände auf Trauben und Most und ihr Einfluß auf die H₂S Bildung. *Wein Wiss.* **6**, 414–420.

White, B. B., and Ough, C. S. (1973) Oxygen uptake studies in grape juice. *Am. J. Enol. Vitic.* **24**, 148–152.

White, K. D. (1984) *Greek and Roman Technology.* Cornell University Press, Ithaca, NY.

Whiting, G. C. (1975) Some biochemical and flavour aspects of lactic acid bacteria in ciders and other alcoholic beverages. In: *Lactic Acid Bacteria in Food and Beverages* (J. G. Carr *et al.*, eds.), pp. 69–85. Academic Press, London.

Wibowo, D., Eschenbruch, R., Davis, C. R., Fleet, G. H., and Lee, T. H. (1985) Occurrence and growth of lactic acid bacteria in wine: A review. *Am. J. Enol. Vitic.* **36**, 302–313.

Wiederkehr, M. (1997) Must heating versus traditional must fermentation. In: *Twenty-sixth Annual Wine Industry Workshop* (T. Henick-Kling, ed.), pp. 18–24. New York State Agricultural Experiment Station, Cornell University, Geneva, NY.

Wightman, J. D., Price, S. F., Watson, B. T., and Wrolstad, R. E. (1997) Some effects of processing enzymes on anthocyanins and phenolics in Pinot noir and Cabernet Sauvignon wines. *Am. J. Enol. Vitic.* **48**, 39–48.

Williams, L. A. (1982) Heat release in alcoholic fermentation: A critical reappraisal. *Am. J. Enol. Vitic.* **33**, 149–153.

Williams, L.A., and Boulton, R. (1983) Modeling and prediction of evaporative ethanol loss during wine fermentations. *Am. J. Enol. Vitic.* **34**, 234–242.

Williams, P. J., Francis, I. L., and Black, S. (1996) Changes in concentration of juice and must glycosides, including flavor precursors. In: *Proceedings of the 4th International Symposium on Cool Climate Viticulture and Enology* (T. Henick-Kling *et al.*, eds.), pp. VI-5–10. New York State Agricultural Experiment Station, Geneva, NY.

Wolf, E., and Benda, I. (1965) Qualität und Resistenz. III. Das Futterwahlvermögen von *Drosophila melanogaster* gegenüber natürlichen Weinhefe-Arten und -Rassen. *Biol. Zentralbl.* **84**, 1–8.

Wolfe, K. W., and Shields, D. S. (1997) Molecular evidence of an ancient duplication of the entire yeast genome. *Nature* **387**, 708–713.

Wong, S., Butler, G., and Wolfe, K. H. (2002) Gene order evolution and paleopolyploidy in hemiascomycete yeasts. *Proc. Natl Acad. Sci.* **99**, 9272–9277.

Zamora, F., Fort, F., Fuguet, J., Bordon, A., and Arola, L. (1996) Influence de certains facteurs lors de la fermentation alcoolique sur le transport de glucose dans les levures. In: *Oenologie 95— 5éme Symposium International d'Oenologie* (A. Lonvaud-Funel, ed.), pp. 167–171. Tec & Doc, Lavoisier, France.

Zavaleta, A. I., Martinex-Murica, A. J., and Rodriguez-Valera, F. (1997) Intraspecific genetic diversity of *Oenococcus oeni* as derived from DNA fingerprinting and sequence analyses. *Appl. Environ. Microbiol.* **64**, 1261–1267.

Zoecklein, B. W., Collins, D., McCarthy, B., and Vaillant, K. (2004) Effect of délestage with seed deportation on Merlot and Cabernet Sauvignon wines. *Am. J. Enol. Vitic.* **55**, 437A.

Zoecklein, B., Fugelsang, K. C., Gump, B. H., and Nury, F. S. (1995) *Wine Analysis and Production.* Chapman-Hall, New York.

Postfermentation Treatments and Related Topics

All wines undergo a period of adjustment (maturation) before bottling. Maturation involves the removal or precipitation of particulate and colloidal material. In addition, the wine experiences a range of physical, chemical, and biological changes that maintain or improve its sensory attributes. Many of these changes occur spontaneously, but can be facilitated by the winemaker. Although undue intervention can disrupt the wine's inherent characteristics, shunning any intervention can be equally detrimental. What is essential is that rational action, not philosophical (or marketing) dictates, take precedence.

Wine Adjustments

Adjustments attempt to correct deficiencies found in the grapes and sensory imbalances that developed during fermentation. In certain jurisdictions, acidity and sweetness adjustments are permitted only before fermentation. This is regrettable, because it is impossible to predict the course of fermentation precisely. Judicious adjustment after vinification can improve the wine so that its geographic and varietal attributes can develop fully.

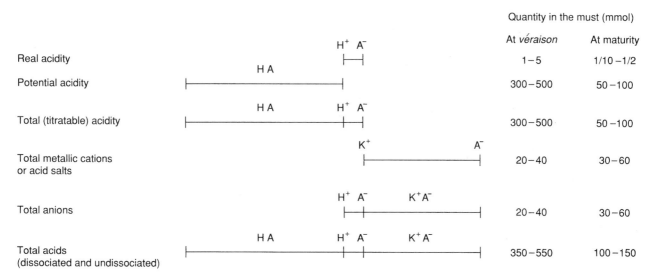

Figure 8.1 Notions characterizing the acid–base equilibrium of a solution. The values given correspond to those found in must: A⁻, acid anion; H⁺, hydrogen ion; HA, undissociated acid; K⁺, potassium ion; K⁺A⁻, undissociated potassium salt. (After Champagnol, 1986, reproduced by permission)

Acidity and pH Adjustment

Theoretically, acidity and pH adjustment could occur at almost any stage during vinification. Nevertheless, postfermentative correction is probably optimal. During fermentation, deacidification often occurs spontaneously, due to acid precipitation or to yeast and bacterial metabolism. In addition, some strains of *Saccharomyces cerevisiae* synthesize significant amounts of malic acid during fermentation (Farris *et al.*, 1989). Thus, the method and extent of deacidification needed are difficult to assess before fermentation ends. However, if the juice is above pH 3.4, some prefermentative lowering of the pH may be advisable to favorably influence fermentation and avoid large adjustments following fermentation, especially with white wines.

Typically, red wines have higher pH values than white wines. Red wines are more frequently produced in warmer regions. Therefore, they tend to have lower malic acid contents at harvest. In addition, more potassium is extracted during the extended maceration of red wines (converting more tartaric acid to its nonionic salt form).

Precise recommendations for optimal acidity are impossible. They reflect stylistic and regional preferences. More fundamentally, acidity and pH are complexly interrelated. The major fixed acids in grapes (tartaric and malic) occur in a dynamic equilibrium of ionized and nonionized states (Fig. 8.1). These include undissociated (nonionized) acids, half-ionized states (with one ionized carboxyl group), fully ionized states (with both carboxyl groups ionized), half-salts (with one carboxyl group associated with a cation), full salts (with both carboxyl groups bound to cations), or as double salts with other acid molecules and cations. The proportion of interconvertible states depends largely on the concentration of the acids and potassium ions. Because of the complexity of the equilibria, precise prediction of the consequences of changing any one of these factors on acidity is impossible. Nevertheless, a range between 0.55 and 0.85% total acidity is generally considered desirable. Red wines are customarily preferred at the lower end of the range, whereas white wines are preferred at the upper end.

Another important aspect of acidity is pH. It represents the proportion of H⁺ to OH⁻ ions in an aqueous solution – the higher the proportion of H⁺ ions, the lower the pH; conversely, the higher the proportion of OH⁻ ions, the higher the pH. Wines vary considerably in pH, with values below 3.1 being perceived as sour, and those above 3.7 being considered flat. White wines are commonly preferred at the lower end of the pH range, whereas red wines are frequently favored in the midrange.

Relatively low pH values in wine are preferred for many reasons. They give wines their fresh taste, improve microbial stability, reduce browning, diminish the need for SO_2, and enhance the production and stability of fruit esters. The concentrations of monoterpenes also may be affected. For example, the concentration of geraniol, citronellol, and nerol may rise at low pH values, whereas those of linalool, α-terpineol, and hotrienol fall. In red wines, color intensity and hue are better at lower pH values.

Because of the importance of pH, the choice of acidity-correction procedure is influenced considerably by how it affects pH. Because tartaric acid is more highly ionized than malic acid, within the usual range of wine pH values, adjusting the concentration of tartaric acid has a greater effect on pH than an equivalent change in the concentration of malic acid (Fig. 8.2).

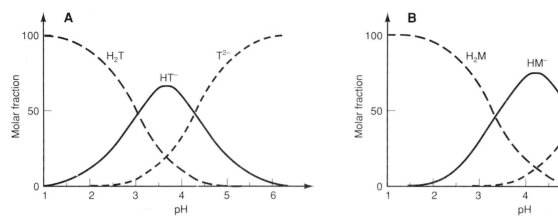

Figure 8.2 Relative concentrations of the three main forms of (**A**) tartaric and (**B**) malic acids as a function of pH. H_2, full acid (unionized); H^-, half acid (half-ionized); $^{2-}$, fully ionized acid. (After Champagnol, 1986, reproduced by permission)

Thus, adjusting the concentration of tartaric acid affects pH more that total acidity. In contrast, adjusting malic acid content affects total acidity more than pH. Which is preferable depends on the rationale for adjustment. Prefermentative adjustments are discussed in Chapter 7.

DEACIDIFICATION

Wine may be deacidified by either physicochemical or biological means. Physicochemical deacidification involves either acid precipitation or column ion-exchange. Biological deacidification usually involves malolactic fermentation (see Chapter 7).

Neutralization can also result from the formation of insoluble half-salts. Deacidification with potassium tartrate by this method is illustrated below.

Of available methods, deacidification with calcium carbonate is probably the most common. Use of potassium carbonate is prohibited in several countries, whereas potassium tartrate is more expensive. Although widely used, calcium carbonate has a number of disadvantages. Its primary drawback is the slow rate at which

Precipitation Precipitation primarily entails the neutralization of tartaric acid; malic acid is less involved due to the higher solubility of its salts. Neutralization occurs when cations (positively charged ions) of a salt exchange with hydrogen ion(s) of an acid. Salt formation can reduce acid solubility, inducing crystallization and precipitation. The removal of the precipitated salt during racking, filtration, or centrifugation makes the reaction irreversible.

To induce the neutralization and precipitation of tartaric acid, finely ground calcium carbonate may be added to the wine. The reaction produces a calcium double salt of tartaric acid.

calcium tartrate precipitates. In addition, formation of calcium malate may produce a salty taste. Furthermore, if tartrate removal is excessive, the resulting increase in pH may leave the wine tasting flat and susceptible to microbial spoilage.

Some of the disadvantages of calcium carbonate addition may be avoided by the use of **double-salt** deacidification. The name refers to the belief that the technique functioned primarily by the formation of an insoluble double calcium salt between malic and tartaric acids. It now appears that very little of the hypothesized double salt forms. Nonetheless, the procedure seems to both speed the precipitation of calcium tartrate and facilitates the partial precipitation of calcium malate (Cole and Boulton, 1989).

The major difference between the single- and double-salt procedures is that the latter involves the addition of calcium carbonate to only a small proportion (~10%) of the wine to be deacidified. Sufficient calcium carbonate is added to raise the pH to above 5.1. This assures adequate dissociation of both malic and tartaric acids (see Fig. 8.2). It also induces the rapid formation and precipitation of the salts. A patented modification of the double-salt procedure (Acidex) incorporates 1% calcium malate-tartrate with the calcium carbonate. The double salt possibly acts as seed crystals, promoting rapid crystallization.

In double-salt procedures, the remainder of the wine is slowly blended back into the treated portion, with vigorous stirring. Subsequent crystal removal occurs by filtration, centrifugation, or settling. Stabilization may take 3 months, during which residual salts precipitate before bottling.

Although precipitation works well with wines of medium to high total acidity (6–9 g/ml) and medium to low pH (<3.5), it can result in an excessive pH rise in wines showing both high acidity (>9 g/ml) and high pH (>3.5). This situation is most common in cool climatic regions where malic acid constitutes the major acid and the potassium content is high. In this situation, column ion exchange may be used (Bonorden *et al.*, 1986), or tartaric acid may be added to the wine before the addition of calcium carbonate in double-salt deacidification (Nagel *et al.*, 1988). Precipitation following neutralization removes the excess potassium with acid salts, and the added tartaric acid lowers the pH to an acceptable value.

Because protective colloids can significantly affect the precipitation of acid salts, it is important to conduct deacidification trials on small samples of the wine. This will establish the amount of calcium carbonate or Acidex required for the desired degree of deacidification.

Ion-exchange Column Ion exchange involves passing the wine through a resin-containing column. During passage, ions in the wine exchange with those in the column. The types of ions replaced can be adjusted by modifying the type of resin and ions present.

For deacidification, the column is packed with an anion-exchange resin. Tartrate ions are commonly exchanged with hydroxyl ions (OH^-), thus removing tartrate from the wine. The hydroxyl ions released from the resin associate with hydrogen ions, forming water. Alternately, malate may be removed by exchange with a tartrate-charged resin. The excess tartaric acid may be subsequently removed by neutralization and precipitation. The major limiting factor in ion exchange use, other than legal restrictions and cost, is its tendency to remove flavorants and color from the wine, reducing wine quality.

Biological Deacidification Biological deacidification, via malolactic fermentation, is possibly the most common means of acidity correction. Because malolactic fermentation can occur before, during, and after alcoholic fermentation, it was discussed in Chapter 7. Alternately, use of malic-acid-degrading strains of *S. cerevisiae* can achieve partial deacidification.

ACIDIFICATION

If wines are too low in acidity, or possess an undesirably high pH, tartaric acid may be added. As an acidulent, tartaric acid has several distinct advantages. These include its fresh crisp taste, high microbial stability, and a dissociation constant (K_a) that allows it to markedly reduce pH. The main disadvantage of tartaric acid use is its expense, especially when added to wines high in potassium content. Crystal formation results in most of the tartaric acid being lost due to precipitation. The addition of citric acid avoids these problems and can assist in preventing ferric *casse*. Nevertheless, the ease with which citric acid is metabolized means that it is microbially unstable. Alternatively, ion exchange may be used to lower pH by exchanging H^+ for the Ca^{2+} or K^+ of tartrate and malate salts.

Sweetening

In the past, stable naturally sweet wines were rare. Most of the sweet wines of antiquity probably contained boiled-down must or honey. Stabilization by the addition of distilled alcohol is a comparatively recent innovation. The stable, naturally sweet table wines of the past few centuries seem to have been produced from highly botrytized grapes (see Chapter 9). In contrast, present-day technology can produce a wide range of sweet wines without recourse to botrytization, baking, or fortification.

Wines may be sweetened with sugar, for example, sparkling wines. However, most still wines possessing a sweet finish are sweetened by the addition of partially fermented or unfermented grape juice, termed **sweet reserve** (*süssreserve*). The base wine is typically fermented dry and sweetened just before bottling. To avoid microbial contamination, both the wine and sweet reserve are sterilized by filtration or pasteurized, and the blend bottled under aseptic conditions, employing sterile bottles and corks.

Various techniques are used in preparing and preserving sweet reserve. One procedure involves separating a small portion of the juice to produce the sweet reserve. Thus, it possesses the same varietal, vintage, and geographical origin as the wine it sweetens. If the sweet reserve is partially fermented, yeast activity may be terminated prematurely by chilling, filtration, centrifugation, or by trapping the carbon dioxide released during fermentation. The pressure buildup stops fermentation. If the sweet reserve is stored as unfermented juice, microbial activity is restricted after clarification by cooling to $-2\,°C$, applying CO_2 pressure, pasteurizing, or sulfiting to above 100 ppm of free SO_2. In the last case, desulfiting is conducted before use by flash heating or sparging with nitrogen gas. If desired, the juice can be concentrated by reverse osmosis or cryoextraction. Heat and vacuum concentration are additional possibilities, but are likely to result in greater flavor modification and fragrance loss.

Dealcoholization

In the past few decades, an increased market for low-alcohol and dealcoholized wines has developed. In addition, delaying harvesting to increase flavor content has resulted in grapes with elevated °Brix values, and corresponding wines with higher alcohol contents. The wines produced may be unbalanced and "hot."

Previously, dealcoholization involved heat-induced alcohol evaporation. Although successful, it generated baked or cooked odors, and drove off important flavorants. Correspondingly, it was appropriate only for inexpensive, low-alcohol wines. With the advent of vacuum distillation, the temperature required could be reduced, thus avoiding heat-generated flavor distortion. It still had the problem that many important volatiles escaped with the alcohol. Although these could be retrieved and added back to the wine, the final product still lacked its original character. Alternative procedures, such as strip-column distillation can lower ethanol contents down to 0.9 g ethanol/liter (Duerr and Cuénat, 1988; Ireton, 1990). Dialysis apparently results in little flavor loss (Wucherpfennig *et al.*, 1986). Pervaporation is another membrane procedure that selectively removes ethanol, permitting the rest of the permeate to be added back to the wine (Takács *et al.*, 2007). Additional potential techniques include the spinning cone column (Rieger, 1994), and volatilization with carbon dioxide (Antonelli *et al.*, 1996; Scott and Cooke, 1995). Both appear to have the advantages of speed and minimal flavor disruption. Nevertheless, the most widely used dealcoholization techniques appear to be vacuum distillation and reverse osmosis, despite their removal of fruity aromatics and the potential accentuation of unpleasant odors (Fischer and Berger, 1996).

When alcoholic wines are treated to bring their alcohol contents down to traditional values, flavor loss is typically relatively minor. Because alcohol content influences wine flavor and balance, alcohol adjustment provides the winemaker with an opportunity (albeit expensive) to tweak the wine's sensory characteristics.

Where the addition of water is permissible in the production of low-alcohol beverages, dilution is the simplest means of dealcoholization. Flavor enhancement, as with wine coolers, offsets flavor dilution.

Flavor Enhancement

Many grape flavorants, notably terpenes, norisoprenoids and volatile phenols, are bound in nonvolatile glycosidic complexes. Consequently, releasing this potential has drawn considerable attention. Glycosidic bonds may be broken by either acidic or enzymic hydrolysis. Because acid-induced hydrolysis is slow, heating the wine to increase the reaction rate has been investigated (Leino *et al.*, 1993). Although successful, it tends to increase the production of methyl disulfide, accentuate terpene oxidation, and promote the hydrolysis of "fruit" esters. These features diminish the floral character of the wine, but enhance oaky, honey, and smoky aspects, attributes typically associated with bottle-aged wine.

Because of the flavor modification associated with heat-induced hydrolysis, enzymic hydrolysis has received most of the attention. Of these, β-glycosidase preparations have been extensively studied. Because flavorants are often bound to variety of sugars, not just glucose, preparations with some α-arabinosidase, α-rhanmosidase, β-xylanosidase, and β-apiosidase activities are preferred. Commercial enzyme preparations are usually derived from filamentous fungi. Their enzymes are relatively insensitive to the acidic conditions in wine, unlike those produced by grape or yeast cells.

When employed, the enzymes are added at the end of fermentation. Glucose is a catalytic inhibitor. Action can be quickly terminated when desired by precipitation with bentonite. More efficient (but expensive) regulation can be achieved with immobilization in a column, through which the wine is passed (Caldini *et al.*, 1994).

Sur lies Maturation

Sur lies maturation is an old procedure enjoying renewed interest and application (Dubourdieu *et al.*, 2000). It has been traditionally used in Burgundian and some Loire white wines. It is now employed fairly extensively worldwide, and occasionally with red wines. In red wines, it can diminish astringency, but also reduce color intensity (Rodríguez *et al.*, 2005).

Sur lies maturation involves leaving the wine in contact with the lees for periods ranging from 3 to 6 months. This usually is in the same barrel in which fermentation occurred. The large surface area/volume ratio of small cooperage favors the diffusion of nutrients and flavorants from yeasts into the wine.

Some of the effects described below occur in all wines, except when racked early and frequently. Nevertheless, the effects are accentuated with the extended lees contact associated with *sur lies* maturation.

During maturation, dead and dying yeast cells begin to autolyze. Although slow under cool storage conditions and within the pH range of wine, autolysis releases many cellular constituents (Charpentier, 2000). Volatile yeast metabolites, such as ethyl octanoate and ethyl decanoate, can add a fruity element to the wine, whereas enzymatic reduction diminishes the sensory impact of carbonyl compounds, such as diacetyl. Lees may also reduce the sensory defect generated by 4-ethylphenols (Chassagne *et al.*, 2005). Susceptibility to oxidative browning may decrease due to the release of various amino acids (increasing the supply of oxidizable substrates). Although yeast hydrolytic enzymes can liberate glycosidically bound flavorants (Zoecklein *et al.*, 1997), they may also degrade other grape aromatics. However, the most significant influence of *sur lies* maturation appears to result from the liberation of mannoproteins (constituents of yeast cell walls).

Mannoproteins have a wide range of effects (Caridi, 2006). These glycoproteins can decolorize wine as well as soften the taste (they complex with and precipitate grape- or oak-derived tannins) (Vidal, S. *et al.*, 2004). Mannoproteins also diminish the likelihood of phenolic pinking (Dubourdieu, 1995). The latter feature is linked with the release of a hydrolytic breakdown product of yeast invertase (Dubourdieu and Moine, 1998a). In addition, mannoproteins minimize haze production from heat-unstable proteins (Dupin, McKinnon *et al.*, 2000), and favor the early completion of malolactic fermentation (Guilloux-Benatier *et al.*, 1995). Improved protein stability is especially valuable in reducing the need for bentonite fining. Mannoproteins also promote tartrate stability (Dubourdieu and Moine, 1998b). This clearly benefits wines consumed early, but its long-term value is less clear. Breakdown of the soluble complexes may eventually lead to in-bottle haze formation and/or tartrate deposition. Mannoproteins also reduce adsorption of aromatic compounds, such as fruit esters, into oak cooperage (Ramirez-Ramirez *et al.*, 2004). Finally, mannoproteins favor the volatilization of some compounds, but bind others (Lubbers *et al.*, 1994; Chalier *et al.*, 2007). Particularly important in this regard may be the absorption of objectionable volatile thiols such as ethanethiol and methanethiol.

To avoid the development of a low redox potential in the lees, and the consequential production of reduced-sulfur off-odors, the wine is periodically stirred (*battonage*). In barrels, mixing limits the development of marked redox potential differences throughout the wine. It also oxidizes hydrogen sulfide. Pressure exerted on the lees in large cooperage appears to promote the synthesis of hydrogen sulfide and several reduced organic sulfur compounds (Lavigne, 1995). Anaerobic conditions in the lees also favor the production of reduced-sulfur compounds. In contrast, lees in barrels appear to absorb or metabolize mercaptans. Nevertheless, *sur lies* maturation increases susceptibility to the development of a sunstruck odor (*goût de lumière*) (La Follette *et al.*, 1993). In addition, the release of significant amounts of glucose during autolysis (Guilloux-Benatier *et al.*, 2001) increases the risks of microbial spoilage. Thus, it is particularly crucial that the cooperage be properly cleaned and sanitized to avoid contamination with yeasts such as *Brettanomyces*. When conducted slowly, *battonage* appears to result in minimal oxygen accumulation in the wine (Castellari *et al.*, 2004). Thus, activation of acetic acid bacteria is minimized (combined with cool storage temperatures). This can avoid the need to add sulfur dioxide that can retard or prevent malolactic fermentation.

A distinctive form of *sur lies* maturation occurs during sparkling wine production. During long contact with the yeasts after the second, in-bottle fermentation, autolysis generates the synthesis of the toasty bouquet that characterizes fine sparkling wines. The release of mannoproteins from cell walls also stabilizes dissolved carbon dioxide, promoting the formation of the long-lasting chains of bubbles. In addition, mannoproteins slow the release of aromatic compounds from the wine (Dufour and Bayonove, 1999), as well as bind potentially undesirable thiols (Tominaga *et al.*, 2003). Another special example of *sur lies* maturation involves the solera aging of *fino* sherries.

Color Adjustment

The bevy of studies on micro-oxygenation with red wines (see below) is one expression of interest in color adjustment. The slow or periodic addition of oxygen favors the polymerization of anthocyanins with tannins. However, some varieties possess low tannin contents, or their tannins are not easily extracted. In these cases, adding enologic tannins may improve color depth and stability. Currently, the results of such investigations have been ambivalent, being encouraging with some cultivars, not with others.

More frequently, though, color adjustment implies its reduction, especially with white wines. All wines can be partially, or completely, decolored by ultrafiltration. Depending on the permeability characteristics of the membrane, ultrafiltration retains macromolecules above a specific size. With membranes of lower cutoff values (~500 Da), ultrafiltration also can remove phenolic pinking. The use of filters, with even lower cutoff values, can produce blush or white wines from red or rosé wines. The major factor limiting the more widespread use of ultrafiltration is its potential for removing important flavorants along with macromolecules.

Adding PVPP (polyvinylpolypyrrolidone) is another procedure used to remove brown or pink pigments (Lamuela-Raventós *et al.*, 2001). By binding tannins into large macromolecular complexes, PVPP facilitates their removal by filtration or centrifugation. Several white wines, such as 'Sauvignon blanc,' have a tendency to turn pinkish within days of oxygen exposure (Simpson, 1977), especially those assiduously protected from oxidation during and subsequent to crushing. Considerable oxygen may be required for development of full pink expression, however (Singleton *et al.*, 1979). Pinking is thought to occur when flavan-3,4-diols (leucoanthocyanins) slowly dehydrate to flavenes under reducing conditions. These can quickly oxidize to their corresponding colored flavylium-forms on exposure to oxygen. The use of moderate levels of sulfur dioxide can be used to limit pinking development (Simpson, 1977).

Other means of color removal involve the addition of casein or special preparations of activated carbon. With activated carbon, the simultaneous removal of aromatic compounds and the occasional donation of off-odors have limited its use. The addition of yeast hulls is also being studied as a means of removing brownish pigments from white wines (Razmkhab *et al.*, 2002).

Blending

Blending wine from different varieties is a long-established procedure. For example, Bordeaux red wines are rarely made from a single variety, up to five being permitted; Châteauneuf-du-Pape permits 13; even more may be blended in the production of *porto*. In the past, the blend often occurred at harvest, as the cultivars were dispersed more or less at random throughout the vineyard. Although rare today, it may still occur in some regions. Where varieties are fermented together (**cofermentation**), it is more common to combine the musts after crushing. Most commonly blending involves only red or white cultivars, though it may include both. Examples of the latter are traditional-style Chianti and some Côte Rôtie wines. The addition of white juice to red must has often been explained as a means of "softening" the wine. While possibly true, it may be more valuable in color enhancement (Gigliotti *et al.*, 1985). If the red wine is low in copigment factors, for which the white variety has an abundant supply, its addition could enhance coloration. Where this is valuable, but dilution with the juice undesired, use of only the skins is an alternative. It is more common,

though, to ferment the must of individual cultivars separately, followed by blending. It has the advantage of permitting selective blending, based on the attributes of the specific wines. Separating grapes into lots at harvest also permits the distinctive qualities of fruit from different sites or maturity grades to be realized. These can be blended later if desired.

The skill and experience of the blender are especially important in the production of fortified and sparkling wines. Without blending, the creation and maintenance of house-styles would be impossible. Computer-aided systems have been proposed to facilitate this important activity (Datta and Nakai, 1992). It is unlikely that they will soon replace the role of the blender.

The production of proprietary table wines also is largely dependent on the judicious combination of diverse wines. Consistency of character typically is more important than the vintage, variety, or vineyard origin. The skill of the blender is often amazing, given the number of wines potentially involved. It also helps that most consumers are ill-adept at remembering subtle sensory differences.

Blending also is used in the production of many premium table wines. In this case, however, the wines typically come from the same geographical region and are often from a single vineyard or holding and vintage. Limitations on blending are usually precisely articulated in the Appellation Control laws – the more prestigious the region, the more restrictive the legislation.

There is little to guide blenders other than past experience. Few studies have investigated the scientific basis of the blender's art. Those that have, have focused on methods predicting color, based on pigmentation of the base wines. Color is particularly significant due to its strong influence on quality perception. Blending diagrams may be developed using colorimeter readings. The diagrams are founded on the reflectance of the wines in the red, green, and blue portions of the visible spectrum. From these data, the lots required to achieve a desired color can be established. Pérez-Magariño and González-San José (2002) have proposed simple absorbance measurements for wineries without the equipment or software necessary to make the complex measurements normally required.

One of the major advantages associated with blending is improved flavor. In a classic study by Singleton and Ough (1962), similar pairs of wines, ranked comparably but recognizably distinct, were reassessed along with a 50:50 blend of the two. In no case was the blend ranked more poorly than the lower ranked of the base wines. More significantly, about 20% of the blends were ranked higher than either of the component wines (Fig. 8.3). Because the relationship between perceived intensity and flavorant concentration is nonlinear,

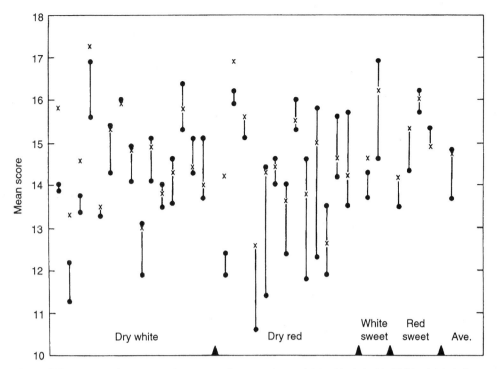

Figure 8.3 Mean quality scores of 34 pairs of wines compared (•) with their 50:50 blend (x), indicating the value of a complex flavor derived from blending. (Based on data from Singleton and Ough, 1962, from Singleton, 1990, reproduced by permission)

blending does not necessarily diminish the sensory characteristics of the individual components.

The origin of the improved sensory quality of blended wines is unknown. It probably relates to the increased flavor subtlety and complexity of the blend. This has been used to commercial advantage when wines of intense flavor have been blended with large volumes of neutral wine. The blend still tends to express the characteristics of the fragrant component. In addition, the negative perception of some off-odors often diminishes with dilution, as they approach their threshold of detection (see Chapter 11).

When blending should occur depends largely on the type and style of wine involved. In sherry production, fractional blending occurs periodically throughout maturation. With sparkling wines, blending occurs in the spring following harvest. At this point the unique features of the wines are often apparent. Red wines are also typically blended in the spring following the vintage. Important in the deliberation is not only the proportional amount of each wine, but also the amount of pressings to use. Wines from poor vintages customarily benefit more from the addition of extra press wine than do wines from good vintages. Press fractions contain a higher proportion of pigment and tannins than the free-run. The addition of pressings can also provide extra body and color to white wines. After blending,

the wine is often aged for several weeks, months or years before bottling.

Wines may not be blended for a number of reasons. Wines produced from grapes of especially high quality are usually kept and bottled separately to retain their distinctive attributes. For wines produced from famous vineyards, blending with wine from other sites, regardless of quality, would prohibit the owner from using the site name. This would significantly reduce the market value of the wine. With famous names, origin can be more important to sales than inherent quality.

Stabilization and Clarification

Stabilization and clarification involve procedures designed to produce a brilliantly clear wine with no flavor faults. Because the procedures can themselves create problems, it is essential that they be used discreetly, and only to the degree necessary to solve specific problems.

Stabilization

TARTRATE AND OTHER CRYSTALLINE SALTS

Tartrate stabilization is one of the facets of wine technology most influenced by consumer perception. The presence of even a few tartrate crystals can too easily be misinterpreted by consumers. As a consequence, considerable effort is expended in avoiding the formation of crystalline deposits in bottled wine. Stabilization is normally achieved by enhancing crystallization, followed by removal. Less frequently, it may be achieved by delaying or inhibiting crystallization.

Potassium Bitartrate Instability Juice is typically supersaturated with potassium bitartrate at crushing. As the alcohol content rises during fermentation, the solubility of bitartrate decreases. This induces the slow precipitation of potassium bitartrate (cream of tartar). Given sufficient time, the salt crystals precipitate spontaneously. In northern regions, low cellar temperatures may induce adequately rapid precipitation. Spontaneous precipitation is seldom satisfactory in warmer areas. Early bottling aggravates the problem. Where spontaneous precipitation is inadequate, refrigeration often achieves rapid and satisfactory bitartrate stability.

Because the rate of bitartrate crystallization is directly dependent on the degree of supersaturation, wines that are only mildly unstable may be insufficiently stabilized by cold treatment. In addition, protective colloids may retard crystallization (Lubbers et al., 1993). Typically, protective colloids have been viewed negatively, as their precipitation after bottling releases tartrates that could subsequently crystallize. Although this may be true for most protective colloids, some mannoproteins appear sufficiently stable to donate tartrate stability in bottled wine (Dubourdieu and Moine, 1998b). Mannoproteins are released during the latter stages of fermentation, and especially during maturation, as yeast cells in the lees autolyze. The addition of yeast-cell-wall enzymic digest (yeast hulls) can promote tartrate stability, without cold or other stabilization treatments.

In the absence of protective colloids, potassium bitartrate exists in a dynamic equilibrium between ionized and salt states.

Under supersaturated conditions, crystals form and eventually reach a critical mass that provokes precipitation. Crystallization continues until an equilibrium develops. If sufficient crystallization and removal occur

before bottling, bitartrate stability is achieved. Because chilling decreases solubility, it speeds crystallization.

In red wines, crystal formation is often associated with yeast cells (about 20% by weight) (Vernhet et al., 1999b). This compares with about 2% in white wines (Vernhet et al., 1999a). Potassium hydrogen tartrate crystals are also associated with smaller amounts of phenolic compounds (notably anthocyanins and tannins in red wines), as well as with polysaccharides, such as rhamnogalacturonans and mannoproteins.

Theoretically, chilling should establish bitartrate stability. However, charged particles in the wine can interfere with crystal initiation and growth. For example, positively charged bitartrate crystals are attracted to negatively charged colloids, blocking growth. The charge on the crystals is created by the tendency of more potassium than bitartrate ions to associate with the crystals early in growth (Rodriguez-Clemente and Correa-Gorospe, 1988). Crystal growth also may be delayed by the binding of bitartrate ions to positively charged proteins. This reduces the amount of free bitartrate and, thereby, the rate of crystallization. Because both bitartrate and potassium ions may bind with tannins, crystallization tends to be delayed more in red than in white wines. The binding of potassium with sulfites is another source of delayed bitartrate stabilization.

For cold stabilization, table wines are routinely chilled to near the wine's freezing point. Five days is usually sufficient at $-5.5°C$, but 2 weeks may be necessary at $-3.9°C$. Fortified wines are customarily chilled to between -7.2 and $-9.4°C$, depending on the alcoholic strength. The stabilization temperature can be estimated using the formula empirically established by Perin (1977).

$$\text{Temperature } (-°C) = (\% \text{ alcohol} \div 2) - 1$$

Direct seeding with potassium bitartrate crystals is occasionally employed to stimulate crystal growth and deposition. Another technique involves filters incorporating seed crystals. The chilled wine is agitated and then passed through the filter. Crystal growth is encouraged by the dense concentration of "seed" nuclei in the filter. The filter acts as a support medium for the crystal nuclei.

At the end of conventional chilling, the wine is filtered or centrifuged to remove the crystals. Crystal removal is performed before the wine warms to ambient temperatures.

Because of the expense of refrigeration, various procedures have been developed to determine the need for cold stabilization. None of the techniques appears to be sufficiently adequate. Potassium conductivity, although valuable, is too complex for regular use in most wineries.

Thus, empirical freeze tests are the most common means of assessing bitartrate stability. For details on the various tests, the reader is directed to Goswell (1981) and Zoecklein *et al.* (1995).

Reverse osmosis is an alternative technique to chilling, agitation, and the addition of nucleation crystals. With the removal of water, the increased bitartrate concentration augments crystallization and precipitation. After crystal removal, the water is added back to reestablish the original balance of the wine.

Electrodialysis is another membrane technique occasionally used for bitartrate stabilization (Uitslag *et al.*, 1996). Electrically charged membranes selectively prevent the passage of ions of the opposite charge. Passing wine between oppositely charged membranes can remove both anions and cations. In practice, though, potassium ions are more rapidly removed than tartrate ions. This not only limits crystallization (requires both ions), but tends to lower the pH. Reducing the potassium content by about 10% seems to be effective in achieving adequate bitartrate stability. Because current membranes are not sufficiently selective, sulfur dioxide and other constituents may also be removed.

Another technique particularly useful for wines with high potassium contents is ion exchange. Passing the wine through a column packed with sodium-containing resin exchanges the sodium for potassium. Sodium bitartrate is more soluble than the potassium salt, and is therefore much less likely to precipitate. Although effective, ion exchange is not the method of choice. Not only is it prohibited in certain jurisdictions, for example member countries of the EU, but it also increases the sodium content of the wine. The high potassium/low sodium content of wine is one of its health benefits.

If the wine is expected to be consumed shortly after bottling, treatment with metatartaric acid is an inexpensive means of establishing short-term tartrate stability. Metatartaric acid is produced by the formation of ester bonds between hydroxyl and acid groups of tartaric acid. The polymer is generated during prolonged heating of tartaric acid at 170°C. When added to wine, metatartaric acid restricts potassium bitartrate crystallization, and interferes with the growth of calcium tartrate crystals. As metatartaric acid slowly hydrolyzes back to tartaric acid, the effect is temporary. At storage temperatures between 12 and 18°C, it may be effective for about 1 year. Because hydrolysis is temperature dependent, the stabilizing action of metatartaric acid quickly disappears above 20°C. If this treatment is used, metatartaric acid is added just before bottling.

Calcium Tartrate Instability Instability caused by calcium tartrate is more difficult to control than that induced by potassium bitartrate. Fortunately, it is less common. Calcium-induced problems usually arise from the excessive use of calcium carbonate in deacidification, but they also may arise from the use of cement cooperage, filter pads, and fining agents.

Several organic acids can significantly influence the crystallization of calcium tartrate (McKinnon *et al.*, 1995). For example, malolactic fermentation removes a major inhibitor of crystallization (malic acid) and, thus, promotes earlier calcium tartrate stability. Conversely, malic acid may be added to sparkling wine (with the *dosage*) to retard calcium tartrate crystallization after bottling. Sparkling wines, because the grapes are pressed whole, contain little polygalacturonic acid, a potent retardant of calcium tartrate crystallization.

Calcium tartrate stabilization is more difficult because precipitation is not activated by chilling. Crystal growth and precipitation occur optimally between 5 and 10°C. It can take months for spontaneous stability to develop. Seeding with calcium tartrate crystals, while deacidifying with calcium carbonate, greatly enhances precipitation (Fig. 8.4). Because the formation of crystal nuclei requires more free energy than crystal growth, seeding circumvents the major limiting factor in the development of stability. A racemic mixture of calcium tartrate seed nuclei, containing both L and D isomers, is preferred. The racemic mixture is about one-eighth as soluble as the naturally occurring L-tartrate salt. This may result from the more favorable (stable) packing of both isomers together in crystals (Brock *et al.*, 1991). The slow conversion of the L form to the D form is one of the major causes of the calcium tartrate formation in bottled wine. Because clarification removes the "seed" crystals that promote crystallization, wine filtration should be delayed until tartrate stability has been achieved. Protective colloids such as soluble proteins

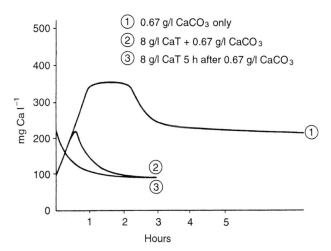

Figure 8.4 Calcium levels after calcium carbonate (CaCO₃) deacidification with and without calcium tartrate (CaT) seeding. (From Neradt, 1984, reproduced by permission)

and tannins can restrict crystal nucleation, but they do not inhibit crystal growth (Postel, 1983).

If protective colloids are a problem, agar may be added. Agar, an algal polysaccharide, tends to neutralize the charge on protective colloids. This eliminates their protective property, allowing colloid-tartrate complexes to dissociate and the colloids to precipitate.

Alternately, calcium content may be directly reduced through ion exchange. Because of the efficiency of ion removal, typically only part of the wine needs to be treated. The treated sample is then mixed with the main volume of the wine. Treating only a small portion of the wine minimizes the flavor loss often associated with ion exchange.

Other treatments that show promise are the addition of stable colloids, such as pectic and alginic acids. These restrict crystallization and keep calcium tartrate in solution (Wucherpfennig *et al.*, 1984).

Other Calcium Salt Instabilities   Occasionally, crystals of calcium oxalate form in wine. The development occurs late, commonly after bottling. The redox potential of most young wines stabilizes the complex formed between oxalic acid and metal ions, such as iron. However, as the redox potential of wine rises during aging, ferrous oxalate changes into the unstable ferric form. After dissociation, oxalic acid may bond with calcium, forming calcium oxalate crystals.

Oxalic acid is commonly derived from grape must, but small amounts may originate from iron-induced structural changes in tartaric acid. Oxalic acid can be removed by blue fining early in maturation (Amerine *et al.*, 1980), but avoiding the development of high calcium levels in the wine is preferable.

Other potential troublesome sources of crystals are saccharic and mucic acids. Both are produced by the pathogen *Botrytis cinerea*. They may form insoluble calcium salts. The addition of calcium carbonate for bitartrate stability often induces their crystallization, precipitation, and separation before bottling.

PROTEIN STABILIZATION

Although a less common cause of wine rejection than crystal formation, protein haze can still cause considerable economic loss. Protein haze results from the clumping of dissolved proteins into light-dispersing particles. Heating accelerates the process, as does reaction with tannins and heavy metals.

The majority of proteins suspended in wine have an isoelectric point (p*I*) above the pH range of wine. The isoelectric point is the pH at which a protein is electrically neutral. Consequently, most soluble proteins in wine possess a net positive charge, generated by ionization of amino groups. The similar charge on proteins slows clumping, while Brownian movement and association (hydration) with water delay settling. In contrast, denaturation coalesces proteins, producing haze.

The proteins primarily involved in haze production are only now being understood. The situation was confusing because protein instability was poorly correlated with total protein content. Two types of proteins appear to be principally involved in haze production (Waters *et al.*, 1996b). Both are pathogenesis-related (PR) proteins (Waters *et al.*, 1996c; Dambrouck *et al.*, 2003). One group consists of thaumatin-like proteins, whereas the other includes a series of chitinases. They often constitute the majority of soluble proteins in pressed juice and wine. Their occurrence typically ranges from 50 to 100 mg/liter, but can vary from 200 to 250 mg/liter in 'Muscat of Alexandria' and 'Sauvignon blanc,' to 62 mg/liter and 31 mg/liter for 'Pinot noir' and 'Shiraz,' respectively (Pocock *et al.*, 2000). Mechanical harvesting, associated with prolonged transport to, or storage at the winery, can permit increased protein extraction and double the bentonite needed for stabilization (from 0.5 g/liter to 1.0 g/liter) (Pocock *et al.*, 1998). Another factor in haze production appears to be sulfate concentration (Pocock *et al.*, 2007).

Although the principal role of these proteins in grapes is to protect the maturing fruit from fungal infection, their production in post-*véraison* healthy fruit suggests a possible developmental role (Tattersall *et al.*, 2001). Despite being termed "unstable," it is their acid stability, resistance to proteolytic action, and minimal bonding with tannins that creates the problem. Their stability permits them to survive throughout fermentation and maturation, denaturing slowly during prolonged in-bottle aging, producing what is termed instability.

These haze-inducing proteins are unrelated to those involved in haze production in beer (Siebert *et al.*, 1996). In beer, proteins possessing a high proline content combine selectively with polyphenolics possessing two- or three-vincinal OH groups (Siebert and Lynn, 1998). Such complexes in wine would precipitate long before bottling.

Yeast mannoproteins and grape arabinogalactan–protein complexes may promote heat-induced protein haze, although specific members in both groups can reduce protein-induced haze formation (Pellerin *et al.*, 1994). Particularly interesting is the action of mannoproteins in reducing the size of haze particles to the threshold level of human detection (Waters *et al.*, 1993; Dupin, Stockdale *et al.*, 2000). Although commercial mannoprotein preparations are available, improved extraction procedures (Dupin, McKinnon *et al.*, 2000), or construction of yeast strains releasing increased amounts of mannoproteins during fermentation have

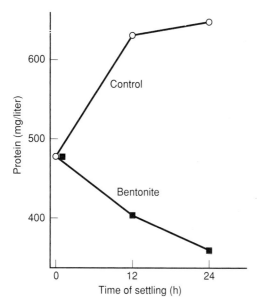

Figure 8.5 Total soluble protein in 'Gewürztraminer' must settled at 20°C for 24h in the presence or absence of 0.5g/liter bentonite. (From Tyson *et al.*, 1982, reproduced by permission)

been proposed (Dupin, McKinnon *et al.*, 2000; Dupin, Stockdale *et al.*, 2000) as alternative solutions.

A number of procedures have been developed to achieve protein stability. The most common involves the addition of bentonite (Fig. 8.5). Because of the abundance of soluble cations associated with bentonite, extensive exchange of ions can occur with ionized protein amino groups. By weakening their association with water, cations make proteins more liable to coalesce and precipitate. Flocculation and precipitation are further enhanced by adsorption onto the negatively charged plates of bentonite. Sodium bentonite is preferred because it separates more readily into individual silicate plates. This generates the largest surface area of any clay and, therefore, the greatest potential for cation exchange and protein adsorption. Regrettably, bentonite generates considerable sediment and associated wine loss. Although much of the wine lost can be recovered by vacuum rotary drum filtration, oxidation during the process can reduce its quality. Despite these limitations, bentonite is still the preferred fining agent for protein stabilization.

Because haze-inducing proteins can be desorbed from bentonite by increasing the pH, there is interest in developing a continuous-flow stabilization system. Bentonite immobilization would not only permit regeneration of the bentonite, but also dramatically reduce the amount of wine lost with currently practiced batch fining. An alternate procedure being investigated is in-line dosing, and bentonite removal with centrifugation (Muhlack *et al.*, 2006).

Other fining agents, such as tannins, are occasionally used in lieu of bentonite. The addition of tannins is ill-advised as they can leave an off-odor and generate an astringent mouth-feel in white wines. However, when immobilized in porous silicon dioxide, tannic acid causes minimal flavor modification or wine loss (Weetal *et al.*, 1984). Kieselsol, a colloidal suspension of silicon dioxide, has occasionally been used to remove proteins. Wines also may be protein-stabilized through heat treatment, followed by filtration or centrifugation.

Ultrafiltration has been investigated as an alternative to bentonite or other types of fining (Hsu *et al.*, 1987). It has the advantage of minimizing wine loss, and the need for a final polishing centrifugation or filtration. Although generally applicable for white wines, ultrafiltration results in excessive color and flavor loss in red wines.

Traditionally, protein stability has been tested by treating wine samples to 80°C for 8 h. Studies by Pocock and Waters (2006) suggest that two hours is fully adequate. In a comparison of various protein stabilization tests, Dubourdieu *et al.* (1988) recommended exposing wine samples to 80°C for just 30 min. After cooling, the sample is observed for signs of haziness, either subjectively (by eye), or objectively (nephelometry or optical density). If found to be unstable, samples of the wine are treated and retested to determine to what degree treatment is required.

POLYSACCHARIDE REMOVAL AND STABILITY

Pectinaceous and other mucilaginous polysaccharides can cause difficulty during filtration, as well as inducing haze formation. Polysaccharides can also act as protective colloids, binding with other suspended materials and slowing or preventing their precipitation. For example, negatively charged pectins collect around positively charged grape solids. In addition, multiple hydrogen bonds formed between water and pectins help these complexes remain in suspension.

Pectin levels can be reduced by the addition of macerating (pectolytic) enzyme preparations. These contain a mixture of enzymes, including a pectin lyase. They split the pectin polymer into simpler, noncolloidal, galacturonic acid subunits. In so doing, positively charged areas of grape colloids are exposed and can bind to the negatively charged surfaces of other colloids. As these complexes increase in mass, they are more likely to precipitate and are more easily removed during fining.

Other grape-derived polysaccharides, such as arabinans, and galactans, have little effect on haze formation or filtration. The same is true for yeast-derived mannans. However, their removal can be beneficial in producing denser lees. This reduces wine loss during racking.

In contrast, β-glucans present in botrytized juice can cause serious filtration problems, even at low

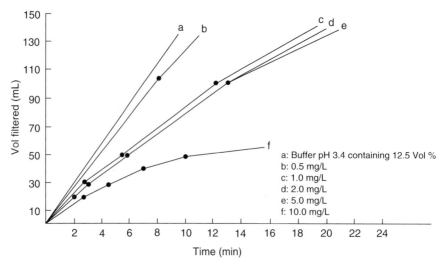

Figure 8.6 Influence of glucan concentration on the filterability of an alcoholic solution through a 0.45-μm membrane. (From Villettaz *et al.*, 1984, reproduced by permission)

concentrations (Fig. 8.6). This is especially serious in highly alcoholic wines, in which ethanol induces aggregation of the glucans. A Kieselsol–gelatin mixture is apparently effective in removing these mucilaginous polymers. Alternatively, the wine may be treated with a formulation of β-glucanases (Villettaz *et al.*, 1984). The enzymes hydrolyze the polymer, destroying both its protective colloidal property and filter-plugging action.

A potential disadvantage of some pectinase preparations is the synthesis of excessive levels of vinylphenols. The production of a phenolic off-odor is associated with the cinnalyl esterase (CE) activity of some *Aspergillus niger* enzyme preparations (Chatonnet, Barbe *et al.*, 1992). The enzymic activity releases *p*-coumaric and ferulic acid. In the presence of some *Saccharomyces cerevisiae* strains, the phenolic acids are transformed into their corresponding vinylphenols (4-vinylphenol and 4-vinylguaiacol). At sufficiently high concentrations they generate a phenolic off-odor. Another disadvantage, specifically with excessive exposure to pectic enzymes, is the production of fine particulate material derived from grape skins. These can cause clarification problems.

Because most commercial enzyme preparations are not highly purified, predicting their effects under specific conditions is difficult (Wightman *et al.*, 1997; Laperche and Görtges, 2001). The only sure way is with individual experimentation under onsite conditions.

TANNIN REMOVAL AND OXIDATIVE *CASSE* (HAZINESS)

Tannins may be directly and indirectly involved in haze development. After exposure to oxygen, tannins oxidize and polymerize into brown, light-diffracting colloids, potentially causing **oxidative *casse***. Shortly after crushing, these reactions normally occur slowly and

nonenzymatically. Depending on the timing and degree of oxidation, tannin oxidation can result in a loss of color intensity, a shift in hue, and an enhancement in long-term color stability. The addition of sulfur dioxide limits oxidation through its antioxidant and antienzymatic properties. However, moldy fruit, contaminated with fungal polyphenol oxidases (laccases), is particularly susceptible to oxidative *casse*. Because laccases are poorly inactivated by sulfur dioxide, pasteurization may be the only convenient means of protecting the juice from oxidative *casse*. Grapes free of fungal infection rarely develop oxidative *casse*. Because *casse* usually develops early during maturation and precipitates before bottling, it does not cause in-bottle clouding.

Chilling wine to achieve bitartrate stability may provoke formation of a protein–tannin haze. Filtration before the wine warms removes the protein–tannin complexes before their dissociation, preventing their reformation post-bottling.

Tannin stability is normally achieved by adding fining agents such as gelatin, egg albumin, or casein. The predominantly positive charge on these proteinaceous agents attracts negatively charged tannins. Their interaction produces large protein–tannin complexes. Their formation is a function of the balance between the potential binding sites of both tannins and proteins. Excess in either tends to diminish binding. Once formed, the complexes may be removed by filtration or centrifugation, if early bottling is desired. Otherwise, adequate spontaneous sedimentation normally occurs during maturation. The removal of excess tannins reduces a major source of astringency, generates a smoother mouth-feel, reduces the likelihood of oxidative *casse*, and limits the formation of sediment following bottling.

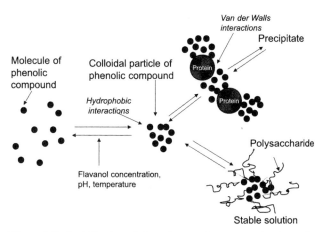

Figure 8.7 Stabilization of phenolic, proteinaceous and polysaccharides (mannoproteins). Protein and phenolic compounds tend to precipitate whereas phenolic and polysaccharides stay in solution. (From Charpentier, 2000, reproduced by permission)

In white wines, addition of PVPP (polyvinylpolypyrrolidone) is a particularly effective means of removing tannin subunits and dimers. Ultrafiltration also may be used to remove excess tannins and other polyphenolic compounds. Ultrafiltration is seldom used with red wines. The filter simultaneously removes important flavorants and anthocyanins.

Additional, but infrequent sources of phenolic instability include oak chips or shavings used to rapidly develop an oak character (Pocock *et al.*, 1984). The accidental incorporation of excessive amounts of leaf material in the grape crush is another atypical source of phenolic instability (Somers and Ziemelis, 1985). Both can generate in-bottle precipitation if the wine is bottled early. These problems can be avoided during maturation by permitting sufficient time for spontaneous precipitation. The instability associated with oak-chip use results from the overextraction of ellagic acid. The **phenolic deposit** produced consists of a fine precipitate of off-white to fawn-colored ellagic acid crystals. A **flavonol haze**, associated with the excessive presence of leaf material during the crushing of white grapes, is produced by the formation of fine, yellow, quercetin crystals (Somers and Ziemelis, 1985). The excessive use of sulfur dioxide has also been associated with cases of phenolic haze in red wines.

Many premium-quality red wines develop a **tannin sediment** during prolonged in-bottle aging. This potential source of haziness is typically not viewed as a fault. Individuals who customarily consume aged wines know its origin. They often consider it to be a quality indicator (incorrectly assuming that fining is inherently prejudicial to quality).

Some of the interactions in wine stabilization are illustrated in Fig. 8.7.

METAL *CASSE* STABILIZATION

A number of heavy metals form insoluble salts and induce additional forms of haziness (*casse*). Although occurring much less frequently than in the past (largely due to the replacement of iron, copper, bronze, and brass fittings with stainless steel), metal *casse* occasionally still causes problems.

The most important metallic ions involved in *casse* formation are iron (Fe^{3+} and Fe^{2+}) and copper (Cu^{2+} and Cu^+). They may be derived from grapes, soil contaminants, fungicidal residues, or winery equipment. Most metal ions so derived are lost during fermentation by coprecipitation with yeast cells. Troublesome concentrations of metal contaminants usually are associated with pickup subsequent to vinification. Corroded stainless steel, improperly soldered joints, unprotected copper or bronze piping, and tap fixtures are the prime sources of contamination. Additional sources may be fining and decoloring agents, such as gelatin, isinglass, activated carbon, and cement cooperage.

Ferric *Casse* Two forms of ferric *casse* are known – white and blue (Fig. 8.8). **White *casse*** is most frequent in white wine. It forms when soluble ferrous phosphate ($FePO_4$) is oxidized to insoluble ferric phosphate ($Fe_3(PO_4)_2$). The white haziness that results may be due solely to ferric phosphate, or to a complex between it and soluble proteins. In red wines, the oxidation of ferrous ions (Fe^{2+}) to the ferric state (Fe^{3+}) can result in the formation of **blue *casse***. Ferric ions form insoluble particles with anthocyanins and tannins. The oxidation of ferrous to ferric ions usually occurs when the wine is exposed to air. In an unstable wine, sufficient oxygen may be absorbed during bottling to induce clouding.

The development of ferric *casse* is dependent on both the wine's metallic content and its redox potential. Its occurrence is also affected by pH, temperature, and the level of certain acids. White *casse* forms only below pH 3.6 and is generally suppressed at cool temperatures. In contrast, blue *casse* is accentuated at cold temperatures. The frequency of white *casse* increases sharply as the iron concentration rises above 15–20 mg/liter. Critical iron concentrations for the production of blue *casse* have been more difficult to estimate. Its occurrence is markedly affected by the wine's phosphate content and traces of copper (1 mg/liter). In addition, citric acid can chelate ferric and ferrous ions, reducing their effective (free) concentration in wine. Correspondingly, citric acid has occasionally been added to wine (~120 mg/liter) to restrict the occurrence of ferric *casse* (Amerine and Joslyn, 1970).

Wines may be directly stabilized against ferric *casse* by iron removal. For example, the addition of phytates, such as calcium phytate, selectively removes iron ions.

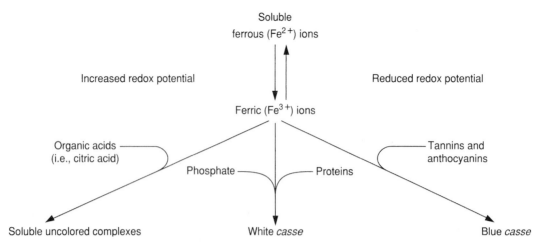

Figure 8.8 Iron-induced *casse* formation.

EDTA (ethylenediaminetetraacetic acid), pectinic acid, and alginic acid can be used to remove iron and copper ions. Removal with ferrocyanide is probably the most efficient method, as it precipitates most metal ions, including iron, copper, lead, zinc, and magnesium. The addition of potassium ferrocyanide is known as **blue fining**. Blue fining is prohibited in many countries and is strictly controlled where permitted. Filtration removes the insoluble metal–ferrocyanide complex.

Ferric *casse* may also be controlled by the addition of agents that limit the flocculation of insoluble ferric complexes. Gum arabic acts in this manner. It functions as a protective colloid, restricting haze formation. Because gum arabic limits the clarification of colloidal material, it can only be safely applied after the wine has undergone all other stabilization procedures.

Copper *Casse* While iron *casse* can form on exposure to oxygen, copper *casse* forms solely in the absence of oxygen. It develops only after bottling and is associated with a decrease in redox potential. Light exposure speeds the reduction of copper, critical in *casse* development. Sulfur dioxide is important, if not essential, as a sulfur source in copper *casse* formation. In a series of incompletely understood reactions, involving the generation of hydrogen sulfide, cupric and cuprous sulfides form. The sulfides produce a fine, reddish-brown deposit, or they flocculate with proteins to form a reddish haze. Copper *casse* is particularly a problem in white wines, but can also cause haziness in rosé wines. Wines with copper contents greater than 0.5 mg/liter are particularly susceptible to copper *casse* development (Langhans and Schlotter, 1985).

MASQUE

Occasionally, a deposit called *masque* forms on the inner surface of sparkling wine bottles. It results from

the deposition of material formed by the interaction of albumin, used as a fining agent, and fatty acids (Maujean *et al.*, 1978). Riddling and disgorging used to eliminate yeast sediment does not remove *masque*. *Masque* is a problem only with traditionally produced (*méthode champenoise*) sparkling wines, in which the wine is sold in the same bottle as used for the second fermentation.

LACQUER-LIKE BOTTLE DEPOSITS

In the 1990s, there was an increase in the worldwide incidence of a lacquer-like deposit in bottles of red wine. It appeared especially in higher-priced wines. The deposit develops within a few years and can cover the entire inner surface of the bottle. It is not associated with a reduction in wine quality, but can impede the sale of wine due to the perception that the wine may be turbid.

The thin, film-like layer results from the deposition of a complex of tannins, anthocyanins and proteins (Waters *et al.*, 1996a). The protein component was unexpected because the high tannin content of red wines has usually been thought to induce the complete precipitation of soluble wine proteins before bottling. Although the mechanism of deposit formation is unknown, several factors can reduce its occurrence. These include the use of bentonite (>50 g/liter) and cold stabilization (at −4 °C for 5 days, followed by centrifugation at −4 °C to remove the insoluble material).

MICROBIAL STABILIZATION

Microbial stability is not necessarily synonymous with microbial sterility. At bottling, wines may contain considerable numbers of viable, but dormant, microorganisms. Under most situations, they provoke no stability or sensory problems.

The simplest procedure for conferring limited microbial stability is racking. Racking removes cells that have fallen out of the wine by flocculation, or have coprecipitated with tannins and proteins. The sediment includes both viable and nonviable microorganisms. The latter slowly undergo autolysis and release nutrients that can favor subsequent microbial growth. Cold temperatures help maintain microbial viability, but retard or prevent growth.

For long-term microbial stability, especially with sweet wines, the addition of antimicrobial compounds or sterilization is required. The antimicrobial agent most frequently used is sulfur dioxide. It may be added at various times during wine production, but almost always after fermentation. Concentrations of 0.8–1.5 mg/liter (molecular) sulfur dioxide inhibits the growth of most yeasts and bacteria. Nevertheless, the precise amounts will depend on the temperature, ethanol content, nutrient availability, concentration of sulfur-binding compounds, the microbial population, and the species and strain involved. For example, spoilage yeasts such as *Saccharomycodes ludwigii*, *Zygosaccharomyces bailii*, and *Brettanomyces* spp. often require >3 ppm molecular SO_2 for limiting the initiation of contamination (Thomas and Davenport, 1985). Current data suggest that standard levels of sulfur dioxide are inadequate to control acetic acid bacteria in wine (see Romano and Suzzi, 1993); thus, the importance of cool storage during maturation. The free sulfur dioxide content should be determined about 24 h after addition – when an equilibrium has developed between the various free- and bound-forms of sulfur dioxide. The proportional molecular SO_2 content can be calculated from the total sulfur dioxide content and the wine's pH. Table 8.1 illustrates the amount of free sulfur dioxide required to have approximately 0.8 mg/liter available within the normal range of wine pH values. Figure 8.9 illustrates the relative toxicity of sulfur dioxide to several wine microbes.

Although generally less effective than sulfur dioxide, sorbic acid (200 mg/l) is effective against several spoilage yeasts. It is often used to control yeast contamination in sweet wines – except where *Zygosaccharomyces bailii* is involved (Rankine and Pilone, 1973). Its effectiveness is largely limited to low pH conditions, where most of the acid exists in its toxic, undissociated state. Because sorbic acid binds SO_2, its addition reduces the activity of sulfur dioxide. Sorbic acid is relatively ineffective in inhibiting bacterial growth. Thus, its use is limited to wines of low pH, where conditions are unfavorable for bacterial activity. In addition, some lactic acid bacteria can metabolize sorbic acid to sorbyl alcohol. Upon esterification with ethanol, it generates 2-ethoxyhexa-3,4-diene. Its accumulation to above threshold values produces an intense geranium-like off-odor.

Benzoic acid and sodium benzoate were once employed as yeast inhibitors, but their general ineffectiveness and taste modification have eliminated their use.

Table 8.1 Amount of free sulfur dioxide required to obtain a concentration of 0.8 ppm molecular SO_2 as a function of pH

pH	Free SO_2
2.8	9.7
2.9	11
3.0	13
3.1	16
3.2	21
3.3	26
3.4	32
3.5	40
3.6	50
3.7	63
3.8	79
3.9	99
4.0	125

Source: Data from Smith, 1982

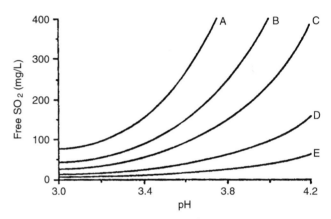

Figure 8.9 The concentration of free and molecular SO_2, over the pH range of wine, necessary to reduce the population of several commonly occurring wine microorganisms by 10^4 cells/ml in 24 h. A, *Lactobacillus plantarum* (4.0 mg/liter); B, *Zygosaccharomyces baillii* (2.5 mg/liter); C, *Lactobacillus plantarum* (1.5 mg/liter) (start); D, *Brettanomyces* spp., *S. cerevisiae* (0.625 mg/liter); and E, *Pichia membranaefaciens* (0.25 mg/liter). (From Sneyd *et al.*, 1993, reproduced by permission)

Dimethyl dicarbonate (DMDC) can effectively sterilize wine, if used just before bottling. DMDC rapidly decomposes to carbon dioxide and methanol, neither leaving a residue nor modifying the wine's sensory attributes (Calisto, 1990). Its action is little affected by pH. Regrettably, its action against *Saccharomyces cerevisiae* and *Oenococcus oeni*, poor solubility, and corrosive nature limit its more general use. In the absence of sulfur dioxide or DMDC, bottled wines can only be securely stabilized against microbial growth by physical means, namely, pasteurization and filter sterilization.

Pasteurization is the older of the two techniques. It has the advantage of promoting protein and copper *casse* stabilization, by denaturing and precipitating colloidal proteins. Although pasteurization may generate increased amounts of protective colloids, cause slight decolorization, and modify wine fragrance, it does not influence the aging process of phenols in wine (Somers and Evans, 1986).

Pasteurization of wine usually occurs for shorter periods or at lower temperatures than typical for products such as milk. This is possible because the low pH and ethanol content of wine markedly depresses the thermal resistance of yeasts and bacteria. Barillère *et al.* (1983) indicate that approximately 3 min at 60 °C should be sufficient for a wine of 11% ethanol. Flash pasteurization at 80 °C usually requires only a few seconds. Sulfur dioxide reduces still further the need for heating. High temperatures markedly increase the proportion of free SO_2 in wine. Although pasteurization destroys most microbes, it does not inactivate the endospores of *Bacillus* species. On rare occasions these bacteria may induce wine spoilage.

Partially because of the complexities of establishing the most appropriate time and temperature conditions for pasteurization, membrane filters have replaced pasteurization in most situations. Filters also result in few physical or chemical disruptions to the sensory characteristics of wine. Membrane filters with a pore size of 0.45 μm or less are adequate for wine sterilization.

Wine sterilization requires the simultaneous use of measures to avoid recontamination. This involves sterilizing all parts of the bottling line and the use of sterile bottles and corks. Sulfur dioxide is commonly added before wines are pasteurized or sterile-filtered to confer protection against oxidation.

OXIDATION CONTROL DURING MATURATION

In most situations, exposure to air during maturation is avoided. Nevertheless, despite a winemaker's best intentions, oxygen exposure does occur. Cellar activities such as tartrate stabilization, refrigeration, and bottle filling all expose wine to air (Vidal *et al.*, 2001; Castellari *et al.*, 2004). Additional oxygen uptake is associated with racking or barrel topping, as well as diffusion through joints, where the head and bilge staves meet (Moutounet *et al.*, 1998). Whether oxygen can permeate directly through barrel staves in significant amounts is a contentious issue, with little solid evidence either way.

Regardless of origin, limited oxygen uptake, either periodically, or slowly over a protracted period, apparently results in no measurable oxygen buildup. The oxygen is rapidly consumed in oxidation reactions. Quinone production is probably the principal oxidation by-product. Polymerization and restructuring regenerate phenol groups capable of consuming additional oxygen (Fig. 6.11). These reactions also favor color stabilization in red wines (by encouraging anthocyanin–tannin

polymerization). In addition, tannin–tannin complexes reduce the bitterness and astringency of tannins, donating a smoother mouth-feel. Furthermore, acetaldehyde produced in the oxidation of ethanol can combine with other wine constituents, notably sulfur dioxide, anthocyanins and tannins. This limits acetaldehyde accumulation, avoiding development of an oxidized attribute.

During maturation, oxidation is limited by restricting air access, as well as by adding SO_2. However, the benefits of limited oxygen ingress for red wine maturation are changing attitudes. Corresponding, there is considerable interest in regulating oxygen uptake. This is especially so for producers who wish to avoid oak flavors, but still desire the perceived benefits of minimal oxygen uptake. This has led to the term **micro-oxygenation**, and procedures for regulating oxygen uptake via inert cooperage (Parish *et al.*, 2000; Lemaire *et al.*, 2002; Jones *et al.*, 2004).

Micro-oxygenation may involve the use of high-density polyethylene cooperage (Flecknoe-Brown, 2005). They can be designed to have a relatively precise oxygen diffusion rate. Oxygen uptake in oak cooperage, though generally thought to be negligible, appears to occur through the ends of the barrel or via the bung hole (if not tightly bunged). The rates most often quoted are estimates, and could vary considerably, due to differences in construction, wood porosity, wood thickness, and the number of repeat uses.

To facilitate micro-oxygenation, silicone tube diffusers may be used to achieve oxidative control in large cooperage. Silicone tubing is about a thousand times more oxygen-permeable than other plastics. Oxygen ingress can be controlled, and kept below the rate of consumption, by adjusting the length, thickness and oxygen pressure in the tubing. Instruments, such as Microdue® or Parsec®, can also supply oxygen at various rates, via microporous diffusers. Castellari *et al.* (2004) consider that about 5 ml O_2/liter/month (at 15–20 °C) is roughly equivalent to barrel uptake, and lower than the rate of oxygen consumption in red wine. Whether these procedures adequately mimic the limited oxidation in barrels is a moot point (du Toit *et al.*, 2006). What micro-oxygenation clearly offers is better control.

After alcoholic (or malolactic) fermentation, sulfur dioxide is frequently added for its joint antimicrobial and antioxidative properties, especially with white wines. Sulfur dioxide reduces the oxidative browning of phenolics, helps convert brown quinones back to colorless diphenols, limits the participation of quinones in further oxidative reactions (by directly binding with quinones), as well as bleaches brown pigments. Sulfur dioxide also appears to protect 3-mercaptohexanol, an important varietal aromatic compound, from oxidative degradation (Blanchard *et al.*, 2004).

The free sulfur dioxide content of wine typically falls shortly after addition. This primarily involves binding with carbonyl compounds existent in the wine. Further, but slow, decreases involve inadvertent oxygen uptake by the wine, irreversible binding to proteins, reduction to hydrogen sulfide, binding with ethanol, and oxidation to sulfate.

Oxygen ingress can favor the development of several types of hazes, notably ferric and oxidative *casse*. Oxygen uptake also reduces the concentration of free sulfur dioxide, resulting in the release of acetaldehyde and the redevelopment of color in bleached anthocyanins. The combined effects of reduced sulfur dioxide content and the ingress of oxygen could favor the reactivation of dormant spoilage microbes in wine. For example, micro-oxygenation has been associated with an increase in *Brettanomyces*, and the presence of a barnyardy/medicinal attribute (du Toit *et al.*, 2006). Although micro-oxygenation is unlikely to lead to vinegarization, the activation of acetic acid bacteria could give the wine a sharp vinegary aspect. Thus, it is important that maturation occurs at cool (12–16 °C) temperatures, to limit the activity of acetic acid bacteria (Vivas *et al.*, 1995).

Fining

Fining is commonly used to accelerate the spontaneous precipitation of suspended material. Fining agents bind to or adsorb particulate matter. The aggregates are generally sufficiently large to precipitate quickly. If not, removal can be achieved by centrifugation or filtration. In addition to facilitating clarification, fining may help stabilize wines against haze formation (by precipitating the compounds involved in haze production), eliminate certain off-odors, and remove excessive amounts of bitter and astringent phenolics. Its effects on removing pesticide residues are often pesticide-specific, as well as significantly affected by pesticide solubility (Ruediger *et al.*, 2004). Although fining may not completely remove undesired compounds, it certainly facilitates negating their sensory impact.

Because fining is an aid to, not a replacement for, spontaneous stabilization, it should be used only to the extent necessary. It is important to avoid sensory disruption by minimizing unnecessary changes to the chemical and physical balance of the wine. Figure 8.10 illustrates the potential of several fining agents to produce aroma changes through the removal of aromatic compounds. Fining should also be conducted as quickly as possible to avoid oxidation.

Tests to determine the need for fining are beyond the scope of this volume, but are detailed in references such as Zoecklein *et al.* (1995). Such testing is essential to

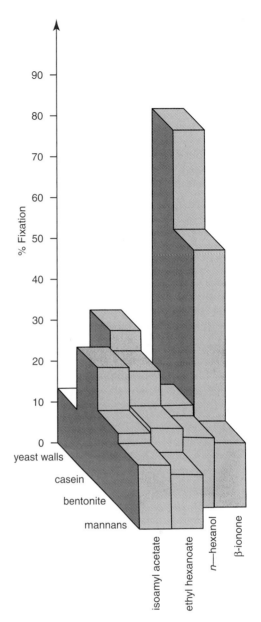

Figure 8.10 Percent removal of several aromatic compounds during fining with bentonite or 1% casein, mannans, or yeast cell walls. (Reprinted with permission from Voilley *et al.*, 1990. Copyright 1990 American Chemical Society)

determine the specific effects of the proposed fining or clarification procedure on the wine involved. It is impossible to precisely predict results. Thus, the information given here is only a guide to their prospective use.

ACTIVATED CARBON (CHARCOAL)

Activated carbon is purified charcoal, treated physically or chemically to generate microfissures that vastly increase its adsorptive surface area. The large surface area (between 500 and 1500 m²/g) and electrical charge effectively adsorb a wide range of polar compounds, especially phenols and their derivatives. Activated carbon is used primarily to decolorize wine or remove off-odors. Different preparations are used for each application. Decolorizing carbons often selectively remove flavonoid monomers and dimers. Larger polymers poorly penetrate the micropores of the carbon fragments (Singleton, 1967). Deodorizing carbons are valuable in removing mercaptan off-odors, but also may remove desirable flavor compounds. Activated carbon also may give the treated wine an atypical odor. Furthermore, activated carbon has an oxidizing property. Although activated carbon can be valuable in some situations, trials using small wine samples are vital to avoiding undesirable, unexpected effects. Typical doses can vary from 2.5 to 50 g/hl.

ALBUMIN

Egg white has long been used in fining wines. The active ingredient is the protein albumin. As with other protein fining agents, albumin is primarily employed to remove excess tannins. The peptide linkages of the albumin form hydrogen bonds with hydroxyl groups on tannins. The opposing charges on the molecules favor the formation of large protein–tannin aggregates. Formerly albumin was added as egg white, which required the addition of sodium chloride to solubilize the albumins. Soluble egg albumin is currently available in pure form.

BENTONITE

Bentonite is a type of montmorillonite clay widely employed as a fining agent. It is used in clarifying juice and wines, in removing heat-unstable proteins, and in limiting the development of copper *casse*. Depending on the objectives, the ability of bentonite to induce partial decolorization and remove nutrients, such as amino acids, is either an advantage or disadvantage. Together with other fining agents, such as tannins and casein, bentonite can speed the settling of particulate matter. It also can correct for the addition of excessive amounts of proteinaceous fining agents by inducing their precipitation. Because bentonite settles out relatively quickly and is easily filtered, it is one of the few fining agents that does not itself potentially create a stability or clarification problem. Bentonite also has, in comparison with other fining agents, a minimal effect on the sensory properties of the treated wine (Fig. 8.10). The major drawbacks of bentonite use are color loss from red wines and a tendency to produce voluminous sediment. The latter can cause considerable wine loss during racking. Correspondingly, small-scale laboratory tests are usually conducted in advance to estimate the minimum amount of bentonite that can achieve the desired results. Details on such tests are given in Weiss *et al.* (2001). Data presented by Lubbers *et al.* (1996) illustrate the complexities of the effects of different bentonites on the

removal of aromatic compounds, and of wine constituents on the aroma absorbency of bentonites.

The bentonite often preferred in the United States is Wyoming bentonite. Because the predominant cation is monovalent (sodium), the particles swell readily in water and separate into separate sheets of aluminasilicate. The sheets are about 1 nm thick and 500 nm wide. The separation of the sheets provides an immense surface area over which cation exchange, adsorption, and hydrogen bonding can occur. When fully expanded (after about 2–3 days in warm water), sodium bentonite has a surface area of about 700–800 m^2/g. Swelling of bentonite before addition to wine significantly improves its efficacy (Marchal *et al.*, 2002). Calcium bentonite is less commonly used. It tends to clump on swelling and provides less surface area for fining. Nevertheless, it has the advantage of producing a heavier sediment that is easier to remove. Calcium bentonite also does not liberate sodium ions into the wine. Details of the physicochemical characteristics of bentonites can be obtained in Marchal *et al.* (1995).

The net negative charge of bentonite attracts positively charged proteins. The immense internal surface area between the individual alumina–silica plates provides abundant sites for attachment (Gougeon *et al.*, 2002). The charged sites on proteins are neutralized by cation exchange, resulting in flocculation and settling as a clay–protein complex. Regrettably, bentonite has little anion-exchange capacity. Therefore, it is relatively ineffective in removing either neutral or negatively charged proteins.

Because the effectiveness of bentonite in removing proteins partially depends on the wine's pH, fining is delayed until after any intended blending is complete. Otherwise, a rise in pH of the blended product could reduce protein solubility and increase the potential for subsequent haze formation.

CASEIN

Casein is the major protein found in milk. In association with sodium or potassium ions, it forms a soluble caseinate that readily dissolves in wine. In wine, the salt dissociates and insoluble caseinate is released. The caseinate adsorbs and removes negatively charged particles as it settles. Casein finds its primary use as a decolorant in white wines. It also has some deodorizing properties.

GELATIN

Gelatin is a soluble albumin-like protein derived from the prolonged boiling of animal tissues (typically bones, skin, and tendons). As a result, the product loses some of its gelling properties, but becomes a more effective fining agent.

Gelatin is employed primarily to remove excess tannins from wines. It is usually added early during maturation. This avoids color loss that would be more pronounced if conducted later (due to the continuing polymerization of anthocyanins with tannins). When gelatin is added to white wine, there is a risk of leaving a gelatin-derived haze. This may be avoided by the simultaneous addition of flavorless tannins, Kieselsol, or other protein-binding agents. These materials favor the formation of the fine meshwork of gelatin fibers that removes tannins and other negatively charged particles. Excessive fining with gelatin can result in undesirable color loss in red wines.

Although the risks are minimal, gelatin use has been mentioned as a possible source of wine contamination with prions associated with Bovine Spongiform Encephalopathy (BSE or mad-cow disease). Wine fined with gelatin derived from infected animal tissue could contain active prion proteins. The internal bondings of this infectious protein are so remarkable that the rendering process used in producing gelatin does not inactivate these infectious agents. Although the actual risk of gelatin's use to human health is unknown, the possibility has prompted the study of substitutes made from plant proteins, such as wheat gluten (Marchal *et al.*, 2002; Fischerleitner *et al.*, 2003). In the United States, most gelatin is derived from pig skins, a source free of BSE.

KIESELSOL

Kieselsol is an aqueous suspension of silicon dioxide. Because it is available in both positively and negatively charged forms, Kieselsol can be formulated to adsorb and remove both positively and negatively charged colloidal material. It is commonly used to remove bitter polyphenolic compounds from white wine. Combined with gelatin, it is effective in clarifying wines containing mucilaginous protective colloids, such as found in botrytized wines. Kieselsol tends to produce a less voluminous sediment than bentonite, removes little color from red wines, and has no tendency to add taste.

ISINGLASS

Isinglass is derived from proteins extracted from the air bladder of fish, notably sturgeons. Similar to most other proteinaceous fining agents, isinglass is primarily used to remove tannins. Because it is less subject to overfining, isinglass requires less added tannin than gelatin to function in fining white wine. Regrettably, it produces a voluminous sediment that tends to plug filters.

POLYVINYLPOLYPYRROLIDONE

Polyvinylpolypyrrolidone (PVPP) is a resinous polymer that acts similarly to proteinaceous fining agents. It is particularly useful in the selective removal of flavans and mono- and dimeric phenolics. As such, PVPP has particular value in diminishing undesirable bitterness. For this, it is usually added relatively early in maturation. It is also efficient in preventing oxidative browning, or removing its brown by-products from white wines after their formation. It functions well at cool temperatures and precipitates spontaneously. Some grades of PVPP can be isolated from the sediment, purified, and reused. Alternately, PVPP may be bonded to a silica support surface, over which the wine is passed in a continuous manner. Regrettably, PVPP, along with charcoal and casein, is particularly effective at removing resveratrol (Castellari *et al.*, 1998), one of the wine components frequently credited with some of the health benefits associated with moderate wine consumption.

TANNIN

Insect galls on oak leaves are the typical source of tannins used in fining. Tannins are commonly combined with gelatin. The tannin–gelatin mixture forms a delicate meshwork that sweeps colloidal proteins out of the wine. Tannins in the mesh join with soluble proteins in the wine to form both weak and strong chemical bonds. The weak bonds involve nonionized carboxyl and hydroxyl groups of the tannins and hydrogen bonds with peptide linkages of the proteins. Strong bonds involve covalent links between tannin quinone groups and protein amino and sulfur groups. The latter form stable links between soluble proteins and the tannin–gelatin meshwork.

COPPER SULFATE

Copper sulfate (or occasionally silver salts) may be used to remove or prevent the accumulation of sulfur off-odors. Such treatment may become more frequent due to the increased adoption of screw cap (ROTE) closures. Screw caps are generally much more effective at excluding oxygen than cork or other closures, potentially increasing the likelihood of the development of reduced-sulfur off-odors. This likelihood can be reduced by the use of various procedures (see Sulfur Off-odors), including copper fining.

Copper fining involves adding up to 0.5 mg/liter of copper sulfate, at least one month before bottling. The copper reacts with hydrogen sulfide and various thiols (notably mercaptans). The odorless copper complexes precipitate readily and are eliminated during racking or filtration. Residual levels of copper should not exceed 0.2 mg/liter (the usual legal limit). If disulfides are also a problem, sulfur dioxide and ascorbic acid (usually 50 mg/liter) may also be added. Sulfite generated from sulfur dioxide binds and eventually splits disulfides, generating thiols that can react with copper (Bobet *et al.*, 1990).

Regrettably, the splitting of disulfides is slow. The ascorbic acid appears to act as an antioxidant, preventing the reoxidation of thiols to disulfides.

Clarification

In contrast to fining, clarification involves only physical means to remove suspended particulate matter. As such, it usually follows fining, although initial clarification often occurs before fermenting white juice. Juice clarification often improves flavor development in white wine. After fermentation, racking removes material that sediments spontaneously, thereby helping prevent microbial spoilage. Subsequently, finer material may be removed by centrifugation or filtration.

RACKING

Until the twentieth century, racking and fining were essentially the only methods available to facilitate wine clarification. Racking can vary from manually decanting wine from barrel to barrel to highly sophisticated, automated, tank-to-tank transfers. In all cases, separation should occur with minimal agitation to avoid resuspending particulate matter. Decanting stops when unavoidable turbulence makes the wine cloudy. The residue may be filtered to retrieve wine otherwise lost with the lees. Racking is generally more effective in clarifying wine matured in small cooperage than in large tanks.

The first racking is conventionally done several weeks after alcoholic fermentation. If malolactic fermentation is desired, racking is delayed until it has come to completion. Racking may also be delayed to permit prolonged lees contact, such as with *sur lies* maturation.

By the first racking, most of the yeast, bacterial, and grape-cell fragments have settled out. Subsequent rackings remove most of the residual microbial population, along with precipitated tannins, pigments, and crystalline material. Later rackings remove sediment generated as a consequence of fining.

If sufficient time is provided, racking and fining can produce stable, crystal clear wines. However, the trend to early bottling, a few weeks or months after fermentation, provides insufficient time for racking and spontaneous precipitation to generate adequate clarification. Consequently, centrifugation and filtration are employed to achieve the required level of clarity.

In addition to aiding clarification, racking plays several additional valuable roles in wine maturation. By removing microbial cells and other sources of nutrients, racking enhances microbial stability. The transfer process also disrupts stratification that may develop within the wine. This is particularly important in large storage tanks, in which stratification can lead to variations in redox potential and rates of maturation throughout the wine. Racking also removes the primary sources of reduced-sulfur taints, such as hydrogen sulfide and mercaptans. These may form under the low-redox conditions that develop in thick layers of lees.

Aeration and escape of CO_2 are also benefits that accrue from racking. Although modest oxygen uptake during racking assists color stability in red wine, its value in white wine maturation is more controversial. As noted, slight aeration benefits white wines *sur lies* matured, but is avoided otherwise. Oxygen exposure can be minimized with automatic pumping systems using carbon dioxide or nitrogen as a blanketing gas. Because nitrogen is lighter than air (in contrast to CO_2), its use in excluding air is largely limited to sealed cooperage. The turbulence generated during pumping and filling helps liberate carbon dioxide that is present in a supersaturated state following fermentation. The escape is essential for the wine to lose its slight *pétillance* before bottling. If necessary, adjustment with additional sulfur dioxide is usually timed to coincide with racking.

The number of rackings recommended varies considerably from region to region, depending on empirically established norms. Cooperage size is also a determining factor. The larger the storage vessel, the more frequently racking is required. This is necessary to avoid the development of a thick sediment layer that is conducive to off-odor production.

The method of racking depends largely on cooperage size and the economics of manual vs. mechanical transfer. Manual draining by gravity, or with a simple hand pump, is adequate where volumes are small and labor costs are low. Hand pumps for transferring wine from one barrel to another have been used at least since the late 1400s, when an illustration of this activity was produced in Nuremberg. For most large wineries, however, manual racking would be prohibitively expensive, both in terms of time and labor. Mechanical pumping is the only reasonable option. Also, if aeration and sulfiting are deemed desirable, they can be controlled more precisely through mechanical than manual racking.

CENTRIFUGATION

Centrifugation employs rotation at high speed to expedite settling. It is equivalent to spontaneous sedimentation, but occurs within minutes rather than months. It often replaces racking when early bottling is desired. Centrifugation also is useful when the wine is heavily laden with particulate matter. Highly turbid wines are prone to off-odor development if they are permitted to clarify spontaneously. Centrifugation is much more efficient in removing large amounts of particulates than are plate filters. Centrifugation also avoids potential health problems (dust and worker allergy) associated with the use and disposal of diatomaceous earth and other filter aids.

Blanketing the wine with an inert gas has minimized a former liability of centrifugation – oxidation. Automation, combined with continuous centrifugation, has improved the efficiency and economy of the process to such an extent that centrifugation is often the preferred clarification technique.

FILTRATION

Filtration involves the physical retention of material on or within a fibrous or porous material. Depending on the pore size, filtration removes coarse particles with diameters larger than $100\,\mu m$ down to molecules and ions with diameters less than $10^{-3}\,\mu m$. However, the greater the retentive property, the greater is the likelihood of plugging. As a consequence, filtration typically is preceded by preliminary clarification, using racking, fining, or centrifugation. This is especially important when employing membrane sterilization or ultrafiltration.

With the development of new filters and support systems, filtration is classified into four categories. Conventional filtration employs depth-type fibrous filters. These remove particles down to about $1\,\mu m$ in diameter. Other filtration techniques involve membranes containing crevices, channels, or pores. Depending on the size range of the perforations, the sieving action is termed microfiltration, ultrafiltration, reverse osmosis or dialysis. Microfiltration and ultrafiltration usually are differentiated on the basis of nominal pore size, $1.0–0.1\,\mu m$ and $0.2–0.05\,\mu m$, respectively. Microfiltration is used primarily to remove fine particles and in sterilization. Ultrafiltration is employed to remove macromolecules and colloidal materials. Reverse osmosis and dialysis are used to remove or concentrate low-molecular-weight molecules or ions. Dialysis involves the same principle (diffusion) as reverse osmosis, but does not use pressure to reverse the direction of flow. Electrodialysis uses an electrical differential across the membrane to influence the flow of charged particles.

Filtration primarily acts by blocking the passage of material larger than the maximum pore size of the filter (Fig. 8.11). However, because material smaller than the smallest perforations may be retained by a filter, other principles are involved. Surface adsorption by electrical attraction can be more important than physical blockage at the lower limit of filtration. Adsorption is generally important with depth filters, but less significant with membrane filters. Conversely, capillary forces may facilitate movement through filters. With depth filters, microbial growth in the filter can result in "grow through." Thus, it is essential that depth filters be frequently cleaned and sterilized, or replaced.

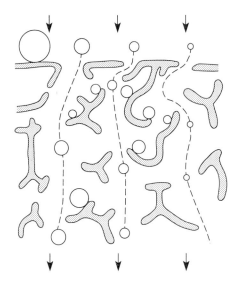

Figure 8.11 Mechanism of membrane filtration. Particles larger than the diameter of the filter pores become trapped at the surface whereas smaller particles pass through or adhere to the filter matrix. (After Helmcke, 1954, from Brock, 1983, reproduced by permission)

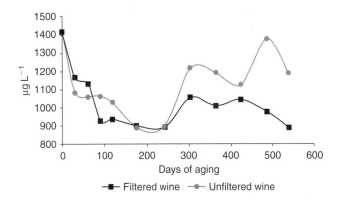

Figure 8.12 Development of total esters (except ethyl lactate) during aging of filtered and unfiltered wine in American and French oak barrels. (From Moreno and Azpilicueta, 2006, reproduced by permission)

Filtration has often been avoided by premium winemakers, supposedly because of flavor loss. Although there is little objective data confirming this view, Ribeiro-Corréa *et al.* (1996) indicate that the contents of several flavorants are reduced. The sensory impact of these influences was, however, not statistically detectable when evaluated by a sensory panel. Interestingly, filtered wines had lower concentrations of esters after maturation in oak than unfiltered wines (Fig. 8.12). The esters most affected where isoamyl acetate, ethyl butyrate and ethyl hexanoate. Whether such changes are sensorially detectable is unknown.

Depth Filters Depth filters are composed of randomly overlapping fibers of relatively inert material.

They may be purchased preformed (**filter pads**) or produced during the filtration process (**filter beds**). Most filters are cellulose-based.

Filter pads come in a broad range of porosities, permitting differential flow rates and selective particle removal. **Tight filters** remove smaller particles, but retain most of the material on the filter surface. Consequently, they tend to plug quickly. In contrast, **loose filters** retain most of the material within the tortuous channels of the pad, plug less quickly, but remove only larger particles. Tight filters are commonly used just before bottling for a final **polishing filtration**.

Filter beds develop as a **filter aid**, suspended in the wine, is progressively deposited on an internal framework during filtration. Filter beds may be employed prior to polishing, sterilization, or ultrafiltration. The filter aid most commonly used has been **diatomaceous earth**. This consists of the remains of countless generations of diatoms – microscopic, unicellular algae that construct their cell walls primarily out of silicon dioxide. Depending on the filtration rate and the particle size to be removed, different formulations of diatomaceous earths are available. Diatomaceous earth is added to the wine during filtration at about 1–1.5 g/liter. Loose cellulose fibers, treated to have a positive charge, may be added to facilitate the adsorption of colloidal materials. **Perlite**, the pulverized remains of heat-treated volcanic glass, has a very fine structure and is occasionally used instead of diatomaceous earth.

Plates possessing a screen of cloth, plastic, or stainless steel are covered with a **precoat** of the filter aid. These are inserted into the framework of a **filter press**. The filter aid is continuously added and mixed with the wine being filtered. Pressure forces the wine through the filter bed (Fig. 8.13). Porous metal or plastic sheets support the filter and provide channels through which the filtered wine escapes. During filtration, the depth of the filter bed grows. The continual addition of filter aid is essential for maintaining a high flow rate at low pressures. Without additional filter aid, the bed would soon plug. Use of higher pressures tends to compact the filter material, aggravating plugging. Choosing the correct grade of filter aid is essential. Particle size affects both flow and plugging rates and, consequently, filtration efficiency. After operation, filtration is temporarily halted to allow removal of the accumulated filter aid and retained material.

Filter beds are usually associated with **plate-and-frame**, **recessed-plate**, or **leaf press** constructions. Plate-and-frame presses consist of alternating precoated plates and frames that provide space for cake development. Recessed-plate presses are similar, but each plate serves both plate and frame functions. However, filter beds can be constructed quite differently. The **rotary vacuum drum** is a prime example (Fig. 8.14). It consists of a large, perforated, cloth-covered, hollow drum. The drum is precoated with 5–10 cm of filter aid, usually diatomaceous earth. Part of the drum is immersed in the wine being filtered. The filter aid is added and kept uniformly dispersed as the wine is drawn through the filter bed into the drum. Shaving the accumulated filter aid and particulate matter off the drum occurs automatically as it rotates. Rotary vacuum drums work particularly well with wines that are highly charged with particulate matter or mucilaginous colloids. Other than the high purchase and operation costs, the major drawback of rotary vacuum filtration is the potential for wine oxidation. Aeration is difficult to limit because part of the drum is raised out of the wine during rotation. Nevertheless, a blanketing atmosphere devoid of oxygen can significantly limit oxidation.

Because filter aids and pads occasionally can be sources of metal and calcium contamination, the material is commonly treated with a tartaric acid wash.

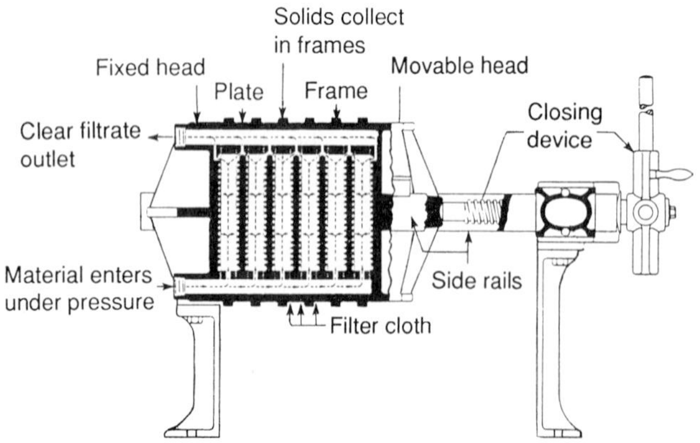

Figure 8.13 Cross-sectional view of a filter press, showing the arrangement of plates, filter cloth, and frames, and the flow of material. (Courtesy of T. Shriver and Co., division of Eimco Inc.)

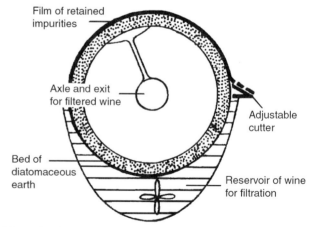

Figure 8.14 Cross-sectional view of a rotary vacuum-drum filtration apparatus.

In addition, the first sample of wine filtered is often kept aside, at least until its freedom from metal contamination, or earthy, paper-like odors has been assessed.

Membrane Filters Membrane filters are constructed out of a wide range of synthetic materials, including cellulose acetate, cellulose nitrate (collodion), polyamide (nylon), polycarbonate, polypropylene, and polytetrafluoroethylene (Teflon). With the exception of polycarbonate filters, most form a complex network of fine, interconnected channels. Polycarbonate (Nuclepore) filters contain cylindrical pores of uniform diameter that pass directly through the filter. Because polycarbonate filters have a small pore-surface area, they are seldom used for wine filtration. In comparison, most other membrane filters contain 50–85% filtering surface. Thus, they have an improved flow rate for the same cutoff point (rated pore size). The applicability of a new inorganic membrane filter (Anapore) with uniform capillary pores and higher flow rates has not been assessed. The recent introduction of sintered, stainless-steel membranes provides both an inert and extremely robust membrane system, combined with high flow-rates.

Because of their small pore-diameter, membrane filters tend to have a slower flow rate than depth filters. They are also more likely to plug, because most of the filtering occurs at the surface. To circumvent rapid plugging, filter holders may be employed that direct the fluid flow parallel to, rather than straight through, the filter (Fig. 8.15).

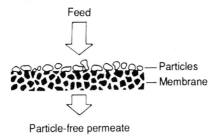

Conventional (dead-end) filtration

Feed

Particles
Membrane

Particle-free permeate

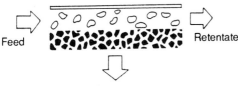

Tangential (cross-flow) filtration

Feed
Retentate

Particle-free permeate

Figure 8.15 Distinction between conventional filtration and tangential (cross-flow) filtration. The washing action of the fluid passing tangentially across the surface of the membrane keeps the filter from becoming clogged. (From Brock, 1983, reproduced by permission)

The parallel system is termed **tangential** or **cross-flow filtration**. The conventional, perpendicular flow is termed **dead-end filtration**. In tangential filtration, wine flow retards suspended material from accumulating on the membrane, causing plugging. Fouling can result from the development of a surface layer, consisting of cell constituents, microbes, and collections of polysaccharides and polyphenolics. Membrane polarity is particularly important to the adherence of polyphenolics (Vernhet and Moutounet, 2002). Short, periodic back-flushing increases the functional life of filters. Tangential filtration can partially eliminate the need for refiltering before wine sterilization. If filtration occurs shortly after the completion of fermentation, the filtrate may be employed as an inoculum for additional alcoholic or malolactic fermentations.

Advanced forms of tangential filtration often employ cylinders of complex internal structure. The external portion, exposed to the wine, contains pores that become progressively smaller toward the center. The central region contains pores of constant diameter, to assure the retention of particles or molecules below a certain size or molecular weight. Cartridge filters contain some of the nonplugging features of depth filters and offer the particle-size retention characteristics of conventional membrane filters. These polypropylene filters are resistant to most chemical reagents. This allows them to be cleaned and used repeatedly. Such developments may reduce, if not eliminate, the need to conduct filterability tests prior to sterile filtration. Filterability tests are discussed in Peleg *et al.* (1979) and de la Garza and Boulton (1984).

Microfiltration is extensively used to sterilize wines. Microfiltration avoids the flavor modification occasionally associated with pasteurization.

Ultrafiltration has been used to a limited extent in protein stabilization of wine. Although it effectively removes most colloidal material from wine, ultrafiltration can also remove important pigments and tannins from red wines (Fig. 8.16). Its use with white wines appears not to produce unacceptable flavor loss (Flores *et al.*, 1991). The benefits and liabilities of ultrafiltration are under active investigation.

Aging

The tendency of wine to improve, or at least change, during aging is one of its most beguiling properties. Regrettably, most wines improve only for a few years before showing irreversible losses in quality. In contrast, red wines produced from varieties such as 'Cabernet Sauvignon,' 'Shiraz,' 'Tempranillo,' 'Nebbiolo,' and 'Pinot noir,' may continue to improve in flavor and subtlety for decades. White wines, produced from varieties such

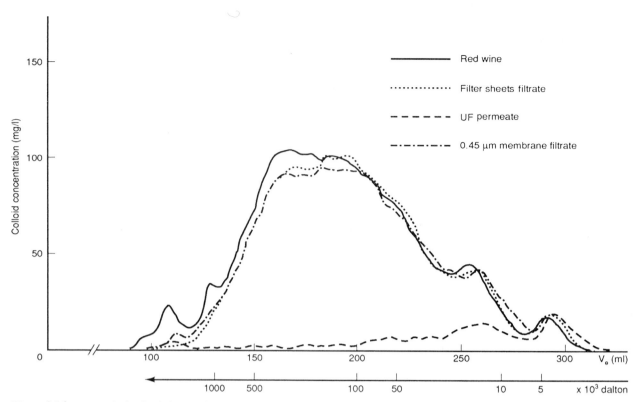

Figure 8.16 Removal of colloids from red wine during various filtration procedures, expressed as elution volumes (V_e in ml) and molecular mass (Da) of the various colloidal fractions. UF, ultrafiltration. (From Cattaruzza *et al.*, 1987, reproduced by permission)

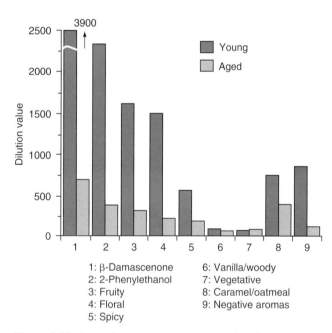

Figure 8.17 Loss of aromatic compounds during the aging of a 'Vidal' wine. (From Chisholm *et al.*, 1995, reproduced by permission)

1: β-Damascenone
2: 2-Phenylethanol
3: Fruity
4: Floral
5: Spicy
6: Vanilla/woody
7: Vegetative
8: Caramel/oatmeal
9: Negative aromas

as 'Riesling,' 'Chardonnay,' 'Sauvignon blanc,' and 'Viura,' also show excellent aging potential.

Quality loss is commonly explained as a dissipation of a fresh, fruity bouquet (Fig. 8.17), along with any aroma donated by the grape variety. Wines noted for continued improvement typically show similar aromatic losses, but they gain in **aged bouquet**. Aging is considered desirable when the development of an aged bouquet, subtle flavor, and smooth texture more than compensate for the fading varietal and fruity character of the wine.

Only since the early 1980s have sufficiently precise analytical tools become available to begin unraveling the nature of wine aging. However, as aging has been studied in only a few wines, caution must be used in generalizing from these findings. More is known about why most wines decline in quality than why some retain or improve in character for decades.

Knowledge of how wines age, and how the effects of aging might be directed is important to all involved or interested in wine. At the very least, quality loss can adversely affect the shelf-life of a wine and the financial return to the producer. On the other hand, the prestige connected with long aging-potential adds greatly to the desirability and appeal of a wine. It also permits consumers to participate in the process, through the conditions and duration of aging they permit. Because the factors affecting aging are poorly understood, a mystique has built up around vineyards and varieties associated with wines that age well.

Aging is occasionally considered to possess two phases. The first, called **maturation**, refers to changes that occur between alcoholic fermentation and bottling. Although maturation frequently lasts from 6 to 24 months, it may continue for decades. During maturation, the wine may undergo malolactic fermentation, be stored in oak cooperage, be racked several times, and be treated to one or more clarification techniques. During racking and clarification, wines may absorb about 40 ml O_2/year, an amount insufficient to give the wine a noticeably oxidized character. Only in some fortified wines is obvious oxidation an important component of maturation.

The second phase of aging commences with bottling. Because this stage occurs essentially in the absence of oxygen, it has been called **reductive aging**. This contrasts with **oxidative aging**, an alternative term for maturation that is used for the aging of some fortified wines.

Effects of Aging

Age-related changes in wine chemistry have long been noted. Initially, these modifications are favorable. They result in the dissipation of the yeasty aspect and spritzy character of newly fermented wines. Subsequently, there is a loss of the fresh fruitiness of the wine. If this is accompanied by the development of an appreciated aged bouquet and smoother mouth-feel, the consequences of aging are highly desirable. To encourage these latter processes, most wine connoisseurs store wine in cool cellars for years to decades. Regrettably, most wines do not age particularly well. Most white wines are recommended to be consumed within a few years of production. Most red wines improve or retain their flavor for little more than 5–10 years. In reality, though, these views reflect professional opinion. It is often thought that most consumers prefer the fresh fruity character of young wines vs. the more general, subtle aspects of an aged bouquet. However, this may simply reflect their disinterest in aging wine, or their acceptance (or insensitivity to) the rough astringency of many young red wines.

Nonenzymatic oxidative reactions produce significant sensory changes during aging. This involves the transfer of an electron (or hydrogen atom) from the oxidized compound to oxygen, or another acceptor. In bottled wine, reactions involving molecular oxygen occur slowly, as oxygen diffuses into the bottle via the cork, or between the cork and the neck. Temperature, pH and the phenolic content significantly affect a wine's oxidative potential. It is estimated that wine can combine with up to about 6 mg/liter O_2 (saturation at 20°C) within a week or less, depending on the wine's phenolic content (Singleton and Cilliers, 1995). Other oxidative reactions (not involving molecular oxygen) occur during wine aging, but their influence on wine

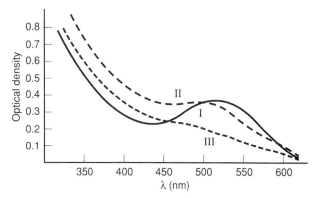

Figure 8.18 Absorption spectra of three red wines of different ages. I, 1 year old; II, 10 years old; III, 50 years old. (From Ribéreau-Gayon, 1986, reproduced by permission)

fragrance and taste are little known. The presence of copper and iron ions are the best known of wine oxidative catalysts. Because the redox potential of wine declines after bottling, reductive reactions are almost undoubtedly involved in wine aging.

As with other aspects of wine chemistry, determining the significance of changes is more difficult than detecting them. To establish their significance, it is necessary to show that the changes detectably impact sensory perception. Because most chemicals occur at concentrations below their sensory threshold, most changes affect neither wine flavor nor the development of an aged bouquet.

APPEARANCE

One of the most obvious changes during aging is a progressive browning. Red wines may initially deepen in color after fermentation, but intensity slowly fades as the tint takes on a ruby and then a brickish hue. These changes result from a disassociation of self-association and copigment anthocyanin complexes (typical of young wines), the formation of new pigments (pyranoanthocyanins, catechinpyrylium, and xanthylium pigments), and the progressive formation of both tannin–tannin and anthocyanin–tannin complexes.

The dynamics of these changes are only beginning to be understood. The significance of polymerization is clear, but little is known about the importance of pigments such as pyranoanthocyanins (see Alcalde-Eon *et al.*, 2006). Polymerization begins during fermentation, constituting about 25–60% of the anthocyanin content within a year (Fig. 6.7). Polymerization is favored by the peroxidation of dihydroxyphenolics (primarily *o*-diphenols) to diquinones. These slowly polymerize into colorless dimers, then yellowish dimers, and finally brownish end products (see Rose and Pilkington, 1989). These changes, originally predicted from a decline in optical density, and a shift in the absorption spectrum (Fig. 8.18), have been confirmed by various methods

(Peng *et al.*, 2002; Remy *et al.*, 2000), including radioactive isotope analysis (Zimman and Waterhouse, 2004). The degree of polymerization is typically measured as a ratio of optical density measurements taken at 520 and 420 nm (Somers and Evans, 1977). High 520/420 nm values indicate a bright-red color, whereas low values indicate a shift to brickish shades. In contrast, white wines darken in color and take on yellow, gold, and finally brownish shades. Browning of white wines is usually assessed by measuring optical density at 420 nm. The color shift results primarily from the accumulation of chromophoric carbonyl compounds. Skouroumounis *et al.* (2003) have developed an *in situ* method to measure the degree of browning in bottled wine. Theoretically, this spectrophotometric technique could be used to remove bottles of wine showing sporadic oxidation before reaching the consumer.

Although less studied in white wine, color change has been extensively investigated in red wines. The initial purplish color, associated with copigmentation, fades as the complexes dissociate and the freed anthocyanins oxidize or progressively polymerize with tannins. Nevertheless, no consensus has been reached about the relative importance (or desirability) of the various polymerization mechanisms involved. Although small amounts of acetaldehyde, produced following oxygen uptake, enhance the polymerization of anthocyanins and flavonoid tannins, polymerization also occurs directly under anaerobic, acid-catalyzed conditions (Somers and Evans, 1986). Because direct polymerization occurs slowly, but continuously throughout the wine, direct polymerization may be the more significant process in bottle aging. Because temperature speeds polymerization, mild heating has been recommended as an alternative to aeration for color stabilization (Somers and Pocock, 1990). Although micro-oxygenation is receiving increased attention, if not carefully controlled, it risks activating dormant acetic acid bacteria, increasing volatile acidity (acetic acid and its esters), aggravating potential problems with *Brettanomyces*, and inducing the precipitation (loss) of polymeric pigments. Red wines are variously considered to benefit from up to about 60 ml O_2 per liter (10 saturation) (Boulton *et al.*, 1996), but show obvious deterioration with more than 25 saturations.

Another source of polymerization involves the catalytic oxidation of tartaric acid to glyoxylic acid. This is activated by metallic ions such as iron (Fulcrand *et al.*, 1997). Glyoxylic acid binds catechins and possibly other phenolics into increasingly long polymers. The reaction can generate both colorless and yellow polyphenols.

Although anthocyanin polymerization is critical to color stabilization, additional mechanisms are involved. One entails a reaction between yeast metabolites, such as pyruvic acid and anthocyanins (Fulcrand *et al.*, 1998). Subsequent structural rearrangement and

dehydration generate a tawny colored product. It is more stable than the original anthocyanin to sulfur dioxide, high temperature, and pH values above 3.5.

Other than phenolic browning reactions, the color shift in white wines is poorly understood. It probably involves structural modification in existing pigments, or synthesis *de novo* by one or more of the following processes: the slow oxidation and polymerization of grape and oak phenols and related compounds;[1] oxidation products derived from added ascorbic acid (vitamin C); metal ion-induced structural modifications in galacturonic acid; ketosamine condensation products produced by Maillard reactions between reducing sugars and amino acids; and sugar caramelization, generating products such as furan-2-aldehyde. The latter two reaction types are particularly likely in sweet wines. Some of these reactions may also participate in the development of a bottle-bouquet. Sulfur dioxide can retard or prevent most of these reactions.

TASTE AND MOUTH-FEEL SENSATIONS

During aging, residual glucose and fructose may react with other compounds and undergo structural rearrangement. Nevertheless, these reactions do not appear to occur to a degree sufficient to affect perceptible sweetness. In contrast, aging can affect acidity, inducing small but perceptible losses. For example, esterification of acids, such as tartaric acid, removes carboxyl groups involved in the sensation of sourness. Upwards of 1.5 g/liter of ethyl bitartrate may form during aging (Edwards *et al.*, 1985). Slow deacidification also can result from the isomerization of the natural L- to the D- form of tartaric acid. The racemic mixture is less soluble than the L-form. This is one of the origins of tartrate instability in wine. Isomerization also results in forming racemic mixtures of L- and D-amino acids (Chaves das Neves *et al.*, 1990). The potential significance of the toxicity of the D-amino acids is unknown. It is probably negligible due to the small amino acid content in wine.

The most significant gustatory changes during aging affect the bitter and astringent sensations of red wines. The best understood of these reactions is the polymerization of tannins and their subunits with themselves, anthocyanins, proteins, and polysaccharides. Autopolymerization tends to induce a progressive decline in bitterness and astringency, due to chemical reactivity changes, or due to precipitation. However, in the early stages of polymerization, there may be an increase in astringency. This results from the greater astringency of medium- versus large-size tannins. The binding of tannins with polysaccharides, peptides or

---

[1] For example, oxidized caffeic acid has a golden color, whereas flavonoid polymers donate a brown coloration.

proteins leads to further reduction in bitter, astringent sensations. Nevertheless, condensed tannins may slowly degrade during aging (Vidal *et al.*, 2002), potentially increasing bitterness. Similarly, the breakdown of hydrolyzable tannins (primarily from oak cooperage) may reduce astringency, but enhance bitterness. Due to our poor understanding of the dynamics of these changes, it is currently impossible to clearly explain in chemical terms the age-related changes in bitterness and astringency experienced by consumers.

FRAGRANCE

Whereas studies on aging in red wines have concentrated primarily on color change, most research on the aging of white wines has focused on fragrance modification. Flavor loss, especially in young white wines, is associated with changes in their ester content. Other known sources of reduced fragrance involve structural rearrangements in terpenes and volatile phenols.

Loss or Modification of Aroma and Fermentation Bouquet Esters produced during fermentation generate much of the fresh, fruity, character of young white wines. The most significant appear to be esters formed between acetic acid and higher alcohols, such as isoamyl and isobutyl acetates. Because yeasts produce and release more of these esters than the equilibrium in wine permits, the esters tend to hydrolyze back to their corresponding acids and alcohols. Thus, the fruity aspect donated by acetate esters tends to fade with time (Fig. 8.19) (Gonzalez Viñas *et al.*, 1996). Nevertheless,

several esters may remain above their detection threshold for several years. Hydrolysis occurs more slowly at higher pHs and at lower temperatures (Marais and Pool, 1980). Hydrolysis of fruit esters also appears to be diminished in the presence of antioxidants, such as sulfur dioxide and various phenolics (caffeic and gallic acids) (Lambropoulos and Roussis, 2007). Sulfur dioxide also retards the loss of varietally important thiols, such as 3-mercaptohexan-1-ol (Blanchard *et al.*, 2004). Conversely, the reaction between sulfur dioxide and damascenone, markedly reduces its impact on the fruity character of several wines (Daniel *et al.*, 2004).

A second major class of esters is based on ethanol and long-chain saturated fatty acids. The level of these esters may decline, remain stable, or increase during aging. Hydrolysis is delayed by the high ethanol content of wine, but is favored by low pH, high temperatures, and increased molecular weight. Those with shorter hydrocarbon chains, such as butanoate, hexanoate, and octanoate ethyl esters, tend to be somewhat fruity in character. As the hydrocarbon chain becomes longer, the odor shifts, becoming soapy and finally lard-like.

A third group of esters, and eventually the most abundant, forms slowly and nonenzymatically between ethanol and organic acids, such as tartaric, malic, lactic, citric and succinic acids. Their formation increases with higher alcohol contents, lower pH values, and at higher temperatures (Shinohara and Shimizu, 1981). The synthesis of diethyl succinate is particularly marked (Fig. 8.19). Both it and methyl succinate appear to be of major significance in the fragrance development of muscadine wines (Lamikanra *et al.*, 1996). However, most ethyl esters of fixed acids probably play little role in bouquet development, due to their low volatility and nondistinctive odor.

Another group of important flavorants that change during aging are terpenes. These are particularly important to the aroma of 'Muscat' and related cultivars. Oxidation of terpenes results in the marked loss of varietal character in these varieties.

The total concentration of monoterpene alcohols falls markedly during aging. The decline in geraniol, linalool, and citronellol is especially marked. For example, the linalool content of 'Riesling' wines can fall by 80%, to below its detection threshold within 3 years. This can result in a noticeable loss in floral character. In contrast, the concentrations of linalool oxides, nerol oxide, hotrienol, and α-terpineol increase. Figure 8.20 presents a proposed reaction scheme for changes in monoterpene alcohols during aging. Most of these derivatives have higher perception thresholds than do their monoterpene progenitors. For example, linalool oxides have flavor thresholds in the 3000–5000 μg/liter range vs. 100 μg/liter for linalool (Rapp, 1988). Oxide terpene derivatives

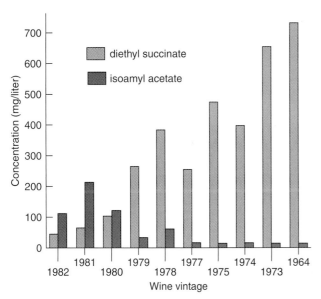

Figure 8.19 Examples of the influence of wine age on the concentration of esters, namely, acetate esters (isoamyl acetate, dark green) and ethanol esters (diethyl succinate, light green). (Data from Rapp and Güntert, 1986)

also have qualitatively different odors. For example, α-terpineol has a musty, pine-like odor, whereas its precursor linalool has a floral aspect. In addition, mixtures may have a qualitative fragrance different from their component compounds. Terpene-related, heterocyclic oxygen compounds also develop during aging, but their sensory significance is unknown.

The role of norisoprenoids in aroma loss has been little studied. However, the concentration of the rose-like β-damascenone declines during aging (Fig. 8.17). The occurrence of other isoprenoid degradation products, such as theaspirane, ionene (a 1,1,6-trimethyl-1,2,3, 4-tetrahydronaphthalene isomer), also appears to decline.

Several volatile phenols, such as vinylphenol and vinylguaiacol, are partially converted to nonvolatile ethyoxyethylphenols during aging (Dugelay *et al.*, 1993). At low concentrations, volatile phenols can enhance wine fragrance, but at higher levels they produce phenolic off-flavors. The concentration of important varietal

aromatics, such as 2-phenylethanol, declines during aging (Fig. 8.17).

In addition to the loss or modification of grape and yeast aromatics, new compounds may be generated. Some of these appear to be the result of oxidation. The best-known of these is acetaldehyde. However, the importance of acetaldehyde to the oxidized odor of white wines has recently come under question (Escudero *et al.*, 2002). In contrast, the production of furfural and hexanal appears to be particularly important to the development of an oxidized odor. Their odor may play a significant role in masking the wine's natural fragrance. Exposure to high temperatures also generates additional compounds (Silva Ferreira *et al.*, 2003). However, the descriptive terms for these compounds appear more related to expressions used to describe an aged bouquet than an oxidized odor. In addition, quinones generated during oxidation may react with amino acids. These reactions can enhance

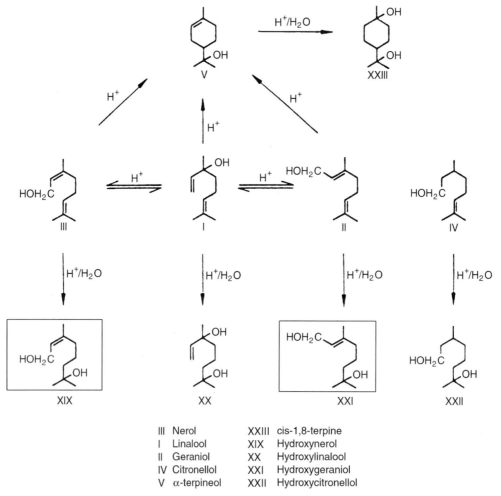

III	Nerol	XXIII	cis-1,8-terpine
I	Linalool	XIX	Hydroxynerol
II	Geraniol	XX	Hydroxylinalool
IV	Citronellol	XXI	Hydroxygeraniol
V	α-terpineol	XXII	Hydroxycitronellol

Figure 8.20 Proposed reaction scheme of monoterpene alcohol evolution during bottle-aging of a white wine. (From Rapp and Mandery, 1986, reproduced by permission)

flavor by generating aldehydes and ketones in Strecker-type degradations.

Although most of the work on wine aroma loss has involved white wines, some studies have been conducted on rosé wines. Like most white wines, rosé wines lose much of their fruity character within 1–2 years. This is correlated with the loss of several volatile thiols (Murat *et al.*, 2003; Murat, 2005). The principal compounds involved are 3-mercaptohexan-1-ol (3-MH) and 3-mercaptohenyl acetate (3-MHA). The concentration of the former (3-MH) can fall to half its initial concentration within one year, while its acetate ester may be no longer detectable. The presence of anthocyanins appears to retard their oxidation and disappearance. The considerably higher antioxidant properties of red wines may, thus, help to explain why rosé wines made from the same cultivars have much shorter shelf-lives. The lack of tannins in rosé wines equally may clarify why they tend to rapidly take on an orangish coloration (oxidation of their slight anthocyanin content). Another compound involved in the fruity character of some rosé wines is phenethyl acetate. The concentration of this fermentation by-product also declines rapidly during aging, further explaining the short shelf-life of rosé wines.

Origin of a Bottle-Aged Bouquet Four groups of compounds are known to be involved in the generation of a bottle-aged bouquet. These include constituents liberated by acid or enzymic hydrolysis from nonvolatile glycosidic conjugates; derived from norisoprenoid precursors and related diterpenes; modified carbohydrates; or generated from reduced-sulfur compounds.

Most flavorants in grapes accumulate as glycosides. This is the basis for the glycosyl-glucose (G-G) analysis of grape quality. The slow release of terpenes from glycosidic linkages under acidic condition may partially offset some of the aroma loss associated with aging. In addition, minor quantities of some monoterpene oxides, such as 2,6,6-trimethyl-2-vinyltetrahydropyran and the 2,2-dimethyl-5-(1-methylpropenyl)tetrahydrofuran isomers, may participate in the cineole-like fragrance of aged 'Riesling' wines (Simpson and Miller, 1983). Understanding the dynamics of their liberation may resolve many of the unexplained phenomena that connoisseurs have subjectively experienced for centuries.

Of isoprenoid degradation products, vitispirane and 1,1,6-trimethyl-1,2-dihydronaphthalene (TDN) appear to be the most important concerning development of an aged bouquet. The two isomers of vitispirane have qualitatively different odors. However, as they occur in concentrations at or below the detection thresholds, their sensory significance of vitispiranes is doubtful. In contrast, the concentration of TDN is clearly sufficient to play a meaningful role in the bouquet development of wines such as 'Riesling.' TDN also has been detected in red wines. Because TDN has a kerosene-like odor, its accumulation at considerably above its threshold value may become undesirable.

Carbohydrate degradation occurs rapidly during the heating of wines, such as madeira and baked sherries. Similar, acid-catalyzed dehydration reactions also occur, but much more slowly, at cellar temperatures. For example, the caramel-like 2-furfural shows a marked increase during aging (Table 8.2). The sensory significance of other decomposition products, such as 2-acetylfuran, ethyl 2-furoate, 5-(hydroxymethyl)-2-furaldehyde, 2-formylpyrrole, and levulinic acid, is unknown. The fruity, slightly pungent ethyl ether, 2-(ethoxymethyl)furan, has been found to form during aging in 'Sangiovese' wines (Bertuccioli and Viani, 1976). This suggests that etherification of Maillard-generated alcohols may play a role in the development of aged bouquets.

The concentration of reduced-sulfur compounds may change during aging. Of these, the most significant is dimethyl sulfide. Its accumulation has occasionally been correlated with the development of a desirable aged bouquet. Spedding and Raut (1982) found that the addition of 20 mg/liter dimethyl sulfide (to wines

Table 8.2 Changes in aroma composition from carbohydrate decomposition during aging of a 'Riesling' wine[a]

	Year					
Substance from carbohydrate degradation	1982	1978	1973	1964	1976 (frozen)	1976 (cellar stored)
2-Furfural	4.1	13.9	39.1	44.6	2.2	27.1
2-Acetylfuran	–	–	0.5	0.6	0.1	0.5
Furan-2-carbonic acid ethyl ester	0.4	0.6	2.4	2.8	0.7	2.0
2-Formylpyrrole	–	2.4	7.5	5.2	0.4	1.9
5-Hydroxymethylfurfural (HMF)	–	–	1.0	2.2	–	0.5

[a] Relative peak height on gas chromatogram (mm).

Source: Data from Rapp and Güntert (1986)

containing 8–15 mg/liter dimethyl sulfide) enhanced the wine's flavor score. Higher concentrations (≥40 mg/liter) were considered detrimental. Occasionally, the production of dimethyl sulfide is so marked at temperatures above 20°C that its presence can mask the varietal character of the wine after several months (Rapp and Marais, 1993). Dimethyl sulfide has also been associated with increased flavor complexity, and the donation of truffle and black olive notes (Segurel et al., 2004).

ADDITIONAL CHANGES

Most wine off-odors do not diminish significantly during aging. Exceptions include hydrogen sulfide and vinylphenols. Hydrogen sulfide tends to oxidize and slow, acid-catalyzed reactions between ethanol and 4-vinylphenol and 4-vinylguaiacol generate 4-(1-ethoxyethyl)-phenol and 4-(1-ethoxyethyl)-guaiacol, respectively (Dugelay et al., 1995). The latter by-products have little if any influence on wine flavor.

A number of age-related changes develop in sherries that generally do not occur in table wines. Notable are increases in the concentrations of aldehydes and acetals. These develop under the oxidizing conditions prevalent during sherry maturation. Similar events also occur in the aging of wood ports, where acetaldehyde reacts with glycerol generating several heterocyclic acetals (da Silva Ferreira et al., 2002). These accumulate to concentrations that could contribute to the sweet, aged bouquet of wood ports. In table wines, the aldehyde content generally declines after bottling. Consequently, aldehyde-derived acetals do not accumulate during aging. An exception involves Tokaji Aszú wines. These are often exposed to oxidizing conditions during maturation (Schreier et al., 1976).

Structural rearrangements of the major fixed acids in wine occur during aging. For example, decarboxylation can convert tartaric acid to oxalic acid and citric acid to citramalic acid. The sensory significance of these changes, if any, is unknown.

During aging, a marked increase occurs in the concentration of abscisic acid. Although abscisic acid is important as a growth regulator in higher plants, its generation in wine is probably purely coincidental. Its presence in wine has no known sensory impact.

The concentrations of several groups of compounds are little affected by aging, notably higher alcohols, their esters, and lactones.

Factors Affecting Aging

In the 1950s, there was active interest in accelerated aging (Singleton, 1962). Subsequently, interest in the topic has waned. Accelerated aging simulates some of the changes of prolonged aging, but generally does not generate the desired subtle complexity of cellar-aged wines. For example, exposure of wine to 45°C for 20 days generates changes in 'Chardonnay' and 'Sémillon' wines equivalent to several years' traditional in-bottle aging (Francis et al., 1993). Generally, though, heating is considered to produce undesirable changes.

OXYGEN

Some wines are much more sensitive to oxidation than others. A detailed explanation of these differences is presently not possible. However, variations in the concentration and types of phenols are undoubtedly involved. For example, the most prevalent phenol in white wines is caftaric acid (an ester between caffeic and tartaric acids). After hydrolysis, caffeic acid (an o-diphenol) can significantly increase oxidative browning (Cillers and Singleton, 1990). This results from polymerization with flavonoids. Red grape varieties particularly susceptible to oxidation, such as 'Grenache,' contain high concentrations of caftaric acid and its derivatives. Another important factor affecting the oxidation potential of wine is pH. As the pH rises, the proportion of phenols in the highly reactive phenolate state increases, enhancing potential oxidation. Finally, the reactivity of the various anthocyanins, differences in their composition, and various condensation products, notably polymerization with tannins, greatly affect susceptibility to oxidation.

Oxygen has usually been thought to disrupt the development of a bottle bouquet. Thus, considerable effort has been expended in protecting bottled wine from oxygen uptake. To assure that bottled wine remains essentially under anaerobic conditions, good-quality closures (corks or pilfer-proof screw caps) are typically used, and oxygen exposure during bottling held to an absolute minimum. Sulfur dioxide is typically added as a protection against the oxidative browning, notably with white wine. Ascorbic acid has often been added for the same purpose. However, its long-term benefits are doubtful (Skouroumounis et al., 2005). In addition, the debate about natural, synthetic and screw cap closures has exposed the degree of our ignorance about the deficit/benefit consequences of minimal, but protracted, oxygen uptake on in-bottle wine development.

The autooxidation of o-diphenols and the accumulation of acetaldehyde to above threshold values has normally been considered the principal undesirable oxidative reactions. Although important in the long-term, Escudero et al. (2002) have shown little if any increase in acetaldehyde content one week after saturation with oxygen. This anomaly may relate to the short duration of the oxidation period, rapid consumption of acetaldehyde, or the low concentration of *o*-diphenols in the white wines studied. The principal descriptors for the treated wines were "cooked vegetables" and "pungent." These attributes were correlated with the presence of furfural and hexanal, respectively. Other aromatic constituents potentially involved were 2-nonenal, 2-octenal, and benzaldehyde. Escudero *et al.* (2000) also associated methional (mercaptopropionaldehyde) with a cooked cabbage odor in oxidized wine. Separate studies with white wines by Silva Ferreira *et al.* (2003) correlated oxidative odors with the production of methional, phenylacetaldehyde, 4,5-dimethyl-3-hydroxy-2(5H)-furanone (sotolon), and trimethyl-1,5-dihydronaphthalene (TDN), particularly at high (45 °C) temperatures and/or low pH values. These compounds were, however, produced in low concentrations at low (15 °C) temperatures, even after 59 days and several saturations with oxygen. It is debatable whether some of the attributes generated by oxidation, reminiscent of honey (phenylacetaldehyde), hay and wood (eugenol), should be considered oxidation off-odors, or the benefits of aging.

Although white wines have a lower content of oxidizable phenolic compounds, they are more susceptible to oxidative browning than red wines. This apparent anomaly results from the ability of many wine phenolics to consume large amounts of oxygen, retarding the undesirable sensory consequences of oxidation. In addition, the paler color and milder flavor of white wines make the consequences of oxidation evident earlier. Factors, such as high pH, cultivation in warm climates, and fungal infection, significantly increase the risk of oxidative browning.

Oxidative browning of bottled wine is universally considered undesirable by wine professionals. Despite this, there is little evidence that most consumers are equally concerned or critical of the visual consequences of oxidative browning. In most cases, unless oxidation is extensive, browning results only in the development of a yellowish or golden tint. Development of obvious browning may require considerable oxygen uptake (Fig. 8.21). In some cultivars, such as 'Sauvignon blanc,' in-bottle oxidation may initially lead to a pinkish blush (pinking) (Singleton *et al.*, 1979).

Browning is usually measured by assessing increased absorption at 420 nm. Nevertheless, this may be inadequate, due to the diverse nature of the pigments involved (Pedretti *et al.*, 2007).

Although much of the attention concerning in-bottle oxidation relates to browning, and the formation of oxidative odors, oxidation may provoke other forms of quality deterioration. This is particularly noticeable with cultivars dependent on monoterpenes for their varietal character. Although their oxidation products are still volatile, they have higher sensory thresholds and possess less appealing odors. Loss of fruitiness also occurs in cultivars not dependent on terpenes for their varietal aroma. An example is the oxidation of varietal thiol flavorants to disulfides. Some ethyl and acetate ester contents also decline rapidly on exposure to air (Roussis *et al.*, 2005).

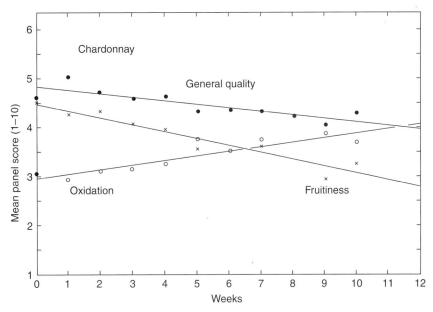

Figure 8.21 Mean panel scores for general quality, fruitiness aroma, and oxidation flavor for a 'Chardonnay' wine exposed weekly to 25.0 ml air of fresh headspace air (~8 ml oxygen per 750 ml bottle). (From Singleton *et al.*, 1979, reproduced by permission)

In bottled wine, oxygen ingress probably involves diffusion through or next to the cork. Oxygen may also dissolve out of cellular voids in the cork into the wine. Additional factors affecting oxidation include the wine's pH, tannin, copper and iron contents. Although oxidation results in the formation of acetaldehyde, its binding with other wine constituents normally keeps its free, volatile content below the detection threshold. Only after considerable oxidation does acetaldehyde accumulate to a point that it could generate an oxidized odor.

One of the principal problems in limiting in-bottle browning is determining the wine's oxidative susceptibility. New laboratory techniques may help in this regard (Oliveira et al., 2002; Palma and Barroso, 2002). They may permit better assessment of the degree of protection required.

In contrast to the aging of table wines, limited oxygen exposure is permitted for many sweet fortified wines. Depending on the method employed, these wines may be characterized by high concentrations of common wine constituents, or relatively unique by-products of oxidative aging. For example, sherries and similar wines are distinguished by the presence of acetaldehyde, acetals, and the lactone, sotolon. Sotolon is thought to be one of the distinctive compounds donating a *rancio* attribute to oxidized red and white sweet wines (Cutzach et al., 1999). The Vins Doux Naturels of Rousillon have also been characterized by the slow development of several unique ethyl esters (Schneider et al., 1998), notably ethyl pyroglutamate, ethyl 2-hydroxyglutarate, and 4-carbethoxy-γ-butyrolactone. They possess honey, chocolate and coconut fragrances. Acetals commonly found in oxidatively aged sweet wines also include compounds such as furfural and 5-(hydroxymethyl)furfural.

TEMPERATURE

To avoid loosening the seal between the cork and the bottle, wine needs to be stored under relatively stable temperature conditions. Rapid temperature changes put pressure on the cork/neck seal by generating sudden fluctuations in wine volume. If this is sufficiently marked, or repeated, rupturing the cork seal would facilitate oxygen ingress. If the wine freezes, the volume increase can be sufficient to force the cork out of the bottle.

Temperature also directly influences the rate and direction of wine aging. Because aging reactions are primarily physicochemical, heat both speeds and activates most of the reactions involved. Thus, cool storage (<10°C) tends to retain the fresh, fruity character of most young wines. For example, the concentrations of fragrant acetate esters, such as isoamyl and hexyl acetates, are stable at 0°C, whereas they rapidly hydrolyze at 30°C (Marais, 1986). In contrast, the formation of less aromatic ethyl esters is rapid at 30°C, but negligible at 0°C. Temperature also has a marked effect on age-induced changes, such as the liberation of norisoprenoid aromatics from their glycosidic precursors (Leino et al., 1993). This may account for some of the increased concentration of TDN in 'Riesling' wine aged at 30 versus 15°C (Marais et al., 1992), as well as the content and types of monoterpene alcohols found in some wines (Rapp and Güntert, 1986). Conversely, other age-related changes are accelerated at cold temperatures. The most well known is the activation of potassium tartrate crystallization.

Heating favors the degradation of carbohydrates to furfurals and pyrroles. Whether a similar activation affects the conversion of norisoprenoid precursors to spiroesters, such as vitispirane and theaspirane, or to hydrocarbons such as TDN and ionene, is unknown. For most wines, prolonged exposure to high temperatures (≥40°C), rapidly show quality deterioration. Carbohydrates in the wine undergo Maillard and thermal degradation reactions, turning brown and producing a baked (madeirized) flavor. Wines also tend to develop a sediment. Even temperatures as low as 30°C show detectable losses in fragrance within a few months.

From the data available, it appears that traditional cellar temperatures (about 10°C) permit the prolonged retention of most fruit esters, while not excessively inhibiting other aging reactions. Nevertheless, temperatures up to 20°C do not appear inimical to the sensory changes desired during aging, at least for red wines. Some of the changes that accrue under different storage temperatures are illustrated in Fig. 8.22.

LIGHT

Exposure to sunlight is generally considered detrimental to wine quality. It can cause heating that may undesirably speed aging, as well as increase wine volume, putting pressure on the cork. In addition, exposure to near-ultraviolet and blue radiation can activate detrimental oxidative reactions. The best known is the shrimp-like or skunky odor of the champagne fault termed light-struck (*goût de lumière*) (Carpentier and Maujean, 1981). Light activates the synthesis of several sulfur compounds, including dimethyl sulfide, dimethyl disulfide, and methyl mercaptan. This synthesis is considered to involve photodegradation of sulfur-containing amino acids, by the activation of riboflavin or its derivatives. D'Auria et al. (2002) have identified an additional source of light-struck off-odors in champagne. In this instance, 2-methylpropanol, and a marked reduction in fruit ester content, appeared to be instrumental. The greater importance of these reactions in sparkling wines may be associated with increased volatilization associated with effervescence.

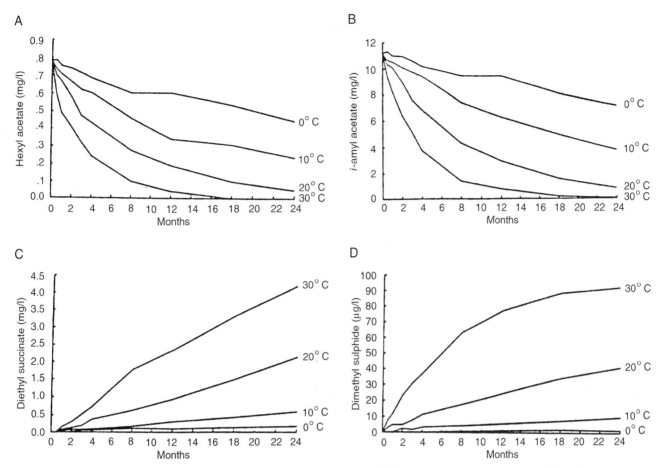

Figure 8.22 Effect of storage temperature and duration on the concentrations of (**A**) hexyl acetate, (**B**) *i*-amyl acetate, (**C**) diethyl succinate, and (**D**) dimethyl sulfide in a 'Colombard' wine. (From Marais, 1986, reproduced by permission)

Light also can promote the production of a potent oxidant, singlet oxygen. The reaction requires both a photosynthesizing pigment and free molecular oxygen. In addition, light exposure can promote the formation of copper *casse*. For these reasons, wine should be stored in darkness whenever possible. Bottling in glass opaque to ultraviolet and blue radiation is also prudent.

VIBRATION

Wines are normally stored in areas free of vibration. Although vibration is commonly considered to disrupt or accelerate aging, no evidence to support this view has been presented. Old claims, concerning the beneficial aspects of vibration, probably refer to facilitated clarification rather than aging.

pH

Wine acidity and pH can affect the rate, as well as the chemical nature of changes that occur during aging. The best understood of these include the effects of pH on red wine color and stability, and phenolic susceptibility

to oxidation. pH also markedly affects the equilibrium between esters and their alcohol and acid precursors (Garofolo and Piracci, 1994).

Rejuvenation of Old Wines

Cuénat and Kobel (1987) have proposed a process for the rejuvenation of wines that have lost their fresh character. It involves dilution of the affected wine with water. The water, and the chemicals presumably involved in the development of the undesired aged character, are subsequently removed by reverse osmosis. Ultrafiltration may have a similarly beneficial effect in rejuvenating old wines.

Aging Potential

Much has been made of a wine's aging potential, or lack thereof. Despite this, wine critics often differ considerably in their advice. For example, French authors often recommend considerably earlier consumption of Bordeaux wines than their British counterparts. Cultural (or habituation) seems to play a part in these differences. Even if wine storage conditions were constant, individual perception is highly variable and strongly influenced by environmental conditions.

How long a wine should be aged primarily depends on personal preference. If a consumer prefers their wine possessing a fresh fruity fragrance and/or showing a distinctive varietal or stylistic character, than aging potential is of little concern. The wine should be drunk within a few years of bottling. This also applies if the consumer enjoys (or is relatively insensitive to) the astringent taste of many young red wines. Preferring coffee strong and black, probably indicates a potential to appreciate strongly astringent red wines. Contrariwise, if one prefers coffee or tea with sugar and cream, it is likely one would prefer wine more mature. Such a consumer may also prefer reds made from 'Pinot noir' or by carbonic maceration. The latter are rarely astringent. Individuals who prefer more subtle flavors are also likely to prefer older wines, despite their having lost much of their original fresh fruity character. If a pleasant bouquet replaces the diminishing varietal character, the wine definitely has aging potential. Whether it is worth the wait depends on one's preference, patience, or wealth.

Probably the most significant factor affecting aging potential is storage temperature. When people ask at what temperature they should store their wines, I half-facetiously ask them how long they expect to live. If you anticipate at least another twenty years, you can probably store with confidence your best wines at 10°C (typically viewed as near ideal). Although no empirical equation predicts how temperature affects maturation rate, a rule of thumb is that aging probably doubles with each 10°C increase. Above 30°C, aging not only occurs much more rapidly, but its nature changes. To most professionals, these changes are not for the better.

For the majority of consumers, though, aging potential has no significance. Most wines are consumed within days, if not hours, of purchase.

Oak and Cooperage

Barrels probably evolved as an extension of skills developed in producing wooden buckets or tanks. Possibly the oldest illustration is a drawing from the tomb of Hesi-Re (ca. 2630–2611 B.C.). It shows a wooden barrel produced from beveled planks, held together by bentwood hoops (Quibell, 1913). It apparently was used as a corn measure. A painting in the tomb of Rekhmire in Thebes (ca. 1400 B.C.) also shows what appear to be barrels, one with straight sides and the other with curved ends (Davies, 1935). Herodotus (ca. 485–425 B.C.)

reports that wine was transported down the Euphrates from Armenia in palm-wood containers. Regrettably, it is unclear whether these had curved sides. Unequivocal written and archeological evidence of wooden cooperage, with hoops, used for wine appear only in Imperial Roman times (Plate 8.1). Pliny the Elder credits peoples from the Alps for their production. Roman barrels were typically longer and thinner than barrels today. They possessed an average diameter/length ratio of about 1:3, in contrast to the more typical current standard of 1:1.4. Examples have been excavated throughout Europe (Ulbert, 1959) and in England (St John Hope and Fox, 1898). Possibly the most well-known illustration appears in the Bayeux Tapestry, chronicling the Norman invasion of England in 1066. Barrels of modern dimensions became standard in the early 1500s. Nevertheless, barrels possessing modern attributes were also apparently produced earlier (Laubenheimer, 1990).

Oak has been used in wine cooperage construction since at least Roman times. Although other types of wood have been used during this period, their use has largely been limited to the construction of large storage cooperage. In Europe, chestnut (*Castanea sativa*) and acacia (*Robinia pseudoacacia*) were employed for this purpose. However, their use, along with oak, has largely been supplanted by inert materials. Similarly, the former use of wooden barrels as the primary container from transporting wine has been supplanted by the glass bottle. Thus, oak use is now principally restricted to the maturation (and occasionally fermentation) of wine. It is particularly popular in the maturation of premium wines. The flavor and occasional slight oxidation provided by in-barrel maturation enhance the character of wines with distinctive varietal aromas.

Oak Species and Wood Properties

Not only does white oak possess the properties required for tight cooperage, but its traditional use has led to an appreciation of (or habituation to) its subtle fragrance. Other woods have either undesirable structural or aromatic characteristics, or have been studied insufficiently to establish their applicability.

Quercus alba, Q. robur, and *Q. sessilis* are the species most commonly used. *Q. alba* and a series of six related white oak species (*Q. bicolor, Q. lyrata, Q. macrocarpa, Q. muehlenbergii, Q. prinus,* and *Q. stelata*) constitute the oaks employed in the construction of American oak cooperage. *Q. alba* provides about 45% of the white oak lumber produced in North America. It has the widest distribution of all American white oak species (Fig. 8.23C) and has the size and structure preferred for select oak lumber. In Oregon, wood from *Q. garryanna* is a new source of oak flavors.

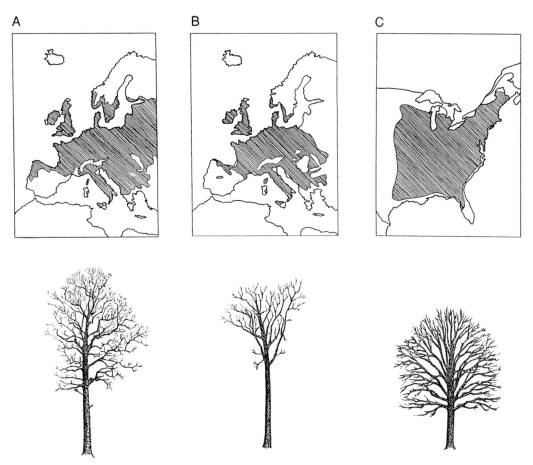

Figure 8.23 Geographic distribution and winter silhouette of (**A**) *Quercus sessilis*; (**B**) *Q. robur*; (**C**) *Q. alba*. (Artwork courtesy of Herman Casteleyn)

In Europe, *Q. robur (Q. pedunculata)* and *Q. sessilis (Q. petraea, Q. sessiliflora)* constitute the primary white oak species employed in cooperage production. Although not genetically isolated, and showing considerable morphologic variability, species identification is generally possible using a combination of features, including leaf and acorn morphology. Differences in chemical composition exist, with *Q. sessilis* tending to possess considerably higher levels of extractable aromatic compounds (oak lactones, eugenol and vanillin), but lower concentrations of ellagitannins, ellagic acid and dry extract than *Q. robur* (Doussot *et al.*, 2000). Nevertheless, individual variation is too extensive to permit unequivocal identification based only on chemical analysis (Mosedale *et al.*, 1998). The same situation exists using wood anatomy (Feuillat *et al.*, 1997).

Both species grow throughout much of Europe (Fig. 8.23A,B). *Quercus robur* does better on deep, rich, moist soils, whereas *Q. sessilis* prefers drier, shallow, hillside soils. Nevertheless, species distribution does not necessarily reflect these ecological preferences. Although *Q. robur* can quickly establish itself in sunlit areas, it is slowly replaced by the more shade-tolerant *Q. sessilis*. In addition, nonselective acorn collection, used in silvicultural plantings, has tended to increase the proportion of *Q. robur*. It is more productive in acorn production.

Staves produced from different American white oak species are almost indistinguishable to the naked eye. The same is true for the two important white oak species in Europe. They can be differentiated with difficulty, and then only with certainty under the microscope. Both genetic and environmental factors often blur the few anatomical features that distinguish the wood of each species (Fletcher, 1978). That *Q. robur* is often considered to possess a coarser grain (more summer wood and larger growth rings) compared to the finer grain of *Q. sessilis* probably reflects as much the growth conditions favored by the two species as genetic factors.

In North America, most of the oak used in barrel construction comes from Kentucky, Missouri, Arkansas, and Michigan. There has been little tendency to separate or distinguish oak coming from different states or sites. In contrast, the identification of oak by origin is traditional in Europe. Geographical designation may indicate the wood's country (French, Russian), region (i.e., Slovenian,

454 8. Postfermentation Treatments and Related Topics

454 / 8. Postfermentation Treatments and Related Topics

 8. Postfermentation Treatments and Related Topics

Limousin), political district (i.e., Vosges, Allier), or forest (i.e., Nevers, Tronçais) origin (Fig. 8.24).

Conditions affecting growth (primarily moisture) also affect wood anatomy and chemistry. Slow growth generally results in the development of less-dense heartwood, due to the higher proportion of large-diameter vessels produced in the spring. In contrast, rapid growth generates wood with a higher portion of small vessels (summer wood). This results from growth continuing into the summer months. The major deposition of tannins occurs some 10–15 years after vessel formation, when the sapwood differentiates into heartwood. Because deposition occurs predominantly in large-diameter spring vessels, the growth rate indirectly affects heartwood chemistry. The phenolics not only contribute significantly to the flavors extracted during in-barrel maturation, but also resist wood rotting.

Due to the higher proportion of large-diameter vessels, slow-grown wood is softer. The lower percentage of cell-wall material in the wood makes it more pliable than oak that grew rapidly. In France, the properties of slowly grown *Q. sessilis*, found in forests such as Nevers and Allier, are commonly preferred for wine maturation. For brandies, the denser, but less aromatic *Q. robur*, found in the Limousin region is preferred. The properties and origin of the wood preferred depend largely on the desired balance between varietal and oak attributes in the finished wine.

In addition to growth-rate induced variations, structural and chemical differences occur throughout the tree. More extractable ellagitannins occur in the heartwood at the base of the tree than near its crown, and in heartwood close to the sapwood. This may reflect increased phenolic deposition as the tree ages, as well as hydrolysis to ellagic acid and oxidative polymerization to less soluble forms. American white oaks tend to have lower levels of extractable ellagitannins than its European counterparts. Significant variations in the concentration of oak lactones and vanillin have also been observed across the grain of wood (Masson *et al.*, 1996). Nevertheless, variation among individual trees is often more marked than average differences between species.

Wood properties reflect both the climatic conditions prevalent in the region, as well as the silvicultural techniques used to maintain forest productivity. A classic example is the denser tree spacing used in Tronçais. This is based on the view that thin annual growth rings (narrow grain), associated with slow growth, generates higher quality wood. In an extensive study of *Quercus robur* and *Q. sessilis*, Doussot *et al.* (2000) found that grain (ring width) was poorly correlated with both extractable ellagitannins and volatile compounds. Ring width (increase) does correlate with the tendency of the wood to shrink more on drying (Vivas, 2001), and reduced oxygen diffusion (Vivas *et al.*, 2003).

Although most chemical variations arise from differences in growth rate and tree age, dissimilarities also originate from genetic divergence. The most significant occur between *Q. alba* (and related species) and European species (*Q. sessilis* and *Q. robur*). American oak appears to possess about 40% of the extractable phenolics of European oak (Singleton *et al.*, 1971). However, not all European oak samples possess high tannin contents (Hoey and Codrington, 1987). In addition, *Q. sessilis* generally contains considerably less extractable phenolics than *Q. robur*. Winemakers desiring higher tannin and phenol levels may choose oak, such as that from Limousin, whereas those preferring lower levels could use *Q. alba*, or the mild *Q. sessilis* of Germany. Those preferring intermediate tannin values may choose one of the forms of *Q. sessilis* grown in France. There is some evidence that wine color stability (higher content of polymeric anthocyanins) develops more rapidly in American than French oak (Pérez-Prieto *et al.*, 2003).

Although significant differences exist in the levels of extractable tannins between American and European oaks, the intensity of oak flavor is similar, albeit different in character (Singleton, 1974). For example, American oak possesses markedly higher levels of isomers of the volatile norisoprenoids, 3,4-dihydro-3-oxoactinidol and oxoedulan derivatives, than Vosges oak (Sefton *et al.*,

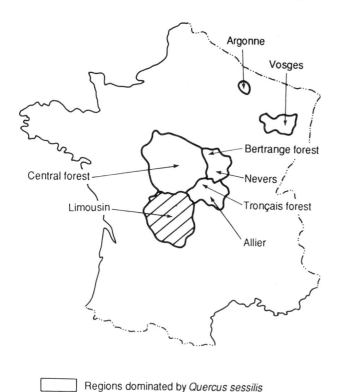

Figure 8.24 Location of the major oak forest in France. Outlined areas are dominated by *Quercus sessilis* and hatched areas by *Quercus robur*. (Modified from Seguin-Moreau, reproduced by permission)

1990). Oak species also differ in oak lactone content (isomers of β-methyl-γ-octalactone), and probably in sesquiterpene, hydrocarbon, and fatty acid concentrations. For example, *Q. alba* has often been found to possess the highest oak lactone content (especially the more aromatic *cis* isomer), whereas *Q. robur* has the lowest. Nevertheless, in a study by Spillman *et al.* (1996), *Q. sessilis* had the highest oak lactone content, whereas *Q. alba* had the lowest. As with extractable tannins, *Q. sessilis* frequently is intermediate in oak lactone content, between *Q. alba* and *Q. robur* (Chatonnet and Dubourdieu, 1998a). *Q. alba* is also distinguishable from European white oaks by the presence of isomers of 3-oxo-*retro*-α-ionol (9-hydroxy-4,6-megastigmadiene-3-one) (Chatonnet and Dubourdieu, 1998a). Differences between specimens of the same species, obtained from different sites, are illustrated in Fig. 8.25.

Habituation to the flavor characteristics of locally or readily available supplies of oak probably explains much of the preferential use of one source over another. Europeans have developed traditional associations with oak derived from particular regions. For example, Spanish vintners customarily prefer American oak cooperage, whereas French producers tend to favor oak derived from their own extensive forests (4.2 million hectares). Intriguingly, the destruction of local oak forests probably explains the nineteenth-century French preference for oak from Russia and Germany, followed by North America, the Austrian Empire, and finally France (Maigne, 1875). French studies support the high quality of Russian oak (Chatonnet, 1998). In another study, eastern European oaks appeared to be intermediate between French and American oak (Prida and Puech, 2006). Although matching oak flavor with the wine remains subjective, progress in oak chemistry may soon facilitate decision-making. Differences in sensory characteristics often can be recognized by trained panels when identical wines are aged similarly in oak of different origin, seasoning, toasting, or production technique. Nevertheless, these influences are frequently incredibly difficult to recognize in nonidentical wines. This indicates that oak extractives are simply another component in the interplay of wine flavors – whose sensory input is often difficult to predict.

With all the sources of variation in oak flavor (Table 8.3), the only way to achieve some degree of standardization is to use barrels incorporating a randomized selection of staves from a relatively common source (standard practice), blend wine matured in a large selection of barrels, and frequently sample to assess that the wines are developing the characteristics desired.

PRIMACY OF OAK

The demands placed on wine cooperage require that the wood possess very specific properties. The wood must be straight-grained, that is, possess vessels and fibers running parallel to the length of the trunk, with no undulating growth patterns or vessel intertwining. In addition, the wood should exhibit both strength and resilience. Structurally, the wood must be free of faults that could make the cooperage leaky. The wood also must be free of pronounced or undesirable odors that could taint the wine. In all these aspects, *Q. alba*, *Q. robur*, and *Q. sessilis* excel. The trees also grow large, straight, and tall. This minimizes wood loss during stave production. Furthermore, white oaks combine two relatively unique quality features, large rays and tyloses. With oak's other characteristics, they make oak the wood of choice in constructing tight cooperage.

All trees produce **rays**, collections of elongated cells positioned radially along the trunk axis. Rays function in conducting water and nutrients between the bark and wood. In oak, the rays are unusually large, being

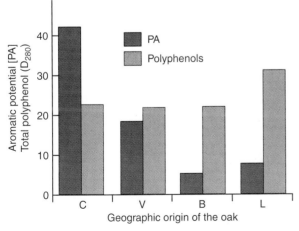

Figure 8.25 Aromatic (dark green) and polyphenolic (light green) profiles of European oak from various regions in France. C, Central group; V, Vosges group; B, Burgundy group; L, Limousin group. Aromatic potential (PA) is based on the concentrations of β-methyloctalactone isomers and eugenol content. (After Chatonnet, 1991, reproduced by permission)

Table 8.3 Summary of sources of variation in oak flavor found in wine

Oak species (coarse/fine grain, tyloses, chemistry, rays)
Geographic origin (rate of growth and ratio of spring to summer wood)
Location along length of tree trunk
Method of drying/seasoning (kiln drying versus the climatic conditions prevalent in the location where external seasoning occurred)
Type of barrel production (steaming versus firing)
Level of toasting
Nature of barrel conditioning prior to use
Size of cooperage, duration, and cellar conditions during maturation
Repeated use (with or without shaving and retoasting)

upward of 15–35 cells thick and 100 or more cells high. In cross-section, the rays resemble elongated lenses. Because staves are split (or sawed) along the radius, the broad surface of the stave runs roughly parallel to the rays. The radial plane becomes the inner and outer surfaces of the cooperage. The high proportion of ray tissue in oak (~28%) and its positioning parallel to the cooperage circumference make rays a major barrier to wine and air diffusion. Wine diffusing into ray cells is deflected along the stave width. Continued lateral flow is limited by nonalignment with the rays in adjacent staves. Wine would have to navigate a very tortuous route past five or more large rays to diffuse out through the sides of a barrel. In practice, wine seldom penetrates more than about 6 mm into oak staves (Singleton, 1974).

Positioning the radial axis of the wood tangential to the barrel circumference has additional benefits in the construction of tight cooperage. The large number of rays permits only minor circumferential swelling. The swelling (~4%) is sufficient, however, to help compress the staves together and seal the joints. Positioning the radial plane of the wood outward also directs the axis of greatest wood expansion (its tangential plane) inward. In this alignment, an expansion of about 7% (Peck, 1957) does not influence barrel tightness. The negligible longitudinal expansion of the staves has no effect on barrel tightness or strength.

The high proportion of rays gives oak much of its flexibility and resilience. Otherwise, the staves would be too tough to be easily bent to form the curved sides (bilge) of the barrel without cracking. The bilge permits full barrels, weighing several hundred kilograms, to be easily rolled.

Oak produces especially large-diameter xylem vessels in the spring. These are large enough to be seen with the naked eye. The vessels allow the rapid flow of water and nutrients up the tree early in the season. However, the vessels could also make barrels excessively porous, permitting wine to seep out through the ends of the staves. In white oak, the vessels become tightly plugged as the sapwood differentiates into heartwood. The plugging results from the expansion of surrounding parenchyma cells into the empty vessels. These ingrowths are called **tyloses** (Fig. 8.26). Tylose production is so extensive that the vessels become essentially impenetrable to

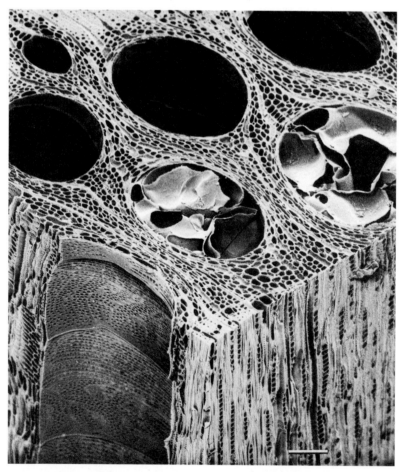

Figure 8.26 Scanning electron micrograph of large vessels of white oak heartwood plugged with tyloses. (Photo courtesy of Dr W. Côté, from Hoadley, 1980)

the movement of liquids or gases. Only heartwood is used in the construction of tight barrels.

The combined effects of rays, tyloses and the placement of the radial plane of the stave tangential to the circumference severely limit the diffusion of air and wine through the wood. With proper construction and presoaking, oak cooperage is essentially an impervious, airtight container.

As sapwood completes its maturation into heartwood, deposition of phenolics kills any remaining living wood cells. The phenolics (primarily ellagitannins) render the heartwood highly resistant to decay. Because mature heartwood contains only dead cells, the lower moisture content makes the lumber less liable to crack or bend on drying. These features give heartwood the final properties required for superior-quality cooperage wood.

Barrel Production

STAVES

For stave production (Fig. 8.27), trees with diameters between 45 and 60 cm are favored (minimum 100–150 years old). Larger trees tend to be reserved for the production of head staves (headings). After felling, the trees may be cut into sections (bolts) equivalent to the stave length desired. They are then split (or sawed) into quarters. The staves are split (or sawed) out of the quarters. In sawing, planks of uniform thickness are cut out, aligning the cuts roughly along the radius (parallel to the rays). In splitting, wedge-shaped planks are removed. Portions too narrow for stave production are removed and discarded. Subsequently, wood is removed from the plank to give it a more uniform width. Because splitting follows the plane of vessel elongation, the sections may be somewhat twisted. Any sapwood associated with a stave piece is removed and discarded. Staves may vary slightly in breadth. Staves with a light pinkish coloration (next to the sapwood) are preferred.

Splitting is generally preferred to sawing because it separates the wood along planes of vessel elongation. This cleaves the staves parallel to the rays. Although oak is "straight-grained," sawing unavoidably cuts across some irregular vessels, increasing surface roughness and potentially enhancing permeability. This is more a problem with European oaks. Their large-diameter spring vessels possess fewer and thinner (more fragile) tyloses than American oak (Chatonnet and Dubourdieu, 1998a). The consequential greater porosity of European oak may also partially explain why the staves release more phenolics than do similarly made American oak staves. Sawing across surface vessels is relatively insignificant with American oak. *Q. alba* possesses sufficiently thick, tightly packed tyloses to make even short, severed vessels liquid and gas impermeable.

With splitting, one side of the staves is shaved obliquely to make the sides parallel. This cuts across wood rays, creating a potential point of leakage. With splitting, only about a quarter of a log (including both sap and heart wood) can be converted into stave wood. Heading pieces are cut out similarly, but are removed from shorter lengths of wood.

Stave length, width, and thickness depend on the volume of wine to be held and the rate of wine maturation desired. To accelerate maturation, barrels constructed of thinner (~2.1 cm), *Château*-style staves may be preferred. For standard barrels, possessing a capacity of 225 liters, staves and headings are roughly 2.7 cm thick.

Once cut, the staves and heading pieces are stacked to dry and season. Natural seasoning for about 3 years is traditional (~1 yr/cm thickness). Stacking each stave row at right-angles favors good air circulation, while close spacing of the stacks diminishes excessively rapid drying (limiting warping or cracking). The stacks are usually dismantled, the staves randomized, and the piles reconstructed every year. This minimizes variation based on positioning within the piles.

It is generally considered that naturally dried oak gives a more pleasant woody, vanilla-like character, whereas kiln drying produces a more aggressive, green, occasionally resinous aspect (Pontallier *et al.*, 1982). The latter may result from a reduction in oak lactone, volatile phenol, fatty acid, and norisoprenoid content, and an enhancement in the concentration of furfural and hydroxymethylfural (Masson *et al.*, 2000). These effects are magnified if drying occurs at higher temperatures and involves green wood (without prior air drying). Kiln drying normally occurs at between 45 and 60 °C. It can rapidly bring newly cut (green) wood down to a desirable moisture content – about 12%. The specific effects of drying method often depend on the species (Chatonnet, 1991). For example, *Q. sessilis* releases more tannins following kiln drying than does *Q. robur*. Although kiln drying decreases the production or release of oak lactones and eugenol, it can increase the availability of *trans*-methyl octalactone from *Q. sessilis*.

With *Q. alba*, natural seasoning increases susceptibility to pyrolytic breakdown, especially cellulosic constituents. The longer the seasoning, the greater the potential production of compounds such as 5-hydroxymethylfurfural, furfural, and 5-methyl furfural (Hale *et al.*, 1999). The effects of air drying are also noticeably influenced by local climatic conditions (Francis *et al.*, 1992; Spillman *et al.*, 2004). Because of the variation in moisture content of air-dried staves, and the potential for undesirable fungal development, it is now common to combine air drying with kiln drying.

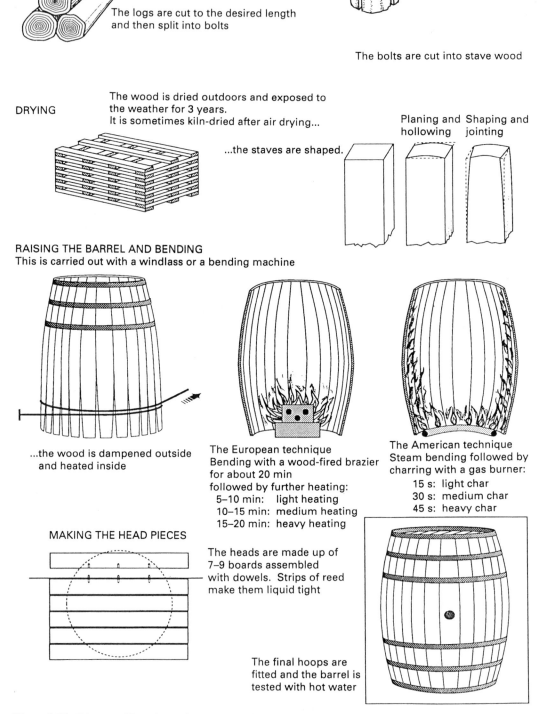

BARREL
MANUFACTURE

SPLITTING

The logs are cut to the desired length
and then split into bolts

The bolts are cut into stave wood

DRYING

The wood is dried outdoors and exposed to
the weather for 3 years.
It is sometimes kiln-dried after air drying...

Planing and   Shaping and
hollowing   jointing

...the staves are shaped.

RAISING THE BARREL AND BENDING
This is carried out with a windlass or a bending machine

...the wood is dampened outside
and heated inside

The European technique
Bending with a wood-fired brazier
for about 20 min
followed by further heating:
5–10 min: light heating
10–15 min: medium heating
15–20 min: heavy heating

The American technique
Steam bending followed by
charring with a gas burner:
15 s: light char
30 s: medium char
45 s: heavy char

MAKING THE HEAD PIECES

The heads are made up of
7–9 boards assembled
with dowels. Strips of reed
make them liquid tight

The final hoops are
fitted and the barrel is
tested with hot water

Figure 8.27   Diagram of barrel manufacture. (Modified from Puech and Moutounet, 1993, reproduced by permission)

Although fungal attack can reduce wood quality, it may also generate some of the benefits of seasoning. Fungal action has the potential to synthesize aromatic aldehydes and lactones from wood lignins (Chen and Chang, 1985). For example, the wood-rotting fungus, *Coriolus versicolor*, produces polyphenol oxidases that can degrade lignins, as well as induce phenolic polymerization. Many fungi have been isolated from the outer few millimeters of staves, but it takes almost a year before penetration of the wood becomes microscopically evident (Vivas *et al.*, 1997). The significance, if any, of the frequent isolation of common saprophytic fungi, such as *Aureobasidium pullans* and *Trichoderma* spp., has not been established.

Polymerization (and reduced solubility) of ellagitannins is a particularly noticeable consequence of natural seasoning (Chatonnet, Boidron *et al.*, 1994b). This undoubtedly affects their extraction by wine. In addition, phenolic oxidation (resulting in the release of peroxide) could favor cellulose hydrolysis (see Evans, 1987). Natural seasoning also can produce changes in the concentration of several oak aromatics. Degradation products of lignin, such as eugenol, vanillin, and syringaldehyde, have been variously found to increase or decrease, notably in the outer portions of staves (Chatonnet, Boidron *et al.*, 1994b). Typically, however, these changes are much less marked than those associated with stave toasting during barrel production. The concentration of the isomers of β-methyl-γ-octalactone also varies during seasoning, increasing or decreasing (Sefton *et al.*, 1993; Chatonnet, Boidron *et al.*, 1994b). A more consistent finding, however, is an increase in the proportion of the more aromatic *cis* isomer of oak lactone. Microbial metabolism could also modify cell-wall constituents. Sugars liberated by cell-wall degradation could increase the furfural content generated during barrel toasting. In addition, the leaching and degradation of phenolic compounds by rain, oxygen, and ultraviolet radiation may be significant. The conversion of the bitter-tasting esculin, to its less-bitter aglycone esculetin, may be another example of how wood character improves with outdoor seasoning.

BARREL ASSEMBLY

In barrel construction, the first step involves checking the staves for knots, cracks, or other structural faults. Once the appropriate number of suitable staves has been assembled, they are **dressed**. Dressing refers to selective shaving in preparation for raising. The first of the dressing procedures, **listing**, tapers the broad ends of the staves to give them their basic shape (Fig. 8.28A). The amount of listing required depends on the desired "height" of the barrel, that is, the length of the staves relative to the maximal circumference of the barrel.

Subsequently, a small amount of wood may be chiseled from the ends (**backing**) and center (**hollowing**) to facilitate bending (Fig. 8.28B,C). Hollowing usually reduces the thickness of the central part of the stave from 27 to 24 mm. The staves are now ready for **jointing**, in which a bevel is planed along the inner edge of the sides of each stave (Fig. 8.28D). Jointing requires considerable skill because the angle changes along the length of the stave. The bevel depends on the barrel height, being maximal at the center (**bilge**) and least at the ends (**heads**). Jointing precision determines the tightness between adjoining staves.

The curved shape of the barrel provides much of its strength. This comes from the engineering principle called the double arch. The sloping sides also provide a point on which the barrel can be pivoted and rolled with comparative ease.

Once dressed, the staves are **raised**. This involves placing the staves (between 28 and 32 for a 225-liter barrel) together in an upright circle. Several temporary hoops, including a trussing **hoop** (runner), help to support the staves (Fig. 8.28E; Plate 8.2). The hoops are forced

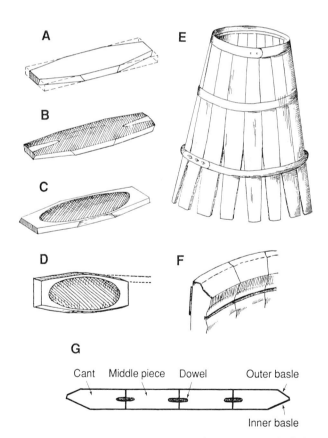

Figure 8.28 Stave preparation and barrel construction: (A) listing; (B) backing; (C) hollowing; (D) jointing; (E) raising; (F) chimed, hoveled, and crozed stave ends; and (G) head cross-section. (After Kilby, 1971, reproduced by permission)

down and begin to force the staves into the curved barrel outline. At this point, the barrel is inverted and placed approximately 5 cm above an open fire (brazier) for softening. The inner and outer stave surfaces are frequently sprayed or splashed with water. Alternatively, the staves may be steamed prior to, or instead of, firing to soften the wood. After sufficient softening, the staves are slowly and periodically pulled together with a windlass (capstan). Positioning temporary hoops holds the staves in place until additional heating (~10–15 min) sets the staves in their curved shape. The firing helps shrink the innermost wood fibers, releasing tension caused by bending. This is termed **setting**.

Additional heating (termed **toasting**) (Plate 8.3) is a comparatively recent innovation in barrel production, not being noted in texts prior to the middle of the nineteenth century. Toasting produces sensory changes in the characteristics of the wood. These result primarily from pyrolysis and thermohydrolysis. Heating may be performed directly over the fire or slightly raised off the floor. The top may be closed with a metal cover or left open. Closed firing requires more frequent moistening, but produces more uniform heating of the barrel's inner surfaces (Chatonnet, 1991; Matricardi and Waterhouse, 1999). Not only does moistening slow the rate of heating, but it also produces steam that promotes the hydrolytic breakdown of hemicelluloses, lignins, and tannins. The inner-surface temperature of the barrel typically reaches 200°C and above. Carbonization (charring) of the wood begins at about 250°C.

The degree and desirability of pyrolysis depend on the style and characteristics of the wine desired. Light toasting, sufficient to ease stave bending (~5 min, inner surface temperatures 100° to 150°C), produces few pyrolytic by-products, leaving the wood with a natural woody aspect (Fig. 8.29). Medium toasting (~15 min, inner surface temperatures >150°C) generates phenolic and furanilic aldehydes (Table 8.4). Phenolic aldehydes, derived from lignins, donate a vanilla roasted character, whereas most furanilic aldehydes, coming from hemicelluloses, may generate a caramel-like aspect. The degradation of hemicelluloses also produces compounds, such as maltol and 2-hydroxy-3-methyl-cyclopentanone. These generate toasty flavors. Heating also may favor the synthesis of oak lactones from precursors (Wilkinson et al., 2004). Medium toasting is generally preferred for Q. sessilis from the forests of central France. Toasting reduces the solubility of oak tannins, particularly useful with European oaks that possesses high levels of soluble ellagitannins. Toasting also activates their degradation, first to ellagic and then gallic acids, and subsequently carbonation. Prolonged exposure (~25 min, inner surface temperatures >200°C) chars the innermost layers of the staves, and destroys or limits the synthesis of phenolic and furanilic aldehydes. These are replaced by volatile phenols, giving the wood a smoky, spicy aspect. Phenolic aldehydes generate guaiacol, 4-methyl guaiacol, and dimethoxy-2,6-phenol, whereas furanilic aldehydes give rise to eugenol and 4-vinylguaiacol. In addition, volatile compounds, such as vanillin and syringaldehyde are progressively destroyed. Heavy toasting also limits the release of phenolic components. Heavy toasting is often preferred for Q. sessilis obtained from southwestern France. With heavy toasting, small (6 μm) fractures, up

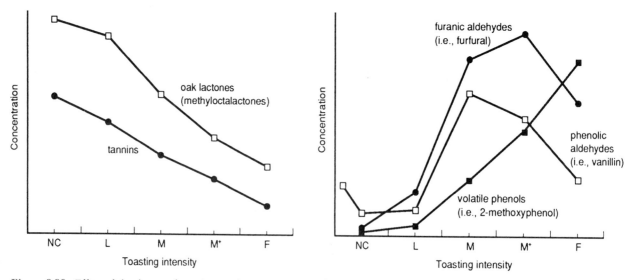

Figure 8.29 Effect of the degree of toasting on the concentration of several compounds extracted from oak barrels constructed from *Quercus sessilis*. NC, not heated; L, light toasting (5 min); M, medium toasting (10 min); M⁺, medium-strong toasting (15 min); F, charring (20 min). (After Chatonnet, 1989, reproduced by permission)

to several millimeters long, appear in the wood. These fissures can penetrate upwards of 600–700 μm into the wood (Hale *et al.*, 1999). These facilitate extraction of constituents from areas less affected by heat. Charring, desirable for Bourbon maturation, is avoided in the production of wine cooperage. Charcoal on the inner surfaces can decolorize red wines, as well as remove desirable flavorants.

Aromatic aldehydes, furfurals, furans, oxygen heterocycles, pyrazines, pyridines, and pyrans are among the many pyrolytic compounds derived from tannins and hemicelluloses. In contrast, volatile phenolics such as guaiacol, 4-methyl guaiacol, vanillin, syringaldehyde, and coniferaldehyde are lignin decomposition products. Toasting also degrades several unsaturated aldehydes, notably (*E*)-2-nonenal, the primary ingredient of the sawdust off-odor occasionally found in wine matured in new oak cooperage (Chatonnet and Dubourdieu, 1998b).

Although barrel manufactures attempt to maintain uniformity in the level of toasting, there are no industry-wide standards for what light, medium, or heavy toasting means. Consequently, there can be considerable variation among manufacturers (Chatonnet *et al.*, 1993b). Significant variation is also detected among barrels assembled by the same producer from a common source of randomly selected staves. The traditional method of toasting over a brassier, burning oak chips, is difficult to control. As well, different portions of the barrel may be exposed to marked temperature variations. Although this increases the incidence of blistering, blisters do not increase the risk of microbial contamination (Vivas, 2001).

The exterior of the fire generates temperatures that can vary from 460 to 600 °C, whereas the interior can vary from 850 to 1000 °C – the wood itself generally does not go much above 200 °C until the later stages of toasting. One technique to reduce the disparity involves a prolonged exposure to a small fire. This produces greater heat penetration, but without the production of dark-colored pyrolytic breakdown products (Hoey and Codrington, 1987). Another potential solution involves the use of several, linear, infrared-heat generators (Chatonnet *et al.*, 1993b). They are arranged in the form of a cone, which is rotated within the barrel. The bottom of the device possesses a crown of water misters, whereas the top possesses a cap that traps the heat. Periodic jets of water provide temporary cooling and generate steam (facilitating heat penetration). Because heating, moistening, and cooling treatments can be programmed, a more standardized product is generated. Alternatively, the wood may be presoaked in hot water. This facilitates bending as well as heat transfer into the wood. Thus, a higher temperature can be used at any toast level, without the risk of charring. This procedure does, however, modify the wood's flavor potential from that of traditional firing, but provides the winemaker with another option in tailoring oak flavor (Schulz, 2004).

Features, such as wood color and temperature control, are too imprecise to be of practical value in defining toast levels (Chatonnet, 1999). Consequently, providing barrels with a consistent toast level is still beyond reach. To date, electronic noses seem to have the greatest potential in providing improved quality control (Chatonnet, 1999). Regrettably, this does not address the other major uncontrolled factor in barrel-to-barrel variation, nonuniformity in wood attributes. Studies by Sauvageot and Feuillat (1999) clearly illustrate the problem. If these features were better understood, variation might be used by winemakers to enhance wine distinctiveness, rather than reduce it.

After setting, and any toasting, the cooper puts a bevel on the inner surface of the stave ends. This is followed by **chiming** – preparing the ends for positioning the **headpieces**. The first task involves planing the ends of the staves (the chime). Shaving the inner edge produces the bevel. Cutting a concave groove slightly below the chime produces the **howel**. A deeper cut into the howel (the **croze**) produces the slot into which the headings fit (Fig. 8.28F).

The outer surface of the barrel is planed to give it a smooth surface, whereas the inner surface is left rough. The rough inner surface aids wine clarification by providing increased surface area for the deposition of suspended particulate matter.

Next, a **bung hole** is bored and enlarged with a special auger to receive a tapered wooden, rubber, or plastic peg. A **tap hole** may also be bored near the end of the central head-stave.

Table 8.4 Aromatic aldehydes produced by toasting or charring oak chips[a]

Product (ppm)	Toasting temperature			
	100 °C	150 °C	200 °C	Charred
Vanillin	1.1	3.8	13.5	2.8
Propiovanillone	0.6	1.1	1.4	0.9
Syringaldehyde	0.1	3.8	32.0	9.2
Acetosyringone	–	0.025	1.5	0.6
Coniferyladehyde	Trace	4.3	24.0	4.8
Vanillic acid	–	1.8	6.1	1.1
Sinapaldehyde	Trace	6.5	60.0	9.0

[a] Oak chips (2%, w/v, in ethanol) were toasted at various temperatures. Charring occurs above 250 °C.

Source: From Nishimura *et al.* (1983), reproduced by permission

If temporary hoops were employed during raising, they are replaced with permanent hoops. For 225-liter barrels, this usually consists of two chime hoops, located just below the heads of the barrel; two bilge hoops, positioned one-third of the way in from the ends; and a set of quarter hoops, placed approximately one-fourth of the way in from the heads. So positioned, the hoops limit the wear on the staves during rolling. At this point, the heading pieces are selected and prepared.

The head consists of several heading pieces, typically between 12 and 16. In contrast to the staves, the joints between the heading are straight, not beveled. In addition, the heading pieces (constituting about 25% of the barrel surface) receive no toasting. Short dowels inserted between each heading piece keep them in alignment (Fig. 8.28G). Caulking with river rushes, called flags, may be used to prevent leakage.

The circular shape of the head is now sawed, in preparation for **cutting** the head. Cutting refers to shaving two bevels, called **basles**, on the upper and lower surfaces of head stave-ends (Fig. 8.28G).

The bottom head is inserted first. Removal of the bottom head hoop allows the head to be forced into the croze. After repositioning the head hoop, the barrel is inverted to remove the opposite head hoop. A heading vice may be screwed into the head and the head lowered sideways into the barrel. The head is pulled up into its groove with the vice. Alternatively, a piece of iron forced in a joint between two staves levers the head into position. The stave alignment of the two heads is positioned perpendicular to one another. This limits the pressure that develops during swelling from acting in the same direction, thus minimizing leakage.

The final task involves hammering the hoops tight. This forces the staves together and closes most cracks. After soaking for about 24 h in water, a well-made barrel becomes leak-proof.

Larger-volume barrels, ovals, and vats are made in essentially the same manner. The primary differences, other than overall size, are stave thickness and degree of curvature. The last affects the need for heating. Typically, large cooperage is not toasted.

COOPERAGE SIZE

Cooperage is manufactured in a wide diversity of sizes, depending on traditional or intended use. In the past, large straight-sided tanks and vats were constructed, possessing capacities greater than 5–10 hl. They often acted as fermentors, as well as storage cooperage following fermentation. Large size minimized oxidation and eased cleansing. However, wooden fermentors and storage tanks have been largely replaced by more durable and easily cleaned tanks, whose standard construction material includes stainless steel, epoxy-lined carbon steel, fiberglass, cement, high-density polyethylene, and flexible plastic storage containers. Wooden cooperage is primarily retained for maturing wine in which an oak aspect is desired. In addition, small cooperage may be used for in-barrel fermentation.

Many of the benefits of barrel use come from their relatively large surface area ($104\,cm^2$/liter for a 225-liter *barrique* vs. $76\,cm^2$/liter for a 500-liter puncheon). Although surface area increases logarithmically with decreasing volume (Singleton, 1974), other factors place practical limits on minimum size. Production economy favors larger size, whereas ease of movement and earlier maturation favors smaller size. A compromise between these opposing factors has led to the widespread adoption of barrels with a capacity of between 200 and 250 liters. Individual regions in Europe often use barrels of a particular capacity. The Bordeaux *barrique* is 225 liters, the Chablis *feuillette* contains 132 liters, the Rhine *doppelohm* holds 300 liters, the sherry butt carries 490.7 liters, and the port pipe possesses 522.5 liters. Premium white wines commonly receive about 3–6 months' maturation in oak, whereas red wines often receive between 18 and 24 months' oak maturation before bottling.

Although much of the literature focuses on the value of maturing wine in small oak cooperage (~225 liters), many fine wines are aged in mid-size to large (>1000 liters) oak cooperage (Plate 8.4). This was characteristic of most European regions until the shift to maturation in small cooperage in the early nineteenth century. Well-sealed tanks restrict oxidation and donate reduced amounts of tannins and oak flavor to the wine. In contrast, the *barrique* has been estimated to allow the ingress of approximately 2–5 mg O_2/liter/year (Ribéreau-Gayon *et al.*, 1976). Large oak cooperage can be used for decades, whereas small cooperage is usually replaced after several uses. Differences in redox potential can develop in large cooperage due to stratification. This requires frequent racking or sampling to assure that hydrogen sulfide and mercaptans do not accumulate to detectable levels.

In one of the few comparative studies on the effects of cooperage size on wine attributes, Pérez-Prieto, de la Hera-Orts *et al.* (2003) found that tasters preferred wine aged in new American oak barrels (220-liter) vs. 500- and 1000-liter oak cooperage. They also found significant differences in the rate and degree of aromatic uptake. This was particularly noticeable with oak lactones and vanillin (Fig. 8.30). The extraction dynamics for different aromatics from oak varies considerably (Spillman, Iland *et al.*, 1998a; Spillman, Pollnitz *et al.*, 1998b), involving both simple extraction, as well as a variety of synthesis and degradative reactions, notably esterification, oxidation, acid hydrolysis and ethanolysis.

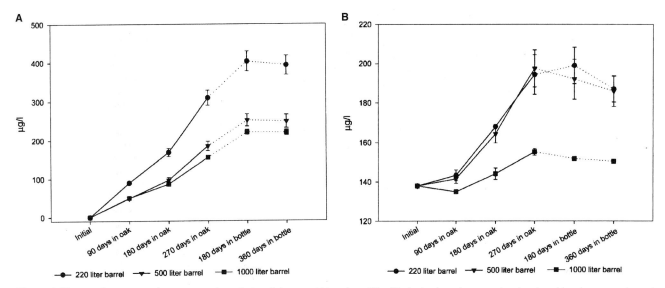

Figure 8.30 Development in the concentration of *cis*-oak lactone (**A**) and vanillin (**B**) during barrel maturation (——) and bottle storage (........) as affected by cooperage size. (From Pérez-Prieto, López-Roca *et al.*, 2003, reproduced by permission. Copyright 2003 American Chemical Society)

Alternatives to the traditional barrel include square oak containers (e.g., Stakvat®). These come with surface areas equivalent to standard barrels, but cost considerably less. They avoid the expense associated with stave bending and barrel raising. They also take less space. Cleaning between fills is easier and new staves can be inserted easily if required. Their use apparently affects wine character by enhancing the extraction of oak lactones, guaiacol, 4-methylguaiacol and eugenol. As with all such changes, their relative desirability depends on the preferences and desires of the winemaker, and the economics of use. Other barrel alternatives are noted below.

CONDITIONING AND CARE

Opinions differ considerably on whether and how to condition new barrels. Furthermore, the need appears to depend on the source and seasoning of the staves. Kiln drying and heavier toasting usually require additional conditioning. Minimal treatment usually involves rinsing and presoaking with cool to lukewarm water – the hotter the water, the more extensive the extraction and loss of important wood flavors (Vivas, 2001). In a comparative study of conditioning techniques, significant differences were noted in the oak flavors subsequently extracted by the wine (Lebrun, 1991). The preferred procedure depended on the type and style of wine. The lighter treatments generally were more appropriate for red wines (benefitting most from the increased intensity and complexity of oak flavors), whereas sparkling wines expressed their subtle features optimally when the base wines were matured in barrels given the strongest conditioning. Nevertheless, the most significant factor in the study was the duration of in-barrel maturation. Additional data can be found in Boidron (1994).

A long-established conditioning procedure is in-barrel fermentation. The constituents most readily extracted are ellagitannins and phenols. After dissolving, they tend to combine with wine constituents, precipitate, and are lost with the lees. Because the desirable flavors in oak dissolve more slowly (for example, oak lactones and aromatics found deeper in the wood), they are not unduly removed by in-barrel fermentation. Although an effective conditioning procedure, the technique is laborious, and the barrels require cleansing before subsequent use in wine maturation. Barrels not subjected to toasting (heated only to facilitate stave bending and setting) may be conditioned with a solution of 1% sodium or potassium carbonate. The alkaline solution accelerates both phenol oxidation and extraction. Subsequently, the barrels are given a thorough rinsing with a 5% solution of citric acid and, finally, a water wash.

In-barrel maturation typically follows an initial clarification of the wine. This minimizes both the adherence of material to the inner surfaces of the barrel and the excessive accumulation of lees. In addition, barrels commonly are racked several times a year to avoid the buildup of a thick sediment layer. These actions decrease the difficult and unpleasant task of barrel cleaning. In addition, potential contamination of the wood with spoilage microorganisms is minimized. However, tradition or personal preference may dictate that the wine be left on the lees for several months. Some vintners believe that the yeast and tartrate coating that develops slow the release of oak flavors.

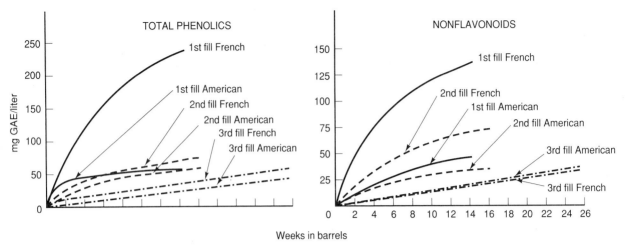

Figure 8.31 Changes in phenolics over time for French and American oak barrels. (From Rous and Alderson, 1983, reproduced by permission)

After use, barrels require cleaning and disinfection. If little precipitate has formed, rinsing with hot water (~80 °C) under high pressure is usually adequate. If a thick layer of tartrates has built up, barrels may require treatment with 0.1 to 1% sodium or potassium carbonate, followed by a thorough rinsing with hot water. After draining, burning a sulfur wick in the barrel usually provides the inner surfaces with sufficient disinfection.

Barrels should be refilled with wine as soon as possible after cleansing and disinfection. If they are left empty for more than a few days, the barrels should be thoroughly drained, sulfited, and tightly bunged. Barrels stored empty for more than 2 months should be filled with an acidified solution at 200 ppm SO_2. Sulfur dioxide inhibits the growth of most microbes, and the water prevents wood shrinkage and cracking (oak is stable only at ≥30% moisture content). Before barrel reuse, the residual sulfur dioxide is removed with several water rinses.

If the cooperage has become contaminated with spoilage microorganisms, various disinfection treatments may be used. Of several procedures compared, only hot water treatment (85–88 °C) for 20 min was effective against acetic acid bacteria (Wilker and Dharmadhikari, 1997). The treatment was more effective prophylactically than as a remedy for extensive contamination. To avoid *Brettanomyces* contamination, Chatonnet *et al.* (1993a) recommend barrel disinfection with at least 7 g SO_2 per barrel and maintenance at 20–25 mg/liter free SO_2. Biofilm production by *Brettanomyces* makes disinfection difficult, especially when the organism becomes established within the wood itself. *Brettanomyces* can survive on constituents found in the wood. Alternatively, ozone generators can sterilize barrels. Ozone (O_3) quickly reverts to molecular oxygen (O_2). However,

Marko *et al.* (2005) found that ozone was only marginally less disruptive to oak aromatic content than hot water. Ozonated water (the form in which it is usually employed) can be used to surface sterilize other types of cooperage, corks, bottles, tubing, and floor surfaces. The major limitation is with standard rubber O-rings, hoses, and filter connections. These must be replaced with ozone-resistant versions.

Treating the outer surfaces of cooperage with 1% rotenone in boiled linseed oil usually controls oak-boring insects, but not fungal growth. Mold growth over the external surface of the barrel may mar its appearance, but does not affect barrel strength, or influence the sensory properties of the wine stored within.

BARREL LIFE SPAN

For certain types of wine, legislation specifies the rate at which barrels must be replaced. In most regions, however, oak use is left to the discretion of the winemaker. Thus, the frequency of reuse can depend as much on economics as on the oak intensity desired. Making these decisions will be facilitated when more is known about the dynamics and sensory impact of oak flavorants.

Figure 8.31 illustrates differences in total and nonflavonoid phenolic extraction from American and French oak. Not surprisingly, the differences are most marked during the first fill. The differences are also more striking with French than American oak. Subsequently, the differences become progressively less marked. Because the rate of nonflavonoid extraction does not drop as rapidly as that of total phenolics, the proportional extraction of nonflavonoids increases with each fill. The sensory significance of this change has not been investigated.

Aromatic extractives, such as furfurals, oak lactones, and phenolic aldehydes, become progressively

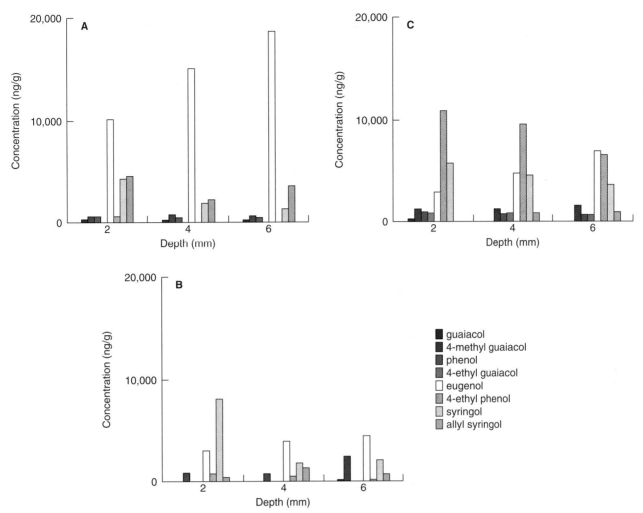

Figure 8.32 Effect of white wine maturation on the level of several volatile phenols present in oak staves at different depths. Oak cooperage used to mature one wine (**A**); three wines (**B**); and five wines (**C**). (After Chatonnet, 1991, reproduced by permission)

exhausted with barrel reuse (Fig. 8.32). In contrast, the extraction of several volatile phenols, with less-pleasant odors, increases (Cerdán *et al.*, 2002). These changes probably reflect chemical extraction in the first instance, and degradative synthesis in the second, or the effects of microbial contamination.

Because wine readily extracts material from oak staves only to a depth of 6–8 mm, shaving off the innermost layers (typically about 4 mm) permits renewed access to oak flavorants. However, as the effect of toasting decreases rapidly away from the innermost surfaces (Hale *et al.*, 1999), shaving exposes wood with different chemical characteristics. Thus, refiring the exposed wood is necessary if reestablishment some of the original toasted aspect is desired. For shaving to be effective, the barrels must be thoroughly cleaned (to remove wine material that may caramelize during firing) and dried (to permit effective planing of the wood and avoid

microbial contamination). Nevertheless, the extractive characteristics are not identical with new barrels, being lower in ellagitannins, furanilic aldehydes, eugenol and vanillin. In addition, there is an increase in the concentration of lignin breakdown products, such as methylphenols (e.g., quaiacol) and dimethoxyphenols (e.g., syringol) (Vivas, 2001). These could donate a burnt aspect to the wine.

To gain some control over the typical between-barrel differences in extractives (see Towey and Waterhouse, 1996), it is common to maintain a constant proportion of new and used barrels (shaved or unshaved). This habit also reduces the intensity of some of the more unpleasant aromatics extracted in higher concentrations from new barrels (Vivas *et al.*, 1995). Blending wine from different barrels before bottling helps generate a relatively uniform oak character in the wine.

Although phenolic extraction does not appear to affect the wood's internal structure (or strength), even after eighty years (Puech, 1984), repeated shaving seriously weakens the barrel's strength.

Chemical Composition of Oak

The major chemical constituents of oak are not markedly different from those of other hardwoods. Because wood consists primarily of dead cells, the major chemical components are cell-wall constituents. In addition, heartwood contains infiltration substances deposited during its differentiation.

CELL-WALL CONSTITUENTS

The cell walls of oak heartwood are composed of about 50% cellulose, 20% hemicellulose, and 30% lignins. Cellulose consists of long fibers of polymerized glucose, locally grouped together in bundles called micelles. The fibers are deposited in different planes, forming interlacings resembling the plies of a tire. These are immersed in a matrix of hemicellulose and lignin polymers. In oak, hemicelluloses are predominantly polymers of xylose. Lignins are complex polymers, formed primarily from oxidative coupling of 4-hydroxyphenylpropanoids.

Cellulose gives wood much of its strength and resilience, whereas lignins limit water permeability and provide much of the wood's structural strength. Hemicelluloses act as binding substances, along with pectins, to hold the cellulose and lignins together.

Because of the high resistance of cellulose to both enzymatic and nonenzymatic degradation, cellulose is unlikely to be involved in the development of oak flavor. However, hemicelluloses slowly hydrolyze after exposure to the acidic conditions of wine, releasing both sugars and acetyl groups. The acetyl groups may be converted to acetic acid during maturation. Hydrolysis is significantly increased during stave toasting. Heating also converts some of the sugars to a wide array of furan aldehydes and ketones, such as furfural, 5-(hydroxymethyl)-2-furaldehyde and 5-methyl-2(3H)-furanone.

Lignins are large, complex, three-dimensionally branched phenylpropanoid polymers. In hardwoods, such as oak, the phenylpropanoid units contain either hydroxyl or methoxyl groups. These form coniferyl (2-methoxyphenol) and sinapyl (2,6-dimethoxyphenol) alcohols, respectively. These polymerize into guaiacol and syringyl lignins. Most lignin polymers contain both phenolic alcohols. A small proportion of lignins, called native lignins, are ethanol-soluble and dissolve readily in wine. Lignins may undergo ethanolysis and be subsequently oxidized to aromatic compounds. Pyrolysis enhances degradation, leading to the production of aromatic phenolic aldehydes, such as syringaldehyde and

vanillin; phenolic alcohols, such as eugenol, quaiacol and syringol; and phenolic ketones, such as acetovanillone and acetosyringone. The hydrophobic nature of lignins limits water penetration and, thus, the denaturing action of hydrolytic enzymes on wall constituents.

CELL-LUMEN CONSTITUENTS

When wood cells die, the cytoplasm degrades, leaving only a central cavity (lumen) formed by the cell wall. Later, when sapwood matures into heartwood, phenolic compounds are deposited in the lumen (up to 7–20 times the level in corresponding sapwood). These are principally hydrolyzable ellagitannins, synthesized by surrounding parenchyma cells. On hydrolysis, ellagitannins yield ellagic and gallic acids. As noted, the tannin content of oak varies considerably, depending on the species, growth conditions, tree age, and position in the trunk. For example, the soluble ellagitannin content of heartwood increases from the inside outward. Insoluble ellagitannins show the reverse trend (Vivas *et al.*, 2004).

Other phenolic compounds extracted from oak heartwood include cinnamic acid derivatives, namely, *p*-coumaric and ferulic acids. Yeasts and lactic acid bacteria may convert these to aromatic phenols, such as 4-ethyl phenol, 4-vinylguaiacol, and 4-ethyl guaiacol (Dubois, 1983). Lyoniresinol is another important phenolic compound derived from oak (Moutounet *et al.*, 1989).

Additional components found in small quantities in oak heartwood include carotenoids, lactones, fats, sterols and resins. Carotenoids (linear polymers of eight isoprene units), such as β-carotene and xanthophyll, may oxidize or degrade during toasting, generating a diversity of aromatic norisoprenoids, terpenes and sesquiterpenes. These could add to the floral or fruity fragrance of wines (Nonier, 2003). One of the breakdown products, 3-oxo-α-ionol, has a high sensory threshold. However, upon dehydration it generates megastigmatrienone isomers, known to be impact compounds in the fragrance of tobacco.

Typical oak flavorants, such as oak lactones are found in lumen of xylem cells. They may also occur as glycosides. This means that, during maturation and in-bottle aging, hydrolysis could slowly increase their sensory impact. This situation also appears valid for eugenol, vanillin and syringaldehyde (Nonier, 2003). Heat-induced hydrolysis may also increase the concentration of octanoic through octadecanoic acids (Chatonnet, 1991).

COMPOUNDS EXTRACTED FROM, AND ABSORBED BY, OAK

The solubility of oak constituents, and their degradation products, varies widely. Extraction of even small amounts of volatile compounds could affect the bouquet, if they possess low thresholds. In contrast, much higher amounts of nonvolatile constituents must be absorbed to influence taste or appearance. For example, wine dissolves about 30% of the tannins from the innermost few millimeters of wood. Because they are largely nonvolatile, this is sufficient to affect color, taste, mouth-feel, and occasionally fragrance. In contrast, wine extracts only about 2% of oak lignins. However, because their breakdown products tend to have low olfactory thresholds, they can easily influence wine fragrance. More than 200 volatile compounds have been identified from oak.

Quantitatively, nonvolatile phenolics are the most important group of oak extractives. Of these, about two-thirds are nonflavonoids. The hydrolyzable tannins (ellagitannins) make up the most significant subgroup of oak nonflavonoids. They can constitute up to 10% of the mass of oak heartwood. Eight ellagitannins have been isolated from oak. The most common of these are the stereoisomers, vescalagin and castalagin, esters of hexahydroxydiphenic acid and glucose. The others are either dimers of these two, or possess a pentose subunit. As the wood ages, polymerization increases. American oaks are characterized by less extractable ellagitannins than their European counterparts (Vivas *et al.*, 1996; Chatonnet and Dubourdieu, 1998a). Although potentially quickly dissolved into wine, the amount of ellagitannins found in wine is low. This partially results from toasting, which hydrolyzes most of the polymers in the surface layers of the wood. Those extracted from the inner layers of oak tend to hydrolyze into their ellagic acid precursors (Vivas *et al.*, 1996). Limited presence also results from binding and precipitation with polysaccharides and proteins. These features, plus their low astringency, probably explain why hydrolyzable tannins seem to have minimal influence on the astringency of white wine matured in oak (Pocock *et al.*, 1994). Lignin-degradation products form the second most important group of extracted phenolics. Their solubilization depends largely on the alcoholic strength and acidity of the wine. Both factors also are involved in the degradation of tannins and lignins to simpler, more soluble compounds.

Oak tannins appear to indirectly play a role in color stability, by promoting anthocyanin polymerization with condensed tannins, through the aegis of acetaldehyde (Vivas and Glories, 1996). Hydrogen peroxide, generated as a result of tannin oxidation, oxidizes ethanol to acetaldehyde. The acetaldehyde produced may also favor the production of stable wine pigments, such as vitisins. Regrettably, the dynamics of the survival of ellagitannins in wine is little known. Thus, the actual significance of ellagitannins to color stabilization

remains unconfirmed. In addition, oak extractives can directly react with anthocyanins and grape tannins. An example is the formation of a red/orange catechin-pyrylium-derived pigment, from the reaction of catechin and sinapaldehyde. Other related pigments are derived from coniferaldehyde-catechin condensation. These have been termed oaklins (Souza *et al.*, 2005).

Although oak tannins do not significantly influence astringency, they can generate bitter tastes (through their breakdown to ellagic acid). Consequently, white wines are usually matured in oak for shorter periods than red wines, and in used barrels or conditioned to release fewer extractable tannins. For red wines, the influence of oak on taste depends on the flavor intensity of the wine – light wines often negatively influenced, whereas full-flavored wines being favorably affected.

Lignin-breakdown products add significantly to the development of an oak bouquet. Lignin degradation involves the action of alcohol, oxygen, and toasting. It is believed that ethanol reacts with certain lignins, forming ethanol–lignins. As these complexes break down, ethanol is released along with the lignin monomers, coniferyl and sinapyl alcohols. These slowly oxidize under acidic conditions, forming phenolic aldehydes, such as coniferaldehyde and vanillin, and sinapaldehyde and syringaldehyde, respectively (Puech, 1987). Above threshold values, they donate woody, vanilla-like odors. Toasting, especially at 200°C, markedly augments their synthesis (Table 8.4). The lower amounts extracted from charred wood probably result from carbonization. Lignin degradation also may generate phenolic acids, such as vanillic and syringic acids, and the coumarin derivatives scopoletin and escutelin. The presence of scopoletin, along with oak lactones, is so characteristic that they are considered diagnostic of oak maturation.

Oak-derived phenols may be modified further by yeast and bacterial metabolism. These changes can influence both volatility and odor quality. For example, the reduction of furfurals to their corresponding alcohols results in a quality shift from almond to hay or verbena-like (Chatonnet, 1991).

Various phenolic and nonphenolic acids have been implicated in the synthesis of esters, acetals, and lactones during oak maturation (Nykänen, 1986). The acids released can lower wine pH and increase acidity. By increasing the proportion of colored anthocyanins, the acids enhance color intensity. The most prevalent acid is acetic acid. It may be formed during the degradation of hemicelluloses (0.1–0.2 g/liter in new barrels), or from the oxidation of acetaldehyde. However, the most significant potential source of acetic acid comes from the activity of acetic acid bacteria, in barrels improperly stored while empty. In addition, oxygen uptake during cellar activities can activate the growth and metabolism of acetic acid bacteria in wine. These generate acetic acid during their metabolism of ethanol and several sugars.

Although compounds such as ethyl phenols are known to be derived from microbial action, most other oak volatiles were thought to be derived by abiotic (physicochemical) reactions. Data from Ferreira *et al.* (2006) suggest that several volatile extractives, such as vanillin, syringaldehyde, and furfuryl alcohol may be also derived from microbial action.

Lignins, tannins, and inorganic salts also influence the poorly understood phenomenon of ethanol–water interactions (D'Angelo *et al.*, 1994). Such interactions are believed to mellow the alcoholic taste of wine and distilled spirits (Nishimura *et al.*, 1983).

Although small amounts of sugar may accumulate due to hemicellulose hydrolysis, it is insufficient to affect taste perception. The simultaneous pyrolytic conversion of some sugars to furfurals and Maillard products appears to be the main sensory significance of

sugar liberation. The furanic by-products are at least partially involved in the toasty, caramel odors of wine matured in medium-toasted barrels.

Oak lactones are the principal volatile constituents in oak, but slowly dissolve into wine. Although often extracted in amounts lower than their threshold values, the presence of oak lactones, notably in red wines, is directly correlated with the presence of flavor characteristics such as berry, vanilla, and coconut (Fig. 8.33). This may partially result from the slow conversion of precursors, such as 3-methyl-4-hydroxyoctanoic acid, to oak lactones during maturation (Wilkinson *et al.*, 2004). Consequently, it can take from 6 months to 1 year for the coconut-like fragrance of oak lactones to become apparent. In most situations, the content of these lactones rises during maturation, notably the more aromatic *cis* isomer. Wines lower in *cis*-oak lactone show more pharmaceutical, clove and hay-like aromas. This suggests that oak lactones may act in additive, synergistic, or suppressive ways in exerting their sensory effects.

Although oak maturation increases the concentration of many important sensory compounds, it also

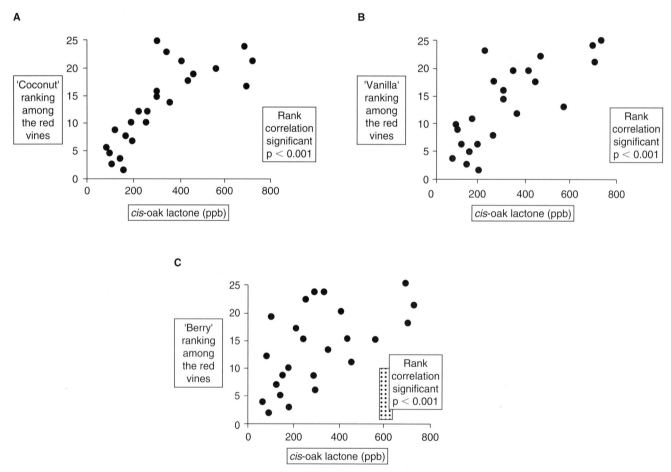

Figure 8.33 Correlation of the perception of several flavor characteristics – (**A**) coconut; (**B**) vanilla, and (**C**) berry – with the presence of *cis*-oak lactones in red wines matured in oak barrels (Spillman *et al.*, 1996).

reduces the concentration of others. For example, the concentration of dimethyl sulfide and dimethyl disulfide decrease during barrel maturation (Nishimura *et al.*, 1983). Methionyl acetate content also decreases, but only if oxygen is also present. The green bean–green chili aspect of some 'Cabernet Sauvignon' wines may dissipate when the wines are matured in oak barrels (Aiken and Noble, 1984).

In addition, oak adsorbs compounds. In absolute amounts, water and alcohol are the most important – estimated at about 5–6 liters in a 225-liter barrel (Chatonnet, 1994). In relative terms, this is of little significance, other than as another origin for barrel ullage. More important in sensory terms is the absorption of wine aromatics. The significance of such losses is only now being assessed. Ramirez-Ramirez *et al.* (2004) have shown that significant amounts of fruit esters (isoamyl acetate, ethyl hexanoate, and ethyl octanoate) can be absorbed by oak from model wines. Because these esters generate a fruity aroma, important in young white wines, their removal could explain why most white wines are not matured in oak cooperage. Although non-polar compounds seem more susceptible to adsorption, polar compounds can be removed, for example benzaldehyde and 2-phenylethanol. Diverse factors, such as ethanol content, yeast autolysate, and filtration can affect absorption (Moreno and Azpilicueta, 2006).

Oxygen Uptake

Slight oxidation is commonly viewed as an important consequence of maturation in oak. Wine placed in well-made barrels, bunged tightly, and rotated, so that wine covers the bung, are generally well protected from oxygen exposure. Moutounet *et al.* (1998) found the ullage oxygen content fell to about 1.8% within 2 months. Concurrently, the carbon dioxide content rose to 28%. Wine exposure to oxygen occurs principally as a consequence of periodic cellaring procedures, such as racking. Air does not usually diffuse into tight barrels in significant quantities. Ribéreau-Gayon (1931) estimated oxygen ingress at about 2–5 ml/liter/yr in tightly bunged, full barrels. This increased to about 15–20 ml/liter/yr in barrels with a typical ullage. The difference appears to relate to slower ingress via staves in contact with wine. Partial drying of the wood above an ullage increases penetration (Vivas *et al.*, 2003). Singleton (1995) argues that most of the oxygen that enters staves likely reacts with phenolic constituents (notably ellagitannins), before traversing the stave. The water and alcohol lost through the surfaces of the barrel are only slowly replaced, resulting in the generation of a partial vacuum. Atmospheric pressure compresses the upper edges of the barrel, frequently

resulting in a stabilization of the vacuum within 5–15 days (Moutounet *et al.*, 1998). Barrels may differ markedly in tightness. The negative pressures observed in barreled wine can vary considerably (Fig. 8.34). This may explain some of the variation in barrel-to-barrel maturation rate. Depending on barrel tightness, temperature, and relative humidity, a barrel may lose from 4 to 10 liters of wine per year. A mathematical model of wine loss may help design air circulation systems in cellars to minimize such losses (Ruiz de Adana *et al.*, 2005).

Evaporative losses tend to be more marked in barrels left with their bungs upright. Despite this, the practice makes topping and sampling considerably easier. Coincidentally, both procedures increase exposure to oxygen. During normal racking, topping, and sampling, oxygen uptake has been variously estimated at between 15 and 40 ml O_2/liter/year. Oxygen access is undoubtedly higher when glass or loosely fitting bungs are used. Under such conditions, the oxygen content in the ullage may remain between 5 and 9% (Moutounet *et al.*, 1998).

Racking can result in up to 6 ml O_2/liter being absorbed per racking. In red wine, the absorbed oxygen is consumed within approximately 6 days at 30°C (Singleton, 1987) – equivalent to 15–20 days at cellar temperature. Most of the oxygen is presumably consumed in the oxidation of *o*-diphenols. Smaller amounts may be involved in the oxidation of ethanol, sulfite, ascorbic acid, ferrous ions, and various other organic constituents.

The slight uptake of oxygen during racking and other processes is generally viewed as desirable, especially for

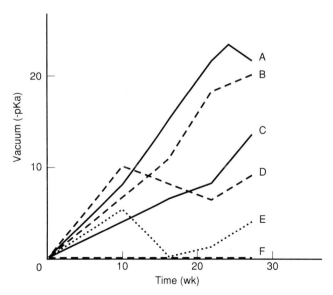

Figure 8.34 Development of a partial vacuum in six wine barrels (A–F) during undisturbed aging of 'Cabernet Sauvignon' wine. (From Peterson, 1976, reproduced by permission)

red wines. It promotes the polymerization of anthocyanins with tannins and, therefore, color stability. In addition, hydrogen sulfide formed in, and released by the lees, is oxidized. This is especially valuable for white wines. The action of yeasts in the lees also speeds oxygen consumption, protecting the wine from excessive oxidation (Pfeifer and Diehl, 1995). Ellagitannins and residual sulfur dioxide provide additional antioxidant protection.

The ullage that develops over wine as liquid escapes through the wood is not a source of spoilage. Its oxygen content is usually low to undetectable. Thus, filling the ullage space (topping) is necessary only if air enters the barrel. Only with prolonged aging, as in brandy, is drying (and shrinkage) of the staves over the ullage likely to be sufficient to generate cracks between (or in) the staves.

In-barrel Fermentation

Most vinifications occur in large tanks or vats. They are easier and more economical to maintain, and can facilitate more uniform fermentation. Therefore, fermentation in-barrel, or in other small-volume cooperage, is normally restricted to situations where the quantities of must are limited, or of unique quality or origin. Although temperatures during in-barrel fermentation may increase more than in cooled tank fermentors, the surface area/volume ratio is often sufficient to avoid serious overheating and stuck fermentation. The probability of stuck fermentation also may be diminished by the uptake of sterols or long-chain unsaturated fatty acids from the wood (Chen, 1970). Sterols are required for the proper maintenance of yeast membranes during and after fermentation. Sugars released during maturation (Nykänen et al., 1985) also may favor malolactic fermentation by providing nutrients for lactic acid bacteria.

In addition to maintaining the individuality of small lots of juice, some winemakers specifically choose in-barrel fermentation for its perceived benefits on wine development. Wine fermented and matured in new oak incorporates less phenolic material than the same wine matured in equivalent barrels after fermentation. This partially results from the coprecipitation of extracted tannins with yeast cells and mannoproteins during and shortly after fermentation (Chatonnet, Dubourdieu et al., 1992). The early extraction and oxidation of readily oxidized tannins help consume oxygen, minimizing wine oxidation. Phenolics also reduce the accumulation of volatile reduced-sulfur compounds (Nishimura et al., 1983). Because the more desirable oak flavors dissolve out more slowly than oak tannins, the wine retains proportionally more oak flavor and fewer harsh tannins.

Other differences have been noted, but their sensory significance is uncertain. For example, the reduction of furfural and 5-(hydroxymethyl)-2-furaldehyde to their less-aromatic corresponding alcohols is enhanced (Marsal and Sarre, 1987). In addition, yeast can metabolize ferulic and p-coumaric acids to aromatic phenols, 4-vinylguaiacol, 4-ethyl guaiacol, and 4-ethyl phenol (Dubois, 1983). These possess spicy, smoky odors. Another significant change is the metabolic conversion of phenolic aldehydes, notably vanillin, to the barely perceptible vanillic alcohol derivative (Chatonnet, Dubourdieu et al., 1992). This partially offsets the accumulation of vanillin extracted from the oak during maturation (Spillman et al., 1997). In-barrel fermentation also increases the level of oak lactones (β-methyl-γ-octalactones), nitrogen compounds, and polysaccharides, primarily those derived from yeast mannoproteins. These polysaccharides may enhance the wine's smooth mouth-feel. In contrast, soluble protein levels may decrease. Finally, yeasts can metabolize furfural to furfurylthiol, a potent coffee-smelling compound (Blanchard et al., 2001). Because furfural is generated during barrel toasting, and is extracted readily during maturation, its presence is more likely to have sensory significance when new barrels are used.

For those desiring the effects associated with in-barrel fermentation, but do not want oak flavors, stainless steel barrels are available. These provide all the perceived benefits, including those of sur lies maturation, but are much easier to clean, and can be reused indefinitely. Any oxygen uptake desired can be strictly controlled.

Advantages and Disadvantages of Oak Cooperage

For premium wines, fermentation and/or aging in oak is often desirable. The expense and effort are justified by the additional flavor complexity obtained. This is especially true for red wines. White wines, generally possessing less flavor, may be overpowered by oak. As a consequence, only a portion of the juice may be in-barrel fermented or barrel matured. Exposure time is also crucial. Short duration tends to extract a higher proportion of woody flavors, with the more appreciated aspects being extracted only with extended contact. The level of toasting and the proportion of new-to-used barrels can also significantly influence the oak character procured (Figs 8.29 and 8.31). For wines of neutral or delicate flavor, exposure to oak is neither cost effective nor necessarily beneficial.

Oak barrels are both costly to purchase and maintain, and new barrels need conditioning prior to use. The tartrates and tannins that accumulate on the inside of the barrel during wine maturation are both difficult and unpleasant to remove. When not containing wine, barrels must be protected from drying and microbial contamination. Off-flavors produced by bacteria and fungi growing on internal surfaces can subsequently taint wine. Examples are corky off-odors (Amon *et al.*, 1987), vinegary taints (predominantly from the metabolism of acetic acid bacteria), and manure or stable notes (due to the enzymatic reduction of vinylphenols to ethyl phenols by *Brettanomyces* spp.).

Because the rate of maturation varies from barrel to barrel, frequent and time-consuming barrel sampling is required to follow the progress of the wine. Racking is more labor intensive and inefficient than its automated equivalent in large cooperage. In addition, considerable economic losses can result from wine evaporation from barrel surfaces. Up to 2–5% of the volume may be lost per year in this way (Swan, 1986). Volume loss is especially marked at warm temperatures. Depending on the relative humidity of the cellars, wine may either increase or decrease in alcoholic strength (Guymon and Crowell, 1977). High relative humidity suppresses water evaporation, but has no influence on alcohol loss. Consequently, the alcoholic strength of wine decreases in humid cellars. Under dry conditions, water evaporates more rapidly than ethanol, resulting in an increase in alcoholic strength. In addition to water and ethanol, small amounts of acetaldehyde, acetal, acetic acid, and ethyl acetate are lost by evaporation from barrel surfaces (Hasuo and Yoshizawa, 1986). In contrast, less volatile and nonvolatile compounds accumulate as a result of the concentrating effect of water and ethanol loss. Relative humidity also influences the types and amounts of phenols extracted. Low relative humidity decreases total phenolic uptake, but increases vanillin synthesis (Hasuo *et al.*, 1983).

Another source of wine loss, associated with in-barrel maturation, results from absorption by the staves. It is estimated that new 225-liter barrels absorb between 5 and 6 liters of wine.

Alternative Sources of Oak Flavor

The addition of oak chips or shavings to wine has been investigated as an economical alternative to barrel aging. Not only does it save by delaying or avoiding the purchase of new barrels, but it can also reduce costs associated with topping and lost wine volume. Values commonly suggested are about 10 g/liter/year for white wines, with more than twice that for red wines. With small oak chips (≤1 mm diameter), about 90% of the extractives are removed within 1 week (Singleton and Draper, 1961). The perception of some aromatic constituents may take longer to appear, partially because they are formed slowly during wine maturation.

The sensory effects obtained from aging on oak chips may differ from those obtained during barrel aging. This may arise from the absence of heat-induced hydrolysis of oak constituents (if the chips are not toasted), reduced oxygen exposure (due to the absence of barrel racking), or differences in microbial modification of oak compounds during seasoning of the wood or during wine maturation. In addition, the surface area-to-volume contact is greater, as is the exposure of the wine to the less-permeable summer wood. An example of these influences may be the increased uptake of oak lactones noted by Swan *et al.* (1997). Increased surface area also greatly facilitates the uptake of ellagitannins, most of which are hydrolyzed on the inner surfaces of barrel staves during toasting. Combining chips with different levels of toasting ideally should reproduce the diversity of extractives obtained from oak barrels (the slow penetration of heat during toasting generating a progressive range of chemical changes across the stave).

When used, chips are usually added during fermentation. This promotes the early precipitation of the extra tannins and phenols, extracted due to the extensive wine/chip contact. In addition, oxygen present in the wood structure is rapidly consumed during fermentation. If the chips are added after fermentation, oxygen uptake can be minimized if the oak is presoaked for a few hours.

Although chips can be added directly to wine, it creates a removal problem. Chips can also clog drains, pumps and filters. This is partially avoided by enclosing the chips in polyester bags. Although facilitating removal, it slows flavor extraction.

Another economic alternative to maturation in oak cooperage is the incorporation of oak strips into large-volume cooperage. This usually involves inserting a series of oak slats (battens) or tubes in a stainless steel holder (Rieger, 1996). The inserts are often derived from used barrels, coming from outer portions of staves unaffected by wine. The sections are usually split in two, essentially doubling the contact surface area. Oak inserts may be exposed to infrared radiation to generate a desired toast level.

An example of the differences that can accrue from various oak treatments is provided in Swan *et al.* (1997). The sensory effects generated may be less complex than those derived from staves (Fig. 8.35). This may be due to the thinness of the slats – thin inserts not possessing the diversity of aromatics found in thicker barrel staves. Conversely, this can be used to advantage, permitting the winemaker to more precisely select the oak attributes (defined by the level of toasting) desired in the wine. Potential problems with oxygen uptake from the wood can be largely avoided by insertion during fermentation or presoaking.

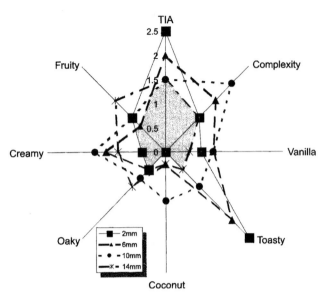

Figure 8.35 Sensory spider diagram for wine from each layer of toasted oak: 'toasty' refers to strong sweet, caramel, vanilla oak attributes, while 'oaky' refers to green, astringent, raw woody attributes. (From Hale *et al.*, 1999, reproduced by permission)

If only oak flavors are desired, an even simpler procedure is the addition of an oak extract (Puech and Moutounet, 1992). Powered oak "flour" is another alternative. When added during fermentation, it tends to settle out with the lees, and is lost during racking.

Previously, European winemakers were prohibited by law from using any of these barrel alternatives. This restriction has been removed, at least for some categories of wine.

Other Cooperage Materials

For both fermentation and storage, cooperage constructed from material other than wood has many advantages. Frequently they are less expensive to maintain. Stainless steel is often preferred, but fiberglass and cement also are extensively used. Because all are impervious to oxygen, wine oxidation is minimized. This preserves the fresh, fruity character, most important to wines designed for early consumption. Stainless steel and fiberglass have the additional benefits of permitting construction in any desired shape. Modern construction materials also facilitate cleaning and dry storage. Gas impermeability permits partial filling, because the ullage can be filled with carbon dioxide or nitrogen (avoiding oxidation). Furthermore, modern construction materials do not modify the wine's fragrance.

Stainless steel is generally the preferred modern cooperage material because of its strength and inertness.

Inertness avoids the need for, and maintenance of coatings of paraffin, glass, or epoxy resin. These are required for cement tanks, otherwise excessive amounts of calcium could seep into the wine. The acidic nature of wine also tends to corrode cement. Stainless steel possesses heat-transfer properties, permitting comparatively easy temperature control during fermentation. Installation is rapid and subsequent movement of the cooperage is possible.

For wine production and storage, stainless steels high in chromium and nickel content are required. A chromium content of between 17 and 18% provides an adequate surface layer of insoluble chromium oxide (e.g., A151 type 304 or 316). It is the chromium oxide that provides most of the anticorrosive properties of stainless steel. Nickel is present in amounts that may vary between 8 and 14%. It facilitates soldering and further enhances corrosion resistance. When wine is stored in stainless steel for only short periods, molybdenum may be omitted from its formulation. Otherwise, molybdenum is required at a concentration of approximately 2–3% (e.g., A151 type 316). It provides protection from corrosion by sulfur dioxide. Titanium may be incorporated because it increases the level of carbon permitted in the finished steel. Titanium also reduces the risk of corrosion next to soldered joints. When added, it is often incorporated at about 0.5%.

With stainless steel, it is important to avoid introducing scratches on the inner surfaces of the tank. Even rinsing with hard water containing minute rust particles can cause damage to polished inner surfaces.

Fiberglass tanks have also become common replacements for wooden cooperage. Fiberglass has the advantages of being less expensive and lighter than stainless steel. However, it possesses less strength, is less conductive to heat, is more porous, and tends to be more difficult to clean (has a rougher surface) than stainless steel. In particular, residual styrene may diffuse into the wine from the polyester resin binding the glass fibers (if the original mixture was improperly formulated). At concentrations above 100 μg/liter, styrene can donate a plastic odor (Anonymous, 1991).

Stainless steel, resin-coated regular steels, and fiberglass have permitted the production of an extensive array of cooperage. Although the containers may be used exclusively for wine maturation, most are designed to facilitate emptying and cleansing after use as fermentors. Thus, they typically possess a slanted floor and exit ports at or near the base. The position of the port (horizontal or vertical) is largely a function of whether cleaning occurs automatically or manually. Other designs may pivot tanks to facilitate emptying, as well as incline a fixed helical blade to facilitate the discharge of pomace or lees.

Cork and Other Bottle Closures

Cork

Cork still remains the bottle closure of choice for most producers. However, cork use in wine preservation predates its use as a bottle closure by more than 2000 years. The ancient Greeks and Romans employed large cork sections to stopper wine amphoras (Frey *et al.*, 1978; Tchernia, 1986). These were often coated with resin to adhere to the neck of the amphora. The seal was protected with a cap of volcanic-tuff cement (*pozzuolana*). The oldest known archaeological evidence of cork as a wine seal comes from an Etruscan amphora unearthed in Tuscany (sixth century B.C.) (Joncheray, 1976). The cork was still in place. There is also evidence that the Greeks stoppered amphoras and wine jars with cork in the sixth century B.C. (Thompson, 1951; Van Buren, 1955). In one example, the cork possessed a small centered hole. It is thought to have been the site for the cord used to extract the cork. Occasionally, amphoras were sealed with terracotta stoppers (Koehler, 1986), or bunged with chopped chaff mixed with mud or other materials. Stoppers made of reeds, tied and glued together, have been found in ancient Egyptian wine amphoras (Hope, 1978). These were covered with a variety of clay caps.

Cork use ceased following the collapse of the Roman Empire, reflecting a decline in amphora use. The reemergence of cork as a wine closure began in the mid-seventeenth century, coincident with the beginnings of industrial-scale glass-bottle manufacture. Fabrication of cork stoppers using knives had already become a thriving industry in France by the mid-1700s (Diderot, 1763). Nevertheless, beer bottles are reported to have been stoppered with cork as far back as late fifteenth-century England (McKearin, 1973). What is uncertain, though, is whether the bottles mentioned were glass or earthenware (a common material for bottle production at the time). By 1615, cork was already being strongly recommended for wine bottles (Dumbrell, 1983). It seems that the seal was obtained by coating the cork with wax or oil. A cord tied around the top of the cork was used to extract the cork. The other end of the cord was attached below the lip of the bottle. The evolution of tightly adhering corks coincided with the production and increased demand for stronger glass bottles. Development of the screwpull (prior to the mid-1680s) suggests that knowledge of cork use was clearly advanced by the late 1600s. Bottles were already being recommended to be laid on their sides (Worlidge, 1676).

As a plant tissue, cork is produced by a layer of cambial cells located in the outer bark (Graça and

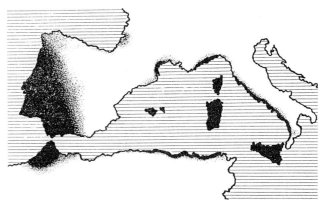

Figure 8.36 Geographical distribution of *Quercus suber*, the cork oak. (From J. V. Vieira Natividade (1956) *Subériculture*. Reproduced by permission of Éditions de l'École nationale du Génie rural des Eaux et des Forêts, Nancy, France)

Pereira, 2004). The cambium produces cork (**phellem**) to the outside and a thinner layer of cells (**phelloderm**) to the inside. Together, these tissues constitute the outer bark. The inner bark, or phloem, consists of cells primarily involved in conducting organic nutrients throughout the plant. In the majority of woody plants, the cork layer is relatively thin. In only a few species is a deep, relatively uniform, cork layer produced. Of these, only the cork oak, *Quercus suber*, produces cork in commercial quantities. Not only does *Q. suber* produce a thick layer of cork, but the cork cambium (phellogen) remains viable throughout the life of the tree. This permits repeat cork harvesting without damaging the tree.

CORK OAK

Quercus suber grows indigenously in a narrow region bordering the western Mediterranean (Fig. 8.36), with most commercial stands located in Portugal and Spain. Portugal produces just above half the world's supply (~200 million kg), with the remainder coming from Spain, Algeria, Morocco, Italy, France, and Greece. Dry, upland sites on rocky soils provide the best areas for quality cork production. Here, the bark is firmer and more resilient. On rich lowland soils, trees produce a thicker, but more spongy cork layer.

The cork oak grows about 16 m high and has a trunk diameter of about 20–60 cm at breast height. Typically, the tree begins to branch out about 4–5 m above ground level. Thus, the lower trunk provides large, clear sections of bark. The lower branches of older trees also may yield bark sections of commercial size. Trees can survive for about 500 years, but their most productive period occurs between the first and second century of growth. As the tree ages, its ability to produce cork slowly declines.

CULTURE AND HARVEST

Most commercial oak stands are of natural origin. However, the selection and planting of superior seedlings occurs in both existing stands and reforestation areas. Pruning helps shape trees for optimal production of quality cork.

About 370,000 tons of cork are harvested yearly (Pereira and Tomé, 2004). Of this, about 51% comes from Portugal, 23% from Spain, with decreasing amounts from other regions. It is estimated that about 60% of the crop is used for bottle closures. Portugal alone annually supplies about 25,000 tons of cork stoppers (Pereira *et al.*, 1994).

When trees reach a diameter of more than 4 cm (after about 20–30 years), cork is stripped from the trunk for the first time. This stimulates new cork growth, and the tree in general. The initial (**virgin**) cork is not used for bottle closures. Its structure is too irregular and porous (Fig. 8.37).

After cork removal, the exposed tissue turns red. Subsequent cork production is more uniform. New cork production is most marked in the first year after stripping, slowing gradually thereafter (Ferreira *et al.*, 2000). Within 7–10 years, the tree may again be ready for stripping. This cork, called **second cork**, also lacks the qualities necessary for good-quality bottle closures. Third and subsequent strippings are referred to as **reproduction cork**. They yield cork suitable for bottle closures. Stripping occurs about every 9 years (the minimum permitted), except in mountainous regions where growth is slow. Here, stripping may occur only every 12–18 years.

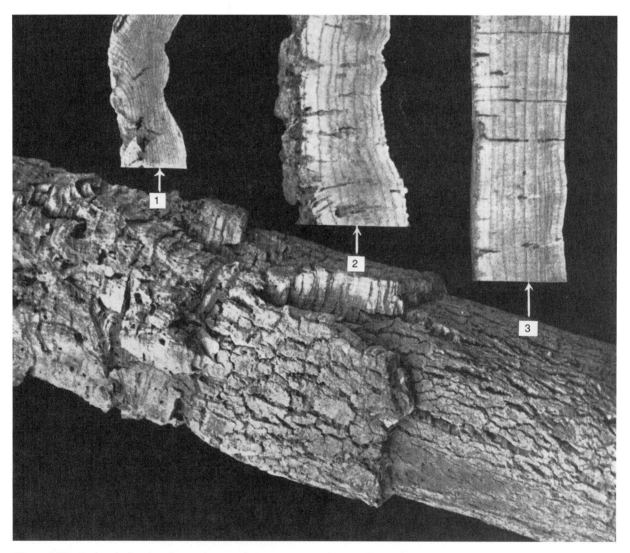

Figure 8.37 Cork oak showing the appearance of virgin (1), second (2), and reproduction cork (3). (From J. V. Vieira Natividade (1956) *Subériculture*. Reproduced by permission of Éditions de l'École nationale du Génie rural des Eaux et des Forêts, Nancy, France)

Stripping usually occurs in late spring to early summer, when the trees are actively growing. During this period, the soft, newly formed cork separates along a line just outside the cork cambium. This greatly facilitates bark removal. Stripping begins with two circumferential cuts with a special ax – one around the base of the tree and the other just below the lowest branches. A subsequent vertical cut connects the initial cuts. A few additional vertical cuts may be made to ease removal and handling. The cork is pried off with the wedge-shaped end of the ax handle. Workers remove the cork from branches, if the diameter is sufficient to yield useful slabs. Because of the damage that can be caused by improper stripping, experienced workers remove the cork. Deep cuts can damage the inner bark, causing permanent scarring that makes subsequent cork removal difficult.

Yield varies widely from tree to tree. Young trees often yield only about 15 kg cork, whereas large trees can produce upward of 200 kg. Bark thickness also differs considerably, depending on growth conditions and tree age. These conditions result in annual growth-ring width varying from 1.5 to 7.0 mm (Pereira *et al.*, 1992). For wine-cork production, the bark needs to be about 4.0 cm thick to yield sheets of usable cork.

After harvesting, laborers bundle and stack the slabs, in preparation for transport to production facilities. Often the slabs are boiled in large vats for 1–1.5 h. This swells and softens the cork, permitting the slabs to flatten. Boiling generally occurs at the beginning of an outdoor seasoning period that can last weeks or months. During seasoning, the cork partially dehydrates and its phenolic content oxidizes.

During storage, the slabs may become covered with a superficial growth of mold, often dominated by *Chrysonila (Monilia) sitophila*. Despite its presence, *C. sitophila* does not participate in production of a corky odor. It is incapable of metabolizing chlorophenol pesticides, such as pentachlorophenol to 2,4,6-trichloroanisole (Silva Pereira, Pires *et al.*, 2000b). In addition, mold growth is typically limited to the outermost 8–15 cell layers (Carriço, 1997). These are usually removed prior to punching out the closures. Only the inner tissues of the cork sheet are used. The relationship between the presence of other surface-growing fungi, such as *Penicillium, Aspergillus,* and *Trichoderma*, to production of 2,4,6-trichloroanisole is unestablished.

Before use at the processing plant, the slabs are again boiled for about 30–60 min. Boiling extracts additional water-soluble compounds and results in tissue swelling and softening. Boiling has little effect in disinfecting cork (Álvarez-Rodríguez *et al.*, 2003), presumably due to cork's insulating properties. The slabs are subsequently stacked for up to 2–3 weeks to reduce their moisture content. Softening facilitates removal of the

outer portion of the cork (**hard back**). It can vary from 1.5 to 3 mm in thickness. The structure of the hard back is too irregular, stiff, and fractured to be used in the production of closures. Finally, boiling tends to equalize elasticity across the cork (Rosa *et al.*, 1990). After removing the hard back, workers trim and sort the slabs into rough grades, based on thickness and surface quality. After grading, the planks are cut transversely into strips, typically between 38 and 45 mm.

CELLULAR STRUCTURE

As cork cells enlarge and mature, they elongate along the radial axis of the trunk. They produce a distinctive cell wall, composed primarily of fatty material. By maturity, the cytoplasm has disintegrated, leaving only the cell wall. Because the typical intercellular cytoplasmic connections (plasmodesmata) become plugged during cell development, each cell acts as a sealed unit. Cork tissue also lacks intercellular spaces, each cell abutting tightly against its neighbors. Cork tissue, especially the valuable reproduction cork, is relatively homogeneous in texture. The only significant disruptions to homogeneity result from the occasional presence of lenticels and the sporadic occurrence of fissures. Lenticels are columns of thin-walled parenchyma cells containing large, irregular, intercellular spaces. They permit gas exchange between the internal tissues of the tree and the exterior. All cork contains lenticels, but the best cork contains small, narrow lenticels, and possesses few fissures.

Individual cork cells resemble elongated prisms in cross-section (Fig. 8.38A). They possess from four to nine sides, with most cells showing a total of seven to nine sides. The cells formed in the spring are larger and thinner-walled than those produced later in the season. The weak early cells often collapse under pressure exerted by

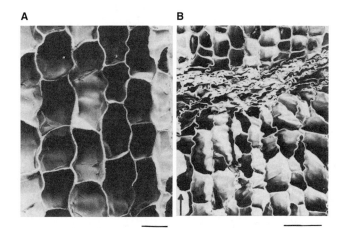

Figure 8.38 Scanning electron micrograph of reproduction cork. (**A**) Radial section; and (**B**) transverse section showing the corrugated appearance and the collapse of cork cells formed in the spring. (From Pereira *et al.*, 1987, reproduced by permission)

subsequent growth of smaller, tougher, cork cells. Even the latter frequently show corrugations in their side walls (Fig. 8.38B). This is caused by their being pressed against the existing outer layers of cork. The highly collapsed, corrugated nature of the early spring cork accounts for the bands (growth rings) that distinguish each year's growth.

Cork that grows slowly has a larger proportion of the smaller, thicker, late-cork cells which have higher mechanical resistance (Pereira *et al.*, 1987). The cork compresses less readily, but is more elastic. The elastic, resilient nature of cork is one of its most important properties. Consequently, slow-grown cork is more valuable for bottle closures than rapidly grown cork (Pereira *et al.*, 1992).

The cell wall of cork tissue shows several unique features. The most notable is the presence of approximately 50 alternating layers of wax and suberin (Fig. 8.39). Both compounds are complex polymers, highly impermeable to liquids, and resistant to the action of acids. About 37% of the wall of reproduction cork consists of suberin (Asensio and Seoane, 1987), a hydrophobic glycerol polyester of long-chain (principally C18 to C22) aliphatic acids, such as ω-hydroxy

acids. These possess ferulic acid as the primary phenolic constituent (Graça and Pereira, 1997; Bento *et al.*, 2001). The wax component, which makes up about 5% of the wall mass, consists primarily of cerin and friedlin, smaller amounts of betulin, and probably fatty acids, such as betulic, cerolic, oxyarachidic, phellonic, oleic, and linoleic acids (Lefebvre, 1988). It is believed that the platelike layers of wax and suberin permit sliding of the layers past one another. Because the wall folds and becomes corrugated under compression, the pressure is absorbed and wall cracking is minimized. Realignment of the wall layers probably explains the elastic return after pressure release.

About 28% of the cell-wall constituents of reproduction cork are guaiacyl lignins, complex water-insoluble polymers composed of phenylpropanoid monomers. Cellulose and related hemicelluloses constitute about 13% of the wall mass. The wall also contains phenolics, such as catechol, orcinol, gallic acid, and tannic acid.

PHYSICOCHEMICAL PROPERTIES

The physicochemical properties of cork make it ideally suited for use as a bottle closure. These include compressibility, resilience, chemical inertness, imperviousness to liquids, and a high coefficient of friction.

Cork is one of the few substances, natural or synthetic, that can be compressed without showing marked lateral (sideway) expansion. In addition, cork shows remarkable resilience on the release of pressure. Cork returns almost immediately to 85% of its original dimensions, and within the next few hours it regains about 98% of its original volume (Fig. 8.40). These properties are

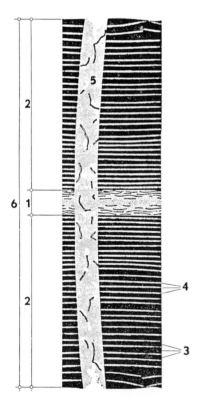

Figure 8.39 Reconstructed electron micrograph through the walls of two cork cells. 1, primary wall; 2, secondary wall; 3, layers of suberin; 4, layers of wax; 5, plasmodesma; and 6, total thickness 1 μm. (From Sitte, 1961 from Honegger, 1966, reproduced by permission)

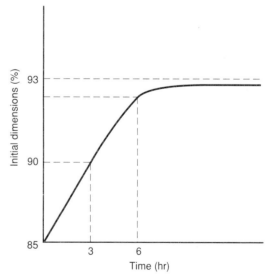

Figure 8.40 Elastic rebound of cork tissue. About 85% of the elastic return is instantaneous. (After Riboulet and Alegoët, 1986, reproduced by permission)

undoubtedly related to the distinctive wall structure and the sealed nature of the cells. The latter gives the cells the property of minute air cushions. Some air is expelled in the early stages of compression, but then gas release ceases until high pressures are reached (Gibson *et al.*, 1981). At that point, the cells rupture and the cell-wall material collapses, destroying its resilient properties.

The ability of cork to spring back to its original shape gives cork much of its sealing properties. Its resilience exerts pressure on the neck of the bottle, providing a long-lasting, tight seal. Eventually, however, resilience declines and adherence to the glass weakens.

Cork moisture content significantly influences resilience and compressibility. Within a moisture range of 5–12%, cork remains sufficiently supple for insertion. Before insertion, corks are held at the lower end of the moisture range, between 5 and 9%. This limits mold growth during storage, but retains sufficient resilience. It also avoids fluid extrusion during insertion.

After insertion, cork absorbs moisture from the wine (Rosa and Fortes, 1993). Although slow, the moisture content can climb to 16% (Jung *et al.*, 1993). Although water uptake is minimal in the radial plane (Rosa and Fortes, 1993), at right-angles to the neck, softening can reduce the cork's adherence to the neck. This feature is less marked with higher quality (less porous) corks.

Cork generally resists microbial attack because of its hydrophobic nature and low nutrient status. Its low moisture content also limits microbial growth. Microbial growth is largely restricted to the outer surfaces of the cork, where it is dependent on wine residues or nutrients available in dust contaminants.

The chemical inertness of cork protects it during prolonged contact with wine. It also means that few breakdown products form and diffuse into wine. Boiling cork after harvest is thought to extract most compounds that might unfavorably affect the sensory properties of a wine. Additional boiling after storage has the advantage of decreasing the concentration of phenolic acids, such as cinnamic acids (which increases during storage). It also augments the content of others, such as vanillin (Mazzoleni *et al.*, 1998).

Impermeability to liquids comes from the tightly packed nature of cork tissue. These provide few channels through which fluids can pass. Cork stoppers of standard lengths present a penetration barrier of from 300 to 500 cells/cm. In addition, the waxy nature of the wall restricts diffusion across or between cells. Although gases, water vapor, and fat-soluble compounds can diffuse into cork, significant movement of these substances through corks in wine bottles is limited. This may result from the pressure created in the gas-fill lumen of cork cells by compression in the neck of the bottle (Casey, 1993).

What oxygen uptake occurs is most marked during the first month, generally slowing considerably thereafter (Fig. 8.41). Although rates of diffusion vary considerably among corks, rates of ingress appear to be in the range of 5–6 mg O_2/L per year. Natural cork is more gas permeable than agglomerate corks (composed of cork particles), but less permeable than most synthetic corks. The uptake of oxygen is probably a significant, but not the sole cause, of sulfur dioxide loss, or oxidative browning during bottle aging. There is considerable controversy as to the relevance (importance) of gas diffusion to development of an aged wine bouquet.

Because cork is slowly permeable to gases, they dry and lose their elasticity if not in contact with the wine. During contact, water vapor diffuses into the cork, condenses and begins to fill some of the cork cells. This not only drastically retards the diffusion of water vapor, but also reduces oxygen ingress. Diffusion occurs many thousands of times more slowly through liquids than gases. As a consequence, water loss from the cork in wine bottles is negligible and oxygen diffusion minimal. Oxygen diffusion may be greater than expected due to excessive porosity (presence of lenticels and hidden fissures).

The hydrophobic nature of most of the cell wall's structure facilitates the absorption/desorption of nonpolar volatile compounds. This is undesirable when the compounds released generate off-odors, such as TCA and quaiacol. Conversely, the absorbency for TCA is such that an uncontaminated cork can remove sufficient TCA from a contaminated wine to eliminate its corked odor (Capone *et al.*, 2003). This property can also remove ('scalp') varietal and age-derived aromatic compounds, such as TDN and long-chain ethyl esters. This could potentially modify the sensory attributes of the wine (Capone *et al.*, 2003). Synthetic stoppers may have even greater absorbent properties for nonpolar compounds than cork, for example removing almost all the TDN and 15–45% of rose oxide.

Cork's high coefficient of friction allows it to adhere tightly, even to smooth surfaces. The cut surfaces of cork cells form microscopic suction cups that adhere tightly to glass. There is also an inelastic loss of energy during compression that increases friction between the cork and the glass surface (Gibson *et al.*, 1981). Such properties, combined with resilience, allow cork to establish a long-lasting, tight seal with the glass within about 8–24 hours.

Over time, cork loses its resilience, strength, and sealing properties. As a consequence, the typical diameter of cork decreases from its original of 24 mm to 20 mm, or less, after several years in-bottle. The reasons for these changes are unclear, but may relate to the slow escape of gas from the cork cells (Casey, 1993), as well

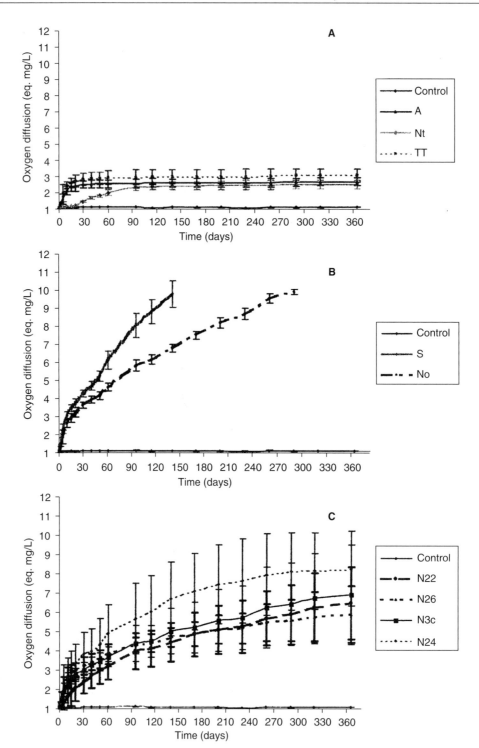

Figure 8.41 Kinetics of oxygen diffusion through each closure type: technical cork closures (**A**); synthetic closures (**B**); natural cork closures (**C**). Commercial wine bottles were stored horizontally over a period of 365 days. A, Agglomerate cork closure; TT, Twin top cork closure; Nt, Neutrocork cork closure; No, Nomacorc closure; S, Supremecorq closure; N22, natural cork, first-grade closure, diameter = 22 mm; N24, natural cork, first-grade closure, diameter = 24 mm; N26, natural cork, first-grade closure, diameter = 25 mm; N3c, natural cork, third-grade colmated. (From Lopes *et al.*, 2005, reproduced by permission. Copyright 2005 American Chemical Society)

as the uptake of moisture. Because of the loss of cork's desirable qualities, wines aged in bottle for long periods may be recorked every 25–30 years. Recorking also permits topping, if the bottles have developed a large headspace.

PRODUCTION OF STOPPERS

Stoppers are produced by placing strips of cork on their side. Cork closures are punched out parallel to the annual growth rings of the bark (Fig. 8.42). This positions the lenticels and any fissures in the cork at right-angles to the length of the stopper. Thus, the major sources of porosity (lenticels and fissures) are positioned such that their ends abut directly against the glass in the bottle. This minimizes their potential to compromise the cork's impermeability.

Trained workers, rather than machines, punch out the stoppers. This permits a better positioning of the cutter to avoid structural faults that occur even in the best cork. Nevertheless, about 75% of the cork slab ends up as residue. This may be used to produce agglomerate corks or other cork products.

If desired, the edges of the stopper may be trimmed at a 45° angle. The process, called **chamfering**, is most commonly used for corks destined to stopper fortified wines. Chamfering eases manual reinsertion of the cork after the bottle is opened. Chamfered corks are routinely bonded to plastic tops to improve grip on the cork. Such corks are called **T-corks**. Chamfered corks are seldom used to seal bottles of table wine as it

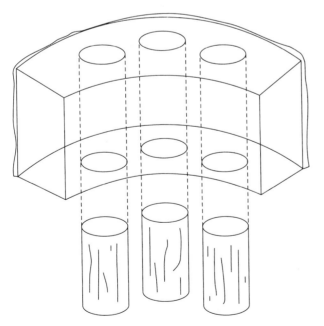

Figure 8.42 Positioning the cutter for the removal of corks from strips of cork tissue. (Redrawn from Riboulet and Alegoët, 1986, reproduced by permission)

reduces the surface area over which the cork adheres to the glass.

After being punched out, corks are treated with a series of washes. Corks have traditionally been rinsed in water to remove debris and cork dust, followed by a soak in a solution of calcium hypochlorite ($Ca(OCl)_2$). This both bleaches and surface-sterilizes the cork. Subsequently, the corks are placed in a bath of dilute oxalic acid. The oxalic acid reacts with residual calcium hypochlorite, liberating chlorine gas and precipitating calcium as an oxalate salt. Oxalic acid also binds with iron particles that may have been left by the cork cutter. If it is not removed, the iron may react with tannins in the cork, forming black ferric tannate spots. After the oxalate wash, corks are given a final water rinse. Some of the calcium oxalate may be purposely left to give the surface a whitish coloring.

Use of the old chlorine treatment is becoming increasingly uncommon. It is thought to be involved in the production of 2,4,6-trichloroanisole (see below). Chlorination may also lead to the production of polychlorinated dibenzo-*p*-dioxins and dibenzofurans (PCDDs and PCDFs). Thankfully, production is minimal. PCDDs and PCDFs intake from wine consumption (not exclusively derived from cork) is estimated to be less than 0.2% of the permissible daily uptake of these substances (Frommberger, 1991). In addition, hypochlorite can corrode cork tissue and oxalic acid can form crystalline deposits, causing leaks by reducing cork adherence to the glass.

The currently preferred bleaching and surface-sterilizing agent is peroxide. This is normally supplied as peracetic acid (CH_3CO_3). On reacting with organic matter, it releases hydrogen peroxide and acetic acid. The bleaching, oxidizing, and sterilizing action of peroxide quickly consumes most of the oxidant. Rinsing or evaporation eliminates the acetic acid. Hydrogen peroxide may also induce polymerization of surface phenols, reducing their potential involvement in the production of TCA. Peroxide is occasionally used alone, but its concentration and exposure time must be carefully monitored. Otherwise, the structural integrity of the cork can be compromised. Excess peroxide can be neutralized by the addition of citric acid.

Microwaving is an alternative, nonchemical treatment. The treatment inactivates both surface and internal microbes and promotes the loss of volatile contaminants.

On request, producers give corks a surface coloration. Although coloring is less frequent than in the past, there is still a demand for corks given a "traditional" color. For example, French wines are occasionally sealed with rose-tinted stoppers, Italian wines are periodically closed with whitened corks, whereas Spanish wines are typically stoppered with uncolored (light-brown) corks.

To bring the moisture content of finished stoppers quickly down to approximately 5–8%, the corks may be placed in centrifugal driers or hot-air tunnels. The corks are next sorted into quality grades. The various grades reflect the presence and size of lenticels, fissures, and other surface structural faults. Even the best-quality bark may contain internal faults undetectable during sorting.

Lower-grade corks are usually treated to a process called *colmatage*. The process fills surface cavities with a mixture of fine cork particles and glue. This normally occurs in a rotary drum in which the corks, cork particles, and glue are tumbled together. After an appropriate period, the corks are removed and rolled to produce a smooth surface. *Colmatage* improves not only the appearance but also the sealing qualities of the cork. Nevertheless, the cork–glue mixture does not possess the elastic, cohesive, and structural properties of natural cork. The treatment is justifiable with corks that are slightly marred with surface imperfections, but not corks so structurally flawed that their sealing properties are severely compromised.

Corks may subsequently be stamped or burnt with marks indicating the winery, producer, wine type, and vintage. Corks may also be coated with paraffin, silicone or both. Paraffin is a petroleum derivative, consisting of a mixture of saturated hydrocarbons roughly 20–50 carbons long. By replacing some of the waxes and oils unavoidably removed during cork processing, paraffin helps reduce penetration of wine into, water loss from, and the extraction of potential taints from the stopper. In addition, paraffin acts as a barrier to oxygen and carbon dioxide passage between the cork and the bottle neck (Keenan *et al.*, 1999). Paraffin is typically applied as an emulsified spray containing water and carriers (such as fatty acids and triethanolamine).

Because paraffin can aggravate cork extraction, it is often overlain with silicone. Silicone is an organosilicon oxide, normally formulated as a methylsilicone polymer (polydimethylsiloxane) (Garcia and Carra, 1996). Silicone eases cork insertion and removal from the bottle. Although compatible with most wines, silicone can disrupt effervescence production (Borges, 1985). Thus, its use is unacceptable with champagne corks. Nevertheless, a slice of silicone, tempered at 200°C for several hours, and attached to the inner surface of a champagne cork, can prevent the diffusion of TCA into the wine for at least one year (Vasserot *et al.*, 2001).

Finally, corks are bundled in large sacks (containing ~10,000 each) or sealed in plastic bags (containing 500–1000 each). Previous use of jute sacks is on the decline. It not only complicates stacking, but it also can result in

quality loss. Baling exposes the corks to contamination from dust, microorganisms, and volatile chemicals. The hydrophobic (nonpolar) nature of cork makes it susceptible to absorbing many odors from the surrounding environment. Known examples of odor contamination include guaiacol (Simpson *et al.*, 1986), naphthalene (Strauss *et al.*, 1985) and TCA. In addition, exposure to high humidity can promote microbial growth, whereas low humidity results in a loss of suppleness.

Storage in polyethylene bags avoids many of these problems, but does not prevent absorption of TCA from the environment (Capone *et al.*, 1999). Bags are commonly packaged in carton boxes for easy storage and shipment. Corks stored at a moisture content of 5–7% usually do not develop a mold growth. Sulfur dioxide is typically added as a prophylactic to limit microbial growth. Its inhibiting action lasts little more than 6 months to 1 year, as sulfur dioxide slowly diffuses through the polyethylene. The alternative of soaking corks in a concentrated solution of sulfur dioxide prior to storage is not recommended. It leads to suberin breakdown in the surface layers of the cork. Also, it can generate sulfurous taints by reacting with other cork constituents.

Because sulfur dioxide, hypochlorite, and peracetic acid disinfect only the outer cork layers, microorganisms found deep in the lenticels, or that have grown into the cork (Fig. 8.43) may remain unaffected. Other treatments offer the possibility of completely sterilizing

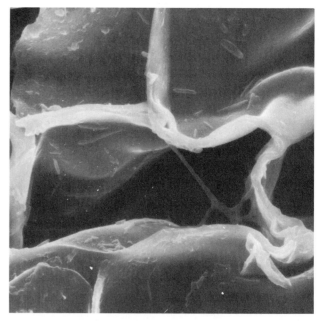

Figure 8.43 Scanning electron micrograph of a block of marbled cork showing the presence of fungal filaments in the tissue. (Photo courtesy of Laboratoire Excell)

corks. The most effective is exposure to gamma radiation. Microwaving also appears to be effective.

In most instances, cork sterilization is unnecessary – most wines and bottles contain viable microorganisms. In addition, sterilization does not destroy existing off-odors in the cork. However, because sweet wines and wines low in alcohol content are particularly susceptible to microbial spoilage, the use of sterilized corks is preferred. The wines are typically sterile-filtered into sterilized bottles to eliminate the likelihood of spoilage.

TECHNICAL AND HYBRID CORK

Cork closures may be derived from granulated cork. These may be untreated (agglomerate cork) or treated to remove TCA or other components (technical cork). In either case, the granules are bound together with a synthetic composite, such as polyurethane. The adhesive usually possesses reactive isocyanide end groups (Six and Feigenbaum, 2003). These react with water molecules in the cork to form the cross-links that bind the cork particles.

Granules are typically produced by grinding the remnants of bark left over from punching out cork stoppers. These are mixed with glue and extruded into long tubular molds. The molds are heated to 95–105 °C to set the glue. The cylinders of agglomerate cork are then removed and cut into stoppers of the desired length. Subsequent treatment is equivalent to that of natural cork. The diameter of agglomerate cork stoppers is slightly less than that of their natural cork equivalents, to adjust for their lower resilience. Lower resilience means that agglomerate cork stoppers are less easily compressed for bottle insertion, and rebound less than the natural-cork equivalents.

The use of agglomerate cork to seal table wines has expanded significantly in recent years. It has been used for decades in closing bottles of sparkling wine. Originally, the corks used to stopper sparkling wines were produced from strips of natural cork. However, because of the larger stopper diameter required for adequate sealing (31 vs. 24 mm), thicker slabs of cork were needed. Because of the cost, corks composed of two, three, or more layers of cork, laminated together, became common. Following the development of better-quality agglomerate cork, it replaced all but two inner disks (*rondelles*) of natural cork. Before insertion, the corks are positioned so that the natural cork will be in contact with the wine.

Rondelles are produced from cork slabs thinner than those required for natural cork stoppers. Top quality cork is steamed, flattened, the hard back removed. Sheets (lamina) of cork (6.4–6.6 mm) are sliced parallel to the surface of the slab. The remainder of the cork is ground into fragments used for agglomeration. Discs of 34.5 mm are bored out at right-angles to the surface. Discs from the layer closest to the outer surface of the slab generate the sections next to the agglomerate portion of the stopper. The inner discs (those in direct contact with the wine) come from the second (innermost) layer. Its smaller cells and thinner growth rings are of lower porosity than the outer layer (Pereira *et al.*, 1996). In larger production facilities, the discs are divided into five quality categories, based on automated image analysis (Lopes and Pereira, 2000). Following gluing the discs together, and to the agglomerate section, the stopper is trimmed and sanded to reach standard dimensions (31.5 × 47 mm). Traditionally, the adhesive has been a casein (gelled in the presence of water and calcium hydroxide). Nevertheless, the present trend is to use polyurethane or polyvinyl adhesives (Six and Feigenbaum, 2003). The familiar mushroom shape of the cork is produced by compressing the top of the cork just after insertion.

Hybrid cork is another alternative to natural cork. For example, Altec® corks are a combination of ground natural cork (with its lignin content removed) and microspheres of a synthetic polymer. The blend is bonded together with polyurethane. The comparative moisture-independence of the cork's elasticity permits stoppered bottles to be stored upright after closure.

Cork Faults

Cork can be both a direct and indirect source of wine faults. Corks can cause leakage, generate deposits, and be a source of off-odors. Corks can also remove (strip) wine flavorants (Capone *et al.*, 2003) and possess an oxidizing potential of uncertain origin (Caloghiris *et al.*, 1997).

LEAKAGE

Leakage around or, rarely, through the cork may have many causes. Incorrect bore size or imperfections in the glass surface may leave or create gaps between the cork and the neck. Improper alignment or compression of the cork during insertion can produce structural faults in an initially flawless cork. Laying the bottle on its side immediately after filling and corking, or rapid temperature changes during storage or shipment, can also induce leakage. Direct leakage due to the cork is typically caused by structural imperfections; however, improper sizing during manufacture also may be a cause.

Cork may show a wide variety of structural imperfections and mechanical faults (Fig. 8.44). Corks containing numerous flaws are customarily avoided by evading poor regions of the bark during punching out, or are rejected during subsequent grading. However, quality

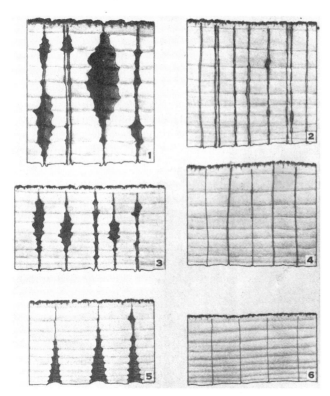

Figure 8.44 Types of porosity associated with the form and size of lenticels in cork. 1 and 3, cavernous; 2, highly porous; 4, moderately porous; 5, conical porous; 6, slightly porous. (From Vieira Natividade, 1956, reproduced by permission)

control systems cannot adequately check every specimen, and some flawed corks are found in all grades.

Lenticels can be the source of large cavernous or conical-shaped crevices in cork. Growth irregularities in the bark or wood may cause splits or fissures in the cork. Less frequently, holes are produced by cork-boring insects, including ants and several moth and beetle larvae. Some of the larvae also may produce holes in corks after insertion. The old practice of dipping the bottle neck in beeswax helped prevent the deposition of insect eggs on the cork. Plastic and metal capsules serve a similar function. Most cork-boring insects are larvae of the common cork moth (*Nemapogon cloacellus*), the grain moth (*N. granellus*), the wine moth (*Oenophila flavum*), the cave moth (*Dryadaula pactolia*), the seed moth (*Hofmannphila pseudopretella*), or the glue moth (*Eudrosis lacteella*) (see Vieira-Natividade, 1956).

Another source of cork rejection involves the presence of *mancha amarela* (yellow stain). The fault occurs sporadically throughout effected cork slabs as yellowish to white spots. These areas are associated with cell degradation by an unknown fungus, suspected to be *Armillariella mellea*. The attack commences with a destruction of the secondary wall of cork cells. It progresses outward and eventually disorganizes the middle lamella (Rocha *et al.*, 1996c). Cell walls become thinner, deformed (corrugated), and separate at the middle lamella. These changes are correlated with degradation of lignin and pectin (Rocha *et al.*, 2000).

Mechanical faults may also result from the development of fibrous tissue in the cork, the growth of fungi (marbled cork) (Fig. 8.43), or insufficient suberification (green cork).

These faults may be the principal cause of one of the major problems with cork – marked variation in oxygen permeability (Caloghiris *et al.*, 1997; Hart and Kleinig, 2005). However, this problem has not consistently been found (Lopes *et al.*, 2005).

Oxygen permeability is not an inherent characteristic of cork, but is thought to be the principal cause of sporadic wine oxidation (Waters and Williams, 1997). This is suspected from the linear relation between SO_2 loss and browning (Waters *et al.*, 1996b). Casey (2003) contends that sporadic wine oxidation is due more to problems associated with cork insertion than the cork itself. This could include variations in fill height, malfunctions in evacuation or gas flushing at bottling, differences in turbulence (and gas uptake) during filling, or residual oxidants in the wine from previous oxygen exposure. In any case, seepage of wine out between the glass and the stopper appears not to be involved. Wines with obvious leakage were no more likely to be oxidized than those with no apparent seepage (Caloghiris *et al.*, 1997). Additional potential sources of oxidation include residual amounts of chlorine or peroxide from disinfectant washes, and improper application of surface agents such as paraffin or silicone. A comparison of methods for measuring headspace oxygen content is given by J.-C. Vidal *et al.* (2004).

DEPOSITS

Cork can occasionally be a source of wine deposits. The most common originates from lenticular dust. Dust also is generated during cork manufacture, but most is removed prior to shipment. Mechanical agitation during transport may loosen more lenticular dust. Coating corks with paraffin or silicone helps to limit, but does not prevent, dust release. If the coating is defective or nonuniform, it can itself generate particulate material. Some coating materials, such as polyvinylidene chloride (PVDC), may become unstable and flake off in contact with highly alcoholic wines. Calcium oxalate crystals, formed after an oxalate wash, occasionally can generate visible deposits in wine. Defective *colmatage* also can produce wine deposits, as can improper cork insertion, by physically damaging the cork.

TAINTS

Cork can give rise to several musty off-odors. Although the odors are often referred to as **corky**, this designation gives a false impression. Sound cork, containing no faults, donates little if any odor. Cork may release several volatile compounds and liberate small amounts of brownish pigments, but none of these have a "corky" smell.

Improper treatment or storage near volatile chemicals can result in the adsorption of off-odors. These can subsequently taint the wine. In this regard, cork can not only release TCA into, but also absorb TCA from wine (Capone *et al.*, 1999). Normally, cork acts as an effective barrier to off-odors originating from a winery or storage environment (Capone *et al.*, 2002).

Cork-derived taints, which are due primarily to microbial growth, are covered later in this chapter. Various authors have estimated that up to 2–6% of all bottled wine may be adversely affected by musty (corky) off-odors. If true, wine producers and retailers are fortunate that the majority of consumers cannot detect, recognize, or do not object markedly to their presence (Prescott *et al.*, 2004). Whether bottles are returned or not, tainted wine may result in a loss of repeat sales.

TCA (2,4,6-trichloroanisole) is the most well known example of a processing-associated cork taint (Pollnitz *et al.*, 1996). TCA produces a marked musty to moldy odor at a few parts per trillion. The threshold rises rapidly with the alcohol content. Consequently, it is seldom detected in distilled beverages such as brandy.

TCA appears to be located primarily in the outer surfaces of contaminated corks (Howland *et al.*, 1997). This finding, combined with the frequent association of hypochlorite treatment with the presence of TCA, suggests that its formation occurs after stoppers have been punched out. If corks are steeped overly long in the bleach solution, chloride ions could diffuse into the cork through lenticels and fissures. Here, chloride could react with phenolic compounds, producing chlorophenols. Subsequent methylation by microorganisms in the cork could produce TCA. Absorption of water vapor by the cork could desorb TCA, permitting it to diffuse into the wine (Casey, 1990). The use of substitute bleaching agents, such as sodium peroxide and peracetic acid, avoids the production of TCA. Nevertheless, they may themselves leave undesirable residues (Fabre, 1989; Puerto, 1992).

Other sources of TCA can originate with the treatment of cork trees, wooden cooperage, or other winery structures with pentachlorophenol (PCP). It is often applied to control insects and wood rot. Several common molds, such as *Penicillium* and *Trichoderma* can metabolize (detoxify) PCP to TCA (Maujean *et al.*, 1985; Coque *et al.*, 2003). PCP also can directly contaminate corks and diffuse into wine.

Although microbial methylation of PCP is frequently implicated in the production of TCA, some white-rot fungi (basidiomycetes), as well as *Penicillium* spp. (Maujean *et al.*, 1985), can directly incorporate chlorine into a benzene ring. Subsequent metabolism could generate TCA (Maarse *et al.*, 1988).

TCA has garnered most of the attention, apparently being the principal cause of corky odors (Sefton and Simpson, 2005). However, there is no established definition for what constitutes a corky odor. Silva Pereira, Figueiredo Marques *et al.* (2000a) have argued that TCA is not the principal cause of corky taints. In an extensive study of some 2400 commercial wines, 145 wines were identified by at least one taster as having a corky odor. About half of these possessed TCA levels below the panel's detection threshold (2 ng/l) (Soleas *et al.*, 2002). In addition, 35% of wines identified as corked possessed TCA contents below analytic detection (0.1 ng/l). None of the 185 wines in the study closed with a screw cap were identified as having a corky taint, whereas about 30% of wines sealed with agglomerate cork were tainted.

The absence of detectable TCA in some wines identified as possessing a corked odor may relate to newly discovered musty-smelling compounds. Chatonnet *et al.* (2004) have isolated 2,4,6-tribromoanisole (TBA) from several corked wines in France. TBA appears to have a similar microbial origin as TCA – methylation of its halophenol precursor, TBP (2,4,6-tribromophenol). The latter is often used as a fire-retardant and wood preservative. As a consequence, it may be found on wooden or wood-based material throughout a winery. The common mold, *Trichoderma longibrachantum*, possesses an o-methyltransferase that can methylate phenols containing fluoro-, chloro- and bromo-substituents (Coque

2,4,6-TCP
2,4,6-trichlorophenol

2,4,6-TCA
2,4,6-trichloroanisole

PCP
(fungicide)
pentachlorophenol

et al., 2003). The conversion of TBP to TBA generates a highly volatile compound (easily contaminating a wine cellar). It adsorbs efficiently into hydrophobic products such as cork, polyethylene, and silicone. Thus, both natural and synthetic corks, the polyethylene liners of screw caps, silicone bungs of barrels, vulcanized rubber gaskets, and polyethylene- or polyester-based winemaking equipment may adsorb significant amounts of TBA. It can subsequently be desorbed into wine. TBA has an extremely low detection threshold, similar to that of TCA (parts per trillion).

In addition to the TBA, Simpson et al. (2004) have isolated and identified another new and potentially significant source of musty odors in cork stoppers. The compound, 2-methoxy-3,5-dimethylpyrazine, is estimated to be the leading cause of moldy odors in Australian wine, after TCA. It has a sensory threshold in the range of TCA and TBA (a few ppt). Currently, the origin of this musty-smelling compound in cork is unknown. Other pyrazines have been hypothesized as potentially being involved in moldy cork odors, but only 2-methoxy-3,5-dimethylpyrazine has thus far been isolated.

Aqueous solutions of sulfur dioxide, often used to surface-sterilize corks, have been implicated in the production of sulfurous off-odors. Sulfur dioxide can react with cork lignins to generate lignosulfurous acid. When leached into wine, lignosulfurous acid may decompose, generating hydrogen sulfide. Reaction with pyrazines could produce musty-smelling thiopyrazines. An additional source of pyrazines, other than that produced by boiling cork slabs, comes from mold growth on and in the cork. Some pyrazines have strong moldy, earthy odors.

Additional cork-associated off-odors are discussed later. Several of the compounds involved may be synthesized by microorganisms on substrates other than cork (i.e., oak staves). Because cork taints may come from a variety of sources, the combined effect of several compounds may be distinct from the odors of each individual component.

Because of the importance of cork-derived off-odors, ingenious procedures have been developed to remove, limit, or prevent their diffusion into wine. These have included autoclaving, pressurized steam, ozone, carbon dioxide, microwaving, as well as coating with silicone or special membranes. Rocha et al. (1996b) studied the potential of autoclaving to remove various musty, moldy odors. Compounds, such as 3-methyl-1-butanol, 1-octanol, and guaiacol, were extracted by the steam, and removed during autoclave venting. Pressurized steam is used commercially by Amorim to remove volatile compounds from cork used to produce technical corks. Ozone can also be used to oxidize (deodorize) cork and

inactivate microbial contaminants. Other procedures used to remove TCA include ultrasound (microwaving), and carbon dioxide in a supercritical state (intermediary between liquid and gaseous phases) (Taylor et al., 2000). Most of these procedures have been most successful with cork particles used to produce technical corks. These extraction procedures may also be combined with encapsulation with a silicone polymer. Silicone polymer membranes can also be formulated to be both highly impermeable to TCA, as well as possess designed degrees of oxygen permeability. This provides the potential for selecting a particular direction for wine aging – once the effects of oxygen permeation are better understood. Such developments bring into question whether cork is actually necessary, if the major protection is provided by a man-made membrane – cork acting almost as an inert filler.

Ingenuity related to TCA problems has not been limited to cork producers. Highly absorbent yeast hulls can be added to contaminated wine to reduce its concentration of TCA and other corky off-odors. Although probably scalping some of the wine desirable aromatics, the wine is saved from rejection. For the consumer, passing wine through a very thin activated-carbon filter system removes TCA and related compounds from the wine. The system involves a hand pump to force the wine out of the bottle and through the filter.

Although these procedures should reduce the incidence of corked wines, it will not eliminate the presence of corky odors in wine. As will be discussed below, corky or moldy odors may affect wine independent of the type of bottle closure. In addition, most of the treatments noted above do not address the equally significant issue of sporadic wine oxidation (Caloghiris et al., 1997). There also appear to be problems associated with the inherent oxidation potential of cork. This may result from a combination of oxygen escaping into the wine from the cork, and residual oxidants (chlorine or peroxide) from bleaching and washing.

Alternative Bottle Closures

Alternatives to cork have been sought for years. This has been spurred by both rising costs and increasing demand for premium quality cork. In addition, some producers believe that the frequency of cork-related faults is on the rise. The reverse seems more likely.

The most successful alternative is the pilfer-proof, roll-on, metal screw cap (Fig. 8.45). Part of its success comes from its not attempting to reproduce the appearance and characteristics of natural cork. Roll-on (RO)[2] closures are generally superior to cork in retaining sulfur dioxide and

---

[2] Alternately referred to as ROTE (roll-on, tamper-evident).

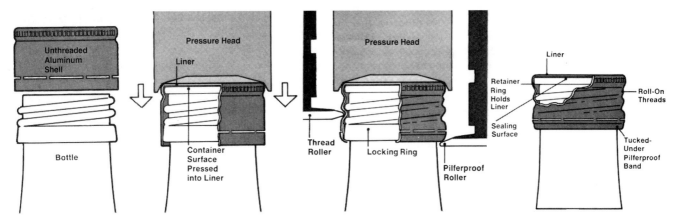

Figure 8.45 Insertion and formation of an Alcoa RO closure. (Courtesy of Alcoa Closure Systems International, Indianapolis, ID)

minimizing oxidation (Caloghiris *et al.*, 1997; Godden *et al.*, 2001). They also maintain the fruitiness of the wine for decades, while being essentially free of TCA tainting. Corked wines found sealed with RO closures probably were contaminated before bottling. Screw-cap closures also obviate the need for lying bottles on their sides, as well as avoiding the nuisance of using a cork-screw. In addition, RO closures do not absorb (scalp) aromatic compounds from the wine, as can cork or synthetic closures (Capone *et al.*, 2003).

The seal is provided by pressure exerted by the cap, forcing the top of the bottle neck against a foam plastic pad. The latter is usually lined with a 19 μm layer of polyvinylidene chloride (Saran®). A 20 μm layer of tin foil acts as a gas barrier. The plastic pad (2 mm) is frequently made of food-grade, high-density polyethylene, such as polytetrafluoroethylene (Teflon®).

The oxygen permeability of RO closures is very low (2–11 μl/month) (Lopes *et al.*, 2006). Although natural cork may vary considerably in oxygen permeability, the best quality corks seem to have permeability attributes similar to those of RO closures (Brajkovich *et al.*, 2005). Lopes *et al.* (2006) note that oxygen ingress is most marked shortly after insertion (20 μl/d), falling to 2–6 μl/d after one month, and declining to about 0.1 μl/d after a year. It appears that most of the oxygen diffuses out of cork cells, with diffusion from the air being low, slow and relatively constant (Lopes *et al.*, 2007).

There is considerable controversy among industry professionals as to the benefits/disadvantages of limited oxygen penetration during bottle aging. Part of this relates to unknown, but potentially beneficial, roles for oxygen during wine aging. Nevertheless, there is growing evidence that both white and red wines age as well under RO closures. For example, 'Riesling' wines retained their fruity character for twenty years in screw-capped bottles. Hart and Kleinig (2005) have shown that even premium quality red wines age well

under RO closures. What is important is that whatever closure is used that it have consistent sealing qualities. All bottles should age equivalently.

The other contentious issue concerning RO closures relates to the possible development of reduced-sulfur odors (Stelzer, 2003). Even if present in some bottles, can consumers recognize it, and if so, will they reject the wine?

One hypothesis is that reduced-sulfur off-odors arise due to the development of excessively low redox potentials in the wine. Were this the case, appropriate adjustment of the oxygen permeability of the pad should correct the problem. As our understanding of the origin of disulfides and other thiols is still rudimentary, it is difficult to explain why limited oxygen access would reduce their accumulation. However, the low redox potential hypothesis has a problem. High-quality corks appear to have oxygen permeability properties similar to those of RO closures. Wines closed with high-quality corks are not known to be characterized by the development of reduced-sulfur odors. However, if reduced-sulfur odors are only an expression of winemaking procedures, as suspected by some, appropriate corrective action in the winery or vineyard should eliminate the problem (see Sulfur Off-odors). The debate is often heated, not only because of a lack of precise information, but also because of the financial stakes involved – for closure producer and winemaker alike.

The historical "disadvantage" to RO use has been its previous association with inexpensive wines. This regrettable perception is rapidly changing, partially due to its widespread adoption in New Zealand, Switzerland, Australia, and its now expanded use in California. Even premium wine producers are selectively choosing RO closures for their best wines.

Most RO closures possess an aluminum casing that can be crimped into the shape of the spiral threads molded onto the bottle neck. An alternative RO closure is composed of molded plastic or metal. It retains the external appearance of the traditional capsule. The threads of the closure are molded onto the inside of the cap. To appeal to producers still wanting a natural cork closure, the top of the plastic closure contains a cork insert anchored to the top of the closure. Unscrewing the cap extracts the cork without the use of a screwpull.

Because of continuing resistance to RO closures in some countries, most cork substitutes have attempted to reproduce the appearance and physical properties of cork. In addition, their synthetic nature makes them amenable to an almost endless range in colors.

The first plastic stoppers used successfully were made of polyethylene. The ease of resealing made them especially appealing for sparkling wines. However, their permeability to volatile compounds has been a major drawback (Skurray *et al.*, 2000). For wines with a rapid turnover, this was relatively insignificant. For premium wines, oxygen permeability precluded their use. Also, their initial use with inexpensive and carbonated wines retarded their adoption when technically improved versions first came on the market.

In contrast to polyethylene, corks made with ethylene vinyl acetate possess most of the appearance and features of cork (Chatonnet *et al.*, 1999). They can also be removed effectively with a cork screw. When injected with air and a hardener (molded), liquid ethylene vinyl acetate forms millions of microscopic gas pockets before hardening. The inflated plastic develops a resilience that permits it to return almost immediately to its original diameter (97%). The plastic stopper regains more than 99% of its initial volume within 1 hour. In addition to molding, synthetic corks may be manufactured by extrusion.

The major disadvantage in most synthetic stoppers has been their less-than-ideal permeability to oxygen (Godden *et al.*, 2001; Lopes *et al.*, 2006). Nevertheless, for wines intended for early consumption (within one to two years), synthetic corks provide adequate oxidation protection. In addition, technical improvements in the past few years appear to have generated synthetic corks with reduced oxygen permeability. Future advances may reduce their aroma scalping potential and provide closures of precise and designed degrees of oxygen permeability.

Ground glass stoppers are a recent entry as a closure option. These have a superficial resemblance to T-corks. As with RO closures, the neck of the bottle must be especially molded to accept the glass stopper.

The crown cap, with a cork or polyethylene liner, formerly so familiar due to its use with soft drinks, has seen little application as a wine closure. The one exception is in the second in-bottle fermentation of sparkling wines. Here, consumer image is not a problem.

The bottle cap is replaced by a traditional cork closure when the wine is disgorged and prepared for shipment. Nevertheless, some producers are starting to seal their finished sparking wines with crown caps.

Another solution to stoppering bottles has been to avoid them altogether. The use of cans, cartons and bag-in-box packaging are existing, if not elegant, solutions. They do appeal to younger drinkers not influenced by the mystique of wine or inclined toward being a connoisseur (at least initially).

Cork Insertion

Various devices are available to insert corks into wine bottles. They all function by compressing the cork to a smaller diameter than the bore of the bottle. After compression, a plunger forces the cork into the neck of the bottle.

Better corking machines apply a uniform pressure over the full length of the cork. This minimizes the production of creases, folds, or puckering that could result in leakage. Standard 24-mm wide corks are compressed to about 14–15 mm before insertion. The plunger is adjusted to assure that, regardless of cork length, the top of the cork rests at, or just below, the lip of the bottle.

Most 750-ml wine bottles are produced to the CETIE (Centre Technique International de l'Embouteillage) standard, with an inner-neck diameter at the mouth of 18.5 ± 0.5 mm, increasing to no more than 21 mm, 4.5 cm below the mouth. An older standard, retained for most sparkling wines, has a bore diameter of 17.5 ± 0.5 mm at the lip. Because most wine corks have a 24-mm diameter, they remain compressed by about 6 mm after insertion. This level of compression is sufficient to generate a constant pressure of approximately 1 to 1.5 kg/cm^2 against the glass (Lefebvre, 1981). For corks of superior quality, a pressure of about 3 kg/cm^2 may be produced. Sweet wines or those containing greater than 1 g/liter CO_2 typically use corks that remain compressed by approximately 7–8 mm against the glass. Sparkling wines commonly use special 30- to 31-mm diameter corks to maintain 12 mm of compression against the neck.

Oversized corks can cause leaks as easily as those that are too narrow. If the cork is too large, creases may occur during insertion; if the cork is too narrow, the seal may be too weak.

Because the diameter of the bore commonly increases down the length of the neck, the desirable cork length depends partially on the extent of the increase. In the CETIE standard, the maximum bore diameter at a depth of 4.5 cm is 21 mm. This means that both medium (44/45-mm) and long (49/50-mm) corks of 25-mm diameter may be held by no more than 4 mm of compression at depths below 4.5 cm. This contrasts with about 6.5 mm compression at the bottle mouth. Deeper than 4.5 cm, the large inner-neck diameter may result in long corks being held only weakly next to the wine. Thus, cork diameter is generally more important to a good seal than length. The advantage of long corks for wines benefitting from long aging comes from factors other than cork contact with the glass.

Cork is noted for its chemical inertness in contact with wine. Nevertheless, prolonged exposure slowly reduces the structural integrity of the cork. Because of cork's low permeability to liquids, corrosion progresses slowly up the cork, from where it contacts the wine. In addition, water uptake makes the cork more supple, weakening its hold on the glass. As the rate of weakening is most rapid in wines with high sugar and alcohol contents, long corks are especially valuable in sealing such wines. Because corrosion affects the substance of the cork, denser cork is preferred if prolonged aging is anticipated. It is estimated that loosening of the cohesive attachment of cork to the glass progress at a rate of about 1.5 mm/year (Guimberteau *et al.*, 1977). The effect is reported to diminish the pressure exerted by the cork against the neck, from its initial value of approximately 100–300 kPa to 80–100 kPa after 2 years, and approximately 50 kPa after 10 years (Lefebvre, 1981). This explains why fine wines are often recorked about every 25 years.

The slight cone shape of the bore (from 18.5 to 21 mm down the neck), the bulge or indentation in the neck about 1.5 cm below the lip, and compression of the cork are all important in limiting movement of the cork in the neck. This is especially important if the wine is exposed to temperature extremes, causing volume changes that can weaken the seal and force the cork out of the bottle.

LEAKAGE CAUSED BY INSERTION PROBLEMS

The piston-like action of cork insertion can compress gases trapped in the neck of the bottle, doubling to quadrupling the gas pressure in the headspace. As soon as the stopper enters the neck, the cork begins to rebound, asserting pressure against the sides of the neck. Full exertion of this elastic pressure takes several hours (Fig. 8.40). Thus, if the bottle is laid on its side, or turned upside down shortly after corking, the pressure exerted by the trapped headspace gases may force a small amount of wine out between the cork and the neck of the bottle. Although seepage is not known to induce wine oxidation (Caloghiris *et al.*, 1997), it can produce a sticky residue that furnishes nutrients for mold growth on the top of the cork.

To avoid seepage, bottles are often left upright for several hours after corking. During this period, entrapped gas can escape or dissolve into the wine, and the internal pressures decline to near atmospheric within a few hours to days. The speed of the decline in headspace pressure depends on the type of cork and the composition of the headspace gases (Fig. 8.46). Temporary upright storage is especially important when the cork is agglomerate (more rigid than natural cork), or if the entrapped gas is air. Air consists of about 78% nitrogen, a gas poorly soluble in wine. If the pressure is not released, nitrogen will continue to exert pressure on the wine and cork after the bottle is placed on its side. Oxygen, the other main atmospheric gas, dissolves quickly in the wine and ceases to exert pressure.

To reduce leakage, bottles may be flushed with carbon dioxide (to remove oxygen and nitrogen) before filling, or be placed under partial vacuum at corking. Partial vacuum (20–80 pKa) minimizes the development of a positive headspace pressure during cork insertion (Casey, 1993).

The use of corking under vacuum or carbon dioxide has benefits beyond reducing the likelihood of leakage. It limits the development of "bottle sickness," presumably by removing the 4–5 mg of oxygen that would otherwise be absorbed from the entrapped air. Both procedures also significantly slow the loss of sulfur dioxide by removing oxygen. For example, De Rosa and Moret (1983) showed that vacuum and flushing, and vacuum application, reduced average SO_2 loss after 12 months from 28 to 16 mg/liter and 5 mg/liter, respectively.

Variation in bottle capacity may be an additional source of seepage problems. If the bottle has a capacity smaller than specified, very little headspace volume may remain after filling. Even with medium-length corks, bottles may possess a headspace volume as little as 1.5 ml after corking. Because wine in a 750-ml bottle can expand by up to 0.23 ml/°C, a rapid rise in temperature can quickly result in a significant increase in the pressure on the cork. More typically, headspace volumes are in the range of 6–9 ml. In such cases, a rapid temperature rise by about 20 °C would be required to produce a doubling of the pressure exerted by the headspace gases on the cork. Leakage becomes probable at internal pressures above 200 kPa (twice atmospheric pressure). At this point, the net outward pressure begins to equal or exceed that exerted by the cork against the glass. This effect would be augmented if the headspace gas contained only nitrogen, or if the wine were sweet or supersaturated with carbon dioxide (Levreau *et al.*, 1977). In such situations, the pressure is likely to remain because nitrogen gas does not effectively dissolve in wine, and wine supersaturated with carbon dioxide absorbs additional CO_2 slowly. Sugar can augment seepage by increasing the capillary action between the cork and the glass. Sugar also increases

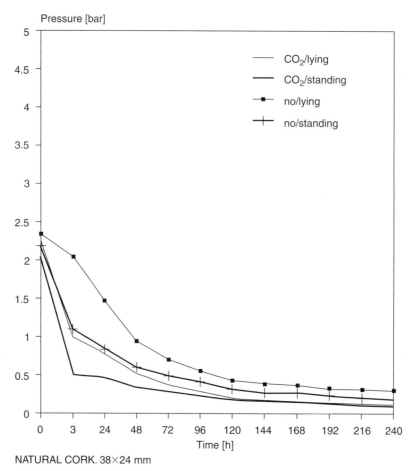

NATURAL CORK. 38×24 mm

Figure 8.46 Change in the inner bottle pressure depending on the position of bottle storage (lying or upright) and with (CO_2) or without (no) carbon dioxide flushing of the bottle prior to filling. (From Jung, 1993, reproduced by permission)

by about 10%, the rate at which wine volume changes with temperature (Levreau *et al.*, 1977).

Finally, cork moisture content influences the likelihood of leakage. At a moisture content between 6 and 9%, cork has sufficient suppleness not to crumble on compression. The lower end of the scale is preferred for fast bottling lines, whereas the higher end is recommended for medium and slow speed bottling lines. Cork within this range of moisture contents also rebounds sufficiently slowly to allow pressurized headspace gases to escape before a tight seal develops. At lower moisture levels, compression and insertion are likely to rupture or crease the cork. At high moisture levels, more gas is likely to be trapped in the headspace (Levreau *et al.*, 1977).

Bottles and Other Containers

Over the years, wine has been stored and transported in a wide variety of vessels. The first storage vessels were probably earthenware or made from animal skins. The interior of the former were probably covered with pitch to make them watertight. Examples of pitched earthenware wine vessels have been discovered in Iran, dating between 5400 and 5000 B.C. (McGovern *et al.*, 1996), and in the tombs of the earliest Egyptian Pharaohs. The use of pitch was carefully described by Pliny in *Historia Naturalis*, and has been frequently confirmed in archeologic finds. Pitch can be prepared from a variety of tree resins, such as pine (*Pinus halepensis*) around the Mediterranean, and terebinth (*Pistacia atlantica*), sandarac (*Tetraclinis articulata*) and myrrh (*Commiphora myrrha*) in the Near East. Nevertheless, impervious amphoras requiring no pitch were eventually made during the Roman era. The addition of potash and firing at an atypically high temperature (between 800 and 1000°C) produced a vitreous glazing. It made the vessel impervious to liquids (Vandiver and Koehler, 1986). Amphoras were the standard storage and transport vessel throughout ancient Greek and Roman times. Their use continued in the Middle East up until at least 625 A.D. (Bass, 1971). They fell into disuse throughout the western Mediterranean after the decline of the Roman Empire. North of the Alps, wood cooperage became the primary wine-storage and -transport vessel. The barrel held its preeminence in Europe until the twentieth century, when the glass bottle replaced wood cooperage as the primary container for aging and transporting wine.

Glass bottles were produced in limited quantities during Roman times. Petronius in his *Satyricon* mentions plaster-sealed glass wine bottles. An example (ca. 200 A.D.) is located in the Wine Museum in Speyer, Germany. The 1.5-liter bottle has shoulders resembling the handles of an amphora. However, Roman glass was generally too fragile for widespread adoption as a wine container.

In Italy, the reintroduction of wine bottles began during the 1500s. However, their fragility meant that the bulbous bottles needed to be protected by a covering of reeds, wicker, or leather. This traditional bottle covering is still used by some Italian producers. Subsequent technical advances, combined with increasing demand, favored the production of strong glass bottles. An edict by James I of England, forbidding the use of wood in glass furnaces, led to wood's substitution by coal. Not only did coal provide a higher and more sustainable heat, but it also supplied sulfur dioxide. A surface reaction between sulfur dioxide and sodium oxide in the glass formed a strengthening layer of sodium sulfate. Prohibition of the sale of bottled wine in 1636 (due to nonstandard volumes) inadvertently increased the demand for bottles. Owners brought their bottles to wholesalers for filling from casks. The need for stoppering necessitated stronger bottles (Dumbrell, 1983). Shortly thereafter, incorporation of iron oxide in the glass formula produced the now famous "black" (dark olive-green) bottle. The advantages of the greatly enhanced strength of these new bottles soon led to the adoption of similar production procedures throughout Europe (Polak, 1975).

Elongation and flattening of the sides, converted the original bulbous form into the now familiar cylindrical shape. This gradual transition occurred roughly between 1720 and 1800 (Dumbrell, 1983). This also corresponds to the discovery of the benefits of bottle-aging wine. These discoveries appear to have been initially associated with port in England and Tokaj in Hungary. During the same period, there was the concurrent evolution of many of the distinctive bottle designs that now characterize particular wine regions. Nevertheless, the extensive use of bottles for wine transport came much later. Bottles were essential only for the transport of sparkling wine and the development of port. Bottles came to supplant barrels as the

principal transport container in the later part of the nineteenth century.

Despite the clarity, inertness, and impervious nature of glass, bottles have their drawbacks. Because of the various shapes, sizes, and colors used in modern commerce, reuse is cost-effective only with a minuscule portion of the market. Bottles can also create a considerable disposal problem. In addition, once opened, wine conservation is difficult. A noticeable loss in quality is usually evident within several hours. Although dispensing systems (Plate 11.2) replace the apportioned wine with an inert gas (nitrogen or argon), they find use principally in restaurants. They are too expensive for the average consumer. As a consequence, new containers, such as bag-in-box or carton packaging, have replaced bottles for many standard wines.

Glass Bottles

Despite the limitations of glass, it has many advantages over other materials in storing wine. Its chemical inertness is especially useful in aging premium wines. Although trace amounts of sodium, chromium, and nickel may dissolve into wine, the levels normally involved are infinitesimal. However, changes in glass formulation may increase the amount of chromium extracted to from 1 to 4 μg/liter (Médina, 1981). Glass is also impermeable to gases and resists all but rough handling. The transparency of most glass to near-ultraviolet and blue radiation is a disadvantage, but can be corrected by the inclusion of metal oxides. The primary disadvantages of glass are its weight and the energy required in its manufacture. This has led to an evaluation of PET (polyethylene terephthalate) plastic for wine bottles.

PRODUCTION

Bottle glass is formed by heating sand (largely silicon dioxide), soda (sodium carbonate), lime (calcium oxide), and small amounts of magnesium and aluminum oxides to approximately 1500°C. During manufacture, an appropriate amount of molten glass (**parison**) is removed from the furnace and placed in the upper end of a rough (blank) bottle mold (Fig. 8.47). Compressed air forces the molten glass down to the bottom of the mold, where the two portions of the neck mold are located. Subsequently, air pressure from the center of the neck mold blows most of the glass back into the configuration of the blank mold. The procedure establishes the finished shape of the neck. The blank mold, containing half of the neck mold, is opened and removed. The outer stiff layer of glass touching the mold maintains the shape of the blank mold during its

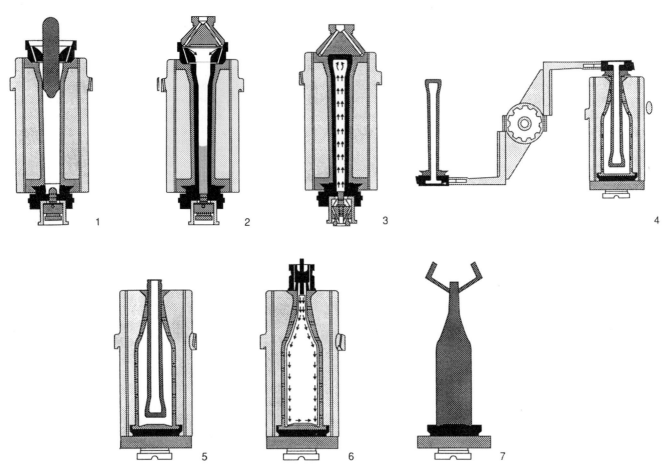

Figure 8.47   Blowing of glass wine bottles. 1, Molten glass (parison) is added to the rough mold; 2, air pressure forces the parison to the base of the mold; 3, air pressure from the mold base forces the still-molten glass into the rough mold shape; 4, the rough mold is removed and transferal of the bottle to the finishing mold; 5, the glass is reheated; 6, air pressure forces the molten glass into the shape of the finishing mold; 7, the finished bottle is removed. (After Riboulet and Alegoët, 1986, reproduced by permission)

transfer to the finishing (blow) mold. The remaining portion of the neck mold (neck ring) is used to raise and invert the bottle, positioning it between the halves of the blow mold. The bottle is released from the neck ring just before closing the halves of the blow mold.

The glass is reheated to bring it back to a moldable temperature. Compressed air, blown in via the neck, drives the glass against the sides of the blow mold. At the same time, a vacuum, created at the base of the mold, removes trapped air. These actions give the bottle its final dimensions. After a short cooling period, to assure retention of the finished shape, the bottle is removed from the blow mold.

During production, various parts of the bottle cool at different rates. This creates structural heterogeneity and areas of weakness. To release these structural tensions, the bottle is annealed at approximately 550°C. After sufficient annealing, the glass is slowly cooled through the annealing range, and then rapidly down to ambient. During annealing, sulfur is typically burned to

produce a thin layer of sodium sulfate on the inner surfaces of the glass. The associated diffusion of sodium ions to the glass surface increases its chemical durability. Alternatively, a thin coating of titanium or other ions may be added. Both procedures harden the surface of the glass and minimize lines of weakness.

Historically, bottles were produced by gathering the parison on the end of a blowpipe. It was rolled (**marvered**) on a flat stone or metal surface to develop its rough shape, as the worker blew into the **blowpipe**. The glass was periodically reheated in the furnace during the process. In early examples, the bottle was left with its inherent, rounded, blown shape. However, bottle shape soon evolved into a mallet form. The flat bottom and more parallel sides allowed the bottle to stand upright more steadily, and facilitated storage on its side.

After establishing the basic bottle shape, but while still malleable, a small dollop of molten glass on a **pontil rod** was attached to the base of the bottle. Once attached, the worker pushed the pontil rod gently into the base, forming an indentation (**punt**). Alternately, the indentation was made with a rounded *mollette*, before the pontil rod was attached at the base. The punt originally had several functions. It held the bottle while the blower cut off the blowpipe from the bottle neck. For the older, onion-shaped bottles, the punt produced a more or less flat bottom, allowing the bottle to sit on a table. The punt also placed the rough edges of the pontil scar out of harms way. The pontil scar was produced when the bottle was snapped off the pontil rod.

With the development of wooden or metal molds, the blower could create a diversity of shapes with more standardized capacities. Subsequently, when the punt had no role in bottle manufacture, the tradition of its presence was incorporated into the mold. The last step in bottle manufacture was to apply a small amount of molten glass around the neck. It was molded to form the lip, often with a pair of bladed forceps. The lip originally acted as an attachment site for the cord that held the cork (for convenient reinsertion). It also provided extra strength to the neck (useful during stoppering). Finally, the bottle was placed in a special annealing oven (**lehr**) to relax stress in the glass. Further details on the history of bottle manufacture can be found in Dumbrell (1983), Jones (1971), and Polak (1975).

SHAPE AND COLOR

Bottles of particular shapes and colors are associated with wines produced in several European regions. These often have been adopted in the New World for wines of analogous style, or to imply similar character. Unique bottle shapes, colors, and markings also are marketing tools used in increasing consumer awareness and recognition of particular wines.

Clear glass is typically produced by adding sufficient magnesium to decolorize iron oxide contaminants that commonly occur in sand. Small amounts of iron, manganese, nickel, and chromium oxides may be added to give the glass a desired color. The yellow to green color of many wine bottles comes primarily from ferric and ferrous oxides (Fe_2O_3 and FeO). The specific shade is influenced by the redox potential, degree of hydration, presence of other metals, and the chemical nature of the glass (Fig. 8.48). Amber is often generated by maintaining the reducing action of sulfur during glass fusion, brown by the addition of manganese and nickel, and emerald green by the incorporation of chromic oxide (CrO_3). Adding chromic oxide under oxidizing conditions or vanadium pentoxide (V_2O_5) greatly increases ultraviolet absorption (Harding, 1972).

An alternative to glass coloring is to apply an outer colored or textured coating. This technique has several advantages, other than the wider range of colors possible. The coating typically provides improved abrasion

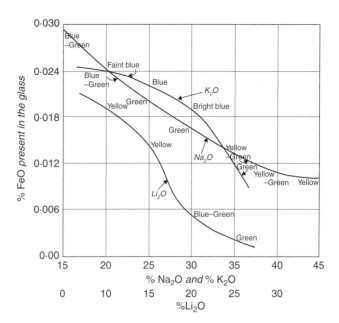

Figure 8.48 The influence of the concentration and nature of the alkaline oxide content of glass on the iron oxide equilibrium in alkali-silica glass of 0.075% Fe_2O_3. (After Densem and Turner, 1938, from Weyl, 1951)

and impact resistance, and facilitates recycling (the coating melts off, leaving clear glass for processing). Even more significant is the development of a coating that absorbs ultraviolet radiation. This now permits the natural color of the wine to be presented, without the risk of light-induced oxidation or spoilage. In addition, the coating can be applied to any shape or bottle size. It is considerably less expensive than changing the color formulation of the bottle glass.

In general, bottle color and shape are more important to marketing than aging, storage or transport. Tradition and image probably explain the inappropriate continued use of ultraviolet-transmissible glass. White wines are the most susceptible to light-induced damage (Macpherson, 1982), but are often the least protected (being commonly sold in clear or light-colored bottles). To partially offset light-activated spoilage, white wines often receive higher doses of sulfur dioxide before bottling.

In contrast, bottle neck design has depended primarily on pragmatic issues, such as the type of closure or reducing the likelihood of dripping. Most table wines can use plain necks, because they require only a simple cork or RO closure. Sparkling wines, however, require a special lip to secure the restraining wire mesh for the stopper.

Bottles of differing filling heights, and permissible capacity variations, are available. Bottles having lower filling heights and smaller volume variation tend to leave more headspace between the wine and the cork.

By providing more space for temperature-induced wine expansion, the bottles are less susceptible to leakage.

WINE PREPARATION

Just prior to bottling, the wine is frequently given a final polishing filtration and a dose of sulfur dioxide. The latter is added primarily for its antioxidative properties (Waters *et al.*, 1996b). For white wines, sulfur dioxide is typically supplied at a concentration sufficient to achieve (after equilibration) a free SO_2 content in the range of 25–40 mg/liter. Where bottles are not flushed with inert gas or filled under vacuum, the headspace may contain up to 5 ml of air. In such situations, an additional 5–6 mg of free SO_2 will be required to bind the presence of approximately 1 mg of oxygen in the headspace. Additional oxygen comes from cellular voids or fissures of the cork. Ascorbic acid has often been added to remove oxygen that may inadvertently be absorbed during bottling. However, a by-product of this reaction is hydrogen peroxide. One of the presumed benefits of joint addition with sulfur dioxide has been the ability of sulfur dioxide to react with hydrogen peroxide, reducing it to water. Nevertheless, Peng *et al.* (1998) found that, with or without sulfur dioxide, ascorbic acid enhanced the browning potential of white wines.

The antioxidant properties possessed by sulfur dioxide include reducing quinones back to diphenols, binding with quinones (preventing their involvement in the formation of brown polymeric pigments), formation of sulfonates with carbonyl compounds (inhibiting their oxidation to brownish pigments as well as producing a nonvolatile complex with acetaldehyde), and bleaching existing pigments.

Browning also appears to involve the oxidation of colorless xanthene to yellowish xanthylium compounds (Bradshaw *et al.*, 2003). In addition, metal ions (e.g., ferric ions) can catalyze catechin dimers, linked through glyoxylic acid, to yellow compounds (Oszmianski *et al.*, 1996). Although most oxidative browning is considered to involve flavonoids, the precise chemical nature of the pigments involved is incompletely understood. Studies by Lutter *et al.* (2007) are beginning this investigation.

FILLING

After a hot-water rinse, bottles are steam-cleaned and allowed to drip dry before filling. Although cleaning is always essential, sterilization is generally unnecessary. Microbial contamination of new bottles is usually negligible and consists mainly of organisms unable to grow in wine. In addition, the existing microbial population of the wine is probably considerably higher than that of the bottle. Thus, only for sweet or low-alcohol wines, or when recycled bottles are being used, is sterilization necessary. The availability of small, more economic ozone producers, should make bottle sterilization both more effective and less expensive.

Various types of automated bottling machines are available (Plate 8.5). Some operate by siphoning or gravity feed, whereas others use pressure or vacuum. Siphoning and gravity feed are the simplest but slowest, whereas pressure and vacuum fillers are more appropriate for rapid, automated filling lines.

Regardless of the machine, precautions must be taken against microbial contamination. The growth of microbes in bottling equipment can quickly contaminate thousands of bottles of wine. Precautions must also be taken to minimize oxidation during filling. This is best achieved by flushing the bottles before filling with carbon dioxide or under a vacuum. Alternately, the headspace may be flushed with carbon dioxide after filling. "Bottle sickness," a mild, temporary form of oxidation, may follow bottling if oxygen remains in the headspace. The volume of oxygen contained in headspace air can amount to eight times that absorbed during filling.

Bag-in-box Containers

Bag-in-box technology has progressed dramatically from its initial start as a means of marketing battery fluid. Subsequent developments have found wide application in the wine industry. Because the bag collapses as the wine is removed, its volume is not replaced with air. This permits wine to be periodically removed over several weeks without a marked loss in quality. This convenience is credited with expanding the wine market in regions without an established wine culture. In some countries, such as Australia, more than half the wine may be sold by the box (Anderson, 1987). Bag-in-box packaging also is ideally suited for "house wine" sold in restaurants, especially the larger 10- to 20-liter sizes.

Bag-in-box packaging protects wine from oxidation for 9 months or longer. This is usually adequate for wines that require no additional aging, and have high turnover rates.

For protection and ease of stacking and storage, the bag is housed in a corrugated or solid fiber box. A handle permits easy transportation. The large surface area of the box provides ample space for marketing information and attracting consumer attention.

No single membrane possesses all the features necessary for a collapsible wine bag. The solution has been to use a two-layered membrane (Webb, 1987). The inner layer, made of a low-density polyester or ethyl vinyl acetate film, provides the necessary protection against flex cracking. The outer layer, typically a metallized polyester laminated to polyethylene, slows the loss of sulfur dioxide and aromatic compounds and limits the inward diffusion of oxygen. Most modern plastic films are inert to wine and do not affect its sensory characteristics. Eliminating amide additives has removed a former source of off-odors.

Taps come in a variety of styles. Each has its potential problems, such as a tendency to leak, relative permeability to oxygen, and expense. The tap is considered the primary source of oxygen ingress in bag-in-box packaging (Armstrong, 1987).

To minimize oxidation, the bags are placed under vacuum before filling, and the headspace is charged with an inert gas after filling. To protect the wine against microbial spoilage and to limit oxidation, the wine is usually adjusted to a final level of 50 mg/liter sulfur dioxide before filling.

Wine Spoilage

Cork-related Problems

The difficulties associated with off-odor identification are well illustrated with cork-related faults. The origins of cork-related problems are diverse and their chemical nature often unknown. Even experienced tasters have great difficulty correctly identifying most faults, especially against the natural aromatic variation of different wines. The situation becomes even more complex if a wine is affected by more than one fault. Combinations of off-odors may influence both the quantitative and qualitative attributes of off-odors. Thus, most off-odors can be identified with confidence only with sophisticated analytic equipment. Although much has been learned, considerably more needs to be known to limit the occurrence of off-odors.

Although several cork-derived taints have a "musty" or "moldy" odor, others do not. Therefore, cork-related taints are usually grouped by presumed origin. Some off-odors come from the adsorption of highly aromatic compounds in the environment. Other taints originate during cork production, notably 2,4,6-trichloroanisole (TCA). Even some glues used in producing sparkling wine corks can generate off-odors (i.e., butyl acrylate) (Brun, 1980). Finally, microbial infestation can be an important source of cork-related taints. Because several of these problems have already been discussed in the section on Cork, consideration here will be limited to microbially induced spoilage.

In a study of tainted wines (Simpson, 1990), the most frequent off-odor compounds detected were TCA (86%), followed by 1-octen-3-one (73%), 2-methyl-isoborneol (41%), guaiacol (30%), 1-octen-3-ol (19%), and geosmin (14%) (Fig. 8.49). Each compound has its own, somewhat distinctive odor, and may not be present at concentrations sufficient for easy recognition.

In about 50% of the wines, TCA was considered the most intense off-odor.

Although TCA is considered the principal corked taint in wine, its musty chlorophenol odor is different from the putrid butyric smell originally ascribed to corked wine (Schanderl, 1971). Its occurrence is apparently now rare. Although of uncertain origin, it has usually been associated with "yellow stain" – an infection thought to be induced by *Armillaria mellea*. Cork infected by *A. mellea* may possess TCA and additional moldy or mushroomy off-odors, uncharacteristic of healthy cork (Rocha *et al.*, 1996a). Conversely, Lefebvre *et al.* (1983) have suggested infection by *Streptomyces*, and its production of guaiacol. Growth of *Streptomyces* spp. on cork surfaces can contaminate stoppers with moldy, smoky, or earthy odors. This probably relates to their synthesis of compounds such as guaiacol, 1-octen-3-ol, 2,5-dimethyl-pyrazine, TCA, and sesquiterpenes.

Moldy odors may originate from bacteria, various molds, and occasionally yeasts growing on cork slabs. Jäger *et al.* (1996) consider contamination, if it occurs, is most likely during seasoning of the bark or storage of finished stoppers in plastic containers. Contamination due to fungal growth after insertion is unlikely. The acidic, low nutrient, and anaerobic conditions of the bottle severely restrict most microbial metabolism.

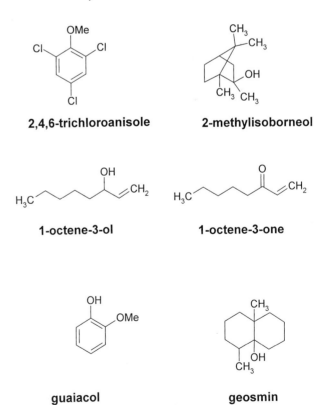

2,4,6-trichloroanisole **2-methylisoborneol**

1-octene-3-ol **1-octene-3-one**

guaiacol **geosmin**

Figure 8.49 Diagrammatic illustration of several major off-odor compounds.

Cork also can be a source of contaminant yeasts and bacteria that can grow in and spoil wine.

Because nonselective media are commonly used when isolating organisms from cork, it is difficult to assess whether the microbes isolated are members of an indigenous cork flora or contaminants. Nonselective culture media favor the growth of fungi that grow rapidly and/or produce large numbers of spores. It may be the equivalent of adding a teaspoon of woodland soil to a pot of sterilized soil to assess the dominant forest species. Regardless of origin or role, several common fungal and bacterial saprophytes are capable of producing moldy- or musty-smelling compounds. Because microorganisms require moist conditions for growth, control is commonly attained by keeping the moisture content of corks below 8%. Sulfur dioxide added to plastic storage bags further minimizes microbial growth. Sterilization by exposure to gamma radiation has been recommended to prevent microbial tainting during storage (Borges, 1985). Valuable as these procedures are, none address the problem of off-odors produced in the bark before harvesting or subsequent seasoning.

Examples of musty- or corky-smelling compounds produced by common fungal saprophytes (such as *Penicillium* and *Aspergillus*) include 3-octanol, as well as mushroom- or metallic-smelling 1-octen-3-one and 1-octen-3-ol. In addition, several *Penicillium* and *Streptomyces* spp. produce musty-smelling sesquiterpenes, such as geosmin (Larsen and Frisvad, 1995). Geosmin has been shown to occur in wines at up to 300 ng/liter (Darriet *et al.*, 2000). Its earthy odor becomes detectable at about 80 ng/liter. Nevertheless, its long-term significance in tainting wine is in doubt due to its short half-life in wine (Amon *et al.*, 1989). The potential role of *Penicillium* in the synthesis of TCA has already been noted. Additional corky taints are suspected to come from the growth of fungi on cork (i.e., *Trichoderma harzianum*) (Brezovesik, 1986). Species of *Streptomyces*, as well as *Bacillus subtilis*, could taint cork by metabolizing vanillin (derived from lignins) to guaiacol (Lefebvre *et al.*, 1983; Álaverz-Rodríguez *et al.*, 2003). Guaiacol possesses a sweet, burnt odor.

Most fungi isolated from cork can also be found in bark taken directly from the tree. In contrast, *Penicillium roquefortii* appears to be a winery contaminant. It has seldom been isolated from corks prior to delivery and storage in wine cellars. It is one of the few organisms that can occasionally grow through cork in bottled wine (Moreau, 1978). Other commonly isolated fungi, such as *Penicillium glabrum*, *P. spinulosum*, and *Aspergillus conicus*, cannot grow in contact with wine. Thus, the lower two-thirds of the cork seldom yields viable fungi after a few months in-bottle (Moreau, 1978). Direct observation of cork from a 61-year-old

bottle showed fungal growth had penetrated only 70% of the cork's length (Jäger *et al.*, 1996).

Among fungi growing on or in cork, few are known to be toxin producers. The major potentially toxigenic species is *P. roquefortii* (Leistner and Eckardt, 1979). Whether the strains of *P. roquefortii* that grow on cork are toxigenic and whether the toxin, if produced, seeps into wine have not been investigated. The major aflatoxin-producing fungus, *Aspergillus flavus*, has not been isolated from cork. Whether *Aspergillus carbonarius* or *A. ochraceus*, the major sources of ochratoxin A, can grow on or in cork appears not to have been investigated.

Fungal growth on the upper surface of corks in bottled wine is favored by moisture retention under unperforated lead capsules. Although growth rarely progresses through the cork, the production of organic acids can speed capsule corrosion. Corrosion eventually can result in contamination of the neck and upper cork surface with soluble lead salts. This knowledge led to replacing lead-tin capsules with those made of aluminum or plastic.

Yeasts have seldom been implicated in cork-derived taints. Exceptions, however, may involve *Rhodotorula* and *Candida*. Both have been isolated from the corks of tainted champagne (Bureau *et al.*, 1974).

Although the importance of cork as a source of spoilage bacteria has been little investigated, the presence of microorganisms on or in cork seems both highly variable and generally low. The most frequently encountered bacterial genus is *Bacillus*. This is not surprising. It produces one of the most highly resistant of dormant structures, endospores. Several *Bacillus* species have been associated with spoilage. *B. polymyxa* has been implicated in the metabolism of glycerol to acrolein, a bitter-tasting compound (Vaughn, 1955), and *B. megaterium* has been observed to produce unsightly deposits in brandy (Murrell and Rankine, 1979).

In addition to cork-related taints, cork can donate a slightly woody character to wine. This is derived from naturally occurring aromatics extracted from cork. More than 80 volatile compounds have been isolated from cork (Boidron *et al.*, 1984). Many of the compounds are similar to those isolated from oak wood used in barrel maturation. This is not surprising. Cork comes from oak, but in this case *Quercus suber*. Nevertheless, the donation of a woody odor from cork appears to be rare. One of the major advantages of cork has been its relative absence of extractable odors.

Yeast-induced Spoilage

A wide variety of yeast species have been implicated in wine spoilage. Even *S. cerevisiae* could be so classed, if it grew where it was unwanted, for example, in bottles of semisweet wine. In addition, epiphytic yeasts such as *Kloeckera apiculata* and *Metschnikowia pulcherrima* can induce wine spoilage. These organisms can donate high contents of aromatics, such as acetic acid, ethyl acetate, diacetyl, and *o*-aminoacetophenone. The latter has frequently been implicated in the generation of an "untypical aged" (UTA) flavor of wines (Sponholz and Hühn, 1996).

Spoilage is commonly associated with the production of off-odors. However, spoilage can simply result from the development of haziness and the deposition of a microbial sediment. The number of cells required to generate haziness varies with the species. With *Brettanomyces*, a distinct haziness is reported to develop at less than 10^2 cells/ml (Edelényi, 1966). In contrast, most yeasts begin to generate a haze at concentrations above 10^5 cells/ml (Hammond, 1976).

Zygosaccharomyces bailii can generate both flocculant and granular deposits (Rankine and Pilone, 1973). It can grow in bottling equipment and, therefore, cloud thousands of bottles. White and rosé wines tend to be more susceptible to attack than red wines. This results from the suppression of *Z. bailii* by the tannins that characterize red wines.

Z. bailii is frequently difficult to control due to its high resistance to yeast inhibitors. For example, it can grow in wine supplemented with 200 mg/liter of either sulfur dioxide, sorbic acid, or diethyl dicarbonate (Rankine and Pilone, 1973). It is also highly tolerant of ethanol, growing in wines at 18% alcohol. Even 1 cell/10 liters may be sufficient to induce spoilage (Davenport, 1982). Along with osmophilic *Kluyveromyces* spp., *Z. bailii* can spoil sweet reserve.

These organisms frequently produce enough acetic acid and higher alcohols to taint the wine. By itself, acetoin produced does not generate a distinct off-odor, because of its high sensory threshold. Nonetheless, its conversion to diacetyl can result in an undesirably intense buttery character. *Z. bailii* also effectively metabolizes malic acid, potentially resulting in an undesired reduction in acidity and rise in pH.

Many yeasts form film-like surface growths on wine under aerobic conditions. These include strains of *Saccharomyces cerevisiae* (*S. prostoserdovii*), *S. bayanus*, and *Z. fermentati*. These are involved in producing the *flor* character of sherries and similarly matured wines. Species of *Candida*, *Pichia*, *Hansenula*, and *Brettanomyces* may occur as minor members in film growths on *flor* sherries, without apparent harm. Alone, however, they usually induce spoilage.

Of the spoilage yeasts, *Brettanomyces* spp. (imperfect state of *Dekkera*) are probably the most notorious (Larue *et al.*, 1991), and controversial (Licker *et al.*,

1999). Surprising to those who detect the malodorous by-products of *Brettanomyces* metabolism, some winemakers appear to appreciate its effects on their wines, considering it part of the *terroir*. These individuals often refer to spicy, smoky, or simply complex aspects, rather than horse sweat or manure. From the point of view of most microbiologists, it is a spoilage organism. Fugelsang and Zoecklein (2003) found no strain that had a positive effect on wine quality, from their perspective.

Brettanomyces contamination is particularly a problem with barrel-matured red wine. Delayed racking can enhance contamination. *Brettanomyces* can be an important member of the lees flora. Its resistance to sulfur dioxide, alcohol and low sugar levels give it great potential to spoil wine.

Both *B. intermedius* and *B. lambicus* produce 2-acetyltetrahydropyridines, compounds that possess mousy odors. These and related compounds (Grbin *et al.*, 1996) are derived from ethyl amino acid precursors (Heresztyn, 1986). These compounds can also be produced by several species of *Lactobacillus*. In addition, "Brett" taints may be associated with the production of above-threshold amounts of isobutyric, isovaleric, and 2-methyl-butyric acids. Licker *et al.* (1999) associated the "Brett" taint primarily with isovaleric acid, and an unknown plastic-smelling compound. Nevertheless, *Brettanomyces* spp. are most well known for their synthesis of several volatile phenolic compounds, notably ethyl phenols (Chatonnet *et al.*, 1997; Licker *et al.*, 1999). In small amounts, ethyl phenols can donate smoky, spicy, phenolic, or medicinal attributes. At higher concentrations, they generate sweaty, leather, barnyardy or manure taints. Most *Brettanomyces* strains possess enzymes that decarboxylate hydroxycinnamic acids, such as *p*-coumaric, caffeic and ferulic acids, to vinylphenols. These are subsequently metabolized to ethyl phenols. Their presence seems to be reduced in the presence of yeast lees. Possibly the lees adsorb some of the ethyl phenols (Guilloux-Benatier *et al.*, 2001). Finally, certain *Brettanomyces* species may generate apple or cider odors, produce high levels of acetic acid, cause haziness, and liberate toxic fatty acids (octanoic and decanoic acids). The toxic fatty acids can accumulate to concentrations sufficient to retard the growth of *Saccharomyces cerevisiae*.

Brettanomyces species can survive, multiply, and contaminate wines from transfer piping, or cooperage that has been insufficiently cleaned and disinfected after use. The yeasts are also effectively transmitted by fruit flies. Their ability to grow on cellobiose, a by-product of toasting during barrel production, means that improperly stored new barrels can also develop *Brettanomyces* contamination.

These organisms are difficult to control due to their insensitivity to sulfur dioxide. They can also survive and multiply on sugar and amino acid (e.g., proline) residues left in tubing or cooperage after use. Thus, control demands prompt cleaning and surface sterilization. Chatonnet *et al.* (1993a) recommend that used barrels be disinfected with at least 7 g/barrel SO_2, and filled barrels be held at 20–25 mg/liter free SO_2. Killer toxins from *Pichia anomala* and *Kluyveromyces wickerhamii* show promise as control agents against *Brettanomyces* (Comitini *et al.*, 2004). Once established, eliminating *Brettanomyces* contamination has been notoriously difficult. Maintaining cellar temperatures at below 12°C (after malolactic fermentation, if desired), maintaining an adequate level of free sulfur dioxide (often considered about 0.8 ppm molecular), and achieving residual sugar levels as low as possible (<1 g/liter) slow its growth in stored wine. Sterile filtration is the only sure means of preventing its activity after bottling.

Compounding problems with its control, *Brettanomyces* grows slowly on culture media. It may take up to two weeks to identify its presence by traditional methods. These methods also limit identification to strains that are readily culturable. Correspondingly, there is much interest in developing molecular methods for rapid identification (Cocolin *et al.*, 2004; Delaherche *et al.*, 2004). In addition, epifluorescence microscopy is a quick method of establishing the wine's microbial status. Their use could permit more rapid application of control measures. However, as noted by Murat *et al.* (2006), the presence of *Brettanomyces* may not be uniform throughout a barrel, often occurring at much higher concentrations in the lees.

Some filamentous fungi, such as *Aureobasidium pullans* and *Exophiala jeanselmei* var. *heteromorpha*, can grow in must, and under suitably aerobic conditions in wine. They often metabolize tartaric acid and can seriously reduce total acidity (Poulard *et al.*, 1983). Must or wine so affected is highly susceptible to further microbial spoilage.

Bacteria-induced Spoilage

LACTIC ACID BACTERIA

Lactic acid bacteria have already been noted in connection with malolactic fermentation. Although lactic acid bacteria are often beneficial, certain strains are spoilage organisms. Spoilage is largely restricted to table wines, but *Lactobacillus hilgardii* can occasionally affect fortified wines (Couto and Hogg, 1994). Spoilage typically occurs under warm conditions, in the presence of insufficient sulfur dioxide, and at pH values higher than 3.5. None of the spoilage problems is induced exclusively by a single species, and the frequency of spoilage strains varies considerably among species.

Tourne is a spoilage problem caused primarily by *Lactobacillus brevis*, although a few strains of *Oenococcus oeni* have been implicated. The primary action is the fermentation of tartaric acid to oxaloacetic acid. Depending on the strain, oxaloacetate is subsequently metabolized to lactic acid, succinic acid, or acetic acid and carbon dioxide. Associated with the rise in pH is the development of a flat taste. Affected red wines usually turn a dull red-brown color, become cloudy, and develop a viscous deposit. Some forms produce an abundance of carbon dioxide, giving the wine an effervescent aspect. In addition to an increase in volatile acidity, other off-odors may develop. These often are characterized as sauerkrauty or mousy.

Amertume is associated with the growth of a few strains of *Lactobacillus brevis* and *L. buchneri*. The strains are characterized by the ability to oxidize glycerol to acrolein, or reduce it to 1,3-propanediol.

$$Glycerol \xrightarrow{\quad H_2O \quad} \text{3-hydroxypropanal}$$

$$\begin{array}{c} \xrightarrow{\quad H_2O \quad} acrolein \\ \xrightarrow[NADH \quad NAD^+]{} \text{1,3-propanediol} \end{array}$$

Acrolein possesses a bitter taste, giving the spoilage its French name, *amertume*. Alternative metabolic routing of glycerol may increase the concentrations of aromatic compounds, such as 2,3-butanediol and acetic acid. As a result of glycerol metabolism, the concentration of glycerol may decrease by 80–90%. In addition, there is a marked accumulation of carbon dioxide and often a doubling of the volatile acidity (Siegrist *et al.*, 1983).

Some heterofermentative strains of lactic acid bacteria, notably *Oenococcus oeni*, induce what is called **mannitic fermentation**. This form of spoilage is characterized by the production of both mannitol and acetic acid.

Ropiness is associated with the synthesis of profuse amounts of mucilaginous polysaccharides (β-1, 3-glucans). Strains of *Oenococcus oeni* and *Pediococcus* are often, but not consistently, associated with ropy wines. In the case of *Pediococcus damnosus*, the property is connected with presence of a plasmid carrying *ropy*[(+)]. The polysaccharides hold the bacteria together in long silky chains. These may appear as floating threads in affected wine. When dispersed, the polysaccharides give the wine an oily look and viscous texture. Although visually unappealing, ropiness is infrequently associated with off-odors or tastes. Real-time PCR detection should facilitate rapid differentiation between the majority of harmless *P. damnosus* strains, from those carrying the *ropy*[(+)] plasmid (Delaherche *et al.*, 2004).

The presence of mousy taints has already been noted in regard to *tourne* and the growth of *Brettanomyces* spp. Mousiness is also associated with the growth of a few strains of *Oenococcus oeni* and *Leuconostoc mesenteroides* (Costello *et al.*, 2001). Mousy odors have been associated with synthesis of several N-heterocycles, notably 2-acetyltetrahydropyridine, 2-ethyltetrahydropyridine, 2-propionyltetrahydropyridines, and 2-acetyl-1-pyrroline (Grbin *et al.*, 1996). The production of 3-acetyl-1-pyrroline has also been associated with mousy taints in several wines (Herderich *et al.*, 1995).

Lactic acid bacteria may produce a geranium-like taint in the presence of sorbic acid (Crowell and Guymon, 1975). Some strains of *Lactobacillus*, and most strains of *Oenococcus oeni*, metabolize sorbic acid to *p*-sorbic alcohol (2,4-hexadienol). Under acidic conditions, 2,4-hexadienol isomerizes to *s*-sorbic alcohol (1,3-hexadienol). It, in turn, reacts with ethanol to form an ester, 2-ethoxyhexa-3,4-diene. It is the latter that generates an intense geranium-like odor.

High volatile acidity is a common feature of most forms of spoilage induced by lactic acid bacteria. Acetic acid can be produced by the metabolism of citric, malic, tartaric, and gluconic acids, as well as hexoses, pentoses, and glycerol. The level of acetic acid synthesis depends on the strain and conditions involved. However, production is very limited in the absence of suitable reducible substances. Thus, acetic acid production is rare under strictly anaerobic conditions.

Species of *Lactobacillus* have also been periodically isolated from fortified wines containing high sugar contents. Stratiotis and Dicks (2002) found that most strains tolerant to high alcohol concentrations (22%) belonged to the species *L. veriforme*.

ACETIC ACID BACTERIA

Acetic acid bacteria were first recognized as causing wine spoilage in the nineteenth century. Their ability to oxidize ethanol to acetic acid both induces wine spoilage and is vital to commercial vinegar production. Although acetic acid synthesis during vinegar production has been intensively investigated, the action of acetic acid bacteria on grapes, and in must and wine, has escaped intensive scrutiny. That acetic acid bacteria could remain viable in wine for years under anaerobic conditions was unexpected. They were thought to be strict aerobes, unable to grow or survive in the absence of oxygen. However, their ability to use hydrogen acceptors other than molecular oxygen suggests that they may show limited metabolic activity under anaerobic conditions. Thus, the role of acetic acid bacteria in all phases of winemaking requires reinvestigation.

It is recognized that acetic acid bacteria form a distinct family of gram-negative, rod-shaped bacteria characterized by the ability to oxidize ethanol to acetic acid. For years, molecular oxygen was thought to be their only acceptable, terminal, respiratory electron acceptor. It is now known that quinones can substitute for oxygen (Aldercreutz, 1986). Thus, acetic acid bacteria may grow in barreled or bottled wine, if acceptable electron acceptors are present. Of even greater practical significance is their ability to grow using traces of oxygen absorbed by wine during clarification and maturation (Joyeux *et al.*, 1984; Millet *et al.*, 1995).

The metabolism of sugar by acetic acid bacteria is atypical in many ways. For example, the pentose phosphate pathway is used exclusively for sugar oxidation to pyruvate, whereas pyruvate oxidation to acetate is by decarboxylation to acetaldehyde, rather than to acetyl-CoA. Sugars also may be oxidized to gluconic and mono- and diketogluconic acids, rather than metabolized to pyruvic acid (Eschenbruch and Dittrich, 1986). Although this property is most commonly associated with *Gluconobacter oxydans*, some strains of *Acetobacter* possess this ability.

In addition to oxidizing ethanol to acetic acid, acetic acid bacteria oxidize other alcohols to their corresponding acids. In addition, they may oxidize polyols to ketones, for example glycerol to dihydroxyacetone.

Of the eight recognized genera of acetic acid bacteria, only *Acetobacter* and *Gluconobacter* commonly occur on grapes or in wine. They can be distinguished both metabolically and by the position of their flagella. Members of the *Acetobacter* have the ability to overoxidize ethanol; that is, they may oxidize ethanol past acetic acid to carbon dioxide and water, via the TCA cycle. Under the alcoholic conditions of wine, however, ethanol overoxidation is suppressed. In contrast, *Gluconobacter* lacks a functional TCA cycle, and cannot oxidize ethanol past acetic acid. The *Gluconobacter* are further characterized by a greater ability to use sugars than *Acetobacter*. Motile forms of both genera can be distinguished by flagellar attachment. *Gluconobacter* has polar flagellation (insertion at the end of the cell), whereas *Acetobacter* has a more uniform (peritrichous) distribution. Of these two genera, only *A. aceti*, *A. pasteurianus*, and *G. oxydans* are commonly found on grapes or in wine. A new species, *Acetobacter oeni*, has recently been isolated from spoiled red wine (Silva *et al.*, 2006).

Although all three main species occur on grapes, and in must and wine, their frequency differs markedly. *Gluconobacter oxydans* is the predominant species on grape surfaces, probably because of its greater ability to metabolize sugars. On healthy fruit, the bacterium commonly occurs at about 10^2 cells/g. On diseased or damaged fruit, this value can rise to 10^6 cells/g (Joyeux et al., 1984). *Acetobacter pasteurianus* is typically present in small numbers, whereas *A. aceti* is only rarely isolated.

During fermentation, the number of viable bacteria tends to decrease, although usually not below 10^2 and 10^3 cells/ml. The most marked change is in the relative proportion of the species. *G. oxydans* falls during fermentation, being replaced by *A. pasteurianus*. Subsequently, the population of *A. pasteurianus* may rise or fall during fermentation and maturation. *A. aceti* tends to become the dominant species after fermentation. Despite this, the population diversity (number of strains) of *A. aceti* declines considerably during fermentation (González et al., 2005). *G. oxydans* tends to disappear entirely during maturation (Fig. 8.50).

Although the viable population of acetic acid bacteria tends to decline during maturation, racking can induce temporary increases. This is probably due to the incorporation of oxygen during racking. Oxygen can participate directly in bacterial respiration, but it also may indirectly generate electron acceptors for respiration, such as quinones.

Spoilage can result from bacterial activity at any stage in wine production. For example, moldy grapes typically have a high population of acetic acid bacteria and can provoke spoilage immediately after crushing. By-products of metabolism, such as acetic acid and ethyl acetate, are retained throughout fermentation and can taint the resulting wine.

Spoilage by acetic acid bacteria during fermentation is rare, largely because most present-day winemaking practices restrict contact with air. Improved forms of pumping over and cooling have eliminated the major sources of must oxidation during fermentation. Also,

a better understanding of stuck fermentation can limit its incidence, permitting the earlier application of techniques that reduce the likelihood of oxidation and microbial spoilage.

Although wine maturation occurs largely under anaerobic conditions, storage in small oak cooperage increases the likelihood of oxygen uptake and activation of bacterial metabolism. Wood cooperage can be a major source of microbial contamination, if improperly stored, cleansed, and disinfected before use. Thus, it is not surprising that red wines have higher levels of volatile acidity (Eglinton and Henschke, 1999).

Alone, the levels of sulfur dioxide commonly maintained in maturing wine are insufficient to inhibit the growth of acetic acid bacteria. Therefore, combinations of techniques such as maintaining or achieving low pH values, minimizing oxygen incorporation, and storing at cool temperatures, along with sulfur dioxide, appear to be the most effective means of limiting the activity of acetic acid bacteria. Spoilage of bottled wine by acetic acid bacteria presumably is limited to situations in which failure of the closure permits seepage of oxygen into the bottle.

The most well-known and serious consequence of spoilage by acetic acid bacteria is the production of high levels of acetic acid (volatile acidity). The recognition threshold for acetic acid is approximately 0.7 g/liter (Amerine and Roessler, 1983). At twice this value, it can give wine an unacceptable vinegary odor and taste. Acetic acid production is more associated with the stationary and decline phases of colony growth than with its log phase (Kösebalaban and Özilgen, 1992).

Although wines mildly contaminated with volatile acidity may be improved by blending with unaffected wine, treating with reverse osmosis (to remove the acetic acid), or blending with grape juice and refermenting (yeasts can metabolize the excess acetic acid) are alternate solutions. However, seriously spoiled wines are fit only for distillation into industrial alcohol, or conversion into wine vinegar.

Under aerobic conditions, acetic acid bacteria do not synthesize noticeable amounts of esters. Although ethyl acetate production is increased at low oxygen levels, most of the ethyl acetate generated during acetic spoilage appears to form from nonenzymatic esterification, or the activity of other contaminant microorganisms. Ethyl acetate may also be metabolized by several microbes. As a consequence, the strong sour vinegary odor of ethyl acetate is not consistently associated with spoilage by acetic acid bacteria (Eschenbruch and Dittrich, 1986). By itself, ethyl acetate possesses an acetone-like odor (nail-polish remover).

Another aromatic compound sporadically associated with spoilage by acetic acid bacteria is acetaldehyde.

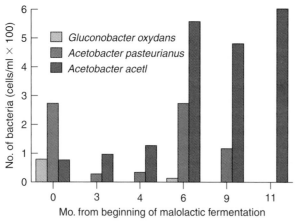

Figure 8.50 Evolution of acetic acid bacteria during malolactic fermentation and maturation in barrel of 'Cabernet Sauvignon' wine. (Data from Joyeux et al., 1984)

Under most circumstances, acetaldehyde is rapidly metabolized to acetic acid and seldom accumulates. However, the enzyme that oxidizes acetaldehyde to acetic acid is sensitive to denaturation by ethanol (Muraoka *et al.*, 1983). As a result, acetaldehyde may accumulate in highly alcoholic wines. Low oxygen tensions also favor the synthesis of acetaldehyde from lactic acid.

Spoilage by acetic acid bacteria generally does not produce a fusel taint. The bacteria oxidize higher alcohols to their corresponding acids. In oxidizing polyols, acetic acid bacteria generate either ketones or sugars. For example, glycerol and sorbitol are metabolized to dihydroxyacetone and sorbose, respectively. The conversion of glycerol to dihydroxyacetone may affect the sensory properties of wine because it has a sweet fragrance and cooling mouth-feel. Dihydroxyacetone may also react with several amino acids to generate a crust-like aroma. Whether this reaction occurs in wine is unknown.

The oxidation of organic acids under acidic conditions appears to be weak. Its only significance in wine spoilage by acetic acid bacteria may be the oxidation of lactic acid to acetaldehyde and acetoin.

Some strains of acetic acid bacteria produce one or more types of polysaccharides from glucose. Such production in grapes may account for some of the difficulties in filtering wines made with infected berries.

In addition to acetic acid, acetic acid bacteria may generate considerable quantities of gluconic and mono- and diketogluconic acids from glucose in grapes. These compounds occur in association with most fungal infections and may be used as indicators of the degree of infection.

OTHER BACTERIA-INDUCED SPOILAGE

Other than that caused by lactic acid and acetic acid bacteria, bacterial spoilage is rare. Nevertheless, its development can have serious financial consequences. The main genus associated with other bacterial spoilage problems is *Bacillus*. Members of the genus are gram-positive, rod-shaped bacteria that commonly produce long-lived, highly resistant endospores. Most species are aerobic, but some are facultatively anaerobic, as well as being acid and alcohol tolerant.

Bacillus polymyxa has been associated with the fermentation of glycerol to acrolein. Other species, including *B. circulans*, *B. coagulans*, *B. pantothenticus*, and *B. subtilis*, have been isolated from spoiled fortified dessert wines (Gini and Vaughn, 1962). *B. megaterium* may even produce sediment in bottled brandy.

Although unlikely to grow directly in wine, species of *Streptomyces* may be involved in tainting wine. Their growth has already been mentioned in regard to the production of off-odors in cork. *Streptomyces* also may grow on the surfaces of unfilled cooperage. Their ability to produce earthy, musty odors is well known. The synthesis of sesquiterpenols, notably geosmin and 2-methylisoborneol, is believed to be the source of the earthy odor of soil. Geosmin may also originate from the joint action of *Botrytis cinerea* and *Penicillium expansum* on infected grapes (La Guerche *et al.*, 2005).

Sulfur Off-odors

In minute quantities, reduced-sulfur compounds may contribute to the varietal fragrance of certain wines, as well as its aromatic complexity. They also can be the source of revolting off-odors. The difference can be a function of concentration. The principal offending compounds are hydrogen sulfide (H_2S), dimethyl disulfide, and mercaptans. Hydrogen sulfide and mercaptans develop putrid odors at concentrations of 1 ppb or less. In contrast, dimethyl disulfide clearly expresses its onion, cooked cabbage and shrimp-like odors only at concentrations some thirty times higher. In all cases, these off-odors have lower thresholds in sparkling and white wines than in red wines. Hydrogen sulfide can be generated throughout fermentation, maturation and bottle aging, depending on conditions, whereas dimethyl disulfide and mercaptans are more frequently formed during maturation and bottle aging.

At much above the sensory threshold, hydrogen sulfide produces a rotten-egg odor. The origin of hydrogen sulfide can be very diverse, due to its pivotal role in sulfur metabolism. Hydrogen sulfide may be generated from sulfur residues on grapes, sulfate found in grape tissue, sulfite (derived from sulfur dioxide addition), and amino acid degradation.

During fermentation, organosulfur fungicides do not appear to be a significant source of hydrogen sulfide, with the possible exception when applied late in the season. The primary source of hydrogen sulfide during vinification was once thought to be elemental sulfur, especially in colloidal form (Fig. 8.51). Stopping application (as a fungicide) at least six weeks prior to harvest limits its potential role in hydrogen sulfide production (Thomas *et al.*, 1993).

Hydrogen sulfide appears to possess two distinct production phases during fermentation. The first occurs during the exponential phase of yeast growth, when demand for sulfur-containing amino acids is high. If these are not present in the must in adequate amounts, yeast cells produce sulfides from intracellular sulfates. Sulfides may also be generated if cells come in contact with sulfur particles. If vitamin co-factors (pantothenate and pyridoxine) and nitrogen precursors of amino acids are also limiting, the sulfides produced are liberated into the must as hydrogen sulfide. Application of diammonium hydrogen phosphate (DAP) tends to limit this release.

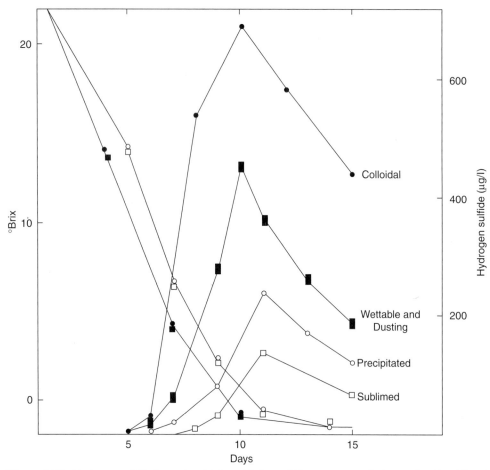

Figure 8.51 Hydrogen sulfide formation (right axis) from different sources of elemental sulfur (50 mg/liter) as sugars are metabolized (left axis) during fermentation. (From Schütz and Kunkee, 1977, reproduced by permission)

Although application of DAP can avoid the early accumulation of hydrogen sulfide, it has little effect on H_2S liberation later in fermentation. The second potential peak for hydrogen sulfide release is correlated with total assimilable nitrogen concentration. Hydrogen sulfide generation is even more closely linked to the ratio of particular amino acids in the must. High concentrations of glutamic acid, alanine, γ-amino butyric acid (GABA), relative to methionine, lysine, arginine and phenylalanine, favor the secretion of hydrogen sulfide.

Additional features involved in limiting hydrogen sulfide synthesis include lower fermentation temperatures, harvesting to possess adequate acidity, reduced levels of suspended-solids (>0.5%), fermentation in small fermentors (slower decline in redox potential), avoiding deficiencies in the vitamins pantothenate and pyridoxine (B_6), and limiting the addition or uptake of sulfur dioxide (such as during racking into newly sulfured barrels). If insufficient acetaldehyde is present to bind sulfur dioxide, other components such as sulfate

may be reduced to hydrogen sulfide. Copper fungicide residues can augment H_2S accumulation.

There have been several attempts to breed yeast strains that do not synthesize hydrogen sulfide. Although some strains are sulfite reductase deficient (Zambonelli *et al.*, 1984), they are dependent on external sources of sulfur-containing amino acids. Because sulfide synthesis is an integral aspect of yeast sulfur metabolism, all yeast strains generate some hydrogen sulfide. Nevertheless, some strains release less hydrogen sulfide than others, for example Pasteur Champagne, Epernay 2 and Prise de Mousse. Because the reduction of elemental sulfur by hydrogen peroxide at the cell-wall surface is nonenzymatic (Wainwright, 1970), this source of H_2S is largely beyond genetic modification.

Despite the inevitable production of hydrogen sulfide during fermentation, most of it is carried off with carbon dioxide. Hydrogen sulfide content also declines during maturation. This may partially involve oxidation to sulfur by hydrogen peroxide (generated during the oxidation of *o*-diphenols). The oxidation of hydrogen sulfide to elemental sulfur may also be coupled with the reduction of sulfur dioxide to sulfate. Racking removes the precipitated sulfur, preventing subsequent reduction back to hydrogen sulfide.

The production of volatile organosulfur compounds, such as mercaptans and dimethyl sulfides, tends to be more intractable. Once formed, they are difficult to remove. It often requires the combined actions of copper fining with the application of sulfur dioxide and ascorbic acid. Although synthesis typically occurs during maturation, compounds such as methylmercaptopropanol may form during fermentation (Lavigne *et al.*, 1992). Cultivar and vineyard conditions play significant roles in their production, as well as high levels of soluble solids and sulfur dioxide.

During maturation, mercaptan synthesis is favored by the development of low redox potentials in the lees (Fig. 8.52). Although hydrogen sulfide theoretically can combine with acetaldehyde, forming ethyl mercaptan (ethanethiol), it apparently does not occur under wine conditions (Bobet, 1987). If present, ethyl mercaptan more likely comes from a slow reaction between diethyl disulfide and sulfite (Bobet *et al.*, 1990). However, hydrogen sulfide and acetaldehyde can generate 1,1-ethanedithiol under wine-like conditions (Rauhut *et al.*, 1993). It possesses a sulfur rubbery note.

The principal source of organosulfur compounds in wine appears to be sulfur-containing amino acids. Methionine can be metabolized by yeasts to methyl mercaptan, while cysteine is a source of dimethyl sulfide (de Mora *et al.*, 1986). Dimethyl sulfide may also be generated when acetic acid bacteria use dimethyl sulfoxide as an electron donor during anaerobic respiration. Additional methyl mercaptan may arise from the hydrolysis of bisdithiocarbamate fungicides. Despite these potential sources, the dynamics and precise origin of most volatile organosulfur compounds in wine remain unclear. This clearly is a topic in need of investigation.

$$CH_3-S\cdot + H\cdot \longrightarrow CH_3-SH \text{ methanethiol}$$

$$2\,CH_3-S\cdot \longrightarrow CH_3-S-S-CH_3 \text{ dimethyl disulfide}$$

The occasional genesis of sulfide off-odors in bottled wine, notably champagne, appears clearer. They form in a complex series of reactions involving methionine, cysteine, riboflavin, and light (Maujean and Seguin, 1983). When photoactivated by light, riboflavin catalyzes the degradation of sulfur-containing amino acids. Various free radicals are formed, some of which combine to form methanethiol and dimethyl disulfide. Hydrogen sulfide also is generated under these conditions. Together, these compounds give rise to a light-struck (*goût de*

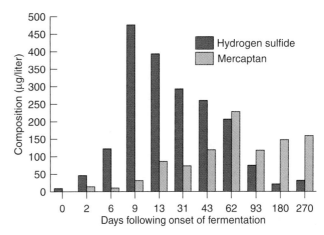

Figure 8.52 Development of reduced-sulfur compounds in wine during fermentation and maturation. The first racking occurred on day 93. (Data from Cantarelli, 1964)

lumière) fault of champagne. The significance and origin of the associated high concentrations of dimethyl sulfide, commonly found in champagne, are unclear. Dimethyl sulfide accumulates regardless of light exposure.

Wines can be protected from this fault by limiting the presence of amino acids and riboflavin, plus bottling in amber or anti-UV glass (to restrict the passage of the blue and UV radiation that catalyzes the reactions). The binding of riboflavin by tannins limits the sensitivity of red wines to light-struck off-odors.

Selectively removing hydrogen sulfide and mercaptans from wine is complicated. Activated carbon absorbs mercaptans, but reduces wine quality by removing essential aromatic compounds. Although mercaptans oxidize readily to disulfides, which have higher sensory thresholds, the reaction slowly reverses under the reducing conditions in bottled wine. Addition of trace amounts of silver chloride (Schneyder, 1965) or copper sulfate (Petrich, 1982) several months before bottling, neutralizes their odor. Removal of their metal precipitates is important as the reactions are reversible. Regrettably, these treatments also remove important thiol aromatics (Hatzidimitriou *et al.*, 1996), important to the varietal aroma of certain *vinifera* cultivars (Tominaga *et al.*, 2000). Copper does not, however, remove disulfides, thioacetic acid esters, or cyclic sulfur-containing off-odors (Rauhut *et al.*, 1993). In addition, care must be taken not to exceed permissible residual levels of copper in wine (0.5 mg/liter in the United States and 0.2 mg/liter in most other wine-producing countries). Ascorbic acid may be added in conjunction with copper sulfate and SO_2 to limit disulfide production. By consuming oxygen, ascorbic acid limits the reoxidation of thiols back to disulfides. The reaction is very slow, though, taking months to complete. Ascorbic acid may also be added to reduce the production of an "untypical aged" flavor (Rauhut *et al.*, 2001).

Metal salts have also been used to unambiguously differentiate between off-odors generated by hydrogen sulfide, mercaptans, or both (Brenner *et al.*, 1954). The laboratory test is based on selective removal by copper and cadmium salts.

An alternative technique showing promise in the removal of mercaptans is the addition of lees (Lavigne, 1998) to affected wines. This can even be the lees that generated the problem. By isolating and gently aerating the lees for approximately 1 day, mercaptans in the lees may disappear. In addition, yeast cell walls appear to bind to various volatile reduced-sulfur compounds, removing them from the wine (Lavigne and Dubourdieu, 1996).

Unacceptably high concentration of hydrogen sulfide can be reduced with aeration. This, however, carries several risks. For white wines, there is the danger of oxidative browning, whereas for red wines, there is the possibility of activating dormant acetic acid bacteria. Oxygen also catalyzes the conversion of mercaptans to disulfides. Although having higher thresholds of detection, in-bottle aging favors the slow reduction of the disulfides back to mercaptans. An alternative procedure for hydrogen sulfide removal is sparging with nitrogen. Its downside is that it also removes desirable aromatics.

Various suggestions have been made to reduce the presence of reduced-sulfur compounds. Principal among these is the use of lower SO_2 additions (≤ 50 mg/liter), low levels of soluble solids (Lavigne *et al.*, 1992), slight oxidation after the first racking (Cuénat *et al.*, 1990), and fermentation at cooler temperatures.

Development of a HPLC method measuring free amino acid levels (Pripis-Nicolau *et al.*, 2001) should facilitate measuring cysteine content of wine. Because cysteine appears to be the primary source of thiol off-odors, assessment of its content may predict the need and degree of protective measures. Figure 8.53 illustrates the considerable variation in the presence of cysteine in the must and wine of several cultivars.

Additional Spoilage Problems

In addition to the light-induced off-odor noted above, light can also cause a fault in Asti Spumante

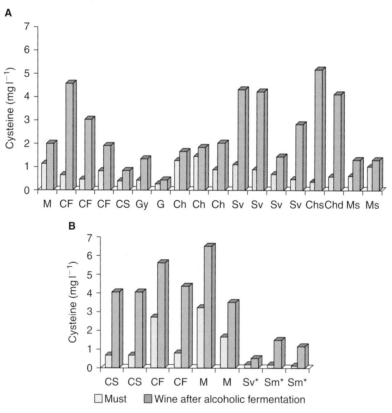

Figure 8.53 Influence of grape variety and alcoholic fermentation on cysteine levels (1999 vintage, *2000 vintage): A, Loire region; B, Bordeaux region. White grape varieties: Ch, 'Chenin'; Chd, 'Chardonnay'; Chs, 'Chasselas'; Ms, 'Muscadet' ('Melon'); Sv, 'Sauvignon'; Sm, 'Sémillon'. Red grape varieties: CF, 'Cabernet Franc'; CS, 'Cabernet Sauvignon'; Gy, 'Gamay'; G, 'Grolleau'; M, 'Merlot'. (From Pripis-Nicolau *et al.*, 2001, copyright Society of Chemical Industry, reproduced by permission John Wiley & Sons Ltd for SCI)

(Di Stefano and Ciolfi, 1985). Exposure to sunlight accelerates the oxidation of aromatic terpenes that give the wine its distinctive aroma. Changes in the composition of esters also can occur. In addition, novel terpenes are generated, notably 3-ethoxy-3,7-dimethyl-1-octen-7-ol. The changes produce an unpleasant taste and fragrance, resembling aromatic herbs and preserved vegetables. In addition, 2-methylpropanol, and a marked reduction in fruit ester content, have been implicated in another light-struck off-odor in sparkling wines (D'Auria *et al.*, 2002).

UNTYPICAL AGED FLAVOR

"Untypical aged" (UTA) is an off-odor that tends to develop about a year after bottling, usually in wines made from grapes having suffered stress during the growing season (Sponholz and Hühn, 1996). The off-odor has usually been associated with the presence of *o*-aminoacetophenone and scatole. The off-odor is characterized by naphthalene, furniture polish, or wet wool attributes. The precursor of *o*-aminoacetophenone appears to be indole-3-acetic acid (IAA), one of the major phytohormones in plants (Hoenicke *et al.*, 2002). Methional (3-methylthio propionaldehyde) may also be associated with UTA (Rauhut *et al.*, 2001).

The occurrence of this off-odor appears to be reduced by the incorporation of thiamine before fermentation, addition of diammonium phosphate (if must assimilable nitrogen is low, or yeast strains requiring high nitrogen contents are used), and the addition of ascorbic acid (150 mg) (Rauhut *et al.*, 2001). Recently, Henick-Kling *et al.* (2005) have questioned the involvement of *o*-aminoacetophenone or scatole in this disorder. They found tridihydronaphthalene, and the loss of volatile terpenes were better indicators of UTA occurrence in New York, at least in association with drought stress just before or during *véraison*. Occasionally the presence of UTA is masked by the simultaneous occurrence of reduced-sulfur odors.

POST-OPENING OXIDATION

Oxidation often causes serious deterioration throughout the winemaking process. Oxidation is also a problem for consumers, partially associated with a rapid loss in freshness after bottle opening. This seemingly results from the degradation of ethyl and acetate esters, and to a lesser extent volatile terpenols (Fig. 8.54). Similar, but less marked aromatic losses were noticed in red wines. These changes presumably reflect oxidative reactions as the addition of antioxidants, such as caffeic acid and N-acetyl-cysteine (Roussis *et al.*, 2005), markedly reduced the loss. The common view that small amounts of acetaldehyde, generated upon exposure to air, mask or dominate the wine's fragrance seems to be in error (Escudero *et al.*, 2002).

Because oxygen uptake is significantly retarded at cool temperatures, and low pH values, the duration and wine/air interface are probably of primary importance in short-term oxidation after opening. For example, the wine/air interface is minimal on opening (about 0.4 cm²/100 ml in a 750 ml bottle). Pouring the wine greatly increases surface-to-air contact. If half the bottle were poured, the surface area/volume ratio of the remaining wine in the bottle increases almost 20 times, to about 7.5 cm²/100 ml. If corked and stored, air in the bottle would contain about 120 mg oxygen, ample to produce significant oxidation in a comparatively short time.

Another factor that has been little investigated is the volatilization of aromatics from the wine. When the bottle is opened, aromatics in the headspace escape from the bottle. The changed equilibrium between aromatics in the wine and the headspace induces further liberation of aromatics. This phenomenon helps maintain the aroma in the glass during tasting, but may depauperate the fragrance of wine left in the bottle.

HEAT

Despite quality deterioration, the specific effects of heat exposure on wine have been little investigated (Beech and Redmond, 1981; Anonymous, 1984). Temperature extremes can induce serious degradation during warehouse storage and on store shelves. Direct exposure to sunlight accentuates the effects of unfavorable store temperatures.

Temperature directly affects the rate of reactions involved in aging, for example, accelerated hydrolysis of aromatic esters and the loss of terpene fragrances. Both changes can result in a decrease in the fruity or varietal character of wine. The synthesis of furfurals from sugars is believed to be partially involved in the development

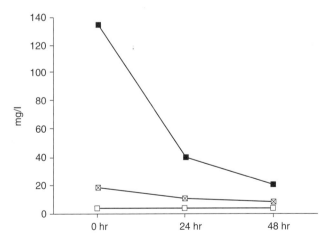

Figure 8.54 Sum of the relative concentrations of volatile ethyl esters (■ —— ■), acetate esters (⊠ —— ⊠), and terpenols (□ —— □) after opening of a bottle of 'Muscat' wine (20 °C). (Data from Roussis *et al.*, 2005)

of a baked character. Exposure to temperatures of more than 50°C quickly accelerates the generation of caramelization products. Similar reactions occur at lower temperatures, but more slowly. These affect both the bouquet and color of wine. Browning in red wines at high temperatures results from the degradation of anthocyanins and the generation of polyphenolic pigments. Subsequent precipitation can lead to sediment formation. Important changes can occur even within a few days at elevated temperatures (Fig. 8.55). Consequently, temperature control is important during all stages of wine transport and storage. Sulfur dioxide (100 mg/liter) apparently minimizes some of the effects (Ough, 1985). This seems unrelated to its antioxidant property.

By affecting wine volume, rapid temperature fluctuation can loosen the seal of a cork, leading to leakage, oxidation, and possible microbial contamination of the wine.

STORAGE ORIENTATION

It is traditional for wine closed with cork to be stored horizontally. This places the cork in direct contact with the wine (preventing drying and shrinkage of the cork). Although the importance of this in limiting wine oxidation is common knowledge, the damage caused

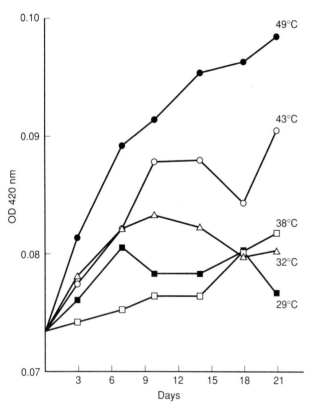

Figure 8.55 Changes in optical density (OD) of white wine with low SO₂ concentrations at five temperatures over a 21-day period. (From Ough, 1985, reproduced by permission)

by upright positioning does not necessarily occur rapidly. For example, Lopes *et al.* (2006) did not find marked differences in oxygen penetration between vertical and horizontal positioning over a two-year period. In addition, Skouroumounis *et al.* (2005) found that it took several years for the effects of upright storage to result in distinct browning and the development of an oxidized odor (cooked vegetable, nutty and sherry-like aspects). In contrast, Mas *et al.* (2002) found vertical positioning distinctly unfavorable to wine development. In addition, Bartowsky *et al.* (2003) found that upright storage made red wine more susceptible to spoilage by *Acetobacter pasteurianus*. Bacterial growth could be detected by the presence of a ring-like growth at the wine-headspace interface prior to opening.

Accidental Contamination

Intentional adulteration of wine has probably occurred since time immemorial. Lead salts were occasionally added during the Middle Ages to "sweeten" highly acidic wine. Laws forbidding such additives were enacted in Europe at least by 1327, when an edict by William, Count of Hennegau, Holland, outlawed the addition of lead sulfate, mercury, and similar compounds to wine (Beckmann, 1846). Concern about accidental contamination is more recent. Only since the 1970s have sufficiently accurate analytic tools become available that can identify many of the contaminants that periodically can taint wine.

Typically, heavy metals occur in wine at considerably below toxic levels. When metals occur at above trace amounts, they usually arise from accidental contamination after fermentation. For example, activated carbon can be an unsuspected source of chromium, calcium, and magnesium; diatomaceous earth may be a source of iron contamination; and bentonite can act as a source of aluminum.

Tin-lined lead capsules may be an occasional source of metal contamination in old wine. Corrosion of the capsule can produce a deposit of soluble lead salts on the lip of the bottle. Although the salts do not diffuse through the cork into the wine (Gulson *et al.*, 1992), failure to adequately cleanse the neck can contaminate the wine during pouring (Sneyd, 1988). In addition, structural faults in the cork could be an infrequent route leading to lead contamination. When wine bottles show obvious corrosion of a lead capsule, the bottle mouth should be thoroughly cleaned and puncture of the cork with a corkscrew avoided. However, the most frequent source of lead contamination in wine appears to be the sporadic use of brass tubing and faucets on winery equipment (Kaufmann, 1998).

Off-odors can arise from a wide variety of unsuspected sources, including blanketing gases, fortifying brandies, oil and refrigerant leaks, absorption of volatile pollutants from the wax coating of cement fermentation and storage tanks, the leaching of chemicals from epoxy paints and resins used on winery equipment, and chlorophenol-treated cooperage (see Strauss *et al.*, 1985).

Naphthalene absorbed by cork from the environment has been mentioned previously. Naphthalene may also come from fiberglass used in the construction of holding tanks, resin used to seal tank joints, compounds used to adhere polyvinyl chloride inlays in bottle caps, and from adsorption onto wax used to coat cement tanks. In the last instance, the naphthalene may come from oil or solvent spills in or near fermentation or storage tanks (Strauss *et al.*, 1985). Naphthalene gives wine a musty note at levels above 0.02 mg/liter. In contrast, the naphthalene-like odor associated with the "untypical aged" flavor (UTA) has usually been associated with the presence of *o*-aminoacetophenone.

Another off-odor occasionally associated with the use of epoxy resins is benzaldehyde. It can produce a bitter almond odor when the concentration reaches 2–3 mg/liter (Brun, 1984). Benzaldehyde is derived from the oxidation of benzyl alcohol. It is occasionally used as a plasticizer in epoxy resins, or found as a contaminant in liquid gelatin (Delfini, 1987). Several microbes can oxidize benzyl alcohol to benzaldehyde. These include *Botrytis cinerea* and the yeasts *Schizosaccharomyces pombe* and *Zygosaccharomyces bailii* (Delfini *et al.*, 1991). The latter two may also synthesize benzyl alcohol and benzoic acid. Various ketone notes may be associated with epoxy resin contaminants. Improperly formulated resins can also be a source of methyl isobutyl ketone, methyl ethyl ketone, acetone, toluene, and xylene (Brun, 1984). In addition, trace amounts of the hardener methylenedianiline, and its monomers bisphenol and epichlorhydrin, may migrate into wine stored in cement and steel tanks coated with some epoxy resins. However, they do not appear to accumulate in concentrations sufficient to affect the wine's sensory attributes (Larroque *et al.*, 1989).

Fiberglass tanks occasionally retain a small amount of free styrene (Wagner *et al.*, 1994). If the styrene migrates into the wine at levels above 0.1–0.2 mg/liter, it can produce a plastic odor and taste (Brun, 1984). Transport tanks lined with polyvinylchloride (PVC) may release butyltins (Forsyth *et al.*, 1992). Butyltin is occasionally used as a PVC stabilizer. The presence of this compound does not affect the sensory qualities of the wine, but does suggest that a food-grade form of PVC was not employed. Butyltin is a member of a group of compounds called organotins. They have diverse uses in industry, including wood preservation, catalysts, and as an antifouling agent in marine paint.

Residues from degreasing solvents and improperly formulated paints can be additional sources of off-odors. When winery equipment is disinfected with hypochlorite, the chlorine may react with degreasing solvents or phenols found in paint (Strauss *et al.*, 1985; Bertrand and Barrios, 1995). As a consequence, one or more chlorophenols may be generated. Wine coming in contact with the affected equipment may show vague medicinal, chemical, or disinfectant odors. Chlorophenols also may be used as preservatives on cooperage oak. These compounds may be microbially metabolized, producing 2,4,6-trichloroanisole and related chloroanisoles. If wine cellars are poorly aerated, these volatile compounds may bind to cooperage surfaces, as well as materials stored therein, such as cork and fining agents (Chatonnet, Guimberteau *et al.*, 1994). These may subsequently taint wine with corky, musty, or moldy off-odors.

Prolonged storage of water in tanks can also lead to the development of strong musty, earthy taints. This is usually associated with the growth of actinomycetes in storage tanks. If the water is subsequently used to dilute distilled spirits in fortifying dessert wines, the wine will develop an undesirable earthy aspect (see Gerber, 1979).

Recently, producers in Ontario and adjacent American states experienced wines tainted with a bell pepper/herbaceous off-odor of methoxypyrazines. These compounds are synthesized in Multicolored Asian lady beetles (*Harmonia axyridis*). As few as one beetle per liter of juice can produce a detectable off-odor (Pickering *et al.*, 2004). At 10 beetles per liter, they seriously distort the wine's varietal character. This equates to about 1200–1500 beetles per 1000 kg of fresh grapes. The off-odor is not dissipated significantly with aging, but can be partially masked with maturation in oak, or reduced by fermentation in the presence of activated charcoal (Pickering *et al.*, 2006). The lady beetles were initially imported and released as a biocontrol agent for important crop pests in the eastern United States. Regrettably, the lady beetle has fared so well that it has itself become an annoying nuisance to home owners, and an occasional headache for wine producers.

The discovery of ochratoxin A, a potential human toxin, has led to an extensive survey of its presence in wine (O'Brien and Dietrich, 2005). The toxin is produced by several hyphomycetes, notably *Aspergillus ochraceus*, as well as a few other related *Aspergillus* and *Penicillium* species. Ochratoxin A primarily occurs in grapes secondarily infected by these saprophytic fungi, or due to their growth on improperly cleaned and maintained winery equipment. The concentration of ochratoxin A is lowered significantly during fining or filtering (Gambuti *et al.*, 2005).

It may also be metabolized during malolactic fermentation. Contamination is usually avoided by proper winery sanitation and the use of healthy grapes. Occasionally, fungicidal sprays (e.g., Switch®) have been used to control the incidence of *Aspergillus* occurring on grapes.

Suggested Readings

General Texts

Amerine, M. A., Berg, H. W., Kunkee, R. E., Ough, C. S., Singleton, V. L., and Webb, A. D. (1980) *The Technology of Wine Making*, 4th edn. Avi, Westport, CT.

Boulton, R. B., Singleton, V. L., Bisson, L. F., and Kunkee, R. E. (1996) *Principles and Practices of Winemaking*. Springer, New York, NY.

Flanzy, C. (ed.) (1998) *Oenologie: Fondements Scientifiques et Technologiques*. Lavoisier, Paris.

Fleet, G. H. (ed.) (1992) *Wine Microbiology and Biotechnology*. Harwood Academic, New York.

Jackson, D., and Schuster, D. (2001) *The Production of Grapes and Wines* In: *Cool Climates*, 3rd edn. Dunmore Press, Auckland, New Zealand.

Ribéreau-Gayon, P., Dubourdieu, D., Glories, Y., and Maujean, A. (2006) *Handbook of Enology* Volume II (*The Chemistry of Wine Stabilization and Treatments*), 2nd edn. John Wiley & Sons, Chichester, UK.

Troost, R. (1988) *Technologie des Weines, Handbuch der Lebensmitteltechnologie*, 2nd edn, Ulmer, Stuttgart, Germany.

Vine, R. P., Bordelon, B., Harkness, E. M., Browning, T., and Wagner, C. (1997) *From Grape Growing to Marketplace*. Chapman & Hall, New York.

Adjustments

Boulton, R. (1985) Acidity modification and stabilization. In: *Proceedings of the International Symposium on Cool Climate Viticulture and Enology* (D. A. Heatherbell *et al.*, eds.), OSU Agricultural Experimental Station Technical Publication No. 7628, pp. 482–495. Oregon State University, Corvallis.

Jones, P. R., Kwiatkowski, M., Skouroumounis, G. K., Francis, I. L., Lattey, K. A., Waters, E. J., Pretorius, I. S., and Høj, P. B. (2004) Exposure of red wine to oxygen post-fermentation – if you can't avoid it, why not control it? *Wine Indust. J.* 19(3), 17–20, 22–24.

Singleton, V. L., and Ough, C. S. (1962) Complexity of flavor and blending of wines. *J. Food Sci.* 12, 189–196.

Stabilization and Clarification

Dickenson, T. C. (1997) *Filters and Filtration Handbook*, 4th edn. Elsevier Science, Oxford, UK.

du Toit, W. J., Marais, J., Pretorius, I. S., and du Toit, M. (2006) Oxygen in must and wine: A review. *S. Afr. J. Enol. Vitic.* 27, 76–94.

Edwards, C. G. (2006) *Illustrated Guide to Microbes and Sediments in Wine, Beer, and Juice*. WineBugs LLC, Pullman, WA.

García-Ruiz, J. M., Alcántara, R., and Martín, J. (1991) Evaluation of wine stability to potassium hydrogen tartrate precipitation. *Am. J. Enol. Vitic.* 42, 336–340.

Marchal, R., Barret, J., and Maujean, A. (1995) Relations entre les caracteristiques physico-chimiques d'une bentonite et son pouvoir d'absorption. *J. Intl. Sci. Vigne Vin* 29, 27–42.

van Rensburg, P., and Pretorius, I. S. (2000) Enzymes in winemaking: Harnessing natural catalysts for efficient biotransformations – a review. *S. Afr. J. Enol. Vitic.* 21 (Special Issue), 52–73.

Waters, E. J., Alexander, G., Muhlack, R., Pocock, K. F., Colby, C., O'Neill, B. K., Høj, P. B., and Jones, P. (2005) Preventing protein haze in bottled white wine. *Aust. J. Grape Wine Res.* 11, 215–225.

Aging

Marais, J. (1986) Effect of storage time and temperature of the volatile composition and quality of South African *Vitis Vinifera* L. cv. Colombar wines. In: *The Shelf Life of Foods and Beverages* (G. Charalambous, ed.), pp. 169–185. Elsevier, Amsterdam.

Rapp, A., and Güntert, M. (1986) Changes in aroma substances during the storage of white wines in bottles. In: *The Shelf Life of Foods and Beverages* (G. Charalambous, ed.), pp. 141–167. Elsevier, Amsterdam.

Rapp, A., and Marais, J. (1993) The shelf life of wine: Changes in aroma substances during storage and ageing of white wines. In: *Shelf Life Studies of Foods and Beverages* (G. Charakanbous, ed.), pp. 891–921. Elsevier, Amsterdam.

Simpson, R. F., and Miller, G. C. (1983) Aroma composition of aged Riesling wines. *Vitis* 22, 1–63.

Oak and Other Cooperage Material

Darlington, P. (1994) Stainless steel in the wine industry: Choosing, using and abusing. *Aust. Grapegrower Winemaker* 365, 17–19.

Feuillat, F., Huber, F., and Keller, R. (1993) La porosité du bois de chêne (*Quercus robur* L.; *Quercus petraea* Liebl.) utilisé en tonnellerie. Relation avec la variabilité de quelques caractéristiques physique et anatomiques du bois. *Rev. Fr. Oenol.* 142, 5–19.

Jarauta, I., Cacho, J., and Ferreira, V. (2005) Concurrent phenomena contributing to the formation of the aroma of wine during aging in oak wood: An analytical study. *J. Agric. Food Chem.* 53, 4166–4177.

Masson, G., Puech, J. L., and Moutounet, M. (1996) Composition chimique du bois de chêne de tonnellerie. *Bull. O.I.V.* 69, 634–657.

Mosedale, J. R., and Puech, J.-L. (2003) Barrels: Wines, spirits and other beverages. In: *Encyclopedia of Food Science, Food Technology and Nutrition* (B. Caballero, L. C. Trugo and P. M. Finglas, eds.), pp. 393–403. Academic Press, London.

Mosedale, J. R., Feuillat, F., Baumes, R., Dupouey, J.-L, and Puech, J.-L. (1998) Variability of wood extractives among *Quercus robur* and *Quercus petraea* trees from mixed stands and their relation to wood anatomy and leaf morphology. *Can. J. Forest. Res.* 28, 994–1006.

Puech, J.-L., Feuillat, F., and Mosedale, J. R. (1999) The tannins of oak heartwood: Structure, properties, and their influence on wine flavor. *Am. J. Enol. Vitic.* 50, 469–478.

Schahinger, G., and Rankine, R. (2005) *Cooperage for Winemakers*, 2nd edn. Ryan, Adelaide, Australia.

Swan, J. S. (2005) Barrel profiling: Helping winemakers attain quality and consistency. *Aust. NZ Grapegrower Winemaker* 498, 72, 74–79.

Vivas, N. (2002) *Manuel de Tonnellerie*. Éditions Fêret, Bordeaux, France.

Waterhouse, A. L., and Caputi, A. (1999) International Symposium on Oak in Winemaking. *Am. J. Enol. Vitic.* 50, 467–546.

Cork, Closures and Bottles

Anderson, T. C. (1987) Understanding and improving bag-in-box technology. In: *Proceedings of the 6th Australian Wine Technical Conference* (T. Lee, ed.), pp. 272–274. Australian Industrial Publishers, Adelaide, Australia.

Butzke, C. E., and Suprenant, A. (1998) *Cork Sensory Quality Control Manual.* University of California Cooperative Extension, Division of Agriculture and Natural Resources, Davis, CA.

Casey, J. A. (1991) The enigmatic properties of cork. *Aust. Grapegrower Winemaker* 328, 83–84, 87–89.

Giles, G. A. (ed.) (1999) *Handbook of Beverage Packaging.* CRC Press, Boca Raton, FL.

Lange, J., and Wyser, Y. (2003) Recent innovation in barrier technologies for plastic packaging – a review. *Packag. Technol. Sci.* 16, 149–158.

Leske, P., and Edlinton, J. (eds.) (1995) *Corks and Closures.* Proceedings of the ASOV Oenology Seminar Australian Society of Viticulture and Oenology, Adelaide, Australia.

Pereira, H. (2007) *Cork Biology, Production and Uses.* Elsevier Press, London.

Riboulet, J. M., and Alegoët, C. (1986) *Aspects pratiques du bouchage des vins.* Collection Avenir Oenologie, Bourgogne, Chaintre, France.

Stelzer, T. (2005) *Taming the Screw: A Manual for Winemaking with Screw Caps.* Winetitles, Adelaide, Australia.

Vieira Natividade, J. (1956) *Subériculture.* Éditions de l'École nationale du Génie Rural des Eaux et Forêts, Nancy, France.

Cork-induced Spoilage

Riboulet, J. M., and Alegoët, C. (1986) *Aspects Pratiques du Bouchage des Vins.* Collection Avenir Oenologie, Bourgogne, Chaintre, France.

Sefton, M. A., and Simpson, R. F. (2005) Compounds causing cork taint and the factors affecting their transfer from natural cork closures to wine – a review. *Aust. J. Grape Wine Res.* 11, 188–200.

Silva Pereira, C., Figueiredo Marques, J. J., and San Romão, M. V. (2000) Cork taint in wine: Scientific knowledge and public perception – a critical review. *Crit. Rev. Microbiol.* 26, 147–162.

Simpson, R. F. (1990) Cork taint in wine: A review of the causes. *Aust. NZ Wine Ind. J.* 5, 286–287, 289, 293–296.

Simpson, R. F., and Sefton, M. A. (2007) Origin and fate of 2,4,6-trichloroanisole in cork bark and wine corks. *Aust. J. Grape Wine Res.* 13, 106–116.

Other Forms of Wine Spoilage

Drysdale, G. S., and Fleet, G. H. (1988) Acetic acid bacteria in winemaking: A review. *Am. J. Enol. Vitic.* 39, 143–154.

du Toit, M., and Pretorius, I. S. (2000) Microbial spoilage and preservation of wine: Using weapons from Nature's own arsenal – a review. *S. Afr. J. Enol. Vitic.* 21 (Special Issue), 74–96.

Fleet, G. (1992) Spoilage yeasts. *Crit. Rev. Biotechnol.* 12, 1–44.

Licker, J. L., Acree, T. E., and Henick-Kling, T. (1999) What is Brett (*Brettanomyces*) flavor? In: *Chemistry of Wine Flavor.* ACS Symposium Series Vol. 714. (A. L. Waterhouse and S. E. Ebeler, eds.), pp. 96–115. Amer. Chem. Soc., Washington, DC.

Limmer, A. (2005) Suggestions for dealing with post-bottling sulfides. *Aust NZ Grapegrower Winemaker* 503, 67, 68–74, 76.

Rapp, A., Pretorius, P., and Kugler, D. (1992) Foreign and undesirable flavours in wine. In: *Off-Flavors in Foods and Beverages* (G. Charalambous, ed.), pp. 485–522. Elsevier, London.

Snowdon, E. M., Bowyer, M. C., Grbin, P. R., and Bowyer, P. K. (2006) Mousy off-flavor: A review. *J. Agric. Food Chem.* 54, 6465–6474.

References

Aiken, J. W., and Noble, A. C. (1984) Comparison of the aromas of oak- and glass-aged wines. *Am. J. Enol. Vitic.* 35, 196–199.

Alcalde-Eon, C., Escribano-Bailón, M. T., Santos-Buelga, C., and Rivas-Gonzalo, J. C. (2006) Changes in the detailed pigment composition of red wine during maturity and ageing. A comprehensive study. *Anal. Chim. Acta.* 563, 238–254.

Aldercreutz, P. (1986) Oxygen supply to immobilized cells, 5. Theoretical calculations and experimental data for the oxidation of glycerol by immobilized *Gluconobacter oxydans* cells with oxygen or *p*-benzoquinone as electron acceptor. *Biotechnol. Bioeng.* 28, 223–232.

Álvarez-Rodríguez, M. L., Belloch, C., Villa, M., Uruburu, F., Larriba, G., and Coque, J. R. (2003) Degradation of vanillic acid and production of guaiacol by microorganisms isolated from cork samples. *FEMS Microbiol. Lett.* 220, 49–55.

Amerine, M. A., and Joslyn, M. A. (1970) *Tables Wines, the Technology of Their Production*, 2nd edn. University of California Press, Berkeley, CA.

Amerine, M. A., and Roessler, E. B. (1983) *Wines – Their Sensory Evaluation.* Freeman, New York.

Amerine, M. A., Berg, H. W., Kunkee, R. E., Ough, C. S., Singleton, V. L., and Webb, A. D. (1980) *The Technology of Wine Making.* Avi, Westport, CT.

Amon, J. M., Simpson, R. F., and Vandepeer, J. M. (1987) A taint in wood-matured wine attributable to microbiological contamination of the oak barrel. *Aust. NZ Wine Ind. J.* 2, 35–37.

Amon, J. M., Vandepeer, J. M., and Simpson, R. F. (1989) Compounds responsible for cork taint in wine. *Aust. NZ Wine Ind. J.* 4, 62–69.

Anderson, T. C. (1987) Understanding and improving bag-in-box technology. In: *Proceedings of the 6th Australian Wine Technical Conference* (T. Lee, ed.), pp. 272–274. Australian Industrial Publishers, Adelaide, Australia.

Anonymous (1984) *Proceedings of the International Symposium on Wine Transportation* (mimeographed). Montreal, Canada.

Anonymous (1991) Wieviel Styrol geben Kunststofftanks ab? *Weinwirtsch. Tech.* 127(6), 23–25.

Antonelli, A., Carnacini, A., Marignetti, N., and Natali, N. (1996) Pilot scale dealcoholization of wine by extraction with solid carbon dioxide. *J. Food Sci.* 61, 970–972.

Armstrong, D. N. (1987) Leakers, equipment failure and quality assurance procedures for defect supervision. In: *Proceedings of the 6th Australian Wine Technical Conference* (T. Lee, ed.), pp. 275–276. Australian Industrial Publishers, Adelaide, Australia.

Asensio, A., and Seoane, E. (1987) Polysaccharides from the cork of *Quercus suber*, I. Holocellulose and cellulose. *J. Nat. Prod.* 50, 811–814.

Barillère, J. M., Bidan, T., and Dubois, C. (1983) Thermorésistance de levures et des bactéries lactique isolées du vin. *Bull. O.I.V.* 56, 327–351.

Bartowsky, E. J., Xia, D., Gibson, R. L., Fleet, G. H., and Henschke, P. A. (2003) Spoilage of bottled red wine by acetic acid bacteria. *Lett. Appl. Microbiol.* 36, 307–314.

Bass, G. F. (1971) A Byzantine trading venture. *Sci. Am.* 225(2), 22–33.

Beckmann, J. (1846) *History of Inventions, Discoveries and Origins* (W. Johnson, trans.), Vol.2, 4th edn. Henry G. Bohn, London.

Beech, F. W., and Redmond, W. J. (eds.) (1981) *Bulk Wine Transport and Storage*. Proceedings of the 5th Wine Subject Day. Long Ashton Research Station, University of Bristol, Long Ashton, UK.

Bento, M. F. S., Pereira, H., Cunha, M. A., Moutinho, A. M. C., van den Berg, K. J., and Boon, J. J. (2001) A study of variability of suberin composition in cork from *Quercus suber* L. using thermally assisted transmethylation GC-MS. *J. Anal. Appl. Pyrolysis* 57, 45–55.

Bertrand, A., and Barrios, M. L. (1995) Contamination des bouchons par les produits de traitements de palletes de stockage des bouchons. *Rev. Fr. Oenologie.* 149, 29.

Bertuccioli, M., and Viani, R. (1976) Red wine aroma: Identification of headspace constituents. *J. Sci. Food Agric.* 27, 1035–1038.

Blanchard, L., Darriet, P., and Dubourdieu, D. (2004) Reactivity of 3-mercaptohexanol in red wine: Impact of oxygen, phenolic fractions, and sulfur dioxide. *Am. J. Enol. Vitic.* 55, 115–120.

Blanchard, L., Tominaga, T., and Dubourdieu, D. (2001) Formation of furfurylthiol exhibiting a strong coffee aroma during oak barrel fermentation from furfural released by toasted staves. *J. Agric. Food Chem.* 49, 4833–4835.

Bobet, R. A. (1987) Interconversion Reactions of Sulfides in Model Solutions. MS thesis, University of California, Davis.

Bobet, R. A., Noble, A. C., and Boulton, R. B. (1990) Kinetics of the ethanethiol and diethyl disulfide interconversion in wine-like solutions. *J. Agric. Food Chem.* 38, 449–452.

Boidron, J. N. (1994) Preparation and maintenance of barrels. In: *The Barrel and the Wine: Scientific Advances of a Traditional Art*, pp. 71–82. Sequin Moreau USA, Napa, CA.

Boidron, J. N., Lefebvre, A., Riboulet, J. M., and Ribéreau-Gayon, P. (1984) Les substances volatiles susceptibles d'être cédées au vin par le bouchon de liège. *Sci. Aliments* 4, 609–616.

Bonorden, W. R., Nagel, C. W., and Powers, J. R. (1986) The adjustment of high pH/high titratable acidity wines by ion exchange. *Am. J. Enol. Vitic.* 37, 143–148.

Borges, M. (1985) New trends in cork treatment and technology. *Beverage Rev.* 5, 15, 16, 19, 21.

Boulton, R. B., Singleton, V. L., Bisson, L. F., and Kunkee, R. E. (1996) *Principles and Practices of Winemaking*, pp. 411–412. Chapman & Hall, New York.

Bradshaw, M. P., Cheynier, V., Scollary, G. R., and Prenzler, P. D. (2003) Defining the ascorbic acid crossover from anti-oxidant to pro-oxidant in a model wine matrix containing (+)-catechin. *J. Agric. Food Chem.* 51, 4126–4132.

Brajkovich, M., Tibbits, N., Peron, G., Lund, C. M., Dykes, S. I., Kilmaretin, P. A., and Nicolau, L. (2005) Effect of screwcap and cork closures on SO$_2$ levels and aromas in a Sauvignon blanc wine. *J. Agric. Food Chem.* 53, 10006–10011.

Brenner, M. W., Owades, J. L., Gutcho, M., and Golyzniak, R. (1954) Determination of volatile sulfur compounds. III. Determination of mercaptans. *Proc. Am. Soc. Brew. Chem.* 88–97.

Brezovesik, L. (1986) Mould fungi as possible cause for corkiness in wine (in Hungarian) *Borgazdasag.* 34, 25–31.

Brock, C. P., Schweizer, W. B., and Dunitz, J. D. (1991) On the validity of Wallach's Rule: On the density and stability of racemic crystals compared with their chiral counterparts. *J. Am. Chem. Soc.* 113, 9811–9820.

Brock, T. D. (1983) *Membrane Filtration. Science Technology.* Science Tech Inc., Madison, WI.

Brun, S. (1980) Pollution du vin par le bouchon et le dispositif de surbouchage. *Rev. Fr. Oenol.* 77, 53–58.

Brun, S. (1984) Toxicology of materials during the transportation of wine. In: *Proceedings of the International Symposium on Wine Transportation* (mimeographed), pp. 72–87. Montreal, Canada.

Bureau, G. M., Charpentier-Massonal, M., and Parsee, M. (1974) Étude des goûts anormaux apportés par le bouchon sur le vin de Champagne. *Rev. Fr. Oenol.* 56, 22–24.

Caldini, C., Bonomi, F., Pifferi, P. G., Lanzarini, G., and Galante, Y. M. (1994) Kinetic and immobilization studies on fungal glycosidases for aroma enhancement in wine. *Enzyme Microb. Technol.* 16, 286–291.

Calisto, M. C. (1990) DMDC's role in bottle stability. *Wines Vines* 71(10), 18–21.

Caloghiris, M., Waters, E. J., and Williams, P. J. (1997) An industry trial provides further evidence for the role of corks in oxidative spoilage of bottled wines. *Aust. J. Grape Wine Res.* 3, 9–17.

Cantarelli, C. (1964) Il difetto da idrogeno solforato dei vini, la natura, le cause, i trattamenti preventivi e di risanamento. *Atti Accad. Ital. Vite Vino Siena* 16, 163–175.

Capone, D., Sefton, M., Pretorius, I., and Høj, P. (2003) Flavour 'scalping' by wine bottle closures – the 'winemaking' continues post vineyard and winery. *Aust. NZ Wine Ind. J.* 18(5), 16, 18–20.

Capone, D. L., Skouroumounis, G. K., Barker, D. A., McLean, H. J., Pollnitz, A. P., and Sefton, M. A. (1999) Absorption of chloroanisoles from wine by corks and by other materials. *Aust. J. Grape Wine Res.* 5, 91–98.

Capone, D. L., Skouroumounis, G. K., and Sefton, M. A. (2002) Permeation of 2,4,6-trichloroanisole through cork closures in wine bottles. *Aust. J. Grape Wine Res.* 8, 196–199.

Caridi, A. (2006) Enological functions of parietal yeast mannoproteins. *Antonie van Leeuwenhoek* 89, 417–422.

Carpentier, N., and Maujean, A. (1981) Light flavours in champagne wines. In: *Flavour '89*. 3rd Weurman Symposium: Proceedings of the International Conference (P. Schreier, ed.), pp. 609–615. de Gruyter, Berlin.

Carriço, S. M. (1997) Estudo da Composição Química da estrutura Celular e dos componentes voláteis da cortiça de *Quercus suber*. PhD Thesis, Universidade de Aveiro, Portugal.

Casey, J. A. (1990) A simple test for cork taint. *Aust. Grapegrower Winemaker* 324, 40.

Casey, J. A. (1993) Cork as a closure material for wine. *Aust. Grapegrower Winemaker* 352, 36–37, 39, 41.

Casey, J. (2003) Controversies about corks. *Aust. NZ Grapegrower Winemaker* 475, 68–70, 72, 74.

Castellari, M., Simonato, B., Tornielli, G. B., Spinelli, P., and Ferrarini, R. (2004) Effects of different enological treatments on dissolved oxygen in wines. *Ital. J. Food Sci.* 16, 387–397.

Castellari, M., Spinabelli, U., Riponi, C., and Amati, A. (1998) Influence of some technological practices on the quality of resveratrol in wine. *Z. Lebensm. Unters. Forsch.* 206, 151–155.

Cattaruzza, A., Peri, C., and Rossi, M. (1987) Ultrafiltration and deep-bed filtration of a red wine: Comparative experiments. *Am. J. Enol. Vitic.* 38, 139–142.

Cerdán, T. G., Mozaz, S. R., and Azpilicueta, C. A. (2002) Volatile composition of aged wine in used barrels of French oak and of American oak. *Food Res. Int.* 35, 603–610.

Chalier, P., Angot, B., Delteil, D., Doco, T., and Gunata, Z. (2007) Interactions between aroma compounds and whole mannoprotein isolated from *Saccharomyces cerevisiae* strains. *Food Chem.* 100, 22–30.

Champagnol, F. (1986) L'acidité des moûts et des vins, 1. Facteurs physico-chimiques et technologiques de variation. *Rev. Fr. Oenol.* 104, 26–30, 51–57.

Charpentier, C. (2000) Yeast autolysis and yeast macromolecules? Their contribution to wine flavor and stability. In: *Proceedings of the ASEV 50th Anniversary Annual Meeting*, Seattle, WA, June 19–23, 2000, pp. 271–277. American Society for Enology and Viticulture, Davis, CA.

Chassagne, D., Guilloux-Benatier, M., Alexandre, H., and Voilley, A. (2005) Sorption of wine volatile phenols by yeast lees. *Food Chem.* 91, 39–44.

Chatonnet, P. (1989) Origines et traitements des bois destinés à l'élevage des vins de qualité. *Rev. Oenologues* 15, 21–25.

Chatonnet, P. (1991) Incidences du Bois de Chêne sur la Composition Chimique et les Qualitiés Organoleptiques des Vins. Applications Technologiques. Thesis, University of Bordeaux II, France.

Chatonnet, P. (1994) Barrel aging of red wines. In: *The Barrel and the Wine*. Scientific Advances of a Traditional Art. Seguin Moreau USA, Inc.

Chatonnet, P. (1998) The characteristics of Russian oak and benefits of its use in wine aging. Data from 1995–1996 experiments. In: *The Barrel and the Wine III: The Taste of Synergy*, pp. 3–17. Seguin Moreau USA, Napa, CA.

Chatonnet, P. (1999) Discrimination and control of toasting intensity and quality of oak wood barrels. *Am. J. Enol. Vitic.* 50, 479–494.

Chatonnet, P., and Dubourdieu, D. (1998a) Comparative study of the characteristics of American white oak (*Quercus alba*) and European oak (*Quercus petraea* and *Q. robur*) for production of barrels used in barrel aging of wines. *Am. J. Enol. Vitic.* 49, 79–85.

Chatonnet, P., and Dubourdieu, D. (1998b) Identification of substances responsible for the 'sawdust' aroma in oak wood. *J. Sci. Food Agric.* 76, 179–188.

Chatonnet, P., Barbe, C., Canal-Llauberes, R.-M., Dubourdieu, D., and Boidron, J. N. (1992) Incidences de certaines préparations pectolytiques sur la teneur en phénols volatils des vins blancs. *J. Intl Sci. Vigne Vin* 26, 253–269.

Chatonnet, P., Boidron, J. N., and Dubourdieu, D. (1993a) Influences des conditions d'élevage et des sulfitage des vins rouges en barriques sur leur teneur en acide acétique et en ethyl-phenols. *J. Int. Sci. Vigne Vin* 27, 277–299.

Chatonnet, P., Boidron, J., and Dubourdieu, D. (1993b) Maîtrise de la chauffe de brûlage en tonnellerie, applications à la vinification et à l'élevage des vins en barriques. *Rev. Fr. Oenol.* 33, 41–58.

Chatonnet, P., Boidron, J. N., Dubourdieu, D., and Pons, M. (1994a) Évolution des composés polyphénoliques du bois de chêne au cours de son séchage, premiers résultats. *J. Intl Sci. Vigne Vin* 28, 337–357.

Chatonnet, P., Boidron, J. N., Dubourdieu, D., and Pons, M. (1994b) Évolution de certains composés volatils du bois chêne au cours de son séchage, premiers résultats. *J. Intl Sci. Vigne Vin* 28, 359–380.

Chatonnet, P., Bonnet, S., Boutou, S., and Labadie, M.-D. (2004) Identification and responsibility of 2,4,6-tribromoanisole in musty, corked odors in wine. *J. Agric. Food Chem.* 52, 1255–1262.

Chatonnet, P., Dubourdieu, D., and Boidron, J. N. (1992) Incidence des conditions de fermentation et d'élevage des vins blanc secs en barriques sur leur composition en substances cédées par le bois de chêne. *Sci. Aliments* 12, 665–685.

Chatonnet, P., Guimberteau, G., Dubourdieu, D., and Boidron, J. N. (1994c) Nature et origin des odeurs de moisi dans les caves, incidences sur la contamination des vins. *J. Intl Sci. Vigne Vin* 28, 131–151.

Chatonnet, P., Labadie, D., and Gubbiotti, M.-C. (1999) Étude comparative des caractéristiques de bouchouns en liège et en matériaux synthétiques – premiere résultats. *Rev. Oenologues* 92, 9–14.

Chatonnet, P., Viala, C., and Dubourdieu, D. (1997) Influence of polyphenolic components of red wines on the microbial syntheses of volatile phenols. *Am. J. Enol. Vitic.* 48, 443–448.

Chaves das Neves, H. J., Vasconcelos, A. M. P., and Costa, M. L. (1990) Racemization of wine free amino acids as function of bottling age. In: *Chirality and Biological Activity* (H. Frank *et al.*, ed.), pp. 137–143. Alan R. Liss, New York.

Chen, C.-L. (1970) Constituents of *Quercus rubra*. *Phytochemistry* 9, 1149.

Chen, C.-L., and Chang, H. M. (1985) Chemistry of lignin biodegradation. In: *Biosynthesis and Biodegradation of Wood Components* (T. Higuchi, ed.), pp. 535–556. Academic Press, New York.

Chisholm, M. G., Guiher, L. A., and Zaczkiewicz, S. M. (1995) Aroma characteristics of aged Vidal blanc wine. *Am. J. Enol. Vitic.* 46, 56–62.

Cillers, J. J. L., and Singleton, V. L. (1990) Nonenzymatic autooxidative reactions of caffeic acid in wine. *Am. J. Enol. Vitic.* 41, 84–86.

Cocolin, L., Rantsiou, K., Iacumin, L., Zironi, R., and Comi, G. (2004) Molecular detection and identification of *Brettanomyces/Dekkera bruxellensis* and *Brettanomyces/Dekkera anomalus* in spoiled wines. *Appl. Environ. Microbiol.* 70, 1347–1355.

Cole, J., and Boulton, R. (1989) A study of calcium salt precipitation in solutions of malic acid and tartaric acid. *Vitis* 28, 177–190.

Comitini, F., Ingeniis De, J., Pepe, L., Mannazzu, I., and Ciani, M. (2004) *Pichia anomala* and *Kluyveromyces wickerhamii* killer toxins as new tools against *Dekkera/Brettanomyces* spoilage yeasts. *FEMS Microbiol. Lett.* 238, 235–240.

Coque, J. R., Álvarez-Rodríguez, M. L., and Larriba, G. (2003) Characterization of an inducible chlorophenol o-methyltransferase from *Trichoderma longibrachiatum* involved in the formation of chloroanisoles and determination of its role in cork taint of wines. *Appl. Environ. Microbiol.* 69, 5089–5095.

Costello, P. J., Lee, T. H., and Henschke, C. (2001) Ability of lactic acid bacteria to produce N-heterocycles causing mousy off-flavour in wine. *J. Grape Wine Res.* 7, 160–167.

Couto, J. A., and Hogg, T. A. (1994) Diversity of ethanol-tolerant lactobacilli isolated from Douro fortified wine: Clustering and identification by numerical analysis of electrophoretic protein profiles. *J. Appl. Bacteriol.* 76, 487–491.

Crowell, E. A., and Guymon, J. F. (1975) Wine constituents arising from sorbic acid addition, and identification of 2-ethoxyhexa-3, 5-diene as source of geranium-like off-odor. *Am. J. Enol. Vitic.* 26, 97–102.

Cuénat, P., and Kobel, D. (1987) La diafiltration en oenologie: Un example d'application à des vins vieux. *Rev. Suisse Vitic. Arboric. Hortic.* 2, 97–103.

Cuénat, P., Zufferey, E., and Kobel, D. (1990) La prévention des odeurs sulfhydriques des vins par la centrifugation après la fermentation alcoolique. *Rev. Suisse Vitic. Arboric. Hortic.* 22, 299–303.

Cutzach, I., Chatonnet, P., and Dubourdieu, D. (1999) Study of the formation mechanisms of some volatile compounds during the aging of sweet fortified wines. *J. Agric. Food Chem.* 47, 2837–2846.

Dambrouck, T., Narchal, R., Marchal-Delahaut, L., Parmentier, M., Maujean, A., and Jeandet, P. (2003) Immunodetection of protein from grapes and yeast in a white wine. *J. Agric. Food Chem.* 51, 2727–2732.

D'Angelo, M., Onori, G., and Santucci, A. (1994) Self-association of monohydric alcohols in water: Compressibility and infrared absorption measurements. *J. Chem. Phys.* 100, 3107.

Daniel, M. A., Elsey, G. M., Capone, D. L., Perkins, M. V., and Sefton, M. A. (2004) Fate of damascenone in wine: The role of SO_2. *J. Agric. Food Chem.* 52, 8127–8131.

Darriet, P., Pons, M., Lamy, S., and Dubourdieu, D. (2000) Identification and quantification of geosmin, and earthy odorant contaminating wines. *J. Agric Food Chem.* 48, 4835–4838.

da Silva Ferreira, A. C., Barbe, J.-C., and Bertrand, A. (2002) Heterocyclic acetals from glycerol and acetaldehyde in port wines: Evolution with aging. *J. Agric. Food Chem.* 50, 2560–2564.

Datta, S., and Nakai, S. (1992) Computer-aided optimization of wine blending. *J. Food Sci.* 57, 178–182.

D'Auria, M., Emanuele, L., Mauriello, G., and Racioppi, R. (2002) On the origin of goût de lumiere in champagne. *J. Photochem. Photobiol A: Chemistry* **158**, 21–26.

Davenport, R. R. (1982) Sample size, product composition and microbial spoilage. In: *Shelf Life – Proceedings of the Seventh Wine Subject Day* (F. W. Beech, ed.), pp. 1–4. Long Ashton Research Station, University of Bristol, Long Ashton, UK.

Davies, N. de G. (1935) *The Tomb of Rekh-Mi-Re at Thebes.* Arno Press, NY (pl. 15).

de la Garza, F., and Boulton, R. (1984) The modeling of wine filtrations. *Am. J. Enol. Vitic.* **35**, 189–195.

Delaherche, A., Claisse, O., and Lonvaud-Funel, A. (2004) Detection and quantification of *Brettanomyces bruxellensis* and 'ropy' *Pediococcus damnosus* in wine by real-time polymerase chain reaction. *J. Appl. Microbiol.* **97**, 910–915.

Delfini, C. (1987) Observations expérimentales sur l'origine et la disparition de l'alcool benzylique et de l'aldéhyde benzoïque dans les moûts et les vins. *Bull. O.I.V.* **60**, 463–473.

Delfini, C., Gaia, P., Bardi, L., Mariscalco, G., Contiero, M., and Pagiara, A. (1991) Production of benzaldehyde, benzyl alcohol and benzoic acid by yeasts and *Botrytis cinerea* isolated from grape musts and wines.*Vitis* **30**, 253–263.

de Mora, S. J., Eschenbruch, R., Knowles., S. J., and Spedding, D. J. (1986) The formation of dimethyl sulfide during fermentation using a wine yeast. *Food Microbiol.* **3**, 27–32.

Densem, N.E., and Turner, W.E.S. (1938) The equilibrium between ferrous and ferric oxides in glasses. *J. Soc. Glass Technol. Trans.* **22**, 372–389.

De Rosa, T., and Moret, I. (1983) Influenza dell'imbottigliamento in ambiente gas inerte sulla conservazione di uno vino blanco tranquillo. *Rev. Viticult. Enol.* **36**, 219–226.

Diderot, M. (1763) Recueil des planches sur les sciences. In: *Encyclopedia.* Diderot, Paris.

Di Stefano, R., and Ciolfi, G. (1985) L'influenza della luce sull'Asti Spumante. *Vini Ital.* **27**, 23–32.

Doussot, F., Pardon, P., Dedier, J., and De Jeso, B. (2000) Individual, species and geographic origin influence on cooperage oak extractible content (*Quercus robur* L. and *Quercus petraea* Liebl.). *Analusis* **28**, 960–965.

Dubois, P. (1983) Volatile phenols in wines. In: *Flavour of Distilled Beverages* (J. R. Piggott, ed.), pp. 110–119. Ellis Horwood, Chichester, UK.

Dubourdieu, D. (1995) Intérêts oenologiques et risques associés à l'élevage des vins blancs sur lies en barriques. *Rev. Fr. Oenol.* **155**, 30–35.

Dubourdieu, D., and Moine, V. (1998a) Recent data on the benefits of aging wine on its lees, I. Molecular interpretation of improvements in protein stability in white wines during aging on the lees. In: *The Barrel and the Wine*, III (*The Taste of Synergy*), pp. 21–25. Seguin Moreau USA, Napa, CA.

Dubourdieu, D., and Moine, V. (1998b) Recent data on the benefits of aging wine on its lees, II. The effect of mannoproteins on wines tartrate stability. In: *The Barrel and the Wine*, III (*The Taste of Synergy*), pp. 27–28. Seguin Moreau USA, Napa, CA.

Dubourdieu, D., Moine-Ledoux, V., Lavigne-Cruège, V., Blanchard, L., and Tominaga, T. (2000) Recent advances in white wine aging: The key role of lees. In: *Proceedings of the ASEV 50th Anniversary Annual Meeting*, Seattle, WA, June 19–23, 2000, pp. 345–352. American Society for Enology and Viticulture, Davis, CA.

Dubourdieu, D., Serrano, M., Vannier, A.C., and Ribéreau-Gayon, P. (1988) Étude comparée des tests de stabilité protéique. *Connaiss. Vigne Vin* **22**, 261–273.

Duerr, P., and Cuénat, P. (1988) Production of dealcoholized wine. In: *Proceedings of the 2nd International Symposium for Cool Climate Viticulture and Oenology* (R. E. Smart *et al.*, eds.),

pp. 363–364. New Zealand Society for Viticulture and Oenology, Auckland, New Zealand.

Dufour, C., and Bayonove, C. L. (1999) Influence of wine structurally different polysaccharides on the volatility of aroma substances in a model system. *J. Agric. Food Chem.* **47**, 671–677.

Dugelay, I., Baumes, R., Gunata, Z., Razungles, R., and Bayonove, C. (1995) Evolution de l'arôme de la conservation du vin, formation de 4-(1-éthoxyéthyl)-phénol et 4-(1-éthoxyéthyl)-gaïacol. *Sci. Alim.* **15**, 423–434.

Dugelay, I., Gunata, Z., Sapis, J.-C., Baumes, R., and Bayonove, C. (1993) Role of cinnamoyl esterase activities from enzyme preparations on the formation of volatile phenols during winemaking. *J. Agr. Food. Chem.* **41**, 2092–2096.

Dumbrell, R. (1983) *Understanding Antique Wine Bottles*, pp. 24, 35–36. Antique Collector's Club, Woodbridge, Suffolk, England.

Dupin, V. S., McKinnon, B. M., Ryan, C., Boulay, M., Markides, A. J., Jones, G. P., Williams, P. J., and Waters, E. J. (2000) *Saccharomyces cerevisiae* mannoproteins that protect wine from protein haze: Their release during fermentation and lees contact and a proposal for their mechanism of action. *J. Agric. Food Chem.* **48**, 3098–3105.

Dupin, V. S., Stockdale, V. J., Williams, P. J., Jones, G. P., Markides, A. J., and Waters, E. J. (2000)*Saccharomyces cerevisiae* mannoproteins that protect wine from protein haze: Evaluation of extraction methods and immunolocalization. *J. Agric. Food Chem.* **48**, 1086–1095.

du Toit, W. J., Lisjak, K., Marais, J., and du Toit, M. (2006) The effect of micro-oxygenation on the phenolic composition, quality and aerobic wine-spoilage microorganisms of different South African red wines. *S. Afr. J. Enol. Vitic.* **27**, 57–67.

Edelényi, M. (1966) Study on the stabilization of sparkling wines (in Hungarian) *Borgazdaság*, **12**, 30–32 (reported in Amerine *et al.*, 1980).

Edwards, T., Singleton, V. L., and Boulton, R. (1985) Formation of ethyl esters of tartaric acid during wine aging: Chemical and sensory effects. *Am. J. Enol. Vitic.* **36**, 118–124.

Eglinton, J. M., and Henschke, P. A. (1999) The occurrence of volatile acidity in Australian wines. *Aust. Grapegrower Winemaker* **426a**, 7–8, 10–12.

Eschenbruch, B., and Dittrich, H. H. (1986) Stoffbildungen von Essigbakterien in bezug auf ihre Bedeutung für die Weinqualität. *Zentralbl. Mikrobiol.* **141**, 279–289.

Escudero, A., Asensio, E., Cacho, J., and Ferreira, V. (2002) Sensory and chemical changes of young white wines stored under oxygen. An assessment of the role played by aldehydes and some other important odorants. *Food Chem.* **77**, 325–331.

Escudero, A., Hernandez-Orte, P., Cacho, J., and Ferreira, V. (2000) Clues about the role of methional as character impact odorant of some oxidized wines. *J. Agric. Food Chem.* **48**, 4268–4272.

Evans, C. S. (1987) Lignin degradation. *Process Biochem.* **22**, 102–105.

Fabre, S. (1989) Bouchons traités au(x) peroxyde(s): détection du pouvoir oxydant et risques d'utilisations pour le vin. *Rev. Oenologues* **15**, 11–15.

Farris, G. A., Fatichenti, F., and Deiana, P. (1989) Incidence de la température et du pH sur la production d'acide malique par *Saccharomyces cerevisiae. J. Intl Sci. Vigne Vin* **23**, 89–93.

Ferreira, A., Lopes, F., and Pereira, H. (2000) Caractérisation de la croissance et de la qualité du liège dans une région de production. *Ann. For. Sci.* **57**, 187–193.

Ferreira, V., Jarauta, I., and Cacho, J. (2006) Physicochemical model to interpret the kinetics of aroma extraction during wine aging in wood. Model limitations suggest the necessary existence of biochemical processes *J. Agric. Food Chem.* **54**, 3047–3054.

Feuillat, F., Dupouey, J.-L., Sciama, D., and Keller, R. (1997) A new attempt at discrimination between *Quercus petraea* and *Quercus robur* based on wood anatomy. *Can. J. For. Res.* **27**, 343–351.

Fischer, U., and Berger, R. (1996) The impact of dealcoholization on the flavor of wines: Correlating sensory and instrumental data by PLS analysis using different transformation techniques. In: *Proceedings of the 4th International Symposium on Cool Climate Viticulture and Enology* (T. Henick-Kling, *et al.*, eds.), pp. VII-13–17. New York State Agricultural Experimental Station, Geneva, NY.

Fischerleitner, E., Wendelin, S., and Eder, R. (2003) Comparative study of vegetable and animal proteins used as fining agents. *Bull. O.I.V.* **76**, 30–52.

Flecknoe-Brown, A. (2005) Oxygen-permeable polyethylene vessels: A new approach to wine maturation. *Aust. NZ Grapegrower Winemaker* **494**, 53–57.

Fletcher, J. (1978) Dating the geographical migration of *Quercus petraea* and *Q. robur* in Holocene times. *Tree-Ring Bull.* **38**, 45–47.

Flores, J. H., Heatherbell, D. A., Henderson, L. A., and McDaniel, M. R. (1991) Ultrafiltration of wine: Effect of ultrafiltration on the aroma and flavor characteristics of White Riesling and Gewürztraminer wines. *Am. J. Enol. Vitic.* **42**, 91–96.

Forsyth, D. S., Weber, D., and Dalglish, K. (1992) Survey of betyltin, cyclohexyltin and phenyltin compounds in Canadian wines. *J. Assoc. Off. Anal. Chem. Intl* **75**, 964–973.

Francis, I. L., Leino, M., Sefton, M. A., and Williams, P. J. (1993) Thermal processing of Chardonnay and Semillon juice and wine – sensory and chemical changes. In: *Proceedings of the 8th Australian Wine Industry Technical Conference* (C. S. Stockley *et al.*, eds.), pp. 158–160. Winetitles, Adelaide, Australia.

Francis, I. L., Sefton, M. A. and Williams, P. J. (1992) A study by sensory descriptive analysis of the effects of oak origin, seasoning, and heating on the aromas of oak models wine extracts. *Am. J. Enol. Vitic.* **43**, 23–30.

Frey, D., Hentschel, F. D., and Keith, D. H. (1978) Deepwater archaeology: The Capistello wreck, excavation, Lipari, Aeolian Islands. *Intl J. Naut. Archaeol. Underwater Explor.* **7**, 279–300.

Frommberger, R. (1991) Cork products – a potential source of polychlorinated dibenzo-*p*-dioxins and polychlorinated dibenzofurans. *Chemosphere* **23**, 133–139.

Fugelsang, K. C., and Zoecklein, B. W. (2003) Population dynamics and effects of *Brettanomyces bruxellensis* strains on Pinot noir (*Vitis Vinifera* L.) wines. *Am. J. Enol. Vitic.* **54**, 294–300.

Fulcrand, H., Benabdeljalil, C., Rigaud, J., Chenyier, V., and Moutounet, M. (1998) A new class of wine pigments generated by reaction between pyruvic acid and grape anthocyanins. *Phytochemistry* **47**, 1401–1407.

Fulcrand, H., Cheynier, V., Oszmianski, J., and Moutounet, M. (1997) The oxidized tartaric acid residue as a new bridge potentially competing with acetaldehyde in flavan-3-ol condensation. *Phytochemistry* **46**, 223–227.

Gambuti, A., Strollo, D., Genovese, A., Ugliano, M., Ritieni, A., and Moio, L. (2005) Influence of enological practices on Ochratoxin A concentration in wine. *Am. J. Enol. Vitic.* **56**, 155–162.

Garcia, C., and Carra, S. (1996) Les silicones: Utilisation dans l'industrie du bouchon de liège. *Rev. Oenologues* **80**, 14–17.

Garofolo, A., and Piracci, A. (1994) Évolution des esters et acids gras pendant la conservation des vins, constantes d'équilibre et énergies d'activation. *Bull. O.I.V.* **67**, 225–245.

Gerber, N. N. (1979) Volatile substance from actinomycetes: Their role in the odor pollution of water. *CRC Crit. Rev. Microbiol.* **7**, 191–214.

Gibson, L. J., Easterling, K. E., and Ashby, M. F. (1981) The structure and mechanics of cork. *Proc. R. Soc. Lond.* A **377**, 99–117.

Gigliotti, A., Bucelli, P.L., and Faviere, V. (1985) Influenza delle uve bianche Trebbiano e Malvasia sul colore del vino Chianti. *VigneVini* **11**, 39–46.

Gini, B., and Vaughn, R. H. (1962) Characteristics of some bacteria associated with the spoilage of California dessert wines. *Am. J. Enol. Vitic.* **13**, 20–31.

Godden, P., Francis, L., Field, J., Gishen, M., Coulter, A., Valente, P., Høj, P., and Robinson, E. (2001) Wine bottle closure: Physical characteristics and effect on composition and sensory properties of a Semillon wine. 1. Performance up to 20 months postbottling. *Aust. J. Grape Wine Res.* **7**, 64–105.

González, A., Hierro, N., Poblet, M., Mas, A., and Guillamón, J. M. (2005) Application of molecular methods to demonstrate species and strain evolution of acetic acid bacteria population during wine production. *Int. J. Food Microbiol.* **102**, 295–304.

Gonzalez Viñas, M. A., Pérez-Coello, M. S., Salvador, M. D., Cabezudo, M. D., and Martín Alvarez, P. J. (1996) Changes in the GC volatiles of young Airen wines during bottle storage. *Food Chem.* **56**, 399–403.

Goswell, R. W. (1981) Tartrate stabilization trials. In: *Quality Control: Proceedings of the Sixth Wine Subject Day* (F. W. Beech and W. J. Redmond, eds.), pp. 62–65. Long Ashton Research Station, University of Bristol, Long Ashton, UK.

Gougeon, R.D., Reinholdt, M., Delmotte, L., Miehé-Brendlé, J., Chézeau, J.-M., Le Dred, R., Marchal, R., and Jeandet, P. (2002) Direct observation of polylysine side-chain interaction with Smectites interlayer surfaces through ^{1}H-^{27}Al heteronuclear correlation NMR spectroscopy. *Langmuir* **18**, 3396–3398.

Graça, J., and Pereira, H. (1997) Cork suberin: A glyceryl based polyester. *Holzforschung* **51**, 225–234.

Graça, J., and Pereira, H. (2004) The periderm development in *Quercus suber*. *IAWA J.* **25**, 325–335.

Grbin, P. R., Costello, P. J., Herderich, M., Markides, A. J., Henschke, P. A., and Lee, T. H. (1996) Developments in the sensory, chemical and microbiological basis of mousy taint in wine. In: *Proceedings of the 9th Australian Wine Industry Technical Conference* (C. S. Stockley, A.N. Sas, R. S. Johnson and T. H. Lee, eds.), pp. 57–61. Winetitles, Adelaide, Australia.

Grew, N. (1681) Musaeum Regalis Societatis. London. (Noted in Watney, B.M. and Babbidge, H. D. (1981) *Corkscrews for Collectors*. Sotheby Parke Bernet, London.)

Guilloux-Benatier, M., Chassagne, D., Alexandre, H., Charpentier, C., and Feuillat, M. (2001) Influence de l'autolyse des levures après fermentation sur le développement de *Brettanomyces/Dekkera* dans le vin. *J. Int. Sci. Vigne Vin* **35**, 157–164.

Guilloux-Benatier, M., Guerreau, J., and Feuillat, M. (1995) Influence of initial colloid content on yeast macromolecule production and on the metabolism of wine microorganisms. *Am. J. Enol. Vitic.* **46**, 486–492.

Guimberteau, G., Lefebvre, A., and Serrano, M. (1977) Le conditionnement en bouteilles. In: *Traité d'Oenologie: Sciences et Techniques du Vin* (J. Ribéreau-Gayon *et al.*, eds.), Vol.4, pp. 579–643. Dunod, Paris.

Gulson, B. L., Lee, T. H., Mizon, K. J., Korsch, M. J., and Eschnauer, H. R. (1992) The application of lead isotope ratios to determine the contribution of the tin-lead to the lead content of wine. *Am. J. Enol. Vitic.* **43**, 180–190.

Guymon, J. F., and Crowell, E. A. (1977) The nature and cause of cap-liquid temperature differences during wine fermentation. *Am. J. Enol. Vitic.* **28**, 74–78.

Hale, M. D., McCafferty, K., Larmie, E., Newton, J., and Swan, J. S. (1999) The influence of oak seasoning and toasting parameters on the composition and quality of wine. *Am. J. Enol. Vitic.* **50**, 495–502.

Hammond, S. M. (1976) Microbial spoilage of wines. In: *Wine Quality – Current Problems and Future Trends* (F. W. Beech *et al.*, eds.), pp. 38–44. Long Ashton Research Station, University of Bristol, Long Ashton, UK.

Harding, F. L. (1972) The development of colors in glass. In: *Introduction to Glass Science* (L. D. Pye *et al.*, eds.), pp. 391–431. Plenum, New York.

Hart, A., and Kleinig, A. (2005) The role of oxygen in the aging of bottled wine. *Aust. NZ Grapegrower Winemaker* **497a**, 79–80, 82–84, 86, 88.

Hasuo, T., and Yoshizawa, K. (1986) Substance change and substance evaporation through the barrel during Whisky ageing. In: *Proceedings of the 2nd Aviemore Conference on Malting, Brewing and Distillation* (I. Campbell and F. G. Priest, eds.), pp. 404–408. Institute of Brewing, London.

Hasuo, T., Saito, K., Terauchi, T., Tadenuma, M., and Sato, S. (1983) Studies on aging of whiskey, III. Influence of environmental humidity on aging of whiskey (in Japanese). *J. Brew. Soc. Jpn* **78**, 966–969.

Hatzidimitriou, E., Bouchilloux, P., Darriet, P., Bugaret, Y., Clerjeau, M., Poupot, C., Medina, B., and Dubourdieu, D. (1996) Incidence d'une protection viticole anticryptpgamique utilisant une formulation cuprique sur le niveau de maturité des raisins et l'arôme variétal des vins de Sauvignon. Bilan de trois années d'expérimentation. *J. Intl Sci. Vigne Vin* **30**, 133–150.

Helmcke, J. G. (1954) Neue Erkenntnisse uber den Aufbau von Membranfiltern. *Kolloid Zeitschift* **135**, 29–43.

Henick-Kling, T., Gerling, C., Martinson, T., Cheng, L., Lakso, A., and Acree, T. (2005) Atypical aging flavor defect in white wines: Sensory description, physiological causes, and flavor chemistry. *Am. J. Enol. Vitic.* **56**, 420A.

Herderich, M., Costello, P. J., Grbin, P. R., and Henschke, P. A. (1995) Occurrence of 2-acetyl-1-pyrroline in mousy wines. *Natl Prod. Lett.* **7**, 129–132.

Heresztyn, T. (1986) Formation of substituted tetrahydopyridines by species of *Brettanomyces* and *Lactobacillus* isolated from mousy wines. *Am. J. Enol. Vitic.* **37**, 127–131.

Hoadley, R. B. (1980) *Understanding Wood.* Taunton Press, Newtown, CT.

Hoenicke, K., Simat, T. J., Steinhart, H., Christoph, N., Geszner, M., and Kohler, H.-J. (2002) 'Untypical aging off-flavor' in wine: Formation of 2-aminoacetophenone and evaluation of its influencing factors. *Anal. Chim. Acta* **458**, 29–37.

Hoey, A. W., and Codrington, J. D. (1987) Oak barrel maturation of table wine. In: *Proceedings of the 6th Australian Wine Technical Conference* (T. Lee, ed.), pp. 261–266. Australian Industrial Publishers, Adelaide, Australia.

Honegger, H. (1966) Qu'est-ce que la liège? La liège sous le microcope electronique. Aide-memoire de lèige 18. Ed. Industries Suisse des Lieges Isolante, Boswill, Switzerland.

Hope, C. (1978) Jar sealing and amphora of the 18th dynasty: A technological study. Excavations at Malkata and the Birket Habu 1971–1974. *Egyptology Today* **5**(2), 1–88.

Howland, P. R., Pollnitz, A. P., Liacopoulos, D., McLean, H. J., and Sefton, M. A. (1997) The location of 2,4,6-trichloroanisole in a batch of contaminated wine corks. *Aust. J. Grape Wine Res.* **3**, 141–145.

Hsu, J., Heatherbell, D. A., Flores, J. H., and Watson, B. T. (1987) Heat-unstable proteins in wine, II. Characterization and removal by ultrafiltration. *Am. J. Enol. Vitic.* **38**, 17–22.

Ireton, D. C. (1990) Spinning cone column: What's it all about? *Wines Vines* **71**(1), 20–21.

Jäger, J., Diekmann, J., Lorenz, D., and Jakob, L. (1996) Cork-borne bacteria and yeasts as potential producers of off-flavours in wine. *Aust. J. Grape Wine Res.* **2**, 35–41.

Joncheray, J.-P. (1976) L'épave grecque, or étrusque, de Bon-Ponté. *Cah. Archéol. Subaquatique* **5**, 5–36.

Jones, O. (1971) Glass bottle push-ups and pontil marks. *Historic. Archaeol.* **5**, 62–73.

Jones, P. R., Kwiatkowski, M. J., Skouroumounis, G. K., Francis, I. L., Lattey, K. A., Waters, E. J., Pretorius, I. S., and Høj, P. B. (2004) Exposure of red wine to oxygen postfermentation – if you can't avoid it, why not control it? *Aust. NZ Wine Ind. J.* **9**, 17–20, 22–24.

Joyeux, A., Lafon-Lafourcade, S., and Ribéreau-Gayon, P. (1984) Evolution of acetic acid bacteria during fermentation and storage of wine. *Appl. Environ. Microbiol.* **48**, 153–156.

Jung, R., Seckler, J., and Zürn, F. (1993) Examination of the corking-procedure concerning cork and bottle. In: *The Cork in Oenology – 5th International Symposium, Pavia, Italy, May 13–14, 1993* (O. Colagrande, ed.), pp. 65–70. Chiriotti Editori, Pinerolo, Italy.

Kaufmann, A. (1998) Lead in wine. *Food Addit. Contam.* **15**, 437–445.

Keenan, C. P., Gözükara, M. Y., Christie, G. B. Y., and Heyes, D. N. (1999) Oxygen permeability of macrocrystalline paraffin wax and relevance to wax coatings on natural corks used as wine bottle closures. *Aust. J. Grape Wine Res.* **5**, 66–70.

Kilby, K. (1971) *The Cooper and His Trade.* John Baker, London. (Republished by Linden, Fresno, CA.)

Koehler, C. G. (1986) Handling of Greek transport amphoras. In: *Recherches sur les Amphores Greques* (J.-Y. Empereur and Y. Garlan, eds.), pp. 49–67. *Bull. Correspondence Hellénique*, suppl. 13. École française d'Anthèns, Paris.

Kösebalaban, F., and Özilgen, M. (1992) Kinetics of wine spoilage by acetic acid bacteria. *J. Chem. Tech. Biotechnol.* **55**, 59–63.

La Follette, G., Stambor, J., and Aiken, J. (1993) Chemical and sensory consideration in *sur lies* production of Chardonnay wines, III. Occurrence of sunstruck flavor. *Wein Wiss* **48**, 208–210.

La Guerche, S., Chamont, S., Blancard, D., Dubourdieu, D., and Darriet, P. (2005) Origin of (−)-geosmin on grapes: On the complementary action of two fungi, *botrytis cinerea* and *Penicillium expansum. Antonie van Leeuwenhock* **88**, 131–139.

Lambropoulos, I., and Roussis, I. G. (2007) Inhibition of the decrease of volatile esters and terpenes during storage of a white wine and a model wine medium by caffeic acid and gallic acid. *Food Res. Intern.* **40**, 176–181.

Lamikanra, O., Grimm, C. C., and Inyang, I. D. (1996) Formation and occurrence of flavor components in Noble muscadine wine. *Food Chem.* **56**, 373–376.

Lamuela-Raventós, R. M., Huix-Blanquera, M., and Waterhouse, A. L. (2001) Treatments for pinking alteration in white wines. *Am. J. Enol. Vitic.* **52**, 156–158.

Langhans, E., and Schlotter, H. A. (1985) Ursachen der Kupfer-Trüng. *Dtsch. Weinbau* **40**, 530–536.

Laperche, S. R., and Görtges, S. (2001) Highly active processing enzyme preparations for red winemaking. *Aust. Grapegrower Winemaker* **449a**, 99–104, 106, 108–111.

Larroque, M., Brun, S., and Blaise, A. (1989) Migration des monomères constitutifs des résines époxydiques utilisées pour revêtir les cuves à vin. *Sci. Aliments* **9**, 517–531.

Larsen, T. O., and Frisvad, J. C. (1995) Characterization of volatile metabolites from 47 *Penicillium* taxa. *Mycol. Res.* **10**, 1153–1166.

Larue, F., Rozes, N., Froudiere, I., Couty, C., and Perreira, G. P. (1991) Influence du développement de *Dekkera/Brettanomyces* dans les moûts et les vins. *J. Intl Sci. Vigne Vin* **25**, 149–165.

Laubenheimer, F. (1990) *Le Temps des amphores en Gaule*, pp. 152–153. Editions Errance, Paris.

Lavigne, V. (1995) Interprétation et prévention des défauts olfactifs de réduction lors de l'élevage sur lies totales. *Rev. Fr. Oenol.* **155**, 36–39.

Lavigne, V. (1998) The aptitude of wine lees or eliminating foul-smelling thiols. In: *The Barrel and the Wine*, III (*The Taste of Synergy*), pp. 31–34. Seguin Moreau USA, Napa, CA.

Lavigne, V., and Dubourdieu, D. (1996) Mise en évidence et interprétation de l'aptitude des lies à éliminer certains thiols volatils du vin. *J. Int. Sci. Vigne Vin* **30**, 201–206.

Lavigne, V., Boidron, J. N., and Dubourdieu, D. (1992) Formation des composés lourds au cours de la vinification des vins blancs secs. *J. Intl Sci. Vigne Vin* **26**, 75–85.

Lebrun, L. (1991) French oak – origin, making, preparation and winemakers' expectations. *Aust. Grapegrower Winemaker* **328**, 143–145, 147–148.

Lefebvre, A. (1981) Le bouchage liège des vins. In: *Actualités Oenologiques et Viticoles* (P. Ribéreau-Gayon and P. Sudraud, eds.), pp. 335–349. Dunod, Paris.

Lefebvre, A. (1988) Le bouchage liege des vins tranquilles. *Connaiss. Vigne Vin*, Numéro hors Séries. 11–35.

Lefebvre, A., Riboulet, J. M., Boidron, J. N., and Ribéreau-Gayon, P. (1983) Incidence des micro-organismes du liège sur les altérations olfactives du vin. *Sci. Aliments* **3**, 265–278.

Leino, M., Francis, I., Kallio, H., and Williams, P.J. (1993) Gas chromotographic headspace analysis of Chardonnay and Sémillon wines after thermal processing. *Z. Lebens.-u Forschung* **197**, 29–33.

Leistner, L., and Eckardt, C. (1979) Vorkommen toxinogener Penicillien bei Fleischerzeugnissen. *Fleischwirtschaft* **59**, 1892–1896.

Lemaire, T., Gilis, J. F., Fort, J.-P., Cucournau, P. (2002) Microoxygenation in extended maceration and early stages of red wine maturation – immediate effects and long term consequences. In: *11th Australian Wine Industry Technical Conference*, October 7–11, 2001, Adelaide, South Australia (R. J. Blair, P. J. Williams and P. B. Høj, eds.), pp. 33–43. Winetitles, Adelaide, Australia.

Lesko, L. H. (1977) *King Tut's Wine Cellar*. Albany Press, Albany, NY.

Levreau, R., Lefebvre, A., Serrano, M., and Ribéreau-Gayon, P. (1977) Études du bouchage liège, I. Rôle des surpressions dans l'apparition des bouteilles couleuses, bouchage sous gas carbonique. *Connaiss. Vigne Vin* **11**, 351–377.

Licker, J. L., Acree, T. E., and Henick-Kling, T. (1999) What is Brett (*Brettanomyces*) flavor? In: *Chemistry of Wine Flavor*. ACS Symposium Series Vol. 714. (A. L. Waterhouse and S. E. Ebeler, eds.), pp. 96–115. American Chemical Society, Washington, DC.

Lopes, F., and Pereira, H. (2000) Definition of quality classes for champagne cork stoppers in the high quality range. *Wood Sci. Technol.* **34**, 3–10.

Lopes, P., Saucier, C., and Glories, Y. (2005) Nondestructive colorimetric method to determine the oxygen diffusion rate through closures used in winemaking. *J. Agric. Food Chem.* **53**, 6967–6973.

Lopes, P., Saucier, C., Teissedre, P. L., and Glories, Y. (2006) Impact of storage position on oxygen ingress through different closures into wine bottles. *J. Agric. Food Chem.* **54**, 6741–6746.

Lopes, P., Saucier, C., Teissedre, P. L., and Glories, Y. (2007) Main routes of oxygen ingress through different closures into wine bottles. *J. Agric. Food Chem.* **55**, 5167–5170.

Lubbers, S., Charpentier, C., and Feuillat, M. (1996) Étude de la rétention de composés d'arôme par les bentonites en moût, vin et milieux modèles. *Vitis* **35**, 59–62.

Lubbers, S., Leger, B., Charpentier, C., and Feuillat, M. (1993) Effet colloide-protecteur d'extraits de parois de levures sur la stabilité tartrique d'une solution hydro-alcoolique modele. *J. Intl Sci. Vigne Vin* **27**, 13–22.

Lubbers, S., Voilley, A., Feuillat, M., and Charpontier, C. (1994) Influence of mannoproteins from yeast on the aroma intensity of a model wine. *Lebensm.-Wiss. u. Technol.* **27**, 108–114.

Lutter, M., Clark, A. C., Prenzier, P. D., and Scollary, G. R. (2007) Oxidation of caffeic acid in a wine-like medium production of dihydrobenzaldenyde and subsequent reactions with (+)-catechin. *Food Chem.* in press.

Maarse, H., Nijssen, L. M., and Angelino, S. A. G. F. (1988) Halogenated phenols and chloranisoles: Occurrence, formation and prevention. In: *Characterization, Production and Application of Food Flavours*. Proceedings of the Second International Wartburg Aroma Symposium 1987, pp. 43–61. Akademie Verlag, Berlin.

Macpherson, C. C. H. (1982) Life on the shelf. In: *Shelf Life – Proceeding of the Seventh Wine Subject Day* (F. W. Beech, ed.), pp. 14–19. Long Ashton Research Station, University of Bristol, Long Ashton, UK.

Maigne, W. (1875) *Nouveau manuel complet du tonnelier et du jaugeage contenant la fabrication des tonneaux de toute dimension des ouves, des foudres, des barils, des seaux et de tous les vaisseaux en bois cerclés suivi du jaugeage de tous les futs* (new edition of text by Paulin-Désormeaux, A. O. and Ott, H.). Encyclopédique de Roret, Paris.

Marais, J. (1986) Effect of storage time and temperature of the volatile composition and quality of South African *Vitis Vinifera* L. cv. Colombar wines. In: *The Shelf Life of Foods and Beverages* (G. Charalambous, ed.), pp. 169–185. Elsevier, Amsterdam.

Marais, J., and Pool, H. J. (1980) Effect of storage time and temperature on the volatile composition and quality of dry white table wines. *Vitis* **19**, 151–164.

Marais, J., van Wyk, C. J., and Rapp, A. (1992) Effect of storage time, temperature and region on the levels of 1,1,6-trimethyl-1, 2-dihydronaphthalene and other volatiles, and on quality of Weisser Riesling wines. *S. Afr. J. Enol. Vitic.* **13**, 33–44.

Marchal, R., Barret, J., and Maujean, A. (1995) Relations entre les caracteristiques physico-chimiques d'une bentonite et son pouvoir d'absorption. *J. Intl Sci. Vigne Vin* **29**, 27–42.

Marchal, R., Marchal-Delahaut, L., Michels, F., Parmentier, M., Lallement, A., and Jeandet, P. (2002) Use of wheat gluten as clarifying agent for musts and white wines. *Am. J. Enol. Vitic.* **53**, 308–314.

Marko, S. D., Dormedy, E. S., Fugelsang, K. C., Dormedy, D. F., Gump, B., and Wample, R. L. (2005) Analysis of oak volatiles by gas chromatography-mass spectrometry after ozone sanitization. *Am. J. Enol. Vitic.* **56**, 46–51.

Marsal, F., and Sarre, C. (1987) Étude par chromatographie en phase gazeuse de substances volatiles issues du bois de chêne. *Connaiss. Vigne Vin* **21**, 71–80.

Mas, A., Puig, J., Llado, N., and Zamora, F. (2002) Sealing and storage position effects on wine evolution. *J. Food Sci.* **67**, 1374–1378.

Masson, E., Baumes, R., Moutounet, M., and Puech, J.-L. (2000) The effect of kiln-drying on the levels of ellagitannins and volatile compounds of European oak (*Quercus petraes* Liebl.) stave wood. *Am. J. Enol. Vitic.* **51**, 201–214.

Masson, G., Puech, J.-L., Baumes, R., and Moutounet, M. (1996) Les extractibles du bois de chêne. Relation entre l'âge et la composition du bois. Impact olfactif de centains composés extraits par le vin. *Rev. Oenologues* **82**, 20–23.

Matricardi, L., and Waterhouse, A. L. (1999) Influence of toasting technique on color and ellagitannins of oak wood in barrel making. *Am J. Enol. Vitic.* **50**, 519–526.

Maujean, A., Haye, B., and Bureau, G. (1978) Étude sur un phénomène de masque observé en Champagne. *Vigneron Champenois* **99**, 308–313.

Maujean, A., Millery, P., and Lemaresquier, H. (1985) Explications biochimiques et métaboliques de la confusion entre goût de bouchon et goût de moisi. *Rev. Fr. Oenol.* **99**, 55–67.

Maujean, A., and Seguin, N. (1983) Contribution à l'étude des goûts de lumière dans les vins de Champagne, 3. Les réactions photochimiques responsables des goûts de lumière dans le vins de Champagne. *Sci. Aliments* **3**, 589–601.

Mazzoleni, V., Caldentey, P., and Silva, A. (1998) Phenolic compounds in cork used for production of wine stoppers as affected by storage and boiling of cork slabs. *Am. J. Enol. Vitic.* **49**, 6–10.

McGovern, P. E., Glusker, D. L., Exner, L. J., and Voigt, M. M. (1996) Neolithic resinated wine. *Nature* 381, 480–481.

McKearin, H. (1973) On stopping, bottling and binning. *Intl Bottler Packer* (April), 47–54.

McKinnon, A. J., Scullery, G. R., Solomon, D. H., and Williams, P. J. (1995) The influence of wine components on the spontaneous precipitation of calcium L(+)-tartrate in a model wine solution. *Am. J. Enol. Vitic.* 46, 509–517.

Médina, B. (1981) Metaux polluants dans les vins. In: *Actualitiés Oenologiques et Viticoles* (P. Ribéreau-Gayon and P. Sudraud, eds.), pp. 361–372. Dunod, Paris.

Millet, V., Vivas, N., and Lonvaud-Funel, A. (1995) The development of the bacterial microflora in red wines during aging in barrels. *J. Sci. Tech. Tonnellerie* 1, 137–150.

Moreau, M. (1978) La mycoflore des bouchons de liège, son évolution au contact du vin: Conséquences possibles du métabolisme des moisissures. *Rev. Mycol.* 42, 155–189.

Moreno, N. J., and Azpilicueta, C. A. (2006) The development of esters in filtered and unfiltered wines that have been aged in oak barrels. *Int. J. Food Sci. Technol.* 41, 155–161.

Mosedale, J. R., Feuillat, F., Baumes, R., Dupouey, J.-L., and Puech, J.-L. (1998) Variability of wood extractives among *Quercus robur* and *Quercus petraea* trees from mixed stands and their relation to wood anatomy and leaf morphology. *Can. J. Forest. Res.* 28, 994–1006.

Moutounet, M., Mazauric, J. P., Saint-Pierre, B., and Hanocq, J. F. (1998) Gaseous exchange in wines stored in barrels. *J. Sci. Tech. Tonnellerie* 4, 131–145.

Moutounet, M., Rabier, P., Puech, J.-L., Verette, E., and Barillère, J. M. (1989) Analysis by HPLC of extractable substances in oak wood, application to a Chardonnay wine. *Sci. Aliments* 9, 35–51.

Muhlack, R., Nordestgaard, S., Waters, E. J., O'Neill, B. K., Lim, A., and Colby, C. B. (2006) In-line dosing for bentonite fining of wine or juice: Contact time, clarification, product recovery and sensory effects. *Aust. J. Grape Wine Res.* 12, 221–234.

Muraoka, H., Watanabe, Y., Ogasawara, N., and Takahashi, H. (1983) Trigger damage by oxygen deficiency to the acid production system during submerged acetic fermentation with *Acetobacter aceti. J. Ferment. Technol.* 61, 89–93.

Murat, M.-L. (2005) Recent findings on rosé wine aromas. Part 1: identifying aromas studying the aromatic potential of grapes and juice. *Aust. NZ Grapegrower Winemaker* 497a, 64–65, 69, 71, 73–74, 76.

Murat, M.-L., Gindreau, E., and Dumeau, F. (2006) Using innovative analytical tools to prevent *Brettanomyces/Dekkera* development and to design preventative winemaking protocols. *Aust. NZ Grapegrower Winemaker* 515, 75–76, 78–82.

Murat, M.-L., Tominaga, T., Saucier, C., Glories, Y., and Dubourdieu, D. (2003) Effect of anthocyanins on stability of a key-odorous compound, 3-mercaptohexan-1-ol, in Bordeaux rosé wines. *Am. J. Enol. Vitic.* 54, 135–138.

Murrell, W. G., and Rankine, B. C. (1979) Isolation and identification of a sporing *Bacillus* from bottled brandy. *Am. J. Enol. Vitic.* 30, 247–249.

Nagel, C. W., Weller, K., and Filiatreau, D. (1988) Adjustment of high pH–high TA musts and wines. In: *Proceedings of the 2nd International Symposium for Cool Climate Viticulture and Oenology* (R. E. Smart *et al.*, eds.), pp. 222–224. New Zealand Society for Viticulture and Oenology, Auckland, New Zealand.

Neradt, F. (1984) Tartrate-stabilization methods. In: *Tartrates and Concentrates: Proceedings of the Eighth Wine Subject Day* (F. W. Beech, ed.), pp. 13–25. Long Ashton Research Station, University of Bristol, Long Ashton, UK.

Nishimura, K., Ohnishi, M., Masuda, M., Koga, K., and Matsuyama, R. (1983) Reactions of wood components during maturation. In: *Flavour of Distilled Beverages: Origin and Development* (J. R. Piggott, ed.), pp. 241–255. Ellis Horwood, Chichester, UK.

Nonier, M. F. (2003) The fruit of oak wood. *Demptos D'Clic* 21, 3–4.

Nykänen, L. (1986) Formation and occurrence of flavor compounds in wine and distilled alcoholic beverages. *Am. J. Enol. Vitic.* 37, 84–96.

Nykänen, L., Nykänen, I., and Moring, M. (1985) Aroma compounds dissolved from oak chips by alcohol. In: *Progress in Flavor Research – Proceeding of the Fourth Weurman Flavour Research Symposium* (J. Adda, ed.), pp. 339–346. Elsevier, Amsterdam.

O'Brien, E., and Dietrich, D. R. (2005) Ochratoxin A: The continuing enigma. *Crit. Rev. Toxicol.* 35, 33–60.

Oliveira, C. M., Silva Ferreira, A. C., Guedes de Pinho, P., and Hogg, T. A. (2002) Development of a potentiometric method to measure the resistance to oxidation of white wines and the antioxidant power of their constituents. *J. Agric Food Chem.* 50, 2121–2124.

Oszmianski, J., Cheynier, V., and Moutounet, M. (1996) Iron-catalysed oxidation of (+)-catechin in model systems. *J. Agric. Food Chem.* 44, 1712–1715.

Ough, C. S. (1985) Some effects of temperature and SO_2 on wine during stimulated transport or storage. *Am. J. Enol. Vitic.* 36, 18–22.

Palma, M., and Barroso, C. G. (2002) Application of a new analytical method to determine the susceptibility of wine to browning. *Eur. Food. Res. Technol.* 214, 441–443.

Parish, M., Wollan, D., and Paul, R. (2000) Micro-oxygenation – a review. *Aust. Grapegrower Winemaker* 438a, 47–50.

Peck, E. C. (1957) How wood shrinks and swells. *For. Prod. J.* 7, 234–244.

Pedretti, F., Barril, C., Clark, A., and Scollary, G. (2007) Yellow, red, brown, pink, orange and blue: Ascorbic acid and white wine. *Aust. NZ Grapegrower Winemaker* 519, 51–54.

Peleg, Y., Brown, R. C., Starcevich, P. W., and Asher, A. (1979) Methods for evaluating the filterability of wine and similar fluids. *Am. J. Enol. Vitic.* 30, 174–178.

Pellerin, P., Waters, E., Brillouet, J.-M., and Moutounet, M. (1994) Effet de polysaccharides sur la formation de trouble proteique dans un vin blanc. *J. Intl Sci. Vigne Vin* 28, 213–225.

Peng, Z., Duncan, B., Pocock, K. F., and Sefton, M.A. (1998) The effect of ascorbic acid on oxidative browning of white wine and model wines. *Aust. J. Grape Wine Res.* 4, 127–135.

Peng, Z., Iland, P. G., Oberholster, A., Sefton, M. A., and Waters, E. J. (2002) Analysis of pigmented polymers in red wine by reverse phase HPLC. *Aust. J. Grape Wine Res.* 8, 70–75.

Pereira, H., and Tomé, M. (2004) In: *Encyclopedia of Forest Science* (J. Burley, J. Evans and J. Youngquist, eds.), pp. 613–620. Elsevier, UK.

Pereira, H., Graça, J., and Baptista, C. (1992) The effect of growth rate on the structure and compressive properties of cork. *IAWA Bull.* 13, 389–396.

Pereira, H., Lopes, F., and Graça, J. (1996) The evaluation of the quality of cork planks by image analysis. *Holzforschung* 50, 111–115.

Pereira, H., Melo, B., and Pinto, R. (1994) Yield and quality in the production of cork stoppers. *Holz Roh- Werkstoff* 51, 301–308.

Pereira, H., Rosa, M. E., and Fortes, M. A. (1987) The cellular structure of cork from *Quercus suber* L. *IAWA Bull.* 8, 213–218.

Pérez-Magariño, S. and González-San José, M. L. (2002) Prediction of red and rosé wine CIELab parameters from simple absorbance measurements. *J. Sci. Food Agric.* 82, 1319–1324.

Pérez-Prieto, L. J., de la Hera-Orts, M. L., López-Roca, J. M., Fernández-Fernández, J. I., and Gómez-Plaza, E. (2003) Oak-matured wines: influence of the characteristics of the barrel on wine colour and sensory characteristics. *J. Sci. Food Agric.* **83**, 1445–1450.

Perin, J. (1977) Compte rendu de quelques essais de réfrigération des vins. *Vigneron Champenois* **98**, 97–101.

Peterson, R. G. (1976) Formation of reduced pressure in barrels during wine aging. *Am. J. Enol. Vitic.* **27**, 80–81.

Petrich, H. (1982) Untersuchungen über den Einfluß des Fungizids N-trichloromethylthio-Phthalimid auf Geruchs- und Geschmacksstoffe von Wienen (Dissertation). F. R. Institute für Lebensmittelchemie, University of Stuttgart, Germany *Vitis Abstr.* **23**, 1 M 13, 1984.

Pfeifer, W., and Diehl, W. (1995) Der Einfluß verschiedener Weinausbaumethoden und -behälter auf die Sauerstoffabbind efähigkeit von Weinen unter besonderer Berüchsichtigung der Barriqueweinbereitung. *Wein Wiss.* **50**, 77–86.

Pickering, G., Lin, J., Reynolds, A., Soleas, G., and Riesen, R. (2006) The evaluation of remedial treatments for wine affected by *Harmonia axyridis*. *Int. J. Food Sci. Technol.* **41**, 77–86.

Pickering, G., Lin, J., Riesen, R., Reynolds, A., Brindle, I., and Soleas, G. (2004) Influence of *Harmonia axyridis* on the sensory properties of white and red wine. *Am. J. Enol. Vitic.* **55**, 153–159.

Pocock, K. F., and Waters, E. J. (2006) Protein haze in bottled white wines: How well do stability tests and bentonite fining trials predict haze formation during storage and transport? *Aust J. Grape Wine Res.* **12**, 212–220.

Pocock, K. F., Alexander, G. M., Hayasaka, Y., Jones, P. R., and Waters, E. J. (2007) Sulfate – a candidate for the missing essential factor that is required for the formation of protein haze in white wine. *J. Agric. Food Chem.* **55**, 1799–1807.

Pocock, K. F., Hayasaka, Y., McCarthy, M. G., and Waters, E. J. (2000) Thaumatin-like proteins and chitinases, the haze-forming proteins of wine, accumulate during ripening of grape (*Vitis Vinifera*) berries and drough stress does not affect the final levels per berry at maturity. *J. Agric. Food Chem.* **48**, 1637–1643.

Pocock, K. F., Hayasaka, Y., Peng, Z., Williams, P. J., and Waters, E. J. (1998) The effect of mechanical harvesting and long-distance transport on the concentration of haze-forming proteins in grape juice. *Aust. J Grape Wine Res.* **4**, 23–29.

Pocock, K. F., Sefton, M. A., and Williams, P. J. (1994) Taste thresholds of phenolic extracts of French and American oakwood: The influence of oak phenols on wine flavor. *Am. J. Enol. Vitic.* **45**, 429–434.

Pocock, K. F., Strauss, C. R., and Somers, T. C. (1984) Ellagic acid deposition in white wines after bottling: A wood-derived instability. *Aust. Grapegrower Winemaker* **244**, 87.

Polak, A. (1975) *Glass: Its Tradition and its Makers.* Putnam, New York.

Pollnitz, A. P., Pardon, K. H., Liacopoulos, D., Skouroumounis, G. K., and Sefton, M. A. (1996) The analysis of 2,4,6-trichloranisole and other chloroanisoles in tainted wines and corks. *Aust. J. Grape Wine Res.* **2**, 184–190.

Pontallier, P., Salagoïty-Auguste, M., and Ribéreau-Gayon, P. (1982) Intervention du bois de chêne dans l'évolution des vins rouges élevés en barriques. *Connaiss. Vigne Vin* **16**, 45–61.

Postel, W. (1983) Solubilité et inhibition du crystallisation de tartrate de calcium dans le vin. *Bull. O.I.V.* **56**, 554–568.

Poulard, A., Leclanche, A., and Kollonkai, A. (1983) Dégradation de l'acide tartrique de moût par une nouvelle espèce: *Exophiala jeanselmei* var. *heteromorpha*. *Vignes Vins* **323**, 33–35.

Prescott, J., Norris, L., Kunst, M., and Kim, S. (2004) Estimating a consumer rejection threshold for cork taint in white wine. *Food Qual. Pref.* **16**, 345–349.

Prida, A., and Puech, J. L. (2006) Influence of geographical origin and botanical species on the content of extractives in American, French, and East European oak woods. *J. Agric. Food Chem.* **54**, 8115–8126.

Pripis-Nicolau, L., de Revel, G., Marchand, S., Beloqui, A. A., and Bertrand, A. (2001) Automated HPLC method for the measurement of free amino acids including cysteine in musts and wines; first applications. *J. Sci. Food Agric.* **81**, 731–738.

Puech, J.-L. (1984) Characteristics of oak wood and biochemical aspects of Armagnac aging. *Am. J. Enol. Vitic.* **35**, 77–81.

Puech, J.-L. (1987) Extraction of phenolic compounds from oak wood in model solutions and evolution of aromatic aldehydes in wines aged in oak barrels. *Am. J. Enol. Vitic.* **38**, 236–238.

Puech, J.-L., and Moutounet, M. (1992) Phenolic compounds in an ethanol-water extract of oak wood and in a brandy. *Lebensm. Wiss. Technol.* **251**, 350–352.

Puech, J.-L., and Moutounet, M. (1993) Barrels. In: *Encyclopedia of Food Science, Food Technology and Nutrition* (R. Macrae *et al.*, eds.), pp. 312–317. Academic Press, London.

Puerto, F. (1992) Traitment des bouchons: Lavage au péroxide contrôlé. *Rev. Oenol.* **64**, 21–26.

Quibell, J. E. (1913) The Tomb of Hesy, Pall Mall, London (reproduced as Figure 53, p. 90 in K. Kilby (1971) *The Cooper and His Trade.* John Baker, London).

Ramirez-Ramirez, G., Chassagne, D., Feuillat, M., Voilley, A., and Charpentier, C. (2004) Effect of wine constituents on aroma compounds sorption by oak wood in a model system. *Am. J. Enol. Vitic.* **55**, 22–26.

Rankine, B. C., and Pilone, D. A. (1973) *Saccharomyces bailii*, a resistant yeast causing serious spoilage of bottled table wine. *Am. J. Enol. Vitic.* **24**, 55–58.

Rapp, A. (1988) Wine aroma substances from gas chromatographic analysis. In: *Wine Analysis* (H. F. Linskens and J. F. Jackson, eds.), pp. 29–66. Springer-Verlag, Berlin.

Rapp, A., and Güntert, M. (1986) Changes in aroma substances during the storage of white wines in bottles. In: *The Shelf Life of Foods and Beverages* (G. Charalambous, ed.), pp. 141–167. Elsevier, Amsterdam.

Rapp, A., and Mandery, H. (1986) Wine aroma. *Experientia* **42**, 873–880.

Rapp, A., and Marais, J. (1993) The shelf life of wine: Changes in aroma substances during storage and ageing of white wines. In: *Shelf Life Studies of Foods and Beverages* (G. Charakanbous, ed.), pp. 891–921. Elsevier, Amsterdam.

Rauhut, D., Kürbel, H., and Dittrich, H. H. (1993) Sulfur compounds and their influence on wine quality. *Wein Wiss.* **48**, 214–218.

Rauhut, D., Shefford, P. G., Roll, C., Kürbel, H., Pour Nikfardjam, M., Loos, U., and Löhnertz, O. (2001) Effect of pre- and/or post-fermentation addition of antioxidants like ascorbic acid or glutathione on fermentation, formation of volatile sulfur compounds and other substances causing off-flavours in wine. In: *Proceeding of the 26th OIV Conference*, Adelaide, 2001, pp. 76–82.

Razmkhab, S., Lopez-Toledano, A., Ortega, J. M., Mayen, M., Merida, J., and Medina, M. (2002) Adsorption of phenolic compounds and browning products in white wines by yeasts and their cell walls. *J. Agric. Food Chem.* **50**, 7432–7437.

Remy, S., Flucrand, H., Labarbe, B., Cheynier, V., and Moutounet, M. (2000) First confirmation in red wine of products resulting from direct anthocyanin-tannin reactions. *J. Sci. Food Agric.* **80**, 745–751.

Ribeiro-Corréa, P., Ricardo-da-Silva, J. M., and Climaco, M. C. (1996) Influence of white wine filtration on flavor quality. In: *Proceedings of the 4th International Symposium on Cool Climate Viticulture and Enology* (T. Henick-Kling, T. E. Wolf, and E. M. Harkness, eds.), pp. VI-80–83. NY State Agricultural Experimental Station, Geneva, New York.

Ribéreau-Gayon, J. (1931) Contribution à l'étude des oxydations et réductions dans les vine. Thesis. Université de Bordeaux.

Ribéreau-Gayon, J., Peynaud, E., Ribéreau-Gayon, P., and Sudraud, P. (eds.) (1976) *Traité d'Oenologie. Sciences et Techniques du Vin*, Vol. 3. Dunod, Paris.

Ribéreau-Gayon, P. (1986) Shelf-life of wine. In: *Handbook of Food and Beverage Stability: Chemical, Biochemical, Microbiological and Nutritional Aspects* (G. Charalambous, ed.), pp. 745–772. Academic Press, Orlando, FL.

Riboulet, J. M., and Alegoët, C. (1986) *Aspects pratiques du bouchage des vins.* Collection Avenir Oenologie, Bourgogne, Chaintre, France.

Rieger, T. (1994) Spinning cone: A tool to remove alcohol and sulfites. *Vineyard Winery Management* 20(4), 29–31.

Rieger, T. (1996) Oak barrel alternatives – care and use of these products of the new age in oak aging. *Vineyard Winery Management* 22(1), 52, 55–56.

Rocha, S., Delgadillo, I., and Correia, A. J. F. (1996a) GC-MS study of volatiles of normal and microbiologically attacked cork from *Quercus suber.* L. *J. Agric. Food Chem.* 44, 865–871.

Rocha, S., Delgadillo, I., and Correia, A. J. F. (1996b) Improvement of the volatile components of cork from *Quercus suber* L. by an autoclaving procedure. *J. Agric. Food Chem.* 44, 872–876.

Rocha, S., Delgadillo, I., and Ferrer Correia, A. J. (1996c) Études des attaques microbiologiques du liège *Quercus suber* L. *Rev. Franç. Oenol.* 161, 31–34.

Rocha, S., Coimbra, M. S., and Delgadillo, I. (2000) Demonstration of pectic polysaccharides in cork cell wall from *Quercus suber* L. *J. Agric. Food Chem.* 48, 2003–2007.

Rodriguez-Clemente, R., and Correa-Gorospe, I. (1988) Structural, morphological and kinetic aspects of potassium hydrogen tartrate precipitation from wines and ethanolic solutions. *Am. J. Enol. Vitic.* 39, 169–179.

Rodríguez, M., Lezáun, J., Canals, R., Llaudy, M. C., Canals, J. M., and Zamora, F. (2005) Influence of the presence of the lees during oak ageing on colour and phenolic compounds composition of red wine. *Food Sci. Tech. Int.* 11, 289–295.

Romano, P., and Suzzi, G. (1993) Sulfur dioxide and wine organisms. In: *Wine: Microbiology and Technology* (G. H. Fleet, ed.), pp. 373–394. Taylor & Francis, London (2002 reprint).

Rosa, M. E., and Fortes, M. A. (1993) Water absorption by cork. *Wood Fiber Sci.* 25, 339–348.

Rosa, M. E., Pereira, H., and Fortes, M. A. (1990) Effects of hot water treatment on the structure and properties of cork. *Wood Fiber Sci.* 22, 149–164.

Rose, A. H., and Pilkington, B. J. (1989) Sulfite. In: *Mechanisms of Action of Food Preservation* (G. W. Gould, ed.), pp. 201–223. Elsevier Applied Science, London.

Rous, C., and Alderson, B. (1983) Phenolic extraction curves for white wine aged in French and American oak barrels. *Am. J. Enol. Vitic.* 34, 211–215.

Roussis, I. G., Lambroooulos, I., and Papadopoulou, D. (2005) Inhibition of the decline of volatile esters and terpenols during oxidative storage of Muscat-white and Xinomavro-red wine by caffeic acid and N-acetyl-cysteine. *Food Chem.* 93, 485–492.

Ruediger, G. A., Pardon, K. H., Asa, A. N., Godden, P. W., and Pollnitz, A. P. (2004) Removal of pesticides from red and white wine by the use of fining and filter agents. *Aust J. Grape Wine Res.* 10, 8–16.

Ruiz de Adana, M., Lopez, L. M., and Sala, J. M. (2005) A Fickian model for calculating wine losses from oak casks depending on conditions in ageing facilities. *Appl. Thermal Engin.* 25, 709–718.

Sauvageot, F., and Feuillat, F. (1999) The influence of oak wood (*Quercus robur* L., *Q. petraea* Liebl.) on the flavor of Burgundy

Pinot noir. An examination of variation among individual trees. *Am. J. Enol. Vitic.* 50, 447–455.

Schanderl, H. (1971) Korkgeschmack von Weinen. *Dtsch. Wein-Zeit.* 107, 333–336.

Schneider, R., Baumes, R., Bayonove, C., and Razungles, A. (1998) Volatile compounds involved in the aroma of sweet fortified wines (Vins Doux Naturels) from Granache noir. *J. Agric. Food Sci.* 46, 3230–3237.

Schneyder, J. (1965) Die Behebung der durch Schwefelwasserstoff und Merkaptane versursachten Geruchsfehler der Weine mit Silberchlorid. *Mitt. Rebe Wein, Ser. A (Klosterneuburg)* 15, 63–65.

Schreier, P., Drawert, F., Kerènyi, Z., and Junker, A. (1976) Gaschromatographisch-massenspektrometrische Untersuchung flüchtiger Inhaltsstoffe des Weines, VI. Aromastoffe in Tokajer Trockenbeerenauslese (Aszu)-Weinen a) Neutralstoffe. *Z. Lebensm. Unters. Forsch.* 161, 249–258.

Schulz, E. (2004) Bending regime influences oak component extraction. *Aust. NZ Grapegrower Winemaker* 486, 66–68.

Schütz, M., and Kunkee, R. E. (1977) Formation of hydrogen sulfide from elemental sulfur during fermentation by wine yeast. *Am. J. Enol. Vitic.* 28, 137–144.

Scott, J. A., and Cooke, D. E. (1995) Continuous gas (CO_2) stripping to remove volatiles from an alcoholic beverage. *J. Amer. Soc. Brew. Chem.* 53, 63–67.

Sefton, M. A., and Simpson, R. E. (2005) Compounds causing cork taint and the factors affecting their transfer from natural cork closures to wine – a review. *Aust. J. Grape Wine Res.* 11, 226–240.

Sefton, M. A., Francis, I. L., Pocock, K. F., and Williams, P. J. (1993) The influence of natural seasoning on the concentration of eugenol, vanillin and *cis* and *trans* β-methyl-γoctalactone extracted from French and American oakwood. *Sci. Alim.* 13, 629–644.

Sefton, M. A., Francis, I. L., and Williams, P. J. (1990) Volatile norisoprenoid compounds as constituents of oak woods used in wine and spirit maturation. *J. Agric. Food Chem.* 38, 2045–2049.

Segurel, M. A., Razungles, A. J., Riou, C., Salles, M., and Baumes, R. L. (2004) Contribution of dimethyl sulfide to the aroma of Syrah and Grenache noir wines and estimation of its potential in grapes of these varieties. *J. Agric. Food Chem.* 52, 7084–7093.

Shinohara, T., and Shimizu, J. (1981) Formation of ethyl esters of main organic acids during aging of wine and indications of aging. *Nippon Nôgeikagaku Kaishi* 55, 679–687.

Siebert, K. J., Carrasco, A., and Lynn, P. Y. (1996) Formation of protein-polyphenol haze in beverages. *J. Agric. Food Chem.* 44, 1997–2005.

Siebert, K. J., and Lynn, P. Y. (1998) Comparison of polyphenol interactions with polyvinylpolypyrrolidone and haze-active protein. *J. Am. Soc. Brew. Chem.* 56, 24–31.

Siegrist, J., Léglise, M., and Lelioux, J. (1983) Caractères analytiques secondaires de quelques vins atteints de la maladie de l'amertume. *Rev. Fr. Oenol.* 23, 47–48.

Silva, L. R., Cleenwerck, I., Rivas, R., Swings, J., Trujillo, M. E., Williams, A., and Velazquez, E. (2006) *Acetobacter oeni* sp. nov., isolated from spoiled red wine. *Int. J. Syst. Evol. Microbiol.* 56, 21–24.

Silva Pereira, C., Figueiredo Marques, J. J., and San Romão, M. V. (2000a) Cork taint in wine: Scientific knowledge and public perception – a critical review. *Crit. Rev. Microbiol.* 26, 147–162.

Silva Ferreira, A. C., Hogg, T., and Guedes de Pinho, P. (2003) Identification of key odorants related to the typical aroma of oxidation-spoiled white wines. *J. Agric. Food Chem.* 51, 1377–2381.

Silva Pereira, C., Pires, A., Valle, M. J., Boas, L., Marques, J. J. F., and San Romão, M. V. (2000b) Role of *Chrysonilia sitophila* in the quality of cork stoppers for sealing wine bottles. *J. Indust. Microbiol. Biotechnol.* 24, 256–261.

Simpson, R. F. (1977) Oxidative pinking in white wines. *Vitis* **16**, 286–294.

Simpson, R. F. (1990) Cork taint in wine: A review of the causes. *Aust. NZ Wine Indust. J.* **5**, 286–287, 289, 293–296.

Simpson, R. F., and Miller, G. C. (1983) Aroma composition of aged Riesling wines.*Vitis* **22**, 51–63.

Simpson, R. F., Amon, J. M., and Daw, A. J. (1986) Off-flavour in wine caused by guaiacol. *Food Technol. Aust.* **38**, 31–33.

Simpson, R. F., Capone, D. L., and Sefton, M. A. (2004) Isolation and identification of 2-methoxy-3,5-dimethylpyrazine, a potent musty compound from wine corks. *J. Agric. Food Chem.* **52**, 5425–5430.

Singleton, V. L. (1962) Aging of wines and other spirituous products, acceleration by physical treatments. *Hilgardia* **32**, 319–373.

Singleton, V. L. (1967) Adsorption of natural phenols from beer and wine. *Tech. Q. Master Brewers Assoc. Am.* **4**(4), 245–253.

Singleton, V. L. (1974) Some aspects of the wooden container as a factor in wine maturation. In: *Chemistry of Winemaking* (A. D. Webb, ed.), pp 254–278. Advances in Chemistry Series No. 137, American Chemical Society, Washington, DC.

Singleton, V. L. (1987) Oxygen with phenols and related reactions in musts, wines, and model systems: Observation and practical implications. *Am. J. Enol. Vitic.* **38**, 69–77.

Singleton, V. L. (1990) An overview of the integration of grape, fermentation, and aging flavours in wines. In: *Proceedings of the 7th Australian Wine Industry Technical Conference* (P. J. Williams *et al.*, eds.), pp 96–106. Winetitles, Adelaide, Australia.

Singleton, V. L. (1995) Maturation of wines and spirits, comparisons, facts, and hypotheses. *Am. J. Enol. Vitic.* **46**, 98–115.

Singleton, V. L., and Cilliers, J. J. L. (1995) Phenolic browning: A perspective from grape and wine research. In: *Enzymatic Browning and Its Prevention* (Y. L. Chang and J. R. Whitaker, eds.), pp. 23–48. ACS Symposium Series No. 600. American Chemical Society, Washington, DC.

Singleton, V. L., and Draper, D. E. (1961) Wood chips and wine treatment: The nature of aqueous alcohol extracts. *Am. J. Enol. Vitic.* **122**, 152–158.

Singleton, V. L., and Ough, C. S. (1962) Complexity of flavor and blending of wines. *J. Food Sci.* **12**, 189–196.

Singleton, V. L., Sullivan, A. R., and Kramer, C. (1971) An analysis of wine to indicate aging in wood or treatment with wood chips or tannic acid. *Am. J. Enol. Vitic.* **22**, 161–166.

Singleton, V. L., Trousdale, E., and Zaya, J. (1979) Oxidation of wines: l. Young white wines periodically exposed to air. *Am. J. Enol. Vitic.* **30**, 49–54.

Sitte, P. (1961) Zum Feinbau der Suberinschichten in Flaschenkork. *Protoplasma* **54**, 555–559.

Six, T., and Feigenbaum, A. (2003) Mechanism of migration from agglomerated cork stoppers. Part 2: Safety assessment criteria of agglomerated cork stoppers from champagne wine cork producers, for users and for control laboratories. *Food Additives Contam.* **20**, 960–971.

Skouroumounis, G. K., Kwiatkowski, M. J., Francis, I. L., Oakey, H., Capone, D. L., Duncan, B., Sefton, M. A., and Waters, E. J. (2005) The impact of closure type and storage conditions on the composition, colour and flavour properties of a Riesling and a wooded Chardonnay wine during five years' storage. *Aust. J. Grape Wine Res.* **11**, 369–384.

Skouroumounis, G. K., Kwiatkowski, M., Sefton, M. A., Gawel, R., and Waters, E. J. (2003) *In situ* measurement of white wine absorbance in clear and in coloured bottles using a modified laboratory spectrophotometer. *Aust J. Grape Wine Res.* **9**, 138–148.

Skurray, G., Castets, E., and Holland, B. (2000) Permeation of vanillin through natural and synthetic corks. *Aust. Grapegrower Winemaker* **438a**, 121–122, 124.

Smith, C. (1982) Enology Briefs Feb/March 1982, University of California, Davis.

Sneyd, T. N. (1988) Tin lead capsules. *Aust. Wine Res. Inst. Tech. Rev.* **56**, 1.

Sneyd, T. N., Leslie, P. A., and Dunsford, P. A. (1993) How much sulfur! In: *Proceedings of the 8th Australian Wine Industry Technical Conference* (C. S. Stockley *et al.*, eds.), pp. 161–166. Winetitles, Adelaide, Australia.

Soleas, G. J., Yan, J., Seaver, T., and Goldberg, D. M. (2002) Method for the gas chromatographic assay with mass selective detection of trichloro compounds in corks and wines applied to elucidate the potential cause of cork taint. *J. Agric. Food Chem.* **50**, 1032–1039.

Somers, T. C., and Evans, M. E. (1977) Spectral evaluation of young red wines, anthocyanin equilibria, total phenolics, free and molecular SO_2 chemical age. *J. Sci. Food. Agric.* **28**, 279–287.

Somers, T. C., and Evans, M. E. (1986) Evolution of red wines, I. Ambient influences on colour composition during early maturation.*Vitis* **25**, 31–39.

Somers, T. C., and Pocock, K. F. (1990) Evolution of red wines. III. Promotion of the maturation phase.*Vitis* **29**, 109–121.

Somers, T. C., and Ziemelis, G. (1985) Flavonol haze in white wines. *Vitis* **24**, 43–50.

Sousa, C., Mateus, N., Perez Alonso, J., Santos Buelga, C., and Freitas, V de (2005) Preliminary study of oaklins, a new class of brick-red catechinpyrylium pigments resulting from the reaction between catechin and wood aldehydes. *J. Agric. Food Chem.* **53**, 9249–9256.

Spedding, D. J., and Raut, P. (1982) The influence of dimethyl sulphide and carbon disulphide in the bouquet of wines.*Vitis* **21**, 240–246.

Spillman, P. J., Pocock, K. F., Gawel, R., and Sefton, M. A. (1996) The influences of oak, coopering heat and microbial activity on oak-derived wine aroma. In: *Proceedings of the 9th Australian Wine Industry Technical Conference* (C. S. Stockley *et al.*, eds.), pp. 66–71. Winetitles, Adelaide, Australia.

Spillman, P. J., Pollnitz, A. P., Liacopoulis, D., Skouroumounis, G. K., and Sefton, M. A. (1997) Accumulation of vanillin during barrel-aging of white, red, and model wines. *J. Agr. Food Chem.* **45**, 2584–2589.

Spillman, P. J., Iland, P. G., and Sefton, M. A. (1998) Accumulation of volatile oak compounds in a model wine stored in American and Limousin oak barrels. *Aust. J. Grape Wine Res.* **4**, 67–73.

Spillman, P. J., Pollnitz, A. P., Liacopoulos, D., Pardon, K. H., and Sefton, M. A. (1998) Formation and degradation of furfuryl alcohol, 5-methylfurfuryl alcohol, vanillyl alcohol, and their ethyl ethers in barrel-aged wines. *J. Agric. Food Chem.* **46**, 657–663.

Spillman, P. J., Sefton, M. A., and Gawel, R. (2004) The effect of oak wood source, location of seasoning and coopering on the composition of volatile compounds in oak-matured wines. *Aust J. Grape Wine Res.* **10**, 216–226.

Sponholz, W. R., and Hühn, T. (1996) Aging of wine: 1,1, 6-Trimethyl-1,2-dihycronaphthalene (TDN) and 2-aminoacetophenone. In: *Proceedings of the 4th International Symposium on Cool Climate Viticulture and Enology* (T. Henick-Kling, *et al.*, eds.), pp. VI-37-57. New York State Agricultural Experimental Station, Geneva, New York.

St John Hope, W. H., and Fox, G. E. (1898) Excavations on the site of the Roman city at Silchester, Hants, in 1897. *Archaeologia* **56**, 103–126.

Stelzer, T. (2003) Uncapping the truth about sulphides: Sulphide management for screw-capped wines. *Aust. NZ Grapegrower Winemaker* **476**, 105–106.

Stratiotis, A. L., and Dicks, L. M. T. (2002) Identification of *Lactobacillus* spp. isolated from different phases during the production of a South African fortified wine. *S. Afr. J. Enol. Vitic.* **23**, 14–21.

Strauss, C. R., Wilson, B., and Williams, P. J. (1985) Taints and off-flavours resulting from contamination of wines: A review of some investigations. *Aust. Grapegrower Winemaker* 256, 20, 22, 24.

Swan, J. S. (1986) Maturation of potable spirits. In: *Handbook of Food and Beverage Stability: Chemical, Biochemical, Microbiological and Nutritional Aspects* (G. Charalambous, ed.), pp. 801–833. Academic Press, Orlando, FL.

Swan, J. S., Newton, J., Larmie, E., and Sayre, R. (1997) Oak and Chardonnay. *Aust. Grapegrower Winemaker* 403, 41–43, 45–46, 48, 50.

Takács, L., Vatai, G., and Korány, K. (2007) Production of alcohol free wine by pervaporation. *J. Food Engin.* 78, 118–125.

Tattersall, D. B., Pocock, K. F., Hayasaka, Y., Adams, K., van Heeswijck, R., Waters, E. J., and Hoj, P. B. (2001) Pathogenesis related proteins – their accumulation in grapes during berry growth and their involvement in white wine heat instability. Current knowledge and future perspectives in relation to winemaking practices. In: *Molecular Biology and Biotechnology of the Grapevine* (K. A. Roubelakis-Angelakis, ed.), pp. 183–201. Kluwer Academic Publ., Netherlands.

Taylor, M. K., Young, T. M., Butzke, C. E., and Ebeler, S. E. (2000) Supercritical fluid extraction of 2,4,6-trichloroanisole from cork stoppers. *J. Agric. Food Chem.* 48, 2208–2211.

Tchernia, A. (1986) *Le Vin de l'Italie Romaine: Essai d'histoire economique d'après les amphores.* École Française de Rome, Rome.

Thomas, D. S., Boulton, R. B., Silacci, M. W., and Gubler, W. D. (1993) The effect of elemental sulfur, yeast strain, and fermentation medium on hydrogen sulfide production during fermentation. *Am. J. Enol. Vitic.* 44, 211–216.

Thomas, D. S., and Davenport, R. R. (1985) *Zygosaccharmyces bailii* – a profile of characteristics and spoilage activities. *Food Microbiol.* 2, 157–169.

Thompson, H. A. (1951) Excavations in the Athenian Agora. *Hesperia* 20, 50–1 (pl. 25a).

Tominaga, T., Baltenweck-Guyot, R., Peyrot des Gachons, C. and Dubourdieu, D. (2000) Contribution of volatile thiols to the aromas of white wines made from several *Vitis Vinifera* grape varieties. *Am. J. Enol. Vitic.* 51, 178–181.

Tominaga, T., Guimbertau, G., and Dubourdieu, D. (2003) Role of certain volatile thiols in the bouquet of aged champagne wines. *J. Agric. Food Chem.* 51, 1016–1020.

Towey, J. P., and Waterhouse, A. L. (1996) Barrel-to-barrel variations of volatile oak extractives in barrel-fermented Chardonnay. *Am. J. Enol. Vitic.* 47, 17–20.

Tyson, P. J., Luis, E. S., and Lee, T. H. (1982) Soluble protein levels in grapes and wine. In: *Grape and Wine Centennial Symposium Proceedings* (A. D. Webb, ed.), pp. 287–290. University of California, Davis.

Uitslag, H., Nguyen, M., and Skurray, G. (1996) Removal of tartrate from wine by electrodialysis. *Aust. Grapegrower Winemaker* 390a, 75–78.

Ulbert, G. (1959) Römische Holzfässer aus Regensburg. *Bayerische Vorgeschichtsblätter* 24, 6–29.

Van Buren, A. W. (1955) Newsletter from Rome. *Am. J. Archaeol.* 59, 303–314.

Vandiver, P., and Koehler, C. G. (1986) Structure, processing, properties, and style of Corinthian amphoras. In: *Ceramics and Civilization*, Vol. 2 (*The Technology and Style of Ceramics*) (W. D. Kingery, ed.), pp. 173–215. American Ceramics Society, Columbus, OH.

Vasserot, Y., Pitois, C., and Jeandet, P. (2001) Protective effect of a composite cork stopper on champagne wine pollution with 2,4,6-trichloroanisole. *Am. J. Enol. Vitic.* 52, 280–281.

Vaughn, R. H. (1955) Bacterial spoilage of wines with special reference to California conditions. *Adv. Food Res.* 7, 67–109.

Vernhet, A., Dupre, K., Boulange-Perermann, L., Cheynier, V., Pellerin, P., and Moutounet, M. (1999a) Composition of tartrate precipitates deposited on stainless steel tanks during the cold stabilization of wines. Part I. *Am. J. Enol. Vitic.* 50, 391–397.

Vernhet, A., Dupre, K., Boulange-Perermann, L., Cheynier, V., Pellerin, P., and Moutounet, M. (1999b) Composition of tartrate precipitates deposited on stainless steel tanks during the cold stabilization of wines. Part I. *Am. J. Enol. Vitic.* 50, 398–403.

Vernhet, A., and Moutounet, M. (2002) Fouling of organic microfiltration membranes by wine constituents: importance, relative impact of wine polysaccharides and polyphenols and incidence of membrane properties. *J. Membrane Sci.* **201**, 103–122.

Vidal, J.-C., Dufourcq, T., Boulet, J.-C., and Moutounet, M. (2001) Les apports d'oxygène au cours des traitments des vins. Bilan des observations sur site, 1ère partie. *Rev. Fr. Oenol.* 190, 24–31.

Vidal, J.-C., Toitot, C., Boulet, J.-Cl., and Moutounet, M. (2004) Comparison of methods for measuring oxygen in the headspace of a bottle of wine. *J. Int. Sci. Vigne Vin* 38, 191–200.

Vidal, S., Cartalade, D., Souquet, J. M., Fulcrand, H., and Cheynier, V. (2002) Changes in proanthocyanidin chain length in wine-like model solutions. *J. Agric. Food Chem.* 50, 2261–2266.

Vidal, S., Francis, L., Williams, P., Kwitkowski, M., Gawel, R., Cheynier, B., and Waters, E. (2004) The mouth-feel properties of polysaccharides and anthocyanins in a wine-like medium. *Food Chem.* 85, 519–525.

Vieira Natividade, J. V. (1956) *Subericulture.* Éditions de l'École nationale du Génie rural des Eaux et Forêts, Nancy, France.

Villettaz, J. C., Steiner, D., and Trogus, H. (1984) The use of a beta glucanase as an enzyme in wine clarification and filtration. *Am. J. Enol. Vitic.* 35, 253–256.

Vivas, N. (2001) Pratiques et recommandations sur la préparation, la mise en service et la conservation de fûts neufs et usagés. *Rev. Oenologues* 28(91), 24–29.

Vivas, N., and Glories, Y. (1996) Role of oak wood ellagitannins in the oxidation process of red wines during aging. *Am. J. Enol. Vitic.* 47, 103–107.

Vivas, N., Amrani-Joutei, K., Glories, Y., Doneche, B., and Brechenmacher, C. (1997) Développement de microorganismes dans le bois de coeur de chêne (*Quercus petraea* Liebl) au cours du séchage naturel à l'air libre. *Ann. Sci. For.* 54, 563–571.

Vivas, N., Debèda, H., Ménil, F., Vivas de Gaulejac, N., and Nonier, M.-F. (2003) Mise en évidence du passage de l'oxygène au travers des douelles consitiuant les barriques par l'utilisation d'un dispositif original de mesure de la porosité du bois. Premieres résultats. *Sci. Alim.* 23, 655–678.

Vivas, N., Glories, Y., Bourgeois, G., and Vitry, C. (1996) The heartwood ellagitannins of different oak (*Quercus* sp.) and chestnut species (*Castanea sativa* Mill.): Quality analysis of red wines aging in barrels. *J. Sci. Tech. Tonnellerie* 2, 51–75.

Vivas, N., Lonvaud-Funel, A., and Glories, Y. (1995) Observations concerning the increase of volatile acidity in red wines whilst ageing in barrels. *J. Sci. Tech. Tonnellerie* 1, 103–122.

Vivas, N., Nonier, M.-F., de Gaulejac, N. V., de Boissel, I. P. (2004) Occurrence and partial characterization of polymeric ellagitannins in *Quercus petraea* Liebl. and *Q. robur* L. wood. *C. R. Chimie* 7, 945–954.

Voilley, A., Lamer, C., Dubois, P., and Feuillat, M. (1990) Influence of macromolecules and treatments on the behavior of aroma compounds in a model wine. *J. Agric. Food Chem.* 38, 248–251.

Wagner, M. S., Jakob, L., Rapp, A., and Niebergall, H. (1994) Untersuchungen zur Migrations aus Tanks aus glasfaserverstärktem Kunststoff in Wein. *Deut. Libens-Rundschau* 90, 218–222.

Wainwright, T. (1970) Hydrogen sulphide production by yeast under conditions of methionine, pantothenate or vitamine B$_6$ deficiency. *J. Gen. Microbiol.* 61, 107–119.

Waters, E. J., and Williams, P. J. (1997) The role of corks in the random oxidation of bottled wines. *Wine Indust. J.* 12, 189–193.

Waters, E. J., Peng, Z., Pocock, K. F., and Williams, P. J. (1996a) Lacquer-like bottle deposits in red wine. In: *Proceedings of the 9th Australian Wine Industry Technical Conference* (C. S. Stockley *et al.*, eds.), pp. 30–32. Winetitles, Adelaide, Australia.

Waters, E. J., Peng, Z., Pocock, K. F., and Williams, P. J. (1996b) The role of corks in oxidative spoilage of white wine. *Aust. J. Grape Wine Res.* 2, 191–197.

Waters, E. J., Shirley, N. J., and Williams, P. J. (1996c) Nuisance proteins of wine are grape pathogenesis-related proteins. *J. Agric. Food Chem.* 44, 3–5.

Waters, E. J., Wallace, W., Tate, M. E., and Williams, P. J. (1993) Isolation and partial characterization of a natural haze protective factor from wine. *J. Agric. Food Chem.* 41, 724–730.

Webb, M. (1987) Cardboard box quality, finish and glues. In: *Proceedings of the 6th Australian Wine Industry Technical Conference* (T. Lee, ed.), pp. 277–280. Australian Industrial Publishers, Adelaide, Australia.

Weetal, H. H., Zelko, J. T., and Bailey, L. F. (1984) A new method for the stabilization of white wine. *Am. J. Enol. Vitic.* 35, 212–215.

Weiss, K. C., Lange, L. W., and Bisson, L. F. (2001) Small-scale fining trials: effect of method of addition on efficiency of bentonite fining. *Am. J. Enol. Vitic.* 52, 275–279.

Weyl, W. A. (1951) *Coloured Glasses.* Society of Glass Technology, Sheffield.

Wightman, J. D., Price, S. F., Watson, B. T., and Wrolstad, R. E. (1997) Some effects of processing enzymes on anthocyanins and phenolics in Pinot noir and Cabernet Sauvignon wines. *Am. J. Enol. Vitic.* 48, 39–48.

Wilker, K. L., and Dharmadhikari, M. R. (1997) Treatment of barrels infected with acetic acid bacteria. *Am. J. Enol. Vitic.* 48, 516–520.

Wilkinson, K. L., Elsey, G. M., Prager, A. P., Pollnitz, A. P., and Sefton, M. S. (2004) Rates of formation of *cis-* and *trans-* oak lactone from 3-methyl-4-hydroxyoctanoic acid. *J. Agric. Food Chem.* 52, 4213–4218.

Worlidge, J. (1676) *Vinetum Britannicum: or a Treatise of Cider, and other Wines and Drinks extracted from Fruits growing in this Kingdom*, pp. 107–108. T. Dring, London.

Wucherpfennig, K., Millies, K. D., and Christmann, M. (1986) Herstellung entalkoholisierter Weine unter besonderer Berüchsichtigung des Dialyseverfahrens. *Weinwirtsch. Tech.* 122, 346–354.

Wucherpfennig, K., Otto, K., and Wittenschläger, L. (1984) Zur Möglichkeit des Einsatzes von Pektin- und Alginsäure zur Entfernung von Erdalkali- und Schwermetallionen aus Weinen. *Wein Wiss.* 39, 132–139.

Zambonelli, C., Soli, M. G., and Guerra, D. (1984) A study of H$_2$S non-producing strains of wine yeasts. *Ann. Microbiol.* 34, 7–15.

Zimman, A., and Waterhouse, A. L. (2004) Incorporation of malvidin-3-glucoside into high molecular weight polyphenols during fermentation and wine aging. *Am. J. Enol. Vitic.* 55, 139–146.

Zoecklein, B. W., Fugelsang, K. C., Gump, B. H., and Nury, F. S. (1995) *Wine Analysis and Production.* Chapman & Hall, New York.

Zoecklein, B. W., Marcy, J. E., and Jasinski, Y. (1997) Effect of fermentation, storage *sur lie* or post-fermentation thermal processing on White Riesling (*Vitis Vinifera* L.) glycoconjugates. *Am. J. Enol. Vitic.* 48, 397–402.

9

Specific and Distinctive Wine Styles

Wine has been produced for millennia, but many modern styles have no ancient equivalent. Wine styles often reflect the unique climatic and politico-economic environment in which they arose. For example, botrytized wines emerged in regions favoring the selective development of noble-rot, sparkling wine evolved in a region (Champagne) unsuitable for standard red wine production, and port arose out of expanded trade between England and Portugal, due to conflicts and trade restrictions with France. Some of these wine styles have spread throughout the world. Others have remained local anomalies. This chapter covers some of the more important and unique wine styles.

Sweet Table Wines

Sweet table wines encompass a wide diversity of styles possessing little in common other than sweetness. They may be white, rosé, or red, and may range from aromatically simple to complexly fragrant. The most famous are those made from noble-rotted (*Botrytis*-infected) grapes.

Botrytized Wines

Wines unintentionally made from grapes infected by *Botrytis cinerea* undoubtedly have been made for centuries. The fungus is omnipresent. It normally produces a destructive bunch rot wherever climatic conditions permit. Wine made from bunch-rotted grapes are unpalatable. Nevertheless, under unique climatic conditions, infected grapes develop what is called 'noble' rot. These grapes beget the most seraphic of white wines.

When noble-rotted grapes were first intentionally used for wine production is unknown. Historical evidence favors the Tokaj region of Hungary, in the mid-sixteenth century (ca. 1560). The production of botrytized wine in Germany is reported to have begun at Schloss Johannisberg (ca. 1750). When the deliberate production of botrytized wines began in France is uncertain, but production appears to have been well established in Sauternes between 1830 and 1850. Botrytized wines are produced throughout much of Europe, wherever conditions are favorable to noble rot. The idea of using *Botrytis*-infected grapes for wine production has been slow to catch on in the New World, but botrytized wine is now produced to a limited degree in Australia, Canada, New Zealand, South Africa, and the United States.

INFECTION

Early in the season, most infections develop from spores produced on overwintered fungal tissue. The initial inoculum probably develops from mycelia on previously diseased tissues, or from resting structures called sclerotia. Infections begin on aborted and senescing flower parts, notably the stamens and petals. Infected flower parts located in the developing cluster may initiate fruit infection later during the season. Although these infections usually cease as hyphae attempt to enter the young green fruit, the mycelium remains viable. Reactivation of latent fruit infections can be important under dry autumn conditions. As the fruit reaches maturity, resistance to fungal growth declines. Under moist conditions, new infections, incited by spores from external sources, are more important than latent infections.

Disease susceptibility and its direction (bunch *vs.* noble rot) depend on several factors. Skin toughness and open fruit clusters reduce disease incidence, whereas heavy rains, protracted moist periods, and shallow rooting increase susceptibility. Conditions such as shallow rooting and high humidity induce berry splitting and favor bunch-rot development. In contrast, noble rot develops late in the season under cyclical conditions of fluctuating humidity (humid nights followed by dry, sunny days). Such conditions permit infection, but limit fungal growth. Fungal metabolism is also modified.

Figure 9.1 Grape-like cluster of *Botrytis cinerea* spores (conidia) produced on the spore-bearing structure, the conidiophore. (Photo courtesy of Dr D. H. Lorenz, Neustadt, Germany)

Depending on the temperature and humidity, new spores (conidia) are produced by fungal hyphae within a few days. The spores are borne on elongated, branched filaments called conidiophores. The shape of the spore clusters (Fig. 9.1) so resembles grape clusters that the scientific name of the fungus comes from Greek – βοτσυς, meaning grape cluster. Because the microclimate of the fruit cluster markedly affects fungal development, various stages of healthy, noble-, and bunch-rotted grapes frequently occur in the same cluster (Plate 9.1).

On infection, several hydrolytic enzymes are released by *Botrytis*. Particularly destructive are the pectolytic enzymes. They degrade the pectinaceous component of the cell wall. The enzymes also cause the collapse and death of affected tissue. With loss of physiological control, the fruit dehydrates under dry conditions. Because the vascular connections with the vine become disrupted as the fruit reaches maturity, the moisture lost is not replaced. Additional water may be lost via evaporation from the conidiophores.

The loss of moisture appears to be the crucial factor in determining the direction of disease progression. Drying retards berry invasion and appears to modify the metabolism of *B. cinerea*. Drying also concentrates the juice, a feature crucial in the development of the wine's sensory properties. Finally, drying limits secondary invasion by bacteria and fungi. Invasion by fungi, including *Penicillium*, *Aspergillus*, and *Mucor*, probably generates most of the moldy off-odors and tastes associated with bunch-rot. For example, *P. frequentans* produces plastic and moldy odors, via the synthesis of styrene (Jouret *et al.*, 1972) and 1-octen-3-ol (Kaminiski *et al.*, 1974), respectively. Saprophytic fungi are commonly present during the development of *Botrytis* bunch-rot. The unpleasant phenolic flavors commonly associated with wines produced from bunch-rotted grapes (Boidron, 1978) is rarely observed in noble-rot wines. This may

Table 9.1 Comparison of juice from healthy and *Botrytis*-infected grapes

Component	'Sauvignon' berries		'Sémillon' berries	
	Healthy	Infected	Healthy	Infected
Fresh weight/100 berries (g)	225	112	202	98
Sugar content (g/liter)	281	326	247	317
Acidity (g/liter)	5.4	5.5	6.0	5.5
Tartaric acid (g/liter)	5.2	1.9	5.3	2.5
Malid acid (g/liter)	4.9	7.4	5.4	7.8
Citric acid (g/liter)	0.3	0.5	0.26	0.34
Gluconic acid (g/liter)	0	1.2	0	2.1
Ammonia (mg/liter)	49	7	165	25
pH	3.4	3.5	3.3	3.6

Source: After Charpentié (1954), from Ribéreau-Gayon *et al.*, 1980, reproduced by permission

result from the low phenol content of the skins and from pressing being conducted without prior crushing.

One of the typical chemical indicators of *Botrytis* infection is the presence of gluconic acid. Although *B. cinerea* produces gluconic acid, the acetic acid bacterium *Gluconobacter oxydans* is particularly active in gluconate synthesis. Because acetic acid bacteria frequently invade grapes infected by *Botrytis*, they are probably responsible for most of the gluconic acid found in diseased grapes (Sponholz and Dittrich, 1985). Acetic acid bacteria also produce acetic acid and ethyl acetate. Thus, they are the most likely source of the elevated concentrations of these compounds in some botrytized wines.

CHEMICAL CHANGES INDUCED BY NOBLE ROTTING

Berry dehydration, combined with the metabolic action of *Botrytis*, are the principal causes of the sensorial effects of noble rotting. Some of the effects of drying simply have concentrating effects, such as the increase in citric acid content (Table 9.1). With other compounds, fungal metabolism is sufficiently active to result in a decrease in concentration, despite the concentrating effect of water loss. This is particularly noticeable with tartaric acid and ammonia (Table 9.1). The selective metabolism of tartaric acid, versus malic acid, is crucial in avoiding a marked drop in pH. In addition, the enhanced acidity, associated with the increased malic acid content, counteracts the cloying effect of the wine's high residual sugar content.

One of the most notable changes during noble rotting is an increase in sugar concentration. This occurs in spite of the metabolism of up to 35–45% of the total sugar

content. The decrease in osmotic potential that results may explain the inability of *B. cinerea* to metabolize a higher proportion of the sugars during infection (Sudraud, 1981). Occasionally, selective glucose metabolism is reflected in an atypically high fructose/glucose ratio.

The production and accumulation of glycerol during infection potentially augment the smooth mouth-feel of botrytized wines. This potential may be enhanced by the simultaneous synthesis of other polyols, such as arabitol, erythritol, mannitol, *myo*-inositol, sorbitol, and xylitol.

A distinctive feature of noble rotting is a loss in varietal aroma. This is particularly noticeable with 'Muscat' cultivars. Aroma loss is explained largely by the destruction of terpenes that give these varieties their distinctive fragrance. Examples are the metabolism of linalool, geraniol, and nerol to less volatile compounds, such as β-pinene, α-terpineol and various furan and pyran oxides (Bock *et al.*, 1986, 1988). These may produce the phenolic and iodine-like odors reported in some botrytized wines (Boidron, 1978). *Botrytis* also produces esterases that can degrade fruit esters that give many white wines their fruity character (Dubourdieu *et al.*, 1983). The significance of these effects depends on their relative importance to wine fragrance. 'Muscat' varieties often lose more character than they gain from *Botrytis*, whereas 'Riesling' and 'Sémillon' generally gain more aromatic complexity than they lose in varietal distinctiveness. Another potential flavorant, seemingly destroyed by *B. cinerea*, is 1,1,6-trimethyl-1,2-dihydronaphthalene (TDN) (Sponholz and Hühn, 1996). At subthreshold levels, it contributes to an aged bouquet, but at above about 20 μg/liter, it can donate a kerosene-like odor.

In addition to destroying certain aromatic compounds, fungal metabolism synthesizes others. The most significant of *Botrytis* flavorants appears to be sotolon. Sotolon has a sweet fragrance and contributes to the characteristic aroma of botrytized wines. In combination with other aromatic compounds found or produced in botrytized wines, sotolon helps give wine a honey-like fragrance (Masuda *et al.*, 1984). Infected grapes also contain the mushroom alcohol, 1-octen-3-ol. Among additional compounds are derivatives generated from grape terpenes. More than 20 terpene derivatives have been isolated from infected grapes (Bock *et al.*, 1985). Additional compounds typical of botrytized wines have been identified by Sarrazin *et al.*, 2007.

Infection by *B. cinerea* not only affects the wine's taste and fragrance, but it also influences the ease of grape picking, modifies yeast and bacterial activity in the juice, and can generate filtration problems. Hydrolytic enzymes from the fungus disrupt the integrity of the grape skin, causing it to split more easily during rainy spells or rupture during harvesting. Surprisingly, infection retards separation of the berry from the pedicel during ripening (Fregoni *et al.*, 1986).

Infection by *Botrytis* may affect the activity of other microorganisms on and in grapes, during fermentation, and in the wine. In the vineyard, infection facilitates secondary invasion by acetic acid bacteria and several saprophytic fungi. Infection also affects the epiphytic yeast flora, both increasing cell numbers and modifying species composition. For example, *Candida stellata*, *Torulaspora delbrueckii*, and *Saccharomyces bayanus* may dominate the yeast flora of infected grapes. In Tokaj, *Candida pulcherrima* was the dominant yeast on botrytized grapes. This shifts to *C. stellata* after collection, transport to, and storage in the winery (Bene and Magyar, 2004). The dependence of *T. delbrueckii* on oxygen availability may partially explain its early demise during fermentation (Mauricio *et al.*, 1991), whereas the high population of *Candida* is probably associated with its marked preferential metabolism of fructose (Mills *et al.*, 2002). In addition, molecular techniques have detected the presence of nonculturable populations of *Candida* and *Hanseniaspora* at the end of fermentation (Mills *et al.*, 2002). *S. bayanus* appears to be particularly able to withstand the high sugar concentration and low nitrogen, sterol, and thiamine conditions found in botrytized juice.

Because thiamine deficiency increases the synthesis of sulfur-binding compounds during fermentation, small amounts of thiamine (0.5 mg/liter) may be added to the juice before fermentation (Dittrich *et al.*, 1975). This avoids increasing the SO_2-binding potential of the wine.

There is considerable variation in the influence of different *Botrytis* strains on yeast growth. Suppression may result from the release of toxic fatty acids by fungal esterases. In addition, some *Botrytis* strains produce compounds that can either stimulate (Minárik *et al.*, 1986) or suppress (Dubourdieu, 1981) yeast metabolism.

Little is known about the specific effects of noble rotting on malolactic fermentation. The presence of high residual sugar and glycerol contents would presumably favor its development. However, the addition of up to 200 or 250 mg/liter SO_2, designed to retain a noticeable residual sugar content, precludes the occurrence of malolactic fermentation. Sulfur dioxide is also required to prevent undesirable microbial activity during maturation and after bottling. Although the addition of sulfur dioxide rapidly reduces the culturable yeast population, a variable portion enters a viable, but nonculturable state (Divol and Lonvaud-Funel, 2005). Depending on the conditions, these cells may reinitiate growth, causing spoilage.

Laccases are one of the most significant enzymes produced by *B. cinerea*. Their function during infection is uncertain, but laccases are suspected to inactivate antifungal phenols, such as pterostilbene and resveratrol in grapes (Pezet *et al.*, 1991). Different laccases are induced by grape juice, gallic acid, and *p*-coumaric acid. In addition, pectin may augment laccase synthesis in the presence of phenolic compounds (Marbach *et al.*, 1985). In wine, laccases oxidize a wide range of important grape phenolics, for example: *p*-, *o*-, and some *m*-diphenols; diquinones; anthocyanins; tannins; and a few nonphenolics such as ascorbic acid. This may partially explain the comparatively high phenolic content of botrytized wines and the atypical absence of hydroxycinnamates. In addition, the oxidation of 2-*S*-glutathionylcaftaric acid may generate much of the golden coloration of botrytized wine (see Macheix *et al.*, 1991).

Unlike grape polyphenol oxidase, laccase is active at wine pH values, and in the presence of typical levels of sulfur dioxide. Concentrations of about 50 mg/liter SO_2 at pH 3.4 are required to inhibit the action of laccase in wine (about 125 mg/liter in must) (Kovač, 1979). In contrast, hydrogen sulfide can completely inhibit laccase activity at contents as low as 1–2.5 mg/liter (added as a sulfide salt) (see Macheix *et al.*, 1991).

The activity and stability of laccase in must and wine have serious consequences for red wines. Because laccase rapidly and irreversibly oxidizes anthocyanins, even low levels of infection can result in considerable color loss and browning. Unlike grape polyphenol oxidases, fungal laccases are little inhibited by oxidized phenolic compounds (Dubernet, 1974). In contrast, the gold color produced from oxidized phenolics in white grapes is considered a positive quality feature. Color development depends largely on the grape variety and the level of infection.

During infection, *Botrytis* synthesizes a series of high-molecular-weight polysaccharides. These form two distinct groups. One group consists primarily of polymers of mannose and galactose, with a small quantity of glucose and rhamnose. They vary between 20,000 and 50,000 Da, and induce increased production of acetic acid and glycerol during fermentation (Dubourdieu, 1981). The other group consists of β-glucans, polymers of glucose ranging from 100,000 to 1,000,000 Da. They have little, if any, affect on yeast metabolism, but form strand-like lineocolloids in the presence of alcohol. They can cause serious plugging problems during clarification. As little as 2–3 mg/liter can seriously retard filtration (Wucherpfennig, 1985). If they contaminate the wine, the addition of β-glucanases can degrade these gel-like compounds. To minimize the release of β-glucans, botrytized grapes are usually harvested manually and pressed slowly. Because the glucans are located predominantly just under the skin, in association with the fungal cells that produce them, gentle harvesting and pressing minimize their dispersion into the released juice.

Botrytis also may induce a form of calcium salt instability. The fungus produces an enzyme that oxidizes galacturonic acid (a breakdown product of pectin) to mucic (galactaric) acid. Mucic acid slowly binds with calcium, forming an insoluble salt. This may produce the sediment that occasionally forms in bottles of botrytized wines.

In addition to the direct and indirect effects of *Botrytis* infection, changes result from the action of secondary invaders, such as the synthesis of gluconic acid by acetic acid bacteria. Although an indicator of *Botrytis* infection, gluconic acid has no known sensory significance. The sweetness of its intramolecular cyclic esters (γ- and δ-gluconolactone) is apparently too slight to be perceptible. However, these lactones, combined with two other bacterial by-products, notably 5-oxofructone and dihydrooxyacetone, constitute the principal SO_2-binding compounds in botrytized must and wine (Barbe *et al.*, 2002). Hydroxypropanedial, another chemical indicator of *Botrytis* (and other fungal) infections (Guillou *et al.*, 1997), can further reduce the antioxidative and antimicrobial effects of sulfur dioxide.

As typical of one of the finest of wines available, botrytized wines cannot be produced inexpensively. Dehydration results in a marked loss in juice volume, there are considerable risks in leaving grapes on the vine to overmature, and fermentation and clarification can be difficult. Nevertheless, their production is one of the crowning achievements of winemaking skill.

TYPES OF BOTRYTIZED WINES

Tokaji Aszú   As noted previously, botrytized wines evolved independently in several European regions. Tokaji may have been the first deliberately produced botrytized wine. Its most famous version, Aszú Eszencia, is derived from juice that spontaneously seeps out of highly botrytized (*aszú*) berries placed in small tubs. Approximately 1–1.5 liters of *eszencia* is derived from about 30 liters of *aszú* grapes. After several weeks, the collected *eszencia* is placed in small wooden barrels for fermentation and maturation (~10°C). Because of the very high sugar content of the juice (occasionally more than 50%), fermentation occurs slowly and often reaches little more than 5–7% alcohol before termination. After fermentation ceases, the bungs are usually left slightly ajar. Oxygen uptake has been thought to be restricted by the growth of a common cellar mold (*Racodium cellare*) on the wine's surface (Sullivan, 1981). However, any velum development is more likely to be of yeast origin. Velum development is no longer favored, except for certain dry Szamorodni styles (Atkin, 2001). After a variable period, the barrels are bunged tight. Up to an additional 20 years of barrel maturation may ensue before bottling.

Table 9.2   Chemical composition of Tokaji wines

Quality grade	Total extract (g/liter)	Extract residue (g/liter)	Sugar content (g/liter)	Ethanol content (%, v/v)
Two *puttonyos*	55	25	30	14
Three *puttonyos*	90	30	60	14
Four *puttonyos*	125	35	90	13
Five *puttonyos*	160	40	120	12
Six *puttonyos*	195	45	150	12
Eszencia	300	50	250	10

Source: After Farkaš (1988), reproduced by permission

In contrast, most Tokaj wines are made from mixing young white wine, or juice derived from healthy grapes, with the juice and pomace ("paste") from crushed *aszú* berries. Legislation concerning the procedure was passed in 1655 (Asvany, 1987). The *aszú* paste may be made from grapes before or after the *eszencia* has drained away. Various categories of *auzú* are produced, based on the amount of *aszú* added to the mixture (3–6 *puttony* – a *puttony* equaling about 28–30 liters). The mixture is placed in open vats for 24–36 h, followed by pressing and transfer to 136-liter barrels (*Gönci*) for fermentation and maturation. The barrels are left partially empty, and the bung left loose for 1–3 months at cool temperatures. This donates an oxidized aspect not found in other botrytized wines. This is attested to by the concentrations of acetaldehyde, acetals, and acetoin in the wine (see Schreier, 1979). The wines are only lightly sulfited. Fermentation continues slowly and may take several weeks or months to finish. The alcohol content may reach as high as 14% (Table 9.2). Occasionally, a distillate made from Tokaj wine may be added to prematurely terminate fermentation to achieve a sweeter finish. The volume lost by evaporation during aging may be replaced with high-strength wine spirits. The wines may be pasteurized before bottling to prevent subsequent fermentation. The principal yeast involved is *Saccharomyces bayanus*. *Candida stellata* and *C. zemplinina* may also be active (Sipiczki, 2003).

German Botrytized Wines   German botrytized wines come in a variety of categories. The basic characteristics are indicated by its Prädikat designation. *Auslesen* wines are derived from specially selected, whole clusters of late-harvested fruit. *Beerenauslesen* (BA) and *trockenbeerenauslesen* (TBA) wines are derived, respectively, from individual berries or dried berries selected from clusters of late-harvested fruit. Although the fruit is typically botrytized, this is not mandatory (Anonymous, 1979). Thus, wines in these categories may or may not

Table 9.3 Composition of some 1971 *Beerenauslesen* and *Trockenbeerenauslesen* wines

Wine type[a]	Total extract (g/liter)	Sugar content (g/liter)	Alcohol content (%, w/v)	Total acidity (g/liter)	Glycerol content (g/liter)	Acetaldehyde content (mg/liter)	Tannin content (mg/liter)	pH
BA	163	74	7.9	8.7	12.0	73	250	3.2
BA	152	103	6.3	9.4	13.6	62	390	3.2
BA	119	78	7.9	7.9	10.9	139	390	3.0
TBA	299	224	5.3	11.4	13.0	56	291	3.6
TBA	303	194	6.4	10.5	40.0	163	446	3.5

[a]BA, *beerenauslese*; TBA, *trockenbeerenauslese*.

Source: Data from Watanabe and Shimazu (1976)

come from botrytized grapes. Each category also is characterized by an increasingly high minimum sugar content in the grapes (see Table 10.1).

BA and TBA juices typically contain more sugar than is converted to alcohol during fermentation. The wines are consequently sweet and low in alcoholic strength, commonly 6–8% (Table 9.3). Auslesen wines may be fermented dry or may retain residual sweetness, depending on the desires of the winemaker.

The other main Prädikat wine categories, namely, *Kabinett* and *Spätlese*, may be derived from botrytized juice, but seldom are. Their sweetness is usually derived from the addition of *süssreserve* – unfermented or partially fermented juice kept aside for sweetening. *Süssreserve* is added just before bottling to the dry wine produced from the majority of the harvest. Various techniques have been used to restrict microbial growth in the *süssreserve*. These include high doses of sulfur dioxide, storage at temperatures near or below freezing, and the application of high carbon dioxide pressures.

Alternatively, sugars may be retained in the fermenting juice by prematurely stopping fermentation. This may be achieved by filtering out the yeasts, by allowing CO_2 pressure to rise in reinforced fermentors, or from the addition of sulfur dioxide. The latter is typical in Sauternes (see below) (Divol *et al.*, 2006).

Because the residual sugar content makes the wine microbially unstable, stringent measures must be taken to avoid microbial spoilage. Sterile filtration of the wine into sterile bottles, sealed with sterile corks, is common. Sterile bottling has supplanted the previous use of high sulfiting during bottling.

French Botrytized Wines In France, the best-known botrytized wines are produced in Sauternes. Over a period of several weeks, noble-rotted grapes may be selectively harvested from clusters in the vineyard. Because of the cost of multiple harvesting, most producers harvest only once and separate the botrytized berries

from the bunches, similar to the procedure followed in Germany. Uninfected grapes are used in the production of dry wine.

Typically, only one sweet style is produced in Sauternes, in contrast to the gradation of botrytized styles produced in Germany. A major stylistic difference between French and German botrytized wines is the alcohol level achieved. French styles commonly exceed 11–13% alcohol, whereas German versions seldom exceed 10%. The other main sensory differences come from the varieties used in their production. For example, the use of 'Sauvignon blanc' and 'Sémillon' explains the presence of a wide range of thiols in Sauternes (Bailly *et al.*, 2006).

Occasionally, sauternes is spoilt by yeast reactivation (Divol *et al.*, 2006). These are strains that went temporarily into a dormant, nonculturable state. They possess characteristics similar to *flor* yeasts: being highly resistant to toxicity of sulfite, acetaldehyde, ethanol, and high concentrations of sugar.

Sweet botrytized wines are produced, in more or less the similar way, in other regions of France. The main locations are Alsace and the Loire Valley.

VARIETIES USED

Many grape varieties are used in the production of botrytized wines. Their appropriateness depends on several factors. Essentially all are white varieties, thus avoiding the brown coloration produced by anthocyanin oxidation. Most varieties mature late, thus, the time of ripening is synchronized to the development of weather conditions suitable for noble-rot development. The cultivars are also relatively thick-skinned. Because of the tissue softening induced by *Botrytis*, harvesting infected varieties with soft skins would be very difficult. Thick-skinned varieties are also less susceptible to splitting and bunch-rot.

'Riesling' and 'Sémillon' are the primary cultivars used in the production of botrytized wines. Both varieties seem ideally suited for the generation of botrytized wines.

The Hungarian variety, 'Furmint,' is the predominant variety used in the production of Tokaji Aszú. Other varieties, such as 'Picolit,' 'Gewürztraminer,' 'Chenin blanc,' and 'Pinot blanc,' are used, depending on tradition and adaptation to local conditions.

INDUCED BOTRYTIZATION

The production of botrytized wines is both a risky and an expensive procedure. Leaving mature grapes on the vine increases the likelihood of bird damage, bunch-rot, and other fruit losses. These dangers may be partially diminished by successive harvesting, but labor costs limit its use to only the most expensive estate-bottled wines. In Germany, most of the crop is usually harvested to produce a nonbotrytized wine. Only a variable portion is left on the vine for noble-rot development or *eiswein* production.

Where climatic conditions are unfavorable for noble-rot development, harvested grapes have occasionally been exposed to conditions that favor its development (Nelson and Amerine, 1957). The fruit is first sprayed with a solution of *B. cinerea* spores. The fruit is subsequently placed on trays and held at about 90–100% humidity for 24–36h at 20–25°C. These conditions permit spore germination and fruit penetration. Subsequently, cool dry air is passed over the fruit to induce partial dehydration and restrict invasion. After 10–14 days, infection has developed sufficiently that the fruit can be pressed and the juice fermented. Induced botrytization has been used successfully, but only on a limited scale in California and Australia.

The inoculation of juice with spores or mycelia of *B. cinerea*, followed by aeration, apparently induces many of the desirable sensory changes that occur in vineyard infections (Watanabe and Shimazu, 1976). Whether the artificial nature of production might limit consumer acceptance is unknown.

Nonbotrytized Sweet Wine

Sweet nonbotrytized wines are produced in most, if not all wine-producing regions. Most have evolved slowly into their present-day forms. On the other hand, several modern versions have developed quickly in response to perceived consumer preferences. Coolers are a prime example.

DRYING

Drying is probably the oldest and simplest procedure used in producing sweet wines. It involves placing fruit on mats or trays in the sun, or shade, to dehydrate. After several weeks or months, the grapes are crushed and the concentrated juice fermented. Variations on the procedure are prevalent throughout southern Europe.

Drying not only results in juice concentration, but also affects grape metabolism. Most of the changes, such as increases in the concentration of abscisic acid (Costantini et al., 2006), may be of little enologic significance. However, activation of lipoxygenases can increase the concentration of hexanal, hex-1-enol and hex-2-enal, whereas alcohol dehydrogenase activation promotes the synthesis of ethanol and acetaldehyde. Additional metabolic changes may augment higher alcohol contents (Bellincontro et al., 2004). Franco et al. (2004) have found that sun-drying is associated with enhanced concentrations of isobutanol, benzyl alcohol, 2-phenyl ethanol, 5-methylfurfural, γ-butyrolactone and γ-hexalactone. Investigation of the sensory consequences of drying is still in its infancy.

Variations in style often depend on the grape variety or varieties used, and the treatments applied before, during, or after fermentation. For example, 'Moscato' grapes in Sicily may be cured in a solution of saltwater and volcanic ash (at least in the past), before being crushed and fermented. For *vin santo*, the juice is placed in small barrels and stored in attics for fermentation (see Stella, 1981). The wines are subsequently aged under fluctuating seasonal temperature extremes for from 2 to 6 years. Whether the progressive dominance of *Metschnikowia pulcherrima* in the grape flora during drying is of significance is unknown (Balloni et al., 1989).

HEATING

Another ancient technique entails concentrating the juice or semisweet wine by gentle heating or boiling. The treatment results in a loss in varietal character, but generates a caramelized or baked odor. The use of heat in the production of madeira is described later in this chapter.

ICEWINE

Juice concentration, as a consequence of freezing, provides the basis for producing icewine (*eiswein*). *Eiswein* is reported to have been first produced in Germany in the late 1700s. For icewine production, grapes are left on the vine until winter temperatures fall to or below −7 to −8°C (Plate 9.2). At this range, most of the water in the fruit freezes, concentrating the remaining liquid. Harvesting and pressing (usually outside) occurs while the grapes are still frozen (Plates 9.3 and 9.4). As a consequence, the concentrate escapes with minimal ice thawing. The ice remains in the press with the seeds and skins. Pressing can take upward of 12h. The sugar concentration is typically so high (often above 35°Brix) that fermentation occurs very slowly and stops prematurely, leaving the wine with a high residual sugar content. Incomplete fermentation probably results from the stresses of the high sugar content, combined with the accumulation of ethanol, acetic acid, and toxic C_8 and C_{10} carboxylic acids. One of the adaptations yeasts make to these unfavorable conditions involves aldehyde dehydrogenase (ALD3) (Pigeau and Inglis, 2005). The enzyme converts acetaldehyde to acetic acid. Acetic acid accumulates partially due to downregulation of enzymes converting acetic acid to acetyl CoA, and its use in fatty acid synthesis. The energy so derived appears to be used to augment glycerol synthesis. Glycerol helps provide osmotolerance by limiting water loss from the cytoplasm.

Because of the difficult fermentation conditions, yeasts are often acclimated before being added to the must (Kontkanen *et al.*, 2004). However, this practice is not universal. Kontkanen *et al.* (2005) have shown that acclimation and nutrient supplementation can significantly modify the flavor characteristics of the wine. Thus, whether, and to what degree, these procedures may be used can give the wine a distinctive character, varying from raisiny, buttery and spicy, to peach/terpene-like, to honey and orange, or pineapple/alcoholic. Choice of yeast strain has also been found to significantly affect the accumulation of acetic acid, glycerol, reduced-sulfur odors, and color (Erasmus *et al.*, 2004). About 0.5 g/liter active dry yeast is normally required to achieve the typical 10% alcohol content.

Prolonged overripening may result in a net loss of some aromatic compounds, but concentration appears to more than make up for the loss. Cold weather conditions result in an absolute loss in acidity (potassium tartrate crystallization), but juice concentration due to ice formation provides sufficient acidity to balance the high residual sugar content of the wine (frequently > 12.5%). The golden color of icewines probably results from the joint effects of juice concentration, caftaric acid oxidation, and the release of catechins on freezing.

Although *eiswein* has been made for at least two centuries in Germany, the worldwide popularity of icewines is comparatively recent. This is correlated with its production in southern Ontario, Canada, and neighboring portions of the United States. The climatic conditions in these regions seem particularly favorable to fine icewine production. Despite its new-found fame, there is a surprising lack of technical information on the processes involved. Thus, little is known about the chemical changes that occur during overripening, pressing, and fermentation, or the microbiology of its production.

One of the major problems connected with icewine production involves protecting the fruit from birds and other predators during the prolonged overmaturation period. Additional problems include molding during rainy spells, and the unpleasant conditions of harvesting and pressing. From 20–60% of the fruit may be lost before sufficient freezing occurs. Difficulties in achieving adequate fermentation are another frequent complication.

In regions where climatic conditions do not permit natural icewine production, cryoextraction can provide its technological equivalent. This has the advantage of permitting the degree of juice concentration to be selected in advance (Chauvet *et al.*, 1986). It also avoids the risks of leaving the fruit on the vine for months, and the difficulties of harvesting and crushing grapes during frigid winter weather. Reverse osmosis has also been used to produce concentrated juices for making icewine-like wines. These techniques, however, seem to lack the flavor changes that develop during long vineyard overripening. They also lack the mystique, often so important in marketing.

ADDITION OF JUICE CONCENTRATE (SWEET RESERVE)

The addition of unfermented grape juice (sweet reserve or *süssreserve*) to dry wine is the most widespread technique for producing sweet wines. It was first perfected in Germany. It has the advantages of losing neither varietal distinctiveness nor producing additional flavors. If the juice is derived from the same grapes used in making the wine, varietal, vintage, and appellation of origin are not compromised. It also may augment varietal distinctiveness, which is occasionally lost during fermentation. Furthermore, the procedure is technologically simpler and more easily controlled than most other sweetening processes.

If storage is short, refrigeration at or just below 0 °C is adequate without sulfiting. However, prolonged storage at low temperatures requires the addition of sulfur dioxide (100 mg/liter), or cross-flow microfiltration to restrict yeast growth. Several spoilage yeasts are known to grow at near 0 °C. Another option is to treat the juice with heavy sulfiting (~1000 mg/liter). Prior to use, the sulfur dioxide is reduced to about 100 mg/liter by flash-heating with steam followed by rapid cooling, heating with nitrogen sparging, or with a spinning cone apparatus.

Red Wine Styles

Recioto-style Wines

In contrast to white wines, few red wines are produced with a sweet finish. Italy is probably the major producer of sweet red wines, usually with moderate petillance. Even fewer wines are made from red grapes infected with *Botrytis cinerea*. Nevertheless, the most famous red wine from Veneto is often made from grapes partially infected with *B. cinerea* (Usseglio-Tomasset *et al.*, 1980). These are the *recioto* (*appassimento*) wines from Valpolicella.

Recioto Valpolicella is made from a blending of musts from 'Corvina,' 'Molinara,' and 'Rondinella' grapes. Similar wines are made in Lombardy from 'Nebbiolo' or 'Groppello.'

Recioto wines develop much of their distinctive fragrance and flavor from the processing of the fruit prior to fermentation. The fragrance contains elements that resemble the sharp phenolic odor of tulip and daffodil blossoms. However, this attribute is limited if the grapes are unaffected by *Botrytis*. The relative occurrence of *B. cinerea* probably depends on conditions during flowering. These significantly affect the incidence of nascent grape infections.

Recioto wines typically have a high alcohol content and a smoother, more harmonious taste than the majority of full-bodied red wines. The wine may be made sweet (*amabile*), dry (*amarone*), or sparkling (*spumante*). Because of the unique fragrance typical of most of *recioto* wines, the process supplies winemakers with an additional means of producing wines with a distinctive character.

PRODUCTION OF *RECIOTO* WINES

Healthy, fully mature clusters, or their most mature portions, are placed in a single layer on trays designed to ease air flow around the fruit. The trays are stacked in rows several meters high, in well-ventilated storage areas or warehouses (Plate 9.5). Natural ventilation may be augmented by fans to keep the relative humidity below 90%. The grapes are left to partially dry for several months under cool ambient temperatures. The fruit is usually turned every few weeks to promote uniform dehydration. Cool temperatures (3–12°C), and humidity levels below 90% are crucial to restricting microbial spoilage.

During the 3–4 months of storage period, the physical and chemical characteristics of the grapes undergo major changes. The most obvious is the 25–40% drop in moisture content. Other significant changes appear to accrue from the action of *B. cinerea*. Fungal growth likely originates from latent fruit infections acquired in the spring during flowering. Under the dry, cool, storage conditions, the fungus reinitiates slow development. Although the color of the grapes changes from bluish purple to pale red, obvious sporulation seldom occurs – unless the humidity becomes high.

The percentage of fruit showing infection usually increases in relation to the duration of storage, the actual percentage depending on the vintage and variety. 'Corvina' seems to be the most susceptible, with up to 45% of the fruit showing infection after 4 months. In contrast, 'Rondinella' may show less than 5% infection after the same period (Usseglio-Tomasset *et al.*, 1980).

As infection progresses, the grapes become flaccid, and the skin loses its strength. The visual and mechanical manifestations of infection are reflected in even more marked chemical alterations (Table 9.4). These changes show the distinctive effects of noble rotting. Although fungal metabolism reduces the sugar content, the greater evaporative water loss results in a marked increase in sugar concentration. The °Brix can rise from 25 to over 40° in heavily noble-rotted grapes. Due to the selective metabolism of glucose, the relative proportion of fructose rises. The grapes also show a marked (10- to 20-fold) increase in gluconic acid and glycerol concentration. Total acidity declines marginally, if at all. Despite dehydration, the tartaric acid concentration remains relatively constant, whereas that of malic acid declines. These data may indicate that selective acid metabolism by *B. cinerea* during storage differs from that occurring in the vineyard. Alternatively, the atypically low malic acid content may reflect the metabolic action of the grapes during prolonged storage.

Browning and loss of color are noticeable, but are less marked than might be expected, considering the infection by *B. cinerea*. This may result from suppression of laccase activity by the high sugar content of the grapes (Doneche, 1991) or skin tannins (Marquette *et al.*, 2003), poorly understood factors limiting laccase activity (Guerzoni *et al.*, 1979), or the resistance of some varieties to infection. However, the most resistant variety, 'Rondinella,' usually constitutes only about 25–30% of the Valpolicella blend.

B. cinerea probably generates the distinctive fragrance of *recioto*-style wines. The fungus has been noted to be

Table 9.4 Changes in must composition during the drying of grapes for *recioto* wines

	Duration of drying (days)				
	0	19	40	73	101
Sugar (g/liter)	181	188	193	204	230
Glucose (g/liter)	92	90	91	95	103
Fructose (g/liter)	89	98	102	109	128
Total extract (g/liter)	205	216	228	232	261
Total acidity (g/liter)	6.9	7.5	6.4	6.5	6.6
pH	3.1	3.0	3.1	3.1	3.2
Volatile acidity (mg/liter)	50	90	90	100	120
Tartaric acid (g/liter)	7.3	7.4	7.7	5.9	7.1
Malic acid (g/liter)	5.7	5.2	3.9	4.4	3.8
Glycerol (g/liter)	0.2	0.7	0.8	2.9	5.1
Gluconic acid (g/liter)	0.2	0.7	1.1	1.0	1.7

Source: Data from Usseglio-Tomasset *et al.* (1980)

associated with the production of a marked oxidized phenolic flavor in red grapes (Boubals, 1982).

Following storage, the grapes are stemmed, crushed, and allowed to ferment under the action of indigenous yeasts. When a noticeable residual sweetness is desired, the must is kept cool (≤12 °C) throughout alcoholic fermentation. For the dry *amarone* style, the must is warmed to, or allowed to rise to, approximately 20 °C during fermentation. The fermentation temperature influences the relative action of different endemic yeast strains. *Saccharomyces uvarum* predominates under cool temperatures, leaving the wine with a perceptibly sweet finish. In contrast, *S. bayanus* metabolizes the fermentable sugars at warmer temperatures, and favors the production of the dry *amarone* style. Fermentation typically lasts approximately 40 days.

Alcoholic fermentation may recommence when ambient temperatures rise in the spring. This may be either prevented by yeast removal (filtration for the *amabile* style) or encouraged (for production of *spumante*-style wines).

Figure 9.2 shows that the concentrations of glycerol, 2,3-butanediol, and gluconic acid are roughly comparable to those expected for highly botrytized wines. The high levels of both glycerol and alcohol contribute to the smooth texture of the wine. The smooth sensation is undoubtedly aided by the limited extraction of tannins during cool fermentation temperatures. Cool temperatures also encourage the production or retention of fragrant esters, augmenting the bouquet of *recioto* wines.

Whether malolactic fermentation plays a significant role in the development of *recioto* wines is unclear. Published data on the chemical composition of *recioto* wines do not suggest its occurrence. Nevertheless, it is reported to be an important factor, even though it may not go to completion (Zapparoli *et al.*, 2003). With inoculation, rates of conversion of malic acid to lactic acid were in the range of 30–60%. It is likely to occur more during the refermentation of sweet *recioto* wines in the spring and summer. Malolactic fermentation may be the source of the carbon dioxide that occasionally gives the *amabile* style a slight effervescence.

Amarone-style wines commonly are aged in oak for several years prior to bottling. This is considered to improve the wine's fragrance and harmony. Old cooperage is preferred, to avoid giving the wine a marked oaky character.

Carbonic Maceration Wines

In its simplest form, carbonic maceration may be as old as winemaking itself. Its involvement was probably widespread until the introduction of mechanical crushers in the nineteenth century. Mechanical crushers permitted complete, rather than partial fruit crushing (treading underfoot). It also improved pigment and tannin extraction. With the advent of mechanical crushers, stemming and crushing became dominant, and considered traditional. The decline in the full or partial use of carbonic maceration also correlates with the increased use of bottles for transport and prolonged aging. Nevertheless, from descriptions of how wines were made at estates, such as Château Lafite, partial carbonic maceration may have still been common during the early 1800s (Henderson, 1824; Cocks, 1846). In only a few regions, notably Beaujolais, has full carbonic maceration remained the dominant technique for red wine production.

Grapes have long been known to metabolize malic acid, especially during ripening under warm conditions. Although carbonic maceration also favors malic acid decarboxylation, the process is not specifically used for deacidification. Its primary advantage has been in the production of early maturing wines. In addition, development of a unique and distinctly fruity aroma resulted in the technique finding a new following. Nevertheless, their image as light quaffable wines has led many critics to malign carbonic maceration wines.

Because of the technique's limited and regional use, carbonic maceration has garnered little attention outside southern France and parts of Italy. With the popularity of *nouveau*-type wines, interest in the technique expanded. The Institut National de la Recherche Agronomique in Montfavet, France has been the primary center for studies on carbonic maceration (Flanzy *et al.*, 1987). The Institute coined the term "carbonic maceration," by which the technique is now known.

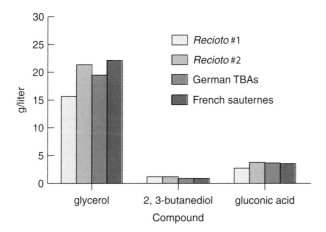

Figure 9.2 Comparison between two *recioto* wines and the averages of three botrytized wines, each from Germany and France for glycerol, 2,3-butanediol, and gluconic acid. (Data for *recioto* wines from Usseglio-Tomasset *et al.*, 1980; those for TBAs and sauternes from Yunome *et al.*, 1981)

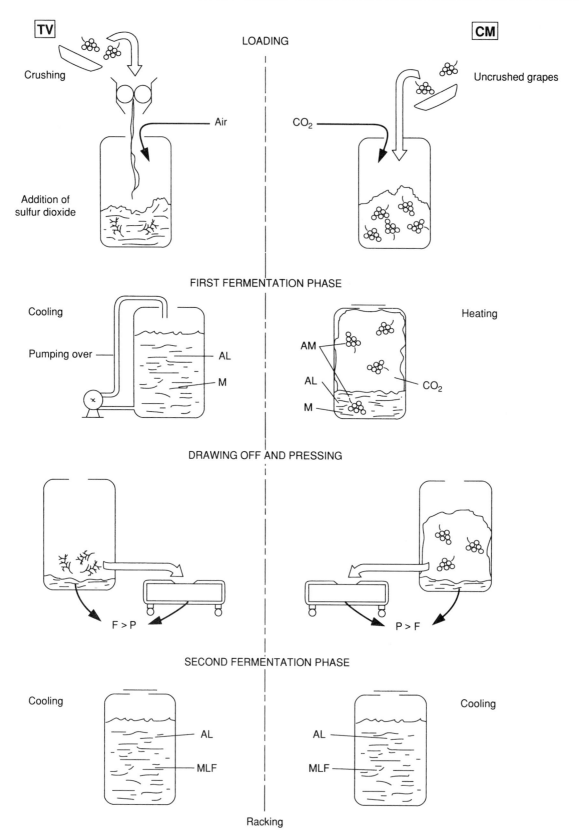

Figure 9.3 Schematic comparison between carbonic maceration (CM) and traditional vinification (TV) with crushing before fermentation. AL, alcoholic fermentation; AM, anaerobic metabolism; F, free-run; M, maceration; MLF, malolactic fermentation; P, press run. (After Flanzy and André, 1973, reproduced by permission)

Figure 9.3 compares carbonic maceration with traditional vinification. Carbonic maceration differs fundamentally from standard vinification in that berries undergo a self-fermentation before yeast and malolactic fermentation. For optimal carbonic maceration, it is essential that the fruit be harvested with minimal breakage.

Carbonic maceration was initially proposed to refer to the anaerobic maceration of whole berries placed in an atmosphere of carbon dioxide. This has subsequently been expanded to include grape-cell alcoholic fermentation, whether or not air is initially removed from the fermentor by flushing with carbon dioxide.

Typically, carbonic maceration takes place in the presence of a small amount of must, released when a portion of the fruit ruptures during loading of the fermentor. This is the case in Beaujolais, France. Thus, berry fermentation typically occurs in association with limited yeast fermentation.

If the fruit is not dumped into a fermentor, anaerobic maceration occurs in the absence of free juice. For example, containers of grapes may be placed in sealed chambers (Fig. 9.4), or wrapped in plastic for carbonic maceration. Pigment extraction during this period is poor. Thus, the juice is left to ferment in contact with the seeds and skins after crushing, until the desired color has been achieved. If, however, deep coloration is not critical, sufficient berries may break open during loading or collapse during maceration to provide adequate pigmentation.

Typically, the whole harvest undergoes carbonic maceration. In some regions, however, the technique may be used for only part of the crop. This may be employed when partial masking a varietal aroma is desired (Fuleki, 1974). If more than 85% of the fruit undergoes carbonic maceration, the 'Cabernet' character of Bordeaux wine is suppressed (Martinière, 1981). Thus, modifying the proportion of the crop left uncrushed can regulate the relative contribution of carbonic maceration vs. the varietal character of the wine. The intensity of the carbonic-maceration aroma may also be modified by adjusting the duration and temperature of the process. Alternatively, the wine may be blended with must vinified by standard procedures.

At the end of carbonic maceration, the grapes are pressed and the juice is allowed to ferment to dryness by yeast action. Malolactic fermentation typically occurs shortly after the termination of alcoholic fermentation. After the completion of malolactic fermentation, the wine typically receives a light dosing with sulfur dioxide (20–50 mg/liter). This helps prevent further microbial action. Racking commonly occurs at the same time. Racking may be delayed for several weeks, however, because some winemakers believe that substances released by yeast autolysis enhance wine flavor.

Although used most extensively in the production of light, fruity red wines, carbonic maceration can yield wines capable of long aging. The procedure has also been used to a limited extent in producing rosé and white wines.

ADVANTAGES AND DISADVANTAGES

Carbonic maceration generates a wine with a unique, fruity aroma. Whether this is due to the liberation of terpenoid aromatics from the grapes (Bitteur *et al.*, 1996; Salinas *et al.*, 1998) is still unclear. The distinctive fragrance produced has been described variously as possessing a kirsch, cherry, or raspberry aspects. Additional descriptors are given in Table 9.5.

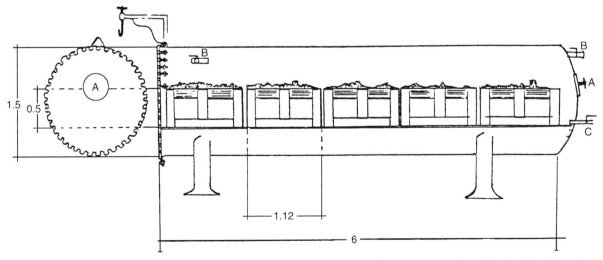

Figure 9.4 Special chamber for conducting carbonic maceration: A, door; B, valve; C, entrance for carbon dioxide. Dimensions are in meters. (From Càstino and Ubigli, 1984, reproduced by permission)

Table 9.5 Sample descriptors used to describe certain carbonic maceration wines

	Full carbonic maceration	Semicarbonic maceration (Beaujolais)
Visual appearance	Ruby red	Ruby red
Fragrance	Kirsch	Hyacinth
	Coffee	Coffee
	English candy	English candy
	Vanilla	Vanilla
	Grilled almonds	Cherry
	Russian leather	Banana
	Resin	Raspberry
Quality	Fine (predominantly vegetal and lactic)	Rich (predominantly winy and phenolic)
Taste	Subtle	Rough
	Buttery	Tannic

Source: After Flanzy *et al.* © INRA, Paris (1987), reproduced by permission

For relatively neutral varieties, such as 'Aramon,' 'Carignan,' and 'Gamay,' carbonic maceration gives an appealing fruitiness that standard vinification does not provide. With some cultivars, such as 'Cabernet Sauvignon,' 'Merlot,' and 'Concord,' carbonic maceration appears to mask the varietal aroma. This may be desirable or not, depending on the appeal of the aroma. With other cultivars, such as 'Syrah' and 'Maréchal Foch,' carbonic maceration has been reported to enhance the complexity of the varietal fragrance. The observed reduction in the herbaceous character of several French-American hybrids may result from the curtailed production of hexan- and hexen-ols (Salinas *et al.*, 1998). Carbonic maceration has even been reported to enhance the varietal aroma of some white grapes (Bénard *et al.*, 1971).

Because standard vinification extracts more tannins than carbonic maceration (Pellegrini *et al.*, 2000), the process may be preferable when used with highly tannic grapes. The potential for deacidification might also partially justify its use with acidic grapes.

Carbonic maceration wines seldom demonstrate a yeasty bouquet after completing vinification. The wine, which also has a smoother taste, can thus be enjoyed sooner. This has financial benefits, because the wines can be bottled and sold within a few weeks of production. Thus, capital is not tied up for years in cellar stock.

Customarily, the carbonic maceration aroma does not improve on aging and fades relatively quickly. Unless a varietal aroma or aged bouquet replaces the fading carbonic maceration fragrance, the wine commonly has a shelf-life of little more than 6 months to 1 year. Carbonic maceration does not in itself limit shelf-life, however. The aging potential depends primarily on the quality and

properties of grapes fermented. Consequently, some carbonic maceration wines show long aging potential, notably those from northern Beaujolais, the Rhône Valley, and Rioja. Prolonged maceration after carbonic maceration favors the extraction of sufficient aromatics, anthocyanin, and tannins to give the wine aging potential.

In the past, the comparative simplicity of carbonic maceration supplemented the benefits of early drinkability. It required neither destemming nor grape treading. Whole grape clusters could simply be loaded into wide shallow vats. Crushing and pressing were easier, because the grapes became weak and flaccid during carbonic maceration.

Some of these advantages are still relevant, for example, early drinkability and the easier pressing of pulpy grapes. However, other aspects of carbonic maceration are incompatible with present-day harvesting and winemaking. In most situations, mechanical harvesting is preferred, due to its cost effectiveness. However, for carbonic maceration, it would rupture more berries than is desirable. In addition, most fermentation tanks are designed for loading with must, not whole grape clusters. Loading into standard fermentors and the cumulative berry mass would result in excessive fruit rupture.

Care must be taken at all stages, from harvesting to loading, to minimize fruit rupture. Broad, shallow (~2.5 m) vats are preferred. They reduce berry rupture as well as ease loading. Vats permit the ready displacement of air with carbon dioxide at the beginning of maceration. Subsequently, the vat opening is covered to restrict air access, while permitting the carbon dioxide generated during berry autofermentation to escape.

In some regions, notably Beaujolais, the juice that slowly accumulates during carbonic maceration is periodically pumped over the fruit. Although frequently practiced, pumping over is not recommended. The procedure increases oxidation and may cause undesirably high concentrations ($\geq$150 mg/liter) of ethyl acetate (Descout, 1986). By temporarily removing the buoyant action of the juice, pumping over induces excessive and premature fruit rupture. The ethanol produced by yeast action in the liberated juice also favors breakage, by weakening and killing skin cells. Rupture curtails both the duration and proportion of the fruit undergoing carbonic maceration.

One of the more serious drawbacks of carbonic maceration is the high demand it places on fermentor capacity. Because the fruit is neither stemmed nor crushed, they displace considerably more volume than must derived from the same amount of fruit. Furthermore, the initial grape-cell fermentation significantly prolongs the fermentation period (Fig. 9.5). Although malolactic fermentation typically commences shortly after yeast fermentation, this does not offset the need for increased fermentor capacity at this critical time during the year.

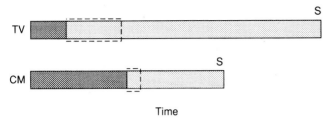

Figure 9.5 Representative duration of vinification by carbonic maceration (CM) and traditional procedures (TV). Dark-shaded bars represent the first fermentation phase, yeast-induced alcoholic fermentation in TV and anaerobic berry metabolism plus alcoholic fermentation of the free-run juice in CM. Light-shaded bars represent the second fermentation phase, the end of alcoholic fermentation and malolactic fermentation. Dashed-line boxes represent the period during which malolactic fermentation occurs. S, biological stability. (After Flanzy *et al.*, 1987, reproduced by permission)

The problem of fermentor capacity can be side-stepped with the use of specially designed storage containers (Fig. 9.4), or by adequately wrapping the fruit in plastic (Rankine *et al.*, 1985). This is easiest when the grapes are left in the same containers in which they were harvested. Either procedure avoids berry rupture, inevitable during vat loading.

PHASE I – CARBONIC MACERATION

Whole-grape Fermentation In the absence of oxygen, grape cells switch from respiratory to fermentative metabolism. This shift is more rapid if air is flushed out with carbon dioxide. Because it is more dense than air, carbon dioxide displaces air in the vat. Carbon dioxide is customarily preferred to nitrogen because of its uniquely desirable properties. CO_2 is readily dissolved by cytoplasm, directly inducing ion leakage from cells (Yurgalevitch and Janes, 1988). It also shifts the equilibria of cellular decarboxylation reactions (Isenberg, 1978). Furthermore, carbon dioxide may accelerate the breakdown of pectins, by inducing the synthesis of grape pectinases.

Alcoholic fermentation in grape cells is similar to that found in yeasts and most other cells. The primary end product is ethanol, with smaller accumulations of glycerol, acetaldehyde, acetic acid, and succinic acid.

Although alcohol dehydrogenase induces the reduction of acetaldehyde to ethanol, its inactivation by ethanol accumulation is insufficient to explain the limited alcoholic fermentation of grape cells (Molina *et al.*, 1986). Instead, enzyme activity probably ceases due to ethanol-induced membrane disruption and cell death (Romieu *et al.*, 1989). The latter would result in the release of organic acids stored in cell vacuoles, lowering the cytoplasmic pH and inhibiting alcohol dehydrogenase activity. Grape ethanol synthesis during carbonic maceration seldom rises above 2%.

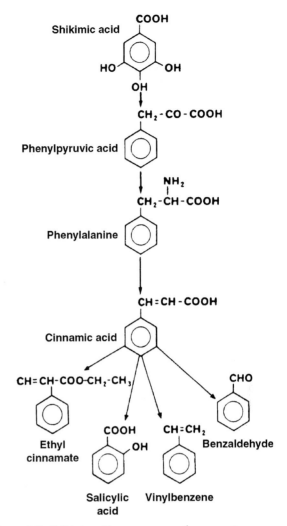

Figure 9.6 Shikimic acid as a precursor for aromatic compounds. (After Flanzy *et al.*, 1987, reproduced by permission)

During grape-cell fermentation, malic acid is metabolized to other acids (primarily oxaloacetic, pyruvic, and succinic acids), as well as ethanol. Depending on the grape variety, and fermentation temperature (Flanzy *et al.*, 1987), upward of 15–60% of the malic acid may be metabolized during carbonic maceration. Significant decarboxylation to lactic acid does not occur. The other major grape acids (tartaric and citric) are occasionally metabolized. Their metabolism appears to depend predominantly on the grape variety.

Associated with grape-cell fermentation is a modified operation of the shikimic acid pathway (Fig. 9.6). Shikimic acid accumulates, along with volatile products of its metabolism, such as ethyl cinnamate, benzaldehyde, vinylbenzene, and salicylic acid. The last is not itself volatile, but it can react with ethanol to form an aromatic ethyl ester. Higher concentrations of ethyl decanoate, eugenol, methyl and ethyl vanillates, ethyl and vinylguaiacols, and ethyl- and vinylphenols develop during carbonic maceration than during standard vinification (Ducruet, 1984). The high ethyl cinnamate and ethyl decanoate levels may be sufficiently distinctive to serve as indicators of carbonic maceration.

The precise chemical nature of the characteristic fragrance of carbonic maceration wines remains unclear. However, some elements of the fragrance have been tentatively ascribed to ethyl cinnamate and benzaldehyde. These may generate some of the characteristic strawberry–raspberry (Versini and Tomasi, 1983) and cherry–kirsch (Ducruet, 1984) fragrances of carbonic maceration.

Low hexyl acetate and hexanol contents are generally associated with carbonic maceration. Without prefermentative crushing, there is less chance for fatty acid oxidation, and the generation of vegetative odors.

One of the distinctive consequences of carbonic maceration is a reduction in the amount of free ammonia and a rise in the concentration of amino acids (Flanzy et al., 1987). Some of the amino acids undoubtedly arise from the enzymatic breakdown of proteins. Others may be biosynthesized from glycolytic or TCA cycle intermediates and ammonia. Although the total concentration of amino acids rises, the content of specific amino acids varies independently. For example, the concentrations of aspartic and glutamic acids fall, due to their metabolism during carbonic maceration (Nicol et al., 1988).

The release of organic nitrogen during maceration helps explain the rapid onset and completion of both alcoholic and malolactic fermentation. Whether the high amino acid content plays a role in the development of the characteristic carbonic maceration bouquet is unknown.

During carbonic maceration, pectins break down in the fruit. Consequently, the attachment of cells to one another weakens, and the pulp loses its solid texture. If the carbon dioxide produced inside the intact fruit escapes, the berries become flaccid. Otherwise, the CO_2 pressure maintains berry shape, but not its strength.

At the beginning of carbonic maceration, the fruit absorbs carbon dioxide from the surrounding environment. The amount dissolved depends on the temperature, varying from approximately 50% of berry volume at 18 °C to 10% at 35 °C. As berries become saturated, carbon dioxide liberated during fermentation begins to be released. The production of carbon dioxide by grapes ceases when the cells die, due to alcohol toxicity or when the energy supply from fermentation is insufficient to sustain cellular integrity.

As grape cells die, the regulation of movement across the membrane ceases. This enables the release of various substances from the cells, notably phenolic compounds. The extraction of phenols is complex and often highly

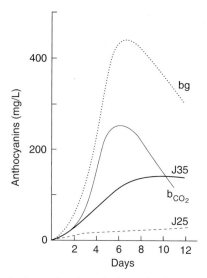

Figure 9.7 Anthocyanin release during carbonic maceration of free-run must (J) and grapes immersed in must (bg) or carbon dioxide (bCO_2); 35, 35 °C; 25, 25 °C. (After Flanzy et al., 1987, reproduced by permission)

specific. The major controlling factors are the temperature and duration of carbonic maceration, as well as the presence of fermenting juice around the fruit.

Anthocyanins are more rapidly and extensively dissolved than tannins. Because high temperatures speed color stability by favoring anthocyanin–tannin polymerization, winemakers prefer a short maceration at temperatures above 30 °C. The extraction of tannins appears to be primarily from the skins, with little coming from the seeds.

Nonflavonoid phenolics are both extracted and structurally modified during maceration. Chlorogenic acid dissolves, and the tartrate esters of p-coumaric and caffeic acids hydrolyze rapidly. As a result, small quantities of free p-coumaric and caffeic acids accumulate.

Submersion of the fruit in fermenting juice, common at the base of the vat, markedly increases anthocyanin and tannin extraction (Fig. 9.7). This presumably is due to the solvent action of the alcohol that accumulates in the fermenting must. As alcohol diffuses into intact grapes, ethanol dissolves phenols in the fruit. Thus, pigment and tannin extraction is not limited just to berries that break open at the bottom of the vat.

Of the factors influencing grape-cell fermentation, temperature is probably the most significant and easily controlled. The initial phase is optimally conducted at between 30 and 32 °C. This shortens the duration of carbonic maceration, promotes grape fermentation, and favors pigment and tannin extraction. To encourage the rapid onset of grape-cell fermentation, the fruit is often picked late in the afternoon on warm sunny days. Alternatively, the fruit may be heated to the desired temperature. Despite this, the preferred initial temperature

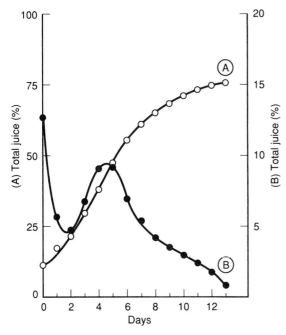

Figure 9.8 Release of free-run must and wine during carbonic maceration: (A), percentage free-run released (left *y*-axis); (B), percentage free-run of total released per day (right *y*-axis). (After André *et al.*, 1967, reproduced by permission)

in Beaujolais is reported to be between 18 and 22 °C (Descout, 1983).

Fermentation of Released Juice In Beaujolais, the fruit is fermented in shallow vats. The breakage and amount of juice released depend on the maturity and health of the fruit, grape variety, tank depth, and mechanism of fruit unloading. To minimize crushing, the tanks are commonly no more than 2.5 m high. Nonetheless, 10–20% of the grapes may break during loading. This level increases during maceration as the berries weaken and the cumulative mass of the fruit ruptures those at the bottom (Fig. 9.8). Pumping over-augments fruit rupture and the amount of free-run liberated. The proportion of juice released by fruit rupture varies widely, but, by the end of maceration, typically reaches 35–55%.

If the juice is low in acidity, tartaric acid may be added to the free-run juice. At this point, the addition of sulfur dioxide usually brings the level of SO_2 in the free-run up to 20–50 mg/liter. Sulfiting is limited to avoid the production of hydrogen sulfide and delay malolactic fermentation. Early completion of malolactic fermentation is essential to the production of *primeur* or *nouveau* wines. Because of early bottling, microbial stability needs to be established within a few weeks of vinification.

Occasionally, chaptalization is conducted before pressing. Normally, though, when the °Brix value is too low, sugar is added after pressing, when alcoholic fermentation is at its apex (Descout, 1983).

Because juice inoculation occurs spontaneously upon rupture, some yeast fermentation occurs concurrently with carbonic maceration. This has a marked effect on the course and duration of grape-cell fermentation. Yeast fermentation has its most marked effect on fruit submerged in the fermenting juice. Even in the absence of released juice, however, the population of yeasts and bacteria on the grapes increases. The grape flora can grow on nutrients already present on the skin. The absence of oxygen is not a limiting factor. The microbial flora is in direct contact with long-chain unsaturated fatty acids needed for growth and membrane production under anaerobic conditions. These compounds are part of the waxy cuticular layer of grape skins.

Few studies on the yeast flora during carbonic maceration have been conducted. *S. cerevisiae* appears to be the dominant species, although *Schizosaccharomyces pombe* may constitute up to 25% of the yeast population (Barre, 1969).

The yeast population reaches approximately 8–12 × 10^7 cells by the time of pressing. Fungicides on the fruit can modify this value significantly, especially because of the small juice volume initially released. Correspondingly, the juice may be inoculated with an active yeast culture to offset potential yeast suppression by fungicide residues.

Thermotolerant yeast strains are required if optimal temperatures for carbonic maceration are used. Nevertheless, it is important to prevent excessive heat buildup. Temperatures above 35 °C can induce yeast death and leave the must open to spoilage yeasts and bacteria.

Yeast inoculation tends to reduce the accumulation of ethyl acetate (Descout, 1986). The origin of this compound is not precisely known. Because its increase is not directly correlated with the simultaneous buildup of acetic acid, it presumably is synthesized directly by grapes or the indigenous flora. Pumping over and periodic chaptalization (frequent but small additions of sugar) encourage ethyl acetate accumulation.

Yeast activity has a considerable influence on the course of carbonic maceration. If carbon dioxide is not employed to displace air from the fermentor, yeast action generates most of the carbon dioxide that eventually blankets the fruit. If the fruit is cool and not heated artificially, yeast metabolism also generates most of the heat that warms the fruit during carbonic maceration. Yeasts quickly convert released sugars to ethanol, in contrast to limited metabolism by whole grapes.

Alcohol vapors generated during carbonic maceration are partially absorbed by the fruit. Not surprisingly, more ethanol diffuses into the fruit submerged

10

in fermenting juice (Fig. 9.9). By acting as a sink for alcohol, intact berries aid yeast fermentation by slowing the accumulation of ethanol in the juice. Malic acid released from broken berries tends to diffuse inward, permitting its continued metabolism by living grape cells. If lactic acid bacteria are active in the juice, the flow of malic acid may reverse. Sugars slowly diffuse out of intact berries, adding to those released by progressive fruit rupture. Sugars provide a continuing nutrient source for yeast metabolism. By the end of carbonic maceration, the sugar content of intact fruit usually has fallen to about 50–70 g/liter (Descout, 1986). Throughout carbonic maceration, nutrients released as berries break open help minimize the accumulation of toxic octanoic and decanoic acids.

Although alcohol accumulation inhibits grape-cell fermentation, it appears to activate production of the aromatic compounds that characterize carbonic-maceration wines (Tesnière *et al.*, 1991).

Winemakers often use the end of carbon dioxide release as an indicator of when carbonic maceration has ceased. Another clue is the drop in juice specific gravity to 1.02 or below. Alternatively, carbonic maceration may

be terminated when juice color or flavor has reached a desirable value.

Maceration typically lasts 6–8 days, but can last up to 2 weeks. Long maceration is more common when there is no simultaneous yeast fermentation. Extended contact between the juice and fruit often leads to the development of a bitter character. This presumably results from extraction of phenolic compounds from the stems.

PHASE II – ALCOHOLIC FERMENTATION

Once the decision has been taken to terminate carbonic maceration, the free-run juice is allowed to escape and the remaining juice extracted from the intact grapes and pomace. If the free-run shows no signs of active malolactic fermentation, it is common to combine all the juice fractions for alcoholic fermentation. However, if the free-run juice is undergoing malolactic fermentation, the press-run juice is usually fermented separately. There is a concern that the higher content of fermentable sugars in the press-run may spur acetic acid production by lactic acid bacteria.

The free- and press-run fractions may also be fermented separately to permit blending based on their respective qualities, and the intentions of the winemaker. Free-run juice, in contrast with traditionally vinified wines, is of lower quality than the press-run fraction. Free-run juice produces wine that is less alcoholic (by 1–2%) than the press-run juice. Free-run wine is also lighter in color, more herbaceous, bitter tasting, and higher in acetaldehyde and 2,3-butanediol contents. The press-run fraction generates wine that is more aromatic, alcoholic, and colored. It also contains most of the esters, fusel alcohols, and aromatic compounds that give carbonic maceration wines their distinctive fragrance. Although the total phenolic contents in both fractions are nearly identical, the specific composition differs. Tannins in the press-run are softer tasting and less bitter than are those in the free-run.

For the lighter *primeur*-style, typical of beaujolais *nouveau*, more free-run wine is used. When a wine of longer aging potential is desired, the blend contains a higher proportion of the press-run fraction.

For the second phase of vinification, a temperature of between 18 and 20°C is generally preferred. It is believed to retain the distinctive fragrance donated by carbonic maceration. If, as usual, the initial phase has taken place or reached temperatures considerably above 18–20°C, cooling is required. Some cooling occurs spontaneously when carbonic maceration comes to completion and during pressing. Nevertheless, additional cooling is often required. Even at 18–20°C, fermentation is tumultuous and customarily complete within 48 h.

In addition to consuming fermentable sugars, yeasts modify the concentration of volatile phenols. Malolactic

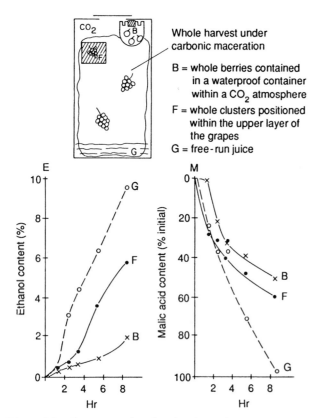

Figure 9.9 Changes in ethanol and malic acid contents in grapes during carbonic maceration in a carbon dioxide atmosphere. (After Flanzy *et al.*, 1987, reproduced by permission)

fermentation further alters the phenolic composition. These effects are more pronounced in carbonic maceration wines than in traditionally produced wines. Both 4-vinylguaiacol and 4-vinylphenol contents increase, whereas 4-ethyl phenol decreases during alcoholic fermentation (Etiévant *et al.*, 1989). Total volatile phenols increase during malolactic fermentation.

Malolactic fermentation typically begins immediately after the completion of alcoholic fermentation, if not before. This is favored by the limited use of sulfur dioxide, storage at warm temperatures, reduced wine acidity, and the ready availability of nitrogenous and other nutrients. If malolactic fermentation is slow to commence, the wine is commonly inoculated with lactic acid bacteria. Inoculation usually involves the addition of wine that has just undergone successful malolactic fermentation. Natural inoculation appears to induce quicker fermentation than commercially available cultures.

AGING

Most carbonic maceration wines are produced for rapid consumption, with only a small proportion vinified for

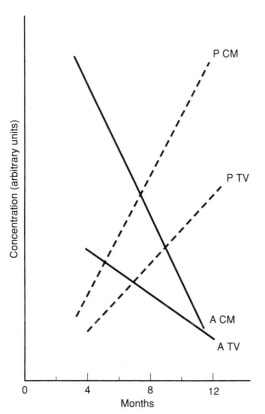

Figure 9.10 Development of aromatic compounds during maturation. 0, end of malolactic fermentation; A, isoamyl acetate and benzaldehyde; CM, carbonic maceration; P, volatile phenols; TV, traditional vinification. (From data published in Etiévant *et al.*, 1989, after Flanzy *et al.*, 1987, reproduced by permission)

extended aging. A few of the changes that can occur during aging are shown in Fig. 9.10. They are similar to those occurring in traditionally vinified wines, but are more pronounced. Subjectively, it is known that the fruity aroma induced by carbonic maceration soon fades. If a desirable varietal aroma or aging bouquet replaces the grape fermentation aroma, the wine is more likely to age well.

Maturation in oak usually has been considered inappropriate for carbonic maceration wines. However, short exposure to oak can add complexity to the wine. Winemakers differ considerably in their opinions concerning whether oak benefits or detracts from the fruity character of wine.

USE WITH ROSÉ AND WHITE WINES

Although carbonic maceration is predominantly used for the production of red wines, rosé and white wines are occasionally vinified using the technique. In the production of rosé wines, grapes are kept from being submerged in free-run juice. This limits pigment and tannin extraction, and maximizes the development of a fruity aroma. If pigment extraction is insufficient, the grapes are crushed and the second phase of fermentation conducted briefly in contact with the seeds and skins. Once sufficient color has been obtained, the fermenting must is pressed to separate the juice from the pomace.

Similarly, white grapes treated by carbonic maceration are kept isolated from fermenting juice. The duration of the process for white wines is commonly shorter than for either red or rosé wine production, often being little more than 48 h. The precise duration and temperature chosen depend on winemaker preferences and how maceration affects the varietal aroma. Carbonic maceration can either suppress or enhance varietal character (Bénard *et al.*, 1971).

An alternative to the typically short carbonic maceration at warm temperatures for white grapes is maceration at 5 °C for approximately 3 days (Montedoro *et al.*, 1974). The procedure favors ester synthesis and retention, as well as reducing phenolic extraction. Centrifugation also may be used to reduce the phenol content and diminish color intensity.

For the second phase of vinification, juice fermentation is conducted without skin contact. If it is low in pH, the wine may be acidified on pressing. Alternatively, the wine may be cooled and sulfited to prevent deacidification by malolactic fermentation.

Occasionally, the harvest is divided into lots – one treated according to standard procedures, the other treated to carbonic maceration. The fractions may be blended together to provide a wine of enriched fragrance and improved acid balance.

Sparkling Wines

Sparkling wine owes much of its development to technical advancements unrelated to the production of the wine itself. These involved the introduction of cork closures and improvements in glass manufacture. The availability of strong glass was a prerequisite for producing bottles able to withstand the high pressures that develop in sparkling wine. The use of coal to fire the glass furnaces permitted the first production of strong glass in England, beginning during the reign of King James I (1603–1625). Similarly, a closure able to withhold the pressure exerted by the carbon dioxide contained was essential. Cork began to seal bottles in the latter part of the 1500s. These developments, along with an atypically long spell of cold weather in Europe (Ladurie, 1971), combined to favor the evolution of sparkling wine. When the still wine from Champagne (shipped in barrels) was bottled in England, it became spritzy in the spring. By 1676, it was sufficiently notorious as to be noted by George Etherege in *The Man of Mode*. Christopher Merret reported to the Royal Society in 1662 that adding sugar promoted effervescence (Stevenson, 2005). Establishing the correct amount of sugar required, and avoiding frequent bottle explosion, took developments in chemistry almost a century to perfect. The result was the conversion of an inferior quality, pale red wine (*vin gris*) into the preeminent celebratory wine. If Dom Perignon (1638–1715) had any role to play in its development, it was possibly in encouraging blending. It offsets the limitations of individual wines, and accentuates their qualities. In addition, it took years to discover how to efficiently remove the yeast sediment that develops after the second, in-bottle fermentation.

The finesse that is now the hallmark of champagne was slow to develop. It became famous in Paris with the jetset of the time in its original sweet version. The taste for dry champagnes appears to have developed as an English predilection, beginning in the 1850s. This penchant quickly spread, becoming the preferred style. The evolution from sweet to dry is reflected in the evolution of champagne terminology, where *doux* is very sweet, *demi-sec* is quite sweet, *sec* is noticeably sweet, *extra sec* is almost dry, and only *brut* actually perceptibly dry. Subsequently, the method spread throughout most of the winemaking world. The twentieth century saw additional improvements, designed primarily to minimize production costs.

A classification of sparkling wines is given in Table 1.2. The three major processes are compared in Figure 9.11.

Sparkling wines derive their carbon dioxide supersaturation from a second alcoholic fermentation. This is typically induced by the addition of yeast and sugar (**tirage**) to dry white wine. Infrequently, the sparkle may come from spontaneous (alcoholic or malolactic) fermentation after bottling. The latter two processes, jointly or combined, were undoubtedly involved in the production of the first sparkling wines. The standard method began to approach its current form when the appropriate tirage formula was discovered.

Although sparkling wines are usually classified by production method (Table 1.2), this is of little practical value to consumers. Wines produced by different techniques often are distinguishable only by close and careful sensory evaluation. More obvious sensory differences develop from the color and aroma of the base wines, the degree of carbon dioxide supersaturation, the duration of in-bottle lees contact, and the sweetness given the finished wine.

In this book, sparkling wines are discussed relative to their method of production. Figure 9.12 outlines the traditional method described below.

Traditional Method

GRAPE CULTIVARS EMPLOYED

Although white or red grapes may be vinified to produce base wine(s), most sparkling wines are white. Thus, if red grapes are used, particular attention must be taken during harvest and pressing to avoid pigment extraction.

In the Champagne region of France, three grape varieties are used – one white ('Chardonnay') and two red ('Pinot noir' and 'Meunier'). Although the varieties may be used separately, most champagnes are produced from a blend of all three. Each variety is deemed to contribute unique qualities to the blend – 'Chardonnay' providing finesse and elegance, 'Pinot noir' donating body, and 'Meunier' giving fruitiness and roundness. Each also is considered to mature at a different rate, with 'Chardonnay' being the slowest and 'Meunier' the most rapid. Thus, 'Meunier' features prominently in nonvintage blends, aged about 1 year in-bottle before disgorging. Conversely, 'Chardonnay' is traditionally an important component in vintage blends aged for at least 3 years before disgorging. The varieties also differ in their tendency to effervesce and develop a *cordon* (ring of bubbles around the glass). 'Pinot noir' has the greatest tendency to generate a stable effervescence, whereas 'Chardonnay' the least (Marchal *et al.*, 2001). Foam stability (which is a function of surfactants surrounding the bubble) has no relationship to the wine's propensity to effervesce.

In other areas of France, regional cultivars, such as 'Chenin blanc' in the Loire Valley, are often the dominant or only varieties employed. Outside France, either indigenous or imported cultivars are used. In Spain, the

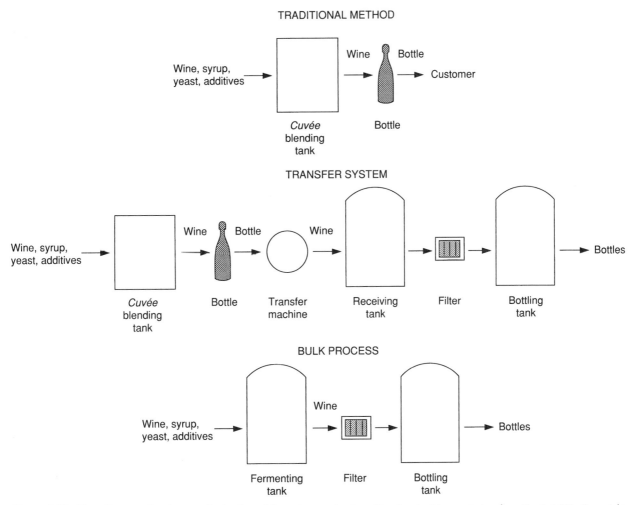

TRADITIONAL METHOD

TRANSFER SYSTEM

BULK PROCESS

Figure 9.11 Flow diagrams for three methods of sparkling-wine production. (Reprinted with permission from Berti, 1981. Copyright 1981 American Chemical Society)

native varieties 'Parellada,' 'Xarel-lo,' and 'Viura' are employed. Each variety is considered to contribute a different important characteristic to the blend – 'Parellada' providing fragrance and softness, 'Viura' donating finesse and elegance, and 'Xarel-lo' imparting strength and a golden color.

HARVESTING

Where the financial return permits, harvesting of both white and red grapes occurs manually. Manual harvesting permits both pre- and post-harvest selection to exclude infected grapes. This is especially critical when red grapes are used. Laccase released by *Botrytis cinerea* can cause serious oxidative browning. *Botrytis* infection also negatively affects foam stability (Marchal *et al.*, 2001). Manual picking minimizes fruit rupture, the release of juice, subsequent oxidation of the juice, and pigment and tannin extraction. However, because of the slowness of manual harvesting, fruit may not

be picked at optimal quality. The inability to harvest quickly can occasionally lead to a considerable quality loss under poor harvest conditions.

Harvesting commonly occurs earlier than is usual for table wine production. This yields grapes higher in total acidity and lower pH. Acidity and pH are important features regulating the freshness desired in sparkling wines. Early picking also assures lower °Brix, yielding wines of reduced alcohol content. An alcohol content of between 9 and 10.5% is preferred for base wines. Slightly immature fruit also have less varietal character, appropriate for the production of most sparkling wines. In addition, the grapes are more likely to be healthy. It is not surprising that Champagne, the most northerly wine-producing region in France, became associated with the evolution of sparkling wine. Grapes from the region routinely yielded poor-colored acidic juice, relatively low in sugar content and varietal aroma – all features now considered desirable in the production of most sparkling wines.

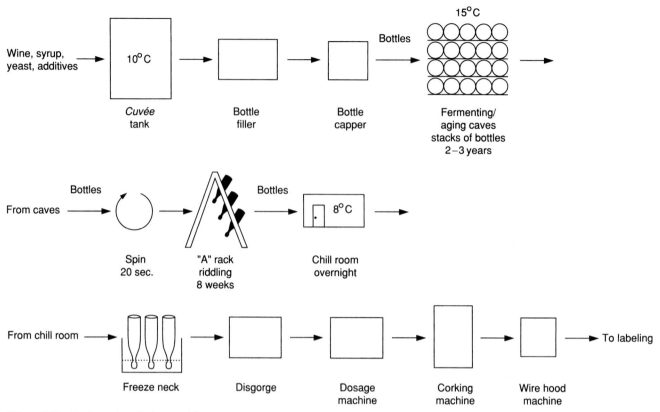

Figure 9.12   Traditional method of sparkling wine production. (Reprinted with permission from Berti, 1981. Copyright 1981 American Chemical Society)

PRESSING

In traditionally produced sparkling wines, the grapes are pressed whole, without prior stemming or crushing. Because this results in the release of juice from the flesh first, the extraction of skin pigments, grape solids, polyphenol oxidases, and potassium is delayed. The juice so liberated is called the *cuvée*. It is higher in acidity and sugar content, with little potential for oxidation. Consequently, little sulfur dioxide or bentonite is required. Intact-grape pressing also limits the extraction of varietal aroma compounds. They could mask the subtle, aged fermentation bouquet so desired in champagne and equivalent sparkling wines. Pressing whole grapes is considered to promote early malolactic fermentation and favors the onset of the second, in-bottle yeast fermentation.

Pressing whole grapes takes considerably longer than conventional pressing. It often takes 2 hours for the release of the *cuvée*, using the large-diameter, vertical presses historically preferred in Champagne. Although the slow liberation exposes the juice to oxidation, the brown pigments that form precipitate during settling or fermentation. This also provides protection against subsequent in-bottle oxidative browning.

Because the grapes are unstemmed and pressed as whole clusters, the stems provide channels for the juice to escape, minimizing the pressure required. The large shallow presses traditionally used provide a large surface area, further minimizing the pressure needed for juice release. Nevertheless, pneumatic horizontal presses have largely replaced traditional presses for all but illustrative purposes. Pneumatic presses generally work well with unstemmed, uncrushed grapes. They also have the distinct advantages of taking up less space, being easier to load and unload, and permitting more efficient pomace crumbling between successive pressings. Whether automated versions of traditional presses can sustain their diminished use is a moot point, even in Champagne.

Approximately 62.5 liters of juice/100 kg grapes is permitted in champagne production. The initial and largest portion (~80%) consists of the *cuvée* (Valade and Blanck, 1989; Fig. 9.13). When juice flow slows to a trickle, pressure is released, the grapes mixed (*retrousse*), and pressure reapplied. This may occur several times during the liberation of the cuvée (~2050 liters from the 4000 kg grapes added to the traditional vertical press). The last fractions used in champagne production

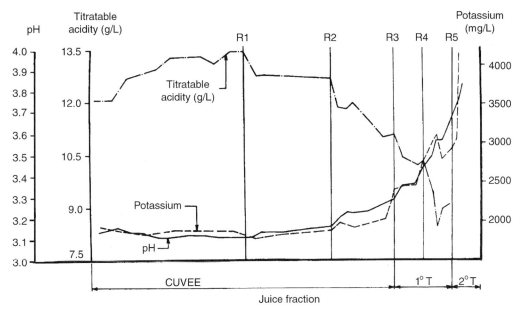

Figure 9.13 Evolution of titratable acidity, pH, and potassium during juice extraction from an automated traditional press in Champagne. R1, R2, R3, R4, and R5 represent the mixings (*retrousses*). The designation given to the various fractions (*cuvée*, 1st *taille*, and 2nd *taille*) are shown on the *x*-axis. (From Dunsford and Sneyd, 1990, reproduced by permission)

are called the *taille*. For their release, pressures exceeding 100 kPa are required. The *première taille* refers to the next 500-liter fraction. Its lower acidity and higher tannin content usually preclude its use in champagne production. The fraction also possesses more fruit flavors. Nevertheless, the *première taille* may be useful in particular proprietary blends. The 2nd *taille* is not used in champagne production.

In much of the New World, quality and pricing constraints, not legislation, are the major factors influencing juice yield. Increasing the volume pressed from the grapes, beyond a particular point, compromises quality. This results as the increased flavor extraction and phenolic content begins to detract from the wine's subtle bouquet. Tannins are particularly undesirable. They can impede the second, in-bottle fermentation and increase the tendency of the wine to gush.

Sulfur dioxide is added to the juice as it comes from the press – 40–60 mg/liter being typical in Champagne. Depending on the quality of the vintage (the maturity and proportion of diseased fruit), sugar, bentonite, charcoal, and pectinase may be added. The sugar is added for chaptalization, whereas bentonite, activated carbon, pectinase, and additional sulfur dioxide may be incorporated to remove glucans, pigments, and inactivate laccases released from diseased fruit. Hyperoxidation has been studied as a substitute for activated carbon in decoloration and astringent phenol removal from must, principally in the *taille* of 'Pinot noir' and 'Pinot Meunier' (Blanck, 1990).

Juice not already cool is routinely chilled down to approximately 10 °C and left to clarify by settling for 12–15 h before fermentation.

As noted, grapes pressed whole liberate fewer solids than those pressed after crushing (0.5 vs. 2–4%) (see Randall, 1987). Where grapes are crushed before pressing, the extra solids are normally removed prior to fermentation via bentonite-facilitated settling, centrifugation, or filtration. The use of peristaltic pumps in transporting the juice to temporary storage tanks minimizes particulate generation following crushing and pressing.

Even under optimal conditions, the juice obtained from red grapes may contain a slight pinkish tinge. The anthocyanins involved usually coprecipitate with yeasts during fermentation or later during fining. Anthocyanase addition, or other forms of decolorization, are customarily unnecessary with juice from healthy grapes.

PRIMARY FERMENTATION

Juice fermentation follows procedures typical of most modern white wines. It usually occurs at approximately 15–18 °C. Lower temperatures are reported to give a grassy odor, whereas higher temperatures yield wines lacking in finesse (Moulin, 1987). Bentonite or casein may be added to aid fermentation and remove excess polyphenolics. Occasionally, a mixture of bentonite, potassium caseinate and microcrystalline cellulose may be substituted with beneficial effect (Puig-Deu *et al.*, 1999). Inoculation with selected yeast strains is almost universal. It helps avoid the production of perceptible amounts of sulfur dioxide, acetaldehyde, acetic acid, or other undesired volatiles synthesized by indigenous yeasts.

If the juice is too low in pH (≤3.0), malolactic deacidification is commonly encouraged. Some producers also believe that malolactic fermentation donates a subtle bouquet they desire. It also reduces excessive acidity, permitting a greater proportion of the wine to be left dry (*brut*). As the bacterial sediment produced is difficult to remove by riddling, it is important that malolactic fermentation be complete before the second, in-bottle, alcoholic fermentation. Malolactic fermentation is encouraged by minimal SO_2 addition and maturation at or above 18°C. Producers also frequently inoculate the wine with a particular strain of *Oenococcus oeni* to encourage rapid onset. If producers wish to avoid malolactic deacidification (to enhance aging potential), the wine is sterile-filtered.

Finally, wines are clarified and cold-stabilized by cultivar, site, and vintage. Maturation may last for several months or years. Aging typically occurs in stainless steel, but occasionally occurs in large or small oak cooperage. Certain producers in Champagne mature some of the base wines on lees under light CO_2 pressure (100–150 kPa) in 1.5-liter bottles (Randall, 1987).

After maturation, the wines are ready for preparation of the *cuvée*. This second use of the term refers to the blend of base wines that will be used in the production of the sparkling wine.

PREPARATION OF THE *CUVÉE*

The blending of wines derived from different sites, varieties, and vintages is one of the hallmarks of sparkling wine production. Because single wines seldom possess the features producers desire, samples from different base wines are combined to obtain a small number of basic blends. Mixing is based solely on sensory evaluation of the wines. Based on the assessment, a formula for the *cuvée* is developed. In addition to improving the quality of the sparkling wine, blending helps negate yearly variations in supply and quality. This is essential in producing the consistency required for proprietary brands.

Because integration can disrupt tartrate equilibrium, the *cuvée* is typically cold-stabilized to reestablish stability. The *cuvée* is loose-filtered cold to remove tartrate crystals that form. Tight filtration is less desirable. It may remove proteins and polysaccharides important in the formation of a fine and stable *mousse*.

TIRAGE

Tirage involves adding a concentrated (50–65%) sucrose solution containing other nutrients to the *cuvée*. The solution is added just before yeast inoculation. The tirage may be made up in water or the *cuvée* itself. When wine is not the solvent, citric acid may be added at 1–1.5% to activate sucrose hydrolysis into glucose and fructose.

Sufficient tirage is added to supply about 24 g sucrose/liter. During fermentation, this produces a pressure considered appropriate for most sparkling wines, namely 600 kPa (~6 atm). Because the pressure exerted by carbon dioxide varies with the temperature and other factors, the concentration of CO_2 is occasionally expressed in terms of mass. Most bottles of sparkling wine contain approximately 15 g CO_2.

If the *cuvée* contains residual fermentable sugars, the amount must be subtracted from the quantity of sucrose added with the tirage. Approximately 4.2 g sugar is required for the generation of 2 g carbon dioxide. During the second, in-bottle fermentation, the alcohol content generally rises 1%.

Thiamine and nitrogen (DAP – diammonium hydrogen phosphate) are often added to the tirage to supply 0.5 and 100 mg/liter, respectively. Thiamine appears to counteract alcohol-induced inhibition of sugar uptake by yeast cells (Bidan *et al.*, 1986). Addition of nitrogen is unnecessary if the concentration of assimilable nitrogen in the *cuvée* is above 15 mg/liter. Nevertheless, it may help suppress the production of hydrogen sulfide. Occasionally, trace amounts of copper salts (≤0.5 mg/liter) are added to further reduce hydrogen sulfide production (Berti, 1981). Some producers incorporate bentonite, casein, gelatin, or isinglass to aid yeast flocculation at the end of fermentation. Evidence suggesting a negative effect of most fining agents on effervescence indicates that their use may be ill-advised (Maujean *et al.*, 1990). Data from Dambrouck *et al.* (2005) suggest that the detrimental effects of bentonite on protein removal (and foam stability) are largely counteracted by the simultaneous addition of casein. There is circumstantial evidence that removal of invertase, a glycoprotein that constitutes 10–20% of wine proteins, may be an important component in foam stability.

If the base wines have not undergone malolactic fermentation, the *cuvée* may be sterile-filtered. Providing a sulfur dioxide content of greater than 10 mg/liter free SO_2 is effective, but less preferable as it may complicate initiation of the second, in-bottle fermentation.

YEASTS AND CULTURE ACCLIMATION

The second fermentation requires inoculation of the *cuvée* with a special yeast strain. These typically are strains of *Saccharomyces bayanus*. Because of the special and exacting conditions that prevail during the second fermentation, yeasts must be capable of commencing fermentation at alcohol contents between 8 and 12%, at temperatures around 10°C, at pH values as low as 2.8, and with free sulfur dioxide contents up to 25 mg/liter. The suppressive influences of pH and sulfur dioxide are

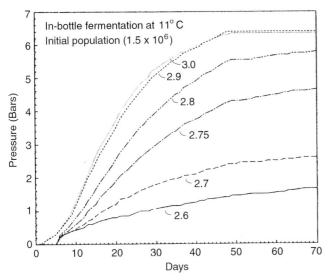

Figure 9.14 Influence of pH on the production of carbon dioxide during the second fermentation in the traditional method of sparkling-wine production. 1 bar = 1 ATM = 101 pKa. (From Bidan *et al.*, 1986, reproduced by permission)

illustrated in Figs 9.14 and 9.15. If cold tolerance is of particular concern, strains of *S. uvarum* may be used. Their suitability may be compromised, however, by their tendency to produce increased amounts of isoamyl and isobutyl alcohol, and quadrupled levels of 2-phenyl alcohol (Massoutier *et al.*, 1998). In addition, *S. uvarum* does not flocculate well. This has recently been overcome by interspecific hybridization, between *S. uvarum* and a flocculent strain of *S. cerevisiae* (Coloretti *et al.*, 2006). In spontaneous second fermentations, strains of *S. cerevisiae* have been isolated (Nadal *et al.*, 1999).

Yeast cells must also flocculate readily to produce a coarse sediment for efficient removal during riddling. Developments in the use of encapsulated yeast may avoid both the need for flocculation and the expense of the riddling–disgorging process (see later).

Yeast cells also must have low tendencies to produce hydrogen sulfide, sulfur dioxide, acetaldehyde, acetic acid, and ethyl acetate. The presence of an active proteolytic ability after fermentation aids amino acid and oligopeptide release during yeast autolysis.

Because of the unfavorable fermentation conditions in the *cuvée*, the yeast inoculum is acclimated before addition. Otherwise, most of the yeast cells die and a prolonged latency results before fermentation commences. Acclimation usually starts with inoculation of a glucose solution at about 20–25 °C. The culture is aerated to assure adequate production of unsaturated fatty acids and sterols (required for cell division and proper membrane function). Once the yeasts are actively growing, the culture may be added to enough *cuvée* to produce a 60:40 mixture. Over the next few days, *cuvée* wine is added

to reach an 80–90% *cuvée* mixture. Simultaneously, the culture is slowly cooled to the desired fermentation temperature (Markides, 1987).

The *cuvée* is inoculated with the acclimated culture to reach a concentration of approximately $3–4 \times 10^6$ cells/ml (2–5% the *cuvée* volume). Higher inoculation levels are thought to increase the likelihood of hydrogen sulfide production, whereas lower levels increase the risk of failed or incomplete fermentation.

SECOND FERMENTATION

Once the *cuvée* has been mixed with the tirage and yeast inoculum, the wine is bottled. In the past, bottles were sealed with a cork stopper, held by a reusable metal clamp called the *agrafe*. This has been replaced with a crown cap possessing a polyethylene bidule. The bidule is an indented plug that helps hold the yeast that collect during riddling and facilitates cleaner disgorging. Crown caps are much less expensive and more easily removed by automated machines. Nevertheless, Vasserot *et al.* (2001) noted that the sealing qualities of crown caps and bottle capacity may favor the excessive accumulation of dimethyl sulfide.

Occasionally 375-ml, 1500-ml, and larger-volume bottles are used, but the 750-ml bottle is standard. Unless a brand-distinctive shape or color is used, the bottle typically has pronounced sloping shoulders and a greenish tint. The glass is thicker than for table wines, to withstand the high pressures that develop during the second fermentation. Special care is taken during annealing to minimize the possibility of bottle explosion.

Filled bottles may be stacked on their sides in large, freestanding piles, in cases, or in specially designed containers ready for mechanical riddling. The wine is kept at a relatively stable temperature, preferably between 10 and 15 °C for the second fermentation. Cooler temperatures may result in premature termination of fermentation, whereas warmer temperatures may result in both a rapid rise in alcohol content and a drop in redox potential. The former may prematurely terminate fermentation and the latter increase hydrogen sulfide production (Markides, 1987). A stable temperature also helps maintain yeast viability under difficult fermentation conditions.

At 11 °C, a common fermentation temperature in Champagne, the second fermentation may last approximately 50 days (Fig. 9.16). During the early stages of fermentation, the yeast population goes through three to four cell divisions, reaching a final concentration of approximately $1–1.5 \times 10^7$ cells/ml. The rate of fermentation is largely dependent on the temperature, pH, and sulfur dioxide content of the *cuvée*.

After fermentation, the bottles may be transferred to a new site for maturation. Storage typically occurs at

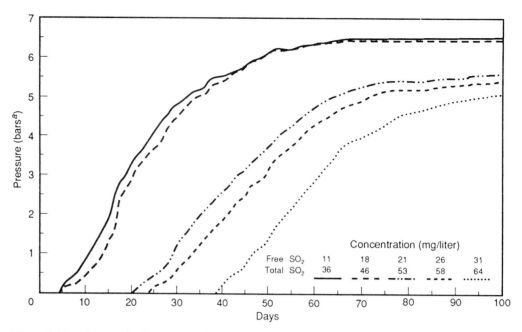

Figure 9.15 Influence of sulfur dioxide concentration on carbon dioxide production during the second fermentation in the traditional method of sparkling wine production. In-bottle fermentation at 11°C started with an initial yeast population of 1.5×10^6 cells/ml; 1 bar = 1 ATM = 101 pKa. (After Bidan *et al.*, 1986, reproduced by permission)

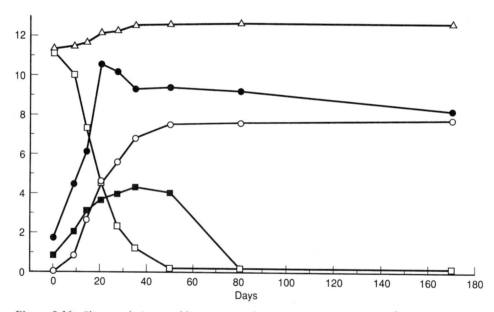

Figure 9.16 Changes during sparkling wine production. ●, total yeasts ($\times 10^6$); ■, viable yeasts ($\times 10^6$); □ sugars/2 (g/liter); ○, pressure (bars); △, ethanol (%). (After Bidan *et al.*, 1986, reproduced by permission)

about 10°C. Yeast contact during maturation lasts a minimum of 9 months, but may continue for 3 or more years, depending on the wine attributes desired. For effervescence production and foam stability, a maturation period of about 18 months seems optimal. These properties appear principally correlated to the accumulation of polysaccharides (Andrés-Lacueva *et al.*,

1997). Further aging seems to result in polysaccharide hydrolysis.

During in-bottle maturation, the number of viable cells drops rapidly. After approximately 80 days, the viable yeast population drops to below 10^6 cells/ml. By the time of disgorgement, normally 9 months to 1 year after tirage, few if any viable cells remain. Even within

6 weeks, the cells show atypical, large, expanded vesicles. By 3 months, the cells become plasmolyzed and most typical membrane-bound organelles have disappeared (Piton *et al.*, 1988). This is associated with equally marked changes in membrane lipid content. Changes in cell-wall structure also occur, notably the disappearance of the innermost layer. The rapid decline in viability contrasts greatly with the slow decline of the yeast population following primary fermentation.

These structural changes are associated with major metabolic perturbations. As the wine becomes depleted in fermentable sugars, the cells begin to metabolize internal energy reserves such as glycogen. As nutrient conditions continue to deteriorate, cells start to show autophagy (Cebollero and Gonzalez, 2006) and begin to die. One of the first indicators of degeneration is the disruption of membrane function and the leakage of cellular nutrients. As cells die (complete about 6 months after tirage), autolysis commences, despite unfavorable low pH and temperature conditions. Autolysis involves the release and activation of cellular hydrolytic enzymes that digest structural components of the cells. Along with cell-wall mucopolysaccharides, several potentially significant flavor enhancers (nucleotides and glutamate) are released. Deletion of the *BCY1* gene from yeast strains appears to speed autolysis (Tabera *et al.*, 2006).

Yeast strain, grape variety, storage conditions, and duration of lees contact all influence the release of nitrogenous compounds from autolyzing yeast cells. Yeast strains differ not only in the amount, but also in the specific amino acids released. It is generally considered that the optimal temperature for lees contact is approximately 10 °C. At higher temperatures, the rate of nitrogen release increases, and the nature of the nitrogenous compounds liberated changes. Temperature also influences the rate and types of changes in the aromatic compounds produced.

The release of amino acids and various oligopeptides during yeast autolysis has frequently been associated with the development of a toasty bouquet. The former could be the source of volatile thiols, such as benzenemethanethiol, 2-furanmethanethiol, and ethyl 3-mercaptopropionate, associated with development of a toasty aspect. These compounds accumulate to above threshold values during aging, especially after disgorging (Tominaga *et al.*, 2003). In addition, amino acids may be precursors for other significant aromatics, such as sotolon (from threonine), ethoxy-5-butyrolactone (from glutamic acid), benzaldehyde (from phenylalanine), and vitispirane (from methionine) (see Bidan *et al.*, 1986). However, changes in their concentration seem not readily correlated with modifications in the content of the pertinent amino acids. Concentration of compounds, such as vitispirane and 1,2-dihydro-1,1,6-trimethylnaphthalene

(TDN), also increase during aging on yeast lees (Francioli *et al.*, 2003). Nucleotides with potential flavor influences also occur in champagne. Whether they accumulate in concentrations sufficient to affect the sensory characteristics of sparkling wine is as yet unknown.

Changes in the concentrations and types of fatty acids and lipids have been noted during lees contact. The level of fatty acids may initially increase, but subsequently decline. Polar lipids decrease in concentration, whereas neutral lipids increase. Such changes may continue for at least 11 years (Troton *et al.*, 1989). The triacylglycerol accumulated may act as an important precursor for aromatic compounds during aging.

Modification in the concentration of esters has been reported during maturation. Similar to the aging of still table wines, most acetate and ethyl esters of fatty acids decline (Francioli *et al.*, 2003), whereas those formed from the major organic acids increase (Silva *et al.*, 1987). Some changes in the volatile composition of champagne are shown in Fig. 9.17. Additional observations are presented in Escudero *et al.* (2000) and Riu-Aumatell *et al.* (2006).

Correlations between bubble size and foam stability with the presence of certain polysaccharides and hydrophobic proteins have been demonstrated (Maujean *et al.*, 1990). Colloidal proteins can increase two to three times in concentration within the first year.

RIDDLING

One of the most involved and expensive procedures in sparkling-wine production involves the removal of

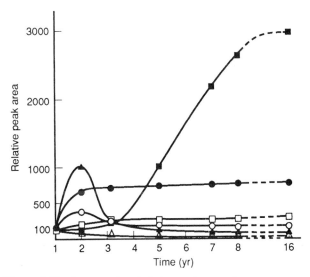

Figure 9.17 Concentration (relative peak area of traces on a gas chromatogram) of several aromatic compounds during the aging of champagne. ■, benzaldehyde; ●, unknown; □, vitispirane; ▲, nerolidol; △, hexyl acetate and isoamyl butyrate; ○, total volatile acidity. (From Loyaux *et al.*, 1981, reproduced by permission)

yeast sediment (lees). The first step entails loosening and suspending the cells in the wine. Subsequent positioning of the bottle moves the lees toward the neck. The agitation involved in the latter process is essential for optimal yeast flocculation (Stratford, 1989).

Historically, riddling was done by hand. It typically took 3–8 weeks, with the bottles positioned neck down in A-racks (*pupitres*). Initially, the bottles were placed about 30° from vertical. Subsequently, the sides of the *pupitres* were moved apart, so that the bottles were about 10–15° from vertical by the end of riddling. By rapid and vigorous twisting of the bottle (about one-eighth of a turn each time), the sediment was dislodged. The bottle was then dropped into the rack, a quarter turn (alternately to the right or left) of its original position. This action was generally repeated at 2-day intervals.

Manual riddling is rapidly disappearing because of its cost, duration and space demands. Automated mechanical riddling is less expensive, takes only about 1 week to 10 days, and requires much less space (Plate 9.6). When fermentation and storage occur in the same container as riddling, bottle handling is also reduced. Various automated riddling systems are available.

DISGORGING, DOSAGE, AND CORKING

At the completion of riddling, the bottles may be left neck down for several weeks in preparation for sediment removal. For disgorging, the bottles are cooled to approximately 7°C, and the necks immersed in an ice bath ($CaCl_2$ or ice–glycol solution, approximately −20°C). This quickly freezes the sediment in the neck. Cooling increases the solubility of carbon dioxide and reduces the likelihood of gushing upon opening. Freezing the yeast plug in the neck facilitates its ejection. Freezing commonly occurs in a trough, while the bottles are being transported through the freezing solution on the way to the disgorging machine.

In rapid succession, the disgorging machine inverts the bottle, removes the cap, allows ejection of the frozen yeast plug, and then covers the mouth of the bottle with a sequence of devices. These prevent further wine escape and adjust the wine level to the desired volume (by either adding or removing wine). Adjustment is often necessary, because the amount of wine lost during disgorging can vary considerably. Furthermore, most sparkling wines have a *dosage liqueur* added before corking.

The *dosage* typically consists of a concentrated sucrose solution (60–70%), dissolved in high-quality aged white wine. Preferably, the dosage wine is the same as the *cuvée* used in the second fermentation. Occasionally, brandy is added to the dosage. A small quantity of sulfur dioxide may be added to prevent subsequent in-bottle fermentation and limit oxidation. The *dosage* is prepared several

weeks in advance to assure that turbidity will not develop. The volume of *dosage* added depends on the sweetness desired and its sugar content.

A few sparkling wines receive no *dosage*. These *nature* (Extra Brut) wines are rare because the *cuvée* seldom has sufficient balance to be harmonious when bone dry. *Brut* wines may be adjusted with *dosage* up to 1.5% sugar. *Extra-sec* wines generally contain between 1.2 and 2% sugar; *sec* wines commonly possess between 2 and 4% sugar; *demi-sec* wines obtain between 3 and 5% sugar; and *doux* styles possess more than 5% sugar. The range of sugar found in each category may vary beyond this, depending on the presence of any residual sugars following the second, in-bottle fermentation.

After volume adjustment and *dosage* are complete, the bottles are sealed with special corks (31 mm in diameter and 48 mm long). They are commonly composed of agglomerate cork, to which two disks of natural cork have been glued. Once the cork is inserted, and just before addition of the wire hood, the upper 10 mm of the cork is compressed into its familiar rounded shape. Once the wire hood has been fastened, the bottle is agitated to disperse the *dosage* throughout the wine. The bottles are stored for 1–3 months to allow the corks to set in the neck. Before setting, cork extraction is particularly difficult. The resting period may occur before or after the bottles are cleaned in preparation for adding the capsule and label. Special glues are commonly used to retard loosening of the label in water.

YEAST ENCLOSURE

Incorporation of yeasts and other microbes into a stable gel matrix is increasingly being used in industrial fermentations. Investigation of this technology for winemaking is comparatively recent (Fumi *et al.*, 1988). By injecting a yeast–gel mixture through fine needles into a fixing agent, small beads of encapsulated yeasts are generated (Fig. 9.18). Each bead contains several hundred yeast cells. Because of the mass of the beads, inversion of the bottles results in rapid settling of the beads to the neck, thus eliminating the need for riddling.

Wines produced and aged with encapsulated yeasts show only subtle chemical differences from their traditionally produced counterparts (Hilge-Rotmann and Rehm, 1990). These differences appear not to influence the sensory properties of the wine.

Transfer Method

The transfer method was developed in the 1940s as a means of avoiding both the expense of manual riddling and the low quality of the wines produced by the bulk method. With advancements in automated riddling and

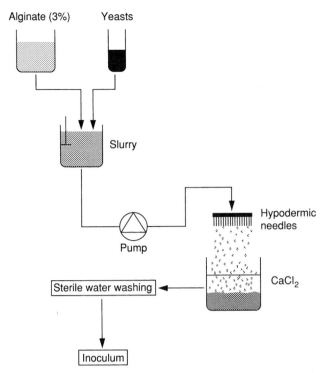

Alginate (3%)   Yeasts

Slurry

Pump

Hypodermic needles

Sterile water washing   CaCl₂

Inoculum

Figure 9.18 Encapsulation of yeasts in alginate. (From Fumi *et al.*, 1988, reproduced by permission)

the developments in yeast encapsulation, most of the advantages of the transfer method have disappeared. Furthermore, advances in the bulk method have eliminated the sources of poor quality that initially plagued the procedure. Because the transfer system is capital-intensive, but does not have the prestige and pricing advantage of the standard method, its continued existence is in doubt. The one remaining advantage of the transfer technique may consist of avoiding bottle-to-bottle variation that can occur with the traditional method.

Preparation of the wine up to riddling is essentially identical to that described for the standard method. Because the wines are not riddled, fining agents need not be added to aid yeast sedimentation. Typically, the bottles are stored neck down in cartons for aging. After aging, the wine is chilled to below 0 °C before discharge. The bottles are opened by a transfer machine and the wine is poured into pressurized receiving tanks. The wine is usually sweetened and sulfited at this stage. Subsequently, the wine is clarified by filtration and decolored if necessary. The wine is typically sterile-filtered just prior to bottling.

Bulk Method

Present-day variations in the bulk method are modifications of the technique initially developed by Charmat

about 1907 (Charmat, 1925). The procedure works well with sweet sparkling wines, accentuating their varietal character. The most well known examples are those produced from 'Muscat' grape varieties. The marked varietal character of 'Muscat' grapes would mask the subtle bouquet generated, at considerable cost, by the standard method.

Occasionally, the wine may be aged on lees for up to 9 months, if a traditional lees-matured attribute is desired. However, because expensive pressurized tanks are tied up for months, many of the economic advantages of the system are lost.

One of the features generally thought to characterize bulk-processed sparkling wine is its poorer effervescence. However, an accurate means of assessing this view has only recently become available (Maujean *et al.*, 1988; Liger-Belair *et al.*, 1999). Objective proof of this assertion still awaits verification.

Fermentation of the juice for base wine production may go to dryness or be terminated prematurely. Frequently, the primary fermentation is stopped at about 6% alcohol to retain sugars for the second fermentation. Termination is either by exposure to cold, followed by yeast removal, or by yeast removal directly. Yeast removal is achieved by a combination of centrifugation and filtration, or by a series of filtrations. Once the *cuvée* has been formulated, the wines are combined with yeast additives (ammonia and vitamins) and sugar, if necessary. The second fermentation takes place in reinforced stainless steel tanks, similar to those employed in the transfer process.

If an extended contact period with the yeasts is desired for bouquet development, the lees are intermittently stirred during this phase. Left undisturbed, a thick layer of yeast cells would form, favoring the generation of reduced-sulfur taints. Stirring also helps to release amino acids from yeast cells that may be involved in evolution of a toasty bouquet. However, stirring also releases fat particles from yeast cells that are not easily removed by filtration. These may interfere with effervescence production (Schanderl, 1965).

At the end of fermentation, or lees contact, the wine is cold-stabilized to precipitate tartrates. Yeast removal involves centrifugation or filtration. It is imperative that these operations be conducted under isobarometric pressure conditions. Otherwise, carbon dioxide may be lost, or gained if the pressurizing gas is carbon dioxide. Sugar and sulfur dioxide contents are adjusted just before sterile filtration and bottling.

Occasionally, still wine may be added to the sparkling wine before final filtration and bottling. This technique may be used to produce wines of reduced carbon dioxide pressure, such as Cold Duck.

Other Methods

A small amount of sparkling wine is produced by the rural or natural method. The primary fermentation is terminated early by repeatedly removing the yeasts by filtration. This also removes essential nutrients from the juice, notably nitrogen. Formerly, fermentation was stopped by repeatedly skimming off the cap of the fermenting juice. After fermentation has ceased, the wine is bottled, and a second in-bottle fermentation slowly converts the residual sugars to carbon dioxide. Yeast removal usually entails manual riddling and disgorging.

Other wines have derived their sparkle from malolactic fermentation. The primary example is Vinho Verde from northern Portugal. The grapes commonly are harvested low in sugar, but high in acidity. They consequently produce wines low in alcohol and high in acidity. The addition of little sulfur dioxide and late racking favored the development of malolactic fermentation. Cool cellar conditions typically resulted in its occurrence in late winter or early spring. Because the wines were kept tightly bunged after fermentation, the small volume of carbon dioxide produced was trapped. The *pétillant* wine that resulted was consumed directly from the barrel. When maturation shifted to large tanks, much of the carbon dioxide liberated by malolactic fermentation escaped from the wine. This was especially marked when the wine was filtered to produce a stable, crystal-clear wine for bottling. Correspondingly, the wine is often carbonated to reintroduce its characteristic *pétillance*. Occasionally, Vinho Verde wines are produced without carbonation or malolactic fermentation when they are low in malic acid content.

In Italy, some red wines become *pétillant* following in-bottle malolactic fermentation. Often the same wine is produced in both still and *spumante* (sparkling) versions.

In the former Soviet Union, sparkling wines were commonly produced in a continuous fermentation process. Though extensively used in Russia, it has been used only comparatively recently outside the former communist state, for example, in Portugal. Multistage, bioreactor, continuous fermentors have also been investigated in Japan (Ogbonna *et al.*, 1989).

Carbonation

The injection of carbon dioxide under pressure is undoubtedly the least expensive method of producing a sparkling wine. It is also the least prestigious. Consequently, carbonation is used only for the least expensive effervescent wines. Nonetheless, the base wine needs to be of good quality. Carbonation can accentuate faults the wine may possess.

Although carbonated wines are generally discounted as unworthy of serious attention, carbonation has the advantage of leaving the aromatic and taste characteristics of the wine unmodified. No secondary microbial activity affects the sensory attributes of the wine.

Production of Rosé and Red Sparkling Wines

Although red grapes are frequently used in the production of sparkling wines, they are typically processed to make a white wine. Only occasionally are grapes fermented on skins used to produce a rosé or light-red wine. The tannins extracted along with the pigments complicate the second fermentation and accentuate gushing. Consequently, the bulk method is preferred. The base wines are almost universally encouraged to undergo malolactic fermentation, prior to the second fermentation to give the wine a smoother taste.

Rosé sparkling wines may be produced from rosé base wines. However, rosé champagnes are typically produced by blending small amounts of red wine into the white *cuvée*. Their pinkish-orange coloration comes partially from the anthocyanin content of the red wine, but may also come from pyranoanthocyanins (Pozo-Bayón *et al.*, 2004). These are cycloaddition products between acetaldehyde, pyruvic acid and phenolics such as 4-vinylphenol, 4-vinylguaiacol and 4-vinylcatechol.

Rosé and red sparkling wines are commonly finished sweet, with low carbon dioxide pressures. They are typically either *pétillant* ($\geqslant 7\,g\ CO_2$/liter) or crackling ($\geqslant 9\,g\ CO_2$/liter). Most white sparkling wines contain at least $12\,g\ CO_2$/liter. The specific carbon dioxide levels applying to each of these terms can vary considerably from jurisdiction to jurisdiction.

Effervescence and Foam Characteristics

Bubble size, foam (*mousse*) characteristics, and the degree, duration, and stability of the effervescence are important aspects in the perceived quality of sparkling wines. Not surprisingly, the origin and factors affecting their development have come under considerable scrutiny (Jordan and Napper, 1987; Liger-Belair *et al.*, 2000, 2001, 2002).

Carbon dioxide may exist in five states in water: microbubbles, dissolved gas, carbonic acid, carbonate ions, and bicarbonate ions. Within the normal pH range of wine, carbon dioxide exists predominantly in the form of dissolved gas, with no carbonated species (Liger-Belair, 2005).

Many factors affect the solubility of carbon dioxide in wine and, therefore, the pressure exerted (Lonvaud-Funel

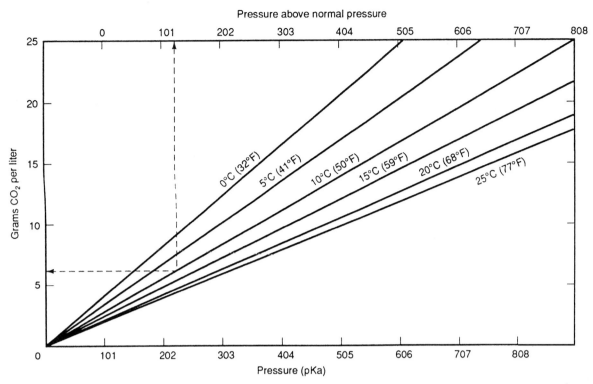

Figure 9.19 Effect of pressure and temperature on the carbon dioxide content of wine. (After Vogt, 1977, reproduced by permission)

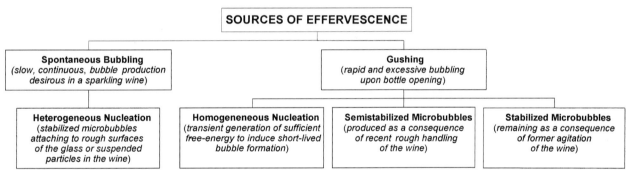

Figure 9.20 Mechanisms of effervescence (CO_2 escape) from sparkling wine.

and Matsumoto, 1979). The most significant factors are the wine's temperature (Fig. 9.19), and its sugar and ethanol contents. Increasing these factors decreases gas solubility and augments the pressure exerted. Once the bottle is opened, the ambient atmospheric pressure becomes a critical factor. The ambient air pressure puts the wine into a supersaturated state that decreases carbon dioxide solubility and promotes bubble nucleation.

Upon opening, the pressure over the wine drops from about 600 to 100 kPa (ambient atmospheric pressure). The partial pressure of carbon dioxide in air is ~0.03 kPa. This decreases carbon dioxide solubility from approximately 12 to 2 g/liter, resulting in the eventual liberation of almost 5 liters of carbon dioxide (from a 750-ml bottle) (Jordan and Napper, 1987). Liger-Belair (2005) estimates that in a champagne flute (100 ml), full degassing would involve the release of about 10 million, 500 μm bubbles. The gas does not escape immediately, because there is insufficient free energy for bubble formation. Most of the carbon dioxide enters a metastable state, from which it is slowly liberated.

Carbon dioxide escapes from the wine via diffusion or bubble formation (Fig. 9.20). The slowest and least significant to the sensory characteristics of sparkling wine is diffusion. Bubble nucleation may arise spontaneously, or through the action of various physical forces. Spontaneous effervescence from nucleation sites is the source for the continuous stream of bubbles so

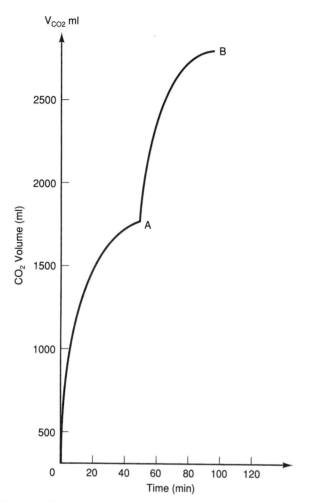

Figure 9.21 Example of the slow effervescence of a sparkling wine upon opening of the bottle (0 to A), and after swirling agitation (A to B). (After Maujean *et al.*, 1988, reproduced by permission)

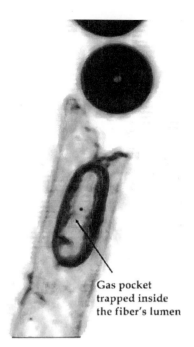

Gas pocket trapped inside the fiber's lumen

Figure 9.22 Close-up of a cellulose fiber acting as a bubble nucleation site. (From Liger-Belair 2005, reproduced by permission. Copyright 2005 American Chemical Society)

desired in sparkling wines. Provoked effervescence is undesirable as it enhances gushing and wine loss.

Spontaneous effervescence results from **heterogeneous nucleation**. This occurs on "imperfections." In a glass of sparkling wine, these appear to be primarily broken ends of cellulose fibers, floating in or adhering to the sides of the glass (Fig. 9.22). These probably come from cloths used to dry the glasses, or dust particles in the air. These possess microscopic cavities that can trap air when wine is poured into the glass. Alternate nucleation sites may originate with gas-filled cavities that form on minute crystals of tartrate suspended in the wine. It is into these nucleation centers that carbon dioxide diffuses. This process accounts for the slow release of almost 60% of the CO_2 over a period of approximately an hour (Fig. 9.21). Even traces of detergent can inhibit this occurrence. When detergent coats nucleation sites, bubble enlargement is prevented.

As carbon dioxide diffuses into entrapped gas pockets (nucleation sites), the nascent bubble enlarges. When the

diameter reaches a critical size (diameters ranging from 14 to 31 μm) (Liger-Belair *et al.*, 2002), buoyancy results in detachment (Fig. 9.23). During ascent, bubble diameter enlarges to more than 600 μm, as additional carbon dioxide diffuses into the bubble. Ascent starts slowly (about 0.2 cm/sec), but rapidly increases as the bubble enlarges – reaching more than 6 cm/sec as it nears the surface. This results in the increasing distance between bubbles as they rise. New bubbles form at a rate of about 15/sec, but this can vary from 1 to 30 per sec.

Gushing (when it occurs) results from a number of different nucleation processes. The mechanical shock of opening or pouring provides sufficient free energy to weaken the bonds between water and carbon dioxide. Disruption of van der Waals forces permits carbon dioxide molecules to form nascent bubbles throughout the wine. The process is called **homogeneous nucleation**. If the bubbles reach a critical size, they incorporate more CO_2 than they lose. They continue to grow and begin their ascent to the surface. Because the energy source for homogeneous nucleation is transient, so is the effervescence it provokes.

Another potential source of gushing comes from **stabilized microbubbles**. These develop from bubbles generated by agitation during handling. Most of the bubbles so formed float to the surface and break. Tiny bubbles may lose carbon dioxide to the wine and dissolve. However, surfactants in the wine may coat the face of these microbubbles, producing a gas-impermeable membrane that

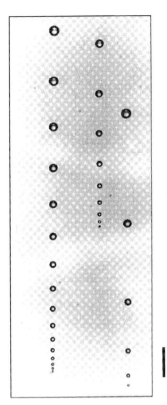

Figure 9.23 Simultaneous formation of distinct bubble chains from a collection of differently sized nucleation sites on the side of a champagne flute (bar = 1 mm). (From Liger-Belair, 2005, reproduced by permission. Copyright 2005 American Chemical Society)

stabilizes the bubble. After the bottle is opened, the bubbles rise to the surface. Gushing from this source takes a few seconds to develop. **Semistabilized microbubbles,** formed shortly after rough handling, may aggravate gushing. They can act as additional sites for bubble growth.

As bubbles reach the surface, they tend to mound in the center or collect around the edge of the glass. They soon burst, due to the combined effects of the wine's alcohol content, various surface-active ingredients that reduce surface tension, liquid drainage from between the bubbles, and insufficient rigidifying agents, such as proteins and glycoproteins. The bubbles tend to implode on rupture. The resulting shock wave propels a minute column of wine up from what was the submerged base of the bubble. The column breaks into a series of microscopic droplets that can be ejected a few centimeters at several m/sec (Liger-Belair *et al.*, 2001). Because several hundred bubbles may burst per second, the surface of the wine is spiked with these thin cone-shaped spires. Their millisecond duration and miniature size make them invisible to the naked eye. They do, however, induce a perceptible touch sensation on the tongue and palate by stimulating nocioreceptors (pain).

Because several aromatic compounds (such as various alcohols, aldehydes and organic acids) adsorb onto the surface of the bubble, they are also ejected into the air at bubble break. These aromatic droplets (or their remnants) can be carried by an air stream into the nose (Liger-Belair *et al.*, 2001). This enhances detection of the subtle fragrance that tends to characterize most sparkling wines. However, as time passes, the accumulation of protein and glycoprotein surfactants on the wine surface modifies and eventually prevents the ejection of aromatic laden droplets (Liger-Belair, 2001).

The formation of durable, continuous chains of small bubbles is an important quality attribute of sparkling wines. The factors that regulate this property are still incompletely understood. Cool temperatures during fermentation and aging, and long contact with the lees are thought to favor the property. Colloidal glycoproteins released during yeast autolysis, notably a portion of mannoproteins (Núñez *et al.*, 2006), appear to be particularly important in the formation of sustained chains of fine bubbles (Senée *et al.*, 1999).

Another property in the perceived quality of sparkling wines is the persistence of a small ring of bubbles (*cordon de mousse*) around the edge of the glass. In contrast to beer, the foam rapidly collapses and must be continuously replenished. The formation and durability of the *mousse* are largely dependent on the nature of the surfactants that decrease surface tension (such as soluble proteins, polyphenols, and polysaccharides), and on the type and number of metallic ions in the wine. The potential for *mousse* formation initially increases after the second fermentation, but may decline thereafter (Andrés-Lacueva *et al.*, 1997). Subsequently, *mousse* stability may again increase as autolysis results in the further degradation of polymeric surfactants.

Gravity tends to remove fluid from between the bubbles, causing them to fuse with one another. Thinning of the fluid layer between the bubbles forces them to assume polyhedral shapes. As a result, uniformity of pressure exerted on the sides of the bubble is lost. This forces fluid into the angled corners of the bubbles and induces further compaction. Carbon dioxide in small bubbles increasingly comes under more pressure than in larger bubbles, promoting CO_2 diffusion from smaller to larger bubbles. As the size of the remaining bubbles enlarges, they become increasingly susceptible to rupture.

The presence of proteinaceous or polysaccharide surfactants can restrict bubble compression. Interaction between surfactants may give a degree of rigidity and elasticity to the *mousse*. Elasticity can absorb the energy of mechanical shocks, limiting fusion, and bubble rupture. Although formation of a *mousse* is desirable, it is also traditional that it be relatively evanescent. The relative absence of stabilizing surfactants limits its duration.

Fortified Wines

Fortified wines are classified together because of their elevated alcohol content. They usually have had wine spirits added at some stage in production. The marked flavor of fortified wines gives the grouping an additional unifying property. Because of flavor intensity, they are seldom consumed with meals, normally being served as aperitifs or dessert wines. Regrettably, several governments also combine them for the purposes of higher taxation.

Most of the well-known fortified wines have developed in the last two to three hundred years. For reasons unknown they have evolved primarily in southern Europe. Examples are sherry (southern Spain), port (Portugal), marsala (Sicily), madeira (Madeira) and vermouth (northern Italy). The production of some versions is discussed below.

Sherry and Sherry-like Wines

Sherry evolved into its near-present-day form in southern Spain, possibly as late as the early 1800s. The details of its development from a young table wine, transported to England in the 1600s, are unclear (Gonzalez Gordon, 1972). The solera system is thought to have originated in the early nineteenth century (Jeffs, 1982). In its current form, only white versions are produced.

In Spain, the designation sherry is used as a geographical appellation. It is restricted to wines produced in and around Jerez de la Frontera in Andalucia. Similar wines produced elsewhere in Spain or the rest of Europe are not permitted to use the sherry appellation. Nevertheless, similar wines may use the stylistic terms *fino*, *amontillado*, and *oloroso*.

Outside Europe, the designation "sherry" is used generically for wines that, to varying degrees, may resemble Spanish sherries. The name of the country or region of origin is typically appended to the term sherry. Such sherries are seldom produced by techniques similar to those employed in Jerez.

Because three distinctly different techniques are used worldwide, each is described separately. These are the traditional **solera** procedure, the **submerged *fino*** technique, and the **baked** method.

SOLERA SYSTEM

The solera system developed in southern Spain as a means of fractional blending. In this expression, younger wine is periodically added and sequentially transferred through a series of casks of older wine. The amount added is equivalent to that removed for transfer to a subsequent series of casks (*criaderas*) containing older wine, or prepared for bottling (Fig. 9.24). Sequential blending and maturation occur in relatively small-volume casks (*butts*). The technique is ideally suited for the production of wine that is both brand-distinctive and consistent from year-to-year.

The frequency and proportion of wine transferred is adjusted to the style desired. The number of *criaderas* is particularly important as it influences the wine's development. For example, *fino* sherries require frequent transfers through many *criadera* stages, whereas *oloroso* sherries develop best with few *criaderas* and infrequent transfers.

These factors also influence the average age of the sherry produced. When a solera system is initiated (a collection of *criaderas*), the average age of the wine rises rapidly (Fig. 9.25). Subsequently, the mean age increases progressively more slowly, finally reaching what approximates a constant age. The plateau is reached more quickly when the frequency and proportion of the wine transferred is increased. The number of *criaderas* in a solera system also influences the rate and maximal age achieved. The greater the number of *criaderas*, the older the stable age finally achieved. The final *criadera* in a solera blending system is termed the *solera*. Formulas for calculating the effects of these factors are discussed in Baker *et al.* (1952).

Spanish sherry is subdivided into three major categories – *fino*, *amontillado*, and *oloroso*. They also may be classified based on where the wines are matured (e.g., Sanlúcar de Barrameda vs. Jerez de la Frontera), by their sensory characteristics (e.g., *palo cortado* vs. *raya oloroso*), or on how they are blended (e.g., cream-type sherries).

BASE WINE PRODUCTION

Whereas the development of a sherry into a *fino* or *oloroso* once seemed arbitrary, it can be largely predicted and directed. Experience has shown that juice derived from grapes grown in cooler vineyards, or in cooler years is more predisposed to becoming a *fino*. Vineyards containing higher proportions of chalk in the soil also tend to favor *fino* development. Gentle grape pressing, and the inclusion of little press-run juice, further predispose evolution toward a *fino*. Conversely, juice derived from grapes ripened under hot conditions, grown on soils containing less chalk, pressed in hydraulic vertical presses, and incorporating press-run fractions generally promote transformation into an *oloroso*. Slightly higher initial phenolic contents are desired in wines designed for *oloroso* production (encouraging oxidation). These tendencies can be particularly enhanced by the level of fortification. Contents of 15 and 18% alcohol favor *fino* or *oloroso* development, respectively. The level of cask (*butt*) filling and maturation temperature also directs the wine's evolution.

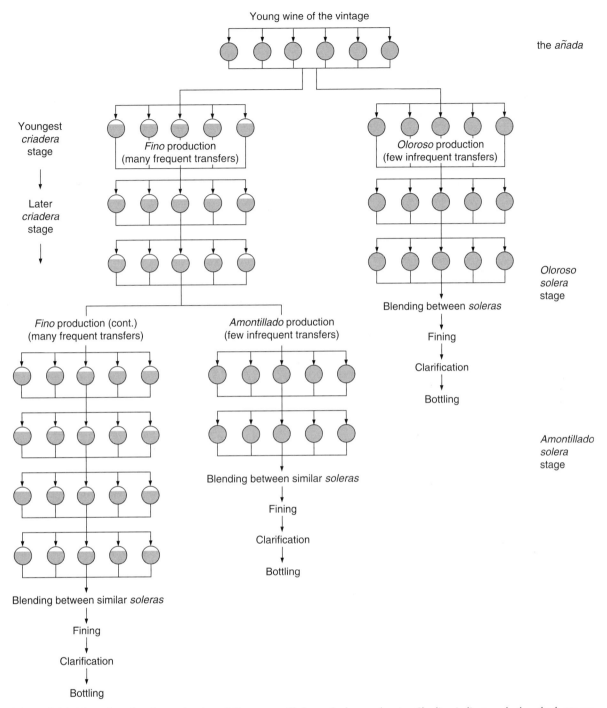

Figure 9.24 Flow chart for the production of *fino*, *amontillado*, and *oloroso* sherries. Shading indicates whether the butts are kept full or partially (~20%) empty. All are produced in a multistage, fractional, blending procedure termed the solera system. The base wine is termed *añada* before entering a solera maturation system. Each of the progressive blending stages is called a *criadera*, the last of which is specifically termed the *solera*

Production of the base wine generally follows standard procedures, except that fermentation occurs between 20 and 27°C – higher than generally preferred elsewhere for white wines. Pressing almost immediately follows crushing to limit tannin extraction. Spontaneous settling for several hours brings the suspended solids content down to 0.5–1%. The effect of suspended solids on the chemical composition of the wine is illustrated in Table 9.6. Increasing tannin content gives a roughness inconsistent with accepted sherry norms. Because

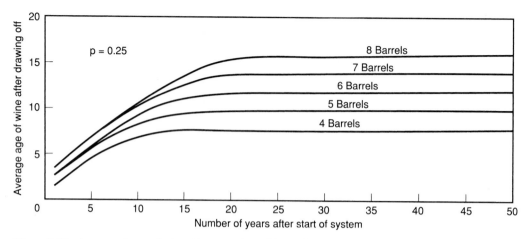

Figure 9.25 Average wine age in the oldest container of four-, five-, six-, seven- and eight-barrel solera systems when 25% of the wine is transferred biennially. (From Baker *et al.*, 1952, reproduced by permission)

Table 9.6 Concentration of several compounds after the fermentation of a must with a different concentration of solids[a,b]

	M1	M2	M3
Volatile acidity (g/L)	0.19	0.25	0.41
Sugars (g/L)	1.7	1.4	1.4
Ethanol (% v/v)	10.6	10.8	10.7
Acetaldehyde (mg/L)	163	56	67
Ethylacetate (mg/L)	23	23	27
Methanol (mg/L)	31	32	36
N-propanol (mg/L)	15	15	16
Isobutanol (mg/L)	27	25	31
Isoamilics (mg/L)	236	189	164
Glycerol (g/L)	7.3	6.5	6.2
Citric acid (g/L)	0.74	0.77	0.64
Malic acid (g/L)	0.07	0.15	0.24
Succinic acid (g/L)	0.52	0.38	0.89
Lactic acid (g/L)	0.11	0.12	0.13

[a] Not decanted (M1, 5 g/liter solids), decanted (M2, 1.1 g/liter solids) and decanted and filtered (M3, 0 g/liter solids)

[b] Analyses correspond to the sixth day after inoculation. The experiment was done per triplicate with a coefficient of variation of <10%.

Source: Martínez *et al.*, 1998, reproduced with permission

the juice often has an undesirably high pH, tartaric acid is commonly added to correct this deficiency. The older procedure (called plastering) involved adding *yeso*, a crude form of gypsum (calcium sulfate). Plastering both lowered the pH and provided a source of sulfate. After conversion to sulfite, it had the advantage of inhibiting the growth of spoilage bacteria, notably *Lactobacillus*

trichodes. Adding sulfur dioxide directly has the same effect, but avoids the addition of calcium and other potential mineral contaminants in *yeso*.

Inoculation with specific yeast strains is still uncommon, with fermentation developing spontaneously from the indigenous grape and winery (*bodega*) flora. If yeast inoculation is employed, it is usually added to one-third of the must. Once fermentation has become turbulent (usually 4–5 days), an equivalent volume of must is added. When this volume is clearly fermenting, the final must portion is added. Esteve-Zarzoso *et al.* (2001) report that the inoculated strain is occasionally replaced by wild strains by the end of fermentation.

To avoid interference with the sherry flavor, the base wine ideally should have little varietal aroma. In Spain, the neutral-flavored 'Palomino' and 'Pedro Ximénez' varieties are preferred.

STYLISTIC FORMS OF JEREZ SHERRY

Fino Fino sherries are the lightest, most subtly flavored sherries. They are also characterized by possession of a *flor* bouquet. This develops from the action of a film of yeast cells (**velum**) that grows on the surface of the wine (Plate 9.7). The film-forming yeasts (*flor*) are typically related to those that induce the original fermentation. If *flor* development does not occur rapidly, an inoculum may be transferred from casks containing an active culture.

After the first racking, the base wine is fortified to bring the alcohol content up to 15–15.5%. Fortification is conducted with a 50:50 blend of rectified (aromatically neutral) wine spirits (~95% ethanol) and aged sherry, called *miteado*. Storage for approximately 3 days permits settling of the cloud that forms, and avoids production of a haze in the young wine. At 15%, the alcohol favors *flor* development as well as restricts the growth of acetic acid bacteria. A velum (pellicle, biofilm) of yeast forms when the elevated alcohol content promotes the production of a hydrophobic cell wall (Alexandre *et al.*, 1999). The change appears to depend on activation and regulation of *FLO11*. The gene encodes production of a hydrophobic glycoprotein (Ishigami *et al.*, 2006). Unlike other microbial biofilms, no protein or polysaccharide extracellular matrix forms between the cells. Low pH and the presence of biotin (Iimura *et al.*, 1980), pantothenate (Martínez *et al.*, 1997c), and phenolic compounds (Cantarelli, 1989) further favor velum formation.

The hydrophobic cell surface favors yeast aggregation. Entrapment of carbon dioxide, generated by yeast metabolism, increases buoyancy (Martínez *et al.*, 1997a; Zara *et al.*, 2005). This permits the aggregated cells to float to the surface, forming a velum. Thus, velum development is considered an adaptive mechanism, whereby starved cells gain access to oxygen, permitting the respiratory metabolism of ethanol and acetaldehyde (Ibeas *et al.*, 1997a). Sulfur dioxide content is commonly adjusted to approximately 100 mg/liter to limit the growth of lactic acid bacteria.

The wine is matured in American oak cooperage. The casks have a capacity of about 490 liters. Typically, they have been used previously to ferment wine. Prior conditioning minimizes oak-flavor extraction that might otherwise mask the *fino* bouquet. Barrels are left with 10–20% ullage to provide sufficient surface for *flor* development.

During initial storage (*añada*), the development of the new wine (*sobretabla*) is periodically checked to assess its development. It remains in the *añada* stage for from one to two years. Flor begins to develop and may soon cover the wine. If *flor* does not form as desired, the wine is either used for the production of another sherry style or distilled.

When wine is removed from the last stage of fractional blending (the *solera*), in preparation for bottling, its volume is replenished from the next oldest *criadera* (Fig. 9.24). This volume is in turn replenished with wine taken from the second oldest *criadera*. This continues sequentially, until the youngest *criadera* is reached. The wine removed from the youngest *criadera* is replenished with *sobretabla* from an *añada*. The wine drawn from each butt is generally blended with wine from other casks in the same *criadera*, before transfer to the next stage.

In the process, about one-quarter of the wine (100 liters) is removed and replenished during each transfer. The transfer frequency depends on development of the wine, as determined by sensory analysis. Typically, transfers occur about twice a year, but may occur more frequently. There generally are four or five *criaderas* in a *fino* solera system. There may, however, be considerably more, especially with Manzanilla *fino* produced in Sanlúcar.

The butts of a *criadera* are arrayed in rows in aboveground buildings called *bodegas* (Plate 9.8). Generally, they are stacked no more than three to four butts high, to avoid structural damage to the cooperage. The different *criaderas* in a particular solera may be housed throughout the *bodega*. Large firms generally have numerous soleras, all in various stages of development. Each style of sherry will usually be a blend of wine from several soleras with different initiation dates.

Wine transfer from *criadera* to *criadera* is being automated. Formerly, transfer was labor-intensive, involving manual siphoning, blending, and subsequent pouring. Previously, several ingenious devices, involving perforated tubes (*rociadors*) and wedge-shaped funnels with an angled spout (*canoas*), were used to minimize disturbance to the yeast film during filling.

Frequent wine transfer is critical to the development and maintenance of an active *flor,* presumably providing nutrients (Berlanga *et al.*, 2004a). Proline is the principal nitrogen source and promotes yeast growth, whereas biotin favors production of a hydrophobic cell wall. Providing a favorable surface area/volume (*SA/V*) ratio is also important. Leaving the butts about 20% empty creates an optimal *SA/V* ratio of about 15 cm^2/liter (Fornachon, 1953). The practice provides both sufficient contact with the primary carbon and energy sources and sufficient oxygen for respiration. The bung hole is left slightly ajar to allow gradual air exchange. In the presence of oxygen, yeast mitochondrial aldehyde dehydrogenase is produced and oxidizes ethanol to acetaldehyde (Millán and Ortega, 1988). Ethanol metabolism is particularly active during the first few months of velum development. The consumption of ethanol by *flor* may require periodic adjustment to maintain its concentration between 15 and 15.5%.

The taxonomic nature of the *flor* population is still in doubt. This may relate as much to diversity from barrel-to-barrel and winery-to-winery, as to changing views of yeast taxonomy. The dominant *flor*-inducing members have been variously identified as strains of *Saccharomyces cerevisiae*, *S. bayanus*, *Torulaspora delbrueckii*, or *Zygosaccharomyces rouxii*. Recent research suggests that unique strains of *S. cerevisiae* constitute the majority of *flor* yeasts (Ibeas *et al.*, 1997b; Esteve-Zarzoso *et al.*, 2004). They appear to possess a particular deletion in their 5.8S ribosomal gene. Their proportions and genetic characteristics change during fermentation, and throughout solera aging. For example, the frequency of cells with functional mitochondria (capable of respiration) is higher in *flor* strains (Martínez *et al.*, 1995). Whether this reflects selection of existing, or mutant strains is unknown. *Flor*

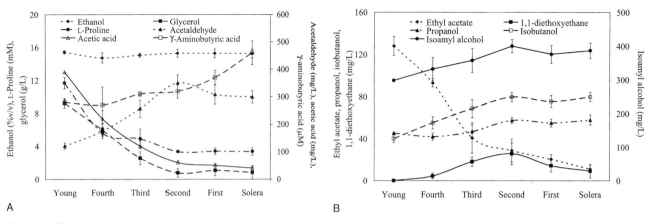

Figure 9.26 Concentration of (**A**) ethanol, L-proline, glycerol, acetaldehyde, acetic acid, and γ-aminobutyric acid; (**B**) ethyl acetate, propanol, isobutanol, 1,1-diethoxyethane, and isoamyl alcohol in young wine and during solera aging. (From Berlanga *et al.*, 2004b, reproduced by permission)

yeasts also have higher expression of *HSP* genes,[1] and correspondingly higher resistance to the toxicity of ethanol and acetaldehyde (Esteve-Zarzoso *et al.*, 2001). They also have greater insensitivity to osmotic stress than fermentative strains. In addition, they overexpress *SSU1*, a gene that encodes for a sulfite pump that translocates sulfite out of the cell. This enhances cellular resistance to sulfur dioxide toxicity. Finally, and most distinctive *flor* strains express *FLO11* (alternately called *MUC1*). This gene encodes for the surface glycoprotein that is responsible for velum development.

Individual *flor* strains tend to vary in their synthesis of volatile compounds, such as esters, higher alcohols, and terpenes. Whether these differences are of sensory significance is unclear (Cabrera *et al.*, 1988) as most yeast vela are a mix of strains and species. Criddle *et al.* (1981) considered that mixed cultures form more uniform pellicles than pure cultures.

Flor yeasts are critical to the development of *fino* sherries. In the absence of fermentable sugars, yeast growth depends on a shift to respiratory metabolism. Formation of a hydrophobic cell wall, and the formation of a film on the wine surface exposes the yeast to the necessary oxygen. As the film grows to cover the wine, diffusion of oxygen into the wine is restricted. Thus, the redox potential of the wine increases, although the wine is seemingly exposed to air. This, plus yeast-induced inhibition of phenol oxidation, probably explains the pale color of wine matured under *flor* (Martínez *et al.*, 1998; Lopez-Toledano *et al.*, 2002).

Flor yeasts partially respire ethanol, glycerol, acetic and several organic acids, producing acetaldehyde and various aromatic metabolic by-products. Examples are 1,1-diethoxyethane, diacetyl, acetoin, 2,3-butanedione, and C_4 organic acids (Cortés *et al.*, 1999). During solera aging, ethanol consumption remains fairly constant (in the range of 5–6 liters per year) (Martínez *et al.*, 1998). In contrast, glycerol consumption rapidly declines as it is consumed (Bravo, 1984). Most of the acetaldehyde generated is respired via the TCA cycle by yeast cells at the surface. Its concentration reaches its highest concentration during the *añada* phase, declines early in solera aging, and then slowly rises again (Martínez *et al.*, 1997b). Fermentation in the lower, submerged portion of the film probably metabolizes residual sugars in the wine.

The accumulation of acetaldehyde (not respired during yeast metabolism) gives sherry its oxidized bouquet. Subsequent reaction of acetaldehyde with ethanol, glycerol, and other polyols generates acetals. Of these, only 1,1-diethoxyethane likely accumulates sufficiently to add a 'green' note to the fragrance of the wine (see Etiévant, 1991). Small amounts of terpenes, such as linalool, *cis*- and *trans*-nerolidol, and *trans,trans*-farnesol are synthesized by *flor* yeasts (Fagan *et al.*, 1981). Several lactones, notably substituted γ-butyrolactones, have been isolated from *fino* sherries. They are generally regarded as important in the development of a *fino* character (Kung *et al.*, 1980). The lactone sotolon is particularly important in contributing to the characteristic walnut-like fragrance of *fino* sherries. Sotolon has also been isolated from *vin jaune*, a sherry-like wine produced in the south of France. It can form due to a slow, abiotic reaction between α-ketoglutaric acid and acetaldehyde (Pham *et al.*, 1995). Nevertheless, the typical *fino* fragrance is derived from the combined effects of several aromatics, including lactones, acetals, terpenes, and aldehydes. Examples of other chemical changes during solera aging are noted in Fig. 9.26.

---

[1] Termed because these genes were first identified as coding for proteins produced as a result of heat shock (heat-shock proteins).

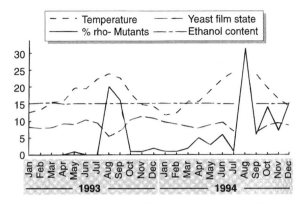

Figure 9.27 Average value of the alcohol content (%vol), temperature (°C), state of the flor (0, deteriorated to 10, optimal), and proportion of *rho* mutants in the yeast film (%), obtained monthly in the studied barrels over two years. (From Ibeas *et al.*, 1997a, reproduced by permission)

In addition to the oxidative metabolism of film yeasts, volatile compounds are lost through the sides and bung hole of the cooperage. Conversely, the evaporation of water from the butts can increase the concentration of various compounds (Martínez de la Ossa *et al.*, 1987). This could potentially increase the alcohol content by about 0.2% v/v per year (Martínez *et al.*, 1998). Water evaporation for barrel surfaces can be reduced by increasing the relative humidity in the *bodega*. This is most simply achieved by sprinkling water on the *bodega* floor.

Although *flor* coverage is commonly complete, yeast activity is not constant. Growth is usually most active in the spring and fall months, when the ambient temperatures in the *bodega* are between 15 and 20 °C. During the winter and summer months, unfavorable temperature conditions slow growth and *flor* coverage may become patchy. This is particularly noticeable when temperatures rise above 22.5 °C (Fig. 9.27). This also correlates with an increase in the number of *rho*° strains (lacking mitochondria).

As the velum thickens, lower sections break off and fall to the bottom. Because of yeast autolysis, sediment accumulates only slowly, and the cooperage seldom needs cleaning. Nutrients released by autolysis are probably important to continued *flor* growth. The substances released also may be important in the development of the typical *fino* bouquet, similar to the situation with sparkling wines.

After maturation is complete, wine removed from one *solera* is usually blended with wines from other *soleras*. Subsequently, the alcohol content is adjusted to 16.5% alcohol, or an amount considered appropriate for the export market. Increasing the alcohol content stops further *flor* activity. During maturation, the malic acid level falls, which may leave the wine with insufficient acidity.

If so, tartaric acid is commonly added. A polishing clarification and cold stabilization prepare the wine for bottling. *Fino* sherries are seldom blended with sweetening or *color* wines. They are sold as dry, pale-colored aperitif wines.

Despite *fino* sherries being considered oxidized (due to the presence of acetaldehyde), they do not possess the oxidized character of white table wines. These fragrant and color attributes develop only after bottling. Browning and development of typical oxidized flavors are accelerated by exposure to light (Benítez *et al.*, 2003).

Amontillado *Amontillado* sherries begin development like a *fino* sherry. Subsequently, the frequency of transfer is slowed, decreasing the rate of nutrient replenishment. This also favors water loss, tending to increase the relative alcohol content. All these features slowly lead to the cessation of *flor* growth. Because an exposed wine surface is no longer required, the butts are usually filled. Without *flor* protection, the wine becomes darker colored and develops a richer, oxidized flavor. There are few *criadera* stages in *amontillado* solera systems, the number depending on the flavor desired. Most *amontillado* soleras are initiated intentionally, rather than occurring by accident.

When drawn from the *solera*, *amontillado* sherries may be sweetened and fortified to meet particular market demands. In Spain, the wine is usually left unmodified. After cold stabilization and a polishing clarification, the wine is ready for bottling. *Amontillado* sherries also may be used in preparing cream-type sherry blends.

Oloroso The first step in the production of an *oloroso* sherry involves fortification of the wine in the *añada* to about 18% alcohol. This inhibits yeast and bacterial growth, and makes *oloroso* maturation less sensitive to temperature fluctuation, as compared to other sherry types. Consequently, the butts are placed in areas of the bodega showing the greatest temperature fluctuation. The butts are commonly filled to about 95% capacity, and irregular topping limits the rate and degree of oxidation. This may partially explain the minimal increase in acetaldehyde content observed during *oloroso* maturation. However, an additional reason may be the conversion of acetaldehyde to acetic acid, and its subsequent esterification with ethanol to ethyl acetate. This is suggested by the progressive increase in the concentration of acetic acid and ethyl acetate during *oloroso* maturation (Martínez de la Ossa *et al.*, 1987).

Because of the longer maturation in oak, the concentration of phenolic compounds is generally higher in *amontillado* and *oloroso* than in *fino* sherries (Estrella *et al.*, 1986). Sugar and alcohol contents also have been noted to rise during the aging of *amontillado* and *oloroso* sherries.

There are typically few *criadera* stages in an *oloroso* solera. Transfer rates are slow, often amounting to only 15% per year. Because fractional blending is limited, the wine shows considerable barrel-to-barrel variation. The cellar master maintains brand consistency through subsequent blending.

Dry *oloroso* sherries are seldom found on the international market. They are usually brought up to about 21% alcohol and blended with sweetening and *color* wines. After clarification and stabilization, they are ready for bottling. *Palo cortado* and *raya* sherries are special *oloroso* sherries. They are more subtle and rougher versions, respectively.

SWEETENING AND *COLOR* WINES

Sweetening is typically achieved by adding one of two special sweetening wines, PX or *mistela*. PX is juice extracted from sun-dried berries (*soleo* process) of the 'Pedro Ximénez' variety. The juice, possessing a marked raisined flavor (Franco *et al.*, 2004), is fortified to about 9% alcohol, allowed to settle, and placed in special soleras for aging. The wine so produced possesses about 40% sugar. In contrast, *mistela* is produced from 'Palomino' grapes, the variety used in sherry production. The free-run juice and first pressing are fortified to about 15% alcohol, allowed to settle, and aged in casks or tanks. *Mistela* is not fractionally blended through a solera system. It generally contains about 16% sugar.

Alternatively, sweetening may be derived from the addition of *color* wine. It is normally obtained from the second pressings of 'Palomino' grapes. Boiling brings the volume down to approximately one-fifth the original volume. The froth that forms during boiling is periodically removed. The product, called *arrope*, is a thick, dark, highly caramelized 70% sugar solution. Addition of *arrope* to fermenting 'Palomino' juice successively slows the rate of fermentation after each addition, until fermentation ceases. It often possesses an alcohol strength of about 8% and contains about 22% sugar. The wine may be raised to about 15% alcohol and be solera-aged. Alternately, *arrope* may be added at the end of fermentation (one part *arrope* to two parts wine). It produces what is called *vino de color*.

EUROPEAN SHERRY-LIKE WINES

The major source of sherry-like wines, other than Jerez, is Montilla-Moriles. It lies about 160 km northeast of Jerez. Its wines were once transported to Jerez for maturation and used in the production of Jerez sherry. This practice is no longer permitted. In Montilla-Moriles, 'Pedro Ximénez' is the predominant cultivar. Grapes of this variety can, without special drying in the sun, yield wines of up to 15.5 to 16% alcohol. Thus, *flor* tends to develop spontaneously without fortification.

Fino sherries are produced from a combination of free-run and first press-run fractions. *Oloroso* sherries are produced primarily from free-run juice plus additional press fractions. Fermentation traditionally occurs in large earthenware vessels (*tinajas*). These possess capacities between 6000 and 9000 liters. They resemble storage vessels, called *pithoi*, used by the ancient Greeks and Romans. The wines are solera-aged, in a procedure analogous to that used in Jerez.

Small amounts of solera-aged sweet wine are also produced in Málaga, about 180 km east of Jerez. Most Málaga wine is produced without solera aging. Those winemakers who use fractional blending employ fermentation procedures distinct from those practiced in Jerez and Montilla. Grapes of the 'Pedro Ximénez' and 'Moscatel' cultivars are placed on mats to dehydrate and overripen in the sun. The grapes are covered at night to prevent the formation of dew on the surface. Juice, fortified to 7% alcohol, may be added to freshly pressed juice before fermentation. The subsequent fermentation is slow and often incomplete. The resulting wine may have an alcohol level of 15–16%, and a residual sugar content of 160–200 g/liter. The wine may be further sweetened with PX and *mistela*. Solera aging, when employed, occurs without the interaction of *flor*, in a manner similar to that of an *oloroso*. The wine may be colored with *sancocho*, a product possessing a specific gravity about 1.24 and similar to the *color* wine of Jerez. Because *sancocho* is concentrated to only one-third of the original volume, it is lighter in color and less caramelized than *color* wine. It is added slowly to the fermenting must to give its distinctive *arrope* flavor and color attributes. The brown pigments are melanoidin compounds (Rivero-Pérez *et al.*, 2002). If aged in an oxidative solera system, the wine develops a complex flavor and an astringent aftertaste.

European sherry-like wines produced outside of Spain include Vernaccia di Oristano and Malvasia di Bosa (Sardinia), as well as *vin jaune* (Jura). The Sardinian wines are *flor*-matured wines made from fully mature grapes of the 'Vernaccia' and 'Malvasia' cultivars, respectively. Natural ripening on the vine commonly produces grapes with sufficient sugar content to yield wines of more than 15% alcohol. Thus, the wines often need no fortification to favor *flor* development.

Vin jaune is usually produced from the 'Savagnin' cultivar, a mild-flavored strain of 'Traminer.' The grapes are harvested late and allowed to dry for several months to develop a high sugar content. The wine is barrel-aged for at least 6 years without racking (Dos Santos *et al.*, 2000). *Flor* development occurs spontaneously, without inoculation or fortification. Without fractional blending, the completeness of the *flor* covering varies with the season.

NON-EUROPEAN SHERRY-LIKE WINES

Solera-aged Sherries Production of solera-aged wine, similar to that practiced in Spain, is uncommon in the New World. The expense of fractional blending and the prolonged maturation undoubtedly explain this situation. Up to 10 times the volume of wine may be maturing as is sold each year using fractional blending.

South African sherries are produced with solera blending, but the details are quite different from those in Spain. 'Palomino' and 'Chenin blanc' ('Steen') are the varieties normally used. The juice is inoculated with selected yeast strains, chosen for their excellent fermentation and film-forming habits.

Wines designed to become *flor*-matured sherries are fortified to 15–15.5% alcohol. They are placed, without clarification, in 450 liter butts for 2–4 years. A 10% ullage provides surface for *flor* development. After the initial maturation, storage of both lees and wine occurs in casks containing about 1500 liters. There are generally two *criaderas* and one *solera* stage. Each stage has only one or two casks. Consequently, little fractional blending occurs between transfers. Wine is generally drawn off in 450-liter lots, equivalent to the contents of *añada* barrels. Due to the proportionally higher evaporation of water through the wood, the alcohol content reaches a level that inhibits *flor* activity. The wine generated is apparently intermediate in character between a *fino* and an *amontillado*.

Wines intended for *oloroso* production are fortified to about 17% alcohol after fermentation. Subsequent storage occurs for about 10 years in butts without fractional blending. Sweetening *mistela*, derived from 'Palomino' or 'Chenin blanc' juice, also is fortified to 17% alcohol and matured for upward of 10 years in oak casks. *Color* wines are produced from *arrobe*, blended into young sherry, and stored in butts for prolonged periods.

In Australia, *flor* sherries are seldom fractionally blended. After fortification, the wine may be inoculated with a film-forming yeast and matured for upward of 2 years in barrels (~275 liters) or cement tanks (~1000 liters). When the desired *flor* character has been reached, the wine is fortified to 18–19% alcohol. Further maturation occurs in oak for 1–3 years.

Submerged-culture Sherries A *flor* procedure, markedly different from that based on the Spanish model described above, has been pioneered in Australia, California, and Canada. It involves a submerged-culture technique. Respiratory growth of the *flor* yeasts is maintained with agitation and aeration throughout the whole volume of the wine.

The base wine is fortified to about 15% alcohol and inoculated with an acclimated culture of *flor* yeast. Optimal growth conditions include a pH of about 3.2, a temperature of 15°C, and SO_2 contents close to 100 mg/liter. Oxygen is provided by bubbling filtered air or oxygen through the wine. The use of porcelain sparging bulbs finely disperses the gas, improving oxygen adsorption and minimizing the loss of aldehydes and other aromatics. The yeasts are kept suspended and highly dispersed by mechanical agitation. The process has the advantage of rapidly producing high levels of acetaldehyde. By adjusting the duration of yeast action, slightly (~200 mg/liter acetaldehyde) to heavily aldehydic wines (>1000 mg/liter) can be obtained.

After *flor* treatment, fortification with relatively neutral spirits raises the alcohol content to 17 to 19%. Fortification appears to intensify the *flor* character. Because the wine generally lacks the complexity and finesse of solera-aged wines, it is customarily used to enhance the complexity of baked sherries (see below), rather than being used alone. The lack of finesse may result from the absence of the reductive phase that occurs under the *flor* growth, plus products released during yeast autolysis.

Baked Sherries Baking has been the most common technique for producing sherries in Canada and the United States. It involves a process that resembles more the production of madeira than it does Jerez sherry. Not surprising, the resulting wines resemble madeira more than Spanish sherry.

Varieties that oxidize fairly readily are preferred in the production of baked sherries. In eastern North America, the variety 'Niagara' has routinely been used, whereas in California, varieties such as 'Thompson seedless,' 'Palomino,' and 'Tokay,' are employed. Both white and red grape varieties may be used. Baking destroys the original color of the wine. Posson (1981) recommends juice possessing a pH no higher than 3.4 for submerged-culture sherries, whereas pH values between 3.4 and 3.6 are optimal for baked sherries.

Slow baking occurs when barrels of wine are exposed to the sun. More rapid baking is achieved in artificially heated rooms. Heating coils also may be inserted directly into wine-storage tanks. Heating is variously provided by passing steam or hot water through the coils. California winemakers appear to prefer baking at 49°C for 4 weeks, rather than the former 10 weeks at 60°C (Posson, 1981).

Heating induces the formation of a wide variety of oxidative and Maillard products, including furfurals, caramelization compounds, and browning by-products. Baking also promotes ethanol oxidation to acetaldehyde (Kundu *et al.*, 1979). Air or oxygen gas may be bubbled through the heated wine to accelerate oxidation.

The desired level of baking may be measured chemically by the production of 5-(hydroxymethyl)-2-furaldehyde, or colorimetrically by the development of brown pigments. Nevertheless, the generally preferred method is by sensory analysis.

After baking, especially by rapid heating, the wine requires maturation to lose some of the resulting strong flavors and rough mouth-feel. Although oak maturation is preferred, used barrels are employed to avoid giving the wine an oaky attribute. Aging may last for from 6 months to more than 3 years.

Baked wines are always finished sweet. The sweetness may come from fortified grape juice added to a base wine. Alternatively, premature termination of fermentation (by fortification) can retain residual sweetness in the base wine.

Porto and Port-like Wines

The beginnings of port development, or *porto* as it is called in Portugal, are unclear. Fortification may have been used as early as 1670. Nevertheless, the practice seems not to have become standard until the mid-eighteenth century. The premature cessation of fermentation by the addition of largely unrectified brandy (*aguardente*) is essential to modern port production. The retention of a high sugar content, and the higher (fusel) alcohols supplied with the brandy during fortification, give port two of its most distinguishing features. The brandy also contributes esters (ethyl hexanoate, ethyl octanoate; ethyl decanoate) and terpenes (α-terpineol, linalool) that donate a fruity, balsamic and spicy aroma (Rogerson and de Freitas, 2002). In addition, wine spirits are rich in aldehyde content, such as acetaldehyde, propionaldehyde, isovaleraldehyde, isobutyraldehyde, benzaldehyde (Pissarra *et al.*, 2005). These not only influence the fragrance, but also contribute to color development in young wines, by participating in the formation of alkyl-linked anthocyanin/tannin polymers. Subsequent aging and blending differentiate the various port styles.

PORTO

Port is primarily a red wine produced in the upper Douro Valley of northern Portugal. Although originating in the upper Douro, the wine is typically transported down river to Oporto for maturation and aging. These processes occur primarily in buildings called *lodges* in Vila Nova de Gala, located at the mouth of the Douro River. A small amount of white port is also produced.

Most port is not vintage dated. Producers blend samples from several vintages and localities to produce brand-named wines of consistent character. After 2–3 years' maturation, the wine is bottled and sold as **ruby** port. Blending small quantities of white port into a ruby port produces most inexpensive brands of **tawny** ports. However, only long aging in oak produces high-quality tawny port. During aging, the bright-red color fades to a tawny hue, and a mild, complex, oxidized character develop. Wines of superior quality from a single vintage, bottled between the second and third years of maturation in cask, become **vintage** port. After long bottle-aging, vintage port develops a distinctive and exquisitely complex fragrance. Wine from a single vintage, aged in cooperage for about 5 years, may be designated **late-bottled vintage** (LBV) port. LBV port matures more rapidly than vintage port, generates no sediment, and is considerably less expensive. A few single-estate (*quinta*) ports are produced, usually from a single vintage. **Vintage-character** ports often are produced from finer-quality ruby ports coming from the Cima Corgo region of the Douro (see Fig. 10.20).

BASE WINE PRODUCTION

Port wine is usually produced from a wide range of grape varieties. There are 28 red and 19 white cultivars authorized in the Douro. Formerly, white and red cultivars were interdispersed in the vineyard, the grapes being harvested, crushed, and vinified together. Currently, plantings are separated, permitting separate harvesting and vinification.

'Touriga Nacional,' 'Mourisco,' 'Mourisco de Semente,' 'Tinta Roriza,' 'Tinta Cão,' and 'Tinta Francisco' are the principal red cultivars. They possess the stable coloration, fruity aromas, and sugar content required to produce good quality port. 'Codega,' 'Malvasia,' and 'Rabigato' are the preferred white varieties.

Formerly, the grapes were vinified on the premises of the vineyard in shallow stone vats called *lagars*. They were granite troughs approximately 3–6 m across and 60 cm deep. Currently, most wine is vinified by regional cooperatives using modern crushing, pressing, and fermenting equipment. Autofermentors or other means are employed to promote early extraction of anthocyanins and tannins. Thermovinification may be used when the crop possesses more the usual level of fungal infection. Little wine is produced by the old foot-treading procedure.

As with most other wine styles, inoculation of the must with a specific yeast strain is becoming standard practice. The primary exception is with treading, in which fermentation by the indigenous flora remains traditional.

A major problem in port production is extracting sufficient anthocyanins to provide an intensely red color before the fermenting must is fortified. Pigment extraction is largely dependent on the heat and ethanol generated during the truncated fermentation period. Because the wine is separated from the pomace when the sugar levels fall to about 14.5°Brix, the opportunity for pigment extraction is short. Extraction is aided by extensive mixing of the juice and pomace during fermentation. Autofermentors achieve this automatically, whereas the long treading, traditionally employed in *lagar* fermentation, required considerable human exertion. The use of deeply pigmented varieties, such as 'Sousão' and 'Tinta Cão,' and the addition of sulfur dioxide (100 mg/liter) further help the release of sufficient anthocyanins.

Stopping fermentation midstream (after about 24–48 h) with brandy retains the wine's high acetaldehyde content present at this stage during fermentation (see Fig. 7.16). This probably aids color stability by favoring the production of anthocyanin–tannin polymers. The high sugar content retained tends to mask the bitterness of tannins, but not their astringency.

After the fermenting must is run off from the skins, it is fortified with wine spirits (77% ethanol). By the time fermentation stops, the must has dropped about another 2°Brix. Thus, the young wine possesses an alcohol concentration of about 19% and retains a sugar content of between 9 and 10%. Figure 9.28 illustrates the relationship between the initial and final °Brix of a wine fortified to 20.5% alcohol, at various °Brix values during fermentation. Fortification prevents the metabolism of lactic acid bacteria, which could produce too much acetic acid.

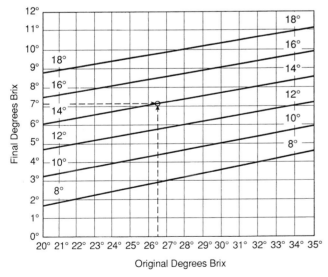

Figure 9.28 Relation between initial and final °Brix of musts fortified to 20.5% ethanol. For example, if the initial reading were 26.5°Brix and the desired final residual value were 7.2°Brix, then fortification should occur at 14°Brix. (From Joslyn and Amerine, 1964, reproduced by permission)

The actual amount of wine spirits required depends on the volume of the fermenting must, its alcohol content at fortification, the alcoholic strength of the wine spirits, and the desired degree of fortification. The proportion of spirit to must can be determined from their respective alcoholic strengths (Joslyn and Amerine, 1964). The proportion of spirit required is calculated by subtracting the alcoholic strength of the must being fortified (i.e., 8%) from the desired alcoholic strength (i.e., 18%). The corresponding must proportion is calculated by subtracting the desired alcoholic strength (i.e., 18%) from that of the fortifying spirit (i.e., 78%). In this example, 10 parts of the spirit (18 − 8 = 10) would be required per 60 parts (78 − 18 = 60) of the fermenting must to achieve the desired 18% alcohol.

The first press fraction from red port is fortified to the same level as the free-run. The press fractions may be kept separate for independent aging or combined immediately with the free-run. Press fractions are an important source of anthocyanins and phenolic flavors.

Previously, white wines were fermented on the skins in a manner similar to that for red port. However, the trend is for a shortened maceration period. As with red ports, most white ports are fortified when half the original sugar content has been fermented. Semidry and dry white ports are fortified later, or when fermentation is complete.

Fortification uses spirits distilled from wine. Unlike most fortifying spirits, what is used in Portugal is not highly rectified. Consequently, it contains many flavorants, notably higher alcohols. This feature gives Portuguese port a distinctive character. It is a property seldom present in non-Portuguese ports. The latter are customarily fortified with highly rectified (neutral) spirits at about 95% alcohol. In contrast, the wine distillate used to fortify port is relatively low in alcohol content (~77%).

To assure the complete termination of fermentation, the wine is thoroughly mixed with the fortifying spirit. Storage occurs in wood or cement cooperage. The first racking usually occurs between November and March. Additional fortification brings the ethanol concentration up to 19–20.5%.

Transportation to the *lodges* in Vila Nova de Gala occurs the following spring. Here, the wine receives most of its aging and blending. In contrast, most white port is aged in the upper Douro.

MATURATION AND BLENDING

Maturation occurs in large wooden or cement tanks, or oak casks of about 525-liter capacity (*pipes*). The type and duration of maturation depend largely on the wine style intended. Racking may vary from quarterly to yearly. Slight fortification after each racking may bring the alcoholic strength up to 21%, and compensate for volume lost via evaporation from the cooperage.

The aging of ruby and tawny ports commonly occurs in *pipes* left partially empty. This enables development of a slightly oxidized character. In contrast, vintage ports are protected from oxygen exposure. Vintage port derives much of its distinctive bouquet from a long, reductive, in-bottle aging.

Because of the large number of producers, blending of individual wines usually begins shortly after transfer to Oporto. As the character of each combination becomes more evident, further mixing reduces the number of blends to a more manageable figure. Blending during the first 2 years is usually confined to wine produced from a single vintage. Later, wines not used in one of the vintage-style ports may be combined fractionally with older wine. For example, in the development of ruby port, wine from reserve blends is commonly added to 2- or 3-year-old ruby ports. A portion of this blend becomes the reserve for the subsequent year. The remainder is used in preparation of the final mix, which may contain several reserve blends, plus optional amounts of sweeter or drier wines. The final blend is left to mature in oak cooperage for several months prior to fining, stabilization, and bottling.

Inexpensive tawny ports are not necessarily older than ruby ports, being produced from the lighter-colored 'Mourisco' cultivar and aged at warmer temperatures. Alternatively, they may be derived from a mixture of ruby and white ports. Quality, long-aged, tawny ports are produced in a manner similar to ruby port, but with extended maturation in *pipes*. White port is not added in the development of aged tawny ports. Tawny ports may indicate the minimum average age of the wine contained – 10, 20, 30, or 40 plus years.

Most ruby and tawny ports are tartrate stabilized by rapidly cooling and holding the wine at −10°C for approximately 2 weeks. The addition of Kieselsol to the cold wine before filtration helps yield a stably clear wine.

Vintage ports are not filtered before bottling. The thick sediment that forms is considered important in the development and aging potential of the wine.

SWEETENING AND BLENDING WINES

During racking, blending, and aging, the sugar content of the port may decline. To bring the sugar content back to the desired level, special sweetening wines may be added. The main sweetening wine is called *jeropiga*. It is port wine fortified to 20% alcohol when a cap begins to form on the fermenting must. Both white (*branca*) and reddish (*loira*) jeropigas are produced. Intensely red (*tinta*) jeropigas, produced with the addition of elderberry juice, are no longer authorized. Juice concentrated under vacuum occasionally may be used for sweetening.

In addition, special wines may be used for coloration. The process for producing these wines is called *repisa*. After half the must is run off in the usual manner, the remaining must is treaded or extensively pumped over to extract additional color.

PORT-LIKE WINES

Many countries produce wines by techniques more or less similar to those used in Portugal. In only a few instances, however, are the wines serious international competitors to *porto*. Australian and South African ports are the primary port alternatives.

In regions where intensely colored varieties are not grown, extracting sufficient pigmentation can be a serious problem. One solution is thermovinification. Various procedures have been used, including exposing the fruit to steam, plunging fruit into boiling water, or heating the juice and pomace. Exposure to steam or boiling water is commonly used in Australia and eastern North America.

Occasionally, the must may be fermented dry before fortification. This improves pigment extraction but can lead to excessive tannin extraction. Sweetening comes from must fortified shortly after fermentation has begun, similar to *jeropiga* production. Alternatively, a must concentrate may be used. If it is important that the end product resemble *porto*, avoiding prolonged heating during must concentration is essential. This is unimportant when baked port is produced.

Baking may take various forms, from storing wine in barrels on the top of wineries to direct heating with oxygen sparging. The duration of baking is generally shorter than that used in producing baked sherries. Baking gives the wine a distinctive oxidized–caramelized bouquet.

Many cultivars are used in producing New World ports. 'Shiraz,' 'Grenache,' and 'Carignan' have often been used in Australia. 'Hermitage' ('Cinsaut') and Portuguese varieties are commonly employed in South Africa. 'Carignan,' 'Petite Sirah,' and 'Zinfandel' are typically used in the cooler regions of California, whereas 'Sousão,' 'Rubired,' and 'Royalty' are preferred in hotter regions. 'Concord' has customarily been used in the eastern parts of Canada and the United States.

AROMATIC CHARACTER OF PORTS

The chemical nature of port fragrance has received comparatively little attention until recently. The common view is that the port-like bouquet comes from the combined effects of many compounds, not a single or a few unique substances (see Williams *et al.*, 1983). As previously noted, higher alcohols derived from fortifying spirits are important in the distinctiveness of *porto*. Ports given extensive wood-aging show high concentrations of diethyl and other succinate esters. These may contribute to the basic port fragrance. Oak lactones (β-methyl-γ-octalactone isomers) and other oxygen heterocycles have also been isolated. Some of the latter are furan derivatives, such as dihydro-2-(3H)-furanone, and may donate a sugary oxidized fragrance. However, the most significant in barrel-aged port appears to be sotolon (Silva Ferreira *et al.*, 2003). The concentration of several norisoprenoids also increases during aging. Examples are β-ionone and β-damascenone in vintage ports, and vitispirane, 2,2,6-trimethylcyclohexanone and TDN in tawny ports (Silva Ferreira and Guedes de Pinho, 2004). Acetals, derived from glycerol and acetaldehyde, also appear to be involved in the flavor of old tawny ports (Silva Ferreira *et al.*, 2002). Esters of 2-phenylethanol may generate part of the fruity, sweet fragrance of ports. Diacetyl contributes to its caramel odor (Rogerson *et al.*, 2001). Many acetals have been isolated from tawny ports, but their contribution to the oxidized character of the wine is unclear.

Madeira

Madeira wine evolved on the island of the same name, some 640 km off the coast of Morocco in the Atlantic Ocean. Madeira is primarily characterized by its distinct baked bouquet. This is obtained by intentional heating of the wine. Subsequent maturation occurs in wooden cooperage for several years.

Heat processing of wine has not been widely adopted in other parts of the world. Outside Madeira, it is most commonly used in North America for the production of baked sherries and some ports. Not surprisingly, such wines resemble madeira more than Spanish sherries or *porto*.

Madeira wines are produced in an incredible range of styles. Some are very sweet; others are almost dry. They range from versions produced from a single grape cultivar, and vintage-dated, to those that are highly blended and carry only a brand name. Some are fractionally blended, using a solera-like system; others are not. Although the variations produce subtle differences in style and character, the predominant factor that distinguishes madeira from most other fortified wines is the exposure to heating, termed *esteufagem*.

BASE WINE PRODUCTION

Better madeiras are produced almost exclusively from white grapes. The preferred varieties are 'Malvasia,' 'Sercial,' 'Verdelho,' and 'Bual de Madeira.' 'Listrão' and two red varieties, 'Tinta Negra Mole' and 'Negra,' are commonly used for inexpensive madeiras. Grapes from better sites and preferred cultivars are crushed, fermented, and stored separately to retain their distinctive attributes, at least until blending. Fermentation typically occurs in large cement fermentors, containing approximately 200–300 hl. Fermentation develops spontaneously from indigenous yeasts. The duration of fermentation depends on the style intended. Very sweet madeiras, commonly called malmsey, are fortified early to retain a high sugar content. Buals (boals) are fortified when about half the sugars have been fermented. Verdelho and especially sercial styles are fermented to or near dryness. Fermentation to dryness may take upward of 4 weeks under the cool conditions prevailing in the wineries. Regrettably, the stylistic names, which are similar, if not identical, to grape varietal names, do not necessarily refer to the grape variety or varieties used in their production. A partial chemical characterization of these styles is provided in Nogueira and Nascimento (1999).

Fortification involves the addition of neutral wine spirits (~95% alcohol). Sufficient spirits are added to raise the alcohol content to 14–18% alcohol. After fortification, clarification occurs with Spanish earth, a form of bentonite. At this point, the wine, called *vinho claro*, is ready for heat processing.

HEAT PROCESSING

If quantities permit, wine from different varieties are separately sealed in large capacity cement tanks. Smaller lots are placed in elongated wooden casks (*charuto*) or shorter casks (*ponche*) for heating. The size and type of cooperage appear to have little influence on the quality of wine produced.

The temperature in the heating room is slowly raised over a period of about 2 weeks (or about 5 °C per day) to a maximum of about 45–50 °C. The wine is customarily exposed to that temperature for at least 3 months. After baking, the wine is slowly cooled to ambient temperatures. Occasionally, cooling is speeded by passing cold water through the heating coils. Additional heating at a cooler temperature may take place in wooden casks positioned directly above the heating rooms.

Alternatively, small lots of wine may be heated in butts stored in non-air-conditioned warehouses. Depending on positioning, the wine is variously heated or cooled for upward of 8 or more years. This old technique is called the *canterio* system.

FURTHER MATURATION

Fining removes most of the heavy brown sediment produced during heating. The use of charcoal achieves any additional decolorization deemed necessary.

Further aging occurs in wood cooperage of differing capacities. Oak is frequently used, but other woods have also been used, such as chestnut, satinwood, and mahogany. The addition of wine spirits supplies the alcohol lost during heating and raises the alcohol content to 18–20%.

Small lots of wine from exceptionally fine vintages may be aged in wood for at least 20 years. After a further 2 years in-bottle, the wine may be called vintage madeira, and the vintage date noted on the label. Such wines are commonly designated as *garrafeira* (or *frasqueira*) wines. Lower-quality madeiras are often aged for only 13 months before being released. Much of this apparently goes into producing madeira sauce. Better-quality madeiras are matured for at least 5 years after baking.

SWEETENING AND BLENDING WINES

Juice from grapes grown on the adjacent island of Porto Santo are commonly used to produce a special sweetening wine called *surdo*. The hotter climate of Porto Santo yields grapes with a higher sugar content than is typical for the main island. The juice is fortified shortly after it begins to ferment. *Surdo* is customarily heated, similar to madeira. Occasionally, however, some may be left unheated. This leaves the *surdo* with a fresh, fruity flavor useful in producing certain proprietary blends. Fortified juice, without fermentation, is called *abafado*.

Coloring wine is produced from must heat-concentrated to about one-third of the original volume. It is dark colored and has a distinct caramelized fragrance.

BLENDING

Wines from different vintages and varieties usually are kept separate for at least the first 2 years of wood maturation. Subsequently, producers begin the process of blending. Further maturation and combination eventually produce the final blend. *Surdo* and coloring wines are added to madeira as required.

CHEMICAL NATURE OF THE BOUQUET

Baking of madeira results in the oxidative production of aldehydes, notably acetaldehyde and acetals. However, production of heterocyclic acetals (*cis*- and *trans*-isomers of dioxanes and dioxolanes) appears to be unaffected by heating (Câmara *et al.*, 2003). Their synthesis was correlated primarily with wine age. A similar finding was found with the synthesis of sotolon (Câmara *et al.*, 2004). As expected, heat breakdown products of sugars, such as furfurals, are common. 5-Ethoxymethyl-2-furfural (derived from 5-hydroxymethyl-2-furfural and 2-furfural) is apparently important in the sweet aroma of madeira wines (Câmara *et al.*, 2005). Many aromatic compounds have been isolated from madeira wines

(Campo *et al.*, 2006). Nonetheless, the chemical nature of some potentially important constituents still awaits discovery.

Vermouth

Various herbs and spices have been added to wine since ancient Greek and Roman times. The production of vermouth, a flavored fortified wine, began in Italy in the 1700s. The German name for the major flavorant, wormwood (*Artemisia absinthium*) *wermut*, is thought to be the etymological origin of vermouth.

In world commerce, vermouths are generally subdivided into sweet Italian and dry French styles. Italian vermouths are usually approximately 16–18% alcohol and may contain up to 4–16% sugar for dry and sweet versions, respectively. French vermouths typically contain 18% alcohol and 4% sugar. The sweetening comes from the addition of *mistelle* – grape juice to which ethanol has been added to bring the alcohol content up to about 18–22%.

The base wine is often a neutral-flavored white wine, although in Italy, the best vermouths are (or were) produced from the aromatic 'Muscato bianco' variety grown in Piedmont. Vinification follows standard procedures. For sweetening, a sugar syrup or grape juice fortified with wine spirits or brandy is added. In the past, a distinct red color was derived from the addition of cochineal (the insect *Dactylopius coccus*). This is apparently no longer permitted – the amber color of red vermouth coming from the addition of caramel.

Upward of fifty herbs and spices may be used in flavoring vermouths. The types and quantities employed in any particular brand are usually a proprietary secret. Examples of a few herbs and spices used are allspice, angelica, anise, bitter almond, chinchona, coriander, juniper, nutmeg, orange peel, and rhubarb. A listing of those potentially used, their Latin names, and primary constituents is given in Joslyn and Amerine (1964).

For the production of Italian Vermouths, extracts are prepared by soaking the herbs and spices (7–11 g/liter) in highly rectified alcohol (~85%). If a darker tawny color is preferred, after the addition of the flavor extract, caramel may be added. Once prepared, the extract is added to the base wine. In France, extraction usually involves soaking the herb and spice mixture (4–8 g/liter) directly in the wine after fortification. To avoid the uptake of undesired herbaceous flavors, extraction usually lasts no more than 2 weeks.

Subsequent to flavorant incorporations, the wine is aged for 4–6 months. During this period, the components blend. Finally, the vermouth is fined, cold-stabilized (at −10°C for 10 days) and filtered. Before bottling, the wine may be sterile-filtered or pasteurized.

Brandy

Brandy is the product of distilled wine aged in small oak cooperage. It is produced in most winemaking regions, but is most well known for two versions produced in southwestern France – cognac and armagnac. Similar beverages, termed *eau-de-vie*, are produced elsewhere in France. If produced from the pomace that remains after fermentation, the product may be given specific regional names, such as *marc* (France), *grappa* (Italy), and *bagaceira* (Portugal).

In comparison with wine, brandy and other distilled spirits have a relatively short history. Arnaud de Villeneuve (Catalonia, Spain) is reported to have distilled wine as early as 1250 (Léauté, 1990). It may also have been produced as early as 1100 in Dalerno, Italy. Nonetheless, distillation of alcoholic beverages throughout Europe began in earnest only in the 1500s. Nevertheless, some commercial distilled-wine beverages apparently began to be produced in Armagnac in the mid-1400s (Bertrand, 2003). Brandy production in Cognac started in the early 1600s. Despite this, credit for the first production of a distilled alcoholic beverage appears to go to the Chinese, some 2000 years ago.

Stills were first extensively used in Europe by alchemists, in their ill-fated attempts to extract precious metals from base metals. These devices were themselves derived from models used, if not developed, by Greco-Egyptians in Hellenistic Egypt (primarily used to concentrate aromatic plant oils). Illustrations of some of these devices from ancient manuscripts are reproduced in Berthelot (1888) and Holmyard (1957). The head of the device was called an *ambix* (Greek). It eventually came to be used for the whole device. The term came into English as alembic (from the Arabic *al-anbïq*), along with importation of still-making skills from North Africa.

BASE WINE PRODUCTION

Production of the base wine follows procedures standard for producing most white wines. An investigation of some of these factors on brandy quality is described in van Jaarsveld *et al.* (2005). Typically, nonaromatic white grape varieties are preferred. The predominant cultivar used in the Cognac region is 'Trebbiano' (called 'Ugni blanc' in France), whereas 'Baco 22A' and 'Ugni blanc' are the standard cultivars used in Armagnac. Some 'Folle blanche' and 'Colombard' are also used. Both standard varieties benefit from being relatively resistant to fungal diseases. 'Baco 22A' has the additional advantage of not requiring grafting to avoid problems with phylloxera. That these cultivars typically do not mature fully in the regions concerned has several advantages. It minimizes aroma development, curtails disease development, retains high acidity (6–10 g/liter),

limits ethanol production (8–11.5%, depending on the region), and facilitates mechanical harvesting. Some of these attributes, undesirable for the production of table wine, are ideal for brandy production.

In California, cultivars without distinctive varietal character, such as 'Chenin blanc,' 'Folle blanche,' 'French Colombard,' 'Palomino,' and 'Thompson Seedless' may be used. In South Africa, 'Colombard' and 'Chenin blanc' are typically preferred. Nevertheless, there is growing interest in other cultivars for brandy production. These permit the production of brandies that generate a distinct regional character. Examples are those produced from 'Gewürztraminer' (rose and tea-like notes), 'Chardonnay' and 'Riesling' (noticeable floral scents) and 'Muscat' (with its distinctive muscat fragrance) (Versini *et al.*, 1993). Typically, only white grapes are used. Short maceration periods avoid the uptake of pectinaceous material from the grapes. On enzymatic breakdown, pectins release methanol.

Spontaneous fermentation involves the action of several yeasts, with *Saccharomyces cerevisiae* initially in the minority. However, it soon comes to dominate fermentation. Occasionally, inoculation with specific strains is employed to minimize the production of undesired odors that could become concentrated in the distillate. In addition, the choice of strain, usually *Saccharomyces bayanus*, can significantly modify the brandy's fragrance. Appropriate strains are those that produce desirable flavorants, such as 2-phenyl alcohol, but synthesize limited amounts of compounds, such as acetic acid and sulfur dioxide (Riponi *et al.*, 1997). The use of neutral-flavored grape varieties avoids concentrating varietal aromas. The retention of high acidity minimizes fungal infection (and associated off-odors such as acetic acid or moldy aspects), as well as the need to add sulfur dioxide. Consequently, the synthesis of acetaldehyde and hydrogen sulfide is minimized. Sulfur dioxide is corrosive to stills and can generate sulfuric acid by reacting with constituents in oak cooperage. In the Cognac and Armagnac regions, the use of sulfur dioxide during wine production is prohibited. Racking off the yeast is uncommon. Usually, some lees are retained or added to the wine for distillation.

DISTILLATION

Distillation preferably occurs shortly after the completion of fermentation. If distillation must be delayed, the wine is stored at a cool temperature. This helps limit the loss of fruit-smelling ethyl and acetate esters (Cantagrel, 2003), reduces the accumulation of ethyl acetate and acetals, and retards both oxidation and microbial spoilage (a risk in the absence of sulfur dioxide). Although accentuating some flavors, such as chocolate and caramel, malolactic fermentation also donates solvent off-odors (du Plessis

Table 9.7 Effect of lees on ester content of brandy distilled in an Alambic still (mg/liter spirit at 70% alcohol)

Constituents	Distillation	
	With few lees	With lees
Ethyl caproate	6.76	8.3
Ethyl caprylate	8.95	23.6
Ethyl caprate	13.8	63.0
Ethyl laurate	12.45	36.2
Ethyl myristate	5.4	9.8
Ethyl palmitate	9.77	13.2
Ethyl palmitoleate	1.44	1.8
Ethyl stearate	0.59	0.61
Ethyl oleate	1.19	1.22
Ethyl linoleate	7.69	9.52
Ethyl linolenate	1.86	2.58
Isoamyl caprylate	0.42	2.48
Isoamyl caprate	1.67	5.76
Isoamyl laurate	0.78	1.83
2-Phenylethyl caprylate	Trace	1.20
2-Phenylethyl caprate	0.25	1.55
Total aromatic esters	73.02	182.65 (= +150%)

Source: From Cantagrel and Vidal, 1993, reproduced by permission

et al., 2004). Consequently, malolactic fermentation is generally discouraged, either by cold storage, or the addition of lysozyme. If malolactic fermentation does occur, distillation is delayed until it comes to completion.

A small fraction of the lees is normally included with the wine for distillation. Lees are the primary source of fatty acid esters (Table 9.7). These high-molecular-weight esters, closely bound to yeast cell membranes, are believed to contribute to the fruitiness of brandy. Fatty acid esters also act as fixing agents, retaining other aromatic compounds. Lees also increase the presence of fragrant, amino acid degradation products.

During heating, aromatic constituents volatilize at different rates and times during distillation. Volatilization depends on the individual vapor pressures of the various constituents, variations in their relative solubility in the two main solvents (water and alcohol), and the changing chemical composition of the wine during distillation (primarily but not exclusively the alcohol/water ratio) (Léauté, 1990). For example, at above 17% alcohol, molecules of ethanol cluster to reduce hydrophobic hydration, forming ethanol-rich regions that reduce the volatility of ethyl esters (Conner *et al.*, 1998). During subsequent cooling and condensation of the escaped vapors, collection of different fractions of the distillate fixes separation of the constituents. The result is the selective concentration (and elimination) of wine constituents, plus the synthesis of heat-generated compounds.

In most areas, speed and economic efficiency favor the use of continuous stills. The major disadvantage of most continuous stills is limited control over separation of the various ingredients. Thus, they often generate a distillate including more wine flavorants and fusel alcohols. **Batch** (**pot** or **Alembic**) **stills** permit more precise separation and the selective inclusion of particular ingredients in the distillate. Their main drawbacks are that the course of distillation is prolonged, is considerably less energy efficient, and requires cleaning between each batch.

Continuous stills come in a variety of forms, possessing one to several columns. Here, only the split- (two-) column still is described. In the split still, wine flows down a spiral tube that coils extensively in the first (rectifying) column. During its descent, the wine is heated by hot distillate ascending the column after escaping from the second (analyzing) column. As the heated wine from the first column is pumped up and released into the top of the second (analyzer) column, it flows down and through a set of perforated plates. Superheated steam ascending in the analyzer column vaporizes the incoming, preheated wine. The vaporized wine is directed into the rectifying column, where it is released at its bottom. As the vapors rise, they are cooled as heat is transferred to fresh wine flowing down, through the coils in the column. Compounds with very low vapor pressures (such as acetaldehyde) are allowed to escape through the top of the column. Ethanol and most aromatics condense in various layers and are separated off near the top of the rectifying column. Compounds with high vapor pressures tend to collect near the bottom of the column. They are collected and a portion incorporated with new wine entering the still for redistillation.

The operation of a batch (pot) still is considerably more complex (Léauté, 1990). Its use in brandy production usually involves double distillation. The Alembic still, employed in the production of cognac, is typical of its type (Prulho, 1993; Fig. 9.29). The boiler is usually onion-shaped with a slightly convex base. They usually have a capacity of up to 30 hl. In the first distillation step, the boiler is partially filled and the base heated over an open flame. The wine reaches a temperature of between 90 and 102 °C in about 90 minutes. The vapors pass into a reflux cap (*chapiteau*) at the top of the still. It has a volume equivalent to about 10–12% of the boiler capacity. Some of the vapors condense here, and return to the boiler for redistillation. The *chapiteau* is about 5 and 8 °C cooler than the boiler. The rest of the vapors pass through a neck to a progressively narrower tubular helix (*serpentine*) that descends through

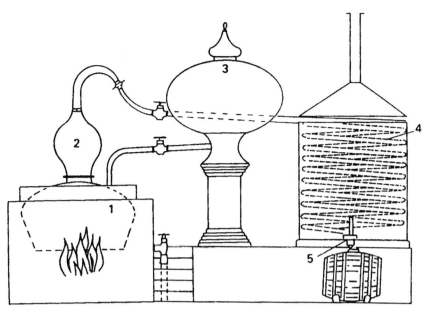

Figure 9.29   Schematic drawing of a batch distillation system: 1, boiler; 2, reflux condenser (*chapiteau*); 3, preheater; 4, cooling coils (*serpentine*); 5, effluent port. (From van den Berg and Maarse, 1993, reproduced by permission)

a water-cooled condenser. Cold water pumped from the bottom cools the hot gases in the *serpentine* and exits at the top. The distillate is filtered, and its temperature, alcohol and volatile content assessed, before being collected at the end of the *serpentine* (hydrometer port). Initially, the upper part of the condenser tube passes through a container, where additional wine samples are preheated to about 50 °C, before being added to the boiler. Boiling continues until the ethanol content of the distillate (tails) approaches 0%.

The first portion of the condensate that collects (~4%) is enriched in acetaldehyde and ethyl acetate. This fraction (**heads**) is often added to a second batch of wine about to be distilled. Subsequently, other highly volatile aromatics, such as ethyl caproate, ethyl caprate, and isoamyl acetate, distill off. These collect primarily in the early portions of the largest distillate fraction, the *brouillis*. Constituents with higher boiling points accumulate in later portions of the *brouillis*. Examples are methanol and higher alcohols (1-propanol, isobutanol, methyl-2-butanol, and methyl-3-butanol). Near the middle and end of distillation, compounds of lower volatility accumulate, notably ethyl lactate, diethyl succinate, acetic acid, and 2-phenyl ethanol (see Léauté, 1990). The *brouillis* is separated and combined with similar fractions from other distillations. Terpenes tend to concentrate at the end of the *brouillis* and in the tails fractions, but some such as geraniol and linalool may distill off early in the heads (Versini *et al.*, 1993). The **tails** fraction (~17%) is concentrated in furfurals. Depending on the preferences of the distiller, the tails

may be separately isolated and added to wine about to be distilled, similar to head fractions. The initial distillation phase often takes about 10 hours.

When sufficient *brouillis* has been collected (about 26–32% ethanol), it is redistilled. The second distillation (*bonne chauffe*) is similar to the first, except that it occurs at a slightly lower temperature. The distillate is usually collected in four fractions. After the heads (~1.5%) has been collected, the next 50% of the distillate (**heart**) is isolated, becoming the nascent brandy. The alcohol content in this fraction is about 70% (67–72%) ethanol. The next two fractions (**heart 2**) (about 40%), and the tails (about 10%) are combined with the heads and added to *brouillis* ready for redistillation. The temperature of the distillate as it runs out of the still is ideally about 18 °C. The second phase of distillation lasts about 14 hours.

The high temperatures to which the wine is exposed (up to 800 °C at the base of the still) induce many Maillard and Strecker degradation reactions. These include reactions between sugars and amino acids, producing heterocyclics, such as furans, pyridines, and pyrazines, as well as aldehydes and acetals, from the degradation of α-amino acids. In addition, heat promotes hydrolytic decomposition of nonvolatile terpene glycosides and polyols to free volatile terpenes (Strauss and Williams, 1983), ketones (such as α- and β-ionones), and norisoprenoids (such as vitispirane and TDN).

Pot stills are typically constructed of copper, being resistant to wine acids, a good conductor of heat, and malleable. However, sulfur dioxide can react with

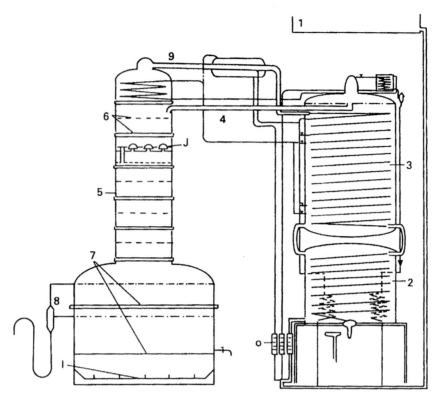

Figure 9.30 Schematic drawing of an Armagnac still: 1, wine container; 2, cooling; 3, wine preheater; 4, wine introduction into the column; 5, distillation column; 6, distillation plate; 7, boiler; 8, exit residue; 9, vapor-transfer pipe. (From van den Berg and Maarse, 1993, reproduced by permission)

copper, corroding the boiler. Thus, stainless steel is occasionally substituted, except for the uppermost portions of the rectifying column. Nevertheless, copper remains the preferred construction material in Cognac and Armagnac. It has the advantage of combining with fatty acids, such as caprylic, caproic, and lauric acids. By forming insoluble constituents (filtered out at the hydrometer port), the contribution of their cheesy or soap-like odors to the distillate is reduced. Copper also fixes hydrogen sulfide typically found in young wines. Where stainless steel is used, copper finings or copper sulfate may be added to the boiler.

In contrast to cognac, armagnac production employs a simple column still possessing 5–15 distillation plates. It differs from most modern stills in that it uses a direct-fired boiler, as in pot stills (Fig. 9.30). At initiation, water is added to the boiler and column. As soon as the water starts to distill, wine is transferred into the preheater. Here, it is heated by exposure to the hot vapors in the condensing coils. The preheated wine (70–85 °C) passes over to the distilling column, where it flows down and over column plates. Alcohol and various aromatics are continuously volatilized as hot vapors rise upward from the boiler. When the wine reaches the bottom, it boils, generating vapors that rise and bubble through the descending wine. To increase contact between rising hot vapors and descending wine, the distillation plates are variously grooved, fitted with mushroom-shaped caps, or possess bell-shaped tunnels. The rising vapors escape into the neck. Here they pass through the preheater before reaching and descending through the condensing coils. A small condenser, usually positioned on top of the distilling column, collects the least volatile constituents. These are usually collected and sent back for redistillation. The condensate that collects at the bottom of the condenser generates a distillate possessing about a 50–54% alcohol content. Armagnac is significantly higher in both total and volatile acidity as well as ester content than cognac. Depending on the distiller, some light lees may be included with the wine for distillation. Their presence supplies the principal source of fatty acids involved in the synthesis of C_8, C_{10}, and C_{12} ethyl esters. They provide a fruity aspect to the distillate. Fatty acids also are the source of methylketones, donating the *rancio* attribute to old, oak-aged brandies.

Adjusting the degree of heating and rate of wine flow are the principal means by which the chemical makeup of the distillate can be regulated. For example, increasing flow rate lowers the temperature, and increases the relative alcohol content, but reduces the concentration of

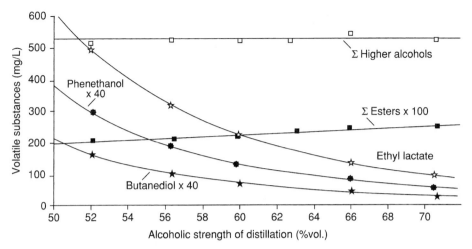

Figure 9.31 Relation between distillation alcoholic strength and content of volatile substances. (From Bertrand, 2003, reproduced by permission)

flavorants, such as phenethyl alcohol, ethyl lactate, and 2,3-butanediol (Fig. 9.31).

Spent wine is drawn off from the lowest level of the still. Periodically, the still needs to be shut down to clean out the sediment that accumulates in the boiler and on the distillation plates. Thus, most armagnac stills function in a semicontinuous manner. This is known as the *système Verdier*.

MATURATION

The second critical step in brandy production involves aging in small oak cooperage. The casks may vary between 200- and 600-liter capacity. Those in Cognac are mostly about 350 liters, while those in Armagnac are between 400 and 420 liters. The preferred source of the oak comes from the adjacent forests of Gascony or Limousin. The wood is more porous and contains more readily extractable tannins. The barrels are typically not filled to capacity, exposing the distillate to oxygen. Oxidation favors the conversion of aldehydes to more pleasant smelling acetals.

It is the aging that converts the sharp roughness of a young brandy distillate into a soft mellow beverage. Maturation typically lasts from 2 to 5 years, but may continue for 20 years or more. During aging, more alcohol than water is lost through the wood, resulting in a drop in the alcohol content. For example, the initial 70%-ethanol content of cognac may fall to about 60% after maturing for some 12 years. Occasionally, an infusion made from oak shavings may be added to provide extra extract, especially where barrel maturation is short. Addition of a sugar syrup (about 6 g/liter) may be used to soften the burning sensation donated by the high alcohol content. Some sugars also accumulate as breakdown products of hemicellulose hydrolysis.

Table 9.8 Example of the effect of aging period on the presence of oak extracts in brandy

Constituent	0.7 years	5 years	13 years
Gallic acid	4.6	9.0	15.3
Vanillic acid	0.3	1.4	2.8
Syringic acid	0.6	2.6	7.0
5-Hydroxymethylfurfural	4.2	4.2	6.3
Furfural	26.8	24.7	21.3
5-Methylfurfural	1.5	1.4	1.6
Vanillin	0.9	4.4	8.8
Syringaldehyde	2.25	8.9	17.6
Coniferaldehyde	3.65	5.9	6.7
Sinapaldehyde	9.45	17.8	17.0

Source: From Cantagrel and Vidal, 1993, reproduced by permission

One of the first noticeable changes during maturation involves the extraction and oxidation of ellagitannins from the cooperage. They generate the typical golden color of brandy. Lignins are degraded and extracted much more slowly (Viriot *et al.*, 1993). The flavorants extracted include oak lactones (β-methyl-γ-octalactones) and the lignin breakdown products, notably vanillin, syringaldehyde, coniferaldehyde, and sinapaldehyde (Puech, 1984; Table 9.8). Ethanolysis is important in lignin degradation.

The nature and amounts of compounds extracted is a function of the duration of in-barrel maturation, the inherent chemistry of the oak, and how these have been modified by wood seasoning and toasting during cooperage construction. In addition to adding flavor, compounds extracted from the wood reduce the volatility of undesirable esters, notably those with longer carbon chains (ethyl octanoate to ethyl hexadecanoate) (Piggott *et al.*, 1992). These produce the undesirable sour, soapy, oily flavors typical of young brandy. Oak extracts reduce the concentration at which ethanol clusters to form pseudo-micelles, normally above 20% (D'Angello *et al.*, 1994). Thus, brandy acts more like a microemulsion than an aqueous ethanol solution. Ethanol soluble compounds, such as higher alcohols and aldehydes, accumulate in these micelles, reducing their volatility and headspace presence (Conner *et al.*, 1998; Escalona *et al.*, 1999).

Although aging in oak is essential for brandy maturation, the use of new oak cooperage is kept to a minimum (often no more than 6 months to a year before transfer to used barrels). There is the desire to avoid the uptake of excessive amounts of tannins and oak flavors (see Calvo *et al.*, 1993). The addition of limited amounts of oak extract may be substituted for long oak aging in inexpensive brandies. This does not, however, provide attributes donated by slow in-barrel oxidation and lignin ethanolysis. In addition, the consequences of acetate esters hydrolysis and the synthesis of fatty acid ethyl esters are less marked when maturation is foreshortened.

Shortly after aging begins, blending with other older brandies commences. The barrels are racked yearly, and the wines from one or more series are mixed before being transferred to barrels for further maturation. As with fortified wines, brand-name identification and distinctiveness are hallmarks of the brandy industry. Nevertheless, some vintage-dated brandies, notably armagnacs, are produced. In Spain, brandies may be aged in a solera system similar to that of sherry (Diez *et al.*, 1985; Quiros Carrasco and Carrascal Garcia, 1993).

During maturation, the alcohol content of the blend is gradually reduced to 40% alcohol (with the addition of distilled water). Caramel may be added to enhance its yellow-gold cast. A final cold treatment and polishing filtration prepare the product for bottling.

The major designations of brandy are based on a combination of the minimum age of the youngest distillate in the blend and the minimum average age of the blend. Respectively, these are Three Stars (2/2 years); VO, VSOP (4/5 years); XO, Extra, Napoleon, Vieille Réserve, Hors d'Age (5/6 years).

Quality in brandy is as difficult to define as in any other grape-derived beverage. By tradition, its characteristics have become associated with moderate levels of higher alcohols, generally in the range of 65–100 mg/liter (pungency), aldehydes and acetals (sharpness), oak lactones (coconut fragrance), phenolic aldehyde derivatives from lignin degradation (vanilla and sweet fragrances), ethyl esters of C_8 to C_{12} fatty acids (fruity/floral notes), the oxidation and transformation of fatty acids into

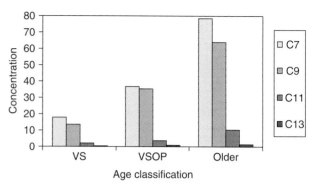

Figure 9.32 Average methylketone concentrations (μg/l) in Cognacs of different age classifications (*n* = 42). (From Watts and Butzke, 2003, copyright Society of Chemical Industry, reproduced by permission John Wiley & Sons Ltd for SCI)

ketones, and heat-derived furans and pyrazines (caramel and roasted notes). Excessive amounts of low volatile constituents, such as ethyl lactate and 2-phenylethanol, tend to donate an atypical heavy flavor, whereas highly volatile constituents provide sharp, irritating notes. Terpenes typically add their particular character to the brandy only when 'Muscat' cultivars are used as the base wine.

Recently, gas chromatography–olfactometry has identified the principal compounds involved in some of the typical flavors of young cognacs. Examples of the odor/chemical associations of compounds present at above detection thresholds are buttery (diacetyl), hay (nerolidol); grass (mainly Z-3-hexen-1-ol); pear and banana (2- and 3-methylbutyl acetates), rose (2-phenylethyl acetate) and lime tree (linalool) (Ferrari *et al.*, 2004). In addition, the *rancio* attribute typical of old (15–20 yr) barrel-aged brandies is primarily due to the presence of methylketones, notably 2-heptanone, 2-nonanone, 2-undecanone and 2-tridecanone (Fig. 9.32). These form from the β-oxidation and decarboxylation of long-chain fatty acids. These are also important to the flavor of blue cheeses. Lignans and lactones derived from the oxidation of oak lignins apparently contribute to the balsamic aspects of the *rancio* attribute (Marche *et al.*, 1975).

Suggested Readings

Sweet Table Wines

Donèche, B. J. (1993) Botrytized wines. In: *Wine Microbiology and Biotechnology* (G. H. Fleet, ed.), pp. 327–351. Taylor & Francis, London, UK.

Ribéreau-Gayon, P. (1988) *Botrytis*: Advantages and disadvantages for producing quality wines. In: *Proceedings of the 2nd International Symposium for Cool Climate Viticulture and Oenology* (R. E. Smart *et al.*, eds.), pp. 319–323. New Zealand Society for Vituculture and Oenology, Auckland, New Zealand.

Carbonic Maceration

Flanzy, C., Flanzy, M., and Bernard, P. (1987) *La Vinification par Macération Carbonique*. Institute National de la Recherche Agronomique, Paris.

Rankine, B. C., Ewart, A. J. W., and Anderson, J. K. (1985) Evaluation of a new technique for winemaking by carbonic maceration. *Aust. Grapegrower Winemaker* **256**, 80–83.

Recioto Process

Usseglio-Tomasset, L., Bosia, P. D., Delfini, C., and Ciolfi, G. (1980) I vini Recioto e Amarone della Valpolicella. *Vini Ital.* **22**, 85–97.

Sparkling Wines

Armstrong, D., Rankine, B., and Linton, G. (1994) *Sparkling Wines – The Technology of Their Production in Australia*. Winetitles, Adelaide, Australia.

Liger-Belair, G. (2004) *Uncorked: The Science of Champagne*. Princeton University Press, Princeton, NJ.

Liger-Belair, G. (2005) The physics and chemistry behind the bubbling properties of champagne and sparkling wines: a state-of-the-art review. *J. Agric. Food Chem.* **53**, 2788–2802.

Markides, A. J. (1987) The microbiology of methode champenoise. In: *Proceedings of the 6th Australian Wine Industry Technical Conference* (T. Lee, ed.), pp. 232–236. Australian Industrial Publishers, Adelaide, Australia.

Moulin, J. P. (1987) Champagne: The method of production and the origin of the quality of this French wine. In: *Proceedings of the 6th Australian Wine Industry Technical Conference* (T. Lee, ed.), pp. 218–223. Australian Industrial Publishers, Adelaide, Australia.

Pernot, N., and Valade, M. (1995) Le pressurage en Champagne, tradition et innovation. *Rev. Fr. Oenol.* **153**, 61–67.

Pool, R., Henick-Kling, T., and Gifford, J. (eds.) (1989) *Production Methods in Champagne*. New York Agricultural Experimental Station, Cornell University, Geneva, NY.

Randall, W. D. (1987) Options for base wine production. In: *Proceedings of the 6th Australian Wine Industry Technical Conference* (T. Lee, ed.), pp. 224–231. Australian Industrial Publishers, Adelaide, Australia.

Fortified Wines

Cristovam, E., and Paterson, A. (2003) Port: The product and its manufacture. In: *Encyclopedia of Food Sciences and Nutrition*, 2nd edn (B. Caballero, L. C. Trugo, and P. M. Finglas, eds.), pp. 4630–4638. Academic Press, Oxford, UK.

Dominquez, M. (1993) *La Viticoltura e la Tecnologia di Produczione de Jerez*. Cámara di CIAA di Venezia, Mestre.

Goswell, R. W. (1986) Microbiology of fortified wines. *Dev. Food Microbiol.* **2**, 1–20.

Liddle, P., and Boero, L. (2003) Vermouth. In: *Encyclopedia of Food Sciences and Nutrition*, 2nd edn (B. Caballero, L. C. Trugo, and P. M. Finglas, eds.), pp. 4630–4638. Academic Press, Oxford, UK.

Martínez, P., Valcárcel, J., Pérez, L., and Binítez, T. (1998) Metabolism of *Saccharomyces cerevisiae* flor yeasts during fermentation and biological aging of fino sherry: By-products and aroma compounds. *Am. J. Enol. Vitic.* **49**, 240–250.

Posson, P. (1981) Production of baked and submerged culture sherry-type wines in California 1960–1980. In: *Wine Production Technology in the United State* (M. A. Amerine, ed.), pp. 143–153. ACS Symposium Series No. 145. American Chemical Society, Washington, DC.

Reader, H. P., and Domingurd, M. (1994) Fortified wines: Sherry, port and madeira. In: *Fermented Beverage Production* (A. G. H. Lea and J. R. Piggott, eds.), pp. 159–203. Blackie, Glasgow, UK.

Brandy

Bertrand, A. (ed.) (1991) *Les Eau-de-vie traditionnelles d'origine viticole*. Lavoisier, Paris.

Bertrand, A. (2003) Brandy and Cognac – Armagnac, brandy and Cognac and their manufacture. In: *Encyclopedia of Food Sciences and Nutrition*, 2nd edn (B. Caballero, L. C. Trugo, and P. M. Finglas, eds.), pp. 584–601. Academic Press, Oxford, UK.

Cantagrel, R. (ed.) (1993) *Premier Symposium Scientifique International de Cognac, Elaboration et Connaissance des Spiritueux*. Lavoisier-Tec & Doc., Paris.

Cantagrel, R. (2003) Brandy and Cognac – chemical composition and analysis of Cognac. In: *Encyclopedia of Food Sciences and Nutrition*, 2nd edn (B. Caballero, L. C. Trugo, and P. M. Finglas, eds.), pp. 601–606. Academic Press, Oxford, UK.

Léaute, R. (1990) Distillation in Alambic. *Am. J. Enol. Vitic.* **41**, 90–103.

Singleton, V. L. (1995) Maturation of wines and spirits, comparisons, facts, and hypotheses. *Am. J. Enol. Vitic.* **46**, 98–115.

Stichlmair, J., and Fair, J. R. (1998) *Distillation: Principles and Practice*. John Wiley, New York.

Swan, J. S. (1986) Maturation of potable spirits. In: *Handbook of Food and Beverage Stability: Chemical, Biochemical, Microbiological, and Nutritional Aspects* (G. Charalambous, ed.), pp. 801–833. Academic Press, Orlando, FL.

References

Alexandre, H., Bertrand, F., and Charpentier, C. (1999) Influence de l'éthanol sur la formation des voiles de levure isolées de vins jaunes. *J. Int. Sci. Vigne Vin* **33**, 25–29.

André, P., Bénard, P., Chambroy, Y., Flanzy, C., and Jouret, C. (1967) Méthode de vinification par macération carbonique, I. Production de jus de goutte en vinification par macération carbonique. *Ann. Technol. Agric.* **16**, 109–116.

Andrés-Lacueva, C., Lamuela-Raventós, R. M., Buxaderas, S., and de la Torre-Boronat, M. D. (1997) Influence of variety and aging on foaming properties of cava (sparkling wine), 2. *J. Agric. Food Chem.* **45**, 2520–2525.

Anonymous (1979) *The Wine Industry in the Federal Republic of Germany*. Evaluation and Information Service for Food, Agriculture and Forestry, Bonn, Germany.

Asvany, A. (1987) Désignation des vins à appellation d'origine de la région de Tokay-Hegyalja. In: *Symposium, Les Appellations d'Origine Historiques*, pp. 187–196. Office International de la Vigne et du Vin, Paris.

Atkin, T. (2001) Tradition and Innovation in the Tokaji Region of Hungary. Masters of Wine Thesis. www. masters-of-wine.org. Assets/files/Student%20pages/Atkin.pdf.

Bailly, S., Jerkovic, V., Marchand-Brynaert, J., and Collin, S. (2006) Aroma extraction dilution analysis of Sauternes wines. Key role of polyfunctional thiols. *J. Agric. Food Chem.* **54**, 7227–7234.

Baker, G. A., Amerine, M. A., and Roessler, E. B. (1952) Theory and application of fractional-blending systems. *Hilgardia* **21**, 383–409.

Balloni, W., Granchi, L., Giovaannetti, L., and Vincenzini, M. (1989) Evolution of yeast flora and lactic acid bacteria in Chianti grapes during the natural drying of Vin Santo production. *Yeast* 5, S37–41.

Barbe, J.-C., de Revel, G., and Bertrand, A. (2002) Gluconic acid, its lactones, and SO$_2$ binding phenomena in musts from botrytized grapes. *J. Agric. Food Chem.* 50, 6408–6412.

Barre, P. (1969) Rendement en levure des jus provenant de baies de raisins placée en atmosphère carbonique. *C. R. Acad. Agric.* 55, 1274–1277.

Bellincontro, A., De Santis, D., Botondi, R., Villa, I., and Mencarelli, F. (2004) Different postharvest dehydration rates affect quality characteristics and volatile compounds of Malvasia, Trebbiano and Sangiovese grapes for wine production. *J. Sci. Food Agric.* 84, 1791–1800.

Bénard, P., Bourzeix, M., Buret, M., Flanzy, C., and Mourgues, J. (1971) Méthode de vinification par macération carbonique. *Ann. Technol. Agric.* 20, 199–215.

Bene, Zs., and Magyar, I. (2004) Characterization of yeast and mould biota of botrytized grapes in Tokaj wine region in the years 2000 and 2001. *Acta Aliment. Pueyo, E.,* 33, 259–267.

Benítez, P., Castro, R., and Varroso, C. G. (2003) Changes in the phenolic and volatile contents of fino sherry wine exposed to ultraviolet and visible radiation during storage. *J Agric. Food Chem.* 51, 6482–6487.

Berlanga, T. M., Peinado, R., Millán, C., Mauricio, J. C., and Ortega, J. M. (2004a) Influence of blending on the content of different compounds in the biological aging of Sherry dry wines. *J. Agric. Food Chem.* 52, 2577–2581.

Berlanga, T. M., Peinado, R., Millán, C., and Ortega, J. M. (2004b) Discriminant analysis of sherry wine during biological aging. *Am. J. Enol. Vitic* 55, 407–411.

Berthelot, M. (1988) *Collection des anciens alchimistes grecs.* Ministère de l'Instruction Public. Paris.

Berti, L. A. (1981) Sparkling wine production in California. In: *Wine Production Technology in the United States* (M. A. Amerine, ed.), pp. 85–121. ACS Symposium Series No. 145, American Chemical Society, Washington, DC.

Bertrand, A. (2003) Brandy and Cognac – Armagnac, brandy and Cognac and their manufacture. In: *Encyclopedia of Food Sciences and Nutrition*, 2nd edn (B. Caballero, L. C. Trugo, and P. M. Finglas, eds.), pp. 584–601. Academic Press, Oxford, UK.

Bidan, P., Feuillat, M., and Moulin, J. P. (1986) Vins mousseux et pétillants, rapport de la France. *Bull. O.I.V.* 59, 565–623.

Bitteur, S., Tesnière, C., Fauconnet, A., Beyonove, C., and Flanzy, C. (1996) Carbonic anaerobiosis of Muscat grape, 2. Changes in the distribution of free and bound terpenols. *Sci. Aliments* 16, 37–48.

Blanck, G. (1990) Utilization de l'hyperoxydation pour la valorisation des moûts de tailles en Champagne. *Rev. Fr. Oenol.* 124, 50–51, 54–55, 57.

Bock, G., Benda, I., and Schreier, P. (1985) Biotransformation of linalool by *Botrytis cinerea. J. Food Sci.* 51, 659–662.

Bock, G., Benda, I., and Schreier, P. (1986) Metabolism of linalool by *Botrytis cinerea.* In: *Biogeneration of Aromas* (T. H. Parliament and R. Crouteau, eds.), pp. 243–253. ACS Symposium Series No. 317, American Chemical Society, Washington, DC.

Bock, G., Benda, I., and Schreier, P. (1988) Microbial transformation of geraniol and nerol by *Botrytis cinerea. Appl. Microbiol. Biotechnol.* 27, 351–357.

Boidron, J. N. (1978) Relation entre les substances terpéniques et la qualité du raisin (rôle du *Botrytis cinerea*). *Ann. Technol. Agric.* 27, 141–145.

Boubals, D. (1982) Progress and problems in the control of fungus diseases of grapevines in Europe. In: *Grape and Wine Centennial Symposium Proceedings*, pp. 39–45. University of California, Davis.

Bravo, F. (1984) Consumo de glicerina por levaduras de flor en vinos finos. *Alimentaria* 156, 19–24.

Cabrera, M. J., Moreno, J., Ortega, J. M., and Medina, M. (1988) Formation of ethanol, higher alcohols, esters, and terpenes by five yeast strains in musts from Pedro Ximénez grapes in various degrees of ripeness. *Am. J. Enol. Vitic.* 39, 283–287.

Calvo, A., Caumeil, M., and Pineau, J. (1993) Extraction des polyphénols et des aldéhydes aromatiques pendant le vieillissement du Cognac, en fonction du titre alcoolique et du degré d'épuisement des fûs. In: *Premier Symposium Scientifique International de Cognac, Elaboration et Connaissance des Spiritueux* (R. Cantagrel, ed.), pp. 562–566. Lavoisier-Tec & Doc., Paris.

Câmara, J. S., Alves, M. A., and Marques, J. C. (2005) Changes in volatile composition of Madeira wines during their oxidative ageing. *Anal. Chim. Acta* 563, 188–197.

Câmara, J. S., Marques, J. C., Alves, A., and Silva Ferreira, A. C. (2003) Heterocyclic acetals in Madeira wines. *Anal. Bioanal. Chem.* 375, 1221–1224.

Câmara, J. S., Marques, J. C., Alves, M. A., and Silva Ferreira, A. C. (2004) 3-Hydroxy-4,5-dimethyl-2,(5H)-furanone levels in fortified Madeira wines: Relationship to sugar content. *J. Agric. Food Chem.* 52, 6765–6769.

Campo, E., Ferreira, V., Escudero, A., Marqués, J. C., and Cacho, J. (2006) Quantitative gas chromatography – olfactometry and chemical quantitative study of the aroma of four Madeira wines. *Anal. Chim. Acta* 563, 180–187.

Cantagrel, R., and Vidal, J. P. (1993) Brandy and Cognac – chemical composition and analysis of Cognac. In: *Encyclopedia of Food Science, Food Technology and Nutrition*, Vol. 1 (R. Macrae *et al.*, eds.), pp. 453–457. Academic Press, London.

Cantagrel, R. (2003) Brandy and Cognac – chemical composition and analysis of Cognac. In: *Encyclopedia of Food Sciences and Nutrition*, 2nd edn (B. Caballero, L. C. Trugo, and P. M. Finglas, eds.), pp. 601–606. Academic Press, Oxford, UK.

Cantarelli, C. (1989) Phenolics and yeast: Remarks concerning fermented beverages. *Yeast* 5, S53–61.

Càstino, M., and Ubigli, M. (1984) Prove di macerazione carbonica con uve Barbera. *Vini Ital.* 26, 7–23.

Cebollero, E., and Gonzalez, R. (2006) Induction of autophagy by second-fermentation yeasts during elaboration of sparkling wines. *Appl Environ Microbiol.* 72, 4121–4127.

Charmat, P. (1925) Ten tank continuous Charmat production of champagne. *Wines Vines* 6(5), 40–41.

Charpentié, Y. (1954) Contribution à l'étude biochemique des facteurs de l'acidité des vins. Doctoral Thesis, University of Bordeaux, France.

Chauvet, S., Sudraud, P., and Jouan, T. (1986) La cryoextraction sélective des moûts. *Rev. Oenologues* 39, 17–22.

Cocks, C. (1846) *Bordeaux: Its Wines and the Claret Country.* Longman, Brown, Green, and Longmans, London.

Coloretti, F., Zambonelli, C., and Tini, V. (2006) Characterization of flocculent *Saccharomyces* interspecific hybrids for the production of sparkling wines. *Food Microbiol.* 23, 672–676.

Conner, J. M., Birkmyre, L., Paterson, A., and Piggott, J. R. (1998) Headspace concentrations of ethyl esters at different alcoholic strengths. *J. Sci. Food Agric.* 77, 121–126.

Cortés, M. B., Moreno, J., Zea, L., Moyano, L., and Median, M. (1999) Response to the aroma fraction in sherry wines subjected to accelerated biological aging. *J. Agric. Food Chem.* 47, 3297–3302.

Costantini, V., Bellincontro, A., De Santis, D., Botondi, R., and Mencarelli, F. (2006) Metabolic changes of Malvasia grapes for wine production during postharvest drying. *J. Agric. Food Chem.* 54, 3334–3340.

Criddle, W. J., Goswell, R. W., and Williams, M. A. (1981) The chemistry of sherry maturation. II. The establishment and operation of a laboratory-scale sherry solera. *Am. J. Enol. Vitic.* 32, 262–267.

Dambrouck, T., Marchal, R., Cilindre, C., Parmentier, M., and Jeandet, P. (2005) Determination of the grape invertase content (using PTA-ELISA) following various fining treatments versus changes in the total protein content of wine. relationships with wine foamability. *J. Agric. Food Chem.* 53, 8782–8789.

D'Angello, M., Onori, G., and Santucci, A. (1994) Self-association of monohydric alcohols in water: compressibility and infrared absorption measurements. *J. Chem. Phys.* 100, 3107–3113.

Descout, J.-J. (1983) Particularités de la vinification en raisins entiers, problèmes posés par la chaptilization. *Rev. Oenologues* 29, 16–19.

Descout, J.-J. (1986) Specificites de la vinification beaujolaise et possibilités d'évolution des productions. *Rev. Fr. Oenol.* 101, 19–26.

Diez, J., Cela, R., and Perez-Bustamante, J. A. (1985) Development of some Sherry brandy aroma components along the solera system. *Am. J. Enol. Vitic.* 36, 86–94.

Dittrich, H. H., Sponholz, W. R., and Göbel, H. G. (1975) Vergleichende Untersuchungen von Mosten und Weinen aus gesunden und aus *Botrytis*-infizierten Traubenbeeren. *Vitis* 13, 336–347.

Divol, B., and Lonvaud-Funel, A. (2005) Evidence for viable but nonculturable yeasts in botrytis-affected wine. *J. Appl. Microbiol.* 99, 85–93.

Divol, B., Miot-Sertier, C., and Lonvaud-Funel, A. (2006) Genetic characterization of strains of *Saccharomyces cerevisiae* responsible for 'refermentation' in *Botrytis*-affected wines. *J. Appl. Microbiol.* 100, 516–526.

Doneche, B. (1991) Influence des sucres sur la laccase de *Botrytis cinerea* dans le cas de la pourriture noble du raisin. *J. Intl Sci. Vigne Vin* 25, 111–115.

Dos Santos, A.-M., Feuillat, M., and Charpentier, C. (2000) Flor yeast metabolism in a model system similar to cellar aging of the French Vin Jaune: Evolution of some by-products, nitrogen compounds and polysaccharides. *Vitis* 39, 129–134.

Dubernet, M. (1974) Recherches sur la Tyrosinase de *Vitis vinifera* et la Laccase de *Botrytis cinerea*, Applications Technologiques. Doctoral Thesis, University of Bordeaux II, France.

Dubourdieu, D. (1981) Les polysaccharides secrétés par *Botrytis cinerea* dans la baie de raisin, leur incidence sur le métabolisme de la levure. In: *Actualités Oenologiques et Viticoles* (P. Ribéreau-Gayon and P. Sudraud, eds.), pp. 224–230. Dunod, Paris.

Dubourdieu, D., Koh, K. H., Bertrand, A., and Ribéreau-Gayon, P. (1983) Mise en évidence d'une activité estérase chez *Botrytis cinerea*, incidence technologique. *C. R. Acad. Sci. Paris. Sér. C* 296, 1025–1028.

Ducruet, V. (1984) Comparison of the headspace volatiles of carbonic maceration and traditional wine. *Lebensm. Wiss. Technol.* 17, 217–221.

Dunsford, P. A., and Sneyd, T. N. (1990) Pressing for quality. In: *Proceedings of the 7th Australian Wine Industry Technical Conference* (P. J. Williams *et al.*, eds.), pp. 89–92. Winetitles, Adelaide, Australia.

du Plessis, H., Steger, C., du Toit, M., and Lambrechts, M. (2004) The impact of malolactic fermentation on South African rebate wine. *Wynboer* **April**, pp. 4 (http://www.wynboer.co.za/recentarticles/0404impact.php3).

Erasmus, D. J., Cliff, M., and van Vuuren, H. J. J. (2004) Impact of yeast strain on the production of acetic acid, glycerol, and the sensory attributes of icewine. *Am. J. Enol. Vitic.* 55, 371–378.

Escalona, H., Piggott, J. R., Conner, J. M., and Paterson, A. (1999) Effects of ethanol strength on the volatility of higher alcohols and aldehydes. *Ital. J. Food Sci.* 11, 241–248.

Escudero, A., Charpentier, M., and Etiévant, P. (2000) Characterization of aged champagne wine aroma by GC-O and descriptive profile analyses. *Sci. Alim.* 20, 331–346.

Esteve-Zarzoso, B., Peris-Torán, M. J., García-Maiquez, E., Uruburu, F., and Querol, A. (2001) Yeast population dynamics during the fermentation and biological aging of sherry wines. *Appl. Environ. Microbiol.* 67, 2056–2061.

Esteve-Zarzoso, B., Fernández-Espinar, M. T., and Querol, A. (2004) Authentication and identification of *Saccharomyces cerevisiae* 'flor' yeast races involved in sherry ageing. *Antonie van Leeuwenhoek* 85, 151–158.

Estrella, M. I., Hernández, T., and Olano, A. (1986) Changes in polyalcohol and phenol compound contents in the aging of sherry wines. *Food Chem.* 20, 137–152.

Etiévant, P. X. (1991) Wine. In: *Volatile Compounds in Foods and Beverages* (H. Maarse, ed.), pp. 483–546. Marcel Dekker, New York.

Etiévant, P. X., Issanchou, S. N., Marie, S., Ducruet, V., and Flanzy, C. (1989) Sensory impact of volatile phenols on red wine aroma: Influence of carbonic maceration and time of storage. *Sci. Aliments* 9, 19–33.

Fagan, G. L., Kepner, R. E., and Webb, A. D. (1981) Production of linalool, *cis*- and *trans*-nerolidol, and *trans*, *trans*-farnesol by *Saccharomyces fermentati* growing as a film on simulated wine. *Vitis* 20, 36–42.

Farkaš, J. (1988) *Technology and Biochemistry of Wine* Vol. 1. Gordon & Breach, New York.

Ferrari, G., Lablanquie, O., Cantagrel, R., Ledauphin, J., Payot, T., Fournier, N., and Guichard, E. (2004) Determination of key odorant compounds in freshly distilled Cognac using GC-O, GC-MS, and sensory evaluation. *J. Agric. Food Chem.* 52, 5670–5676.

Flanzy, C., Flanzy, M., and Bernard, P. (1987) *La Vinification par macération carbonique.* Institute National de la Recherche Agronomique, Paris.

Flanzy, M., and André, P. (eds.) (1973) La Vinification par macération carbonique. Étude No. 56. Éditions S.E.I. Centre Nationale de la Recherche Agronomique, Versailles, France.

Fornachon, J. C. M. (1953) *Studies on the Sherry Flor.* Australian Wine Board, Adelaide, Australia.

Francioli, S., Torrens, J., Riu-Aumatell, M., and López-Tamames, E. (2003) Volatile compounds by SPME-GC as age markers of sparkling wines. *Am. J. Enol. Vitic.* 54, 158–162.

Franco, M., Peinado, R. A., Medina, M., and Moreno, J. (2004) Off-vine grape drying effect on volatile compounds and aromatic series in must from Pedro Ximénez grape variety. *J. Agric. Food Chem.* 52, 3905–3910.

Fregoni, M., Iacono, F., and Zamboni, M. (1986) Influence du *Botrytis cinerea* sur les caractéristiques physico-chimiques du raisin. *Bull. O.I.V.* 59, 995–1013.

Fuleki, T. (1974) Application of carbonic maceration to change the bouquet and flavor characteristics of red table wines made from Concord grapes. *Can. Inst. Food Sci. Techol. J.* 7, 269–273.

Fumi, M. D., Trioli, G., Colombi, M. G., and Colagrande, O. (1988) Immobilization of *Saccharomyces cerevisiae* in calcium alginate gel and its application to bottle-fermented sparkling wine production. *Am. J. Enol. Vitic.* 39, 267–272.

Gonzalez Gordon, M. M. (1972) *Sherry. The Noble Wine.* Cassell, London.

Guerzoni, M. E., Zironi, R., Flori, P., and Bisiach, M. (1979) Influence du degré d'infection par *Botrytis cinerea* sur les caractéristiques du moût et du vin. *Vitivinicoltura* 11, 9–14.

Guillou, I., Bertrand, A., De Revel, G., and Barbe, J. C. (1997) Occurrence of hydroxypropanedial in certain musts and wines. *J. Agric. Food Chem.* 45, 3382–3386.

Henderson, A. (1824) *The History of Ancient and Modern Wines*, p. 182. Baldwin, Craddock & Joy, London.

Hilge-Rotmann, B., and Rehm, H.-J. (1990) Comparison of fermentation properties and specific enzyme activities of free and calcium-alginate entrapped *Saccharomyces cerevisiae*. *Appl. Microbiol. Biotech.* 33, 54–58.

Holmyard, E. J. (1957) Alchemical equipment. In: *A History of Technology* Vol. II (*The Mediterranean Civilization and the Middle Ages, c. 700 B.C. to c. A.D. 1500*) (R. J. Singer, E. J. Holmyard, and A.R. Hall, eds.), pp. 731–752 (Figure 660).Clarendon Press, Oxford.

Ibeas, J. I., Lozano, I., Perdigones, F., and Jimenez, J. (1997a) Effects of ethanol and temperature on the biological aging of sherry wines. *Am. J. Enol. Vitic.* 48, 71–74.

Ibeas, J. I., Lozano, I., Perdigones, F., and Jimenez, J. (1997b) Dynamics of flor yeast populations during the biological aging of sherry wines. *Am. J. Enol. Vitic.* 48, 75–70.

Iimura, J., Hara, S., and Otsuka, K. (1980) Cell surface hydrophobicity as a pellicle formation factor in film strain of *Saccharomyces*. *Agric. Biol. Chem.* 44, 1215–1222.

Isenberg, F. M. R. (1978) Controlled atmosphere storage of vegetables. *Hortic. Rev.* 1, 337–394.

Ishigami, M., Nakagawa, Y., Hayakawa, M., and Iimura, Y. (2006) *FLO11* is the primary factor in flor formation caused by cell surface hydrophobicity in wild-type flor yeast. *Biosci. Biotechnol. Biochem.* 70, 660–666.

Jeffs, J. (1982) *Sherry*. Faber and Faber, London.

Jordan, A. D., and Napper, D. H. (1987) Some aspects of the physical chemistry of bubble and foam phenomena in sparkling wine. In: *Proceedings of the 6th Australian Wine Industry Technical Conference* (T. Lee, ed.), pp. 237–246. Australian Industrial Publishers, Adelaide, Australia.

Joslyn, M. A., and Amerine, M. A. (1964) *Dessert, Appetizer and Related Flavored Wines*. University of California Press, Berkeley, CA.

Jouret, C., Moutounet, M., and Dubois, P. (1972) Formation du vinylbenzène lors de la fermentation alcoolique du raisin. *Ann. Technol. Agric.* 21, 69–72.

Kaminiski, E., Stawicki, S., and Wasowicz, E. (1974) Volatile flavour compounds produced by moulds of *Aspergillus, Penicillium* and *Fungi Imperfecti. Appl. Microbiol.* 27, 1001–1004.

Kontkanen, D., Inglis, D. L., Pickering, G. J., and Reynolds, A. (2004) Effect of yeast inoculation rate, acclimatization, and nutrient addition on icewine fermentation. *Am. J. Enol. Vitic* 55, 363–370.

Kontkanen, D., Pickering, G. J., Reynolds, A., and Inglis, D. L. (2005) Impact of yeast conditioning on the sensory profile of Vidal icewine. *Am. J. Enol. Vitic.* 56, 298A.

Kovač, V. (1979) Étude de l'inactivation des oxydases du raisin par des moyens chimiques. *Bull. O.I.V.* 52, 809–826.

Kundu, B. S., Bardiya, M. C., and Tauro, P. (1979) Sun-baked sherry. *Process Biochem.* 14(4), 14–16.

Kung, M. S., Russell, G. F., Stackler, B., and Webb, A. D. (1980) Concentration changes in some volatiles through six stages of a Spanish-style solera. *Am. J. Enol. Vitic.* 31, 187–191.

Ladurie, E. L. (1971) *Times of Feast, Times of Famine: A History of Climate since the Year 1000*. Doubleday, Garden City, NY.

Léauté, R. (1990) Distillation in alambic. *Am. J. Enol. Vitic.* 41, 90–103.

Liger-Belair, G. (2005) The physics and chemistry behind the bubbling properties of champagne and sparkling wines: a state-of-the-art review. *J. Agric. Food Chem.* 53, 2788–2802.

Liger-Belair, G. (2001) Une première approche des processus physiochimiques liés à l'effervescence des vins de Champagne. Thesis, Laboratoire d'Oenologie UFR Sciences Exactes, Reims, France.

Liger-Belair, G., Marchal, R., and Jeandet, P. (2002) Close-up on bubble nucleation in a glass of champagne. *Am. J. Enol. Vitic.* 53, 151–153.

Liger-Belair, G., Lemaresquier, H., Robillard, B., Duteurtre, B., and Jeandet, P. (2001) The secrets of fizz in champagne wines: A phenomenological study. *Am. J. Enol. Vitic.* 52, 88–92.

Liger-Belair, G., Marchal, R., Robillard, B., Dambrouck, T., Maujean, A., Vignes-Adler, M., and Jeandet, P. (2000) On the velocity of expanding spherical gas bubbles rising in line in supersaturated hydroalcoholic solutions: Application to bubble trains in carbonated beverages. *Langmuir* 16, 1889–1895.

Liger-Belair, G., Marchal, R., Robillard, B., Vignes-Adler, G., Maujean, A., and Jeandet, P. (1999) Study of effervescence in a glass of Champagne: Frequencies of bubble formation, growth rates, and velocities of rising bubbles. *Am. J. Enol. Vitic.* 50, 317–323.

Lonvaud-Funel, A., and Matsumoto, N. (1979) Le coeffecent de solubilité du gaz carbonique dans les vins. *Vitis* 18, 137–147.

Lopez-Toledano, A., Mayen, M., Merida, J., and Medina, M. (2002) Yeast-induced inhibition of (+)-catechin and (−)-epicatechin degradation in model solutions. *J. Agric. Food Chem.* 50, 1631–1635.

Loyaux, D., Roger, S., and Adda, J. (1981) The evolution of champagne volatiles during aging. *J. Sci. Food Agric.* 32, 1254–1258.

Macheix, J. J., Sapis, J. C., and Fleuriet, A. (1991) Phenolic compounds and polyphenoloxidases in relation to browning grapes and wines. *Crit. Rev. Food Sci. Nutr.* 30, 441–486.

Marbach, I., Harrel, E., and Mayer, A. M. (1985) Pectin, a second inducer for laccase production by *Botrytis cinerea*. *Phytochemistry* 24, 2559–2561.

Marchal, R., Tabary, I., Valade, M., Moncomble, D., Viaux, L., Robillard, B., and Jeandet, P. (2001) Effects of *Botrytis cinerea* infection on champagne wine foaming properties. *J. Sci. Food Agric.* 81, 1371–1378.

Marche, M., Joseph, E., Goizet, A., and Audebert, J. (1975) Etude théorique sur le Cognac, sa composition et son vieillissement naturel en futs de chêne. *Rev. Fr. Oenol.* 57, 3–108.

Markides, A. J. (1987) The microbiology of methode champenoise. In: *Proceedings of the 6th Australian Wine Industry Technical Conference* (T. Lee, ed.), pp. 232–236. Australian Industrial Publishers, Adelaide, Australia.

Marquette, B., Dumeau, F., Murat, M.-L., and Daviaud, F. (2003) Mise en évidence de l'inhibition de l'activité laccase de *Botrytis cinerea* par l'utilisation de tanins en vinification. *Rev. Oenolog.* 30, 23–26.

Martin, B., Etiévant, P. X., Le Quéré, J. L., and Schlich, P. (1992) More clues about sensory impact of sotolon in some *flor* sherry wines. *J. Agric. Food Chem.* 40, 475–478.

Martínez, P., Codón, A. C., Pérez, L., and Benítez, T. (1995) Physiological and molecular characterization of flor yeasts: polymorphism of populations. *Yeast* 11, 1399–1411.

Martínez, P., Pérez Rodríguez, P., and Benítez, T. (1997a) Velum formation by flor yeasts isolated from sherry wine. *Am. J. Enol. Vitic.* 48, 55–62.

Martínez, P., Perez Rodríguez, L., and Bénitez, T. (1997b) Evolution of flor yeast population during the biological aging of fino sherry wine. *Am. J. Enol. Vitic.* 48, 160–168.

Martínez, P., Perez Rodríguez, L., and Bénitez, T. (1997c) Factors which affect velum formation by flor yeasts isolated from Sherry wine. *System. Appl. Microbiol.* 20, 154–157.

Martínez, P., Valcárcel, J., Pérez, L., and Binítez, T. (1998) Metabolism of *Saccharomyces cerevisiae* flor yeasts during fermentation and biological aging of fino sherry: By-products and aroma compounds. *Am. J. Enol. Vitic.* 49, 240–250.

Martínez de la Ossa, E., Pérez, L., and Caro, I. (1987) Variations of the major volatiles through aging of sherry. *Am. J. Enol. Vitic.* 38, 293–297.

Martinière, P. (1981) Thermovinification et vinification par macération carbonique en bordelais. In: *Actualités Oenologiques et Viticoles* (P. Ribéreau-Gayon and P. Sudraud, eds.), pp. 303–310. Dunod, Paris.

Massoutier, C., Alexandre, H., Feuillat, M., and Charpentier, C. (1998) Isolation and characterization of cryotolerant *Saccharomyces* strains. *Vitis* 37, 55–59.

Masuda, M., Okawa, E., Nishimura, K., and Yunome, H. (1984) Identification of 4,5-dimethyl-3-hydroxy-2(5H)-furanone (Sotolon) and ethyl 9-hydroxynonanoate in botrytised wine and evaluation of the roles of compounds characteristic of it. *Agric. Biol. Chem.* 48, 2707–2710.

Maujean, A., Gomerieux, T., and Garnier, J. M. (1988) Étude de la tenue et de la qualité de mousse des vins effervescents. *Bull. O.I.V.* 61, 24–35.

Maujean, A., Poinsaut, P., Dantan, H., Brissonnet, F., and Cossiez, E. (1990) Étude de la tenue et de la qualité de mousse des vins effervescents, II. Mise au point d"une technique de mesure de la moussabilité, de la tenue et de la stabilité de la mousse des vins effervescents. *Bull. O.I.V.* 63, 405–427.

Mauricio, J. C., Guijo, S., and Ortega, J. M. (1991) Relationship between phospholipid and sterol contents in *Saccharomyces cerevisiae* and *Torulaspora delbrueckii* and their fermentation activity in grape musts. *Am. J. Enol. Vitic.* 42, 301–308.

Millán, C., and Ortega, J. M. (1988) Production of ethanol, acetaldehyde and acetic acid in wine by various yeast races: Role of alcohol and aldehyde dehydrogenase. *Am. J. Enol. Vitic.* 39, 107–112.

Mills, D. A., Johannsen, E. A., and Cocolin, L. (2002) Yeast diversity and persistence in *Botrytis*-infected wine fermentations. *Appl. Environ. Microbiol.* 68, 4884–4893.

Minárik, E., Kubalová, V., and Šilhárová, Z. (1986) Further knowledge on the influences of yeast starter amount and the activator *Botrytis cinerea* on the course of fermentation of musts under unfavourable conditions (in Polish). *Kvasný Prům.* 28, 58–61.

Molina, I., Nicolas, M., and Crouzet, J. (1986) Grape alcohol dehydrogenase. I. Isolation and characterization. *Am. J. Enol. Vitic.* 37, 169–173.

Montedoro, G., Fantozzi, P., and Bertuccioli, M. (1974) Essais de vinification de raisins blancs avec macération à basse température et macération carbonique. *Ann. Technol. Agric.* 23, 75–95.

Moulin, J. P. (1987) Champagne: The method of production and the origin of the quality of this French wine. In: *Proceedings of the 6th Australian Wine Industry Technical Conference* (T. Lee, ed.), pp. 218–223. Australian Industrial Publishers, Adelaide, Australia.

Nadal, D., Carro, D., Fernándes-Larrea, J., and Piña, B. (1999) Analysis and dynamics of the chromosomal complements of wild sparkling-wine yeast strains. *Appl. Environ. Microbiol.* 65, 1688–1695.

Nelson, K. E., and Amerine, M. A. (1957) The use of *Botrytis cinerea* Pers. in the production of sweet table wines. *Hilgardia* 26, 521–563.

Nicol, M.-Z., Romieu, C., and Flanzy, C. (1988) Catabolisme de l'aspartate et du glutamate dans les baies de raisin en anaérobiose. *Sci. Aliments* 8, 51–65.

Nogueira, J. M. F., and Nascimento, A. M. D. (1999) Analytical characterization of Madeira wine. *J. Agric. Food Chem.* 47, 566–575.

Núñez, Y. P., Carrascosa, A. V., González, R., Polo, M. C., and Martínez–Rodríguez, A. (2006) Isolation and characterization of a thermally extracted yeast cell wall fraction potentially useful for improving the foaming properties of sparkling wines. *J. Agric. Food Chem.* 54, 7898–7903.

Ogbonna, J. C., Amano, Y., Nakamura, K., Yokotsuka, K., Shimazu, Y., Watanabe, M., and Hara, S. (1989) A multistage bioreactor with replaceable bioplates for continuous wine fermentation. *Am. J. Enol. Vitic.* 40, 292–298.

Pellegrini, N., Simonetti, P., Gardana, C., Brenna, O., Brighenti, F., and Pietta, P. (2000) Polyphenol content and total antioxidant activity of *vini novelli* (young red wines) *J. Agric. Food Chem.* 48, 732–735.

Pezet, R., Pont, V., and Hoang-Van, K. (1991) Evidence for oxidative detoxification of pterostilbene and resveratrol by laccase-like stilbene oxidase produced by *Botrytis cinerea. Physiol. Molec. Plant Pathol.* 39, 441–450.

Pham, T. T., Guichard, E., Schlich, P., and Charpentier, C. (1995) Optimal conditions for the formation of sotolon from α-ketobutyric acid in the French Vin Jaune. *J. Agric. Food Chem.* 43, 2616–2619.

Pigeau, G. M., and Inglis, D. L. (2005) Upregulation of ALD3 and GPD1 in *Saccharomyces cerevisiae* during icewine fermentation. *J. Appl. Microbiol.* 99, 112–125.

Piggott, J. R., Conner, J. M., Clyne, J., and Paterson, A. (1992) The influence of non-volatile constituents on the extraction of ethyl esters from brandies. *J. Sci. Food Agric.* 59, 477–482.

Pissarra, J., Lourenco, S., Machado, J. M., Mateus, N., Guimaraens, D., and de Freitas, V. (2005) Contribution and importance of wine spirit to the port wine final quality – initial approach. *J. Sci. Food Agric.* 85, 1091–1097.

Piton, F., Charpentier, M., and Troton, D. (1988) Cell wall and lipid changes in *Saccharomyces cerevisiae* during aging of champagne wine. *Am. J. Enol. Vitic.* 39, 221–226.

Posson, P. (1981) Production of baked and submerged culture Sherry-type wines in California 1960–1980. In: *Wine Production Technology in the United States* (M. A. Amerine, ed.), pp. 143–153. ACS Symposium Series No. 145, American Chemical Society, Washington, DC.

Pozo-Bayón, M.A., Monagas, M., Polo, M. C., and Gómex-Cordovés, C. (2004) Occurrence of pyranoanthocyanins in sparkling wines manufactured with red grape varieties. *J. Agric. Food Chem.* 52, 1300–1306.

Prulho, R. (1993) La distillerie et son environnnement. In: *Premier Symposium Scientifique International de Cognac, Elaboration et Connaissance des Spiritueux* (R. Cantagrel, ed.), pp. 245–256. Lavoisier-Tec & Doc., Paris.

Puech, J.-L. (1984) Characteristics of oak wood and biochemical aspects of Armagnac aging. *Am. J. Enol. Vitic.* 35, 77–81.

Puig-Deu, M., López-Tamames, E., Buxaderas, S., and Torre-Boronat, M. C. (1999) Quality of base and sparkling wines as influenced by the type of fining agent added pre-fermentation. *Food Chem.* 66, 35–42.

Quiros Carrasco, J. M., and Carrascal Garcia, V. (1993) Ageing brandy de Jerez by the solera system. In: *Premier Symposium Scientifique International de Cognac, Elaboration et Connaissance des Spiritueux* (R. Cantagrel, ed.), pp. 603–609. Lavoisier-Tec & Doc., Paris.

Randall, W. D. (1987) Options for base wine production. In: *Proceedings of the 6th Australian Wine Industry Technical Conference* (T. Lee, ed.), pp. 224–231. Australian Industrial Publishers, Adelaide, Australia.

Rankine, B. C., Ewart, A. J. W., and Anderson, J. K. (1985) Evaluation of a new technique for winemaking by carbonic maceration. *Aust. Grapegrower Winemaker* 256, 80–83.

Ribéreau-Gayon, J., Ribéreau-Gayon, P., and Seguin, G. (1980) *Botrytis cinerea* in enology. In: *The Biology of Botrytis* (J. R. Coley-Smith, K. Verhoeff, and W. R. Jarvis, eds.), pp. 251–274. Academic Press, London.

Riponi, C., Carnacini, A., Antonelli, A., Castellari, L., and Zambonelli, C. (1997) Influence of yeast strain on the composition of wines for the production of brandy. *J. Wine Res.* 8, 41–50.

Riu-Aumatell, M., Bosch-Fusté, J., López-Tamames, E., and Buxaderas, S. (2006) Development of volatile compounds of cava (Spanish sparkling wine) during long ageing time in contact with lees. *Food Chem.* 95, 237–242.

Rivero-Pérez, M. D., Pérez-Magariño, S., and González-San José, M. L. (2002) Role of melanoidins in sweet wines. *Anal. Chim. Acta* 458, 169–175.

Rogerson, F. S., and de Freitas, V. (2002) Fortification spirit, a contributor to the aroma complexity of Porto. *J. Food Sci.* 67, 564–569.

Rogerson, F. S. S., Castro, H., Fortunato, N., Azevedo, Z., Macedo, A., and de Freitas, V. P. A. (2001) Chemicals with sweet aroma descriptors found in Portuguese wines from the Douro region: 2,6,6-trimethylcyclohex-2-ene-1,4-dione and diacetyl. *J. Agric. Food Chem.* **49**, 263–269.

Romieu, C., Sauvage, F. X., Robin, J. P., and Flanzy, C. (1989) Évolution de diverses activités enzymatiques au cours du métabolisme anaérobie de la baie de raisin. *Connaiss. Vigne Vin* **23**, 165–173.

Salinas, M. R., Alonso, G. L., Pardo, F., and Bayonove, C. (1998) Free and bound volatiles of Monastrell wines. *Sci. Alim.* **18**, 223–231.

Sarrazin, E., Dubourdieu, D., and Darriet, P. (2007) Characterization of key-aroma compounds of botrytized wines, influence of grape botrytization. *Food Chem.* **103**, 536–545.

Schanderl, H. (1965) Über die Entstehung voh Hefefett bei der Schaumweingärung. *Mitteil Klostern.* **15**, 1–13.

Schreier, P. (1979) Flavor composition of wines: A review. *Crit. Rev. Food Sci. Nutr.* **12**, 59–111.

Senée, J., Robillard, B., and Vignes-Adler, M. (1999) Films and foams of Champagne wines. *Food Hydrocolloid.* **13**, 15–26.

Silva, A., Fumi, M. D., Montesissa, G., Colombi, M., and Colagrande, O. (1987) Incidence de la conservation en présence de levures sur la composition des vins mousseux. *Connaiss. Vigne Vin* **21**, 141–162.

Silva Ferreira, A. C., and Guedes de Pinho, P. G. (2004) Nor-isoprenoids profile during port wine aging – influence of some technological parameters. *Anal. Chim. Acta* **513**, 169–176.

Silva Ferreira, A. C., Barbe, J.-C., and Bertrand, A. (2002) Heterocyclic acetals from glycerol and acetaldehyde in port wines: evolution with aging. *J. Agric. Food Chem.* **50**, 2560–2564.

Silva Ferreira, A. C., Barbe, J. C., and Bertrand, A. (2003) 3-Hydroxy-4,5-dimethyl-2(5H)-furanone: a key odorant of the typical aroma of oxidative aged Port wine. *J. Agric. Food Chem.* **51**, 4356–4363.

Sipiczki, M. (2003) *Candida zemplinina* sp. Nov., an osmotolerant and psychrotolerant yeast that ferments sweet botrytized wines. *Int. J. Syst. Evol. Microbiol.*, **53**, 2079–2083.

Sponholz, W. R., and Dittrich, H. H. (1985) Über die Herkunft von Gluconsäure, 2-und 5-Oxo-Gluconsäure sowie Glucuron- und Galacturonsäure in Mosten und Weinen.*Vitis* **24**, 51–58.

Sponholz, W. R., and Hühn, T. (1996) Aging of wine: 1,1,6-Trimethyl-1,2-dihycronaphthalene (TDN) and 2-aminoacetophenone. In: *Proceedings of the 4th International Symposium on Cool Climate Viticulture and Enology* (T. Henick-Kling *et al.*, eds.), pp. VI-37–57. New York State Agricultural Experimental Station, Geneva, NY.

Stella, C. (1981) Tecnologia di produzione dei vini bianchi passiti. *Q. Vitic. Enol. Univ. Torino* **5**, 47–58.

Stevenson, T. (2005) *The Sotheby's Wine Encyclopedia*, 4th edn, p. 175. Dorling Kindersley, London.

Stratford, M. (1989) Yeast flocculation – the influence of agitation. *Yeast* **5**, S97–103.

Strauss, C. R., and Williams, P. J. (1983) The effect of distillation on grape flavour components. In: *Flavour of Distilled Beverages* (J. R. Piggott, ed.), pp. 120–133. Ellis Horwood, Chichester, UK.

Sudraud, P. (1981) Problèmes de stabilisation des vins provenant de raisins atteints de pourriture noble. In: *Actualités Oenologiques et Viticoles* (P. Ribéreau-Gayon and P. Sudraud, eds.), pp. 278–289. Dunod, Paris.

Sullivan, C. L. (1981) Tokaj. *Vintage* **10**(3), 29–35.

Tabera, L., Munoz, R., and Gonzalez, R. (2006) Deletion of *BCY1* from the *Saccharomyces cerevisiae* genome is semidominant and induces autolytic phenotypes suitable for improvement of sparkling wines. *Appl. Environ. Microbiol.* **72**, 2351–2358.

Tesnière, C., Nicol, M.-Z., Romieu, C., and Flanzy, C. (1991) Effect of increasing exogenous ethanol on the anaerobic metabolism of grape berries. *Sci. Aliment* **11**, 111–124.

Tominaga, T., Guimbertau, G., and Dubourdieu, D. (2003) Role of certain volatile thiols in the bouquet of aged champagne wines. *J. Agric. Food Chem.* **51**, 1016–1020.

Troton, D., Piton, F., Charpentier, M., and Duteurtre, B. (1989) Changes of the lipid composition of *Saccharomyces cerevisiae* during aging of Champagne yeast. *Yeast* **5**,141–143.

Usseglio-Tomasset, L., Bosia, P. D., Delfini, C., and Ciolfi, G. (1980) I vini Recioto e Amarone della Valpolicella. *Vini Ital.* **22**, 85–97.

Valade, M., and Blanck, G. (1989) Évolution des paramètres analytiques au cours du pressurage en Champagne. *Rev. Fr. Oenol.* **118**, 23–27.

van den Berg, F., and Maarse, H. (1993) Brandy and Cognac – brandy and its manufacture. In: *Encyclopedia of Food Science, Food Technology and Nutrition* (R. Macrae *et al.*, eds.), pp. 450–453. Academic Press, London.

van Jaarsveld, F. P., Blom, M., Hattingh, S., and Marais, J. (2005) Effect of juice turbidity and yeast lees content on brandy base wine and unmatured pot-still brandy quality. *S. Afr. J. Enol. Vitic.* **26**, 116–130.

Vasserot, Y., Jacopin, C., and Jeandet, P. (2001) Effect of bottle capacity and bottle-cap permeability to oxygen on dimethylsulfide formation in champagne wines during aging on the lees. *Am. J. Enol. Vitic.* **52**, 54–55.

Versini, G., and Tomasi, T. (1983) Confronto tra i componenti volatili dei vini rossi ottenuti con macerazione tradizionale e macerazione carbonica. Importanza differenziante del cinnamato di etile. *Enotecnico* **19**, 595–600.

Versini, G., Inama, S., and Pilzer, B. (1993) Aroma characteristics of Gewürztraminer grape distillate and grappa in relation to the varietal aroma distribution in berry parts and in comparison with other nonvarietal distillates. In: *Premier Symposium Scientifique International de Cognac, Elaboration et Connaissance des Spiritueux* (R. Cantagrel, ed.), pp. 69–76. Lavoisier-Tec & Doc., Paris.

Viriot, C., Scalbert, A., Lapierre, C., and Moutounet, M. (1993) Ellagitannins and lignins in aging of spirits on oak barrels. *J. Agric. Food. Chem.* **41**, 1872–1879.

Vogt, E. (1977) *Der Wein, seine Bereitung, Behandlung und Untersuchung*, 7th edn. Verlag Eugen Ulmer, Stuttgart, Germany.

Watanabe, M., and Shimazu, Y. (1976) Application of *Botrytis cinerea* for wine making. *J. Ferment. Technol.* **54**, 471–478.

Watts, V. A. and Butzke, C. E. (2003) Analysis of microvolatiles in brandy: relationship between methylketone concentration and Cognac age. *J. Sci. Food Agric.* **83**, 1143–1149.

Williams, A. A., Lewis, M. J., and May, H. V. (1983) The volatile flavor components of commercial port wines. *J. Sci. Food Agric.* **34**, 311–319.

Wucherpfennig, K. (1985) The influence of *Botrytis cinerea* on the filterability of wine. *Wynboer Tegnies* **13**, 18–19.

Yunome, H., Zenibayashi, Y., and Date, M. (1981) Characteristics components of botrytised wines – sugars, alcohols, organic acids, and other factors. *J. Ferment. Technol.* **59**, 169–175.

Yurgalevitch, C. M., and Janes, H. W. (1988) Carbon dioxide enrichment of the root zone of tomato seedlings. *J. Hortic. Sci.* **63**, 265–270.

Zapparoli, G., Torriani, S., Malacrino, P., Suzzi, G., and Dellaglio, F. (2003) Interactions between *Saccharomyces* and *Oenococcus oeni* strains from Amarone wine affect malolactic fermentation and wine composition. *Vitis* **42**, 107–108.

Zara, S., Bakalinsky, A. T., Zara, G., Pirino, G., Demontis, M. A., and Budroni, M. (2005) FLO11-based model for air–liquid interfacial biofilm formation by *Saccharomyces cerevisiae*. *Appl. Environ. Microbiol.* **71**, 2934–2939.

10

Wine Laws, Authentication, and Geography

Appellation Control Laws

Basic Concepts and Significance

Governments typically legislate the health and safety of merchandise. They also often dictate quality standards. Wines are no exception. Possibly because of their diversity, wines are subject to more regulations than most other commodities. Statutes may cover aspects ranging from how grapes are grown, to when and where wine may be consumed. Because of this complexity, only those aspects dealing with geographic origin and style are discussed here. These features are of more general interest since they reflect cultural differences on how wine quality is perceived and on what attributes it is based.

Of wine regulations, the laws affecting naming by geographic origin (*appellation contrôlée*) are the most well known. The limits of designated regions often correspond to the boundaries of specific geologic and/or geographic features. This reflects the common belief that soil type and climate generate conditions that give regional wines distinctive attributes. This opinion is so widely held that most countries have, or are developing, Appellation Control (AC) laws. Although the legislation attempts to prevent the inappropriate (or fraudulent) use of regional names, the marketing advantage obtained by restricting regional name use appears to be a prime factor promoting their establishment. A corollary of the belief that geographic features establish potential wine quality is that the smaller and more unique the region, the more likely the wines are to be distinctive. More importantly, this tends to decrease availability and enhance prestige, potent attributes for increasing a wine's price and profitability.

Although soil type is often used to define the geographic boundaries of an AC region, there is little evidence that soil type itself significantly influences grape or wine characteristics (see Chapter 5). What is more important is the regional mesoclimate and vineyard microclimate, of which the soil structure (not type) is a significant component.

An additional and significant element of many AC laws (notably European) is the regulation of grape varietal use. Such legislation is appropriate where the intention is to maintain a traditional style central to existing commercial success. Nevertheless, strict adherence to such regulations slows and can inhibit stylistic evolution and improvement.

Many AC laws also regulate viticultural and enologic practice. Some, such as the ban on irrigation are unjustified, although, admittedly they are under review. The standard argument is that irrigation increases yield, thereby reducing grape quality. This is valid only if irrigation is used to excess. Amazingly, the same regions that are opposed to irrigation still permit increased yield in better years. Thus, the genuine nature of their concern about overcropping is questionable. Developments in vine training permit simultaneous improvements in yield, without detrimental effects on grape quality (see Chapter 4). The greatest damage caused by the zealous imposition of AC regulations is the stifling of grape grower and winemaker initiative. Legislation ideally should be based on verifiable principles, such as yield per meter canopy (Intrieri and Filippetti, 2000). Increased yield, associated with canopy management to maintain adequate leaf area/fruit ratios, are unaccompanied with noticeable changes in wine quality (Reynolds, 2000). In countries such as the United States, legislation is limited to aspects of geographic and varietal authenticity. This is wise, because a fully objective assessment of wine excellence is impossible. Thus, governments are ill-advised to imply or give anything more than minimum quality endorsements.

Although Spain was the originator of the Appellation Control concept, France instituted the first national AC system in the 1930s. All wine-producing members of the European Economic Community (EEC) now possess their own AC laws, within the framework of a general set of EEC regulations. Outside Europe, South Africa has established the most comprehensive set of Appellation Control laws.

In their simplest form, Appellation Control laws apply only to geographic origin. More complex forms often dictate grape varietal use, prohibit certain vineyard practices, limit maximum yield, designate required production procedures, and specify maturation conditions. Inherent in these laws is the desire to assure both quality and stylistic authenticity. This is often misinterpreted as guaranteeing quality. Wine excellence can never be guaranteed. Quality can easily be compromised by conditions beyond the control of the producer, shipper, or country. The best that can be achieved is to assess representative samples for freedom from fault before shipping. How specific bottles of wine may improve or deteriorate after leaving the winery cannot be predicted, regulated, or controlled.

Although sensory evaluations are to a large degree subjective, and therefore largely unquantifiable, they have greater significance than objective chemical analyses. Germany possesses a system whereby all wines must pass a sensory evaluation test to warrant a QbA and QmP designation. In a few countries, certain quality designations require yearly sensory evaluation. Examples are the DOCG and VQA designations of Italy and Canada, respectively. French AOC wines require periodic sensory evaluation, but not all wines are sampled annually.

Although Appellation Control laws probably have helped maintain or raise average quality standards, the importance given ranking by wine critics suggests a lack of confidence in AC "guarantees" of quality. Similarly, the popularity of vintage guides also attests to a failure of Appellation Control designations to adequately indicate excellence.

As noted, an aspect of AC legislation is the implication that the smaller the region, the greater the potential sensory distinctiveness of the wine. This has led to the mistaken view that blending wines, from different sites or regions, is inimical to quality. The fallacy of this opinion is obvious from the quality of sherry, port, champagne and cognac. In these regions, house style (proprietary labeling) replaces vineyard location as the major indicator of character and quality. The skillful combination of wines from different sites often enhances the desirable features of the component wines (Fig. 8.3). The value of blending is also evident in many Australian wines, where blending from different vineyards and regions is common. Their best-known prestigious wine, Penfolds Grange, is a blended wine.

Although Appellation Control laws promote authenticity in geographic labeling, they can disadvantage wines of equal or better quality produced in regions not possessing an established prestige. By promoting the sale of wine from well-known regions, AC laws tend to increase their profitability and, thereby, the use of costly practices that may enhance quality. Conversely, they also may promote the continued use of inefficient and outmoded technologies. As with other products, success tends to breed success. For example, the land values of Burgundian vineyards markedly reflect the relative importance ascribed to their AC location (Fig. 10.1). Regrettably, success may also lead to arrogance and a failure to improve.

Some European AC laws tend to fix the regions in which particular grape varieties are cultivated. This is reflected in the highly localized distribution of many grape varieties in Europe. Lamentably, the same ordinances retard the adoption of practices that could enhance wine excellence. Outside Europe, legislation regulating viti- and vinicultural practices is less repressive, and allows greater flexibility for experimentation and change. The longer growing season and more frequent sunny conditions have facilitated many economic modifications to traditional viticultural practice in regions like Australia, California, and Chile, with an enhancement in grape quality. Viticultural flexibility is also particularly useful in the New World. It permitted rapid accommodation to frequent or marked changes in consumer taste. In addition, winemaking styles are not so inexorably linked with particular regions that stylistic continuity must be maintained for market acceptance.

Although views on what constitutes consumer quality or desirability may fluctuate quickly, legislative change seems inexorably slow. As a consequence, traditional varieties accepted at the inception of regulations may not possess the properties later considered desirable (Pouget, 1988). This has led several producers to dissociate

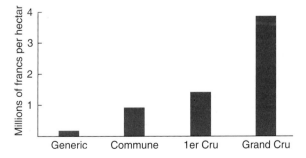

Figure 10.1 Average land values for vineyards carrying the Côte d'Or appellations for white wine. (From Moran, 1988, reproduced by permission)

themselves from the legal constraints, and accept the "penalty" of using the lowest official wine designation. Many of the best-known Italian wine producers have chosen this route to institute changes they felt compelled to improve quality. If the producers have sufficient reputation and skill, their wines often gain a prestige above those following the officially prescribed procedures. Innovation, although not always successful, is surely the route to further advances. It certainly has paid off for Australian, Californian, and Chilean producers.

Appellation and related control laws also regulate the use of terms placed on wine labels. In addition to geographic origin, labels often show the vintage, cultivar(s) used, and whether the wine is estate-bottled. Regrettably, consumers may assume greater precision than is warranted. For example, geographic designation often does not require that 100% of the wine come from the named appellation. Equally, the regulations defining vintage designation, varietal composition, and estate bottling often vary from jurisdiction to jurisdiction. For example, varietal designation in the United States generally requires that more than 75% of the wine come from the named cultivar. In Oregon, however, state regulations require 90%. In the eastern United States, wine made from *Vitis labrusca* cultivars can be varietally labeled as long as they possess 51% of the stated variety. Although this may be considered deception by some, the information provided gives the knowledgeable consumer useful information, without being needlessly precise.

In most jurisdictions, wines varietally designated must contain wine more than 75% derived from the named cultivar. This is usually more than sufficient to assure that the wine reflects the essential character of the cultivar. Flexibility is useful in permitting adjustments to maintain or improve wine character. This is well known in Bordeaux, where wines from related varieties are often blended with 'Cabernet Sauvignon.' The proportions typically vary from year to year. Excellence should be the ultimate goal of winemaker and consumer alike. Label data are there as a consumer guide, not to provide technical production details.

Geographic Expression

FRANCE

As the first country to establish a national set of Appellation Control laws, its approach has, for better or worse, been a model for most equivalent legislation. The legislation took 30 years (1905–1935) to approach its near-present-day expression. Fine-tuning and subsequent integration with EEC regulations have not modified the basic spirit or tenets of the laws. The statutes were designed to assure the geographic authenticity, traditional character, and reputation of the region's wines.

Part of the goal has involved intense legal efforts to inhibit the use of French geographic names on wines produced elsewhere. In this regard, France has been largely successful. The restriction of European geographic names to wines produced within the officially designated region has become part of the mandate of EEC Appellation Control regulators. This has unintentionally benefitted the countries that formerly used European appellations. Eliminating European regional names on New World wines not only avoids promoting consumer recognition of these appellations, but also permits the wines to become known and valued for what they are.

The French AC legislation established in 1935 recognized only a single category, the *Appellation d'Origine Contrôlée* designation (AOC). Subsequently, categories designating less-distinguished viticultural regions have been delimited. This has led to the creation of the *Vins Délimités de Qualité Supérieure* (VDQS), *Vins de Pays* and *Vins de Table* categories.

The least regulated and uniform category is the *Vins de Table* designation. A similar category is found in all EEC countries. The French and EEC "table wine" category should not be confused with the common English distinction between table and fortified wines. The EEC designation refers to wines coming from one or several EEC member countries. If the wine comes from a single country, it may originate from one or more regions not possessing, or using, an official AC designation. Only one country, or a general EEC origin, may be specified on the label. Beyond limits on generally permitted grape varieties, minimum alcoholic strength, and acidity level, there are few viti- or vinicultural restrictions on the production or characteristics of the wine.

The *Vins de Pays* category was first created in 1968. It is similar in concept to the *Vins de Table* designation, except for the permission to designate the wine's regional origin. The label also may state the vintage date and varietal origin of the wine. The regulatory freedom of the category has permitted some serious and dedicated producers to innovate and produce wines of a quality superior to most AOC wines.

The VDQS category, established in 1949, is supposed to be phased out, in conformance with the two basic categories of wines recognized by EEC regulations, namely Table and VQPRD (*Vins de Qualité produits dans les Régions Déterminées*). Most VDQS (*Vins Délimités de Qualité Supérieure*) regions are being upgraded to the AOC category (the VQPRD version in France), with a slight reduction in delimited area. The VDQS category covers wines produced in regions without a long tradition of renown. Otherwise, production

regulations are similar to those of AOC regions. There are currently only 33 VDQS regions, in comparison to 470 AOC designated areas. Of the latter, many are subdivisions of larger AOC regions. For example, the Beaujolais region possesses a general Beaujolais AOC, plus Beaujolais Superiéur, Beaujolais-Villages and several individual Beaujolais village AOCs. The latter generally mention only village name, such as Morgon, Juliénas, St-Amour, etc.

AOC-designated regions are those considered to possess a long history of producing fine wines. To maintain the traditional character of these regional wines, laws regulate aspects of varietal use, grape growth, wine production, and maturation. Because most wine styles have evolved in specific regions, the protection of the character and associated geographic name has been a cornerstone of French AOC (and European AC) legislation.

A consistent feature of French AC classification is its geographic subdivision into smaller and smaller regions. This is based on the assumption that wines having their origin from smaller regions will be more consistent and distinct, and "better" than those produced (blended) from larger regions. This hierarchy of geographical regions has occasionally been taken down to single vineyards. In most cases, however, it only goes down to single parishes or townships (communes). As the area becomes smaller, the restrictions imposed by the dictates tend to become more rigid and limiting. For example, wine with a general Burgundy AC designation may be produced from vineyards having yields typically up to 45 hl/ha. In contrast, wine with regional or village AC designations, such as Nuits Saint-Georges or Pommard, must come from vineyards having yields usually no greater than 35 hl/ha. Finally, individual vineyard AC designations, such as Clos Vougeot and Chambertin, have their yield tentatively capped at 30 hl/ha. These maxima may be increased (by decree) if the vintage is considered to be plentiful and of above-average quality. In association with restricted yield, there is also a requirement for higher alcohol contents (reflecting higher °Brix values usually correlated with greater grape maturity and flavor). In the example given above, the minimum alcohol values for the wines must be at or above 9.0, 10.5, and 11.5%, respectively.

A humorous, but illustrative, example of a hypothetical AOC designation is given in Fig. 10.2. Wine from within the basic enclosed region (solid line) would have the right to the *Celliers* (fermentation room) appellation. Wine from Château Cider, located outside the region, could be designated only as a *Vin de Pays*. The Celliers AOC is subdivided (dashed line) into three AOCs – *Carafe* (decanter), *Charnu* (full-bodied), and *Bouchon* (stopper). The appellation regions are traversed by the *rivière Levure* (yeast river). The Charnu appellation

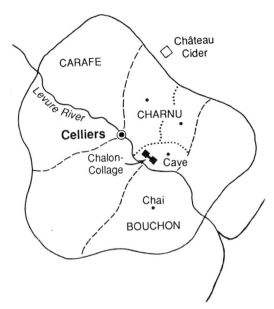

Figure 10.2 Hypothetical AOC region according to *Appellation d'Origine Contrôlée* laws in France. (From de Blij, 1983, reproduced by permission)

Grand Cru Classé

(*Premier Grand Cru*)
(*Deuxième Grand Cru*)
(*Troisième Grand Cru*)
(*Quatrième Grand Cru*)
(*Cinquième Grand Cru*)

Cru Exceptionnel

Cru Bourgeois Supérieur

Cru Bourgeois

Figure 10.3 The *cru classé* system of the Haut Médoc (Bordeaux), initially prepared for the Paris World Exhibition of 1855.

itself possesses several smaller AOC designations (dotted lines). *Appellation Cave Contrôlée*, the most central appellation, is located close to the town of *Cave* (cellar). The highest AOC ranking is accorded the vineyard called *Chalon-Collage* (dragnet-glue).

An important notion in the development of AOC legislation has been the concept of *terroir* (local soil and climatic conditions, occasionally considered to include regional viniviticultural tradition) (Laville, 1990). The geographic component of this concept has been codified in the AC control system, where distinctive geologic and/or soil conditions usually define the geographic limits of each designation.

The significance given to local conditions probably gave birth to an even earlier hierarchical system, the *cru classé* system. It coexists with AC designations in a few AOC regions. The *cru classé* system ranks, in a variable number of superimposed categories, either vineyard sites or wineries. In this instance, sites are ranked solely on historical perceptions of quality. As with AC regulations, higher positioning in the *cru classé* system may entail more stringent demands on viti- and vinicultural practices. In most cases, sensory tests are not required to confirm the relative quality of the wines, nor are periodic readjustments of the system mandated (except in St Émilion). In some regions, such as Alsace, producers have the option of participating or remaining outside the system.

Somewhat confusingly, the meanings of the terms used in *cru* classification can vary from region to region. For example, Premier Grand Cru in Bordeaux is equivalent

to Grand Cru in Burgundy – with Premier Cru being a level below Grand Cru in Burgundy. In addition, a *cru* in Burgundy can refer to a defined piece of real estate, whereas in Bordeaux it is the winery (not the property) that is codified. Thus, the defined limits of the associated vineyard(s) in Bordeaux have changed over time. These differences in the *cru classé* system reflect their separate and independent origins. The newest version is in Alsace. The oldest is the Médoc system, codified in 1855 (Fig. 10.3). It has had one modification since – when a Deuxième Grand Cru was upgraded to Premier Grand Cru.

Wines produced within AOC regions are frequently required to undergo both chemical and sensory testing by winemakers from the affected region. However, the tests are designed only to remove faulty or atypical wines, not to assign a quality ranking (Marquet, 1987). Because tasting usually occurs a few months after fermentation, they give little indication of the potential of the wine to develop. In addition, not all wines receive yearly governmental sensory evaluation.

Another French concept, that of *terroir,* has been enthusiastically embraced by most viticultural regions. It essentially signifies that local soil and climate conditions influence the sensory characteristics of the wine. That vineyard microclimate influences grape development is beyond question (Plate 4.19). In that regard, every vineyard site has a *terroir,* and, as Hugh Johnson (1994) has noted, so does every front- and backyard. In general, the best *terroirs* (sites) are those where the ripening curves for sugar, acid, flavor, color and tannins occur "ideally" and uniformly (Kosuge, 1999). In some regions, such as Saint-Émilion, this situation may be primarily controlled by factors that influence vine water status (van Leeuwen *et al.*, 2004). In other regions, different factor(s) may be the overriding controllers of quality.

What is regrettable is how the term has been misappropriated to claim almost mystical powers for certain plots of land. Too often, *terroir* has been used to imply a sensory distinctness to the wine that, if present, is far too nebulous for the vast majority of (if any) humans to detect (see Chapter 11, Odor Perception). To avoid the term's emotional and occasional mythological overtones, my preference is to use the acronym SAM (soil and atmospheric microclimate). For marketing purposes, however, this acronym has none of the "romance" inherent in a foreign term.

The implications that particular *terroirs* donate unique sensorial attributes has also been misused in marketing wine of particular regions or sites. Not surprisingly, other regions want the same commercial advantages. Thus, the increased interest and desire to define appellations with supposedly distinctive SAMs.

Determination of the soil/atmospheric features that truly influence a grape's ripening potential merit serious study. This is called **precision viticulture** (Chapter 4). The potential benefits go far beyond those associated with marketing. By defining the features that donate micro-climatic distinction, the viticultural factors that limit wine potential may be discovered, permitting them to be diminished. In addition, precision viticulture should permit accentuation of the best attributes of a prove-nance. In the long run, detection of a region's deficits and assets may be the most beneficial aspect of the cur-rent emphasis on *terroir*. Quality products are eventu-ally recognized as such. There will then be no need to resort to the mystique of *terroir*.

As will be noted later, given sufficient analysis, regional wines usually can be differentiated chemically. However, this does not indicate that the wines can be identified by human sensory skill. Evidence suggests the opposite. Individuals rarely, if ever, consistently detect or recog-nize such differences. Those most likely to detect such differences (experienced wine judges and winemakers) frequently do surprisingly poorly (Winton *et al.*, 1975; Noble *et al.*, 1984; Morrot, 2004). Although the dis-tinctiveness of regional varietal wines is often consid-ered *ipso facto*, rigorous investigations of these claims are rare. One example is the study of regional varietal wines in California (McCloskey *et al.*, 1996). Although regional wines could be distinguished by descriptive analysis under laboratory conditions, whether consum-ers under real-life conditions could make such distinc-tions is a moot point. Other examples are Douglas *et al.* (2001) and Schlosser *et al.* (2005).

GERMANY

The German Appellation Control system is conceptually different from its French counterpart. German legislation separates its geographic and quality aspects. This prob-ably reflects their traditional interest in distinguishing

Naturweins, wines made from fully mature grapes without chaptalization.

Unlike the French system, German law designates all its major wine-producing regions. The regions are sub-divided by size, but without any official implication that wines produced in smaller regions are necessarily better or more distinctive than wines from larger regions. The largest appellations are called *bestimmte Anbaugebiete* (designated growing regions). These are divided into one or more areas designated as *Bereich*. Further divisions are based on either group vineyard sites (*Grosslage*) or individual vineyard sites (*Einzellage*). The latter divi-sions are usually associated with the name of the clos-est village or suburb (*Ortsteil*). This greatly facilitates the identification of the exact geographic origin of the wine. Regrettably, there is no simple way of distinguish-ing between *grosslage* and *einzellage* sites. This may be intentional, to avoid giving a marketing advantage to either designation, but it is a disservice to the consumer looking for wine from an individual vineyard site.

Although differences in maximum allowable yield per hectare occur among regions, the differences are particularly marked between the various *Qualitätswein* (VQPRD) levels. It is primarily this feature that dif-ferentiates the German VQPRD system from other equivalents. Also, samples of every *Qualitätswein* must pass several tests. If the wine fails to achieve a mini-mum grade, it may be declassified to a *Landwein*. This is the German equivalent of *Vins de Pays* in France. Geographic origin may be no more precise than one of several specified regional names. The basic EEC table wine designation, equivalent to *Vins de Table* in France, is termed *Tafelwein*.

Within the *Qualitätswein* designation, Germany rec-ognizes seven categories. The least demanding, in terms of grape maturity, is the QbA (*Qualitätswein bestim-mter Anbaugebietes*) category. It has more stringent yield/hectare limits than *Landweins*, but is less restric-tive than the higher categories, grouped under the QmP (*Qualitätswein mit Prädikat*) designation. The maxi-mum yield/hectare varies depending on the region and cultivar used. In addition, QbA juice may be chaptal-ized, whereas QmP juice cannot.

QmP (*Prädikat*) wines are subdivided into six cat-egories, based largely on the degree of grape ripeness at harvest. Although this system suggests the poten-tial sweetness level of the wine, it does not necessarily relate to the overall quality of the wine. Each designated wine region (*Anbaugebiete*) establishes for the cultivars growing in its region an official harvest-commencement date. The dates vary yearly, depending on the prevail-ing climatic conditions. Once harvesting of the *kabinett* quality is complete, and after a specified period, permis-sion must be requested to gather late-harvest grapes.

Table 10.1 Starting soluble solids and alcohol potential of juice in the Rheingau and Mosel–Saar–Ruwer regions of Germany for *Tafelwein, Qualitätswein* and *Qualitätswein mit Prädikat* wines

	Starting soluble solids (Oechsle)/alcohol potential(%)						
Cultivar	Table wine	QbA wine	Kabinett wine	Spätlese wine	Auslese wine	Beerenauslese wine	Trockenbeerenauslese wine
Rheingau							
'Riesling'	44/5	60/7.5	73/9.5	85/11.4	95/13.0	125/17.7	150/21.5
Other white grapes	44/5	60/7.5	73/9.5	85/11.4	100/13.8	125/17.7	150/21.5
Weissherbst varieties	44/5	60/7.5	78/10.3	85/11.4	105/14.5	125/17.7	150/21.5
'Pinot noir'	44/5	68/8.8	80/10.6	90/12.2	105/14.5	125/17.7	150/21.5
Red cultivars	44/5	60/7.5	80/10.6	90/12.2	105/14.5	125/17.7	150/21.5
Mosel-Saar-Ruwer							
'Riesling'	44/5	57/7	70/9.1	76/10.0	83/11.1	110/15.3	150/21.5
'Elbling'	44/5	57/7	70/9.1	80/10.6	83/11.9	110/15.3	150/21.5
Other varieties	45/5	60/7.5	73/9.5	80/10.6	88/11.9	110/15.3	150/21.5

Source: From Anonymous, 1979, reproduced by permission

Typically, the grapes must be inspected before crushing. Each late-harvest designation must be made from grapes possessing specific minimum Oechsle (sugar content) values (Table 10.1). Thus, depending on the maturity, late-harvested grapes may produce *spätlese, auslese, beerenauslese* (BA), and *trockenbeerenauslese* (TBA) wines. *Spätlese* is the basic late-harvest category. *Auslese* wines come from individually selected grape clusters, possessing overripe fruit. *Beerenauslese* wines come from individually selected grapes that are over-ripe and preferably botrytized. *Trockenbeerenauslese* wines come from individually selected grapes that have shriveled on the vine, and are usually botrytized. A special late-harvest category is the *eiswein*. For *eiswein* production, grapes must be picked and crushed frozen (below −6° to −7°C), and must possess Oechsle values equivalent to that of a *beerenauslese*.

Before the producer can use any of the designations, samples must pass both chemical and sensory evaluation tests, administered by regional authorities. The sensory evaluation assesses whether the wine is typical of the cultivar indicated (if specified) and the desired QmP category. Failure to reach established norms for each category may prevent the use of the cultivar name and result in a demotion to a lower category, or declassification to a *Landwein*. During the 1970s, when the system was first introduced, approximately 1.5 and 3% of tested wines were demoted and declassified, respectively. Wine passing both the chemical and sensory analyses receive an official control number (*Amtliche Prüfungsnummer, A.P.Nr.*) for a specified amount of wine. Two bottles of each sample tested is retained for 2–3 years.

Individual producers may band together, imposing more restrictive regulations to obtain a higher market profile. The most famous of these is the VDP (*Verein Deutscher Prädikatsweingüter*). It uses an eagle symbol on the label and neck capsule. Increased profile also may be achieved by winning quality seals at regional and national competitions.

Because quality designations are associated with grape maturity, which can vary from year to year, Germany possesses no official hierarchical ranking of vineyard sites or regions.

ITALY

The Italian AC system bears greater resemblance to its French than to its German counterpart. The table wine category possesses two levels, namely the *Vino da Tavola* and more newly established *Indicazione Geografica Tipica* (IGT) categories. *Vino da Tavola* refers to wines possessing only a general regional designation, wines from areas not currently possessing an official appellation, or wines made in contravention of local appellation regulations. Because some of the most prestigious wine producers in Italy disagree with the restrictions imposed by existing local AC regulations, some of the finest Italian wines carry the lowest official AC designation. Although this has increased the prestige of Italian wine, it has usually been at the expense of endemic cultivars. The impact would have been broader if the acclaim had been achieved without using foreign cultivars. IGT wines possess the right to use specific geographic designations, mention cultivar names, and state year of production.

Italy possesses two categories of VQPRD wines. The first, introduced in 1966, is the DOC (*Denominazione di Origine Controllata*). It covers about 10–12% of all Italian wine production. Although resembling the French

AOC, the DOC differs in several respects. For example, there is no specific division into smaller, more restrictive designations. DOC designations may be variously based on regional names with no varietal indication (i.e., Barolo, Chianti, and Soave), geographic locations, combined with varietal and stylistic names (i.e., Grave del Friuli Pinot Grigio and Recioto della Valpolicella Amarone), or traditional, but nongeographic expressions (i.e., Lacryma Christi del Vesuvio, Est! Est!! Est!!!).

Since 1982, selected DOC designations have been raised to the DOCG (*Denominazione di Origine Controllata e Garantita*) category. Wines in this class must be submitted to several sensory analyses at various points during production and maturation, in addition to standard chemical analyses.

One of the unique features of the Italian AC system is the potential pyramidal interconnection of the categories. Wines in the DOCG category are recommended not to exceed 55 hl/ha for red and rosé wines (60 hl/ha for white wines). If these values are exceeded by 20%, the wine is demoted to the next lower (DOC) category. DOC regions are recommended to not exceed 70 hl/ha for red and rosé wines (85 hl/ha for white wines). Exceeding this limit by 20% demotes the wine to the IGT classification. Official recommendations for IGT regions are no more than 100 hl/ha for red and rosé wines (115 hl/ha for white wines). Overproduction by a further 20% declassifies the wine to the *Vino da Tavola* category. The law provides only recommended maximum production levels. More precise limits can be set by committees associated with individual AC regions. Additional information on the AC system may be found in a general discussion by Fregoni (1992).

For consumers, possibly the most serious deficiency in the Italian AC system is the potential for multiple styles within the single DOC. In some regions, for example, there may be dry, semidry, sweet, sparkling or *passito* styles. Admittedly, these features are noted on the label. However, without careful inspection of the label, the purchaser may unintentionally buy an undesired style. Another deficiency is that the province from which the wine comes is rarely, if every, noted. This does not facilitate consumers connecting the wine with a specific region. Most consumers correlate their wine knowledge with regions that can be easily visualized on a map. France, to its credit (and profit), has largely avoided both these deficits.

SOUTH AFRICA

The South African Appellation Control system reflects many aspects found in European legislation, but has incorporated distinctive concepts. The system recognizes five levels of geographic designation. These progress from Regional Viticultural Designations to Areas, Districts, Ward (vineyard groupings), and finally Estates (individual vineyards). Most vineyards have not been granted officially designated status.

Wines produced within the appellations may be submitted for official evaluation and certification. Passing the certification tests permits the producer to apply an official neck seal from the Wine and Spirit Board. The neck seal possesses a certification number and bands that note the vintage, varietal origin, and whether the wine was estate-bottled. Furthermore, if the wine achieves an especially high ranking during sensory evaluation, the rarely used designation "Superior" appears on the seal.

UNITED STATES

Since 1978, the BATF (renamed the Alcohol and Tobacco Tax and Trade Bureau – TTB) has been empowered to designate viticultural area appellations in the United States. There had been previous regulations affecting the use of geographic names, but they had involved only country, state, or county designations. Approved Viticultural Areas (AVA) need not reflect existing state or county boundaries. There are at least 130 designated AVAs. The largest (Texas Hill Country) encompasses nearly 4 million hectares, whereas the smallest (Cole Ranch, CA) involves only some 60.7 hectares.

For the designation of an AVA, a proposal must be submitted to the TTB. Various types of supporting documentation must accompany the proposal. These include data establishing the local or national recognition of the named area; historical or present-day evidence for the proposed boundaries; climatologic, geologic, and topographic features distinguishing the region from surrounding areas; and precise boundaries noted on topographic maps from the US Geologic Survey. Geographic size is not a critical factor. Existing viticultural areas may have smaller appellations designated within their borders, and larger appellations may be formed from part or all of several existing appellations. There is no intent that size should imply anything about potential wine quality. There is, of course, the veiled assumption (or hope) that the regions so designated will produce wines that have a perceptible regional character. In almost all cases, this is unsupported by objective sensory data. One notable exception comes from the work of Arrhenius *et al.* (1996). Descriptive sensory analysis of 'Chardonnay' wines from four California viticultural areas (from a single vintage) supported the contention that the wines from these regions were statistically distinguishable (Fig. 10.4).

The designation of an AVA imposes no special regulations on cultivar use, viticultural practices, or winemaking procedures. It does contain the 85% rule, however. This stipulates that at least 85% of the juice must come

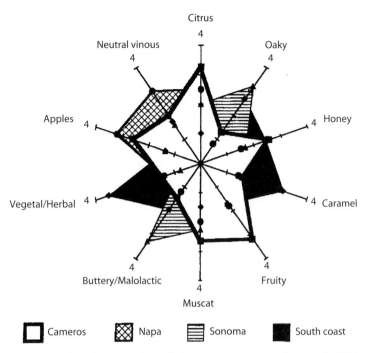

Figure 10.4 Polar plot derived from 10 aroma attributes of 1991 Chardonnay wines, showing regional variation within four viticultural areas of California. (From Arrhenius *et al.*, 1996, reproduced by permission)

from vines grown in the designated viticultural area (in contrast to only 75% for single-county or -state designations, with some exceptions such as 85% in Texas and 100% in California). In addition, the mention of a single variety on the label requires that at least 75% of the wine comes from the mentioned cultivar. The major exception relates to wines produced from *V. labrusca* cultivars or hybrids. In this instance, only 51% of the juice must come from the named cultivar. Thus, the American expression is intended primarily to regulate the authenticity and distinctiveness of the wine's geographic origin, with no implication concerning quality. This is a wise move. That is a job for wine critics, and consumers to accept or reject as they see fit.

Although foreign geographic names such as Chablis and Burgundy still appear on some American wines, producers of premium wines scrupulously avoid their use. Varietal name and vineyard location have proved a more effective means of obtaining recognition and market position. The primary exceptions are the appellations sherry, port, and champagne. For sherry and port, the terms have been used as both generic and regional designations. Although English possesses a legitimate generic term for champagne-like wines – sparkling – this has not achieved the prestige or market recognition it deserves. This situation will continue as long as producers promote recognition of the champagne appellation on their labels. The inability to use an established

geographic wine designation need not limit marketability as some fear. The absence of the sherry designation on similar wines, such as Montilla or Château-Chalon, does not seem to have limited their acceptance or marketability.

CANADA

Only one province in Canada possesses a legislated set of Appellation Control laws – British Columbia. Although the other major wine-producing province, Ontario, possesses a similar set of regulations, they are voluntary and administered by an independent body, the Vintners Quality Alliance (VQA). In the administration of the regulations, the Alliance is aided by provincial government agencies.

The systems in both provinces have regulations affecting the cultivars permitted, and they require wines to pass minimum sensory analysis standards to earn the right to use the VQA designation. Although both provinces have designated several viticultural areas, there is no stated or implied superiority to smaller appellations. The same is true for single-vineyard and estate bottling. Nevertheless, more stringent regulations affecting cultivar use apply to the regional than to province-wide appellations in Ontario. For example, estate-bottled wine must be 100% from grapes owned or regulated by the winery, whereas vineyard-labeled wines must be 100% from grapes grown on the vineyard designated.

AUSTRALIA

Although Australia appears philosophically opposed to an Appellation Control system, based on the European formula, it has established the boundaries of several regions, termed Geographical Indications (GI). To designate one of these regions on the label, at least 85% of the grapes used in making the wine must have been grown in the named region. In addition, the Hunter Valley region has a voluntary accreditation system. It is unique in its simplicity and degree of governmental support. Wines may be awarded one of two seals. These distinguish wines of superior quality, based on maturity (Classic) and potential to improve (Benchmark). Each year's winners are noted in a public ceremony and tasting at Parliament House in Sydney. Open endorsement by the state government is accredited with improving the profile of the quality wines in the region.

Detection of Wine Misrepresentation and Adulteration

In the previous section, laws designed to assure the authenticity of wine origin were discussed. The effectiveness of any regulation depends largely on the willingness and ability of enforcing agencies to assess compliance. Enforcement is a political–economic decision beyond the scope of this book. However, the technical ability to assess compliance is within the realm of science and is an appropriate topic for discussion.

Establishing conformity with many of the viticultural constraints of AC legislation is comparatively simple. Features such as use of irrigation, cultivars grown, training systems employed, or percentage of land planted can be directly assessed with on-site inspection. Yield per hectare, total soluble solids, and harvest date often can be checked against cellar records, simple chemical analyses, or corroborative testimony. In the past, however, the validation of the geographic and varietal nature of wine located in a winery, or the degree of chaptalization and fortification, was often impossible. Sensory analysis is insufficiently precise and cannot be automated.

Detection of contravention of several AC regulations has depended on the development of modern analytical tools, including scintillation counters, mass spectrometry, nuclear magnetic resonance (NMR), and the latest forms of gas and liquid chromatography. Because of the large amount of data that can be collected, advanced statistical analysis such as Principal Component Analysis may be required to detect pertinent information. Such an analysis can generate the equivalent of a fingerprint for each wine category or variety. The detection of sophisticated or toxic adulteration may require even more sophisticated instrumentation. Because of the incredible range of adulteration possible, it is usually necessary to suspect a priori the duplicity involved.

Although most modern wine falsification is nontoxic, it can do irreparable harm to the reputation and sales of a producer or region. The image of wine as a natural and wholesome beverage is central to most wine marketing. Geographic and varietal identity is also important because many consumers believe that it is almost synonymous with quality. For premium wines, the prestige associated with the appellation can be as or more important than the sensory quality of the wine. Lack of character is too often dismissed, and explained as a sign of finesse. Thus, the more famous the region, the more critical it becomes to assure consumers of authenticity of origin.

In addition to marketing concerns, circumvention of AC restrictions places conscientious producers at an economic disadvantage. Regional governments are also concerned because they have a financial stake in avoiding tax revenue loss associated with wine scandals.

The methods used to detect violations of Appellation Control statutes depend on the type of doctoring suspected. Because the instrumentation is expensive and requires considerable skill in its use and interpretation, tests cannot be used routinely. Their use is justified only when falsification is suspected. Nevertheless, the ability of instrumentation to detect fraudulent activity is impressive.

Validation of Geographic Origin

The major discriminant procedures are based on differences in the isotopes of hydrogen (H), carbon (C), and oxygen (O). The proportion of strontium isotopes ($^{87}Sr/^{86}Sr$) also appears to have potential value (Almeida and Vasconcelos, 2004). In addition, features as diverse as differences in the local distributions of microelements (Baxter et al., 1997), amino acid production (Tusseau et al., 1993), and distinctive phenol and aromatic characteristics of grape varieties (Etiévant et al., 1988; Rapp et al., 1993; Martí et al., 2004) have potential in detecting noncompliance with AC regulations. For example, differences in the volatile constituents of 'Syrah' wines have been used to successfully characterize wines produced in the different regions of the Rhône Valley (Vernin et al., 1989).

Isotopes of an element possess similar chemical properties, but differ in mass. The mass differences come from the number of neutrons present in the nucleus. For example, the most common isotope of hydrogen possesses a single neutron (^1H), whereas deuterium (^2H) and tritium (^3H) contain two and three neutrons, respectively. Although seven carbon isotopes are known, only three are sufficiently common to be of practical use.

The most common carbon isotope, representing about 99% of the total, possesses 12 neutrons (^{12}C). Carbon-13 (^{13}C) contains 13 neutrons, and carbon-14 (^{14}C) possesses 14 neutrons. The two important oxygen isotopes (^{16}O) and (^{18}O) possess 16 and 18 neutrons, respectively. Although most of the isotopes are stable, some, such as carbon-14, are unstable and release radioactive particles as they decay.

The value of isotopes in wine authentication comes from the influence of biotic and environmental factors on isotope distribution and relative proportion. Enzymes and biological processes such as transpiration may discriminate among isotopes. This typically shows as a differential favoring the lightest isotope. In addition, enzymes in different organisms, or diverse enzymes in the same organism, may show characteristic and divergent levels of preferential isotope use. Thus, the proportions of isotopes may differ from compound to compound, from variety to variety, from species to species, from region to region, from year to year, and with or without irrigation. Differences in specific isotopic proportion are measured relative to the deviation ($\delta D\permil$) from an international standard (sample ratio minus the standard ratio, divided by the standard, times 1000).

Because enzymic and climatic factors differentiate among isotopes, distinctive and stable variations among similar wines produced in different locations may be precisely and reproducibly detected. For example, regional precipitation and temperature differences affect the rate of evapotranspiration, and thus influence the $^{2}H/^{1}H$ ratio of the water and ethanol of wines produced at particular sites (Martin, 1988). Local environmental and biotic factors can also generate distinctive deviations in the $^{18}O/^{16}O$ ratio of water, and the $^{13}C/^{12}C$ ratio of grape constituents (Fig. 10.5). These differences may be sufficiently marked to allow tentative identification of the wine's country (Fig. 10.6A), region (Fig. 10.6B), and vintage (Fig. 10.6C) origin. They also may permit the identification and proportion of unauthorized juice or wine added to a particular wine.

Not only is the isotopic signature of a wine influenced by the regional climate, it also is affected by the grape cultivar, as well as the yeasts and bacteria involved in fermentation. These factors can either enhance or diminish the distinctive isotopic ratio produced by local weather conditions. Even juice concentration is likely to affect the isotopic ratio. Thus, precise information on standard wine production procedures in the region is required when assessing fraudulent adulteration. Authenticated examples

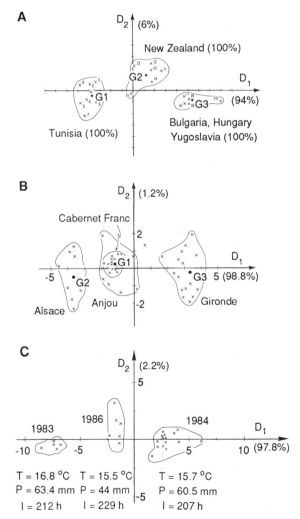

Figure 10.6 Discrimination of wines produced from different (**A**) countries, (**B**) geo-climatic regions, and (**C**) 'Cabernet Sauvignon' vintages from Saumur-Champigny. T, temperature; P, precipitation; I, incident radiation. (After Martin, 1988, reproduced by permission)

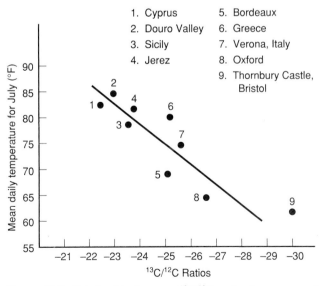

Figure 10.5 Relationship between $^{13}C/^{12}C$ ratios in grapes and temperature for selected growing areas in 1980. (From Stacey, 1984, reproduced by permission)

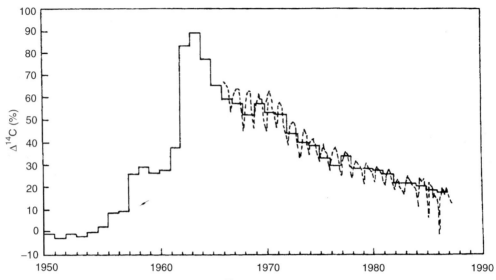

Figure 10.7 Thermonuclear bomb-generated ^{14}C variation in samples of annually produced Georgian wines and in monthly samples of atmospheric CO_2 in Bratislava. (From Burchuladze *et al.*, 1989, reproduced by permission)

of the wine in question are needed to recognize the effects of viti- and vinicultural practices on isotopic ratios.

Under most circumstances, the ^{14}C content of wine provides little useful information concerning origin. The rate of carbon-14 decay in organic compounds only permits precise measurement of time units greater than 80 years. However, the atmospheric testing of nuclear weapons during the late 1950s and early 1960s raised the relatively stable concentration of carbon-14 in the atmosphere. Neutrons released by these tests increased the amount of ^{14}C normally generated by the collision of solar cosmic rays with atmospheric carbon. This significant, but temporary, atmospheric perturbation is reflected in the ^{14}C level of wines produced during that period, and subsequently (Fig. 10.7). Additional data on the ^{14}C levels in wines up to 1999 is provided in Zoppi *et al.* (2004). From 1956 on, the ^{14}C content of a wine permits one or more potential vintage dates to be proposed. Any measurement may indicate two or occasionally several possible vintage dates. Precise vintage pinpointing requires additional information. However, as yearly differences are becoming less marked, new isotopic criteria are being developed for dating, based on the stable isotopes of water and carbon dioxide (Martin *et al.*, 1998).

Validation of Conformity to Wine Production Regulations

Although radiocarbon (^{14}C) dating has limited value in authenticating wine origin, it can be very useful in detecting the presence of synthetic additives. Synthetic organic compounds often are derived from fossil fuels, and may be either nature identical or artificial. The former are chemically and isotopically identical to naturally occurring compounds. Artificial compounds may be either isomers of natural compounds or unrelated compounds possessing sensory properties similar to natural compounds, for example the sugar substitute saccharine. Although additives may not be dangerous to human health, their addition to wine may be illegal. Synthetic compounds usually can be recognized as such, due to their fossil-fuel derivation. Because of the age of fossil fuels, compounds derived from them possess no detectable ^{14}C. Organic compounds lose detectable carbon radioactivity after approximately 50,000 years. Thus, a port fortified with synthetic alcohol, rather than with grape spirits, could be recognized by appearing to be several thousand years old, according to ^{14}C measurement of its ethanol content. The synthetic alcohol would lower the average ^{14}C age of the ethanol content of the wine. In addition, nature-identical compounds often can be differentiated from those found in wine by their different stereoisomeric (enantiomeric) composition (see Ebeler *et al.*, 2001).

In the past, ^{14}C dating could detect only the addition of major organic wine constituents. However, the combination of liquid/liquid extraction and gas chromatography, with improvements in ^{14}C assessment, can permit differentiation between even milligram samples of natural versus synthetic compounds (McCallum *et al.*, 1986). The use of carbonation versus secondary yeast fermentation can be detected with ^{14}C analysis, if the carbon dioxide is derived from a fossil fuel source.

In most countries, adding water to wine is illegal. However, assessment of compliance has been fraught with difficulties (Ribéreau-Gayon *et al.*, 1982). Of the older techniques, a depression in the relatively stable magnesium content of wine has been the most accepted indicator of dilution (Robertson and Rush, 1979). However, this difference could be corrected by adding the requisite amount of magnesium. Modifications in the $^2H/^1H$ and $^{18}O/^{16}O$ ratios of water have been investigated as more accurate dilution indicators. Even the groundwater from a vineyard site may possess a detectably different isotopic balance, compared to the water content of the wine. These ratios may be modified during transpirational loss, selective use in photosynthesis, and involvement in cellular hydrolytic reactions. In addition, water deficit and rootstock variety can adjust the $^{18}O/^{16}O$ ratio of water extracted from soil (Tardaguila *et al.*, 1997).

The addition of sugar to must is forbidden in most jurisdictions. In others (i.e., Germany), its use in chaptalization is regulated, but permissible only in the lower-quality categories of wine. In France, it is permitted, but under strict regulation. Sugar is added either as sucrose or invert sugar (enzymatically or acid hydrolyzed sucrose). In either case, the sugar is quickly assimilated and converted to ethanol. Thus, the detection of chaptalization requires a means of distinguishing between grape- and nongrape-derived ethanol. This is most conveniently achieved by measuring the $^{13}C/^{12}C$ ratio of the alcohol. Ethanol derived from different sugar sources normally can be differentiated. The addition of any commercially available sugar would shift the $^{13}C/^{12}C$ ratio of the ethanol produced during fermentation. The level of discrimination would depend on the amount of sugar added and the source of the sugar. The detection of chaptalization down to 0.3% is possible (Martin, 1988). Measurement of modifications in the $^2H/^1H$ ratio of ethanol can provide additional data on the degree of chaptalization (Giraudon, 1994; Fauhl and Wittkowski, 2000). An additional indicator of chaptalization with invert sugar is 2-hydroacetylfuran (Rapp *et al.*, 1983). The compound appears not to occur in either must or wine without the addition of acid-hydrolyzed cane sugar.

The $^{13}C/^{12}C$ ratio also can be used to detect the addition of other plant- and microbial-derived constituents, such as glycerol and organic acids. Each compound possesses a $^{13}C/^{12}C$ ratio characteristic of its biological origin. Although the differences are most marked among plants possessing distinctive photosynthetic processes (C_3, C_4, and CAM pathways), differences also occur among different plant species, and in the same species in different climatic regions. A complicating factor, however, is the modification of the ratio due to action of metabolic enzymes during fermentation (Weber *et al.*, 1997). This is less of a problem if addition occurs after fermentation.

Adulteration with most nongrape pigments, such as cochineal, poke, or orseille, can be readily detected by the distinctive chemical properties of the pigments. Even the addition of oenocyamine, a color extract from grapes, can be determined. High-performance liquid chromatography permits the separation and quantitative measurement of the many anthocyanins present in red wine. A pattern atypical of the cultivar and year would indicate the likelihood of color addition, or the incorporation of must or wine from unauthorized grape varieties. Chromatography has been used for years in Europe to verify the presence or absence of must or wine from non-*vinifera* cultivars in Appellation Control wines. Anthocyanins from other *Vitis* species, or from interspecific hybrids with *V. vinifera*, typically possess diglucoside anthocyanins not found in *V. vinifera* grapes. However, some complex *V. vinifera* interspecific hybrids do not synthesize diglucoside anthocyanins (van Buren *et al.*, 1970; Guzun *et al.*, 1990). In addition, the absence of particular anthocyanins may be used to partially verify the varietal origin of a wine. For example, 'Pinot noir' grapes and wine do not possess acylated anthocyanins (Rankine *et al.*, 1958).

Other harmless, but unauthorized, adulterants that are relatively easily detected in wine are sorbitol and apple juice (Burda and Collins, 1991). Sorbitol may be added with the intention of giving the wine a smoother mouth-feel. The adulteration of wine with glycerol, for the same purpose, is more difficult to detect, but is possible with multielement stable-isotope-ratio analysis (Roßmann *et al.*, 1998), or the presence of 3-methoxy-1,2-propanediol and/or cyclic diglycerols (Lampe *et al.*, 1997). These compounds occur in commercial sources of glycerol, but are absent in wine. Use of apple juice (or cider) as an inexpensive extender in wine production can be determined by measuring the concentrations of chlorogenic acid and sorbitol. Sorbitol and chlorogenic acid do not occur naturally in wine at concentrations greater than 1 g/liter and 1 mg/liter, respectively (Pocock and Somers, 1989).

Although instrumental analysis will not stop wine adulteration, it can make detection and conviction more certain. By making the avoidance of detection very complex and costly, adulteration may become unprofitable.

World Wine Regions

In the remainder of the chapter, the distinctive climatic, soil, varietal, viticultural, and enologic features of the various wine regions of the world are highlighted.

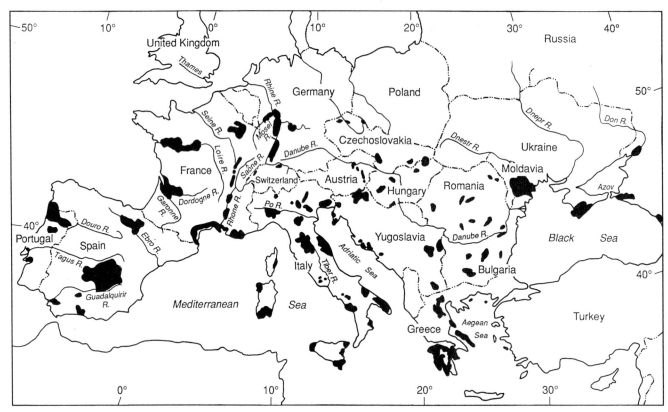

Figure 10.8 Map showing the approximate location of some of the wine regions of European countries.

Western Europe constitutes the most significant region, both in terms of vineyard area and quantity of production. Thus, more detail is available about European wine regions than about other viticultural areas. Nevertheless, where possible, other regions are covered similarly.

A map indicating some of the major European wine regions is given in Fig. 10.8. The majority of the production is centered in southern portions of western Europe, around the Mediterranean Sea. The region is characterized by cool moist winters and warm (to hot), relatively dry summers. Viticulture has traditionally been dryland farming, without irrigation. Grape hectarage and wine production data for various viticultural regions are listed in Table 10.2. These figures are down considerably from their peak in the late 1970s, when total vineyard area was about 10,200,000 ha and production in the range of 320,000,000 hl. World figures are currently about 7,960,000 ha and 265,000,000 hl. For additional data on wine and grape production, and consumption statistics see OIV (2005).

Western Europe

The importance of western Europe in wine production is a reflection of both its physical and social climate. Richness in natural resources combined with historical

and geographic factors to favor industrialization and the accumulation of free capital – conditions conducive to developing an expanding and increasingly discriminating class of wine connoisseurs. This became more important as transport improved and vineyard proximity to the consuming public became less significant. Increased profit both permitted and spurred the use of more costly and complex practices that could enhance grape and wine quality. The improvements complemented the inherent climatic advantages of the region. Technical developments also permitted the potential preservation and development of the wine's finer characteristics. Consequently, distinctive regional features became more pronounced as a social class willing and desirous to appreciate these aspects grew. In contrast, repeated invasions, limited natural resources, and the periodic imposition of Islamic rule largely prevented an equivalent development occurring in eastern Europe.

Although western Europe is not inherently more suited to grape growing and winemaking than some other European regions, or parts of the world, it has evolved wines well suited to highlight the qualities of indigenous cultivars and to take advantage of local climatic conditions. Cultural tastes throughout the world have developed largely as a reflection of European precepts. In addition, climatic vicissitudes have been used

Table 10.2 Comparison of viticultural surface areas and wine production[a]

Country	Wine production (10^3 hl)	Vineyard area (10^3 ha)	Total land area (10^6 ha)	Wine yield per total surface area (ha/10^3 ha)	Vineyard area per surface area (ha/10^3 ha)	Yield per vineyard hectarage (hl/ha)
France	46,360	887	55	893	16.1	52.2
Italy	44,086	868	30	1,469	28.9	50.8
Spain	42,802	1,207	51	839	23.7	35.5
United States	20,770	415	963	22	0.4	50.0
Argentina	13,225	211	278	48	0.8	62.7
China	11,600	453	960	12	37.8	25.6
Australia	10,194	157	768	13	0.2	64.9
South Africa	8,853	132	122	73	1.1	67.0
Germany	8,191	102	36	228	2.8	80.3
Portugal	7,340	249	9	815	27.6	29.5
Chile	6,682	185	76	88	2.4	36.1
Roumania	5,555	239	24	231	9.9	23.2
Russia	4,530	70	1,707	3	0.04	64.7
Hungary	3,880	88	9	431	9.8	44.1
Greece	3,799	130	13	292	10.0	29.2
Moldova	3,215	148	3	946	43.5	21.7
Brazil	2,620	72	851	3	0.08	36.4
Austria	2,526	48	8	316	6.0	52.6
Ukraine	2,380	99	60	40	1.7	24.0
Bulgaira	2,314	99	11	210	9.0	23.4
Croatia	1,768	61	6	294	10.0	30.0
Serbia	1,734	71	9	193	7.9	24.4
Mexico	1,096	42	196	6	0.2	26.1
Switzerland	967	15	4	242	3.8	64.5
Japan	938	21	38	25	0.6	44.7
Uruguay	837	10	18	47	0.6	83.7
Slovenia	671	17	2	336	8.5	39.5
New Zealand	550	19	27	20	0.7	28.9
Canada	359	11	997	0.4	0.01	32.6
United Kingdom	15	1	24	0.6	0.04	15.0

[a]Data for wine yield and vineyard area are from the OIV (2005). Note that the production figures per hectare have not been adjusted to account for the proportion of the harvest used as table grapes, in raisin production, or wine produced from imported grapes and must. These data were unavailable for most countries.

to create the mystique of vintage and regional variability, often important in marketing fine (at least expensive) wines.

CLIMATE

The climate of western Europe is significantly modified by the influence of adjacent bodies of water, namely the Atlantic Ocean, and the Mediterranean and Baltic seas. Slow heating and cooling of these large water bodies, combined with predominant westerly winds, retard rapid and marked seasonal changes in temperature over much of western Europe. The general east–west orientation of European mountain ranges does not impede the moderating influence of wind flow off these large bodies of water onto the landmass. These influences are accentuated by the warming action of the Gulf Stream during the winter months (Fig. 10.9), and its cooling action during the summer months. Consequently, Europe's climate is milder than

latitude alone would suggest. This permits commercial grape culture at latitudes higher than on any other continent.

In western Europe, the limit of commercial wine production angles northward from the Loire Valley (47°30′ N) up into the Rhine Valley (51°N) (Fig. 10.10). The increasingly northern limit for grape culture, as one passes eastward through France, results from the progressive overland warming of the predominant cool winds coming off the Atlantic. Nevertheless, the northernmost viticultural regions of Germany hug river valleys, and the most favored sites occur on south-facing slopes. East of the Rhine and its tributaries, the northern limit of grape cultivation turns slowly southward, as the moderating effect of the Gulf Stream and surrounding seas diminish. Here, the frigid winter cold of the continental climate becomes more limiting than reduced light intensity or accumulated heat units of these northern latitudes. Thus, the limit of viticulture moves south

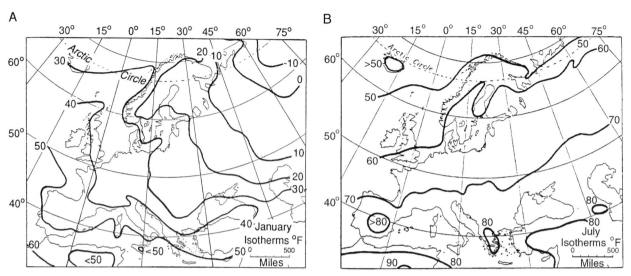

Figure 10.9 January (**A**) and July (**B**) isotherms (°F) of Europe. Note the marked warming and cooling seasonal influence of the Gulf Stream on the western portion of Europe. (From Shackleton, 1958, reproduced by permission)

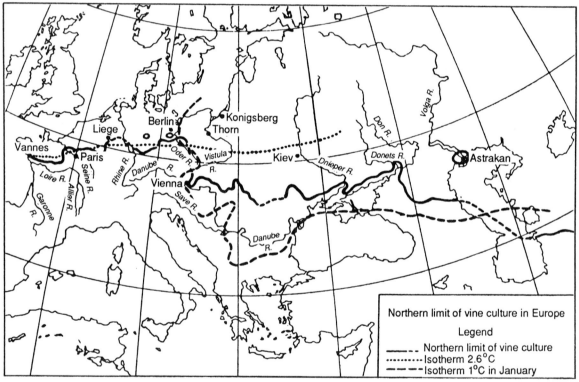

Figure 10.10 Northern limit of most grapevine culture in continental Europe. (After Branas *et al.*, 1946, reproduced by permission)

through Czechoslovakia and Romania, to just north of the Black Sea.

The southern extent of commercial viticulture is limited by warm winters that ineffectively break bud dormancy, and the increasing arid climate.

Although western regions of Europe share many common cultural, historical, and climatic features related to their proximity to several seas and the Atlantic Ocean, there are equally important differences. For grape-growing and winemaking practices, climatic differences are the most significant. Climatically, Europe may be divided into several broad zones. The area encompassing most of the western portion of southern France possesses both mild winters and summers.

Table 10.3 Climatic change spanning the major regions supporting commercial planting of 'Pinot noir'

Region	Latitude	Frost-free period (Celsius degree-days)	Average temperature of warmest month (°C)	Must minimal potential alcohol required (%)	Maximum permissible chaptalization (% potential alcohol increase)
Ahr	50°50′N	887	17.8	7.5	3.5
Champagne	49°20′N	988	18.3	8.5	2.5
Kairserstuhl	48°10′N	1045	18.8	8.9	2.5
Mâcon	46°20′N	1223	20.0	10.0	2.0

Source: After Becker, 1977, reproduced by permission

More precipitation falls during the autumn and winter months than the rest of the year. Nevertheless, adequate precipitation for vine development usually occurs throughout the growing season. The climates of north-central and western France and southwestern Germany possess several common features. Although the growing season is shorter, abundant spring and summer rain and longer day-length spur growth. Dry, sunny autumn days favor fruit maturation. Austria and Hungary possess similar climatic conditions, but tend to have higher humidities and colder winters. Southeastern France and much of Italy, Spain, Portugal, and Greece possess a Mediterranean climate characterized by dry hot summers, with most rainfall occurring during the mild winter months.

Geographic and topographical features can significantly modify local climatic conditions. For example, Table 10.3 shows the effect of latitude on the climate of regions spanning the major north–south European limits of 'Pinot noir' cultivation. The influence is reflected in an increase in permitted chaptalization as one progresses northward. The vineyard altitude also is critical; lower altitudes are usually preferable at high latitudes, and higher altitudes desired at lower latitudes. In general, the average annual temperature (isotherm) decreases by about 0.5 °C/100 m elevation in altitude, or degree-latitude increase (Hopkin's bioclimatic law). In practice, however, these preferences may be significantly modified by regional factors, such as site openness, valley width, or nearness to water bodies (Seeman *et al.*, 1979).

In addition to affecting the distribution of grape culture, climate also has a deciding influence on the potential for fine wine production. For example, most of the well-known wine regions of Europe are in mild to cool climatic zones. The absence of hot weather during ripening favors the retention of grape acidity. This gives the resulting wine a fresh taste and helps restrict microbial spoilage. Cool harvest conditions also promote the development or retention of varietal flavors and minimize overheating during fermentation. In addition,

storage in cool cellars represses microbial growth that could induce spoilage. Finally, cool conditions have required that most vineyards be situated on south- or west-facing slopes, to obtain sufficient heat and light exposure. This incidentally has positioned vineyards on less-fertile, but better-drained sites. These features have restrained excessive vine vigor, while promoting fruit ripening and providing a degree of frost protection.

In contrast, the hot conditions typical of southern regions favor acid metabolism and a rise in juice pH. In addition to producing a flat taste, the low acidity makes the wines more susceptible to oxidation and microbial spoilage. Although only approximately 0.004% of grape phenols are in a readily oxidized state at pH 3.5 (Cilliers and Singleton, 1990), they are so unstable that oxidative reactions occur readily. Even minor increases in pH can significantly increase the tendency of wine to oxidize. Thus, protection from oxidation tends to be more critical in warm areas than in cooler regions. In addition, grapes tend to accumulate higher sugar contents under warm conditions. These increase the likelihood of premature cessation of fermentation, along with harvesting and fermentation under warm conditions. By retaining fermentable sugars, the wine is much more susceptible to undesirable forms of malolactic fermentation and microbial spoilage. Warm cellar conditions further enhance the likelihood of spoilage.

Although advancements in viticulture and enology have increased the potential to produce a wider range of wine styles, prevailing conditions had a decisive effect on the evolution of regional wine styles. Cool climates favored the production of fruity, tart, white wines. Such wines normally have been consumed alone as sipping wines, before or after meals. More alcoholic white wines functioned primarily as food beverages. In warmer regions, red wines have tended to predominate. Here, the higher sugar content of the grapes permitted the production of full-bodied wines, well suited to consumption with meals. In hot Mediterranean regions, high-sugar and low-acid grapes favored the production of alcoholic wines that tended to oxidize readily.

These features encouraged the development of oxidized, or sweet, artificially flavored, high-alcoholic wines, appropriate for use as aperitifs or dessert wines.

Nevertheless, present-day viticultural and enologic techniques now permit the production of almost any wine style in southern regions. Equivalent techniques have not allowed the reverse situation to develop in northern regions.

CULTIVARS

In Europe, notably in France, the cultivation of grape varieties has tended to be highly localized (Fig. 10.11). This has given rise to the view that cultivar distribution reflects a conscious selection of cultivars, particularly suited to local climatic conditions. At sites where religious orders have produced wine for centuries, empirical trials may have found grape cultivars especially suited to the local climate. In most localities, however, wine was consumed within the year of its production, a situation incompatible with assessing aging potential (the *sine qua non* of wine quality). Also, varieties were commonly planted more or less at random within vineyards, as well as harvested and vinified together. Thus, assessment of the relative quality of one cultivar versus another would have been essentially impossible. Finally, there is little solid evidence documenting the continued cultivation of specific varieties in particular regions. Important exceptions are 'Riesling' and 'Pinot noir,' for which information may go back to

the 1400s. Going beyond this requires a leap of faith. Even in Bordeaux, 'Cabernet Sauvignon' is reported to have been rare, or nonexistent, until the early 1800s (Penning-Rowsell, 1985). More likely, unintentional selection resulted among vines of local origin, derived from indigenous strains of *V. vinifera*, or accidental crossing with imported cultivars. Most selection would have been for obvious traits, such as compatibility with local climatic and soil conditions, higher sugar content, adequate acidity, and aroma. Subtleties such as aging potential, development of delicate bouquets, and complexity would have been selected more by accident than design. Only since the 1700s have conditions become more conducive to the intentional selection of premium-quality cultivars.

VITICULTURE

In Europe, traditional cultivation has ranged from dense plantings, with about 5,000 to 10,000 vines/ hectare, to interplanting with field crops with trees as supports. In densely planted vineyards, each vine occupied about $1\,m^2$, resulting in intervine competition and restrained vigor. This had the effect of reducing the level of pruning and manual labor required. Spacing was also compatible with the use of single horse-drawn equipment. Restrained vigor also resulted from the confinement of most vineyards to poor soils, where cereal and other food crops would not grow well. This applied equally to sloped sites, which generally were (and are) ill-suited to annual crop production. The relatively dry summer months, combined with dryland farming, equally tended to result in an early termination of vegetative growth. Combined with traditional pruning, fruit production was limited, but promoted early maturity. This had the distinct advantage of permitting an early harvest, important in regions where the onset of cold, rainy, autumn weather could ruin the crop. Hedging, to allow easier access to the vines, directed nutrient supplies to the fruit. As well, hedging tended to limit fruit shading, thus promoting improved coloration and flavor development.

The previous relegation of most grape culture to poorer or sloped agricultural sites, and the advantages of dense plantings, has regrettably led to the erroneous view that these conditions were necessary for grape quality. That these conditions reduce individual vine size and productivity, and favor fruit maturation, is not in question. However, as noted in Chapter 4, new training systems involving canopy management permit the cultivation of widely spaced vines on rich soils without sacrificing fruit quality. In addition, fruit yield is increased and mechanization facilitated. Some of these techniques are being slowly integrated into standard European vineyard practice.

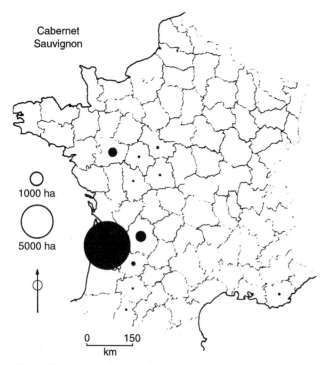

Figure 10.11 Distribution of 'Cabernet Sauvignon' by department in France. (From Moran, 1988, reproduced by permission)

ENOLOGY

In Europe, more variation is probably found in wine-making procedures than in viticultural practice. The differences reflect the wine styles that have evolved in response to climatic or marketing conditions. In addition, major developments have occurred since the mid-nineteenth century. Previously, long-aging wines were matured in large-capacity wood cooperage, including both white and red wines. Furthermore, red and occasionally white wines were produced by fermenting the juice with the seeds and skins for up to several weeks. The resulting wines were often partially oxidized and possessed higher volatile acidity than now considered acceptable (Sudraud, 1978). Current trends have been to reduce the maceration period for both white and red wines. In addition, shorter maturation in wood is also favored. Premium white wines may receive up to 6 months' maturation in small oak cooperage, whereas premium reds often receive up to 2 years of in-barrel aging. Limited oak exposure may be used to preserve the fresh aroma of the wine. Greater emphasis is placed on the development of a reductive in-bottle aging bouquet than in the past.

CENTRAL WESTERN EUROPE

France   France has a diverse topography with few homogeneous regions. Most agricultural regions of similar geographic character are comparatively small and specialized in crop production. This disparity is reflected in the country's localized cultivar plantings and regional wine styles. Although no one soil type or geologic origin distinguishes French vineyards, several regions possess calcareous soils or cover chalky substrata.

Although the effect of Appellation Control laws has tended to stabilize cultivar plantings in AOC- and VDQS-regions, marked changes in the varietal composition have occurred in nondesignated regions. For example, the proportion of French-American hybrid varieties has dropped from about 30% in 1958 to less than 5% by 1988 (Boursiquot, 1990). It is supposedly to go to 0% by 2010. Remarkable in terms of worldwide trends is the increase in red cultivar plantings. In France, the hectarage of red *V. vinifera* cultivars has grown by 9% since 1958, whereas the planting of white cultivars has declined by 7%. White cultivars cover only about 30% of French vineyard area. Of these, about 40% of the yield is used in the production of brandies, notably cognac and armagnac.

Over the period 1958–1988, there was an increase in the cultivation of premium red cultivars (Fig. 10.12). Several well-known white cultivars lost ground (Fig. 10.13), whereas others came close to extinction.

For example, 'Sémillon' and 'Chenin blanc' plantings declined by about 50%, and 'Viognier' fell to only 82 ha (Boursiquot, 1990) – subsequently, 'Viognier' cultivation has rebounded to an estimated 1500 ha (Bonfiglioli *et al.*, 1999). 'Chardonnay' and 'Sauvignon blanc' are two of the few famous white cultivars to

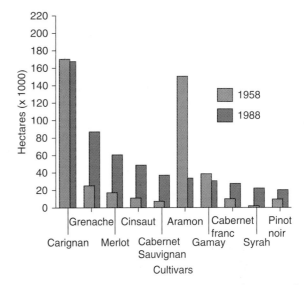

Figure 10.12 Planting of the 10 most common red *Vitis vinifera* cultivars in France in 1958 and 1988. (Data from Cadastre Viticole, IVCC (Institut des Vins de Consommation Courante), Recensement Agricole, and SCEES–INSEE (Service Central des Enquêtes et Études Statistiques–Institut National de Statistique et des Études Economiques), reported in Boursiquot, 1990)

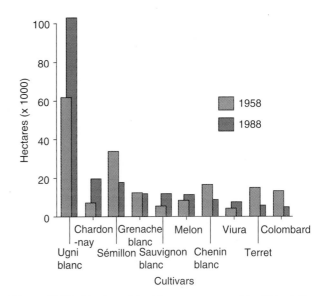

Figure 10.13 Planting of the 10 most common white *Vitis vinifera* cultivars in France in 1958 and 1988. (Data from Cadastre Viticole, IVCC (Institut des Vins de Consommation Courante), Recensement Agricole, and SCEES–INSEE (Service Central des Enquêtes et Études Statistiques–Institut National de Statistique et des Études Economiques), reported in Boursiquot, 1990)

see cultivation increases. The most marked increase in cultivar coverage has occurred with 'Syrah,' whose hectarage expanded about 16-fold. Other red varieties with significantly expanded plantings are 'Cabernet Sauvignon,' 'Cabernet franc,' 'Merlot,' and 'Pinot noir.'

Although France is particularly famous for one of its sparkling wines (champagne), the vast majority of French wines are still. As in other major wine-producing and -consuming countries, most French wines are red. Only a few sweet or fortified wines are produced.

Although wine is produced commercially in most regions of France, only a few regions are widely represented in world trade, notably Alsace, Bordeaux, Burgundy, Champagne, Loire, and the Rhône. These regions are briefly discussed here, along with the less prestigious, but most important wine-producing region (in terms of volume), the Midi (Languedoc-Roussillon).

Alsace   Alsace is one of the most culturally distinctive regions of France. This reflects its German-French heritage and alternate French and German nationality. Not surprisingly, wines from Alsace bear a varietal resemblance to German wines, but a stylistic similarity to French wines. It is also the one "French" region where varietal origin is typically and prominently displayed on the label.

Alsatian vineyards run from north to south, along the eastern side of the Vosges Mountains (47°50′ to 49°00′ N). The vineyard region possesses three structurally distinctive zones. The zone running along the edge of the mountains has excellent drainage, and benefits from solar warming of its shallow, rocky, siliceous soil. The foothill region is predominantly calcareous and generally possesses the best microclimate for grape cultivation. The soils of the plains are of more recent alluvial origin and possess excellent water-retention properties. Most vineyards occur at an altitude ranging between 170 and 360 m.

Alsace produces predominantly dry white wines, although some sweet and sparkling wines are produced. Yearly production averages about 0.9 million hl. Predominant cultivars are 'Gewürztraminer' (20%), 'Pinot blanc' (18%), 'Riesling' (20%), and 'Silvaner' (21%), with small plantings of 'Chasselas,' 'Muscat,' 'Auxerrois,' 'Pinot gris,' and 'Pinot noir.'

Because of the cool climate, grapes are often harvested high in acidity and the wines treated to promote malolactic fermentation. Local strains of lactic acid bacteria may be involved in generating some of the distinctive regional flavor found in many Alsatian wines.

Bordeaux   Bordeaux is the largest of the famous French viticultural regions (44°20′ to 45°30′ N). It runs southeast for about 150 km along the banks of the Gironde, Dordogne, and Garonne Rivers. The vineyards cover about 112,000 ha, and annual production averages more than 6 million hl.

The Bordeaux region, located near the mouth of the Aquitaine Basin, is trisected by the junction of the Dordogne and Garonne Rivers to form the Gironde. These zones are divided into some 30 variously sized AOC areas. The best-known are those on the western banks of the Gironde (Haut Médoc) and the Garonne (Graves and Sauternes), and the eastern bank of the Dordogne (Pomerol and Saint-Émilion). Although Bordeaux is best known for its red wines, about 40% is white. White wines are produced primarily in Graves and Entre-Deux-Mers. The best vineyard sites are generally on shallow slopes or alluvial terraces adjacent to the Gironde, or low-lying regions along the Garonne and Dordogne Rivers at altitudes between 15 and 120 m.

Geologically, Bordeaux shows relatively little diversity. The bedrock is predominantly composed of Tertiary marls or sandstone, intermixed with limestone inclusions. The substrata are usually covered by alluvial deposits of gravel and sand of Quaternary origin, topped with silt. The soils are generally poor in humus and exchangeable cations. This is partially offset at better sites by the soil depth, often 3 m or more. Deep soils also provide vines with access to water during periods of drought and good drainage in heavy rains.

The presence of extensive forests to the east and south protects Bordeaux vineyards from direct exposure to cool winds off the Atlantic. Nevertheless, proximity to the ocean and rivers provides some protection from rapid temperature changes, but limits summer warmth. Wet autumns occasionally cause difficulty during harvest, causing fruit to split, encouraging bunch rot, and diluting the sugar and flavor content of the grapes. Soil depth and drainage, along with local microclimate, appear to be more significant to quality than soil type or geologic origin (Seguin, 1986).

Unlike the wines of some French viticultural regions, Bordeaux wines typically are blends of wines produced from two or more cultivars. Depending on the AOC, the predominant cultivar can vary. In the Haut Médoc, the prevailing variety in red wines is 'Cabernet Sauvignon,' whereas in Pomerol, it is 'Merlot.' This partially results from 'Cabernet Sauvignon' grapes being more herbaceous on the clay soils of Pomerol, but less so on the sandy and gravelly soils of the Haut-Médoc. Maturing somewhat earlier and reaching high °Brix, 'Merlot' is also more forgiving of the slower warming of the clay soils in Pomerol.

The presence and percentage of each cultivar in a vineyard often varies considerably among estates. In addition, the blend usually changes from year to year. This permits the winemaker to compensate for deficiencies in the character of the different varietal base wines. Wine not incorporated into the final blend may be bottled under an alternate (second) label, or used in the makeup of a general blend. The standard red cultivars are 'Cabernet Sauvignon,' 'Merlot,' 'Cabernet franc,' 'Petit Verdot,' and 'Malbec.' The first three constitute about 90% of red cultivar planting in Bordeaux. Although 'Cabernet Sauvignon' is the best-known Bordeaux grape, and grown particularly in the Haut-Médoc, the related 'Merlot' constitutes about 60% of the red grapes grown in Bordeaux. 'Petit Verdot' and 'Malbec' constitute only a small proportion of the hectarage.

White bordeaux also tends to be a blend of wines produced from two or more cultivars. In most areas, the predominant cultivar is 'Sauvignon blanc,' with 'Sémillon' coming second. The 2:1 proportion of these cultivars is reversed in Sauternes and Barsac, in which sweet, occasionally botrytized, wines are produced. In contrast to German botrytized wines, Bordeaux botrytized wines are usually high in alcohol content (14–15%). Other permitted white cultivars are 'Muscadelle,' 'Ugni blanc,' and 'Colombard.'

Because of good harbor facilities, proximity to the climate-moderating ocean, and a long-established association with discriminating wine-importing countries, Bordeaux was well positioned to capitalize on the benefits of many winemaking developments. It was one of the first regions to initiate the modern practice of estate bottling and in-bottle aging. It also influenced the shift from tank to barrel maturation of wine. Except for some white wines, Bordeaux wines are tank- or vat-fermented, rather than in-barrel fermented, as in Burgundy. This situation reflects the relatively large size of many Bordeaux estates (*châteaux*), in contrast to Burgundy. Most Bordeaux vineyards cover 5–20 ha, with some encompassing 40–80 ha.

Burgundy Burgundy is often considered to include several regions beyond the strict confines of Burgundy proper (the Côte d'Or). The ancillary regions include Chablis, to the northwest, and the more southern areas of Challonais, Mâconnais, and Beaujolais. The total wine production in all Burgundian regions averages about 1.5 million hl/year.

For all its fame, the Côte d'Or region consists of only a narrow strip of land, seldom more than 2 km wide. The strip runs about 50 km in a northeasterly direction from Chagny to Dijon (46°50′ to 47°20′ N), along the western edge of the broad Saône Valley. Although the vineyards are in a river valley, the Saône River is too distant (≥ 20 km) to have a significant effect on vineyard microclimate. A major physical feature favoring viticulture in the region is the southeasterly inclination of the valley wall. The porous soil structure and 5–20% slope promote good drainage and favor early-spring warming of the soil. Sites located partially up the slope, at an altitude between 250 and 300 m, are generally preferred (Fig. 10.14). Wine produced from grapes grown on higher ground or on the alluvial soils of the valley floor are generally regarded as inferior.

The Côte d'Or is divided into two subregions, the northern Côte de Nuits and the southern Côte de Beaune. Although there are exceptions, the Côte de Nuits is known more for its red wines, whereas the Côte de Beaune is more renowned for its white wines. This difference is commonly ascribed to the steeper slopes and limestone-based soils of the Côte de Nuits versus the shallower slopes and marly clays of the Côte de Beaune.

The predominant cultivars planted in the Côte d'Or are both early maturing – 'Pinot noir' and 'Chardonnay.' These cultivars produce some of the best-known wines in the world. The cool climate slows ripening, a factor often considered to limit the loss of important varietal flavors, especially 'Pinot noir.'

'Pinot noir' can produce delicately fragrant, subtle, smooth wines of great quality under ideal conditions. Regrettably, optimal conditions occur rarely, even in Burgundy. Climatic factors appear far more significant to the flavor characteristics of 'Pinot noir' wines than other varieties (Miranda-Lopez *et al.*, 1992). In addition, there is the notorious clonal diversity of 'Pinot noir.' Further complicating an already difficult situation is the multiple ownership of most of the vineyards. Individual owners frequently possess a few rows of vines at numerous sites scattered throughout the region. Thus, grapes are fermented in small lots (to maintain site identity) by producers whose technical skill and equipment are highly variable. The wines are usually fermented and matured in older, small, oak cooperage. New oak is not considered the quality feature here that it is in Bordeaux.

Although 'Pinot noir' matures early, it is not intensely pigmented. Thus, to improve color extraction, part of the crop may be subjected to thermovinification. However, the trend is to return to a cold maceration period prior to fermentation. Frequent punching down of the cap during fermentation is usually necessary. Because of the onerous nature of punching down, considerable interest has been shown in the use of small-capacity (~50-hl) rotary fermentors. They frequently and automatically mix the pomace and fermenting juice.

White wine is produced primarily from 'Chardonnay' grapes, although some comes from 'Aligoté.' If so, its presence must be designated on the label. Most white wines are fermented in barrels or small tanks. In the region of Macon, about 5–10% of the 'Chardonnay' clones possess a muscat character. These are considered

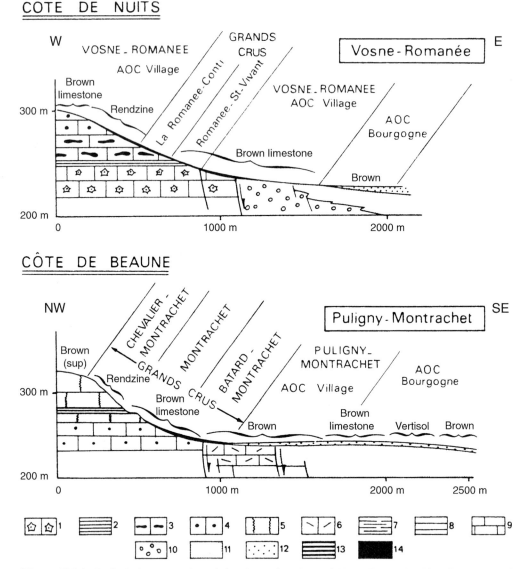

Figure 10.14 Geological cross-section of the vineyard regions of Vosne-Romanée (Côte de Nuits) and Puligny-Montrachet (Côte de Beaune) in Burgundy. (From Mériaux *et al.*, 1981, reproduced by permission.) The Jurassic structure of the Côte: 1, crinoidal limestone (Bajocain); 2, Ostrea acuminata marl (Upper Bajocain); 3, Prémeaux stone; 4, white oolite, oolite and bioclastic limestone (Middle Bathonian); 5, Comblanchien limestone with, in the south, Pholasomya bellona marl (Middle Bathonian); 6, Dalle nacrée (pearly slab) (Callovian); 7, limestone marl and Pernand marl (Middle Oxfordian); 8, Pommard marl (Upper Oxfordian); Nantoux limestone (Upper Oxfordian); 13, marly intercalations between 4 and 5. Infilling of the Bresse Graben: 10, conglomerate, limestone, and clay (Oligocene); 11, marl, sand, and gravel (plio-Pleistocene). Cover: 12, alluvials and silt cover; 14, colluvials and limestone scree.

to give the region's wines its distinctive aroma. Some rare, pink-colored 'Chardonnay' clones are grown in the northern part of the Côte d'Or.

Because of the cool climate, chaptalization is commonly required to reach the alcohol content considered typical (12–13%) for Burgundian wines. Malolactic fermentation is promoted for its beneficial deacidification effect. As a consequence, the wines usually are racked infrequently, and the associated long contact with the

lees tends to influence the character of Burgundian wines. It is thought by some that the accumulation of yeasts and tartrate on the insides of the barrel limits excessive uptake of an oak flavor from the cooperage.

Chablis is a delimited region some 120 km northwest of the Côte d'Or (47°48′ to 47°55′ N), just east of Auxerre. The region is characterized by a marly subsoil topped by a limestone- and flint-based clay. Sites located on well-exposed slopes (15–20%) are preferred

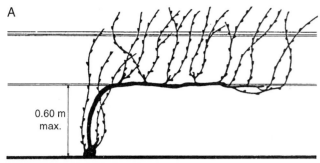

Pinot noir vine before being pruned in the
Cordon de Royat training system

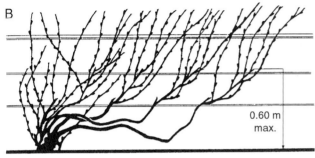

Chardonnay vine before being pruned in the
Chablis training system

Figure 10.15 Training of vines low to the ground in Chablis to provide protection from frost damage in late spring or early fall. (A) 'Pinot noir' before pruning to Cordon de Royat training, (B) 'Chardonnay' before pruning to Chablis training. (From Dovaz, 1983, reproduced by permission)

to achieve better sun exposure and drainage. This is especially important because the region frequently suffers killing late-spring frosts. To further enhance protection against frost damage, the vines are trained low to the ground. *Cordon de Royat* and short-trunk double *Guyot* training systems are common (Fig. 10.15). Shoot growth seldom reaches more than 1.5 m above ground level. Thus, the vines remain close to heat radiated from the soil. The wines typically have little (subtle) fragrance and are more acidic than their equivalent wines from central Burgundy. 'Chardonnay' is the only authorized cultivar in Chablis. Yield varies from approximately 50,000 to 100,000 hl from 1500 ha planted with vines.

Beaujolais is the most southerly region in Burgundy (45°50′ to 46°10′N). It runs approximately 70 km as a broad strip of hilly land from just north of Lyon to just south of Mâcon. Most vineyards are located on slopes that are part of the eastern edge of the Massif Central. Here, the subsoil is deep and derived from granite and schists. The soil has considerable clay content and may be admixed with calcareous and black-shale deposits.

The most distinctive feature of Beaujolais has been its retention of an old production technique, currently called carbonic maceration (see Chapter 9).

The procedure can generate wines that are pleasantly drinkable within a few weeks of production. It also results in the synthesis of a distinctively fresh, fruity fragrance. The light style generated by carbonic maceration became very popular in the late 1960s, primarily as beaujolais *nouveau*. The red cultivar grown in the region, 'Gamay noir,' responds well to carbonic maceration. Nevertheless, the technique can also yield wines that age well. These come predominantly from several villages in the northern portion of Beaujolais. Possibly to distance themselves stylistically from *nouveau* wines, most producers from these villages (*crus*) avoid mentioning Beaujolais on their labels. Beaujolais produces approximately 1 million hl wine/year, with more than 60% going into beaujolais *nouveau*.

Champagne   Champagne is probably the best-known French wine. So many producers in other parts of the world use the term in a semigeneric manner that "champagne" has become identified with, if not considered synonymous with, sparkling wines in the minds of most consumers.

The designated region of Champagne is quite large, constituting about 30,000 ha (3% of French vineyard hectarage). The annual production of approximately 2 million hl is largely, but not exclusively, used for the production of sparkling wine. Most of the region lies east-northeast of Paris, spanning out equally on both sides of the Marne River for about 120 km. The other main section lies to the southeast, in the Aube *département*. Nevertheless, the greatest concentration of vineyards (~50%) and the best sites occur within the vicinity of Épernay (49°02′N). Here lie two prominences (*falaises*) that rise above the valley floor. The Montagne de Reims creates steep south- and east-facing slopes along the Marne River, and more gentle slopes northward toward Reims, some 6 km away. The Côte des Blancs, just south of Épernay, provides steeply sloped vineyard sites facing eastward. Soil cover is shallow (15–90 cm) and overlies a hard bedrock of chalk. Because of the slope, the topsoil needs to be periodically restored.

All three authorized grape cultivars are planted in the Épernay region, but the pattern of distribution varies among regions. The north and northeastern slopes of the Montagne de Reims are planted almost exclusively with 'Pinot noir,' whereas along the eastern and southern inclines both 'Pinot noir' and 'Chardonnay' are cultivated. On the eastern ascent of the Côte des Blancs essentially only 'Chardonnay' is grown. The best vineyards tend to lie between 140 and 170 m altitude (about halfway up the slopes) and possess eastern to southern orientations. 'Pinot Meunier' may be grown on the *falaises* as well, but it is primarily cultivated along the Marne Valley and other delimited regions.

In the valley, soils are more fertile and less calcareous than those of the *falaises*, but the area is more susceptible to frost damage. Although the cultivar is less preferred, about 48% of Champagne plantings are 'Pinot Meunier,' with the rest divided about equally between 'Pinot noir' and 'Chardonnay.'

In Champagne, the northernmost French vineyard region, the vines are trained low to the ground. As previously noted, the best sites are on slopes that direct the flow of cold air away from the vines and out onto the valley floor. The inclination of the sites also can provide conditions that enhance spring and fall warming. Surprisingly, some excellent 'Pinot noir' vineyards face north. Although the slopes tend to be shallow, solar warming will still be less than on level sites. In Champagne, however, good coloration and phenol synthesis are not essential to wine quality. In contrast, low color content simplifies the extraction of uncolored juice from the grapes. In addition, delayed fruit ripening probably aids the harvesting of healthy grapes – a prerequisite for producing white wines from red grapes. 'Pinot noir' becomes very susceptible to bunch rot at maturity. With maturation delayed, the fruit remains relatively resistant to *Botrytis* infection. This is especially important because precipitation (that favors infection) occurs predominantly in the late summer and fall.

That grapes may be harvested somewhat immature is not the disadvantage it would be with other wine styles. It actually favors the production of wines low in alcohol content ($\sim$9%). This facilitates the initiation of the second in-bottle fermentation, so integral to champagne production. If the acidity is excessive, deacidification can be achieved with malolactic fermentation or blending. Although training vines close to the ground makes mechanical harvesting impossible, this is acceptable because red varieties must be handpicked to avoid berry rupture. Mechanical harvesting would unavoidably result in some berry rupture and the associated diffusion of pigment into the juice. Manual harvesting also permits the removal of diseased fruit by hand.

Although regions are ranked relative to the potential quality of the fruit produced, champagnes are rarely vineyard-designated. Champagnes usually are blended from wines from different sites and vintages, to generate consistent house styles. Each champagne firm (house) creates its own proprietary style(s). The procedure also helps cushion variations in annual yield and quality, and tends to stabilize prices. In exceptional years, vintage champagnes may be produced. In vintage champagnes, at least 80% of the wine must come from the indicated vintage.

Loire The Loire marks the northern boundary of commercial viticulture in western France ($\sim$47°N), a full degree latitude south of Champagne. This apparent anomaly results from proximity and access of the Loire Valley to cooling winds off the Atlantic Ocean. The region consists of several distinct subregions, stretching from the mouth of the Loire River near Nantes to Pouilly sur-Loire, some 450 km upstream. Most regions specialize in varietal wines produced from one or a few grape cultivars. Loire vineyards cover some 61,000 ha and annually produce approximately 3.5 million hl wine.

Nearest the Atlantic Ocean is the Pays Nantais. It produces white wines from the 'Muscadet' ('Melon') and 'Gros Plant' varieties. About 100 km upstream is Anjou-Saumur. Here, 'Chenin blanc' is the dominant cultivar. Although 'Chenin blanc' usually produces dry white wines, noble-rotted fruit produce sweet wines high in alcohol content ($\sim$14%). Rosés are also a regional speciality, coming from the 'Cabernet franc' and 'Groslot' varieties. In the central district of Touraine, light-red wines are derived from 'Cabernet franc' and 'Cabernet Sauvignon' (Chinon and Bourgueil) and white wines from 'Chenin blanc' (Vouvray). In Vouvray, dry, sweet, (botrytized), and sparkling wines are also produced. The best-known upper-Loire appellations are Sancerre and Pouilly sur-Loire. Their wines come primarily from 'Sauvignon blanc,' although some wine also is made from 'Chasselas.'

In the Loire Valley, vineyard slope becomes significant only in the upper reaches of the river, around Sancerre and Pouilly sur-Loire. Here, chalk cliffs rise to an altitude of 350 m. In most regions, moisture retention, soil depth, and drainage are the most significant factors influencing the microclimate (Jourjon *et al.*, 1991).

Southern France Progressing south from the union of the Saône and Rhône Rivers, just below Beaujolais, the climate progressively takes on a Mediterranean character. Total precipitation declines, and peak rainfall shifts from the summer to winter months. The average temperature also rises considerably. Here, red grapes consistently develop full pigmentation. Not surprisingly, red wines are the dominant type produced. Because of the longer growing season, cultivars adapted to such conditions predominate.

'Syrah' is generally acknowledged to be the finest red cultivar in southern France. 'Syrah' has shown a marked increase in cultivation, following a decline associated with and following the phylloxera devastation of the 1870s. A similar fate befell many other cultivars, such as 'Roussanne,' 'Marsanne,' 'Viognier,' and 'Mataro' (Mourvèdre). The fruitfulness of French-American hybrids, developed initially to avoid the expense of grafting in phylloxera control, induced further displacement of the indigenous varieties. The shift of vine culture from cool highland slopes to the hotter rich plains resulted in overcropping and a reduction in wine quality. Thus, wines from southern regions, ranked as highly as Bordeaux in the mid-1800s, are little known today.

Generally, the best-known regions are those in the upper Rhône Valley. In regions such as the Côte Rôtie (45°30′ N) and Hermitage (45°N), the best sites are on steep slopes. In the Côte Rôtie, 'Syrah' is often cofermented with 10% 'Viognier,' to add the distinctive fruitiness of the white cultivar. In some areas, the slopes are terraced, for example, Condrieu. One exception is Châteauneuf-du-Pape (44°05′ N). It is in the center of the lower Rhône Valley and situated on shallow slopes. The Rhône Valley possesses approximately 38,000 ha of vines and produces about 2 million hl wine/year.

In the upper Rhône, most of the wines are produced from a single cultivar. Progressing southward, the tendency shifts to the blending of several to many cultivars. Also, the predominant cultivar changes from 'Syrah' in the upper Rhône, to 'Grenache' in the lower Rhône, to 'Carignan' in the Midi (primarily Languedoc and Roussillon).

With a tendency for long, hot, dry summers in the south, vegetative growth ceases early, producing short sturdy shoots. This, and the value of fruit shading, have promoted the continued use of the *Goblet* training system. The bushy form developed has been thought to help minimize water loss by ground shading. However, data from van Zyl and van Huyssteen (1980) counter this view. The system also obviates the need for a trellis and yearly shoot positioning. In addition, the short vine stature and sturdy shoots are less vulnerable to the strong, southerly, *mistral* winds, common in the lower Rhône and Rhône Delta (Fig. 10.16). Windbreaks are a common feature in the area.

Although the upper Rhône Valley, and to a lesser extent the lower Rhône, produces several wines of international reputation, this is rare in the Midi. Only some sweet wines, such as Banyuls, appear to have gained a small international clientele. Much of the nearly 33 million hl wine production is sold in bulk or converted to industrial alcohol. Grapes are the single most important crop in the region.

Improvement in wine quality in the Midi will depend on planting better cultivars, eliminating overcropping, and adopting mechanized winemaking equipment and techniques. However, this is difficult to achieve in an economically depressed area, where most vineyards are small and too often owned by poorly trained producers, and the prominence of a mentality accustomed to subsidized prices for wine destined largely for distillation into industrial alcohol.

Germany Germany's reputation for quality wines far exceeds its significance in terms of quantity. Germany

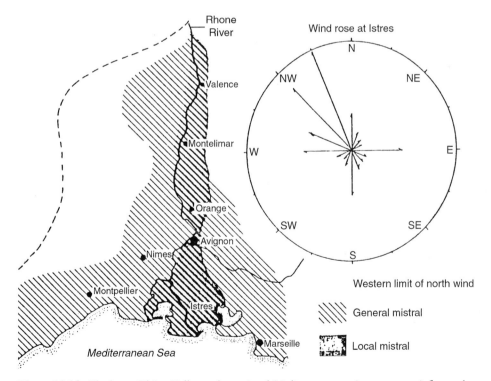

Figure 10.16 The lower Rhône Valley and associated Mediterranean regions prone to influence by strong *mistral* winds. (From Rougetet, 1930, reproduced by permission)

produces only about 3% of the world's supply. However, much of the international repute comes from a small quantity of botrytized and drier 'Riesling' Prädikat wines. Nevertheless, much fine wine comes from the lesser-known cultivar 'Muller-Thurgau.' The reputation of German wines, despite the high latitude, partially reflects the high technical skill of its grape growers and winemakers, and the assistance provided by the many excellent research facilities throughout the country.

The high latitude of vineyard regions in Germany (47°40′ to 50°40′ N), and resulting cool climate and relatively short growing season, favor the retention of fruit flavors and a refreshing acidity. The "liability" of low °Brix levels has been turned into an asset by producing naturally light, low-alcohol wines (7.5–9.5%). With the advent of sterile filtration in the late 1930s, crisp semi-sweet wines with fresh fruity and floral fragrances could be produced in quantity. These wines ideally fit the role they have often played in Germany, namely, light sipping wines consumed before or after meals. The botrytized specialty wines have for centuries been favorite dessert replacements. As befits their use, more than 85% of all German wines are white. Even the red wines come in a light style, often more resembling a rosé than a standard red wine. Although most German wines generally are not considered ideal meal accompaniments, several producers are developing dry wines to meet the growing demand for this style in Germany.

German viticultural regions reflect the typical European regional specialization with particular cultivars. Nevertheless, in only a few regions does one cultivar predominate. In addition, modern cultivars are grown extensively. For example, 'Müller-Thurgau' is the most extensively grown German cultivar (24%), ahead of the more well-known 'Riesling' (21%). Nearly half the vineyards are planted with varieties developed in ongoing German grape-breeding programs. Their earlier maturity, higher yield, and floral fragrance have made them valuable in producing wines at the northern limit of commercial viticulture. Both new cultivars and clonal selection of established varieties have played a significant role in raising vineyard productivity, without resulting in a loss in wine quality. Of red cultivars, the most frequently planted are 'Spätburgunder' ('Pinot noir') and 'Portugieser.'

One of the most distinctive features of German viticulture is the high proportion of vineyards on slopes (Fig. 10.17). This has meant that viticulture usually has not competed with other crops for land. Most of the famous sites are on valley walls, unsuitable for other crop cultivation. Although steep inclinations may produce favorable microclimates for grape growth, they are often incompatible with mechanized vineyard activities. Thus, to facilitate viticulture, vineyard consolidation has been encouraged in several regions, as well as structural modification to produce terraces suitable for mechanization (Luft and Morgenschweis, 1984).

Formerly, vines were trained on short trunks (10–30 cm), as is still common in northern France. This has changed to trunks between 50 and 80 cm high. Not only is cultivation and harvesting easier with taller trunks, but the vines are less susceptible to disease, due to better air circulation and surface drying. Trellising also helps position shoot growth and leaf production for optimal light exposure. This is important because long summer days (≥16.5 h) and abundant precipitation promote rapid development of a large photosynthetic assimilation area. In addition, by locating

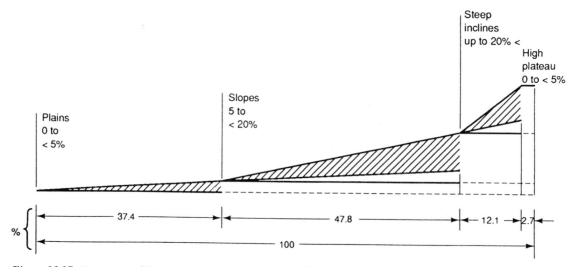

Figure 10.17 Proportion of German vineyards on land within different slope ranges in 1977. Note: Angles of 10° and 20° correspond to inclines of 18 and 37%, respectively. (From Anonymous, 1979, reproduced by permission)

shoot-bearing wood low on the vine, and directing shoot growth upward, grapes are kept as close as possible to heat radiated from the soil. Only on particularly steep slopes, notably along the Mosel–Saar–Ruwer, are vines still trained to individual stakes with two arched canes.

On all but the steepest inclines, narrow-gauge tractors permit most vineyard activities to be mechanized. Small tractors are required to pass between the densely planted vines.

Although German vineyard regions have a cool temperate climate, the southernmost viticultural portions of Baden possess warmer but wetter conditions than elsewhere in Germany. They occur on the windward side of the Black Forest, across from Alsace. The Baden region consists of a series of noncontiguous areas spanning 400 km, from the northern shores of Lake Constance (Bodensee) (47°40′ N) to the Tauber River, just south of Würzburg (49°44′ N). Most of the vineyards occur in the 130-km stretch between Freiburg and Baden-Baden. The most favored sites are located on the volcanic slopes of the Kaiserstuhl and Tuniberg.

'Spätburgunder' ('Pinot noir') is often grown on well-drained south-facing slopes of Baden, and produces about 90% of Germany's 'Spätburgunder' wines. On heavier loamy soils, 'Müller-Thurgau,' 'Ruländer,' and 'Gutendal' ('Chasselas') tend to predominate. 'Gutendal' is especially well adapted to the more humid portions of the region. Wine production is largely under the control of several large, skilled cooperatives. The region has been particularly active in vineyard consolidation and terrace construction.

Directly north of Alsace, on the western side of the Rhine Valley, is the Rheinpfalz. It continues as the Rheinhessen to where the Rhine River turns westward at Mainz. Combined, these regions possess almost one-half of all German vineyards (48,000 ha), and produce most of the wine exported from Germany. The region consists largely of rolling fertile land at the northern end of the Vosges (Haardt) Mountains. *Liebfraumilch* is the best-known regionally exported wine. The regions span latitudes from 49° to 50° N.

As with other German wine regions, the best vineyard sites occupy south- and east-facing slopes, and are planted primarily with 'Riesling' vines. Occasional black-basalt outcrops, as found around Deidesheim, are thought to improve the local microclimate and provide extra potassium. The longer growing cycle of 'Riesling' demands the warmest, sunniest microclimates to reach full maturity. In the Rheinpfalz, 'Riesling' constitutes the second most commonly cultivated variety (17%), whereas in the Rheinhessen it covers only 7% of vineyard hectarage. In both regions, 'Müller-Thurgau' is the dominant variety, covering about 22% of vineyard sites. Many new cultivars are grown and add fragrance to the majority

of wines produced. Soil type varies widely over both regions. Combined, the regions are up to 30 km wide and together stretch about 150 km in length.

At Mainz, the Rhine River turns and flows southwest for about 25 km until it turns northwest again past Rüdesheim (~50° N). Along the northern slope of the valley are some of the most famous vineyard sites in Germany. The region, called the Rheingau, possesses approximately 3000 ha of vines. The soil type varies along the length of the region and up the slope. Along the riverbanks are alluvial sediments. Further up, the soil becomes more clayey, finally changes to a brown loess. The soil is generally deep, well-drained, and calcareous. Although soil structure is important, factors such as slope aspect and inclination, wind direction and frequency, and sun exposure are of equal or greater importance. Because the river widens to 800 m in the Rheingau section, it significantly moderates the climate. Mists that arise from the river on cool autumn evenings, combined with dry sunny days, favor the development of noble-rotted grapes (*edelfäule*). This development is crucial to the production of the most prestigious wines of the region.

'Riesling' is the predominant cultivar grown in the Rheingau (82%), followed by 'Spätburgunder' (6.7%). The latter is cultivated almost exclusively on the steep slopes between Rüdesheim and Assmannshausen, where the course of the river turns northward again. On less-favored sites, new cultivars such as 'Müller-Thurgau,' 'Ehrenfelser,' and 'Kerner' may be grown.

Vineyards continue to be planted along the slopes of the Rhine and its tributaries up to Bonn, as the river continues its flow toward the Atlantic. Red wines from 'Spätburgunder' and 'Portugieser' are produced along one of the tributaries, the Ahr. Heat retained by the volcanic slate and tufa of the steep slopes may be crucial in permitting red wine production in this northerly location (50°34′ N). However, most of the region, called the Mittelrhein, cultivates 'Riesling' along the steep banks. As, in the more southerly regions of the Nahe and Neckar valleys, most of the wine is sold locally and seldom enters export channels.

A major artery joining the Rhine at Koblenz is the Mosel River. Along the banks of the Mosel are vineyards with a reputation as high as those of the Rheingau. The Mosel–Saar–Ruwer region stretches from the French border to Koblenz (49°30′ to 50°18′ N) and has 13,000 ha of vines. Although the overall orientation is northwest, the Mosel snakes extensively for 65 km along its middle section. This section possesses many steep slopes (Plate 10.1) with southern aspects between the Eifel and Hunsruck mountains. This stretch of the river, and its tributaries, the Saar and Ruwer, are planted almost exclusively with 'Riesling' vines. The central section, called the Mittelmosel, is blessed with a slate-based soil that favors heat retention. Thus, not only does the soil possess excellent drainage, but it also provides frost protection and promotes grape maturation. The Mittelmosel contains most of the region's most famous vineyards.

The final German river valley particularly associated with viticulture is the Main. The Main flows westward and joins the Rhine at Mainz, the beginning of the Rheingau. Vineyards in the Main Valley are upstream and centered around Würzburg (~49°N). The region, called Franken, is noted for 'Silvaner' wines and its squat green bottle, the *bocksbeutel*. Franken is the only German region that does not use slim bottles for wine. Although best known for 'Silvaner' wines, Franken's has a continental climate more suited to the cultivation of the newer, earlier maturing cultivars, such as 'Müller-Thurgau.' The latter covers almost half of the 5500 vineyard hectarage in Franken, whereas 'Silvaner' covers only about 20%.

Switzerland Although a small country, Switzerland has vineyards covering approximately 15,000 ha, with an annual wine production of close to 1.0 million hl. The ethnic diversity and distribution of its inhabitants are reflected in the wines it produces. Along the northern border, Switzerland produces light wines resembling those in neighboring Baden. In the southwest, wines are more alcoholic, similar to those across the frontier in France. In the southeast, the Italian-speaking region specializes in producing full-bodied red wines.

Although the southern vineyard regions of Switzerland are parallel to the Côte de Nuits (~46°30′N), the altitude, between 400 and 800 m, produces a cooler climate than its similar latitude with Burgundy might suggest. Consequently, most vineyards are on south-facing slopes, adjacent to lakes or river valleys. The southwestern vineyards of the Vaud and Valais hug the northern slopes of Lake Geneva and the head of the Rhône River, whereas those of the west occur along the eastern slopes of Lake Neuchâtel and associated tributaries. In northern Switzerland, the vineyards congregate around the Rhein, Rheintal, and Thurtal river valleys and the south-facing slopes of Lake Zürich. In the southeast (Ticino), vines embrace the slopes around Lakes Lugano, Maggiore, and their tributaries. All the regions are associated with headwaters of three great viticultural river valleys, the Rhine, the Rhône, and the Po. As in Germany, the use of steep slopes has avoided competition with most other agricultural use.

Because of frequent strong winds, the shoots are tied early and topped to promote stout cane growth in most Swiss vineyard regions. Abundant rainfall has fostered the use of open canopy configurations to favor foliage

drying and minimize fungal infection. Soil management typically employs the use of cover crops or mulches to minimize the need for herbicide use and for erosion control. There is increasing interest in the use of several French-American hybrids. Their greater disease resistance is compatible with the growing popularity of organically produced wines (Basler and Wiederkehr, 1996).

The moist cool climate usually necessitates chaptalization if normal alcohol levels are to be achieved. Malolactic fermentation is used widely to reduce excessive wine acidity. The short growing season requires the cultivation of early-maturing varieties, such as 'Müller-Thurgau' in the north and 'Chasselas doré' in the southwest. There has been a recent trend to replace 'Chasselas' with other indigenous cultivars, such as 'Petit Arvine,' 'Cornalin,' Armine,' and 'Humagne Rouge.' Although most cultivars are white, there is a shift toward red wine production. The demand has resulted in a marked increase in the planting of 'Pinot noir' in the north and both 'Pinot noir' and 'Gamay' in western regions. In the southeast, 'Merlot' is the dominant cultivar.

Czech Republic and Slovakia The former Czechoslovakia covers a primarily mountainous region, divided along most of its length by the Carpathian Mountains to the east and the Moravian Heights and Sumava in the west. The latter separate the small northern vineyard area of Bohemia (50°33′N) from the warmer, more protected, southern regions of Moravia and Slovakia (~48°N).

In Bohemia, the vineyards are located on slopes along the Elbe River Valley, about 40 km northwest of Prague. Low altitude and the use of a variety of rootstocks differentially affecting bud break help to cushion the influence of the cool and variable climate of the region (Hubáčková and Hubáček, 1984). The predominant cultivars are 'Riesling,' 'Traminer,' 'Müller-Thurgau,' and 'Chardonnay.' The small size of the viticultural region, and its proximity to Germany, probably explains the resemblance of the cultivars and wine styles to those of its neighbor to the west.

In south-central portion of the Czech Republic, most vineyard sites are positioned on the slopes of the Moravia River valley, north of Bratislava. The Moravian vineyards (12,000 ha) cultivate primarily white varieties, such as 'Riesling,' 'Traminer,' 'Grüner Veltliner,' and 'Müller-Thurgau,' similar to their Austrian neighbors. Small amounts of red wine are produced from 'Portugieser,' 'Limberger,' and 'Vavřinecké.'

Most of Slovakia's vineyards (35,000 ha) adjoin the Danube basin along the southern border of the country. In common with adjacent viticultural regions, most of Slovakia's wines are white. Although many western European cultivars are grown, the junction of the region with eastern Europe is equally reflected in the cultivation of eastern varieties. This is particularly marked in the most easterly viticultural region. The area is an extension of the Tokaj region of Hungary. Here, the varieties 'Furmint' and 'Hárslevelű' are the dominant cultivars. The wines are identical in style to those of the adjoining Hungarian region.

Austria   The mountains of the region restrict the 48,000 ha of Austrian vineyards to the eastern flanks of the country. Much of the wine-growing area is in low-lying regions along the Danube River and Danube basin to the north (~48 °N) and the Hungarian basin to the east. A small section, Steiermark, is located across from Slovenia (~46°50′ N).

The best-known viticultural regions are the Burgenland (Rust-Neusiedler-See), Gumpoldskirchen, and Wachau. The Burgenland vineyards adjacent to the Neusiedler See (47°50′ N) occur on rolling slopes at an altitude between 150 and 250 m. The shallow, 30-km-long lake creates autumn evening mists that, combined with sunny days, promote the development of highly botrytized grapes and lusciously sweet wines. Here, 'Riesling,' 'Müller-Thurgau,' 'Muscat Ottonel,' 'Weissburgunder' ('Pinot blanc'), 'Ruländer,' and 'Traminer' predominate. Vineyards located north and south of Vienna produce light dry wines. These are derived primarily from the popular Austrian variety 'Grüner Veltliner.' Other varieties grown are 'Riesling,' 'Traminer,' and in Gumpoldskirchen, south of Vienna, the local specialities 'Zierfändler' and 'Rotgipfler.' Most of the vineyards in the latter region are positioned on southeastern slopes at an altitude between 200 and 400 m. About 65 km west of Vienna is the Wachau region. Vineyards in Wachau lie on steep south-facing slopes of the Danube, just west of Krems. They occur at altitudes of between 200 and 300 m. The predominant cultivar is 'Grüner Veltliner.'

Although most Austrian wine is white, some red wine is produced. The most important red cultivar is the new variety 'Zweigelt' (Mayer, 1990). Although Austria annually produces approximately 2.5 million hl wine, only a small proportion is exported. Most wine production is designed for early and local consumption. As in Germany, the wines usually have the varietal origin and grape maturity noted on the label. The designations from *Qualitätswein* to *Trockenbeerenauslese* are the same, except that the requirements for total soluble solids (Oechsle) are higher than in Germany.

United Kingdom   Although the United Kingdom is not a viticulturally important region on the world scene, UK vineyards indicate that commercial wine production is possible up to 54°45′ N. The extension of wine production further north than German vineyards results from both the moderating influence of the Gulf Stream and the cultivation of early-maturing varieties, such as 'Müller-Thurgau' and 'Seyval blanc.' Other potentially valuable cultivars are 'Auxerrois,' 'Chardonnay,' 'Madeleine Angevine,' and several of the new German cultivars. Improved viticultural and enologic practices also have played an important role in reestablishing winemaking in England. The cooling of the European climate in the thirteenth century (Le Roy Ladurie, 1971), combined with the joint rule of England and Bordeaux, probably explain the demise of English viticulture during that period.

SOUTHERN EUROPE

Although French and German immigrants were seminal in the development of the wine industry in the New World, the basic practices used in Central Europe were originally introduced from Southern Europe. During the ancient Classical period, Greece and Italy were considered, albeit by themselves, to be the most famous wine countries. With the fall of these civilizations, economic activity languished for almost a millennium. The result was a marked reduction in both the quantity and quality of wine production throughout Europe.

At the end of the medieval period, the rebirth of centralized governments and improved economic conditions north of the Alps favored the renewed production of fine wine. In contrast, grapes continued to be grown admixed with other crops in much of southern Europe, often trained up trees in fields amongst other crops (Fregoni, 1991). Without monoculture, insufficient time could be devoted to controlling yield and optimizing fruit excellence. The subsistence economy in much of southern Europe at the time was incompatible with the exacting demands of quality viticulture. In addition, the absence of a large, discriminating, middle class resulted in the persistence of low-quality wines. Only in the northern parts of Italy, associated with the Renaissance, did fine wine production again develop. Regrettably, political division, associated with the rise of aggressive nation-states in Central Europe, resulted in a subsequent return to stagnation in quality wine production. In Greece, repeated invasion by oppressive foreign regimes severely restricted the redevelopment of a vibrant wine industry, as well as that in much of eastern Europe. Although Spain and Portugal did not experience similar events, their initial colonial prowess did not translate into lasting economic benefits at home. The absence of a growing, prosperous, middle class may explain why wine skills remained comparatively simple in most of southern Europe until the last few decades.

Mediterranean wines tended to be alcoholic, oxidized, and low in total acidity, but high in volatile acidity. Inadequate temperature control often resulted in wines high in residual sugar content and susceptible to spoilage. In contrast, wines produced north of the Alps tended to be less alcoholic, fresh, dry, flavorful, and microbially stable.

With the growth and spread of technical skill, the standard of winemaking has improved dramatically throughout southern Europe. Regrettably, implementation of advances was limited because the economic return was, until recently, insufficient to support major improvements. Better-known regions, producing wines able to command higher prices, were those best able to benefit from technical advances.

Italy   Italy and France often exchange top ranking as the world's major wine-producing country. Annual production has fallen considerably over the past 20 years, and now stands at about 45–50 million hl, approximately one-fifth of the world's production. Italy's vineyard area is about the same as France (868,000 vs. 887,000 ha), but more table grapes are grown in Italy than in France (7% vs. 2%).

Italian vineyards are predominantly exposed to a Mediterranean climate, receiving most of the rain during the winter months. Nevertheless, nearly one-third of Italy (the Po River valley and the Italian Alps) possesses a more continental climate, without a distinct dry summer season. Also, the Apennines running down the middle of the Italian Peninsula, provide a cooler, more moist climate along leeward slopes of the mountains. The mountains in Italy produce considerable variation in yearly precipitation, from more than 170 cm on some slopes to less than 40 cm in southern areas. The predominantly north–south axis of Italy results in coverage of over 10° latitude, from 46°40′ N, parallel with Burgundy, to 36°30′ N, parallel with the northern tip of Africa. Nonetheless, a common feature of the country, apart from mountainous regions, is a mean July temperature of between 21 and 24 °C. Combined with a wide range of soil types and exposures, Italy possesses an incredible variety of viticultural microclimates.

In addition, Italy possesses one of the largest and oldest collections of grape varieties known, plus about a 2600-year-old history of wine production. Thus, it is not surprising that Italy produces an incredible range of wine styles. Many regions produce several white and red wines in dry, sweet, and slightly sparkling (*frizzante*) versions. Although this may give confusion to those outside the region, it supplies a diverse range of styles desired by the local clientele. Many regions also produce limited quantities of wine using distinctive, often ancient, techniques not found elsewhere.

The diversity of wine styles found in many Italian regions probably has hindered their acceptance internationally. Also confusing to many consumers is a lack of consistency in wine designation. In some regions, the wines are varietally designated, whereas in others by geographic origin, producer, or by mythological names. This situation may be explained by the long division of Italy into many separate city-states, duchies, kingdoms, and papal dominions. The difficulty of land transport within Italy, until comparatively recently, impeded wine shipments throughout the country, and the development of a national wine-designation system.

For many centuries, poverty in most Italian regions resulted in a system of subsistence farming based on polyculture. Only in a few regions, such as northwestern Italy and Castelli Romani in central Italy, was pure viticulture practiced. Today, the old polyculture has essentially vanished, and Italian viticulture is similar to that in other parts of the world. There is, however, a remarkable range of training systems in practice, many used only locally. In several areas, high pergola or tendone training is used, either to favor light and air exposure for disease control, or to limit sun- and wind-burning of the fruit. Low training is more common in the hotter, drier south. This may minimize water stress and promote sugar accumulation, desirable in the production of sweet fortified wines. Irrigation may be practiced in southern regions, where protracted periods of drought occur during the hot summer months.

In most regions, enologic practice is now modern and has had a profound effect on improving wine quality. Modernization is also affecting a shift from the traditional long maturation in large wooden cooperage (≥ 5 years) to shorter aging in small cooperage (≤ 2 years). There is considerable debate concerning the relative merits of oak flavor in Italian wine. Nevertheless, several unique and distinctive winemaking styles are practiced. Regrettably, most have been little investigated scientifically. Hopefully, they will be studied adequately before they potentially fall victim to standardized winemaking procedures.

Northern Italy   Much of the Italian wine sold internationally, except for that sold in bulk, comes from northern Italy. The regions involved are situated on the slopes of the Italian Alps and the North Italian Plain (Fig. 10.18). These regions possess a mild continental climate, without a distinct drought period. The arch of Alps that forms Italy's northern frontier usually protects the region from cold weather systems coming from the north, east, and west.

The most northerly region is Trentino-Alto Adige. Although the vineyard area covers only about 13,000 ha, and annually produces approximately 1.5 million hl

Figure 10.18 The major natural regions of Italy. (From Shackleton, 1958, reproduced by permission)

wine, production in Trentino-Alto Adige constitutes almost 35% of Italy's total bottled-wine exports. The region also leads in the proportion of DOC appellations. Unlike most Italian wines, the label usually indicates the name of the grape variety used, followed by its regional origin.

Trentino-Alto Adige contains a wide diversity of climatic and soil conditions. The climate ranges from Alpine in the north to subcontinental, and finally sub-Mediterranean along the coast. Soils equally vary considerably, depending largely on the slope and altitude. In the valley bottom, the soil is generally alluvial and deep, shifting to sandy clay-loam on lower slopes. On the steeper, upper slopes, the soil is shallow with low water-holding capacity.

The vineyards in the northern half of Trentino-Alto Adige often line the narrow portion of the Adige River valley on steep slopes at altitudes between 450 and 600 m. The region called Alto Adige (South Tyrol) stretches 30 km, both north and south from Bolzano (46°31′N). The considerable German-speaking population of the region is reflected in the style and care with which the wines are produced. The region produces many white and red wines. Whites may be produced from cultivars such as 'Riesling,' 'Traminer,' 'Silvaner,'

'Pinot blanc,' 'Chardonnay,' and 'Sauvignon blanc.' Red wines may be produced from distinctively local cultivars such as 'Schiava' and 'Lagrein,' or French cultivars, such as 'Pinot noir' and 'Merlot.'

In contrast to Alto Adige, Trentino covers a 60-km strip of gravelly alluvial soil on the valley floor that widens 20 km north of Trento (46°04′N). Here the vineyards are no higher than 200 m in altitude. The most famous wine of the region comes from the local red cultivar 'Teroldego.' The vines are trained on supports resembling an inverted L (*pergola trentino*) to increase canopy exposure to light and air. Many different white and red wines are produced, primarily from the same varieties cultivated in Alto Adige. In addition to standard table wines, the region produces a *vin santo* from the local white cultivar 'Nosiola,' and considerable quantities of dry sparkling wine (*spumante*) from 'Pinot noir' and 'Chardonnay.'

Veneto is the major wine-producing region adjacent to Trentino, where the Alps taper off into foothills and the broad Po Valley. The vineyard area in Veneto covers some 80,000 ha and annually produces approximately 8 million hl wine. Internationally, the best-known wines come from hilly country above Verona (45°28′N), where the Adige River turns eastward. The dry white wine Soave comes primarily from the 'Garganega' grape, cultivated on the slopes east of the city. The vineyards producing the grapes used in Valpolicella are situated north and east of Verona, whereas those involved in making Bardolino are further west, along the eastern shore of Lake Garda. Both red wines are produced primarily from 'Corvina' grapes, with slightly different proportions of 'Molinara' and 'Rondinella' involved. 'Negrara' also may be part of the Bardolino blend. Grapes for Valpolicella are generally grown on higher ground (200–500 m) than those for Bardolino (50–200 m), and produce a darker red wine. The most distinctive Valpolicellas come from specially selected, and partially dried grape clusters, using the *recioto* process (see Chapter 9). During vinification, fermentation may be stopped prematurely to retain a detectable sweetness, or continued to dryness to produce an *amarone*.

Two other regions in Veneto are also fairly well known outside Italy. These are Breganze, 50 km northeast of Verona, and Conegliano, 50 km north of Venice. Although Breganze produces both red and white wines, it is most famous for a sweet white wine made from the local cultivar 'Vespaiolo.' The grapes are processed similarly to those used to produce red *recioto* wines. In Conegliano, another local white cultivar, 'Prosecco,' is used to produce still, *frizzante*, and *spumante* wines. The *frizzante* and *spumante* wines may contain some 'Pinot bianco' and 'Pinot grigio.' In addition to regional cultivars, Conegliano also cultivates several widely grown Italian and French varieties.

Although fine wines are produced in Friuli-Venezia Giulia and Lombardy, provinces to the east and west of Veneto, respectively, Piedmont in the northwest is the most internationally recognized region. The largest and best-known of Piedmontese wine areas are centered around Asti (44°54′N), in the Monferrato Hills. This subalpine region receives abundant rainfall, with moderate precipitation peaks in the spring and autumn. The latter is often associated with foggy autumn days at lower altitudes. The area is famous for full-bodied reds, sweet spumante, and bittersweet vermouths. As a whole, Piedmont possesses a vineyard area of about 60,000 ha (mostly on hillsides) and yields almost 4 million hl wine/year.

The most famous red wines of the region come from the 'Nebbiolo' grape, grown north and south of Alba (44°41′N). They are especially associated with the villages of Barbaresco and Barolo. In both appellation regions, the best sites are located on higher portions of either east- or south-facing slopes. Those associated with Barolo generally are at a higher altitude, and less likely to be fog-covered in the fall than the slopes around Barbaresco. 'Nebbiolo' is grown in other areas of Piedmont, notably the northwest corner, and in northern Lombardy. In both areas, local regional names such as 'Spanna' and 'Chiavennasca' are used. Many 'Nebbiolo' wines are named after the town from which they come, without mention of the cultivar.

Lombardy's 'Nebbiolo' region (Valtellina) occurs on steep, south-facing slopes lining the Adda River, where it leaves the Alps and flows westward into Lake Como. The vineyards along this 50-km strip often occur on 1–2 m wide, constructed terraces. Owing to the strong westerly winds, the canes of the vines are often intertwined. In addition, the vines are trained low to gain extra warmth from the soil and garner protection from winter storms. *Recioto*-like wines are occasionally produced in Valtellina, where they are called Sfursat or Sforzato.

Most Piedmontese wines are produced from indigenous grape varieties. Most are red, such as 'Barbera' (about 50% of all production), 'Dolcetto,' 'Bonarda,' 'Grignolino,' and 'Freisa,' although some whites are grown, such as 'Moscato bianco,' 'Cortese,' 'Chardonnay,' and 'Arneis.' 'Moscato bianco' is used primarily in the production of the sweet, aromatic, sparkling wine Asti Spumante. 'Cortese' is commonly used to produce a dry still white wine in southern Piedmont, but it also is employed in the production of some *frizzante* and *spumante* wines.

Pure varietal wines, rather than blends, are the standard in Piedmont. Modern fermentation techniques are common. However, for some wines – notably those from 'Nebbiolo' – aging in large oak casks for several years is still common. Extensive experimentation with barrel aging is in progress. Carbonic maceration is used to a limited extent to produce *vino novello* wines, similar to those of beaujolais *nouveau*.

Piedmont is also the center of vermouth production in Italy. The fortified herb-flavored aperitif was initially produced from locally grown 'Moscato bianco' grapes. However, most 'Moscato' grapes are now used in the production of Asti Spumante. The majority of the wine used in producing vermouth currently comes from further south.

Central Italy Although wine is produced in all Italian provinces, chianti is the wine most commonly associated with Italy in the minds of wine consumers. Chianti is Tuscany's (and Italy's) largest appellation. It includes seven separate subregions that constitute about 70% of the 83,000 ha of Tuscan vines. These cover a 180 km-wide area in central Tuscany. The most famous section, Chianti Classico, incorporates the central hilly region between Florence and Siena (~43°30′N). Vineyards grow mostly 'Sangiovese.' 'Sangiovese' makes up between 75 and 90% of the blend. 'Canaiolo' and 'Colorino' are the two other red varieties. Wine from two white cultivars ('Trebbiano' and 'Malvasia') may also be employed. These are added within specified limits to make chianti. The must from white cultivars improves color development by supplying missing phenolic compounds for copigmentation.

The total region encompassed by Chianti includes a wide range of soils, from clay to gravel, although most of the region is calcareous. Slopes in this predominantly hilly zone vary considerably in aspect and inclination, with altitudes extending almost from 200 to 500 m. These geographic differences, combined with the flexibility in varietal content and occurrence of many clones, confer on chianti the potential for as much variation as bordeaux. There is a tendency to increase vine density (from the standard 2000–2500 vines per hectare to some 5000 vines per hectare) and to shift from the outdated Guyot system to spur-pruned cordons.

Although most versions of chianti are made using standard vinification techniques, several winemakers are returning to the ancient *governo* process. It is particularly advantageous when making light, early-drinking wines (Bucelli, 1991) – those features that initially made chianti famous. A similar technique is also used by several producers in Verona. In the *governo* process, from 3 to 10% of the grape harvest is kept aside. During a 2-month storage period, the grapes undergo slow partial drying. In addition, there is a change in the grape yeast flora. Whereas the population of apiculate yeasts, such as *Kloeckera apiculata*, declines markedly, the proportion of *Saccharomyces cerevisiae* increases (Messini *et al.*, 1990). After storage, the grapes are crushed and allowed to commence fermentation. At this point, the fermenting must is added to wine previously made from the main portion of the crop. The cellar may be heated to facilitate the slow refermentation of the wine–must mixture. The second yeast fermentation, induced primarily by *S. cerevisiae*, donates a light *frizzante* that enhances the early drinkability of the wine. The process also delays the onset, if not the eventual occurrence, of malolactic fermentation.

In addition to chianti, several other Tuscan red wines are made from the 'Sangiovese' grape. These wines may involve the use of non-Italian grape varieties, such as the inclusion of 'Cabernet Sauvignon' in Carmignano, or the production of a pure varietal wine, such as Brunello di Montalcino. Although 'Sangiovese' is the most common name for the variety, vernacular names such as 'Brunello' and 'Prugnolo' are regionally used. Superior clones, such as 'Sangioveto,' have also been individually named.

On the eastern side of central Italy are several wine-producing regions. The most well known, due to its expensive exports of Lambrusco, is Emilia-Romagna. Although shunned by critics, the bubbly, semisweet, cherry-fruit-like fragrance of this red wine is highly appreciated by millions of consumers. A well-made wine, without fault, that sells well is a success. Not everyone wants to drink a liquid art object. The wine is made from the 'Lambrusco' cultivar.

Besides 'Lambrusco,' Emilia-Romagna, Marche and Abruzzi possess a wide range of indigenous cultivars. Their potential is largely unknown. Until vintners with sufficient capital seriously study their potential, they will continue to languish in possibly undeserved ignominy. All one has to think of are the difficulties producers still have with 'Pinot noir.' How would it be regarded had it occurred in the backwaters of some inland valley of Italy, rather under the hands of rich monasteries in Burgundy?

Southern Italy Southern Italy produces most of the country's wine, much of it going into inexpensive Euroblends or distilled into industrial alcohol. The application of contemporary viticultural and enologic practices is gradually increasing the production of better wines, bottled and sold under their own name. Examples of local cultivars producing fine red wines are 'Aglianico' in Basilicata and Campania, 'Gaglioppo' in Calabria, 'Negro Amaro,' 'Malvasia nera' and 'Primitivo' in Apulia, 'Cannonau' in Sardinia, and 'Nerello Mascalese' in Sicily. Several indigenous cultivars, such as 'Greco,' 'Fiano,' and 'Torbato,' also generate interesting dry white wines. It is far better that these regions concentrate on the qualities of local cultivars than produce another variation-on-a-theme of 'Cabernet Sauvignon,' 'Merlot,' or 'Chardonnay.'

Reflecting the hot dry summers in southern Italy, sweet and fortified wines have long been the best wines of the region. These can vary from dessert wines made from 'Malvasia,' 'Moscato,' and 'Aleatico' grapes, to sherry-like wines from 'Vernaccia di Oristano,' to Marsala produced in Sicily.

Spain Along with Portugal, Spain forms the most westerly of the three major Mediterranean peninsulas, the Iberian Peninsula. It is the largest and consists primarily of a mountain-ribbed plateau averaging 670 m in altitude. The altitude and the mountain ranges along the northern and northwestern edges produce a long rain shadow over most of the plateau (Meseta). With its Mediterranean climate, most of Spain experiences hot, arid to semiarid conditions throughout the summer.

Climate conditions have markedly influenced Spanish viticultural practice. Because of a shortage of irrigation water, most vines are trained with several low trunks or arms (≤ 50 cm), each pruned to two short spurs. Where irrigation is feasible, recent experiments have shown that yield can be increased while maintaining grape quality (Esteban *et al.*, 1999). The short, unstaked bushy vines may limit water demand and shade the immediate soil, where most surface roots are located. Vine density is kept as low as 1200 vines/ha in the driest areas of the Meseta (La Mancha) to assure sufficient moisture during the long hot summers. Daytime maxima of 40–44 °C are common in the region. In areas with more adequate and uniform precipitation, such as Rioja, planting densities average between 3000 and 3600 vines/ha. In most of Europe, planting densities of 5000 to 10,000 vines/ha are not uncommon. Thus, although Spain possesses more vineyard area than any other country (~1.2 million ha), it ranks third in wine production (~43 million hl/year). Its yield per hectare is about one-third that of France, currently averaging about 36 hl/ha.

In the northern two-thirds of Spain, table wines are the primary vinous product, whereas in the southern third (south of La Mancha), fortified sweet or sherry-like wines are produced. In the past, wine production in most regions was unpretentious. Although the wines were acceptable as an inexpensive local beverage, they had little appeal outside Spain, except for use in blending. They often supplied the deep color and alcohol lacking in many French wines, a practice now banned except for Euroblend wines. Reminders of ancient practices are evident in the continuing but diminishing use of *tinajas*. These large-volume amphora-like clay fermentors are occasionally employed in southern Spain

(Valdepeñas and Montilla-Moriles). The Penedés region south of Barcelona is a major center for enologic and viticultural innovation. Nevertheless, traditional centers of wine excellence, such as Rioja in the north and the Sherry region in the south, are involved in considerable experimentation and modernization. Modern trends also affect grape and wine production in the major producing areas of central Spain, and have begun to provide wines that can compete internationally under their own appellations.

Although the reputation of fine Spanish wines is based largely on indigenous cultivars, such as 'Tempranillo,' 'Viura,' and 'Palomino,' most wines are produced from the varieties 'Airen' and 'Garnacha.' So widely are these grown that they were and may still be the two most extensively cultivated varieties in the world.

Rioja Rioja has a long tradition of producing the finest table wines in Spain. The region spans a broad 120-km section of the Ebro River valley and its tributaries from northwest of Haro (42°35′N) to east of Alfaro (42°08′N). The upper (western) Alta and Alavesa regions are predominantly hilly and possess vineyards mostly on slopes rising nearly 300 m above the valley floor. The altitude of the valley floor drops steadily eastward, from 480 m in the west to about 300 m at Alfaro. The lower (eastern) Baja region generally possesses a

rolling landscape with many vineyards on level expanses of valley floor.

The south-facing slopes of the northern Sierra Cantábrica possess a primarily stony, calcareous, clayey soil, whereas the south side of the valley and the north-facing slopes of the southern Sierra de la Demanda possess a calcareous subsoil, overlaid by stony ferruginous clay or alluvial silt. The parallel sets of mountain ridges that run east and west along the valley help shelter the region from the cold north winds and hot blasts off the Meseta to the south.

The western portions of Rioja generally yield more delicately flavored wines than the eastern region. This probably results from the cooler climate generated by the greater altitude. The altitude also produces a higher and more uniform annual precipitation (Fig. 10.19). In contrast, the eastern Baja region possesses a more distinctly Mediterranean climate, with precipitation averaging about 60% that of Haro in Rioja Alta. Differences in varietal composition in the regions also affect wine characteristics. In the upper Rioja, 'Tempranillo' is the predominant red cultivar, whereas in the Baja, 'Garnacha' ('Grenache') is dominant. Soil and microclimate differences between the southern and northern slopes in the upper Rioja may explain the presence of more 'Graciano,' 'Mazuelo' ('Carignan'), and 'Garnacha' in the Alta and more 'Viura' in the Alavesa.

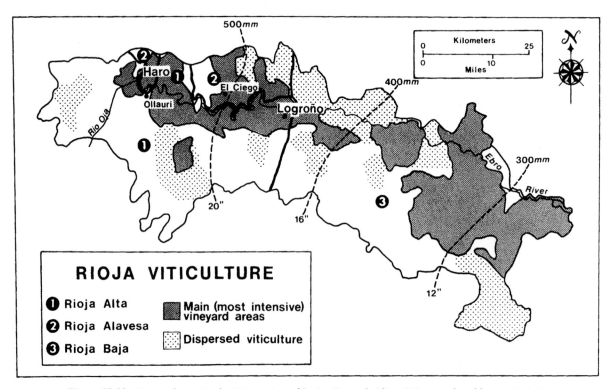

Figure 10.19 Vineyard areas in the Rioja region of Spain. (From de Blij, 1983, reproduced by permission)

The wines of Rioja are generally blends of wines from several cultivars and vineyards. 'Tempranillo' is the predominant cultivar in red wines, with various amounts derived from the other red cultivars and occasionally the white 'Viura.' Most traditional white wines are blends of 'Viura' and 'Malvasia,' with small quantities of 'Garnacha blanco.' Modern white wines are nearly pure 'Viura.'

In the past, vineyards were planted in the proportions desired for the finished wine. The grapes were also picked and fermented together. The present trend is to separate the different cultivars in the vineyard, to permit each variety to be picked at its optimal maturity. The separate fermentation of each variety also permits blending based on the properties of each wine.

In the blend, each cultivar is considered to add a component deficient in the others. For red wines, 'Tempranillo' provides the acid balance, aging potential, and traditional distinctive fragrance, 'Graciano' donates additional subtle flavors, 'Mazuelo' adds color and tannins, whereas 'Garnacha' confers alcoholic strength. With white Riojas, 'Viura' provides a light fruitiness and resistance to oxidation, whereas 'Malvasia' donates fragrance, color, and body.

Vinification procedures often vary considerably among *bodegas* (wineries). For red wines, traditional procedures involve a form of carbonic maceration (see Chapter 9). Whole grape clusters are placed in *lagars* (open concrete or stone vats) for fermentation. Large *bodegas* may employ a semicarbonic maceration process, where the grapes are stemmed before being added uncrushed to fermentation vats. The recent tendency is to employ what is now considered standard degrees of stemming and crushing, before fermentation of the must in wooden vats or stainless steel tanks. The latter more easily permit cooling to keep the fermenting must below 25°C.

Traditionally produced, white wines involve maceration on the skins for several hours before pressing. Usually fermentation occurs at warm temperatures (25–28 °C). Modern white wines are separated from the skins shortly after crushing and fermented at cool temperatures (14–18 °C).

The preferred method of maturation varies considerably. The traditional procedure for both white and red wines is prolonged lees contact, associated with infrequent racking, and several years in oak barrels (225 liters). American oak barrels are preferred and used repeatedly to avoid a new-oak flavor. During aging, the wines mellow, acquire a distinctive vanilla flavor, and develop a complex, slightly oxidized bouquet. In contrast, the present-day trend is to use shorter aging, a higher proportion of new oak barrels, and proportionately longer in-bottle aging before release from the winery.

Penedés Catalonia in northeastern Spain possesses several viticultural regions along its coast. In the past, Catalonia was known largely for its alcoholic, dark-red wines added to weaker French wine. It also produced a special, sweet, oxidized red wine that often exhibited a *rancio* odor – Priorato from Tarragona. However, Penedés, located just southwest of Barcelona, is now the most productive and important wine-producing region in Catalonia.

Despite its relatively small size, Penedés (~41°20′ N) annually produces about 1.5 million hl wine from 25,000 ha of vines. In the late 1800s, a transformation began that has made Penedés the world's largest producer of sparkling (*cava*) wine. In addition, innovations begun by Bodega Torres in the 1960s have resulted in a break with tradition. Considerable commercial success followed the introduction of cold fermentation for white wines. Red wines with limited barrel maturation also are produced in increasing quantities. Penedés now produces a greater range of wines, from both indigenous and foreign cultivars, than any other Spanish region.

Although Penedés consists of a narrow strip of land along the Mediterranean coast, it shows considerable climatic variation across its 30-km width. Penedés changes westward from a hot coastal zone to cool slopes in the Catalonian Mountains. Most of the traditional red cultivars, such as 'Monastrell,' 'Ull de Llebre' ('Tempranillo'), 'Cariñena,' and 'Garnacha,' are grown on the coastal plain. However, most newer vineyards are located along the more temperate central zone, at an altitude of 200 m. The majority of grapes are traditional cultivars, such as 'Xarel-lo' and 'Macabeo' ('Viura'). Nevertheless, several foreign red cultivars are grown in the region, notably 'Cabernet Sauvignon.' The *cava*-producing facilities use most of the local white grape harvest. The mountainous foothills further west have an even cooler, more moist environment. This favors the growth of local cultivars, such as 'Parellada,' and foreign varieties, such as 'Chardonnay,' 'Sauvignon blanc,' 'Riesling,' and 'Pinot noir.' This region possesses slopes ranging in altitude from 500 to 800 m above sea level.

Although training still favors *Goblet* systems, commonly used throughout Spain, the trunks are often higher, with the first branches allowed to originate at 50 cm above-ground level. Nevertheless, most foreign cultivars are trained using trellising systems common throughout most of western central Europe. Because of the above-average precipitation in much of Penedés, vine density is higher (4000–5000 vines/ha) than usual in Spain. Yield per hectare correspondingly is considerably above the national average.

Sherry Sherry is produced in southwestern Spain, where the lowlands of the Guadalquivir River Valley meet the Atlantic Ocean. The region encompasses an area 60 km in diameter, centered around Jerez de la Frontera (36°42′N). The preferred sites are mostly aggregated in the north and northwest, from around Jerez to the coastal town of Sanlúcar de Barrameda. The area contains 19,000 ha of vines and annually produces approximately 1.2 million hl wine.

The most significant factor influencing the quality and stylistic features of sherry is the form of processing that follows fermentation (see Chapter 9). Nonetheless, microclimatic features influence the quality of the base wine used in sherry production.

One of the more significant vineyard factors involves the soil's modification of the vine microclimate. *Albariza* sites, possessing a high chalk content (30–60%), exhibit several desirable properties. The soils are very porous and permit the rapid uptake of the winter rains. Under the hot summer sun, the soil forms a hard, noncracking crust. This enhances water available by restricting evaporation from the soil surface. Because the water-holding capacity of the soil is only 35% by weight, soil depth is important to sustained water supply throughout the typically long summer drought. Level land and low hills tend to limit excessive water loss by drainage. Because of the low latitude (36°42′N), vineyard slope and aspect are relatively unimportant. Proximity to the Atlantic Ocean has an important moderating influence on the climate. As a consequence, both winters and summers are comparatively mild, with the maximum temperatures rarely rising above 37°C (vs. 45°C and above on the Meseta). Ocean breezes also increase the region's relative humidity.

Although *albariza* soils generally possess desirable properties, they also have drawbacks. It has low fertility and susceptibility to nematode infestation limits fruit yield.

The predominant cultivar grown is the *fino* clone of 'Palomino.' It has the advantage of higher yield than standard clones. Other cultivars, such as 'Pedro Ximénez' and 'Moscatel,' are grown in limited amounts, especially on poorer sites. The latter is used only for a varietally designated sweet wine. The main advantages of 'Palomino' are its tough skin, disease resistance, and low varietal aroma. The disadvantages of late maturity are partially offset by the heat and light reflected from the white *albariza* soil.

Portugal Portugal produces an amazing range of wines for a comparatively small country. In addition, wine production is largely limited to the northern half of the country. Nonetheless, it annually produces about 7.5 million hl wine from 250,000 ha of vines. Thus, Portugal ranks among the top 10 wine-producing countries. Among the wines produced are possibly the world's most

popular wine (Mateus rosé), some of the most aromatically intricate wines (*porto*), distinctive and delicately effervescent white wines (vinho verde), dark tannic long-aging red wines (bairrada and dão), and a complex baked wine (madeira, from the island of the same name). All of these are produced almost exclusively from indigenous Portuguese cultivars, rarely planted elsewhere, even in neighboring Spain. Although producing some of the most skillfully blended wines for mass distribution, Portugal has retained much of its vinous heritage unaffected by outside influences.

The Upper Douro The eastern portion of the Douro River Valley is the origin of Portugal's most prestigious wine, *porto*. Port is produced in what was one of the most inaccessible and inhospitable regions of Portugal. The delimited port region stretches nearly 120 km along the banks of the Douro and its tributaries, from Barqueiros in the west to Barca d'Alva near the Spanish border. The region roughly parallels 41°N latitude. Although the present area has expanded extensively eastward from its original delimitation in 1761 (Fig. 10.20), most of the 34,000 ha of vines are still centered around Pêso da Régua. The region produces about 2 million hl wine annually, of which only about 40% is used in port production.

Because of the steep mountainous terrain through which the Douro River passes, slopes along the valley may possess inclinations of 60°. Because of the steep slopes and prevailing dry conditions, little soil has accumulated. Thus, most vineyards are a series of constructed terraces held by stone retaining walls (Plate 10.2). The soil itself consists primarily of schist. Although high in potassium (~12%), the rocky soil is low in phosphorus and organic material. These deficiencies were once offset by the addition of manure, usually from the animals that used to work the vineyards. Because most old terraces are too steep to permit mechanization, many new and some old terraces have been graded to permit partial mechanization.

As a wine style, port evolved as British shippers catered to an increasingly discriminating market at home. Thus, quality very early became an important aspect in port ontogeny. This culminated in a highly detailed and critical analysis of the vineyards. Each vineyard site is assigned to one of six quality categories, based on features affecting microclimate, soil conditions, and vine characteristics. Of a maximum of 1680 points, about two-thirds reflect environmental features, whereas the remaining third concerns viticultural practices. More points may be subtracted for negative influences than are granted for desirable features.

Many of the factors indicated in Table 10.4 reflect the importance assigned to regional temperature and

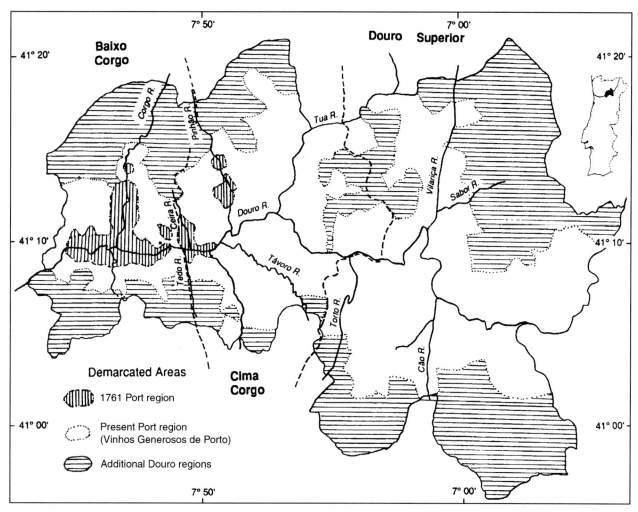

Figure 10.20 Map showing present-day (cross-hatched regions) and 1761 boundaries of the port wine area (within the larger Upper Douro), indicating the major subdivisions of the region. (Data from Stanislawski, 1970, and Instituto de Vinho do Porto)

moisture conditions. For example, the Upper Douro is divided into five geographic subregions, of initially increasing rank moving eastward. This mirrors the higher temperatures and lower precipitation in the upper regions of the valley. The most easterly subregion, beyond the watershed dividing the Tua and Vilariça rivers, often receives insufficient rainfall to offset the baking summer drought. Because the grapes may raisin before harvest, the wines rarely possess the characteristics deemed appropriate for port production. The bias for low altitude in site ranking also reflects the general desirability of warmer, drier conditions near the valley floor, versus the cooler, wetter conditions of the upper slopes. Schistose soils are preferred to granitic soils, possibly because of the fractured structure of the former. Schist more readily permits rain and root penetration, features vital to a steady and sufficient supply of water throughout the summer drought. The narrow sections of fertile alluvial deposits on the valley floor are given 600

demerit points, almost assuring that grapes grown there will be excluded from use in port production. The aspect and degree of slope are calculated into the ranking, but they are considered of minor importance.

Of viticultural features, the training system is considered of the greatest importance, earning up to 12.1% of possible points. The highest number of points is assigned for single-cane *Guyot* training. It has the advantage of restricting vine vigor on nutrient poor soils. Arbors are discouraged by being penalized 500 demerit points. In contrast, cultivar composition is awarded only a maximum of 150 points. Although some cultivars, such as 'Touriga Nacional,' 'Tinta Francisca,' 'Tinta Cão,' 'Tinta Roriz,' 'Mourisco,' and 'Bastardo,' are considered finer than others, the granting of only 6.1% of points for cultivar composition is revealing. It acknowledges the need for the grower to be able to adjust vineyard cultivar composition to the demanding and variable conditions of the region and site. It also acknowledges the

Table 10.4 Apportionment of points to features considered important to the quality of grapes for port elaboration

Trait	Penalty points	Award points	Spread	Percentage of spread
Primary importance				
Low productivity	−900	150	1050	20.6
Altitude	−900	240	1140	22.4
Physical nature of soil	−400	100	500	9.8
Locality	−50	600	650	12.8
Type of training	−500	100	600	11.8
Subtotal	−2750	1190	3940	77.4
Secondary importance				
Cultivars used	−400	200	600	11.8
Degree of slope	−1	101	102	2.0
Subtotal	−401	301	702	13.8
Tertiary importance				
Site aspect	−30	100	130	2.6
Vine spacing	−50	50	100	2.0
Soil texture	0	80	80	1.6
Vine age	0	60	60	1.2
Shelter	0	80	80	1.6
Subtotal	−80	370	450	8.8
Total	−3231	1861	5092	100.0

Source: Data from Instituto do Vinho do Porto, Ministerío da Agricultura, Porto, 1992

contribution of the varietal mix to the final quality of port. Although almost as many white cultivars are permitted as red, most ports are red. Thus, although some white port is produced, most white grapes are used in the production of table wines.

Depending on the ranking of the vineyard, and the market for port, port officials set the quantity of wine permitted from each vineyard category. For example, vineyards in category A normally can sell up to 700 liters/100 vines to port shippers, whereas those in category F seldom can sell any wine for port production. Each ranking, commencing at 1200 points for category A, is separated by a 200-point differential. This explains why little more than 40% of the wine produced in the Upper Douro is used for port elaboration. The remainder usually goes into Douro table wine, or the production of *aguardente*, the distilled wine spirits used to fortify port.

Vinho Verde Vinho Verde is the largest appellation in Portugal, located just west and north of the port region. It produces about 15% of the country's wine. The region's five subregions, crossed by five of the countries largest rivers, lie between 41 and 42 °N latitude. Although the borders of the Vinho Verde appellation extend beyond the limits of Minho province in northwestern Portugal, most of the vineyards lie within

its boundaries. Vineyards cover an estimated 71,000 ha and yearly produce about 2 million hl wine.

In contrast to the Upper Douro, where schistose soils are predominant, granite forms the soil substratum in Minho. Although rich in potassium, the soil is deficient in phosphorus. Depending on the region, vineyards receive between 120 and 160 cm precipitation per year. The weather becomes progressively drier and warmer as one moves from the eastern highlands to the narrow western coastal belt. Although the rainfall is higher than typical for the Iberian Peninsula, the distribution is distinctly Mediterranean, peaking in the winter and being minimal during the summer.

Legal impositions, designed to prevent competition with Upper Douro producers and increase in population density, led to the removal of most vines from fertile sites during the mid-1700s. Vines were largely relegated to polyculture, in association with field crops on terraced sites, or were planted to grow up and between trees along roadsides and in fields. The latter method, although complicating grape harvest, freed land for food crop production. Unintentionally, the change in viticultural practice may have favored the production of the wine style for which the region is now known. Because the growth habit limited sugar accumulation and malic acid respiration, grapes were often harvested high in acidity and low in sugar content. In contrast, grapes trained on the horizontal T trellises being used in newer vineyards are more likely to reach standard sugar and acidity levels by harvest.

Due to cool winter conditions, malolactic fermentation generally occurred in late winter or spring in unheated wine cellars. Because the wine historically was stored in sealed barrels or casks, the carbon dioxide released by malolactic fermentation was trapped in the wine. The slight fizz, combined with the low alcohol content (~8–9.5%), produced a light, refreshingly tart wine. Aromatic substances released during malolactic fermentation also may contribute to the distinctive fragrance of the wine. Under the demands of commercial success, large producers rarely depend on malolactic fermentation for the *pétillance*. Instead, they use carbonation.

As in port elaboration, most vinho verdes are blends of wines derived from several grape cultivars. Only rarely are varietal vinho verdes produced. An exception is the vinho verde produced from 'Alvarinho' grapes in Monção, across from Spanish Galicia. The wine is usually less effervescent than most vinho verdes and has traditionally been trained on low arbors.

Although most vinho verdes are red (~70%), the white version is almost exclusively the style exported. Thus, most exported vinho verde comes from the northern Monção and southern Penafiel regions, areas specializing in cultivating white varieties. 'Alvarinho' cultivation is mainly centered in Monção, whereas other white varieties, such as 'Loureiro,' 'Trajudura,' 'Azal,' 'Avesso,' 'Bataco,' and 'Pedernã,' occur throughout Minho.

Setubal Setubal is an oxidized dessert wine coming from a delimited region south of Lisbon, in the Setubal Peninsula. The vineyards are planted primarily on lime soils on north-facing hillsides. This provides some protection from the hot drying sun and permits more flavor development in the primary cultivar grown, 'Muscat of Alexandria.' Other cultivars grown in smaller amounts are the local white cultivars 'Arinto,' 'Boais,' 'Rabo de Ovelha,' 'Ovelha,' 'Roupeiro,' and 'Tamarez.' These constitute no more than 30% of the wine going into the production of setubal. Condensed humidity from the Tagus River Basin is considered important in diminishing vine water demand in the hot Mediterranean climate.

Vinification occurs traditionally for a white wine, followed by fortification (up to 18% with wine spirits) and contact with the pomace for approximately 5 months. The wine is subsequently aged for many years in oak cooperage. The initial sugar content of the wine is about 12%. This slowly increases during aging due to evaporative loss. This is estimated at close to 4%/year (Franco and Singleton, 1984). The content of nonflavonoid phenolics also increases, due both to the concentrating effect of vapor loss and prolonged maturation in oak. Most of the wine's characteristic orange-brown color probably comes from the oxidation of these phenolics.

Madeira Grape cultivation occurs over much of the island of Madeira, but is principally concentrated in the south. Madeira is the largest and most significant of three volcanic islands in the Madeira Archipelago, situated about 640 km off the western coast of Morocco. The latitude of 32°40′0″N gives Madeira a subtropical climate, moderated by the surrounding Atlantic Ocean. Its ancient volcanic peak, rising to a height of 1800 m, induces sufficient precipitation to favor luxurious plant growth, but also produces a humid climate. This resulted in decimation of the vine population when powdery mildew and phylloxera reached the island in the late 1800s. This partially explains the displacement of many traditional grape varieties with French-American hybrids, such as 'Jacquez.' With modern chemical control and phylloxera-resistant rootstock, traditional cultivars are regaining vineyard area lost over a century ago. The smaller island of Porto Santo, without significant highlands, is considerably hotter and drier. No grapes appear to be grown on the smallest island of Desertas.

The steep slopes of Madeira have required the construction of a tiered series of narrow terraces up the slopes of the volcano. The volcanic ash that forms most of the soil is clay-like and rich in potassium, phosphorus, and nitrogen. Although much of the island is cultivated, vineyards are largely restricted to the lower and mid altitudes. Of traditional varieties, 'Malmsey' ('Malvasia') is concentrated closest to the shore, generally at altitudes up to 300 m; 'Verdelho' is usually cultivated between 300 and 600 m; 'Bual' is planted between 400 and 1000 m, but down to the coast on the north side; 'Tinta Negra' is generally grown between 300 and 1000 m; and 'Sercial' cultivation is limited mostly to between 800 and 1100 m.

Greece Greece occupies the most easterly of the three major Mediterranean peninsulas. Although Greece produces approximately 3.8 million hl wine from 130,000 ha of grapes, its major importance comes from the role it played in the ancient dispersal of winemaking throughout the western Mediterranean.

In ancient times, highly regarded wines were produced in northern Greece, notably Thrace and the island of Thásos (40°41′N), and especially the Aegean islands of Khios (38°20′N) and Lemnos (39°58′N) off the coast of Turkey. Today, only Samos, a 'Muscat'-based dessert wine from the island of Samothráki (40°23′N), reflects the former vinous glory of the Aegean Islands.

Some Greek wines (*retsina*) still use an ancient wine preservation technique – the addition of resin (principally 1,1-dimethyl hydrazine). Its addition during fermentation gives the wine a terpene-like character. The preferred resin comes from the Aleppo pine (*Pinus halepensis*). It grows south of Athens in central Greece. Various pitch resins, such as terebinth (*Pistacia atlantica*), sandarac (*Tetraclinis articulata*) and myrrh (*Commiphora myrrha*) were also used to flavor and preserve wines in antiquity (Hayes, 1951). Most retsinas (~85%) are produced from the 'Savatiano' and 'Rhoditis' cultivars, which produce white and rosé *retsinas*, respectively. Although *retsinas* have maintained broad popularity in Greece, the appreciation of this wine style has not spread significantly. The other distinctive Greek wine commonly seen internationally is a red fortified dessert wine made from 'Mavrodaphne.' The appellation has the same name, Mavrodaphne. It comes from Patras (38°15′N), on the northern coast of Peloponnesus.

Similar to Portugal, Greece produces the vast majority of its wines from indigenous, possibly ancient, varieties. Thus, Greece retains a wealth of grape varieties whose merits in most cases are largely untapped and unknown internationally. For example, 'Rhoditis' can produce a delicate and uniquely flavored dry white wine. The loss of political independence and its economic base for almost two millennia, combined with prolonged Ottoman oppression, robbed Greece of the potential vinous excellence inherent in its climate and early winemaking expertise.

EASTERN EUROPE

The viticulturally important countries of eastern Europe include Hungary, Romania, Bulgaria, the states that formerly comprised Yugoslavia, and the western regions of the former Soviet Union. The area incorporates two major arched mountain chains, the Carpathian and Dinaric. Together, they form the Hungarian and Romano-Bulgarian basins. Unlike the Alps, the mountains seldom rise above 1500 m, and have rounded, tree-covered tops. The region also includes the lowlands north of and between the Black and Caspian Seas.

With the exception of the Dalmatian coast, the entire region has cold winters. Hot summers occur everywhere except in the high mountains. Along the Dalmatian coast, precipitation averages over 100 cm per year, and summer drought is experienced. Moving eastward, rainfall declines to less than 50 cm north of the Black Sea, but occurs primarily in the spring and summer months.

Because the region has served as a gateway between western Europe and Asia, it has borne the brunt of repeated incursions from both east and west. Repeated invasions frequently forced much of the population out of the plains, to seek refuge in the mountains. The vinous effect was a severe retardation in viticultural development and wine production for centuries. The least affected was Hungary, the region most distant from Turkey. It was the last country invaded by the Ottomans, and the first free from its rule.

The lack of significant resources of coal, iron, and water power for rapid industrial growth also favored the continuance of subsistence farming. It persisted throughout much of the region, far longer than in other parts of Europe, and still exists in some regions. Wines produced locally functioned as an unsophisticated but safe food beverage.

In the past, the comparative isolation from western Europe resulted in wine production being based on indigenous cultivars. Thus, eastern Europe contains a wonderfully complex collection of varieties. How many diamonds-in-the-rough occur among these varieties is little known, at least outside of their particular regions. Among the vast collection of cultivars, it is reasonable to expect some to be of equal quality to those in western Europe. It is hoped that their cultivation will not be supplanted by western cultivars in the rush to appease the international bias for recognized cultivar names.

The world already is more than amply supplied with 'Cabernet Sauvignon,' 'Chardonnay,' and 'Riesling' wines. The area also possesses the largest assortment of freely growing wild *V. vinifera*. These may constitute a valuable genetic resource for future cultivar improvement.

Because of the major vineyard-expansion program following the Second World War, eastern Europe had become by the early 1990s a major wine producer. The former Soviet Union was fifth in world ranking, with 18 million hl, whereas Romania, Hungary, and the former Yugoslavia were in tenth, eleventh, and twelfth position, respectively. This situation has changed considerably since.

Hungary Hungarian vineyards cover only 2% of the cultivated land mass and are dispersed throughout the country. The total vineyard area is 88,000 ha. It produces approximately 4 million hl wine annually.

The northern latitude (45°50′ to 48°40′N) and cold winters are reflected in the primary cultivation of white cultivars. However, the warm summers often permit the grapes to develop high °Brix values, permitting the frequent production of slightly sweet white wines high in alcohol content (13–14%). These apparently suit the Hungarian preference for spicy food. Although most wines are intended to be consumed with meals, some dessert wines are produced. The most famous of these is the wine from Tokaj.

The majority of wine (60–70%) comes from the sandy Hungarian basin in the south-central region of the country. Most of the production is consumed locally. Higher-quality wine, coming from the slopes bordering the basin, is usually exported. Examples of some of the better sites are Pécs and Vilány from the southwest, Lake Balaton to the west, Sopron and Mór from the northwest, and Eger and Tokaj in the northeast.

Tokaji comes from an ancient volcanic region in the hilly, northeast corner of the country, across from eastern Slovakia (Tokaj-Hegyalja). Vineyards occur on the sandy loam of southeast-facing slopes bounded by the Szerencs and Bodrog rivers (48°06′N). They cover about 275 km² at altitudes roughly between 100 and 300 m. The region is characterized by an extended warm autumn associated with humid evenings. The wine usually comes from a blending of several cultivars. The dominant varieties are 'Furmint' (60–70%) and 'Hárslevelü' (15–20%), with small plantings of 'Sárga Muskotály' ('Yellow Muscat;' 'Muscat Lunel') and 'Zéta.' Additional cultivars possible still grown are 'Leányka,' 'Traminer,' and 'Wälschriesling.' Although the sweet, botrytized *aszú* styles are the most renowned (see Chapter 9), dry and sweet versions (Szamorodni) are made from nonbotrytized grapes.

One of the red wines frequently seen internationally is Egri Bikavér. The wine comes from Eger (47°53′N), about 100 km southwest of Tokaj and northeast of Budapest. It is produced from a blending of several varieties, including 'Kadarka,' the most significant Hungarian red cultivar. Also potentially included are 'Merlot,' 'Pinot noir,' and 'Oporto.'

Other wines occasionally seen internationally come from the northwestern slopes of Lake Balaton (46°46′ to 47°00′ N). In the central zone, vineyards are scattered over the steeper hills that rise between 150 and 200 m above the shoreline. The lake moderates rapid climatic change and reflects light up into the canopy. Some protection from north winds is derived from the Bakony Forest, which reaches a maximum altitude of 700 m. The sandy slopes drain well and warm early in the season. The wines are almost exclusively white, alcoholic (13–15%) and frequently semisweet. Several endemic cultivars are grown, such as 'Kéknyelü' and 'Zöldszilváni,' as well as foreign cultivars, such as 'Pinot gris' ('Szürkebarát'), 'Muscat Ottonel,' and 'Wälschriesling' ('Olaszrizling'). Wines coming from the plains are produced from a wide variety of cultivars, the most common being 'Olaszrizling' and 'Ezerjo.'

Republics of the Former Yugoslavia Combined, the states that formerly composed Yugoslavia show a superficial resemblance to Italy, their neighbor across the Adriatic Sea. Both are elongated along a northwest–southeast axis, have mountainous regions along the northern frontiers, and are divided along the length by mountain ranges. However, the extended eastern connection (41° to 46°50′ N) of the region with the land mass of Eurasia permits the ready access of continental climatic influences. Thus, most of the area experiences cold snowy winters and hot moist summers. The Mediterranean influence is limited to the western (Dalmatian) coastline, between the Adriatic Sea and the Dinaric Mountains. Even here, rainfall is higher than is typical for most Mediterranean regions.

Except where the Hungarian Basin extends into northeastern Croatia, there is very little lowland. Most of the region is mountainous and generally above 400 m in altitude, with extensive areas above 1000 m. Because of this feature, most vineyards are arranged around the edges of the former country, namely, the Dalmatian coast and associated islands, the Hungarian basin in the northeast, and the Morava-Vardar corridor in the southeast. Combined, the Yugoslav states used to produce about 5 million hl wine annually (now down to 3.5 million hl). Because of the highly diverse climatic, topographic, and ethnic divisions, it is not surprising that the wine styles the region produces are equally diverse.

Partially due to the proximity of Slovenia to Austria and Italy, and its once being a part of Austria, its wines and grape varieties are similar to those of its neighbors. 'Graševina' ('Wälschriesling'), 'Silvaner,' 'Sauvignon blanc,' 'Traminer,' 'Šipon' ('Furmint'), and the indigenous 'Plavać' are the most common varieties. The Adriatic moderates the alpine climate, and the region benefits from mild winters and temperate summers. The best-known wine region is situated around Ljutomer in the northeast (46°25′ N).

The wines of Croatia fall into two groups, those from the Hungarian basin and the mountains north of Zagreb, and those from the Dalmatian coast. The highlands in the northeast, adjacent to Slovenia, continue to show an Austrian influence. The dominance of white cultivars continues in the Hungarian basin lowlands. However, on the plains the prevalence of 'Wälschriesling' reflects a similar dominance of the cultivar in the neighboring region of Hungary. In contrast, the Dalmatian coastal region produces predominantly red wines. It cultivates primarily indigenous varieties. The most distinctive of these is the dark red 'Plavać mali.' The most extensively grown red cultivar in much of the region is 'Prokupac.' Endemic white cultivars favored are 'Plavaać,' 'Maraština,' 'Grk,' 'Vugava,' and 'Zilavka.' The Istrian Peninsula in the northeast reflects in its varietal plantings a strong historical association with Italy.

Serbia has been the main region for bulk-wine production. Most of the production is red and comes from 'Prokupac.' 'Smederevka' is the main white variety. However, there is a trend to replace native cultivars with those from western Europe, presumably to gain easier acceptance in foreign markets.

Although Macedonian vineyards were decimated during the Ottoman domination, and subsequently by phylloxera in the last century, they were reestablished after the Second World War. Most of the cultivars chosen were domestic cultivars, including such red varieties as 'Prokupac,' 'Kadarka,' and 'Stanusina.' There also are smaller plantings of the white cultivars 'Žilavartea' and 'Smederevka.'

Romania Romania is another eastern European country that has most of its vineyards distributed around its periphery. The eastern curved arch of the Carpathian Mountains divides the country in two. Mountains reach up to 2400 m and enclose the central Transylvanian plateau (~600 m).

The 239,000 ha of vineyard in Romania are divided equally between red and white cultivars. From these, the country produces approximately 5.5 million hl wine annually. Extensive plantings of both native and western European cultivars occurred following the Second World War. The better, local white cultivars appear to be 'Fetească alba,' 'Grasă de Cotnari,' 'Tămîioasa romînească' ('Muscat blanc'), and 'Frîncuşa.' The most well known red varieties are 'Fetească neagră' and 'Babeasca neagră.' As with other indigenous cultivars, their evaluation elsewhere might expand the variety and interest of wines produced worldwide. This will also require a change in the eurocentric bias of wine critics and producers.

Few Romanian wines are well known internationally. This partially relates to most trade having been conducted with its eastern European neighbors. The most highly regarded of Romanian wines, at least historically, are those coming from around Cotnari and Grasă (47°27′ N), along the lower slopes of the Carpathian Mountains in the northeast.

Bulgaria As in so many other eastern European countries, Ottoman domination brought wine production in Bulgaria to a virtual halt for several centuries. Viticulture was reestablished following the First World War, but the vineyards were extensively destroyed during the Second World War. The major replanting that followed, combined with an emphasis on export to Western countries, help explain the extensive use of western cultivars. Bulgaria currently exports over 60% of its wine. Production often reaches 2.5 million hl, from 100,000 ha of vines.

Because the winery facilities also had to be reconstructed, the wine industry is comparatively up-to-date. The emphasis on quantity, however, has been at the expense of fine-wine production. The vineyards lie largely between latitudes 41°45′ and 43°40′ N, on land ideally suited to viticulture.

Of the native cultivars, approximately 8000 ha are devoted to 'Dimiat,' a grape commonly used for the production of sweet wines. The Georgian cultivar 'Rkátsiteli' is the dominant white cultivar. It is used extensively in wines sold in eastern Europe. Native red cultivars are 'Pamid,' 'Shiroka Melnishka Losa,' and 'Mavrud.'

Russia, Moldova, Ukraine and other Former Soviet Union States The states that formed the western portion of the Soviet Union lie in an immense plain. The east European (Russian) plain stretches from the Caucasus 2400 km north to the Arctic Ocean and an equal distance eastward to the Ural Mountains. Without moderation by large bodies of water, or mountains to deflect air flow, continental weather systems move largely unimpeded over the region. Thus, summers may be very hot and winters bitterly cold, with marked and rapid changes in temperature throughout much of the year. Precipitation falls off moving eastward, especially north of the Black Sea. These climatic influences limit commercial viticulture primarily to the more moderate climates north of the Black Sea, Moldova, Georgia, and the southern region of the Ukraine and Russia. Other

Russian vineyards, and those of Azerbaijan, are found along the eastern edges of the Caspian Sea.

Due to a decision taken in the early 1950s, the Soviet Union embarked on a massive vineyard-expansion program. Vineyard area increased from nearly 400,000 to over 1.1 million ha by the 1970s. This propelled the Soviet Union from tenth position, in terms of vineyard hectarage, to second in importance worldwide. The annual yield of about 18 million hl placed the region fifth globally in terms of wine production. The disparity between vineyard area and wine production reflected the hectarage devoted to table grape and raisin production.

About 50% of the vineyard area was located in the European section of the country, near the Black and Caspian seas; 30% was positioned in the Transcaucasian zone between the Black and Caspian seas, with the remainder occurring in south central Asia. In the Asian region, the production of raisins and table grapes is the main viticultural activity.

Some 70 foreign cultivars are grown throughout the former Soviet Union. Nevertheless, they constitute only approximately 30% of plantings. Some 100 indigenous varieties occupy most of the remaining hectarage. Over 200 additional varieties are cultivated, but only in limited quantities. The most extensively grown white cultivar was 'Rkátsiteli,' which used to cover 250,000 ha. This made 'Rkátsiteli' second only to 'Airen' as the most extensively grown white grape variety. Other commonly grown local cultivars are 'Mtsvane,' originally coming from Georgia, and 'Fetească.' Popular red cultivars are 'Saperavi,' 'Khindogny,' and 'Tsimyansky.' Many western cultivars, such as 'Traminer,' 'Riesling,' 'Aligoté,' and 'Cabernet Sauvignon,' are grown in Moldova and the southern Ukraine. There is renewed interest in the cultivation of local varieties in the Crimea. These include 'Kefessia,' 'Soldaia,' 'Savy Pandas,' 'White Kokur,' and 'Jevet Kara' (Rybintsev, 1995).

Although the Soviet Union was a major wine-producing region, only about 2.5% of the annual production was exported. Vineyard coverage has declined considerably from its high in the early 1980s, due both to policies taken by the Soviet government (to curtail alcoholism), and the economic turmoil following the transition to capitalism. Of the now-independent states, Moldova has the largest production, consistent with its possessing the most extensive vineyard area, approximately 150,000 ha. Vineyard area in Azerbaijan has fallen precipitously, from about 130,000 ha in the early 1990s to 8000 ha. The Ukraine, including the Crimea, possesses about 100,000 ha of vines, whereas the vineyard area in Russia covers almost 70,000 ha. Although possessing smaller total vineyard areas, Georgia (64,000 ha) and

Armenia (13,000 ha) are important in the production of fine wines. In these regions, vineyards may date back to the first attempts to establish viticulture five to seven millennia ago.

Unlike the wines of most wine-producing regions, about three-quarters of all wine produced in the former Soviet Union possess a distinctly sweet taste (over 15% sugar). Sparkling wines are especially appreciated and constitute 10% of the total production (~256 million bottles in 1989). To economize production costs, a continuous fermentation system was developed. Much of the production of sparkling wine is centered close to the Black Sea, around Krasnodar in the Kubar Valley (45°03′N), and near Rostov-na-Donu along the Don River (47°16′N) in southwestern Russia.

One of the major factors limiting viticulture in this region is its continental climate. The bitterly cold winters require that the vines in about 50% of the area's vineyards be laid down and covered with soil each winter. The annual practice is not only labor-intensive, but exposes the vines to mechanical damage and additional disease problems. Therefore, an extensive breeding program has been in progress for several decades to increase cultivar hardiness. Central to success has been the incorporation of frost resistance from *V. amurensis*. Improved cultivars, such as 'Burmunk,' 'Mertsavani,' 'Karmreni,' and 'Nerkarat' are able to survive adequately without burial during the winter. The development of high trunks has also proven beneficial in raising buds above the coldest zone near ground level. Restricting irrigation late in the season and applying cryoprotectants, such as mirval, migugen, and krezatsin, has further improved bud survival (Kirillov *et al.*, 1987).

North Africa and the Near East

In both North Africa and the Near East, wine production has been declining for decades. This has been particularly marked in Moslem countries. The overthrow of French colonial rule was followed by the strict imposition of religious restrictions against wine consumption. For example, wine production in Algeria dropped to 3% of its preindependence (1962) value. Even in Israel, where wine has religious significance, recent production is about 40% of the 1971–1975 level.

The regions of Israel and Lebanon are potentially capable of producing fine wines: for example, the wines from Château Musar in the Bekaa Valley of Lebanon. However, most wines have been either excessively sweet or flat and unbalanced, at least to those accustomed to European wines.

Because wine consumption can be a serious felony in Islamic countries, most cultivated grapevines are table or raisin varieties. The rapid loss of acidity in such grapes during maturation is appropriate for a fresh fruit crop, or for raisining, but it is unacceptable if wine is the intended product. The colossal cluster of grapes, reportedly carried back to Moses from Canaan (Numbers 13:23), is indicative of the agricultural fertility of the region, not the quality of its wine grapes.

Far East

Both China and Japan have been repeatedly introduced to winemaking during their long history. In addition, *V. vinifera* varieties have been cultivated in these countries for several centuries. However, wine did not become part of their cultural fabric. Various cultural and genetic hypotheses have been presented to explain this phenomenon, but none seem adequate. In Japan, the general inappropriateness of the climate probably has been the principal reason. In China, however, several regions are suitable to the cultivation of *V. vinifera*. Regardless of the reason(s), viticulture, and especially winemaking, attracted limited interest in the Far East until recently.

CHINA

In the past few decades, vineyard hectarage in China has expanded extensively, from 143,000 ha in the mid-1980s, to about 450,000 ha in 2003. Wine production has jumped from about 2,700,000 to 11,600,000 hl during the same period. By 2005, some 47,000 ha were dedicated to the production of 'western-style' winemaking (Gastin, 2006). This involves the cultivation of varieties such as 'Cabernet Sauvignon,' 'Merlot,' 'Cabernet Franc,' 'Riesling' and 'Chardonnay.'

Surprisingly, the majority of Chinese vineyards occur in one of the most rigorous climatic regions of the country, the far northwestern province of Xinjiang Uygur (~40°15′N). Its location in central Asia, north of Tibet, exposes the vines to extremes of drought, summer heat, and frigid cold. The grapes used to be almost exclusively used for raisin production, but most of the extensive new plantings are for wine production. Although rainfall is low, irrigation based on water from the adjacent Tian Shan mountain range is potential. The soils are principally sandy loam covering granite.

Limited cultivation of table and wine grapes also occurs in the east, north of the Yangtze River (~30°30′N). Monsoon rains and summer heat south of the Yangtze are unfavorable to most grape cultivars, whereas north of 35° latitude, cold winters usually require *V. vinifera* vines to be covered with soil during the winter. However, the alkaline nature of the soil makes it suitable for growing *V. vinifera* on its own root system.

This is possible because phylloxera is of limited occurrence in China.

The cultivation of *V. labrusca* cultivars and hybrids occurs in the north-central portions of Manchuria (~44°N). Here, abundant rainfall produces more acidic soils, suitable for *V. labrusca* cultivars. However, use of cold-hardy *V. rupestris* or *V. amurensis* rootstock is generally necessary to limit frost-induced root damage during the frigid winters. Temperatures can frequently dip to −30° to −40°C. The indigenous *V. amurensis* is the most cold-tolerant of *Vitis* species, surviving temperatures down to −50°C without significant harm. Several pure *V. amurensis* cultivars are grown, such as 'Tonghua' and 'Changbeisan,' but their unisexual habit makes yield erratic. The bisexual cultivar 'Shuanqing' is a major improvement. However, *V. vinifera* hybrids, such as 'Beichum' and 'Gongniang,' possessing cold-hardiness derived from *V. amurensis,* are more popular. Older *V. vinifera* varieties still widely cultivated are 'Longyan,' 'Niunai,' and 'Wuhebai' ('Sultana').

Due to the cold sensitivity of most cultivars, the vines are trained for easy removal from the trellis for winter burial. In hilly terrain, trellising has usually been on sloping elongated pergolas. On level ground, fan training has been common. Both systems use multiple cordons or bearing shoots to maintain sufficient wood subtlety to permit the annual lying down and raising of the vines. A unique adaptation of existing systems, incorporating concepts of canopy management, is the single-Dragon training system using a vertical T trellis (Fig. 10.21).

The most important winemaking area of China is the Liaoning Peninsula in Shandong province (~37°N). Shandong possesses over 14,000 ha of vines. The area experiences both maritime and continental climatic influences. About 12,700 ha of vines are cultivated in the coastal region of the Bo Hai Bay, and west and north of Beijing (~39°55′N). The other main regions include 11,500 ha in Henan province (33° to 34°N), and the adjoining northern portions of Jiangsu and Anhai, as well as 3,600 ha in Manchuria (42°30′ to 46°N).

Traditionally, Chinese wines have been fortified and sweet. Often, only 30% of the content has been grape-derived (Hua, 1990). The government has begun to improve quality standards, but the required purchase of local grapes, regardless of maturity, keeps wine quality minimal. Joint ventures with European firms have more freedom in choosing grapes, and wines up to international standards are being produced.

JAPAN

Although winemaking experience in Japan goes back at least to the eighth century, social acceptance of wine has been slow. Viticulture occurs sporadically throughout the main island of Honshu and the northern island of Hokkaido. Nevertheless, activity is largely concentrated around Kofu (35°41′N), 100 km southwest of Tokyo. The drier foothills climate of Mount Fuji provides the most suitable conditions in Japan for cultivating European *V. vinifera* varieties.

The major problems facing Japanese viticulture are the monsoon rains (often occur during flowering and harvesting), and the cold winters. To counteract these undesirable conditions, the vines are trained high on vertically branched pergolas. Sloped sites are preferred. These provide both better drainage and sun exposure. It also avoids competition for the limited supply of level arable land.

Of *V. vinifera* cultivars, the indigenous 'Koshu' variety is the most well adapted to Japanese conditions. The other dominant cultivars are *V. labrusca* hybrids, such as 'Delaware,' 'Campbell's Early,' 'Neo-Muscat,' and 'Muscat Bailey A.' Both disease resistance and acid-soil tolerance make *V. labrusca* hybrids more suitable for cultivation than most *V. vinifera* varieties. Japan produces approximately 940,000 hl wine from 21,000 vineyard ha. Wine production is about 1/100 that of the volume of sake, produced from fermented rice.

Australia and New Zealand

AUSTRALIA

Although Australia is now one of the top 10 wine-producing countries, the quality of its wines thrust Australia into world significance decades earlier. The range of climatic conditions, from the cool moist highlands of Tasmania to the hot arid conditions of the Murray Valley, creates many opportunities for

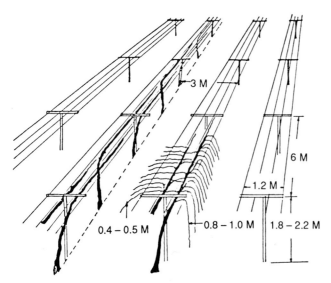

Figure 10.21 Diagram of the single-Dragon training of vines in China. (By Xiu Deren from Luo, 1986, reproduced by permission)

producing distinctive wines. The judicious selection of premium European cultivars has given even the simplest Australian wines a quality seldom found elsewhere.

For over 100 years, local preference for sweet fortified wines dictated wine production in Australia. However, consumer preference shifted dramatically during the 1960s and 1970s to dry table wines. This led to major changes in both viticultural and enologic practice, as well as expansion of grape growing into cooler regions of the country. Fortified wines now constitute less than 2% of Australia's wine production.

Viticulture is concentrated principally in the south-eastern portion of the continent (Fig. 10.22). The region forms a triangle from the Clare Valley, north of Adelaide (South Australia), to Muswellbrook in the Hunter Valley (New South Wales), and south to Geelong below Melbourne (Victoria). The region incorporates most of the eastern Australian continent between the 10 and 20°C annual isotherms (Fig. 1.1). However, the latitude range (32° to 38°S) is equivalent to that between southern Spain and Madeira. Although much of Australia is hot and arid, southern regions are significantly influenced by cold Antarctic currents flowing eastward below the continent. The high-pressure systems that prevail over the South Indian Ocean generate westerly winds that cool southern and western coastal regions. In addition, the southeast trade winds and low-pressure weather systems coming down the eastern Australian coast provide associated coastal regions with precious moisture. Thus, Australian viticultural regions are influenced by a range of climatic conditions from Mediterranean to temperate maritime.

The mid-latitudes and generally dry climate of Southern Australia provide longer growing seasons, with higher light intensities than are typical of Europe. This means that planting on sloped sites is of less significance than in much of Europe. Also, the vineyards are rarely subject to frost. Rainfall in coastal regions is well distributed throughout the year, and usually adequate for viticulture. However, precipitation rapidly declines toward the arid interior. Most Australian vineyards lie in the transition zone between the coast and interior, where irrigation is typically essential.

With the exception of some regions in Victoria and adjacent New South Wales, grafting for phylloxera control is unnecessary (Ruhl, 1990). Although phylloxera

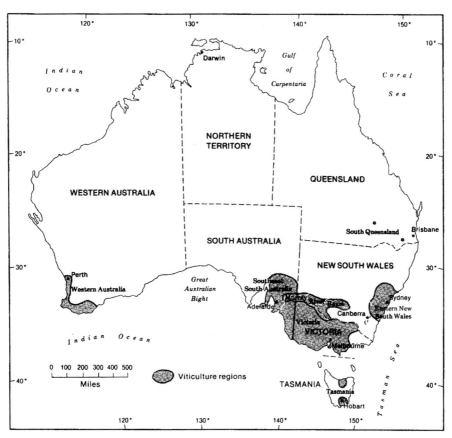

Figure 10.22 Wine-producing regions of Australia. (From de Blij, 1985, reproduced by permission)

was accidentally introduced into Australia in the nineteenth century, it has not spread widely. Otherwise, grafting is limited to mainly nematode-infested vineyards. 'Ramsey,' which possesses both nematode and drought tolerance, is the most extensively used rootstock in Australia.

Because of the former importance of sweet fortified wines and brandy in Australia, grape growing became, and still remains, centered along the Murray and Murrumbidgee river valleys. This region contains contiguous areas of New South Wales, South Australia, and Victoria. With the shift to dry wines, replanting focused on aromatically distinctive cultivars, such as 'Chardonnay,' 'Riesling,' and 'Traminer,' and away from neutral-flavored varieties such as 'Sultana' and 'Trebbiano.' Even 'Cabernet Sauvignon' has joined 'Shiraz' ('Syrah') as the most important red varieties grown in these hot regions. The proportion of premium cultivars in the cooler viticultural regions to the south is even more marked.

Although varieties such as 'Cabernet Sauvignon,' 'Pinot noir,' and 'Chardonnay' are being more extensively planted, 'Shiraz' remains the main red variety cultivated in Australia. It constitutes about 40% of red cultivar hectarage, followed by 'Cabernet Sauvignon' (30%), 'Merlot' (11%) and 'Pinot noir' (4%). 'Sémillon' has long been the major white variety cultivated in the Hunter Valley of New South Wales, although it constitutes only 10% of Australia's white grape hectarage. The most widely cultivated white variety is 'Chardonnay' (40%). Country wide, 'Riesling' and 'Sauvignon blanc' occupy 7% and 5% of white cultivar hectarage, respectively.

Amazing, from a European perspective, is the diversity of climatic regimes in which European cultivars excel. In the moderate climate of northeastern Victoria (Rutherglen), flinty chablis-like wines have been made from 'Pedro Ximénez.' Fine 'Cabernet Sauvignon' and 'Chardonnay' wines are regularly produced in the hot, arid climate of the Murray River Valley. Excellent 'Marsanne' wines are produced from the cool Yarra Valley in Victoria to the subtropical Hunter Valley in New South Wales. Even award-winning 'Pinot noir' wines have come from the Hunter Valley. Thus, Australia illustrates how viticultural and enologic practice can contradict the commonly assumed limitations of macroclimate.

Australia also illustrates that irrigation does not necessarily lead to reduced grape and wine quality. On the contrary, irrigation provides the grower with the potential of controlling vine vigor and ripening. Furthermore, Australia has amply demonstrated that low yield is not an *a priori* condition for high grape and wine quality.

One of the more distinctive characteristics of winemaking in Australia is the dominance of a few major producers. They not only possess major holdings in widely dispersed wine regions, but they also produce a full range of wines, from bag-in-box to prestigious estate-bottled wines. The relaxed attitude about wine culture is reflected in the extensive use of bag-in-box containers.

The extensive use of rotary fermentors has permitted the production of intensely colored and flavorful red wines, without the excessive extraction of tannic substances. The technique is applicable to both early-maturing and long-aging, premium wines. The adoption of refrigeration to control fermentation temperature has also been a critical element in Australia's success in producing superior table wines. Another aspect in the triumphant conversion from fortified to table wines has been the addition of tartaric acid. This is frequently required in hot, irrigated regions, where acidity falls excessively during ripening.

Australia is particularly unique in its general disinterest in Appellation Control laws. This permits producers to blend wines from different vineyards or regions, to maximize expression of their individual qualities. Even the wine often accorded premier ranking among Australian red wines, Penfolds Grange, comes from a blend of several wines from different vineyards in South Australia. For decades, blends, designated by bin number, rather than vineyard names, have become the quality hallmark of several producers. In Australia, blending has been raised to an art frequently associated with the production of champagne and port. Smaller producers of premium wines, as in other parts of the world, try to accentuate regional distinctiveness rather than ideal harmony. It is to the credit of Australians that both views are accepted and appreciated equally.

Other expressions of Australian inventiveness are the novel training systems and attention to harvest criteria. Australia, and its distant neighbor New Zealand, have both generated new training systems. These have been designed to improve early productivity (Tatura), increase yield (Lincoln), improve canopy exposure (RT2T), or achieve better pruning economy (minimal pruning). In determining the harvest date, there is greater concern about fruit flavor than in most countries. Combined with advanced enologic practices, both light aromatic 'Traminer' and smooth, dark, full-flavored 'Shiraz' wines can be produced from grapes grown in hot arid climates.

The 67,000 ha of vines in South Australia yield about 6.5 million hl wine annually. This constitutes about 50% of all Australian wine. The majority of vineyards occur on irrigated lands associated with the Murray Valley ($\sim$34°10′S), some 250 km northeast of Adelaide. The best-known wine region in South Australia is situated 55 km northeast of Adelaide, in the 30-km-long

Barossa Valley (~34°35′S). Cooling sea breezes and cool nights help to moderate the effect of the warm climate on acidity loss during ripening. Irrigation supplies the moisture not supplied by sufficient rainfall. Although most of the vines grow on the valley floor, vineyards have been increasingly planted up the western slopes of the valley. Here, the altitude of the Mount Lofty Range retards ripening, and is thought to enhance flavor development. Vineyards of the Eden Valley occur on the eastern slopes of the mountain. Smaller, but increasingly important, regions occur in the Clare Valley northeast of Adelaide, the Southern Vales regions south of the city, and the milder regions of Padthaway (36°30′S) and Coonawarra (37°15′S) in the southeastern corner of the state. Coonawarra has garnered a reputation that rivals the most famous vineyards of France.

New South Wales is the second most important Australian state in terms of wine production (32%). It produces about 4.6 million hl wine from 36,000 ha of vines. As in South Australia, the largest production comes from the irrigated regions along the Murray and Murrumbidgee Valleys. Nevertheless, the fame of the state rests with its Hunter Valley vineyards. The original area is centered just northeast of Cessnock (32°58′S). Although initially famous for 'Shiraz' and 'Sémillon,' it has developed a reputation for excellent 'Chardonnay' and 'Cabernet Sauvignon' wines. Breezes off the South Pacific bring precipitation and cooling to the subtropical latitude. Cloud cover during the hottest part of the day also tends to moderate the heat and intensity of the sun. Frequent drying winds from the interior help to offset the disease-favoring humidity of the sea breezes. Although rains brought by Pacific breezes obviate the need for irrigation, they occasionally produce problems during harvest, making the quality of Hunter Valley wine one of the most variable in Australia. Further inland, to the northwest, is the Muswellbrook region (32°15′S). Its drier climate usually makes irrigation necessary. Further inland again, in the high valleys of the Great Dividing Range, are the vineyards of the Mudgee region.

Victoria is the third major wine-producing Australian state. It produces about 15% of Australia's wine. Recently it has regained some of its former viticultural importance. As with its neighboring states, much wine comes from irrigated vineyards along the Murray Valley. Nevertheless, particular interest has been given to plantings in the cooler maritime south. Examples are the Yarra Valley (37°50′S), just east of Melbourne, Geelong (38°06′S), 70 km southwest of the city, and Drumborg, in the southwestern corner, near Portland (38°20′S).

About 2000 km to the west are the vineyard regions of Western Australia. Most of its vineyard area is situated northeast of Perth (31°45′S) in the arid Swan Valley. However, cooler regions in the southwestern corner of the state are favored for making premium table wine. In the Margaret River district, westerly winds off the Indian Ocean provide both cooling and limited summer rains. The abundant sunshine and typically dry harvest conditions are ideal for the production of intensely colored, flavorful red wines. However, due to the closeness to the ocean, spring gale-force winds can inflict serious damage to varieties that undergo early break bud. There also can be considerable salt transport associated with the winds. The other prime viticultural region occurs around Mount Barker, in the southeast corner of the state. Vineyard altitude typically provides climatic moderation and uniform year-round precipitation. With the exception of Swan Valley, and a few other sites where nematodes are a problem, grafting is unnecessary and most grapevines are own-rooted.

The coolest of the Australian vineyards are located on the southern island of Tasmania. Its climate is primarily maritime, with moderate seasonal temperature variation. This can give problems with nonuniform bud break and fruit ripening. The vineyard area, while small (~1000 ha), is expanding. Most of the vines are planted around the capital, Hobart (37°50′S). It is favorably influenced by the warming of the Derwent Estuary and adjacent bays. Another region is in the northern portion of the island, near Launceston (37°50′S). It is the warmest and sunniest region of the island.

NEW ZEALAND

New Zealand (Fig. 10.23) was the last region in the Southern Hemisphere to see a major expansion in its wine industry. Although New Zealand has produced wine for more than 160 years, the industry underwent most of its development and considerable expansion from the 1970s onwards. In 1965, vineyard coverage approximated about 300 ha. By 2006, vineyard area had expanded to about 23,000 ha (a doubling since 2001) (Anonymous, 2006). During the same period, wine production went from 533,000 hl to 1,330,000 hl. Although the North Island initially possessed more vineyard area, plantings in the South Island are now more extensive. The South Island possesses about two-thirds of the vineyard area, with the Marlborough area alone growing about 52% of the country's vines. The latitude and position of the islands, and their distance from any large land mass, provide a moderate to cool, but variable maritime climate. The central ranges of mountains that run the length of both main islands generate marked contrasts between the higher rainfall, cloudy, windward west, and the milder, sunnier, leeward side.

Most of the vineyards in the North Island occur on the eastern side of the mountains that divide the island. They possess a drier, sunny climate. In contrast, the western portion of the island may receive up to

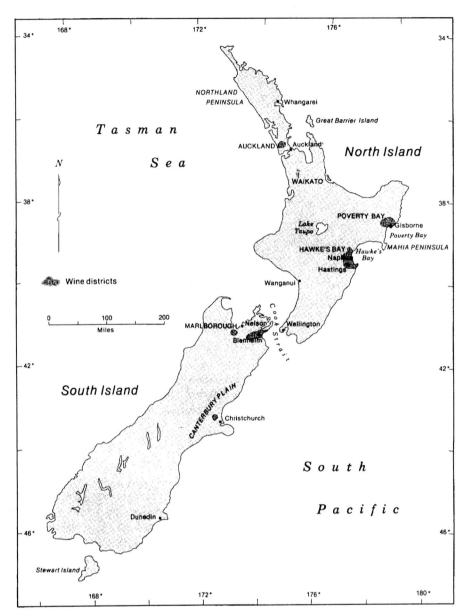

Figure 10.23 Wine-producing regions of New Zealand. (From de Blij, 1985, reproduced by permission)

200 cm rain/year. Combined with fertile volcanic soils, vines in the Auckland region tend to grow very vigorously, produce dense canopies, and be particularly susceptible to fungal diseases. These conditions prompted the development of new training systems, such as the TK2T and RT2T. With enhanced canopy openness, disease control was facilitated and fruit quality increased, without negating the natural benefits of the region's fertile soils. Ground covers further help restrain excessive vigor and enhance evapotranspiration, minimizing some of the potential dangers of the region's high rainfall.

The Hawke's Bay region, around Napier (39°30′ S), has about half the rainfall of the Auckland area. It possesses about 60% of the North Island's vineyard area of 7700 ha. Second in importance in regard to size is the Gisborne area (38°40′ S), in the northeastern portion of the island. The west-coast vineyard regions are located north and south of Auckland (~37° S), while the Wairarapa region is located at the southern tip of the island (41°15′ S).

The newer vineyard regions are primarily located on the South Island. The largest region is in the northeastern tip of Marlborough district (Cloudy Bay), around

Blenheim (41°30′S). Sheltering provided by the enclosing branches of the Spencer Mountains limits annual precipitation to almost 50 cm/year. This provides needed sunshine and warm long summers to fully ripen the grapes. Basalt gravelly soil acts as a useful heat sink (and source at night). Otago in the southeastern, cooler portion of the island (~46°S) is its second most significant region, with some 1253 ha of vines. Additional vines are grown on the Canterbury Plain, west of Christchurch (43°30′S). The South Island has become world famous for its superior 'Sauvignon blanc' and exquisite 'Pinot noir' wines. 'Pinot noir' constitutes about 78% of the planting in the Otago region.

The distinctiveness of Marlborough and Wairarapa 'Sauvignon blanc' wines appears to derive, at least partially, from the concentrations of 3-mercaptohexyl acetate and 3-mercaptohexan-1-ol. These compounds, plus higher contents in 2-methoxy-3-isobutyl pyrazine, distinguish the wines from their counterparts in the United States, France, and South Africa (Nicolau *et al.*, 2006). Data from Masneuf-Pomarède *et al.* (2006), on the effect of yeast strain and fermentation temperature, suggest that these factors do not explain the observed differences in volatile thiols noted above. The differences may result from climatic differences, such as the frequency of cool nights during the growing season in Marlborough and Wairarapa.

Because of the disease severity, associated with the original vineyard regions in New Zealand, French-American hybrids covered most of the vineyards. With improved chemical control and a shift toward the making of table wines, *Vitis vinifera* cultivars replaced French-American hybrids. The most widely planted cultivar used to be 'Müller Thurgau,' but it has fallen to about 1% of white cultivar plantings. 'Sauvignon blanc' is now the most cultivated variety (40%) followed by 'Pinot noir' (18%) and 'Chardonnay' (17%).

South Africa

South Africa has a long history of grape culture and winemaking, dating back to 1655. It initially became famous for Constantia, a sweet fortified wine. It was once one of the most sought-after wines in Europe. The specialization in fortified wine production remained up until recently, with about half of South African wines becoming sherry and port, or being distilled into brandy. The recent shift to table wine production, so pronounced in most New World countries, was slower to express itself in South Africa.

South Africa ranks third, after Australia and Argentina, as the largest wine producer in the Southern Hemisphere. The 132,000 ha of vines annually yield approximately 9 million hl wine. Viticultural activity is largely concentrated in the southwestern coastal region of Cape Province (Fig. 10.24). This area, spanning 31°00′ to 34°15′S, is equivalent to the latitude spread from southern California (Los Angeles) to the Gulf of California. The other viticultural regions are centered around Upington (28°25′S), along the Orange River Valley, and in the Douglas region in central South Africa.

Given the subtropical latitude, the early excellence of South African fortified wines is not surprising. Most of the coastal Cape region is influenced by a Mediterranean climate, with its limited rainfall occurring primarily in the winter months. The climate is relatively stable and endures few erratic variations. Consequently, South African vintages are almost consistently good to excellent.

Although not possessing the largest vineyard area (~17,500 ha), Stellenbosch has the highest concentration of grape cultivation of any district in South Africa. The central location of the city of Stellenbosch, in the L-shaped viticultural region of the Cape, has made it central to viticultural and enologic research in South Africa. The adjacent Cape Peninsula contains what remains of the famous Constantia vineyards. The Constantia district is permitted only a single agricultural activity, viticulture.

Stellenbosch is only 40 km east of Cape Town (33°48′S). Its position on the north shore of False Bay shields it from direct exposure to Atlantic influences. Nonetheless, the cooling effect of the Atlantic Ocean reaches the vineyards and slows vine growth. This postpones ripening into the milder autumn, and reduces excessive vineyard evapotranspiration. Mountain chains to the north and east of Stellenbosch favor cloud and rain formation. These conditions result in 20–25% of the 50 cm annual precipitation falling during the growing season. Nevertheless, some of the finest sites still require periodic irrigation. The famed red wines of the region generally come from vines grown on the moister west-facing slopes of Mt Simonsberg and the Stellenbosch Mountains. White cultivars tend to do best on the sandy soils of the western lowlands. 'Cabernet Sauvignon' appears to do better in the warm climate of South Africa than in equivalent climates in Europe.

Viticulture becomes progressively less important both north and east of Stellenbosch. Grape culture also declines through a complex patchwork of mountains and valleys into the *highveld* interior. This reflects the decline in precipitation and increase in temperature in the associated regions. One exception is the most southerly extension of the Cape, the Overberg district. Here, conditions are cooler than in Stellenbosch, but high

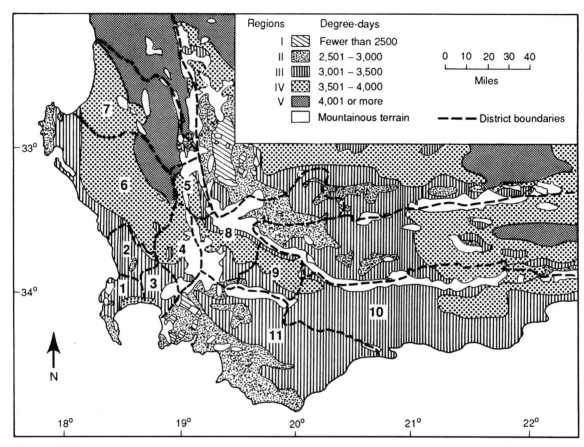

Figure 10.24 Mountainous and Fahrenheit degree-day climatic regions superimposed on viticultural districts of Cape Province in South Africa: (1) Constantia, (2) Durbanville, (3) Stellenbosch, (4) Paarl, (5) Tulbagh, (6) Swartland, (7) Piketberg, (8) Worcester, (9) Robertson, (10) Swellendam, and (11) Overberg. (After Le Roux, 1974, reproduced by permission)

winds and sand dunes hampers viticulture throughout much of the district.

The L-shaped stretch of coastline viticulture, extending north and east of Stellenbosch, also shows a shift from table to fortified and distilling wine production. This change is already apparent on entering the Paarl district, just north of Mt Simonsberg. Although the Paarl district is larger and possesses more vineyard area (18,000 ha), the vineyards are less concentrated than those of Stellenbosch. Even larger in size is the adjacent Worcester district to the east. It has close to 19,500 vineyard ha, located largely in the western region. Winds funneling through the valley provide some cooling and extra moisture, especially to vineyards on higher slopes. The western region annually collects about 75 cm of rain, whereas the eastern portion receives only 25 cm. Shadows cast by mountain peaks over 850 m high limit the duration of exposure to the hot subtropical sun. Vines producing table wines usually benefit from east-facing slopes, whereas those producing fortified wines are favored by cultivation on west-facing slopes. Similar, but drier, is the adjacent Tulbagh district to the north of

Paarl. Further west again, and along the Atlantic coast, is Swartland. Cool sea breezes and heavy dew help partially compensate for the low annual rainfall (25 cm). Additional districts extend viticulture northward and eastward to form a region up to 130 km deep, 250 km to the north and 400 km to the east of Cape Town.

One of the more important viticultural problems in South Africa is the high proportion of acidic soils. It is estimated that 70% of the vines of the Cape grow in soils below pH 5. Toxic levels of available aluminum probably explain much of the observed poor root growth in acidic soils. This influence may be counteracted by the incorporation of lime to raise the soil pH. Liming also probably helps root growth by improving the soil structure. The use of acid-tolerant rootstocks can further counter the detrimental effects of acidic soils.

Although the production of different wine styles has not changed as dramatically as in other non-European countries, the varietal composition of South African vineyards has changed considerably since the 1960s. The most marked transformation has been the striking increase in 'Steen' ('Chenin blanc'), which covers about 20% of the vineyard hectarage in South Africa. 'Colombard' makes up another 11% coverage, followed by 'Chardonnay' and 'Sauvignon blanc,' at about 7% each. A corresponding decline has occurred in the cultivation of 'Hermitage' ('Cinsaut'). Other cultivars with decreased use have been 'Palomino,' 'Green Grape' ('Sémillon'), 'Hanepoot' ('Muscat of Alexandria'), and the local cultivar 'Pinotage.' Although white cultivars still make up 55% of vineyard coverage, red varieties are on the increase. For example, 'Cabernet Sauvignon,' 'Shiraz,' 'Pinotage' and 'Merlot' now constitute 15%, 9%, 7% and 7% of vineyard plantings, respectively.

South America

South America has been associated with viticulture and winemaking almost since its discovery and colonization by the Spanish. Nevertheless, the emergence of South America as an important wine-producing region has been comparatively recent. Figure 10.25 shows some of the main wine-producing areas.

CHILE

Chile is unique among world nations in having such an extensive coastline. Although spanning almost 40° of latitude, from the Peruvian border (17°24′ S) to Cape Horn (56°S), it is only an average of 180 km wide. Along the 4500-km coastline, only the zone between 32° and 38°S is amenable to viticulture and premium wine production. Chilean vineyard area (185,000 ha) and yearly wine production (~7 million hl) are considerably less than that of its eastern neighbor, Argentina. Within this region, however, conditions are more favorable for premium wine production than in Argentina.

The best viticultural regions in Chile lie at latitudes roughly equivalent to those of southern California in the Northern Hemisphere. The cool Humboldt Current, and the altitude of the vineyards in the central valley (*Nucleo Central*), provide a temperate Mediterranean climate. Precipitation increases rapidly along this 600-km stretch from north of Santiago, down to the Bío-Bío River, near Concepción. Further south, the climate becomes maritime, without a summer drought period. At the northern end of the *Nucleo Central,* the annual precipitation averages 25 cm, whereas at the southern end near Concepción, it averages more than 75 cm annually.

Of the several viticultural zones in Chile, the most highly regarded section is termed the *Regadio*. It encompasses central Chile, from north of the Aconcagua River to south of the Maule River. Except for the Aconcagua region, the best sites occur within the broad *Nucleo Central,* formed by the coastal *cordillera* and the Andes

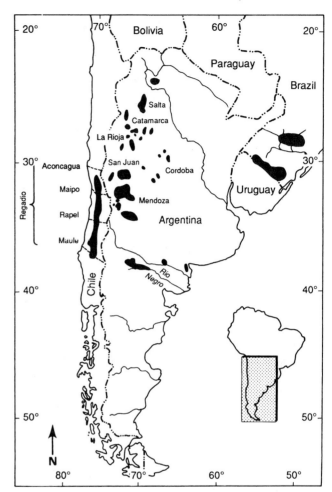

Figure 10.25 Main wine-producing regions of Chile, Argentina, and Brazil. (After de Blij, 1985, reproduced by permission)

to the east. Most of the vineyards are associated with river valleys that cross the *cordillera*. From north to south, they include the Maipo, Cachopoal, Tinguiririca, Lontué, and Maule rivers. The *Regadio* is subdivided from north to south into the Aconcagua, Maipo, Rapel, and Maule regions.

Within the *Regadio*, sites north and south of Santiago (33°26′ S) in the Maule River Valley are considered superior. The regional average rainfall of around 40 cm provides moisture to the deep loamy–gravelly soils. This, combined with irrigation water from the Maipo River, provide adequate moisture for vine growth and grape ripening. The gentle rolling landscape facilitates excellent drainage. Prevalent dry sunny conditions and moderate temperatures help limit disease development. The calcareous layer in the soil favors root growth. Chile is the major wine-producing country still unafflicted by phylloxera, and most cultivars can be grown on their own root system. The winter season is sufficiently cool to satisfy bud-dormancy requirements and permit bud break in the spring. Maipo's moderately stable climate usually provides conditions optimal for producing fine-quality fruit. The adjacent Rapel region, south of Maipo, is slightly cooler and yearly receives between 50 and 60 cm of rain. It also produces excellent-quality fruit and wine. Annually, the regions produce about 90,000 and 230,000 hl wine, respectively.

Although most grape varieties in Chile are red (75%), the cultivars grown in the central Regadio regions are different from the 'País'[1] variety that dominates regions such as Bío-Bío. 'Cabernet Sauvignon' makes up 55% and 70% of the red cultivars in the Maipo and Rapel regions, and 47% of all red cultivars grown in Chile. In the Maipo region, 'Merlot' is the second most widely grown red variety, whereas in the Rapel region, the rare old Bordeaux cultivar 'Carmenère' is third in red coverage. Of white varieties, 'Chardonnay' is by far the dominant cultivar, with 'Sauvignon blanc' a distant second. The predominance of French cultivars in a region colonized by Spanish immigrants reflects the effect of Silvestre Ochagavia, an influential viticulturist and politician in the 1850s.

Of the other two regions in the Regadio zone, Maule, south of Rapel, is the largest. It possesses about 50,000 ha of vines, in contrast to the 11,000 and 33,000 ha in Maipo and Rapel. It generates about 50% of Chile's table wines. Some sites are as fine as those further north, but others possess a more unstable, moist environment. This is reflected in the increased importance of the 'País' variety (23%). Nevertheless, the conditions are still favorable for premium white cultivars. 'Sauvignon blanc' constitutes about 50% of plantings, with 'Chardonnay' and 'Sémillon' at 25% and 10%, respectively.

In the northern Regadio region of Aconcagua, red wine production is dominated by 'Cabernet Sauvignon,' 'Merlot' and 'Pinot noir,' whereas 'Chardonnay' and 'Sauvignon blanc' are the dominant white cultivars. However, the warmer climate is reflected in the increased cultivation of varieties used in the production of Pisco brandy.

The shift of cultivars to those used in Pisco production is particularly noticeable in the adjacent northern Pisquera viticultural zone. Here, a variety of 'Muscat' cultivars and 'País' constitute the majority of white and red cultivars, respectively. Table grapes also increase in importance in the Pisquera zone. The primary deficit of the zone is neither its hot climate nor dry environment, but the erratic and violent storms that can periodically ravage the area.

---

[1] Synonym for 'Criollo' in Argentina and Mexico, and 'Mission' in California.

South of the *Regadio*, one enters a transition zone where irrigation becomes progressively unnecessary. In the Secano zone, south of the Bío-Bío River, the increasing rainfall and cool climate make viticulture increasingly problematic. Nevertheless, the zone still has an extensive vineyard area, covering 14,000 ha.

ARGENTINA

Grape growing and winemaking have been conducted in Argentina since the mid-1500s, as in Chile. However, the growth of the Argentinian wine industry into the fifth largest in the world occurred only in the twentieth century. The vineyard area of 211,000 ha and annual wine production of approximately 13 million hl make it the largest wine producer in the Southern Hemisphere, though this is being challenged by Australia. Viticulture constitutes the third largest industry in Argentina. As in Russia, most of the production is consumed locally. Argentina used to export only between 1.5 and 3% of the annual production. By 2003, this had risen to 14%. This does not include the export of concentrated grape juice for fermentation in other countries. Were this included, it would add significantly to its contribution to the world's wine supply.

Vineyard regions in Argentina occur almost exclusively in the rain shadow of the Andes, along the western border of the country. Wines are grown from south of the province of Jujuy (24°S) to along the Negro River (40°S). However, the major concentration of vineyards occurs in the province of Mendoza (~32–36°S).

The location of Mendoza in the lee of the Andes gives it an arid climate, annually receiving about 20 cm rain. Irrigation of the region's extensive vineyards is possible only because of the relative flatness of the land, and the ready availability of river and artesian water (Fig. 10.26). The deep, loosely compacted soils permit good drainage, water retention, and root penetration. Irrigation costs are partially offset by the savings derived from the disease-limiting dry air. The clear skies and altitude (~500–750 m) of the region generate day–night temperature fluctuations of up to 25 °C. Thus, the heat summation of some 1900–2100 Celsius degree-days gives the impression that the region is cooler than it is. To avoid excessive acid loss during grape ripening, it is usual to harvest early and prevent malolactic fermentation. Although the summers are hot, the winter period is adequate to permit bud break in the spring.

Regional differences in latitude and altitude across Mendoza produce significant differences in fruit ripening and quality. The basins of the Tunuyána and Mendoza rivers, southwest of Mendoza (32°54'S) are currently considered the best viticultural areas. Regional differences are partially reflected in cultivar distributions.

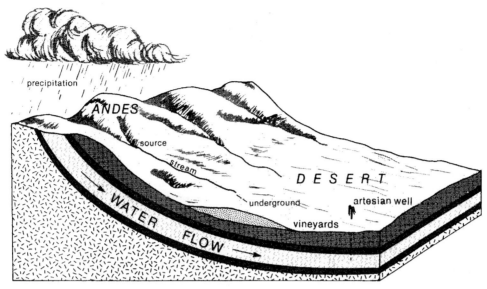

Figure 10.26 Source of the artesian water (from the eastern flank of the Andes) essential for irrigation of vineyards of Mendoza, Argentina. (From de Blij, 1985, reproduced by permission)

'Cabernet Sauvignon' and 'Malbec' are more common in the northern portions of Mendoza, whereas 'Tempranillo' and 'Sémillon' are more frequently cultivated in central regions. Nearly 50% of all cultivars are premium red varieties, and 20% are white. The remainder consists of varieties established in Argentina for centuries, such as 'Criolla'[2] and 'Cereza.'

The province of San Juan, to the north of Mendoza, ranks second in terms of quantity of wine produced. It has 58,000 ha of vines and annually yields approximately 5 million hl wine. San Juan also produces much of the wine used in sherry and brandy production, as well as most of the exported concentrated grape juice. The San Juan area is slightly hotter and drier than Mendoza, but more frequently subjected to strong desiccating winds from the Andes. Consequently, pergola-training systems are commonly used to protect the grapes from intense sun and wind exposure. The same training system has been used in Mendoza for hail protection.

In San Juan, premium red cultivars, such as 'Barbera,' 'Nebbiolo,' and 'Malbec' cover only about 10% of the vineyard area. White varieties, such as 'Pedro Ximénez,' 'Torrontés,' and 'Muscat,' and the reds 'Criolla' and 'Cereza' are the dominant cultivars planted. 'Torrontés' is the second most cultivated white cultivar. It appears to be derived from an indigenous crossing between 'Muscat of Alexandria' and 'Criolla' (Agüero *et al.*,

2003). Additional details are available in This *et al.* (2006). A genetic connection with 'Torrontés' in Spain seems unlikely, as vines with that name appear to consist of a collection of distinct varieties, different from those bearing the name in Argentina (Borrego *et al.*, 2002).

Of the other provinces that produce wine, only the south central Rio Negro is of considerable importance. The more southerly latitude (~38°S) provides Rio Negro with a cooler climate. Thus, grapes mature more slowly and generally develop a better acid/sugar balance during ripening.

In Rio Negro, the important viticultural area occurs on the broad flood plain of the Negro River, east of the junction of the Limay and Neuquén rivers. Because the annual precipitation averages 20 cm, irrigation is essential. Textural soil differences may generate the features that distinguish wines produced on opposite sides of the river. The north side possesses more sand and gravel, whereas the southern portion has finer, more fertile soils.

Argentinian vineyards have a much wider diversity of cultivars than neighboring Chile. The original European settlers brought Spanish varieties and techniques that still dominate the wine industry. French cultivars came later, primarily via Chile. Finally, Italian varieties came with the influx of Italian immigrants, beginning in the middle of the nineteenth century.

As noted, the winemaking procedures are predominantly Spanish. Thus, the finer wines are given several years of aging in casks before bottling. With red wines, the results meet with widespread approval, both in Argentina and abroad. The slightly oxidized character

---

[2] Originating from Spain (Tapia *et al.*, 2007) and called 'Mission' in Mexico and California; País in Chile, and 'Rosa del Peru' or 'Negra corriente' in Peru.

given white wine by prolonged aging in oak cooperage is appreciated in Argentina, but has not developed an equivalent following in foreign markets.

BRAZIL

Although not widely recognized as a wine-producing nation, Brazil possesses approximately 72,000 ha of vines and annually produces approximately 3 million hl wine. Of the several states involved, only the southernmost Rio Grande do Sul is of considerable significance. It contains about 70% of Brazil's vineyard area. In addition, 6000 ha of grapes are grown in tropical Brazil, largely in the Saõ Francisco River valley, in the Petrolina region (~9°2′S). Most of the production (two crops per year) is grown as a fresh fruit crop, with only about 10% being used in wine production.

Most of the vineyards are congregated north of the Jacui River (~29°30′S), 120 km northwest of Porto Alegre. The other main region in Rio Grande do Sul lies along the border with Uruguay (~31°S). Both regions have moist warm summers and mild winters. The moist climate generally is unfavorable to the cultivation of *Vitis vinifera*. Nevertheless, modern chemical disease control has permitted the expansion of *V. vinifera* cultivation in Brazil and neighboring Uruguay. The predominant red cultivars are either Italian, such as 'Barbera,' 'Bonarda,' or 'Nebbiolo,' or French, such as 'Cabernet Franc' or 'Merlot.' 'Muscat' varieties, along with some 'Trebbiano' and 'Sémillon,' are the main white cultivars. Nevertheless, almost 80% of all cultivars are *V. labrusca* varieties or French-American hybrids. Important cultivars are 'Isabella,' 'Dutchess,' 'Niagara,' 'Delaware,' 'Concord,' and Seibel hybrids. Their main advantage in the humid climate of Brazil is their greater disease resistance.

URUGUAY

As with most other South American countries, Uruguay has a long history of grape production, commencing with the arrival of Spanish colonizers. Nevertheless, production has been limited and consumed almost exclusively locally. Most of the vineyards are located near Montevideo, in the southern part of the country. However, other vineyard regions are located throughout much of the southwestern portions of Uruguay. Vineyards cover about 10,000 ha, of which about half are *Vitis vinifera* cultivars. The most important red *vinifera* cultivar is 'Tannat,' amounting to about 32% of the wine grape coverage. In Latin America, this variety from southern France is almost exclusively grown in Uruguay. Its coverage is superseded only by 'Muscat Hamburg.' 'Isabella' and French-American hybrids are grown primarily for domestic wine production. Wine production has remained relatively stable over the past decade or so, being in the range of about 850,000 hl.

North America

The North American market has experienced the same dramatic shift in consumer preference noted previously in Australia and New Zealand. In addition to a move away from fortified wines, white wines became the preference with the majority of consumers. More recently, red wines have shown a return to favor, presumably due to their supposed superior health benefits. These changes have prompted considerable adjustment in viticultural practice, cultivar planting, and winemaking. They also have spawned the creation of an increasing number of wineries specializing in premium wines. Even more important in the long-run has been the greater communication, cooperation, and integration of views between grape growers and winemakers.

These developments have also spurred legal changes that have induced many grape growers to start their own wineries. Consequently, the North American industry has begun to resemble that of Europe. The effect has been an improvement in the level of grape and wine quality. Nevertheless, North American winery conglomerates still retain a dominant position in wine production. Although continuing to produce fortified wines, they have adopted and championed technological advancements that have made North America one of the best quality and price wine regions of the world.

UNITED STATES

The experimental and technological innovation generated by the rapid adjustment of winemaking to new preferences thrust the United States into the forefront of wine research. This has occurred despite its production of somewhat less than half the wine volume generated by any one of the three major wine-producing countries, Italy, France, and Spain. US production stands above 21 million hl; vineyards cover approximately 415,000 ha.

California Although wine is made in nearly every state in the Union, California is the major producer. Despite the growing importance of other states, California continues to produce over 90% of American wine.

Since the shift toward table wines, the focus of enologic activity has moved from the Central Valley to the numerous valleys that directly open to the Pacific Ocean. Nevertheless, about 60% of Californian vines grow in the southern portion of the Central Valley, the San Joaquin. In addition, nearly 70% of Californian wines come from this strip of land, 650 km long and up to 150 km wide. The Central Valley lies approximately between latitudes 35° and 38°N. The reliably warm Mediterranean climate, rich soils, and flat landscape are ideal for most forms of agriculture. The high Coastal

Range effectively separates the valley from moisture-carrying sea breezes.

Although hot and arid, the San Joaquin Valley provides the table wines most Americans consume. This has been possible with practices such as early harvesting, in-field crushing, cool fermentation, prevention of malolactic fermentation, and protection from oxidation. These procedures are comparable to those used in similar regions in Australia, Chile, and South Africa. New cultivars specifically bred for Central Valley conditions helped its transformation into a producer of consistently good, inexpensive, table wines. Particularly valuable were 'Rubired,' 'Ruby Cabernet,' and 'Emerald Riesling.' However, around Lodi (38°07′N), cultivars such as 'Chenin blanc,' 'Colombard,' 'Chardonnay,' 'Zinfandel,' and 'Cabernet Sauvignon' are grown extensively. Here, the opening of the Central Valley to the Pacific produces cooler nights, which favor acid retention and flavor development in premium cultivars. The southern portion of the San Joaquin Valley remains the center for sherry and port production. It also is the main location of the prominent raisin and table grape industries. These constitute 10% and 30%, respectively, of vineyard production. Average rainfall declines in the valley, from 45 cm/year around Sacramento (38°35′N), to 13 cm at the southern tip. Because most of the precipitation comes during the winter, irrigation is essential in most years.

Napa is the best-known coastal valley, located northeast of San Francisco (37°45′N). It is typical of several valleys affected by an influx of cool, moist air from the Pacific Ocean. This results as rising air currents, generated by heating of interior parts of the valley, draw in cooler air from the ocean. This often is associated with the development of fog at the mouth of the valley. As the air is heated, the humidity drops. This influence is less marked along the valley walls. Precipitation is typically higher on the western wall and further up the valley (Fig. 10.27).

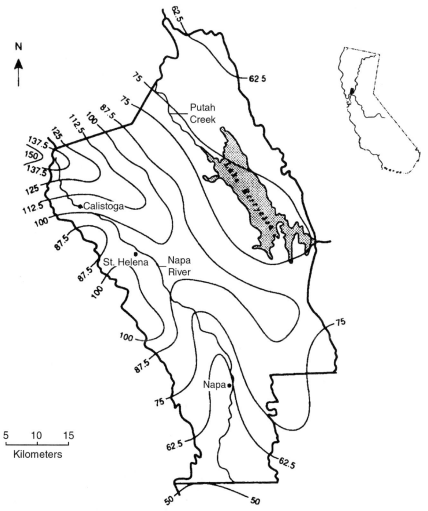

Figure 10.27 Mean annual precipitation in Napa County. (From US Department of Agriculture, 1978)

In addition to the declining precipitation up the valleys, there is a similar reduction moving down the coast. For example, average precipitation decreases from near 100 cm/year in Mendocino (39°18′ N), to roughly 45 cm around Santa Barbara (34°26′ N), and just over 20 cm in San Diego (32°43′ N). Because most of the rain comes during the winter, coastal-valley viticulture typically requires irrigation. Although irrigation increases the cost of production, it provides an opportunity to regulate vine growth and limit several diseases and pests. For example, California is generally unaffected by downy mildew, black rot, and some grape and berry moths. The absence of the winged stage of phylloxera may partially explain the freedom of most central-coastal vineyards from phylloxera, years after its accidental introduction into the state.

The Sonoma Valley, adjacent to Napa, is only slightly less well known. Both are about equal in vineyard area and wine production. The vineyards in the northern region of Mendocino County are more dispersed and scattered through many valleys. The Monterey region, south of San Francisco, has also become an important viticultural area. Other regions, although important for quality, are minor in terms of production. Several regions possess viticultural areas possessing a reputation and consumer following exceeding that of the county name. Examples are the Alexander Valley in Sonoma and the Santa Ynez Valley in Santa Barbara.

Although viticulture is practiced over nearly 6° of latitude in California, temperature changes are often more marked along the length of the coastal valleys than down its coastline. For example, several coastal valleys pass from heat summation Region I (cool) to Region III (mild) along their length (see Fig. 5.7). This change can occur over a distance of less than 40 km. Temperature regimes near the opening of coastal valleys can be similar to those found at a latitude 13° further north in Europe. The influences of altitude up valley slopes further enhances short-distance climatic diversity. However, valleys in California possess longer growing seasons, have milder winters, and experience higher light intensities than many of their European counterparts. The greater proximity of the vineyard regions to the coast partially explains the apparent anomaly that some of the coolest viticultural regions of California are in the south (San Luis Obispo and Santa Barbara, ~35°N).

The combination of diverse temperature regimes with local differences in moisture and soil conditions have provided California with a remarkably varied range of juxtaposed growing conditions. Consequently, cultivars typically separated by hundreds of kilometers in Europe may grow within sight of one another in California. This has influenced the production of stylistically different wines within kilometers of one another. It also has fostered an acceptance of a much wider range of varietal expressions than in Europe. The absence of a traditional style for particular regions in California has left winemakers free to experiment and create their own distinctive wines. The rejection of the view that grape varieties succeed only under a limited range of climatic conditions advances the development of new and better regional and varietal wines.

The shift in consumer preference to dry white wines that occurred in the 1980s had marked effects on Californian viticulture. One solution was the production of blush and "white" wines from excess red grape capacity. Another solution was the grafting over to, or replanting with, white cultivars. This was particularly marked with 'Chardonnay.' It still remains the most cultivated wine cultivar (17%). The next most grown white variety is 'Colombard' (7%), most of which is found in the San Joaquin Valley. Other important white cultivars grown extensively in California are 'Chenin blanc,' 'Sauvignon blanc,' and 'Riesling.'

Another indicator of present trends is the increased importance of 'Cabernet Sauvignon.' Following 'Chardonnay,' 'Cabernet Sauvignon' is the most common cultivar (12.5%). Coverage by 'Zinfandel' is almost the same as 'Merlot,' about 10%. The enigmatic 'Pinot noir' constitutes about 2% of Californian wine grape hectarage. Other red cultivars include 'Rubired,' 'Grenache,' 'Barbera,' 'Sangiovese,' and 'Carignan.' Planting of vines has also increased fairly steadily since 1995.

The proportion of premium cultivars grown in coastal valley vineyards is significantly higher than for the state as a whole. This reflects the cooler environment of the coastal valleys and greater similarity of the climate to the European homeland of the cultivars.

As in Australia, extensive research has been directed toward improving grape quality. One aspect of this research has been the isolation of clones with distinctive flavor characteristics. Some producers are combining the must or wine from several clones to enhance wine complexity (Long, 1987). Such activity may produce, under precise control, some of the clonal diversity that has tended to exist in many European vineyards.

One of the few disappointing aspects of wine production in California, and in most other New World countries, is the inordinate space given a few grape varieties. Although several Spanish and Italian cultivars are grown widely, they are seldom used in varietally designated wines. Although essential to the quality of inexpensive wines, their anonymity prevents their receiving the recognition they deserve. This is probably one of the unfortunate legacies of the prevalent British view that French wines, or at least their cultivars, are superior to all others. Thus, as usual, it will depend on a few, dedicated, skilled visionaries to convince consumers and wine critics alike of the merits of other grape cultivars. Consumers deserve more variation in sensory stimulation than can be provided by the principal, so-called 'premium,' cultivars grown worldwide.

Pacific Northwest Of the two states separating California and Canada, Washington is the more significant in wine production. It possesses about 12,000 ha devoted to wine production, and equivalent coverage for juice production. This qualifies it as the third most important wine-producing state (2.6%). In comparison, Oregon comes in fourth in wine production from 5600 ha.

In Washington, the primary vineyard area is situated in the south-central region, approximately between 45° and 48°N latitude. The region encompasses the connecting valleys of the Columbia River and its tributaries, the Yakima and Snake Rivers. It is bounded on all sides by mountains – the Rocky Mountains to the east, the Cascades to the west, the Okanagan highlands to the north, and the Blue to the south. The mountains provide protection both from cold north and east winds and from moisture-laden winds from the west. The area possesses a dry, sunny climate, with much of the limited precipitation coming in winter as snow. The summers are warm with cool nights. The rapid decline in temperature in mid-September helps retain fruit acidity. Heat summation varies from 1220 to 1500 Celsius degree-days (Regions I to II). The soils are primarily sandy loams of various depths and are commonly underlain by a calcareous hardpan, typical of most dryland regions. Irrigation is typically necessary. Cold winter temperatures occasionally reach −25°C and the region may experience marked and precipitous temperature drops. Sites midway up slopes provide the optimal frost protection by draining cool air away, while avoiding the cold at the slope apex. Cool evenings, combined with stable sunny conditions throughout the growing season, favor full ripening, excellent color production, and the retention of ample acidity. South-facing orientations are preferred because of the extra spring and fall light received.

In the past, much of the grape culture in central Washington was destined for use in juice production, notably from 'Concord.' Since the late 1960s, plantings of *Vitis vinifera* cultivars have increased considerably. In areas experiencing the coldest winters, varieties possessing cold tolerance, such as 'Riesling,' 'Gewürztraminer,' 'Pinot noir,' and 'Chardonnay' are cultivated. 'Cabernet Sauvignon' and other longer-season cultivars grow better on warmer sites. Because phylloxera is absent, a major cost saving is achieved by avoiding the necessity of grafting.

A second viticultural region occurs in the area surrounding Seattle and extending to the Canadian border (~47–49°N). The temperate maritime climate of the Puget Sound area is markedly different from the arid conditions of the central and eastern parts of the state. Although freezing is seldom experienced, the cool moist climate (heat summation of 850–1050 Celsius degree-days) retards growth and grape ripening. Thus, short-season varieties such as 'Müller-Thurgau,' 'Madeleine Angevine,' and 'Okanagan Riesling' are preferred. The vineyard area is highly dispersed and small.

In Oregon, commercial viticulture exists in four regions – the southern portion of the Columbia River valley across from Washington, and a string of three valleys formed between the low Coast Range and Cascade Mountains. Of the valleys, the largest and most significant is the Willamette. It extends southward from the Washington border for about 280 km and can be 100 km wide. It spans a latitude between 44° and 46°N. The Umpqua Valley lies south of the Willamette Valley, from which it is separated by a semimountainous divide. An even smaller region situated within the Klamath Mountains is the Rogue Valley. It lies close to the Californian border.

The Willamette Valley receives on average 80–120 cm rain/year, but only about one-third falls during the growing season. Long sunny days during the spring and summer tend to compensate for the moderate to cool temperatures (heat summation of 1050–1250 Celsius degree-days), with from 165 to 210 frost-free days. South-facing slopes, between 120–220 m altitude, are preferred for the extra light and heat received. As with other cool climatic regions, the mid-slope region provides optimal frost protection, consistent with heat gain. Deep soils are preferred to provide protection against occasional summer droughts. Excellent drainage is necessary, however, to avoid excessive early shoot vigor in such sites.

The preferred cultivar is clearly 'Pinot noir.' It covers over 55% of the vineyard hectarage. 'Pinot gris' comes in a distant second, followed by 'Chardonnay,' and 'Riesling.' As the varietal coverage suggests, Oregon has been particularly successful with 'Pinot noir.' The wine is produced in a manner similar to that in Burgundy. This may involve early harvesting to retain sufficient acidity, even though it requires chaptalization. During fermentation, some whole clusters may be incorporated with the must, and the temperatures permitted to rise to 29°C. Fermentation usually is preferred in small cooperage, with gentle punching down rather than pumping over. Malolactic fermentation is favored by infrequent racking and the minimal addition of sulfur dioxide. There is considerable interest in the use of special strains of lactic acid bacteria to enhance flavor complexity during malolactic fermentation. Furthermore, some vineyards are planted at high densities (10,000 vines/ha). The narrow rows require the use of special machinery designed to work under such conditions (Adelsheim, 1988).

Oregon has the advantages of a climate that accentuates the varietal character of 'Pinot noir,' and a fairly consistent concept among producers on how the wine should be made. The combination has helped the state achieve rapid international recognition for its 'Pinot noir' wines. It is a good example of how a small region can quickly establish an identity in the crowded and competitive world wine market.

East of the Rockies The most significant wine-producing region after California is New York state. New York annually produces approximately 1 million hl wine from about 12,500 vineyard ha. This constitutes about 5% of America's wine production. Most eastern states, as well as those in the south and southwest, produce wine. Although the wines are of considerable local interest and pride, the production capacity is small and the wines are seldom found outside their home state. Following Washington and Oregon, New Jersey, Florida, Kentucky, Vermont, Texas and Ohio produce the most wine.

The northeastern states receive precipitation throughout the year. During the winter, snow cover often provides needed frost protection. However, high humidity during the growing season favors several severe fungal pathogens, notably downy mildew, and both black and bunch rots. In addition, the region is endemically infested with phylloxera and several insect pests absent west of the Rockies.

The southern and coastal states seldom suffer from vine-damaging frost conditions, but they are more humid. Humid conditions demand greater fungicide use, or require the cultivation of varieties derived from indigenous *Vitis* species, or varieties containing resistance genes. The bacterium *Xylella fastidiosa*, the causal agent of Pierce's disease, severely limits cultivation of *V. vinifera* cultivars in the southern and coastal states.

The central portion of the United States has little viticultural activity. Most of that is located either in the southwestern states of Texas, New Mexico, and Arizona, or in the east-central states of Michigan and Ohio. Viticulture in Wisconsin usually requires that the vines be laid down and covered with soil each winter. States like Colorado have abundant sunshine, low relative humidity, marked day/night temperature fluctuations, but frequently experience bitterly cold winters. Throughout the northern states, vines are commonly trained with two or more trunks. This minimizes the damage caused by death of one trunk by crown gall, Eutypa dieback, or other related problems.

One of the most distinctive features of the eastern wine industry is the tremendous diversity of cultivars. Because of repeated failures of *V. vinifera* to survive east of the Rockies, the industry developed using *V. labrusca* cultivars and hybrids in the northeast, *V. aestivalis* cultivars in the Midwest, and *V. rotundifolia* cultivars in the southeast. *V. labrusca* cultivars, such as 'Concord,' 'Niagara,' 'Catawba,' 'Isabella,' and 'Ives' are well adapted to the climate and prevalent pathogens and pests of the northeast. They produce good fortified and sparkling wines, which were the staple of the wine industry for more than a century. In the Midwest (Arkansas, Illinois, Kentucky, Missouri) and mid-Atlantic states (Maryland and Virginia) cultivars such as 'Norton' and 'Cynthiana' are popular (Dami *et al.*, 2001). This derives primarily from their cold-hardiness and withstanding the typical hot humid summers. In the South, the predominant cultivars are muscadine varieties.

With an increase in the popularity of dry wines, the northeastern states and adjacent Canada began to explore the use of French-American hybrids. These possessed many of the winemaking properties of *V. vinifera*, combined with some of the disease resistance of one or more indigenous American *Vitis* species. The success of the trials added cultivars like 'Maréchal Foch,' 'Baco noir,' 'Vidal blanc,' 'Seyval blanc,' and 'Aurora' to the list of commonly grown cultivars. Although widely cultivated, 'Aurora' plantings are on a decline. They are being replaced by varieties with better winemaking characteristics. Breeding programs in both New York state and the neighboring province of Ontario have produced new *V. vinifera* hybrids, such as 'Cayuga White' and 'Ventura,' respectively. In the Midwest, varieties such as the Swenson hybrid 'Frontenac' have proven successful and popular under severe winter conditions. In the southeast, renewed breeding work has generated new muscadine cultivars, such as 'Carlos,' 'Noble,' and 'Magnolia.' Further south, the cultivars 'Stover' and 'Suwannee' have been bred to the climate and disease situation in Florida.

To these new varieties are now being added *vinifera* cultivars from Europe. With appropriate rootstocks, pesticide application, site selection, training, and winter protection, European cultivars can now survive and prosper after more than two centuries of failure. Although the success is gratifying, the long-term benefit of *V. vinifera* cultivars to the wine industry in the eastern and central states is unclear. One short-term effect has been to direct interest away from hybrids inherently more suited to the local climate. The hybrid cultivars could give regional wines a distinctiveness that European cultivars cannot. If as much attention were given these cultivars, they would undoubtedly contribute significantly to the repute of the wines of eastern North America. However, wine critics and many producers are intent on trying to turn eastern North America into another Europe. This is not to imply that producing wines of superior caliber from French-American or newer hybrids will necessarily be any easier than with other grape varieties. They tend to have problems associated with excessive secondary shoot production, pronounced vegetative flavors (especially in young wine), and lower tannin levels. Nevertheless, the acceptance and prestige, derived from successfully cultivating familiar European varieties, may eventually encourage young winemakers to branch out and produce distinctively regional wines, possessing new and interesting flavors.

Wine production in New York is largely centered in the Finger Lakes region (~42°25′ to 42°50′N), nearly 100 km south of Lake Ontario. The vineyards are primarily situated on slopes adjacent to a series of narrow elongated lakes oriented north and south. Thus, the vines receive either an eastern or western exposure. Although such sites receive less light than south-facing slopes, this can be beneficial in delaying the loss of insulating snow cover and retarding premature bud burst in the spring. The lakes moderate temperature fluctuations, a feature especially important in the late winter and spring, when the vines are losing their cold acclimation. The marked slopes also facilitate drainage, early warming of the soils, and direct cold air away from the vines. The lakes act as a heat source and sink, helping to prolong autumn warmth and favor fruit ripening.

During the summer, cloud cover and precipitation can reduce the temperature maxima in the Finger Lakes. These factors also limit solar intensity and increase humidity. The Geneva Double Curtain (GDC) training system, developed in Geneva, New York, was designed primarily to counteract these effects (see Chapter 4). By opening the canopy and allowing the shoots to grow pendulously, the system greatly improved the degree and uniformity of fruit exposure to air and sun. In addition to increased fruit quality and health, vineyard yield was boosted due to improvement in leaf photosynthetic efficiency. In wine production, the region is predominantly a producer of white wine. 'Cayuga White' and 'Seyval blanc' are the most cultivated French-American cultivars, followed by 'Vignobles' and 'Vidal blanc.' There has also been considerable success with several *V. vinifera* cultivars, notably 'Riesling,' 'Chardonnay,' as well as 'Cabernet Sauvignon.'

Other increasingly significant wine regions in the Empire State occur about 150 km north of New York City, along the Hudson River, and on the northeastern branch of Long Island, east of New York City. The latter is somewhat unique in its almost exclusive use of *V. vinifera* cultivars, with about a 50:50 split between red and white varieties. The maritime climate of Long Island extends the growing season and raises the average winter temperature. However, the additional cloud cover delays ripening.

In Ohio, the vineyards are concentrated along the southern side and islands of Lake Erie. These are desirably close to the large population center of Cleveland. Similarly, the majority of Michigan vineyards are located near Chicago, along the southeastern portion of Lake Michigan.

In Virginia, most vineyards are situated in close proximity to Washington, D.C. This permits consumers to quickly reach farm wineries by car. Although avoiding the colder climate of more northern states, Virginia suffers erratic winter temperatures. These occasionally cause severe bud damage to *V. vinifera* and French-American hybrid cultivars. Otherwise, Virginia possesses a desirably mild climate, and occurs north of the natural distribution of Pierce's disease. Considerable interest has been shown in the viticultural potential of Virginia by Californian, New York, Canadian, and European wine enterprises.

Florida is the largest wine producer of southern states. However, because of its size and diverse climatic regions, Texas may soon become the major wine-producing area of the southeastern and Gulf States. The south Plains region around Lubbock possesses cool nights at its elevation of approximately 1000 m. The limited precipitation in this semiarid region is concentrated in the spring and early summer months, ideal for vine growth and to limit the need for irrigation. The deep, well-drained, rich soils permit excellent root penetration, favoring good nutrition and minimizing water stress. Premium French and German varieties are those preferred for cultivation.

CANADA

Although the wine industry of Canada is small on a world scale, it has considerable regional economic significance. Most Canadian production is located in southwestern Ontario and south-central British Columbia. The wine industry was forced into a major restructuring due to a General Agreement on Tariffs and Trade (GATT) ruling in 1987, and implementation of the Free-Trade Agreement with the United States in 1989. The lack of protective tariffs has made most wines based on *Vitis labrusca* and French-American hybrids unprofitable. For example, vineyard hectarage in British Columbia shrank from a high of 1375 ha in 1988 to 460 ha in 1989. The vineyards that have remained are planted primarily with *V. vinifera* cultivars, or have been replanted with them. This is reflected in an almost doubling of the *vinifera* hectarage between 1993 and

2005. French-American hybrids now cover less than one-quarter of the vineyard hectarage. In Ontario, the larger fresh fruit and juice market partially cushioned the effects of these changes. Nevertheless, the shift to *V. vinifera* from French-American hybrids and *V. labrusca* is also a major trend in Ontario vineyards. In both provinces, there is a marked preference for white cultivars, notably 'Riesling,' 'Chardonnay,' and 'Gewürztraminer.' For red wines, 'Cabernet Sauvignon,' 'Cabernet franc,' and 'Pinot noir' are favored.

Ontario possesses the largest vineyard area devoted to wine production (about 5700 ha). Most of this is located along the southwestern edge of Lake Ontario, between Hamilton and Niagara Falls (~43°N). The soils are deep, fertile, silt clay to sandy loam, underlaid by shale. Precipitation is relatively uniform throughout the year and averages 80 cm. The region is bounded on the south by the Niagara Escarpment, a prominent geologic feature that markedly affects the climate of the region (Fig. 5.12). Sites on 4–10% slopes, 4–6 km from the lake, are the most favored. Cold air drainage draws warmer air down from the temperature inversion layer that often develops during calm cold nights. The vines also are sufficiently far from the lake to avoid a marked chilling by the flow of cool air off the water during the summer. Although the north-facing slopes of the Escarpment limit sun exposure, it also delays bud burst, further minimizing the likelihood of frost damage in the spring. It shortens the growing season, however. The location of the vineyards between Lake Ontario and Lake Erie greatly cushions the effect of the otherwise continental climate. Long mild autumns usually supply ample time for the ripening of most short-season *V. vinifera* cultivars. Nevertheless, the region often experiences a hard freeze in late November or December. Although normally undesirable, it has favored the production of icewines, a style that is becoming an Ontario speciality. The other main region for wine production in Ontario is along the northern portion of Lake Erie.

On Canada's west coast, the vineyards (about 2700 ha) are considerably further north than those in Ontario. The vines in British Columbia grow primarily between 49° and 50°N, versus 43°N in Ontario. Nevertheless, dry, sunny conditions, cold protection provided by the surrounding mountains, and the influence of the Pacific coast provides the Okanagan Valley with a moderate climate. Semiarid conditions make irrigation necessary in most locations. The vineyards are located primarily on slopes and plateaus lining a series of narrow elongated lakes in the Okanagan River Valley. Vines also grow along the Similkameen River, a tributary joining the Okanagan, about 25 km from the US border. Although the southern portion of the Okanagan Valley is warmer, its drier environment provides less snow cover. Thus, the vines are about as vulnerable to cold damage as in the northern portions of the valley. Nonetheless, about 60% of the grape production occurs in the south, between Penticton (49°30′N) and the border with Washington state (49°00′N).

Nova Scotia, on the eastern coast of Canada, possesses several vineyards and local wineries (~45°N). Because the province is almost entirely surrounded by water, the vines are exposed to a maritime climate. Both the Bay of Fundy and Atlantic Ocean moderate continental influences from the west. These same influences also retard early bud break. The typically long autumn helps to compensate for delayed spring growth, permitting the ripening of short-season cultivars. In addition to familiar French-American hybrids, the region also grows several new German *Vitis vinifera* cultivars. Particularly interesting is the cultivation of *V. vinifera* [×] *V. amurensis* cultivars from Russia, notably 'Michurinetz' and 'Severnyi.'

Québec has a small wine industry (~100 h) located largely in the southwestern portion of the province, adjacent to Vermont and New York. Because of the short growing season and cold climate, most cultivation involves French-American hybrids, notably 'Seyval blanc' and some local hybrids. Some red varieties are grown and produce light-red to rosé wines. Even with these varieties, the vines must be cropped low and protected by partial burial during the winter. This keeps production low and adds to production costs. Judicious selection of cultivars, site, slope, drainage and wind breaks is critical to improving the chances of commercial viability in such marginal viticultural regions.

Suggested Readings

Introductory Books

Clarke, O. (1995) *Wine Atlas – Wines and Wine Regions of the World.* Little-Brown, Boston, MA.

Johnson, H., and Robinson, J. (2001) *The World Atlas of Wine*, 5th edn. Simon & Schuster, New York.

Stevenson, T. (2005) *The Sotheby's Wine Encyclopedia*, 4th edn. Dorling Kindersley, London.

Appellation Control Laws

Dickenson, J. P. (1992) 'Aristocratic' versus 'popular' wine: Trends in the geography of viticulture. In: *Viticulture in Geographic Perspective: Proceedings of the 1991 Miami AAG Symposium* (H. J. de Blij, ed.), pp. 23–43. Miami Geographical Society, Coral Gables, FL.

Moran, W. (1988) The wine appellation, environmental description or economic device? In: *Proceedings of the 2nd International Symposium for Cool Climate Viticulture and Oenology* (R. E. Smart *et al.*, eds.), pp. 356–360. New Zealand Society for Viticulture and Oenology, Auckland, New Zealand.

O. I. V. (1987) *Symposium Les Appellations d'Origine Historiques.* Office International de la Vigne et du Vin, Paris.

Detection of Wine Misrepresentation and Adulteration

Capron, X., Smeyers-Verbeke, J., and Massart, D. L. (2007) Multivariate determination of the geographical origin of wines from four different countries. *Food Chem.* **101**, 1585–1597.

Kelly, S., Heaton, K., and Hoogewerff, J. (2005) Tracing the geographical origin of food: the application of multi-element and multi-isotope analysis. *Trends Food Sci. Technol.* **16**, 555–567.

Martin, G. J., and Martin, M. L. (1988) The site-specific natural isotope fractionation-NMR method applied to the study of wines. In: *Wine Analysis* (H. F. Linskens and J. F. Jackson, eds.), pp. 258–275. Springer-Verlag, Berlin.

Martin, G. J., Thibault, J.-N., and Bertrand, M.-J. (1995) Spatial and temporal dependence of the ^{13}C and ^{14}C isotopes of wine ethanols. *Radiocarbon* **37**, 943–954.

Médina, B. (1996) Wine authenticity. In: *Food Authentication* (P. R. Ashurst and Dennis, M. J., eds.), pp. 61–107. Blackie Academic, London.

Geographical Regions

General

de Blij, H. J. (1983) *Wine: A Geographic Appreciation.* Rowman & Allanheld, Totowa, NJ.

de Blij, H. J. (1985) *Wine Regions of the Southern Hemisphere.* Rowman & Allanheld, Totowa, NJ.

de Blij, H. J. (ed.) (1992) *Viticulture in Geographical Perspective, Proceedings of the 1991 Miami AAG Symposium.* University of Miami, Coral Gables, FL.

Dickenson, J., and Unwin, T. (1992) *Viticulture in Colonial Latin America, Essays on Alcohol, the Vine and Wine in Spanish America and Brazil.* Institute of Latin American Studies, University of Liverpool, Liverpool, UK.

Fegan, P. W. (2004) *The Vineyard Handbook: Appellation, Maps and Statistics.* Chicago Wine School Inc., Chigago, IL.

Gladstone, J. (1992) *Viticulture and Environment.* Winetitles, Adelaide, Australia.

Prescott, J. A. (1965) The climatology of the vine (*Vitis vinifera* L.), the cool limits of cultivation. *Trans. R. Soc. South Aust.* **89**, 5–22.

Argentina

Catania, C., and de del Monte, S. A. (1986) Détermination des aptitudes oenologiques des différents cépages dans la républic Argentine. In: *19th International Viticulture and Oenology Congress,* Vol. 2, pp. 17–33. Office International de la Vigne et du Vin, Paris.

Australia

Coombe, B. G., and Dry, P. T. (eds.) (1988) *Viticulture,* Vol. I (*Resources in Australia*). Australian Industrial Publishers, Adelaide, South Australia.

Helm, F. K., and Cambourne, B. (1988) The influence of climatic variability on the production of quality wines in the Canberra district of South Eastern Australia. In: *Proceedings of the 2nd International Symposium for Cool Climate Viticulture and Oenology* (R. E. Smart, *et al.*, eds.), pp. 17–20. New Zealand Society for Viticulture and Oenology, Auckland, New Zealand.

Brazil

Lakatos, T. (1995) Grape growing and winemaking in Tropical Brazil. *J. Small Fruit Vitic.* **3**, 3–14.

Canada

Anonymous (1984) *Atlas of Suitable Grape Growing Locations in the Okanagan and Similkameen Valleys of British Columbia.* Agriculture Canada & Association of B. C. Grape Growers, Kelowna, Canada.

Deshaies, L., and Dubois, J.-M. M. (eds.) (1993) Vins et vignoles artisanaux au Québec. *Géographes* **4**, 5–110.

Sayed, H. (1992) *Vineyard Site Suitability in Ontario.* Ministry of Agriculture and Food Ontario & Agriculture Canada, Queen's Printer for Ontario, Toronto, Canada.

Chile

Santibañez, M. M. F., Diaz, F., Gaete, C., and Daneri, D. (1986) Bases climatiques pour le zonage de la région viti-vinicole chilienne. In: *19th International Viticulture and Oenology Congress,* Vol. 2, pp. 93–124. Office International de la Vigne et du Vin, Paris.

China

Hua, L. (1990) Les indications géographiques des vins et leur protection en Chine. *Bull. O.I.V.* **63**, 282–287.

Huang, S. B. (1990) Agroclimatology of the major fruit production in China: A review of current practice. *Agr. Forest Met.* **53**, 125–142.

England

Pearkes, G. (1984) England. In: *International Symposium on Cool Climate Viticulture and Enology* (D. A. Heatherbell *et al.*, eds.), OSU Agriculture Experimental Station Technical Publication No. 7628, pp. 347–357. Oregon State University, Corvallis, OR.

France

Boursiquot, J. M. (1990) Évolution de l'encépagement du vignoble français au cours des trente dernières années. *Prog. Agric. Vitic.* **107**, 15–20.

duPuy, P. (1984) Endeavors to produce quality in the wines of Burgundy: Control of soil, clonal selection, maturity, processing techniques and commercial practices. In: *International Symposium on Cool Climate Viticulture and Enology* (D. A. Heatherbell *et al.*, eds.), OSU Agriculture Experimental Station Technical Publication No. 7628, pp. 292–314. Oregon State University, Corvallis, OR.

Pomeral, C. (ed.) (1989) *The Wines and Winelands of France, Geological Journeys.* Éditions BRGM, Orléans, France.

Wilson, J. T. (1998) *Terroir: The Role of Geology, Climate and Culture in the Making of French Wines.* University of California Press Ltd., Berkeley, CA.

Germany

Anonymous (1979) *The Wine Industry in the Federal Republic of Germany.* Evaluation and Information Services for Food, Agriculture and Forestry, Bonn, Germany.

Wahl, K. (1988) Climate and soil effects on grapevine and wine. The situation on the northern borders of viticulture – the example Franconia. In: *Proceedings of the 2nd International Symposium for Cool Climate Viticulture and Oenology* (R. E. Smart *et al.*, eds.), pp. 1–5. New Zealand Society for Viticulture and Oenology, Auckland, New Zealand.

Italy

Anderson, B. (1980) *Vino – the Wines and Wine Makers of Italy.* Little-Brown, Boston, MA.

Hazan, V. (1982) *Italian Wine.* Alfred A. Knopf, New York.

India

Shanmugavelu, K. G. (1989) *Viticulture in India.* Agro Botanical Publishers, Bikaner, India.

New Zealand

Eschenbruch, R. (1984) New Zealand. In: *International Symposium on Cool Climate Viticulture and Enology* (D. A. Heatherbell *et al.*, eds.), OSU Agriculture Experimental Station Technical Publication No. 7628, pp. 345–346. Oregon State University, Corvallis, OR.

Portugal

Gonçalves, F. E. (1984) *Portugal – A Wine Country.* Editora Portuguesa de Livros Técnicos e Científicos, Lisbon.

Stanislawski, D. (1970) *Landscapes of Bacchus, the Vine in Portugal.* University of Texas Press, Austin.

Russia

Batukaev, A. (1995) La situation de la viticulture dans la Fédération de Russie. *Bull. O.I.V.* 68, 615–623.

Titov, A. P., and Djeneev, S. U. (1991) Viticulture et oenologie de l'U.S.S.R. *Bull. O.I.V.* 64, 575–583.

Spain

García de Luján y Gil de Bernabé, A. (1997) *La Viticultura del Jerez.* Ediciones Mondi-Prensa. Madrid.

Ruiz Hernández, M. (1985) *Elementos Técnicos para Elaboraciones de Vino de Rioja.* Estación de Viticultura y Enología, Consejería de Agricultura, Haro, Spain.

United States

Baxevanis, J. J. (1992) *The Wine Regions of America, Geographical Reflections and Appraisals.* Vinifera Wine Growers Journal, Stroudsburg, PA.

Elliott-Fisk, D. L., and Noble, A. C. (1992) *Environments in Napa Valley, California, and their influence on Cabernet Sauvignon*

wine flavors. In:*Viticulture in Geographic Perspective, Proceeding of the 1991 Miami AAG Symposium* (H. J. de Blij, ed.), pp. 45–71. Miami Geographical Society, Coral Gables, FL.

Morton, L. T. (1985) *Winegrowing in Eastern America.* Cornell University Press, Ithaca, NY.

Olien, W. C. (1990) The Muscadine grape: Botany, viticulture, history and current industry. *HortScience* 25, 732–739.

Templer, O. W. (1992) The southern High Plains: focal point of the Texan growing industry. In: *Viticulture in Geographic Perspective*: *Proceeding of the 1991 Miami AAG Symposium* (H. J. de Blij, ed.), pp. 97–110. Miami Geographical Society, Coral Gables, FL.

Uruguay

Carrau, F. M. (1997) The emergence of a new Uruguayan wine industry. *J. Wine Res.* 8, 179–185.

References

Adelsheim, D. (1988) Oregon experiences with Pinot noir and Chardonnay. In: *Proceedings of the 2nd International Symposium for Cool Climate Viticulture and Oenology* (R. E. Smart *et al.*, eds.), pp. 264–269. New Zealand Society for Viticulture and Oenology, Auckland, New Zealand.

Agüero, C. B., Rodríguez, J. G., Martínez, L. E., Dangle, G. S., and Meredith, C. P. (2003) Identity and parentage of Torrontés cultivars in Argentina. *Am. J. Enol. Vitic.* 54, 318–321.

Almeida, C. M. R., and Vasconcelos, T. S. D. (2004) Does the winemaking process influence the wine $^{87}Sr/^{86}Sr$? A case study. *Food Chem.* 85, 7–12.

Anonymous (1979) *The Wine Industry in the Federal Republic of Germany.* Evaluation and Information Service for Food, Agriculture and Forestry, Bonn, Germany.

Anonymous (2006) New Zealand Winegrowers Statistical Annual 2006. http://www.nzwine.com/statistics.

Arrhenius, S. P., McCloskey, L. P., and Sylvan, M. (1996) Chemical markers for aroma of *Vitis vinifera* var. Chardonnay regional wines. *J. Agric. Food. Chem.* 44, 1085–1090.

Basler, P., and Wiederkehr, M. (1996) Considering cultivar selection as a main element in sustainable viticulture. In: *Proceedings of the 4th International Symposium on Cool Climate Viticulture and Enology* (T. Henick-Kling *et al.*, eds.), pp. I-5–9. NY State Agricultural Experimental Station, Geneva, New York.

Baxter, M. J., Crews, H. M., Dennis, M. J., Goodall, I., and Anderson, D. (1997) The determination of the authenticity of wine from its trace element composition. *Food Chem.* 60, 443–450.

Becker, N. (1977) The influence of geographical and topographical factors on the quality grape crop. In: *International Symposium on Quality in the Vintage*, pp. 169–180. Oenology and Viticulture Research Institute, Stellenbosch, South Africa.

Bonfiglioli, R. G., Habili, N., Green, M., Schliefert, L. F., and Symons, R. H. (1999) The hidden problem – *Rugose* wood associated viruses in Australian viticulture. *Aust. Grapegrower Winemaker* 420, 9–13.

Borrego, J., de Andrés, M. T., Gómex, J. L., and Ibáñex, J. (2002) Genetic study of Malvasia and Torrontes groups through molecular markers. *Am. J. Enol. Vitic.* 53, 125–130.

Boursiquot, J. M. (1990) Évolution de l'encépagement du vignoble français au cours des trente dernières années. *Prog. Agric. Vitic.* 107, 15–20.

Branas, J., Bernon, G., and Levadoux, L. (1946) *Eléments de Viticulture Générale*. L'École Nationale de Agriculture, Montpellier, France.

Bucelli, P. (1991) Il governo del vino all'uso del Chianti, note storiche e aspetti tecnici. *Vini d'Italia* 33(4), 63–70.

Burchuladze, A. A., Chudý, M., Eristavi, I. V., Pagava, S. V., Povinec, P., Šivo, A., and Togonidze, G. I. (1989) Anthropogenic ^{14}C variations in atmospheric CO_2 and wines. *Radiocarbon*. 31, 771–776.

Burda, K., and Collins, M. (1991) Adulteration of wine with sorbitol and apple juice. *J. Food Protect*. 54, 381–382.

Cilliers, J. J. L., and Singleton, V. L. (1990) Nonenzymatic autooxidative reactions of caffeic acid in wine. *Am. J. Enol. Vitic*. 41, 84–86.

Dami, I., Eberle, P., and Brown, M. R. (2001) *Illinois Vineyard Survey in 2000*. Southern Illinois University, Carbondale, IL.

de Blij, H. J. (1983) *Wine: A Geographic Appreciation*. Rowman & Allanheld, Totowa, NJ.

de Blij, H. J. (1985) *Wine Regions of the Southern Hemisphere*. Rowman and Allanheld, Totowa, NJ.

Douglas, D., Cliff, M. A., and Reynolds, A. G. (2001) Canadian terroir: sensory characterization of Riesling in the Niagara Peninsula. *Food Res. Int*. 34, 559–563.

Dovaz, M. (1983) *L'Encyclopédie des Vins de Champagne*. Julliard, Paris.

Ebeler, S. E., Sun, G. M., Datta, M., Stremple, P., and Vickers, A. K. (2001) Solid-phase microextraction for the enantiomeric analysis of flavors in beverages. *J. AOAC Int*. 84, 479–485.

Esteban, M. A., Villanueva, M. J., and Lissarrague, J. R. (1999) Effect of irrigation on changes in berry composition of Tempranillo during maturation. Sugars, organic acids, and mineral elements. *Am. J. Enol. Vitic*. 50, 418–434.

Etiévant, P., Schlich, P., Bertrand, A., Symonds, P., and Bouvier, J.-C. (1988) Varietal and geographic classification of French red wines in terms of pigments and flavonoid compounds. *J. Sci. Food Agric*. 42, 39–54.

Fauhl, C., and Wittkowski, R. (2000) Oenological influences on the D/H ratios of wine ethanol. *J. Agric. Food Chem*. 48, 3979–3984.

Franco, D. S., and Singleton, V. L. (1984) The changes in certain components of Setubal wines during aging. *Am. J. Enol. Vitic*. 35, 146–150.

Fregoni, M. (1991) *Origines de la Vigne et de la Viticulture*. Musumeci Editeur, Quart (Vale d'Aosta), Italy.

Fregoni, M. (1992) The new millennium opens for the wines of Italy. *Ital. Wines Spirits* 16(1), 9–18.

Gastin, D. (2006) The grape wall of China. *Aust NZ Grapegrower Winemaker* 508, 72, 74, 76.

Giraudon, S. (1994) Détection de la chaptalisation des vins: Constitution d'une base de données. *J. Intl. Sci. Vigne Vin* 28, 47–55.

Guzun, N. I., Nedov, P. N., Usatov, V. T., Kostrikin, I. A., and Meleshko, L. F. (1990) Grape selection for resistance to biotic and abiotic environmental factors. In: *Proceedings of the 5th International Symposium on Grape Breeding*, pp. 219–222 (Special issue of *Vitis*). St Martin, Pfalz, Germany.

Hayes, W. C. (1951) Inscriptions from the palace of Amenhotep III. *J. Near East. Stud*. 10, 35–40, 82–104.

Hua, L. (1990) Les indications géographiques des vins et leur protection en Chine. *Bull. O.I.V*. 63, 282–287.

Hubáčková, M., and Hubáček, V. (1984) Frost resistance of grapevine buds on different rootstocks (in Czech). *Vinohrad* 22, 55–56.

Intrieri, C., and Filippetti, I. (2000) Planting density and physiological balance: comparing approaches to European viticulture in the 21st century. In: *Proceedings of the ASEV 50th Anniversary Annual Meeting*, Seattle, Washington, June 19–23, 2000, pp. 296–308. American Society for Enology and Viticulture, Davis, CA.

Johnson, H. (1994) *The World Atlas of Wine*, p. 22. Simon & Schuster, New York.

Jourjon, F., Morlat, R., and Seguin, G. (1991) Caractérisation des terroirs viticoles de la moyenne vallée de la Loire. Parcelles expérimentales, climat, sols et alimentation en eau de la vigne. *J. Intl Sci. Vigne Vin* 25, 179–202.

Kirillov, A. F., Levit, T. K., Skurtul, A. M., Koz'mik, R. A., Grozova, V. M., Syli, V. N., Khanin, Y. D., Baryshok, V. P., Semenova, N. V., and Voronkov, M. G. (1987) Enhancement of the cold hardiness in grapevine as affected by cryoprotectors (in Russian). *Sadovod. Vinograd. Mold*. 42, 36–39.

Kosuge, B. (1999) Tannins, color and flavor maturity in California Pinot noir – the winemaker's perspective. In: *RAVE 99 2nd Joint Burgundy–California–Oregon Winemaking Symposium*, pp. 30–32 (C. Butzke, P. Durnad, M. Feuillat, and B. Watson, Coordinators), Department of Viticulture and Enology, University of California, Davis. Lampe, U., Kreisel, A., Burkhard, A., Bebiolka, H., Brzezina, T., and Dunkel, K. (1997) Zum Nachweis eines Glycerinzusatzes zu Wein. *Deut. Lebens.-Rundschau* 93, 103–110.

Laville, P. (1990) Le terroir, un concept indispensable à l'élaboration et à la protection des appellations d'origine comme à la gestion des vignobles: le cas de la France. *Bull. O.I.V*. 63, 217–241.

Le Roux, E. (1974) Kilmaatstreek Indeling van die Suidwes Kaaplandse Wynbougebiede. Master's thesis, University of Stellenbosch, South Africa.

Le Roy Ladurie, E. (1971) *Times of Feast, Times of Famine. A History of Climate since the Year 1000*. Doubleday, Garden City, NY.

Long, Z. R. (1987) Manipulation of grape flavour in the vineyard, California, north coast region. In: *Proceedings of the 6th Australian Wine Industry Technical Conference* (T. Lee, ed.), pp. 82–88. Australian Industrial Publishers, Adelaide, Australia.

Luft, G., and Morgenschweis, G. (1984) Zur Problematik großterrassierter Flurbereinigung im Weinbaugebiet des Kaiserstuhls. *Z. Kulturtech. Flurbereinig*. 25, 138–148.

Luo, G. (1986) Dragon system of training and pruning in China's viticulture. *Am. J. Enol. Vitic*. 37, 152–157.

Marquet, P. (1987) L'importance des facteurs naturels et humains dans le développement des appellations d'origine françaises. In: *Symposium Les Appellations d'Origine Historiques*, pp. 33–41. Office International de la Vigne et du Vin, Paris.

Martí, M. P., Busto, O., and Guasch, J. (2004) Application of a headspace mass spectrometry system to the differentiation and classification of wines according to their origin, variety and aging. *J. Chromatography A* 1057, 211–217.

Martin, G. J. (1988) Les applications de la mesure par résonance magnétique nucléaire du fractionnement isotopique naturel spécifique (RMN-Fins) en viticulture et en oenologie. *Rev. Fr. Oenol*. 114, 23–24, 53–60.

Martin, G. J., Nicol, L., Naulet, N., and Martin, M. L. (1998) New isotopic criteria for the short-term dating of brandies and spirits. *J. Sci. Food Agric*. 77, 153–160.

Masneuf-Pomarède, I., Mansour, C., Murat, M. L., Tominaga, T., and Dubourdieu, D. (2006) Influence of fermentation temperature on volatile thiols concentrations in Sauvignon blanc wines. *Int. J. Food Microbiol*. 108, 385–390.

Mayer, G. (1990) Results of cross-breeding. In: *Proceedings of the 5th International Symposium on Grape Breeding*, pp. 148 (special issue of *Vitis*). St Martin, Pfalz, Germany.

McCallum, N. K., Rothbaum, H. P., and Otlet, R. L. (1986) Detection of adulteration of wine with volatile compounds. *Food Technol Aust*. 38, 318–319.

McCloskey, L.P., Sylvan, M., and Arrhenius, S.P. (1996) Descriptive analysis for wine quality experts determining appellations by Chardonnay wine aroma. *J. Sensory Stud*. 11: 49–67.

Mériaux, S., Chrétien, J, Vermi, P., and Leneuf, N. (1981) La Côte Viticole, ses sols et ses crus. *Bull. Sci. Bourgogne* **34**,17–40.

Messini, A., Vincenzini, M., and Materassi, R. (1990) Evolution of yeasts and lactic acid bacteria in Chianti wines as affected by the refermentation according to Tuscan usage. *Ann. Microbiol.* **40**, 111–119.

Miranda-Lopez, R., Libbey, L. M., Watson, B. T., and McDaniel, M. R. (1992) Odor analysis of Pinot noir wines from grapes of different maturities by a gas-chromatographic-olfactometery technique (Osme). *J. Food Sci.* **57**, 985–993.

Moran, W. (1988) The wine appellation, environmental description or economic device? In: *Proceedings of the 2nd International Symposium for Cool Climate Viticulture and Oenology* (R. E. Smart *et al.*, eds.), pp. 356–360. New Zealand Society for Viticulture and Oenology, Auckland, New Zealand.

Morrot, G. (2004) Cognition et vin. *Rev. Oenologues* **111**, 11–15.

Nicolau, L., Benkwitz, F., Tominaga, T. (2006) Characterising the aroma of New Zealand Sauvignon Blanc. *Aust. NZ Grapegrower Winemaker* **509**, 46–49.

Noble, A. C., Williams, A. A., and Langron, S. P. (1984) Descriptive analysis and quality ratings of 1976 wines from four Bordeaux communes. *J. Sci. Food Agric.* **35**, 88–98.

OIV (2005) Situation and statistics of the world vitiviniculture sector. http://news.reseau-concept.net/images/oiv_uk/Client/Stat_2003_EN.pdf

Penning-Rowsell, E. (1985) *The Wines of Bordeaux*, 5th edn. Wine Appreciation Guild, San Francisco, CA.

Pocock, K. F., and Somers, C. (1989) Detection of wine adulteration by cider. *Aust. NZ Wine Ind. J.* **4**, 302–303.

Pouget, R. (1988) L'encépagement des vignobles français d'appellation d'Origine contrôlée: historique et possibilités d'évolution. *Bull. O.I.V.* **61**, 185–195.

Rankine, B. C., Kepner, R. E., and Webb, A. D. (1958) Comparison of anthocyanin pigments of vinifera grapes. *Am. J. Enol. Vitic.* **9**, 105–110.

Rapp, A., Mandery, H., and Heimann, W. (1983) Flüchtige Inhaltsstoffe aus Flüssigzucker. *Vitis* **22**, 387–394.

Rapp, A., Volkmann, C., and Niebergall, H. (1993) Untersuchung flüchtiger Inhaltsstoffe des Weinaromas, Beitrag zur Sortencharakterisierung von Riesling und Neuzüchtungen mit Riesling-Abstammung.*Vitis* **32**, 171–178.

Reynolds, A. G. (2000) Impact of trellis/training systems and cultural practices on production efficiency, fruit composition, and vine balance. In: *Proceedings of the ASEV 50th Anniversary Annual Meeting*, Seattle, Washington June 19–23, 2000, pp. 309–317. American Society for Enology and Viticulture, Davis, CA.

Ribéreau-Gayon, J., Peynaud, E., Sudraud, P., and Ribéreau-Gayon, P. (1982) Recherche des fraudes. In: *Traité d'Oenologie. Sciences et Techniques du Vin* (J. Ribéreau-Gayon *et al.*, eds.), 2nd edn, Vol. I, pp. 575–617. Dunod, Paris.

Robertson, J. M., and Rush, G. M. (1979) Chemical criteria for the detection of winemaking faults in red wine. *Food Technol. NZ* **14**, 3–11.

Roßmann, A., Schmidt, H.-L., Hermann, A., and Ristow, R. (1998) Multielement stable isotope ratio analysis of glycerol to determine its origin in wine. *Z. Lebensm. Unters. Forsch.* **207**, 237–243.

Rougetet, E. (1930) Le mistral dans les plaines du Rhône Moyen entre Bas-Dauphiné et Province. *Météorologie* **6**, 341–385.

Ruhl, E. H. (1990) Better rootstocks for wine grape production. *Aust. Grapegrower Winemaker* **304**, 113–115.

Rybintsev, V. (1995) Viticulture and oenology in the Ukraine: A survey of their history, ecology and economic development. *J. Wine Res.* **6**, 35–47.

Schlosser, J., Reynolds, A. G., King, M., and Cliff, M. (2005) Canadian terroir: sensory characterization of Chardonnay in the Niagara Peninsula. *Food Res. Int.* **38**, 11–18.

Seeman, J., Chirkov, Y. I., Lomas, J., and Primault, B. (1979) *Agrometeorology*. Springer-Verlag, Berlin.

Seguin, G. (1986) Terroirs and pedology of wine growing. *Experientia* **42**, 861–873.

Shackleton, M. R. (1958) *Europe: A Regional Geography*, 6th edn. Longmans, London.

Stacey, R. J. (1984) Isotopic analysis, its application in the wine trade. In: *Tartrates and Concentrates – Proceedings of the Eighth Wine Subject Day* (F. W. Beech, ed.), pp. 45–51. Long Ashton Research Station, University of Briston, Long Ashton, UK.

Stanislawski, D. (1970) *Landscapes of Bacchus, the Vine in Portugal*. University of Texas Press, Austin, TX.

Sudraud, P. (1978) Evolution des taux d'acidité volatile depuis le début du siècle. *Ann. Technol. Agric.* **27**, 349–350.

Tapia, A. M., Cabezas, J. A., Cabello, F., Lacombe, T., Martínez-Zapater, J. M., Hinrichsen, P., and Cervera, M. T. (2007) Determining the Spanish origin of representative ancient American grapevine varieties. *Am. J. Enol. Vitic.* **58**, 242–251.

Tardaguila, J., Bertamini, M., Reniero, F., and Versini, G. (1997) Oxygen isotope composition of must-water in grapevine: effects of water deficit and rootstock. *Aust. J. Grape Wine Res.* **3**, 84–89.

This, P., Lacombe, T., and Thomas, M. R. (2006) Historical origins and genetic diversity of wine grapes. *Trends Genet.* **22**, 511–519.

Tusseau, D., Valade, M., Virion, M. C., and Moncomble, D. (1993) Controle de l'origine et de la nature des vins de champagne. *Vigneron Champenois* **114**(4), 8–21.

US Department of Agriculture (1978) *Soil Survey of Napa County, California*. Soil Conservation Service, Napa, CA.

van Buren, J. P., Bertino, J. J., Einset, J., Remaily, G. W., and Robinson, W. B. (1970) A comparative study of the anthocyanin pigment composition in wines derived from hybrid grapes. *Am. J. Enol. Vitic.* **21**, 117–130.

van Leeuwen, C., Friant, P., Choné, X., Tregoat, O., Koundouras, S., and Dubourdieu, D. (2004) Influence of climate, soil and cultivar on terroir. *Am. J. Enol. Vitic.* **55**, 207–217.

van Zyl, J. L., and van Huyssteen, L. (1980) Comparative studies on wine grapes on different trellising systems, I. Consumptive water use. *S. Afr. J. Enol. Vitic.* **1**, 7–14.

Vernin, G., Boniface, C., and Metzger, J. (1989) Les constituants volatils de l'arôme de la Syrah, classification statistique des vins de Syran d'origine differente. *Rev. Fr. Oenol.* **29**, 12–20.

Weber, D., Roßmann, A., Schwarz, S., and Schmidt, H. L. (1997) Correlations of carbon isotope ratios of wine ingredients for the improved detection of xterterations, I. Organic acids and ethanol. *Food Res. Technol.* **205**, 158–164.

Winton, W., Ough, C. S., and Singleton, V. L. (1975) Relative distinctiveness of varietal wines estimated by the ability of trained panelists to name the grape variety correctly. *Am. J. Enol. Vitic.* **26**, 5–11.

Zoppi, U., Skopec, Z., Skopec, J., Jones, G., Fink, D., Hua, Q., Jacobsen, G., Tuniz, C., and Williams, A. (2004) Forensic applications of ^{14}C bomb-pulse dating. *Nucl. Inst. Meth. Phys. Res. B.* **223–224**, 770–775.

11

Sensory Perception and Wine Assessment

Visual Sensations

Color

The visual characteristics of a wine depend on how its chemical and particulate nature transmit, absorb, and reflect visible radiation. Although some of these characteristics can be accurately measured with a spectrophotometer (Fig. 8.18), the relevance of the data to human color perception is far from direct. Spectrophotometric measurements assess the intensity of individual wavelengths, whereas the eye responds to the reflective and transmissive properties of light and its relative brightness. This involves a combination of reactions from two types of receptor neurons (cones and rods), as well as several other receptors. The cones adjust quickly to changing light intensities and quality. They generate color and high resolution vision. Cones exist in three forms, L, M, and S. They respond differentially, but over a broad range, to portions of the visible spectrum.

They respectively have peak sensitivities in the yellow, green, and violet. In contrast, the rods adjust more slowly to adjusting light levels, are receptive to low intensity light of medium and short wavelengths, are primarily responsible for motion detection, and generate night vision. The perception of color is complicated as it involves interpretation not only of impulses from the cones, but also information from other receptors concerning light intensity and contrast between colored regions. Consequently, there is no simple relationship between spectrophotometric measurements and human color perception (see Kaiser, 1996).

The CIE (Commission Internationale de l'Eclairage) system is frequently used to measure wine color (CIE, 1986). A simple method, proposed by the OIV (1990), measures light transmittance at four wavelengths (445, 495, 550, and 625 nm). From these data, tristimulus values (X, Y, and Z) are derived. Tristimulus colorimeters, in contrast to spectrophotometers, directly correlate these values in terms of color, hue and depth. While adequate with lighter wines, serious deviations develop with deeply colored wines. Several researchers have proposed changes to improve its applicability to wine (Ayala *et al.*, 1997). Continuing difficulties in measuring color variability are discussed in Huertas *et al.* (2003) and Pridmore *et al.* (2005).

Other than the pleasure color can give, observation of a wine's color yields little precise information to the consumer or judge. It provides only a rough indication of grape pigmentation, the potential duration of skin contact, probable wine age, and the presence or absence of a few wine faults. Even here, caution is required to avoid being influenced unjustly, especially if wines of differing ages, or winemaking procedures, are assessed together. Nevertheless, color often influences the perception of wine quality (Tromp and van Wyk, 1977; Williams *et al.*, 1984). For red wines, there may be justification for this association. Assessed wine quality is often correlated to color density (Iland and Marquis, 1993) and hue (the proportion of "ionized" anthocyanins and other pigments) (Somers and Evans, 1974; Bucelli and Gigliotti, 1993). Color can indicate if the wine was well made (at an appropriate pH, low in SO_2, and at an adequate ethanol concentration), and suggest high flavor (varietal aromatics, located primarily in the skins, are more likely to be extracted in wines left sufficiently long on the skins). Nevertheless, the interaction between wine color (and total soluble solids) and perceived wine quality is not simple (Gishen *et al.*, 2002). Because wine color can so influence perception, wines are occasionally sampled in black wine glasses, or under red lights, to negate the potential of color to unduly bias wine evaluation.

The influence of color on perception has been long known. One of the first studies on this phenomenon

was conducted by André *et al.* (1970). In the investigation, rosé wines were ranked by color alone, tasted with the color visible, and tasted blind. The results from the first two conditions were essentially identical. However, when visual cues were absent, ranking was markedly different. The preferred wines, when color was visible, were the least preferred when color was hidden. More recently, the influence of color on perception has been demonstrated by Morrot *et al.* (2001). In this study, adding anthocyanin pigments to a white wine induced tasters to describe the wine in terms typically used to describe red wines. The biasing influence of color on wine perception has been studied under nonlaboratory conditions by Delwiche (2003). Parr *et al.* (2003) found the influence of color was less marked on "expert" tasters than on "novice" tasters, at least when the disparity between color and other sensory attributes was obvious to the expert tasters.

The depth of color of a sample apparently enhances the intensity with which its odors are perceived, regardless of the appropriateness of the color (Zellner and Whitten, 1999). The biasing effect of color has also been demonstrated on the neuronal level. Österbauer *et al.* (2005) have shown that simultaneously showing a color, typically associated with a particular fragrance (for example, red with strawberry), enhances the response in the orbitofrontal complex. This is the cerebral region known to integrate sensory impulses. In contrast, an inappropriate color (blue with strawberry) depresses the response in the orbitofrontal complex. This influence also appears to apply for the anterior insula cortex, principally associated with primary taste sensations.

These influences are examples of the phenomenon where the *context* of an experience strongly influences the interpretation of sensory inputs (see Palmer, 1999). Thus, with experience, people tend to learn associations between a variety of sensory inputs, be they from fruit, a location, or wine.

Young dry white wines generally range from nearly colorless to pale straw colored. A more obvious yellow tint may be considered suspicious, unless it is associated with long maceration or maturation in oak cooperage. Prolonged maturation in oak can donate a yellow to golden color. Sweet white wines may vary from a pale straw to yellow-gold (for botrytized wines). Sherries vary from pale straw to golden-brown, depending on the style. Rosé wines are expected to be pale pink, without shades of blue. Hints of brown, purple, or orange usually indicate oxidation. Red wines vary from deep purple to pale tawny red, depending on age, variety and wine style. Initially, most red wines have a purplish-red hue. Varieties such as 'Gamay' and 'Pinot noir' seldom yield wines with deep colors, usually possessing a ruby color. More intensely pigmented varieties, such as

'Nebbiolo' and 'Cabernet Sauvignon,' may remain deep red for decades. Red ports, depending on style, may be deep red, ruby, or tawny.

Eventually, wines take on brownish hues. This shift is measured spectrophotometrically as the ratio of absorption at 420 and 520 nm (E_{420}/E_{520}) (Somers and Evans, 1977). Although an increase in absorption at 420 nm is an indicator of aging, it can also be a sign of oxidation or heating. Therefore, wine age, type, and style must be known before interpreting the meaning and significance of a brownish hue. Brown (dark orange) shades are acceptable only if associated with the development of a desirable processing or aged bouquet. The heating of madeira, which gives the wine its brownish coloration and baked bouquet, is an example of process-produced browning. Because many wines fail to develop a desirable aged bouquet, brown casts are typically an indicator of a wine "past its peak."

Because wine color can markedly influence the perception of its taste and odor, whether the color should be masked during a tasting depends on its intent. If the purpose is to assess consumer acceptance, then it is important that its biasing influence on sensory evaluation be present. If, however, the intent is to assess how some aspect of production affects a specific taste or olfactory attribute, then the color should be masked.

Clarity

In contrast to the complexity of interpreting the significance of wine color, haziness is always considered a fault. With current quality control procedures, consumers have come to expect perfectly clear wines. Thus, considerable effort is expended in producing wines stable in terms of clarity (see Chapter 8).

Most wines are initially supersaturated with tartrate salts. During maturation, physicochemical isomerization reduces tartrate solubility, and cool storage enhances crystallization. Crusty, flake-like crystals are usually potassium bitartrate, whereas fine crystals are typically calcium tartrate (Lüthi and Vetsch, 1981). Additional crystalline deposits may consist of calcium malate, calcium oxalate, calcium sulfate, and calcium mucate. Consumers occasionally mistake crystalline wine deposits for glass fragments.

Another potential source of haziness is the resuspension of sediment. Sediment occurs most frequently in older red wines, and may consist of one or more of polymerized and precipitated anthocyanins, tannins, proteins, tartrate crystals, fining agents, and cell fragments. The presence of sediment often is considered a sign of quality by many wine connoisseurs. To others, it is an indication of improper clarification or stabilization.

Depending on the chemical composition, sediment may have a bitter or chalky taste.

Casse is an infrequent cause of haziness, resulting from a reaction between metallic ions and soluble proteins. As the components of *casse* coalesce and reach colloidal size, a milky haze develops. Although unacceptable, *casse* does not affect the taste or aromatic character of the wine.

Microbial spoilage may be an additional source of haziness. Although either bacteria or yeasts may be involved, bacteria are the most frequent causal agents. For example, some lactic acid bacteria form long macroscopic filaments, producing a condition called **ropiness**. Disruption of the filaments generates turbidity and an oily texture. The condition may be associated with the occurrence of off-odors.

When oxygen has access to wine, microaerobic yeasts and acetic acid bacteria may grow in or on the wine. Occasionally, they produce a thick film on the wine's surface. It generates variously sized particles when disrupted. Such growths taint the wine with off-odors and off-tastes.

Viscosity

Although viscosity is often mentioned at tastings, perceptible increases usually occur only when the sugar or alcohol contents are high (Burns and Noble, 1985), or in cases of wine showing ropiness. The glycerol content apparently needs to be high ($\geqslant 25$ g/liter) to affect perceived viscosity (Noble and Bursick, 1984). Sugar (15 g fructose/5 g glucose) generates about the same degree of perceived viscosity as does 25 g glycerol (Nurgel and Pickering, 2005). At these concentrations, viscosity values of about 1.5 cP (mPa) become perceptible. At these and higher values, viscosity reduces the perceived intensity of astringency and sourness (Smith and Noble, 1998).

Spritz (Effervescence)

In sparkling wines, numerous chains of bubbles are an important quality feature (see Chapter 9). In this instance, effervescent is associated with a second, intended, in-bottle, yeast fermentation.

Still wines may occasionally contain sufficient carbon dioxide to produce bubbles along the sides and bottom of the glass. This usually is caused by the wine being bottled before excess dissolved carbon dioxide has had time to escape following fermentation. Occasionally, bubbles may result from the metabolism of contaminant microbes after bottling.

Tears

Tears ("legs") is another phenomenon given undue attention by consumers. Tears form after wine is swirled in the glass, and a film of wine coats the inner surfaces. Because ethanol evaporates from the film more rapidly than from the main volume of wine, the surface tension on the sides of the glass increases, relative to that in the bowl. As water molecules in the film pull closer together, due to increased water activity, droplets begin to form. As their size increases, drops start to sag, producing "arches." Finally, the drops slide down, forming the tears. When the drops reach the surface of the wine in the bowl, fluid is lost, and the drops pull back.

Once formed, tears continue to develop as long as alcohol evaporation pulls up sufficient wine to offset the action of gravity pulling the film downward. Cooling generated by alcohol evaporation further helps generate convection currents that draw wine up the glass (Neogi, 1985). Thus, factors affecting the rate of evaporation, such as temperature, alcohol content, and the liquid–air interface, influence tears formation. Contrary to popular belief, glycerol neither significantly affects, nor is required for tears formation. The movement of wine up the sides of the glass can be demonstrated by adding food coloring, or nonwettable powder, to wine after tears have begun to form.

Taste and Mouth-feel

Taste and mouth-feel are perceptions derived from two distinct sets of chemoreceptors in the mouth. Taste is initiated by specialized receptor neurons located in taste buds. They generate at least five basic tastes – sweet, umami, bitter, sour, salty. Mouth-feel is activated by free nerve endings, and gives rise to the sensations of astringency, dryness, viscosity, heat, coolness, prickling, and pain.

Taste

Taste buds are located primarily on the tongue, but may also occur on the soft palate, pharynx, epiglottis, larynx, and upper portions of the esophagus. On the tongue, taste buds form depressions on the sides of raised growths called papillae. Individual taste buds resemble pear-shaped objects, possessing up to 50 neuroepithelial cells (Fig. 11.1B). Individual receptor cells remain active for only approximately 10 days, before being replaced by differentiating adjacent epithelial cells. Each receptor (gustatory) cell terminates in a receptive dendrite or several microvilli. These project into the oral cavity. Impulses initiated from the receptive endings pass down to the cell body to connect with one of several cranial nerves enervating the oral cavity. Nerve stimulation not only generates impulses sent to the brain, but also maintains the integrity of the taste buds. The distribution pattern of cranial nerves in the tongue partially reflects the differential sensitivity of areas of the tongue to taste substances (Fig. 11.2).

Although taste buds have a basic structural similarity, they are found in three morphologically distinct types of papillae (Fig. 11.1A). Fungiform papillae occur primarily on the anterior two-thirds of the tongue. They are the most significant type of papillae relative to taste

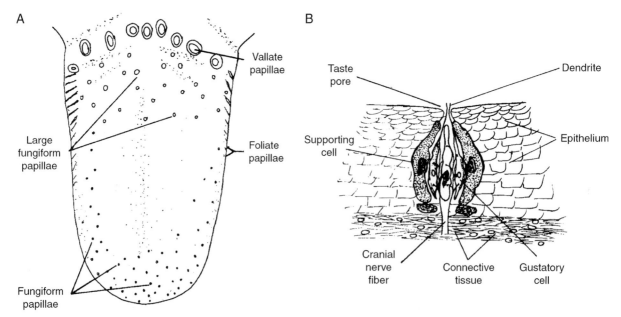

Figure 11.1 Taste receptors. (A) Location of the types of papillae on the tongue and (B) structure of taste buds that line the inner cavities of several types of papillae.

acuity. Taste sensitivity has been directly correlated to their density on the tongue (Zuniga *et al.*, 1993). A few, large, circumvallate papillae develop along a V-shaped zone across the back of the tongue. Foliate papillae are restricted to two sets of parallel ridges between folds along posterior margins of the tongue. The filiform papillae, a fourth type of papilla, are the most common, but contain no taste buds. Their tapering, fibrous extensions give the tongue its characteristic texture.

Four cranial nerves enervate the taste and mouth-feel receptor cells of the oral cavity. Free nerve endings, extension of the trigeminal nerve, also enervate the nasal mucosa. The other cranial nerves enervate specific groups of taste buds. The geniculate ganglion contains neurons that enter taste buds of the fungiform papillae and the frontal region of the soft palate. The petrous ganglion services the taste buds of foliate and circumvallate papillae, posterior portions of the palate, tonsils, and fauces. Finally, the nodose ganglion branches into the taste buds of the epiglottis, larynx, and upper reaches of the esophagus. It appears that the topographical distribution of these cranial nerves reflects some of the differential sensitivity of different areas of the mouth to sapid (taste) substances.

As noted, five major taste perceptions are recognized – sweet, umami, bitter, sour, salty. Some researchers have proposed an expansion to recognize the taste of free fatty acids (Gilbertson *et al.*, 1997). Other apparent taste sensations, such as metallic, may simply be poorly recognized aspects of olfaction (Hettinger *et al.*, 1990; Lawless, 2004). Psychophysical, neurophysiological, cytological, biochemical, and genetic techniques have been used to clarify the precise nature of taste perception.

Sensitivity to the various taste modalities is associated with specific receptor proteins or their combinations (Gilbertson and Boughter, 2003). Specific taste neurons produce only one or a select pair of receptor proteins and, thus, produce impulses corresponding usually to only one or a few tastant categories. Response to modalities such as sweet, umami, and bitter are associated with a group of about 30 related *TAS* genes. Sour and salty sensations are associated with an unrelated group of genes that encode for ion channels. They respond principally to either H^+ ions or metallic ions, notably Na^+.

Some studies have noted that receptors of similar sensitivity tend to be grouped together (Scott and Giza, 1987). Nevertheless, individual receptor neurons appear to react, although differentially, to more than one sapid category (Scott and Chang, 1984). In addition, individual taste buds generally respond to more than one type of taste compound (Beidler and Tonosaki, 1985). These factors may partially explain imprecise taste localization with complex solutions such as wine.

SWEET, UMAMI, AND BITTER TASTES

Although seemingly unrelated, these sensations are mechanistically related, being partially dependent on van der Waals forces. The receptor genes (*TAS*) fall into two subgroups –*TAS1R* and *TAS2R*. They code for proteins possessing seven transmembrane domains, with active sites on the extracellular surface. The *TAS1R*[1] group consists of three genes that encode proteins sensitive to sweet and/or savory (umami) tastants. Expression of only *TAS1R3* generates low level response to sugar, whereas joint expression of *TAS1R2* with *TAS1R3* provides high sensitivity to sugars and artificial sweeteners. The proteins transcribed by *TAS1R2* and *TAS1R3* form a dimer that is the principal sweetness receptor. Various allelic forms of *TAS1R3* significantly affect the response to various sweet tastants (monosaccharides, polysaccharides, sweeteners and sweet-tasting amino acids), and individual sensitivity to sapid substances. In addition, each receptor possesses several distinct sites (cavities) that can react with one or more sweet substances (see Temussi, 2007). In contrast, joint expression

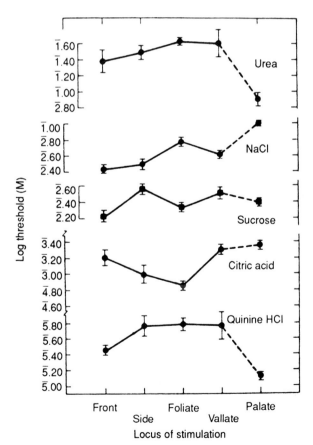

Figure 11.2 Threshold for five compounds as a function of locus on the tongue and soft palate. (From Collings, 1974, by permission)

---

[1] Alternately referred to as T1R.

of *TAS1R1* and *TAS1R3* permits formation of a protein dimer that reacts with L-amino acids (for example, monosodium glutamate – MSG). The latter permits detection of the umami attribute (Matsunami and Amrein, 2004).

The *TAS2R* group consists of some 25 genes. These code for receptors that react with a wide range of bitter tastants (Meyerhof *et al.*, 2005). They occur primarily on taste buds at the back of the tongue and palate, but sporadically elsewhere in the mouth. Individual receptor cells may occasionally coexpress several *TAS2R* gene transcripts. These presumably respond to several classes of bitter compounds. All cells expressing *TAS2R* genes also express the gustatory G protein, β-gustaducin, important in the sensation of bitterness.

Each TAS receptor protein is associated with GTP (guanosine triphosphate). Correspondingly, they are often referred to as G-proteins. Reaction with the tastant indirectly activates depolarization of the cell membrane, releasing neurotransmitters from the receptor axon.

Slight structural changes in many sweet- and bitter-tasting compounds can change their taste quality from sweet to bitter, or vice versa. Bitter- and sweet-tasting compounds can also mask perception of each other's intensity, without modifying their individual sensory modality.

Glucose and fructose are the primary sources of sweet sensations in wine, with fructose being sweeter. The perception of sweetness may be enhanced by the glycerol and ethanol content of the wine.

Flavonoid phenolics are the primary bitter compounds in wines, with tannin monomers (catechins) being the principal active components (Kielhorn and Thorngate, 1999). In tannic red wines, the bitterness of tannins is often confused with (Lee and Lawless, 1991) or masked by (Arnold and Noble, 1978) the astringent sensation of tannins. During aging, wine often develops a smoother taste because tannins polymerize and may precipitate, resulting in an eventual decline in bitterness and astringency. However, if smaller phenolics remain in solution, or larger tannins hydrolyze, perceived bitterness may increase with age.

Other bitter compounds occasionally found in wine are glycosides, terpenes, and alkaloids. Naringin is one of the few bitter glycosides occurring in wine. Bitter terpenes rarely occur in wine, except when pine resin is added (e.g., *retsina*). Similarly, bitter alkaloids rarely occur in wine, except if they come from herbs and barks used in flavoring wines such as vermouth.

SOUR AND SALTY TASTES

Sourness and saltiness are commonly called the electrolytic tastes, because small soluble inorganic cations (positively charged ions) induce changes in membrane potential of the receptor cell. These activate the release of neurotransmitters from axons that induce the firing of associated nerve fibers. Sourness is induced primarily by H^+ ions, whereas saltiness is activated by metal and metalloid cations.

Present evidence suggests that both sensations are regulated by a pair of related genes. Acid detection is based on an ion-sensing channel controlled by *ASIC2* (Gilbertson and Boughter, 2003). It is particularly sensitive to H^+ ions. Salt detection appears to be associated with a related gene, *ENaC*, that encodes another ion-sensitive channel. It is primarily responsive to metal or metalloid ions, notably sodium.

Because the tendency of acids to dissociate into ions is influenced by pH, it significantly affects perceived sourness. Undissociated acid molecules are relatively inactive in stimulating receptor neurons, but may indirectly affect perceived acidity (Ganzevles and Kroeze, 1987). The major acids affecting wine sourness are tartaric, malic, and lactic acids. These acids also induce astringency, presumably by denaturing saliva proteins (Sowalsky and Noble, 1998). Additional acids occur in wine, but, with the exception of acetic acid, they do not occur at sufficient concentrations to influence wine sourness.

Salts also dissociate into positively and negatively charged ions. Salt cations are typically a metal ion, for example, K^+ and Ca^{2+}, whereas anions may be either inorganic or organic, such as Cl^- and bitartrate, respectively. As with sourness, salt perception is not solely influenced by the activating salt cation. The tendency of a salt to ionize affects perceived saltiness, as does the size of the associated anion. For example, large organic anions suppress the sensation of saltiness, as well as delay reaction time (Delwiche *et al.*, 1999). Because the major salts in wine possess large organic anions (i.e., tartrates and bitartrates), and dissociate poorly at the pH values of wine, their common cations (K^+ and Ca^{2+}) do not actively stimulate salt receptors. In addition, the comparative scarcity of Na^+ in wine, the primary cation inducing saltiness, is a contributing factor in explaining the relative absence of salty sensations in wine.

Factors Influencing Taste Perception

Many factors affect a person's ability to detect and identify taste sensations. These may be divided into four categories – physical, chemical, biological, and psychological.

Interest in the effect of temperature on taste has existed since the mid-1800s. Regardless, the effect of temperature on perception is still unclear. Perception is considered to be optimal at normal mouth temperature. For example, cooling reduces sensitivity to sugars and bitter alkaloids (Green and Frankmann, 1987). Nevertheless, low temperatures appear to enhance the perception of bitterness (and astringency). This contradiction may relate to the different receptors involved in alkaloid vs. phenolic-induced bitterness.

Another important physicochemical factor affecting taste perception is pH. Through its effect on organic and amino acid ionization, and on their salts, pH influences the solubility, shape, and biological activity of proteins. The modification of receptor-molecule shape on the gustatory neurons could markedly affect taste sensitivity.

Sapid substances not only directly stimulate receptor neurons, but may also influence the perception of other tastants. For example, mixtures of different sugars suppress the perception of sweetness, especially at high concentrations (McBride and Finlay, 1990). Suppression also occurs among members of different sapid categories, for example salt (from food) or sugars (in port) suppress the bitter aspect of tannins, whereas acidity diminishes the perception of sweetness (and vice versa). Although suppression of perception is more common, ethanol both enhances the sweetness of sugars and the bitterness of alkaloids. Acidity enhances the bitterness and astringency of tannins. In contrast, ethanol suppresses the sourness of some acids and the astringency of tannins.

Sapid substances often have more than one sensory modality. For example, procyanidins may be both bitter and astringent; glucose can be sweet and mildly tart; potassium salts are salty and bitter; and alcohol possesses a sweet taste, as well as generating burning and weight sensations. In heterogeneous mixtures, these **side-tastes** can significantly affect overall taste perception (Kroeze, 1982). The intensity of a mixture generally reflects the intensity of the dominant component, not an integration of the separate intensities of its individual components (McBride and Finlay, 1990). The origin of these interactions may be various and complex (Avenet and Lindemann, 1989).

The interaction of sapid compounds on wine perception is further complicated by its changing chemical nature in the mouth. Wine stimulates salivary flow (Fig. 11.3), which both dilutes and modifies wine chemistry. The proline-rich proteins (PRP) of saliva, which make up approximately 70% of saliva proteins, effectively bind tannins. The histatins in saliva also effectively complex with wine polyphenolics (Wróblewski *et al.*, 2001). Because saliva chemistry can change temporally, and often differs between individuals, it is difficult to generalize on the effect of saliva on taste.

Several studies have noted a loss in sensory acuity with age (Bartoshuk *et al.*, 1986; Stevens and Cain, 1993). There is also a general reduction in the number of both taste buds and sensory receptors per taste bud. Nevertheless, age-related sensory loss is not known to seriously limit wine-tasting ability. Certain medications also reduce taste sensitivity (see Schiffman, 1983),

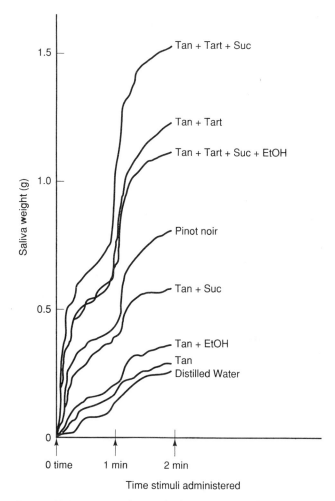

Figure 11.3 Amount of parotid saliva secreted in response to tasting 'Pinot noir' wine and selected constituents of the wine, singly and in combination. EtOH, ethanol; Suc, sucrose; Tan, tannin; Tart, tartaric acid. (From Hyde and Pangborn, 1978, by permission)

generate their own tastes, and produce a variety of taste distortions (Doty and Bromley, 2004). In addition, taste perception can be disrupted by various environmental influences. A common example, but unrelated to wine, is the disruption of taste perception by sodium lauryl sulfate (sodium dodecyl sulfate), a standard ingredient in most toothpastes (DeSimone *et al.*, 1980).

Acuity loss generally is detected as an increase in the detection threshold – the lowest concentration at which a substance can be detected. Chronic oral and dental ailments may create lingering mouth tastes, complicating discrimination at low concentrations (Bartoshuk *et al.*, 1986). This could explain why detection thresholds are usually higher in elderly people with natural dentition than those with dentures. Acuity loss also appears to depress the ability to identify sapid substances in mixtures (Stevens and Cain, 1993).

Although recessive genetic traits can produce specific taste deficiencies, subtle variations in taste acuity are more common (Fig 11.4). Subjective responses to these sensations can also be marked (Fig. 11.5). Cultural influences, such as family upbringing, or peer pressure, can override some of the genetic underpinnings of personal preference (Barker, 1982; Mennella *et al.*, 2005). Acuity also varies over time. For example, sensitivity to the bitter tastant, phenylthiocarbamide (PTC), may vary by a factor of 100 over several days (Blakeslee and Salmon, 1935).

Taste **adaptation** is a transient loss in acuity to a tastant associated with extended exposure. At moderate levels, adaptation can become complete. Consequently, it is usually recommended that wine tasters cleanse the palate between samples. **Cross-adaptation** is the effect of adaptation to one compound reducing sensitivity to another.

Color not only influences the perception of wine quality (Tromp and van Wyk, 1977; Pokorný *et al.*, 1998), but also affects taste perception (Maga, 1974; Clydesdale *et al.*, 1992). Most of the data suggest that these influences are learned associations (see Clydesdale *et al.*, 1992).

Most people are aware that many volatile compounds possess a taste modality (Enns and Hornung, 1985;

Frank and Byram, 1988). For example, several ethyl esters and furanones are commonly described as possessing a sweet aspect. In some instances, as with butyl acetate, these cross-modalities are actual – the compound stimulates both olfactory and gustatory or trigeminal receptors. In most cases, though, the taste qualities of odorants are learned associations. These apparently develop in the orbitofrontal cortex (Rolls *et al.*, 1998; Prescott *et al.*, 2004). This phenomenon also appears to have a cultural element, expressed in differences in odor/taste judgments (Chrea *et al.*, 2004). Real (and even suggested) fragrances have been shown to modify taste perception (Djordjevic *et al.*, 2004). Where there is a conflict between expected gustatory and olfactory sensations, the olfactory stimulus appears to have priority, and represses "aberrant" taste perceptions (Murphy *et al.*, 1977; Djordjevic *et al.*, 2004).

Mouth-feel

Mouth-feel refers to sapid sensations activated by free nerve endings of the trigeminal nerve. The distribution of free nerve endings throughout the oral cavity generates

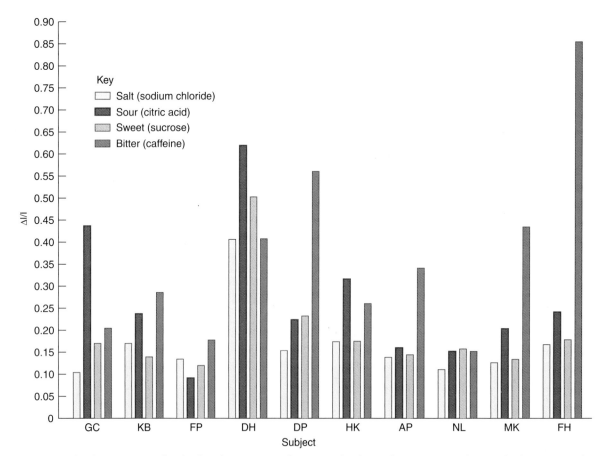

Figure 11.4 The sensitivities for the four basic taste qualities in each of 10 subjects. (From Schutz and Pilgrim, 1957, by permission)

diffuse, poorly localized sensations. In wine, mouth-feel includes the perceptions of astringency, temperature, prickling, body, and burning. They derive from the stimulation of one or more of the (at least) four types of trigeminal receptors. These are mechanoreceptors (touch), thermoreceptors (heat and cold), nocireceptors (pain), and proprioreceptors (movement and position).

ASTRINGENCY

Astringency refers to a complex of sensations inducing a variety of dry, puckery, dust-in-the-mouth sensations. In red wines, these perceptions are activated primarily by flavonoid tannins, extracted from grape seeds and skins. Anthocyanins can enhance the astringency induced by procyanidins, but do not directly contribute to bitterness (Brossaud *et al.*, 2001). White wines show less astringency, because they generally possess low concentrations of phenolic compounds. When astringency is detected in white wines, it is usually due to high acidity.

Although astringency may be confused with bitterness (Lee and Lawless, 1991), and both may be induced by related compounds, they are distinct sensations. The similar nature of their response curves also contributes to potential confusion. Both perceptions develop comparatively slowly, and possess lingering aftertastes (Figs 11.6 and 11.7). At high concentrations, the astringency of condensed tannins may partially mask the perception of bitterness (Arnold and Noble, 1978). When demanded, trained tasters often indicate that they can differentiate between these sensations. How effectively they succeed is unestablished.

Astringency in wine is normally ascribed to the binding and precipitation of salivary proteins and glycoproteins by phenolic compounds (Haslam and Lilley, 1988). Although flavonoid monomers and dimers do not effectively precipitate proteins, they may provoke astringency by structural protein deformation, possibly those that are constituents of the epithelial membrane of the oral cavity. This may explain why astringency, induced by catechins and their dimers,

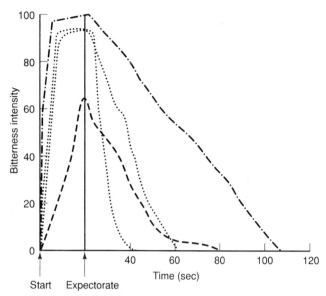

Figure 11.6 Individual time-intensity curves for four judges in response to bitterness of 15 ppm quinine in distilled water. (From E. J. Leach and A. C. Noble (1986) Comparison of bitterness of caffeine and quinine by a time-intensity procedure. *Chem. Senses* **11**, 329–345. By permission of Oxford University Press)

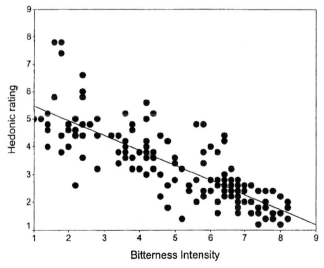

Figure 11.5 Inverse relationship between perceived bitterness and rated acceptability of PROP (6-*n*-propylthiouracil). (From Drewnowski *et al.*, 2001, reproduced by permission)

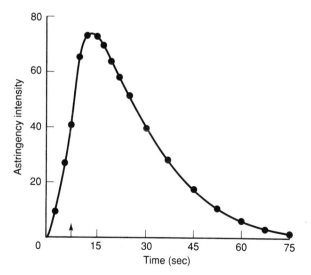

Figure 11.7 Average time-intensity curve for astringency of 500 mg/liter tannic acid in white wine. The sample was held in the mouth for 5 s (↑) and then expectorated. (From Guinard *et al.*, 1986a, reproduced by permission)

correlates more with the quantity remaining in solution than that which precipitates saliva proteins (Kallithraka *et al.*, 2000).

The main reaction between flavonoid phenolics and proteins involves -NH$_2$ and -SH groups of proteins and *o*-quinone groups of tannins (Haslam *et al.*, 1992). Other tannin–protein reactions are known (see Guinard *et al.*, 1986b), but apparently are of little significance in wine. In the binding of tannins with one or more proteins, their mass, shape, and electrical properties change, leading to precipitation.

An important factor influencing the perceived astringency of white wine is the typically low pH of the wine (Fig. 11.8). Hydrogen ions in the wine can affect protein hydration, as well as phenolic and protein ionization, thereby influencing both direct precipitation and via phenol–protein bonding. Precipitation of salivary proteins has been correlated with astringency and related sensations (Thomas and Lawless, 1995). Acidity (Peleg *et al.*, 1998) and ethanol (Vidal, Courcoux *et al.*, 2004) enhance the perception of astringency.

As noted above, the dry mouth-feel of astringency is thought to derive partially from precipitated salivary proteins (mostly proline- and histidine-rich proteins), histatins and glycoproteins (notably α-amylase). Their precipitation coats the teeth and oral cavity. On the teeth, the coating produces a rough texture. On the mucous epithelium, precipitated proteins and tannins force water away from the cell surface, simulating dryness. Reactions with

cell-membrane glycoproteins and phospholipids of the mucous epithelium may be even more important than those with salivary proteins. This may explain why astringency increases with repeated exposure to tannins (Guinard *et al.*, 1986a), but is unrelated to saliva flow. Astringency may result from reversible malfunctioning of the cell membrane, such as disruption of catecholamine methylation. In addition, the relatedness of certain tannin constituents to adrenaline and noradrenaline could stimulate localized blood vessel constriction, further enhancing a dry, puckery, mouth-feeling.

Astringency is one of the slowest in-mouth sensations to express itself. With tannic acid, maximal perception develops within approximately 15 s (Fig. 11.7). This is similar to the intensity response to the astringency of red wines. The subsequent response decline occurs more slowly. Different tasters have similar but nonidentical response curves to various tannins.

That the intensity and duration of an astringent response often increases during repeated sampling has been already noted. This phenomenon is less likely to occur when wine is consumed with food, due to the reaction between food proteins and tannins. However, if wines are tasted in quick succession, and without adequate palate cleansing, the increase in apparent astringency can produce tasting **sequence errors**. Sequence errors are differences in perception due to the order in which objects are tasted. Although tannins stimulate the secretion of saliva (Fig. 11.3), production is insufficient to restrict an increase in perceived astringency.

One of the most important factors influencing astringency is the molecular size of the phenolic. Catechin monomers bond weakly to proline-rich salivary proteins, but not detectably to α-amylase (de Freitas and Mateus, 2003). Bonding roughly increases with molecular size (polymerization). However, at above 3400 Da, the molecules lose conformational flexibility, and steric hindrance limits the availability of binding sites. Correspondingly, the polymerization that occurs during aging may result in a decline in red wine astringency.

Although bonding with saliva proteins and their precipitation have been the most studied molecular aspect of astringency, this, in itself, does not explain all aspects of astringency. As noted, procyanidins and their monomers precipitate saliva proteins poorly, but generate astringency. In addition, several authors have described distinct modalities of astringency (Gawel *et al.*, 2000, 2001). These include features described as rough, grainy, puckery, and dry. For example, oligomeric procyanidins were less drying and grainy than larger polymers, while increased galloylation was associated with the coarse aspect of astringency (Vidal *et al.*, 2003).

Polymerization between catechins and anthocyanins (either direct or ethyl-linked) also affects perceived

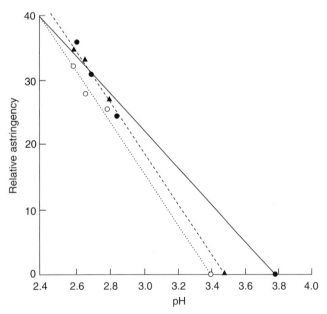

Figure 11.8 Effect of pH on perceived intensity of astringency in model wine solutions at three tannic acid concentrations, ● ——, 0.5 g/l; ▲ - - - -, 1 g/l; ○ . . . , 2 g/l. (From Guinard *et al.*, 1986b, reproduced by permission)

astringency. Tentative evidence suggests that pigmented condensation products may partially explain the decrease in astringency typically associated with red wine aging. This may result from the formation of nonastringent complexes, as well as a reduction in self-polymerization of catechins into astringent procyanidins (Vidal, Francis, Noble *et al.*, 2004).

BURNING

Wines high in ethanol content produce a burning sensation in the mouth, especially noticeable at the back of the throat. Some phenolics also produce a peppery burning sensation, as do high sugar contents. These perceptions probably result from the activation of polymodal nocireceptors on the tongue and palate. These neurons possess vanilloid receptors (TRPV1) that act as an integrator of many noxious stimuli (heat, acids, and chemicals, such as capsaicin). These receptors can generate either heat or pain sensations. Most intense sensations stimulate nocireceptors.

TEMPERATURE

The cool mouth-feel produced by chilled sparkling or dry white wine adds an element of interest and pleasure to wines of subtle flavor. Cool temperatures also help retain the effervescence of sparkling wines. In contrast, red wines typically are served at room temperature. This preference may be based on a reduction in the perception of bitterness and astringency, and an increased volatility of wine aromatics. Nevertheless, the preferred serving temperature of wine may reflect custom, as much as any other factor (Zellner *et al.*, 1988). This is suggested by the apparent nineteenth-century preference for drinking red Bordeaux wines cold (Saintsbury, 1920).

PRICKLING

Bubbles bursting in the mouth produce a prickling, tingling, occasionally burning/painful sensation. These are associated with stimulation of trigeminal nerve endings. The feeling is elicited by wines containing more than 3–5‰ carbon dioxide. The sensation appears partially related to bubble size and temperature, and is more pronounced at cold temperatures. Carbon dioxide modifies the perception of sapid compounds (Cowart, 1998), and significantly enhances the perception of cold in the mouth (Green, 1992). Carbon dioxide also may suppress odor perception (Cain and Murphy, 1980).

BODY (WEIGHT)

Although "body" is a desirable aspect in most wines, the precise origin of the sensation remains unclear, despite attempts to quantify it (Bertuccioli and Ferrari, 1999). Although fullness in the mouth is roughly correlated with the wine's sugar content, alcohol content is of little significance (Pickering *et al.*, 1998). Glycerol can increase the perception of body, but is not present in sufficient quantity to influence this sensation, except in some sweet wines. In contrast, acidity reduces the perception of body. Less recognized is the importance of grape and yeast polysaccharides (Vidal, Francis, Williams *et al.*, 2004). The principal yeast polysaccharides are mannoproteins, whereas the principal grape polysaccharides are arabinogalactan-proteins and rhamnogalacturonans. All increase the perception of body. The rhamnogalacturonan fraction may also reduce the astringency induced by organic acids (Vidal, Francis, Williams *et al.*, 2004). The perception of fullness also appears to involve aspects of wine fragrance, notably its intensity.

METALLIC

A metallic sensation is occasionally detected in dry wines. It is especially noticeable as an aftertaste in some sparkling wines. The origin of this sensation is unestablished. It could be induced by iron and copper ions. However, the concentrations required are normally well above those found in wine (>20 and 2 mg/liter, respectively). Oct-1-en-3-one has been associated with a metallic sensation in dairy products, and acetamides can have metallic aspects (Rapp, 1987). In addition, some reduced sulfur compounds are considered to have a metallic odor, for example 2-methyltetrahydrothiophen-3-one and ethyl-3-methylthiopropionate. These data support the contention of Hettinger *et al.* (1990) that metallic "tastes," are, in reality, misinterpreted olfactory sensations.

Taste and Mouth-feel Sensations in Wine Tasting

To distinguish between the various taste and mouth-feel sensations, tasters may concentrate sequentially on the sensations of sweetness, sourness, bitterness, astringency, and balance. Their temporal responses can be useful in confirming the identification of taste sensations (Kuznicki and Turner, 1986). The localization of the sensations in the mouth and on the tongue also can be useful in affirming taste characterization. Balance is a summary perception, derived from the interaction of sapid and mouth-feel sensations.

Sweetness is usually the most rapidly detected taste sensation. Sensitivity to sweetness appears to be optimal at the tip of the tongue (Fig. 11.2). It also tends to be the first taste sensation to show adaptation. The intensity of its perception is reduced in the presence of significant concentrations of wine acids and phenolics.

Sourness is also detected rapidly. The rate of adaptation to sourness may be slower, and may generate a lingering aftertaste. Acid detection is commonly strongest along the sides of the tongue. This varies considerably among individuals, with some people detecting sourness more distinctly on the back of the lips, or insides of the cheeks. Strongly acidic wines can induce astringency, giving the teeth a rough feel. Both the sour and astringent aspects of markedly acidic wine may be decreased by the presence of perceptible levels of viscosity (Smith and Noble, 1998).

The detection of bitterness usually follows sweetness and sourness. It typically takes several seconds to express itself. Peak intensity may not be reached for 10–15 s (Fig. 11.6). After expectoration, the sensation declines gradually, and may linger for several minutes. Most bitter-tasting compounds in wine, primarily phenolics, are perceived at the back-central portion of the tongue. In contrast, bitter alkaloids are perceived primarily on the soft palate and the front of the tongue (Boudreau *et al.*, 1979). Because bitter sensations at the back of the tongue develop more rapidly, and most bitter sensations in wine are induced by phenolic compounds, bitter sensations in wine are perceived most commonly at the back of the tongue (McBurney, 1978). The bitterness of a wine is more difficult to assess accurately when the wine is astringent. High levels of astringency may partially mask the perception of bitterness. A high sugar content also reduces the perception of bitterness.

Astringency is often the last sensation detected. It often takes 15 or more seconds for its perceived intensity to develop fully (Fig. 11.7). After expectoration, the sensation slowly declines over a period of several minutes. Astringency is poorly localized because it is sensed by free nerve endings distributed throughout the mouth. Because the perceived intensity and duration of astringency increase with repeat sampling, some judges recommend that astringency be assessed with the first taste. This would give a perception more closely approximating the astringency detected on consuming the wine with food. Others consider that the assessment of astringency should occur only after several samplings, when the mollifying affects of saliva have diminished.

The increase in perceived astringency that can occur when tasting a series of wines (Guinard *et al.*, 1986a) could seriously affect the validity of a wine's assessment. This is especially true with red wines, for which the first wine in a series often appears the smoothest. A similar situation could occur in a series of dry wines, as well as making a sweeter wine appear overly sweet. These influences are sufficiently well known that tastings are organized to avoid the comparison of wines of markedly different character. However, design errors can still have significant effects on well-conceived comparative tastings. The effect of **sequence error** may be partially offset in group tasting by arranging each taster to sample the wines in random order. In addition, lingering taste effects can be minimized by assuring that adequate palate cleansing occurs after each wine sample.

Although the number of in-mouth sensations is limited, they are particularly important to consumer acceptance. Unlike professionals, consumers seldom dote on the wine's fragrance. Thus, in-mouth sensations are far more important to overall perception. Nevertheless, even for connoisseurs, one of the ultimate tests of greatness is the holistic impression of mouth-feel and balance. These are phenomena principally associated with joint gustatory and tactile sensations. Producing a wine with a fine, complex, and interesting fragrance is often a significant challenge for the winemaker. Assuring that the wine also possesses a rich, full and balanced in-mouth sensation is the ultimate achievement.

Odor

The Olfactory System

NASAL PASSAGES

Olfactory tissue occurs as two small patches in the upper portions of the nasal passages (Fig. 11.9). Volatile compounds may reach the olfactory epithelium either directly, via the nostrils (orthonasal), or indirectly from the back of the throat (retronasal). The latter route is especially important in the generation of flavor – the combined sensation derived from gustatory, tactile and olfactory stimuli.

The nasal passage is bilaterally divided into right and left segments by a central septum. The receptors in each cavity send signals to the corresponding halves of the olfactory bulb, located directly above them, at the base of the skull. Because impulses from both olfactory bulbs subsequently connect, via the anterior olfactory nuclei, both hemispheres of the brain are involved in odor processing.

Each nasal cavity is further, but incompletely, subdivided transversely by three outgrowths, the turbinate bones (Zhao *et al.*, 2004). These increase the contact area between the air and the epithelial linings of the nasal passages. Although inducing turbulence, warming, and cleaning of the air, the folds restrict air flow past the recessed olfactory regions. It is estimated that in ordinary breathing, only about 10% of the inhaled air moves past the olfactory epithelium (Hahn *et al.*, 1993). At high rates of air intake, the value may increase to approximately 20%. Although higher flow rates may enhance odor perception slightly, the vigor

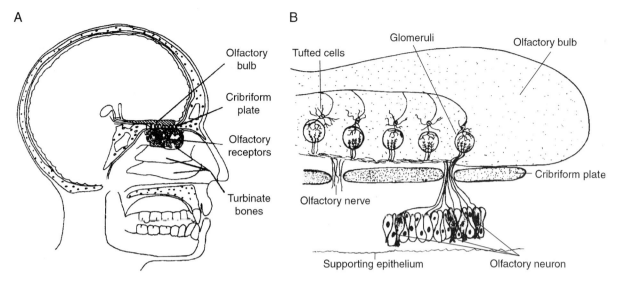

Figure 11.9 Receptors for olfaction. (**A**) Location of the olfaction region in the nasal cavity and (**B**) enlarged section showing the olfactory neurons (receptors) and their connections to the olfactory bulb.

of breathing apparently does not affect perceived odor intensity (Laing, 1983). The usual recommendation to take short whiffs of a wine during tasting probably has more to do with avoiding odor adaptation than enhancing odor perception.

Surfaces of the olfactory regions (olfactory cilia) are coated with a highly organized, extracellular, mucous matrix. A major component of the matrix is olfactomedin. The protein is thought to be a neurotrophic factor, affecting the growth and differentiation of olfactory neurons. Odorant-binding proteins (OBP) are another group of proteins typically found in the mucous matrix. These lipophilic carriers are also produced by lateral nasal glands, and are secreted near the tip of the nose. They are thought to bind odorants and assist their transport across the mucous to receptor sites on olfactory neuron cilia (Nespoulous *et al.*, 2004). This may explain the concentration of odorant molecules in the nasal mucus, at up to 1000 to 10,000 times above their gaseous concentration (Senf *et al.*, 1980). Alternately, OBPs may function to inactivate or remove odorants from olfactory receptor sites.

Only a fraction of the aromatic molecules that reach the olfactory patches is absorbed into the mucus that coats the epithelium. Of these, only a proportion is likely to diffuse through the mucus and reach reactive sites on the olfactory receptor neurons. In some animals, high concentrations of cytochrome-dependent oxygenases accumulate in the olfactory mucus (Dahl, 1988). These enzymes may interact with olfactory UDP glucuronosyl transferase to catalyze a wide range of reactions. These have been hypothesized to terminate olfactory activation (Lazard *et al.*, 1991).

OLFACTORY EPITHELIUM, RECEPTOR NEURONS, AND CONNECTION WITH THE BRAIN

The olfactory epithelium consists of a thin layer of tissue covering an area of almost $2.5\,cm^2$ on each side of the nasal septum. Each region contains approximately 10 million receptor neurons, plus associated supporting and basal cells (Fig. 11.10). Receptor neurons are specialized nerve cells that respond to aromatic compounds; supporting cells (and the glands underlying the epithelium) produce a special mucus and several classes of odorant-binding proteins that coat the olfactory epithelium. In humans, each receptor neuron expresses one of about 340 odorant receptor genes[2] (Malnic *et al.*, 2004). Basal cells differentiate into receptor neurons, as older receptor neurons degenerate. Receptor neurons remain active for a variable period (possibly up to 1 year), before they degenerate and are replaced. Differentiating basal cells produce extensions that grow through openings in the skull (cribriform plate) to connect with the olfactory bulb at the base of the skull. These nonmyelinated extensions (axons) associate into bundles as they pass through the cribriform plate. In humans, olfactory and gustatory neurons are the only nerve cells known to regenerate regularly. Supporting cells electrically isolate adjacent receptor cells and help maintain normal function.

As with gustatory cells in the mouth, olfactory neurons show a common cellular structure. Odor quality, the distinctive aromatic character of any odorant, is

---

[2] There are also 297 OR pseudogenes – inactive gene sequences having similarity to and having probably been derived from active equivalents.

not associated with an obvious morphological neuronal differentiation. As receptor neurons differentiate, they send a dendrite down to the surface of the olfactory epithelium (Fig. 11.11A). Here they form a swelling called the olfactory knob (Fig. 11.11B). From this emanate a variable number of long (1–2 μm) hairlike projections (cilia). These are thought to markedly increase

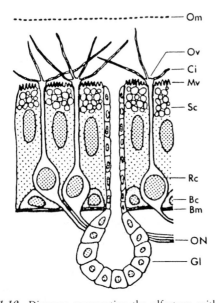

Figure 11.10 Diagram representing the olfactory epithelium. Bc, basal cell; Bm, basement membrane; Ci, cilium; Gl, olfactory gland; Mv, microvilli; Om, mucous surface; ON, olfactory nerve; Ov, vesicle; Rc, receptor cell; Sc, supporting cell. (From Takagi, 1978, reproduced by permission)

the surface contact between odorants and the receptor membrane. Odorant molecules are thought to bind to one (or more) olfactory receptors (OR) proteins. These are located in and transverse the receptor cell membrane. ORs are a class of G-protein-coupled complexes that, on activation, stimulate adenylate cyclase to liberate cyclic AMP. It, in turn, activates the opening of ion channels in the membrane, resulting in an influx of Na^+ and Ca^{2+}. This initiates membrane depolarization and impulse transmission along the nerve fiber.

Odor quality is thought to arise from the differential sensitivity of one or more types of receptor neurons to aromatic compounds. This reflects the presence of a unique family of proteins produced by the olfactory epithelium (Buck and Axel, 1991). About 340 unique olfactory receptor (OR) proteins are produced. However, only one type is produced per olfactory neuron. OR proteins possess several regions that span the membrane, the outer portions of which bind with odorants. Each OR bears several variable regions that generate the uniqueness required for receptor specificity. Although some odorants activate only a single type of OR, most odorants activate several different ORs (Malnic *et al.*, 1999). Thus, the olfactory system seems to encode odor identity by a combination of stimuli, usually from several receptors. This is equivalent to a chord played on the piano. Odor qualities may also be derived from the duration and number of receptors stimulated. The selective production of subclass(es) of basal cells, expressing particular *OR* genes, may explain

Figure 11.11 Scanning electron micrographs of the human olfactory mucosal surface (**A**) and olfactory dendritic knobs and cilia (**B**). (Photos courtesy of Drs Richard M. Costanzo and Edward E. Morrison, Virginia Commonwealth University)

the increased sensitivity of some individuals to an odorant upon repeat exposure (Wysocki *et al.*, 1989). This property is apparently a general characteristic of women (during the childbearing years) (Fig. 11.12). It is less marked in men.

After stimulation, the electrical impulse from an olfactory neuron rapidly travels along the filamentous extensions of the cell to the olfactory bulb. In the olfactory bulb, bundles of receptor axons terminate in spherical regions called glomeruli. It appears that axons for olfactory receptors, responding to the same or related odorants, connect with a particular glomerulus (Tozaki *et al.*, 2004). Within the glomeruli, the axons synapse with one or more of several types of nerve cells (mitral and tufted cells) in the olfactory bulb (Fig. 11.9).

The olfactory bulb is a small bilaterally lobed portion of the brain that collects and edits the information received from olfactory receptors in the nasal passages. From here, impulses are sent, via the lateral olfactory tract, to the hypothalamus and several higher centers in the brain. Feedback impulses may also pass downward to the olfactory bulb and regulate its response to incoming signals. The interaction between the olfactory bulb and other centers of the brain, notably the orbitofrontal cortex, is of great importance. This has special significance to the perception of flavor, as taste, touch,

odor and visual centers of the brain interact within the orbitofrontal cortex, and can influence each other's perception (Fig. 11.13).

The integration of taste and odor stimuli may explain why identifying constituents of taste–odor mixtures is poorer than taste mixtures alone (Laing *et al.*, 2002). Even more fascinating may be the significance of visual clues to olfactory sensations, explaining why a white wine (colored red) is described in terms appropriate for a red wine (Fig. 11.14).

Odorants and Olfactory Stimulation

There is no precise definition of what constitutes an olfactory compound. Based on human perception, there are thousands of olfactory substances, spanning a huge range of chemical groups. For air-breathing animals, an odorant must be volatile (pass into a gaseous phase at ambient temperatures). Although this places limitations on odorant molecular size (≤ 300 Da), low molecular mass implies neither volatility nor aromaticity. Most aromatic compounds have strongly hydrophobic (fat-soluble) and weakly polar (water-soluble) sites. They also tend to bind weakly with cellular constituents, and dissociate readily. Volatility in wine may be influenced by the presence of the other constituents in wine, such as sugars (Sorrentino *et al.*, 1986), ethanol (Fischer *et al.*, 1996), oils (Roberts and Acree, 1995; Roberts *et al.*, 2003), and macromolecules (Voilley *et al.*, 1991). Volatility for compounds that ionize is a function of the proportion in its molecular form. Only in the molecular state is a compound, such as acetic acid, volatile. In addition, volatility may be affected by the redox potential of the wine. Low redox potentials, for example, increase the volatility of reduced-sulfur compounds, such as hydrogen sulfide and mercaptans.

Although the chemical nature of odor quality has been studied for decades, no general theory has found widespread acceptance. Present thought favors the view that several molecular properties may be involved in olfactory stimulation and the perception of odor quality. These include electrostatic attraction, hydrophobic bonds, van der Waals forces, hydrogen bonding, and dipole–dipole interactions. Small structural modifications, such as found in stereoisomers, can markedly affect perceived intensity and quality. For example, the D- and L-carvone stereoisomers possess spearmint-like and caraway-like qualities, respectively.

Compounds possessing similar odors, and belonging to the same chemical group, appear to show competitive inhibition. This phenomenon, called **cross-adaptation**, suppresses the perception of an aromatic compound by prior exposure to a related odorant. Mixtures of odorants with markedly different modalities generally retain their distinct and separate qualities, but occasionally

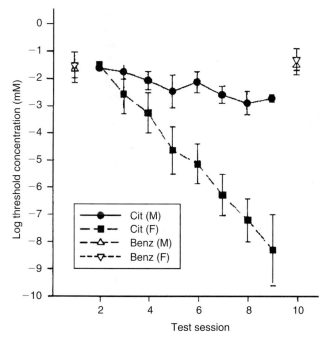

Figure 11.12 Gender (F, female; M, male) effects of repeated test exposure to citralva (orange–lemon fragrance) (Cit). Benzaldehyde (Benz) tested as a control at the beginning and end of the eight sessions. (Reprinted by permission from Macmillan Publishers Ltd: *Nature Neuroscience* 5, 199–200. Dalton *et al.*, 2002)

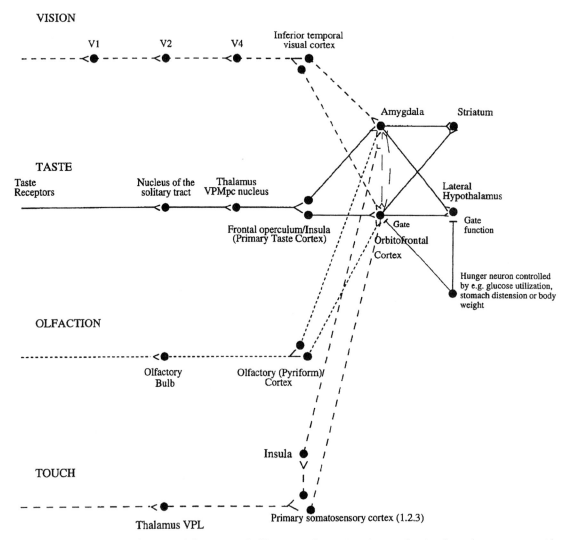

Figure 11.13 Schematic diagram of the taste and olfactory pathways in primates showing how they converge with each other and with visual pathways. The gate functions shown refer to the finding that the responses of taste neurons in the orbitofrontal cortex and the lateral hypothalamus are modulated by hunger, VPMpc, ventralposteromedial thalamic nucleus; V1, V2 and V4, visual cortical areas. (From Rolls, 2001, reproduced by permission)

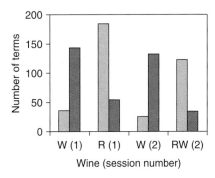

Figure 11.14 Frequency distribution of olfactory terms representative of red/dark objects (light green bars) and yellow/light objects (dark green bars) used to describe two paired tastings of a white and a red wine (W: 'Sémillon'/'Sauvignon blanc;' R: 'Cabernet Sauvignon;' and RW: sample W colored red with anthocyanins). (From Morrot *et al.*, 2001, reproduced by permission)

mixtures produce a unitary impression unrelated to their individual components (Laing and Panhuber, 1978). Conversely, single compounds may be recognized as possessing several distinguishable attributes. For example, dihydromyrcenol has both woody and citrus attributes (Lawless, 1992). Whether these differences arise from having already come to associate certain receptor stimulation patterns with separate fruits, beverages, locations, etc. is unknown, but probable. In addition, mixtures of aromatics each occurring at below their threshold, may act synergistically, promoting their mutual perception (Selfridge and Amerine, 1978). Occasionally, both synergistic and suppressive effects may be found. Piggott and Findlay (1984) found both reactions with different pairs of esters, with even opposite effects at different concentrations for some pairs.

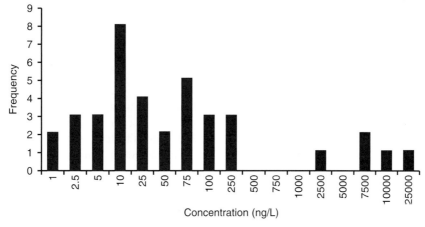

Figure 11.15 Frequency distribution for individual 2,4,6-TCA thresholds in 'Sauvignon blanc' wine. (From Suprenant and Butzke, 1996, reproduced by permission)

These diverse reactions, combined with variability of individual sensitivity and idiosyncratic term use, help to explain the divergence of opinion frequently expressed at wine tastings.

Sensations from the Trigeminal Nerve

Free nerve endings of the trigeminal nerve in the nose originate from the same cranial nerve that enervates the oral cavity. In the nose, the nerve endings are scattered throughout the nasal epithelium, except the olfactory epithelia. They respond to a wide range of pungent and irritant chemicals, often at very low concentrations. At higher concentrations, the receptors respond to most odorant molecules (Cain, 1974; Ohloff, 1994). Vanillin is an apparent exception.

Most pungent chemicals react nonspecifically with protein sulfhydryl (-SH) groups, or break protein disulfide (-S-S-) bridges (see Cain, 1985). The resultant reversible structural changes in membrane proteins may stimulate firing of free nerve endings. Most pungent compounds apparently have a net positive charge, whereas putrid compounds commonly possess a net negative charge (Amoore *et al.*, 1964).

As already noted, most aromatic compounds can stimulate trigeminal nerve fibers. This stimulation is referred to as the **common chemical sense**. In the nose, it may be expressed variously as an irritation, burning, stinging, tingling, or pain. Volatile compounds that are strongly hydrophobic may dissolve into the lipid component of the cell membrane, disrupting cell permeability and inducing nerve firing (Cain, 1985). Unlike the reduced excitability of olfactory neurons after prolonged odorant exposure, free nerve endings are less susceptible to adaptation (Cain, 1976). The stimulation of free nerve endings by an odorant can also modify

odor quality. For example, a small amount of sulfur dioxide can be pleasing, but at high concentrations, it becomes an overpowering irritant. In addition, hydrogen sulfide contributes to a yeasty bouquet, and adds an aspect of fruitiness to wine at low concentrations (~1 μg/liter). At slightly higher concentrations, it produces a revolting rotten egg odor (MacRostie, 1974). At high concentrations, most fragrant compounds lose any pleasantness they might have had at lower concentrations.

Odor Perception

Individual variation in odorant perception has long been known. What is new is our increasing understanding of the extent of this variation (Pangborn, 1981; Stevens *et al.*, 1984). Variation can affect the ability to detect, identify, and measure the intensity of odors, as well as our emotional response to odors. Thresholds, the amount of compound required to produce a positive response half of the time, have been assessed by a variety of procedures. Procedural and related issues are covered in detail in Bi and Ennis (1998), O'Mahony and Rousseau (2002), and Walker *et al.* (2003).

The **detection threshold** is the concentration at which the presence of a substance becomes noticeable. Human sensitivity to odorants varies over 10 orders of magnitude, from ethane at 2×10^{-2} M, to mercaptans at 10^{-10} to 10^{-12} M. Even sensitivity to the same or chemically related compounds can show tremendous variation. For example, the detection thresholds for TCA (Fig. 11.15) and pyrazines (Seifert *et al.*, 1970) span 4 and 9 orders of magnitude, respectively. TCA is a frequent source of corked odors in wine, and certain pyrazines contribute to the bell pepper and moldy aspects of some wines. Examples of the detection thresholds of a variety of important aromatic compounds in wines are given in Francis and Newton (2005).

When the detection threshold of an individual is markedly below normal, the condition is called **anosmia**. Anosmia can be general, or may affect only a small range of related compounds (Amoore, 1977). The occurrence of specific anosmias varies widely in the population. For example, it is estimated that about 3% of the human population is anosmic to isovalerate (sweaty), whereas 47% is anosmic to 5α-androst-16-en-3-one (urinous). The genetic and physiologic origin of most specific anosmias is unknown.

Hyperosmia, the detection of odors at abnormally low concentrations, is little understood. One of the most intriguing accounts of hyperosmia relates to a 3-week episode of suddenly being able to recognize people and objects by their odor (Sachs, 1985). Also unclear is the origin of the normally limited olfactory skills of humans, compared to most animals. The limited skill may arise from the comparatively small size of the human olfactory epithelium, the olfactory bulbs, and the orbitofrontal complex of the brain. For example, the total olfactory epithelium in dogs can be up to $15 \, cm^2$, compared with about $2–5 \, cm^2$ in humans. Comparative measurements of the odor thresholds in dogs and humans indicate that dogs possess thresholds approximately 100 times lower than humans (Moulton *et al.*, 1960). There is also marked degeneration in the human genome associated with odor receptors. For example, humans possess about 340 functional *OR* genes, vs. approximately 920 for mice and more than 970 in dogs (Olender *et al.*, 2004).

The detection threshold may be temporarily influenced by the presence of other volatile substances. As mentioned previously, at subthreshold concentrations, two or more compounds may act synergistically (or suppressively). Another aspect of mixture interaction relates to how the solute affects volatility, and therefore threshold values. Figure 11.16 shows how low ethanol concentrations affect the volatilization of several esters, perceived fruitiness being maximal at 0.75% alcohol. This could be important to a wine's finish. In addition, the rapid evaporation of alcohol from wine coating the sides of a glass (following swirling) would enhance the liberation of esters. At higher concentrations (up to 5%), ethanol reduces perceived fruitiness.

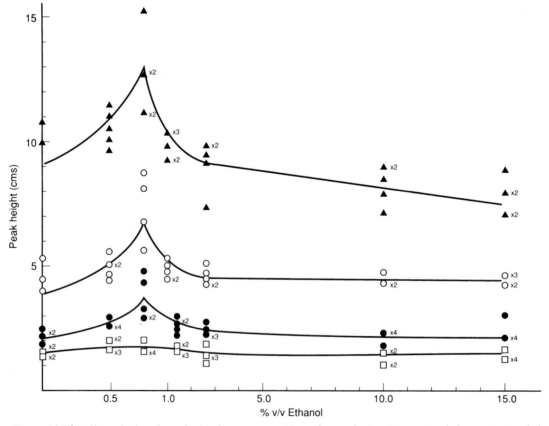

Figure 11.16 Effect of ethanol on the headspace composition of a synthetic mixture. ●, ethyl acetate; ○, ethyl butrate; ▲, 3-methylbutyl acetate; □, 2-methylbutanol. (From A. A. Williams and P. R. Rosser (1981) Aroma enhancement effect of ethanol. *Chem. Senses* 6, 149–153. By permission of Oxford University Press)

Guth (1997) also found that alcohol reduced the perception of several esters and higher alcohols (Table 11.1), while augmenting acidic and astringent tastes. Guth and Sies (2002) concluded that the suppression of fruitiness results more from diminished perception than volatilization. However, other researchers have found that increasing alcohol content progressively decreases the volatility of several higher alcohol and aldehydes (Escalona *et al.*, 1999) and ethyl esters (Conner *et al.*, 1998). There are anecdotal accounts that slight modifications in alcohol content can significantly influence wine flavor. Alcohol concentrations associated with improved flavor are called "sweet spots."

Other wine constituents can also influence the release of volatiles. For example, grape and yeast polysaccharides influence the volatility of esters, higher alcohols and diacetyl (Dufour and Bayonove, 1999a); flavonoids weakly affect the release of various esters and aldehydes (Dufour and Bayonove, 1999b) and other volatiles (Aronson and Ebeler, 2004); whereas anthocyanins bind with volatile phenolics, such as vanillin, reducing their sensory impact (Dufour and Sauvaitre, 2000). Such effects could easily affect the development of wine fragrance after pouring, as well as its finish in the mouth.

In a study using the buccal odor screening system (BOSS), Buettner (2004) found that the finish of two 'Chardonnay' wines differed with regard to their respective rates of odor loss from the oral cavity (fruity/floral aspects diminishing more quickly than oak attributes). In contrast, no differences in odorant release timing were noted. Those that persisted the longest became increasingly important to the perceived finish.

Saliva is an additional factor potentially affecting the release and perception of aromatic compounds. An indirect effect already noted results from ethanol dilution in the mouth. However, saliva contains several enzymes that could modify the chemical nature (and volatility) of wine aromatics. For example, saliva can modify retronasal odor by degrading volatile esters and thiols (Buettner, 2002a), as well as reduce aldehydes to their corresponding alcohols (Buettner, 2002b). Although not marked within the normal time frame of a tasting, such effects might marginally affect the wine's finish and aftertaste. Because saliva chemistry is not constant, even in the same individual, saliva variability may be another source of idiosyncratic differences in perceptive abilities.

The **recognition threshold** refers to the minimum concentration at which an aromatic compound can be correctly identified. The recognition threshold is typically higher than the detection threshold. It is generally acknowledged that people have considerable difficulty in correctly identifying odors in the absence of visual clues (Engen, 1987). Nevertheless, it is often thought that expert tasters and perfumers have superior odor acuity. Although this may be true, winemakers often fail to recognize their own wine in blind tastings, and experienced wine tasters frequently misidentify the varietal and geographical origin of wines (Winton *et al.*, 1975; Noble, Williams *et al.*, 1984; Morrot, 2004). In this regard, it is important to distinguish between wine *differentiation* (by direct comparison) and wine *recognition* (in isolation). The first is direct comparison, the second demands precise recall of differences minutes, hours, or days later.

Odors commonly are organized into groups based on origin, such as floral, smoky, and resinous. Nevertheless, odor memories appear to be arranged according to the events with which they have been associated, not conceptual groupings. The more significant an event, the more intense and stable the association. Engen (1987) views this memory pattern as equivalent to the nonscalar use of words by young children. Children tend to categorize objects and events functionally, rather than in terms of abstract concepts: for example, a chair is something on which one sits vs. a type of furniture. If odor memories are associated in categories related to experience, then it is not surprising that prior knowledge of a wine's origin can bias a taster's sensory perception (see Herz, 2003). This view may help to explain why it is so difficult to use unfamiliar terms for familiar odors. The language describing wine fragrance is relatively impoverished, and relies heavily on objects or events associated with aromatic experiences. The words people use to describe wine often say more about the taster and their emotional response to the wine than its sensory attributes (Lehrer, 1975; Dürr, 1985). At best, this expresses feelings in a poetic manner; at worst, it is used to impress or intimidate. The difficulty of correctly naming familiar odors has been called the "tip-of-the-nose" phenomenon by Lawless and Engen (1977). The phenomenon is often experienced when tasting wines blind. Suggestions can

Table 11.1 Effect of ethanol on the odor threshold of some wine aromatics in air (ethanol in the gas phase 55.6 mg/liter)

Compound	Odor threshold (ng/l)		
	Without ethanol (*a*)	With ethanol (*b*)	Factor *b/a*
Ethyl isobutanoate	0.3	38	127
Ethyl butanoate	2.5	200	80
Ethyl hexanoate	9	90	10
Methylpropanol	640	200 000	312
3-Methylbutanol	125	6300	50

Source: From Grosch, 2001, based on data from Guth, reproduced by permission

improve identification (de Wijk and Cain, 1994), but also can unduly skew opinions during a tasting.

The perceived intensity of aromatic compounds, compared to their detection or recognition thresholds, often varies considerably. For example, compounds such as hydrogen sulfide or mercaptans are perceived as intense, even at their recognition thresholds. In addition, the rate of change in perceived intensity varies widely among compounds. For example, a threefold increase in perceived intensity was correlated with a 25-fold increase in the concentration of propanol, whereas a 100-fold concentration increase was required to have a similar increase in perceived intensity for amyl butyrate (Cain, 1978). A rapid increase in perceived intensity is characteristic of most off-odors.

Factors Affecting Olfactory Perception

There are small sex-related differences in olfactory acuity. Women are generally more sensitive to, and more skilled at, identifying odors than men (Doty *et al.*, 1984; Choudhury *et al.*, 2003). There are also sex-related (or experience-related) differences in the types of odors identified. In addition, women between puberty and menopause experience changes in olfactory discrimination that are correlated with cyclical changes in hormone levels (see Doty, 1986). However, the most significant factor may be the remarkable genetic diversity between individuals in their olfactory receptor (*OR*) genes (Menashe *et al.*, 2003). Each person tested had a distinctive pattern of *OR* genes. Their results also indicate that distinct differences may exist among ethnic groups concerning olfactory sensitivity.

Age also affects olfactory acuity, by elevating both detection and recognition thresholds (Stevens *et al.*, 1984; Cowart, 1989). Although these effects may begin at age 20, they tend to increase markedly only during a person's 70s and 80s (Fig. 11.17). Considerable diversity in individual sensitivity occurs in all age groups.

The reduction in the turnover rate of olfactory receptor neurons may partially explain age-related olfactory loss (Doty and Snow, 1988). However, the degeneration of the olfactory bulb, and nerve connections in the rhinencephalon (olfactory cortex), may be more important. Olfactory regions frequently degenerate earlier than other parts of the brain (Schiffman *et al.*, 1979). This may account for smell often being the first of the chemical senses to show age-related loss. Although wine-judging ability may decline with age, experience and mental concentration may compensate for the sensory loss.

Nasal and sinus infections may accelerate certain degenerative changes, and upper respiratory tract infections can negatively affect perception long after the infection has passed. Short-term effects involve a massive increase

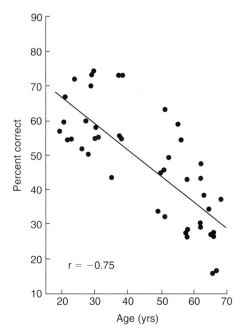

Figure 11.17 Odor identification ability for a battery of 80 common odors, plotted as a function of the age of the subject. (From Murphy, 1995, reproduced by permission)

in mucus secretion, which both blocks air access to the olfactory regions and retards diffusion to the receptor neurons. A loss of olfactory ability is also associated with several major diseases, such as polio, meningitis, and osteomyelitis. These may destroy the olfactory nerve and cause generalized anosmia. In addition, some genetic defects are associated with generalized anosmia, such as Kallmann's syndrome. Certain medications and illicit drugs, such as cocaine, disrupt the olfactory epithelium and diminish the sense of smell (see Schiffman, 1983).

It is commonly believed that hunger increases olfactory sensitivity and, conversely, that satiation lowers it. This view is supported by a report that hunger and thirst increased the general reactiveness of the olfactory bulb and cerebral cortex (Freeman, 1991). Nevertheless, another study showed an increase in olfactory acuity following food intake (Berg *et al.*, 1963).

Smoking produces long-term, but potentially reversible, impairment of olfactory discrimination (Frye *et al.*, 1990). Thus, the perception of wine to smokers and nonsmokers likely differs. However, smoking has not prevented some individuals from becoming highly skilled winemakers and cellar masters. Nevertheless, the short-term impairment of olfactory skill has made smoking in tasting rooms unacceptable.

Adaptation is an additional source of altered olfactory perception. Adaption may result from either a loss in the excitability of olfactory receptors, or a decline in sensitivity of interpretive centers in the brain. Thus, after

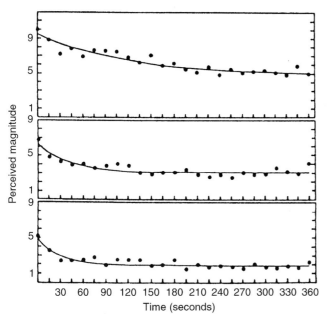

Figure 11.18 Adaptation in perceived magnitude of *n*-butyl acetate at 0.8 mg/liter (*bottom*), 2.7 mg/liter (*middle*), and 18.6 mg/liter (*top*). (From Cain, 1974, reproduced by permission)

continued exposure, the apparent intensity of an odorant may fall both rapidly and exponentially (Fig. 11.18). It appears that the more intense the odor, the longer it takes for adaptation or habituation to develop.

The effect of adaptation is very important in wine tasting. Because adaptation develops rapidly (often within a few seconds), the perceived fragrance of a wine often changes quickly. Thus, wine tasters are counseled to take only short whiffs of wine. An interval of 30 s to 1 min usually allows reestablishment of normal acuity. Recovery seems to follow a curve similar to, but the inverse of, adaptation (Ekman *et al.*, 1967). However, with aromatically complex wines, such as vintage ports, it can be beneficial to continuously smell the wine for an extended period. The progressive adaptation successively reveals different components of the wine's complex fragrance, and may present new and pleasurable experiences. This view is supported by one of the few studies of adaptation to odorant mixtures (de Wijk, 1989).

Mixtures of odorants, each below their individual detection thresholds, often act in an additive manner. That is, two related compounds at half their individual detection thresholds often become detectable when combined (Patterson *et al.*, 1993). This may be of importance in wine assessment (in which hundreds of compounds occur below their detection or recognition thresholds). Probably of equal importance, but little investigated, are the roles of odor masking, and cross-adaptation. For example, odorants formed during roasting mask the presence of methoxypyrazines in coffee (Czerny and Grosch, 2000) and "Brett" taints suppress the perception of fruit aromas in wine (Licker *et al.*, 1999). Data from Livermore and Laing (1996) demonstrate that even professional perfumers and flavorists fail to identify more than three or four components in a complex aromatic mixture. People also have difficulty recognizing when single components have been omitted from an odor mixture (Laska and Hudson, 1982). These situations may result from identification of odors based on the modulated response of several different receptors, with any one receptor reacting to more than one odorant (Malnic *et al.*, 2004). In other words, odor identification may resemble the pattern memory used in facial or object recognition. If so, not all qualities of an odorant are necessary for recognition. However, stimulation of critical receptors by several odorants may make individual identification difficult to impossible (Jinks and Laing, 2001). In addition, the identification of varietal or regional wine styles may be based on the combined impression, not on the remembrance of distinct aromatic sensations (much as a series of musical chords). This could explain why experts may describe wines using markedly different terms, but still equally acknowledge its varietal or stylistic origin.

Odor and taste interactions can be as complex as those of odorant mixtures. Odors can induce taste sensations and modify the perceived intensity of tastants, whereas sapid substances can affect the perception of odorants. For example, sugar can increase perceived fruitiness (von Sydow *et al.*, 1974). Tastants and odorants may also have mutual effects. For example, subthreshold mixtures (e.g., benzaldehyde and saccharin) may permit both to be detected (Dalton *et al.*, 2000).

Psychological factors also play an important role in odor response. For example, color (Iland and Marquis, 1993; Zellner and Whitten, 1999), as well as appropriate visual images (Sakai *et al.*, 2005), enhance perceived odor intensity. As already noted (Fig. 11.14), adding tasteless anthocyanins to a white wine can induce tasters to expect (and perceive?) flavors typical of red wines. Another example of tasting conditions affecting perception is provided by Brochet and Morrot (1999). Providing access to knowledge that indicated the apparent prestige of the wines tasted markedly affected the terms used to describe the wines. Pejorative terms were used when the wines were thought to be *vin de table*. In contrast, positive attributes were ascribed to the same wine when it was thought to be a *grand cru classé*.

Experience tends to generate idiotypic memories, against which new examples are compared (see Hughson and Boakes, 2002). In contrast, novice tasters use general sensory attributes, such as sweetness, on which to base assessment (Solomon, 1988).

Odor Assessment in Wine Tasting

Contemporary analytical instrumentation is far superior to human sensory acuity in chemical measurement and identification. In addition, "electronic noses" associated with computer neural networks can learn characteristic odor patterns (Buratti *et al.*, 2004; Martí *et al.*, 2005). These instruments may possess an array of sensors to different aromatic compounds or be connected to a mass spectrometer. They have been used in routine assays of off-odors in several foods and beverages. Electronic noses would also be useful in the preparation and verification of odor samples used by sensory panels. Electronic tongues are also being investigated. Nevertheless, the need for skilled tasters will remain into the foreseeable future.

In wine tasting, consumers are primarily interested in the positive, pleasure-giving aspects of wine attributes. Regrettably, little progress has been made in describing varietal, regional, or stylistic features in meaningful terms. Although suggestions for a practical range of wine descriptors have been proposed, only a small set of descriptive terms are usually necessary to discriminate a particular style (Lawless, 1984).

Wine fragrance is commonly subdivided into two categories, aroma and bouquet – their differentiation being based solely on origin. **Aroma** refers to odorants, or their precursors, derived from grapes. Although usually applied to compounds that give certain grape varieties their distinctive fragrance, it can also involve odorants that develop in grapes due to excessive sun exposure, disease, or overripening. There is no evidence supporting the common belief that grapes derive specific flavors from the soil in which they grow, as implied by the terms "flinty", "chalky" or "goût de terroir." The gun flint attribute is apparently derived primarily from the production of benzenemethanethiol (Tominaga *et al.*, 2003), presumably from the breakdown of cysteine during or after fermentation. Other related thiols may also contribute to this odor.

The other major category, **bouquet**, refers to aromatic compounds that develop during fermentation, processing, and aging. **Fermentative bouquets** include aromatic compounds derived from yeast (alcoholic), bacterial (malolactic) and grape-cell (carbonic maceration) fermentation. **Processing bouquets** refer to odorants derived from procedures such as the addition of brandy (port), baking (madeira), fractional blending (sherry), yeast autolysis (sparkling wines), and maturation in oak cooperage. **Aging bouquets** refer to the fragrant compounds that develop during wine maturation and aging.

Although the subdivision of wine fragrance into aroma and bouquet is common, it is difficult to use precisely. Similar or identical aromatics may be derived from grape,

yeast or bacterial metabolism, or from strictly organic chemical reactions, for example, acetic acid. In addition, it is often only with long experience and after assessing wines made by excellent wine-producing techniques that varietal aromas can be recognized.

An improvement in wine terminology would be beneficial even at the consumer level. It could help focus attention on the aromatic characteristics of wine that enhance appreciation. Vital to this goal is the availability of representative odor samples. Published lists are available (see Meilgaard *et al.*, 1982; Noble, Arnold *et al.*, 1984; Jackson, 2002). Regrettably, samples are a nuisance to prepare, maintain, and standardize. Even analytically pure chemicals can contain contaminants that can alter both the perceived intensity and quality of the odor. Microencapsulated (scratch-and-sniff) samples of representative wine flavors and off-odors would be convenient and efficient, but are not commercially available.

More is known about the chemical nature of wine faults than about the positive fragrant attributes of wine. The section that follows briefly summarizes the characteristics of several important off-odors in wine. A discussion of the volatile compounds important in wine fragrance and the origin of off-odors is given in Chapters 6 and 8. Directions for preparing faulty samples for training tasters are provided by Jackson (2002) and Meilgaard *et al.* (1982).

Off-odors

The quick and accurate identification of off-odors is vital to winemaker and wine merchant alike. For the winemaker, early remedial action often can correct the situation before the fault becomes serious or irreversible. For the wine merchant, avoiding losses associated with faulty wines can improve the profit margin. Consumers should also know more about wine faults, so that rejection, if justified, is based on genuine faults, not unfamiliarity, the presence of bitartrate crystals, or the wine being "too" dry (often incorrectly called "vinegary").

There is no precise definition of what makes a wine faulty – human perception is too variable. In addition, it is the vinous equivalent of incorrect grammar, and therefore open to interpretation. Furthermore, compounds that produce off-odors at certain concentrations are often deemed desirable at low concentrations, at which they may produce interesting and subtle nuances. In addition, faults in one wine may not be undesirable in another: for example, the complex oxidized bouquets of sherries, the fusel odors of *porto* and the baked character of madeiras. Some faults, such as a barnyardy odor, generated by ethyl phenols, may be considered pleasingly "rustic," or part of the *terroir* character of certain wines. The evident presence of ethyl acetate in the aroma of a wine is also usually considered a fault. However, in expensive Sauternes, it appears to be acceptable (or ignored). Even with oak, noticeable oakiness is a fault to some, whereas it a prized aspect for others. Nevertheless, there is general agreement among most wine professionals as to what constitutes an aromatic fault in table wines. In contrast, there is much less agreement on taste and mouth-feel faults.

Once one of the most frequently occurring wine faults, marked oxidation is now comparatively rare in commercial table wines. In extreme cases, it produces a flat, acetaldehyde off-odor. More frequently, it is expressed as a loss of freshness, or the development of odors characterized variously as cooked vegetable, cabbage, or simply pungent (Escudero *et al.*, 2000; Silva Ferreira *et al.*, 2003). In white wines, it is also associated with a "brownish" tint.

Although usually thought to result from oxygen ingress into the bottle, oxidation may occur in the absence of molecular oxygen – being catalyzed by metal ions, especially in the presence of light. Wine supplied in bag-in-box containers often oxidizes within a year of filling. This is considered to occur due to oxygen penetration around the spigot.

Factors influencing the tendency of bottled wine to oxidize are its phenolic content, notably *o*-diphenols, copper and iron contents, level of free sulfur dioxide; pH; and the temperature and light conditions during storage.

Significant flavor changes also develop upon bottle opening. Within the normal time frame of meal consumption, this is undetectable (Russell *et al.*, 2005). However, after several hours, the wine begins to lose its original character. The escape and oxidation of ethyl and acetate esters, and to a lesser extent volatile terpenols, may be involved in these changes (Fig. 8.54).

Presence of an ethyl acetate off-odor is less common than in the past (Sudraud, 1978). At concentrations below 50 mg/liter, ethyl acetate can add a subtle fragrance. However, at about 100 mg/liter, it begins to have a negative effect. This may result from its masking of the fragrance of fruit-smelling esters (Piggott and Findlay, 1984). At above 150 mg/liter, its own acetone (Cutex-like) odor becomes marked. It is usually more readily obvious in white than red wines.

Although ethyl acetate is produced early in fermentation, the concentration usually falls below the recognition threshold by the end of fermentation. A concentration of ethyl acetate sufficiently high to generate an off-odor usually results from the metabolism of acetic acid bacteria. An infrequent source of undesirable levels of ethyl acetate comes from the metabolism of *Hansenula* during spontaneous fermentation.

The metabolism of acetic acid bacteria also results in the accumulation of acetic acid (volatile acidity) to detectable levels. The thresholds of detection and recognition are approximately 100 times higher those of ethyl acetate. Vinegary wines typically are sharply acidic, with an irritating odor derived from the combined effects of acetic acid and ethyl acetate. Ideally, the concentration of acetic acid should not exceed 0.7 g/liter.

Although the benefits of sulfur dioxide are multiple, excessive addition can produce an irritating, burnt-match odor. This fault usually dissipates rapidly if the wine is swirled in the glass. However, because sulfur dioxide can initiate asthmatic attacks in sensitive individuals (Taylor *et al.*, 1986), there is a concerted effort worldwide to minimize its use.

A geranium-like odor can develop from the use of another wine preservative, sorbate. People hypersensitive to sorbate detect a butter-like odor when it is used. However, the most important off-odor potentially associated with sorbate use is a geranium-like off-odor. The sharp penetrating odor is produced by 2-ethoxyhexa-3,5-diene. It forms as a consequence of sorbate metabolism by certain lactic acid bacteria (see Chapter 8).

During fermentation, yeasts produce limited amounts of higher (fusel) alcohols. At concentrations close to their detection thresholds, they can add complexity to a wine's fragrance. If the alcohols accumulate to levels greater than about 300 mg/liter, they become a negative quality factor in most wines. However, detectable levels are an expected and characteristic feature of Portuguese ports (*porto*). The elevated fusel aspect comes from the addition of largely unrectified wine spirits.

Diacetyl usually is found in low concentrations, as a result of yeast metabolism, or as an oxidation by-product of oak constituents. Nevertheless, when present in amounts sufficient to affect a wine's flavor, its occurrence is usually associated with malolactic fermentation. Diacetyl is typically considered to possess a buttery aroma. Surprisingly, this attribute can be present despite diacetyl occurring at below threshold values (Bartowsky *et al.*, 2002). This presumably results from additive effects with other wine components. Typically considered desirable by most tasters, it is highly disagreeable to others. Bertrand *et al.* (1984) considered people fall into two distinct groups, based on their response to diacetyl. This may be due to its association with trace contaminant(s) possessing a vile odor. This attribute is presumably undetected to those responding positively to diacetyl.

Wines may express corked or moldy odors, due to the presence of a variety of compounds (see Chapter 8). Of these, the most fully documented is 2,4,6-trichloroanisole (TCA). It usually develops as a consequence of fungal growth on or in cork, following the use of PCP (a pentachlorophenol fungicide) on cork trees, or the bleaching of stoppers with chlorine. It produces a distinctive chlorophenol odor at a few parts per trillion. Other corky off-odors may come from the presence of 2,4,6-tribromoanisoles or 2-methoxy-3,5-dimethylpyrazine (see Chapter 8). Another moldy off-odor occasionally contaminating wine derives from the growth of filamentous bacteria, such as *Streptomyces*. Additional moldy odors can result from the production of guaiacol and geosmin by fungi, such as *Penicillium* and *Aspergillus*. Although most moldy (corky) taints come from cork, oak cooperage and infected grapes are also potential sources. It is critical that cooperage be properly treated to restrict microbial growth on its inner surfaces during storage.

Some fortified and dessert wines are purposely heated to over 45°C for several months. Under such conditions, the wine develops a distinctive, baked, caramel odor. Although characteristic and expected in wines such as madeira, a baked odor is a negative feature in table wines. In table wines, a baked odor is usually indicative of excessive heat exposure during transit or storage.

Hydrogen sulfide and mercaptans may be produced in wine during fermentation or aging. Their presence may be undetected, because they frequently occur at levels below their recognition threshold. Although hydrogen sulfide usually can be eliminated by mild aeration in the glass, sulfide by-products such as mercaptans are more intractable. Their removal usually requires the addition of trace amounts of silver chloride or copper sulfate. Mercaptans impart off-odors reminiscent of farmyard manure or rotten onions. Disulfides are formed under similar reductive conditions and generate cooked-cabbage to shrimp-like odors. Related compounds, such as 2-mercaptoethanol and 4-(methylthio)butanol, produce intense barnyardy and chive–garlic odors, respectively.

Light-struck (*goût de lumière*) refers to a reduced-sulfur odor that can develop in wine during exposure to light (see Chapter 8). This fault is but one of several undesirable consequences of wine exposure to light.

Several herbaceous off-odors have been detected in wines. The best known are associated with leaf (C_6) aldehydes and alcohols. They are derived from the oxidation of grape lipids. Fruit shading and maceration, particular climates, and some strains of lactic acid bacteria can influence the development of vegetable off-odors. Exposure to light can also induce a "strange, vegetable-like" off-odor in Asti Spumante (Di Stefano and Ciolfi, 1985). Its formation is inhibited by the presence of sorbate. Depending on one's reaction to the bell-pepper aroma of 2-methoxy-3-isobutylpyrazine, a characteristic fragrance in most 'Cabernet' cultivars, it is an off-odor or an enticing aroma compound.

A mousy taint occasionally noticed in wine is associated with the metabolism of spoilage microbes, principally species of *Lactobacillus* and *Brettanomyces*. The odor is caused by several tetrahydropyridines. Because they are not readily volatile at wine pH, their presence is seldom detected on smelling the wine. Their odors become evident on tasting or after swallowing (Grbin and Henschke, 2000). Winemakers often put a small amount of wine on their hand and use the "palm and sniff" technique for quick detection. Sensitivity to this taint can vary by 2 orders of magnitude (Grbin *et al.*, 1996).

Bitter-almond odors in wine may have several origins. One involves residual ferrocyanides following "blue fining." The decomposition of ferrocyanides can release small quantities of hydrogen cyanide, which possesses a bitter-almond odor. A more common source of bitter-almond odors is the microbial conversion of benzyl alcohol to benzaldehyde. The precursor, benzyl alcohol, may come from gelatin used as a fining agent, from grapes, or cement cooperage covered with or containing epoxy resins.

Additional off-odors include raisined (use of sun-dried grapes), cooked (wines fermented at high temperatures), stemmy (presence of green grape stems during fermentation), and rancio[3] (old oxidized red and white wines) (Cutzach *et al.*, 1999). Other off-odors, depending on one's subjective response, may be unique aroma compounds of particular cultivars, for example the foxy and strawberry-like aspects of some *Vitis labrusca* hybrids, and the methoxyisobutylpyrazines of several *V. vinifera* cultivars. Off-odors of unknown chemical nature that are occasionally noted are rubbery (possibly associated with reduced sulfur compounds), weedy, and earthy (*goût de terroir*).[4]

Wine Assessment and Sensory Analysis

Wine assessment and sensory analysis cover various aspects of wine evaluation. Examples include preference determination, assessment of specific attributes, and the development of flavor profiles.

The intent of a tasting profoundly affects both its design and analysis. Wine-society tastings often involve wines whose origin and price are known in advance. Analysis, if any, involves little more than ranking the wines in order of preference. In regional and international tastings, wines usually are grouped by region, variety, and style. Simple numerical averaging of the scores is used

---

[3] *Rancio* has different meanings, and chemical origins, when used in reference to dessert wines or old brandies. In wines, it apparently is primarily due to the presence of high concentrations of sotolon, whereas in brandies, it is largely due to methylketones.

[4] Exactly what is meant by this term is not clear, but it presumably does not imply an actual soil-like smell. Most of the odor of soil comes from compounds such as geosmin. These largely originate from the growth of actinomycetes in the soil.

to develop a ranking, usually without analysis of significance. Tastings intended to assess vini- and viticultural practices require more exacting conditions, including appropriate experimental design for the legitimate application of statistical tests. These can estimate the degree and significance of unavoidable taster variation and unsuspected interactions that might obscure or compromise valid conclusions. Computers have made sophisticated statistical procedures readily available.

Conditions for Sensory Analysis

TASTING ROOM

Ideal lighting is still considered to be natural north luminance. However, under most tasting situations this is impossible. In addition, the light source may be less important than previously thought (Brou *et al.*, 1986). Any bright, white light source probably is acceptable, although full-spectrum fluorescent lighting is preferable. In situations in which wines of different hues must be tasted together, it can be advantageous to have the option of using red light to disguise the color. The use of red or black wine glasses is an alternative technique. Under most situations, however, use of white tabletops or countertops, and white or neutral-colored walls, facilitate color differentiation.

Tasting rooms must be adequately air-conditioned, both for taster comfort and to limit the development of a background odor. Covers on wine glasses also help to limit the escape and accumulation of wine odors in tasting rooms. Watch glasses are commonly used for this purpose, but plastic petri-dish bottoms are superior. Because the bottom may fit snugly over the glass mouth, holding the cover during swirling is unnecessary. The use of dentist-type sinks (Plate 11.1), or cuspidors with tops, at each tasting station further minimizes odor buildup.

Tasting stations should be physically isolated (cubicles) to limit taster interaction (Fig. 11.19). Silence also prevents among-taster influence and facilitates concentration. Where tasters cannot be physically separated, the order of wine presented to each taster should be varied to negate taster interaction.

NUMBER OF WINES

The number of wines adequately evaluated per session depends on the level of assessment required for each sample. If the rejection of faulty samples is the only intent, 20–50 wines can easily be assessed at one sitting. However, if the wines are similar and must be compared critically, five to six wines is a reasonable limit. The wines should be tasted at a relaxed pace to avoid odor adaptation, or an increase in perceived astringency. Frequent breaks are desirable if wines are

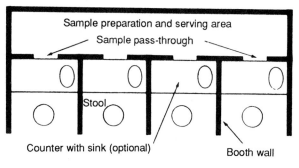

Floor plan

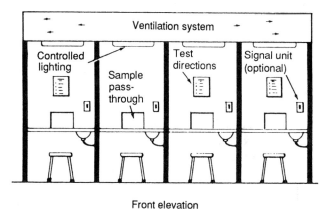

Front elevation

Figure 11.19 Diagram of the booth area in a sensory-evaluation facility (not drawn to scale). (From Stone and Sidel, 1985, by permission)

assessed critically. Detailed written analyses will often require 15 min or more per wine. The evaluation of each wine may be spread over 30 min to observe fragrance development and duration.

PRESENTATION OF SAMPLES

Glasses   Glass shape is well known to affect wine perception. This effect has partially spawned the production of wine glasses, supposedly accentuating the properties of particular wines. Only recently have researchers studied the actual significance of glass shape to wine assessment (Cliff, 2001; Delwiche and Pelchat, 2002; Hummel *et al.*, 2003; Russell *et al.*, 2005).

As expected, shape affects wine perception, but the differences are relatively small. Of those tested, the International Standards Organization (ISO) wine-tasting glass was found to be fully adequate for both red and white wines. It also enhanced color discrimination by maximizing wine depth relative to volume (Cliff, 2001).

The ISO glass (Fig. 11.20) expresses the basic requirements of a wine-tasting glass. Its bowl is broader at the base than the top, and the glass is clear and colorless. Its sloping sides permit vigorous swirling. For swirling, the glass should be no more than one-third full. Its thin crystal construction is also esthetically pleasing. Noncrystal

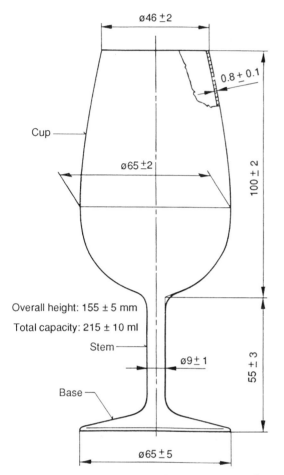

ø46 ±2

0.8 ± 0.1

Cup

ø65 ±2

100 ± 2

Overall height: 155 ± 5 mm

Total capacity: 215 ± 10 ml

Stem

ø9 ± 1

55 ± 3

Base

ø65 ± 5

Figure 11.20 ISO wine-tasting glass; dimensions are in millimeters. (Courtesy of International Standards Organization, Geneva, Switzerland)

versions are available (e.g., Durand Viticole), or of very similar shape (Libby Citation No. 8470). For sparkling wines, flute-shaped glasses are preferred to permit adequate viewing of the wine's effervescence. There is no published evidence supporting claims that particular shapes uniquely enhance the attributes of specific table or fortified wines. Although glass shape modifies the dynamics of aroma release from wine, appearance and feel may be of greater significance to a consumer's appreciation of wine than its chemical sensory attributes.

Despite which shape is chosen, it is important that all glasses in a tasting be identical. They also need to be filled to the same level. This permits each wine to be sampled under equivalent conditions. Between 30 and 50 ml is adequate for most analyses. Not only are small volumes economic, but they facilitate holding the wine at a steep angle (for viewing color and clarity) and allow vigorous swirling (to enhance the release of aromatics).

It is particularly important that glasses be properly washed, rinsed and stored between tastings. Residual

odors can easily distort a wine's fragrance, and traces of oily residues or detergent readily suppress the effervescence of sparkling wines. Storage in cardboard boxes or painted cabinetry can quickly contaminate the glasses with alien odors.

Temperature   There is general agreement that most red wines taste best between 18 and 20 °C. Young carbonic maceration wines, such as most *nouveau* wines, are preferred at between 14 and 16 °C. With white wines, there is less agreement, some preferring 11–13 °C, whereas others suggest 16 °C. Generally, the sweeter the wine, the cooler the optimal temperature. There also is divergence in opinion concerning the ideal temperature for sparkling wines, varying between 8 and 13 °C. Sweet fortified wines are commonly served at about 18 °C, whereas dry fortified wines, notably *fino* sherries, are taken cool (14 °C) to cold (8 °C).

In large public tastings, it may be impossible to serve wines at the "optimal" temperature. If they are served at or above room temperature (~20 °C), off-odors may be accentuated. If the wine temperature can be controlled, it might be preferable to obtain a consensus from the judges concerning the serving temperature. If the wines are to be assessed over an extended period, it is better to present the wines cooler than ideal, so that the wine passes through the preferred temperature range during the tasting.

Wine Identity   Wine identity should be withheld at all but consumer or informal tastings. This is facilitated by decanting into carafes or prepouring into glasses (covered to prevent aroma loss). This removes clues as to origin, such as bottle size, shape, or color. A simpler, but less effective technique, used at informal tasting, is covering the bottle with a paper bag.

Information provided to tasters depends on the purpose of the tasting. If the wines are to be judged relative to a particular style, variety, or region, this clearly needs to be known. However, if wines from various regions, varieties, or styles are being assessed together, it may be inappropriate for the tasters to know this in advance. If simple hedonic preferences are being assessed, information concerning origin and price probably is best concealed. Double-blind tests may be useful in critical tastings, in which neither the tasters nor the pourers know the origin or nature of the wines being assessed.

Breathing   Opening bottles in advance to "breathe" is unnecessary and undesirable. The limited wine/air interface generated by cork removal is minimal. Even decanting, which exposes the wine to more oxygen, does not rapidly generate noticeable changes. However, the equilibrium between weakly bound aromatics in the wine and the headspace shifts. This can result in the release of aromatics – what is termed the wine's "opening." It is not infrequently noted in finer quality, young wines. This is a fascinating phenomenon. It should not be wasted by having it occur unnoticed in a decanter. It is best observed in the glass as the wine is periodically swirled. With very old wines, though, the fragrance may be so evanescent that it may dissipate completely with a few minutes. In this case, tasting should commence as soon as possible after decanting (to separate the wine from any sediment).

Presentation Sequence   To avoid unintended or perceived ranking by the presentation sequence, a two-digit code for each wine may be used. Amerine and Roessler (1983) suggest numbers from 14 to 99, to avoid psychological biases associated with lower numbers. Code numbers are used in a random order.

If different groups of wines are tasted, the standard serving recommendations of white before red, dry before sweet, and young before old are appropriate. If possible, each set of wines should differ from the previous one, to help maintain interest, and minimize fatigue throughout the tasting session.

Time of Day   It is common to hold technical tasting in the late morning or afternoon. This is based partially on the view that sensory acuity is optimal when a subject is hungry. Although cyclical changes in sensory acuity occur throughout the day, variation from individual to individual makes designing tasting around this feature of dubious value.

Replicates   Replicates seldom are incorporated into the protocol of a tasting because of the extra time and expense involved. If the tasters have established records of consistency, there may be little need for replication. Nevertheless, duplicate bottles should be available to substitute for faulty wines. The presence of an off-odor will severely depreciate its quality ranking (Bett and Johnsen, 1996).

Wine Score Cards

Although many score cards are available, few have been studied sufficiently to establish how quickly they come to be used consistently. This obviously is critical if important decisions are based on the results. Generally, the more detailed the score cards, the slower is the development of consistent use (Ough and Winton, 1976). Thus, the score cards should be as simple as possible, compatible with the intent of the tasting. The incorporation of unnecessary detail reduces consistency of use and, therefore, the potential value of

Table 11.2   The modified Davis score card for wine grading

Characteristic	Maximum points
Appearance	2
Cloudy 0, clear 1, brilliant 2	
Color	2
Distinctly off 0, slightly off 1, correct 2	
Aroma and bouquet	4
Vinous 1, distinct but not varietal 2, varietal 3	
Subtract 1 or 2 for off-odors, add 1 for bottle bouquet	
Vinegary	2
Obvious 0, slight 1, none 2	
Total acidity	2
Distinctly high or low 0, slightly high or low 1, normal 2	
Sweetness	1
Too high or too low 0, normal 1	
Body	1
Too high or low 0, normal 1	
Flavor	2
Distinctly abnormal 0, slightly abnormal 1, normal 2	
Bitterness	2
Distinctly high 0, slightly high 1, normal 2	
General quality	2
Lacking 0, slight 1, impressive 2	

Source: From Amerine and Singleton, 1977, by permission

the data obtained. Conversely, insufficient choice may result in **halo-dumping**. This refers to use of an existing, but unrelated category, to register an important perception not recorded elsewhere (Lawless and Clark, 1992; Clark and Lawless, 1994).

Possibly the most widely used scoring system is the modified Davis score card (Table 11.2). It was developed to compare young table wines as a means of focusing on areas for improvement. Thus, it has several weaknesses when applied to fine wines of equal and high quality (those showing few, if any, faults). In addition, it contains aspects that are inappropriate for, or lacks critical features essential to, particular wine styles. The lack of a site for comments on effervescence is an obvious example relative to sparkling wines. Finally, the Davis score card often does not reflect features considered central to present-day winemaking styles (Winiarski *et al.*, 1996).

If ranking and detailed sensory analyses are desired, two independent scoring systems may be valuable (Table 11.2 and Fig. 11.21). Not only does this simplify assessment, but it may also avoid a **halo effect**, in which one assessment prejudices another (Lawless and Clark, 1992). For example, astringent bitter wines might be scored poorly on overall quality and drinkability, but be rated more leniently when marked separately on its various sensory attributes. Separating the

Sample Number: _____	Wine Category: _____	E x c e p t i o n a l	V e r y G o o d	A b o v e A v e r a g e	A v e r a g e	B e l o w A v e r a g e	P o o r	F a u l t y	Comments
Visual	Clarity								
Odor	Intensity								
	Duration								
	Quality								
Flavor	Intensity								
	Duration								
	Quality								
Finish	Duration								
	Quality								
Conclusion									

Figure 11.21 Hedonic wine-tasting sheet for still wines.

two assessments temporally decreases the likelihood of one assessment influencing the other. This view is consistent with studies that indicate that total taste intensity reflects the intensity of the strongest component, whereas the perception of individual components may be differentially influenced by interaction among the components (McBride and Finlay, 1990). If the panel members are sufficiently experienced, discrimination among wines may be as accurate with simple hedonic scales as with detailed scales (Lawless *et al.*, 1997).

No one score card is universally applicable to all wines. For example, effervescence is very important in assessing sparkling wines, but irrelevant for still wines. In addition, authors in different countries seem to rate features differently. For example, out of 20, taste and smell are given 9 and 4 marks, respectively by Johnson (1985) in England, whereas taste, smell, and flavor are assessed 6, 6, and 2 marks, respectively, by Amerine and Singleton (1977) in the United States. For the descriptive analysis of the wines (see later), score cards specific of each wine type are required.

An example of a simple hedonic tasting sheet is given in Fig. 11.21. More detailed sensory analysis forms can be found in Jackson (2002). An example of a form specifically designed for sparkling wines is found in Anonymous (1994).

Number of Tasters

If wines are to be sampled repeatedly, ideally the same tasters should be present at all tastings. However, in most tasting situations, it may be impossible or unnecessary for all tasters present at every tasting. Usually, a nucleus of 12–15 tasters is assembled so that at least 10 tasters are present at any one tasting. If continual monitoring of the tasters is performed, this number should assure that an adequate selection of tasters will be "in form" to permit a statistically valid data analysis. The exclusion of results from individuals who are assessed to be temporarily "out of form" (see later) should be done prior to data analysis.

Because individual tasters perceive tastes and odors differently, they also may diverge considerably in their concepts of wine quality. Therefore, the number of tasters should be sufficiently large to buffer individual idiosyncrasies, or the tasters should be trained and selected for the particular skills necessary. Which approach is preferable will depend on the purpose and nature of the tasting. When the panel is intended to function as a precise sensory instrument, the members' sensory acuity must be up to the task. Detailed descriptive analysis requires extensive and specific training, whereas consumer-acceptance testing should involve no prior coaching.

In expansive public tastings, the number of tasters usually is correspondingly large. Individual tasters can taste only a limited number of wines accurately and must be asked to assess only those wines within their range of expertise. For example, tasters having little experience with dry sherries cannot be expected to evaluate these wines adequately.

When the purpose of the tasting is to establish consumer preference, a representative cross-section of the target group is essential. Marked divergence can exist both within and among consumer subgroups. For example, sweetness and freedom from bitterness and astringency are particularly important to members of less-affluent social groups and infrequent wine drinkers (Williams, 1982). Woodiness and spiciness have often been considered to be disliked by infrequent wine drinkers, but appreciated by most wine experts. However, this generality probably glosses over important subtleties. This is suggested by divergent conclusions concerning the acceptance of oak flavor in studies by Hersleth *et al.* (2003) and Binders *et al.* (2004). The importance of color and aroma to general acceptability increases with consumer age and the frequency of wine consumption. Not surprisingly, differences in experience also influence the terms used by people to describe wine (Dürr, 1984; Solomon, 1990).

Tasters

Training

In the past, wine assessment was done primarily by winemakers or wholesalers. As all individuals possess biases and genetic idiosyncrasies, most wine assessment is now done by teams of trained tasters. This has required the preparation of more tasters than generated by the former, relatively informal, in-house approach. There also is a desire to have standardization in the instruction and assessment of sensory skills.

Sensory training often consists of an extensive series of tastings of a wide variety of wines. This rapidly gives trainees a basic experience on which to build. Because the instructional rational and availability of wines vary widely, specific suggestions for appropriate wines would be inappropriate and inutile.

For economy and convenience, aspects of varietal grape aromas may be simulated by producing standard odor samples. These have the advantage of being continuously available as a reference. Standards may be prepared and stored in small sample bottles under wax or in wine. Examples can be found in Williams (1978), Noble *et al.* (1987) and Jackson (2002).

Although training improves the consistency with which descriptive terms are used, it may not eliminate idiosyncrasies in term use (Lawless, 1984). Thus, unambiguous

identification of odors in wine frequently requires the use of several tasters (Clapperton, 1978). The need for developing consistent term use depends on the purpose of the tasting. Achieving this goal may be even more difficult than suspected if Parr *et al.* (2004) are correct. They contend that perceptive skill (ability to recognize odors on repeat exposure) is poorly correlated with linguistic ability (correct identification of an odorant).

In addition to correctly recognizing varietal odors, the identification of odor faults is a vital component of taster training. In the past, faulty samples usually were obtained from wineries, but samples prepared in the laboratory are preferable. They can be presented in any wine and at any desired concentration (Jackson, 2002).

Training usually includes gustatory samples, prepared in either aqueous solutions or wine. As with odor training, testing allows the student to discover personal idiosyncrasies. Sample preparation is described in Marcus (1974) and Jackson (2002).

When selecting tasters for training, it is more important to select for motivation and ability to learn, than initial skill. Motivation is critical to both learning and consistent attendance at tastings. Because initial skill in recognizing odors usually reflects previous exposure (Cain, 1979), not innate ability, measures of learning ability are more important in screening potential tasters (Stahl and Einstein, 1973). In addition, some people initially anosmic to a compound develop the ability to smell the compound after repeated exposure (Wysocki *et al.*, 1989). Enhanced sensitivity can also result following repeat exposure in individuals of normal sensory acuity (Stevens and O'Connell, 1995). However, decreases in thresholds, averaging 5 orders of magnitude, appear to be restricted to women during their reproductive years (Dalton *et al.*, 2002).

One of the multiple problems in assessing tasting ability is assuring that samples assessed over several days are identical (avoiding between sample differences). This can be partially offset by the use of a dispensing machine (Plate 11.2). Samples can be removed over several days without the main sample being exposed to oxygen.

Basic screening tests are usually designed to eliminate tasters with insufficient sensory acuity. Subsequently, the ability to develop identification and differentiation skills, and consistent use of important sensory terms, are assessed. Basic screening tests are discussed in Amerine and Roessler (1983), Basker (1988), and discrimination tests by Jackson (2002).

Measuring Tasting Acuity and Consistency

Assessing tasting acuity has two primary objectives – to measure skills learned during training and monitor consistency during and between tasting sessions.

Depending on the purpose of instruction, various skills are required. In descriptive sensory analysis, odor and taste acuity are essential, whereas in quality evaluation, discrimination among subtle differences is more important. However, regardless of the task, the correct and consistent use of terms is essential. Because tests of consistency require repeat sampling, the tests may involve only olfactory sensations. Considerable economy is achieved by having all participants smell the same samples. Taste and mouth-feel acuity are the only tests requiring literal tasting. Tests ideally should be conducted over several days (or weeks). Acuity often expresses considerable daily variation.

The assessment of performance with a series of simple tests before or during each tasting session is valuable for several reasons. Because individual acuity varies, data from tasters having "off" days can be removed before the analysis of the tasting results. Furthermore, the tests show whether individual tasters are using terms consistently and if they require a refresher course.

Brien *et al.* (1987) distinguish five aspects of taster consistency. Discrimination is defined as a measure of the ability to distinguish among wines of distinct character; stability refers to reproducibility of scoring results for similar wines from tasting to tasting; reliability assesses the reproducibility of score differences between replicate sets of the same wines; variability gauges the range of scores between replicate wine samples; and agreement evaluates the scoring differences among tasters. Of these measures, two require identical wines be sampled repeatedly, either on one occasion (reliability), or on separate occasions (variability).

Measures of discrimination, stability, and variability are derived from the analyses of variance among scores in successive tastings. Measures of agreement and reliability are derived from correlation matrices obtained from the scoring results. For recent views on these and other measures of panel reliability see Cicchetti (2004), Huon de Kermadec and Pagès (2005), and Latrielle *et al.* (2006).

Although analyses of consistency are useful, caution must be used in their interpretation. Although a high degree of agreement may appear desirable, it may indicate uniform lack of skill, or an inadequate reflection of normal variability in perceptive ability. The latter is especially important if the results are expected to reflect consumer opinion. Also, measures of consistency that require replicate tastings may be invalid if the samples are not actually identical. In addition, reliability may be affected by the number of replicates, improving as the tasters learn from repeat sampling. Finally, variability may be higher with experienced tasters than with inexperienced tasters. This may simply result from the experienced taster developing the confidence to use a wider range of marks.

Wine-tasting Technique

There is no single procedure appropriate for all tasting situations. What is required depends on the conditions and purpose of the tasting. Tasting during a meal is typically simple and short – most of the attention being directed to the meal and the social nature of the occasion. In contrast, critical tastings require strict silence, absence of any distracting odors, and presentation of the wine in black glasses. Normally, though, every aspect of a wine's potential quality is assessed. The procedure described below focuses sequentially on separate wine attributes. Because complete assessment can take up to 30 minutes, several wines are usually assessed together.

Appearance

Except in special situations, visual attributes are judged first. To improve light transmission, the glass is tilted against a bright, white background (35–45°). This produces a curved meniscus of varying depths, through which the wine's color and clarity may be assessed.

Ideally, the appearance should give a sense of pleasure. More significantly, it tends to provide clues as to the wine's sensory attributes, including potential faults. These should be interpreted with caution, to avoid allowing visual clues from unduly biasing one's opinions.

CLARITY

Wine should be brilliantly clear. Haziness in barrel samples is of minor concern, because it is eliminated before bottling. Cloudiness in bottled wine is another matter, despite its limited association with modified taste or aromatic attributes. Because most sources of cloudiness are understood and controllable, haziness in commercially bottled wine is rare. The major exception involves some well-aged red wines. Careful pouring or prior decanting can avoid suspending the sediment in the wine on serving.

COLOR

The two most significant features of a wine's color are its hue and depth. Hue denotes its shade or tint, whereas depth refers to the relative brightness of the color. Both aspects can provide clues as to grape maturity, duration of skin contact, cooperage use, and wine age. Immature white grapes yield almost colorless wines, whereas fully to over-mature grapes generate yellowish wines. Increased grape maturity often enhances the color intensity of red wine. The extent to which these tendencies are realized depends on the duration of skin contact. Maturation in oak cooperage speeds age-related color changes, but temporarily enhances color depth. During aging, golden tints in white wines increase, whereas red wines lose color density. Eventually, all wines take on tawny to brown shades.

Because many factors affect wine color, it is impossible to be dogmatic about the significance of a particular color. If the origin, style, and age of the samples are known, color can indicate the "correctness" of the wine. An atypical color can be a sign of several faults. The less known about a wine, the less significant color becomes in assessing quality.

Tilting the glass has the advantage of creating a gradation of wine depths. Viewed against a bright background, a range of color characteristics is observed. The meniscus provides one of the better measures of relative wine age. A purplish to mauve edge is an indicator of youthfulness in a red wine. A brickish tint along the rim is often the first sign of aging. In contrast, observing wine from the top is the best means of judging color depth.

The most difficult task associated with color assessment is expressing the impressions meaningfully. There is no accepted terminology for wine colors. In addition, color terms frequently are used inconsistently. Until a practical standard is developed, use of a few, simple, relatively self-explanatory terms is preferable. Terms such as purple, ruby, red, brick and tawny for red wines, and straw, yellow, gold and amber for white wines, combined with qualifiers such as pale, light, medium and dark are probably sufficient and adequate.

VISCOSITY

Viscosity refers to the resistance of wine to flow. Typically, detectable differences in viscosity can be found only in dessert or highly alcoholic wines. Because these differences are minor and of diverse origin, they are of little diagnostic value.

SPRITZ

Spritz refers to the bubbles that may form on the sides of a glass. If present, they usually develop along the bottom and sides of glass. Alternately, they may be detected as a slight effervescence in the wine, or detected as a prickling on the tongue. Active and continuous bubble formation is expected only in sparkling wines. In the latter instance, the size, number, and duration of the bubbles are important quality attributes.

When present, slight bubbling usually is an indicator of early bottling, before excess dissolved carbon dioxide in the wine has escaped. Infrequently, a slight spritz may result from post-bottling malolactic fermentation. In either case, it is of no significance.

TEARS

Tears (rivulets, legs) develop and flow down the sides of glass following swirling. They are little more than a crude indicator of alcohol content. Other than for the intrigue or visual amusement they may provide, tears are sensory trivia.

Orthonasal Odor

Tasters are often counseled to smell the wine before swirling. This assessment exposes the taster to the wine's most ethereal fragrance attributes. Subsequent assessments follow swirling.

Learning to effectively swirl the wine usually takes some practice. Until comfortable with the process, it is best to start by slowly rotating the base of glass on a level surface. Most of the action involves a cyclical arm movement at the shoulder, while the wrist remains stationary. As one becomes familiar with the action, shift to a wrist-induced swirling and slowly lift the glass. Some connoisseurs hold the glass by the edge of the base. While adequate, it is an affectation.

Swirling increases air–wine contact, facilitating the liberation of aromatic compounds. The incurved sides of tulip-shaped glasses permit not only vigorous rotation, but also slightly concentrate the released aromatics. Whiffs are taken at the rim of the glass and then in the bowl. This permits sensation of the fragrance at different concentrations, potentially generating distinct perceptions. Considerable concentration is usually required in detecting and recognizing varietal, stylistic, or regional attributes. It frequently requires repeat attempts and a process of elimination. As the primary source of a wine's unique character, study of its fragrance warrants the attention it demands.

Occasionally, glass or plastic covers are placed over the mouth of the glass. Covers serve two purposes. With highly fragrant wines, it limits aromatic contamination of the surroundings. Such contamination can seriously complicate the assessment of less aromatic wines. The primary function, though, is to permit especially vigorous swirling of the wine (if held on tightly with the index finger). This can be valuable when the wines are aromatically mild.

No special method of inhalation is required when assessing wine (Laing, 1986; Fig. 11.22). Breathing in for about 2 seconds seems fully adequate. Longer periods generally lead to adaptation and loss of sensitivity. Although the wine should be smelled several to multiple times, each sniff should be separated by about 30–60 seconds. Olfactory receptors take about this long to reestablish their intrinsic sensitivity. In addition, measurements of the rate of wine volatilization suggest that the headspace takes about 15 seconds for replenishment (Fischer *et al.*, 1996). In comparative tastings, each wine should be sampled in sequence. This avoids odor fatigue from sampling the same wine over a short period.

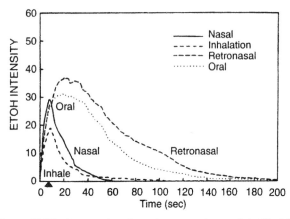

Figure 11.22 Average time–intensity curves for nasal (sniff), inhalation (inhale), retronasal (sip), and oral (sip, nose plugged) response for ethanol intensify of 10% v/v ethanol. (From Lee, 1989, reproduced by permission)

Ideally, assessment of olfactory features should be spread over about 30 minutes. This period is necessary to evaluate features such as duration and development. Duration refers to how long the fragrance lasts and development denotes changes in its intensity and modality. Both are highly regarded attributes, and particularly important in premium wines. The higher costs of these wines are justifiable only if accompanied with exceptional sensory endowments.

Regardless of the technique employed, recording one's impressions clearly and precisely is important. This is difficult for everyone, possibly because we are not systematically trained from an early age to develop verbal–olfactory associations. The common difficulty in recalling odor names has been aptly dubbed "the-tip-of-the-nose" phenomenon by Lawless and Engen (1977). For this purpose, fragrance and off-odor charts are often provided (Noble *et al.*, 1987; Jackson, 2002).

Stress on descriptive terms is too often misinterpreted, especially by consumers. Charts are intended to guide as well as encourage attention to a wine's aromatic attributes. Once people recognize the importance of examining the wine's olfactory attributes, search for descriptive terms often becomes irrelevant, and may become counter productive. Most people have legitimate difficulty in adequately describing olfactory sensations. It is more valuable for consumers to concentrate on recognizing the differences that exemplify varietal characteristics, production styles, and wine age. Except for educational purposes, lexicons of descriptive terms are best left for the purposes for which they were initially intended – descriptive sensory analysis.

Impressions (both positive and negative) should be recorded on some form of chart. Multiple forms have been generated over the years, depending on the purpose.

For a range of examples see Anonymous (1994) and Jackson (2002).

In addition to verbal descriptions, the dynamics of temporal odor changes can be represented graphically as a line drawn on a time/intensity axis. Both qualitative and intensity transformations can be clearly and rapidly expressed. This is especially useful when time is at a premium. The technique also helps focus the taster's attention on the dynamic nature of wine quality. Most tasting charts give a more static impression.

In-mouth Sensations

TASTE AND MOUTH-FEEL

After an initial assessment of fragrance, attention turns to taste and mouth-feel. About a 6–8 ml sample is sipped. As far as feasible, the volume of each sample should be kept equivalent of valid comparison among samples. Actively moving the jaw, or rolling with the tongue, brings wine in contact with all sensory regions in the mouth.

The first taste sensations recognized are those of sweetness and sourness. Sweetness (if detectable) is generally most noticeable at the tip of the tongue. In contrast, sourness is more evident along the sides of the tongue and insides of the cheek. The sharp aspect of acidity typically lingers considerably longer than that of sweetness. Response to bitterness is slower to develop, and may take upward of 15 seconds or more before reaching its peak, usually in the central, posterior portion of the tongue. Thus, it is important to retain the wine in the mouth for at least 15 seconds, preferably longer. During this period, the taster concentrates on mouth-feel sensations, such as the rough, dry, dust-in-the-mouth sensations of astringency, and any burning or prickling sensations. These and other tactile sensations are dispersed throughout the mouth, without specific localization. Subsequently, focus turns to integration of these perceptions on the overall in-mouth sensation.

As indicated, the temporal sequence of detection helps confirm specific taste sensations (Kuznicki and Turner, 1986). However, the duration of these sensations is not particularly diagnostic. Persistence reflects more tastant concentration and maximum perceived intensity than its category (Robichaud and Noble, 1990). Although significant in some critical tastings, the purpose of noting sapid sensations is not so much to record individual sensations as to concentrate on how they interact to generate overall perceptions, such as balance, flavor, and body.

There are differing opinions on whether taste and mouth-feel should be assessed with the first sip, or during subsequent samplings. Tannins react with proteins in the mouth, diminishing their potential bitter and

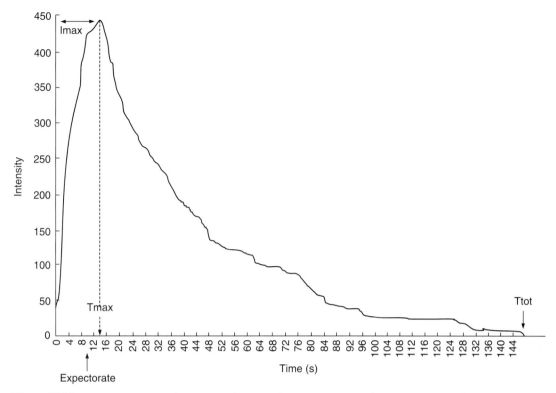

Figure 11.23 Average T-1 curve for retronasal (in-mouth) aroma intensity for response to vanillin and limonene. Tmax (time to maximum intensity), Imax (maximum intensity), and Ttot (total duration) values are indicated. (From Mialon and Ebeler, 1997, reproduced with permission)

astringent sensations. These reactions probably explain why red wines are usually less bitter and astringent on the first than subsequent samplings. The first taste more closely simulates the perception generated when wine is taken with food. If this is an important aspect to assess, it is essential that the tasting progress slowly. This permits stimulated salivary production to compensate for its dilution during the tasting.

In professional tastings, sampling typically occurs in fairly rapid succession. This can lead to "carry-over" effects, where subsequent wines appear more astringent and/or bitter. To avoid this, water, bread or unsalted crackers are commonly available to cleanse the palate. Various recommendations have come out of studies on these and other palate cleansers (Brannan *et al.*, 2001; Colonna *et al.*, 2004; Ross *et al.*, 2006). In some, pectin (1 g/litre) was found superior, in others unsalted crackers. They do concur, though, in noting that water is a poor palate cleanser.

RETRONASAL ODOR

To enhance the in-mouth detection of fragrance, tasters frequently aspirate the wine. This involves tightening the jaws, pulling the lips slightly ajar, and drawing air through the wine. Alternately, some tasters purse the lips before aspirating the wine. Either procedure

increases volatilization – analogous to swirling wine in the glass. Although less effective, vigorous agitation of the wine in the mouth has a similar effect. The liberated aromatic compounds flow up into the nasal passages, producing what is called **retronasal olfaction** (Fig. 11.23). The combination of retronasal olfaction with taste and mouth-feel generates the perception called **flavor**. In addition, the volatility of wine constituents can be markedly different in-mouth than in-glass (Diaz, 2004). This results from factors such as dilution in, and modification by, saliva, and the changed temperature. These perceptions should be recorded quickly as they are often evanescent and change unpredictably.

Some tasters complete their assessment of a wine's fragrance with a prolonged aspiration. Following inhalation, the wine is swallowed, and the vapors slowly exhaled through the nose. Any aromatic sensations detected are termed the **after-smell**. While occasionally informative, it typically is of value only with highly aromatic wines such as ports.

Following assessment, the wine is either swallowed or expectorated. In wine appreciation courses, wine societies, and the like, the samples are typically consumed. Because the number of wines being tasted is often small, and assessment is not critical, any minor effect of alcohol consumption on tasting skill is insignificant. However, twenty or more wines may be sampled in competitions or technical tastings. Consequently, consumption must be assiduously avoided. Scholten (1987) has shown that expectoration avoids significant amounts of alcohol accumulating in the blood.

Finish

Finish refers to the aromatic and sapid sensations that linger following swallowing or expectoration. It is the vinous equivalent of a sunset. Typically, the longer the finish, the more highly rated the wine. Some tasters consider its duration a major indicator of quality. Its measure may be formalized in the term *caudalie*. One *caudalie* represents the duration of the finish for one second. Fruity–floral essences, associated with refreshing acidity, epitomize most superior white wines; while complex berry fragrances combined with flavorful tannins exemplify the best red wines. Fortified wines, possessing more intense flavors, have a very long finish. Exceptions to the generally desirable nature of a protracted finish are features such as a persistent metallic aspect, or excessively acidic, bitter and astringent sensations.

The finish may be affected by features such as the volatility and polarity of individual aromatic compounds, and how these properties are affected by the wine matrix and conditions in the mouth (see Buettner, 2004). Matrix features include aspects such as the changing alcohol content of wine residues in the mouth, or the presence of binding compounds such as mannoproteins. Another significant factor may be the action of salivary enzymes on chemical persistence or food constituents.

Assessment of Overall Quality

After focusing on the sensory aspects of individual sensations, attention shifts to integrating the sensations. This may involve aspects of conformity with, and distinctiveness within, regional standards; the development, duration, and complexity of the fragrance; and uniqueness of the tasting experience.

Many of the terms used for overall quality have been borrowed from the art world. As such, they are very subjective, varying considerably from individual to individual. Despite this drawback, most professional tasters tend to agree in general on the relative application of these terms. For wine, **complexity** refers to the presence of many, distinctive, aromatic elements, rather than one or a few easily recognizable odors. **Balance** (harmony) denotes the perceptive equilibrium of all olfactory and sapid sensations, where individual perceptions do not dominate. At its simplest, it applies to the acid/sugar interaction in the mouth, but involves all sapid constituents.

However, the true core of this attribute involves the integration of olfactory and sapid sensations. Balance is lost when excessive astringency reduces appreciation of the jammy fruitiness of a red wine, or by insufficient fragrance or acidity in a sweet wine. Occasionally, individual aspects may be sufficiently intense to give the impression that balance is on the brink of collapse. In this situation, the near-imbalance can donate a **nervous** aspect that can be fascinating. **Development** designates changes in the aromatic character that occur throughout the sampling period. Ideally, these changes maintain interest and keep drawing the taster's attention back to its latest transmutation. **Duration** refers to how long the fragrance retains a unique character, before losing its individuality, and simply becoming vinous. **Interest** is the combined influences of the previous factors on retaining the taster's attention. Implied, but often not specifically stated, is the requirement for both **power** and **elegance** in the wine's sensory characteristics. Without these attributes, attractiveness is short-lived. If the overall sensation is sufficiently remarkable, the experience becomes unforgettable, an attribute Amerine and Roessler (1983) call **memorableness**. This feature is particularly important in the training of tasters and directing future expectations.

Most European authorities feel that quality should be assessed only within regional appellations, counseling against comparative tastings across regions or grape varieties. Although these restrictions make tastings simpler, they negate much of their value in promoting quality improvement. When tasting concentrates on artistic quality, rather than stylistic purity, comparative tasting can be especially revealing. Comparative tastings are more popular in the UK and New World, where artistic merit tends to be considered more highly than regional "purity."

Wine Terminology

Lehrer (1983) notes that scientific writing is judged by its success in clear, critical communication. Neither of these attributes characterizes the majority of wine descriptions (Brochet and Dubourdieu, 2001). Most expressions evoke an image, usually a holistic emotional response or aromatic/taste resemblance. Wine writers tend to use language as much to entertain as to inform. Precise terms, such as bitterness and astringency are seldom employed, presumably because they possess negative connotations. Instead, where sensory terminology is used, only positive aspects are noted, as with expression such as "a lot of character" or "a long aftertaste" (Lesschaeve, unpublished observations). Similar findings were noted by Lehrer (1975). Interestingly, Bastian *et al.* (2005) found that tasting notes on back-labels are often considered useless by consumers. They can even be discouraging, especially when consumers find themselves unable to detect the flavors or nuances so lovingly described.

Because tasting notes usually reflect subjective reaction to the wine, tasters often develop their own idiosyncratic lexicon. These expressions have personal meaning, but rarely accurately describe the wine's sensory attributes. Frequently, they cannot even be employed by the user to identify the wine in a subsequent tasting. At best, descriptors may express overall quality, attributes typical of white vs. red wines, or features considered typical of particular wines. The latter reflects the norms attributed to specific regional, stylistic, or varietal wines. Ranking typically concentrates on how well a particular wine expresses features preferred or expected by the taster.

Memory aids, in the form of flavor charts can be useful learning tools. These were first developed by Meilgaard *et al.* (1979) for beer. They have subsequently been prepared for wine (Noble, Arnold *et al.*, 1984; Jackson, 2002), brandy (Jolly and Hattingh, 2001), whiskey (Lee *et al.*, 2001), as well as mouth-feel sensations (Gawel *et al.*, 2000). Regrettably, the popularization of aroma charts has led to the faulty impression that descriptive terms are what serious consumers should look for. The descriptors are noted primarily to prod the memory. Only rarely are the terms accurate representations.

Statistical and Descriptive Analysis of Tasting Results

Simple Tests

For most tastings, simple statistical tests are usually adequate in assessing whether tasters can distinguish any, or all, of the sampled wines. One measure of significance is based on the range of scores for each wine and the cumulative score range. An example is given below (Table 11.3).

For the wines to be considered distinguishable from one another, the range in scores must be greater than the statistic given in Appendix 11.1, multiplied by the sum of the score ranges for individual wines. In this example, the pertinent statistic for five tasters and five wines is 0.81, for significance at a 5% level. For the tasters to be considered capable of distinguishing among the wines, the range of total scores must be greater than the product of the statistic (0.81) and by the sum of score ranges (13) [0.81 × 13 = 10.5]. Because the range of total scores (11) is greater than the calculated product (10.5), the tasters are considered able to distinguish differences among the wines.

To determine which wines were distinguished, the second (lower) statistic in Appendix 11.1 (0.56 in this instance) is multiplied by the sum of score ranges (13) to produce the product (7.3). When the difference between the total scores of any pair of wines is greater than the calculated product (7.3), the wines may be considered significantly different. Table 11.4 shows that Wine 1 was distinguishable from Wines 3 and 5, but not from Wines 2 and 4, whereas Wines 2, 3, 4, and 5 were indistinguishable from one another.

Caution must be exercised in interpreting results of such tests. For example, Wine 1 might have been faulty, and had an appropriate sample been substituted, none of the wines might have been considered significantly different. In addition, there is no means of determining whether the tasters were scoring consistently (a single test). Thus, had more competent tasters been involved, significant differences might have been detected. Even if one or two tasters were "out of form," incorrect conclusions might be drawn. Conclusions can be no more valid than the quality of the data on which they are based.

Numerous examples of this and other statistical techniques are given by Amerine and Roessler (1983). Their book is still an excellent primer for those wishing details on the use of statistics in wine analysis.

Table 11.3 Hypothetical scores of five tasters for five wines

	Judging results				
	Wine 1	Wine 2	Wine 3	Wine 4	Wine 5
Taster 1	5	6	7	5	8
Taster 2	6	6	8	6	5
Taster 3	4	7	7	4	6
Taster 4	5	4	6	6	6
Taster 5	4	6	7	7	7
Total scores	24	29	35	28	32
Range (max − min)	2	3	2	3	3
Sum of score ranges = (2 + 3 + 2 + 3 + 3) = 13					
Range of total scores = (35 − 24) = 11					

Table 11.4 Difference between the sum of scores of pairs of wines[a]

	Wine 1	Wine 2	Wine 3	Wine 4	Wine 5
Wine 1	–				
Wine 2	5	–			
Wine 3	11[a]	6	–		
Wine 4	4	1	7	–	
Wine 5	8[a]	3	3	4	–

[a]Values are significant at a 5% probability.

Analysis of Variance

For a more detailed evaluation, analysis of variance (ANOVA) may be used. Although ANOVA techniques are more complicated, computers have made them readily available. Direct electronic incorporation of data further eases the analysis of large amounts of data. This has developed to the point that complete computer programs for the sensory analysis of foods and beverages are available (i.e., Compusense *five*®). They can be adjusted to suit special needs of the user.

Analysis of variance can assess not only whether any two or more wines are detectably different, but also whether the tasters are scoring differently. In addition, the analysis permits evaluation of significant interaction among factors in a tasting. Furthermore, it can provide measures of taster discrimination, stability, and variability.

Multivariate Analysis and Descriptive Analysis of Wine

Another powerful statistical tool is the application of multivariate analysis (Zervos and Albert, 1992; Kaufmann, 1997). It has the potential to isolate features that distinguish wines made from specific varieties (Guinard and Cliff, 1987), from within particular regions (Williams *et al.*, 1982), or made by particular processes.

To use the technique, prospective tasters undergo a screening, followed by a sampling of a series of wines representing the variety, region, or style investigated. During the tastings, potential judges work toward a consensus on the choice of descriptive terms that adequately represent the distinctive features of the wines. Subsequently, wine samples are judged using these descriptors to assess their adequacy, and assess whether the number of terms can be further consolidated. The analysis of consistent and correct term use is typically performed, and inconsistent or divergent tasters are eliminated before formal tasting begins.

In most studies, there is an attempt to associate features such as varietal or geographic origin with chemical or sensory characteristics, or investigate the influence of a particular technique on a wine's fragrance (Fig. 11.24).

Because of the extensive training and discussion required before conducting descriptive sensory analysis, some researchers have questioned the polarization of views involved (Myers and Lamm, 1975). Further questions about the appropriateness of descriptive analysis relate to the tendency of people to be highly idiosyncratic in term use (Lawless, 1984) and whether "correct" sets of descriptors are possible (Solomon, 1991). Although selectively reducing panel variation makes obtaining statically valid results more likely, the data reflect only the views of that specific subset of individuals. Whether or not this is important depends on the purpose of the analysis. If intended to represent the views of the general tasting public, or even special subgroups of consumers, the results may be significant but invalid.

To avoid such concerns, Williams and Langron (1983) propose that tasters be allowed to use their own vocabulary to describe wine appearance, aroma, taste, and flavor. A scale is used to measure the intensity of each attribute, and these are subjected to a multidimensional mathematical model (Procrustes analysis). It adjusts individual results so that they can be compared and assessed statistically (Oreskovich *et al.*, 1991; Dijksterhuis, 1996). Although this technique avoids some of the problems of descriptive sensory analyses, Procrustes analysis assumes that tasters experience the same sensations. Psychophysical tests suggest that this is not necessarily correct and that people may perceive sensory inputs both quantitatively and qualitatively differently. This could undermine the potential applicability of the data.

Of particular interest is the combination of chemical analysis, descriptive sensory analysis, and consumer subgroup preferences. Such combinations may permit the correlation of preference data with particular aromatic and sapid substances (Herraiz and Cabezudo, 1980/81; Williams *et al.*, 1982; Williams, 1984). If the chemical nature of consumer preferences could be defined, it would allow a more precise selection and blending of wines for particular consumer groups. Such designing of wines may not be consistent with the romantic image

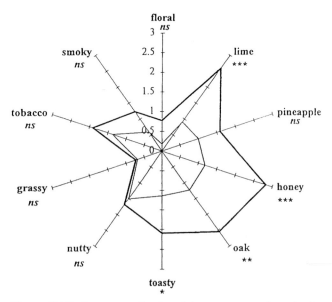

Figure 11.24 Sensory profile plot of the mean aroma intensity for attributes of a wine made from 'Sémillon' juice containing glycosides (——) and having had the glycosides removed before fermentation (——). (From Williams and Francis, 1996, reproduced by permission)

cultivated by boutique wineries, wine merchants and journalists, but providing the right wine at the right price can be decidedly profitable!

Although designing wines for a particular market can be profitable, it is happily not the only means of providing consumers with what they want. As Vernon Singleton (1976) has said:

> *Wine is, and must remain I feel, one of the few products with almost unlimited diversity ... keeping the consumer forever intrigued, amused, pleased, and never bored.*

Objective Wine Analysis

Most of the chapter has dealt with the human assessment of wine, its complexities, and limitations. Nevertheless, surprisingly good correlation has been obtained between perceived quality in red wine and certain aspects of its phenolic content (Fig. 11.25; Somers, 1998). These data indicate that color density, measured as the sum of absorbency (extinction) at 420 and 520 nm, correlate highly with quality rating.

Extinction values at 420 and 520 nm were chosen because they change the most during aging. The quality correlation is apparently not due to visually perceptible differences in color depth. Judges were inconsistent in recognizing color differences in the wines, and only 3 out of 20 marks used to rank quality were assigned to color (Somers, 1975). Color density is also closely linked to the total phenolic extract in red wines (as measured by UV absorbency (280 nm). It, unlike color

Table 11.5 Direct comparison of phenolic levels in high volume and ultra premium wines (mg/liter)

Compound	HV weighted average	Ultra-premium average	% Difference	*t*-test result
cis-Caftaric acid	11.31	9.84	−13	0.837
trans-Caftaric acid	20.12	37.38	86	0.093
cis-Coutaric acid	5.02	5.14	2	0.869
trans-Coutaric acid	10.89	18.69	72	0.271
Caffeic acid	7.07	19.24	172	0.00131
Coumaric acid	7.50	13.22	76	0.413
Total cinnamates	**61.91**	**103.5**	**67**	**0.083**
Gallic acid	25.97	43.09	66	0.014
Syringic acid	4.23	5.73	35	0.930
Total gallates	**30.2**	**48.82**	**62**	**0.018**
Flavonol 1	16.19	48.09	197	0.0078
Flavonol 2	10.55	35.86	240	0.0083
Flavonol 3	7.44	20.46	175	0.0032
Myricitin	9.00	45.06	401	0.0000017
Quercetin	9.61	53.01	452	0.00033
Total flavonols	**52.79**	**202.5**	**284**	**0.00026**
Procyanidin B3	24.02	28.63	19	0.512
Catechin	26.95	33.00	22	0.453
Procyanidin B2	19.40	–	NA	
Epicatechin	22.62	21.37	−6	0.318
Total flavan-3-ols	**73.59**	**83.00**	**13**	**0.929**
Malvidin-3-glu	66.49	74.01	11	
Malvidin-3-glu-ac	17.19	19.99	16	
Malvidin-3-glu-*p*-coum	5.12	8.61	68	
Total malvidin derivatives	**88.8**	**100.9**	**14**	

Source: From Ritchey and Waterhouse, 1999, reproduced by permission

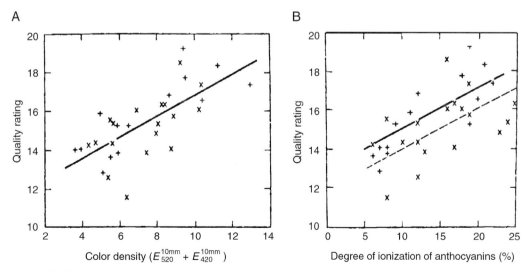

Figure 11.25 Relation between quality rating and wine color density (**A**) and quality rating and degree of anthocyanin ionization (**B**) of 'Cabernet Sauvignon' (+) and 'Shiraz' (×) wines. (From Somers and Evans, 1974, *J. Sci. Food Agric.* **26**, 1369–1379. Copyright SCI, reproduced with permission)

density, remains relatively constant during aging (maturation in oak typically will add only ~5% to the total phenolic content of the wine). Another parameter that correlates well with perceived quality is the proportion of colored anthocyanins (primarily in the red ionized flavylium state). Regrettably, these studies have been conducted only with the wines of two deeply pigmented varieties, 'Cabernet Sauvignon' and 'Shiraz.' Studies with white grape wines have shown no relationship between phenolic content and assessed quality (Somers and Pocock, 1991).

Data from Ritchey and Waterhouse (1999) provide a different and fascinating approach in wine quality. They compared the phenolic chemistry of high volume with ultra premium 'Cabernet Sauvignon' wines. Their results are summarized in Table 11.5. They show that the concentration of most phenolic groups in ultra premium wines are markedly higher than that in high volume wines. This is particularly noticeable for flavonols (increase of about 280%); less so with cinnamates and gallates, showing increases of 60–70%. Ultra premium wines were also higher in alcohol content (12.3 vs. 14.1), lower in residual sugar content, as well as malic acid content.

These findings are no more likely to replace human tasters than electronic noses, but the intricate links between grape composition and wine characteristics are under investigation. This has particular applicability to understanding how microclimate, grape culture, and wine production enhance (and diminish) wine quality. Human appreciation will always remain the most important indicator of wine quality, but its imprecision makes isolating those chemical factors essential to quality perception particularly vexing.

Appendix 11.1

Multipliers for Estimating Significance of Difference by Range: One-Way Classification, 5% Level[a,b]

Number of judges	Number of wines								
	2	3	4	5	6	7	8	9	10
2	3.43	2.35	1.74	1.39	1.15	0.99	0.87	0.77	0.70
	3.43	1.76	1.18	0.88	0.70	0.58	0.50	0.44	0.39
3	1.90	1.44	1.14	0.94	0.80	0.70	0.62	0.56	0.51
	1.90	1.14	0.81	0.63	0.52	0.44	0.38	0.33	0.30
4	1.62	1.25	1.01	0.84	0.72	0.63	0.57	0.51	0.47
	1.62	1.02	0.74	0.58	0.48	0.40	0.35	0.31	0.28
5	1.53	1.19	0.96	0.81	0.70	0.61	0.55	0.50	0.45
	1.52	0.98	0.72	0.56	0.47	0.40	0.34	0.30	0.27
6	1.50	1.17	0.95	0.80	0.69	0.61	0.55	0.49	0.45
	1.50	0.96	0.71	0.56	0.46	0.40	0.34	0.30	0.27
7	1.49	1.17	0.95	0.80	0.69	0.61	0.55	0.50	0.45
	1.49	0.96	0.71	0.56	0.47	0.40	0.35	0.31	0.28
8	1.49	1.18	0.96	0.81	0.70	0.62	0.55	0.50	0.46
	1.49	0.97	0.72	0.57	0.47	0.41	0.35	0.31	0.28
9	1.50	1.19	0.97	0.82	0.71	0.62	0.56	0.51	0.47
	1.50	0.98	0.73	0.58	0.48	0.41	0.36	0.31	0.28
10	1.52	1.20	0.98	0.83	0.72	0.63	0.57	0.52	0.47
	1.52	0.99	0.74	0.59	0.49	0.42	0.37	0.32	0.29
11	1.54	1.22	0.99	0.84	0.73	0.64	0.58	0.52	0.48
	1.54	0.99	0.75	0.60	0.49	0.42	0.37	0.32	0.29
12	1.56	1.23	1.01	0.85	0.74	0.65	0.58	0.53	0.49
	1.56	1.00	0.75	0.60	0.50	0.43	0.38	0.32	0.30

[a]Entries in the table are to be multiplied by the sum of ranged within wines. The upper value must be exceeded by the range in wine totals to indicate significance. If significance is indicated, the lower value must be exceeded by pairs of wine totals to indicate a significance between individual wines.

[b]After T. E. Kurtz, R. F. Link, J. W. Tukey, and D. L. Wallace (1965). Short-cut multiple comparisons for balanced single and double classifications: Part I, Results. *Technometrics* 7, 95–165. Reprinted with permission from *Technometrics*. Copyright (1965) by the American Statistical Association and the American Society for Quality Control. All rights reserved. From Amerine and Roessler (1983).

Suggested Readings

Visual Sensations

Clydesdale, F. M. (1993) Color as a factor in food choice. *Crit. Rev. Food Sci. Nutr.* **33**, 83–101.

Kaiser, P. K., and Boynton, R. M. (1996) *Human Color Vision*, 2nd edn. Optical Society of America, Washington, DC.

Schwartz, S. H. (2004) *Visual Perception*, 3rd edn. McGraw-Hill Medical, New York.

Walker, J. (1983) What causes the tears that form on the inside of a glass of wine? *Sci. Am.* **248**, 162–169.

Taste and In-Mouth Sensations

Beauchamp, G. K., and Bartoshuk, L. (1997) *Tasting and Smelling.* Academic Press, San Diego, CA.

Clifford, M. N. (1997) Astringency. In: *Phytochemistry of fruits and vegetables.* Proceedings of the Phytochenical Society of Europe (F. S. Tomás-Barberán and R. J. Robins, eds.), pp. 87–107. Clarendon Press, Oxford, UK.

Doty, R. L. (ed.) (1995) *Handbook of Olfaction and Gustation.* Marcel Dekker, New York.

Drayna, D. (2005) Human taste genetics. *Annu. Rev. Genomics Hum. Genet.* **6**, 217–235.

Francis, I. L., Gawel, R., Iland, P. G., Vidal, S., Cheynier, V., Guyot, S., Kwiatkowski, M. J., and Waters, E. J. (2002) Characterising mouth-feel properties of red wines. In: *Proceeding of the 11th Australian Wine Industry Technical Conference* (R. Blair, P. Williams and P. Høj, eds.), pp. 123–127. Australian Wine Industry Technical Conference Inc., Urrbrae, Australia.

Given, P., and Paredes, D. (eds.) (2002) *Chemistry of Taste. Mechanisms, Behaviors and Mimics.* ASC Symposium Series No. 825. American Chemical Society Publication, Washington, DC.

Kim, U.-K., Breslin, P. A. S., Reed, D., and Drayna, D. (2004) Genetics of human taste perception. *J. Dent. Res.* **83**, 448–453.

Schifferstein, H. N. J. (1996) Cognitive factors affecting taste intensity judgements. *Food Qual. Pref.* **7**, 167–176.

Stevens, J. C. (1996) Detection of tastes in mixture with other tastes: Issues of masking and aging. *Chem. Senses* **21**, 211–221.

Swiegers, J. H., Chambers, P. J., and Pretorius, I. S. (2005) Olfaction and taste: Human perception, physiology and genetics. In: *AWRI Advances in Wine Science* (R. J. Blain, M. E. Francis and I. S. Pretorius, eds.), pp. 213–217. The Australian Wine Research Institute, Glen Osmond, SA, Australia.

Olfactory Sensations

Beauchamp, G. K., and Bartoshuk, L. (1997) *Tasting and Smelling.* Academic Press, San Diego, CA.

Doty, R. L. (ed.) (1995) *Handbook of Olfaction and Gustation.* Marcel Dekker, New York.

Firestein, S. (2001) How the olfactory system makes sense of scents. *Nature* **413**, 211–218.

Grosch, W. (2001) Evaluation of the key odorants of foods by dilution experiments, aroma models and omission. *Chem. Senses* **26**, 533–545.

Malnic, B., Godfrey, P. A., and Buck, L. B. (2004) The human olfactory receptor gene family. *Proc. Natl Acad. Sci.* **101**, 2584–2589.

Murphy, C. (1995) Age-associated differences in memory for odors. In: *Memory of Odors* (F. R. Schab and R. G. Crowder, eds.), pp. 109–131. Lawrence Erlbaum, Mahwah, NJ.

Prescott, J. (2004) Psychological processes in flavour perception. In: *Flavour Perception* (A. J. Taylor and D. Roberts, eds.), pp. 256–278. Blackwell, Oxford.

Rolls, E. T. (2004) The functions of the orbitofrontal cortex. *Brain Cognit.* **55**, 11–29.

Ronnett, G. V., and Moon, C. (2002) G proteins and olfactory signal transduction. *Annu. Rev. Physiol.* **64**, 189–222.

Schab, F. R., and Crowder, R. G. (1995) *Memory of Odors.* Lawrence Erlbaum, New York.

Swiegers, J. H., Chambers, P. J., and Pretorius, I. S. (2005) Olfaction and taste: Human perception, physiology and genetics. In: *AWRI Advances in Wine Science* (R. J. Blain, M. E. Francis and I. S. Pretorius, eds.), pp. 213–217. The Australian Wine Research Institute, Glen Osmond, SA, Australia.

Taylor, A. J., and Roberts, D. D. (eds.) (2004) *Flavor Perception.* Blackwell, Oxford.

Sensory Analysis

Amerine, M. A., and Roessler, E. B. (1983) *Wines – Their Sensory Evaluation*, 2nd edn. Freeman, San Francisco, CA.

Durier, C., Monod, H., and Bruetschy, A. (1997) Design and analysis of factorial sensory experiments with carry-over effects. *Food Qual. Pref.* **8**, 141–149.

Earthy, P. J., MacFie, H. J. H., and Hederley, D. (1997) Effect of question order on sensory perception and preference in central location trials. *J. Sens. Stud.* **12**, 215–238.

Jackson, R. S. (2002) *Wine Tasting. A Professional Handbook.* Academic Press, London.

Lawless, H. T., and Heymann, H. (1998) *Sensory Evaluation of Food: Principles and Practices.* Kluwer Academic Press, Norwell, MA.

Lehrer, A. (1983) *Wine and Conversation.* Indiana University Press, Bloomington, IN.

Meilgaard, M. C., Civille, G. V., and Carr, B. T. (1999) *Sensory Evaluation Techniques*, 3rd edn. CRC Press. Boca Raton, FL.

Peynaud, E. (1987) *The Taste of Wine.* MacDonald, London.

Stone, H., and Sidel, J. (2004) *Sensory Evaluation Practices*, 3rd edn. Academic Press, Orlando, FL.

References

Amerine, M. A., and Roessler, E. B. (1983) *Wines, Their Sensory Evaluation*, 2nd edn. Freeman, San Francisco, CA.

Amerine, M. A., and Singleton, V. L. (1977) *Wine – An Introduction.* University of California Press, Berkeley, CA.

Amoore, J. E. (1977) Specific anosmia and the concept of primary odors. *Chem. Senses Flavours* **2**, 267–281.

Amoore, J. E., Johnson, J. W., Jr., and Rubin, M. (1964) The stereochemical theory of odor. *Sci. Am.* **210**(2), 42–49.

André, P., Aubert, S., and Pelisse, C. (1970) Contribution aux études sur les vins rosés meridionaux. I. La couleur. Influence sur la degustation. *Ann. Technol. Agric.* **19**, 323–340.

Anonymous (1994) Resolution Eno 2/94, OIV standard for international wine competitions. *Bull. O.I.V.* **67**, 558–597.

Arnold, R. A., and Noble, A. C. (1978) Bitterness and astringency of grape seed phenolics in a model wine solution. *Am. J. Enol. Vitic.* **29**, 150–152.

Aronson, J., and Ebeler, S. E. (2004) Effect of polyphenol compounds on the headspace volatility of flavors. *Am. J. Enol. Vitic.* **55**, 13–21.

Avenet, P., and Lindemann, B. (1989) Perspective of taste reception. *J. Membr. Biol.* **112**, 1–8.

Ayala, F., Echávarri, J. F., and Negueruela, A. I. (1997) A new simplified method for measuring the color of wines. II. White wines and brandies. *Am. J. Enol. Vitic.* **48**, 364–369.

Barker, L. M. (ed.) (1982) *The Psychobiology of Human Food Selection.* AVI, Westport, CT.

Bartoshuk, L. M., Rifkin, B., Marks, L. E., and Bars, P. (1986) Taste and aging. *J. Gerontol.* **41**, 51–57.

Bartowsky, E. J., Francis, I. L., Bellon, J. R., and Henschke, P. A. (2002) Is buttery aroma perception in wines predictable from the diacetyl concentration? *Aust. J. Grape Wine Res.* **8**, 180–185.

Basker, D. (1988) Assessor selection, procedures and results. In: *Applied Sensory Analysis of Foods* (H. Moskowitz, ed.), Vol. 1, pp. 125–143. CRC Press, Boca Raton, FL.

Bastian, S., Bruwer, J., Alant, K., and Li, E. (2005) Wine consumers and makers: are they speaking the same language? *Aust. NZ Grapegrower Winemaker* **496**, 80–84.

Beidler, L. M., and Tonosaki, K. (1985) Multiple sweet receptor sites and taste theory. In: *Taste, Olfaction and the Central Nervous System* (D. W. Pfaff, ed.), pp. 47–64. Rockefeller University Press, New York.

Berg, H. W., Pangborn, R. M., Roessler, E. B., and Webb, A. D. (1963) Influence of hunger on olfactory acuity. *Nature* **197**, 108.

Bertrand, A., Zmirou-Bonnamour, C., and Lonvaud-Funel, A. (1984) Aroma compounds formed in malolactic fermentation. Alko Symposium on Flavor Research of Alcoholic Beverages (L. Nykänen and P. Lehtonen, eds). *Foundation Biotech. Indust. Ferm.* **3**, 39–49.

Bertuccioli, M., and Ferrari, S. (1999) *Laboratory Experience on the Influence of Yeast on Mouthfeel.* Les Entretiens Scientidiques Lallemand, Montreal, CA.

Bett, K. L., and Johnsen, P. B. (1996) Challenges of evaluating sensory attributes in the presence of off-flavors (geosmin and 2-methylisoborneol) *J. Sens. Stud.* **11**, 1–18.

Bi, J., and Ennis, D. M. (1998) Sensory thresholds: concepts and methods. *J. Sens. Stud.* **13**, 133–148.

Binders, G., Pintzler, S., Schröder, J., Schmarr, H. G., and Fischer, U. (2004) Influence of German oak chips on red wine. *Am. J. Enol. Vitic.* **55**, 323A.

Blakeslee, A. F., and Salmon, T. N. (1935) Genetics of sensory thresholds, individual taste reactions for different substances. *Proc. Natl Acad. Sci. U.S.A.*, **21**, 84–90.

Boudreau, J. C., Oravec, J., Hoang, N. K., and White, T. D. (1979) Taste and the taste of foods. In: *Food Taste Chemistry* (J. C. Boudreau, ed.), pp. 1–30. American Chemical Society, Washington, DC.

Brannan, G. D., Setser, C. S., and Kemp, K. E. (2001) Effectiveness of rinses in alleviating bitterness and astringency residuals in model solutions. *J. Sens. Stud.* **16**, 261–275.

Brien, C. J., May, P., and Mayo, O. (1987) Analysis of judge performance in wine-quality evaluations. *J. Food Sci.* **52**, 1273–1279.

Brochet, F., and Dubourdieu, D. (2001) Wine descriptive language supports cognitive specificity of chemical senses. *Brain Lang.* **77**, 187–196.

Brochet, F., and Morrot, G. (1999) Influence du contexte sure la perception du vin. Implications cognitives et méthodologiques. *J. Int. Sci. Vigne Vin* **33**, 187–192.

Brossaud, F., Cheynier, V., and Noble, A. C. (2001) Bitterness and astringency of grape and wine polyphenols. *J. Grape Wine Res.* **7**, 33–39.

Brou, P., Sciascia, T. R., Linden, L., and Lettvin, J. Y. (1986) The colors of things. *Sci. Am.* **255**(3), 84–91.

Bucelli, P., and Gigliotti, A. (1993) Importanza di alcuni parametri analatici nella valutazione dell'attitudine all'invecchiamento dei vini. *Enotecnico* **29**(5), 75–84.

Buck, L., and Axel, R. (1991) A novel multigene family may encode odorant receptors, a molecular basis for odor recognition. *Cell* **65**, 175–187.

Buettner, A. (2002a) Influence of human salivary enzymes on odorant concentration changes occurring *in vivo*. 1. Esters and thiols. *J. Agric. Food Chem.* **50**, 3283–3289.

Buettner, A. (2002b) Influence of human saliva on odorant concentrations. 2. Aldehydes, alcohols, 3-alkyl-2-methoxypyrazines, methoxyphenols, and 3-hydroxy-4,5-dimethyl-2(5H)-fruanone. *J. Agric. Food Chem.* **50**, 7105–7110.

Buettner, A. (2004) Investigation of potent odorants and afterodor development in two Chardonnay wines using the buccal odor screening system (BOSS) *J. Agric. Food Chem.* **52**, 2339–2346.

Buratti, S., Benedetti, S., Scampicchio, M., and Pangerod, E. C. (2004) Characterization and classification of Italian Barbera wines by using an electronic nose and an amperometric electronic tongue. *Anal. Chim. Acta* **525**, 133–139.

Burns, D. J. W., and Noble, A. C. (1985) Evaluation of the separate contributions of viscosity and sweetness to perceived viscosity, sweetness and bitterness of vermouth. *J. Texture Stud.* **16**, 365–381.

Cain, W. S. (1974) Perception of odour intensity and the time-course of olfactory adaption. *Trans. Am. Soc. Heating Refrigeration Air-Conditioning Eng.* **80**, 53–75.

Cain, W. S. (1976) Olfaction and the common chemical sense: Some psychophysical contrasts. *Sens. Processes* **1**, 57.

Cain, W. S. (1978) The odoriferous environment and the application and the application of olfactory research. In: *Handbook of Perception: Tasting and Smelling* (E. C. Carterette and P. M. Friedman, eds.), Vol. 6A, pp. 197–229. Academic Press, New York.

Cain, W. S. (1979) To know with the nose: Keys to odor identification. *Science* **203**, 467–469.

Cain, W. S. (1985) Chemical sensation: Olfaction. In: *Nutrition in Oral Health and Disease* (R. L. Pollack and E. Cravats, eds.), pp. 68–83. Lee & Febiger, Philadelphia, PA.

Cain, W. S., and Murphy, C. L. (1980) Interaction between chemoreceptive modalities of odour and irritation. *Nature* **284**, 255–257.

Choudhury, E. S., Moberg, P., and Doty, R. L. (2003) Influences of age and sex on a microencapsulated odor memory test. *Chem. Senses* **28**, 799–805.

Chrea, C., Valentin, D., Sulmont-Rossé, C., Mai, H. L., Nguyen, D. H., and Adbi, H. (2004) Culture and odor categorization: Agreement between cultures depends upon the odors. *Food Qual. Pref.* **15**, 669–679.

Cicchetti, D. (2004) On designing experiments and analysing data to assess the reliability and accuracy of blind wine tastings. *J. Wine Res.* **15**, 221–226.

CIE (1986) *Colorimetry*, 2nd edn. Central Bureau of the Commission Internationale de l'Eclairage, Vienna.

Clapperton, J. F. (1978) Sensory characterization of the flavour of beer. In: *Progress in Flavour Research* (D. G. Land and H. E. Nursten, eds.), pp. 1–20. Applied Science Publishers, London.

Clark, C. C., and Lawless, H. T. (1994) Limiting response alternatives in time-intensity scaling. An examination of the halo-dumping effect. *Chem. Senses* **19**, 583–594.

Cliff, M. A. (2001) Influence of wine glass shape on perceived aroma and colour intensity in wines. *J. Wine Res.* **12**, 39–46.

Clydesdale, F. M., Gover, R., Philipsen, D. H., and Fugardi, C. (1992) The effect of color on thirst quenching, sweetness, acceptability and flavor intensity in fruit punch flavored beverages. *J. Food Quality* **15**, 19–38.

Collings, V. B. (1974) Human taste response as a function of locus of stimulation on the tongue and soft palate. *Percept. Psychophys.* **16**, 169–174.

Colonna, A. E., Adams, D. O., and Noble, A. C. (2004) Comparison of procedures for reducing astringency carry-over effects in evaluation of red wines. *Aust J. Grape Wine Res.* **10**, 26–31.

Conner, J. M., Birkmyre, A., Paterson, A., and Piggott, J. R. (1998) Headspace concentrations of ethyl esters at different alcoholic strengths. *J. Sci. Food Agric.* **77**, 671–677.

Cowart, B. J. (1989) Relationship between taste and smell across the adult life span. *Ann. NY Acad. Sci.* **561**, 39–55.

Cowart, B. J. (1998) The addition of CO_2 to traditional taste solutions alters taste quality. *Chem. Senses* **23**, 397–402.

Cutzach, I., Chatonnet, P., and Dubourdieu, D. (1999) Study of the formation mechanisms of some volatile compounds during the aging of sweet fortified wines. *J. Agric. Food Chem.* **47**, 2837–2846.

Czerny, M., and Grosch, W. (2000) Potent odorants of raw Arabica coffee. Their changes during roasting. *J. Agric. Food Chem.* **48**, 868–872.

Dahl, A. R. (1988) The effect of cytochrome P-450 dependent metabolism and other enzyme activities on olfaction. In: *Molecular Neurobiology of the Olfactory System* (F. L. Margolis and T. V. Getchell, eds.), pp. 51–70. Plenum, New York.

Dalton, P., Doolittle, N., and Breslin, P. A. S. (2002) Gender-specific induction of enhanced sensitivity to odors. *Nature Neurosci.* **5**, 199–200.

Dalton, P., Doolittle, N., Nagata, H., and Breslin, P. A. S. (2000) The merging of the senses: integration of subthreshold taste and smell. *Nature Neurosci.* **3**, 431–432.

de Freitas, V., and Mateus, N. (2003) Nephelometric study of salivary protein-tannin aggregates. *J. Sci. Food Agric.* **82**, 113–19.

Delwiche, J. F. (2003) Impact of color on perceived wine flavor. *Foods Food Ingred. J. Jpn* **208**, 349–352.

Delwiche, J. F., Halpern, B. P., and Desimone, J. A. (1999) Anion size of sodium salts and simple taste reaction times. *Physiol. Behav.* **66**, 27–32.

Delwiche, J. F., and Pelchat, M. L. (2002) Influence of glass shape on wine aroma. *J. Sens. Stud.* **17**, 19–28.

DeSimone, J. A., Heck, G. L., and Bartoshuk, L. M. (1980) Surface active taste modifiers, a comparison of the physical and psychophysical properties of gymnemic acid and sodium lauryl sulfate. *Chem. Senses* **5**, 317–330.

de Wijk, R. A. (1989) Temporal factors in human olfactory perception. Doctoral thesis, University of Utrecht, The Netherlands. Cited in Cometto-Muñiz, J. E., and Cain, W. S. (1995) Olfactory adaptation. In: *Handbook of Olfaction and Gustation* (R. L. Doty, ed.), pp. 257–281. Marcel Dekker, New York.

de Wijk, R. A., and Cain, W. S. (1994) Odor quality: Discrimination versus free and cued identification. *Percept. Psychophys.* **56**, 12–18.

Diaz, M. E. (2004) Comparison between orthonasal and retronasal flavour perception at different concentrations. *Flavour Fragr.* **19**, 499–504.

Dijksterhuis, G. (1996) Procrustes analysis in sensory research. In: *Multivariate Analysis of Data in Sensory Science. Data Handling in Science and Technology* (T. Naes and E. Risvik, eds.), Vol. 16, pp. 185–220. Elsevier Science, Amsterdam.

Di Stefano, R., and Ciolfi, G. (1985) L'influenza dell luce sull'Asti Spumante. *Vini Ital.* **27**, 23–32.

Djordjevic, J., Zatorre, R. J., and Jones-Gotman, M. (2004) Effects of perceived and imagined odors on taste detection. *Chem. Senses* **29**, 199–208.

Doty, R. L. (1986) Reproductive endocrine influences upon olfactory perception, a current perspective. *J. Chem. Ecol.* **12**, 497–511.

Doty, R. L., and Bromley, S. M. (2004) Effects of drugs on olfaction and taste. *Otolaryngol. Clin. N. Am.* **37**, 1229–1254.

Doty, R. L., and Snow, J. B. Jr. (1988) Age-related alterations in olfactory structure and function. In: *Molecular Neurobiology of the Olfactory System* (F. L. Margolis and T. V. Getchell, eds.), pp. 355–374. Plenum, New York.

Doty, R. L., Shaman, P., Applebaum, S. L., Giberson, R., Siksorski, L., and Rosenberg, L. (1984) Smell identification ability changes with age. *Science* **226**, 1441–1443.

Drewnowski, A., Henderson, A. S., and Barratt-Fornell, A. (2001) Genetic taste markers and food preferences. *Drug Metab. Disposit.* **29**, 535–538.

Dufour, C. and Bayonove, C. L. (1999a) Influence of wine structurally different polysaccharides on the volatility of aroma substances in a model system. *J. Agric. Food Chem.* **47**, 671–677.

Dufour, C., and Bayonove, C. L. (1999b) Interactions between wine polyphenols and aroma substances. An insight at the molecular level. *J. Agric. Food Chem.* **47**, 678–684.

Dufour, C., and Sauvaitre, I. (2000) Interactions between anthocyanins and aroma substances in a model system. Effect on the flavor of grape-derived beverages. *J. Agric. Food Chem.* **48**, 1784–1788.

Dürr, P. (1984) Sensory analysis as a research tool. *Found. Biotech. Ind. Ferment. Res.* **3**, 313–322.

Dürr, P. (1985) Gedanken zur Weinsprache. *Alimentia* **6**, 155–157.

Ekman, G., Berglund, B., Berglund, U., and Lindvall, T. (1967) Perceived intensity of odor as a function of time of adaptation. *Scand. J. Psychol.* **8**, 177–186.

Engen, T. (1987) Remembering odors and their names. *Am. Sci.* **75**, 497–503.

Enns, M. P., and Hornung, D. E. (1985) Contributions of smell and taste to overall intensity. *Chem. Senses* **10**, 357–366.

Escalona, H., Piggott, J. R., Conner, J. M., and Paterson, A. (1999) Effects of ethanol strength on the volatility of higher alcohols and aldehydes. *Ital. J. Food Sci.* **11**, 241–248.

Escudero, A., Hernandez-Orte, P., Cacho, J., and Ferreira, V. (2000) Clues about the role of methional as character impact odorant of some oxidized wines. *J. Agric. Food Chem.* **48**, 4268–4272.

Fischer, C., Fischer, U., and Jakob, L. (1996) Impact of matrix variables, ethanol, sugar, glycerol, pH and temperature on the partition coefficients of aroma compounds in wine and their kinetics of volatization. In: *Proceedings of the 4th International Symposium on Cool Climate Viticulture and Enology* (T. Henick-Kling *et al.*, eds.), pp. VII-42–46. New York State Agricultural Experimental Station, Geneva, NY.

Francis, I. L., and Newton, J. L. (2005) Determining wine aroma from compositional data. *Aust. J. Grape Wine Res.* **11**, 114–126.

Frank, R. A., and Byram, J. (1988) Taste-smell interactions are tastant and odorant dependant. *Chem. Senses* **13**, 445–455.

Freeman, W. J. (1991) The physiology of perception. *Sci. Am.* **264** (2), 78–85.

Frye, R. E., Schwartz, B. S., and Doty, R. L. (1990) Dose-related effects of cigarette smoking on olfactory function. *J. Am. Med. Assoc.* **263**, 1233–1236.

Ganzevles, P. G. J., and Kroeze, J. H. A. (1987) The sour taste of acids, the hydrogen ion and the undissociated acid as sour agents. *Chem. Senses* **12**, 563–576.

Gawel, R., Oberholster, A., and Francis, I. L. (2000) A 'Mouth-feel Wheel': terminology for communicating the mouth-feel characteristics of red wine. *Aust. J. Grape Wine Res.* **6**, 203–207.

Gawel, R., Iland, P. G., and Francis, I. L. (2001) Characterizing the astringency of red wine: a case study. *Food Qual. Pref.* **12**, 83–94.

Gilbertson, T. A., and Boughter, J. D. Jr. (2003) Taste transduction: appetizing times in gustation. *NeuroReport* **14**, 905–911.

Gilbertson, T.A., Fontenot, D. T., Liu, L., Zhang, H., and Monrot, W. T. (1997) Fatty acid modulation of K^+ channels in taste receptor cells: gustatory cues for dietary fat. *Am. J. Physiol.* **272**, C1203–1210.

Gishen, M., Iland, P. G., Dambergs, R. G., Esler, M. B., Francis, I. L., Kambouris, A., Johnstone, R. S., and Høj, P. B. (2002) Objective measures of grape and wine quality. In: *11th Australian Wine Industry Technical Conference*, Oct. 7–11, 2001, Adelaide, South Australia. (R. J. Blair, P. J. Williams and P. B. Høj, eds.), pp. 188–194. Winetitles, Adelaide, Australia.

Grbin, P. R., and Henschke, P. A. (2000) Mousy off-flavour production in grape juice and wine by *Dekkera* and *Brettanomyces* yeasts. *Aust. J. Grape Wine Res.* **6**, 255–262.

Grbin, P. R., Costello, P. J., Herderich, M., Markides, A. J., Henschke, P. A., and Lee, T. H. (1996) Developments in the sensory, chemical and microbiological basis of mousy taint in wine. In: *Proceedings of the 9th Australian Wine Industry Technical Conference*. Adelaide, 16–19 July, 1996 (C. S. Stockley, A. N. Sas, R. S. Johnson and T. H. Lee, eds.), pp. 57–61. Winetitles, Adelaide, Australia.

Green, B. G. (1992) The effects of temperature and concentration on the perceived intensity and quality of carbonation. *Chem. Senses* **17**, 435–450.

Green, B. G., and Frankmann, S. P. (1987) The effect of cooling the tongue on the perceived intensity of taste. *Chem. Senses* **12**, 609–619.

Grosch, W. (2001) Evaluation of the key odorants of foods by dilution experiments, aroma models and omission. *Chem. Senses* **26**, 533–545.

Guinard, J., and Cliff, M. (1987) Descriptive analysis of Pinot noir wines from Carneros, Napa, and Sonoma. *Am. J. Enol. Vitic.* **38**, 211–215.

Guinard, J., Pangborn, R. M., and Lewis, M. J. (1986a) The time-course of astringency in wine upon repeated ingestion. *Am. J. Enol. Vitic.* **37**, 184–189.

Guinard, J., Pangborn, R. M., and Lewis, M. J. (1986b) Preliminary studies on acidity-astringency interactions in model solutions and wines. *J. Sci. Food Agric.* **37**, 811–817.

Guth, H. (1997) Objectification of white wine aromas. Thesis, Technical University, Munich (in German).

Guth, H., and Sies, A. (2002) Flavor of wines: towards an understanding by reconstitution experiments and an analysis of ethanol's effect on odour activity of key compounds. In: *Proceedings of the 11th Australian Wine Industry Technical Conference* (R. J. Blair, P. J. Williams, and P. B. Høj, eds.), pp. 128–139. Australian Wine Industry Technical Conference Inc., Urrbrae, Australia.

Hahn, I., Scherer, P. W., and Mozell, M. M. (1993) Velocity profiles measured for airflow through a large-scale model of the human nasal cavity. *J. Appl. Physiol.* **75**, 2273–2287.

Haslam, E., and Lilley, T. H. (1988) Natural astringency in foods. A molecular interpretation. *Crit. Rev. Food Sci. Nutr.* **27**, 1–40.

Haslam, E., Lilley, T. H., Warminski, E., Liao, H., Cai, Y., Martin, R., Gaffney, S. H., Goulding, P. N., and Luck, G. (1992) Polyphenol complexation, a study in molecular recognition. In: *Phenolic Compounds in Food and Their Effects on Health. 1: Analysis, Occurrence, and Chemistry* (C.-T. Ho et al., eds.), pp. 8–50. ACS Symposium Series No. 506. American Chemical Society, Washington, DC.

Herraiz, J., and Cabezudo, M. D. (1980/81) Sensory profile of wines, quality index. *Process Biochem.* **16**, 16–19, 43.

Hersleth, M., Mevik, B.-H., Naes, T., and Guinard, J.-X. (2003) Effect of contextual factors on liking for wine – use of robust design methodology. *Food Qual. Pref.* **14**, 615–622.

Herz, R. S. (2003) The effect of verbal context on olfactory perception. *J. Exp. Psychol. Gen.* **132**, 595–606.

Hettinger, T. P., Myers, W. E., and Frank, M. E. (1990) Role of olfaction in perception of non-traditional 'taste' stimuli. *Chem. Senses* **15**, 755–760.

Huertas, R., Yebra, Y., Pérez, M. M., Melgosa, M., and Negueruela, A. I. (2003) Color variability for a wine sample poured into a standard glass wine sampler. *Color Res. Appl.* **28**, 473–479.

Hughson, A. L., and Boakes, R. A. (2002) The knowing nose: the role of knowledge in wine expertise. *Food Qual. Pref.* **13**, 463–472.

Hummel, T., Delwiche, J. F., Schmidt, C., and Hüttenbrink, K.-B. (2003) Effects of the form of glasses on the perception of wine flavors: a study in untrained subjects. *Appetite* **41**, 197–202.

Huon de Kermadec, F., and Pagès, J. (2005) Methodology to analyse rank and carry-over effects. *Food Qual. Pref.* **16**, 600–607.

Hyde, R. J., and Pangborn, R. M. (1978) Parotid salivation in response to tasting wine. *Am. J. Enol. Vitic.* **29**, 87–91.

Iland, P. G., and Marquis, N. (1993) Pinot noir – viticultural directions for improving fruit quality. In: *Proceedings of the 8th Australian Wine Industry Technical Conference* (P. J. Williams et al., eds.), pp. 98–100. Winetitles, Adelaide, Australia.

Jackson, R. S. (2002) *Wine Tasting. A Professional Handbook.* Academic Press, London.

Jinks, A., and Laing, D. G. (2001) The analysis of odor mixtures by humans: evidence for a configurational process. *Physiol Behav.* **72**, 51–63.

Johnson, H. (1985) *The World Atlas of Wine.* Simon & Schuster, New York.

Jolly, N. P., and Hattingh, S. (2001) A brandy aroma wheel for South African brandy. *S. Afr. J. Enol. Vitic.* **22**, 16–21.

Kaiser, P. K. (1996) *The Joy of Visual Perception. A Web Book.* http://www.yorku.ca/.

Kallithraka, S., Bakker, J., and Clifford, M. N. (2000) Interaction of (+)-catechin, (−)-epicatechin, procyanidin B2 and procyanidin C1 with pooled human saliva in vitro. *J. Sci. Food Agric.* **81**, 261–268.

Kaufmann, A. (1997) Multivariate statistics as a classification tool in the food laboratory. *J. AOAC Intl* **80**, 665–675.

Kielhorn, S., and Thorngate, J. H. (1999) Oral sensations associated with the flavan-3-ols (+)-catechin and (−)-epicatechin. *Food Qual. Pref.* **10**, 109–116.

Kroeze, J. H. A. (1982) The relationship between the side taste of masking stimuli and masking in binary mixtures. *Chem. Senses* **7**, 23–37.

Kurtz, T. E., Link, T. E., Tukey, R. F., and Wallace, D. L. (1965) Short-cut multiple comparisons for balanced single and double classifications. *Technometrics* **7**, 95–165.

Kuznicki, J. T., and Turner, L. S. (1986) Reaction time in the perceptual processing of taste quality. *Chem. Senses* **11**, 183–201.

Laing, D. G. (1983) Natural sniffing gives optimum odour perception for humans. *Perception* **12**, 99–117.

Laing, D. G. (1986) Optimum perception of odours by humans. *Proceedings of the 7th World Clean Air Congress*, Vol. 4, pp. 110–117. Clear Air Society of Australia and New Zealand.

Laing, D. G., and Panhuber, H. (1978) Application of anatomical and psychophysical methods to studies of odour interactions. In: *Progress in Flavour Research.* Proceedings, 2nd Weurman Flavour Research Symposium (D. G. Land and H. E. Nursten, eds.), pp. 27–47. Applied Science, London.

Laing, D. G., Link, C., Jinks, A. L., and Hutchinson, I. (2002) The limited capacity of humans to identify the components of taste mixtures and taste-odour mixtures. *Perception* **31**, 617–635.

Laska, M. and Hudson, R. (1992) Ability to discriminate between related odor mixtures. *Chem. Senses* **17**, 403–415.

Latreille, J., Mauger, E., Ambroisine, L., Tenenhaus, M., Vincent, M., Navarro, S., and Guinot, C. (2006) Measurement of the reliability of sensory panel performances. *Food Qual. Pref.* **17**, 369–375.

Lawless, H. T. (1984) Flavor description of white wine by expert and nonexpert wine consumers. *J. Food Sci.* **49**, 120–123.

Lawless, H. T. (1992) Unexpected congruence of odor quality and intensity rating. *Chem. Senses* **17**, 657–658.

Lawless, H. T., and Clark, C. C. (1992) Psychological biases in time-intensity scaling. *Food Technol.* **46** (11), 81, 84–86, 90.

Lawless, H. T., and Engen, T. (1977) Associations of odors, interference, mnemonics and verbal labeling. *J. Exp. Psychol. Human Learn. Mem.* 3, 52–59.

Lawless, H. T., Liu, Y.-F., and Goldwyn, C. (1997) Evaluation of wine quality using a small-panel hedonic scaling method. *J. Sens. Stud.* 12, 317–332.

Lawless, H. T., Schlake, S., Smythe, J., Lim, J., Yang, H., Chapman, K., and Bolton, B. (2004) Metallic taste and retronasal smell. *Chem. Senses* 29, 25–33.

Lazard, D., Zupko, K., Poria, Y., Nef, P., Lazarovits, J., Horn, S., Khen, M., and Lancet, D. (1991) Odorant signal termination by olfactory UDP glucuronosyl transferase. *Nature* 349(6312), 790–793.

Leach, E. J., and Noble, A. C. (1986) Comparison of bitterness of caffeine and quinine by a time-intensity procedure. *Chem. Senses* 11, 339–345.

Lee, C. B., and Lawless, H. T. (1991) Time-course of astringent sensations. *Chem. Senses* 16, 225–238.

Lee, K. (1989) Perception of irritation from ethanol, capsaicin and cinnamyl aldehyde via nasal, oral and retronasal pathways. MS thesis, University of California, Davis. (Reproduced in Noble, A. C. (1995) Application of time-intensity procedures for the evaluation of taste and mouthfeel. *Am. J. Enol. Vitic.* 46, 128–133.)

Lee, K.-Y. M., Paterson, A., and Piggott, J. R. (2001) Origins of flavour in whiskies and a revised flavour wheel: a review. *J. Inst. Brew.* 107, 287–313.

Lehrer, A. (1975) Talking about wine. *Language* 51, 901–923.

Lehrer, A. (1983) *Wine and Conversation.* Indiana University Press, Bloomington, IN.

Licker, J. L., Acree, T. E., and Henick-Kling, T. (1999) What is Brett (*Brettanomyces*) flavor? In: *Chemistry of Wine Flavor.* ACS Symposium Series Vol. 714. (A. L. Waterhouse and S. E. Ebeler, eds.), pp. 96–115. Amer. Chem. Soc., Washington, DC.

Livermore, A., and Laing, D. G. (1996) Influence of training and experience on the perception of multicomponent odor mixtures. *J. Exp. Psychol. Hum. Percept. Perform.* 22, 267–277.

Lüthi, H., and Vetsch, U. (1981) *Practical Microscopic Evaluation of Wines and Fruit Juices.* Heller Chemie und Verwaltsingsgesellschaft mbH, Schwäbisch Hall, Germany.

MacRostie, S. W. (1974) Electrode measurement of hydrogen sulfide in wine. M.S. thesis, University California, Davis.

Maga, J. A. (1974) Influence of color on taste thresholds. *Chem. Senses Flavor* 1, 115–119.

Malnic, B., Godfrey, P. A., and Buck, L. B. (2004) The human olfactory receptor gene family. *Proc. Natl Acad. Sci.* 101, 2584–2589.

Malnic, B., Hirono, J., Sato, T., and Buck, L.B. (1999) Combinatorial receptor codes for odors. *Cell* 96, 713–723.

Marcus, I. H. (1974) *How to Test and Improve Your Wine Judging Ability.* Wine Publishers, Berkeley, CA.

Martí, M. P., Boqué, R., Busto, O., and Guasch, J. (2005) Electronic noses in the quality control of alcoholic beverages. *TRAC – Trends Anal. Chem.* 24, 57–66.

Matsunami, H., and Amrein, H. (2004) Taste perception: How to make a gourmet mouse. *Curr. Biol.* 14, R118–R120.

McBride, R. L., and Finlay, D. C. (1990) Perceptual integration of tertiary taste mixtures. *Percept. Psychophys.* 48, 326–336.

McBurney, D. H. (1978) Psychological dimensions and perceptual analyses of taste. In: *Handbook of Perception* (E. C. Carterette and M. P. Friedman, eds.), Vol. 6A, pp. 125–155. Academic Press, New York.

Meilgaard, M. C., Dalgliesch, C. E., and Clapperton, J. F. (1979) Beer flavor terminology. *J. Am. Soc. Brew. Chem.* 37, 47–52.

Meilgaard, M. C., Reid, D. C., and Wyborski, K. A. (1982) Reference standards for beer flavor terminology system. *J. Am. Soc. Brew. Chem.* 40, 119–128.

Menashe, I., Man, O., Lancet, D., and Gilad, Y. (2003) Different noses for different people. *Nature Genetics* 34, 143–144.

Mennella, J. A., Pepino, M. Y., and Reed, D. R. (2005) Genetic and environmental determinants of bitter perception and sweet preferences. *Pediatrics.* 115, e216–22.

Meyerhof, W., Behrens, M., Brockhoff, A., Bufe, B., and Kuhn, C. (2005) Human bitter taste perception. *Chem. Senses* 30, i14–i15.

Mialon, V. S., and Ebeler, S. E. (1997) Time-intensity measurement of matrix effects on retronasal aroma perception. *J. Sensory Stud.* 12, 303–316.

Morrot, G. (2004) Cognition et vin. *Rev. Oenologues* 111, 11–15.

Morrot, G., Brochet, F., and Dubourdieu, D. (2001) The color of odors. *Brain Lang.* 79, 309–320.

Moulton, D. E., Ashton, E. H., and Eayrs, J. T. (1960) Studies in olfactory acuity, 4. Relative detectability of *n*-aliphatic acids by the dog. *Anim. Behav.* 8, 117–128.

Murphy, C. (1995) Age-associated differences in memory for odors. In: *Memory of Odors* (F. R. Schab and R. G. Crowder, eds.), pp. 109–131. Lawrence Erlbaum, Mahwah, NJ.

Murphy, C., Cain, W. S., and Bartoshuk, L. M. (1977) Mutual action of taste and olfaction. *Sens. Processes* 1, 204–211.

Myers, D. G., and Lamm, H. (1975) The polarizing effect of group discussion. *Am. Sci.* 63, 297–303.

Neogi, P. (1985) Tears-of-wine and related phenomena. *J. Colloid Interface Sci.* 105, 94–101.

Nespoulous, C., Briand, L., Delage, M. M, Tran, V., and Pernollet, J. C. (2004) Odorant binding and conformational changes of a rat odorant-binding protein. *Chem Senses* 29, 189–198.

Noble, A. C., and Bursick, G. F. (1984) The contribution of glycerol to perceived viscosity and sweetness in white wine. *Am. J. Enol. Vitic.* 35, 110–112.

Noble, A. C., Arnold, R. A., Masuda, B. M., Pecore, S. D., Schmidt, J. O., and Stern, P. M. (1984) Progress towards a standardized system of wine aroma terminology. *Am. J. Enol. Vitic.* 35, 107–109.

Noble, A. C., Williams, A. A., and Langron, S. P. (1984) Descriptive analysis and quality ratings of 1976 wines from four Bordeaux communes. *J. Sci. Food Agric.* 35, 88–98.

Noble, A. C., Arnold, R. A., Buechsenstein, J., Leach, E. J., Schmidt, J. O., and Stern, P. M. (1987) Modification of a standardized system of wine aroma terminology. *Am. J. Enol. Vitic.* 38, 143–146.

Nurgel, C., and Pickering, G. (2005) Contribution of glycerol, ethanol and sugar to the perception of viscosity and density elicited by model white wines. *J. Texture Studies* 36, 303–323.

Ohloff, G. (1994) *Scents and Fragrances*, p. 6. Springer Verlag, Berlin.

OIV (1990) *Récueil de Méthodes Internationales d'Analyse des Vine et des Moûts.* Office International de la Vigne et du Vin, Paris.

Olender, T., Fuchs, T., Linhart, C., Shamir, R., Adams, M., Kalush, F., Khen, M., and Lancet, D. (2004) The canine olfactory subgenome. *Genomics* 83, 361–372.

O'Mahony, M., and Rousseau, B. (2002) Discrimination testing: a few ideas, old and new. *Food Qual. Pref.* 14, 157–164.

Oreskovich, D. C., Klein, B. P., and Sutherland, J. W. (1991) Procrustes analysis and its applications to free-choice and other sensory profiling. In: *Sensory Science Theory and Application in Foods* (H. T. Lawless and B. P. Klein, eds.), pp. 353–393. Marcel Dekker, New York.

Österbauer, R. A., Matthews, P. M., Jenkinson, M., Beckmann, C. F., Hansen, P. C., and Calvert, G. A. (2005) Color of scents: chromatic stimuli modulate odor responses in the human brain. *J. Neurophysiol.* 93, 3434–3441.

Ough, C. S., and Winton, W. A. (1976) An evaluation of the Davis Wine Score Card and individual expert panel members. *Am. J. Enol. Vitic.* 27, 136–144.

Palmer, S. E. (1999) *Vision Science: From Photons to Phenomenology.* MIT Press, Cambridge, MA.

Pangborn, R. M. (1981) Individuality in responses to sensory stimuli. In: *Criteria of Food Acceptance* (J. Solms and R. L. Hall, eds.), pp. 177–219. Forster, Zurich.

Parr, W. V., White, K. G., and Heatherbell, D. (2003) The nose knows: Influence of colour on perception of wine aroma. *J. Wine Res.* **14**, 99–121.

Parr, W. V., White, K. G., and Heatherbell, D. (2004) Exploring the nature of wine expertise: What underlies wine expert's olfactory recognition memory advantage? *Food Qual. Pref.* **15**, 411–420.

Patterson, M. Q., Stevens, J. C., Cain, W. S., and Cometto-Muniz, J. E. (1993) Detection thresholds for an olfactory mixture and its three constituent compounds. *Chem. Senses* **18**, 723–734.

Peleg, H., Bodine, K., and Noble, A. C. (1998) The influence of acid on astringency of alum and phenolic compounds. *Chem. Senses* **23**, 371–378.

Pickering, G. J., Heatherbell, D. A., Vanhaenena, L. P., and Barnes, M. F. (1998) The effect of ethanol concentration on the temporal perception of viscosity and density in white wine. *Am. J. Enol. Vitic.* **49**, 306–318.

Piggott, J. R., and Findlay, A. J. F. (1984) Detection thresholds of ester mixtures. In: Proceedings of the Alko Symposium on Flavour Research of Alcoholic Beverages, Helsinki 1984 (L. Nykänen and P. Lehtonen, eds.). *Foundation Biotech. Indust. Ferm.* **3**, 189–197.

Pokorný, J., Filipü, M., and Pudil, F. (1998) Prediction of odour and flavour acceptances of white wines on the basis of their colour. *Nahrung* **42**, 412–415.

Prescott, J., Johnson, V., and Francis, J. (2004) Odor-taste interactions: Effects of attentional strategies during exposure. *Chem. Senses* **29**, 331–340.

Pridmore, R. W., Huertas, R., Melgosa, M., and Negueruela, A. I. (2005) Discussion on perceived and measured wine color. *Color Res. Appl.* **30**, 146–152.

Rapp, A. (1987) Veränderung der Aromastoffe während der Flaschenlagerung von Weißweinen. In: *Primo Simposio Internazionale: Le Sostanze Aromatiche dell'Uva e del Vino*, pp. 286–296.

Ritchey, J. G., and Waterhouse, A. L. (1999) A standard red wine: Monomeric phenolic analysis of commercial Cabernet Sauvignon wines. *Am. J. Enol. Vitic.* **50**, 91–100.

Roberts, D. D., and Acree, T. E. (1995) Simulation of retronasal aroma using a modified headspace technique: investigating the effects of saliva, temperature, shearing, and oil on flavor release. *J. Agric. Food Chem.* **43**, 2179–2186.

Roberts, D. D., Pollien, P., Antille, N., Lindinger, C., and Yeretzian, C. (2003) Comparison of nosespace, headspace, and sensory intensity ratings for the evaluation of flavor absorption by fat. *J. Agric. Food Chem.* **51**, 3636–3642.

Robichaud, J. L., and Noble, A. C. (1990) Astringency and bitterness of selected phenolics in wine. *J. Sci. Food Agric.* **53**, 343–353.

Rolls, E. T. (2001) The rules of formation of the olfactory representations found in the orbitofrontal cortex olfactory areas in primates. *Chem. Senses* **26**, 595–604.

Rolls, E. T., Critchley, H. D., Browning, A., Hernadi, I. (1998) The neurophysiology of taste and olfaction in primates, and umami flavor. *Ann. NY Acad. Sci.* **855**, 426–437.

Ross, C. F., Hinken, C., and Weller, K. (2006) Efficacy of rinse procedures for reducing astringency carryover in red wine upon repeated ingestion. *Am. J. Enol. Vitic.* **57**, 394A.

Russell, K., Zivanovic, S., Morris, W. C., Penfield, M., and Weiss, J. (2005) The effect of glass shape on the concentration of polyphenolic compounds and perception of Merlot wine. *J. Food Qual.* **28**, 377–385.

Sachs, O. (1985) The dog beneath the skin. In: *The Man Who Mistook His Wife for a Hat*, pp. 149–153. Duckworth, London.

Saintsbury, G. (1920) Notes on a Cellar-Book. Macmillan, London.

Sakai, N., Imada, S., Saito, S., Kobayakawa, T., and Deguchi, Y. (2005) The effect of visual images on perception of odors. *Chem. Senses* **30** (Suppl. 1), 1244–1245.

Schiffman, S. S. (1983) Taste and smell in disease. *N. Engl. J. Med.* **308**, 1275–1279.

Schiffman, S., Orlandi, M., and Erickson, R. P. (1979) Changes in taste and smell with age, biological aspects. In: *Sensory Systems and Communication in the Elderly* (J. M. Ordy and K. Brizzee, eds.), pp. 247–268. Raven, New York.

Scholten, P. (1987) How much do judges absorb? *Wines Vines* **69**(3), 23–24.

Schutz, H. G., and Pilgrim, F. J. (1957) Differential sensitivity in gustation. *J. Exp. Psychol.* **54**, 41–48.

Scott, T. R., and Chang, G. T. (1984) The state of gustatory neural coding. *Chem. Senses* **8**, 297–313.

Scott, T. R., and Giza, B. K. (1987) Neurophysiological aspects of sweetness. In: *Sweetness* (J. Dobbing, ed.), pp. 15–32. Springer-Verlag, London.

Seifert, R. M., Buttery, R. G., Guadagni, D. G., Black, D. R., and Harris, J. G. (1970) Synthesis of some 2–methoxy-3–alkylpyrazines with strong bell pepper-like odors. *J. Agric. Food Chem.* **18**, 246–249.

Selfridge, T. B., and Amerine, M. A. (1978) Odor thresholds and interactions of ethyl acetate and diacetyl in an artificial wine medium. *Am. J. Enol. Vitic.* **29**, 1–6.

Senf, W., Menco, B. P. M., Punter, P. H., and Duyvesteyn, P. (1980) Determination of odour affinities bases on the dose-response relationships of the frog's electro-olfactogram. *Experientia* **36**, 213–215.

Silva Ferreira, A. C., Hogg, T., and Guedes de Pinho, P. (2003) Identification of key odorants related to the typical aroma of oxidation-spoiled white wines. *J. Agric. Food Chem.* **51**, 1377–2381.

Singleton, V. L. (1976) Wine aging and its future. First Walter and Carew Reynell Memorial Lecture, pp. 1–39. Tanunda Institute, Roseworthy Agricultural College, Australia.

Smith, A. K., and Noble, A. C. (1998) Effects of increased viscosity on the sourness and astringency of aluminum sulfate and citric acid. *Food Qual. Pref.* **9**, 139–144.

Solomon, G. E. A. (1988) Great expectorations: The psychology of novice and expert wine talk. Doctoral thesis. Harvard University (cited in Hughson and Boakes, 2002).

Solomon, G. E. A. (1990) Psychology of novice and expert wine talk. *Am. J. Psychol.* **103**, 495–517.

Solomon, G. E. A. (1991) Language and categorization in wine expertise. In: *Sensory Science Theory and Applications in Foods* (H. T. Lawless and B. P. Klein, eds.), pp. 269–294. Marcel Dekker, New York.

Somers, T. C. (1975) In search of quality for red wines. *Food Technol. Australia* **27**, 49–56.

Somers, T. C. (1998) *The Wine Spectrum.* Winetitles, Adelaide, Australia.

Somers, T. C., and Evans, M. E. (1974) Wine quality: Correlations with colour density and anthocyanin equilibria in a group of young red wine. *J. Sci. Food Agric.* **25**, 1369–1379.

Somers, T. C., and Evans, M. E. (1977) Spectral evaluation of young red wines, anthocyanin equilibria, total phenolics, free and molecular SO_2 chemical age. *J. Sci. Food Agric.* **28**, 279–287.

Somers, T. C., and Pocock, K. F. (1991) Phenolic assessment of white musts: Varietal differences in free-run juices and pressings. *Vitis* **26**, 189–201.

Sorrentino, F., Voilley, A., and Richon, D. (1986) Activity coefficients of aroma compounds in model food systems. *AIChE J.* **32**, 1988–1993.

Sowalsky, R. A., and Noble, A. C. (1998) Comparison of the effects of concentration, pH and anion species on astringency and sourness of organic acids. *Chem. Senses* **23**, 343–349.

Stahl, W. H., and Einstein, M. A. (1973) Sensory testing methods. In: *Encyclopedia of Industrial Chemical Analysis* (F. D. Snell and L. S. Ettre, eds.), Vol. 17, pp. 608–644. John Wiley, New York.

Stevens, D. A., and O'Connell, R. J. (1995) Enhanced sensitivity to androstenone following regular exposure to pemenone. *Chem. Senses* 20, 413–420.

Stevens, J. C., Bartoshuk, L. M., and Cain, W. S. (1984) Chemical senses and aging: Taste *versus* smell. *Chem. Senses* 9, 167–179.

Stevens, J. C., and Cain, W. C. (1993) Changes in taste and flavor in aging. *Crit. Rev. Food Sci. Nutr.* 33, 27–37.

Stevens, J. C., Cain, W. S., and Burke, R. J. (1998) Variability of olfactory thresholds. *Chem. Senses* 13, 643–653.

Stone, H., and Sidel, J. (1985) *Sensory Evaluation Practices.* Academic Press, Orlando, FL.

Sudraud, P. (1978) Évolution des taux d'acidité volatile depuis le début du siècle. *Ann. Technol. Agric.* 27, 349–350.

Suprenant, A., and Butzke, C. E. (1996) Implications of odor threshold variations on sensory quality control of cork stoppers. In: *Proceedings of the 4th International Symposium on Cool Climate Viticulture and Enology* (T. Henick-Kling *et al.*, eds.), pp. VII-70–74. New York State Agricultural Experimental Station, Geneva, NY.

Takagi, S. F. (1978) Biophysics of smell. In: *Handbook of Perception* (E. C. Carterette and M. P. Friedman, eds.), Vol. 6A, pp. 233–243. Academic Press, New York.

Taylor, S. L., Higley, N. A., and Bush, R. K. (1986) Sulfites in foods: Uses, analytical methods, residues, fate, exposure assessment, metabolism, toxicity, hypersensitivity. *Adv. Food Res.* 30, 1–76.

Temussi, P. (2007) The sweet taste receptor: a single receptor with multiple sites and modes of interaction. *Adv. Food Nutr. Res.* 53, 199–239.

Thomas, C. J. C., and Lawless, H. T. (1995) Astringent subqualities in acids. *Chem Senses* 20, 593–600.

Tominaga, T., Guimbertau, G., and Dubourdieu, D. (2003) Contribution of benzenemethanethiol to smoky aroma of certain *Vitis vinifera* L. wines. *J. Agric. Food Chem.* 51, 1373–1376.

Tozaki, H., Tanaka, S., and Hirata, T. (2004) Theoretical consideration of olfactory axon projection with an activity – dependent neural network model. *Mol. Cell Neurosci.* 26, 503–517.

Tromp, A., and van Wyk, C. J. (1977) The influence of colour on the assessment of red wine quality. In: *Proceedings of the South African Society for Enology and Viticulture*, pp. 107–117.

Vidal, S., Francis, L., Guyot, S., Marnet, N., Kwiatkowski, M., Gawel, R., Cheynier, V., and Waters, E. J. (2003) The mouth-feel properties of grape and apple proanthocyanidins in a wine-like medium. *J. Sci. Food Agric.* 83, 564–573.

Vidal, S., Courcoux, P., Francis, L., Kwiatkowski, M., Gawel, R., Williams, P., Waters, E., and Cheynier, V. (2004) Use of an experimental design approach for evaluation of key wine components on mouth-feel perception. *Food Qual. Pref.* 15, 209–217.

Vidal, S., Francis, L., Noble, A., Kwiatkowski, M., Cheynier, V., and Waters, E. (2004) Taste and mouth-feel properties of different types of tannin-like polyphenolic compounds and anthocyanins in wine. *Anal. Chim. Acta* 513, 57–65.

Vidal, S., Francis, L., Williams, P., Kwiatkowski, M., Gawel, R., Cheynier, V., and Waters, E. (2004) The mouth-feel properties of polysaccharides and anthocyanins in a wine like medium. *Food Chem.* 85, 519–525.

Voilley, A., Beghin, V., Charpentier, C., and Peyron, D. (1991) Interactions between aroma substances and macromolecules in a model wine. *Lebensm. Wiss. Technol.* 24, 469–472.

von Sydow, E., Moskowitz, H., Jacobs, H., and Meiselman, H. (1974) Odor taste interaction in fruit juices. *Lebens. Wiss. Technol.* 7, 9–16.

Walker, J. C., Hall, S. B., Walker, D. B., Kendal-Reed, M. S., Hood, A. F., and Niu, X.-F. (2003) Human odor detectability: New methodology used to determine threshold and variation. *Chem. Senses* 28, 817–826.

Williams, A. A. (1978) The flavour profile assessment procedure. In: *Sensory Evaluation – Proceeding of the Fourth Wine Subject Day*, pp. 41–56. Long Ashton Research Station, University of Bristol, Long Ashton, UK.

Williams, A. A. (1982) Recent developments in the field of wine flavour research. *J. Inst. Brew.* 88, 43–53.

Williams, A. A. (1984) Measuring the competitiveness of wines. In: *Tartrates and Concentrates – Proceeding of the Eighth Wine Subject Day Symposium* (F. W. Beech, ed.), pp. 3–12. Long Ashton Research Station, University of Bristol, Long Ashton, UK.

Williams, A. A., and Langron, S. P. (1983) A new approach to sensory profile analysis. In: *Flavor of Distilled Beverages: Origin and Development* (J. R. Piggott, ed.), pp. 219–224. Ellis Horwood, Chichester, UK.

Williams, A. A., Bains, C. R., and Arnold, G. M. (1982) Towards the objective assessment of sensory quality in less expensive red wines. In: *Grape and Wine Centennial Symposium Proceedings* (A. D. Webb, ed.), pp. 322–329. University of California, Davis.

Williams, A. A., Langron, S. P., and Noble, A. C. (1984) Influence of appearance on the assessment of aroma in Bordeaux wines by trained assessors. *J. Inst. Brew.* 90, 250–253.

Williams, A. A., and Rosser, P. R. (1981) Aroma enhancing effects of ethanol. *Chem. Senses* 6, 149–153.

Williams, P. J., and Francis, I. L. (1996) Sensory analysis and quantitative determination of grape glycosides – the contribution of these data to winemaking and viticulture. In: *Biotechnologically Improved Foods Flavors* (G. R. Takeoka *et al.*, eds.), pp. 124–133. ACS Symposium Series 637. American Chemical Society, Washington, DC.

Winiarski, W., Winiarski, J., Silacci, M., and Painter, B. (1996) The Davis 20-point scale: How does it score today. *Wines Vines* 77, 50–53.

Winton, W., Ough, C. S., and Singleton, V. L. (1975) Relative distinctiveness of varietal wines estimated by the ability of trained panelists to name the grape variety correctly. *Am. J. Enol. Vitic.* 26, 5–11.

Wróblewski, K., Muhandiram, R., Chakrabartty, A., and Bennick, A. (2001) The molecular interaction of human salivary histatins with polyphenolic compounds. *Eur. J. Biochem.* 268, 4384–4397.

Wysocki, C. J., Dorries, K. M., and Beauchamp, G. K. (1989) Ability to perceive androsterone can be acquired by ostensibly anosmic people. *Proc. Natl Acad. Sci. USA* 86, 7976–7978.

Zellner, D. A., and Whitten, L. A. (1999) The effect of color intensity and appropriateness on color-induced odor enhancement. *Am. J. Psychol.* 112, 585–604.

Zellner, D. A., Stewart, W. F., Rozin, P., and Brown, J. M. (1988) Effect of temperature and expectations on liking for beverages. *Physiol. Behav.* 44, 61–68.

Zervos, C. and Albert, R.H. (1992) Chemometrics: The use of multivariate methods for the determination and characterization of off-flavors. In: *Off-Flavors in Foods and Beverages* (G. Charalambous, ed.), pp. 669–742. Elsevier Science, Amsterdam.

Zhao, K., Scherer, P. W., Hajiloo, S. A., and Dalton, P. (2004) Effect of anatomy on human nasal air flow and odorant transport patters: Implications for olfaction. *Chem. Senses* 29, 365–379.

Zuniga, J. R., Davies, S. H., Englehardt, R. A., Miller, I. J., Jr., Schiffman, S. S., and Phillips, C. (1993) Taste performance on the anterior human tongue varies with fungiform taste bud density. *Chem. Senses* 18, 449–460.

12

Wine and Health

Introduction

The contrasting social and antisocial effects of alcohol consumption must have become evident shortly after the discovery of winemaking. Time has only augmented our understanding of the multifaceted nature of this Dr Jekyll–Mr Hyde phenomenon. It is clear that excessive alcohol consumption, both acute and chronic, can have devastating effects on physical and mental well-being. Excessive ethanol consumption can cause cirrhosis of the liver, increase the likelihood of hypertension and stroke, favor the development of breast and digestive tract cancers, and induce fetal alcohol syndrome. Many of these effects may stem from the activation of free-radical damage induced by high alcohol intake (Meagher et al., 1999). Because the problems associated with alcoholism (Abrams et al., 1987; Schmitz and Gray, 1998), and the cerebral chemical changes associated with addiction (Nestler and Malenka, 2004; Heinz, 2006) have been well documented, they will not be discussed here. On the other hand, it is becoming equally clear that moderate wine consumption (250–300 ml/day – about one-third of a 750-ml bottle) has distinct health benefits. Multiple epidemiological studies suggest that daily, moderate, alcohol consumption (Thun *et al.*, 1997; Doll *et al.*, 2005), and especially wine (Grønbæk *et al.*, 2000; Renaud *et al.*, 2004) is associated with a reduction in all-cause mortality. This is expressed in a U-shaped curve, with increased mortality being associated with both excess alcohol intake and abstinence. This is particularly evident in the reduced incidence of cardiovascular disease in moderate alcohol consumers. In addition, it reduces the likelihood of type 2 diabetes, combats hypertension, and reduces the frequency of certain cancers and several other diseases. These epidemiological correlations are being supported by *in vivo* studies that provide molecular explanations for these associations. The principal element missing in confirming a causal relationship involves detailed information on the dynamics of absorption, metabolism, and elimination of the proposed active ingredients.

Faced with a chemical and beverage that can be not only salubrious but also addictive, the fluctuations in society's attitude toward alcohol are not surprising (Musto, 1996; Pittman, 1996; Vallee, 1998). Thankfully for those in the wine industry, wine drinkers are less likely to show those alcohol-related problems that have given alcohol a bad reputation (Smart and Walsh, 1999). In addition, wine has a higher social image than other alcoholic beverages (Klein and Pittman, 1990).

The use of wine as a medicine, or as a carrier for medications, has a long history. It goes back at least to the ancient Egyptians (Lucia, 1963). Ancient Greek and Roman society used wine extensively in herbal infusions. This practice continued largely unabated until the beginning of the twentieth century. The excessive abuse of distilled alcoholic beverages, combined with religious and political conservatism, created a backlash against all beverages containing alcohol, notably in North America. Alcohol was viewed as an agent of corruption to be annihilated. Following the failure of Prohibition, humans themselves, not alcohol, came to be viewed as the source of evil. Alcoholism is now appropriately viewed as a genuine disease, possessing a complex etiology (Nurnberger and Bierut, 2007), with both genetic and environmental aspects. Thus, the social climate is changing, and the legitimate use of wine in medicine and health is being investigated seriously.

Nevertheless, it is unlikely that doctors will soon be prescribing wine for its health benefits. Too often, people have a difficulty recognizing the limits of rational use. Even dietary flavonoid supplements (one of the benefits of wine consumption) may be detrimental if taken in excess (Skibola and Smith, 2003).

Metabolism of Alcohol

Alcohol is the primary by-product of fermentative metabolism in many organisms. Ethanol is also an energy source for an even larger number of species. Thus, it is not surprising that enzymes involved in ethanol oxidation are found in most life forms, including humans.

In humans, ethanol enters the bloodstream either via the consumption of beverages containing alcohol, or from ethanol synthesized by the bacterial flora in the intestinal tract. When the concentration of alcohol is low, most of it is metabolized in the liver before it enters the systemic blood flow. Most of the blood supply from the digestive tract passes through the liver before dispersing to the rest of the body.

The liver metabolizes about 95% of alcohol in the plasma, at about 15 ml/h. The rest of the alcohol tends to be lost in the breath, or secreted in the urine and other bodily fluids. The liver possesses two enzymic pathways for ethanol metabolism. The primary (constitutive) mechanism involves the oxidation of ethanol to acetaldehyde, through the action of cytoplasmic alcohol dehydrogenases (ADHs). Of the seven known *ADH* genes, three function in the liver. The others act in the gastric epithelium and other tissues. Subsequent oxidation converts acetaldehyde to acetic acid. This occurs principally under the action of mitochondrial acetaldehyde dehydrogenase (ALDH2). Cytoplasmic acetaldehyde dehydrogenase (ALDH1) is less active. The acetic acid may be released into the blood or converted to acetyl CoA. From this point, metabolism may flow along any of the standard biochemical pathways (Fig. 7.15).

As noted, alcohol metabolizing enzymes frequently occur in allelic forms (isozymes). Their relative occurrence also tends to vary among ethnic groups. Some of the isozymes possess distinct physiological attributes. For example, *ADH1B*1* codes for an isozyme that oxidizes ethanol slowly, whereas *ADH1B*3* encodes for a highly active version (about 30 times more efficient) (Thomasson *et al.*, 1995). These allelic variants are common among African Americans. Fast and slow acting isozymes of ADH1C are common in Europeans. Rapid alcohol oxidation may donate a degree of protection against alcoholism, by quickly converting ethanol to acetaldehyde. Such individuals tend to drink less and show more rapid and marked negative consequences to alcohol consumption than do those metabolizing alcohol more slowly (Wall *et al.*, 2005). However, this probably reduces some of the health benefits of ethanol consumption. The effect of mutants of mitochondrial acetaldehyde dehydrogenase gene (*ALDH2*) is discussed below.

A second metabolic routing of ethanol metabolism occurs only when the blood alcohol concentration is high. It entails an inducible pathway involving micro-somal cytochrome P4502E1. It oxidizes ethanol to acetaldehyde, using molecular oxygen rather than NAD^+. The activation of the microsomal oxidation pathway has the undesirable side-effect of generating free oxygen radicals (Meagher *et al.*, 1999). The free-radical activity can remain long after alcohol intake has ceased. Although most free oxygen radicals are inactivated by glutathione, superoxide dismutase, and catalase, long-term exposure to trace amounts of oxygen radicals may induce the slow, progressive accumulation of irreparable cellular damage. Subsequent oxidation of acetaldehyde, generated by the microsomal pathway, to acetic acid is identical to that derived by alcohol dehydrogenase.

The metabolism of ethanol to acetate (acetic acid) has the advantage that tissue cells can regulate its transport. This is not true for ethanol, which can diffuse freely across cellular membranes. Control of transport is central to proper cellular function.

Physiological Actions

The ability of ethanol to displace water, and its unregulated passage cross cell membranes, explains much of alcohol's cytoplasmic toxicity. In addition, its oxidation to acetaldehyde is more rapid than the subsequent oxidation to acetate. Thus, acetaldehyde may accumulate in the blood and other bodily fluids. This is often viewed as an important contributor to the toxicity associated with excessive alcohol consumption. Differentiating between the direct and indirect toxic effects of excessive ethanol intake has proven difficult.

One of the first physiologic effects of alcohol consumption is a suppression of higher brain function. This is most noticeable in enhanced sociability – by blocking social inhibitions. For others, it quickly induces drowsiness (Stone, 1980). This probably explains why taking a small amount of wine (90–180 ml) before bed often helps people, notably the elderly, suffering from insomnia (Kastenbaum, 1982). Half a glass of wine provides the benefits of sleep induction, without causing agitation and sleep apnea – often associated with greater alcohol consumption. The effect on sleep may arise from alcohol facilitating the transmission of inhibitory γ-aminobutyric acid (GABA), while suppressing the action of excitatory glutamate receptors (Haddad, 2004). GABA and glutamate are estimated to be involved in about 80% of neuronal circuitry in the brain.

Another effect on brain function results from a reduction in hormonal secretion – notably vasopressin. As a consequence, urine production increases, producing the diuretic effect frequently associated with alcohol consumption. Less well known is how alcohol acts as a crucial regulator of the hypothalamic–pituitary–adrenal axis, modulating the release of hormones such as adrenocorticotropic hormone (ACTH) and corticosterone (Haddad, 2004).

Although alcohol has a general depressive action on brain function, the levels of some brain modulators show transitory increases. Examples are serotonin and histamine. The latter may activate a cascade of reactions leading to headache production.

Another of the multiple influences of alcohol is the conversion of hepatic glycogen to sugar. This results in a short-lived increase in plasma glucose content. This, in turn, can cause glucose loss in the urine, as well as an increase in insulin release by the pancreas. Both result in a drop in blood sugar content. If sufficiently marked, hypoglycemia results. This apparently causes the temporary weakness often associated with excessive alcohol intake.

In addition to the direct effects of ethanol, the accumulation of acetaldehyde may have several undesirable consequences. It may be involved in much of the chronic damage associated with alcoholism. At low rates of alcohol intake, acetaldehyde metabolism is sufficiently rapid to limit its accumulation and liberation from the liver. At higher concentrations, acetaldehyde rapidly consumes glutathione reserves in the liver – an crucial cellular antioxidant. This coincides with activation of the microsomal ethanol oxidation pathway that generates toxic free-oxygen radicals. In the absence of sufficient glutathione, free-oxygen radicals can accumulate, disrupting mitochondrial function. Elsewhere in the body, acetaldehyde can bind with proteins and cellular constituents, forming stable complexes (Niemela and Parkkila, 2004). These can lead to the production of immunogenic determinants, which can stimulate antibody production against acetaldehyde adducts. This may generate some of the chronic tissue damage associated with alcohol abuse (Niemela and Israel, 1992). The binding of acetaldehyde to the plasma membrane of red blood cells is known to increase their rigidity. By limiting their ability to squeeze through the narrowest capillaries, oxygen supply to tissue cells may be restricted. This could participate in suppressed brain function. Nerve cells show only respiratory (oxygen-dependent) metabolism. It is estimated that the brain consumes up to 20% of the blood's oxygen supply.

Although ethanol and acetaldehyde can produce severe, progressive, and long-term damage to various organs, and incite alcohol dependence, these consequences are absent when alcohol consumption is moderate and taken with meals. As the sections below demonstrate, moderate, daily, wine consumption has clear health benefits for the majority of people.

Food Value

Wine's major nutritional value comes from the rapidly metabolized, caloric value of its ethanol content. Alcohol does not need to be digested, and can be absorbed directly through the intestinal wall. In rural viticultural areas, wine historically functioned as a major source of metabolic energy for the adult population. Wine in those regions was a food.

Wine contains small quantities of several vitamins, notably the B vitamins, such as B_1 (thiamine), B_2 (riboflavin), and B_{12} (cobalamin). However, wine is virtually devoid of vitamins A, C, D, and K. In excess, ethanol can impair vitamin uptake. Wine contains various minerals in readily available forms, especially potassium and iron (in the ferrous state). Nevertheless, excessive alcohol consumption can disturb the uptake of calcium, magnesium, selenium, and zinc, and increase the excretion of zinc via the kidneys. The low sodium/high potassium content of wine makes it one of the more effective sources of potassium for individuals on diuretics. Although wine contains soluble dietary fiber, especially red wines (Díaz-Rubio and Saura-Calixto, 2006), it is insufficient to contribute significantly to the daily recommended fiber content in the human diet.

Digestion

Wine has several direct and indirect effects on food digestion. The phenolic (Hyde and Pangborn, 1978) and alcohol (Martin and Pangborn, 1971) contents of wine activate the release of saliva. In addition, wine promotes the release of gastrin as well as gastric juices. Succinic acid is the principal wine ingredient activating the release of gastric juices (Teyssen *et al.*, 1999). It does not, however, activate gastrin release. The substance(s) involved in stimulating gastrin secretion are unknown. Wine also significantly delays gastric emptying, both on an empty stomach (Franke *et al.*, 2004), or when consumed with food (Benini *et al.*, 2003). The latter favors digestion by extending acid hydrolysis. It also promotes inactivation of pathogenic food contaminants. In addition, wine slows plasma glucose uptake, independent of any insulin response (Benini *et al.*, 2003). Furthermore, at the levels found in most table wines, ethanol activates the release of bile in the intestines. Wine acids and aromatics also induce the same effects. In contrast, the high alcohol content of distilled beverages can suppress digestive juice flow, the release of bile, and induce stomach spasms.

Despite the general beneficial effects of alcohol on digestion, the phenolic content of red wine may retard digestion. For example, tannins and phenolic acids interfere with the action of certain digestive enzymes, such as α-amylase, lipase and trypsin (Rohn *et al.*, 2002). Digestion may be further delayed by polymerization between tannins and food proteins. Although potentially delaying digestion in the small intestine, it continues in the colon. In contrast, some phenolics, such as quercetin, resveratrol, catechin, and epigallocatechin gallate, promote pepsin-activated protein breakdown (Tagliazucchi *et al.*, 2005).

The phenolic content of wine also decreases the intestinal absorption of iron and copper. Although this may be undesirable by limiting the bioavailability of iron under deficiency conditions, it has the benefit of reducing the formation of toxic lipid hydroperoxides during digestion in the intestines. The antioxidant effect of polyphenolics also applies to hydroperoxide generated in the stomach (Kanner and Lapidot, 2001).

Wine also has a cultural/psychologic effect on digestion. The association of wine with refined eating promotes slower food consumption, permitting biofeedback mechanisms to induce satiety and regulate food intake. In addition, wine consumption can promote a more relaxed lifestyle, something increasingly valuable in our overly compulsive society. Whether this explains the improved appetite of many elderly and anorectic patients when wine is taken with the meal is unknown.

The activation of gastric juice release not only aids food digestion, but also inactivates enzymes involved in ulceration. Even more significant may be the antibiotic action of wine constituents on *Helicobacterium pylori* (Fugelsang and Muller, 1996). *H. pylori* is considered the primary cause of stomach ulceration. Moderate wine consumption appears have a prophylactic effect limiting ulcer initiation (Brenner *et al.*, 1997). The bacterium is also implicated in gastritis, vitamin B_{12} malabsorption, and gastric adenocarcinoma.

Wine may further aid human sustenance by increasing nutrient uptake. Congeners combine with metallic ions, vitamins, and fatty acids, facilitating their transport across the intestinal wall.

Finally, consuming wine with food slows the rate of alcohol uptake in the blood (Fig. 12.1). In the absence of food, about 80% of alcohol absorption occurs through the intestinal wall. This value increases in the presence of food. By retarding gastric emptying, food consumption slows the transfer of alcohol into the intestine. This extends the period over which the liver can metabolize ethanol. The result is a reduction in the maximal alcohol content reached in the blood. However, taking sparkling wine on an empty stomach increases short-term alcohol uptake by about 35% (Ridout *et al.*, 2003). Although the cause of the difference is unknown, it is suspected that carbon dioxide relaxes the pyloric sphincter,

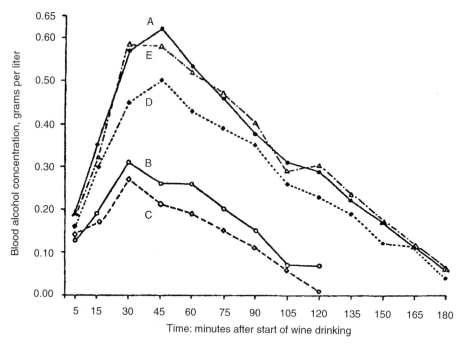

Figure 12.1 Blood alcohol concentrations after wine drinking in a single dose. A, fasting; B, during a meal; C, 2 hours after a meal; D, 4 hours after a meal; E, 6 hours after a meal. (From Serianni *et al.*, 1953, reproduced by permission)

allowing earlier transfer of fluids from the stomach into the duodenum.

The rate of alcohol metabolism differs considerably among individuals, with rates commonly varying between 90 and 130 mg/kg/h. The hormonal and nutritional state of the individual can affect the rate of ethanol metabolism.

Phenolic Bioavailability

To fully understand the significance of wine phenolics to health, it will be necessary to know the dynamics of their uptake, concentration, metabolism, and elimination. Such data are just now starting to become available.

In the mouth, mid-sized flavonoid polymers often bond to salivary proteins, forming stable complexes (De Freitas and Mateus, 2003; Pizarro and Lissi, 2003). This significantly limits their uptake in the stomach and small intestine. Passage through the stomach does not modify the majority of wine phenolics. Nevertheless, flavonoids such as anthocyanins quickly traverse the stomach and pass into the blood (Passamonti *et al.*, 2003). They are also effectively translocated across the wall of the small intestine (Talavéra *et al.*, 2005). Phenolic acids, such as caffeic acid (Simonetti *et al.*, 2001) and resveratrol (Soleas *et al.*, 2001) also readily pass into the plasma via the intestinal tract. In contrast, polymers tend to remain in the intestine until degraded by bacteria in the colon (Scalbert *et al.*, 2002). A small portion of their breakdown products, primarily phenolic acids, are subsequently absorbed into the blood via the colon (Ward *et al.*, 2004).

Studies on the bioavailability of flavonoids in the blood are in their infancy (Williamson and Manach, 2005). Although many flavonoids are quickly absorbed into the plasma, most appear to be rapidly conjugated – being methylated, sulfated, transformed to glucuronsides, or otherwise metabolized (see Williams *et al.*, 2004). These transformations could significantly affect their antioxidant and other properties, as well as their ability to move into tissues. Most of these metabolites still retain a reducing phenolic group, and, thus, may possess antioxidant properties. Nevertheless, there is growing evidence that phenolic metabolites act primarily as signaling molecules, notably in oxygen-stress-related pathways (Williams *et al.*, 2004). Smaller amounts of chemical are needed for signaling reactions, than direct antioxidant effects. This might explain the discrepancy between the low levels of free phenolics in the plasma and their apparent effects in the body. Future studies will have to investigate the efficacy of phenolic metabolites and conjugated complexes, at concentrations found in the plasma. Their efficacy at binding to, or translocation into tissue cells, is little known. Survival of most of these constituents in the plasma is comparatively short (a few hours). Breakdown products of phenolic metabolism rapidly appear in urine shortly after uptake in the plasma.

Presence in the plasma probably permits their diffusion into most body tissues. This does not apply to the brain. Except where there are specific transport proteins, most compounds above a molecular weight of 500 Da are excluded by the blood–brain barrier. This barrier exists due to tight connections between the endothelial cells that makeup the lining of cerebral capillaries. This prevents the diffusion of molecules between vascular endothelial cells that is typical elsewhere in the body. However, with anthocyanins (Passamonti *et al.*, 2005) and simple flavonols (Youdim *et al.*, 2004), access to the brain occurs within minutes of consumption.

Antimicrobial Effects

The prophylactic action of wine against gastrointestinal infections has been known for millennia, long before the microbial nature of infectious diseases was suspected. This action is complex and not fully understood.

The antimicrobial effect of alcohol was discovered in the late 1800s. Nevertheless, alcohol is not particularly antimicrobial at the concentrations found in wine (optimal at about 70%). Thus, the antibiotic action of wine results primarily from other constituents, probably its phenolic content. Modification of anthocyanins during fermentation increases their toxicity to viruses, protozoans, and bacteria. Other phenolic compounds commonly found in red wines are also bacteriostatic and fungistatic. For example, *p*-coumaric acid is particularly active against gram-positive bacteria, such as *Staphylococcus* and *Streptococcus*, whereas other phenols inhibit gram-negative bacteria, for example *Escherichia*, *Shigella*, *Proteus*, and *Vibrio* (Masquelier, 1988). The latter cause serious forms of diarrhea and dysentery. Despite wine being more effective than antimicrobial agents, such as bismuth salicylate (Weisse *et al.*, 1995), full action may take several hours (Møretrø and Daeschel, 2004; Dolara *et al.*, 2005). In most instances, the mechanism by which phenolics have their action is unknown. However, in the case of quercetin, the effect may be partially attributed to its inhibition of DNA gyrase, whereas with epigallocatechin, disruption of cell membrane function appears central to its antibiotic action. The low pH and presence of various organic acids appear to accentuate the antimicrobial action of both wine phenolics and ethanol. It is not without good reason that Roman armies added wine or vinegar to their drinking water. Diarrhetic soldiers do not win wars.

Wine is also active against several viruses, including the herpes simplex virus, poliovirus, hepatitis A virus, as well

as rhinoviruses and coronaviruses. The effect on the latter two groups appears reflected in the reduced incidence of the common cold in moderate alcohol consumers (Cohen *et al.*, 1993), particularly those drinking red wines (Takkouche *et al.*, 2002). If you have to gargle, port is certainly one of the more pleasant options available.

Cardiovascular Disease

The most clearly established benefit of moderate alcohol consumption, notably wine, relates to a nearly 30–35% reduction in death rate due to cardiovascular disease (Klatsky *et al.*, 1974, 2003; Renaud and de Lorgeril, 1992). Figure 12.2 provides an example of such results. Alcohol consumption also decreases the likelihood of intermittent claudication (pain or cramping in the calf of the leg). Claudication is a common indicator of peripheral arterial disease. Recent studies have confirmed that incidental factors, such as gender, race, lifestyle, educational level, etc. do not affect the results (for example, Mukamal *et al.*, 2006). Studies have also demonstrated that daily consumption of alcohol significantly reduces the incidence of other forms of cardiovascular disease, such as hypertension (Keil *et al.*, 1998), heart attack (Gaziano *et al.*, 1999), stroke (Truelsen *et al.*, 1998; Hillbom, 1999), and peripheral arterial disease (Camargo *et al.*, 1997). Those who consume wine moderately live, on average, 2.5–3.5 years longer than teetotalers, and considerably longer than heavy drinkers. The prime area of contention is the degree to which these benefits accrue from the effects of ethanol *vs.* phenolic constituents (Rimm *et al.*, 1996).

Atherosclerosis is the principal cause of most cardiovascular disease (Libby, 2001). It apparently results from chronic injury to the arteries (Fig. 12.3). Although associated with several independent factors, most damage develops as a result of the oxidation of lipids in a special subgroup of cholesterol–apoproteins complexes, the low-density lipoproteins (LDLs). Because of the hydrophobic nature of cholesterol and triglycerides, their transfer via the blood requires a special transport vehicle. As illustrated in Fig. 12.4, lipoprotein complexes consist of an outer membrane of phospholipids, in which apoproteins and free cholesterol occur. They enclose a hydrophobic core possessing numerous triglycerides and cholesteryl esters. The specific apoproteins in the complex regulate the metabolism of the associated lipids.

Normally, LDLs function in supplying cholesterol for cellular membrane repair and the synthesis of steroids. However, in high concentrations, they may accumulate in the artery wall. If they remain there for an extended period, their lipid content tends to become oxidized. In an oxidized form, lipids are cytotoxic and indirectly

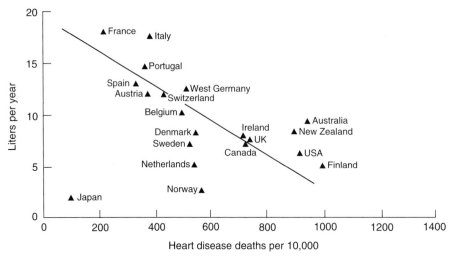

Figure 12.2 Relationship of per capita alcohol consumption with 1972 heart disease death rates in men aged 55–64 in 20 countries. (From La Porte *et al.*, 1980, reproduced by permission)

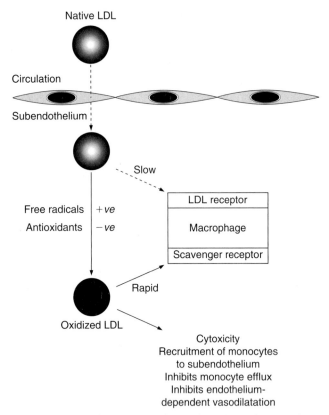

Figure 12.3 The oxidative-modification hypothesis of arteriosclerosis. (From Maxwell, 1997, reproduced by permission)

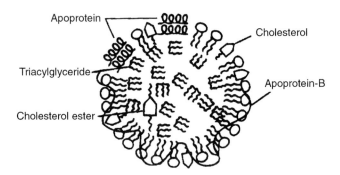

Figure 12.4 General structure of a triglyceride-rich lipoprotein. (From Walzem and Hansen, 1996, reproduced by permission)

irritate the artery wall. As a consequence, special adhesion proteins attach to the artery wall. Monocytes and helper T-cells of the immune system bond to these proteins. In addition, affected endothelial cells may secrete compounds, such as endothelin-1. Endothelin-1 activates the migration of monocytes and T-cells into the artery wall. Procyanidins, principally found in red wines, are particularly effective in suppressing the production of endothelin-1 (Corder *et al.*, 2001). In the layer just underneath the endothelial lining (intima), monocytes mature into macrophages. Both macrophages and T-cells may release cytokines that further activate the immune system, involving localized inflammation. Activated macrophages tend to engulf oxidized LDLs. However, as the LDLs are not degraded, their progressive accumulation gives the macrophage the appearance of being full of bubbles. This has given rise to the term foam cells. They are the first clear evidence of the beginning of localized arterial swelling (plaques). Occasionally plaques enlarge inward, but more frequently they bulge outward into the surrounding tissue. Action of immune cells in the plaque also induces migration of smooth muscle cells from the artery wall into the intima. Here they proliferate and produce collagen, forming a fibrous cap over the plaque. Additional LDLs slowly collect, provoking further rounds of inflammation and enlargement of the plaque. These accretions may develop their own vasculature, becoming fibrous and inelastic. As the plaques enlarge, they may produce irregular protrusions into and block the artery lumen.

Even without restricting blood flow, plaques set the stage for platelet aggregation, clot formation (thrombus) and the blockage that can precipitate a heart attack or stroke. In the later phases of plaque formation, unknown factors enhance inflammatory changes in the plaque. These disrupt the integrity of the cap. For example, collagenases secreted by macrophages inhibit collagen synthesis by smooth muscle cells. Sudden rupture of a plaque permits blood infiltration into the plaque. Because plaques contain potent blood clotting factors, thrombus development is almost instantaneous. It is currently thought that plaque rupture is the principal factor inducting thrombus formation, and precipitating a heart attack, stroke, or other cardiovascular trauma.

If the risk factors of atherosclerosis, such as smoking, high blood pressure, high dietary sources of cholesterol, and possibly infection by pathogens such as *Chlamydia pneumoniae* and cytomegalovirus (CMV) are eliminated, atherosclerosis appears to be at least partially reversible. Part of the reversal process involves the action of high-density lipoproteins (HDLs). Of the two principal forms, ethanol augments the presence of HDL_3, whereas exercise increases the level of HDL_2. Either form tends to remove cholesterol from the arteries, transferring it to the liver for metabolism. HDLs also appear to interfere with LDL oxidation. The effect of ethanol on HDL concentration appears to be independent of beverage type (van der Gaag *et al.*, 2001). The slower the rate of LDL turnover, the greater the likelihood of oxidation (Walzem *et al.*, 1995).

The beneficial effect of moderate alcohol consumption on the HDL/LDL ratio is now clearly established. Less well understood is its effect in lowering the concentration of C-reactive protein (CRP) (Levitan *et al.*, 2005). CRP is an indicator of inflammation.

Moderate alcohol consumption also reduces the incidence of another risk factor for cardiovascular disease – type 2 diabetes. Chronically high values of circulatory glucose, associated with type 2 diabetes, appear to generate high plasma triglyceride and LDL levels. The beneficial effects of alcohol on glucose and insulin metabolism appear not to occur if intake is not coincident with meal consumption (Augustin *et al.*, 2004). Phytoestrogens, such as resveratrol, have a similar effect in reducing triglyceride and LDL contents in the circulatory system (see Bisson *et al.*, 1995).

Another of alcohol's beneficial influences involves disruption of events leading to clot formation. Platelets are less 'sticky' in the presence of alcohol, thus, less likely

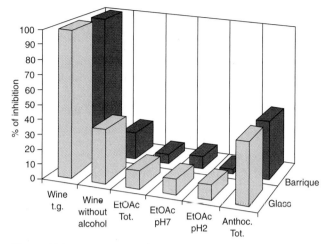

Figure 12.5 In activation of platelet aggregation induced by several red wine fractions (barrel- or bottle-aged, their dealcoholized versions, and total ethanol at pH 7 and 2) and anthocyanin extracts from the wines. (From Baldi *et al.*, 1997, reproduced by permission)

to aggregate to form a clot. Alcohol also increases the level of prostacyclin (interferes with clotting) and raises the level of plasminogen activator (a clot-dissolving enzyme). Clots adhering or becoming stuck to the roughened surfaces of narrowed atherosclerotic vessels may block blood flow. The oxygen deficiency and cell death that result are central to the damage caused by a heart attack or stroke. Thus, it is not surprising that inhibitors of platelet aggregation reduce the frequency of these cardiovascular crises. It is the rationale for recommending the daily consumption of acetylsalicylic acid (ASA) (an inhibitor of platelet aggregation). Ethanol (Renaud and Ruf, 1996), as well as wine phenolics, such as resveratrol and anthocyanins, have similar effects (Fig. 12.5).

An additional example of the importance of ethanol in cardiovascular disease is the correlation between alcohol dehydrogenase (ADH) genotype and the incidence of myocardial infarction. Those individuals homozygous for ADH1C*2 (slow metabolizers of ethanol) are significantly less likely to have a heart attack than heterozygous individuals, and even less likely than homozygous individuals for ADH1C*1 (fast metabolizers of ethanol) (Hines *et al.*, 2001).

Individually, many phenolics, such as resveratrol, catechin, epicatechin, and quercetin, have inhibitory effects on platelet aggregation (Keli *et al.*, 1994). Nevertheless, the combined effect of several phenolics is superior to single compounds (Wallerath *et al.*, 2005). The action partially results from the enhanced synthesis and release of nitric oxide by endothelial cells. Nitric oxide induces vasodilation (by relaxing vascular smooth muscle), and limits platelet adhesion to blood vessel endothelia. Indicative of the complexities of such interactions is the observation that flavonoids may also inactivate nitric oxide (Verhagen *et al.*, 1997). In addition, nitric oxide, notably as peroxynitrile, oxidizes LDLs.

In addition to effects on platelet aggregation, phenolic wine constituents bind directly with LDLs (limiting their oxidation), indirectly reduce macrophage-mediated oxidation of LDLs, and preserve the action of paraoxonase (further protecting LDLs from oxidation) (Aviram and Fuhrman, 2002). Furthermore, red wine phenolics limit the migration of smooth muscle cells into the intima of artery walls. These influences probably explain some of the added benefits of wine versus other alcoholic beverages in reducing the incidence and severity of cardiovascular disease. Although flavonoids tend to suppress inflammation, conflicting observations put the clinical significance of their anti-inflammatory action to atherosclerosis in question.

Red wines usually have been credited with superior benefits to white wines, relative to cardiovascular disease. This probably results from their higher flavonoid concentration. This view is supported from studies where white wine has shown the same effects as red wine when supplemented with grape polyphenolics (Fuhrman *et al.*, 2001). Nevertheless, prolonged skin contact, or choice of particular cultivars, can enhance the presence of phenolic acids in white wines. Common phenolics in white wine, such as caffeic and coumaric acids, as well as flavonols, such as quercetin, are well-known, potent antioxidants.

The low sodium content of wine is an incidental benefit. It may permit wine consumption by those on a low-sodium diet, for example heart attack victims. The high potassium to sodium ratio of wine (20 : 1) is a further benefit.

Antioxidant Effects

Wine phenolics are not only important antioxidants in wine preservation, but also play a role as antioxidants in the human body. By limiting LDL peroxidation (Maxwell *et al.*, 1994; Rice-Evans *et al.*, 1996), they restrict one of the critical early stages of atherosclerosis. This probably results from the inhibition of lipoxygenases, as well as the scavenging of free oxygen radicals, such as superoxide and hydroxyl radicals, and by chelating iron and copper (involved in radical formation) (Morel *et al.*, 1994; Rice-Evans *et al.*, 1996). In addition, tannin subunits (catechins and epicatechins) appear to protect other cellular components from oxidation. Other antioxidants of importance in the human diet are vitamins E and C (tocopherol and ascorbic acid), β-carotene, and selenium. The occurrence of tocopherol in the precursor of LDLs may provide a natural, but short-term, protection from oxidation.

One of the antioxidants relatively unique to wine is resveratrol. It is a phenolic (stilbene) compound produced in response to fungal attack or other stresses. Other plants producing resveratrol include mulberries, blueberries, peas, and peanuts. It has greater antioxidant action than dietary antioxidants such as vitamin E and ascorbic acid (Frankel *et al.*, 1993). There is also direct evidence that resveratrol can enter the blood system at levels sufficient to suppress cyclooxygenase (COX) and 5-lipoxygenase pathways. These are involved in the synthesis of proinflammatory mediators (Bertelli, 1998). In addition, resveratrol activates proteins involved in nerve cell differentiation, synaptic plasticity (important in learning), and neuronal survival (Tredici *et al.*, 1999).

Additional potent antioxidants in wine are flavonols, such as quercetin, and flavonoid tannin subunits (Miller and Rice-Evans, 1995). Flavonoids have been shown to possess various mechanisms of action, some directly quenching free radicals, others increasing the level of endogenous antioxidants, such as glutathione, whereas others prevent the influx of calcium ions associated with oxidative stress (Ishige *et al.*, 2001). They may also be more effective antioxidants, as well as occurring at higher concentrations, than resveratrol. Their content depends partially on the duration of skin contact and the type of fining. For example, PVPP (polyvinylpolypyrrolidone) markedly reduces quercetin content (Fluss *et al.*, 1990).

Alternately, de la Torre *et al.* (2006) have suggested that ethanol itself may modify human metabolism, increasing the synthesis of hydroxytyrosol. The latter is a well-known antioxidant phenolic found in olive oil. Consumption of red wine, with a concentration of 0.35 mg hydroxytyrosol, increased the concentration of hydroxytyrosol in the blood more than administration of olive oil, at a concentration of 1.7 mg hydroxytyrosol.

Vision

Many of the beneficial influences of alcohol and wine consumption show a U-shaped curve. This also applies to its effect on age-related macular degeneration (Obisesan *et al.*, 1998; Fraser-Bell *et al.*, 2006). The disease expresses itself as a progressive degeneration of the central region of the retina (macula) that leads to blurred or distorted vision. It results as a consequence of local atherosclerosis that deprives the retina of oxygen and nutrients. It is the leading cause of blindness in adults over the age of 65. A similar relationship has been found for cataract development. In both situations, wine antioxidants are suspected to be the active protective agent. At higher rates of intake, ethanol promotes a prooxidant action that could negate the benefits of wine antioxidants.

Neurodegenerative Diseases

Alzheimer's is one of the most investigated of neurodegenerative diseases, affecting approximately 15 million people worldwide. It is not surprising that researchers have investigated whether wine consumption reduces the incidence of neurodegenerative diseases affected by oxidative stress (Barnham *et al.*, 2004). A pattern appears to apply here, as with so many other health-related benefits of wine and alcohol consumption. Moderate intake is beneficial, whereas heavier consumption or abstinence is prejudicial.

Alzheimer's disease is associated with the accumulation of extracellular amyloid β-peptide (plaque), and the formation of intracellular neurofibrillar tangles containing tau-protein. The latter supports microtubule cytoplasmic structures. Many *in vitro* studies have shown that antioxidant compounds, such as vitamin E, protect neurons from β-amyloid accumulation. Tannic acid has also been shown to inhibit the formation of, and destabilize preexisting β-amyloid fibrils (Ono *et al.*, 2004), whereas resveratrol promotes the degradation of amyloid β-peptides (Marambaud *et al.*, 2005). Wine consumption is also linked in epidemiological studies to a reduction in the incidence of Alzheimer's disease (Truelsen *et al.*, 2002; Letenneur, 2004; Luchsinger *et al.*, 2004). Even mild cognitive impairment and the progression of idiopathic dementia may be reduced with moderate alcohol consumption (Solfrizzi *et al.*, 2007). Like other health benefits, these finding may not, in themselves, justify wine consumption, but they are encouraging to those who choose wine as part of their preferred lifestyle.

Osteoporosis

Age-related loss in bone mass affects both sexes, but is particularly prominent in postmenopausal women. Many risk factors, dietary influences, and hormonal supplements can affect its occurrence and severity. Of these factors, moderate alcohol consumption has been found to favor bone retention (Ganry *et al.*, 2000; Ilich *et al.*, 2002). These studies did not separate between various alcoholic beverages, thus, whether bone retention is due to enhanced calcium uptake associated with alcohol consumption (Ilich *et al.*, 2002), the phytoestrogen effects of phenolics, such as resveratrol, or some other influence is unknown.

Gout

In the 1800s, there were many reports linking gout with wine consumption, especially port. Gout is caused by the localized accumulation of uric acid crystals in the synovium of joints. Their presence stimulates the synthesis and release of humoral and cellular inflammatory mediators (Choi *et al.*, 2005). Gout is often associated with reduced excretion of uric acid in the kidneys. Mutations in the gene that encodes urease, the enzyme that metabolizes uric acid to allantoin (a soluble by-product), is often involved in gout.

Dietary predisposing factors include red meat, seafood, and beer – presumably due to the increased availability of purines, the principal source of uric acid. Alcohol consumption may occasionally aggravate gout by increasing lactic acid synthesis. It, in turn, favors uric acid reabsorption by the kidneys. Despite this, wine consumption has not been found to be associated with an increased risk for gout (Choi *et al.*, 2004).

Medical historians suspect the gout–port connection in the nineteenth century was associated with lead-induced kidney damage (Yu, 1983; Emsley, 1986/1987). Samples of port from the nineteenth century show high lead contents. Lead contamination probably came from its uptake from stills, used in making the brandy added in port production. In addition, the former use of pewter and lead-glazed drinking cups, and prolonged storage of port in lead crystal decanters, could have further augmented the lead content (Falcone, 1991).

Arthritis

A number of drugs used in treating arthritis have a tendency to irritate the lining of the stomach. This side-effect may be counteracted by the mildly acidic, dilute alcohol content of table wines. Other beneficial effects connected with moderate wine consumption may accrue from its mildly diuretic and muscle-relaxant properties. The diuretic action of wine can help reduce water retention and minimize joint swelling. Wine can also directly reduce muscle spasms and the stiffness associated with arthritis. The antiinflammatory influences of wine phenolics, notably resveratrol, may also play a role in diminishing the suffering associated with arthritis.

Diabetes

Wine consumption has been shown to attenuate insulin-resistance in type 2 diabetes (Napoli *et al.*, 2005). This may result from wine phenolics quenching oxygen radicals, thought to be pivotal in the damage associated with type 2 diabetes. Type 2 diabetes appears to result when body cells do not respond properly to the presence of insulin. Moderate alcohol consumption is also associated with reduced incidence of this form of diabetes

(Dixon *et al.*, 2001). The incidence of metabolic syndrome X is also lower in wine drinkers (Rosell *et al.*, 2003). These effects may be due to one or more of the following: the influences of alcohol on metabolism; the antidiabetic properties of the element vanadium (of which wine is a significant source) (Brichard and Henquin, 1995; Teissèdre *et al*, 1996); the hypoglycemic and hypolipidemic effects of phenolics like resveratrol (Su *et al.*, 2006); or through some effect on endothelial nitric oxidase synthase (Leighton *et al.*, 2006). Moderate consumption of dry wine has no adverse effect on sugar control in diabetic patients (Gin *et al.*, 1992; Bell, 1996).

Goitre

In an epidemiological study, Knudsen *et al.* (2001) found a strong link between alcohol consumption and a reduced prevalence of goitre and solitary thyroid nodules. This effect may result from some unknown protective effect of ethanol.

Kidney Stones

Drinking water has long been associated with reducing the development of kidney stones. Increased urine production helps prevent the crystallization of calcium oxalate in the kidneys. What is new is the observation that wine consumption further reduces the production of these painful and dangerous inclusions (Curhan *et al.*, 1998).

Cancer

The consumption of moderate amounts of wine has not been shown to increase the incidence of most cancers. In contrast, increased consumption tends to increase the risk of certain cancers (see Ebeler and Weber, 1996).

In certain instances, these influences may derive from indirect effects on carcinogens, such as ethyl carbamate. At the concentrations typically found in table wine, though, ethanol diminishes the carcinogenesis of ethyl carbamate. Certain wine phenolics can be protective, whereas others can be mutagenic, especially at high concentrations. For example, quercetin can induce mutations in laboratory tissue culture, but is a potent anticarcinogen in whole-animal studies (Fazal *et al.*, 1990). This apparent anomaly may result from differences in the concentrations of quercetin used, and the low level of metal ions and free oxygen found in the body (vs. tissue culture). In addition, phenolics

may detoxify the small quantities of nitrites commonly found in food. However, in the presence of high nitrite concentrations (a preservative found in smoked and pickled foods), nitrites are converted into diazophenols (Weisburger, 1991). These can induce oral and stomach cancers.

Several phenolics can limit or prevent cancer development through a diversity of effects, such as DNA repair, carcinogen detoxification, enhanced apoptosis (programmed cell death), and disrupted cell division (Hou, 2003; Aggarwal *et al.*, 2004). For example, resveratrol induces the redistribution of the Fas receptor. It is a site for the attachment of TNF (tumor necrosis factor) in colon cancer. Its action is part of a sequence that leads to the death of the affected cancer cells (Delmas *et al.*, 2003). Resveratrol is also well known as an inhibitor of angiogenesis – the production of new vasculature essential for tumor growth. Other effects of resveratrol include inhibition of cyclooxygenase-2 (Subbaramaiah *et al.*, 1998) and P450 1A1 (Chun *et al.*, 1999). Cyclooxygenase-2 is thought to be involved in carcinogenesis, whereas P450 1A1 is an important hydroxylase. It can convert several environmental toxicants and procarcinogens into active carcinogens.

Flavones and flavonols strongly restrict the action of the common dietary carcinogens, heterocyclic amines (Kanazawa *et al.*, 1998). It is estimated that these compounds, produced during cooking, are consumed at a rate of approximately 0.4–16 μg per day (Wakabayashi *et al.*, 1992). The antiallergic and antiinflammatory properties of flavonoid phenolics probably also contribute to the anticancer aspects of these flavonoids (see Middleton, 1998).

The major exception to the general benefit of moderate wine consumption against cancer may be breast cancer (Viel *et al.*, 1997). This correlation is more evident in those with the ADH1C*1 (fast metabolizers of ethanol to acetaldehyde, a known carcinogen) (Terry *et al.*, 2006). However, findings from the long-duration Framingham Study indicate no relationship between moderate alcohol consumption and the incidence of breast cancer (Zhang *et al.*, 1999). High rates of wine consumption have been linked with an increased incidence of mouth and throat cancers (Barra *et al.*, 1990). Ethanol itself is not carcinogenic, but can enhance the transforming effect of some carcinogens.

Allergies and Hypersensitivity

Alcoholic beverages may induce a wide diversity of allergic and allergic-like reactions. In sensitive individuals, these may express as rhinitis, itching, facial swelling, headache, cough, or asthma. Occasionally, ethanol plays a role in these responses, for example the flushing reaction of many Asians. Nevertheless, sulfur dioxide is potentially the most significant wine irritant. A small proportion of wine-sensitive asthmatics may experience bronchial constriction on exposure to sulfite (Dahl *et al.*, 1986). In a study by Vally and Thompson (2001), wine containing 300 ppm sulfite induced a rapid drop in forced expiratory volume, reaching a maximal decline within about 5 min. Recovery took between 15 and 60 min. The same individuals did not respond to wine containing 20, 75, or 150 ppm sulfite. Why sensitive asthmatics episodically react to wines with low SO_2 contents may be related to changes in the state of their asthma control. Surprisingly, red wines tend to provoke more asthma problems than white wines, even though red wines typically have lower sulfur dioxide contents than white wines.

The speed of the reaction to sulfite suggests some malfunction in the amount of glutathione in lung tissue, or the activity of glutathione *S*-transferase reducing sulfite to glutathione *S*-sulfonate. Normally, sulfite is rapidly converted to sulfate by sulfite oxidase in the blood. However, low levels of this enzyme could permit sulfite to persist, provoking a heightened response in hypersensitive individuals.

At potentially greater risk are individuals afflicted with a rare autosomal genetic disease caused by a deficiency in sulfite oxidase (Shih *et al.*, 1977; Crawhall, 1985). Affected individuals must live on a very restricted diet, low in sulfur-containing proteins. It is estimated that the synthesis of sulfite, associated with normal food metabolism, generates approximately 2.4 g sulfite/day. The sulfites in wine contribute only marginally to this amount. Because of the gravity of sulfite oxidase deficiency, most affected people die before reaching adulthood.

An intriguing allergy-like reaction is a rapid facial and neck flushing (cutaneous erythema) shortly after alcohol consumption. Other symptoms often include peripheral vasodilation, elevated heart rate, nausea, abdominal discomfort, and broncho-constriction. This reaction is found in up to 50% of eastern Asians. The syndrome is associated with a malfunctional form of mitochondrial acetaldehyde dehydrogenase (ALDH2) (Enomoto *et al.*, 1991). This is the principal enzyme oxidizing acetaldehyde to acetic acid. A malfunctional alcohol dehydrogenase allele (ADH1B) may also play a contributing role in the flushing reaction (Takeshita *et al.*, 2001).

Elevated levels of acetaldehyde are thought to cause flushing by activating the localized release of histamine from mast cells. Histamine induces vasodilation and the associated influx of blood that appears as a reddening of the associated tissue. The connection between acetaldehyde and histamine is supported by the action

of antihistamines in reducing this flushing, if taken in advance of an alcohol challenge (Miller *et al.*, 1988).

Antihistamines can also diminish the rhinitis that may be associated with the consumption of red wine (Andersson *et al.*, 2003). In addition, antihistamines counteract the bronchoconstriction in individuals showing histamine intolerance. Histamine intolerance presumably relates to reduced activity of diamine oxidase (Wantke *et al.*, 1996).

It has been suggested that the *ALDH2* mutant in eastern Asians noted above may reflect an evolutionary selective adaptation to the endemic occurrence of hepatitis B in that region of Asia (Lin and Cheng, 2002). The resulting avoidance of alcohol would avert any synergism with hepatitis B-induced liver damage. Onset of a similar collection of unpleasant symptoms, associated with the abnormal accumulation of plasma acetaldehyde, is often used in the treatment of alcoholism. Disulfiram (Antabuse) is a potent inhibitor of ALDH.

Facial flushing, concomitant with alcohol consumption, but devoid of other symptoms, is occasionally experienced by Caucasians. Whether this is related to an ALDH malfunction is unclear. However, this possibility is supported by its suppression by acetylsalicylic acid (aspirin), if taken before an alcohol challenge (Truitt *et al.*, 1987). An alternative proposal is that the flushing results from a direct, cutaneous, alcohol-induced vasodilation.

Idiosyncratic allergic and other immune hypersensitive responses to wine are difficult to predict or diagnose. Reactions may include the induction of headaches, nausea, vomiting, general malaise, or a combination of these. In a few instances, IgE-related anaphylaxis reactions have been reported to grape PR proteins (endochitinase and thaumatin) (Pastorello *et al.*, 2003). The reactions may involve urticaria/angiodema (red patches or wheals on the skin/swelling), and occasionally shock. Residual amounts of fining agents, such as egg whites, have also been implicated in some allergic reactions (Marinkovich, 1982). However, in a double-blind, placebo-controlled trial, wines fined with egg white, isinglass, or non-grape derived tannins presented "an extremely low risk of anaphylaxis" to egg-, fish-, or peanut-allergic consumers (Rolland *et al.*, 2006). In an ELISA study, only egg white and lysozyme could be detected in wine samples (Weber *et al.*, 2007). Nevertheless, with more than 800 compounds potentially occurring in wine, it is not surprising that some individuals may occasionally show some form of adverse reaction to specific wines or wine types.

In addition to physiological reactions to wine constituents, there is a wide range of equally important psychological responses (Rozin and Tuorila, 1993), both positive and negative. Traumatic memories associated with the first exposure to, or excessive consumption of, a particular beverage can create an association that lasts a lifetime. Other people have come to associate certain products with social groups, lifestyles, or behaviors. Such attitudes can make the beverage unacceptable.

Headaches

People occasionally avoid wine consumption because it induces headaches. This situation has regrettably seen little study. Central to progress is effective differentiation of the multiplicity of headache phenomena and their specific etiologies.

One of the most severe headache syndromes potentially associated with wine consumption is the **migraine**. Migraines may be induced by a wide range of environmental stimuli, possibly because migraines are themselves a complex of etiologically distinct events. The dilation of blood vessels in the brain, as a result of histamine release, can be the common element in many instances of headache development. When red vs. white wines were discovered to contain higher concentrations of biogenic amines, such as histamine and tyramine, there was the initial assumption that they were the culprits. However, it was later realized that the normal levels of histamine found in red wines are below those that generally trigger a migraine. In addition, double-blind studies have seemingly exonerated histamine in red-wine-induced migraine headaches (Masyczek and Ough, 1983). Nevertheless, alcohol can suppress the action of diamine oxidase, an important enzyme of the small intestine that inactivates histamine and other biogenic amines (Jarisch and Wantke, 1996). Thus, in individuals with histamine intolerance, sufficient histamine may enter the blood system to provoke a vascular headache. However, this view does not correlate with the observation that spirits and sparkling wine were more frequently associated with migraine attacks than other alcoholic beverages (Nicolodi and Sicuteri, 1999). Both spirits and sparkling wines are low in histamine content.

Although the biogenic amine content of wine appears to be insufficient to cause migraines in most individuals, the phenolic content of wine can induce headaches. Correspondingly, red wines are more frequently associated with headache production than white wines. On average red wines contain about 1200 mg/liter vs. 200 mg/liter for whites. Some phenolics suppress the action of platelet phenolsulfotransferase (PST) (Jones *et al.*, 1995). Individuals having low levels of platelet-bound PST are apparently more susceptible to migraine headaches (Alam *et al.*, 1997). Suppression of PST results in reduced sulfation (detoxification) of a variety of endogenous and xenobiotic ("foreign") compounds, including biogenic amines and phenolics. Without inactivation, biogenic amines can activate the liberation of

5-hydroxytryptamine (5-HT, or serotonin), an important neurotransmitter in the brain. 5-HT also promotes platelet aggregation and the dilation of small blood vessels in the brain. The pressure so created can cause intercranial pain, and the instigation of a migraine (Pattichis *et al.*, 1995). People prone to migraine headaches also may show abnormal and cyclical patterns in platelet sensitivity to 5-HT release (Jones *et al.*, 1982; Peatfield *et al.*, 1995). This may be involved in why wine consumption is not consistently linked to headache induction. Small phenolic components in wine also prolong the action of potent hormones and nerve transmitters, such as histamine, serotonin, dopamine, adrenalin and noradrenaline. These could affect headache severity and other allergic reactions.

In the treatment of the possibly closely related **cluster-headache syndrome**, small doses of lithium may be preventive (Steiner *et al.*, 1997). Because some red wines have higher than average lithium contents, they may prevent, rather than induce this type of headache.

Another recognized headache syndrome is the **red wine headache** (Kaufman, 1986). It may develop within minutes of consuming red wine and is often dose-related. The headache reaches its first peak within approximately 2 hours, tends to fade, but returns roughly 8 hours later in a more intense form. The headache seems related to the release of type E prostaglandins, important chemicals involved in dilating blood vessels. If this association is correct, the inhibition of prostaglandin synthesis by acetylsalicylic acid, acetaminophen and ibuprofen could explain how these medications prevent headache development, when taken prior to wine consumption (Kaufman, 1992).

A distinct wine-related headache has been dubbed the **red head** (Goldberg, 1981). It develops within an hour of waking, after drinking no more than two glasses of red wine the previous evening. The headache is associated with nausea. The headache is particularly severe when reclining. Although the headache is relieved somewhat by standing, this itself exacerbates the nausea. The headache usually lasts a few hours before dissipating. A similar phenomenon has been reported with some white wines, or mixtures of white wine, taken alone or with coffee or chocolates. Its chemical cause is unknown (Kaufman, 1986).

In most instances, headaches associated with wine consumption are assumed to be induced by tannins. However, tannins (polyphenolics) are poorly absorbed in the digestive tract. In contrast, their monomers are readily absorbed. This may explain why aged red wines (in which most tannins occur as large polymers) tend to be less associated with headache induction than their younger counterparts. A classic example is the ease with which the youngest of all red wines, Beaujolais *nouveau*, produces headaches in those prone to their occurrence. Large tannin polymers remain largely unmodified until entering the colon, where bacteria metabolize them to low-molecular-weight phenolics (Déprez *et al.*, 2000). Because this can take up to two days, they cannot be involved in headache induction. However, headache induction may result from monomeric phenolics, such as caffeic acid and catechins. Depending on their metabolism in the plasma, phenolics may be detoxified (*o*-methylated or sulfated), or made more 'toxic' (oxidized to *o*-quinones). *o*-Quinones can inhibit the action of the enzyme catechol-O-methyltransferase (COMT). By so doing, the neurotransmitter dopamine is not broken down, and the availability of μ-opinoid (painkilling) receptors is restricted. Consequently, the perception of pain associated with cerebral blood vessel dilation may be enhanced.

Resveratrol, a phenolic found in higher concentration in red than white wines, inhibits the expression of cyclooxygenases. Cyclooxygenase is involved in the synthesis of prostaglandins (Jang and Pezzuto, 1998), dilators of cerebral blood vessels. This is another example of where some wine phenolics may counter, rather than induce headache development. In contrast, ethanol tends to elevate the concentration of prostaglandins (Parantainen, 1983).

The ability of some yeast strains to produce prostaglandins (Botha *et al.*, 1992) introduces the intriguing possibility that prostaglandins may occur as constituents in wine. If produced in sufficient quantity, yeast-derived prostaglandins could be another, or supplemental agent, involved in headache development. Yeast-derived prostaglandins could also theoretically provoke inflammatory lung problems such as asthma.

Although red wines are generally more associated with headache production than white wine, some headaches are exclusively associated with white table wines. Its characteristics and etiology are even less well understood than those evoked by red wines. In some individuals, this situation may be associated with a sensitivity to sulfites, which are generally found in higher concentrations in white wines than red wines, especially when young.

One of the most recognized alcohol-related headache phenomena is that associated with binge drinking – the **hangover** (veisalgia) (Wiese *et al.*, 2000). Although not consistently associated with a headache, it is frequently an accompanying phenomenon. A hangover is characterized with tremulousness, palpitations, tachycardia, sweating, loss of appetite, anxiety, nausea, and possibly vomiting and amnesia. When a headache accompanies the hangover, it possesses symptoms similar to a migraine. The headache is global, more frequently located anteriorly, and associated with heavy, pulse-synchronous throbbing. It usually starts a few (>3) hours after the cessation of drinking, when the blood alcohol level is declining and other hangover symptoms have already developed (Sjaastad and Bakketeig, 2004). Duration is seldom more than 12 hours.

Despite its all-too-frequent occurrence, the causal mechanism(s) remain unclear. Various compounds have been implicated, notably ethanol (and its primary breakdown products, acetaldehyde and acetic acid), methanol (through its metabolic by-products, formaldehyde and formic acid), and various congeners. None of these has been adequately established as individually or collectively being the principal causal agent(s).

Despite the absence of clear causal relationships, ethanol is typically viewed as the principal perpetrator. This view is supported by the physiologic effect of ethanol on the pituitary gland. Ethanol limits production of the hormone vasopressin. The result is a reduction in water reabsorption by the kidneys (increased urination), resulting in partial tissue dehydration. Contraction of the membrane covering the brain (the dura) could pull on fibers attaching the dura to the skull, causing pain sensors to discharge. Alternately, the diuretic effect of ethanol could result in electrolytic imbalance. The resultant disruption of normal nerve and muscle function could theoretically induce symptoms such as headache, nausea and fatigue. Ethanol can also induce the breakdown of glycogen in the liver. The resulting influx of glucose is eliminated in the urine, producing hypoglycemia and a feeling of weakness. Finally, the breakdown of ethanol via the hepatic microsomal pathway increases the release of free radicals in the blood, causing cellular damage and a diverse range of symptoms. It is said that:

Wine hath drowned more men than the sea.

Because glutathione is very important in the inactivation of free radicals, taking an amino acid supplement, N-acetyl-cysteine (NAC), has been suggested as a partial remedy. NAC is rich in cysteine, an amino acid that forms the core of glutathione. In addition, glutathione facilitates the conversion of acetaldehyde to acetic acid and its subsequent metabolism, as well as binding irreversibly with acetaldehyde.

The accumulation of acetaldehyde has also been proposed as a major activator of hangover initiation. For example, acetaldehyde can disrupt membrane function (partially by interfering with the action of cytochrome P-450 oxidase), and consequently cerebral neurotransmitter action. Commercial products such as Hangover Helper™ and Rebound™ have been developed to counter the effects of acetaldehyde. In addition, congeners (such as fusel alcohols and methanol) may exacerbate the effects of ethanol and acetaldehyde. Although the methanol content of wine is particularly low, its metabolism by ADH to formaldehyde, and subsequently formic acid, could be a significant factor with distilled beverages. The product Chaser™ has been developed as a means of limiting the uptake of such congeners. Its formulation of activated calcium carbonate and vegetable carbon is thought to bind congeners in the stomach, preventing their uptake in the blood.

Hangovers are also associated with disregulation of cytokine pathways (Kim *et al.*, 2003), as well as increased levels of C-reactive protein in the plasma (Wiese *et al.*, 2004).

Some purported remedies, such as artichoke extract, have not stood up to rigorous clinical trial (Pittler *et al.*, 2003), but others, such as an extract of *Opuntia fiscus-indica* (Prickly Pear), apparently reduces the severity of some hangover symptoms (Wiese *et al.*, 2004). It is thought to work as an inflammatory mediator. Pyritinol (a vitamin B_6 derivative) has also been reported to reduce some hangover symptoms (Khan *et al.*, 1973). Regrettably, there is no known universally effective treatment for a hangover. Although time is the only sure cure, avoidance is preferable.

Taking wine with meals is probably the best-known and reliable preventive, combined with limited consumption. Food delays the movement of alcohol into the intestinal tract, where some 80% of the alcohol is absorbed. Because the uptake is slowed, absorption more evenly matches the body's ability to metabolize alcohol. It also delays the uptake of phenolic compounds and diminishes their maximum concentration in the blood.

Dental Erosion

Wine tasting is not normally considered hazardous to one's health. However, recent studies have found that dental erosion can be a risk (Mok *et al.*, 2001; Mandel, 2005; Chikte *et al.*, 2006). Damage results from the frequent and extended exposure to wine acids. Removal of calcium softens the enamel, which becomes susceptible to erosion by masticatory forces and tooth brushing. Demineralization commences at about a pH of 5.7. This is typically not a problem for the consumer who takes wine with meals. Food and saliva secretion limit, if not prevent, demineralization of tooth enamel.

After many years, professional wine tasters may experience tooth disfiguration, affecting both tooth shape and size. Cupping, a depression in the enamel, exposing dentine at the tip of molar cusps, is a frequent clinical sign. Erosion can also contribute to severe root abrasion at the gum line. Protection is partially achieved by rinsing the mouth with an alkaline mouthwash after tasting, application of a fluoride gel (such as APF), and refraining from tooth brushing for at least one hour after tasting.

The delay permits minerals in the saliva to rebind with the enamel.

Fetal Alcohol Syndrome

Fetal Alcohol Syndrome (FAS) refers to a set of phenomena including suppressed growth, mild mental retardation, and subtle facial abnormalities (Wattendorf and Muenke, 2005). It was first described in 1973 and appeared most markedly in the children of alcoholic mothers. They also tended to be heavy smokers, use illicit drugs, consume large amounts of coffee, have poor nutrition, or show a combination of these (see Scholten, 1982; Whitten, 1996).

It is suspected that alcohol is the principal cause, although the concentration and timing associated with FAS are still uncertain. Acetaldehyde accumulation may also be involved. In addition, even more subtle effects have now been detected, giving rise to the acronym FASD (fetal alcohol spectrum disorders). Because the consequences may be lifelong, it is now generally recommended that pregnant women, or women wishing to become pregnant, refrain from alcohol consumption during this period. Whether total abstinence is fully warranted is unknown, but erring on the side of caution is certainly judicious.

Contraindications

The most important contraindication relates to those with a past history of alcohol abuse. For the majority of the adult population (except pregnant women), moderate wine consumption appears to have considerable health benefits. Nevertheless, there are several situations in which wine consumption, even in moderate amounts, can complicate or diminish the effectiveness of disease treatment.

1. The acidic nature of wine can aggravate the inflammation and slow the natural healing of ulcers in the mouth, throat, stomach, and intestinal tract. Other constituents in wine may also be detrimental in this regard. Thus, all beverages containing alcohol are usually contraindicated in cases of gastritis, gastric cancer, and bleeding in the upper digestive tract. Nevertheless, the prophylactic action of red wine against *Helicobacterium pylori* and the suppression of histamine production by the gastric mucosa (Masquelier, 1986) may require a reconsideration of the old prohibition in mild cases. In the presence of pancreatitis, alcohol is absolutely contraindicated.

2. Wine, along with other alcoholic beverages, may provoke gastroesophageal (acid) reflux in individual prone to this problem.

3. In liver disease, the consumption of wine is normally contraindicated. The presence of alcohol puts additional stress on an already weakened vital organ. Chronic alcohol abuse can lead to cirrhosis of the liver.

4. In acute kidney infection, wine should be avoided. The consumption of alcohol increases the burden on an organ essential to eliminating toxic metabolic wastes.

5. In prostatitis or genitourinary infections, the consumption of alcohol can complicate matters. The diuretic action of wine may increase the frequency of urination or, conversely, it may induce highly painful urinary retention.

6. In epilepsy, the consumption of even moderate amounts of wine may increase the frequency of seizures.

7. In patients about to undergo surgery, the effect of alcohol on reducing platelet aggregation is undesirable. Thus, it is recommended that patients terminate any alcohol (as well as aspirin) consumption before surgery. This avoids increasing the incidence of intra- and postoperative bleeding (Wolfort *et al.*, 1996).

The consumption of alcohol is also ill advised when eating certain mushrooms. The most well-known example is the antabuse reaction associated with simultaneous consumption with *Coprinus atramentarius* (Inky Cap). Another mushroom generating the same response is *Boletus luridus* (Budmiger and Kocher, 1982). The antabuse reaction derives its name from the trade name of disulfiram, a medication used in the treatment of alcoholism. It functions as an inhibitor of acetaldehyde dehydrogenase. Even when small amounts of alcohol are consumed along with disulfiram, it induces a very unnerving reaction. This may include flushing, sweating, weakness, vertigo, blurred vision, difficulty breathing, nausea, chest pain, palpitation, and tachycardia. In severe cases, the reaction can provoke acute congestive heart failure, convulsion, and death. Simultaneous consumption of alcoholic beverages while using certain drugs (e.g., cephalosporins, griseofulvin, chloramphenicol, sulfonylurea, metronidazole) can produce similar symptoms in sensitive individuals.

Wine and Medications

In addition to the antabuse reaction just noted, consumption of alcohol can generate various unpleasant to dangerous reactions. Regrettably, most of the literature relating to alcohol–drug interactions comes from studies on alcoholics or binge drinkers. This limits their potential applicability under conditions of moderate consumption with meals. Nevertheless, even small amounts of alcohol may cause loss of muscle control in people taking tricyclic antidepressants. In addition, red

wines can reduce the effectiveness of MAO (monoamine oxidase) inhibitors used in controlling hypertension. The long-term use of acetaminophen can enhance alcohol-induced kidney damage.

Other contraindications involve the intensification of the effects of barbiturates and narcotics. In combination with certain antidiabetic agents, such as tolbutamide and chlorpropamide, alcohol can cause dizziness, hot flushes, and nausea. Mild reactions may occur with a wide range of other medications, such as sulfanilamide, isoniazid, and aminopyrine. Details can be found in Adams (1995), Fraser (1997), and Weathermon and Crabb (1999).

Suggested Readings

Baur, J. A., and Sinclair, D. A. (2006) Therapeutic potential of resveratrol: the *in vivo* evidence. *Nat. Rev. Drug Discov.* 5, 493–506.

Bauza, T., Blaise, A., Teissidre, P. L., Cabanis, J. C., Kanny, G., and Moneret-Vautin, D. A. (1995) Biogenic amines. Metabolism and toxicity. *Bull. OIV* 68, 42–67.

Cushnie, T. P., and Lamb, A. J. (2005) Antimicrobial activity of flavonoids. *Int. J. Antimicrob. Agents* 26, 343–356.

da Luz, P. L., and Coimbra, S. R. (2004) Wine, alcohol and atherosclerosis: clinical evidences and mechanisms. *Brazil. J. Med. Biol. Res.* 37, 1275–1295.

Das, D. K., and Ursini, F. (eds.) (2003) Alcohol and Wine in Health and Disease. *Ann. NY Acad. Sci.* 957, 350.

de la Lastra, C. A., and Villegas, I. (2005) Resveratrol as an anti-inflammatory and anti-aging agent: mechanisms and clinical implications. *Mol. Nutr. Food Res.* 49, 405–430.

Dulak, J. (2005) Nutraceuticals as anti-angiogenic agents: hopes and reality. *J. Physiol. Pharmacol.* 56 (Suppl. 1), 51–69.

German, J. B., and Walzem, R. L. (2000) The health benefits of wine. *Annu. Rev. Nutr.* 20, 561–593.

Gershwin, M. E., Ough, C., Bock, A., Fletcher, M. P., Nagy, S. M., and Tuft, D. S. (1985) Adverse reactions to wine. *J. Allergy Clin. Immunol.* 75, 411–420.

Iriti, M., and Faoro, F. (2001) Plant defense and human nutrition: phenylpropanoids on the menu. *Curr. Topics Nutraceut. Res.* 2, 47–65.

Libby, P. (2002) Atherosclerosis: The new view. *Sci. Am.* 286(5), 47–55.

Lieber, C. S. (2000) Alcohol: Its metabolism and interaction with nutrients. *Annu. Rev. Nutr.* 20, 395–430.

Lucia, S. P. (1963) A History of Wine as Therapy. Lippincott, Philadelphia, PA.

Ramassamy, C. (2006) Emerging role of polyphenolic compounds in the treatment of neurodegenerative diseases: a review of their intracellular targets. *Eur. J. Pharmacol.* 545, 51–64.

Rice-Evans, C. A., Miller, N. J., and Paganga, G. (1996) Structure-antioxidant activity relationships of flavonoids and phenolic acids. *Free Radic. Biol. Med.* 20, 933–956.

Sandler, M., and Pinder, R. (eds.) (2003) *Wine: A Scientific Exploration*. Taylor & Francis, London.

Stockley, C. S., and Høj, P. B. (2005) Better wine for better health: Faction or fiction. *Aust. J. Grape Wine Res.* 11, 127–138.

Stoclet, J.-C., Chaigneau, T., Ndiaye, M., Oak, M.-H., Bedoui, J. E., Chaigneau, M., and Schini-Kerth, V. B. (2004) Vascular protection by dietary polyphenols. *Eur. J. Pharmacol.* 500, 299–313.

Vally, H., and Thompson, P. J. (2003) Allergic and asthmatic reactions to alcoholic drinks. *Addiction Biol.* 8, 3–11.

Waterhouse, A. L., and Rantz, J. M. (eds.) (1996) *Wine in Context: Nutrition, Physiology, Policy, Proceedings of the Symposium on Wine and Health.* American Society for Enology and Viticulture, Davis, CA.

Watkins, T. R. (ed.) (1997) *Wine – Nutritional and Therapeutic Benefits.* ACS Symposium Series No. 661. American Chemical Society, Washington, DC.

References

Abrams, A., Aronson, M. D., Delbanco, T., and Barnes, H. N. (eds.) (1987) *Alcoholism.* Springer-Verlag, New York.

Adams, W. L. (1995) Interactions between alcohol and other drugs. *Int. J. Addict.* **30**, 1903–1923.

Aggarwal, B. B., Bhardwaj, A., Aggarwal, R. S., Seeram, N. P., Shishodia, S., and Takada, Y. (2004) Role of resveratrol in prevention and therapy of cancer: preclinical and clinical studies. *Anticancer Res.* **24**(5A), 2783–2840.

Alam, Z., Coombes, N., Waring, R. H., Williams, A. C., and Steventon, G. B. (1997) Platelet sulphotransferase activity, plasma sulfate levels, and sulphation capacity in patients with migraine and tension headache. *Cephalalgia* **17**, 761–764.

Andersson, M., Persson, C. G. A., Persson, G. G., Svensson, C., Cervin-Hoberg, C., and Greiff, L. (2003) Effects of loratadine in red wine-induced symptoms and signs of rhinitis. *Acta Otoloryngol.* **123**, 1087–1093.

Augustin, L. S. A., Gallus, S., Tavani, A., Bosetti, C., Negri, E., and La Vecchia, C. (2004) Alcohol consumption and acute myocardial infarction: a benefit of alcohol consumed with meals? *Epidemiology* **15**, 767–769.

Aviram, M., and Fuhrman, B. (2002) Wine flavonoids protect against LDL oxidation and atherosclerosis. *Ann. NY Acad. Sci.* **957**, 146–161.

Baldi, A., Romani, A., Mulinacci, N., Vincieri, F. F., and Ghiselli, A. (1997) The relative antioxidant potencies of some polyphenols in grapes and wines. In: *Wine: Nutritional and Therapeutic Benefits* (T. R. Watkins, ed.), pp. 166–179. ACS Symposium Series No. 661, American Chemical Society, Washington, DC.

Barnham, K. J., Masters, C. L., and Bush, A. I. (2004) Neurodegenerative disease and oxidative stress. *Nat. Rev. Drug Discov.* **3**, 205–214.

Barra, S., Franceschi, S., Negri, E., Talamini, R., and La Vecchia, C. (1990) Type of alcoholic beverage and cancer of the oral cavity, pharynx and oesophagus in an Italian area with high wine consumption. *Intl J. Cancer* **46**, 1017–1020.

Bell, D. S. H. (1996) Alcohol and the NIDDM patient. *Diabetes Care* **19**, 509–513.

Benini, L., Salandini, L., Rigon, G., Tacchella, N., Brighenti, F., and Vantini, I. (2003) Effect of red wine, minor constituents, and alcohol on the gastric emptying and the metabolic effects of a solid digestible meal. *Gut* **52** (Suppl. 1), pA79–pA80.

Bertelli, A. A. E. (1998) Modulatory effect of resveratrol, a natural phytoalexin, on endothelial adhesion molecules and intracellular signal transduction. *Pharmaceut. Biol.* **36** (Suppl.), 44–52.

Bisson, L. F., Butzke, C. E., and Ebeler, S. E. (1995) The role of moderate ethanol consumption in health and human nutrition. *Am. J. Enol. Vitic.* **46**, 449–462.

Botha, A., Kock, J. L. F., Coetzee, D. J., van Dyk, M. S., van der Berg, L., and Botes, P. J. (1992) Yeast eicosanoids, IV. Evidence for prostaglandin production during ascosporogenesis by *Dipodascopsis tothii. Syst. Appl. Microbiol.* **15**, 159–163.

Brenner, H., Rothenbacher, D., Bode, G., and Adler, G. (1997) Relation of smoking, alcohol and coffee consumption to active *Helicobacterium pylori* infection: Cross sectional study. *Br. Med. J.* **315**, 1489–1492.

Brichard, S. M., and Henquin, J.-C. (1995) The role of vanadium in the management of diabetes. *Trends Pharmaceut. Sci.* **16**, 265–270.

Budmiger, H., and Kocher, F. (1982) *Boletus luridus* and alcohol. Case report (in German). *Schweiz. Med. Wochenschr.* **112**, 1179–1181.

Camargo, C. A., Jr., Stampfer, M. J., Glynn, R. J., Gaziano, J. M., Manson, J. E., Goldhaber, S. Z., and Hennekens, C. H. (1997) Prospective study of moderate alcohol consumption and risk of peripheral arterial disease in US male physicians. *Circulation* **95**, 577–580.

Chikte, U. M., Naidoo, S., Kolze, T. J., and Grobler, S. R. (2005) Patterns of tooth surface loss among winemakers. *S. Afr. Dental J.* **9**, 370–374.

Choi, H. K., Atkinson, K., Karlson, E. W., Willett, W., and Curhan, G. (2004) Alcohol intake and risk of incident gout in men: a prospective study. *Lancet* **363**, 1277–1281.

Choi, H. K., Mount, D. B., and Reginato, A. M. (2005) Pathogenesis of gout. *Ann. Intern. Med.* **143**, 499–516.

Chun, Y. J., Kim, M. Y., and Guengerich, F. P. (1999) Resveratrol is a selective human cytochrome P450 1A1 inhibitor. *Biochem. Biophys. Res. Commun.* **262**, 20–24.

Cohen, S., Tyrrell, D. A. J., Russell, M. A. H., Jarvis, M. J., and Smith, A. P. (1993) Smoking, alcohol consumption, and susceptibility to the common cold. *Am. J. Publ. Health* **83**, 1277–1283.

Corder, R., Douthwaite, J. A., Lees, D. M., Khan, N. Q., Viseu Dos Santos, A. C., Wood, E. G., and Carrier, M. J. (2001) Endothelin-1 synthesis reduced by red wine. *Nature* **414**(6866), 863–864.

Crawhall, J. C. (1985) A review of the clinical presentation and laboratory findings of two uncommon hereditary disorders of sulfur amino acid metabolism, β-mercaptolactate cysteine disulfideuria and sulfite oxidase deficiency. *Clin. Biochem.* **18**, 139–142.

Curhan, G. C., Willett, W. C., Speizer, F. E., and Stampfer, M. J. (1998) Beverage use and risk for kidney stones in women. *Ann. Intern. Med.* **128**, 534–540.

Dahl, R., Henriksen, J. M., and Harving, H. (1986) Red wine asthma: A controlled study. *J. Allergy Clin. Immunol.* **78**, 1126–1129.

De Freitas, V., and Mateus, N. (2003) Nephelometric study of salivary protein-tannin aggregates. *J. Sci. Food Agric.* **82**, 113–119.

de la Torre, R., Covas, M. I., Pujadas, M. A., Fito, M., and Farre, M. (2006) Is dopamine behind the health benefits of red wine? *Eur. J. Nutr.* **45**, 307–310.

Delmas, D., Rébés, C., Lacours, S., Filomenko, R., Athias, A., Cambert, P., Cherkaoui-Malki, M., Jannin, B., Dubrez-Daloz, L., Latruffe, N., and Solary, E. (2003) Resveratrol-induced apoptosis is associated with Fas redistribution in the rafts and the formation of a death-inducing signaling complex in colon cancer cells. *J. Biol. Chem.* **278**, 41482–41490.

Déprez, S., Brezillon, C., Rabot, S., Philippe, C., Mila, I., Lapierre, C., and Scalbert, A. (2000) Polymeric proanthocyanidins are catabolized by human colonic microflora into low-molecular-weight phenolic acids. *J. Nutr.* **130**, 2733–2738.

Díaz-Rubio, M. E., and Saura-Calixto, F. (2006) Dietary fiber in wine. *Am. J. Enol. Vitic.* **57**, 69–72.

Dixon, J. B., Dixon, M. F., and O'Brien, P. E. (2001) Alcohol consumption in the severely obese: relationship with the metabolic syndrome. *Obesity Res.* **10**, 245–252.

Dolara, P., Arrigucci, S., Cassetta, M. I., Fallani, S., and Novelli, A. (2005) Inhibitory activity of diluted wine on bacterial growth: the secret of water purification in antiquity. *Int. J. Antimicrob. Agents* **26**, 338–341.

Doll, R., Peto, R., Boreham, J., and Sutherland, I. (2005) Mortality in relation to alcohol consumption: a prospective study among male British doctors. *Int. J. Epidemiol.* **34**, 199–204.

Ebeler, S. E., and Weber, M. A. (1996) Wine and cancer. In: *Wine in Context: Nutrition, Physiology, Policy, Proceedings of the Symposium on Wine and Health* (A. L. Waterhouse and J. M. Rantz, eds.), pp. 16–18. American Society for Enology and Viticulture, Davis, CA.

Emsley, J. (1986/1987) When the Empire struck lead. *New Scientist.* **112**(1581), 64–67.

Enomoto, N., Takada, A., and Date, T. (1991) Genotyping of the aldehyde dehydrogenase 2(ALDH2) gene using the polymerase chain reaction: evidence for single point mutation in the ALDH2 gene for ALDH2-deficiency. *Gastroenterol. Jpn* **26**, 440–447.

Falcone, F. (1991) Migration of lead into alcoholic beverages during storage in lead crystal decanters. *J. Food Protect.* **54**, 378–380.

Fazal, F., Rahman, A., Greensill, J., Ainley, K., Hasi, S. M., and Parish, J. H. (1990) Strand scission in DNA by quercetin and Cu(II): Identifi-cation of free radical intermediates and biological consequences of scission. *Carcinogenesis* **11**, 2005–2008.

Fluss, L., Hguyen, T., Ginther, G. C., and Leighton, T. (1990) Reduction in the direct-acting mutagenic activity of red wine by treatment with polyvinylpolypyrrolidone. *J. Wine Res.* **1**, 35–43.

Franke, A., Teyssen, S., Harder, H., and Singer, M. V. (2004) Effects of ethanol and some alcoholic beverages on gastric emptying in humans. *Scand. J. Gastroenterol.* **39**, 638–645.

Frankel, E. N., Waterhouse, A. L., and Kinsella, J. E. (1993) Inhibition of human LDL oxidation by resveratrol. *Lancet* **341**, 1103–1104.

Fraser, A. G. (1997) Pharmacokinetic interactions between alcohol and other drugs. *Clin. Pharmacokinet.* **33**, 79–90.

Fraser-Bell, S., Wu, J., Klein, R., Azen, S. P., and Varma, R. (2006) Smoking, alcohol intake, estrogen use, and age-related macular degeneration in Latinos: the Los Angeles Latino Eye Study. *Am. J. Ophthalmol.* **141**, 79–87.

Fugelsang, K. C., and Muller, C. J. (1996) The *in vitro* effect of red wine on *Helicobacterium pylori*. In: *Wine in Context: Nutrition, Physiology, Policy.* Proceedings of the Symposium on Wine and Health (A. L. Waterhouse and J. M. Rantz, eds.), pp. 43–45. American Society for Enology and Viticulture, Davis, CA.

Fuhrman, B., Volkova, N., Suraski, A., and Aviram, M. (2001) White wine with red wine-like properties: increased extraction of grape skin polyphenols improves the antioxidant capacity of the derived white wine. *J. Agric. Food Chem.* **49**, 3164–3168.

Ganry, O., Baudoin, C., and Fardellone, P. (2000) Effect of alcohol intake on bone mineral density in elderly women: The EPIDOS Study. Epidemiologie de l'Osteoporose. *Am. J. Epidemiol.* **151**, 773–780.

Gaziano, J. M., Hennekens, C. H., Godfried, S. L., Sesso, H. D., Glynn, R. J., Breslow, J. L., and Buring, J. E. (1999) Type of alcoholic beverage and risk of myocardial infarction. *Am. J. Cardiol.* **83**, 52–57.

Gin, H., Morlat, P., Ragnaud, J. M., and Aubertin, J. (1992) Short-term effect of red wine (consumed during meals) on insulin requirement of glucose tolerance in diabetic patients. *Diabetes Care* **15**, 546–548.

Goldberg, D. (1981) Red head. *Lancet* **8227**, 1003.

Grønbæk, M., Becker, U., Johansen, D., Gottschau, A., Schnohr, P., Hein, H. O., Jensen, G., and Sorensen, T. I. (2000) Type of alcohol consumed and mortality from all causes, coronary heart disease, and cancer. *Ann. Intern. Med.* **133**, 411–419.

Haddad, J. J. (2004) Alcoholism and neuro-immune–endocrine interactions: physiochemical aspects. *Biochem. Biophys. Res. Commun.* **323**, 361–371.

Heinz, A. (2006) Staying sober. *Sci. Am. Mind* **17**, 57–61.

Hillbom, M. (1999) Oxidants, antioxidants, alcohol and stroke. *Front. Biosci.* **4**, 67–71.

Hines, L. M., Stampfer, M. J., Ma, J., Gaziano, J. M., Ridker, P. M., Hankinson, S. E., Sacks, F., Rimm, E. B., and Hunter, D.J. (2001) Genetic variation in alcohol dehydrogenase and the beneficial effect of moderate alcohol consumption on myocardial infarction. *N. Engl. J. Med.* **344**, 549–555.

Hou, D.-X. (2003) Potential mechanism of cancer chemoprevention by anthocyanins. *Curr. Molec. Med.* **3**, 149–159.

Hyde, R. J., and Pangborn, R. M. (1978) Parotid salivation in response to tasting wine. *Am. J. Enol. Vitic.* **29**, 87–91.

Ilich, J. Z., Brownbill, R. A., Tamborini, L., and Crncevic-Orlic, Z. (2002) To drink or not to drink: how are alcohol, caffeine and past smoking related to bone mineral density in elderly women? *J. Am. Coll. Nutr.* **21**, 536–544.

Ishige, K., Schubert, D., and Sagara, Y. (2001) Flavonoids protect neuronal cells from oxidative stress by three distinct mechanisms. *Free Radic. Biol. Med.* **30**, 433–446.

Jang, M., and Pezzuto, J. M. (1998) Resveratrol blocks eicosanoid production and chemically-induced cellular transformation: Implications for cancer chemoprevention. *Pharmaceut. Biol.* **36** (Supp.), 28–34.

Jarisch, R., and Wantke, F. (1996). Wine and headache. *Intl. Arch. Allergy Immunol.* **110**, 7–12.

Jones, A. L., Roberts, R. C., Colvin, D. W., Rubin, G. L., and Coughtrie, M. W. H. (1995) Reduced platelet phenolsulphotransferase activity towards dopamine and 5-hydroxytryptamine in migraine. *Eur. J. Clin. Pharmacol.* **49**, 109–114.

Jones, R. J., Forsythe, H. M., and Amess, J. A. (1982) Platelet aggregation in migraine patients during the headache-free interval. *Adv. Neurol.* **33**, 275–278.

Kanazawa, K., Yamashita, T., Ashida, H., and Danno, G. (1998) Antimutigenicity of flavones and flavonols to heterocyclic amines by specific and strong inhibition of the cytochrome P450 1A family. *Biosci. Biotechnol. Biochem.* **62**, 970–977.

Kanner, J., and Lapidot, T. (2001) The stomach as a bioreactor: dietary lipid peroxidation in the gastric fluid and the effects of plant-derived antioxidants. *Free Radic. Biol. Med.* **31**, 1388–1395.

Kastenbaum, R. (1982) Wine and the elderly person. In: *Proceedings of the Wine, Health and Society. A Symposium*, pp. 87–95. GRT Books, Oakland, CA.

Kaufman, H. S. (1986) The red wine headache: A pilot study of a specific syndrome. *Immunol. Allergy Prac.* **8**, 279–284.

Kaufman, H. S. (1992) The red wine headache and prostaglandin synthetase inhibitors: A blind controlled study. *J. Wine Res.* **3**, 43–46.

Keil, U., Liese, A., Filipiak, B., Swales, J. D., and Grobbee, D. E. (1998) Alcohol, blood pressure and hypertension. *Novartis Found. Symp.* **216**, 125–144 (discussion 144–151).

Keli, S. O., Hertog, M. G. L., Feskens, E. J. M., and Kromhout, D. (1994) Dietary flavonoids, antioxidant vitamins and the incidence of stroke: the Zutphen Study. *Arch. Intern. Med.* **154**, 637–642.

Khan, M. A., Jensen, K., and Krogh, H. J. (1973) Alcohol-induced hangover. A double-blind comparison of pyritinol and placebo in preventing hangover symptoms. *Q. J. Stud. Alcohol.* **34**, 1195–1201.

Kim, D.-J., Kim, W., Yoon, S.-J., Choi, B.-M., Kim, J.-S., Go, H. J., Kim, Y.-K., and Jeong, J. (2003) Effects of alcohol hangover on cytokine production in healthy subjects. *Alcohol* **31**, 167–170.

Klatsky, A. L., Friedman, G. D., and Siegelaub, A. B. (1974) Alcohol consumption before myocardial infarction: Results from the Kaiser-Permanente epidemiologic study of myocardial infarction. *Ann. Intern. Med.* **81**, 294–301.

Klatsky, A. L., Friedman, G. D., Armstrong, M. A., and Kipp, H. (2003) Wine, liquor, beer, and mortality. *Am. J. Epidemiol.* **15**, 585–595.

Klein, H., and Pittman, D. (1990) Drinker prototypes in American Society. *J. Substance Abuse* **2**, 299–316.

Knudsen, N., Bülow, I., Laurberg, P., Perrild, H., Ovesen, L., and Jørgensen, T. (2001) Alcohol consumption is associated with reduced prevalence of goitre and solitary thyroid nodules. *Clin. Endocrinol.* **55**, 41–46.

La Porte, R. E., Cresanta, J. L., and Kuller, L. H. (1980) The relationship of alcohol consumption to arteriosclerotic heart disease. *Prev. Med.* **9**, 22–40.

Leighton, F., Miranda-Rottmann, S., and Urquiaga, I. (2006) A central role of eNOS in the protective effect of wine against metabolic syndrome. *Cell Biochem. Funct.* **24**, 291–298.

Letenneur, L. (2004) Risk of dementia and alcohol and wine consumption: a review of recent results. *Biol Res.* **37**, 189–193.

Levitan, E. B., Ridker, P. M., Manson, J. E., Stampfer, M. J., Buring, J. E., Cook, N. R., and Liu, S. (2005) Association between consumption of beer, wine, and liquor and plasma concentration of high-sensitivity C-reactive protein in women aged 39 to 89 years. *Am. J. Cardiol.* **96**, 83–88.

Libby, P. (2001) Atherosclerosis: A new view. *Sci. Am.* **286**(5), 46–55.

Lin, Y., and Cheng, T. (2002) Why can't Chinese Han drink alcohol? Hepatis B virus infection and the evolution of acetaldehyde dehydrogenase deficiency. *Med. Hypotheses* **59**, 204.

Luchsinger, J. A., Tang, M. X., Siddiqui, M., Shea, S., and Mayeux, R. (2004) Alcohol intake and risk of dementia. *J. Am. Geriatr. Soc.* **52**, 540–546.

Lucia, S.P. (1963) *A History of Wine as Therapy*. Lippincott, Philadelphia, PA.

Mandel, L. (2005) Dental erosion due to wine consumption. *J. Am. Dent. Assoc.* **136**, 71–75.

Marambaud, P., Zhao, H. and Davies, P. (2005) Resveratrol promotes clearance of Alzheimer's disease amyloid-beta peptides *J. Biol. Chem.* **280**, 37377–37382.

Marinkovich, V. A. (1982) Allergic symptoms from fining agents used in winemaking. In: *Proceedings of Wine, Health and Society. A Symposium*, pp. 119–124. GRT Books, Oakland, CA.

Martin, S., and Pangborn, R. M. (1971) Human parotid secretion in response to ethyl alcohol. *J. Dental Res.* **50**, 485–490.

Masquelier, J. (1986) Azione portettrice del vino sull'ulcera gastrica. *Indust. Bevande* **81**, 13–16.

Masquelier, J. (1988) Effets physiologiques du vin. Sa part dans l'alcoolisme. *Bull. O.I.V.* **61**, 555–577.

Masyczek, R., and Ough, C. S. (1983) The red wine reaction syndrome. *Am. J. Enol. Vitic.* **32**, 260–264.

Maxwell, S. R. J. (1997) Wine antioxidants and their impact on antioxidant activity *in vivo*. In: *Wine: Nutritional and Therapeutic Benefits* (T. R. Watkins, ed.), pp. 150–165. ACS Symposium Series No. 661, American Chemical Society, Washington, DC.

Maxwell, S. R. J., Cruickshank, A., and Thorpe, G. H. G. (1994) Red wine and antioxidant activity in serum. *Lancet* **334**, 193–194.

Meagher, E. A., Barry, O. P., Burke, A., Lucey, M. R., Lawson, J. A., Rokach, J., and FitzGerald, G. A. (1999) Alcohol-induced generation of lipid peroxidation products in humans. *J. Clin. Invest.* **104**, 805–813.

Middleton, E. Jr. (1998) Effect of plant flavonoids on immune and inflammatory cell function. *Adv. Exp. Med. Biol.* **439**, 175–182.

Miller, N. J., and Rice-Evans, C. A. (1995) Antioxidant activity of resveratrol in red wine. *Clin. Chem.* **41**, 1789.

Miller, N. S., Goodwin, D. W., Jones, F. C., Gabrielle, W. F., Pardo, M. P., Anand, M. M., and Hall, T. B. (1988) Antihistamine blockade of alcohol-induced flushing in orientals. *J. Stud. Alcohol* **49**, 16–20.

Mok, T. B., McIntyre, J., and Hunt, D. (2001) Dental erosion: *In vitro* model of a wine assessor's erosion. *Aust. Dental J.* **46**, 263–268.

Morel, I., Lescoat, G., Cillard, P., and Cillard, J. (1994) Role of flavonoids and iron chelation in antioxidant action. *Methods Enzymol.* **234**, 437–443.

Møretrø, M and Daeschel, M. A. (2004) Wine is bactericidal to foodborne pathogens. *J. Food Science* **69**, M 251–257.

Mukamal, K. J., Chiuve, S. E., and Rimm, E. B. (2006) Alcohol consumption and risk for coronary heart disease in men with healthy lifestyles. *Arch. Intern. Med.* **166**, 2145–2150.

Musto, D. F. (1996) Alcohol in American History. *Sci. Am.* **274**(4), 78–83.

Napoli, R., Cozzolino, D., Guardasole, V., Angelini, V., Zarra, E., Matarazzo, M., Cittadini, A., Sacca, L., and Torella, R. (2005) Red wine consumption improves insulin resistance but not endothelial function in type 2 diabetic patients. *Metabolism* **54**, 306–313.

Nestler, E. J., and Malenka, R. C. (2004) The addicted brain. *Sci. Am.* **290**(3), 78–85.

Nicolodi, M., and Sicuteri, F. (1999) Wine and migraine: compatibility or incompatibility? *Drugs Exp. Clin. Res.* **25**, 147–153.

Niemela, O., and Israel, Y. (1992) Hemoglobin-acetaldehyde adducts in human alcohol abusers. *Lab. Invest.* **67**, 246–252.

Niemela, O., and Parkkila, S. (2004) Alcoholic macrocytosis – is there a role for acetaldehyde and adducts? *Addict. Biol.* **9**, 3–10.

Nurnberger, J. I., and Bierut, L. J. (2007) Seeking the connections: alcoholism and our genes. *Sci. Am.* **296**(4), 46–53.

Obisesan, T. O., Hirsch, R., Kosoko, O., Carlson, L., and Parrott, M. (1998) Moderate wine consumption is associated with decreased odds of developing age-related macular degeneration in NHANES-1. *J. Am. Geriat. Soc.* **46**, 1–7.

Ono, K., Hasegawa, K., Naiki, H., and Yamada, M. (2004) Anitamyloidogenic activity of tannic acid and its activity to destabilize Alzheimer's beta-amyloid fibrils *in vitro*. *Biochem. Biophys. Acta* **1690**, 193–202.

Parantainen, J. (1983) Prostaglandins in alcohol intolerance and hangover. *Drug Alcohol Depend.* **11**, 239–248.

Passamonti, S., Vrhovsek, U., Vanzo, A., and Mattivi, F. (2003) The stomach as a site for anthocyanins absorption from food. *FEBS Lett.* **544**, 210–213.

Passamonti, S., Vrhovsek, U., Vanzo, A., and Mattivi, F. (2005) Fast access of some grape pigments to the brain. *J. Agric. Food Chem.* **53**, 7029–7034.

Pastorello, E. A., Farioli, L., Pravettoni, V., Ortolani, C., Fortunato, D., Giuffrida, M. G., Garoffo, L. P., Calamari, A. M., Brenna, O., and Conti, A. (2003) Identification of grape and wine allergens as an endochitinase 4, a lipid-transfer protein, and a thaumatin. *J. Allergy Clin. Immunol.* **111**, 350–359.

Pattichis, K., Louca, L. L., Jarman, J., Sandler, M., and Glover, V. (1995) 5-Hydroxytryptamine release from platelets by different red wines: Implications for migraine. *Eur. J. Pharmacol.* **292**, 173–177.

Peatfield, R. C., Hussain, N., Glover, V. A. S., and Sandler, M. (1995) Prostacyclin, tyramine and red wine. In: *Experimental Headache Models* (J. Olesen and M. A. Moskowitz, eds.), pp. 267–276. Lippincott-Raven, Philadelphia, PA.

Pittler, M. A., White, A. R., Stevinson, C., and Ernst, E. (2003) Effectiveness of artichoke extract in preventing alcohol-induced hangovers: a randomized controlled trial. *Can. Med. Assoc. J.* **169**, 1269–1273.

Pittman, D. J. (1996) Cross-cultural aspects of drinking, alcohol abuse and alcoholism. In: *Wine in Context: Nutrition, Physiology, Policy: Proceedings of the Symposium on Wine and Health* (A. L. Waterhouse and J. M. Rantz, eds.), pp. 1–5. American Society for Enology and Viticulture, Davis, CA.

Pizarro, M., and Lissi, E. (2003) Red wine antioxidants. Evaluation of their hydrophobicity and binding extent to salivary proteins. *J. Chil. Chem. Soc.* **48**(3), 57–59.

Renaud, S., and de Lorgeril, M. (1992) Wine alcohol, platelets and the French paradox for coronary heart disease. *Lancet* **339**, 1523–1526.

Renaud, S. C., and Ruf, J.-C. (1996) Effects of alcohol on platelet functions. *Clin. Chem. Acta* **246**, 77–89.

Renaud, S., Lanzmann-Petithory, D., Gueguen, R., and Conard, P. (2004) Alcohol and mortality from all causes. *Biol. Res.* **37**, 183–187.

Rice-Evans, C. A., Miller, N. J., and Paganga, G. (1996) Structure-antioxidant activity relationships of flavonoids and phenolic acids. *Free Radical Biol. Med.* **20**, 933–956.

Ridout, F., Gould, S., Nunes, C., and Hindmarch, I. (2003) The effects of carbon dioxide in champagne on psychometric performance and blood-alcohol concentration. *Alcohol Alcohol* **38**, 381–385.

Rimm, E. B., Klatsky, A., Grobbee, D., and Stampfer, M. J. (1996) Review of moderate alcohol consumption and reduced risk of coronary heart disease: Is the effect due to beer, wine, or spirits? *Br. Med. J.* **312**, 731–736.

Rohn, S., Rawel, H. M., Kroll, J. (2002) Inhibitory effects of plant phenols on the activity of selected enzymes. *J. Agric. Food Chem.* **50**, 3566–3571.

Rolland, J. M., Apostolou, E., Deckert, K., de Leon, M. P., Douglass, J. A., Glaspole, I. N., Bailey, M., Stockley, C. S., and O'Hehir, R. E. (2006) Potential food allergens in wine: double-blind, placebo-controlled trial and basophil activation analysis. *Nutrition.* **22**, 882–888.

Rosell, M., de Faire, U., and Hellenius, M. L. (2003) Low prevalence of the metabolic syndrome in wine drinkers – is it the alcohol beverage or the lifestyle? *Eur. J. Clin. Nutr.* **57**, 227–234.

Scalbert, A., Morand, C., Manach, C., and Remesy, C. (2002) Absorption and metabolism of polyphenols in the gut and impact on health. *Biomed. Pharmacother.* **56**, 276–282.

Schmitz, C. M., and Gray, R. A. (1998) *Alcoholism: The Health and Social Consequences of Alcohol Use.* Pierian Press, Ann Arbor, MI.

Scholten, P. (1982) Moderate drinking in pregnancy. In: *Proceedings of Wine, Health and Society. A Symposium*, pp. 71–86. GRT Books, Oakland, CA.

Serianni, E., Cannizzaro, M., and Mariani, A. (1953) Blood alcohol concentrations resulting from wine drinking timed according to the dietary habits of Italians. *Quart. J. Stud. Alcohol* **14**, 165–173.

Shih, V. E., Abroms, I. F., Johnson, J. L., Carney, M., Mandell, R., Robb, R. M., Cloherty, J. P., and Rajagopalan, K. V. (1977) Sulfite oxidase deficiency. *New Engl. J. Med.* **297**, 1022–1028.

Simonetti, P., Gardana, C., Pietta, P. (2001) Plasma levels of caffeic acid and antioxidant status after red wine intake. *J. Agric. Food Chem.* **49**, 5964–5968.

Singleton, V. L., Trousdale, E., and Zaya, J. (1979) Oxidation of wines. l. Young white wines periodically exposed to air. *Am. J. Enol. Vitic.* **30**, 49–54.

Sjaastad, O., and Bakketeig, L. S. (2004) Hangover headache: various manifestations and proposal for criteria. Vågå study of headache epidemiology. *J. Headache Pain* **5**, 230–236.

Skibola, C. F., and Smith, M. T. (2003) Potential health impacts of excessive flavonoid intake. *Free Rad. Biol. Med.* **29**, 375–383.

Smart, R. G., and Walsh, G. (1999) Heavy drinking and problems among wine drinkers. *J. Stud. Alcohol* **60**, 467–471.

Soleas, G. J., Yan, J., and Goldberg, D. M. (2001) Ultrasensitive assay for three polyphenols (catechin, quercetin and resveratrol) and their conjugates in biological fluids utilizing gas chromatography with mass selective detection. *J. Chromatogr. B Biomed. Sci. Appl.* **757**, 161–172.

Solfrizzi, V., D'Introno, A., Colacicco, A. M., Capurso, C., Del Parigi, A., Baldassarre, G., Scapicchio, P., Scafato, E., Amodio, M., Capurso, A., and Panza, F. (2007) Alcohol consumption, mild cognitive imparment, and progression to dementia. *Neurology* **68**, 1790–1799.

Steiner, T. J., Hering, R., Couturier, E. G., Davies, P. T., and Whitmarsh, T. E. (1997) Double-blind placebo-controlled trial of lithium in episodic cluster headache. *Cephalalgia* **17**, 673–675.

Stone, B. M. (1980) Sleep and low doses of alcohol. *Electroencephalogr. Clin. Neurophysiol.* **48**, 706–709.

Su, H. C., Hung, L. M., and Chen, J. K. (2006) Resveratrol, a red wine antioxidant, possesses an insulin-like effect in streptozotocin-induced diabetic rats. *Am. J. Physiol. Endocrinol. Metab.* **290**, E1339–1346.

Subbaramaiah, K., Michaluart, P., Chung, W. J., and Dannenberg, A. J. (1998) Resveratrol inhibits the expression of cyclooxygenase-2 in human mammary and oral epithelial cells. *Pharmaceut. Biol.* **36** (Suppl.), 35–43.

Tagliazucchi, D., Verzelloni, E., and Conte, A. (2005) Effect of some phenolic compounds and beverages on pepsin activity during simulated gastric digestion. *J. Agric. Food Chem.* **53**, 8706–8713.

Takeshita, T., Mao, Z. Q., and Morimoto, K. (1996) The contribution of polymorphism in the alcohol dehydrogenase bets subunit to alcohol sensitivity in a Japanese population. *Hum. Genet.* **97**, 409–413.

Takeshita, T., Yang, X., and Morimoto, K. (2001) Association of the ADH2 genotypes with skin responses after ethanol exposure in Japanese male university students. *Alcohol Clin. Exp. Res.* **25**, 1264–1269.

Takkouche, B., Requeira-Mendez, C., Garcia-Closas, R., Figueiras, A., Gestal-Otero, J. J., and Hernan, M. A. (2002) Intake of wine, beer, and spirits and the risk of clinical common cold. *Am. J. Epidemiol.* **155**, 853–858.

Talavéra, S., Felgines, C., Texier, O., Besson, C., Manach, C., Lamaison, J.-L., and Rémésy, C. (2005) Anthocyanins are efficiently absorbed from the small intestine in rats. *J. Nutr.* **134**, 2275–2279.

Teissèdre, P. L., Cros, G., Krosniak, M., Portet, K., Serrano, J. J., and Cabanis, J. C. (1996) Contribution to wine in vanadium dietary intake: Geographical origin has a significant impact on wine vanadium levels. In: *Metal Ions in Biology and Medicine* (P. Collery *et al.*, eds.), Vol. 4, pp. 183–185. John Libbey Eurotext, Paris.

Terry, M. B., Gammon, M. D., Zhang, F. F., Knight, J. A., Wang, Q., Britton, J. A., Teitelbaum, S. L., Neugut, A. I., and Santella, R. M. (2006) ADH3 genotype, alcohol intake, and breast cancer risk. *Carcinogenesis* **27**, 840–847.

Teyssen, S., González-Calero, G., Schimiczek, M., and Singer, M. V. (1999) Maleic acid and succinic acid in fermented alcoholic beverages are the stimulants of gastric acid secretion. *J. Clin. Invest.* **103**, 707–713.

Thomasson, H. R., Beard, J. D., and Li, T. K. (1995) ADH2 gene polymorphism are determinants of alcohol pharmacokinetics. *Alcohol Clin. Exp. Res.* **19**, 1494–1499.

Thun, M. J., Peto, R., Lopez, A. D., Monaco, J. H., Henley, S. J., Heath, C. W. Jr., and Doll, R. (1997) Alcohol consumption and mortality among middle-aged and elderly US adults. *N. Engl. J. Med.* **337**, 1705–1714.

Tredici, G., Miloso, M., Nicolini, G., Galbiati, S., Cavaletti, G., and Bertelli, A. (1999) Resveratrol, MAP kinases and neuronal cells: might wine be a neuroprotectant? *Drugs Exptl Clin. Res.* **25**, 99–103.

Truelsen, T., Grønbaek, M., Schnohr, P., and Boysen, G. (1998) Intake of beer, wine, and spirits and risk of stroke: The Copenhagen city heart study. *Stroke* **29**, 2467–2472.

Truelsen, T., Thudium, D., and Grønbaek, M. (2002) Amount and type of alcohol and risk of dementia: the Copenhagen City Heart Study. *Neurology* **59**, 1313–1319.

Truitt, E. B., Jr, Gaynor, C. R., and Mehl, D. L. (1987) Aspirin attenuation of alcohol-induced flushing and intoxication in oriental and occidental subjects. *Alcohol Alcohol* Suppl.1, 595–599.

Vallee, B. L. (1998) Alcohol in the western world. *Sci. Am.* **278**(6), 80–85.

Vally, H., and Thompson, P. J. (2001) Role of sulfite additives in wine induced asthma: single dose and cumulative dose studies. *Thorax* **56**, 763–769.

van der Gaag, M. S., van Tol, A., Vermunt, S. H. D., Scheek, L. M., Schaafsma, G., and Hendriks, H. F. J. (2001) Alcohol consumption stimulates early steps in reverse cholesterol transport. *J. Lipid Res.* **42**, 2077–2083.

Verhagen, J. V., Haenen, G. R. M. M., and Bast, A. (1997) Nitric oxide radical scavenging by wines. *J. Agric. Food Chem.* **45**, 3733–3734.

Viel, J. F., Perarnau, J. M., Challier, B., and Faivrenappez, I. (1997) Alcoholic calories, red wine consumption and breast cancer among premenopausal women. *Europ. J. Epidemiol.* **13**, 639–643.

Wakabayashi, M., Nagao, M., Esumi, H., and Sugimura, T. (1992) Food-derived mutagens and carcinogens. *Cancer Res.* **52**, 2092s–2098s.

Wall, T. L., Shea, S. H., Luczak, S. E., Cook, T. A. R., and Carr, L. G. (2005) Genetic association of alcohol dehydrogenase with alcohol use disorders and endophenotypes in white college students. *J. Abnorm. Psychol.* **114**, 456–465.

Wallerath, T., Li, H., Godtel-Ambrust, U., Schwarz, P. M., and Forstermann, U. (2005) A blend of polyphenolic compounds explains the stimulatory effect of red wine on human endothelial NO synthase. *Nitric Oxide* **12**, 97–104.

Walzem, R. L., and Hansen, R. J. (1996) Alterosclerotic cardiovascular disease and antioxidants. In: *Wine in Context: Nutrition, Physiology, Policy. Proceedings of the Symposium on Wine and Health* (A. L. Waterhouse and J. M. Rantz, eds.), pp. 6–12. American Society for Enology and Viticulture, Davis, CA.

Walzem, R. L., Watkins, S., Frankel, E. N., Hansen, R. J., and German, J. B. (1995) Older plasma lipoproteins are more susceptible to oxidation: A linking mechanism for the lipid and oxidation theories of atherosclerotic cardiovascular disease. *Proceedings of the Natl Acad. Sci. USA.* **92**, 7460–7464.

Wantke, F., Hemmer, W., Haglmuller, T., Gotz, M., and Jarisch, R. (1996) Histamine in wine – bronchoconstriction after a double-blind placebo-controlled red wine provocation test. *Intl Arch. Allegery Immunol.* **110**, 397–400.

Ward, N. C., Croft, K. D., Puddey, I. B., and Hodgson, J. M. (2004) Supplementation with grape seed polyphenols results in increased urinary excretion of 3-hydroxyphenylpropionic acid, an important metabolite of proanthocyanidins in humans. *J. Agric. Food Chem.* **52**, 5545–5549.

Wattendorf, D. J., and Muenke, M. (2005) Fetal alcohol spectrum disorders. *Am. Fam. Physician.* **72**, 279–282, 285.

Weathermon, R., and Crabb, D. W. (1999) Alcohol and medication interactions. *Alcohol Res. Health* **23**, 40–54.

Weber, P., Steinhart, H., and Paschke, A. (2007) Investigation of the allergenic potential of wines fined with various proteinogenic fining agents by ELISA. *J. Agric. Food. Chem.* **55**, 3127–3133.

Weisburger, J. H. (1991) Nutritional approach to cancer prevention with emphasis on vitamins, antioxidants, and carotenoids. *Am. J. Clin. Nutr.* **53** (Suppl.), 226S–237S.

Weisse, M. E., Eberly, B., and Person, D. A. (1995) Wine as a digestive aid: Comparative antimicrobial effects of bismuth salicylate and red and white wine. *Br. Med. J.* **311**, 1657–1660.

Whitten, D. (1996) Fetal alcohol risk: A current perspective. In: *Wine in Context: Nutrition, Physiology, Policy. Proceedings of the Symposium on Wine and Health* (A. L. Waterhouse and J. M. Rantz, eds.), pp. 46–49. American Society for Enology and Viticulture, Davis, CA.

Wiese, J.G., Shlipak, M. G., and Browner, W. S. (2000) The alcohol hangover. *Ann. Intern. Med.* **132**, 897–902.

Wiese, J., McPherson, S., Odden, M. C., and Shlipak, M. G. (2004) Effect of *Opuntia fiscus indica* on symptoms of the alcohol hangover. *Arch. Intern. Med.* **164**, 1334–1340.

Williams, R. J., Spencer, J. P., and Rice-Evans, C. (2004) Flavonoids: antioxidants or signalling molecules? *Free Radic. Biol. Med.* **36**, 838–849.

Williamson, G., and Manach, C. (2005) Bioavailability and bioefficacy of polyphenols in humans. II. Review of 93 intervention studies. *Am. J. Clin. Nutr.* **81**, 243S–255S.

Wolfort, F. G., Pan, D., and Gee, J. (1996) Alcohol and preoperative management. *Plastic Reconstruct. Surg.* **98**, 1306–1309.

Youdim, K. A., Qaiser, M. Z., Begley, D. J., Rice-Evans, C. A., and Abbott, N. J. (2004) Flavonoid permeability across an *in situ* model of the blood–brain barrier. *Free Radic. Biol. Med.* **36**, 592–604.

Yu, T. (1983) Lead nephropathy and gout. *Am. J. Kidney Dis.* **11**, 555–558.

Zhang, Y., Kreger, B. E., Dorgan, J. F., Splansky, G. L., Cupples, L. A., and Ellison, R. C. (1999) Alcohol consumption and risk of breast cancer: The Framingham Study revisited. *Am. J. Epidemiol.* **149**, 93–101.

Glossary

Acclimation, physiological changes that adapt cells and tissues to environmental stress.

Acetic acid bacteria, aerobic bacteria that frequently cause wine spoilage through the production of acetic acid and ethyl acetate.

Acidity, the concentration of nonvolatile organic acids in must or wine, or the perception of acids in the mouth.

After-smell, the fragrance that lingers in the mouth after swallowing wine.

After-taste, the lingering taste perception in the mouth after wine has been swallowed.

Aging, changes in wine chemistry that occur after bottling; occasionally includes maturation.

Agglomerate cork, cork reconstituted from fragments of cork bark glued together.

Alambic still, a type of pot still used in brandy production.

Amelioration, the dilution of must with water to reduce the relative concentration of grape constituents, plus the addition of sucrose to compensate for the reduction in sugar content.

Ampelography, the identification of grape cultivars based on the vegetative characteristics of the vine.

Amphora, an elongated ceramic container historically used for the storage and transport of wine and other products.

Amplified Fragment Length Polymorphism (AFLP), a technique used in DNA fingerprinting that uses specific segments of labeled DNA.

Analgen, an embryonic inflorescence.

Aneuploidy, the presence or absence of one or more chromosomes from the normal genomic complement of chromosomes.

Anthesis, the stage in anther development when they rupture and pollen is released.

Anthocyanin, flavonoid pigments that generate the red to purple color of red grapes and wine (*see* Table 6.6).

Antioxidant, a compound that reacts readily with oxygen, limiting the oxidation of other wine or cellular constituents (or the reaction with toxic oxygen radicals).

Apoplast, the free space outside the plasma membrane of plant cells (cell wall and intercellular spaces).

Appellation Control (AC), a set of laws that regulate the use of geographic names to wines produced in the region; may also legislate aspects, such as permitted grape varieties, viticultural practices and wine-production techniques.

Arm, a short branch more than 1 year old growing from the trunk (Fig. 4.6).

Aroma, the fragrant perception that is derived from aromatic grape constituents.

Aromatic, lipid-soluble compounds sufficiently volatile to stimulate the olfactory receptors in the nose.

Aspect, the north–south–east–west orientation of a slope.

Astringency, a dry, puckery, dust-in-the-mouth touch sensation induced primarily by tannins and acids.

Autofermentor, a fermentor design that automatically induces periodic punching down of the cap of seeds and skins.

Autolysis, the breakdown of cell constituents by the activation of cellular hydrolytic enzymes.

Auxins, a group of growth regulators that influences many aspects of plant growth and development, notably through the stimulation of cell division and enlargement.

Available water, the portion of water held in soil that can be taken up through plant roots.

Axil, the upper angle formed by leaves where they attach to a shoot.

Axillary bud, a bud that forms in an axil of a leaf.

Bacteriophage, a virus that infects bacteria.

Baking, the heating used in processing wines such as madeira to obtain their distinctive bouquet.

Balanced pruning, a pruning technique in which the extent of pruning is related to the vigor of the vine, based on the weight of pruned wood removed.

Base bud, a bud that forms in the axil of a bract (scale-like leaf) at the base of a cane.

Base wine, a wine used as part of a blend in the production of certain wine styles (i.e., champagnes and sherries) or brandy.

Barrel, wood cooperage with arched sides used to mature and occasionally ferment wine; commonly occurring in volumes between 200 and 300 liter (depending on the wine region).

Barrel fermentation, wine fermentation conducted in-barrel.

Bearing shoot, a growing stem that possesses two (one to four) fruit clusters.

Bearing wood, a stem more than 1 year old from which bearing shoots grow.

Bench grafting, indoor grafting of the scion to a rootstock (usually in a greenhouse or nursery).

Bentonite, a formulation of montmorillonite clay used in juice clarification and wine fining.

Berry, a fleshy fruit derived from several fused ovaries.

Bilateral cordon, two horizontal extensions of the vine trunk aligned in opposite directions, but parallel to the vine row.

Biogenic amine, short amino acid derivatives that act as inter-cellular mediators animals, for example histamine.

Biological control, the use of parasites and predators in the control of pests, diseases, and weeds.

Bleeding, the extrusion of plant sap from pruning cuts in the early spring.

Blending, the mixing of wines from different grape varieties, vintages, vineyards or a combination of these to enhance wine quality (or at least to minimize the failing of the individual wines).

Bloom, the matte-like appearance of grapes due to the deposition of waxy plates on the surface.

Blue fining, the addition of ferrocyanide in removing metal ions to limit the formation of *casse* and metal-catalyzed oxidation of the wine.

Body, a perception of weight in the mouth produced by the major organic constituents in wine, ethanol, sugars, glycerol and tannins; synonymous with weight.

Botrytized wine, wine produced from grape that have undergone partial or complete noble rotting by *Botrytis cinerea*.

Bottle sickness, the temporary change in fragrance characteristics of wine shortly after bottling; usually interpreted as due to one or more unknown oxidation reactions.

Bottle stink, a nineteenth-century term referring to an off-odor that occasionally dissipated after decanting.

Bouquet, the fragrant sensation in wine derived from aromatics produced during fermentation, maturation or aging.

Brandy, the product derived from distilled wine aged in small oak cooperage.

Breathing, a term that refers to either the exposure of wine to air shortly following the opening or the decanting of bottled wine.

Brettanomyces, a yeast that typically is viewed as a spoilage organism, but is occasionally espoused by some as a source of distinctive flavors.

Brix (*Brix*), an indirect (specific gravity) measure of the total soluble solids in grape juice or wine; typically sugar content in juice and (by adjustment to alcohol content) in wine.

Brouillis, the major middle fraction collected from a pot still and used for redistillation into brandy.

Browning, an undesired increase in the brownish cast of a wine due to the oxidation of phenolic compounds.

Bud, one or more embryonic shoots protected in a series of modified leaves called bud scales.

Bud break, when buds begin to swell and grow.

Bung, the closure used to seal the bung hole through which wine may be sampled or added.

Butt, a type of oak barrel (~500 liter) used in the solera maturation of sherry.

Calcareous soil, soil possessing a high proportion of calcium carbonate (often with magnesium carbonate), frequently found in Europe.

Callus, undifferentiated parenchyma tissue that forms in response to damage, either as a result of wounding (i.e., grafting or pruning) or excised tissue placed on culture media.

Calyptra, the apically fused petals of the grape flower.

Cambium, a laterally positioned meristem – the vascular cambium produces new xylem and phloem tissue in the shoot and root, whereas the cork cambium produces cork tissue.

Cane, the vine shoot from the period it matures (turns brown and woody) until the end of the second year of growth (*see* Fig. 4.9).

Canopy, the foliage cover of the vine.

Cap, the collection of seeds and skins that forms on the top of fermenting must; an alternate name for the fused petals of the grape flower (calyptra).

Capacity, the productivity of the whole vine; typically applied to the fruit-bearing component.

Capillary water, the water that is held or moves up microscopic channels in soil.

Carbonation, the incorporation of carbon dioxide into a beverage under high pressure.

Carbonic maceration, the intracellular fermentation of grape cells that may precede yeast fermentation; used in the production of beaujolais-like wines.

Carboxylic acid, a hydrocarbon possessing an acid grouping.

Casse, haziness caused by the production of complexes between or metal salts with some organic acids, proteins, or phenolic components.

Catch wire, one or more trellis wires to which young shoots may be attached and moved to position the shoots in a particular training system.

Catechin, the major type of monomeric flavonoid tannin; isolated primarily from grape seeds; consists of two optical isomers of flavan-3-ols.

Cation, an positively (+) charged ion.

Caudalie, after swallowing, the unit of flavor duration (seconds).

cDNA, a chemically synthesized strand of DNA (deoxyribonucleic acid) complementary to a segment of RNA (usually a mRNA transcript of a structural gene).

Chalasa, the region of the seed where it was attached to the ovary wall of the fruit (*see* Fig. 2.3C).

Chamfered, a cork with beveled edges; used for T-corks that close sherry and port bottles.

Champagne, a region in northern France famous for its sparkling wines; (lower case), the sparkling wine produced in the Champagne region of France (used in some New World countries as a generic term for sparkling wines).

Chaptalization, the addition of sugar to grape must to increase the alcohol content of wine, initially proposed as a means of limiting wine spoilage in poor vintage years.

Chelator, heterocyclic organic compounds that form weak bonds with soluble ions; important in maintaining a supply of inorganic nutrients for roots in soil.

Chimera, an organism containing one or more genetically different cell lines in its tissues.

Chlorosis, the loss of chlorophyll in young plant tissue, allowing the yellow color of the carotenoids present to become apparent; develops during leaf senescence or earlier under environmental stress (e.g., disease and iron deficiency).

Clarification, the reduction of the concentration of suspended particles in must or wine by physical or physicochemical processes.

Clonal selection, the comparison, selection, and propagation of one or more clones of a variety possessing particular characteristics.

Clone, a population of vines derived vegetatively from a single cell or plant; initially genetically identical.

Closure, an object whose function is to seal the opening of a bottle, barrel, or other container.

Cluster thinning, the removal of a portion of the fruit clusters after fruit set.

Cold stabilization, a process in which the wine is chilled to below 0°C to promote the formation and precipitation of salt crystals (notably tartrates), and then filtered to remove them.

Colloid, macromolecules (i.e., soluble proteins), or molecular complexes (i.e., clay) dispersed in a medium by random, heat-induced (Brownian) molecular movement.

Color density, the sum of the absorbency of a wine at 420 and 520 nm ($E_{420} + E_{520}$).

Color stability, the long-term retention of a wine's young color; favored by low pH, oxygen exclusion, and (for red wines) anthocyanin polymerization with tannins.

Compatibility, the ability of a scion and rootstock to form a functional long-term union (graft); a genetic system that regulates self-fertility in plants.

Condensed tannins, covalently bonded polymers of flavonoid phenolic; they do not hydrolyze readily.

Compound bud, the mature axillary bud that survives the winter; typically it possesses three immature buds in different states of development.

Congener, compounds that influence the sensory quality (or intestinal uptake) of related substances; usually refers to alcohols other than ethanol in wines or distilled beverages.

Cooperage, a large container in which wine is fermented or matured.

Copigmentation, the weak association (stacking) of anthocyanins in complexes in association with other compounds that can significantly change in hue created by the pigments.

Cordon, an arm or trunk extension positioned horizontally or at an angle to the main axis of the trunk.

Cordon training, any of the training systems in which the bearing wood comes from a cordon.

Cork, the outer bark of woody stems; for wine, the bark of the cork oak (*Quercus suber* or related species); used in the production of stoppers for wine bottles.

Corky, a moldy off-odor most commonly associated with the presence of above-threshold values of 2,4,6-trichloroanisole.

Coulure, a diverse collection of environmentally induced disturbances that result in abnormally poor berry development, due either to excessive drop of flowers (shelling), immature berries (shatter), or failure of fertilized fruit to develop.

Count bud, the readily visible buds on a dormant cane (not including the small base buds); used in determining where to prune canes to regulate vine yield.

Criadera, a collection of barrels (butts having a capacity of about 490 l) constituting any one of the stages of the fractional blending (solera) system used in sherry maturation.

Crown cap, a lined metal closure used to seal bottles; the edges are crimped around the lip of the bottle.

Crushing, the forceful rupture of grapes to allow juice release and initiate the liberation of compounds from the seeds and skins.

Cryoextraction, the selective partial freezing of grapes to allow a predetermined level of juice concentration (via water removal as ice), followed by cold pressing.

Cultivar, a clone or series of related clones propagated vegetatively from a single parent plant (monoclonal origin) or several genetically similar parents (polyclonal origin); synonymous with variety.

Cuvée, the blend of wines in which the second carbon dioxide-producing fermentation occurs during the production of sparkling wines.

Deacidification, the reduction in total (titratable) acidity by any of a series of biological, physical or physicochemical means.

Dealcoholization, the reduction in alcohol content either by limiting ethanol production or by removing ethanol after its formation.

Degorgement, the removal of yeast from a sparkling wine after completion of the second carbon dioxide-producing fermentation.

Degree-days, a measure of climate suitability based on the sum of values obtained by multiplying the average daily temperature (−10 for °Celsius) by the number of days with that temperature, in months with an overall average temperature above 10 °C.

Dejuicer, a device used to allow the escape of juice from crushed grapes before pressing begins.

Dense planting, planting a vineyard such that there are approximately 4000–5000 vines per hectare (occasionally more), with rows spaced approximately 1 meter apart.

Destemming, the removal of grapes from fruit clusters, usually concomitant with crushing.

Devigoration, any procedure designed to reduce vine vigor (i.e., dense planting, root trimming, or grafting to a low-vigor rootstock).

Diaphragm, the woody tissue that may transverse the node in shoots (*see* Fig. 2.3A).

Diatomaceous earth, geologic deposits of the cell-wall remains of centric diatoms; occasionally used as a filter aid.

Dioecious, possessing only functional female or male flowers.

o-Diphenol, a phenolic compound possessing two adjacently located phenolic (-OH) groups; also referred to as vincinal diphenols or catechols.

Diploid, possessing two copies of each chromosomes; normally designated as 2N (N refers to the genome – a single copy of each chromosome).

Direct producer, a cultivar derived from backcrossing hybrids between *Vitis vinifera* and *V. riparia*, *V. rupestris*, or *V. aestivalis* with other *V. vinifera* cultivars (synonymous with French-American hybrid).

Distillation, the process of heating wine to remove and concentrate its more volatile aromatic ingredients.

Divided canopy, a technique in which a single dense canopy of foliage is separated into two or more thinner adjacent canopies to achieve better light and air exposure.

Dosage, the small quantity of aged wine or brandy, usually sweetened, added to sparkling wine immediately after disgorging.

Dry, having no perceived sweetness, in wine.

ELISA, an enzyme-enhanced form of serology; based on an antibody produced against a specific antigenic complex; often used to detect the presence of the antigen source (i.e., a virus) in an organism.

Ellagitannins, the primary hydrolyzable tannins in wine; derived principally from wood cooperage used to mature wine; polymers of ellagic acid or ellagic and gallic acids.

Embryo rescue, the isolation and cultivation of an immature seed embryo to a seedling on culture media (used when the embryo would abort otherwise).

Encapsulation, the entrapment of microbial cells in a gelatinous matrix to limit cell growth while permitting continued cell metabolism.

Enology, the scientific investigation of wine production.

Enzyme, an organic molecule (usually a protein) that acts as a catalyst in one (or several similar) chemical reactions.

Fatty acid, a long, straight hydrocarbon possessing a carbonyl (acid) group at one end.

Fermentation, an energy-yielding form of metabolism in which the electron donors and terminal electron acceptors are organic compounds (oxygen is not required).

Fermentor, cooperage in which fermentation occurs.

Fertigation, the joint application of inorganic nutrients along with irrigation water.

Fertilization, union of a pollen tube nucleus with an egg nucleus in the ovary of a flower; the addition of inorganic or organic nutrients to soil to improve plant growth.

Field capacity, the maximum amount of water that can be retained by a soil against the force of gravity.

Field grafting, promoting the union of a scion with a rootstock, conducted in a vineyard.

Film yeast, any yeast that can develop a hydrophobic wall, resulting in the formation of a floating colony (pellicle) on wine.

Filter aid, material suspended in wine prior to filtering that settles and becomes part of the filter bed of a depth filter.

Fining, the addition of particulate (i.e., bentonite) or macromolecules (i.e., gelatin) to remove excess tannins, dissolved proteins, or other colloidal material.

Fixed acidity, the organic acid content of wine that cannot readily be volatilized (removed) by steam distillation.

Flavonoid, a phenolic compound based on two phenols bonded by a pyran carbon ring; in wine they primarily come from the grape seeds and skins (*see* Table 6.2).

Flavor, the combined sensation of taste, touch, and odor of food and beverages in the mouth.

Flocculation, the process by which some yeast strains adhere to one another, especially as the cells shift from the exponential growth to stationary growth phase.

Flor, the pellicle of film yeasts that forms on the surface of maturing *fino* sherry.

Foliar feeding, the application of a dilute solution of inorganic nutrients to the foliage.

Fortification, the addition of wine spirits to arrest fermentation, increase alcohol content, or influence the course of wine development.

Foxy, an aroma associated with wine produced from some *Vitis labrusca* cultivars.

Fragrance, the aromatic aspect of wine.

Free-run, the portion of juice or wine that can flow away from the seeds and skins (pomace) of crushed grapes under the influence of gravity.

French-American hybrids, cultivars derived from backcrossing hybrids between *Vitis vinifera* and *V. riparia*, *V. rupestris*, or *V. aestivalis* with other *V. vinifera* cultivars (synonymous with direct producer).

Fruit bud, a bud that contains embryonic flower clusters.

Fruit ester, a low-molecular-weight (1- to 10-carbon) fatty acid ester that possesses a fruit-like fragrance.

Fruit set, the period when flower fertilization occurs and the berries begin to develop.

Fungicide, a chemical used to kill or inhibit spore germination, or eradicate existing fungal tissue.

Fusel alcohol, a short-chain (3- to 5-carbon) alcohol possessing a pronounced fusel or petroleum odor (*see* Higher alcohol).

Genetic engineering, a set of techniques designed to introduce genes from one organism into another which could not occur by sexual means.

Glycerol, an essential component of cell-membrane lipids; may contribute to the smooth mouth-feel and viscosity of some wines.

Glycoside, an organic compound linked glycosidicly (C–O–C) to one or more sugars.

Governo, the technique of adding must from partially dried grapes to wine to encourage a second fermentation; formerly common in the production of Chianti wines.

Grafting, promoting the union of a shoot (scion) with a root-possessing shoot (rootstock) such that their cambial tissues are adjacent to one another.

Green grafting, grafting of actively growing scions and rootstocks.

Hard back, the outermost layer of cork that forms after the periodic stripping of bark; removed from harvested cork slabs because it is too rigid and inelastic for use in stopper production.

Hardpan, a dense layer of clay or carbonate salts that may form in the subsoil, restricting both root penetration and water drainage (if it is of clay, synonymous with claypan).

Head, the apex of the trunk from which arms or cordons originate.

Heads, the first fraction derived from the pot distillation of wine (primarily acetaldehyde).

Headspace, the volume of gas left in a container after filling and attaching the closure.

Heartwood, the inner portion of the trunk of some hardwoods where phenolic compounds have been deposited; cell function ceases and the wood darkens.

Heat Summation Units, a numeric means of comparing viticultural regions; based on a summation of the degree-days per year.

Hectarage, a metric unit of area ($10,000 \, m^2$), equivalent to 2.37 acres.

Hectoliter, a metric unit of volume (100 liters) – equivalent to 26.3 US gal.

Hedging, pruning of the growing vine in early summer to limit further vegetative growth, to produce vine rows resembling a straight-sided hedge.

Hemicelluloses, large polysaccharides that form a variable portion of plant cell walls; as source of sugars they may caramelize during the toasting of oak barrels.

Herbaceous, describing an odor induced by the presence of above-threshold levels of several hexanols and hexanals, or certain pyrazines.

Herbicide, a chemical applied to kill or limit weed growth.

Hermaphrodite, a flower possessing both male and female functional parts.

Hexaploid, an organism that possesses six sets of chromosomes; these may be multiples of a single set of chromosomes or a combination of the chromosome sets from two or three different organisms.

Higher alcohol, a straight-chain hydrocarbon (3- to 8-carbon) possessing a single alcohol (-OH) group; occasionally considered synonymous with fusel alcohols.

Hybrid, the progeny of a cross between two species, or between cultivars of a single species.

Hydrolyzable tannin, a polymer of ellagic, gallic, or ellagic and gallic acid esters with glucose; in wine, it primarily comes from oak cooperage and is a polymer of ellagic acid esters (ellagitannins).

Hyperoxidation, the pumping (sparging) of juice or must with oxygen, ostensively to oxidize readily oxidizable phenols, promote their subsequent precipitation during fermentation and enhance color stabilization in red wines.

Hypodermis, a layer of parenchyma cells located just underneath the epidermis of the grape berry; it possesses much of the flavor of the skin as well as the anthocyanins of most red grapes.

Ice-nucleating bacteria, members of the surface flora (notably *Pseudomonas*) that promote the formation of ice crystals on plants.

Icewine, produced from the concentrated juice extracted from grapes left on the vine until they freeze when temperatures fall below $-7\ °C$.

Inbreeding depression, the progressive loss of vegetative vigor and sexual fertility that results when cross-fertilized plants are repeated inbred.

Inclination, the angle formed by sloped terrain relative to the horizontal plane.

Induced fermentation, inoculation of must with a known yeast strain ($\geq 10^5$ cells/ml); this strain typically will suppress yeasts derived from the grapes or winery equipment and conducts the alcoholic fermentation.

Inflorescence, a cluster of flowers on a common stalk.

Inoculation, the addition of living microbes, such as yeasts and bacteria, to must or wine.

Integrated Pest Management (IMP), a coordinated approach to the control of diseases, pests, and weeds to minimize the application and negative interaction between control measures while maximizing their effectiveness.

Internode, the interval along a shoot between two adjacent nodes.

Invertase, an enzyme that hydrolyzes sucrose into its two component fermentable sugars, glucose and fructose.

Ion exchange column, a column packed with a resin than can exchange its ions (i.e., Na^+) for similarly charged ions (i.e., H^+) of a solution passed through the column.

Isomers, two or more compounds possessing the same molecular formula, but differing in at least one chemical or physical property.

Isotopes, variants of a chemical element that differ in mass due to the number of neutrons present in the nucleus of the atom.

Killer Factor, one of a series of proteins, produced by carrier strains of yeast, that can kill related strains or occasionally unrelated yeasts, filamentous fungi or bacteria.

Laccase, a fungal polyphenol oxidase that is relatively resistance to inactivation by sulfur dioxide; it can oxidize a wide range of mono- and di-phenols.

Lactic acid bacteria, a group of strictly fermentative bacteria that can induce malolactic fermentation or various forms of wine spoilage.

Layering, a method of vegetative reproduction by placing a trailing cane in the ground; once it has taken root, the young vine can be severed from the mother plant.

Lees, sediment that forms during and after fermentation; it includes material such as dead and dying yeasts and bacteria, grape cell remains, seeds, tartrate salts and precipitated tannins.

Leg, *see* Tears

Lenticel, region in bark containing radially elongated cells and gaps designed to ease gas exchange between living cells of the shoot, stem, or trunk and the atmosphere.

Light-struck, a light-induced off-odor associated with the production of methanthiol and dimethyl disulfide.

Lignin, a large polymer of hydroxycinnamyl alcohols that characterizes woody cell walls; it is the source of important aromatic phenolics extracted during wine maturation in oak barrels.

Maceration, the enzymatic breakdown of grape cell constituents following crushing while the juice remains in contact with the seeds and skins of the fruit.

Macroclimate, the climatic conditions that characterize large geographic regions (i.e., continental or maritime).

Maillard product, the product of nonenzymatic reactions between reducing sugars and amine compounds (i.e., amino acids and proteins); which produce polymeric brown pigments and caramel-like aromatics.

Malic acid, one of the two major organic acids in grapes and wine.

Malolactic Fermentation, the decarboxylation of malic to lactic acid by several lactic acid bacteria and some yeast strains; a biologic form of wine deacidification.

Material-Other-than-Grapes (MOG), non-grape material that is collected during harvesting.

Maturation, the period between the end of fermentation and bottling, during which the wine is clarified, stabilized and possibly held in oak cooperage.

Mercaptan, a short-chain hydrocarbon containing a thiol group (-SH); it is highly volatile, possessing strong, putrid odors.

Meristem, the embryonic tissue that gives rise to cells that differentiate in the mature tissues of plants.

Metallic, aluminum-like taste; occasionally an aromatic sensation mistaken as a perception on the tongue.

Metabisulfite, a salt, usually potassium, added to must or wine to supply sulfur dioxide.

Metabolite, a by-product of a biochemical pathway (i.e., ethanol via fermentation).

Microbe, a microscopic organism that in wine may be either a bacterium, yeast, or filamentous fungus.

Microclimate, the climatic conditions in the immediate vicinity of a plant or other object.

Micronutrient, an inorganic nutrient required in trace amounts (e.g., boron, chlorine, copper, iron, manganese, molybdenum, or zinc).

Minimal pruning, a pruning technique in which a trained vine is no longer pruned, other than to remove shoot segments that may interfere with vineyard management; the vines self-prune as nonwoody shoot segments die and become detached by natural physical forces.

Mouth-feel, the touch sensation (i.e., astringency, prickling, and viscosity) perceived in the mouth via the trigeminal nerve.

Muscadine, a general term referring to grapes in the subgenus *Muscadinia*.

Mushroom alcohol, an aromatic alcohol (1-octen-3-ol) produced by many filamentous fungi; it possesses a mushroom-like fragrance.

Must, the juice, seeds, and skins of crushed grapes before, and occasionally after, fermentation has become apparent.

Mycorrhiza, the symbiotic association between young roots and a fungus; the fungus enhances water and phosphorus uptake by the root and may limit attack by root pathogens.

Nature identical, identical to compounds produced by organisms, but synthesized organically in a laboratory.

Nectary, a series of swollen regions at the base of the ovary of grape flowers; it is thought to produce an insect-attracting scent, at least in unisexually female flowers of wild grapevines.

Nematodes, a group of roundworms frequently found in decomposing vegetation; some are vine pathogens.

Noble rot, the limited infection of ripening grapes by *Botrytis cinerea* under a repeating cycle of moist nights and dry sunny days.

Node, the enlarged region of a shoot where leaves, tendrils, flower clusters and buds originate.

Nonflavonoid, a phenolic compound or its derivative, based on a phenol possessing a 1- or 3-carbon side chain (*see* Table 6.2).

Norisoprenoid, an aromatic breakdown product of carotenoids, which may contribute significantly to the aged bouquet of some white wines.

Oak, the wood used in wine cooperage, derived from several North American and European white oak species – notably *Quercus alba*, *Q. robur*, and *Q. sessilis*.

Oak lactones, a pair of optical isomers found in oak that contribute to the characteristic flavor of wine matured in oak cooperage.

Oenoccus oeni, the primary bacterium inducing beneficial forms of malolactic fermentation.

Off-odor, a fragrant or pungent compound that is generally considered undesirable.

Off-taste, an imbalance in taste sensations that is generally considered undesirable.

Oidium, an old name for downy mildew of grapes caused by *Plasmopara viticola*.

Old wood, the portion of a vine's shoot system that is older than two years.

Organic viticulture, the cultivation of grapes without the use of laboratory-synthesized pesticides.

Osmoticum, a substance accumulated by cells that limits the net water loss from cells in concentrated solutions, such as grape juice.

Ovary, the swollen base of the pistil that bears the ovules (egg-bearing tissue) and differentiates into the fruit.

Overcropping, the vine bearing more fruit than it can fully ripen.

Overmature, juice concentration and change in flavor associated with fruit having been left on the vine (or in storage) for several week or months after ripening.

Oxidation, a reaction in which a compound loses an electron (or hydrogen atom) and becomes oxidized; only occasionally is molecular oxygen directly involved in most wine oxidations as the hydrogen acceptor (usually hydrogen peroxide or various oxygen radicals), although it is the principal initiator of a chain of oxidation-reduction reactions in wine; in more restricted senses, oxidation may be used in referring to the browning of white and red wines, to the development of a pungent, cooked vegetable off-odor in bottled wines, to the development of a distinct aldehyde odor in sherries, the development of "bottle sickness", or the fragrance loss after bottle opening.

Oxidized, having lost an electron (or hydrogen atom); in wine, having been exposed to sufficient oxygen to have generated above-threshold concentrations of acetaldehyde and develop a brownish cast.

Parenchyma, the major cell type found in young plant tissues.

Pasteurization, the process of heat treating a beverage or food to inactivate potential human pathogens or reduce the number of potential spoilage microbes.

Pectins, a series of gel-like galacturonic acid polymers important in holding plant cells together; frequently they release methanol on degradation under the affect of certain pectinases.

Pedicel, the short extension of the rachis that hold berries to the cluster (*see* Fig. 3.16).

Pellicle, the layer of yeast cells that may form at the top of fermenting juice or wine; it is induced by the increased hydrophobic nature of the cell wall resulting from the changed and increased protein portion of the wall.

Pergola, a horizontal trellis supported by vertical columns, on which vines may be trained.

Pericarp, the skin and pulp of the grape; it develops from the ovary wall (*see* Fig. 3.23).

Petiole, the stalk of a leaf that connects the leaf blade to the shoot.

Petiole analysis, the collection and assessment of petioles for the presence of mineral deficiencies or toxicities.

Pétillance, oral sensation generated by a slight amount of carbon dioxide (about 200 kPa to 2 atm).

pH, the relative concentration of H^+ ions to OH^- ions in an aqueous solution; it significantly affects the taste, color, and microbial stability of wine.

Phenolic compounds, compounds containing one or more benzene-ring structures and at least one hydroxyl (OH^-) group.

Pheromone, a volatile compound released by an organism (i.e., an insect) that affects the behavior or physiology of other members of the same species.

Photoperiod, the duration of darkness (night length) per day which regulates many physiological and developmental phases of plant growth (i.e., flowering).

Photosynthate, a general term for the organic compounds derived directly from photosynthesis.

Photosynthesis, the chain of reactions that trap light energy and use it to fix carbon dioxide gas in organic compounds.

Phylloxera, an indigenous aphid-like insect that endemically parasitized eastern North American grapevines, and to which the roots of *Vitis vinifera* are particularly vulnerable.

Pinching, the removal of 7–15 cm of the shoot tip of a vine; it is used to limit wind damage and in training the vine.

Pith, the inner soft spongy tissue of shoots.

Plasmids, self-replicating DNA segments located in the cytoplasm that are unessential for cell function; they are frequently used in transferring genes in genetic engineering.

Plastering, an old acidification technique in which plaster of Paris is added to the juice; formerly used in sherry production.

Pneumatic press, a device that uses the application of air pressure in an elongated rubber tube to gently release juice (or wine) from crushed grapes (or fermented red must).

Polishing filtration, a final prebottling filtration to assure a stable crystal-clear wine.

Polymerization, any of a series of reactions that induce the bonding of identical or related compounds into macromolecules.

Polyphenol oxidases, enzymes that catalyze the oxidation of phenolic compounds; grape enzymes oxidize monophenols whereas fungal enzymes oxidize mono- and di-phenols.

Polyvinylpolypyrrolidine (PVPP), a reusable fining agent.

Pomace, the solid component of crushed grapes or fermenting red must (seeds and skins).

Press, a device used to separate juice (or wine) from crushed grapes (or fermented red must).

Press cake, the solid remains after pressing crushed grapes or fermented red must (primarily seeds and skins).

Press-run, the juice (or wine) released by pressing the must (after the free-run has escaped).

Primary bud, the largest and central bud of a compound bud; usually develops into the fruit-bearing shoot the subsequent spring.

Procyanidins, oligomers of flavanols (catechin and epicatechin) and occasionally gallic acid; they induce bitter and astringent sensations and promote color stability (synonymous with grape tannin).

Protective colloids, the dispersed polysaccharides or dissolved proteins that limit the crystallization, nucleation, and precipitation of other compounds in wine.

Pruning, severing the connection of canes, shoots, leaves, fruit clusters and roots from the vine.

Pulp, the central and major fleshy portion of grapes (synonymous with flesh).

Pumping over, the transfer of wine to flow through, or submerge, the cap of seeds and skins that forms on the surface of fermenting red must.

Punching down, the periodic mechanical submergence of the cap of seeds and skins into the fermenting juice.

Pyrazines, a group of aromatic nitrogen-containing cyclic compounds; they are the source of important aroma compounds in 'Cabernet Sauvignon' and 'Sauvignon blanc'.

Quality, the property of wine showing marked aromatic and flavor complexity, subtlety, harmony, and development, associated with a distinct aroma and aged bouquet; in aromatic compounds, quality signifies the subjective similarity to a known flavor or aroma, for example apple-like (in contrast to the intensity of the sensation).

Rachis, the central stalk of a grape cluster (*see* Fig. 3.14).

Racking, the separation of wine from the lees by transferring wine from one cooperage to another.

Raisined, a flavor term to describe wines made from grapes possessing a significant number of sun-dried fruit.

Random Amplified Polymorphic DNA (RAPD), a technique of DNA fingerprinting using labeled segments of DNA isolated and copied with a DNA polymerase.

Recioto, a wine made from grapes slowly partially dried and allowed to overmature under cool conditions for several months; primarily used in the Veneto and Lombardy provinces of Italy.

Rectified, distilled to separate and purify an alcoholic beverage into its distinctive aromatic constituents.

Redox balance, the relative proportion of the oxidized and reduced forms of the electron carriers NAD and NADP involved in cellular oxidation and reduction reactions.

Redox potential, the relative potential of compounds to be oxidized (lose electrons) or be reduced (receive electrons), e.g., oxygen is considered to have a high redox potential (receiving electrons as it is reduced to water, whereas phenols are considered to have a low redox potential (losing electrons as they are oxidized to quinones).

Reduction, a reaction in which a compound gains an electron (or hydrogen atom) and becomes reduced.

Refractometer, a hand-held device used to measure the sugar content of grapes; the measurement is based on the proportional diffraction of light by the sugar in grape juice.

Residual sugar, the sugar content that remains in wine after fermentation is complete; in a dry wine this primarily involves the nonfermentable sugars arabinose and rhamnose.

Restriction Fragment Length Polymorphism (RFLP), a technique used in DNA fingerprinting that uses specific segments of labeled DNA.

Resveratrol, a stilbene compound produced by grapes in response to environmental stress, notably to attack by pathogenic fungi; an important antioxidant involved in some of the beneficial health consequences of moderate wine consumption.

Reverse osmosis, the concentration of a liquid by the selective loss of water through a differentially permeable membrane.

Ripening, the physiological changes that occur as grapes increase in sugar content, decrease in acidity, and accumulate flavor and mature color.

Rondelle, circular section of cork cut out tangentially to the bark surface; two copies are glued to the base of agglomerate champagne corks.

Root pruning, a pruning technique in which a portion of the root system of a vine is severed.

Rootstock, the lower section of a grafted vine that serves to develop the root system.

Ropiness, the presence of long visible chains of spoilage lactic acid bacteria in wine; the large amount of gelatinous polysaccharides that are produced noticeably increases the wine's viscosity.

Rosé, a wine possessing sufficient anthocyanins to give the wine a light-pinkish color.

Saccharomyces cerevisiae, the primary wine yeast; if it does not initiate fermentation, it typically soon dominates and completes alcoholic fermentation.

Salinization, the accumulation of salts in the soil; this process is typical in hot regions of low rainfall and poor drainage.

SAM, the acronym for soil and atmospheric microclimate – equivalent to *terroir*, but without mystical connotations with which *terroir* is so often associated.

Scion, the upper vegetative and fruit-bearing portion of a grafted vine.

Second crop, the fruit produced if lateral shoots develop and bear grape clusters.

Sediment, the material that accumulates on the sides or bottom of bottled wine; it usually consists of precipitated tartrate crystals and tannin complexes.

Sensory evaluation, the scientific evaluation of the quality and characteristics of a food or beverage.

Shatter, the fall of unfertilized flowers and young fruit about a week after bloom.

Shoot, the vegetative stem of the current year that bears the leaves, tendrils, and fruit.

Skin contact, the stage between crushing the grapes and pressing of the must; it is associated with maceration.

Sodic, referring to soils high in exchangeable sodium ions, low total ionic content, and with a pH greater than 8.5.

Soil structure, the association (aggregate) arrangement of the inorganic and organic content of soil.

Soil texture, the relative proportion of the sand, silt, and clay particles in soil.

Solera system, the fractional blending system used primarily in Spain for the maturation of sherry.

Somatic mutation, a change in the structure (and possible function) of genes in vegetative (nonreproductive) cells.

Sparkling wine, a wine that has undergone a second fermentation to generate a high carbon dioxide content.

Spontaneous fermentation, alcoholic fermentation induced by the yeast inoculum that occurs on the grapes or that is picked up from winery equipment.

Springwood, the wood produced shortly after growth begins in the spring; it tends to possess a larger proportion of wide water-conduction vessels.

Spur, a short cane possessing two to four count buds (*see* Fig. 4.9).

Stabilization, the process of treating a wine to make it microbially stable and crystal clear for many years after bottling.

Stave, a shaped strip of wood that forms part of a barrel or other type of cooperage.

Stoma (*pl.* Stomata), the gap in the epidermis of green tissue surrounded by guard cells that helps to regulate the exchange of water vapor and gases with the atmosphere.

Stuck fermentation, fermentation that stops prematurely, leaving residual fermentable sugars.

Sucker, a shoot that originates from the trunk just above or below ground level (*see* Fig. 4.9).

Suckering, the process of removing suckers from grapevines.

Sulfiting, the addition of sulfur dioxide to whole grapes, crushed grapes, or wine.

Sulfur dioxide, the major antimicrobial and antioxidant agent used in wine production; it is also a by-product of yeast metabolism.

Summerwood, the small-vesseled wood produced during the summer.

Sunflecks, the shafts of sunlight that periodically penetrate the canopy as wind moves the leaves.

Surface sterilant, an agent that kills microbes and inactivates viruses on the surface of an object.

Surfactant, a compound that reduces the surface tension and eases the spread of fluids over a surface.

Sur lies, the technique of permitting the uptake of flavorants from autolyzing yeast cells in the lees by delaying racking; it requires frequent agitation to avoid the generation of reduced sulfur off-odors in the lees.

Suspended solids, microscopic particulate matter that is held temporarily in suspension following crushing of the grapes.

Sweet reserve, the portion of the juice that is kept unfermented and added to the wine produced from the main portion of the juice to give the finished wine a sweet, fresh-fruit aspect.

Systemic, of a chemical, absorbed and potentially translocated throughout plant tissues; of pathogens, occurring throughout most tissues of a plant.

Tails, the last fraction extracted during the pot distillation of wine.

Tank, a fermentor or storage cooperage that possesses a sealed top (vs. a vat).

Tannins, polymeric phenolic compounds that can tan (precipitate proteins) in leather; in wine they contribute to bitter and astringent sensations, promote color stability, and are potent antioxidants.

Tartaric acid, one of the two major organic acids in grapes and wine; because it is relatively resistant to microbial breakdown, tartaric acid provides much of the stable acidity to wine.

Taste buds, specialized regions on the tongue that possess receptor cells for taste sensation.

TCA cycle, the metabolic cycle that generates most of the reduced NAD in respiration; also an important source of components for conversion to amino acids (also called the Krebs or citric acid cycle).

Tears, the droplets that slide down the sides of a swirled glass of wine; they form as alcohol evaporates and the increased surface tension of the film pulls the fluid on the glass together.

Temperature inversion, an atmospheric condition during which a cold layer of air traps warm air close to the ground; it restricts air flow and can result in the accumulation of air pollutants.

Temperature isotherm, a line on a map that connects regions with similar average temperatures.

Tendril, a twining modified shoot that originates in leaf axils.

Terpenes, polymers of isoprene structure important in generating may floral and fruit fragrances.

Terroir, the combined influences of vineyard atmospheric, soil, and cultural conditions on vine growth and fruit ripening; the term is often misused in an attempt to justify the supposedly unique quality of wines from certain vineyard sites.

Thermotherapy, the process of heat treating shoots or canes to eliminate infection with certain systemic bacterial, mycoplasmal, and viral pathogens.

Thermovinification, the process of heat treating grapes or must to enhance the extraction of anthocyanins; if they are heated above 60°C it can also inactivate laccases that may be present due to fungal infection.

Tirage, a solution of sugar and nutrients added to the *cuvée* to provide the ingredients required for the second fermentation in sparkling wine production.

Tissue culture, cells grown in a liquid or on a semisolid nutrient medium.

Tolerance, the ability to withstand exposure (or infection) to a toxic (or pathogenic) agent without marked damage; the maximum amount of toxic residue permitted in a food or beverage.

Training system, the pruning and placement of a vine to facilitate long-term vine health while optimizing crop yield, easing harvest, and reducing production costs.

Transpiration, the loss of water vapor from the surfaces of a plant, primarily through the stomata.

Treading, the crushing of grapes with the feet.

Trellis, the support structure on which vines may be trained.

Triangle test, a method designed to determine statistically whether two samples can be distinguished.

2,4,6-Trichloroanisole (TCA), the compound most frequently associated with a corky off-odor in wine.

Tricolpate, the three oblong-to-elliptical thin areas in a pollen grain through which pollen germ tubes may rupture.

Trigeminal nerve, the fifth cranial nerve, two branches of which carry impulses from the nose and mouth; it gives rise to the sensations of astringency, heat, body, prickling, and pain in wine.

Trunk, the vertical wood stem of a vine up to the origin of the branches (*see* Fig. 4.9).

Tuberosity, a gall-like swelling formed on mature vine roots as a consequence of the feeding of phylloxera.

Tunica, the outer embryonic layers of the shoot that divide primarily perpendicular to the surface; T1 gives rise to the epidermis whereas T2 provides the outer cortex and generative (sex) cells.

Tylose, the outgrowth of parenchyma cells that can plug the cental cavity of xylem vessels in wood.

Ullage, the gaseous volume that forms over wine in oak cooperage, initially as wine absorbs into the wood (up to 5–6 liters in a 225-liter barrel) and as water and alcohol slowly evaporate from the barrel surface.

Undercanopy, the layer(s) of leaves that occur beneath the uppermost foliage.

Variety, a clone, or series of related clones, propagated vegetatively from a single parent plant (monoclonal origin) or several genetically similar parents (polyclonal origin) (synonymous with cultivar).

Vat, a fermentor with an open top (vs. tank).

Vegetative propagation, a reproduction technique that does not involve sexual fertilization and seed production (i.e., tissue culture or rooted cuttings).

Vein, the bundle of vascular tissue that extends into leaves and fruit from the stem.

Véraison, the beginning of the last growth phase of grapes, when the green color begins to fade and the pulp starts to soften.

Vermouth, a flavored fortified wine characteristically produced in sweet or bitter styles.

Vigor, the rate or duration of shoot growth, or the weight of shoot growth; vines showing high vigor are usually characterized by large leaf size, long internodes, thick canes, active lateral growth, and a yield/pruning weight ratio significantly less than 6–7 : 1.

Vintage, the year in which the fruit develops and the grapes are fermented into wine.

Viroid, an infectious nucleic acid that can cause disease.

Virus, an infectious nucleic acid enclosed in a proteinaceous layer that can cause disease.

Viscosity, the perception of the resistance of wine to flow; a smooth, velvety mouth-feel.

Vinification, wine production.

Viticulture, the scientific investigation of grape cultivation.

Vitis vinifera, the primary grape species cultivated and used as a source of wine, table grapes, and raisin grapes.

Volatile, evaporating into the air (typically at room temperature).

Volatile acids, organic acids that can be readily removed by steam distillation, almost exclusively acetic acid in wine.

Weight, a sensory perception of the organic content of a wine, notably alcohol, sugar, glycerol, and tannins (synonymous with body).

Wild yeast, species or strains of yeast that occur on grape surfaces or are contaminants on winery equipment.

Wine spirit, distilled wine used to fortify wines such as sherry and port; it may be highly rectified to produce a neutral-flavored source of high-strength alcohol.

Wing, the smaller side-cluster of grapes produced on the lateral branch of the main cluster (*see* Fig. 3.16).

Xylem, the major woody tissue of stems, which contains the water- and mineral-conducting cells of plants.

Yeast, a primarily unicellular fungus that divides by budding (or fission) and possesses a glucan or mannoprotein cell wall (budding yeast may contain some chitin around the bud scar).

Yeast hulls (ghosts), the partially purified cell-wall remains of yeasts.

Yield, the fruit crop per area planted; variously calculated in terms of fresh weight (i.e., tons/acre) or wine production (hectoliters/hectare); rough equivalencies among measures are:

$$1 \text{ ton/ha} = \sim 7 \text{ hl/ha} = \sim 10 \text{ quintal/ha}$$
$$1 \text{ ton/acre} = \sim 17 \text{ hl/ha} = \sim 25 \text{ quintal/ha}$$
$$1 \text{ ton/acre} = 2.5 \text{ ton/ha}$$

Index

Abscisic acid:
    dormancy, 57
    growth regulator, 79, 80, 81, 85, 87, 207
    in wine, 448
    water stress, 65, 67, 151, 154,
Abscission:
    fruit, 72, 78, 80, 81, 154, 207
    leaf, 63, 190
    root, 123
Acetals, 301, 448, 450, 524, 556, 563, 565, 567, 569, 571
Acetaldehyde, 254–255
    affected by:
        aging, 446, 448
        carbonic maceration, 533, 536
        fermentation temperature, 384
        sulfur dioxide, 312–315, 335
    affecting:
        color stability, 291–94,
        phenol polymerization, 283, 284, 297, 298
    botrytized wines, 524, 525
    brandy, 565–567
    chemistry, 300

    off-odor,
        bottle sickness, 337
        oxidized, 434, 435, 663
    oxidation, 298, 313, 450, 503
    port, 375, 560, 561, 563
    madeira, 564
    metabolism by:
        acetic acid bacteria, 497, 498–99
        lactic acid bacteria, 391, 392
        yeasts, 315, 347, 358–62
    sherry, 554–557, 559
    tokaji, 524
    toxicity, 305, 688, 97, 699, 700
2-Acetyltetrahydropyridine, 273, 306, 495, 497
Acetamides, 272, 307, 651
Acetic acid, 279–280; *see also* Volatile acidity:
    affect on:
        malolactic fermentation, 400
        yeast fermentation, 387
    esterification of, 301–03
    in phenol synthesis, 286
    metabolism by:
        flor yeasts, 556

        yeasts during fermentation, 359
    off-odor, *see* Volatile acidity:
    sensory effect, 279, 663
    source:
        acetic acid bacteria, 423, 435, 444, 497–99
        botrytized wines, 523
        brandy, 567
        carbonic maceration, 535, 536
        icewine, 527
        infected fruit, 183, 387, 498, 565, 522
        lactic acid bacteria, 387, 390, 392, 401, 496, 536
        oak cooperage, 464, 466, 467, 468
        peracetic acid, 479, 480, 483
        sherry maturation, 556, 557
        yeast:
        during fermentation, 347, 359, 361, 362, 367
        spoilage, 380, 495, 496
        strain variability, 368, 369
    synthesis affected by:
        concurrent malolactic fermentation, 402

Food Science
and Technology

International Series

Maynard A. Amerine, Rose Marie Pangborn, and Edward B. Roessler, *Principles of Sensory Evaluation of Food*. 1965.

Martin Glicksman, *Gum Technology in the Food Industry*. 1970.

Maynard A. Joslyn, *Methods in Food Analysis*, second edition. 1970.

C. R. Stumbo, *Thermobacteriology in Food Processing*, second edition. 1973.

Aaron M. Altschul (ed.), *New Protein Foods*: Volume 1, *Technology, Part A*—1974. Volume 2, *Technology, Part B*—1976. Volume 3, *Animal Protein Supplies, Part A*—1978. Volume 4, *Animal Protein Supplies, Part B*—1981. Volume 5, *Seed Storage Proteins*—1985.

S. A. Goldblith, L. Rey, and W. W. Rothmayr, *Freeze Drying and Advanced Food Technology*. 1975.

R. B. Duckworth (ed.), *Water Relations of Food*. 1975.

John A. Troller and J. H. B. Christian, *Water Activity and Food*. 1978.

A. E. Bender, *Food Processing and Nutrition*. 1978.

D. R. Osborne and P. Voogt, *The Analysis of Nutrients in Foods*. 1978.

Marcel Loncin and R. L. Merson, *Food Engineering: Principles and Selected Applications*. 1979.

J. G. Vaughan (ed.), *Food Microscopy*. 1979.

J. R. A. Pollock (ed.), *Brewing Science*, Volume 1—1979. Volume 2—1980. Volume 3—1987.

J. Christopher Bauernfeind (ed.), *Carotenoids as Colorants and Vitamin A Precursors: Technological and Nutritional Applications*. 1981.

Pericles Markakis (ed.), *Anthocyanins as Food Colors*. 1982.

George F. Stewart and Maynard A. Amerine (eds), *Introduction to Food Science and Technology*, second edition. 1982.

Malcolm C. Bourne, *Food Texture and Viscosity: Concept and Measurement*. 1982.

Hector A. Iglesias and Jorge Chirife, *Handbook of Food Isotherms: Water Sorption Parameters for Food and Food Components*. 1982.

Colin Dennis (ed.), *Post-Harvest Pathology of Fruits and Vegetables*. 1983.

P. J. Barnes (ed.), *Lipids in Cereal Technology*. 1983.

David Pimentel and Carl W. Hall (eds), *Food and Energy Resources*. 1984.

Joe M. Regenstein and Carrie E. Regenstein, *Food Protein Chemistry: An Introduction for Food Scientists*. 1984.

Maximo C. Gacula, Jr., and Jagbir Singh, *Statistical Methods in Food and Consumer Research*. 1984.

Fergus M. Clydesdale and Kathryn L. Wiemer (eds), *Iron Fortification of Foods*. 1985.

Robert V. Decareau, *Microwaves in the Food Processing Industry*. 1985.

S. M. Herschdoerfer (ed.), *Quality Control in the Food Industry*, second edition. Volume 1—1985. Volume 2—1985. Volume 3—1986. Volume 4—1987.

F. E. Cunningham and N. A. Cox (eds), *Microbiology of Poultry Meat Products*. 1987.

Walter M. Urbain, *Food Irradiation*. 1986.

Peter J. Bechtel, *Muscle as Food*. 1986.

H. W.-S. Chan, *Autoxidation of Unsaturated Lipids*. 1986.

Chester O. McCorkle, Jr., *Economics of Food Processing in the United States*. 1987.

Jethro Japtiani, Harvey T. Chan, Jr., and William S. Sakai, *Tropical Fruit Processing*. 1987.

J. Solms, D. A. Booth, R. M. Dangborn, and O. Raunhardt, *Food Acceptance and Nutrition*. 1987.

R. Macrae, *HPLC in Food Analysis*, second edition. 1988.

A. M. Pearson and R. B. Young, *Muscle and Meat Biochemistry*. 1989.

Marjorie P. Penfield and Ada Marie Campbell, *Experimental Food Science*, third edition. 1990.

Leroy C. Blankenship, *Colonization Control of Human Bacterial Enteropathogens in Poultry*. 1991.

Yeshajahu Pomeranz, *Functional Properties of Food Components*, second edition. 1991.

Reginald H. Walter, *The Chemistry and Technology of Pectin*. 1991.

Herbert Stone and Joel L. Sidel, *Sensory Evaluation Practices*, second edition. 1993.

Robert L. Shewfelt and Stanley E. Prussia, *Postharvest Handling: A Systems Approach*. 1993.

R. Paul Singh and Dennis R. Heldman, *Introduction to Food Engineering*, second edition. 1993.

Tilak Nagodawithana and Gerald Reed, *Enzymes in Food Processing*, third edition. 1993.

Dallas G. Hoover and Larry R. Steenson, *Bacteriocins*. 1993.

Takayaki Shibamoto and Leonard Bjeldanes, *Introduction to Food Toxicology*. 1993.

John A. Troller, *Sanitation in Food Processing*, second edition. 1993.

Ronald S. Jackson, *Wine Science: Principles and Applications*. 1994.

Harold D. Hafs and Robert G. Zimbelman, *Low-fat Meats*. 1994.

Lance G. Phillips, Dana M. Whitehead, and John Kinsella, *Structure-Function Properties of Food Proteins*. 1994.

Robert G. Jensen, *Handbook of Milk Composition*. 1995.

Yrjö H. Roos, *Phase Transitions in Foods*. 1995.

Reginald H. Walter, *Polysaccharide Dispersions*. 1997.

Gustavo V. Barbosa-Cánovas, M. Marcela Góngora-Nieto, Usha R. Pothakamury, and Barry G. Swanson, *Preservation of Foods with Pulsed Electric Fields*. 1999.

Ronald S. Jackson, *Wine Science: Principles, Practice, Perception*, second edition. 2000.

R. Paul Singh and Dennis R. Heldman, *Introduction to Food Engineering*, third edition. 2001.

Ronald S. Jackson, *Wine Tasting: A Professional Handbook*. 2002.

Malcolm C. Bourne, *Food Texture and Viscosity: Concept and Measurement*, second edition. 2002.

Benjamin Caballero and Barry M. Popkin (eds), *The Nutrition Transition: Diet and Disease in the Developing World*. 2002.

Dean O. Cliver and Hans P. Riemann (eds), *Foodborne Diseases*, second edition. 2002.

Martin Kohlmeier, *Nutrient Metabolism*, 2003.

Herbert Stone and Joel L. Sidel, *Sensory Evaluation Practices*, third edition. 2004.

Jung H. Han, *Innovations in Food Packaging*. 2005.

Da-Wen Sun, *Emerging Technologies for Food Processing*. 2005.

Hans Riemann and Dean Cliver (eds), *Foodborne Infections and Intoxications*, third edition. 2006.

Ioannis S. Arvanitoyannis, *Waste Management for the Food Industries*. 2008.

Ronald S. Jackson, *Wine Science: Principles and Applications*, third edition. 2008.

Da-Wen Sun, *Computer Vision Technology for Food Quality Evaluation*. 2008.

Kenneth David, *What Can Nanotechnology Learn From Biotechnology?* 2008.

Elke Arendt, *Gluten-Free Cereal Products and Beverages*. 2008.